Lecture Notes in Computer S

Commenced Publication in 1973
Founding and Former Series Editors:
Gerhard Goos, Juris Hartmanis, and Jan van Leeuwen

D1757050

Editorial Board

Amit Sheth   Steffen Staab   Mike Dean
Massimo Paolucci   Diana Maynard
Timothy Finin   Krishnaprasad Thirunarayan (Eds.)

The Semantic Web – ISWC 2008

7th International Semantic Web Conference, ISWC 2008
Karlsruhe, Germany, October 26-30, 2008
Proceedings

 Springer

Volume Editors

Amit Sheth
Krishnaprasad Thirunarayan
Wright State University, Dayton, OH 45435-0001, USA
E-mail: {amit.sheth,t.k.prasad}@wright.edu

Steffen Staab
University of Koblenz, 56016 Koblenz, Germany
E-mail: staab@uni-koblenz.de

Mike Dean
BBN Technologies, Ann Arbor, MI 48103, USA
E-mail: mdean@bbn.com

Massimo Paolucci
DOCOMO Euro-Labs, 80687 Munich, Germany
E-mail: paolucci@docomolab-euro.com

Diana Maynard
University of Sheffield, Sheffield S1 4DP, UK
E-mail: d.maynard@dcs.shef.ac.uk

Timothy Finin
University of Maryland Baltimore County, Baltimore, MD 21250, USA
E-mail: finin@umbc.edu

The picture of Karlsruhe Castle on the cover of this volume was kindly provided by
Monika Müller-Gmelin from the *Bildstelle Stadtplanungsamt Karlsruhe*

Library of Congress Control Number: 2008937502

CR Subject Classification (1998): H.4, H.3, C.2, H.5, F.3, I.2, K.4

LNCS Sublibrary: SL 3 – Information Systems and Application, incl. Internet/Web
and HCI

ISSN   0302-9743
ISBN-10   3-540-88563-3 Springer Berlin Heidelberg New York
ISBN-13   978-3-540-88563-4 Springer Berlin Heidelberg New York

Springer is a part of Springer Science+Business Media

springer.com

© Springer-Verlag Berlin Heidelberg 2008
Printed in Germany

Typesetting: Camera-ready by author, data conversion by Scientific Publishing Services, Chennai, India
Printed on acid-free paper   SPIN: 12548668   06/3180   5 4 3 2 1 0

Preface

The Web is a global information space consisting of linked documents and linked data. As the Web continues to grow and new technologies, modes of interaction, and applications are being developed, the task of the Semantic Web is to unlock the power of information available on the Web into a common semantic information space and to make it available for sharing and processing by automated tools as well as by people. Right now, the publication of large datasets on the Web, the opening of data access interfaces, and the encoding of the semantics of the data extend the current human-centric Web. Now, the Semantic Web community is tackling the challenges of how to create and manage Semantic Web content, how to make Semantic Web applications robust and scalable, and how to organize and integrate information from different sources for novel uses. To foster the exchange of ideas and collaboration, the International Semantic Web Conference brings together researchers and practitioners in relevant disciplines such as artificial intelligence, databases, social networks, distributed computing, Web engineering, information systems, natural language processing, soft computing, and human–computer interaction.

This volume contains the main proceedings of ISWC 2008, which we are excited to offer to the growing community of researchers and practitioners of the Semantic Web. We got a tremendous response to our call for research papers from a truly international community of researchers and practitioners from 41 countries submitting 261 papers. Each paper received an average of 3.25 reviews as well as a recommendation by one of the Vice Chairs who read the papers under investigation as well as the comments made by PC members. Based on a first round of reviews, authors had the opportunity to rebut leading to further discussions among the reviewers and—where needed—to additional reviews. Reviews for all papers with marginal chances of acceptance were discussed in the Programme Committee meeting attended by the Vice Chairs.

As the Semantic Web field develops we have observed the existence of a stable set of subjects relevant to the Semantic Web, such as reasoning (19)[1], knowledge representation (14), knowledge management (12), querying (9), applications (8), semantic Web languages (7), ontology mapping (6), ontology modelling (6), data integration (6), and semantic services (5). Some of the paper topics substantiate the role of the Semantic Web as being at the intersection of several technologies, e.g., collaboration and cooperation (5), interacting with Semantic Web data (5), human-computer interaction (4), information extraction (4), content creation and annotation (4), uncertainty (3), database technology (3), social networks

---

[1] Numbers in parentheses indicate the frequency of this topic among the set of accepted papers as given by the authors at the time of submission. Many papers addressed multiple topics.

(3), data mining and machine learning (3), semantic search (3), information retrieval (3), Semantic Wikis (3), social processes (2), peer-to-peer (2), personal information management (2), visualization (2), multimedia (1), grid (1), semantic desktop (1), trust (1), and middleware (1). Eventually, new areas that are core to the Semantic Web field gain prominence as data and ontologies become more widespread on the Semantic Web, e.g., ontology evaluation (4), ontology reuse (4), searching and ranking ontologies (3), ontology extraction (2), and ontology evolution (1).

Overall, as the field matures, ISWC Programme Committee members have adopted high expectations as to what constitutes high-quality Semantic Web research and what must be delivered in terms of theory, practice and/or evaluation in order to be accepted in the research track. Correspondingly, the Programme Committee accepted only 43 papers (i.e., 16.7%); three of the submissions were accepted for the in-use track after further discussion with its Track Chairs.

The Semantic Web In-Use Track received 26 submissions, each of which was reviewed by 3 members of the In-Use Track Programme Committee. We accepted 11 papers, along with 3 papers referred from the Research Track. Submissions came from both research and commercial organizations, reflecting the increased adoption and use of Semantic Web technologies. Traditional business, Web, and medical applications were joined with home automation, context-aware mobile computing, and satellite control. Papers also addressed deployment, scalability, and explanations.

This year ISWC 2008 hosted, for the fourth consecutive year, a doctoral consortium for Ph.D. students within the Semantic Web community, giving them the opportunity to discuss in detail their research topics and plans, and to get extensive feedback from leading scientists in the field. This year, in order to minimize overlap with other events and to increase attendance, the consortium was held on the day before the main conference. There were 39 submissions in total, which is approximately a 33% increase over 2007 and the highest ever submission rate to the ISWC DC. Submissions were reviewed by a panel of experienced researchers. The acceptances comprised 7 papers and 12 posters, which were presented in a full-day session. Each student was also assigned a mentor who led the discussions following the presentation of the work, and provided more detailed feedback and comments, focusing on the PhD proposal itself and presentation style as well as on the actual work presented. The mentors were drawn from the set of reviewers and comprised some leading researchers in the field—both from academia and from industry.

A unique aspect of the International Semantic Web Conferences is the Semantic Web Challenge. This is a competition in which participants from both academia and industry are encouraged to show how Semantic Web techniques can provide useful or interesting applications to end-users. This year the Semantic Web Challenge was organized by Jim Hendler and Peter Mika. It was further extended to include the Billion Triple Challenge. Here the focus was not so much on having a sleek and handsome application, but rather on managing a huge mass of heterogeneous data — semantic data, microformat data or data scraped from syntactic sources that one finds out there on the Semantic Web.

Keynote Talks from prominent scientists and managers further enriched ISWC 2008: Ramesh Jain, an eminent figure in the field of multimedia and beyond, gave a talk indicating the importance of semantics in the field of experiential computing. Stefan Decker, one of the founding members of the Semantic Web field in research and practice, presented his ideas about further developing the Semantic Web in order to give the common user the power of the Semantic Web at his fingertips. Finally, John Giannandrea considered the Semantic Web from the point of view of a business person. As co-founder and CTO of MetaWeb technologies, he explained how the Semantic Web helps his customers, showing the importance of thinking out of the box in order to exploit the strength of Semantic Web technologies. In addition, there were seven invited talks from industry that focused on the development and application of Semantic Web Technology along with a panel titled "An OWL 2 Far?" moderated by Peter F. Patel Schneider.

The conference was enlivened by a highly attractive Poster and Demo Session organized by Chris Bizer and Anupam Joshi and a large Tutorial Program supervised by Lalana Kagal and David Martin and including 11 unique events to learn more about current Semantic Web technologies. A great deal of excitement was generated by 13 workshops, which were selected from 22 high-quality proposals under the careful supervision of Melliyal Annamalai and Daniel Olmedilla.

The final day of the conference included a Lightning Talk session, where ISWC attendees could submit one slide and get five minutes of attention from a broad audience to report on what they learned and liked or disliked and how they see the Semantic Web continuing to evolve.

We are much indebted to Krishnaprasad Thirunarayan, Proceedings Chair, who provided invaluable support in compiling the printed proceedings. We also offer many thanks to Richard Cyganiak and Knud Möller, Meta data Co-chairs, for their expert coordination of the prodution of the semantic mark-up associated with contributions to the conference.

The meeting would not have been possible without the tireless work of the Local Organization Chair, Rudi Studer, the Local Organizing Committee including Anne Eberhardt, Holger Lewen and York Sure and their team from Karlsruhe.

We thank them all for providing excellent local arrangements. We would also like to thank the generous contribution from our sponsors and the fine work of the Sponsorship Chairs, John Domingue and Benjamin Grosof, and Li Ding, Publicity Chair. Finally, we are indebted to Andrei Voronkov and his team for providing the sophisticated yet free service of EasyChair, and to the team from Springer for being most helpful with publishing the proceedings.

October 2008

Amit Sheth
Steffen Staab
Diana Maynard
Mike Dean
Massimo Paolucci
Tim Finin

Organization

General Chair

Tim Finin   University of Maryland Baltimore Country

Programme Chairs

Amit Sheth   Wright State University
Steffen Staab   Universität Koblenz Landau

Semantic Web In Use Chairs

Mike Dean   BBN
Massimo Paolucci   DOCOMO Euro-Labs

Semantic Web Challenge Chairs

Jim Hendler   Rensselaer Polytechnic Institute
Peter Mika   Yahoo! Research

Doctoral Consortium Chair

Diana Maynard   University of Sheffield

Proceedings Chair

Krishnaprasad   Wright State University
Thirunarayan

Local Chair

Rudi Studer   Universität Karlsruhe

Local Organizing Committee

Anne Eberhardt   Universität Karlsruhe
Holger Lewen   Universität Karlsruhe
York Sure   SAP Research Karlsruhe

Workshop Chairs

Melliyal Annamalai   Oracle
Daniel Olmedilla   Hannover University

Tutorial Chairs

Lalana Kagal   MIT
David Martin   SRI

Poster and Demos Chairs

Chris Bizer   Freie Universität Berlin
Anupam Joshi   University of Maryland Baltimore County

Sponsor Chairs

John Domingue   The Open University
Benjamin Grosof   Vulcan Inc.

Metadata Chairs

Richard Cyganiak   DERI/Freie Universität Berlin
Knud Möller   DERI

Publicity Chair

Li Ping   Rensselaer Polytechnic Institute

Fellowship Chair

Joel Sachs   University of Maryland Baltimore County

Vice Chairs - Research Track

Abraham Bernstein   Wolfgang Nejdl
Vassilis Christophides   Natasha Noy
Thomas Eiter   Uli Sattler
Yolanda Gil   Guus Schreiber
Vasant Honavar   Luciano Serafini
Anupam Joshi   Umberto Straccia
David Karger   Gerd Stumme
Craig Knoblock   Frank van Harmelen
Riichiro Mizoguchi   Kunal Verma
Enrico Motta

Programme Committee - Research Track

Karl Aberer
Mark Ackerman
Harith Alani
Boanerges Aleman-Meza
José Julio Alferes
Dean Allemang
José Luis Ambite
Anupriya Ankolekar
Grigoris Antoniou
Kemafor Anyanwu
Marcelo Arenas
Lora Aroyo
Naveen Ashish
Uwe Assmann
Yannis Avrithis
Wolf-Tilo Balke
Leopoldo Bertossi
Olivier Bodenreider
Paolo Bouquet
François Bry
Christoph Bussler
Vinay Chaudhri
Paul - Alexandru Chirita
Philipp Cimiano
Oscar Corcho
Isabel Cruz
Bernardo Cuenca Grau
Carlos Damasio
David De Roure
Mike Dean
Stefan Decker
Paola Di Maio
Tharam Dillon
John Domingue
Marlon Dumas
Martin Dzbor
Jérôme Euzenat
Gerhard Friedrich
Aldo Gangemi
Chiara Ghidini
C. Lee Giles
Fausto Giunchiglia
Carole Goble

Jennifer Golbeck
Marko Grobelnik
Daniel Gruhl
Ramanathan Guha
Asunción Gómez-Pérez
Peter Haase
Siegfried Handschuh
Kevin Hass
Manfred Hauswirth
Jeff Heflin
Pascal Hitzler
Andreas Hotho
Jane Hunter
David Huynh
Eero Hyvönen
Vipul Kashyap
Michael Kifer
Jihie Kim
Hong-Gee Kim
Matthias Klusch
Mieczyslaw Kokar
Yiannis Kompatsiaris
Manolis Koubarakis
Georg Lausen
Kristina Lerman
Bertram Ludaescher
Thomas Lukasiewicz
Carsten Lutz
Alexander Löser
Jan Maluszynski
Tiziana Margaria
Trevor Martin
David Martin
Wolfgang May
Pankaj Mehra
Peter Mika
John Miller
Dunja Mladenic
Ralf Moeller
Premand Mohan
Boris Motik
Daniel Oberle
Jeff Z. Pan

Programme Committee - Semantic Web in Use

Programme Committee - Doctoral Consortium

Aldo Gangemi
Marko Grobelnik
Peter Haase
Siegfried Handschuh
Michael Hausenblas
Conor Hayes
Natasha Noy
Horacio Saggion
Guus Schreiber
Elena Simperl
York Sure
Holger Wache

External Reviewers

Faisal Alkhateeb
Ricardo Amador
Renzo Angles
Manuel Atencia
Grigori Babitski
Claudio Baldassarre
Sruthi Bandhakavi
Andreas Bartho
Sean Bechhofer
Khalid Belhajjame
Dominik Benz
Veli Bicer
Stephan Bloehdorn
Eva Blomqvist
Jürgen Bock
Uldis Bojars
Nicolas Bonvin
Stefano Bortoli
Amancio Bouza
Shawn Bowers
Saartje Brockmans
Adriana Budura
Alexandros Chortaras
David Corsar
James Cunningham
Richard Cyganiak
Maciej Dabrowski
Brian Davis
Jan Dedek
Marco de Gemmis

Geeth de Mel
Renaud Delbru
Stefan Dietze
Jiri Dokulil
Christian Drumm
Alistair Duke
Georges Dupret
Alan Eckhardt
Savitha Emani
Irma Sofia Espinosa Peraldi
Nicola Fanizzi
Bettina Fazzinga
Blaz Fortuna
Ronan Fox
Thomas Franz
Yusuke Fukazawa
George Georgakopoulos
Fausto Giunchiglia
Birte Glimm
Tim Glover
Carole Goble
Karthik Gomadam
Miha Grcar
Gunnar Grimnes
Paul Groth
Tudor Groza
Slawomir Grzonkowski
Alessio Gugliotta
Jun Han
Ramaswamy Hariharan

Ragib Hasan
Kevin Hass
Jakob Henriksson
Raphael Hoffmann
Aidan Hogan
Thomas Hornung
Matthew Horridge
David Huynh
Luigi Iannone
Robert Jäschke
Bo-Yeong Kang
Zoi Kaoudi
Patrick Kapahnke
Alissa Kaplunova
Malte Kiesel
Jihie Kim
Nick Kings
Joachim Kleb
Alexander Kleiner
Pavel Klinov
Mitch Kokar
Beate Krause
Markus Krötzsch
Joey Sik-Chun Lam
Sébastien Laborie
Dave Lambert
Christoph Lange
Jiwen Li
Yuan-Fang Li
Nuno Lopes
Nikos Loutas
Alessandra Martello
Paolo Massa
Christian Meilicke
Anousha Mesbah Shoulami
Martin Michalowski
Matthew Michelson
Zoltan Miklos
Iris Miliaraki
Manuel Moeller
Fergal Monaghan
Mauricio Monsalve
Michael Mrissa
Sergio Munoz-Venegas
Eetu Mäkelä

Knud Möller
Takefumi Naganuma
Rammohan Narendula
Stefan Nesbigall
Spiros Nikolopoulos
Andriy Nikolov
Barry Norton
Vit Novacek
Kieron O'Hara
Kemafor Ogan
Carlos Pedrinaci
Paul Peitz
Rafael Penaloza
Jorge Perez
Jan Polowinski
Livia Predoiu
Guilin Qi
Angus Roberts
Marco Rospocher
Sujith Ravi
Quentin Reul
Christoph Ringelstein
Brahmananda Sapkota
Saket Sathe
Simon Scerri
Simon Schenk
Anne Schlicht
Florian Schmedding
Michael Schmidt
Joo Seco
Nigam Shah
Erin Shaw
Kostyantyn Shchekotykhin
Rob Shearer
Amandeep Sidhu
Heiko Stoermer
Giorgos Stoilos
Mari Carmen Suárez
   de Figueroa Baonza
Martin Szomszor
Stuart Taylor
Arash Termehchy
VinhTuan Thai
David Thau
Yannis Theodoridis

Sponsors

Platinum Sponsors

- Ontoprise

Gold Sponsors

- BBN
- eyeworkers
- Microsoft
- NeOn
- SAP Research
- Vulcan

Silver Sponsors

- ACTIVE
- ADUNA
- Saltlux
- SUPER
- X-Media
- Yahoo

Table of Contents

4. Non-standard Reasoning with Ontologies

5. Semantic Retrieval

6. OWL

11. Semantic Web Services

12. Semantic Social Networks

13. Rules and Relatedness

Semantic Web in Use Track

1. Knowledge Management

2. Business Applications

3. Applications from Home to Space

4. Services and Infrastructure

Doctoral Consortium Track

Involving Domain Experts in Authoring OWL Ontologies[*]

Vania Dimitrova[1], Ronald Denaux[1], Glen Hart[2], Catherine Dolbear[2],
Ian Holt[2], and Anthony G. Cohn[1]

[1] School of Computing, University of Leeds, Woodhouse Lane, Leeds, LS2 9JT, UK
[2] Ordnance Survey Research, Romsey Rd, Southampton, SO16 4GU, UK
{vania,rdenaux,agc}@comp.leeds.ac.uk
{Glen.Hart,Catherine.Dolbear,Ian.Holt}@ordnancesurvey.co.uk

Abstract. The process of authoring ontologies requires the active involvement of domain experts who should lead the process, as well as providing the relevant conceptual knowledge. However, most domain experts lack knowledge modelling skills and find it hard to follow logical notations in OWL. This paper presents ROO, a tool that facilitates domain experts' definition of ontologies in OWL by allowing them to author the ontology in a controlled natural language called Rabbit. ROO *guides* users through the ontology construction process by following a methodology geared towards domain experts' involvement in ontology authoring, and exploiting intelligent user interfaces techniques. An evaluation study has been conducted comparing ROO against another popular ontology authoring tool. Participants were asked to create ontologies based on hydrology and environment modelling scenarios related to real tasks at the mapping agency of Great Britain. The study is discussed, focusing on the usability and usefulness of the tool, and the quality of the resultant ontologies.

Keywords: Ontology Authoring, Controlled Natural Language Interfaces, Evaluation of Ontology Building Tools, Geographical Ontologies.

1 Introduction

The need to construct ontologies – ranging from small domain ontologies to large ontologies linked to legacy datasets– hinders the ability and willingness of organisations to apply Semantic Web (SW) technologies to large-scale data integration and sharing initiatives [1,7,9]. This is due to the time and effort required to create ontologies [1,19]. Most ontology construction tools aggravate the situation because they are designed to be used by specialists with appropriate knowledge engineering and logic skills, but who may lack the necessary domain expertise to create the relevant ontologies. At present, it is knowledge engineers who usually drive the ontology authoring process, which creates an extra layer of bureaucracy in the development cycle [19].

---

[*] The work reported here is part of a research project, called Confluence, funded by the Ordnance Survey and conducted by an interdisciplinary team from the University of Leeds and Ordnance Survey. The main goal of the project is the development of the ontology construction tool ROO, presented in this paper.

A. Sheth et al. (Eds.): ISWC 2008, LNCS 5318, pp. 1 – 16, 2008.
© Springer-Verlag Berlin Heidelberg 2008

Furthermore, this knowledge engineer led approach can hinder the ontology construction process because the domain expert and domain knowledge may become secondary to the process of efficient knowledge modelling. This is especially true where the domain expert has no understanding of the languages and tools used to construct the ontology. The development of approaches that facilitate the engagement of domain experts in the ontology construction process can lead to a step change in the deployment of the Semantic Web in the public and industrial sector.

Such an approach, drawn upon extensive experience in creating topographic ontologies at Ordnance Survey, the mapping agency of Great Britain, is described here. Ordnance Survey is developing a topographic domain ontology to empower the integration and reuse of their heterogeneous topographic data sets with third party data [9]. At the heart of Ordnance Survey's ontology development process is the *active involvement of domain experts* [20]. They construct *conceptual ontologies* that record domain knowledge in a human readable form with appropriate formality using a controlled language, Rabbit[1] [14], that is translated into OWL DL [8].

The paper presents ROO (Rabbit to OWL Ontology authoring), a user-friendly tool that guides the authoring of a conceptual ontology which is then converted to a logical ontology in OWL. The *distinctive characteristics* of our approach are: (a) catering for the needs of domain experts without knowledge engineering skills; (b) exploiting techniques from intelligent user interfaces to assist the ontology construction process by following an ontology authoring methodology (the current implementation follows the methodology used at Ordnance Survey for developing several large ontologies with the active involvement of domain experts [20]); (c) providing an intuitive interface to enter knowledge constructs in Rabbit. We describe an experimental study that examines the degree to which domain experts (i.e. not knowledge engineers) can build ontologies[2] following real scenarios based on work at Ordnance Survey.

An analysis of related work (§2) positions ROO in the relevant SW research. §3 presents the ROO tool and gives illustrative examples of user interaction taken from an experimental study reported in §4. §5 discusses the findings of the study, and outline implications for SW research.

2   Related Work

Recent developments of ontology authoring tools are increasingly recognising the need to cater for users without knowledge engineering skills. Controlled language (CL) interfaces have been provided for entering knowledge constructs in an intuitive way close to Natural Language (NL) interface (see [11,23] for recent reviews). ROO builds on the strengths and minimises the usability limitations of existing CL tools. Positive usability aspects have been followed in the design of ROO, such as: *look*

---

[1] Named after Rabbit in *Winnie the Pooh*, who is actually cleverer than Owl.

[2] Our expectation is not that domain experts will be able to completely author large complex ontologies without assistance (although this might be for small ontologies), but to establish that they can actively participate in the authoring process and construct significant portions of the ontology themselves. This means that domain experts can capture much of the ontology in a form that can be manipulated by knowledge engineers, who can in turn concentrate on the "hard modelling".

ahead to provide suggestions by guessing what constructs the users might enter [24]; showing the parsed structure to help the user recognise correct sentence patterns ([10,21,26]); providing a flexible way to parse English sentences using robust language technologies [8,11,24]; automatically translating to OWL ([17,4,11]); using templates to facilitate the knowledge entering process [22,24]; maintaining a text-based glossary describing parsed concepts and relationships [26]; and distributing the CL tool as a Protégé plug-in [10]. At the same time, we have tried to minimise the negative usability issues exhibited in existing CL tools, such as reliance on the user having knowledge engineering skills to perform ontology authoring (all existing tools suffer from this to an extent) and lack of immediate feedback and meaningful error messages [10,11,26].

Although the goal of CL tools is to assist in entering knowledge constructs, the existing tools focus solely on the CL aspect - they *do not aim to provide assistance for the whole ontology construction process*. In this vein, the HALO project[3] makes an important contribution by offering holistic and intuitive support at all stages of ontology authoring [2]. This key design principle is also followed in ROO. HALO focuses on providing advanced functionality based on the state-of-the-art SW technologies, e.g. sophisticated NL parsing of source documents, graphical interface for entering ontology constructs and rule-based queries. In contrast, ROO offers simpler functionality and follows the Ordnance Survey's practice in ontology construction when taking design decisions. For example, we do not use information extraction techniques to pull out domain concepts from documents, as domain experts normally know what the key concepts are. Our experience shows that the major challenge is to perform abstraction and to a lesser degree reformulation (from NL to CL) and to formulate ontology constructs in a CL, which is the main focus in ROO. It provides *intelligent support* for ontology definition by offering *proactive guidance* based on monitoring domain experts' activities when performing ontology construction steps. Essentially, certain knowledge engineering expertise has been *embedded into* ROO to compensate for the lack of such skills in domain experts. This ensures rigour and effectiveness of the ontology development process, and can lead to better quality ontologies (ontology "quality" is described further in §4.3). Furthermore, ROO aims to improve users' understanding of the knowledge engineering process, and to gradually develop their ontology modelling skills. The study presented in this paper is an initial examination of some of these assumptions.

3   The ROO Tool

The design of ROO takes into account factors that may hinder the involvement of domain experts in the ontology authoring process. As identified through Ordnance Survey's experience in ontology construction, they are: the need to follow a systematic methodology for capturing the knowledge of domain experts; the difficulty in expressing knowledge constructs in a formal language; and the need to cater for the lack of knowledge engineering skills in domain experts.

ROO follows the main steps in Kanga, the Ordnance Survey's methodology for involving domain experts in the authoring of conceptual ontologies [20]. It includes

---

[3] www.projecthalo.com

the following steps: (a) identify the scope, purpose and other requirements of the ontology; (b) gather sources of knowledge (e.g. documents and external ontologies); (c) define lists of concepts, relationship and instances supplied with NL descriptions; (d) formalise *core* concepts and their relations in structured English sentences; (e) generate the OWL ontology. Once step (a) is complete, steps (b)-(d) are performed iteratively by domain experts, while step (e) is performed by ROO automatically. Note that the focus in Kanga is to capture domain experts' knowledge and encode it in OWL, so it can be further examined, validated and improved by knowledge engineers who can use ROO in combination with other ontology engineering tools, for example querying tools [3,18,28,31].

The formalisation, step (d), uses a controlled natural language, called Rabbit, developed in response to a need for domain experts to be able to understand and author ontologies [14]. Rabbit covers every construct in OWL 1.1 [14,8], allowing domain experts to express sufficient detail to describe the domain.

ROO[4] is an open source tool distributed as a Protégé 4 plugin [6]. ROO extends the Protégé 4 user interface by simplifying it as much as possible[5] - hiding advanced options from the user and using what we believe to be less-confusing terminology (e.g. instead of 'classes and properties', ROO shows 'concepts' and 'relations').

In order to explain the services provided by the

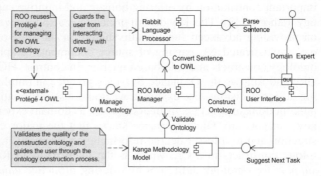

Fig. 1. UML 2.0 component diagram shows the architectural elements, interfaces and inter-element connections in ROO

Rabbit *Language Processor*, the Kanga *Methodology Model* and the ROO *Model Manager* (see Fig. 1), we show two typical user interactions with the system and explain how they are handled by ROO. The examples are taken from the experimental study described in §4.

Domain experts edit the ontology using Rabbit sentences instead of directly editing OWL or the Manchester Syntax. Fig. 2 depicts how a domain expert enters sentences in ROO using the Rabbit *editor*. The user has entered two Rabbit sentences defining the concept river. The first one (Every river transports freshwater) is a valid Rabbit pattern but uses the concept freshwater which is not defined in the ontology. The Rabbit *Language Processor* recognises that freshwater is likely to be a domain concept and composes a corresponding error message. The user has typed the second sentence (Every river flows into one or more of a sea, a lake, or a river) while looking into the

---

[4] ROO is built as part of the Confluence project. http://sourceforge.net/projects/confluence
[5] The default Protégé 4 GUI components are still available for the more advance users, but are not used as a default in the ROO application.

existing **Rabbit** patterns (shown by clicking on the **Rabbit** *patterns* tab). However, the **Rabbit** pattern for *non-exclusive OR* is applied wrongly – instead of commas the user should have used or[6], and the sentence uses a relationship flows into which is not defined in the ontology. Corresponding error messages to help the user are generated, as shown in Fig. 2. The user then corrects the errors by adding the missing concept and relationship and correcting the **Rabbit** pattern. Every time the user makes changes, the input is re-parsed and, if necessary, error messages are generated accordingly. When the input does not contain errors, the user confirms the sentence. It is then translated into OWL by the **Rabbit** *Language Processor*, then validated by the **Kanga** *Methodology Manager*[7], and added to the ontology by the ROO *Model Manager*.

Domain experts can also ask a "guide dog" in ROO to *suggest tasks*, which is a "wizard"-like feature which monitors the state of the ontology and the user's activities, and suggests the most appropriate actions. Fig. 3 shows how the system handles these requests. The user has already entered several concepts to the ontology. The user then asks for a next task. The **Kanga** *Methodology Model* then derives a list of possible tasks and sorts them according to the current ontology state and the user's recent activity. In Fig. 3, the user is prompted to enter **Rabbit** sentences for the concept freshwater which was created with the previous concept definition of river (see Figure 2) but did not yet give **Rabbit** definitions for it. Other task suggestions include reminding the user to enter missing natural language descriptions or pointing at other previously entered concepts which lack **Rabbit** definitions.

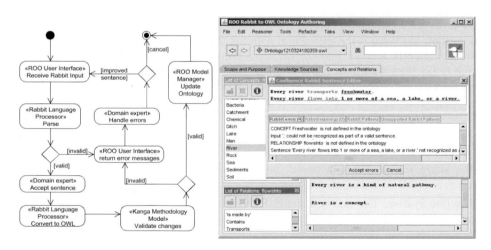

Fig. 2. State chart and screenshot showing how a **Rabbit** sentence is handled by ROO. The parsed syntax elements are highlighted, and possible errors/suggestions are reported to the user.

---

[6] The correct **Rabbit** pattern is: Every river flows into 1 or more of a sea or a lake or a river.

[7] This includes checks whether the input is appropriate to the current stage of the ontology construction; e.g. scope and purpose must be enterered before later stages can commence; **Rabbit** definitions require existence of NL descriptions.

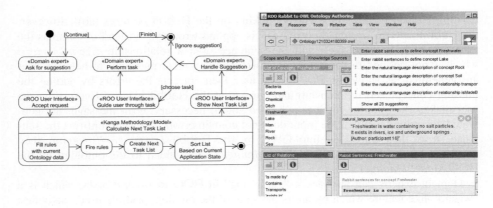

Fig. 3. State chart and screenshot depicting how ROO handles the suggestion of next tasks

The development of ROO has been guided by regular usability tests with potential users – domain experts in different domains. This has led to a fairly robust version that has been evaluated following real scenarios at Ordnance Survey. This evaluation is discussed in the rest of the paper.

4   Experimental Study

To assess the effectiveness of ROO, we conducted an experimental study following the criteria for evaluating ontology tools in [15]. The study addressed three groups of questions: (1) What is the interaction with the tool like? How usable is the tool? Can domain experts without knowledge engineering skills create OWL ontologies with ROO? (2) How well does ROO facilitate the ontology construction process? Do users develop ontology modelling skills as a result of the assistance the tool provides? (3) What is the quality of the resultant ontologies produced with ROO? Is the quality influenced by assistance provided by the tool?

4.1   Experimental Design

The study followed a *task-based*, *between-subjects* experimental methodology to compare ROO with a baseline system.

Baseline System. The study compares ROO with a similar tool that allows the user to author in a CL. From the available CL tools for ontology authoring, ACEView for Protégé [16] was chosen since the user interaction with it is the closest to the user interaction with ROO: both tools extend Protégé as plug-ins, support text input in a CL compatible with OWL-DL, provide error messages for sentence composition, and produce an ontology in OWL[8]. The main difference between ROO and ACEView is that ROO offers assistance with the whole ontology authoring process (§3).

---

[8] The other available CL ontology authoring tools are CLONE [11] and PENG [16]. They were used during a pilot but discarded for the actual study. CLONE is more suitable for users with some knowledge engineering skills, while the users in our study did not have such skills. The interaction with PENG is pattern-based and is notably different from the ROO interface.

Participants. The study involved 16 volunteers from the departments of Geography (8 students) and Earth and Environment (8 students) at the University of Leeds. The participants were chosen to closely resemble domain experts who may perform ontology modelling tasks at Ordnance Survey (Hydrology) or the Environment Agency for England and Wales (Flooding and Water Pollution). The main requirement for attending the study was to have knowledge and experience (confirmed with the modules attended and practical work done) in Hydrology, for Geography students, and Flooding and Water Pollution, for Environmental Studies students. In each domain, 4 participants used ACEView and 4 used ROO; this was assigned on a random basis. None of the participants was familiar with ontologies or ontology construction tools. They had not heard of RDF or OWL. None had previous background in encoding knowledge and for most participants "structuring knowledge" meant writing reports/essays in a structured way.

Scenarios. The study involved two ontology authoring scenarios.

Scenario 1 [Geography participants]: This scenario resembles ontology modelling tasks performed by domain experts at Ordnance Survey to describe geographical features whose spatial representations are included in Ordnance Survey's OS Master-Map®[9]. The participants were asked to describe several hydrology concepts: River, River Stretch, River Bank, Ditch, Catch Drain, Balancing Pond, Canal and Reservoir. These concepts are included in a large Hydrology ontology[10] defined by Ordnance Survey. The Geography participants were familiar with OS MasterMap®, which is used at the School of Geography at Leeds University.

Scenario 2 [Environmental Studies participants]: This scenario resembles ontology modelling tasks performed by domain experts at one of Ordnance Survey's customers –the Environment Agency of England and Wales– who can use OS MasterMap® for flooding and water pollution analysis. The participants were asked to describe: River, Catchment, Flood Plain, Ditch, Water Pollution, Sediments, Colloids, Land Use and Diffuse Pollution. These concepts were selected from a list derived by an Ordnance Survey researcher interviewing an expert from the Environment Agency as part of a project to scope a semantic data integration scenario. Many of these concepts required references to hydrology features from OS MasterMap® but the participants were unaware of this. None of the Environment subjects had knowledge of OS MasterMap®. Ontologies for geography and environment were also produced by Ordnance Survey and were used as comparators with the ontologies produced by the participants.

Procedure and Materials[11]. Depending on their background, the participants were sent the corresponding list of concepts, and were asked to prepare brief textual descriptions for these concepts by using specialised dictionaries or other sources. Each session was conducted individually and lasted 2 hours. It included several steps.

Pre-study questionnaire [20 min] included a brief introduction to the study and several questions to test the participants' ontology modelling background.

---

[9] OS MasterMap® www.ordnancesurvey.co.uk/osmastermap/ is a nationally contiguous vector map containing more than 450 million individual features down to street, address and individual building level, spatial data to approximately 10cm accuracy.

[10] www.ordnancesurvey.co.uk/ontology

[11] All materials are available from www.comp.leeds.ac.uk/confluence/study.html

Introduction to the scenario and training with the ontology authoring tool [10 min] was given to each participant by an experimenter, describing the main parts of the interface and entering of several definitions from a Building and Places[12] ontology. The examples used for the ACEView and ROO sessions were similar (the differences came from the CL and the errors given by each tool). The training with ROO also required entering the ontology's scope and purpose and knowledge sources.

Interaction with the tool [60 min] The participants had to use the tool allocated to them to describe the concepts following the descriptions they had prepared. Each session was monitored by an experimenter who provided some general help when the participants got stuck with the language. Help materials with printed examples of the corresponding CL were provided. The interactions were logged and video recorded. The experimenters kept notes of the user interaction.

Post-study questionnaire [20 min] included checking the participants' ontology modelling background (repeating questions from the pre-study questionnaire); a usability questionnaire using a seven-point Likert scale; and open questions about benefits, drawbacks, and future improvement of the tool used.

General impression and clarification [10 min] included a brief interview with each participant about their general impression of the CL used, interaction with the tool, and any additional aspects the participants wished to mention.

Data Collected. The following data was collected during the study: (a) *Questionnaires* – used for examining the usability of each tool and examining possible changes in the participants' understanding of ontology modelling; (b) *Log data, video records of the sessions, and experimenter's notes* – used for clarifying aspects of the interaction with each tool; (c) *Resultant OWL ontologies* – the quality of these ontologies was analysed following the O2 framework [12]. The data was analysed quantitatively and qualitatively. The quantitative analysis used Mann-Whitney U test[13] for discrete measurements and t-test for interval data.

4.2 Comparing the Interaction with ROO and ACEView

Interaction Patterns. Both tools have fairly simple interfaces and were easy to use. The first quarter of the interaction was usually slower as the participants had to learn to formulate sentences in the corresponding CL. During this time, the definition of the first concept `river` (common for both scenarios) was completed. Both tools offer a tab to show the *CL errors*, this was used extensively. Initially, most users did not realise that the error messages refer to incorrect CL grammar that the computer could not parse or translate into a logical form, rather than incorrect domain facts. From the second quarter, the users established a routine to describe a concept, including:

1. *Check the NL description for the currently entered concept and identify a statement with knowledge to be encoded.* The ACEView users had a printout of the descriptions they had prepared, while the ROO users followed the NL descriptions the tool prompted them to enter.

---

[12] www.ordnancesurvey.co.uk/ontology.

[13] Mann-Whitney U test is a powerful nonparametric test used as an alternative to the parametric t-test to compare two independent samples [27]. It is often used when the measurement is weaker than interval scaling or the parametric assumptions are not met.

2. *Look for a CL pattern that matches the NL statement.* The ACEView users used only the printed list of CL examples provided, ROO users could, in addition, see the available patterns within the tool, and they gradually moved to using this;

3. *(Re)Formulate the NL statement in a CL pattern.* This usually involved simplifying the constructs or taking away unnecessary detail, e.g. simple patterns were easily created, more complex patterns were normally not written correctly in the first instance and required several iterations and checking the system feedback.

4. *Check for error messages* – if there are no error messages, continue with another NL statement (i.e. go to step 1). When there are error messages, the users would usually repeat steps 2-4. Some participants would be persist, reformulating the CL statement until there were no errors (and it was translated to OWL), while others would continue and leave the CL statement with errors (i.e. not encoded in OWL).

For both tools, the users were occupied mostly with steps 3 and 4 and would often refer to step 2 for a quick check. Two of the eight ACEView users entered sentences to describe all concepts from the given list (see scenarios), while none of the ROO users managed to complete the descriptions; in most cases the last two concepts were not defined. Table 1 summarises the main interaction problems.

Usability. Table 2 summarises the findings from the usability questionnaire. For both tools, the users were positive. ROO was found to be significantly less frustrating than ACEView, which may be due to the much more intuitive interface, much less confusing error messages, and the help offered from the "guide dog". The messages in ROO were more helpful, the tool was less complex than ACEView, and users would be more willing to use ROO again (note the very low significance).

Ontology Modelling Skills. The answers to six ontology modelling questions (covering the main steps and building blocks in conceptual models, definition of ontology, concepts, and relations) in the pre- and post-study questionnaires were compared by

Table 1. Summary of the main interaction problems identified in the study

Problem	Tool	Explanation
Error messages lack detail.	ACEView ROO	When the CL pattern entered was not recognised, the users would not always get informative error messages. In such cases, the users had to guess what may be misleading, e.g. ACEView: `The sentence is not correct ACE syntax.` ROO: `Sentence is not recognised as correct Rabbit sentence.`
Error messages confusing.	ACEView ROO	When the user entered sentences which could not be recognised, they sometimes received error messages that were misleading. ACEView messages included ??? to indicate unrecognised parts in the sentence or referred to grammatical constructs which some users found hard to follow. ROO gave at times misleading suggestions when the sentence was unrecognised.
Dealing with adjectives and compound noun phrases	ACEView ROO	Recognising a concept which includes a compound noun phrase (e.g. adjective-noun) can be a challenging problem. ACEView users often received the message "`adjectives are not supported`", in which case they had to use hyphenation (see above problem). ROO parses for compound noun phrases and in most cases could make helpful suggestions about what the concept might be, e.g. `natural waterway`, `man-made feature`. However, when the compound nouns were not recognised and this led to confusing error messages, e.g. `natural body of water` was not recognised as a possible concept.
Dealing with a specialised vocabulary	ACEView ROO	The parsers in both tools could not recognise some specialised vocabulary which did not allow entering certain concepts, such as: ACEView: `sediment`, `irritation`; ROO: `watershed`. ACEView deals with this by pre-entering classes. However, it would be hard to predict in advance what phrases a user may enter. A more flexible way would be to allow the user to enter a phrase which should be added to the vocabulary used by the NL parser.
Next task suggestion not always useful	ROO	On several occasions, users ignored the task suggestions and commented that not all of them were useful. E.g. ROO suggested that the participant enter definitions of secondary concepts, such as `man` or `bacteria` The Kanga methodology discerns between core concepts and secondary concepts. Only core concepts need to be formalised. However, the current ROO tool does not discriminate between core and secondary concepts yet.

Table 2. Summary of the comparison of the usability of both tools (post-study questionnaire)

Question (1-Strongly disagree; 4-Neutral; 7-Strongly agree)	ROO median	ACEView median	U (Mann-Whitney, 1-tail)	p	Significance
The error messages helped me write CL sentences	5	4.5	16.5	p≤0.1	LOW:
The error messages were confusing	2	4.5	11.5	p≤0.025	YES
The guide dog was helpful	5	—	—	—	—
The guide dog suggestions were not easy to understand	2	—	—	—	—
I did not follow the suggestions from guide dog	4	—	—	—	—
The interaction was demanding	3	4	39	p>0.1	NO
I had no idea what I was doing	2	1.5	16	p>0.1	NO
It took me too long to compose what I wanted	4	3	21	p>0.1	NO
The interaction was intuitive	5	3.5	11.5	p≤0.025	YES
The feedback was prompt and timely	5	4.5	24	p>0.1	NO
It was clear to me what to do in this tool	5	4.5	24	p>0.1	NO
The tool was frustrating	3	5	5.5	p≤0.01	YES (HIGH)
The tool was unnecessary complex	2.5	3.5	18	p≤0.1	LOW
I'd like to use the tool again	5	4	18.5	p≤0.1	LOW

to examine whether the users' ontology modelling skills had changed as a result of the interaction with the tool. Two evaluators with a sound ontology background worked independently and marked the users' answers. The following scheme was applied to each question: -1 (the understanding has worsened, e.g. because the user was confused); 0 (no change to the user's understanding on the questions), +1 (correct aspects are added but gaps exist), +2 (the understanding is improved, and now is correct and complete). The marker compared their results and the discrepancies were clarified in a discussion. The maximum score, if a user had not had any ontology modelling knowledge and has become an expert, would have been 12, while the worst score meaning a user was an expert and became totally confused would have been -6.

The **ROO** users scored significantly higher than the ACE users - ACEView score mean 0.38, STDEV 2.97; ROO score mean 5, STDEV 2.78; U (Mann-Whitney)=8.5, p≤0.01. This shows that the users' understanding in ontology modelling improves significantly more when using **ROO** than when using ACEView.

4.3 Quality of the Resultant Ontologies

The resultant ontologies were analysed following the ontology evaluation framework in [12] considering structural, functional, and usability ontology measures.

Ontology Structural Measures. Since the size of the ontologies is limited, we have used fairly simple structural metrics based on [29], calculated by Protégé 4[14].

There are no significant differences in the structural characteristics of the ontologies created, with exception to annotations per entity, as shown in Table 3.

Table 3. Summary of ontology structural measures

	Average Class Count	Average Object Property Count	Average Properties Relative to number of Classes	Average Annotations per Entity	Average Subclass Axiom per Class (Inheritance Richness)
ROO	21.875	8.250	0.367	2.625	0.634
ACE	28.125	11.875	0.420	0.582	0.877
p (t-test)			0.263	0.000	0.095
U (Mann-Whitney)	19.5	21.5			
p (Mann-Whitney)	0.104	0.147			

---

[14] We also attempted deeper graph-based structural metrics with the Protégé 3 plugin OntoCAT [5] but it could not properly analyse the produced ontologies due to version compatibility.

The results show that ontologies built with ROO have a *significantly better readability* than ontologies built with ACEView. Both systems store the entered sentences as annotations in the ontology. Since both Rabbit and ACE are quite readable for humans, these annotations can be used to understand the meaning of the OWL entities. The main reason why ROO ontologies are more readable is that ROO encourages users to provide additionally natural language descriptions for both concepts and relationships. When Rabbit sentences are translated and new classes and properties are added to the ontology, an appropriate rdf:comment is added containing the Rabbit sentence, with an rdf:label containing the Rabbit concept name. In contrast, ACEView does not add annotations when classes or properties are added.

We measured inheritance richness based on OntoQA[29]. ACEView ontologies had higher inheritance richness (Table 3), i.e. the classes built with ACEView had more connections to other classes. However, the functional measures (see Table 4 below) indicate that ACEView ontologies were more tangled than ROO ontologies. Domain experts seemed slightly more productive using ACEView than using ROO but the Mann-Whitney U-test does not provide conclusive significance.

Ontology Functional Measures. A domain expert who is also a knowledge engineer[15] at Ordnance Survey produced two benchmark ontologies to quantify the fitness-for-purpose of the participants' ontologies. A scoring system was devised:

+1 point for each axiom produced by the participant ontology that exactly matched[16] an axiom from the benchmark ontology;

+1 point for each additional valid axiom, i.e. axioms that were considered to be valid even though an equivalent did not exist in the benchmark;

-1 point deducted for each axiom in the benchmark but absent the user's ontology;

-1 point deducted for any axiom containing a modelling error.

The participants did not define axioms for all the concepts they were given. Where this was the case, we did not count any metrics for that concept for that participant. We only scored against axioms belonging to the concepts in the concept list given to the participants. The total score for each ontology was therefore the sum of the points added or deducted.

Table 4. Summary of the scores from the functional analysis of the resultant ontologies

Scenario	ROO (mean)	ACEView (mean)	U (p)
Geography	1.25	-3.5	3.5 (p>0.1)
Environment	3.75	-5	0 (p≤0.025)
Combined	2.5	-4.25	9 (p≤0.1)

Subjectively, the ACEView ontologies appeared to be more complete, whereas the ROO ontologies appeared to be better structured and with fewer modelling errors.

The data for each set of ontologies was analysed statistically using the Mann Whitney U test (Table 4). At a 95% confidence level this indicates that there is no significant difference between the sets of data collected for the geography ontologies but that ROO out-performs ACEView with respect to the environmental ontologies and overall (geography and environment combined). The weakest participant by far was a

---

[15] We were lucky that such an expert existed, making it possible to examine in depth the functional dimensions of the ontology.

[16] Some interpretation was required owing to variances in terminology.

ROO geographer who despite only recording axioms for three concepts achieved a negative overall score, but this alone would not have accounted for the overall differences even given the small sample sizes.

ACEView users tended to describe more concepts and add more axioms (Table 4). This applied to both the "in scope" concepts and also those out of scope. Some of the latter group were secondary concepts necessary to define the core concepts – for example `water body` used to super class `river` and `reservoir`. But others were irrelevant clutter, such as Scotland, and it was not clear why they were added.

ACEView users did better than ROO in getting exact axiom matches with the benchmark ontologies (with a mean that was 1.5 matches higher per person). They also had a higher mean for providing additional axioms, with an average of three more per person. However, ACEView users did very much worse when it came to the number of errors they made, that is the number of axioms that were deemed to be incorrect, averaging 8 errors per person more that ROO users. Even taking into account that ACEView users enter more axioms proportionately they enter 0.4 errors per axiom, compared to 0.13 errors for ROO users. Erroneous axioms were not included in the other axiom counts. If included, it would show that ACEView users are even more prolific – it seems to be a case of quantity over quality. Table 5 summarises the modelling problems that occurred.

Ontology Usability. None of the ontologies as produced would have been usable without modification. This is unsurprising given the fact that the users were essentially untrained in the language and knowledge modelling techniques. No user produced an ontology that provided a complete description of the concepts, but again this is unsurprising given the experience levels and time available. In simple terms the ROO ontologies were less complete, containing fewer concepts and fewer

Table 5. Types of modelling problems found with the functional analysis of ontologies

Problem	Tool	Explanation
Multiple tangled inheritance	ACEView ROO (much less frequently)	This was a very common error in ACE ontologies. In the worst case `Drainage` had five separate immediate simple super classes: `Artificial Object`, `Depression`, `Drainage`, `Long Trench` and `Narrow Trench`. An error was scored for each extra entanglement so in the case above a score of 4 would have been recorded. The axioms would have been included in the overall total of axioms. Although also occurring in ROO ontologies, the rate and degree of multiple inheritance was much lower.
Definition of an instance instead of a class	ACEView ROO	There were a number of occasions where a class was recorded as an instance. ACEView example: in one ontology `Flood-Plain` is declared to be an individual of class `sediment-deposition`. In examining the ACE log file the first mention of `flood-plain` is the sentence: `Flood-plain borders a river`. There is no use of every in the sentence so ACE assumes `Flood-Plain` is an individual, and so records the assertion `Flood-plain` is an individual of the anonymous class "`borders some River`". The next correct sentence: `Flood-plain is a sediment-deposition` adds `Flood-plain` as an individual of the class `sediment-deposition`. ROO example: user entered `Flood Plain is a Land Area` rather than `Flood Plain is a kind of` Land Area.
Generation of 'random' individuals	ACEView	ACEView also appears to generate "random" individuals. For example the sentence: `Scotland contains a farm and contains a forest and contains a reservoir.` Generates three individuals. It is probable that what the user meant was that `Scotland` (also an individual) contains *some* farms, forests and reservoirs. What is even less clear is why the user felt it necessary to add this out of scope information at all.
Repeated Knowledge	ACEView ROO (much less frequently)	In a number of cases ACEView users tended to enter axioms that were similar to axioms already entered. An example is: `Every flood-plain experiences flooding` and `Every flood-plain experiences periodic-flooding`. Such repetitiveness also occurred in the ROO ontologies, but much less frequently.

axioms. However, the greater number of modelling errors in the ACEView ontologies, combined with the amount of unnecessary clutter in terms of out-of-scope concepts and axioms would indicate that it would take longer to get them to a usable state. ROO ontologies were certainly better annotated and this helped significantly in terms of evaluating the usability of ontologies for a certain purpose.

5 Discussion and Conclusions

To the best of our knowledge, the study presented here is the first attempt to evaluate how domain experts without knowledge engineering skills can use CL-based tools to complete ontology modelling tasks close to real scenarios (existing studies have either used people with knowledge engineering skills and simple tasks [11] or looked into recognising CL constructs [14]). The results enable us to address key questions concerning the authoring of ontologies where a domain expert takes a central role: Can we use CL to involve domain experts in ontology construction? To what degree can a tool support help the authoring process and substitute for a knowledge engineer? What further support is needed?

Involvement of Domain Experts. Accepting that the users who participated in our study had minimal training in the languages and the tools, it is fair to conclude from the resultant ontologies that domain experts alone, even with tool assistance, would be unable to author anything more than simple ontologies without some formal training. Nevertheless, almost a quarter of the participants entered axioms that matched roughly 50% of the axioms in the benchmark ontologies. This would indicate that with even a minimal amount of training these domain experts could become quite competent as authors. It is always likely that for complex ontologies knowledge engineering skills will be required. However, if the domain expert is able to author most of the ontology, they will be more easily able to engage with the knowledge engineer who can then express the more difficult aspects. Furthermore, the study indicated that if methodical, intelligent support for ontology authoring is *embedded* in the authoring tool, domain experts can gain an understanding of the ontology modelling process, that can gradually lead to the development of knowledge engineering skills.

The study confirmed that domain experts are able to start authoring relatively quickly and without the need to learn obscure terminology and esoteric languages such as OWL. In fact, it is unlikely that the study would have been possible if OWL had been used rather than Rabbit (or ACE) given the need to provide training in OWL. That no real training was provided to participants is, at the very least, indicative of the benefits to domain experts in using intuitive CL interfaces. We are confident that a central involvement by domain experts in the authoring process is not possible if the only way of expressing the ontology is in a logic-based language expressed using esoteric terms and symbols, without a lengthy process of turning the domain expert into a fully fledged knowledge engineer, something that few domain experts have the time or inclination to do.

Existing Tool Support. The various processes involved in authoring an ontology include: (a) identification of concepts and relationships (classes and properties); (b) development of an overall structure for the ontology; (c) capturing of axioms for each concept; (d) development of patterns to express certain model constructs; (e) optimisation and rationalisation; (f) testing and validation; (g) documentation. This list is not exhaustive, nor does it attempt to imply a priority of one process over another. ROO and ACEView currently provide degrees of support for (a), (c), (d) and (g). The study gives strong evidence that offering intuitive error messages, making users aware of the knowledge constructs they are creating, and offering methodical guidance can have a positive effect on the usability and efficacy of ontology construction tools. It also indicates that this additional functionality tailored to domain experts (as in ROO) can have impact on the quality of the resultant ontologies - domain experts make fewer errors, detect unwanted concepts and relationships, avoid repetition, and document the ontology more consistently and in more detail.

Required Tool Support. The interaction with both tools suggests that additional support should be provided. This may have implications for ontology authoring in general, including the newly emerging collaborative ontology editing environments [21] where support is even more critical. Patterns of modelling errors can be recognised and pointed out with the error message or the task guidance (the guide dog in ROO). For instance, definition of an instance instead of a class can be detected based on the CL pattern (e.g. is a vs is a kind of in Rabbit), as the error can turn the OWL ontology from OWL-DL to OWL Full; likely repetition or redundancy can be recognised by using synonyms (e.g. is part of, consists of, contains, comprises) and indicated in a 'warning' message; both multiple tangled inheritance and isolated classes can be detected with structural analysis and warnings generated or advice given. The study also indicated that flexible CL parsing should be provided, such as recognising similarity between NL and CL sentences (e.g. no need to ask the user to specify a determiner, as in ACEView, as this is not normally needed in a correct NL sentence; missing 'Every' can be spotted easily and pointed out in a meaningful error message); recognising compound noun phrases and the underlying structure (e.g. the parsers can recognise that natural body of water may require two concepts linked with subsumption, so the user may be asked whether natural body of water is a kind of body of water); or enabling the users to add missing specialised terminology (e.g. sediment) that can then be considered by the parser in future sentences.

Although there is evidence that the guidance offered in ROO is beneficial, it has to be improved further. For instance, the suggestions should take into account the current task better to avoid distracting and confusing the user (e.g. a task context could be retained in ROO and only activities/concepts relevant to that context would be suggested). The ontology status should be better monitored more closely and potential limitations pointed out (e.g. some of the structural metrics can indicate unpopulated parts of the ontology). Lastly, more proactive help should be offered (instead of waiting for the user to click on the guide dog, certain suggestions could be brought to the user's attention automatically). The study confirmed that systematic support based on an ontology methodology is beneficial. The current implementation of ROO can be considered as a proof of concept that a methodology can be embedded in the planning process. An interesting research question would be to define ontology construction methodologies explicitly, e.g. by using an ontology and rules. For instance, ROO could be easily adapted to work with methodologies which Kanga is similar to, e.g. Uschold and King's method [30] or METHONTOLOGY [13]. It would then be possible to choose the most appropriate methodology for the current ontology authoring task, or to compare the effect of different methodologies.

At the time of writing **ROO** implements only the core **Rabbit** constructs. We intend to complete all **Rabbit** constructs and implement some of the additional support outlined above. This will give us a much more robust and usable tool, that can then be the basis for a larger user study in real settings, facilitating further examination of the extent to which domain experts can be involved in ontology authoring.

Acknowledgements. The authors would like to thank Kaarel Kaljurand, for kindly providing us with the ACEView tool. Thanks go to Paula Engelbrecht for sharing her material on the Environment Agency and Ilaria Corda for helping with the initial design of ROO. Special thanks go to the participants in the experimental study.

References

1. Alani, H., Dupplaw, D., Sheridan, J., O'Hara, K., Darlington, J., Shadbolt, N., Tullo, C.: Unlocking the Potential of Public Sector Information with Semantic Web Technology. In: Aberer, K., Choi, K.-S., Noy, N., Allemang, D., Lee, K.-I., Nixon, L., Golbeck, J., Mika, P., Maynard, D., Mizoguchi, R., Schreiber, G., Cudré-Mauroux, P. (eds.) ASWC 2007 and ISWC 2007. LNCS, vol. 4825. Springer, Heidelberg (2007)
2. Angele, J., Moench, E., Oppermann, H., Staab, S., Wenke, D.: Ontology-Based Query and Answering in Chemistry: OntoNova @ Project Halo. In: Fensel, D., Sycara, K.P., Mylopoulos, J. (eds.) ISWC 2003. LNCS, vol. 2870. Springer, Heidelberg (2003)
3. Cimiano, P.: ORAKEL: A natural language interface to an F-logic knowledge base. In: Proceedings of Natural Language Processing and Information Systems, pp. 401–406 (2004)
4. Cregan, A., Meyer, T.: Sydney OWL Syntax - towards a Controlled Natural Language Syntax for OWL 1.1. In: Proceedings of OWLED 2007 (2007)
5. Cross, V., Pal, V.: An ontology analysis tool. International Journal of General Systems 37(1), 17–44 (2008)
6. Denaux, R., Holt, I., Dimitrova, V., Dolbear, C., Cohn, A.G.: Supporting the construction of conceptual ontologies with the ROO tool. In: OWLED 2008 (2008)
7. Dimitrov, D.A., Heflin, J., Qasem, A., Wang, N.: Information Integration Via an End-to-End Distributed Semantic Web System. In: Cruz, I., Decker, S., Allemang, D., Preist, C., Schwabe, D., Mika, P., Uschold, M., Aroyo, L.M. (eds.) ISWC 2006. LNCS, vol. 4273, pp. 764–777. Springer, Heidelberg (2006)
8. Dolbear, C., Hart, G., Goodwin, J., Zhou, S., Kovacs, K.: The Rabbit language: description, syntax and conversion to OWL. Ordnance Survey Research Labs Techn. Rep. IRI-0004 (2007)
9. Dolbear, C., Hart, G.: Combining spatial and semantic queries into spatial databases. In: Cruz, I., Decker, S., Allemang, D., Preist, C., Schwabe, D., Mika, P., Uschold, M., Aroyo, L.M. (eds.) ISWC 2006. LNCS, vol. 4273. Springer, Heidelberg (2006)
10. Fuchs, N.E., Kaljurand, K., Schneider, G.: Attempto Controlled English meets the challenges of knowledge representation, reasoning, interoperability and user interfaces. In: Proceedings of FLAIRS 2006 (2006)
11. Funk, A., Tablan, V., Bontcheva, K., Cunningham, H., Davis, B., Handschuh, S.: CLOnE: Controlled Language for Ontology Editing. In: Aberer, K., Choi, K.-S., Noy, N., Allemang, D., Lee, K.-I., Nixon, L., Golbeck, J., Mika, P., Maynard, D., Mizoguchi, R., Schreiber, G., Cudré-Mauroux, P. (eds.) ASWC 2007 and ISWC 2007. LNCS, vol. 4825. Springer, Heidelberg (2007)

12. Gangemi, Catenacci, C., Ciaramita, M., Lehmann, J.: Modelling ontology evaluation and validation. In: Sure, Y., Domingue, J. (eds.) ESWC 2006. LNCS, vol. 4011. Springer, Heidelberg (2006)
13. Gomez-Perez, A., Fernandez-Lopez, M., Juristo, N.: Methontology: from ontological art toward ontological engineering. In: Proceedings of AAAI 1997 Spring Symposium Series on Ontological Engineering, pp. 33–40 (1997)
14. Hart, G., Johnson, M., Dolbear, C.: Rabbit: Developing a Control Natural Language for Authoring Ontologies. In: Bechhofer, S., Hauswirth, M., Hoffmann, J., Koubarakis, M. (eds.) ESWC 2008. LNCS, vol. 5021, Springer, Heidelberg (2008)
15. Hartman, J., Spyns, P., Giboin, A., Maynard, D., Cuel, R., Suárez-Figueroa, M., Sure, Y.: Methods for ontology evaluation, Knowledge Web Deliverable, D1.2.3 (2004)
16. Kaljuran, K.: Attempto Controlled English as a Semantic Web Language. PhD thesis, Faculty of Mathematics and Computer Science, University of Tartu (2007)
17. Kaljurand, K., Fuchs, N.E.: Bidirectional mapping between OWL DL and Attempto Controlled English. In: Workshop on Principles and Practice of Semantic Web Reasoning (2006)
18. Kaufmann, E., Bernstein, A., Fischer, L.: NLP-Reduce: A "naive" but domain-independent natural language interface for querying ontologies. In: Franconi, E., Kifer, M., May, W. (eds.) ESWC 2007. LNCS, vol. 4519. Springer, Heidelberg (2007)
19. Klischewski, R.: Ontologies for e-document management in public administration. Business Process Management Journal 12, 34–47 (2006)
20. Kovacs, K., Dolbear, C., Hart, G., Goodwin, J., Mizen, H.: A Methodology for Building Conceptual Domain Ontologies. Ordnance Survey Research Labs Techn. Report IRI-0002 (2006)
21. Kuhn, T.: AceWiki: A Natural and Expressive Semantic Wiki. In: Proceedings of Workshop on Semantic Web User Interaction, held at CHI 2008 (2008)
22. Lopez, V., Sabou, M., Motta, E.: PowerMap: Mapping the Semantic Web on the Fly. In: Sure, Y., Domingue, J. (eds.) ESWC 2006. LNCS, vol. 4011. Springer, Heidelberg (2006)
23. Schwitter, R., Kaljurand, K., Cregan, A., Dolbear, C., Hart, G.: A comparison of three controlled natural languages for OWL 1.1. In: Proc. of OWLED 2008 workshop (2008)
24. Schwitter, R.: A. Ljungberg and D Hood, ECOLE - A Look-ahead Editor for a Controlled Language. In: Proc. of EAMT-CLAW 2003, pp. 141–150 (2003)
25. Schwitter, R.: English as a formal specification language. In: Hameurlain, A., Cicchetti, R., Traunmüller, R. (eds.) DEXA 2002. LNCS, vol. 2453. Springer, Heidelberg (2002)
26. Schwitter, R.: Representing Knowledge in Controlled Natural Language: A Case Study. In: Negoita, M.G.R., Howlett, J., Jain, L.C. (eds.) KES 2004. LNCS (LNAI), vol. 3213, pp. 711–717. Springer, Heidelberg (2004)
27. Siegel, S., Castellan, J.N.: Nonparametric Statistics for the Behavioral Scences, 2nd edn. McGraw-Hill, New York (1988)
28. Tablan, V., Damljanovic, D., Bontcheva, K.: A natural language query interface to structured information. In: Bechhofer, S., Hauswirth, M., Hoffmann, J., Koubarakis, M. (eds.) ESWC 2008. LNCS, vol. 5021. Springer, Heidelberg (2008)
29. Tartir, S., Arpinar, I.B., Moore, M., Sheth, A.P., Aleman-Meza, B.: Ontoqa:Metric-based ontology quality analysis. In: Proc. of W. on Knowledge Acquisition from Distributed, Autonomous, Semantically Heterogeneous Data and Knowledge Sources (2006)
30. Uschold, U., King, M.: Towards and methodology for building ontologies. In: Workshop on Basic Ontological Issues in Knowledge Sharing, held at IJCAI 1995 (1995)
31. Wang, C., Xiong, M., Zhou, Q., Yu, Y.: PANTO: A Portable Natural Language Interface to Ontologies. In: Franconi, E., Kifer, M., May, W. (eds.) ESWC 2007. LNCS, vol. 4519. Springer, Heidelberg (2007)

Supporting Collaborative Ontology Development in Protégé

Tania Tudorache, Natalya F. Noy, Samson Tu, and Mark A. Musen

Stanford University, Stanford, CA 94305, US
{tudorache,noy,tu,musen}@stanford.edu

Abstract. Ontologies are becoming so large in their coverage that no single person or a small group of people can develop them effectively and ontology development becomes a community-based enterprise. In this paper, we discuss requirements for supporting collaborative ontology development and present Collaborative Protégé—a tool that supports many of these requirements, such as discussions integrated with ontology-editing process, chats, and annotations of changes and ontology components. We have evaluated Collaborative Protégé in the context of ontology development in an ongoing large-scale biomedical project that actively uses ontologies at the VA Palo Alto Healthcare System. Users have found the new tool effective as an environment for carrying out discussions and for recording references for the information sources and design rationale.

1   Ontology Development Becomes Collaborative

Recent developments are dramatically changing the way that scientists are building ontologies. First, as ontologies are becoming commonplace within many scientific domains, such as biomedicine, they are being developed collaboratively by increasingly large groups of scientists. Second, ontologies are becoming so large in their coverage (e.g., NCI Thesaurus with 80K concepts) that no one user or small group of people can develop them effectively. Hence, organizations such as the NCI Center for Bioinformatics "outsource" some of their ontology development to the scientific community at large. Third, in the last one or two years, many users have become quite familiar and comfortable with the concept of user-contributed content, both in their personal and professional lives (cf. Web 2.0). Thus, domain experts need tools that would support collaborative ontology development and would include collaboration as an integral part of the ontology development itself.

Researchers are only now beginning to develop such tools. Last year, tool developers were invited to contribute their tools for collaborative construction of structured knowledge (which included not only ontologies, but also any structured data) to the CKC Challenge, which brought together developers and users in order to examine the state-of-the-art and to understand the requirements for new tools [10]. In general, the participants in the CKC Challenge agreed on several key points. First, the notion of collaborative development of ontologies and most of the tool support was in its infancy. Second, the spectrum of tools even in the relatively small set of the challenge participants (from tools to organize tags in a hierarchy to full-fledged ontology editors) demonstrated that no single tool is likely to fill the niche completely. Third, the

A. Sheth et al. (Eds.): ISWC 2008, LNCS 5318, pp. 17–32, 2008.

requirements for such tools to support collaborative development in any specific setting were still poorly understood. The challenge participants started identifying these requirements. Starting with the initial set of requirements identified as the result of the CKC workshop, we continued the requirements-gathering phase in the context of extending the Protégé ontology editor to support collaborative ontology development. To gather specific requirements, we conducted interviews with representatives of several groups that currently use Protégé for ontology development and that were trying to adopt a more formal process for development. These projects included the development of the NCI Thesaurus [17], the ontologies for the ATHENA-DSS project at the VA Palo Alto Healthcare System [7], the Ontology of Biomedical Investigations (OBI) [2], the RadLex ontology for annotating radiological images [14], and many others. As the result of this process, we collected a set of requirements for an ontology editor supporting collaboration. Note that we focused on the projects that need a full-fledged ontology editor and where ontologies are fairly rich in structure and large in size. For example, the NCI Thesaurus is an OWL DL ontology with more than 80K classes, several thousand of which are defined classes. Both RadLex and ATHENA-DSS ontologies are frame-based ontologies that use different types of constraints on properties extensively. We then developed Collaborative Protégé by extending the Protégé tool with a set of features to support these requirements. We have performed the formative evaluation of Collaborative Protégé in several different projects in order to evaluate the usability of the tool and to understand what users like and do not like about it, how they use it, and what other features they need to support their work.

More specifically, this paper makes the following contributions:

- We identify a set of requirements for developing expressive ontologies and knowledge bases collaboratively (Section 2).
- We present Collaborative Protégé—an ontology editor that supports collaboration through integration of features such as discussions, chats, and annotations in the ontology editor (Sections 4, 5, and 6).
- We perform the formative evaluation of Collaborative Protégé in the context of representing formally clinical practice guidelines in the ATHENA-DSS project (Sections 7, 8, and 9).

2   Requirements for Support of Collaborative Ontology Development

We have identified our requirements for tool support for collaborative ontology development through interviews with many institutional Protégé users. The requirements that we identified significantly extend the set of requirements from the CKC workshop, and focus on the requirements of ontology developers for domains such as biomedicine. These developers are usually domain experts rather than knowledge engineers.

In most of these projects, users have already used Protégé in a client–server mode that enabled distributed users to edit the ontology simultaneously, immediately seeing the changes that others make. Thus, we focused on the features that would explicitly support collaboration. Furthermore, by the nature of projects already having chosen

Protégé for their ontology development, most of them had to work with expressive ontologies and knowledge bases. In some cases, users worked collaboratively to extend the ontologies themselves (e.g., the NCI Thesaurus or OBI), and in others they additionally used an expressive ontology to create a knowledge base of classes and instances (e.g., ATHENA-DSS). The overarching theme of these interviews was the disconnect between the produced ontology on the one hand and all the thought and discussion that went into producing this artifact on the other hand. The former was captured in Protégé, but the latter was captured in myriads of email messages, forum posts, phone conversations, and MS Access databases. When someone browsed the ontology, it was often impossible to understand the rationale that went into the design decisions, to find which references were relevant, to find the external resources that informed the modeling decisions. Conversely, when developers read a mailing list post discussing a modeling issue, they do not see the context for that post.

The specific requirements for supporting collaborative ontology development that our users identified included the following:

Integration of discussions and annotations in ontology development. Almost by definition, an ontology is an artifact that requires its authors to reach consensus. At the same time, our experience demonstrates that developing an ontology is not a straightforward task and the developers can disagree on the best way to model concepts in the ontology or, in fact, on which concepts to model. Thus, tools that support discussion, such as forums and chats, are essential. However, these discussions happen in email messages and similar venues and are completely separated from the resulting artifact. For example, one of our collaborators (the OBI developers) reported recently that they found themselves in a heated discussion on the definition of a specific term (namely, analyte), something they thought they have resolved several months before. However, that discussion was not captured in the class definition and was not available when the question arose again. In fact, linking interactions among users and their comments and annotations directly to the artifacts they are producing, carries several advantages. First, when browsing the ontology later, developers and users can understand better why certain design decisions were made, what alternatives were considered, which group advocated a certain position, and so on. Second, when carrying out the discussion itself, if it is integrated in the development tool, the participants can immediately see the context for the components being discussed, they can examine the corresponding definitions and relations. Thus, the requirement for integrating discussion tools into ontology development environment is two-fold: Make the discussions accessible from the ontology components that are being discussed and make the ontology components accessible when one examines or writes a discussion message.

Support for various levels of expressiveness. The projects that use Protégé for collaborative development have rather expressive ontologies. For instance, one often comes across defined classes, complex restrictions, with intersections, in class definitions for the NCI Thesaurus. Thus, in the settings of these biomedical projects that heavily rely on ontologies, the collaborative version of the tools must ultimately have the same expressive power as a stand-alone ontology editor. It must support editing both ontology classes and instances.

User management and provenance of information. With multiple authors contributing to the ontology and the corresponding discussion, it is critical for users to understand where information is coming from. Thus, users must be able to see who makes specific changes and when, who creates a new proposal for change, who votes on it, and so on. This information must also be searchable. One must be able to find all changes or comments made by a specific user, or all recent changes and comments.

Scalability, reliability, and robustness. The traditional requirements of using tools in production systems include scalability (both in the size of ontologies and in the number of users), reliability (domain experts cannot afford to loose their data), and robustness (ontology-development tools should be no less robust than other tools that domain experts use). While several prototypes of collaborative tools have appeared recently, our experience shows that domain experts are usually reluctant to try a new tool until they are convinced the tools is ready to be used in production environment. Ontology development is not their primary task and they need tools that would help them perform this task quickly and reliably.

Access control. We often hear from our users who develop ontologies collaboratively that one of the features that all ontology-development tools largely lack today is access control. Today, for the most part, any user with writing privileges can edit anything in an ontology. However, users need to have more fine-grained control, particularly in the development of large ontologies. For example, users with expertise in an area represented by some part of an ontology should be able to edit that part, but may be able only to browse other parts or link to them. In fact, many ontology-development projects today maintain separation between what different users can do: For instance, some users can make proposals for changes but not make the changes themselves; others can comment on these proposal, but not create new ones; another group of users can affect the changes in the ontology based on the discussion; yet others can perform quality control by reviewing and approving the changes. We need to extend access-control policies with a more detailed model of user roles and privileges [4]. Because in ontologies concept definitions are often intertwined and a change in one part can affect definitions in another part, making such separation is far from trivial.

Workflow support. Many collaborative development projects have specific workflows associated with making changes. For example, there is a formal workflow for development of ontologies for the Food and Agriculture Organization (FAO) of the United Nations in the NeOn project [9]. The DILIGENT methodology for collaborative development [19], which focuses on formalizing the argumentation process, has been used in several European projects. A workflow specification may include different tasks that editors are charged with; the process for proposing a change and reaching consensus; roles that different users play, and so on. We are only beginning to understand different workflow models that collaborative ontology development requires [5]. Flexible support for these workflows must be an integral part of tools for collaborative development.

Synchronous and asynchronous access to shared ontologies. Depending on the size of the group and the complexity of the ontology, users might prefer synchronous or asynchronous editing [16]. In some of the projects we studied, users wanted to have their changes seen by everyone as soon as they make them, without the additional step of "checking in" their changes. In other cases, users preferred to have their own "sandbox" to test out the changes they are proposing before sharing them with everyone.

The core Protégé system supports some of the requirements listed here. Specifically, Protégé provides support for various levels of expressiveness, user management and provenance information, access control, and synchronous access to ontologies. It also addresses the requirement for scalability, reliability, and robustness. We describe work-flow support elsewhere [15]. In this paper, we focus on the support for integration of discussion and annotations with ontology-development environment.

3   Related Work

A number of ontology editors support some aspects of collaborative development. For instance, OntoWiki [1] is a web-based ontology and instance editor that provides such capabilities as history of changes and ratings of ontology components. OntoWiki provides different views on instance data (e.g., a map view for geographical data or a calendar view for data containing dates). OntoWiki focuses on instance acquisition and provides only rudimentary capabilities for ontology editing. The Hozo ontology editor [18] enables asynchronous development of ontologies that are subdivided into multiple inter-connected modules. A developer checks out and locks a specific module, edits it locally, and then checks it back in. If the ontology is not modularized, however, a developer must lock the whole ontology preventing others from editing it while he makes his change—an approach that may not be practical in many circumstances.

Several wiki-based environments support editing ontologies and instance data. The adaptation of the wiki environments that are particularly suited for ontology editing usually support a specific editing workflow. For example, a LexWiki platform developed at the Mayo Clinic, which is based on Semantic MediaWiki, currently is at the core of community-based development of BiomedGT.[1] BiomedGT is a terminology from the NCI Center for Bioinformatics (the same group that develops the NCI Thesaurus). The goal of BiomedGT is to enable the wider biomedical research community to participate directly and collaboratively in extending and refining the terminology. LexWiki enables users to browse an ontology, to make comments or to propose changes to (usually text-based) definitions. The BiomedGT curators with the privileges to make changes then open this annotated ontology in Protégé and perform the actual edits there. Wikis provide a natural forum for discussions, and the provenance information for suggested changes is easy to archive. Wikis, however, are not intended for ontology development and users cannot easily edit class definitions using this kind of framework. For example, in BiomedGT, curators must switch to Protégé to make the actual changes.

The coefficientMakna and Cicero tools (also based on wikis) implement the DILI-GENT methodology for collaborative development [3, 19]. The DILIGENT workflow focuses on the process of argumentation. The users discuss *issues*, which are usually specified at the ontology level (e.g., how should a particular classification be structured). The users present their arguments, suggest alternatives, agree and disagree with one another, and vote on the resolution. The editing environment explicitly supports these steps.

---

[1] http://biomedgt.org

Tools such as BiomedGT, Cicero, and coefficientMakna are designed to support specific workflows and could potentially work very well in the projects that use that specific workflow. The wiki-based tools have a simple interface that is best suited for making simple changes to the ontology. Wikis provide a natural forum for discussions, and the provenance information for suggested changes is easy to archive. However, these tools inherently cannot address the requirement of supporting ontology editing that conforms to a different workflow than the one for which they were designed. In the development of Collaborative Protégé, one of our goals is to make as few assumptions as possible about the editorial workflow that users will have and to develop mechanisms to make the tools customizable for different workflows.[2] Furthermore, these implementations do not provide structured access-control mechanisms.

4 Architecture of Collaborative Protégé

Our laboratory has developed Protégé—a widely used open-source ontology and knowledge base editor [6, 13]. At the time of this writing, Protégé has more than 100,000 registered users. Users can build ontologies in Protégé using different representation formalism ranging from *Frames*, to *RDF(S)* and *OWL*, and store them in file or database backends. Protégé is both robust and scalable and is being used in production environment by many government and industrial groups. The *ontology and knowledge base API* and *the plugin architecture* – one of the most successful features of Protégé, allow other developers to implement their own custom extensions that can be used either in the Protégé user interface or as part of other applications.

Protégé can be run as a standalone application, or in a client–server setting. In the client–server mode, ontologies are stored on a central Protégé server. Users access the ontologies on the server to browse and edit them through desktop or web Protégé clients. The client–server mode uses the Remote Method Invocation (RMI) mechanism of Java.

We have developed Collaborative Protégé as an extension to the client–server Protégé. Collaborative Protégé enables users who develop an ontology collaboratively to hold discussions, chat, annotate ontology components and changes—all as an integral part of the ontology-development process. The key feature of Collaborative Protégé is the ability to create annotations. In this context, **annotations** are typed comments (e.g. example, proposal, question, etc.) attached to ontology components, or to the descriptions of ontology changes, or to other annotations. We define the structure of the annotations in the *Changes and Annotations ontology* (ChAO), which we describe in Section 5.

Figure 1 gives an overview of the main components of Collaborative Protégé. The Protégé server has an **ontology repository** that contains all the ontologies that Protégé clients can edit in the collaborative mode. The repository has ChAO knowledge bases (instances of the ChAO classes) for each of the domain ontologies in the repository. These instances represent the changes and the annotations for the corresponding ontology. Several related domain ontologies can share the same ChAO knowledge base. For example, in Figure 1, the *ATHENA-DSS* and the *Guideline* ontologies share the same ChAO knowledge base, while the NCI Thesaurus has its own ChAO knowledge base.

---

[2] We are currently working on adding customizable workflow support for Collaborative Protégé, but this work is outside of the scope of this paper.

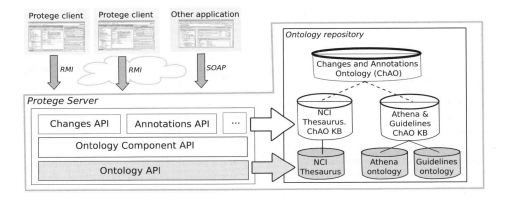

Fig. 1. The client–server architecture of Collaborative Protégé. The users work in Protégé clients or in other Protégé-based applications. All the changes made by a user in a client are sent to the server, and are immediately propagated to all other clients. The server has an ontology repository and several APIs to support the collaborative functionalities. Each domain ontology in the server repository has a *Changes and Annotations knowledge base* (ChAO KB) associated with it. This knowledge base contains instances of the ChAO ontology that describe the changes and annotations for the specific domain ontology.

When a user edits the domain ontology in the Protégé client, each change that the user performs, is sent to the server. The server then performs several actions: (a) updates the central (server-side) ontology; (b) pushes the change to the other clients so that other Protégé users can see them immediately; and (c) creates one or several ChAO instances that represent the change [11]. The server also pushes the changes in the ChAO knowledge bases to the Protégé clients. When users create an annotation in the Protégé client, the Protégé server adds the corresponding instances to the ChAO knowledge base.

The server also provides several layered Java APIs for accessing the collaborative features. The **Changes API** provides methods for getting the structured log of ontology changes, to get detailed information about a change (like author and date of the change), and transactions – changes that are composed of several atomic changes, which are executed together as one single change. The **Annotations API** provides methods for adding annotations to ontology components and changes, for accessing the meta-data of an annotation (e.g. provenance information), to get the discussion threads, and so on. The **Ontology Components API** has common methods for both the Changes and the Annotations API and supports the access to the ontology components (e.g. classes, properties, individuals) stored as instances in the ChAO knowledge bases. The **Ontology API** has methods for accessing and changing the content of the ontologies and knowledge bases. It also provides support for transactions, caching, for multiple back-ends and support for the client-server architecture. The layered APIs can be used by other applications to access all domain ontologies as well as the collaborative information from the ChAO knowledge bases stored on the server side.

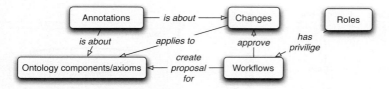

Fig. 2. Representation modules for collaborative ontology development. The *Ontology com-ponents* module represents the ontology elements. The *Changes* module captures declarative rep-resentations of changes to these elements. The *Annotations* module represents different types of annotations users can make about ontology elements and changes. The *Workflows* module repre-sents activities and tasks in collaborative ontology development. The arrows in the diagram are labeled with sample relationships that may exist between classes in one ontology and another.

5   Ontologies for Supporting the Collaborative Development

Collaborative Protégé uses a set of ontology modules to drive the collaborative devel-opment process (Figure 2) .

The Roles module describes the users, roles, operations and policies that apply to a certain ontology. The Protégé server uses the *Roles* module for checking the users cre-dentials at login time, and for determining whether a user is allowed to perform a certain operation based on the policies attached to an ontology instance. A user is represented as an instance of the User class and can play several roles (instances of Group class). For example, a user Ricardo can play the role of software developer and of editor. New roles can be easily added by creating new instances of Role, if a certain project requires them. To each ontology instance we associate a set of policies that define what opera-tions are allowed for a role. For example, the NCI Thesaurus would be a represented as an instance of the Project class and would have associated to it a set of policy instances. One of the policies would allow editors to change the ontology. Because Ri-cardo is an editor, he will be allowed to write to the ontology, while for non-editor users the write access will be denied.

The Workflows module provides a formal language for describing workflows for col-laborative ontology development. The Workflow class represents the workflow object. Each instance of this class describes a workflow (e.g., an approval workflow or a vot-ing workflow). Each workflow is associated with a set of initialization parameters, a workflow target, a partially ordered set of activities or states. For example, a workflow for a change proposal can be attached to a particular class in an ontology and would guide the flow of operations in the collaborative platform (e.g. first, start a proposal, then users votes, then count votes, then take a decision, etc.). We envision that future versions of Collaborative Protégé will provide flexible workflow support that would al-low us just by changing a workflow description in the *Workflow* module to regenerate the collaborative platform to use the new workflow description.

The Ontology Components module provides a meta-language for describing represen-tational entities in different ontology languages. For example, it contains classes such as `Class`, `Property`, and `Instance`. An instance of a `Class` represents a reified object for a real class in an ontology (e.g. in the ATHENA-DSS ontology, we would have an instance of `Class`, called `Guideline`). *The Ontology Components* module provides classes for representing entities in OWL, RDF(S) and Frames. Collaborative Protégé uses this ontology, when users add comments to ontology components and also for change tracking. For example, if the user adds a comment to the `Guideline` class, the annotation instance will be attached to the corresponding `Class` instance (`Guideline`) in *the Ontology Components* module. This instance also references all the changes made to that class, and all other comments and annotations that users have attached to the class. For future versions, we are considering integrating the *Ontology Metadata Vocabulary* (OMV) [12] for the representation of OWL language constructs.

The Annotations module represents the different types of annotations that users make. The annotation types are extensions of the Annotea [8] annotations and contain concepts such as Comment, Question, Advice, Example, and so on. Each comment or annotation is linked to one or several ontology elements, or changes, which are represented in the ontologies describing `Ontology components` and `Ontology changes` [11]. If users need a new annotation type, they can simply extend this ontology by creating a new subclass of the `Annotation` class. In fact, users in our evaluation (Section 7) found this feature critical.

The Changes module contains classes representing different types of changes that can occur in an ontology. For example, an instance of the class `Class_Created` will represent a class creation event that references the `Class` instance from *the Ontology Components* module corresponding to the new class in the domain ontology. One of the challenges that we are facing is that each ontology language has its own types of changes. For example, in a Frames ontology, changing the domain of a slot will be recorded as a domain change event, while in OWL, the real change would actually be a remove and add domain axiom for a certain property. We plan to address this issue by defining a common layer for changes such as creating a class or adding a subclass and then creating subontologies for changes that are unique to each of the languages.

These service ontologies reference the components in the domain ontology. However, note that the domain ontology does not have references to the annotations, changes, and so on. Thus, the developers have the choice of whether or not to make their annotations public when they publish the ontology itself.

6   User Interface

The user interface of Collaborative Protégé (Figure 3) is implemented as a graphical extension of Protégé. Panel A in Figure 3 shows the class tree, Panel B shows the selected class information (in this case `Gene_Product`)—just like in the original Protégé user interface, while panel C displays the *collaborative tabs*. Each of the collaborative tabs supports one of the several collaboration features. For example, in the *Annotations* tab, the user can add comments to ontology components; in the *Changes* tab, the user may see the change history of the selected class and comment on a change; in the *Search* tab,

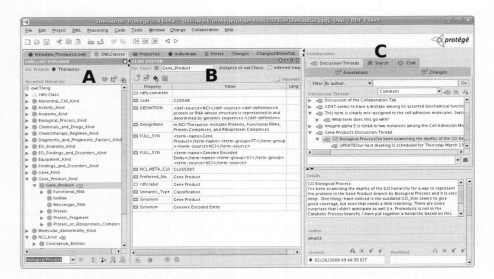

Fig. 3. The Protégé user interface, with the Collaborative Protégé plug-in. This screen capture shows the OWL Classes tab, in which the user edits and browses the classes that describe a domain ontology – here the NCI Thesaurus. Panel A shows the class tree; panel B displays the form for entering and viewing the description of the selected class Gene_Product, as a collection of attributes; and panel C shows the discussion among users about this class.

the user can search all annotations on different criteria; in the *Chat* tab, the user may discuss with other online users, and so on.

The **Annotations tab** is the default tab that users see when logging into Collaborative Protégé. The Annotations tab shows the annotations that are attached to the selected class in the tree (it also works for properties and individuals). The small callout icon shown in the class tree (Figure 3) next to the class name, indicates that the class has annotations. The lower part of the Annotations Tab shows the details of the selected annotation (e.g. the author, creation date, annotated entity, etc.). The annotations shown in the user interface are instances of the Annotation class. The user can create annotations of specific type (for example, Comment, Question, Example, Proposal, etc.). These types are defined in the *Annotations ontology* as subclasses of the Annotation class. Users can also reply to existing comments or notes—creating discussion threads related to a specific entity (Figure 3). The user may filter the displayed annotations by using one of the filtering criteria available at the top of the Annotations Tab. For example, she can filter by author, date, type and body of annotation.

Because the user interface takes the annotation types from the *Annotations ontology*—they are subclasses of the Annotation class—users can create their own types of annotation. To create a new annotation type, the user can edit the *Annotations ontology* itself, add the new type as the subclass of the Annotation class, define any additional properties that this custom-tailored annotation type should have, and the new annotation type will be available for use in Collaborative Protégé. In fact, in our evaluation (Section 7) users have defined their custom annotation type.

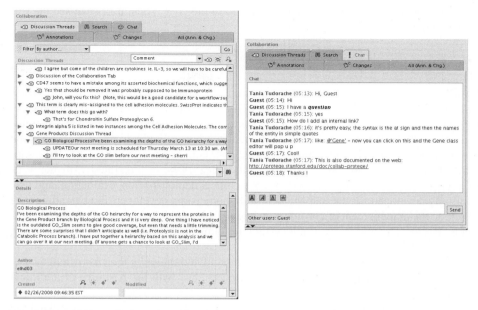

Fig. 4. Two of the collaborative tabs. The left screenshot shows the *Discussions Thread tab* where users can add comments on the ontology. The right screenshot is *the Chat Tab*, which allows users to chat and exchange internal and external links.

The **Discussion Thread tab** has a similar user interface and features as the Annotation tab (Figure 4). However, the annotations from the Discussion Thread tab are not attached to a particular ontology component, as the other annotations, but refer to the ontology itself. For example, users may discuss modeling patterns, or naming conventions that are broader in scope and that should apply to the whole ontology, rather than to individual ontology components.

The **Changes Tab** shows a chronological list of all the changes for the selected ontology component. For each change, the tab shows the change details (e.g. author, date, sub-changes, etc.). Users may also comment and have discussion threads related to a certain change as also shown in our example.

Users may also search all annotations based on different criteria in the **Search Tab**. For example, a user can search for all annotations of type `Comment` that have been made by an author `eldh` between `05/14/2007` and `05/14/2008`. The search result will show all the annotations that satisfy the criteria and will provide direct access to the annotated ontology elements or changes.

One of the popular features of Collaborative Protégé is the **Chat Tab** (Figure 4). Users connected to the Protégé server can exchange live messages. The chat panel supports HTML formatting of the message, such as bold, italics, highlight. One feature that sets the Collaborative Protégé chat functionality apart from other chat clients is the support for sending internal and external links. An internal link points to an ontology component. In the example in Figure 4, one of the users sends an internal link to the `Gene` class. The other user who is receiving the message can click on the internal link and see the definition for the class mentioned in the chat. Thus, users can see the full context of the discussion in the chat.

7   Evaluation

We have performed the formative evaluation of Collaborative Protégé in the context of the ATHENA-DSS project. ATHENA (Assessment and Treatment for Healthcare: EvideNcebased Automation) [7] is a clinical decision-support system that generates guideline-based recommendations for the management of patients suffering from some clinical conditions. The system, developed as a collaboration between VA Palo Alto Healthcare System and Stanford University since 1998, is integrated with the VA's Computerized Patient Record System for a clinical demonstration, evaluation, and use. Initially developed for the management of hypertension, developers are extending it to include the management of chronic pain and diabetes, and the screening of chronic kidney disease. The end-users of the system are clinicians who are making decisions on the management of care for patients.

ATHENA-DSS developers use Protégé to build and maintain their knowledge base. The team of clinicians and knowledge engineers start with the narrative of a clinical guideline and distill this narrative into a set of related Protégé classes and instances that represent the guideline formally. Currently, the developers use an MS Access database to save the recommendation text and the associated annotations that they create. Thus, the information is spread across different tools and it is not linked. As the developers formalize medical concepts, such as diseases and drugs, and instantiate guideline recommendations as parts of flow-chart-like clinical algorithms, they have to work closely with one another, making sure that they do not overwrite one another's work. As the knowledge bases evolve, the developers have to ensure that the recommendations and annotations in the MS Access databases and Protégé knowledge bases are in synch.

As Collaborative Protégé became available, the team of one clinician and two knowledge engineers evaluated it over the period of one month. The three users actively used the tool during the evaluation period. They had access to the web pages that briefly describe the tool[3] but they did not have any training on how to use Collaborative Protégé. They were experienced users of the regular Protégé tool.

After the evaluation period, we conducted extensive interviews with the users to gauge their level of satisfaction with the tool, to understand how they used the it, to learn which features they liked and did not like, and to get new feature requests from them. In addition, we examined the annotations and the changes that the developers produced during the evaluation period to determine how they used the annotation and discussion feature, what was the nature of their posts, and how much of their time spent with the system was spent on collaboration activities compared to modeling activities.

8   Results

During the evaluation period, the developers entered 22 comments. All comments were comments on instances. There were three short discussion threads. We observed two

---

[3] http://protege.stanford.edu/doc/collab-protege/

main uses for the comments in this project. First, the developers used the discussion feature to ask each other questions. For instance, the clinicians described some modeling problems and asked the knowledge engineers for the best ways to model the situation. Conversely, the knowledge engineers asked about some clinical concepts that they needed to represent.

Each clinical guideline has a narrative description and a set of qualitative parameters. The ATHENA-DSS developers represent each guideline as classes and instances in the ATHENA-DSS knowledge base. The developers found that annotations provided a good way to record the narrative and the parameters of the original guideline and to link them to the ontology components that represent the guideline. In a sense, the information about the original guideline provided the background information for ontology components, and annotations were a natural way to represent this link. The ATHENA-DSS developers currently store the information on the original guidelines in an MS Access database and they wanted this information to be accessible during ontology browsing. Because the reference guideline contains not only text, but a number of additional fields, we used the flexible design of Collaborative Protégé to produce a custom-tailored annotation type for ATHENA-DSS. We created a subclass of the `Annotatation` class, a `GuidelineComment` class. This subclass contained the fields specific to that type of annotation, such as quality of evidence and recommendation code. Because the Collaborative Protégé implementation simply displays the subclasses of the `Annotation` class as its available annotation types, we did not need to change any code to display the custom-tailored annotation. The ATHENA-DSS developers found this flexibility to be a particularly useful feature. They reported that they are now considering porting all the annotations from the MS Access databases to Collaborative Protégé as annotations. They cited several advantages of this approach in our interviews: First, they will be able to stay within one environment and not have to maintain the synchronization between the two sources. Second, they can see the reference source immediately as they browse the instances and can understand why the guideline was modeled the way it was. After we provided them with the new annotation type, about 25% of their comments were of this type.

In general, the members of the ATHENA-DSS team found Collaborative Protégé "very useful." They appreciated that the knowledge engineers could see the questions from the clinician in context of where the question was asked (rather than in an email, detached from the ontology). As one of the participants told us "It's just there, at the point where the problem is."

The ATHENA-DSS developers did not use the chat feature, mainly because they were never on-line at the same time. Another group that is currently evaluating Collaborative Protégé (the editors of the the NCI Thesaurus) found the chat to be one of the more useful features. The main difference between the two groups is that the second group is much larger and ontology development is their primary task. Thus, most editors are on-line editing the ontology during their workday.

In our interviews, the ATHENA-DSS developers indicated other potential uses that they see for the annotation features. These uses included recording detailed design rationale, having one developer explain to the others how he is approaching a specific modeling problem in the context of the ontology, and having developers educate new users on the structure and intricacies of the ontology.

9 Discussion and Future Work

The analysis of the results, even from this fairly small evaluation period, points to several issues. First, users found Collaborative Protégé useful and did not require any special training to use it. We know that they did not find or use all the features that were available, and we expect that they would use the collaboration features even more extensively after a short training session (or with better documentation).

Second, the innovative use of Collaborative Protégé features points to the versatility of the tool. In fact, some of these use prompted us to consider new features. For example, we might link the tool to an issue-tracker system, to enable users to see which task assignments have been made as part of the discussion, and to track their progress.

Third, the flexibility of the tool and the ease of extending it with new annotation types proved crucial in the ATHENA-DSS project. We envision that other users will create their own annotation types, with properties that are relevant in their settings.

One of the surprising findings for us (which we also observed in other settings) was that users do not add annotations to changes, but annotate only ontology components (in this case, instances). Even the rationale for changes themselves is recorded at the level of the ontology component, not the change or a group of changes. This observation suggests that users think in terms of ontology components rather than changes, even as they are closely involved in ontology editing.

In Collaborative Protégé, facilities for reaching consensus, recording design rationale, and noting outstanding issues are an integral part of the process of ontology browsing and editing. As users examine, say, a class in the ontology, they can immediately see all the discussion and questions pertaining to this class, whether there was any contention in its definition, alternatives that the authors considered. An editor, when coming upon a class that, he feels, must be changed, can post a request immediately, in the context of this class. This dual advantage of *context-sensitivity and archival* character of annotations adds the greatest value to Collaborative Protégé compared to discussion lists and issue trackers that are not integrated with an ontology environment.

Our infrastructure and the use of ontologies to represent many of the components that drive our software, enables other developers to reuse these components easily. Specifically, while Collaborative Protégé uses all the service ontologies described in Section 5, the service ontologies themselves are not specific to Protégé. We expect that other developers will reuse the ontologies in their tools, thus providing interoperability between the tools. For instance, different tools can implement their own mechanism for supporting or displaying discussions. If they use the same annotation ontology, then annotations created in one of the tools can be visible in the other tool.

There are many outstanding issues, however, that we must address in order to support truly collaborative ontology development.

In our original model, each annotation annotates a single object: a single class in the ontology, a single instances, a single other annotation. However, in the ATHENA-DSS use case a single guideline description could refer to different concepts such as hypertension and diabetes. Thus, there must be a way of associating an annotation to several different objects. We do not currently have such support in the user interface. However, because annotations are simply instances, the `annotates` property can have more than one value and thus reference more than one object.

While we have a set of annotation types for proposals and voting, we do not have any workflow support for it. Our users (in ATHENA-DSS, and other projects) indicated that the proposals feature would be much more useful with such workflow support. For instance, when someone initiates a new round of voting, a workflow engine might inform other users that they are expected to vote, can tally the votes or wait for a certain period of time to elapse, and can produce the voting result.

Currently, Collaborative Protégé has only simple support for different user roles. In the future, we plan to adopt a policy mechanism that would enable us to describe privileges of users with different roles at different levels of granularity. For example, not all users in a project may have the privileges to create change proposals or to comment on the propsals. Some users may be able to edit only a part of the ontology. We plan to analyze the different scenarios and workflows that the biomedical ontology-development projects employ and add flexible support for roles and policies in future versions.

Finally, as we studied the different workflows that the projects described in the introduction to this paper used, one thing became clear: Developers of biomedical ontologies need tools that are flexible enough to work with different workflows. For instance, a group of users working together on developing an ontology in the context of a specific project will have different requirements compared to an open community developing a lightweight taxonomy that anyone can edit. In some cases, tools should support specific protocols for making changes, where some users can propose changes, others can discuss and vote on them, and only users with special status can actually perform the changes. At the other end of the spectrum are settings where anyone can make any changes immediately. Thus, tools need to support different mechanisms for building consensus, depending on whether the environment is more open or more controlled.

We are currently evaluating Collaborative Protégé in several other settings: the development of the NCI Thesaurus, the development of the Software Resource Ontology to be used by the NIH Roadmap's NCBCs, the development of the 11th revision of the International Classification of Diseases (ICD-11) at the World Health Organization, and other projects. These projects are all active ongoing projects and have different scope, workflow, the number of contributors, and so on. We expect to these evaluation to produce additional requirements for the tools and also to demonstrate innovative uses of the capabilities that we described here.

Acknowledgments

This work was supported in part by a contract from the U.S. National Cancer Institute. Protégé is a national resource supported by grant LM007885 from the United States National Library of Medicine. Initial development of ATHENA-DSS for diabetes mellitus is supported by the Palo Alto Institute for Research and Education at VA Palo Alto Health Care System. Views expressed are those of the authors and not necessarily those of the Department of Veterans Affairs. We are indebted to Susana Martins, Martha Michel, and Mary Goldstein of the VA Palo Alto Healthcare System for their help with the evaluation and for their insightful feedback on the tool.

References

1. Auer, S., Dietzold, S., Riechert, T.: OntoWiki–a tool for social, semantic collaboration. In: Cruz, I., Decker, S., Allemang, D., Preist, C., Schwabe, D., Mika, P., Uschold, M., Aroyo, L.M. (eds.) ISWC 2006. LNCS, vol. 4273. Springer, Heidelberg (2006)
2. OBI Consortium, http://obi.sourceforge.net/
3. Dellschaft, K., Engelbrecht, H., Barreto, J.M., Rutenbeck, S., Staab, S.: Cicero: Tracking design rationale in collaborative ontology engineering (2008)
4. Finin, T., Joshi, A., Kagal, L., Niu, J., Sandhu, R., Winsborough, W., Thuraisingham, B.: Rowlbac: Role based access control in owl. In: ACM Symposium on Access Control Models and Technologies (SACMAT 2008), Colorado, US (2008)
5. Gangemi, A., Lehmann, J., Presutti, V., Nissim, M., Catenacci, C.: C-ODO: an OWL meta-model for collaborative ontology design. In: Workshop on Social and Collaborative Construction of Structured Knowledge at WWW 2007, Banff, Canada (2007)
6. Gennari, J., Musen, M.A., Fergerson, R.W., Grosso, W.E., Crubézy, M., Eriksson, H., Noy, N.F., Tu, S.W.: The evolution of Protégé: An environment for knowledge-based systems development. International Journal of Human-Computer Interaction 58(1) (2003)
7. Goldstein, M.K., et al.: Translating research into practice: organizational issues in implementing automated decision support for hypertension in three medical centers. Journal of the American Medical Informatics Association 11(5), 368–376 (2004)
8. Kahan, J., Koivunen, M.-R.: Annotea: an open RDF infrastructure for shared web annotations. In: The 10th International World Wide Web Conference, pp. 623–632 (2001)
9. Muñoz García, O., Gómez-Pérez, A., Iglesias-Sucasas, M., Kim, S.: A Workflow for the Networked Ontologies Lifecycle: A Case Study in FAO of the UN. In: Borrajo, D., Castillo, L., Corchado, J.M. (eds.) CAEPIA 2007. LNCS (LNAI), vol. 4788, pp. 200–209. Springer, Heidelberg (2007)
10. Noy, N.F., Chugh, A., Alani, H.: The CKC Challenge: Exploring tools for collaborative knowledge construction. IEEE Intelligent Systems 23(1), 64–68 (2008)
11. Noy, N.F., Chugh, A., Liu, W., Musen, M.A.: A framework for ontology evolution in collaborative environments. In: Cruz, I., Decker, S., Allemang, D., Preist, C., Schwabe, D., Mika, P., Uschold, M., Aroyo, L.M. (eds.) ISWC 2006. LNCS, vol. 4273. Springer, Heidelberg (2006)
12. Palma, R., Hartmann, J., Haase, P.: OMV: Ontology Metadata Vocabulary for the Semantic Web. Technical report (2008), http://ontoware.org/projects/omv/
13. Protégé, http://protege.stanford.edu/
14. Rubin, D.L., Noy, N.F., Musen, M.A.: Protégé: A tool for managing and using terminology in radiology applications. Journal of Digital Imaging (2007)
15. Sebastian, A., Noy, N.F., Tudorache, T., Musen, M.A.: A generic ontology for collaborative ontology-development workflows. In: The 16th International Conference on Knowledge Engineering and Knowledge Management (EKAW 2008), Catania, Italy. Springer, Heidelberg (2008)
16. Seidenberg, J., Rector, A.: The state of multi-user ontology engineering. In: The 2nd International Workshop on Modular Ontologies at KCAP 2007, Whistler, BC, Canada (2007)
17. Sioutos, N., de Coronado, S., Haber, M., Hartel, F., Shaiu, W., Wright, L.: NCI Thesaurus: A semantic model integrating cancer-related clinical and molecular information. Journal of Biomedical Informatics 40(1), 30–43 (2007)
18. Sunagawa, E., Kozaki, K., Kitamura, Y., Mizoguchi, R.: An environment for distributed ontology development based on dependency management. In: Fensel, D., Sycara, K.P., Mylopoulos, J. (eds.) ISWC 2003. LNCS, vol. 2870. Springer, Heidelberg (2003)
19. Tempich, C., Simperl, E., Luczak, M., Studer, R., Pinto, H.S.: Argumentation-based ontology engineering. IEEE Intelligent Systems 22(6), 52–59 (2007)

Identifying Potentially Important Concepts and Relations in an Ontology*

Gang Wu[1,2], Juanzi Li[1], Ling Feng[1], and Kehong Wang[1]

[1] Department of Computer Science, Tsinghua University, Beijing 100084, P.R. China
[2] Department of Computer Science, Southeastern University,
Nanjing 210000, P.R. China
wug@keg.cs.tsinghua.edu.cn, ljz@keg.cs.tsinghua.edu.cn,
fengling@tsinghua.edu.cn, wkh@keg.cs.tsinghua.edu.cn

Abstract. More and more ontologies have been published and used widely on the web. In order to make good use of an ontology, especially a new and complex ontology, we need methods to help understand it first. Identifying potentially important concepts and relations in an ontology is an intuitive but challenging method. In this paper, we first define four features for potentially important concepts and relation from the ontological structural point of view. Then a simple yet effective Concept-And-Relation-Ranking (CARRank) algorithm is proposed to simultaneously rank the importance of concepts and relations. Different from the traditional ranking methods, the importance of concepts and the weights of relations reinforce one another in CARRank in an iterative manner. Such an iterative process is proved to be convergent both in principle and by experiments. Our experimental results show that CARRank has a similar convergent speed as the PageRank-like algorithms, but a more reasonable ranking result.

1 Introduction

Ontology provides Artificial Intelligence and Web communities the remarkable capability of specifying shared conceptualization explicitly and formally. A diversity of ontologies have been widely used as the bases of semantic representation in many applications such as knowledge bases, multi-agents and the Semantic Web. As the amount, scale, and complexity of ontologies are increasing rapidly, it requires more efforts for ontologists and domain experts to understand them. Hence, *Ontology Understanding*, the process of getting familiar with an ontology [4], has to seek helps from computer intelligence.

The state-of-the-art ontology engineering projects, like IsaViz, Ontoviz, and Jambalaya, use information visualization techniques to represent ontologies. They have the ability to help humans understand and navigate in complex information spaces [9]. However, for a complex ontology, graphically presenting all concepts

---

* This work is supported by the National Natural Science Foundation of China under Grant No.90604025 and the Major State Basic Research Development Program of China (973 Program) under Grant No.2003CB317007 and No.2007CB310803.

and relations indistinctively makes above tools generate unreadable visualization results. Users who are unfamiliar with the ontology will probably get lost in such a maze.

To resolve the problem, some researchers have proposed approaches by drawing users' attention to those potentially *important* (or alternatively *interesting*) concepts within one ontology. They calculate the importance of concepts either by tracking the user's browsing activities [7], or according to the concept hierarchy [20]. These solutions are straightforward. While more detailed information about ontology structure, like the correlation between concepts and relations, is not explored. In some other studies, traditional link analysis ranking algorithms on Web pages and objects are employed to rank the importance of concepts [3], and even the importance of relations [8, 17]. These solutions need the help of additional statistic information or time-consuming machine learning schemes.

In this paper, we propose a simple yet effective algorithm, named Concept And Relation Ranking (CARRank), for identifying potentially important concepts and relations in an ontology. By efficiently ranking the importance of concepts and relations simultaneously, CARRank can find out which concepts and relations might be the ones the ontology creator would like to suggest to users for further consideration. In this way, CARRank can promote the usability for ontology understanding. Users can even outline an interested *sub-scope* of an ontology, of which important parts are taken out. Although CARRank is rather an automatic ranking algorithm than a specific visualization approach, it can be easily integrated into the existing ontology visualization tools to provide a novel perspective. Main contributions of this paper include:

1) To make good use of ontology structural information, we give a graph representation of ontology which makes it easy for applying link analysis ranking algorithms while preserves the semantics expressed by RDF-based ontology languages.

2) To determine the potentially important concepts and relations in an ontology, we introduce an importance ranking model. The model tries to imitate the creation process of an ontology from the ontological structural point of view by defining four representative features.

3) To calculate the importance of concepts and relations, we propose an efficient algorithm according to the model, named CARRank. The difference between CARRank and existing PageRank-like algorithms is two-fold. Firstly, with this algorithm, the importance of vertices (i.e. concepts) and the weights of edges (i.e. relations) reinforce one another in an iterative process. Such a dynamic computation on edges weights as well as vertices importance has never been studied previously. Secondly, the directions of *walk* for the algorithms are opposed, which makes CARRank more suitable for supporting ontology understanding. CARRank is proved to be convergent, and thus is universal for simultaneously ranking vertices importance and edges weights in arbitrary directed labeled graph.

4) Experiments are conducted to demonstrate the effectiveness and efficiency of the approach to support understanding of ontologies.

The remainder of the paper is organized as follows. We review the closely related work in Section 2, and present our CARRank model in Section 3. We then bring forward the CARRank algorithm in Section 4. Experimental results are shown in Section 5. The final section is about the conclusion and discussion.

2   Related Work

Cognitive Support for Ontology Understanding. The DIaMOND project [7] and the holistic "imaging" ontology [20] are two most related studies. DIaMOND [7] is a plug-in for Protégé[1] to help users find concepts of interest within an ontology. By tracking user's navigation activities on an ontology, it continuously calculates the degree of interest for each concept. The navigation overhead can thus be reduced by drawing user's attention to the highlighted concepts of high interest degrees. The degree calculation of this method is user-specific. In [20], authors exploited degrees of interest of concepts as a filter for labeling important concepts in a large scale ontology. Its degree calculation is holistically based on concept hierarchy without considering non-subsumption relations between concepts. Our work differs from these approaches. First, we think that the importance measurement of a concept should take into account the contributions from all the other concepts in the ontology through relations including both subsumption and non-subsumption ones. Second, relations between concepts are also helpful for ontology understanding.

Ontology Ranking in the Semantic Web. OntoSelect [6], OntoKhoj [18], and AKTiveRank [1] are three approaches that were developed to select (or rank) one or more ontologies that satisfy certain criteria [19], with an ontology document as the ranking granularity. The first two approaches relied on the *popularity*, which assumed that ontologies referenced by many ontologies are more popular, while the third one considered several structural evaluation metrics, including Density (DEM), Betweenness (BEM), Semantic Similarity (SSM), and Class Match measure (CMM). Although AKTiveRank does not intend to rank the importance of concepts or relations in an ontology, the above complex networks analysis metrics it employs are useful for reference in this work. According to the pre-existing statistic information on instances, Swoogle [8] could enable both document level and term level ranking, including the class-property relationship ranking.

Compared with this line of research, our study aims to finding out potentially important information in a given ontology, so the granularity of output is concept and relation, rather than a whole ontology. Besides, the method can evaluate the importance of general relations of concepts, as well as concepts themselves. Furthermore, no prior knowledge or user interaction is required, which may be more applicable in dealing with new ontologies. Table 1 lists some of the differences.

Ranking Algorithms. In ranking Web pages, hyperlink is the only relation to be considered. PageRank [5] pointed out that a good *authority* page is the one pointed to by many good authorities. The evaluation is performed in a *random surfer* manner over all pages on the Web graph. Unlike PageRank, HITS [13]

---

[1] http://protege.stanford.edu/

Table 1. Related Work in the Semantic Web

	Concept Rank	Relation Rank	Ranking Methods
CARRank	✓	✓	CARRank
DIaMOND	✓	-	Tracking users' navigation
[20]	✓	-	Concept hierarchy
OntoSelect	-	-	PageRank-like
OntoKhoj	-	-	PageRank-like
AKTiveRank	-	-	CMM+DEM+SSM+BEM
Swoogle	✓	✓	PageRank-like

exploited a mutual reinforcing relationship between *hub* pages and authority pages within a subgraph of the Web. By extension of PageRank and HITS, Reverse PageRank [10] was investigated as a reasonable approach to *browse* the Web, which reverses the direction of all hyperlinks before applying PageRank. In this study, we browse an ontology in a similar manner to Reverse PageRank.

Apart from the hyperlink relation, there exist more edge types in an ontology, such as property-of, subclass-superclass, etc. The edge type is an important factor in determining the importance of vertices. This was addressed recently in a series of object-level link analysis ranking algorithms. In the field of database, ObjectRank [3] applied link analysis methods to rank the importance of database objects and tuples. Different weights are set according to link types either manually or by statistic information. PopRank [17] is a machine learning approach to automatically assign the weights and rank the importance of Web objects. These weight assignment approaches are not applicable for ontology understanding where absence of priori knowledge is fairly common. We attempt to resolve it by evaluating the weights simultaneously in the ranking process according to only the mutually reinforcing relationship between concepts and relations.

3   CARRank Model

3.1   Ontology Graph

Before any link analysis could be performed, an ontology should be represented as a graph. As an ontology defines the concepts and the relations between them in certain domain [16, 11], it is suggested to model a concept as a vertex and a relation as a directed edge linking two concepts. We call such constructed graph the *ontology graph*.

Definition 1. *Given an ontology $\mathcal{O}$, the **ontology graph** $\mathcal{G} = (\mathcal{V}, \mathcal{E}, l_{\mathcal{V}}, l_{\mathcal{E}})$ of $\mathcal{O}$ is a directed labeled graph. $\mathcal{V}$ is a set of nodes representing all concepts in $\mathcal{O}$. $\mathcal{E}$ is a set of directed edges representing all relations in $\mathcal{O}$. $l_{\mathcal{V}}$ and $l_{\mathcal{E}}$ are labeling functions on $\mathcal{V}$ and $\mathcal{E}$ respectively.*

Definition 1 is a representation of an ontology at the syntactic level. Its semantic capabilities will be presented in section 3.4.

The ontology graph illustrated in Figure 1 is our running example. It describes concepts and relations in an open software project domain, especially the relationships between developers and projects.

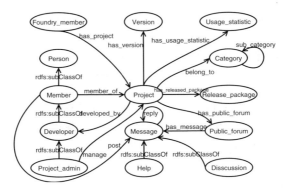

Fig. 1. The running example

3.2   Mapping RDF-Based Ontology to Ontology Graph

In practice, the most important ontology languages in the Semantic Web are RDF Schema (RDFS) and OWL. In these languages, an ontology is expressed as a set of triples. A triple $(s, p, o) \in (U \cup B) \times U \times (U \cup B \cup L)$ is called an RDF triple where U, B, and L are infinite sets of URI references, blank nodes, and literals respectively. Here, s is called the *subject*, p the *predicate*, and o the *object* of the triple. A set of such RDF triples is defined as an *RDF graph* [21], and represented as a directed labeled graph as shown in Definition 2. We will use the RDF graph to refer to both a set of RDF triples and its directed labeled graph representation throughout the rest of this paper.

Definition 2. *Let T be a set of RDF triples. The **directed labeled graph representation** of T is $G = (V, E, l_V, l_E)$, where*

$$V = \{v_x | x \in subject(T) \cup object(T)\}$$
$$E = \{e_{s,p,o} | (s, p, o) \in T\}$$
$$l_V(v_x) = \begin{cases} (x, d_x) & \text{if } x \text{ is literal } (d_x \text{ is datatype identifier}) \\ x & \text{else} \end{cases}$$
$$from(e_{s,p,o}) = v_s, \ to(e_{s,p,o}) = v_o, \ \text{and} \ l_E(e_{s,p,o}) = p$$

V is the set of vertices in G. E is the set of directed edges. l_V and l_E are labeling functions on V and E. subject(T) and object(T) are used to achieve all the subjects and the objects in T. Function from() and to() return the starting and ending vertex of an edge.

However, for the same ontology, an RDF graph and an ontology graph are unequal. Suppose an ontology consists of a relation "manage" linking from "Project_Admin" to "Project". The ontology graph is shown in Figure 2. To express the same semantics, an RDF graph needs two triples (manage, rdfs:domain, Project_Admin) and (manage, rdfs:range, Project) as shown in Figure 3.

The difference lies in that, for an ontology, a relation does not exist as a directed edge but a vertex in an RDF graph. A relation is associated with a concept

Fig. 2. An ontology graph representation   **Fig. 3.** An RDF graph representation

by the semantics of rdfs:domain or rdfs:range (the concept is named *domain* or *range* accordingly). Such indirect relationships will hinder the importance propagation during the ranking, because there is no path between the domain and the range. Hence, we propose a map function ω to map an RDF graph to an ontology graph in Definition 3.

Definition 3. *Let* $G = (V, E, l_V, l_E)$ *be the RDF graph of an ontology* $\mathcal{O}$. *We define a map* $\omega : G \to \mathcal{G}$ *as follows:* $\omega(G) = (\mathcal{V}, \mathcal{E}, l_{\mathcal{V}}, l_{\mathcal{E}})$ *where,*

$$\mathcal{V} = V, \quad l_{\mathcal{V}} = l_V,$$
$$\mathcal{E} = \{e_{s,p,o}|e_{s,p,o} \in E \wedge l_E(e_{s,p,o}) \neq \textsf{rdfs:domain} \wedge l_E(e_{s,p,o}) \neq \textsf{rdfs:range}\}$$
$$\cup E_{DR} \cup E_D \cup E_R,$$
$$E_{DR} = \{e_{s,p,o}|\exists e_{p,\textsf{rdfs:domain},s} \in E \wedge \exists e_{p,\textsf{rdfs:range},o} \in E\},$$
$$E_D = \{e_{s,p,\textsf{keg:Sink}}|\exists e_{p,\textsf{rdfs:domain},s} \in E \wedge \not\exists e_{p,\textsf{rdfs:range},o} \in E\},$$
$$E_R = \{e_{\textsf{keg:Source},p,o}|\exists e_{p,\textsf{rdfs:range},o} \in E \wedge \not\exists e_{p,\textsf{rdfs:domain},s} \in E\},$$
$$\forall e_{s,p,o} \in \mathcal{E}, \ \textsf{from}(e_{s,p,o}) = v_s, \ \textsf{to}(e_{s,p,o}) = v_o, and \ l_{\mathcal{E}}(e_{s,p,o}) = p$$

Here, keg:Source *and* keg:Sink *are defined to be the virtual domain and range of those relations having no domain or range defined explicitly.*

Each edge in the output ontology graph is an RDF triple. Therefore the same relation can be distinguished between different domain concepts and range concepts. The map removes those edges taking rdfs:domain or rdfs:range as their labels, while adds new labeled edges to directly link the domains to the ranges according to the rules in Definition 3. In this way, $\omega(G)$ presents an ontology graph that preserves the semantics of G and makes it easy for ranking. Thus, the RDF graph in Figure 3 can be mapped to the ontology graph in Figure 2. In fact, our running example shown in Figure 1 is mapped from a real ontology[2].

3.3   Model Description

The creation of an ontology is a composition process where the creator operates with a set of concepts and relations. Hence, the ontology could be considered as the image of the creator's own understanding of the knowledge, just like a literary work to its author. This phenomenon of human consciousness can be best explained with William James' famous *stream of consciousness* theory [12]. He observed that human consciousness has a composite structure including *substantive parts* (thought or idea) and *transitive parts* (fringe or penumbra), and keeps moving from thought to thought. Transitive parts play an important role in controlling the orderly advance of consciousness from one thought to another.

---

[2] http://keg.cs.tsinghua.edu.cn/project/software.owl

By analogizing concepts and relations to substantive parts and transitive parts, the creation of an ontology could be described as drifting on the stream of the creator's consciousness of the domain knowledge from one concept to another via a particular relation. The initially created concept has a certain possibility of being one of the creator's emphasis (suggestions to users). For the concepts to be suggested, the creator would always like to create more relations to describe its relationships with other concepts. Consequently, ontology users will implicitly follow the creator's stream of consciousness for understanding the ontology.

We characterize four features for potentially important concepts and relations which drive the drift on the stream of consciousness. It turns out to be our model for Concepts And Relations Ranking (the **CARRank** model):

1. A concept is more important if there are more relations starting from the concept.

2. A concept is more important if there is a relation starting from the concept to a more important concept.

3. A concept is more important if it has a higher relation weight to any other concept.

4. A relation weight is higher if it starts from a more important concept.

There are three meanings here. First, it explains what is *important* (or alternatively *interesting*). In this paper, term *importance* is used as a metric for measuring the extent that the ontology creator suggests a concept or relation to users. Second, a concept is regarded as a source that owns a set of relations related to other concepts. We refer to this character as the *hub* like that in HITS [13]. Finally, concepts and relations exhibit a mutually reinforcing relationship.

In our running example, concepts "Project", "Project_admin" and "Developer" are more attractive because they either have abundant relations to other concepts (e.g. "Project"), or locate deeply in the subsumption hierarchy (e.g. "Project_admin"), or have a relation to other attractive concept (e.g. "Developer"). Accordingly, relation "manage" between "Project" and "Project_admin" becomes more meaningful. These observations coincide with the creator's comment that declares to emphasize the relationship between developers and projects. Our inquiry to the creator about the design process is answered as follows: First defined the concept "Project" with some decorative literals such as "Version" and "Usage_statistic". Next, provided another concept "Developer" to complement the description of "Project" through a relation "developed_by" from "Project" to "Developer". Then, a hierarchy was built about "Developer" from "Person" to "Project_admin". The process continued until all information was included.

3.4 Semantic Abilities

By using ω mapping, any RDF-based ontology, like RDF Schema, DAML+OIL, and OWL (including three increasingly-expressive sublanguages: OWL Lite, OWL DL, and OWL Full), can be ranked with the **CARRank** model. In the section of experiments, we will further analyze the ranking results of **CARRank** for the same ontology in three languages with different expressive powers.

Furthermore, CARRank even has the ability to support axioms expressed as rules, e.g. SWRL [22] rules, because there exists RDF-compatible model-theoretic semantics [15] of SWRL by which we can interpret SWRL rules in the framework of RDF graphs. In a broad sense, any inference scheme for ontology is supported by CARRank, if it is resolvable on the level of RDF graphs.

Moreover, since a relation is represented as a vertex in an RDF graph, and then kept in the ontology graph after ω mapping, the hierarchies and properties of relations will also impact the global importance of these relations. That means if there is a deeper hierarchy or more properties for a specific relation, the importance of that relation is higher. Here, whereas we only concern about the comparison locally among relations starting from the same concepts rather than globally among all relations, because the importance may be quite different when associated with different concepts.

Finally, since ontology understanding is affected by many factors, here the importance only means some *potential* to be important in our context.

4   CARRank Algorithm

Definition 4. *Suppose an ontology graph $\mathcal{G}$ has $|\mathcal{V}| = n \geq 1$ concepts $v_1, ..., v_n \in \mathcal{V}$. The **adjacency matrix representation** of $\mathcal{G}$, $\mathbf{A} = (a_{i,j})$, is a $n \times n$ matrix where $1 \leq i, j, k \leq n$ and*

$$a_{i,j} = \begin{cases} 1 & \text{if } \exists e_{i,k,j} \in \mathcal{E}, \\ 0 & \text{otherwise.} \end{cases} \tag{1}$$

*Let $\mathrm{w}(v_i, v_j)$ be a relation weight function, and $w_{i,j} = \mathrm{w}(v_i, v_j)$ be the weight of all relations from v_i to v_j. The **relation weight matrix representation** of $\mathcal{G}$, $\mathbf{W} = (w_{i,j})$, is a $n \times n$ matrix where $1 \leq i, j, k \leq n$, and*

$$\begin{cases} 0 < w_{i,j} \leq 1 & \text{if } \exists e_{i,k,j} \in \mathcal{E}, \\ w_{i,j} = 0 & \text{otherwise.} \end{cases} \tag{2}$$

Definition 5. *For any concept $v_i \in \mathcal{V}$, **the forward concepts** of v_i are defined as $F_{v_i} = \{v_j | v_j \in \mathcal{V} \wedge \exists e_{i,k,j} \in \mathcal{E}\}$, and **the backward concepts** of v_i are defined as $B_{v_i} = \{v_j | v_j \in \mathcal{V} \wedge \exists e_{j,k,i} \in \mathcal{E}\}$.*

Definition 6. *Suppose an ontology graph $\mathcal{G}$ has $|\mathcal{V}| = n \geq 1$ concepts $v_1, ..., v_n$. Let $\mathrm{r}(v_i)$ be an importance function on $\mathcal{V}$, and $r_i = \mathrm{r}(v_i)$ be the importance value of v_i where $0 \leq r_i \leq 1$, $\sum r_i = 1$, and $\mathbf{W} = (w_{i,j})$ be the relation weight matrix. We call $\mathbf{R} = (r_1, ..., r_n)$ the ontology graph $\mathcal{G}$'s **concept importance vector** , and $\mathbf{L}_i = (r_1 w_{i,1}, \cdots, r_n w_{i,n})$ the concept v_i's **relation importance vector** .*

It is possible that there exists more than one relation from concept v_i to concept v_j. Therefore, $r_j w_{i,j}$ is the total importance value of all the relations from concept v_i to concept v_j. Suppose there are $m > 0$ such relations, $e_{i,k_1,j}, ..., e_{i,k_m,j}$. We define the importance of individual relation $e_{i,k_l,j}$ to be $\frac{r_j w_{i,j}}{m}$ for any $1 \leq l \leq m$.

Since a concept, like a hub according to the first two features of our model, sinks the importance of other concepts, the computation for the importance is totally the reverse of the process in PageRank. In fact, CARRank traces the stream of consciousness reversely similar to the idea of Reverse PageRank [10]. The difference is that it updates the weight of relations during the iteration according to the last two features of the model. Given an ontology graph $\mathcal{G} = (\mathcal{V}, \mathcal{E}, l_{\mathcal{V}}, l_{\mathcal{E}})$, after k ($k = 0, 1, 2, ...$) iterations, the importance of a concept $s \in \mathcal{V}$ and the weight of relation(s) from s to another concept $t \in \mathcal{V}$ are written as $r_{k+1}(s)$ and $w_{k+1}(s, t)$ respectively. They are recursively evaluated in Equations 3 and 4.

$$w_{k+1}(s, t) = \frac{r_k(s)}{\displaystyle\sum_{t_i \in B_t} r_k(t_i)} \tag{3}$$

$$r_{k+1}(s) = \frac{1 - \alpha}{|\mathcal{V}|} + \alpha \sum_{t_i \in F_s} r_k(t_i) w_{k+1}(s, t_i) \tag{4}$$

Like PageRank-like algorithms, we use a damping factor $0 < \alpha < 1$ as the probability at which CARRank will get bored of reversely tracing the stream of consciousness and begin looking for another concept on the ontology graph.

Equations 3 and 4 reflect the features of our potentially important concepts and relations model. Equation 3 formalizes the last feature, which computes the weight of relation(s) starting from concept s to concept t at the $(k+1)$th iteration. The weight is in proportional to the importance of s and in the inverse ratio of the sum of all importance of t's backward concepts at the kth iteration. Therefore, an important concept will increase the weight of those relations starting from itself. Equation 4 formalizes the first three features, which compute the importance of concept s at the $(k + 1)$th iteration. The importance consists of two parts. One is contributed by all the importance of s's forward concepts and the weight of relations from s to the forward concepts with probability α. The other is contributed by some independent jump probabilities (here is $\frac{1}{|\mathcal{V}|}$) when CARRank leaves the current stream of consciousness with probability $1 - \alpha$.

For any initial distribution of concept importance vector $\mathbf{R_0} = (r_1^0, r_2^0, ..., r_n^0)$, we have proved[3] that the iterative sequence $\{\mathbf{R}_k \mid k = 0, 1, 2, ...\}$ will converge to $\mathbf{R}^*$ which is the solution of this non-linear equations, i.e. the final result of concept importance vector. Correspondingly, $\mathbf{W}^*$ is the final result of the relation weight matrix. In numerical analysis, it is reasonable to take $\mathbf{R}_{k+1}$ as the approximation of $\mathbf{R}^*$ and stop the iterative process, if the difference between two successive iterations $\| \mathbf{R}_{k+1} - \mathbf{R}_k \|$ is small enough. Thus ranking the importance of the concepts is performed by sorting the entries in $\mathbf{R}^*$. With a slight effort, ranking the importance of the relations related to certain concept is performed by sorting the entries in the relation importance vector which is computed with $\mathbf{W}^*$ and $\mathbf{R}^*$.

---

[3] For the details of the proof, see our technical report [24]. The proof indicates that CARRank is a flexible algorithm for evaluating the importance of vertices and edges simultaneously in any kind of directed graph.

Let $\mathbf{A}$ be the adjacency matrix representation of an ontology graph[4], and $\mathbf{S}$ be the initial concept importance vector. In terms of Equation 3, 4 and the above descriptions, we present the CARRank algorithm as follows.

```
CARRANK(A, S)
 1    R₀ ← S, W₀ ← 0, k ← 0
 2    repeat
 3          Σ ← ARₖ
 4          for i ← 1, 2, ..., n
 5              do for j ← 1, 2, ..., n
 6                  do if σᵏᵢ,ⱼ ≠ 0
 7                      then wᵏ⁺¹ᵢ,ⱼ ← rᵏᵢ / σᵏᵢ,ⱼ
 8          Rₖ₊₁ ← Wₖ₊₁Rₖ
 9          d ← ‖Rₖ‖₁ − ‖Rₖ‖₁
10          Rₖ₊₁ ← Rₖ₊₁ + dE
11          δ ← ‖Rₖ₊₁ − Rₖ‖₁
12          k ← k + 1
13    until δ < ε
14    return (Wₖ, Rₖ)
```

The algorithm consists of two parts, the update of the relation weight matrix (line 3 to 7) and the update of the concept importance vector (line 8 to 10). $\sigma^k_{i,j}$ is the sum of ranks of concepts which are i's backward concepts at step k. Damping factor α in Equation 4 is represented in vector as $\mathbf{E}$ where $\|\mathbf{E}\|_1 = \alpha$. Ignoring the differences in concepts, $\mathbf{E}$ is usually a uniform distribution. Threshold $0 < \varepsilon < 1$ controls the termination of the iteration. The algorithm returns $\mathbf{R}_k$ and $\mathbf{W}_k$ as the limits of the concept importance vector and the relation weight matrix.

5   Experiments

We study the feasibility of CARRank from three aspects: ranking qualities, semantic abilities, and efficiencies.

5.1   Experimental Settings

Evaluation Metrics. The metric for measuring the efficiency of ranking algorithms is the number of iterations k that minimizes the difference between two successive iterations $\| \mathbf{R}_{k+1} - \mathbf{R}_k \|$ to a given threshold ε. A smaller k indicates a faster convergence.

In order to measure the quality of concepts ranking results, we employ a variant first 20 precision metric [14]. The improved first 20 precision, $P@20 = \frac{n_{1\sim3}\times20 + n_{4\sim10}\times17 + n_{11\sim20}\times10}{279}$, assigns different weights for the first 3, the next 7, and the last 10 results to increase the value for ranking effectiveness.

Similarly, we define $PR = \frac{\sum_{c\in C_{1\sim20}} \frac{m_c}{5}}{|C_{1\sim20}|}$ to measure the quality of relation ranking results, where $C_{1\sim20}$ is the relevant concepts in the first 20 most important concepts, and m_c is the count of relevant relations in the first 5 most important relations starting from concept c.

---

[4] $\mathbf{A}$ is obtained by parsing an ontology file into an RDF graph, and mapping it to an ontology graph, and finally constructed according to Definition 4.

A higher value of $\widetilde{P@20}$ or PR means a better quality of ranking the importance of concepts or relations.

Ranking Methods. Most of the related work in Section 2 are not specific for ontology understanding as shown in Table 1. Appropriate modifications are made in order to make them comparable. **1)** We choose the standard PageRank(**PR**) algorithm [5] on behalf of those PageRank-like algorithms. **2)** We extract the importance based labeling method from [20] which represents the methods that only consider concept hierarchy(**CH**). **3)** AKTiveRank [1] algorithm is modified by only considering the aggregation of density and betweenness measures (**DEM+BEM**) for each concept as the importance. CMM and SSM are irrelevant to the task of ontology understanding.

Experimental Environments. The experiments were carried out on a Windows 2003 Server with two Dual-Core Intel Xeon processors (2.8 GHz) and 3GB memory. For some ranking methods, let damping factor $\alpha = 0.85$, and threshold $\varepsilon = 1 \times 10^{-6}$ by default.

5.2   Ranking Qualities

To evaluate our proposed approach, we tried to collect representative ontologies and their accurate answers (a list of ranked concepts and relations) as possible as we could. In this experiment, four representative ontologies from the SchemaWeb[5] dataset are selected as shown in Table 2. "OWL" is a well-known meta ontology. "Software Project" is a full version of our running example which has a small number of concepts and relations, while, "Copyright Ontology" and "Travel Ontology" are more complex.

Table 2. Four ontologies

	Concept#	Property#	URL
OWL	17	24	http://www.w3.org/2002/07/owl.rdf
Software Project	14	84	http://keg.cs.tsinghua.edu.cn/persons/tj/ontology/software.owl
Copyright Ontology	98	46	http://rhizomik.net/ontologies/2006/01/copyrightonto.owl
Travel Ontology	84	211	http://learn.tsinghua.edu.cn:8080/2003214945/travelontology.owl

We take the ontology creators' feedback to the ranking task as the reference answers. We sent emails to the four contact creators, and got three ranks (for Software Project, Copyright Ontology, and Travel Ontology) and one suggestion (the creator of OWL recommended [23] as his answer) back in their replies. In our inquiry email, the following ranking instruction is described:

For each ontology file, list top 20 (or as many as you like) important concepts (with URI) of your ontology in your mind. And for each top concept, please give top 5 (or as many as you like) important relations (with URI) for that concept.

---

[5] http://www.schemaweb.info/

Table 3. The importance of concepts – Software ontology

Rank	Reference Answer	PageRank	DEM+BEM	CARRank	User Study
1	Project	*Message*	*Project*	*Project*	*Project*
2	Member	has_usage_statistics	*Usage_statistics*	*Usage_statistics*	*Category*
3	Developer	statistics_bugs	*Developer*	Statistic_record	*Message*
4	Category	statistic_record_support	Statistic_record	*Developer*	Discussion
5	Public_forum	*Member*	*Member*	*Category*	Help
6	LastestNew	*Project*	*Message*	Release_package	Person
7	Message	*Developer*	*Public_forums*	*Member*	*Member*
8	Version	*Category*	Person	*Message*	*Developer*
9	homepage	super_category	*Category*	Help	Project_admin
10	Usage_statistics	page_views	Project_admin	*Public_forums*	*Public_forums*

Table 4. The importance of relations – Software ontology

Top 5 Concepts		Reference Answer	Ranking results CARRank	User Study
Project	1	title	has_usage_statistics	*project_homepage*
	2	summary	developed_by	*title*
	3	activity_ranking	belong_to_category	*activity_ranking*
	4	project_homepage	translations	has_public_forum
	5	project_of_statistic	intended_audience	has_usage_statisitics
Member	1	login_name	post_message	person_name
	2	publicly_displayed_name	*site_member_since*	
	3	email_address	*login_name*	
	4	user_id	*email_address*	
	5	site_member_since	*publicly_displayed_name*	
Developer	1	skills	member_of_project	person_name
	2	project_role	*project_role*	
	3		*skills*	
	4		*user_id*	
	5			
Category	1	hasProject	*hasProject*	*super_category*
	2	category_name	*sub_category*	*sub_category*
	3	super_category	*super_category*	*category_name*
	4	sub_category	*category_name*	*hasProject*
	5			
Public_Forum	1	hasMessage	*hasMessage*	*hasMessage*
	2	belong_to_project		
	3	project_of_forum		
	4			
	5			

With these reference answers, we compare CARRank with the four other ranking methods mentioned above and a user study. The user study was conducted on 5 volunteers whose research interests include the Semantic Web. We provided each volunteer the four ontologies that they never knew about before, in their original file formats, e.g. RDF or OWL. And then, for each ontology, volunteers were required to independently give the top 20 important concepts and the top 5 important relations for each top concept as their own ranking results. In this way, given one of the four ontologies, for each volunteer, we can computed a $P@20$ value and a PR values according to his/her ranking results. The arithmetic means on five $P@20$ values and five PR values are used to represent the corresponding metrics of the user study.

Table 3 and 4 present the comparisons on concepts and relations ranking for a full version of our running example. Here, we choose one of the five ranking results collected in the user study which has the highest $P@20$ value.

Items listed in italic bold font are relevant ranking results. In Table 3, there are 5 relevant items in the first 10 ranking results for PageRank, 7 for DEM+BEM, 7 for CARRank, and 6 for the user study. Obviously, CARRank and DEM+BEM both have better ranking qualities than the user study. It means that they can somewhat support the ontology understanding. It also shows that PageRank is not a proper method in ranking the importance of concepts with less relevant results than the user study. Both CARRank and DEM+BEM rank concept "Project" the first place. The major difference of their results is that DEM+BEM considers "Person" and "Project_admin", while CARRank considers "Help" and "Release package". However, "Person" is relatively not important in this ontology because it is a base class of "Developer" and "Member" in the class hierarchy and rarely instantiated. PageRank fails in ranking "Project" the first place, which greatly lower its ranking qualities.

As the other four ranking methods do not directly support to rank the importance of relations, Table 4 only gives the comparisons of CARRank and the user study. It lists the first 5 relations (if available) starting from each concept of the first 5 concepts in the reference answers[6]. Apparently, CARRank can better reflect the importance of relations except for the concept "Project", since its ranking results are closer to the reference answers most of the time. For concept "Project", several owl:DatatypeProperty type relations, e.g. "title", "summary", "activity_ranking", and "project_homepage", are given in the reference answers. Such relations usually link to those simple data type values which have no outgoing edges hence very low importance as concepts. Therefore, according to Equation 3, owl:DatatypeProperty type relations are assigned low importance. We believe that it is beyond the scope of link analysis ranking algorithms.

We further examine the quality of ranking results with $\widetilde{P@20}$ and PR. The comparisons are illustrated in Figure 4 and Table 5. CARRank has some affirmative ability for helping ontology understanding, because it obtained a better result than the user study did. Though the precision of CARRank for "Software" is only about 4 percentage higher than that of users' decision, the degree of the support will be amplified along with the increase of the ontology's scale and complexity as shown in Figure 4. We find users can hardly decide the top important

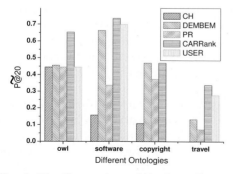

Table 5. The Comparison of Ranking Relations

	CARRank	User
copyright	0.06	0
software	0.586	0.562

Fig. 4. The Comparison of Ranking Concepts

---

[6] In fact, every concept listed in Table 4 has more than five relations except "Public_Forum". However, the creator could not provide us more relations than the reference answers.

concepts for "Copyright Ontology" for its complexity. Obviously, CARRank is helpful in this case. Another interesting observation is that our algorithm is also effective to those meta ontologies like "OWL".

5.3   Comparison of Semantic Abilities

To exhibit the semantic abilities of CARRank, we generate three variations of FOAF ontology[7], i.e. OWL-Full, OWL-DL, and OWL-Lite, with a tool named foaf_cleaner [2]. Then, CARRank is applied on the three versions of FOAF and the original FOAF. Results are shown in Table 5.

	Original	OWL-Full	OWL-DL	OWL-Lite
Person	1	1	1	1
Document	2	2	2	2
Organization	3	3	3	5
Project	4	4	4	4
Agent	5	5	5	3
OnlineEcommerceAccount	6	6	6	7
OnlineChatAccount	7	7	7	8
OnlineGamingAccount	8	8	8	9
OnlineAccount	9	9	9	10
PersonalProfileDocument	10	10	10	11
Image	11	11	11	6
Group	12	12	12	12
Pearson Correlation Coefficients		1.0	1.0	0.867

Fig. 5. Top 10 Concepts for FOAF

Table 6. Top 10 Concepts for CYC

Rank	Concepts
1	RNAPolymerase
2	ExtensionOf-C-Regular
3	ClosedUnderGeneralizations-Classical
4	NetworkPortNumber
5	SimpleWord
6	GLFGraph
7	BrigadeOrRegimentSized
8	BrigadeOrRegimentSized
9	ExtensionOf-K-Normal
10	GLFAnalysisDiagramGraph

There are totally 12 concepts involved. The values in the first two columns are the concepts and their ranks produced by applying CARRank on the original FOAF ontology. The values in the last three columns are the ranks for the three versions. We use the Pearson Correlation Coefficient to measure the similarity of ranking results between one OWL version and the original version. The ranking results for the OWL-Full and OWL-DL are the same as that for the original one, though owl:imports of the OWL and RDFS ontologies are removed from the original, and owl:InverseFunctionalProperty on owl:DatatypeProperty is removed from OWL-Full. The only affection happens to the ranking results of OWL-Lite when owl:disjointWith is removed from OWL-DL. However, the similarity is still over 85%. This indicates that CARRank can capture most of the semantics even when the language expressive power changes.

Another challenge for semantic abilities of CARRank is to rank large scale ontologies, e.g. CYC[8] (23.7MB). Large scale ontologies are always developed collaboratively by many creators for a long time. Because of the limitations of individual creator and the limitation of the time, a global design intention may be unstable or even inconsistent. The interesting ranking results of CYC are listed in Table 6. There are 30432 classes and properties defined with 254371 RDF triples. It seems that CARRank ranks higher some abstract concepts for

---

[7] http://xmlns.com/foaf/spec/
[8] http://www.cyc.com/2004/06/04/cyc

their complicated class hierarchy constructed with rdfs:subClassOf. Although it is hard to determine the quality of ranking results for such large scale ontology, we still suggest to use CARRank to periodically rank the concepts during its composition in order to discover early the deviation of design intention.

5.4   Efficiencies

Convergence Comparison. Figure 6 presents the comparisons among PageRank, Reverse PageRank, and CARRank. Rankings are performed on "Relationship"[9] ontology which has 169 vertices and 252 directed labeled edges in its ontology graph. Obviously, CARRank and Reverse PageRank have conformable convergent speed because both consider the hub score instead of authority score. The only difference is that the additional time spent on updating the relation weight matrix makes CARRank a little slower than Reverse PageRank.

On the other hand, the convergent speed of both CARRank and Reverse PageRank are quite different from that of PageRank. The reason is that PageRank considers authority score instead of hub score. Therefore, the convergent speed may be various with respect to the topological structure of the ontology graph. In Figure 6 the convergent speed of PageRank is much faster. However, take "UNSPSC"[10] ontology on SchemaWeb for another example. There are 19600 vertices and 29386 directed labeled edges. As shown in Figure 7, CARRank and Reverse PageRank express the same convergent speed and converge to the threshold early than PageRank. In any case, the convergent speed is acceptable for CARRank.

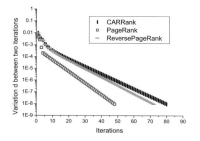
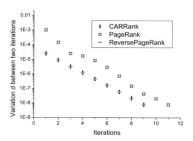

Fig. 6. Convergence ("Relationship")   **Fig. 7.** Convergence ("UNSPSC")

6   Conclusion and Discussion

CARRank is a simple yet effective algorithm for identifying potentially important concepts and relations in an ontology. The experimental results show the feasibility of CARRank from the ranking qualities and the semantic abilities.

---

[9] http://purl.org/vocab/relationship/
[10] http://www.ksl.stanford.edu/projects/DAML/UNSPSC.daml

Although ontology understanding means much more than our proposed solution. we expect **CARRank** to be a preliminary step towards identifying potentially important concepts and relations user-independently. In addition, we also agree that being user-independent may not meet all the needs of application. Fortunately, **CARRank** can be personalize by letting user provide a sub-graph of the ontology which mainly contains the concepts and relations concerned about. It would be interesting to explore the ranking based on users' tasks and needs in the future work.

Acknowledgments

We would like to thank all the ontology creators who contributed their ranking results, and all the reviewers for their constructive comments and suggestions.

References

[1] Alani, H., Brewster, C., Shadbolt, N.: Ranking ontologies with aktiverank. In: Cruz, I., Decker, S., Allemang, D., Preist, C., Schwabe, D., Mika, P., Uschold, M., Aroyo, L.M. (eds.) ISWC 2006. LNCS, vol. 4273. Springer, Heidelberg (2006)

[2] Alford, R.: Using FOAF and OWL (July 2005), http://www.mindswap.org/2005/foaf_cleaner/

[3] Balmin, A., Hristidis, V., Papakonstantinou, Y.: Objectrank: Authority-based keyword search in databases. In: VLDB, pp. 564–575 (2004)

[4] Bontas, E.P., Mochol, M.: Towards a cost estimation model for ontology engineering. In: Berliner XML Tage, pp. 153–160 (2005)

[5] Brin, S., Page, L.: The anatomy of a large-scale hypertextual web search engine. Computer Networks 30(1-7), 107–117 (1998)

[6] Buitelaar, P., Eigner, T., Declerck, T.: Ontoselect: A dynamic ontology library with support for ontology selection. In: The Demo Session at the ISWC (2004)

[7] d'Entremont, T., Storey, M.-A.: Using a degree-of-interest model for adaptive visualizations in protégé. In: 9th International Protégé Conference (2006)

[8] Ding, L., Pan, R., Finin, T., Joshi, A., Peng, Y., Kolari, P.: Finding and Ranking Knowledge on the Semantic Web. In: Gil, Y., Motta, E., Benjamins, V.R., Musen, M.A. (eds.) ISWC 2005. LNCS, vol. 3729, pp. 156–170. Springer, Heidelberg (2005)

[9] Ernst, N.A., Storey, M.-A., Allen, P.: Cognitive support for ontology modeling. Int. J. Hum.-Comput. Stud. 62(5), 553–577 (2005)

[10] Fogaras, D.: Where to start browsing the web? In: Böhme, T., Heyer, G., Unger, H. (eds.) IICS 2003. LNCS, vol. 2877, pp. 65–79. Springer, Heidelberg (2003)

[11] Gruber, T.R.: What is an ontology (December 2001)

[12] James, W.: The principles of psychology. Harvard (1890)

[13] Kleinberg, J.M.: Authoritative sources in a hyperlinked environment. J. ACM 46(5), 604–632 (1999)

[14] Leighton, H.V., Srivastava, J.: First 20 precision among world wide web search services (search engines). Journal of the American Society for Information Science 50(10), 870–881 (1999)

[15] Mei, J., Boley, H.: Interpreting swrl rules in rdf graphs. Electr. Notes Theor. Comput. Sci. 151(2), 53–69 (2006)

[16] Neches, R., Fikes, R., Finin, T., Gruber, T., Patil, R., Senator, T., Swartout, W.R.: Enabling technology for knowledge sharing. AI Mag. 12(3), 36–56 (1991)

[17] Nie, Z., Zhang, Y., Wen, J.-R., Ma, W.-Y.: Object-level ranking: bringing order to web objects. In: WWW, pp. 567–574 (2005)

[18] Patel, C., Supekar, K., Lee, Y., Park, E.K.: Ontokhoj: a semantic web portal for ontology searching, ranking and classification. In: WIDM, pp. 58–61 (2003)

[19] Sabou, M., Lopez, V., Motta, E.: Ontology selection for the real semantic web: How to cover the queens birthday dinner? In: Managing Knowledge in a World of Networks. LNCS, pp. 96–111. Springer, Heidelberg (2006)

[20] Tu, K., Xiong, M., Zhang, L., Zhu, H., Zhang, J., Yu, Y.: Towards imaging large-scale ontologies for quick understanding and analysis. In: Gil, Y., Motta, E., Benjamins, V.R., Musen, M.A. (eds.) ISWC 2005. LNCS, vol. 3729. Springer, Heidelberg (2005)

[21] W3C. Resource Description Framework (RDF): Concepts and Abstract Syntax (2004), http://www.w3.org/TR/rdf-concepts/

[22] W3C. SWRL: A Semantic Web Rule Language Combining OWL and RuleML (2004), http://www.w3.org/Submission/SWRL/

[23] Wang, T.D., Parsia, B., Hendler, J.: A survey of the web ontology landscape. In: Cruz, I., Decker, S., Allemang, D., Preist, C., Schwabe, D., Mika, P., Uschold, M., Aroyo, L.M. (eds.) ISWC 2006. LNCS, vol. 4273. Springer, Heidelberg (2006)

[24] Wu, G.: Understanding an ontology by ranking its concepts and relations. Technical report, Tsinghua University (January 2008), http://166.111.68.66/persons/gangwu/publications/kegtr-carrank.pdf

RoundTrip Ontology Authoring

Brian Davis[1], Ahmad Ali Iqbal[1,3], Adam Funk[2], Valentin Tablan[2],
Kalina Bontcheva[2], Hamish Cunningham[2], and Siegfried Handschuh[1]

[1] Digital Enterprise Research Institute, Galway, Ireland
[2] University of Sheffield, UK
[3] University of New South Wales, Australia

Abstract. Controlled Language (CL) for Ontology Editing tools offer
an attractive alternative for naive users wishing to create ontologies, but
they are still required to spend time learning the correct syntactic structures and vocabulary in order to use the Controlled Language properly.
This paper extends previous work (CLOnE) which uses standard NLP
tools to process the language and manipulate an ontology. Here we also
generate text in the CL from an existing ontology using template-based
(or shallow) Natural Language Generation (NLG). The text generator
and the CLOnE authoring process combine to form a RoundTrip Ontology Authoring environment: one can start with an existing imported
ontology or one originally produced using CLOnE, (re)produce the Controlled Language, modify or edit the text as required and then turn
the text back into the ontology in the CLOnE environment. Building
on previous methodology we undertook an evaluation, comparing the
RoundTrip Ontology Authoring process with a well-known ontology editor; where previous work required a CL reference manual with several
examples in order to use the controlled language, the use of NLG reduces
this learning curve for users and improves on existing results for basic
ontology editing tasks.

1 Introduction

Formal data representation can be a significant deterrent for non-expert users
or small organisations seeking to create ontologies and subsequently benefit
from adopting semantic technologies. Existing ontology authoring tools such
as Protégé[1] attempt to resolve this, but they often require specialist skills in
ontology engineering on the part of the user. This is even more exasperating
for domain specialists, such as clinicians, business analysts, legal experts, etc.
Such professionals cannot be expected to train themselves to comprehend Semantic Web formalisms and the process of knowledge gathering; involving both
a domain expert and an ontology engineer can be time-consuming and costly.
Controlled languages for knowledge creation and management offer an attractive
alternative for naive users wishing to develop small to medium sized ontologies
or a first draft ontology which can subsequently be post-edited by the Ontology

---

[1] http://protege.stanford.edu

A. Sheth et al. (Eds.): ISWC 2008, LNCS 5318, pp. 50–65, 2008.

Engineer. In previous work[1], we presented CLOnE - *Controlled Language for Ontology Editing* which allows naive users to design, create, and manage information spaces without knowledge of complicated standards (such as XML, RDF and OWL) or ontology engineering tools. CLOnE's components are based on GATE's existing tools for IE (Information Extraction) and NLP (Natural Language Processing) [2].

The CLOnE system was evaluated using a *repeated-measures, task-based* methodology in comparison with a standard ontology editor – Protégé. CLOnE performed favourably with test users in comparison to Protégé. Despite the benefits of applying Controlled Language Technology to Ontology Engineering, a frequent criticism against its adoption, is the learning curve associated with following the correct syntactic structures and/or terminology in order to use the Controlled Language properly. Adhering to a controlled language can be, for some naive users, time consuming and annoying. These difficulties are related to the *habitability* problem, whereby users do not really know what commands they can or cannot specify to the NLI (Natural Language Interface) [3]. Where the CLOnE system uses natural language analysis to unambiguously parse CLOnE in order to create and populate an ontology, the reverse of this process, NLG (Natural Language Generation), involves the generation of the CLOnE language from an existing ontology. The text generator and CLOnE authoring processes combine to form a RoundTrip Ontology Authoring(ROA) environment: a user can start with an existing imported ontology or one originally produced using CLOnE, (re)produce the Controlled Language using the text generator, modify or edit the text as required and subsequently parse the text back into the ontology using the CLOnE environment. The process can be repeated as necessary until the required result is obtained. Building on previous methodology [1], we undertook a repeated-measures, task-based evaluation, comparing the RoundTrip Ontology Authoring process with Protégé. Where previous work required a reference guide in order to use the controlled language, the substitution of NLG can reduce the learning curve for users, while simultaneously improving upon existing results for basic Ontology editing tasks. The remainder of this paper is organized as follows: Section 2 discusses the design and implementation of the ROA pipeline focusing on the NLG component - the ROA text generator, Section 3 presents our evaluation and discusses our quantitative findings. Section 4 discusses related work. Finally, Section 5 and Section 6 offer conclusions and future work.

2   Design and Implementation

In this section, we describe the overall architecture of the Round Trip Ontology Authoring (ROA) pipeline which is implemented in GATE [2]. We discuss briefly extensions to existing CLOnE components of ROA, but focus the attention of this section towards describing the CLOnE text generator, the algorithm used and the XML configuration file containing templates needed to configure the controlled language output of the generator.

2.1   RoundTrip Ontology Authoring (ROA) and CLOnE

ROA builds on and extends the existing advantages of the CLOnE software and input language, which are described below:

1. ROA requires only one interpreter or runtime environment, the Java 1.6 JRE.
2. ROA like CLOnE uses a sub-language of English.
3. As far as possible, CLOnE is grammatically lax; in particular it does not matter whether the input is singular or plural (or even in grammatical agreement).
4. ROA can be compact; the user can create any number of classes or instances in one sentence.
5. ROA is more flexible and easier to learn by using simple examples of how to edit the controlled language generated by the text generator in order to modify the Ontology. It reduces the need to learn the Controlled Language by following examples, style guides or CLOnE syntactic rules. Instead, a user can create or modify various classes and instances in one (generated) sentence or (using simple copy and paste) create new properties between new or existing classes and instances.
6. The CLOnE grammar within ROA has been extended to handle simple verbs and phrasal verbs.
7. Like CLOnE any valid sentence of ROA can be unambiguously parsed.
8. The advantage of the GATE Ontology API allows users to import existing Ontologies for generation, subsequent editing in ROA and export the result to different Ontology formats.
9. SimpleNLG[2] has been added into the ROA text generator to lexicalize unseen properties.

   Procedurally, CLOnE's analysis consists of the ROA pipeline of processing resources (PRs) shown in Figure 1 (left dotted box). This pipeline starts with a series of fairly standard GATE NLP tools which add linguistic annotations and annotation features to the document. These are followed by three PRs developed particularly for CLOnE: the gazetteer of keywords and phrases fixed in the controlled language and two JAPE[3] transducers which identify quoted and unquoted chunks. Names enclosed in pairs of single or double quotation marks can include reserved words, punctuation, prepositions and determiners, which are excluded from unquoted chunks in order to keep the syntax unambiguous. The last stage of analysis, the CLOnE JAPE transducer, refers to the existing ontology in several ways in order to interpret the input sentences. Table 1 below provides an excerpt of the grammar rules of the CLOnE language. We refer the reader to [1,4] for additional rules and examples.

---

[2] http://www.csd.abdn.ac.uk/~ereiter/simplenlg/
[3] GATE provides the *JAPE* (Java Annotation Pattern Engine) language for matching regular expressions over annotations, adding additional annotations to matched spans, and manipulating the match patterns with Java code.

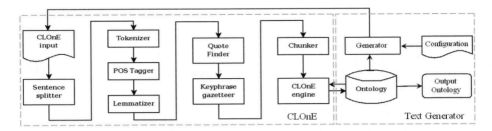

Fig. 1. The ROA RoundTrip Ontology Authoring pipeline

Table 1. Excerpt of CLOnE grammar with examples

Sentence Pattern	Example	Usage
Forget everything.	Forget everything.	Clear the whole ontology corpus to start with the new ontology.
(Forget that) There is/are <CLASSES>.	There are researchers, universities and conferences.	Create or delete (new) classes.
(Forget that) <INSTANCES> is a/are <CLASS>.	Ahmad Ali Iqbal and Brian Davis are 'Ph.D. Scholar'.	Create (or delete) instances of the class.
(Forget that) <SUBCLASSES> is/are a type/types of <SUPERCLASS>.	'Ph.D. Scholar' is a type of Student.	Make subclass(es) of an existing super-class. 'Forget that' only unlinks the the subclass-superclass relationship.
(Forget that) <CLASSES/INSTANCES> <VERB PROPERTY> <CLASSES/INSTANCES>.	Professor supervises student.	Create the property of the form Domain_verb_Range either between two classes or instances.

2.2 Text Generation of CLOnE

The text generation component in Figure 1 (right dotted box) displayed in the ROA pipeline is essentially an Ontology Verbalizer. Unlike some NLG systems, the communicative goal of the text generator is not to construct tailored reports for specific content within the knowledge base or to respond to user specific queries. Hence no specific content selection subtask or "choice" is performed since our goal is to describe and present the Ontology in textual form as unambiguous subset of English - the CLOnE language for reading, editing and amendment. We select the following content from the Ontology: top level classes, subclasses, instances, class properties, their respective domain and ranges and instance properties. The text generator is configured using an XML file, whereby text templates are instantiated and filled by the values from the Ontology. This

```
56  <!-- Template for all the other top classes -->
57  <template>
58      <in>
59          <triple id="t1">
60              <property ns="rdf" name="type"/>
61              <object ns="owl" name="Class"/>
62          </triple>
63          <triple id="t1.subject">
64              <subject ref="t1.subject"/>
65              <property ns="rdfs" name="subClassOf"/>
66              <object name="Entity"/>
67          </triple>
68      </in>
69      <out>
70          <singular>
71              <phrase>There are
72              <ref ref="t1.subject" number="plural"/>.
73              </phrase>
74          </singular>
75          <plural>
76              <phrase>There are
77              <ref ref="t1.subject" number="plural"/>.
78              </phrase>
79          </plural>
80      </out>
81      <ignoreIf>
82      </ignoreIf>
83  </template>
```

```
166  <!-- Template for defining class properties -->
167  <template>
168      <in>
169          <triple id="t1">
170              <property ns="rdf" name="type"/>
171              <object ns="rdf" name="Property"/>
172          </triple>
173
174          <triple id="t2">
175              <subject ref="t1.subject"/>
176              <property ns="rdfs" name="domain"/>
177          </triple>
178
179          <triple id="t3">
180              <subject ref="t1.subject"/>
181              <property ns="rdfs" name="range"/>
182          </triple>
183      </in>
184      <out>
185          <singular>
186              <phrase>
187              <ref ref="t2.object" number="singular"/>
188              <ref ref="t1.subject" number="singular"/>
189              <ref ref="t3.object" number="singular"/>.
190              </phrase>
191          </singular>
192          <plural>
193              <phrase>
194              <ref ref="t2.object" number="singular"/>
195              <ref ref="t1.subject" number="singular"/>
196              <ref ref="t3.object" number="plural"/>.
197              </phrase>
198          </plural>
199      </out>
200  </template>
```

Fig. 2. Example of a generation template

file is decoupled from the text generator PR. Examples of two templates used to generate top level classes and class properties are displayed in Figure 2. The text generator (See **Generator** in Figure 1) is realised as a GATE PR and consists of **three** stages:

Stage 1 within the text generator converts the input ontology into an internal GATE ontological resource and flattens it into RDF style triples. This is executed in a breadth-first manner—so lists are created where super-classes always precede their corresponding subclasses—in the following order: top-level classes, subclasses, instances, class properties, and instance properties.

Stage 2 matches generation templates from the configuration file (See Figure 2) with the triples list derived from the Ontology in Stage 1. A generation template has three components: (1) an `in` element containing a list of triple specifications, (2) an `out` element containing phrases that are generated when a successful match has occurred and (3) an optional `ignoreiIf` element for additional triple specifications that cause a match specified in the `in` element to be ignored if the conditions are satisfied. The triple specifications contained within the `in` portion of the template can have subject, property and object XML elements. The triple specifications act as restrictions or conditions, such that an input triple generated from the Ontology must match this template. If more than one triple is included in the `in` element they are considered as a conjunction of restrictions, hence the template will only match if one or more actual triples for all triple specifications within the `in` element are found. One triple can reference another, i.e., a specification can constrain a second triple to have the same object as the subject of the first triple. Only backward referencing is permitted since the triples are matched in a top down fashion according to their textual ordering. An example of referencing can be seen in line 188 of the `out` element of the template shown in Figure 2 for generating class properties.

In **Stage 3** the `out` section of the template describes how text is generated from a successful match. It contains phrase templates that have text elements and references to values matched within the `in` elements. Phrases are divided into singular and plural forms. Plural variants are executed when several triples are grouped together to generate a single sentence (Sentence Aggregation) based on a list of Ontology objects (i.e., `There are Conferences, Students and Universities`). Text elements within a template are simply copied into the output while reference values are replaced with actual values based on matching triple specifications. We also added a small degree of lexicalization into the Text Generator PR, whereby, for example, an unseen property, which is treated as a verb is inflected correctly for surface realisation i.e. `study` and `studies`. This involves a small amount of dictionary look-up using the SimpleNLG Library to obtain the third person singular inflection `studies` from `study` to produce `Brian Davis studies at NUIG`. The `out` elements of the generation template also provide several phrase templates for the singular and plural sections. These are applied in rotation to prevent tedious and repetitive output.

Stage 2 also groups matches together into sets that can be expressed together in a plural form. For this to proceed, the required condition is that the difference between matches, occurs in only one of the references used in the phrase templates, i.e., if singular variants would only differ by one value. A specialized generation template with no `in` restrictions is also included in the configuration file. This allows for the production of text where there are no specific input triple dependencies.

3   Evaluation

3.1   Methodology

Our methodology is deliberately based on the criteria previously used to evaluate CLOnE [1,4], so that we can fairly compare the earlier results using the CLOnE software with the newer RoundTrip Ontology Authoring(ROA) process. The methodology involves a *repeated-measures, task-based* evaluation: each subject carries out a similar list of tasks on both tools being compared. Unlike our previous experiment, the CLOnE reference guide list and examples are withheld from the test users, so that we can measure the benefits of substituting the text generator for the reference guide and determine its impact on the learning process and usability of CLOnE. Furthermore, we used a larger sample size and more controls for bias. All evaluation material and data are available online for inspection, including the CLOnE evaluation results for comparison[4]. The evaluation contained the following:

---

[4] `http://smile.deri.ie/evaluation/2008/ROA`

- A pre-test questionnaire asking each subject to test their degree of knowledge with respect to ontologies, the Semantic Web, Protégé and Controlled Languages. It was scored by assigning each answer a value from 0 to 2 and scaling the total to obtain a score of 0–100.
- A short document introducing Ontologies, the same 'quick start' Protégé instructions as used in [4] (partly inspired by Protégé's *Ontology 101* documentation [5]), and an example of editing CLOnE text derived from the text generator. The CLOnE reference guide and detailed grammar examples used in for the previous experiment [4] were withheld. Subjects were allowed to refer to an example of how to edit generated Controlled Language but did *not* have access to CLOnE reference guide.
- A post-test questionnaire for each tool, based on the *System Usability Scale* (SUS), which also produces a score of 0–100 to compare with previous results [6].
- A comparative questionnaire similar to the one used in [4] was applied to measure each user's preference for one of the two tools. It is scored similarly to SUS so that 0 would indicate a total preference for Protégé, 100 would indicate a total preference for ROA, and 50 would result from marking all the questions *neutral*. Subjects were also given the opportunity to make comments and suggestions.
- Two equivalent lists of ontology-editing tasks, each consisting of the following subtasks:
  - creating two subclasses of existing classes,
  - creating two instances of different classes, and
  - either (A) creating a property between two classes and defining a property between two instances, or (B) extending properties between two pairs of instances.

For both task lists, an initial ontology was created using CLOnE. The same ontology was loaded into Protégé for both tasks and the text generator was executed to provide a textual representation of the ontology for editing purposes(see Figure 3), again for both tasks.

For example, Task List A is as follows.

- Create a subclass *Institute* of *University*.
- Create a subclass *Workshop* of *Conference*.
- Create an instance *International Semantic Web Conference* of class *Conference*.
- Create an instance *DERI* of class *Institute*.
- Create a property that *Senior Researchers supervise Student*.
- Define a property that *Siegfried Handschuh supervises Brian Davis*.

3.2   Sample Quality

We recruited 20 volunteers from the Digital Enterprise Research Institute, Galway[5]. The sample size ($n = 20$) satisfies the requirements for reliable SUS

---

[5] http://www.deri.ie

Fig. 3. Text Generated by ROA

evaluations [7]. We recruited subjects with an industrial background (**I**) and participants with a research background (**R**). See (in Table 5) for details. In addition we attempted to control bias by selecting volunteers who were either:

- Research Assistants/Programmers/Post-Doctoral Researchers with an industrial background either returning (or new) to Academic Research respectively(**I**),
- Postgraduate Students who were new to the Semantic Web and unfamiliar with Ontology Engineering(**R**),
- Researchers from the E-learning and Sensor Networks lab but not from the Semantic Web Cluster(**R**),
- Researchers with no background in Natural Language Processing or Ontology Engineering(**R**) or
- Industrial Collaborators (**I**).

In all cases, we tried to ensure that participants had limited or no knowledge of GATE or Protégé. First, subjects were asked to complete the pre-test questionnaire, then they were permitted time to read the Protégé manual and Text Generator examples, and lastly they were asked to carry out each of the two task lists with one of the two tools. (Half the users carried out task list A with ROA and then task list B with Protégé; the others carried out A with Protégé and then B with ROA.) Each user's time for each task list was recorded. After each task list the user completed the SUS questionnaire for the specific tool used, and finally the comparative questionnaire. Comments and feedback were also recorded on the questionnaire forms.

3.3   Quantitative Findings

Table 2 summarizes the main measures obtained from our evaluation. We used SPSS[6] to generate all our statistical results. In particular the mean ROA SUS

---

[6] SPSS 2.0, http://www.spss.com

Table 2. Summary of the questionnaire scores

Measure	min	mean	median	max
Pre-test scores	17	42	42	75
ROA SUS rating	48	74	70	100
Protégé SUS rating	10	41	41	85
R/P Preference	40	72	79	95

Table 3. Confidence intervals (95%) for the SUS scores

Tool	Confidence intervals		
	Task list A	Task list B	Combined
Protégé	28–55	29–51	32–49
ROA	63–77	69–84	68–79

Table 4. Correlation coefficients

Measure	Measure	Pearson's	Spearman's	Correlation
Pre-test	ROA time	-0.41	-0.21	weak −
Pre-test	Protégé time	-0.28	-0.35	none
Pre-test	ROA SUS	-0.02	-0.00	none
Pre-test	Protégé SUS	-0.32	-0.29	weak −
ROA time	Protégé time	0.53	0.58	+
ROA time	ROA SUS	-0.65	-0.52	−
Protégé time	Protégé SUS	0.53	0.56	+
ROA time	Protégé SUS	-0.14	-0.10	none
Protégé time	ROA SUS	-0.02	-0.09	none
ROA SUS	Protégé SUS	0.04	-0.01	none
ROA SUS	R/P Preference	0.58	0.56	+
Protégé SUS	R/P Preference	-0.01	0.10	none

score is above the baseline of 65–70% while the mean SUS score for Protégé is well below the baseline [8]. In the ROA/Protégé Preference (R/P Preference) scores, based on the comparative questionnaires, we note that the scores also favour on average ROA over Protégé. Confidence intervals are displayed in Table 3.[7]

We also generated Pearson's and Spearman's correlations coefficients [9,10]. Table 4 displays the coefficients. In particular, we note the following results.

− The pre-test score has a weak negative correlations the with ROA task time.
− There are no correlations with pre-test score and the ROA SUS score.
− The pre-test score has a weak negative correlation with the Protégé SUS score.
− There are no correlations with pre-test score and the Protégé time.

---

[7] A data sample's *95% confidence interval* is a range 95% likely to contain the mean of the whole population that the sample represents [9].

- In previous results in comparing CLOnE and Protégé, the task times for both tools were more positively correlated with each other while in the case of ROA and Protégé, there correlation has being weakened by a significant 32% of its original value (of 78%) reported for CLOnE [1], indicating that the users tended not spend the equivalent time completing both ROA and Protégé tasks.
- There is a moderate correlation with Protégé task time and Protégé SUS scores.
- There is a strong negative correlation of -0.65 between the ROA task time and the ROA SUS scores. Our previous work reported no correlation between the CLOnE task time and CLOnE SUS time. A strong negative or inverse correlation implies that users who spent less time completing a task using ROA tended to produce high usability scores - favouring ROA. More importantly, we noted that the associated probability reported by SPSS, was less then the typical 5% cut-off point used in social sciences. This implies there is a 5% chance that the true population coefficient is very unlikely to be 0 (no relationship). Conversely, one can infer statistically that for 19 out of 20 (95%)users, with little or no experience in either NLP or Protégé who favour RoundTrip Ontology Authoring over Protégé also tend to spend less time completing Ontology editing tasks.
- The R/P Preference score correlates moderately with the ROA SUS score, similar to previous results, but no longer retains a significant inverse correlation with the Protégé SUS score. The reader should note the R/P Preference scores favour ROA over Protégé.

We also varied the tool order evenly among our sample. As noted previously in [1], once again the SUS scores have differed slightly according to tool order (as indicated in Table 3). Previous SUS scores for Protégé tended to be slightly lower for B than for A, which we believe may have resulted from the subjects' decrease in interest as the evaluation progressed. While in previous results there was a decrease in SUS scores for CLOnE (yet still well above the SUS baseline), in the case of ROA however, the SUS scores increased for task B (see Table 3), implying that if waning interest was a factor in the decrease in SUS scores for CLOnE, it does not appear to be the case for ROA. What is of additional interest is that group **I**, subjects with industrial background scored on average 10% higher for both ROA SUS and ROA/Protégé, which implies that Industrial collaborators or professionals with an Industrial background favoured a natural language interface over a standard Ontology Editor even more than Researchers.

3.4   User Feedback

The test users also provided several suggestions/comments about ROA.

- "RoundTrip Ontology Authoring becomes much easier, once the rules are learnt". (This is very interesting considering that no syntax rules, extended examples or restricted vocabulary list were provided).

- Use of inverted commas should be used only once and afterwards, if same the class /instance is reused, the system should automatically recognise it as the previous word.
- Many users suggested displaying the ontology pane on the right hand side of the text pane, where test users edit the text instead of moving between two separate panes.
- Some users suggested dynamic ontology generation, once a user finishes typing a sentence, the changes should be displayed automatically in the ontology pane.
- Similar suggestions to the previous evaluation were provided for user auto-completion, syntax highlighting, options about available classes, instances or property names and keywords should be displayed, a similar concept to modern Word Processor or programming IDEs such as eclipse.
- Some test users with an industrial background demonstrated concern regarding scalability and ROA using with a larger business related ontology and suggest capabilities for verbalizing a portion of the ontology tree within the Ontology viewer, using text generation for subsequent editing.
- Some test users appreciated the singular/plural forms and sentence handling of ROA (e.g., study, studies).

Table 5. Groups of subjects by source and tool order

Source	Tool order		Total
	PR	RP	
R Researcher	5	7	12
I Industry	5	3	8
Total	10	10	20

Table 6. Comparison of the two sources of subjects

Measure	Group	min	mean	median	max
Pre-test	R	17	38	38	58
	I	17	47	50	75
ROA SUS	R	48	69	70	82
	I	65	80	80	100
Protégé SUS	R	10	30	28	52
	I	12	48	49	85
R/P Preference	R	40	68	72	88
	I	65	78	78	95

4   Related Work

"Controlled Natural Languages (CL)s are subsets of natural language whose grammars and dictionaries have been restricted in order to reduce or eliminate both ambiguity and complexity"[11]. CLs were later developed specifically for computational treatment and have subsequently evolved into many variations and flavours such as Smart's Plain English Program (PEP), White's International Language for Serving and Maintenance (ILSAM) [12] and Simplified English.[8] They have also found favour in large multi-national corporations, usually within the context of machine translation and machine-aided translation of user documentation [11,12].

The application of CLs for ontology authoring and instance population is an active research area. *Attempto Controlled English*[9] (ACE) [13], is a popular CL for ontology authoring. It is a subset of standard English designed for knowledge representation and technical specifications, and is constrained to be unambiguously machine-readable into DRS - Discourse Representation Structure. ACE OWL, a sublanguage of ACE, proposes a means of writing formal, simultaneously human- and machine-readable summaries of scientific papers [14,15]. Similar to RoundTrip Ontology Authoring, ACE OWL also aims to provide reversibility (translating OWL DL into ACE). The application NLG, for the purposes editing existing ACE text, is mentioned in [16]. The paper discusses the implementation of the shallow NLG system - an OWL Verbalizer, focusing primarily on the OWL to ACE rewrite rules, however no evaluation or quantitative data are provided in attempt to measure the impact of NLG in the authoring process. Furthermore OWL's *allValuesFrom* must be translated into a construction which can be rather difficult for humans to read. A partial implementation is however available for public testing[10].

Another well-known implementation which employs the use of NLG to aid the knowledge creation process is WYSIWYM (*What you see is what you meant*). It involves direct knowledge editing with natural language directed feedback. A domain expert can edit a knowledge based reliably by interacting with natural language menu choices and the subsequently generated feedback, which can then be extended or re-edited using the menu options. The work is conceptually similar to RoundTrip Ontology Authoring, however the natural language generation occurs as a feedback to guide the user during the editing process as opposed to providing an initial summary in Controlled Language for editing. A usability evaluation is provided in [17], in the context of knowledge creation, partly based on IBM heuristic evaluations[11], but no specific quantitative data that we are aware of, is presented. However, evaluation results are available for the MILE (Maritime Information and Legal Explanation) application, which used WYSIWYM, but in the context of query formulation for the CLIME[12] project, of which the outcome was favourable [17].

Similar to WYSIWYM is *GINO* (Guided Input Natural Language Ontology Editor) provides a guided, controlled NLI (natural language interface) for

---

[8] http://www.simplifiedenglish-aecma.org/Simplified_English.htm
[9] http://www.ifi.unizh.ch/attempto/
[10] http://attempto.ifi.uzh.ch/site/tools/
[11] http://www-03.ibm.com/able/resources/uebeforeyoubegin.html
[12] CLIME, Cooperative Legal Information Management and Explanation, Esprit Project EP25414.

domain-independent ontology editing for the Semantic Web. GINO incrementally parses the input not only to warn the user as soon as possible about errors but also to offer the user (through the GUI) suggested completions of words and sentences—similarly to the "code assist" feature of Eclipse[13] and other development environments. GINO translates the completed sentence into triples (for altering the ontology) or SPARQL[14] queries and passes them to the Jena Semantic Web framework. Although the guided interface facilitates input, the sentences are quite verbose and do not allow for aggregation. A full textual description of the Ontology is not realized as is the case of the CLOnE text generator [18]. Furthermore, similar, to our evaluation, a small usability evaluation was conducted using SUS [6], however the sample set of six was too small to infer any statistically significant results [7]. In addition, GINO was not compared to any existing Ontology editor during the evaluation. Finally, [19] presents an Ontology based Controlled Natural Language Editor, similar to GINO, which uses a CFG (Context-free grammar) with lexical dependencies - CFG-DL to generate RDF triples. To our knowledge the system ports only to RDF and does not cater for other Ontology languages. Furthermore no quantitative user evaluation is provided.

Other related work involves the application of Controlled Languages for Ontology or knowledge base querying, which represent a different task than that of knowledge creation and editing but are worth mentioning for completeness sake. Most notably *AquaLog*[15] is an ontology-driven, portable Question-Answering (QA) system designed to provide a natural language query interface to semantic mark-up stored in a knowledge base. PowerAqua [20] extends AquaLog, allowing for an open domain question-answering for the semantic web. The system dynamically locates and combines information from multiple domains.

5   Conclusion and Discussion

The main research goal of this paper is to assess the effect of introducing Natural Language Generation (NLG) into the CLOnE Ontology authoring process to facilitate RoundTrip Ontology Authoring. The underlying basis of our research problem is the *habitability* problem (See Section 1): How can we reduce the learning curve associated with Controlled Languages? And how can we ensure their uptake as a Natural Language Interface (NLI)? Our contribution is empirical evidence to support the advantages of combining of NLG with ontology authoring, a process known as RoundTrip Ontology Authoring (ROA).

The reader should note, that we compared Protégé with ROA, because Protégé is the standard tool for ontology authoring. Previous work [1] compared CLOnE with Protégé. Hence, in order to compare ROA with CLOnE, it

---

[13] http://www.eclipse.org/

[14] http://www.w3.org/TR/rdf-sparql-query/

[15] http://kmi.open.ac.uk/technologies/aqualog/

was necessary to repeat the experiment and use Protégé as the baseline. We make no claims that Protégé should be replaced with ROA, the point is that ROA can allow for the creation of a quick easy first draft of a complex Ontology by domain experts or the creation of small to medium sized Ontologies by novice users. Domain experts are not Ontology Engineers. Furthermore, a large percentage of an initial Ontology would naturally consists of taxonomic relations and simple properties/relations.

Our user evaluation consistently indicated that our subjects found ROA (and continue to find CLOnE) significantly more usable and preferable than Protégé for simple Ontology editing tasks. In addition our evaluation differs, in that we implemented more tighter restrictions during our selection process, to ensure that users had no background in NLP or Ontology engineering. Furthermore, 40% of our subjects with an industrial background, tended to score ROA 10% higher then Researchers indicating that a NLI to a Ontology Editor might be a preferred option for Ontology development within industry.

In detail, this evaluation differs from previous work [1] by two important factors: (1) we *excluded* the CLOnE reference manual from the training material provided in the previous evaluation; and (2) we introduced a Text Generator, verbalizing CLOnE text from a given populated Ontology and asked users to edit the Ontology, using the generated CLOnE text based on an example provided. We observed two new significant improvements in our results: (1) the previous evaluation indicated a strong correlation between CLOnE task times and Protégé task times, this correlation has significantly weaken by 32% between ROA and Protégé task times. Hence, where users previously required the equivalent time to implement tasks both in CLOnE and Protégé, this is no longer the case with ROA (the difference being the text generator); and (2) our previous evaluation indicated no correlation between either CLOnE/Protégé task times and their respective SUS scores. However, with ROA, we can now infer that 95% of the total population of naive users, who favour RoundTrip Ontology Authoring over Protégé, would also tend to spend less time completing Ontology editing tasks. We suspect that this is due to the reduced learning curve caused by the text generator. Furthermore, ROA tended to retain user interest, which CLOnE did not. We suspect that the absence of the need to refer to the CL reference guide was a factor in this. While Protégé is intended for more sophisticated knowledge engineering work, this is not the case for ROA. Scalability, both in performance and usage, was also an issue raised by our test subjects. From a performance perspective, when loading large Ontologies, we do not forsee any major issues as ROA is currently being ported to the newest release of GATE which contains a completely new Ontology API that utilises the power of OWLIM - OWL in Memory, a high performance semantic repository developed at Ontotext[16]. Finally, from a user perspective, authoring memory frequently used in translation memory systems or text generation of selective portions of the Ontology (using a Visual Resource) could significantly aid the navigation and authoring of large Ontologies.

---

[16] http://www.ontotext.com/owlim/

6 Continuing and Future Work

Several interesting and useful suggestions for improvements to ROA were made, many of which were already under development within the Nepomuk[17] (The Social Semantic Desktop) project. ROA has been ported to a Nepomuk-KDE [18] application, Semn[19] for Semantic Notetaking and will be also be targeted towards the task of semi-automatic semantic annotation. Furthermore, the ROA text generator was recently used in KnowledgeWeb[20] for the verbalization of suggestions for semi-automatic ontology integration. Finally, ROA is being applied within the EPSRC-funded Easy project to create a controlled natural language interface for editing IT authorization policies (access to network resources such as directories and printers) stored as Ontologies.

Acknowledgements

This research has been partially supported by the following grants: KnowledgeWeb (EU Network of Excellence IST-2004-507482), TAO (EU FP6 project IST-2004-026460), SEKT (EU FP6 project IST-IP-2003-506826, Líon (Science Foundation Ireland project SFI/02/CE1/1131) and NEPOMUK (EU project FP6-027705).

References

1. Funk, A., Tablan, V., Bontcheva, K., Cunningham, H., Davis, B., Handschuh, S.: Clone: Controlled language for ontology editing. In: ASWC 2007 and ISWC 2007. LNCS, vol. 4825, pp. 142–155. Springer, Heidelberg (2007)
2. Cunningham, H., Maynard, D., Bontcheva, K., Tablan, V.: GATE: A Framework and Graphical Development Environment for Robust NLP Tools and Applications. In: Proceedings of the 40th Anniversary Meeting of the Association for Computational Linguistics (ACL 2002) (2002)
3. Thompson, C.W., Pazandak, P., Tennant, H.R.: Talk to your semantic web. IEEE Internet Computing 9(6), 75–78 (2005)
4. Funk, A., Davis, B., Tablan, V., Bontcheva, K., Cunningham, H.: Controlled language IE components version 2. Deliverable D2.2.2, SEKT (2006)
5. Noy, N.F., McGuinness, D.L.: Ontology development 101: A guide to creating your first ontology. Technical Report KSL-01-05, Stanford Knowledge Systems Laboratory (March 2001)
6. Brooke, J.: SUS: a "quick and dirty" usability scale. In: Jordan, P., Thomas, B., Weerdmeester, B., McClelland, A. (eds.) Usability Evaluation in Industry, Taylor and Francis, London (1996)
7. Tullis, T.S., Stetson, J.N.: A comparison of questionnaires for assessing website usability. In: Usability Professionals' Association Conference, Minneapolis, Minnesota (June 2004)

---

[17] http://nepomuk.semanticdesktop.org/xwiki/
[18] http://nepomuk-kde.semanticdesktop.org/xwiki/bin/view/Main/WebHome
[19] http://smile.deri.ie/projects/semn/
[20] http://knowledgeweb.semanticweb.org/

8. Bailey, B.: Getting the complete picture with usability testing. Usability updates newsletter, U.S. Department of Health and Human Services (March 2006)
9. Phillips, J.L.: How to Think about Statistics. W.H. Freeman and Company, New York (1996)
10. Connolly, T.G., Sluckin, W.: An Introduction to Statistics for the Social Sciences, 3rd edn. Macmillan, Basingstoke (1971)
11. Schwitter, R.: Controlled natural languages. Technical report, Centre for Language Technology, Macquarie University (June 2007)
12. Adriaens, G., Schreurs, D.: From COGRAM to ALCOGRAM: Toward a controlled English grammar checker. In: Conference on Computational Linguistics (COLING 1992), Nantes, France, pp. 595–601 (1992)
13. Fuchs, N., Schwitter, R.: Attempto Controlled English (ACE). In: CLAW 1996. Proceedings of the First International Workshop on Controlled Language Applications, Leuven, Belgium (1996)
14. Kaljurand, K., Fuchs, N.E.: Bidirectional mapping between OWL DL and Attempto Controlled English. In: Fourth Workshop on Principles and Practice of Semantic Web Reasoning, Budva, Montenegro (June 2006)
15. Kuhn, T.: Attempto Controlled English as ontology language. In: Bry, F., Schwertel, U. (eds.) REWERSE Annual Meeting 2006 (March 2006)
16. Kaljurand, K., Fuchs, N.: Verbalizing OWL in Attempto Controlled English. In: Proceedings of OWL: Experiences and Directions, OWLED 2007 (2007)
17. Piwek, P.: Requirements definition, validation, verification and evaluation of the clime interface and language processing technology. Technical report, ITRI-University of Brighton (2002)
18. Bernstein, A., Kaufmann, E.: GINO—a guided input natural language ontology editor. In: Cruz, I., Decker, S., Allemang, D., Preist, C., Schwabe, D., Mika, P., Uschold, M., Aroyo, L.M. (eds.) ISWC 2006. LNCS, vol. 4273. Springer, Heidelberg (2006)
19. Namgoong, H., Kim, H.: Ontology-based controlled natural language editor using cfg with lexical dependency. In: Aberer, K., Choi, K.-S., Noy, N., Allemang, D., Lee, K.-I., Nixon, L., Golbeck, J., Mika, P., Maynard, D., Mizoguchi, R., Schreiber, G., Cudré-Mauroux, P. (eds.) ASWC 2007 and ISWC 2007. LNCS, vol. 4825, pp. 353–366. Springer, Heidelberg (2007)
20. Lopez, V., Motta, E., Uren, V.: Poweraqua: Fishing the semantic web. In: Sure, Y., Domingue, J. (eds.) ESWC 2006. LNCS, vol. 4011, pp. 393–410. Springer, Heidelberg (2006)

nSPARQL: A Navigational Language for RDF

Jorge Pérez[1], Marcelo Arenas[1], and Claudio Gutierrez[2]

[1] Pontificia Universidad Católica de Chile
[2] Universidad de Chile

Abstract. Navigational features have been largely recognized as fundamental for graph database query languages. This fact has motivated several authors to propose RDF query languages with navigational capabilities. In particular, we have argued in a previous paper that *nested regular expressions* are appropriate to navigate RDF data, and we have proposed the nSPARQL query language for RDF, that uses nested regular expressions as building blocks. In this paper, we study some of the fundamental properties of nSPARQL concerning expressiveness and complexity of evaluation. Regarding expressiveness, we show that nSPARQL is expressive enough to answer queries considering the semantics of the RDFS vocabulary by directly traversing the input graph. We also show that nesting is necessary to obtain this last result, and we study the expressiveness of the combination of nested regular expressions and SPARQL operators. Regarding complexity of evaluation, we prove that the evaluation of a nested regular expression E over an RDF graph G can be computed in time $O(|G| \cdot |E|)$.

1 Introduction

The Resource Description Framework (RDF) [8,14] is the W3C recommendation data model for the representation of information about resources on the Web. The RDF specification includes a set of reserved keywords with its own semantics, the RDFS vocabulary. This vocabulary is designed to describe special relationships between resources like typing and inheritance of classes and properties [8]. As with any data structure designed to model information, a natural question that arises is what the desiderata are for an RDF query language. Among the multiple design issues to be considered, it has been largely recognized that navigational capabilities are of fundamental importance for data models with explicit tree or graph structure (like XML and RDF).

Recently, the W3C Working Group issued the specification of a query language for RDF, called SPARQL [20], which is a W3C recommendation since January 2008. SPARQL is designed much in the spirit of classical relational languages such as SQL. It has been noted that, although RDF is a directed labeled graph data format, SPARQL only provides limited navigational functionalities. This is more notorious when one considers the RDFS vocabulary (which current SPARQL specification does not cover), where testing conditions like being a subclass of or a subproperty of naturally requires navigating the RDF data. A good illustration of this is shown by the following query, which cannot be expressed in SPARQL without some navigational capabilities. Consider the RDF graph shown in Fig. 1. This graph stores information about cities, transportation services between cities, and further relationships among those transportation

A. Sheth et al. (Eds.): ISWC 2008, LNCS 5318, pp. 66–81, 2008.

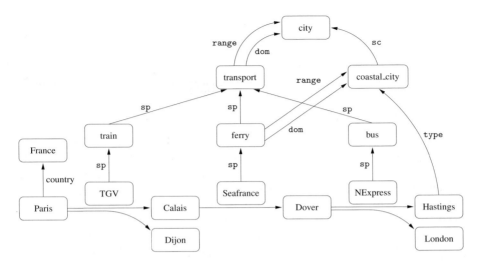

Fig. 1. An RDF graph storing information about transportation services between cities

services (in the form of RDFS annotations). For instance, in the graph we have that a
"Seafrance" service is a subproperty of a "ferry" service, which in turn is a subproperty
of a general "transport" service. Assume that we want to test whether a pair of cities
A and B are connected by a sequence of transportation services, but without knowing
in advance what services provide those connections. We can answer such a query by
testing whether there is a path connecting A and B in the graph, such that every edge
in that path is connected with "transport" by following a sequence of subproperty relationships. For instance, for "Paris" and "Calais" the condition holds, since "Paris" is
connected with "Calais" by an edge with label "TGV", and "TGV" is a subproperty
of "train", which in turn is a subproperty of "transport". Notice that the condition also
holds for "Paris" and "Dover".

Driven by these considerations, we introduced in [7] the language nSPARQL, that incorporates navigational capabilities to a fragment of SPARQL. The main goal of [7] was
not to formally study nSPARQL, but instead to provide evidence that the navigational
capabilities of nSPARQL can be used to pose many interesting and natural queries over
RDF data. Our goal in this paper is to formally study some fundamental properties of
nSPARQL. The first of these fundamental questions is whether the navigational capabilities of nSPARQL can be implemented efficiently. In this paper, we show that this is
indeed the case. More precisely, the building blocks of nSPARQL patterns are *nested
regular expressions*, which specify how to navigate RDF data. Thus, we show in this
paper that nested regular expressions can be evaluated efficiently; if the appropriate data
structure is used to store RDF graphs, the evaluation of a nested regular expression E
over an RDF graph G can be computed in time $O(|G| \cdot |E|)$.

The second fundamental question about nSPARQL is how expressive is the language.
In this paper, we first show that nSPARQL is expressive enough to capture the deductive
rules of RDFS. Evaluating queries which involve the RDFS vocabulary is challenging,
and there is not yet consensus in the Semantic Web community on how to define a query

language for RDFS. In this respect, we show that the RDFS evaluation of an important fragment of SPARQL can be obtained by posing nSPARQL queries that directly traverse the input RDF data. It should be noticed that nested regular expressions are used in nSPARQL to encode the inference rules of RDFS. Thus, a second natural question about nSPARQL is whether these expressions are necessary to obtain this result. In this paper, we show that nesting is indeed necessary to deal with the semantics of RDFS. More precisely, we show that regular expressions alone are not enough to obtain the RDFS evaluation of some queries by simply navigating RDF data.

Finally, we also consider the question of whether the SPARQL operators add expressive power to nSPARQL. Given that nested regular expressions are a powerful navigational tool, one may wonder whether the SPARQL operators can be somehow represented by using these expressions. Or even if this is not the case, one may wonder whether there exist natural queries that can be expressed in nSPARQL, which cannot be expressed by using only nested regular expressions. In our last result, we show that this is the case. More precisely, we prove that there are simple and natural queries that can be expressed in nSPARQL and cannot be expressed by using only nested regular expressions.

Organization of the paper. In Section 2, we introduce some basic notions about RDF and RDFS. In Section 3, we define the notion of nested regular expression, and prove that these expressions can be evaluated efficiently. In Section 4, we define the language nSPARQL, and study the expressiveness of this language. Concluding remarks and related work are given in Section 5.

2   Preliminaries

RDF is a graph data format for the representation of information in the Web. An RDF statement is a *subject-predicate-object* structure, called RDF *triple*, intended to describe resources and properties of those resources. For the sake of simplicity, we assume that RDF data is composed only by elements from an infinite set U of IRIs[1]. More formally, an RDF triple is a tuple $(s, p, o) \in U \times U \times U$, where s is the *subject*, p the *predicate* and o the *object*. An RDF graph is a finite set of RDF triples. Moreover, we denote by $\mathrm{voc}(G)$ the elements from U that are mentioned in G.

Figure 1 shows an RDF graph that stores information about transportation services between cities. In this figure, a triple (s, p, o) is depicted as an edge $s \xrightarrow{p} o$, that is, s and o are represented as nodes and p is represented as an edge label. For example, (Paris, TGV, Calais) is a triple in the graph that states that TGV provides a transportation service from Paris to Calais. Notice that an RDF graph is not a standard labeled graph as its set of edge labels may have a nonempty intersection with its set of nodes. For instance, in the RDF graph in Fig. 1, TGV is simultaneously acting as a node and as an edge label.

The RDF specification includes a set of reserved words (reserved elements from U) with predefined semantics, the RDFS vocabulary (RDF Schema [8]). This set of

---

[1] In this paper, we do not consider anonymous resources called blank nodes in the RDF data model, that is, our study focuses on *ground* RDF graphs. We neither make a special distinction between IRIs and Literals.

Table 1. RDFS inference rules

1. Subproperty:	*2. Subclass:*	*3. Typing:*
(a) $\dfrac{(\mathcal{A},\text{sp},\mathcal{B})\;(\mathcal{B},\text{sp},\mathcal{C})}{(\mathcal{A},\text{sp},\mathcal{C})}$	(a) $\dfrac{(\mathcal{A},\text{sc},\mathcal{B})\;(\mathcal{B},\text{sc},\mathcal{C})}{(\mathcal{A},\text{sc},\mathcal{C})}$	(a) $\dfrac{(\mathcal{A},\text{dom},\mathcal{B})\;(\mathcal{X},\mathcal{A},\mathcal{Y})}{(\mathcal{X},\text{type},\mathcal{B})}$
(b) $\dfrac{(\mathcal{A},\text{sp},\mathcal{B})\;(\mathcal{X},\mathcal{A},\mathcal{Y})}{(\mathcal{X},\mathcal{B},\mathcal{Y})}$	(b) $\dfrac{(\mathcal{A},\text{sc},\mathcal{B})\;(\mathcal{X},\text{type},\mathcal{A})}{(\mathcal{X},\text{type},\mathcal{B})}$	(b) $\dfrac{(\mathcal{A},\text{range},\mathcal{B})\;(\mathcal{X},\mathcal{A},\mathcal{Y})}{(\mathcal{Y},\text{type},\mathcal{B})}$

reserved words is designed to deal with inheritance of classes and properties, as well as typing, among other features [8]. In this paper, we consider the subset of the RDFS vocabulary composed by rdfs:subClassOf, rdfs:subPropertyOf, rdfs:range, rdfs:domain and rdf:type, which are denoted by sc, sp, range, dom and type, respectively. This fragment of RDFS was considered in [17]. In that paper, the authors provide a formal semantics for it, and also show that this fragment is well-behaved as the remaining RDFS vocabulary does not interfere with the semantics of this fragment. The semantics proposed in [17] was shown to be equivalent to the full RDFS semantics when one focuses on the mentioned fragment.

We use the system of rules in Tab. 1. This system was proved in [17] to be sound and complete for the inference problem for RDFS in the presence of sc, sp, range, dom and type, under some mild assumptions (see [17] for further details). In every rule, letters $\mathcal{A}$, $\mathcal{B}$, $\mathcal{C}$, $\mathcal{X}$, and $\mathcal{Y}$, stand for *variables* to be replaced by actual terms. More formally, an *instantiation* of a rule is a replacement of the variables occurring in the triples of the rule by elements of U. An *application* of a rule to a graph G is defined as follows. Given a rule r, if there is an instantiation $\frac{R}{R'}$ of r such that $R \subseteq G$, then the graph $G' = G \cup R'$ is the result of an application of r to G. We say that a triple t is *deduced from* G, if there exists a graph G' such that $t \in G'$ and G' is obtained from G by successively applying the rules in Tab. 1.

Example 1. Let G be the RDF graph in Fig. 1. This graph contains RDFS annotations for transportation services. For instance, (Seafrance, sp, ferry) states that Seafrance is a subproperty of ferry. Thus, we know that there is a ferry going from Calais to Dover since (Calais, Seafrance, Dover) is in G. This conclusion can be obtained by a single application of rule (1b) to triples (Calais, Seafrance, Dover) and (Seafrance, sp, ferry), from which we deduce triple (Calais, ferry, Dover). Moreover, by applying the rule (3b) to this last triple and (ferry, range, coastal_city), we deduce triple (Dover, type, coastal_city) and, thus, we conclude that Dover is a coastal city.   □

3   Nested Regular Expressions for RDF Data

Navigating graphs is done usually by using an operator *next*, which allows one to move from one node to an adjacent one in a graph. In our setting, we have RDF "graphs", which are sets of triples, not classical graphs. In particular, instead of classical edges (pair of nodes), we have directed triples of nodes (*hyperedges*). Hence, a language for navigating RDF graphs should be able to deal with this type of objects. In [7], we

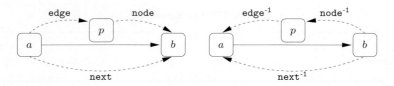

Fig. 2. Forward and backward axes for an RDF triple (a, p, b)

introduce the notion of *nested regular expression* to navigate through an RDF graph. This notion takes into account the special features of the RDF data model. In particular, nested regular expressions use three different *navigation axes* to move through an RDF triple. These axes are shown in Fig. 2 (together with their inverses).

A navigation axis allows one to move one step forward (or backward) in an RDF graph. Thus, a sequence of these axes defines a path in an RDF graph, and one can use classical regular expressions over these axes to define a set of paths that can be used in a query. An additional axis self is used not to actually navigate, but instead to test the label of a specific node in a path. The language also allows *nested expressions* that can be used to test for the existence of certain paths starting at any axis. The following grammar defines the syntax of nested regular expressions:

$$exp \ := \ \text{axis} \ | \ \text{axis::}a \ (a \in U) \ | \ \text{axis::}[exp] \ | \ exp/exp \ | \ exp|exp \ | \ exp^* \quad (1)$$

where axis $\in \{\text{self, next, next}^{-1}, \text{edge, edge}^{-1}, \text{node, node}^{-1}\}$.

Before introducing the formal semantics of nested regular expressions, we give some intuition about how these expressions are evaluated in an RDF graph. The most natural navigation axis is next::a, with a an arbitrary element from U. Given an RDF graph G, the expression next::a is interpreted as the a-*neighbor* relation in G, that is, the pairs of nodes (x, y) such that $(x, a, y) \in G$. Given that in the RDF data model a node can also be the label of an edge, the language allows us to navigate from a node to one of its leaving edges by using the edge axis. More formally, the interpretation of edge::a is the pairs of nodes (x, y) such that $(x, y, a) \in G$. The nesting construction $[exp]$ is used to check for the existence of a path defined by expression exp. For instance, when evaluating nested expression next::$[exp]$ in a graph G, we retrieve the pairs of nodes (x, y) such that there exists z with $(x, z, y) \in G$, and such that there is a path in G that follows expression exp starting in z.

The evaluation of a nested regular expression exp in a graph G is formally defined as a binary relation $[\![exp]\!]_G$, denoting the pairs of nodes (x, y) such that y is reachable from x in G by following a path that conforms to exp. The formal semantics of the language is shown in Tab. 2. In this table, G is an RDF graph, $a \in U$, voc(G) is the set of all the elements from U that are mentioned in G, and exp, exp_1, exp_2 are nested regular expressions.

As is customary for regular expressions, given a nested regular expression exp, we use exp^+ as a shortcut for exp^*/exp. The following is a simple example of the evaluation of a nested regular expression. We present more involved examples when introducing the nSPARQL language.

Table 2. Formal semantics of nested regular expressions

$[\![\mathtt{self}]\!]_G = \{(x,x) \mid x \in \mathrm{voc}(G)\}$

$[\![\mathtt{self::}a]\!]_G = \{(a,a)\}$

$[\![\mathtt{next}]\!]_G = \{(x,y) \mid \text{there exists } z \text{ s.t. } (x,z,y) \in G\}$

$[\![\mathtt{next::}a]\!]_G = \{(x,y) \mid (x,a,y) \in G\}$

$[\![\mathtt{edge}]\!]_G = \{(x,y) \mid \text{there exists } z \text{ s.t. } (x,y,z) \in G\}$

$[\![\mathtt{edge::}a]\!]_G = \{(x,y) \mid (x,y,a) \in G\}$

$[\![\mathtt{node}]\!]_G = \{(x,y) \mid \text{there exists } z \text{ s.t. } (z,x,y) \in G\}$

$[\![\mathtt{node::}a]\!]_G = \{(x,y) \mid (a,x,y) \in G\}$

$[\![\mathtt{axis}^{-1}]\!]_G = \{(x,y) \mid (y,x) \in [\![\mathtt{axis}]\!]_G\} \quad \text{with axis} \in \{\mathtt{next}, \mathtt{node}, \mathtt{edge}\}$

$[\![\mathtt{axis}^{-1}\mathtt{::}a]\!]_G = \{(x,y) \mid (y,x) \in [\![\mathtt{axis::}a]\!]_G\} \quad \text{with axis} \in \{\mathtt{next}, \mathtt{node}, \mathtt{edge}\}$

$[\![exp_1/exp_2]\!]_G = \{(x,y) \mid \text{there exists } z \text{ s.t. } (x,z) \in [\![exp_1]\!]_G \text{ and } (z,y) \in [\![exp_2]\!]_G\}$

$[\![exp_1|exp_2]\!]_G = [\![exp_1]\!]_G \cup [\![exp_2]\!]_G$

$[\![exp^*]\!]_G = [\![\mathtt{self}]\!]_G \cup [\![exp]\!]_G \cup [\![exp/exp]\!]_G \cup [\![exp/exp/exp]\!]_G \cup \cdots$

$[\![\mathtt{self::}[exp]]\!]_G = \{(x,x) \mid x \in \mathrm{voc}(G) \text{ and there exists } z \text{ s.t. } (x,z) \in [\![exp]\!]_G\}$

$[\![\mathtt{next::}[exp]]\!]_G = \{(x,y) \mid \text{there exist } z, w \text{ s.t. } (x,z,y) \in G \text{ and } (z,w) \in [\![exp]\!]_G\}$

$[\![\mathtt{edge::}[exp]]\!]_G = \{(x,y) \mid \text{there exist } z, w \text{ s.t. } (x,y,z) \in G \text{ and } (z,w) \in [\![exp]\!]_G\}$

$[\![\mathtt{node::}[exp]]\!]_G = \{(x,y) \mid \text{there exist } z, w \text{ s.t. } (z,x,y) \in G \text{ and } (z,w) \in [\![exp]\!]_G\}$

$[\![\mathtt{axis}^{-1}\mathtt{::}[exp]]\!]_G = \{(x,y) \mid (y,x) \in [\![\mathtt{axis::}[exp]]\!]_G\} \quad \text{with axis} \in \{\mathtt{next}, \mathtt{node}, \mathtt{edge}\}$

Example 2. Let G be the graph in Fig. 1, and consider expression $exp_1 = \mathtt{next::}[\mathtt{next::sp/self::train}]$. The nested expression $[\mathtt{next::sp/self::train}]$ performs an existential test; it defines the set of nodes z in G such that there exists a path from z that follows an edge labeled sp and reaches a node labeled train. There is a single such node in G, namely TGV. Restricted to graph G, expression exp_1 is equivalent to $\mathtt{next::TGV}$ and, thus, it defines the pairs of nodes that are connected by an edge labeled TGV. Hence, the evaluation of exp_1 in G is $[\![exp_1]\!]_G = \{(\text{Paris, Calais}), (\text{Paris, Dijon})\}$. □

In the following section, we introduce the language nSPARQL that combines the operators of SPARQL with the navigational capabilities of nested regular expressions. But before introducing this language, we show that nested regular expressions can be evaluated efficiently, which is an essential requirement if one wants to use nSPARQL for web-scale applications.

3.1 Complexity of Evaluating Nested Regular Expressions

In this section, we study the complexity of evaluating nested regular expressions over RDF graphs. We present an algorithm for this problem that works in time proportional to the size of the input graph times the size of the expression being evaluated. As is customary when studying the complexity of the evaluation problem for a query language (cf. [21]), we consider its associated decision problem. For nested regular expressions, this problem is defined as:

PROBLEM   : Evaluation problem for nested regular expressions.
INPUT   : An RDF graph G, a nested regular expression exp, and a pair (a,b).
QUESTION : Is $(a,b) \in [\![exp]\!]_G$?

We assume that an RDF graph G is stored as an adjacency list that makes explicit the navigation axes (and their inverses). Thus, every $u \in \text{voc}(G)$ is associated with a list of pairs $\alpha(u)$, where every pair contains a navigation axis and the destination node. For instance, if (s, p, o) is a triple in G, then $(\texttt{next::}p, o) \in \alpha(s)$ and $(\texttt{edge}^{-1}\texttt{::}o, s) \in \alpha(p)$. Moreover, we assume that $(\texttt{self::}u, u) \in \alpha(u)$ for every $u \in \text{voc}(G)$. Notice that if the number of triples in G is N, then the adjacency list representation uses space $O(N)$. Thus, when measuring the size of G, we use $|G|$ to denote the size of its adjacency list representation. We further assume that given an element $u \in \text{voc}(G)$, we can access its associated list $\alpha(u)$ in time $O(1)$. This is a standard assumption for graph data-structures in a RAM model.

In this section, we assume some familiarity with automata theory. Recall that given a regular expression r, one can construct in linear time a nondeterministic finite automaton with ε-transitions $\mathcal{A}_r$ that accepts the language generated by r.

A key idea in the algorithm introduced in this section is to associate to each nested regular expression a nondeterministic finite automaton with ε-transitions (ε-NFA). Given a nested regular expression exp, we recursively define the set of *depth-0 terms* of exp, denoted by $\mathbf{D}_0(exp)$, as follows:

$\mathbf{D}_0(exp) = \{exp\}$ if exp is either axis, or axis::a, or axis::$[exp']$,
$\mathbf{D}_0(exp_1/exp_2) = \mathbf{D}_0(exp_1|exp_2) = \mathbf{D}_0(exp_1) \cup \mathbf{D}_0(exp_2)$,
$\mathbf{D}_0(exp^*) = \mathbf{D}_0(exp)$,

where axis $\in \{\texttt{self}, \texttt{next}, \texttt{next}^{-1}, \texttt{edge}, \texttt{edge}^{-1}, \texttt{node}, \texttt{node}^{-1}\}$. For instance, for the nested expression:

$$exp = \texttt{next::}a/(\texttt{next::}[\texttt{next::}a/\texttt{self::}b])^*/(\texttt{next::}[\texttt{node::}b] \mid \texttt{next::}a)^+,$$

we have $\mathbf{D}_0(exp) = \{\texttt{next::}a, \texttt{next::}[\texttt{next::}a/\texttt{self::}b], \texttt{next::}[\texttt{node::}b]\}$. Notice that a nested regular expression exp can be viewed as a classical regular expression over alphabet $\mathbf{D}_0(exp)$. We denote by $\mathcal{A}_{exp}$ the ε-NFA that accepts the language generated by the regular expression exp over alphabet $\mathbf{D}_0(exp)$.

The algorithm for the evaluation of nested regular expressions is similar to the algorithms for the evaluation of some temporal logics [11] and propositional dynamic logic [1]. Given an RDF graph G and a nested regular expression exp, it proceeds by recursively labeling every node u of G with a set $\text{label}(u)$ of nested expressions. Initially, $\text{label}(u)$ is the empty set. Then at the end of the execution of the algorithm, it holds that $exp \in \text{label}(u)$ if and only if there exists z such that $(u, z) \in [\![exp]\!]_G$. In the algorithm, we use the product automaton $G \times \mathcal{A}_{exp}$, which is constructed as follows. Let Q be the set of states of $\mathcal{A}_{exp}$, and $\delta : Q \times (\mathbf{D}_0(exp) \cup \{\varepsilon\}) \to 2^Q$ the transition function of $\mathcal{A}_{exp}$. The set of states of $G \times \mathcal{A}_{exp}$ is $\text{voc}(G) \times Q$, and its transition function $\delta' : (\text{voc}(G) \times Q) \times (\mathbf{D}_0(exp) \cup \{\varepsilon\}) \to 2^{\text{voc}(G) \times Q}$ is defined as follows. For every $(u, p) \in \text{voc}(G) \times Q$ and $s \in \mathbf{D}_0(exp)$, we have that $(v, q) \in \delta'((u, p), s)$ if and only if $q \in \delta(p, s)$ and one of the following cases hold:

- $s = $ axis and there exists a such that $(\text{axis::}a, v) \in \alpha(u)$,
- $s = $ axis::a and $(\text{axis::}a, v) \in \alpha(u)$,
- $s = $ axis::$[exp]$ and there exists b such that $(\text{axis::}b, v) \in \alpha(u)$ and $exp \in \text{label}(b)$,

where axis $\in \{\texttt{self}, \texttt{next}, \texttt{next}^{-1}, \texttt{edge}, \texttt{edge}^{-1}, \texttt{node}, \texttt{node}^{-1}\}$. Additionally, if $q \in \delta(p, \varepsilon)$ we have that $(u, q) \in \delta'((u, p), \varepsilon)$ for every $u \in \text{voc}(G)$. That is, $G \times \mathcal{A}_{exp}$ is the standard product automaton if G is viewed as an NFA over alphabet $\mathbf{D}_0(exp)$. It is straightforward to prove that $G \times \mathcal{A}_{exp}$ can be constructed in time $O(|G| \cdot |\mathcal{A}_{exp}|)$.

Now we have all the necessary ingredients to present the algorithm for the evaluation problem for nested regular expressions. This algorithm is split in two procedures: LABEL labels G according to nested expression exp as explained above, and EVAL returns YES if $(a, b) \in [\![exp]\!]_G$ and NO otherwise.

LABEL(G, exp):
1. **for each** axis::$[exp'] \in \mathbf{D}_0(exp)$ **do**
2.   call LABEL(G, exp')
3. construct $\mathcal{A}_{exp}$, and assume that q_0 is its initial state and F is its set of final states
4. construct $G \times \mathcal{A}_{exp}$
5. **for each** state (u, q_0) that reaches a state (v, q_f) in $G \times \mathcal{A}_{exp}$, with $q_f \in F$ **do**
6.   label(u) := label$(u) \cup \{exp\}$

EVAL$(G, exp, (a, b))$:
1. **for each** $u \in \text{voc}(G)$ **do**
2.   label(u) := $\emptyset$
3. call LABEL(G, exp)
4. construct $\mathcal{A}_{exp}$, and assume that q_0 is its initial state and F is its set of final states
5. construct $G \times \mathcal{A}_{exp}$
6. **if** a state (b, q_f), with $q_f \in F$, is reachable from (a, q_0) in $G \times \mathcal{A}_{exp}$
7.   **then return** YES
8.   **else return** NO

It is not difficult to see that these procedures work in time $O(|G| \cdot |exp|)$. Just observe that step 5 of procedure LABEL and step 6 of procedure EVAL, can be done in time linear in the size of $G \times \mathcal{A}_{exp}$ by traversing $G \times \mathcal{A}_{exp}$ in a depth first search manner.

Theorem 1. *Procedure* EVAL *solves the evaluation problem for nested regular expressions in time* $O(|G| \cdot |exp|)$.

4   The Navigational Language nSPARQL

In this section, we introduce the language nSPARQL, and we formally study its expressiveness. nSPARQL is essentially obtained by using triple patterns with nested regular expressions in the predicate position, plus SPARQL operators AND, OPT, UNION, and FILTER. Before formally introducing nSPARQL, we recall the necessary definitions about SPARQL.

SPARQL [20] is the standard language for querying RDF data. We use here the algebraic formalization introduced in [19]. Assume the existence of an infinite set V of variables disjoint from U. A SPARQL graph pattern is defined as follows:

-   A tuple from $(U \cup V) \times (U \cup V) \times (U \cup V)$ is a graph pattern (a *triple pattern*).
-   If P_1 and P_2 are graph patterns, then $(P_1 \text{ AND } P_2)$, $(P_1 \text{ OPT } P_2)$, and $(P_1 \text{ UNION } P_2)$ are graph patterns.

- If P is a graph pattern and R is a SPARQL *built-in* condition, then the expression $(P$ FILTER $R)$ is a graph pattern.

A SPARQL *built-in* condition is a Boolean combination of terms constructed by using equality $(=)$ among elements in $U \cup V$, and the unary predicate bound over variables.

To define the semantics of SPARQL graph patterns, we need to introduce some terminology. A *mapping* μ from V to U is a partial function $\mu : V \rightarrow U$. For a triple pattern t, we denote by $\mu(t)$ the triple obtained by replacing the variables in t according to μ. The domain of μ, denoted by $\mathrm{dom}(\mu)$, is the subset of V where μ is defined. Two mappings μ_1 and μ_2 are *compatible* if for every $x \in \mathrm{dom}(\mu_1) \cap \mathrm{dom}(\mu_2)$, it is the case that $\mu_1(x) = \mu_2(x)$, i.e. when $\mu_1 \cup \mu_2$ is also a mapping. Let Ω_1 and Ω_2 be sets of mappings. We define the join, the union, the difference, and the left-outer join between Ω_1 and Ω_2 as:

$$\Omega_1 \bowtie \Omega_2 = \{\mu_1 \cup \mu_2 \mid \mu_1 \in \Omega_1, \mu_2 \in \Omega_2 \text{ are compatible mappings}\},$$
$$\Omega_1 \cup \Omega_2 = \{\mu \mid \mu \in \Omega_1 \text{ or } \mu \in \Omega_2\},$$
$$\Omega_1 \smallsetminus \Omega_2 = \{\mu \in \Omega_1 \mid \text{ for all } \mu' \in \Omega_2, \mu \text{ and } \mu' \text{ are not compatible}\},$$
$$\Omega_1 \bowtie\hspace{-0.9em}\bowtie\, \Omega_2 = (\Omega_1 \bowtie \Omega_2) \cup (\Omega_1 \smallsetminus \Omega_2).$$

The *evaluation* of a graph pattern over an RDF graph G, denoted by $[\![\cdot]\!]_G$, is defined recursively as follows:

- $[\![t]\!]_G = \{\mu \mid \mathrm{dom}(\mu) = \mathrm{var}(t) \text{ and } \mu(t) \in G\}$, where $\mathrm{var}(t)$ is the set of variables occurring in t.
- $[\![(P_1 \text{ AND } P_2)]\!]_G = [\![P_1]\!]_G \bowtie [\![P_2]\!]_G$, $[\![(P_1 \text{ UNION } P_2)]\!]_G = [\![P_1]\!]_G \cup [\![P_2]\!]_G$, and $[\![(P_1 \text{ OPT } P_2)]\!]_G = [\![P_1]\!]_G \bowtie\hspace{-0.9em}\bowtie\, [\![P_2]\!]_G$.

The semantics of FILTER expressions goes as follows. Given a mapping μ and a built-in condition R, we say that μ satisfies R, denoted by $\mu \models R$, if (we omit the usual rules for Boolean operators):

- R is bound($?X$) and $?X \in \mathrm{dom}(\mu)$;
- R is $?X = c$, where $c \in U$, $?X \in \mathrm{dom}(\mu)$ and $\mu(?X) = c$;
- R is $?X = ?Y$, $?X \in \mathrm{dom}(\mu)$, $?Y \in \mathrm{dom}(\mu)$ and $\mu(?X) = \mu(?Y)$.

Then $[\![(P \text{ FILTER } R)]\!]_G = \{\mu \in [\![P]\!]_G \mid \mu \models R\}$.

It was shown in [19], among other algebraic properties, that AND and UNION are associative and commutative, thus permitting us to avoid parenthesis when writing sequences of either AND operators or UNION operators.

Now we formally define the language *nested* SPARQL (or just nSPARQL), by considering triples with nested regular expressions in the predicate position. A *nested-regular-expression triple* (or just nre-triple) is a tuple t of the form (x, exp, y), where $x, y \in U \cup V$ and exp is a nested regular expression. nSPARQL patterns are recursively defined from nre-triples:

- An nre-triple is an nSPARQL pattern.
- If P_1 and P_2 are nSPARQL patterns and R is a built-in condition, then $(P_1 \text{ AND } P_2)$, $(P_1 \text{ OPT } P_2)$, $(P_1 \text{ UNION } P_2)$, and $(P_1 \text{ FILTER } R)$ are nSPARQL patterns.

To define the semantics of nSPARQL, we just need to define the semantics of nre-triples. The evaluation of an nre-triple $t = (?X, exp, ?Y)$ over an RDF graph G is defined as the following set of mappings:

$$[\![t]\!]_G = \{\mu \mid \mathrm{dom}(\mu) = \{?X, ?Y\} \text{ and } (\mu(?X), \mu(?Y)) \in [\![exp]\!]_G\}.$$

Similarly, the evaluation of an nre-triple $t = (?X, exp, a)$ over an RDF graph G, where $a \in U$, is defined as $\{\mu \mid \mathrm{dom}(\mu) = \{?X\} \text{ and } (\mu(?X), a) \in [\![exp]\!]_G\}$, and likewise for $(a, exp, ?X)$ and (a, exp, b) with $b \in U$.

Notice that every SPARQL triple $(?X, p, ?Y)$ with $p \in U$ is equivalent to (has the same evaluation of) nSPARQL triple $(?X, \mathtt{next}::p, ?Y)$. Also notice that, since variables are not allowed in nested regular expressions, the occurrence of variables in the predicate position of triple patterns is forbidden in nSPARQL. Nevertheless, every SPARQL triple of the form $(?X, ?Y, a)$, with $a \in U$, is equivalent to nSPARQL pattern $(?X, \mathtt{edge}::a, ?Y)$. Similarly, the triple $(a, ?X, ?Y)$ is equivalent to $(?X, \mathtt{node}::a, ?Y)$. Thus, what we are loosing in nSPARQL is only the possibility of using variables in the three positions of a triple pattern.

As pointed out in the introduction, it has been largely recognized that navigational capabilities are fundamental for graph databases query languages. However, although RDF is a directed labeled graph data format, SPARQL only provides limited navigational functionalities. In [7], we introduced nSPARQL as a way to overcome this limitation. The main goal of [7] was not to formally study nSPARQL, but instead to provide evidence that the navigational capabilities of nSPARQL can be used to pose many interesting and natural queries over RDF data. Our goal in this paper is to formally justify nSPARQL. In particular, we have already shown that nested regular expressions can be evaluated efficiently, which is an essential requirement if one wants to use nSPARQL for web-scale applications. In this section, we study some fundamental properties related to the expressiveness of nSPARQL. But before doing that, we provide some additional examples of queries that are likely to occur in the Semantic Web, but cannot be expressed in SPARQL without using nested regular expressions.

Example 3. Let G be the RDF graph of Fig. 1 and P_1 the following pattern:

$$P_1 = (?X, (\mathtt{next}::\mathrm{TGV} \mid \mathtt{next}::\mathrm{Seafrance})^+, \mathrm{Dover}) \text{ AND } (?X, \mathtt{next}::\mathrm{country}, ?Y)$$

Pattern P_1 retrieves cities, and the country where they are located, such that there is a way to travel from those cities to Dover using either TGV or Seafrance in every direct trip. The evaluation of P_1 over G is $\{\{?X \to \mathrm{Paris}, ?Y \to \mathrm{France}\}\}$. Notice that although there is a direct way to travel from Calais to Dover using Seafrance, Calais does not appear in the result since there is no information in G about the country where Calais is located. We can relax this last restriction by using the OPT operator:

$$P_2 = (?X, (\mathtt{next}::\mathrm{TGV} \mid \mathtt{next}::\mathrm{Seafrance})^+, \mathrm{Dover}) \text{ OPT } (?X, \mathtt{next}::\mathrm{country}, ?Y)$$

Then we have that $[\![P_2]\!]_G = \{\{?X \to \mathrm{Paris}, ?Y \to \mathrm{France}\}, \{?X \to \mathrm{Calais}\}\}$.   □

Example 4. Assume that we want to obtain the pairs of cities $(?X, ?Y)$ such that there is a way to travel from $?X$ to $?Y$ by using either Seafrance or NExpress, with an intermediate stop in a city that has a direct NExpress trip to London. Consider nested expression:

$$exp_1 = (\texttt{next::Seafrance} \mid \texttt{next::NExpress})^+/$$
$$\texttt{self::}[\texttt{next::NExpress}/\texttt{self::London}]/(\texttt{next::Seafrance} \mid \texttt{next::NExpress})^+$$

Then pattern $P = (?X, exp_1, ?Y)$ answers our initial query. Notice that expression $\texttt{self::}[\texttt{next::NExpress}/\texttt{self::London}]$ is used to perform the intermediate existential test of having a direct NExpress trip to London. □

Example 5. Let G be the graph in Fig. 1 and P_1 the following pattern:

$$P_1 = (?X, \texttt{next::}[(\texttt{next::sp})^*/\texttt{self::transport}], ?Y). \qquad (2)$$

Pattern P_1 defines the pairs of cities $(?X, ?Y)$ such that, there exists a triple $(?X, p, ?Y)$ in the graph and a path from p to transport where every edge has label sp. Thus, nested expression $[(\texttt{next::sp})^*/\texttt{self::transport}]$ is used to emulate the process of inference in RDFS; it retrieves all the nodes that are *sub-properties* of transport (rule (1a) in Tab. 1). Therefore, pattern P_1 retrieves the pairs of cities that are connected by a direct transportation service, which could be a train, ferry, bus, etc. In general, if we want to obtain the pairs of cities such that there is a way to travel from one city to another, we can use the following nSPARQL pattern:

$$P_2 = (?X, (\texttt{next::}[(\texttt{next::sp})^*/\texttt{self::transport}])^+, ?Y). \qquad (3)$$

In this section, we formally prove that (2) and (3) cannot be expressed without using nested expressions of the form $\texttt{axis::}[exp]$. □

4.1 On RDFS and nSPARQL

We claimed in [7] that the language of nested regular expressions is powerful enough to deal with the predefined semantics of RDFS. In this section, we formally prove this fact. More precisely, we show that if one wants to answer a SPARQL query P according to the semantics of RDFS, then one can rewrite P into an nSPARQL query Q such that Q retrieves the answer to P by directly traversing the input graph. We also show that the nesting operation is crucial for this result.

SPARQL follows a *subgraph-matching* approach, and thus, a SPARQL query treats RDFS vocabulary without considering its predefined semantics. We are interested in defining the semantics of SPARQL over RDFS, that is, taking into account not only the explicit RDF triples of a graph G, but also the triples that can be derived from G according to the semantics of RDFS. Let the *closure* of an RDF graph G, denoted by $\text{cl}(G)$, be the graph obtained from G by successively applying the rules in Tab. 1 until the graph does not change. The most direct way of defining a semantics for the RDFS evaluation of SPARQL patterns is by considering not the original graph but its closure. The theoretical formalization of such an approach was studied in [12]. The following definition formalizes this notion.

Definition 1. *Given a SPARQL graph pattern P, the RDFS evaluation of P over G, denoted by $[\![P]\!]_G^{\text{rdfs}}$, is defined as the set of mappings $[\![P]\!]_{\text{cl}(G)}$, that is, as the evaluation of P over the closure of G.*

Regular expressions alone are not enough. Regular expressions are the most common way of giving navigational capabilities to query languages over graph databases [5], and recently to query languages over RDF graphs [3,16,6]. Our language not only allows regular expressions over navigational axes but also nesting of those regular expressions. In our setting, regular expressions are obtained by forbidding the nesting operator and, thus, they are generated by the following grammar:

$$exp \; := \; \text{axis} \; | \; \text{axis}::a \; (a \in U) \; | \; exp/exp \; | \; exp|exp \; | \; exp^* \qquad (4)$$

where axis $\in \{\text{self}, \text{next}, \text{next}^{-1}, \text{edge}, \text{edge}^{-1}, \text{node}, \text{node}^{-1}\}$. Let *regular* SPARQL (or just rSPARQL) be the language obtained from nSPARQL by restricting nre-triples to contain in the predicate position only regular expressions (generated by grammar (4)). Notice that rSPARQL is a fragment of nSPARQL and, thus, the semantics for rSPARQL is inherited from nSPARQL.

Our next result shows that regular expressions are not enough to obtain the RDFS evaluation of some simple SPARQL patterns by directly traversing RDF graphs. In fact, the following theorem shows that there is a SPARQL triple pattern whose RDFS evaluation cannot be obtained by any rSPARQL pattern.

Theorem 2. *Let* $p \in U \smallsetminus \{\text{sp}, \text{sc}, \text{type}, \text{dom}, \text{range}\}$ *and consider triple pattern* $(?X, p, ?Y)$. *There is no* rSPARQL *pattern* Q *such that* $[\![(?X, p, ?Y)]\!]_G^{\text{rdfs}} = [\![Q]\!]_G$ *for every RDF graph* G.

nSPARQL and RDFS evaluation. In this section, we show that if a SPARQL pattern P is constructed by using triple patterns having at least one position with a non-variable element, then the RDFS evaluation of P can be obtained by directly traversing the input graph with an nSPARQL pattern. More precisely, consider the following *translation* function from elements in U to nested regular expressions:

$$
\begin{aligned}
trans(\text{sc}) \;\; &= (\text{next}::\text{sc})^+ \\
trans(\text{sp}) \;\; &= (\text{next}::\text{sp})^+ \\
trans(\text{dom}) \;\; &= \text{next}::\text{dom} \\
trans(\text{range}) &= \text{next}::\text{range} \\
trans(\text{type}) \;\; &= (\; \text{next}::\text{type}/(\text{next}::\text{sc})^* \; | \\
& \qquad \text{edge}/(\text{next}::\text{sp})^*/\text{next}::\text{dom}/(\text{next}::\text{sc})^* \; | \\
& \qquad \text{node}^{-1}/(\text{next}::\text{sp})^*/\text{next}::\text{range}/(\text{next}::\text{sc})^* \;) \\
trans(p) \;\; &= \text{next}::[(\text{next}::\text{sp})^*/\text{self}::p] \;\; \text{for } p \notin \{\text{sc}, \text{sp}, \text{range}, \text{dom}, \text{type}\}.
\end{aligned}
$$

Notice that we have implicitly used this translation function in Example 5.

Lemma 1. *Let* (x, a, y) *be a SPARQL triple pattern with* $x, y \in U \cup V$ *and* $a \in U$, *then* $[\![(x, a, y)]\!]_G^{\text{rdfs}} = [\![(x, trans(a), y)]\!]_G$ *for every RDF graph* G.

That is, given an RDF graph G and a triple pattern t not containing a variable in the predicate position, it is possible to obtain the RDFS evaluation of t over G by navigating G through a nested regular expression.

Suppose now that we have a SPARQL triple pattern t with a variable in the predicate position, but such that the subject and object of t are not both variables. We show how to construct an nSPARQL pattern P_t such that $[\![t]\!]_G^{\text{rdfs}} = [\![P_t]\!]_G$. Assume that $t = (x, ?Y, a)$ with $x \in U \cup V$, $?Y \in V$, and $a \in U$, that is, t does not contain a variable in the object position. Consider for every $p \in \{\text{sc}, \text{sp}, \text{dom}, \text{range}, \text{type}\}$, the pattern $P_{t,p}$ defined as $((x, trans(p), a)$ AND $(?Y, \text{self}::p, ?Y))$. Then define then pattern P_t as follows:

$$P_t = ((x, \text{edge}::a/(\text{next}::sp)^*, ?Y) \text{ UNION } P_{t,\text{sc}} \text{ UNION } P_{t,\text{sp}} \text{ UNION }$$
$$P_{t,\text{dom}} \text{ UNION } P_{t,\text{range}} \text{ UNION } P_{t,\text{type}}).$$

We can similarly define pattern P_t for a triple pattern $t = (a, ?Y, x)$, where $a \in U$, $?Y \in V$ and $x \in U \cup V$. Thus, we have the following result.

Lemma 2. *Let* $t = (x, ?Y, z)$ *be a triple pattern such that* $?Y \in V$, *and* $x \notin V$ *or* $z \notin V$. *Then* $[\![t]\!]_G^{\text{rdfs}} = [\![P_t]\!]_G$ *for every RDF graph G.*

Let $\mathcal{T}$ be the set of triple patterns of the form (x, y, z) such that $x \notin V$ or $y \notin V$ or $z \notin V$. We have translated every triple pattern $t \in \mathcal{T}$ into an nSPARQL pattern P_t such that $[\![t]\!]_G^{\text{rdfs}} = [\![P_t]\!]_G$. Moreover, for every triple pattern t, its translation is of size linear in the size of t. Given that the semantics of SPARQL is defined from the evaluation of triple patterns, we can state the following result.

Theorem 3. *Let* P *be a SPARQL pattern constructed from triple patterns in* $\mathcal{T}$. *Then there exists an nSPARQL pattern* Q *such that* $[\![P]\!]_G^{\text{rdfs}} = [\![Q]\!]_G$ *for every RDF graph G. Moreover, the size of* Q *is linear in the size of* P.

The following example shows that one can combine the translation function presented in this section with nested regular expression patterns to obtain more expressive queries that take into account the RDFS semantics.

Example 6. Let G be the RDF graph shown in Fig. 1. Assume that one wants to retrieve the pairs of cities such that there is a way of traveling (by using any transportation service) between those cities, and such that every stop in the trip is a coastal city. The following nSPARQL pattern answers this query:

$$P = (?X, (trans(\text{transport})/\text{self}::[trans(\text{type})/\text{self}::\text{coastal\_city}])^+, ?Y). \qquad \square$$

Notice that Theorems 2 and 3 imply that nSPARQL is strictly more expressive than rSPARQL. We state this result in the following corollary.

Corollary 1. *There exists an nSPARQL pattern that is not equivalent to any rSPARQL pattern.*

4.2   On the Expressiveness of the SPARQL Operators in nSPARQL

Clearly, nested regular expressions add expressive power to SPARQL. The opposite question is whether using SPARQL operators in nSPARQL patterns add expressive power to the language. Next we show that this is indeed the case. In particular, we show that there are simple and natural queries that can be expressed by using nSPARQL features and that cannot be simulated by using only nested regular expressions. Let us present the intuition of this result with an example.

Example 7. Let G be the RDF graph shown in Fig. 1. Assume that one wants to retrieve from G the cities $?X$ such that there exists exactly one city that can be reached from $?X$ by using a direct Seafrance service. The following nSPARQL pattern answers this query:

$$[\ (?X, \text{next::Seafrance}/\text{next}^{-1}, ?X)$$
$$\text{OPT}\ (\ (\ (?X, \text{next::Seafrance}, ?Y)\ \text{AND}\ (?X, \text{next::Seafrance}, ?Z)\ )$$
$$\text{FILTER}\ \neg ?Y =?Z\ )\ ]\ \ \text{FILTER}\ \neg\, \text{bound}(?Y)$$

The first nre-triple $(?X, \text{next::Seafrance}/\text{next}^{-1}, ?X)$ retrieves the cities $?X$ that are connected with some other city by a Seafrance service. The optional part obtains additional information for those cities $?X$ that are connected with at least two different cities by a Seafrance service. Finally, the pattern filters out those cities for which no optional information was added (by using $\neg\, \text{bound}(?Y)$). That is, only the cities $?X$ that are connected with exactly one city by a Seafrance service remains in the evaluation. If we evaluate the above pattern over G, we obtain a single mapping μ such that $\text{dom}(\mu) = \{?X\}$ and $\mu(?X) = \text{Calais}$. □

The nSPARQL pattern in the above example is essentially counting (up to a fixed threshold) the cities that are connected with $?X$ by a Seafrance service. In the next result, we show that some counting capabilities cannot be obtained by using nSPARQL patterns without considering the OPT operator, even if we combine nested regular expressions by using the operators AND, UNION and FILTER. The query used in the proof is similar to that of Example 7. It retrieves the nodes $?X$ for which there exists at least two different nodes connected with $?X$.

Theorem 4. *There is an* nSPARQL *pattern that is not equivalent to any* nSPARQL *pattern that uses only* AND, UNION, *and* FILTER *operators.*

5   Related Work and Concluding Remarks

Related work. The language of nested regular expressions has been motivated by some features of query languages for graphs and trees, namely, XPath [10], temporal logics [11] and propositional dynamic logic [1]. In fact, nested regular expressions are constructed by borrowing the notions of *branching* and navigation axes from XPath [10], and adding them to regular expressions over RDF graphs. The algorithm that we present in Section 3.1 is motivated by standard algorithms for some temporal logics [11] and propositional dynamic logic [1].

Regarding languages with navigational capabilities for querying RDF graphs, several proposals can be found in the literature [18,3,16,6,4,2]. Nevertheless, none of these languages is motivated by the necessity to evaluate queries over RDFS, and none of them is comparable in expressiveness and complexity of evaluation with the language that we study in this paper. Probably the first language for RDF with navigational capabilities was Versa [18], whose motivation was to use XPath over the XML serialization of RDF graphs. Kochut et al. [16] propose SPARQLeR, an extension of SPARQL that works with *path variables* that represent paths between nodes in a graph. This language also allows to check whether a path conforms to a regular expression. Anyanwu et al. [6] propose a language called SPARQ2L. The authors further investigate the implementation of a query evaluation mechanism for SPARQ2L with emphasis in some secondary memory issues. The language PSPARQL was proposed by Alkhateeb et al. in [3]. PSPARQL extends SPARQL by allowing regular expressions in triple patterns. The same authors propose a further extension of PSPARQL called CPSPARQL [4] that allows constraints over regular expressions. CPSPARQL also allows variables inside regular expressions, thus permitting to retrieve data *along* the traversed paths. In [3,4], the authors study some theoretical aspects of (C)PSPARQL.

Alkhateeb has recently shown [2] that PSPARQL, that is, the full SPARQL language extended with regular expressions, can be used to encode RDFS inference. Although PSPARQL [2] and the language rSPARQL that we present in Section 4.1 are similar, when defining rSPARQL we use a fragment of SPARQL, namely, the graph pattern matching facility without solution modifiers like projection. Alkhateeb's encoding [2] needs the projection operator, and in particular, extra variables (not needed in the output solution) appearing in the predicate position of triple patterns. This feature is not allowed in the fragment that we use to construct languages rSPARQL and nSPARQL. Although PSPARQL could be used to answer some RDFS queries, the additional abilities needed in PSPARQL come with an associated complexity impact in the evaluation problem for the conjunctive fragment, namely, NP-completeness [2]. By using the results in [19] and the complexity of the evaluation problem for nested regular expressions, it is easy to show that the complexity of the evaluation problem for the conjunctive fragment of nSPARQL is polynomial.

Evaluating queries which involve RDFS vocabulary is challenging, and there is not yet consensus in the Semantic Web community on how to define a query language for RDFS. Nevertheless, there have been several proposals and implementations of query languages for RDF data with RDFS vocabulary, e.g. [15,9,13,12]. It would be interesting to compare these approaches with the process of answering a SPARQL query under the RDFS semantics by first compiling it into an nSPARQL query.

Concluding Remarks. In this paper, we have started the formal study of nested regular expressions and the language nSPARQL, that we proposed in [7]. We have shown that nested regular expressions admit a very efficient evaluation method, that justifies its use in practice. We further showed that the language nSPARQL is expressive enough to be used for querying and navigating RDF data. In particular, we proved that besides capturing the semantics of RDFS, nSPARQL provides some other interesting features that allows users to pose natural and interesting queries.

Acknowledgments. We are grateful to the anonymous referees for their helpful comments. The authors were supported by: Arenas - Fondecyt grant 1070732; Gutierrez - Fondecyt grant 1070348; Pérez - Conicyt Ph.D. Scholarship; Arenas, Gutierrez and Pérez - grant P04-067-F from the Millennium Nucleus Center for Web Research.

References

1. Alechina, N., Immerman, N.: Reachability Logic: An Efficient Fragment of Transitive Closure Logic. Logic Journal of the IGPL 8(3), 325–338 (2000)
2. Alkhateeb, F.: Querying RDF(S) with Regular Expressions. PhD Thesis, Université Joseph Fourier, Grenoble (FR) (2008)
3. Alkhateeb, F., Baget, J., Euzenat, J.: RDF with regular expressions. Research Report 6191, INRIA (2007)
4. Alkhateeb, F., Baget, J., Euzenat, J.: Constrained regular expressions in SPARQL. In: SWWS 2008, pp. 91–99 (2008)
5. Angles, R., Gutierrez, C.: Survey of graph database models. ACM Comput. Surv. 40(1), 1–39 (2008)
6. Anyanwu, K., Maduko, A., Sheth, A.: SPARQ2L: Towards Support for Subgraph Extraction Queries in RDF Databases. In: WWW 2007, pp. 797–806 (2007)
7. Arenas, M., Gutierrez, C., Pérez, J.: An Extension of SPARQL for RDFS. In: SWDB-ODBIS 2007, pp. 1–20 (2007)
8. Brickley, D., Guha, R.V.: RDF Vocabulary Description Language 1.0: RDF Schema. W3C Recommendation (February 2004), http://www.w3.org/TR/rdf-schema/
9. Broekstra, J., Kampman, A., van Harmelen, F.: Sesame: A generic architecture for storing and querying RDF and RDF schema. In: Horrocks, I., Hendler, J. (eds.) ISWC 2002. LNCS, vol. 2342, pp. 54–68. Springer, Heidelberg (2002)
10. Clark, J., DeRose, S.: XML Path Language (XPath). W3C Recommendation (November 1999), http://www.w3.org/TR/xpath
11. Clarke, E., Grumberg, O., Peled, D.: Model Checking. The MIT Press, Cambridge (2000)
12. Gutierrez, C., Hurtado, C., Mendelzon, A.: Foundations of Semantic Web Databases. In: PODS 2004, pp. 95–106 (2004)
13. Harris, S., Gibbins, N.: 3store: Efficient bulk RDF storage. In: PSSS 2003, pp. 1–15 (2003)
14. Hayes, P.: RDF Semantics. W3C Recommendation (February 2004), http://www.w3.org/TR/rdf-mt/
15. Karvounarakis, G., Alexaki, S., Christophides, V., Plexousakis, D., Scholl, M.: RQL: a declarative query language for RDF. In: WWW 2002, pp. 592–603 (2002)
16. Kochut, K., Janik, M.: SPARQLeR: Extended SPARQL for Semantic Association Discovery. In: Franconi, E., Kifer, M., May, W. (eds.) ESWC 2007. LNCS, vol. 4519, pp. 145–159. Springer, Heidelberg (2007)
17. Muñoz, S., Pérez, J., Gutierrez, C.: Minimal Deductive Systems for RDF. In: Franconi, E., Kifer, M., May, W. (eds.) ESWC 2007. LNCS, vol. 4519, pp. 53–67. Springer, Heidelberg (2007)
18. Olson, M., Ogbuji, U.: The Versa Specification, http://uche.ogbuji.net/tech/rdf/versa/etc/versa-1.0.xml
19. Pérez, J., Arenas, M., Gutierrez, C.: Semantics and Complexity of SPARQL. In: Cruz, I., Decker, S., Allemang, D., Preist, C., Schwabe, D., Mika, P., Uschold, M., Aroyo, L.M. (eds.) ISWC 2006. LNCS, vol. 4273, pp. 30–43. Springer, Heidelberg (2006)
20. Prud'hommeaux, E., Seaborne, A.: SPARQL Query Language for RDF. W3C Recommendation (January 2008), http://www.w3.org/TR/rdf-sparql-query/
21. Vardi, M.Y.: The Complexity of Relational Query Languages (Extended Abstract). In: STOC 1982, pp. 137–146 (1982)

An Experimental Comparison of RDF Data Management Approaches in a SPARQL Benchmark Scenario

Michael Schmidt[1,*], Thomas Hornung[1], Norbert Küchlin[1], Georg Lausen[1],
and Christoph Pinkel[2]

[1] Freiburg University, Georges-Köhler-Allee 51, 79106 Freiburg, Germany
{mschmidt,hornungt,kuechlin,lausen}@informatik.uni-freiburg.de
[2] MTC Infomedia OHG, Kaiserstr. 26, 66121 Saarbrücken, Germany
c.pinkel@mtc-infomedia.de

Abstract. Efficient RDF data management is one of the cornerstones in realizing the Semantic Web vision. In the past, different RDF storage strategies have been proposed, ranging from simple triple stores to more advanced techniques like clustering or vertical partitioning on the predicates. We present an experimental comparison of existing storage strategies on top of the SP²Bench SPARQL performance benchmark suite and put the results into context by comparing them to a purely relational model of the benchmark scenario. We observe that (1) in terms of performance and scalability, a simple triple store built on top of a column-store DBMS is competitive to the vertically partitioned approach when choosing a physical (predicate, subject, object) sort order, (2) in our scenario with real-world queries, none of the approaches scales to documents containing tens of millions of RDF triples, and (3) none of the approaches can compete with a purely relational model. We conclude that future research is necessary to further bring forward RDF data management.

1 Introduction

The Resource Description Framework [1] (RDF) is a standard format for encoding machine-readable information in the Semantic Web. RDF databases are collections of so-called "triples of knowledge", where each triple is of the form (*subject,predicate,object*) and models the binary relation *predicate* between the *subject* and the *object*. For instance, the triple (*Journal1,issued,"1940"*) might be used to encode that the entity *Journal1* has been *issued* in year *1940*. By interpreting each triple as a graph edge from a *subject* to an *object* node with label *predicate*, RDF databases can be seen as labeled directed graphs.

To facilitate RDF data access, the W3C has standardized the SPARQL [2] query language, which bases upon a powerful graph pattern matching facility. Its very basic construct are simple triple graph patterns, which, during query evaluation, are matched against components in the RDF graph. In addition, different SPARQL operators can be used to compose more advanced graph patterns.

---

* The work of this author was funded by DFG, grant GRK 806/2.

A. Sheth et al. (Eds.): ISWC 2008, LNCS 5318, pp. 82–97, 2008.

An efficient RDF storage scheme should support fast evaluation of such graph patterns and scale to RDF databases comprising millions (or even billions) of triples, as they are commonly encountered in today's RDF application scenarios (e.g., [3,4]). The straightforward relational implementation, namely a single `Triples` relation with three columns *subject*, *predicate*, and *object* that holds all RDF triples, seems not very promising: The basic problem with this approach is that the evaluation of composed graph patterns typically requires a large amount of expensive self-joins on this (possibly large) table. For instance, the query "Return the year of publication of *Journal1 (1940)*" might be expressed in SQL as follows (for readability, we use shortened versions of the RDF URIs).

```
SELECT T3.object AS yr
FROM Triples T1 JOIN Triples T2 ON T1.subject=T2.subject
     JOIN Triples T3 ON T1.subject=T3.subject                         (1)
WHERE T1.predicate='type' AND T1.object='Journal' AND T2.predicate='title'
      AND T2.object='Journal 1 (1940)' AND T3.predicate='issued'
```

The `Triples` table access `T1` and the associated WHERE-conditions extract all *Journal* entities, `T2` fixes the title, and `T3` extracts the year of publication. We observe that even this rather simple query requires two *subject-subject* self-joins over the `Triples` table. Practical queries may involve much more self-joins.

To overcome this deficiency, other physical organization techniques for RDF have been proposed [5,6,7,8,9,10,11]. One notable idea is to cluster RDF data, i.e. to group entities that are similar in structure [9,10] and store them in flattened tables that contain all the shared properties. While this may significantly reduce the amount of joins in queries, it works out only for well-structured data. However, one strength of RDF is that it offers excellent support for scenarios with poorly structured information, where clustering is not a feasible solution.

A conceptually simpler idea is to set up one table for each unique *predicate* in the data [5,11], which can be seen as full *vertical partitioning* on the predicates. Each such predicate table consists of two columns (*subject*, *object*) and contains all *subject-object* pairs linked through the respective predicate. Data is then distributed across several smaller tables and, when the predicate is fixed, joins do not involve the whole set of triples. By physically sorting data on the *subject* column, *subject-subject* joins between two tables, a very frequent operation, can be realized in linear time (w.r.t. the size of the tables) by merging their subject columns [11]. In such a scenario, the query from above might be formulated as,

```
SELECT DI.object AS yr
FROM type TY JOIN title TI ON TY.subject=DT.subject
     JOIN issued IS ON TY.subject=IS.subject                          (2)
WHERE TY.object='bench:Journal' AND TI.object='Journal 1 (1940)'
```

where `type`, `title`, and `issued` denote the corresponding predicates tables. Predicate selection now is implicit by the choice of the predicate table (i.e., no longer encoded in the WHERE-clause) and, given that the *subject*-column is sorted, both joins might be efficiently implemented as linear merge joins.

In the experiments in [11] on top of the Barton library data [12], vertical partitioning turns out to be clearly favorable to the triple table scheme and always competitive to clustering. Although the scenario is a reasonable choice that illustrates many advantages of vertical partitioning, several issues remain open. One point is that, in the partitioned scenario, efficient *subject-subject* merge joins on the predicate tables (which are possible whenever predicates are fixed) are a key to performance. However, when physically sorting table `Triples` by (*predicate, subject, object*), linear merge joins might also apply in a triple store.

A study of the Barton benchmark shows that one query (out of seven) requires no join on the triple (resp., predicate) table(s), and each two involve (a) a single *subject-subject* join, (b) two *subject-subject* joins, and (c) one *subject-subject* plus one *subject-object* join. Thus, none involves more than two joins. The simplicity of these join patterns to a certain degree contrasts with the Introduction of [11], where the authors state that "almost all interesting queries involve many self-joins" and motivate vertical partitioning using a five-way self-join query. We agree that real-world queries often involve complex join-patterns and see an urgent need for reevaluating the vertical approach in a more challenging scenario.

To this end, we present an experimental comparison of the triple and vertically partitioned scheme on top of the the SP^{2}Bench SPARQL benchmark [13]. The SP^{2}Bench queries implement meaningful requests in the DBLP scenario [14] and have been designed to test challenging situations that may arise in the context of SPARQL and Semantic Web data. In contrast to the Barton queries, they contain no aggregation, due to missing SPARQL language support. But except for this construct, they cover a much wider range of operator constellations, RDF data access paths, join patterns, and advanced features (e.g., OPTIONAL clauses, solution modifiers). The queries for the vertical and the triple store are obtained from a methodical SPARQL-to-SQL translation and reflect these characteristics.

To put our analysis into context, we consider two more scenarios. First, we test the *Sesame* SPARQL engine [15] as a representative SPARQL processor that relies on a native RDF store. Second, we translate the SP^{2}Bench scenario into a purely relational scheme, thus comparing the current state-of-the-art in RDF data management against established relational database technologies.

Contributions. Among others, our experiments show that (1) when triple tables are physically sorted by (*predicate, subject, object*), efficient merge joins can be exploited (just like in the vertical scheme) and the triple table approach becomes more competitive, (2) for the challenging SP^{2}Bench queries neither the vertical nor the triple scheme shows a good overall performance, and (3) while both schemes typically outperform the *Sesame* SPARQL engine, the purely relational encoding is almost always at least one order of magnitude faster. We conclude that there is an urgent need for future research in this area.

Related Work. An experimental comparison of the triple table and a vertically partitioned scheme has been provided in [5]. Among others, the authors note the additional costs of predicate table unions in the vertical scenario, which will be discussed later in this paper. Nevertheless, the setting in [5] differs in several aspects, e.g. in the vertically partitioned scheme the RDF schema layer was stored in separate tables and physical sorting on the *subject*-column (to allow for *subject-subject* merge joins), a central topic in our analysis, was not tested.

We point the interested reader to the experimental comparison of the triple and vertical storage scheme in [16]. This work has been developed independently from us. It presents a reevaluation of the experiments from [11] and, in this line, identifies situations where vertical partitioning is an insufficient solution. Several findings there are similar to our results. While the latter experiments are carried out in the Barton scenario (like the original experiments in [11]), we go one step further, i.e. perform tests in a different scenario and put the results into context by comparing them to a purely relational scheme, as well as a SPARQL engine.

The Berlin SPARQL Benchmark [17] is settled in an e-commerce scenario and strictly use-case driven. In contrast, the language-specific SP^{2}Bench suite used in this work covers a broader range of SPARQL/RDF constructs and, for this reason, is preferable for testing the generality of RDF storage schemes.

Structure. In the next section we summarize important characteristics of the SP^{2}Bench SPARQL performance benchmark [13], to facilitate the interpretation of the benchmark results. In Section 3 we then sketch the tested storage schemes and the methodical query translation into these scenarios. Finally, Section 4 contains the in-depth discussion of our experiments and a conclusion. In the remainder, we assume the reader to be familiar with RDF [1] and SPARQL [2].

2   The SP^{2}Bench Scenario

SP^{2}Bench [13] is settled in the DBLP [14] bibliographic scenario. Central to the benchmark is a data generator for creating DBLP-like RDF documents, which mirror characteristics and relations found in the original DBLP data. It relies on natural function families to capture social-world aspects encountered in the DBLP data, e.g. the citation system is modeled by powerlaw distributions, while limited growth functions approximate the number of publications per year. Supplementary, the SP^{2}Bench suite provides a set of meaningful SPARQL queries, covering a variety of SPARQL operator constellations and data access patterns.

According to DBLP, the SP^{2}Bench generator creates nine distinct types of bibliographic entities, namely ARTICLE, JOURNAL, INPROCEEDINGS, PROCEEDINGS, BOOK, INCOLLECTION, PHDThESIS, MASTERSThESIS, and WWW documents, where each document is represented by a unique URI. In addition, there are persons that act as authors or editors. They are modeled by blank nodes.

Each document (resp., person) is described by a set of properties, such as *dc:title*, *dc:creator* (i.e., the author), or *swrc:isbn*. Outgoing citations are expressed through predicate *dcterms:references*, which points to a blank node of type *rdf:Bag* (a standard RDF container class) that links to the set of all document URIs referenced by the respective document. Attribute *dcterms:partOf* links inproceedings to the proceedings they appeared in; similarly, *swrc:journal* connects articles to journals. Several properties (e.g., *dc:creator*) are multi-valued.

The first part of Table 1 lists the number of document class instances of type INPROCEEDINGS, PROCEEDINGS, ARTICLE, JOURNAL, INCOLLECTION, and the

Table 1. Key characteristics of documents generated by the SP^{2}Bench generator

#triples	#Inpr.	#Proc.	#Art.	#Journ.	#Inc.	#Oth.	#auth./#dist.	#prop.	file size	year
10k	169	6	916	25	18	0	1.5k/0.9k	23+34	1.0MB	1955
50k	1.4k	37	4.0k	104	56	0	6.8k/4.1k	23+34	5.1MB	1967
250k	9.2k	213	17.1k	439	173	39	34.5k/20.0k	23+43	26MB	1979
1M	43.5k	903	56.9k	1.4k	442	551	151.0k/82.1k	23+44	106MB	1989
5M	255.2k	4.7k	207.8k	4.6k	1.4k	1.4k	898.0k/429.6k	23+52	533MB	2001
25M	1.5M	24.4k	642.8k	11.7k	4.5k	2.4k	5.4M/2.1M	25+52	2.7GB	2015

remaining types *#Oth.* (BOOK, WWW, PHD- and MASTERSTHESIS) for generated documents up to $25M$ RDF triples. ARTICLE and INPROCEEDINGS documents clearly dominate. The total number of authors (i.e., triples with predicate *dc:creator*) increases slightly super-linear to the total number of documents. This reflects the increasing average number of authors per paper in DBLP over time.

The table also lists the number *#prop.* of distinct properties. This value $x + y$ splits into x "standard" attribute properties and y bag membership properties *rdf:_1*, ..., *rdf:_y*, where y depends on the maximum-sized reference list in the data. We observe that larger documents contain larger reference lists, and hence more distinct properties. As discussed later, this might complicate data processing in the vertically partitioned scenario. Finally, we list the physical size of the RDF file (in NTriples format) and the year up to which data was generated.

To support queries that access an author with fixed characteristics, the documents contain a special author, named after the mathematician Paul Erdös, who gets assigned 10 publications and 2 editor activities in-between 1940–1996. As an example, $Q8$ (Appendix A) extracts all persons with *Erdös Number* 1 or 2.[1]

3   The Benchmark Scenarios

We now describe the four benchmark scenarios in detail. The first system under consideration is (1) the *Sesame* [15] SPARQL engine. *Sesame* constitutes a query engine that, like the other three scenarios, relies on a physical DB backend. It is among the fastest SPARQL engines that have been tested in the context of the SP^{2}Bench benchmark (cf. [13]) and has been chosen as a representative for the class of SPARQL engines. The remaining scenarios are (2) the triple table approach, (3) the vertically partitioned approach as described in [11], and (4) a purely relational DBLP model. They are all implemented on top of a relational DBMS. Accordingly, a translation of the SP^2*Bench* SPARQL queries into SQL is required. We will sketch the detailed settings and our methodical query translation approaches for scenarios (2)-(4) in the remainder of this section. The resulting SQL queries are available online[2]; still, to be self-contained we will summarize their key characteristics when discussing the results in Section 4.

According to [11], to reach best performance all relational schemes should be implemented on top of a column-store DBMS, which stores data physically

---

[1] See http://www.oakland.edu/enp/.

[2] http://dbis.informatik.uni-freiburg.de/index.php?project=SP2B/translations.html

by column rather than row (see [11] for the advantages of column-oriented systems in the RDF scenario). The C-Store research prototype [18] used in [11] misses several SQL features that are essential for the SP²Bench queries (e.g. left joins), so we fall back on the *MonetDB* [19] column-store, a complete, industrial-strength relational DBMS. We note that MonetDB differs from C-Store in several aspects. First, data processing in MonetDB is memory-based while it is disk-based in C-Store. Moreover, C-Store exhibits a carefully optimized merge-join implementation (on top of run-length encoded data) and makes heavy use of this operation. Although we observe that MonetDB uses merge joins less frequently (cf. Section 4), the system is known for its performance and has recently been shown to be competitive to C-Store in the Barton Library RDF scenario [16].

3.1   The Triple Table Storage Scheme

In the *triple table* scheme a single table `Triples`(*subject, predicate,object*) holds all RDF triples. Methodical translations of SPARQL into this scheme have been proposed in [20,21,22]. The idea is to evaluate triple patterns separately against table `Triples`, then combining them according to the SPARQL operators in the query. Typically, SPARQL operator AND is expressed by a relational join, UNION by a SQL union, FILTER clauses result in WHERE-conditions, and OPTIONAL is modeled by a left outer join. For instance, SPARQL query $Q1$ (Appendix A) translates into query (1) from the Introduction (prefixes and data types are omitted). Observe that $Q1$ connects three patterns through two AND operators (denoted as "."), resulting in two SQL joins. The patterns are connected through variable *?journal* in *subject* position, so both are *subject-subject* joins. We emphasize that, although queries were translated manually, the scheme is very close to the approaches used by SPARQL engines that build on the relational model.

Dictionary Encoding. URIs and Literals tend to be long strings; they might blow up relational tables and make joins expensive. Therefore, we store integer keys instead of the string value, while keeping the key-value mapping in a `Dictionary`(*ID,val*) table (cf. [15,23,24,11]). Note that dictionary encoding implies additional joins with the `Dictionary` table in the translated queries.

Implementation. We sort data physically by (*predicate, subject, object*) rather than (*subject, predicate, object*). While this contrasts with the experiments in [11], we will show that this sort order makes the triple approach more competitive, because fast linear merge joins across property tables in the vertical scenario can now be realized by corresponding merge joins in the triple scenario.

We note that indexing in *MonetDB* differs from conventional DBMS; it interprets INDEX statements as advices, feeling free to ignore them and create its own indices.[3] Though, we issue a secondary BTree index for all remaining permutations of the *subject*, *predicate*, and *object* columns. The `Dictionary` table is physically sorted by *ID* and we request a secondary index on column *val*.

---

[3] See http://monetdb.cwi.nl/projects/monetdb/SQL/Documentation/Indexes.html.

3.2   The Vertically Partitioned Storage Scheme

The vertically partitioned relational store maintains one two-column table with schema (*subject*, *object*) for each unique predicate in the data. The query translation for the vertical scenario is similar to the triple table translation. The translation of SPARQL query $Q1$ into this scenario is exemplarily shown in the Introduction, query (2). Here, data is extracted from the predicate tables, so predicate value restrictions in the WHERE-clause are no longer necessary.

One major problem in the vertical scheme arises when predicates in queries are not fixed (i.e., when SPARQL variables occur in predicate position). Then, information cannot be extracted from a single predicate table, but queries must compute the union over *all* these tables. As discussed in Section 2 (Table 1), in our scenario the number of distinct properties (and hence, predicate tables) increases with document size. Consequently, such queries require more unions on large documents. This illustrates a basic drawback of the vertical approach: Query translation depends on the structure of the data and, what is even more urgent, queries may require a large number of unions over the predicate tables.

Implementation. We sort the predicate tables physically on (*subject*, *object*) and issue an additional secondary BTree index on columns (*object*, *subject*). Dictionary encoding is implemented analogously to the triple scheme.

3.3   The Purely Relational Scheme

We started from scratch and developed an Entity Relationship Model (ERM) of DBLP. Using ERM translation techniques, we end up with the following tables, where primary keys are underlined and foreign keys are marked by prefix "*fk_*".

- Document(*ID,address,booktitle,isbn,. . .,stringid,title,volume*)
- Document_homepage(*fk_document,homepage*)
- Document_seeAlso(*fk_document,seeAlso*)
- Venue(*ID,fk_document,fk_venue_type*)
- Publication(*ID,chapter,fk_document,fk_publication_type,fk_venue,pages*)
- Publication_cdrom(*fk_publication,cdrom*)
- Abstract(*fk_publication,txt*)
- PublicationType(*ID,name*) and VenueType(*ID,name*)
- Person(*ID,name,stringid*)
- Author(*fk_person,fk_publication*) and Editor(*fk_document,fk_person*)
- Reference(*fk_from,fk_to*)

The scheme distinguishes between venues (i.e., JOURNAL and PROCEEDINGS) and publications (such as ARTICLE, INPROCEEDINGS, or BOOK). The dictionary tables PublicationType and VenueType contain integer *ID*s for the respective venue and publication classes. Table Document constitutes a base table for both document types, containing properties that are common to both venues and publications. Supplementary, Venue and Publication store the properties that are specific for the respective type. For instance, if a new BOOK document is inserted, its base properties are stored in table Document, while publication-type specific properties (e.g., *chapter*) are stored in table `Publication`. The entries are linked through foreign key `Publication`.*fk_document*; the type (in this case BOOK) is fixed by linking `Publication`.*fk_publication_type* to the BOOK ID in `PublicationType`. Properties *foaf:homepage*, *rdf:seeAlso*, and *bench:cdrom* are multi-valued in the SP^{2}Bench scenario, so they are stored in the separate tables `Document_homepage`, `Document_seeAlso`, and `Publication_cdrom`. We use a distinguished `Abstract` table for the larger-than-average abstract strings.

Finally, there is one table `Person` that stores person information, two tables `Author` and `Editor` that store the author and editor activity of persons, and a table `Reference` that contains all references between documents.

Implementation. The scheme was implemented in *MonetDB* exactly as described above, using the specified `PRIMARY` and `FOREIGN KEY` constraints, without additional indices. In the sense of a relational schema we omit prefix definitions (such as "rdf:", "dc:"). The data was translated using a conversion script.

4 Experimental Results

Setting. The experiments were carried out on a Desktop PC running ubuntu v7.10 gutsy Linux, with Intel Core2 Duo E6400 2.13GHz CPU and 3GB DDR2 667 MHz nonECC physical memory. We used a 250GB Hitachi P7K500 SATA-II hard drive with 8MB Cache. The relational schemes were executed with *MonetDB* mserver v5.5.0, using the (more efficient) algebra frontend (flag "-G").

As discussed in Section 3, we tested (1) the *Sesame* v2.0 engine *SP* (coupled with its native storage layer, providing all possible combinations of indices) and three *MonetDB* scenarios, namely (2) the triple store *TR*, (3) the vertically partitioned store *VP*, and (4) the purely relational scheme *RS*. We report on user (`usr`), system (`sys`), and elapsed time (`total`). While `usr` and `sys` were extracted from the */proc* file system, elapsed time was measured through a timer. *MonetDB* follows a client-server architecture and we provide the sum of the `usr` and `sys` times of the client and server processes. Note that the experiments were run on a DuoCore CPU, where the linux kernel sums up `usr` and `sys` of the individual processor units, so `usr+sys` might be greater than `total`.

For all scenarios we carried out three runs over all queries on documents of $10k$, $50k$, $250k$, $1M$, $5M$, and $25M$ triples, setting a 30 minutes timeout and $2GB$ memory limit (using *ulimit*) per query. As our primary interest is the basic performance of the approaches (rather than caching or learning strategies), we performed *cold* runs, i.e. destroyed the database in-between each two consecutive runs and always restarted it before evaluating a query. We provide average times and omit the deviation from the average (which was always negligible).

Discussion of the Benchmark Results. All results were verified by comparing the outcome of the engines among each other (where possible). Table 2 summarizes the query result sizes and the physical DB sizes for each scenario on all documents. The *VP* scheme requires less disk space than *TR* for large documents, since predicates are not explicitly stored for each triple. For *Sesame*, indices

Table 2. Query result sizes on documents up to $25M$ triples and physical DB size

	Q1	Q2	Q3a	Q3b	Q3c	Q4	Q5a/b	Q6	Q7	Q8	Q9	Q10	Q11	SP	TR	VP	RS
			Number of query results for individual queries											Phys. DB size (MB)			
$10k$	1	147	846	9	0	23.2k	155	229	0	184	4	166	10	3	3	6	4
$50k$	1	965	3.6k	25	0	104.7k	1.1k	1.8k	2	264	4	307	10	14	5	8	5
$250k$	1	6.2k	15.9k	127	0	542.8k	6.9k	12.1k	62	332	4	452	10	69	18	20	13
$1M$	1	32.8k	52.7k	379	0	2.6M	35.2k	62.8k	292	400	4	572	10	277	63	58	42
$5M$	1	248.7k	192.4k	1.3k	0	18.4M	210.7k	417.6k	1.2k	493	4	656	10	1376	404	271	195
$25M$	1	1.9M	594.9k	4.1k	0	n/a	696.7k	1.9M	5.1k	493	4	656	10	6928	2395	1168	913

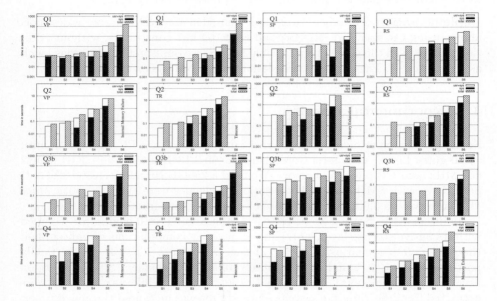

Fig. 1. Results on S1=10k, S2=50k, S3=250k, S4=1M, S5=5M, and S6=25M triples

occupy more than half of the required space. In RS there is no redundancy, no dictionary encoding, and no prefixes are stored, so least space is required.

The query execution times are shown in Figures 1, 2, and 3 (the y-axes are always in log scale). Please note that the individual plots scale differently.

Q1. Return the year of publication of "Journal 1 (1940)".

This simple query returns exactly one result on all documents. The TR and VP translations are shown in the Introduction. The RS query joins tables Venue, Document, and VenueType on the connecting foreign keys and then filters for VenueType.*name*="Journal" and Document.*title*="Journal 1 (1940)".

We observe that both the TR and VP scenario scale well for documents up to $5M$ triples, but total time explodes for $25M$ triples. The gap between total and usr+sys for $25M$ indicates that much time is spent in waiting for data being read from or written to disk, which is caused by query execution plans (QEPs) that

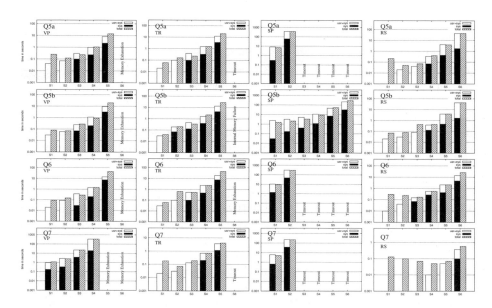

Fig. 2. Results on S1=10k, S2=50k, S3=250k, S4=1M, S5=5M, and S6=25M triples

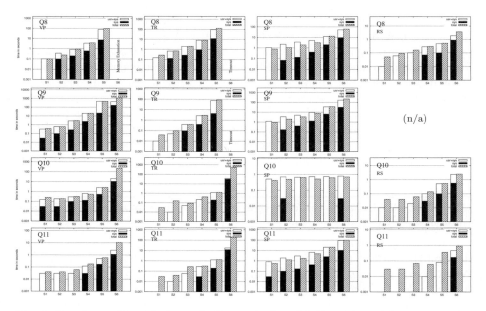

Fig. 3. Results on S1=10k, S2=50k, S3=250k, S4=1M, S5=5M, and S6=25M triples

involve expensive fetch joins, instead of efficient *subject-subject* merge joins. We claim that using merge joins would be more efficient here. Due to this deficiency, both *Sesame* and the *RS* scenario outperform the *TR* and *VP* schemes.

Q2. Extract all inproceedings with properties *dc:creator*, *bench:booktitle*, *dc:title*, *swrc:pages*, *dcterms:partOf*, *rdfs:seeAlso*, *foaf:homepage*, *dcterms:issued*, **and optionally** *bench:abstract*, **including these properties.**

Q2 implements a star-join-like graph pattern. Result size grows with document size (cf. Table 2) and the solution modifier ORDER BY forces result ordering. The nine outer SPARQL triple patterns translate into nine predicate (triple) table accesses in the *VP* (*TR*) scenario, connected through eight *subject-subject* joins, due to variable *?inproc*. The OPTIONAL clause causes an additional left outer join. The *RS* query gathers all relevant information from tables `Document`, `Publication`, `PublicationType`, `Author`, `Person`, `Document_seeAlso`, `Venue`, and `Document_homepage`, and also contains a left outer join with table `Abstract`.

Like for *Q1*, the *subject-subject* joins should be realized by merge joins in the *TR* and *VP* scenario, but MonetDB chooses QEPs that mostly use fetch joins, involving merge joins only in few cases. These fetch joins consume the major part of execution time. Lastly, none of both schemes succeeds for the 25*M* triples document. *Sesame* is about one order of magnitudes slower. The *RS* scheme requires less joins and is significantly faster than the other approaches.

Q3abc. Select all articles with property (a) *swrc:pages*, **(b)** *swrc:month*, **or (c)** *swrc:isbn*.

We restrict on a discussion of *Q3b*, as the results for *Q3a* and *Q3c* are similar. As explained in [13], the FILTER in *Q3b* selects about 0.65% of all articles. The *TR* translation contains a *subject-subject* join on table `Triples` and a WHERE value-restrictions for predicate *swrc:month*. Although variable *?property* occurs in *predicate* position, we chose a *VP* translation that does not compute the union of all predicate tables, but operates directly on the table for predicate *swrc:month*, which is implicitly fixed by the FILTER. The *RS* translation is straightforward.

The *VP* approach is a little faster than *TR*, because it operates on top of the *swrc:month* predicate table, instead of the full triples table. The query contains only one *subject-subject* join, and we observe that the *VP* and *TR* approaches explode for the 25*M* document, again due to expensive fetch joins (cf. *Q1*, *Q2*). *Sesame* is competitive and scales even better, while *RS* shows best performance.

Q4. Select all distinct pairs of article author names for authors that have published in the same journal.

Q4 contains a long graph chain, i.e. variables *?name1* and *?name2* are linked through the articles that different authors have published in the same journal. When translated into *TR* and *VP*, the chain is mapped to a series of *subject-subject*, *subject-object*, and *object-object* joins. The *RS* query gathers all articles and their authors from the relevant tables twice and joins them on `Venue`.*ID*.

As apparent from Table 2, the query computes very large results. Due to the *subject-object* and *object-object* joins, the *TR* and *VP* scenarios have to compute many expensive (non-merge) joins, which makes the approaches scale poorly. *Sesame* is one order of magnitude slower. In contrast, *RS* involves simpler joins (e.g., efficient joins on foreign keys) and shows the best performance.

Q5ab. Return the names of all persons that occur as author of at least one inproceeding and at least one article.

Q5a joins authors implicitly on author names (through the FILTER condition), while *Q5b* explicitly joins on variable *?person*. Although in general not equivalent, the one-to-one mapping between authors and their names in SP^{2}Bench implies equivalence of *Q5a* and *Q5b*. All translations share these join characteristics, i.e. all translations of *Q5a* model the join by an equality condition in the SQL WHERE-clause, whereas translations of *Q5b* contain an explicit SQL JOIN.

Sesame scales bad for *Q5a*, probably due to the implicit join (it performs much better for *Q5b*). In the SQL scenarios there are no big differences between implicit and explicit joins; such situations are resolved by relational optimizers.

Q6. Return, for each year, the set of all publications authored by persons that have not published in years before.

Q6 implements closed world negation (CWN), expressed through a combination of operators OPTIONAL, FILTER, and BOUND. The block outside the OPTIONAL computes *all* publications and the inner one constitutes earlier publications from authors that appear outside. The outer FILTER then retains all publications for which *?author2* is unbound, i.e. those from newcomers. In the *TR* and *VP* translation, a left outer join is used to connect the outer to the inner part. The *RS* query extracts, for each year, all publications and their authors, and uses a SQL NOT EXISTS clause to filter away authors without prior publications.

One problem in the *TR* and *VP* queries is the left join on top of a less-than comparison, which complicates the search for an efficient QEP. In addition, both queries contain each two *subject-object* joins on the left and on the right side of the left outer join. Ultimately, both scale poorly. Also *Sesame* scales very bad. In contrast, the purely relational encoding is elegant and much more efficient.

Q7. Return the titles of all papers that have been cited at least once, but not by any paper that has not been cited itself.

This query implements a double-CWN scenario. Due to the nested OPTIONAL clauses, the *TR* and *VP* translations involve two nested left outer joins with join-intensive subexpressions. The *VP* translation is complicated by three unions of all predicate tables, caused by the SPARQL variables *?member2*, *?member3*, and *?member4* in predicate position. When encoding them at the bottom of the evaluator tree, the whole query builds upon these unions and the benefit of sorted and indexed predicate tables gets lost. We tested different versions of the query and decided for the most performant (out of the tested variants), where we pulled off the outermost union, thus computing the union of subexpressions rather than individual tables. The *RS* query uses two nested SQL NOT IN-clauses to express double negation. We could have used nested NOT EXISTS-clauses instead (cf. *Q6*), but decided to vary, to test the impact of both operators.

Due to the unbound predicates, the *VP* approach has severe problems in evaluating this query and behaves worse than the *TR* scheme. This illustrates the disadvantages of the vertical approach in scenarios where unbound predicates occur. *Sesame* also behaves very bad, while the nested NOT IN-clause in *RS*, a common construct in relational queries, constitutes the only practical solution.

Q8. Compute authors that have published with Paul Erdoes or with an author that has published with Paul Erdoes.

Q8 contains a SPARQL UNION operator, so all translations contain a SQL union. The *TR* and *VP* versions of this query are straightforward. The *RS* translation separately retrieves persons that have published with Paul Erdoes and persons that have published with one of its coauthors (each from the **Author** and the **Person** table), and afterwards computes the union of both person sets.

Again, the *TR* scenario turns out to be competitive to *VP*, but both schemes fail to find an efficient QEP for large documents, due to the *subject-object* and *object-object* joins and the additional non-equality WHERE-condition over the *subject* and *object* columns. The *Sesame* engine scales surprisingly well for this query, but is still one order of magnitude slower than the relational scheme.

Q9. Return incoming and outgoing properties of persons.

Both parts of the union in *Q9* contain a fully unbound triple pattern, which selects all RDF database triples. The *TR* translation is straightforward. Concerning the unbound *?predicate* variable, we again pulled off the union of the predicate tables in the *VP* scenario, thus computing the same query separately for each predicate table and building the union of the results afterwards. As discussed in *Q7*, this was more efficient than the union at the bottom of the operator tree. The result size is always 4 (the first part constitutes properties *dc:creator* and *swrc:editor*, and the second one *rdf:type* and *foaf:name*). A meaningful *RS* translation of this query, which accesses schema information, is not possible: In *RS*, the properties are encoded as (fixed) table attributes names.[4]

Although a little bit slower than the *TR* approach for small documents, *VP* succeeds in evaluating the 25*M* triple document. Though, both approaches seem to have problems with the unbound triple pattern and scale poorly. *Sesame*'s native store offers better support, but is still far from being performant.

Q10. Return all subjects that stand in any direct relation with Paul Erdoes. In our scenario the query can be reformulated as "Return publications and venues in which Paul Erdoes is involved as author or editor, respectively".

Q10 implements an *object* bound-only RDF access path. The *TR* and *RS* translations are standard. Due to the unbound variable *?predicate*, the *VP* query involves a union of the predicate tables. As for *Q9*, the implementation of this union on top of the operator tree turned out to be the most performant solution.

Recalling that "Paul Erdoes" is active between 1940 and 1996, the result size has an upper bound (cf. Table 2 for the 5*M* and 25*M* documents). *VP* and *TR* show very similar behavior. As illustrated by the results of *Sesame*, this query can be realized in constant time (with an appropriate index). The index

---

[4] A lookup query for fixed values in the DBMS system catalog is not very interesting.

selection strategy of MonetDB in *TR* and *VP* is clearly suboptimal. *RS* scales much better, but (in contrast to *Sesame*) still depends on the document size.

Q11. Return (up to) 10 electronic edition URLs starting from the 51st publication, in lexicographical order.

*Q*11 focuses on the combination of solution modifiers ORDER BY, LIMIT, and OFFSET, which arguably remains the key challenge in all three translations.

The *VP* query operates solely on the predicate table for *rdfs:seeAlso* and, consequently, is a little faster than *TR*. *Sesame* scales superlinearly and is slower than both. Once more, *RS* dominates in terms of performance and scalability.

Conclusion. Our results bring many interesting findings. First, the MonetDB optimizer often produced suboptimal QEPs in the *VP* and *TR* scenario (e.g., for *Q*1, *Q*2, and *Q*3*b* not all *subject-subject* join patterns were realized by merge joins). This shows that relational optimizers may have problems to cope with the specific challenges that arise in the context of RDF. Developers should be aware of this when implementing RDF schemes on top of relational systems.

Using the SP^{2}Bench queries we have identified limitations of the vertical approach. We observe performance bottlenecks in complex scenarios with unbound predicates (e.g., *Q*7), for challenging operator constellations (e.g., CWN-queries *Q*6, *Q*7), and identified queries with many non-*subject-subject* joins as a serious weakness of the *VP* scheme. While the latter weakness has been noted before in [11], our experiments reveal the whole extent of this problem. The materialization of path expressions might improve the performance of such queries [11], but comes with additional costs (e.g., disk space), and is not a general solution.

Another finding is that a triple store with physical (*predicate,subject,object*) sort order is more competitive to the vertical scheme, and might even outperform it for queries (e.g., *Q*7) with unbound predicates (cf. [16]). This relativizes the results from [11], where the triple store was implemented with (*subject, predicate, object*) sort order and only tested in combination with a row-store DBMS.

Finally, none of the tested RDF schemes was competitive to a comparable purely relational encoding. Although relational schemata are domain-specific and, in this regard, optimized for the underlying scenario, we observed a gap of at least one order of magnitude for almost all queries already on small documents, typically increasing with document size. We therefore are convinced that there is still room for optimization in RDF storage schemes, to reduce the gap between RDF and relational data processing and bring forward the Semantic Web vision.

Acknowledgment. The authors thank the MonetDB team for its support in setting up MonetDB and interesting discussions on RDF storage technologies.

References

1. W3C: Resource Description Framework (RDF), http://www.w3.org/RDF/
2. W3C: SPARQL Query Language, http://www.w3.org/TR/rdf-sparql-query/

3. Bizer, C., Cyganiak, R.: D2R Server – Publishing the DBLP Bibliography Database (2007), http://www4.wiwiss.fu-berlin.de/dblp/
4. Tauberer, J.: U.S. Census RDF Data, http://www.rdfabout.com/demo/census/
5. Alexaki, S., Christophides, V., Karvounarakis, G., Plexousakis, D.: On Storing Voluminous RDF Descriptions: The case of Web Portal Catalogs. In: WebDB (2001)
6. Broekstra, J., Kampman, A., van Harmelen, F.: Sesame: A Generic Architecture for Storing and Querying RDF and RDF Schema. In: Horrocks, I., Hendler, J. (eds.) ISWC 2002. LNCS, vol. 2342, pp. 54–68. Springer, Heidelberg (2002)
7. Bonstrom, V., Hinze, A., Schweppe, H.: Storing RDF as a Graph. In: Web Congress, pp. 27–36 (2003)
8. Theoharis, Y., Christophides, V., Karvounarakis, G.: Benchmarking RDF Representations of RDF/S Stores. In: Gil, Y., Motta, E., Benjamins, V.R., Musen, M.A. (eds.) ISWC 2005. LNCS, vol. 3729, pp. 685–701. Springer, Heidelberg (2005)
9. Chong, E.I., Das, S., Eadon, G., Srinivasan, J.: An Efficient SQL-based RDF Querying Scheme. In: VLDB, pp. 1216–1227 (2005)
10. Wilkinson, K.: Jena Property Table Implementation. In: International Workshop on Scalable Semantic Web Knowledge Base, pp. 35–46 (2006)
11. Abadi, D.J., Marcus, A., Madden, S., Hollenbach, K.J.: Scalable Semantic Web Data Management Using Vertical Partitioning. In: VLDB, pp. 411–422 (2007)
12. Abadi, D.J., Marcus, A., Madden, S., Hollenbach, K.J.: Using the Barton libraries dataset as an RDF benchmark. Technical report, MIT-CSAIL-TR-2007-036, MIT (2007)
13. Schmidt, M., Hornung, T., Lausen, G., Pinkel, C.: SP^{2}Bench: A SPARQL Performance Benchmark. Technical report, arXiv:0806.4627v1 cs.DB (2008)
14. Ley, M.: DBLP Database, http://www.informatik.uni-trier.de/~ley/db/
15. openRDF.org: Home of Sesame, http://www.openrdf.org/documentation.jsp
16. Sidirourgos, L., Goncalves, R., Kersten, M., Nes, N., Manegold, S.: Column-store Support for RDF Data Management: not all swans are white. In: VLDB (2008)
17. Bizer, C., Schultz, A.: The Berlin SPARQL Benchmark, http://www4.wiwiss.fu-berlin.de/bizer/BerlinSPARQLBenchmark/
18. Stonebraker, M., et al.: C-store: a Column-oriented DBMS. In: VLDB, pp. 553–564 (2005)
19. CWI Amsterdam: MonetDB, http://monetdb.cwi.nl/
20. Chebotko, A., Lu, S., Yamil, H.M., Fotouhi, F.: Semantics Preserving SPARQL-to-SQL Query Translation for Optional Graph Patterns. Technical report, TR-DB-052006-CLJF (2006)
21. Cyganiac, R.: A Relational Algebra for SPARQL. Technical report, HP Bristol
22. Harris, S.: SPARQL Query Processing with Conventional Relational Database Systems. In: SSWS (2005)
23. SourceForge: Jena2, http://jena.sourceforge.net/DB/index.html
24. Harris, S., Gibbins, N.: 3store: Efficient Bulk RDF Storage. In: PSSS (2003)

A SP²Bench SPARQL Benchmark Queries

```
SELECT ?yr                                              Q1
WHERE {
  ?journal rdf:type bench:Journal.
  ?journal dc:title "Journal 1 (1940)"^^xsd:string.
  ?journal dcterms:issued ?yr }
```

```
SELECT ?inproc ?author ?booktitle ?title              Q2
       ?proc ?ee ?page ?url ?yr ?abstract
WHERE {
  ?inproc rdf:type bench:Inproceedings.
  ?inproc dc:creator ?author.
  ?inproc bench:booktitle ?booktitle.
  ?inproc dc:title ?title.
  ?inproc dcterms:partOf ?proc.
  ?inproc rdfs:seeAlso ?ee.
  ?inproc swrc:pages ?page.
  ?inproc foaf:homepage ?url.
  ?inproc dcterms:issued ?yr
  OPTIONAL { ?inproc bench:abstract ?abstract }
} ORDER BY ?yr
```

```
(a) SELECT ?article                                    Q3
    WHERE {
      ?article rdf:type bench:Article.
      ?article ?property ?value
      FILTER (?property=swrc:pages) }
(b) Q3a, but "swrc:month" instead of "swrc:pages"
(c) Q3a, but "swrc:isbn" instead of "swrc:pages"
```

```
SELECT DISTINCT ?name1 ?name2                          Q4
WHERE {
  ?article1 rdf:type bench:Article.
  ?article2 rdf:type bench:Article.
  ?article1 dc:creator ?author1.
  ?author1 foaf:name ?name1.
  ?article2 dc:creator ?author2.
  ?author2 foaf:name ?name2.
  ?article1 swrc:journal ?journal.
  ?article2 swrc:journal ?journal.
  FILTER (?name1<?name2) }
```

```
(a) SELECT DISTINCT ?person ?name                      Q5
    WHERE {
      ?article rdf:type bench:Article.
      ?article dc:creator ?person.
      ?inproc rdf:type bench:Inproceedings.
      ?inproc dc:creator ?person2.
      ?person foaf:name ?name.
      ?person2 foaf:name ?name2
      FILTER(?name=?name2)
    }
(b) SELECT DISTINCT ?person ?name
    WHERE {
      ?article rdf:type bench:Article.
      ?article dc:creator ?person.
      ?inproc rdf:type bench:Inproceedings.
      ?inproc dc:creator ?person.
      ?person foaf:name ?name
    }
```

```
SELECT ?yr ?name ?doc                                  Q6
WHERE {
  ?class rdfs:subClassOf foaf:Document.
  ?doc rdf:type ?class.
  ?doc dcterms:issued ?yr.
  ?doc dc:creator ?author.
  ?author foaf:name ?name
  OPTIONAL {
    ?class2 rdfs:subClassOf foaf:Document.
    ?doc2 rdf:type ?class2.
    ?doc2 dcterms:issued ?yr2.
    ?doc2 dc:creator ?author2
    FILTER (?author=?author2 && ?yr2<?yr)
  } FILTER (!bound(?author2))
}
```

```
SELECT DISTINCT ?title                                 Q7
WHERE {
  ?class rdfs:subClassOf foaf:Document.
  ?doc rdf:type ?class.
  ?doc dc:title ?title.
  ?bag2 ?member2 ?doc.
  ?doc2 dcterms:references ?bag2
  OPTIONAL {
    ?class3 rdfs:subClassOf foaf:Document.
    ?doc3 rdf:type ?class3.
    ?doc3 dcterms:references ?bag3.
    ?bag3 ?member3 ?doc
    OPTIONAL {
      ?class4 rdfs:subClassOf foaf:Document.
      ?doc4 rdf:type ?class4.
      ?doc4 dcterms:references ?bag4.
      ?bag4 ?member4 ?doc3
    } FILTER (!bound(?doc4))
  } FILTER (!bound(?doc3))
}
```

```
SELECT DISTINCT ?name                                  Q8
WHERE {
  ?erdoes rdf:type foaf:Person.
  ?erdoes foaf:name "Paul Erdoes"^^xsd:string.
  {
    ?doc dc:creator ?erdoes.
    ?doc dc:creator ?author.
    ?doc2 dc:creator ?author.
    ?doc2 dc:creator ?author2.
    ?author2 foaf:name ?name
    FILTER (?author!=?erdoes &&
            ?doc2!=?doc &&
            ?author2!=?erdoes &&
            ?author2!=?author)
  } UNION {
    ?doc dc:creator ?erdoes.
    ?doc dc:creator ?author.
    ?author foaf:name ?name
    FILTER (?author!=?erdoes)
  }
}
```

```
SELECT DISTINCT ?predicate                             Q9
WHERE {
  {
    ?person rdf:type foaf:Person.
    ?subject ?predicate ?person
  } UNION {
    ?person rdf:type foaf:Person.
    ?person ?predicate ?object
  }
}
```

```
SELECT ?subj ?pred                                     Q10
WHERE {
  ?subj ?pred person:Paul_Erdoes
}
```

```
SELECT ?ee                                             Q11
WHERE {
  ?publication rdfs:seeAlso ?ee
} ORDER BY ?ee
LIMIT 10
OFFSET 50
```

Anytime Query Answering in RDF through Evolutionary Algorithms

Eyal Oren, Christophe Guéret, and Stefan Schlobach

Vrije Universiteit Amsterdam, de Boelelaan 1081a, Amsterdam, The Netherlands

Abstract. We present a technique for answering queries over RDF data through an evolutionary search algorithm, using fingerprinting and Bloom filters for rapid approximate evaluation of generated solutions. Our evolutionary approach has several advantages compared to traditional database-style query answering. First, the result quality increases monotonically and converges with each evolution, offering "anytime" behaviour with arbitrary trade-off between computation time and query results; in addition, the level of approximation can be tuned by varying the size of the Bloom filters. Secondly, through Bloom filter compression we can fit large graphs in main memory, reducing the need for disk I/O during query evaluation. Finally, since the individuals evolve independently, parallel execution is straightforward. We present our prototype that evaluates basic SPARQL queries over arbitrary RDF graphs and show initial results over large datasets.

1   Introduction

Almost ten years after its birth as a W3C recommendation, RDF is now used to represent data in an uncountable variety of applications. Together with other (almost) standards, such as RDF schema, OWL, or SPARQL, we now have widely accepted formalisms for the Semantic Web. For all their success there remains a strange discrepancy between the type of representation and retrieval mechanisms and the type of knowledge and data that they are meant to represent. Looking, for example, at SPARQL as a query-language for RDF we have a database-style query language which returns perfect answers on finite repositories. However, the Semantic Web is intrinsically imperfect, too large to represent entirely, with errors, incompleteness, misrepresentations, omissions, ambiguity and so forth.

In this paper we introduce a novel method to query RDF datasets with SPARQL, which is scalable, but which might produce imperfect, approximate answers; first, as a method to deal with *ever larger datasets* such as the billion triples made available for the Semantic Web challenge[1], and secondly, as a method to retrieve an almost correct *approximate answer quickly*. Given the imprecise nature and the size of the Semantic Web, we believe that approximation will be useful in many applications and even essential for others.

---

[1] http://challenge.semanticweb.org

A. Sheth et al. (Eds.): ISWC 2008, LNCS 5318, pp. 98–113, 2008.

1.1   Method

Our method is based on the application of evolutionary techniques in searching for an assignment that validates entailment between a graph representing a query and a data graph. More concretely, we encode a query as a set of triple-constraints with variables where a perfect solution is, as usually, an assignment which maps nodes from the domain of the graph to each variable in such a way that the instantiated constraints are all in the data graph.

To find such an assignment we do not apply exhaustive search on a pre-computed index as is commonly done, but instead evolve the solutions through standard evolutionary methods of mutation and crossover, guided by the number of satisfied constraints as our fitness function. To efficiently calculate this fitness function we represent the original graph data using Bloom filters [4], an efficient and space-reduced data representation for set membership. With each evolutionary step, we converge closer to a solution to our query.

This method is approximate in two ways: Bloom filters are unsound and may lead to false positives. However, the confidence level of the filter can be tuned by increasing the size of the filter (space-correctness trade-off). Secondly, and more importantly, our evolution process might not reach 100% correctness, i.e. solutions may still contain unsatisfied constraints; again, the approximation level may be tuned by longer evolution cycles (time-correctness trade-off). For both sources of approximation we provide a formal model to estimate the probability of correctness of our answer.

The advantage of our method is that its behaviour is intrinsically any-time: our evolutionary algorithm, for which we will demonstrate convergence, can be stopped at any time and produce a meaningful result.

1.2   Research Questions

When presenting a new, approximate, method for querying potentially huge Semantic Web data, two types of questions arise: can we compute useful answers in an any-time manner? And secondly, how scalable is our approach when it comes to runtime and representation size? In the following, we will address these questions:

1. As the main requirement for *any-time* algorithms: does our evolutionary strategy evolve monotonically, i.e. can we expect the next result in an iteration to be at least as good as the previous one.
2. How does *query time* relate to the prospected quality of the answers, and how does query-time compares to traditional approaches?

1.3   What to Expect from This Paper?

This paper introduces a new method for querying RDF repositories for approximate answers, using Bloom filters for fast approximate access to the triple graph and evolutionary methods for searching an (almost) optimal solution. We have implemented this idea and evaluated it on a number of real-life data-sets.

However, our implementation itself is only preliminary and unoptimised, using a fairly standard evolution strategy and a relatively simple fitness function. Therefore, this paper should be read mostly as a proof of concept, where even a rather naive implementation indicates that of our idea can have significant impact as a new querying paradigm for the Semantic Web.

The paper is structured as follows: in Section 2 we give the necessary background to make the paper self-contained. Section 3 introduces our instance of the RDF querying problem formally, before we give details of our evolutionary querying method in Section 4. Section 5 presents our prototype implementation and 6 presents our initial experimental results.

2   Background

Before outlining our approach, we briefly present an overview of evolutionary algorithms. We also discuss existing approaches for querying of RDF data, mostly based on database techniques, and related work in approximate query answering.

2.1   Evolutionary Algorithms

The evolutionary computing paradigm [6] consists of a number of algorithms such as genetic algorithms, evolutionary programming, and others, that are all based on natural selection and genetic inheritance; these algorithms can be used for optimisation problems or for modelling and simulating complex systems.

In this paper we use an evolutionary algorithm, based on the general idea of a population of individuals that evolve in some environment. Each individual is a candidate solution to the problem at hand. The environment is the solution space in which the individual competes with its siblings, based on survival of the fittest. Starting with an initial population, individuals recombine and mutate to produce an offspring. During each iteration of the algorithm, the current individuals are evaluated against a fitness function, the worst performing are removed and replaced by new individuals. Finally, when a stop criterion is satisfied (eg. minimal fitness or maximum number of generations), the best individuals are presented as final solutions. Many variations on this basic evolutionary schema are possible; our particular strategy will be presented in Section 4.

2.2   RDF Query Answering

Existing RDF stores such as Sesame [5] or YARS [10] mostly employ standard database techniques for query answering. Some stores represent triples directly in relational tables, possibly with some optimised partitioning or storage scheme [1, 16, 18]. Others re-implement these well-known database techniques on their own representation [10]. Generally speaking, all systems construct partial indices for simple triple patterns such as $(?s, p, o)$ and $(s, p, ?o)$ during loading time. During query execution single patterns can be answered with direct index lookups, while joins require some nested loops [14], assigning one value at a time for each variable and backtracking when encountering wrong paths. With such loop joins, and in the absence of special path indices, additional query clauses lead to exponential runtime. In contrast, in our approach, additional clauses make the problem *easier* instead of harder, since individuals can be more easily distinguished and have more variation between their fitness values.

2.3   Approximate Query Answering

Generally speaking, when querying a dataset, three kinds of approximations can be made: one can approximate the query, one can approximate the dataset, and one can approximate the reasoning strategy (e.g. returning partial matches).

As an example of the first strategy, Stuckenschmidt and van Harmelen [17] present an approximation technique that first relaxes all query constraints and then stepwise restores them, leading to monotonically improving query results; each approximate query returns a (ever smaller) superset of the original query results and is complete with increasing correctness. The last strategy, approximating the reasoning process, has been investigated for RDF by Kiefer *et al.* [11]. They introduce a similarity join extension to SPARQL; during query answering, potential assignments to join-variables are compared using user-specified similarity functions. The second strategy, approximating the dataset, e.g., through random sampling, is often applied when dealing with very large datasets.

In comparison, our method can be seen as an approximation of the dataset, but not only by random sampling. We have two sources of approximate answers: first, the evolution can be stopped at any point without all constraints necessarily satisfied. Found results are then incorrect to some degree (since some constraints are not satisfied) and may also be incomplete (since some possibilities would not have been explored).

3   Problem Description

In this section we give the necessary, and standard, formal definitions for the problem we address (see also [12, 13]). We also introduce the motivating example used in the following section

Given three infinite sets I, B and L called respectively URI references, blank nodes and literals, an *RDF triple* (s, p, o) is an element of $(I \cup B) \times I \times (I \cup B \cup L)$. Here, s is called the subject, p the predicate, and o the object of the triple. An *RDF graph* (or graph or dataset) is then a set of RDF triples. In this paper we only consider basic SPARQL queries using so called *graph patterns*, which are subsets of $(I \cup L \cup V) \times (I \cup V) \times (I \cup L \cup V)$, where V is a set of variables (disjoint from $U \cup I \cup B$).[2] Whenever, in the remainder of the paper, we discuss SPARQL queries, we will refer to the sublanguage language of graph patterns.

---

[2] An extension to complex query expressions with the usual algebraic operators such as UNION, FILTER, OPTIONAL etc. is conceptually straightforward, and will be considered in more detail in future research.

We define the semantics of a query through a mapping μ which is a partial function $\mu : V \rightarrow U \cup I \cup B$. For a triple pattern t, $\mu(t)$ is the triple obtained when the variables in t are replaced according to μ.

The set of solutions to a query G over a data-set D is now defined as follows: let D be an RDF data-set over $U \cup I \cup B$, and $G =$ a graph pattern. Then we say that a mapping μ is *a solution* for G in D if, and only if, $\mu \in \bigcap_{t \in G} \{\mu \mid dom(\mu) = var(t)$ and $\mu(t) \in D\}$, where $var(t)$ is the set of variables occurring in t.

In the following we will call the graph pattern our *query*, and a solution for G in D an *assignment*. Furthermore, we will refer to a triple pattern within our query as a *constraint*.

3.1 Approximation through Constraint Violation

Based on our definition of query answering, we can now define our notion of "approximation". An approximate solution is a variable assignment for which not all constraints are satisfied, ie. for which not all constraints, after substitution, appear in the original set of triples. To quantify the level of approximation, we therefore count the number of unsatisfied query clauses: the more clauses satisfied, the better the approximation.

Formally, we say that a mapping μ is *an approximate solution* for G in D if, and only if, $\mu \in \{\mu \mid dom(\mu) = var(t)$ and $\mu(t) \in D\}$ for some $t \in G$. To refine the notion of approximation, we have to take the number of satisfied query triple patterns into account, as a solution is of course *better* the more triple patters are satisfied. More concretely, we define the trust in our approximations based on an ordering using the number of violations of constraints t in G.

3.2 Approximation through Unsound Look-Up

On top of the notion of approximation by ignoring some triple patterns in the query graph, we also introduce approximation by using an unsound method for checking whether a mapping μ is indeed a solution to a query G for a graph D. The reason for this is that Bloom filters are fast but unsound lookup mechanisms. As shown in Equation 1, the probability of false positives (because of hash collisions) depends on the number k of hash functions used, the bitsize m of the Bloom filter, and the number n of elements inserted into the filter. During loading time, if a particular confidence level is required, we can tune the size of the Bloom filter; alternatively, with a given filter and domain size, we can estimate the confidence of false positives in the answers using the same equation.

$$confidence = 1 - p_{collision} = 1 - (1 - e^{-\frac{kn}{m}})^k \qquad (1)$$

3.3 Motivating Example

A short snippet of RDF, taken from the SwetoDblp dataset [2] of CS publications, is shown in Listing 1.1. It states that the "Principles of Database Systems" book was written by some unnamed blank node, whose first element is Jeff Ullman, with a homepage at Stanford. All authors in the SwetoDblp dataset are RDF sequences (ordered lists), although in this particular case that sequence has only one member. We will reuse this example and this dataset throughout the rest of the paper.

An example SPARQL query that could be executed over the SwetoDblp dataset is shown in Listing 1.2, with namespace declarations removed for brevity. The query selects the titles of all books in the dataset. A more extensive query is shown in Listing 1.3, which selects the first author of each publication in a conference proceedings, limiting the number of results. Here the "[ ]" brackets indicate traversal of blank nodes, the ";" indicates repetition of the previous subject, and "rdf:_1" is a RDF predicate for the first position in a list.

Listing 1.1. RDF snippet from SwetoDBLP dataset

```
<Ullman88> rdf:type opus:Book .
<Ullman88> rdfs:label "Principles␣of␣Database␣and␣Knowledge-Base␣Systems" .
<Ullman88> opus:author _:b1 .
_:b1 rdf:_1 dblp:ullman .
dblp:ullman foaf:homepage <http://www-db.stanford.edu/~ullman/> .
```

Listing 1.2. SPARQL query for book title

```
SELECT ?title WHERE {
  ?publication rdf:type opus:Book .
  ?publication rdfs:label ?title .
}
```

Listing 1.3. SPARQL query for publication title and first author

```
SELECT ?author ?title WHERE {
  [ rdf:type opus:Article_in_Proceedings ;
    rdfs:label ?title ;
    opus:author [ rdf:_1 [foaf:name ?author ]]
  ] .
} LIMIT 1
```

4   Method

In this section, we present the details of our evolutionary technique. We explain how we represent the RDF input data and the SPARQL query as an evolutionary problem, we present a fitness function for our candidate solutions, and we explain the overall evolution strategy. The advantage of an evolutionary algorithm is that each generated individual contains a complete assignment for all variables, and we verify each complete assignment as a whole. Since our tasks is *verifying* solutions instead of generating them, Bloom filters are very useful, since they do not allow lookups but only membership testing.

In the rest of the section, we explain our technique using the SPARQL query shown earlier in Listing 1.2, which selects all publications and their titles in the SwetoDblp dataset.

Table 1. Translation of SPARQL query into constraints

	Constraint	Filter name
❶	`?publication rdf:type opus:Book`	*spo*
❷	`?publication rdf:type`	*sp*
❸	~~`rdf:type opus:Book`~~	~~*po*~~
❹	`?publication opus:Book`	*so*
❺	`?publication rdfs:label ?title`	*spo*
❻	`?publication rdfs:label`	*sp*
❼	`rdfs:label ?title`	*po*
❽	`?publication ?title`	*so*

4.1   Encoding

To setup our evolutionary algorithm, we need to choose a representation for the query (constraints) and for the individuals (solutions).

Constraints. The graph patterns of our SPARQL query is translated into constraints that will be verified against the populated Bloom filters. We use four Bloom filters (*spo*, *sp*, *so*, *po*) to check both complete and partial triple assignments (to have more fine-grained fitness levels in the individuals).

An example translation is shown in Table 1, listing the constraints for the query shown earlier in Listing 1.2. Constraints 1–4 are generated from the first WHERE clause (`?publication rdf:type opus:Book`), the next ones correspond to the second clause (`?publication rdfs:label ?title`). Constraints using only ground terms, like the third one in our example, are discarded. The user is warned if the constraint was unsatisfied. Otherwise, this operation is silent.

Splitting the triples into more fine-grained (e.g. binary) constraints allows us to define a fitness function with better predictive power.

Individuals. Each individual is a fully instantiated solution to our problem, ie. an assignment for all variables. Therefore, the encoding template for the individuals is the set of terms (both ground term and free variables) defined by the query, as shown in Figure 1. Each individual consists of a set of variable assignments, assigning one domain element to each query term. Each variable assignment can be seen as a gene, and together they form the individual's chromosome.

?publication	ground1	ground2	ground3	?title

Fig. 1. Encoding template for individuals

The domain of candidates depends on the usage of the variable. In total, we have seven domains of candidate assignments: s, p, o, sp, so, po, spo. During graph parsing we populate the three domains s, p and o with nodes occurring at subject, predicate and object position. Then, for each variable, its domain will be set to the intersection of its position in the query clauses. Ground terms in the query are bound to a special domain, containing their only (already known) value, as shown in Table 2.

Table 2. Variables and corresponding domain snippets

variable	domain
?publication	s: <Ullman88>, _:b1, dblp:ullman
ground1	rdf:type
ground2	opus:Book
ground3	rdfs:label
?title	o: <http:/...>, _:b1, dblp:ullman, "Principles...", opus:Book

Moreover, we use a dictionary encoding for all nodes in the dataset. Only these dictionary keys (integers) are used during computation, requiring very little memory space. However, for the sake of readability, nodes values will be used instead of their keys in all the following examples.

4.2   Fitness Evaluation

Next, we establish a metric for the quality of individuals: a fitness function. This function should be designed in such a way that individuals closer to the optimal solution can be identified by the system. For our application, an optimal solution consist of a valid variable assignment.

A candidate solution is optimal if it satisfies all constraints. The quality of our individual is therefore related to the number of constraint that they do violate. To illustrate the fitness, we consider the candidate solution shown in Table 3(a). To evaluate the fitness of this individual, the query instantiated with the variable assignment corresponding to the individual is checked against all relevant corresponding Bloom filters. For each (possibly binary) constraint that is not present in a filter, the involved variables are penalised by one point, as shown in Table 3(b). Table 3(c) shows the complete fitness evaluation for this individual; the individual violated the two constraints in several manners, leading to a total fitness of 8 (lower is better). In addition to this overall fitness, we will also use the individual score per variable later to determine how to control mutation.

4.3   Evolution Process

The evolution process consists of four operators: parent selection, recombination (crossover), mutation and survivor selection. We now describe our implemented choice for each of these operators.

Parent selection. Evolution loops create new individuals and destroy previous ones. The parent selection operator is aimed at selecting from the current population the individuals that will be allowed to mate and create offspring. Selection is commonly aimed at the best individuals. The underlying assumption of this selection pressure is that mating two good individuals will lead to better results than combining two bad individuals.

Several parent selection schemes can be used. We employ a tournament-based selection, in which two individuals are randomly picked from the population, the best one is kept as the first parent. This process is repeated to get more parents.

Table 3. Evaluation of a candidate solution

(a) Candidate solution

dblp:ullman	rdf:type	opus:Book	rdfs:label	"Principles..."

(b) Evaluating the individual

	Constraint	Filter	Test result
❶	*dblp:ullman* rdf:type opus:Book	*spo*	*false*
❷	*dblp:ullman* rdf:type	*sp*	*false*
❹	*dblp:ullman* opus:Book	*so*	*false*
❺	*dblp:ullman* rdfs:label *"Principles..."*	*spo*	*false*
❻	*dblp:ullman* rdfs:label	*sp*	*false*
❼	rdfs:label *"Principles..."*	*po*	*true*
❽	*dblp:ullman* *"Principles..."*	*so*	*false*

(c) Summing constraint violations

variables	?publication	?title
violation	❶ ❷ ❹ ❺ ❻ ❽	❺ ❽

Table 4. One-point crossover operator process

(a) Selection of random pivot gene

dblp:ullman	rdf:type	opus:Book	rdfs:label	"Principles..."

<Ullman88>	rdf:type	opus:Book	rdfs:label	_:b1

(b) Creation of two children

dblp:ullman	rdf:type	opus:Book	rdfs:label	_:b1

<Ullman88>	rdf:type	opus:Book	rdfs:label	"Principles..."

Recombination. Recombination acts as exploration during the search process. This operator is aimed at creating new individuals in unexplored regions of the search space. Its operation takes two parents and combines them into two children. After various experiments, we opted for a classical one-point crossover operator, in which one pivot gene is randomly selected and the parts around it are swapped between the parents, demonstrated in Table 4.

Mutation. As compared with the recombination operator whose objective is to do "big jumps" in the search space, the mutation operator is meant to explore the neighbourhood of an individual. A slight modification is applied to one or more genes. This perturbation is commonly referred to as an exploitation scheme.

Table 5. Mutation operator process

(a) The gene responsible for the highest number of errors is selected

dblp:ullman	rdf:type	opus:Book	rdfs:label	"Principles..."

6	0	0	0	2

(b) and a new value is randomly assigned

<Ullman88>	rdf:type	opus:Book	rdfs:label	"Principles..."

In a standard genetic algorithm, mutation is a blind operator. The gene to modify is randomly selected and the mutation is applied. After some experimentation, we instead designed a mutation operator which is biased towards mutating badly performing genes, based on the score per variables computed during fitness evaluation. In case of a tie between two or more genes, a random selection is performed among them, as shown in Table 5.

Such a mutation operator improves the convergence speed of the population by identifying the most problematic variables. However, such a greedy strategy may lead to local optimums, without reaching proper global optimums. To reduce the risk of premature convergence, we therefore also apply a blind random mutation, after our optimised local search. This mutation is applied randomly, with low probability, to one gene, randomly assigning a new value to it.

Survivor Selection. At this point of the evolution, we have both a parent population and an offspring population (created by the parents). During the survivor selection phase we select the individuals to keep for the next evolution round. After experimenting with several possible strategies, we chose a generational selection: at the end of each evolutionary cycle, the parent population is discarded and replaced by its offspring.

5 Prototype Implementation

We implemented our technique into an initial prototype in C++, using the Open Beagle framework [7] for the evolutionary computing and Redland [3] for the RDF graph parsing and SPARQL query parsing. As is commonly done, we split the problem into a parsing and a querying phase, each with their own executable. The prototype is open-source and available from http://eardf.few.vu.nl.

During parsing, we fill the Bloom filters with the triples found in the RDF input graph and collect the candidate assignments for each triple position. To reduce the memory requirements in the Bloom filter and to increase the speed of the fitness calculation, we construct a dictionary of all nodes in the input graph. The dictionary maps each distinct node (URI, blank node and literal) to some index number; internally, only these indices are used (the chromosomes are simply a list of index numbers); when outputting the final results we transform the solution indices back into the original node. To reduce the size of the dictionary, we compress all nodes using the zlib library[3].

We use Redland to parse the RDF into streams of triples. For each node in each triple, we retrieve or construct its corresponding dictionary index. We then insert the triple into the relevant Bloom filters, substituting all nodes by their index number, and collect the candidates for each domain (s, p, o, sp, so, po, and spo). After parsing, we serialise the dictionary, domains, and Bloom filters.

When querying a parsed RDF graph, we load the previously generated Bloom filters, domains, and dictionary. We then parse the SPARQL query (also using Redland) and transform all WHERE clauses into constraints on the evolutionary problem. We also transform all ground terms in the query into variables and problem constraints, but with domains that contain only a single element (the ground term), so the individuals have no choice in the assignment of that variable. We then start the evolutionary process; when it finishes, we rewrite the found solution to the SPARQL result format using the dictionary.

6   Evaluation

We evaluate our technique on several datasets, with several different queries. Since our implementation is very basic without any optimisations, the absolute loading and runtime numbers are not that meaningful. Instead we focus on the curve of the graph, and especially on the first question: do we evolve monotonically towards a better solution.

6.1   Experimental Setup

We used three publicly available datasets: LUBM, an automatically generated dataset targeted towards OWL reasoning [9], FOAF, a publicly available collection of FOAF profiles, and DBLP, an extract from the SwetoDblp collection mentioned earlier. From the SwetoDblp dataset we extracted two subsets, one containing 5000 triples and one containing 500.000 triples. All these datasets are available at http://eardf.few.vu.nl.

On each dataset, we evaluated one query. For the DBLP datasets, we used one of the benchmark queries proposed by [15], shown in Listing 1.4. For the FOAF dataset, we query for the name, work homepage, and publications of all people, as shown in Listing 1.6. For LUBM, we used the standard LUBM query #2, shown in Listing 1.5. This query relies on OWL semantics and is not satisfiable under simple RDF entailment; still, we will see that our technique manages to converge towards an approximate answer.

All experiments were run using 100–200 individuals, over 500 generations, and each experiment was repeated 100 times. We measured data loading time, query execution time, and average and best fitness of the population in each

---

[3] http://zlib.net

Listing 1.4. Query on DBLP dataset

```
SELECT ?author ?art1 ?art2 ?proctitle ?year
WHERE {
    ?a1 rdf:type opus:Article_in_Proceedings;
         opus:author [ rdf:_1 ?au ] ;
         opus:isIncludedIn ?proc ;
         rdfs:label ?art1.
    ?a2 rdf:type opus:Article_in_Proceedings;
         opus:author [ rdf:_1 ?au ] ;
         opus:isIncludedIn ?proc ;
         rdfs:label ?art2.
    ?au foaf:name ?author .
    ?proc rdfs:label ?proctitle ;
              opus:year ?year .
} LIMIT 1
```

Listing 1.5. Query on LUBM dataset

```
SELECT ?student, ?univ, ?dept
WHERE {
    ?student rdf:type ub:GraduateStudent .
    ?univ rdf:type ub:University .
    ?dept rdf:type ub:Department .
    ?student ub:memberOf ?dept .
    ?dept ub:subOrganizationOf ?univ .
    ?student ub:undergraduateDegreeFrom ?univ
}
```

generation. The experiments were performed on a 64-bit 2GHz Intel Core2 Duo machine with 2Gb of RAM, running Linux kernel 2.6.24.

6.2   Evaluation Results

Figure 2 demonstrates that on all datasets, for all queries, our technique converges towards a complete solution. For each dataset, we show the best fitness in each generation, averaged over the 100 different runs. One should note, that even if we do not reach perfect fitness in the allocated evolution time, the solution are typically very close to perfection. Since we have several Bloom filters and since queries contain many clauses with several variables, a difference in fitness between 0 and n points is often caused by only one or two wrong assignments. Establishing a direct relation between the usefulness of a candidate solution and its fitness value is a tricky task. In absence of a gold standard with "most useful" ranked answers, fitness values are meaningfull essentially when compared pair-wise.

Table 6 shows the average query times for these queries. These times include de-serialising the Bloom filters, the domains and the dictionary, and the evolu-

Listing 1.6. Query on FOAF dataset

```
WHERE {
    ?person foaf:name ?name .
    ?person foaf:workplaceHomepage ?work .
    ?person foaf:publications ?pubs .
}
```

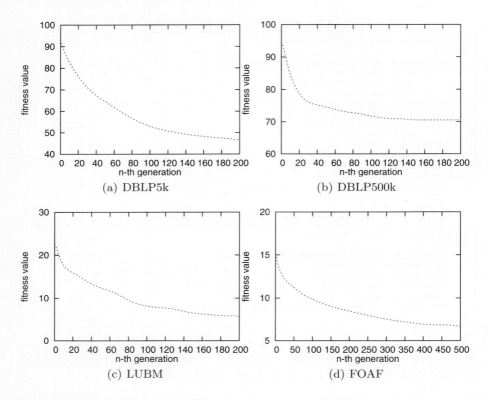

Fig. 2. Evolution of best fitness in the population for different datasets

Table 6. Average query execution time

dataset	nr. triples	nr. variables	runtime
LUBM	8502	3	3.60s
FOAF	21163	4	2.31s
DBLP5k	5000	9	6.76s
DBLP500k	500000	9	38.74s

tion of 100 individuals for 500 generations. In absolute terms, these times are still one order of magnitude slower than existing systems (we have compared with Sesame2) but we have much room for optimisation, both in the implemented code and in the evolution strategy. One interesting option is to use the parallel (distributed) execution extension of the Beagle framework, which allows sets of individuals to evolve separately on distributed machines, especially since memory usage during our evolution is minimal. Note that, due to our unoptimised implementation, most of the querying time is actually spent to de-serialise the previously parsed information; the actual evolution is almost constant with the size of the dataset. This remarks also applies to memory usage. As we expected,

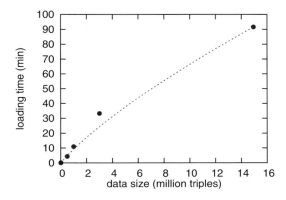

Fig. 3. Data loading time for different data sizes

each of the Bloom filters only requires very little memory to reach an acceptable confidence rate and the size of the individuals during evolution remains small, as it only depends on the number of variables in the query.

Figure 3 shows how loading times relate to the size of the datasets. In absolute terms our loading times are in the same order of magnitude as Sesame2, presumably because we, on the one hand, do not construct any indices but only construct the dictionary and populate our Bloom filters, while our implementation on the other hand is still unoptimised. In general, the loading times seems to grow linearly with the number of triples in the dataset [4], presumably because most time is spent in computing the hashes before Bloom filter insertion, which needs to be repeated for each triple in the graph.

7   Conclusion

We have introduced a novel method for querying RDF datasets. In contrast to traditional database-oriented techniques, our method is not focused on finding perfect solutions but rather on finding *good enough* solutions. Given the imprecise nature and the size of the Semantic Web, we believe that such approximations are useful in many applications.

We generate different solutions using an evolutionary algorithm; to enable fast computation of the fitness of solutions, we verify assignments using Bloom filters containing a compressed representation of the data graph. Our evolutionary approach features anytime and approximate answering, and we have demonstrated that even with a rather straightforward evolutionary strategy our solutions improve monotonically with each generation. This answers our first research question positively.

The prototype used for this paper and the results for small datasets should be seen as a proof of concept. Our first experiments confirm our intuition, showing

---

[4] Our current prototype was able to load, but not able to proceed, a 15M triples dataset.

that it is indeed possible to construct query solutions "from scratch" for RDF datasets, guided by the estimated quality of variable assignments.

The answer to the second research question is less easily given, as the comparison is intrinsically unfair as, on the one hand, our method is still unoptimised and, on the other hand, produces approximate results. However, initial experiments indicate that the acual costs in runtime for the evolution part is constant, and that the low memory requirement will indeed reduce the number of I/O operations.

Using our rather unoptimised implementation as a baseline, we are currently improving the evolutionary operators to increase convergence speed and efficiency (i.e. converging to useful results). We are also improving the code for loading data, and repeat run our experiments on bigger datasets and more complex queries. An insight in this ongoing work can be found in a follow-up paper focused on scalability [8], featuring a new and improved implementation with promising performance and scalability results.

References

[1] Abadi, D.J., Marcus, A., Madden, S., Hollenbach, K.J.: Scalable semantic web data management using vertical partitioning. In: Proceedings of the International Conference on Very Large Data Bases (VLDB), pp. 411–422 (2007)

[2] Aleman-Meza, B., Hakimpour, F., Arpinar, I., Sheth, A.: SwetoDblp ontology of computer science publications. Journal of Web Semantics 5(3), 151–155 (2007)

[3] Beckett, D.: The design and implementation of the Redland RDF application framework. Computer Networks 39(5), 577–588 (2002)

[4] Bloom, B.H.: Space/time trade-offs in hash coding with allowable errors. Communications of the ACM 13(7), 422–426 (1970)

[5] Broekstra, J., Kampman, A., van Harmelen, F.: Sesame: A generic architecture for storing and querying RDF and RDF Schema. In: Horrocks, I., Hendler, J. (eds.) ISWC 2002. LNCS, vol. 2342, pp. 54–68. Springer, Heidelberg (2002)

[6] Eiben, A.E., Smith, J.E.: Introduction to Evolutionary Computing. Springer, Berlin (2003)

[7] Gagné, C., Parizeau, M.: Genericity in evolutionary computation software tools: Principles and case-study. International Journal on Artificial Intelligence Tools 15, 173–194 (2006)

[8] Guéret, C., Oren, E., Schlobach, S., Schut, M.: An evolutionary perspective on approximate RDF query answering. In: Proceedings of the International Conference on Scalable Uncertainty Management (2008)

[9] Guo, Y., Pan, Z., Heflin, J.: LUBM: A benchmark for OWL knowledge base systems. Journal of Web Semantics 3, 158–182 (2005)

[10] Harth, A., Decker, S.: Optimized index structures for querying RDF from the web. In: Proceedings of the Latin-American Web Congress (LA-Web), pp. 71–80 (2005)

[11] Kiefer, C., Bernstein, A., Stocker, M.: The fundamentals of iSPARQL: A virtual triple approach for similarity-based semantic web tasks. In: Aberer, K., Choi, K.-S., Noy, N., Allemang, D., Lee, K.-I., Nixon, L., Golbeck, J., Mika, P., Maynard, D., Mizoguchi, R., Schreiber, G., Cudré-Mauroux, P. (eds.) ASWC 2007 and ISWC 2007. LNCS, vol. 4825, pp. 295–309. Springer, Heidelberg (2007)

[12] Muñoz, S., Pérez, J., Gutierrez, C.: Minimal deductive systems for RDF. In: Franconi, E., Kifer, M., May, W. (eds.) ESWC 2007. LNCS, vol. 4519. Springer, Heidelberg (2007)

[13] Pérez, J., Arenas, M., Gutierrez, C.: Semantics and complexity of SPARQL. In: Cruz, I., Decker, S., Allemang, D., Preist, C., Schwabe, D., Mika, P., Uschold, M., Aroyo, L.M. (eds.) ISWC 2006. LNCS, vol. 4273. Springer, Heidelberg (2006)

[14] Selinger, P.G., Astrahan, M.M., Chamberlin, D.D., Lorie, R.A., et al.: Access path selection in a relational database management system. In: Proceedings of the ACM SIGMOD International Conference on Management of Data (1979)

[15] Shvila, M., Jelinek, I.: Benchmarking RDF production tools. In: Wagner, R., Revell, N., Pernul, G. (eds.) DEXA 2007. LNCS, vol. 4653, pp. 700–709. Springer, Heidelberg (2007)

[16] Sintek, M., Kiesel, M.: RDFBroker: A signature-based high-performance RDF store. In: Sure, Y., Domingue, J. (eds.) ESWC 2006. LNCS, vol. 4011, pp. 363–377. Springer, Heidelberg (2006)

[17] Stuckenschmidt, H., van Harmelen, F.: Approximating terminological queries. In: Andreasen, T., Motro, A., Christiansen, H., Larsen, H.L. (eds.) FQAS 2002. LNCS (LNAI), vol. 2522, pp. 329–343. Springer, Heidelberg (2002)

[18] Wilkinson, K., Sayers, C., Kuno, H.A., Reynolds, D.: Efficient RDF storage and retrieval in Jena2. In: Proceedings of the International Workshop on Semantic Web and Databases (SWDB) (2003)

The Expressive Power of SPARQL

Renzo Angles and Claudio Gutierrez

Department of Computer Science, Universidad de Chile
{rangles,cgutierr}@dcc.uchile.cl

Abstract. This paper studies the expressive power of SPARQL. The main result is that SPARQL and non-recursive safe Datalog with negation have equivalent expressive power, and hence, by classical results, SPARQL is equivalent from an expressiveness point of view to Relational Algebra. We present explicit generic rules of the transformations in both directions. Among other findings of the paper are the proof that negation can be simulated in SPARQL, that non-safe filters are superfluous, and that current SPARQL W3C semantics can be simplified to a standard compositional one.

1 Introduction

Determining the expressive power of a query language is crucial for understanding its capabilities and complexity, that is, what queries a user is able to pose, and how complex the evaluation of queries is, issues that are central considerations to take into account when designing a query language.

SPARQL, the query language for RDF, has recently become a W3C recommendation [9]. In the RDF Data Access Working Group (WG) were it was designed, expressiveness concerns generated ample debate. Many of them remained open due to lack of understanding of the theoretical expressive power of the language.

This paper studies in depth the expressive power of SPARQL. A first issue addressed is the incorporation of negation. The W3C specification of SPARQL provides explicit operators for join and union of graph patterns, even for specifying optional graph patterns, but it does not define explicitly the difference of graph patterns. Although intuitively it can be emulated via a combination of optional patterns and filter conditions (like negation as failure in logic programming), we show that there are several non-trivial issues to be addressed if one likes to define the difference of patterns inside the language.

A second expressiveness issue refers to graph patterns with non-safe filter, i.e., graph patterns $(P \text{ FILTER } C)$ for which there are variables in C not present in P. It turns out that these type of patterns, which have non-desirable properties, can be simulated by safe ones (i.e., patterns where every variable occurring in C also occurs in P). This simple result has important consequences for defining a clean semantics, in particular a compositional and context-free one.

A. Sheth et al. (Eds.): ISWC 2008, LNCS 5318, pp. 114–129, 2008.

A third topic of concern was the presence of non desirable features in the W3C semantics like its operational character. We show that the W3C specification of the semantics of SPARQL is equivalent to a well behaved and studied compositional semantics for SPARQL, which we will denote in this paper SPARQL$_C$ [6].

Using the above results, we are able to determine the expressive power of SPARQL. We prove that SPARQL$_C$ and non-recursive safe Datalog with negation (nr-Datalog$^\neg$) are equivalent in their expressive power. For this, first we show that SPARQL$_C$ is contained in nr-Datalog$^\neg$ by defining transformations (for databases, queries, and solutions) from SPARQL$_C$ to nr-Datalog$^\neg$, and we prove that the result of evaluating a SPARQL$_C$ query is equivalent, via the transformations, to the result of evaluating (in nr-Datalog$^\neg$) the transformed query. Second, we show that nr-Datalog$^\neg$ is contained in SPARQL$_C$ using a similar approach. It is important to remark that the transformations used are explicit and simple, and in all steps bag semantics is considered.

Finally, and by far, the most important result of the paper is the proof that SPARQL has the same expressive power of Relational Algebra under bag semantics (which is the one of SPARQL). This follows from the well known fact that Relational Algebra has the same expressive power as nr-Datalog$^\neg$ [1].

The paper is organized as follows. In Section 2 we present preliminary material. Section 3 presents the study of negation. Section 4 studies non-safe filter patterns. Section 5 proves that the W3C specification of SPARQL and SPARQL$_C$ are equivalent. Section 6 proves that SPARQL$_C$ and nr-Datalog$^\neg$ have the same expressive power. Section 7 presents the conclusions.

Related Work. The W3C recommendation SPARQL is from January 2008. Hence, it is no surprise that little work has been done in the formal study of its expressive power. Several conjectures were raised during the WG sessions [1]. Furche et al. [3] surveyed expressive features of query languages for RDF (including old versions of SPARQL) in order to compare them systematically. But there is no particular analysis of the expressive power of SPARQL.

Cyganiak [2] presented a translation of SPARQL into Relational Algebra considering only a core fragment of SPARQL. His work is extremely useful to implement and optimize SPARQL in SQL engines. At the level of analysis of expressive issues it presented a list of problems that should be solved (many of which still persist), like the filter scope problem and the nested optional problem.

Polleres [8] proved the inclusion of the fragment of SPARQL patterns with safe filters into Datalog by giving a precise and correct set of rules. Schenk [10] proposed a formal semantics for SPARQL based on Datalog, but concentrated on complexity more than expressiveness issues. Both works do not consider bag semantics of SPARQL in their translations.

The work of Perez et al. [6] and the technical report [7], that gave the formal basis for SPARQL$_C$ compositional semantics, addressed several expressiveness issues, but no systematic study of the expressive power of SPARQL was done.

---

[1] See `http://lists.w3.org/Archives/Public/public-rdf-dawg-comments/`, especially the years 2006 and 2007.

2   Preliminaries

2.1   RDF and Datasets

Assume there are pairwise disjoint infinite sets I, B, L (IRIs, Blank nodes, and RDF literals respectively). We denote by T the union $I \cup B \cup L$ (RDF *terms*). A tuple $(v_1, v_2, v_3) \in (I \cup B) \times I \times T$ is called an *RDF triple*, where v_1 is the *subject*, v_2 the *predicate*, and v_3 the *object*. An *RDF Graph* [4] (just graph from now on) is a set of RDF triples. Given a graph G, term(G) denotes the set of elements of T occurring in G and blank(G) denotes the set of blank nodes in G. The *union* of graphs, $G_1 \cup G_2$, is the set theoretical union of their sets of triples.

An *RDF dataset* D is a set $\{G_0, \langle u_1, G_1 \rangle, \ldots, \langle u_n, G_n \rangle\}$ where each G_i is a graph and each u_j is an IRI. G_0 is called the *default graph* of D and it is denoted dg(D). Each pair $\langle u_i, G_i \rangle$ is called a *named graph*; define name$(G_i)_D = u_i$ and gr$(u_i)_D = G_i$. We denote by term(D) the set of terms occurring in the graphs of D. The set of IRIs $\{u_1, \ldots, u_n\}$ is denoted names(D). Every dataset satisfies that: (i) it always contains one default graph (which could be empty); (ii) there may be no named graphs; (iii) each u_j is distinct; and (iv) blank$(G_i) \cap$ blank$(G_j) = \emptyset$ for $i \neq j$. Finally, the *active graph* of D is the graph G_i used for querying D.

2.2   SPARQL

A SPARQL query is syntactically represented by a block consisting of a *query form* (SELECT, CONSTRUCT or DESCRIBE), zero o more *dataset clauses* (FROM and FROM NAMED), a WHERE *clause*, and possibly *solution modifiers* (e.g. DISTINCT). The WHERE clause provides a *graph pattern* to match against the RDF dataset constructed from the dataset clauses.

There are two formalizations of SPARQL which will be used throughout this study: SPARQL$_{\text{WG}}$, the W3C recommendation language SPARQL [9] and SPARQL$_{\text{C}}$, the formalization of SPARQL given in [6]. We will need some general definitions before describe briefly both languages.

Assume the existence of an infinite set V of variables disjoint from T. We denote by var(α) the set of variables occurring in the structure α. A tuple from $(I \cup L \cup V) \times (I \cup L \cup V) \times (I \cup V)$ is called a *triple pattern*. A *basic graph pattern* is a finite set of triple patterns.

A *filter constraint* is defined recursively as follows: (i) if $?X, ?Y \in V$ and $u \in I \cup L$ then $?X = u$, $?X = ?Y$, bound($?X$), isIRI($?X$), isLiteral($?X$), and isBlank($?X$) are *atomic filter constraints*[2]; (ii) if C_1 and C_2 are filter constraints then $(\neg C_1)$, $(C_1 \wedge C_2)$, and $(C_1 \vee C_2)$ are *complex filter constraints*.

A *mapping* μ is a partial function $\mu : V \to T$. The domain of μ, dom(μ), is the subset of V where μ is defined. The *empty mapping* μ_0 is a mapping such that dom(μ_0) = $\emptyset$. Two mappings μ_1, μ_2 are *compatible*, denoted $\mu_1 \sim \mu_2$, when for all $?X \in$ dom(μ_1) $\cap$ dom(μ_2) it satisfies that $\mu_1(?X) = \mu_2(?X)$, i.e., when $\mu_1 \cup \mu_2$ is also a mapping. The expression $\mu_{?X \to v}$ denote a mapping such that dom(μ) = $\{?X\}$ and $\mu(?X) = v$.

---

[2] For a complete list of atomic filter constraints see [9].

The evaluation of a filter constraint C against a mapping μ, denoted $\mu(C)$, is defined in a three value logic with values $\{true, false, error\}$ as follows:

- If C is an atomic filter constraint, excluding bound($\cdot$), and var(C) $\not\subseteq$ dom(μ), then $\mu(C) = error$; else if C is $?X = u$ and $\mu(?X) = u$, or if C is $?X = ?Y$ and $\mu(?X) = \mu(?Y)$, or if C is isIRI($?X$) and $\mu(?X) \in I$, if C is isLiteral($?X$) and $\mu(?X) \in L$, if C is isBlank($?X$) and $\mu(?X) \in B$, then $\mu(C) = true$; otherwise $\mu(C) = false$.
- If C is bound($?X$) then $\mu(C) = true$ if $?X \in$ dom(μ) else $\mu(C) = false$.[3]
- If C is $(\neg C_1)$ then $\mu(C) = true$ when $\mu(C_1) = false$; $\mu(C) = false$ when $\mu(C_1) = true$; and $\mu(C) = error$ when $\mu(C_1) = error$.
- If C is $(C_1 \vee C_2)$ then $\mu(C) = true$ if either $\mu(C_1) = true$ or $\mu(C_2) = true$; $\mu(C) = false$ if $\mu(C_1) = false$ and $\mu(C_2) = false$; otherwise $\mu(C) = error$.
- If C is $(C_1 \wedge C_2)$ then $\mu(C) = true$ if $\mu(C_1) = true$ and $\mu(C_2) = true$; $\mu(C) = false$ if either $\mu(C_1) = false$ or $\mu(C_2) = false$; otherwise $\mu(C) = error$.

A mapping μ satisfies a filter constraint C, denoted $\mu \models C$, iff $\mu(C) = true$. Consider the following operations between two sets of mappings Ω_1, Ω_2:

$$\Omega_1 \bowtie \Omega_2 = \{\mu_1 \cup \mu_2 \mid \mu_1 \in \Omega_1, \mu_2 \in \Omega_2 \text{ and } \mu_1 \sim \mu_2\}$$
$$\Omega_1 \cup \Omega_2 = \{\mu \mid \mu \in \Omega_1 \text{ or } \mu \in \Omega_2\}$$
$$\Omega_1 \setminus \Omega_2 = \{\mu_1 \in \Omega_1 \mid \text{ for all } \mu_2 \in \Omega_2, \mu_1 \text{ and } \mu_2 \text{ are not compatible }\}$$
$$\Omega_1 \setminus_C \Omega_2 = \{\mu_1 \in \Omega_1 \mid \text{ for all } \mu_2 \in \Omega_2, \mu_1 \text{ and } \mu_2 \text{ are not compatible }\} \cup$$
$$\{\mu_1 \in \Omega_1 \mid \text{ for all } \mu_2 \in \Omega_2 \text{ such that } \mu_1 \sim \mu_2, (\mu_1 \cup \mu_2) \not\models C\}$$
$$\Omega_1 \bowtie\!\!\!\!\!\!\!\!\!\!\!\!\!\!\! \bowtie\ \Omega_2 = (\Omega_1 \bowtie \Omega_2) \cup (\Omega_1 \setminus \Omega_2)$$
$$\Omega_1 \bowtie\!\!\!\!\!\!\!\!\!\!\!\!\!\!\! \bowtie\ {}_C\Omega_2 = \{\mu \mid \mu \in (\Omega_1 \bowtie \Omega_2) \text{ and } \mu \models C\} \cup (\Omega_1 \setminus_C \Omega_2)$$

Syntax and Semantics of SPARQL$_\text{C}$.

A SPARQL$_\text{C}$ graph pattern P is defined recursively by the following grammar:

```
P  ::= t | "(" GP ")"
GP ::= P "AND" P | P "UNION" P | P "OPT" P | P "FILTER" C |
       n "GRAPH" P
```

where t denotes a triple pattern, C denotes a filter constraint, and $\mathbf{n} \in I \cup V$.

The evaluation of a SPARQL$_\text{C}$ graph pattern P over an RDF dataset D having active graph G, denoted $[\![P]\!]_G^D$, is defined recursively as follows:

- if P is a triple pattern t, $[\![P]\!]_G^D = \{\mu \mid$ dom(μ) = var(t) and $\mu(t) \in G\}$ where $\mu(t)$ is the triple obtained by replacing the variables in t according to mapping μ.
- if P is a complex graph pattern then $[\![P]\!]_G^D$ is defined as given in Table 1.

Syntax and Semantics of SPARQL$_\text{WG}$.

A SPARQL$_\text{WG}$ graph pattern GroupGP is defined by the following grammar[4]:

```
GroupGP        ::= "{" TB? ((GPNotTriples | Filter) "."? TB?)* "}"
GPNotTriples   ::= OptionalGP | GroupOrUnionGP | GraphGP
```

---

[3] Functions invoked with an argument of the wrong type are evaluated to $error$.

[4] http://www.w3.org/TR/rdf-sparql-query/#grammar. We use GP and TB to abbreviate GraphPattern and TriplesBlock respectively

Table 1. Semantics of SPARQL$_C$ graph patterns. P_1, P_2 are SPARQL$_C$ graph patterns, C is a filter constraint, $u \in I$ and $?X \in V$.

Graph pattern P	Evaluation $[\![P]\!]_G^D$
$(P_1 \text{ AND } P_2)$	$[\![P_1]\!]_G^D \bowtie [\![P_2]\!]_G^D$
$(P_1 \text{ OPT } P_2)$	$[\![P_1]\!]_G^D \mathbin{\rlap{$\bowtie$}{\,\,\,\supset}} [\![P_2]\!]_G^D$
$(P_1 \text{ UNION } P_2)$	$[\![P_1]\!]_G^D \cup [\![P_2]\!]_G^D$
$(P_1 \text{ FILTER } C)$	$\{\mu \mid \mu \in [\![P_1]\!]_G^D \text{ and } \mu \models C\}$
$(u \text{ GRAPH } P_1)$	$[\![P_1]\!]_{\mathrm{gr}(u)_D}^D$
$(?X \text{ GRAPH } P_1)$	$\bigcup_{v \in \mathrm{names}(D)}([\![P_1]\!]_{\mathrm{gr}(v)_D}^D \bowtie \{\mu_{?X \to v}\})$

```
OptionalGP      ::= "OPTIONAL" GroupGP
GraphGP         ::= "GRAPH" VarOrIRIref GroupGP
GroupOrUnionGP ::= GroupGP ( "UNION" GroupGP )*
Filter          ::= "FILTER" Constraint
```

where `TB` denotes a basic graph pattern (a set of triple patterns), `VarOrIRIref` denotes a term in the set $I \cup V$ and `Constraint` denotes a filter constraint. Note that the operator `{A . B}` represents the AND but it has not fixed arity.

The evaluation of a SPARQL$_{\mathrm{WG}}$ graph pattern `GroupGP` is defined by a series of steps, starting by transforming `GroupGP`, via a function T, into an intermediate algebra expression E (with operators BGP, Join, Union, LeftJoin, Graph and Filter), and finally evaluating E on an RDF dataset D.

The transformation $T(\texttt{GroupGP})$ is given by Algorithm 1. The evaluation of E over an RDF dataset D having active graph G, which we will denote $\langle\!\langle E \rangle\!\rangle_G^D$ (originally denoted $\mathrm{eval}(D(G), E)$ in [9]), is defined recursively as follows:

- if E is BGP(TB), $\langle\!\langle E \rangle\!\rangle_G^D = \{\mu \mid \mathrm{dom}(\mu) = \mathrm{var}(E) \text{ and } \mu(E) \subseteq G\}$ where $\mu(E)$ is the set of triples obtained by replacing the variables in the triple patterns of TB according to mapping μ.
- if E is a complex expression then $\langle\!\langle P \rangle\!\rangle_G^D$ is defined as given in Table 2.

Note 1. In the definition of graph patterns, we avoided blank nodes, because this restriction does not diminish the generality of our study. In fact, each SPARQL query Q can be simulated by a SPARQL query Q' without blank nodes in its pattern. It follows from the definitions of RDF instance mapping, solution mapping, and the order of evaluation of solution modifiers (see [9]), that if Q is a query with graph pattern P, and Q' is the same query where each blank node b in P has been replaced by a fresh variable $?X_b$ then Q and Q' give the same results. (Note that, if Q has the query form SELECT or DESCRIBE, the "$*$" parameter is –according to the specification of SPARQL– an abbreviation for all variables occurring in the pattern. In this case the query Q' should explicit in the SELECT clause all variables of the original pattern P.)

Note 2. SPARQL$_C$ follows a compositional semantics, whereas SPARQL$_{\mathrm{WG}}$ follows a mixture of compositional and operational semantics where the meaning of certain patterns depends on their context, e.g., lines 7 and 8 in algorithm 1.

Algorithm 1. Transformation of $\text{SPARQL}_{\text{WG}}$ patterns into algebra expressions.

1: // Input: a $\text{SPARQL}_{\text{WG}}$ graph pattern GroupGP
2: // Output: an algebra expression $E = T(\texttt{GroupGP})$
3: $E \leftarrow$ empty pattern; $FS \leftarrow \emptyset$
4: **for** each syntactic form f in GroupGP **do**
5:     **if** f is TB **then** $E \leftarrow \text{Join}(E, \text{BGP}(\texttt{TB}))$
6:     **if** f is OPTIONAL $\texttt{GroupGP}_1$ **then**
7:         **if** $T(\texttt{GroupGP}_1)$ is $\text{Filter}(F, E')$ **then** $E \leftarrow \text{LeftJoin}(E, E', F)$
8:         **else** $E \leftarrow \text{LeftJoin}(E, T(\texttt{GroupGP}_1), true)$
9:     **if** f is $\texttt{GroupGP}_1$ UNION $\cdots$ UNION $\texttt{GroupGP}_n$ **then**
10:        **if** $n > 1$ **then**
11:            $E' \leftarrow \text{Union}(\cdots(\text{Union}(T(\texttt{GroupGP}_1), T(\texttt{GroupGP}_2)))\cdots), T(\texttt{GroupGP}_n))$
12:        **else** $E' \leftarrow T(\texttt{GroupGP}_1)$
13:        $E \leftarrow \text{Join}(E, E')$
14:    **end if**
15:    **if** f is GRAPH VarOrIRIref $\texttt{GroupGP}_1$ **then**
16:        $E \leftarrow join(E, \text{Graph}(\text{VarOrIRIref}, T(\texttt{GroupGP}_1)))$
17:    **if** f is FILTER constraint **then** $FS \leftarrow (FS \wedge \textbf{constraint})$
18: **end for**
19: **if** $FS \neq \emptyset$ **then** $E \leftarrow \text{Filter}(FS, E)$
20: **return** E

Table 2. Semantics of $\text{SPARQL}_{\text{WG}}$ graph patterns. A pattern GroupGP is transformed into an algebra expression E using algorithm 1. Then E is evaluated as the table shows. E_1 and E_2 are algebra expressions, C is a filter constraint, $u \in I$ and $?X \in V$.

Algebra Expression E	Evaluation $\langle\!\langle E \rangle\!\rangle_G^D$
$\text{Join}(E_1, E_2)$	$\langle\!\langle E_1 \rangle\!\rangle_G^D \bowtie \langle\!\langle E_2 \rangle\!\rangle_G^D$
$\text{LeftJoin}(E_1, E_2, C)$	$\langle\!\langle E_1 \rangle\!\rangle_G^D \;{}_{\displaystyle \bowtie}\, _C \langle\!\langle E_2 \rangle\!\rangle_G^D$
$\text{Union}(E_1, E_2)$	$\langle\!\langle E_1 \rangle\!\rangle_G^D \cup \langle\!\langle E_2 \rangle\!\rangle_G^D$
$\text{Filter}(C, E_1)$	$\{\, \mu \mid \mu \in \langle\!\langle E_1 \rangle\!\rangle_G^D \text{ and } \mu \models C\}$
$\text{Graph}(u, E_1)$	$\langle\!\langle E_1 \rangle\!\rangle_{\text{gr}(u)_D}^D$
$\text{Graph}(?X, E_1)$	$\bigcup_{v \,\in\, \text{names}(D)} (\langle\!\langle E_1 \rangle\!\rangle_{\text{gr}(v)_D}^D \bowtie \{\mu_{?X \rightarrow v}\})$

Note 3. In this paper we will follow the simpler syntax of SPARQL_{C}, better suited to do formal analysis and processing than the syntax presented by $\text{SPARQL}_{\text{WG}}$. There is an easy and intuitive way of translating back and forth between both syntax formalisms, which we will not detail here.

2.3 Datalog

We will briefly review notions of Datalog (For further details and proofs see [1,5]).

A *term* is either a variable or a constant. An *atom* is either a *predicate formula* $p(x_1, ..., x_n)$ where p is a predicate name and each x_i is a term, or an *equality formula* $t_1 = t_2$ where t_1 and t_2 are terms. A *literal* is either an atom (a *positive literal* L) or the negation of an atom (a *negative literal* $\neg L$).

A Datalog *rule* is an expression $H \leftarrow B$ where H is a positive literal called the *head* [5] of the rule and B is a set of literals called the *body*. A rule is *ground* if it does not have any variables. A ground rule with empty body is called a *fact*.

A *Datalog program* Π is a finite set of Datalog rules. The set of facts occurring in Π, denoted facts(Π), is called the *initial database* of Π. A predicate is *extensional* in Π if it occurs only in facts(Π), otherwise it is called *intensional*.

A Datalog program is *non-recursive* and *safe* if it does not contain any predicate that is recursive in the program and it can only generate a finite number of answers. In what follows, we only consider non-recursive and safe programs.

A *substitution* θ is a set of assignments $\{x_1/t_1, \ldots, x_n/t_n\}$ where each x_i is a variable and each t_i is a term. Given a rule r, we denote by $\theta(r)$ the rule resulting of substituting the variable x_i for the term t_i in each literal of r.

The *meaning* of a Datalog program Π, denoted facts$^*(\Pi)$, is the database resulting from adding to the initial database of Π as many new facts of the form $\theta(L)$ as possible, where θ is a substitution that makes a rule r in Π true and L is the head of r. Then the rules are applied repeatedly and new facts are added to the database until this iteration stabilizes, i.e., until a *fixpoint* is reached.

A *Datalog query* Q is a pair (Π, L) where Π is a Datalog program and L is a positive (goal) literal. The *answer* to Q over database $D = $ facts(Π), denoted ans$_d(Q, D)$ is defined as the set of substitutions $\{\theta \mid \theta(L) \in \text{facts}^*(\Pi)\}$.

2.4   Comparing Expressive Power of Languages

By the *expressive power* of a query language, we understand the set of all queries expressible in that language [1,5]. In order to determine the expressive power of a query language L, usually one chooses a well-studied query language L' and compares L and L' in their expressive power. Two query languages have the same expressive power if they express exactly the same set of queries.

A given query language is defined as a quadruple $(\mathcal{Q}, \mathcal{D}, \mathcal{S}, \text{eval})$, where $\mathcal{Q}$ is a set of queries, $\mathcal{D}$ is a set of databases, $\mathcal{S}$ is a set of solutions, and eval $: \mathcal{Q} \times \mathcal{D} \rightarrow \mathcal{S}$ is the evaluation function. The evaluation of a query $Q \in \mathcal{Q}$ on a database $D \in \mathcal{D}$ is denoted eval(Q, D). Two queries $Q_1, Q_2 \in \mathcal{Q}$ are *equivalent*, denoted $Q_1 \equiv Q_2$, if eval(Q_1, D) = eval(Q_2, D) for every $D \in \mathcal{D}$.

Let $L_1 = (\mathcal{Q}_1, \mathcal{D}_1, \mathcal{S}_1, \text{eval}_1)$ and $L_2 = (\mathcal{Q}_2, \mathcal{D}_2, \mathcal{S}_2, \text{eval}_2)$ be two query languages. We say that L_1 is *contained* in L_2 if and only if there are bijective data transformations $\mathcal{T}_D : \mathcal{D}_1 \rightarrow \mathcal{D}_2$ and $\mathcal{T}_S : \mathcal{S}_1 \rightarrow \mathcal{S}_2$, and query transformation $\mathcal{T}_Q : \mathcal{Q}_1 \rightarrow \mathcal{Q}_2$, such that for all $Q \in \mathcal{Q}_1$ and $D \in \mathcal{D}_1$ it satisfies that $\mathcal{T}_S(\text{eval}_1(Q, D)) = \text{eval}_2(\mathcal{T}_Q(Q), \mathcal{T}_D(D))$. We say that L_1 and L_2 are *equivalent* if and only if L_1 is contained in L_2 and L_2 is contained in L_1. (Note that if L_1 and L_2 are subsets of a language L, then $\mathcal{T}_D$, $\mathcal{T}_S$ and $\mathcal{T}_Q$ are the identity.)

---

[5] We may assume that all heads of rules have only variables by adding the corresponding equality formula to its body.

3   Expressing Difference of Patterns in SPARQL$_{\mathrm{WG}}$

The SPARQL$_{\mathrm{WG}}$ specification indicates that it is possible to test if a graph pattern does not match a dataset, via a combination of optional patterns and filter conditions (like negation as failure in logic programming)([9] Sec. 11.4.1). In this section we analyze in depth the scope and limitations of this approach.

We will introduce a syntax for the "difference" of two graph patterns P_1 and P_2, denoted $(P_1 \text{ MINUS } P_2)$, with the intended informal meaning: "the set of mappings that match P_1 and does not match P_2". Formally:

Definition 1. *Let P_1, P_2 be graph patterns and D be a dataset with active graph G. Then $\langle\!\langle(P_1 \text{ MINUS } P_2)\rangle\!\rangle^D_G = \langle\!\langle P_1 \rangle\!\rangle^D_G \setminus \langle\!\langle P_2 \rangle\!\rangle^D_G$.*

A *naive implementation* of the MINUS operator in terms of the other operators would be the graph pattern $((P_1 \text{ OPT } P_2) \text{ FILTER } C)$ where C is the filter constraint
$(\neg \text{ bound}(?X))$ for some variable $?X \in \text{var}(P_2) \setminus \text{var}(P_1)$. This means that for each mapping $\mu \in \langle\!\langle(P_1 \text{ OPT } P_2)\rangle\!\rangle^D_G$ at least one variable $?X$ occurring in P_2, but not occurring in P_1, does not match (i.e., $?X$ is unbounded). There are two problems with this solution:

- Variable $?X$ cannot be an arbitrary variable. For example, P_2 could be in turn an optional pattern $(P_3 \text{ OPT } P_4)$ where only variables in P_3 are relevant.
- If $\text{var}(P_2) \setminus \text{var}(P_1) = \emptyset$ there is no variable $?X$ to check unboundedness.

The above two problems motivate the introduction of the notions of non-optional variables and copy patterns.

The set of *non-optional variables* of a graph pattern P, denoted $\text{nov}(P)$, is a subset of the variables of P defined recursively as follows: $\text{nov}(P) = \text{var}(P)$ when P is a basic graph pattern; if P is either $(P_1 \text{ AND } P_2)$ or $(P_1 \text{ UNION } P_2)$ then $\text{nov}(P) = \text{nov}(P_1) \cup \text{nov}(P_2)$; if P is $(P_1 \text{ OPT } P_2)$ then $\text{nov}(P) = \text{nov}(P_1)$; if P is $(n \text{ GRAPH } P_1)$ then either $\text{nov}(P) = \text{nov}(P_1)$ when $n \in I$ or $\text{nov}(P) = \text{nov}(P_1) \cup \{n\}$ when $n \in V$; and $\text{nov}(P_1 \text{ FILTER } C) = \text{nov}(P_1)$. Intuitively $\text{nov}(P)$ contains the variables that necessarily must be bounded in any mapping of P.

Let $\phi : V \to V$ be a variable-renaming function. Given a graph pattern P, a *copy pattern* $\phi(P)$ is an isomorphic copy of P whose variables have been renamed according to ϕ and satisfying that $\text{var}(P) \cap \text{var}(\phi(P)) = \emptyset$.

Theorem 1. *Let P_1 and P_2 be graph patterns. Then:*

$$(P_1 \text{ MINUS } P_2) \equiv ((P_1 \text{ OPT}((P_2 \text{ AND } \phi(P_2)) \text{ FILTER } C_1)) \text{ FILTER } C_2) \quad (1)$$

where:

- C_1 *is the filter constraint* $(?X_1 = ?X'_1 \wedge \cdots \wedge ?X_n = ?X'_n)$ *where* $?X_i \in \text{var}(P_2)$ *and* $?X'_i = \phi(?X_i)$ *for* $1 \leq i \leq n$.
- C_2 *is the filter constraint* $(\neg \text{ bound}(?X'))$ *for some* $?X' \in \text{nov}(\phi(P_2))$.

Note 4 (Why the copy pattern $\phi(P)$ is necessary?).

Consider the naive implementation of difference of patterns, that is the graph pattern $((P_1 \text{ OPT } P_2) \text{ FILTER } C)$ where C is the filter constraint $(\neg \text{bound}(?X))$ for some $?X \in \text{var}(P_2) \setminus \text{var}(P_1)$. Note that such implementation would fail when $\text{var}(P_2) \setminus \text{var}(P_1) = \emptyset$, because *there exist no variables* to check unboundedness.

To solve this problem, P_2 is replaced by $((P_2 \text{ AND } \phi(P_2)) \text{ FILTER } C_1)$ where $\phi(P_2)$ is a copy of P_2 whose variables have been renamed and whose relations of equality with the original ones are in condition C_1. Then we can use some variable from $\phi(P_2)$ to check if the graph pattern P_2 does not match. The copy pattern ensure that there will exist a variable to check unboundedness.

Note 5 (Why non-optional variables?). Consider the graph pattern

$$P = ((?X, \text{name}, ?N) \text{ MINUS}((?X, \text{knows}, ?Y) \text{ OPT}(?Y, \text{mail}, ?Z))).$$

The naive implementation of P would be the graph pattern

$$P' = ((P_1 \text{ OPT } P_2) \text{ FILTER}(\neg \text{bound}(?Z))),$$

where $P_1 = (?X, \text{name}, ?N)$, $P_2 = ((?X, \text{knows}, ?Y) \text{ OPT}(?Y, \text{mail}, ?Z))$ and $?Z$ is the variable selected to check unboundedness. (Note that variable $?Y$ could also have been selected because $?Y \in \text{var}(P_2) \setminus \text{var}(P_1)$.)

Note that the evaluation of graph pattern P' differs from that of pattern P. To see the problem recall the informal semantics: a mapping μ matches the pattern P if and only if μ matches P_1 and μ does not match P_2. This latter condition means: it is false that every variable in P_2 (but not in P_1) is bounded. But to say "every variable" is not correct in this context, because P_2 contains the optional pattern $(?Y, \text{mail}, ?Z)$, and its variables could be unbounded for some valid solutions of P_2. The problem is produced by the expression $(\neg \text{bound}(?Z))$, because the bounding state of variable $?Z$ introduces noise when testing if pattern P_2 gets matched.

Now, if we ensure the selection of a "non-optional variable" to check unboundedness when transforming P, we have that $?Y$ is the unique non-optional variable occurring in P_2 but not occurring in P_1, i.e., variable $?Y$ works exactly as the test to check if a mapping matching P_1 matches P_2 as well. Hence, instead of P', the graph pattern

$$P'' = ((P_1 \text{ OPT } P_2) \text{ FILTER}(\neg \text{bound}(?Y)))$$

is the one that expresses faithfully the graph pattern $(P_1 \text{ MINUS } P_2)$, and in fact, the evaluation of P'' gives exactly the same set of mappings as P.

4   Avoiding Unsafe Patterns in SPARQL_{WG}

One influential point in the evaluation of patterns in $\text{SPARQL}_{\text{WG}}$ is the behavior of *filters*. What is the scope of a filter? What is the meaning of a filter having variables that do not occur in the graph pattern to be filtered?

It was proposed in [6] that for reasons of simplicity for the user and cleanness of the semantics, the scope of filters should be the expression which they filter, and free variables should be disallowed in the filter condition. Formally, a graph pattern of the form $(P \, \text{FILTER} \, C)$ is said to be *safe* if $\text{var}(C) \subseteq \text{var}(P)$. In [6] only safe filter patterns were allowed in the syntax, and hence the scope of the filter C is the pattern P which defines the filter condition. This approach is further supported by the fact that non-safe filters are rare in practice.

The WG decided to follow a different approach, and defined the scope of a filter condition C to be a case-by-case and context-dependent feature:

1. The scope of a filter is defined as follows: a filter "is a restriction on solutions over the whole group in which the filter appears".
2. There is one exception, though, when filters combine with optionals. If a filter expression C belongs to the group graph pattern of an optional, the scope of C is local to the group where the optional belongs to. This is reflected in lines 7 and 8 of Algorithm 1.

The complexities that this approach brings were recognized in the discussion of the WG, and can be witnessed by the reader by following the evaluation of patterns in $\text{SPARQL}_{\text{WG}}$.

Let $\text{SPARQL}_{\text{WG}}^{\text{Safe}}$ be the subset of queries of $\text{SPARQL}_{\text{WG}}$ having only filter-safe patterns. In what follows, we will show that, in $\text{SPARQL}_{\text{WG}}$, non-safe filters are superfluous, and hence its non-standard and case-by-case semantics can be avoided. In fact, we will prove that non-safe filters do not add expressive power to the language, or in other words, that $\text{SPARQL}_{\text{WG}}$ and $\text{SPARQL}_{\text{WG}}^{\text{Safe}}$ have the same expressive power, that is, for each pattern P there is a filter-safe pattern P' which computes exactly the same mappings as P.

The transformation $\text{safe}(P)$ is given by Algorithm 2. This algorithm works as the identity for most patterns. The key part is the treatment of patterns which combine filters and optionals. Line 9 is exactly the codification of the WG evaluation of filters inside optionals. For non-safe filters (see lines 15-20), it replaces each atomic filter condition C', where a free variable occurs, by either an expression *false* when C' is bound$(\cdot)$; or an expression bound(a) otherwise. (note that bound(a) is evaluated to a logical value of error because a is a constant.)

Note 6 (On Algorithm 2). The expression in line 9 must be refined for bag semantics to the expression:

$$P' \leftarrow (((\text{safe}((P_1 \, \text{AND} \, P_3) \, \text{FILTER} \, C) \, \text{UNION} \, (\text{safe}(P_1) \, \text{MINUS} \, \text{safe}(P_3)))$$
$$\text{UNION} \, (\text{safe}(P_1) \, \text{MINUS}(\text{safe}(P_1) \, \text{MINUS} \, \text{safe}(P_3))))$$
$$\text{MINUS} \, \text{safe}((P_1 \, \text{AND} \, P_3) \, \text{FILTER} \, C))$$

Lemma 1. *For every pattern P, the pattern* $\text{safe}(P)$ *defined by Algorithm 2 is filter-safe and it holds* $\langle\!\langle P \rangle\!\rangle = \langle\!\langle \text{safe}(P) \rangle\!\rangle$.

Thus we proved:

Theorem 2. *$SPARQL_{WG}$ and $SPARQL_{WG}^{Safe}$ have the same expressive power.*

Algorithm 2. Transformation of a general graph pattern into a safe pattern.

1: // Input: a SPARQL$_{\mathrm{WG}}$ graph pattern P
2: // Output: a safe graph pattern $P' \leftarrow \mathrm{safe}(P)$
3: $P' \leftarrow \emptyset$
4: **if** P is $(P_1 \,\mathrm{AND}\, P_2)$ **then** $P' \leftarrow (\mathrm{safe}(P_1) \,\mathrm{AND}\, \mathrm{safe}(P_2))$
5: **if** P is $(P_1 \,\mathrm{UNION}\, P_2)$ **then** $P' \leftarrow (\mathrm{safe}(P_1) \,\mathrm{UNION}\, \mathrm{safe}(P_2))$
6: **if** P is $(n \,\mathrm{GRAPH}\, P_1)$ **then** $P' \leftarrow (n \,\mathrm{GRAPH}\; \mathrm{safe}(P_1))$
7: **if** P is $(P_1 \,\mathrm{OPT}\, P_2)$ **then**
8:     **if** P_2 is $(P_3 \,\mathrm{FILTER}\, C)$ **then**
9:       $P' \leftarrow (\mathrm{safe}(P_1) \,\mathrm{OPT}(\mathrm{safe}((P_1 \,\mathrm{AND}\, P_3) \,\mathrm{FILTER}\, C)))$
10:    **else** $P' \leftarrow (\mathrm{safe}(P_1) \,\mathrm{OPT}\, \mathrm{safe}(P_2))$
11: **end if**
12: **if** P is $(P_1 \,\mathrm{FILTER}\, C)$ **then**
13:    **if** $\mathrm{var}(C) \subseteq \mathrm{var}(\mathrm{safe}(P_1))$ **then** $P' \leftarrow (\mathrm{safe}(P_1) \,\mathrm{FILTER}\, C)$
14:    **else**
15:      **for all** $?X \in \mathrm{var}(C)$ and $?X \notin \mathrm{var}(\mathrm{safe}(P_1))$ **do**
16:        **for all** atomic filter constraint C' in C
17:          **if** C' is $(?X = u)$ or $(?X = ?Y)$ or $\mathrm{isIRI}(?X)$ or $\mathrm{isBlank}(?X)$ or $\mathrm{isLiteral}(?X)$
18:             Replace in C the constraint C' by $\mathrm{bound}(a)$ //where a is a constant
19:         **else if** C' is $\mathrm{bound}(?X)$ **then**
20:             Replace in C the constraint C' by *false*
21:        **end for**
22:      **end for**
23:      $P' \leftarrow (\mathrm{safe}(P_1) \,\mathrm{FILTER}\, C)$
24:    **end if**
25: **end if**
26: **return** P'

---

5   Expressive Power of SPARQL$_{\mathrm{WG}}$ is Equivalent to SPARQL$_{\mathrm{C}}$

As we have been showing, the semantics that the WG gave to SPARQL departed in some aspects from a compositional semantics. We also indicated that there is an alternative formalization, with a standard compositional semantics, which was called SPARQL$_{\mathrm{C}}$ [6].

The good news is that, albeit apparent differences, these languages are equivalent in expressive power, that is, they compute the same class of queries.

Theorem 3. *SPARQL$_{WG}^{Safe}$ is equivalent to SPARQL$_C$ under bag semantics.*

The proof of this theorem is an induction on the structure of patterns. The only non-evident case is the particular evaluation of filters inside optionals where the semantics of SPARQL$_{\mathrm{WG}}^{\mathrm{Safe}}$ and SPARQL$_{\mathrm{C}}$ differ. Specifically, given a graph pattern $P = (P_1 \,\mathrm{OPT}(P_2 \,\mathrm{FILTER}\, C))$, we have that SPARQL$_{\mathrm{WG}}^{\mathrm{Safe}}$ evaluates the algebra expression $\mathrm{LeftJoin}(P_1, P_2, C)$, whereas SPARQL$_{\mathrm{C}}$ evaluates P to the expression $[\![P_1]\!] \bowtie [\![P_2 \,\mathrm{FILTER}\, C]\!]$, which is the same as the SPARQL$_{\mathrm{WG}}$ algebra expression $\mathrm{LeftJoin}(P_1, \mathrm{Filter}(C, P_2), true)$.

6   Expressive Power of SPARQL$_C$

In this section we study the expressive power of SPARQL$_C$ by comparing it against non recursive safe Datalog with negation (just Datalog from now on).

Note that because SPARQL$_C$ and Datalog programs have different type of input and output formats, we have to normalize them to be able to do the comparison. Following definitions in section 2.4, let $L_s = (\mathcal{Q}_s, \mathcal{D}_s, \mathcal{S}_s, \text{ans}_s)$ be the SPARQL$_C$ language, and $L_d = (\mathcal{Q}_d, \mathcal{D}_d, \mathcal{S}_d, \text{ans}_d)$ be the Datalog language.

In this comparison we restrict the notion of $SPARQL_C$ Query to a pair (P, D) where P is a graph pattern and D is an RDF dataset.

6.1   From SPARQL$_C$ to Datalog

To prove that L_s is contained in L_d, we define transformations $\mathcal{T}_Q : \mathcal{Q}_s \rightarrow \mathcal{Q}_d$, $\mathcal{T}_D : \mathcal{D}_s \rightarrow \mathcal{D}_d$, and $\mathcal{T}_S : \mathcal{S}_s \rightarrow \mathcal{S}_d$. That is, $\mathcal{T}_Q$ transforms a SPARQL$_C$ query into a Datalog query, $\mathcal{T}_D$ transforms an RDF dataset into a set of Datalog facts, and $\mathcal{T}_S$ transforms a set of SPARQL$_C$ mappings into a set of Datalog substitutions.

RDF datasets as Datalog facts. Given a dataset D, the transformation $\mathcal{T}_D(D)$ works as follows: each term t in D is encoded by a fact $iri(t)$, $blank(t)$ or $literal(t)$ when t is an IRI, a blank node or a literal respectively; the set of terms in D is defined by the set of rules $term(X) \leftarrow iri(X)$, $term(X) \leftarrow blank(X)$, and $term(X) \leftarrow literal(X)$; the fact $Null(null)$ encodes the *null* value [6]; each triple (v_1, v_2, v_3) in the default graph of D is encoded by a fact $triple(g_0, v_1, v_2, v_3)$; each named graph $\langle u, G \rangle$ in D is encoded by a fact $graph(u)$ and each triple in G is encoded by a fact $triple(u, v_1, v_2, v_3)$.

SPARQL$_C$ mappings as Datalog substitutions. Given a graph pattern P, a dataset D with default graph G, and the set of mappings $\Omega = [\![P]\!]_G^D$. The transformation $\mathcal{T}_S(\Omega)$ returns a set of substitutions defined as follows: for each mapping $\mu \in \Omega$ there exists a substitution $\theta \in \mathcal{T}_S(\Omega)$ satisfying that, for each $x \in \text{var}(P)$ there exists $x/t \in \theta$ such that $t = \mu(x)$ when $\mu(x)$ is bounded and $t = null$ otherwise.

Graph patterns as Datalog rules. Let P be a graph pattern to be evaluated against an RDF graph identified by g which occurs in dataset D. We denote by $\delta(P, g)_D$ the function which transforms P into a set of Datalog rules. Table 3 shows the transformation rules defined by the function $\delta(P, g)_D$. The notion of compatible mappings is implemented by the rules:

$comp(X, X, X) \leftarrow term(X)$, $comp(X, null, X) \leftarrow term(X)$,

$comp(null, X, X) \leftarrow term(X)$ and $comp(X, X, X) \leftarrow Null(X)$.

Let $?X, ?Y \in V$ and $u \in I \cup L$. An atomic filter condition C is encoded by a literal L as follows: if C is either $(?X = u)$ or $(?X=?Y)$ then L is C; if C is $(\text{isIRI}(?X))$ then L is $iri(?X)$; if C is $(\text{isLiteral}(?X))$ then L is $literal(?X)$; if C is $(\text{isBlank}(?X))$ then L is $blank(?X)$; if C is $(\text{bound}(?X))$ then L is $\neg Null(?X)$.

The transformation follows essentially the intuitive transformation presented by Polleres [8] with the improvement of the necessary code to support faithful

---

[6] We use the term null to represent an unbounded value.

Table 3. Transforming SPARQL$_C$ graph patterns into Datalog Rules. D is a dataset having active graph identified by g. $\overline{\mathrm{var}}(P)$ denotes the tuple of variables obtained from a lexicographical ordering of the variables in the graph pattern P. Each p_i is a predicate identifying the graph pattern P_i. If L is a literal, then $\nu_j(L)$ denotes a copy of L with its variables renamed according to a variable renaming function $\nu_j : V \rightarrow V$. *cond* is a literal encoding the filter condition C. Each P_{1i} is a copy of P_1 and $u_i \in \mathrm{names}(D)$. $P_3 = (P_1 \mathrm{\ AND\ } P_2)$, $P_4 = (P_1 \mathrm{\ FILTER\ } C_1)$ and $P_5 = (P_1 \mathrm{\ FILTER\ } C_2)$.

Pattern P	$\delta(P, g)_D$
(x_1, x_2, x_3)	$p(\overline{\mathrm{var}}(P)) \leftarrow triple(g, x_1, x_2, x_3)$
$(P_1 \mathrm{\ AND\ } P_2)$	$p(\overline{\mathrm{var}}(P)) \leftarrow \nu_1(p_1(\overline{\mathrm{var}}(P_1))) \ \wedge \ \nu_2(p_2(\overline{\mathrm{var}}(P_2)))$ $\bigwedge_{x \in \mathrm{var}(P_1) \cap \mathrm{var}(P_2)} comp(\nu_1(x), \nu_2(x), x),$ $\delta(P_1, g)_D \ , \ \delta(P_2, g)_D$ $\mathrm{dom}(\nu_1) = \mathrm{dom}(\nu_2) = \mathrm{var}(P_1) \cap \mathrm{var}(P_2), \mathrm{range}(\nu_1) \cap \mathrm{range}(\nu_2) = \emptyset.$
$(P_1 \mathrm{\ UNION\ } P_2)$	$p(\overline{\mathrm{var}}(P)) \leftarrow p_1(\overline{\mathrm{var}}(P_1)) \bigwedge_{x \in \mathrm{var}(P_2) \wedge x \notin \mathrm{var}(P_1)} Null(x),$ $p(\overline{\mathrm{var}}(P)) \leftarrow p_2(\overline{\mathrm{var}}(P_2)) \bigwedge_{x \in \mathrm{var}(P_1) \wedge x \notin \mathrm{var}(P_2)} Null(x),$ $\delta(P_1, g)_D \ , \ \delta(P_2, g)_D$
$(P_1 \mathrm{\ OPT\ } P_2)$	$p(\overline{\mathrm{var}}(P)) \leftarrow p_1(\overline{\mathrm{var}}(P_1)) \wedge \neg p_1'(\overline{\mathrm{var}}(P_1)) \bigwedge_{x \in \mathrm{var}(P_2) \wedge x \notin \mathrm{var}(P_1)} Null(x),$ $p(\overline{\mathrm{var}}(P)) \leftarrow p_3(\overline{\mathrm{var}}(P_3)),$ $p_1'(\overline{\mathrm{var}}(P_1)) \leftarrow p_3(\overline{\mathrm{var}}(P_3)),$ $\delta(P_1, g)_D \ , \ \delta(P_2, g)_D \ , \ \delta(P_3, g)_D$
$(u \mathrm{\ GRAPH\ } P_1)$ and $u \in I$	$p(\overline{\mathrm{var}}(P)) \leftarrow p_1(\overline{\mathrm{var}}(P_1)),$ $\delta(P_1, u)_D$
$(?X \mathrm{\ GRAPH\ } P_1)$ and $?X \in V$	$p(\overline{\mathrm{var}}(P)) \leftarrow p_{11}(\overline{\mathrm{var}}(P_{11})) \wedge graph(?X) \ \wedge \ ?X = u_1,$ $\delta(P_{11}, u_1)_D,$ $\ldots$ $p(\overline{\mathrm{var}}(P)) \leftarrow p_{1n}(\overline{\mathrm{var}}(P_{1n})) \wedge graph(?X) \ \wedge \ ?X = u_n,$ $\delta(P_{1n}, u_n)_D$
$(P_1 \mathrm{\ FILTER\ } C)$ C is atomic	$p(\overline{\mathrm{var}}(P)) \leftarrow p_1(\overline{\mathrm{var}}(P_1)) \wedge cond$ $\delta(P_1, g)_D$
$(P_1 \mathrm{\ FILTER\ } C)$ C is $(\neg(C_1))$	$p(\overline{\mathrm{var}}(P)) \leftarrow p_1(\overline{\mathrm{var}}(P_1)) \wedge \neg p_4(\overline{\mathrm{var}}(P_1)),$ $\delta(P_1, g)_D \ , \ \delta(P_4, g)_D$
$(P_1 \mathrm{\ FILTER\ } C)$ C is $(C_1 \wedge C_2)$	$p(\overline{\mathrm{var}}(P)) \leftarrow p_1(\overline{\mathrm{var}}(P_1)) \wedge \neg p'(\overline{\mathrm{var}}(P_1)),$ $p'(\overline{\mathrm{var}}(P_1)) \leftarrow p_1(\overline{\mathrm{var}}(P_1)) \wedge \neg p''(\overline{\mathrm{var}}(P_1)),$ $p''(\overline{\mathrm{var}}(P_1)) \leftarrow p_4(\overline{\mathrm{var}}(P_1)) \wedge p_5(\overline{\mathrm{var}}(P_1)),$ $\delta(P_4, g)_D \ , \ \delta(P_5, g)_D$
$(P_1 \mathrm{\ FILTER\ } C)$ C is $(C_1 \vee C_2)$	$p(\overline{\mathrm{var}}(P)) \leftarrow p_1(\overline{\mathrm{var}}(P_1)) \wedge \neg p'(\overline{\mathrm{var}}(P_1)),$ $p'(\overline{\mathrm{var}}(P_1)) \leftarrow p_1(\overline{\mathrm{var}}(P_1)) \wedge \neg p''(\overline{\mathrm{var}}(P_1))$ $p''(\overline{\mathrm{var}}(P_1)) \leftarrow p_4(\overline{\mathrm{var}}(P_1)),$ $p''(\overline{\mathrm{var}}(P_1)) \leftarrow p_5(\overline{\mathrm{var}}(P_1)),$ $\delta(P_4, g)_D \ , \ \delta(P_5, g)_D$

translation of bag semantics. Specifically, we changed the transformations for complex filter expressions by simulating them with double negation.

SPARQL$_C$ queries as Datalog queries. Given a graph pattern P, a dataset D with default graph G, and the SPARQL$_C$ query $Q = (P, D)$. The function $T_Q(Q)$ returns the Datalog query $(\Pi, p(\overline{\mathrm{var}}(P)))$ where Π is the Datalog program $T_D(D) \cup \delta(P, g_0)_D$, g_0 identifies the default graph G, and p is the goal literal related to P.

The following theorem states that the above transformations work well.

Theorem 4. *SPARQL$_C$ is contained in non-recursive safe Datalog with negation.*

6.2   From Datalog to SPARQL$_C$

To prove that L_d is contained in L_s, we define transformations $T_Q' : Q_d \to Q_s$, $T_D' : D_d \to D_s$, and $T_S' : S_d \to S_s$. That is, T_Q' transforms a Datalog query into an SPARQL$_C$ query, T_D' transforms a set of Datalog facts into an RDF dataset, and T_S' transforms a set of Datalog substitutions into a set of SPARQL$_C$ mappings.

Datalog facts as an RDF Dataset. Given a Datalog fact $f = p(c_1, ..., c_n)$, consider that $\mathrm{desc}(f) = \{ (\_:b, \mathrm{predicate}, p), (\_:b, \mathrm{rdf}:\_1, c_1), ..., (\_:b, \mathrm{rdf}:\_n, c_n) \}$, where $\_:b$ is a fresh blank node. Given a set of Datalog facts F, we have that $T_D'(F)$ returns an RDF dataset with default graph $\{\mathrm{desc}(f) \mid f \in F\}$, where $\mathrm{blank}(\mathrm{desc}(f_i)) \cap \mathrm{blank}(\mathrm{desc}(f_j)) = \emptyset$ for each $f_i, f_j \in F$ with $i \neq j$.

Datalog substitutions as SPARQL$_C$ mappings. Given a set of substitutions Θ, the transformation $T_S'(\Theta)$ returns a set of mappings defined as follows: for each substitution $\theta \in \Theta$ there exists a mapping $\mu \in T_S'(\Theta)$ satisfying that, if $x/t \in \theta$ then $x \in \mathrm{dom}(\mu)$ and $\mu(x) = t$.

Datalog rules as SPARQL$_C$ graph patterns. Let Π be a Datalog program, and L be a literal $p(x_1, ..., x_n)$ where p is a predicate in Π and each x_i is a variable. We define the function $\mathrm{gp}(L)_\Pi$ which returns a graph pattern encoding the program (Π, L), that is, the fragment of the program Π used for evaluating literal L.

The translation works intuitively as follows:

(a) If predicate p is extensional, then $\mathrm{gp}(L)_\Pi$ returns the graph pattern
$$((?Y, \mathrm{predicate}, p) \,\mathrm{AND}\, (?Y, \mathrm{rdf}:\_1, x_1) \,\mathrm{AND} \cdots \mathrm{AND}\, (?Y, \mathrm{rdf}\_n, x_n)),$$
where $?Y$ is a fresh variable.

(b) If predicate p is intensional, then for each rule in Π of the form
$$L \leftarrow p_1 \wedge \cdots \wedge p_s \wedge \neg q_1 \wedge \cdots \wedge \neg q_t \wedge L_1^{eq} \wedge \cdots \wedge L_u^{eq},$$
where L_k^{eq} are literals of the form $t_1 = t_2$ or $\neg(t_1 = t_2)$, we have have that $\mathrm{gp}(L)_\Pi$ returns a graph pattern with the structure

$$(((\cdots ((\mathrm{gp}(p_1)_\Pi \,\mathrm{AND} \cdots \mathrm{AND}\, \mathrm{gp}(p_s)_\Pi)$$
$$\mathrm{MINUS}\, \mathrm{gp}(q_1)) \cdots) \,\mathrm{MINUS}\, \mathrm{gp}(q_t))$$
$$\mathrm{FILTER}(L_1^{eq} \wedge \cdots \wedge L_u^{eq})). \quad (2)$$

The formal definition of $\mathrm{gp}(L)_\Pi$ is Algorithm 3.

Algorithm 3. Transformation of Datalog rules into SPARQL$_C$ graph patterns

---

1: //Input: a literal $L = p(x_1, \ldots, x_n)$ and a Datalog program Π
2: //Output: a SPARQL$_C$ graph pattern $P = \mathrm{gp}(L)_\Pi$
3: $P \leftarrow \emptyset$
4: **if** predicate p is extensional in Π **then**
5:     Let ?Y be a fresh variable
6:     $P \leftarrow ((?Y, \mathrm{predicate}, p) \, \mathrm{AND}(?Y, \mathrm{rdf:\_1}, x_1) \, \mathrm{AND} \cdots \mathrm{AND}(?Y, \mathrm{rdf\_n}, x_n))$
7: **else if** predicate p is intensional in Π **then**
8:     **for** each rule $r \in \Pi$ with head $p(x'_1, \ldots, x'_n)$ **do**
9:         $P' \leftarrow \emptyset$
10:        $C \leftarrow \emptyset$
11:        Let $r' = \nu(r)$ where ν is a substitution such that $\nu(x'_i) = x_i$
12:        **for** each positive literal $q(y_1, \ldots, y_m)$ in the body of r' **do**
13:            **if** $P' = \emptyset$ **then** $P' \leftarrow \mathrm{gp}(q)_\Pi$
14:            **else** $P' \leftarrow (P' \, \mathrm{AND} \, \mathrm{gp}(q)_\Pi)$
15:        **end for**
16:        **for** each negative literal $\neg q(y_1, \ldots, y_m)$ in the body of r' **do**
17:            $P' \leftarrow (P' \, \mathrm{MINUS} \, \mathrm{gp}(q))$
18:        **end for**
19:        **for** each equality formula $t_1 = t_2$ in r' **do**
20:            **if** $C = \emptyset$ **then** $C \leftarrow (t_1 = t_2)$
21:            **else** $C \leftarrow C \wedge (t_1 = t_2)$
22:        **end for**
23:        **for** each negative literal $\neg(t_1 = t_2)$ in r' **do**
24:            **if** $C = \emptyset$ **then** $C \leftarrow \neg(t_1 = t_2)$
25:            **else** $C \leftarrow C \wedge \neg(t_1 = t_2)$
26:        **end for**
27:        **if** $C \neq \emptyset$ **then** $P' \leftarrow (P' \, \mathrm{FILTER} \, C)$
28:        **if** $P = \emptyset$ **then** $P \leftarrow P'$
29:        **else** $P \leftarrow (P \, \mathrm{UNION} \, P')$
30:    **end for**
31: **end if**
32: **return** P

Datalog queries as SPARQL$_C$ queries. Given a Datalog program Π, a literal $L = p(x_1, \ldots, x_n)$, and the Datalog query $Q = (\Pi, L)$. The function $\mathcal{T}'_Q(Q)$ returns the SPARQL$_C$ query (P, D) where P is the graph pattern $\mathrm{gp}(L)_\Pi$ and D is an RDF dataset with default graph $\mathcal{T}'_D(\mathrm{facts}(\Pi))$.

The following theorem states that the above transformations work well.

Theorem 5. *nr-Datalog$^\neg$ is contained in SPARQL$_C$.*

7   Conclusions

We have studied the expressive power of SPARQL. Among the most important findings are the definition of negation, the proof that non-safe filter patterns are superfluous, the proof of the equivalence between SPARQL$_{WG}$ and SPARQL$_C$.

From these results we can state the most relevant result of the paper:

Theorem 6 (main). $SPARQL_{WG}$ has the same expressive power as Relational Algebra under bag semantics.

This result follows from the well known fact (for example, see [1] and [5]) that relational algebra and non-recursive safe Datalog with negation have the same expressive power, and from theorems 2, 3, 4 and 5.

Relational Algebra is probably one of the most studied query languages, and has become a favorite by theoreticians because of a proper balance between expressiveness and complexity. The result that SPARQL is equivalent in its expressive power to Relational Algebra, has important implications which are not discussed in this paper. Some examples are the translation of some results from Relational Algebra into SPARQL, and the settlement of several open questions about expressiveness of SPARQL, e.g., the expressive power added by the operator *bound* in combination with optional patterns. Future work includes the development of the manifold consequences implied by the Main Theorem.

Acknowledgments. R. Angles was supported by Mecesup project No. UCH0109. R. Angles and C. Gutierrez were supported by FONDECYT project No. 1070348. The authors wish to thank the reviewers for their comments.

References

1. Abiteboul, S., Hull, R., Vianu, V.: Foundations of Databases. Addison-Wesley, Reading (1995)
2. Cyganiak, R.: A relational algebra for sparql. Technical Report HPL-2005-170, HP Labs (2005)
3. Furche, T., Linse, B., Bry, F., Plexousakis, D., Gottlob, G.: RDF Querying: Language Constructs and Evaluation Methods Compared. In: Barahona, P., Bry, F., Franconi, E., Henze, N., Sattler, U. (eds.) Reasoning Web 2006. LNCS, vol. 4126, pp. 1–52. Springer, Heidelberg (2006)
4. Klyne, G., Carroll, J.: Resource Description Framework (RDF) Concepts and Abstract Syntax (February 2004),
   http://www.w3.org/TR/2004/REC-rdf-concepts-20040210/
5. Levene, M., Loizou, G.: A Guided Tour of Relational Databases and Beyond. Springer, Heidelberg (1999)
6. Pérez, J., Arenas, M., Gutierrez, C.: Semantics and Complexity of SPARQL. In: Cruz, I., Decker, S., Allemang, D., Preist, C., Schwabe, D., Mika, P., Uschold, M., Aroyo, L.M. (eds.) ISWC 2006. LNCS, vol. 4273, pp. 30–43. Springer, Heidelberg (2006)
7. Pérez, J., Arenas, M., Gutierrez, C.: Semantics of SPARQL. Technical Report TR/DCC-2006-17, Department of Computer Science, Universidad de Chile (2006)
8. Polleres, A.: From SPARQL to rules (and back). In: Proceedings of the 16th International World Wide Web Conference (WWW), pp. 787–796. ACM, New York (2007)
9. Prud'hommeaux, E., Seaborne, A.: SPARQL Query Language for RDF (January 2008), http://www.w3.org/TR/2008/REC-rdf-sparql-query-20080115/
10. Schenk, S.: A sparql semantics based on datalog. In: Hertzberg, J., Beetz, M., Englert, R. (eds.) KI 2007. LNCS (LNAI), vol. 4667, pp. 160–174. Springer, Heidelberg (2007)

Integrating Object-Oriented and Ontological Representations: A Case Study in Java and OWL

Colin Puleston, Bijan Parsia, James Cunningham, and Alan Rector

School of Computer Science, University of Manchester, United Kingdom

Abstract. The Web Ontology Language (OWL) provides a modelling paradigm that is especially well suited for developing models of large, structurally complex domains such as those found in Health Care and the Life Sciences. OWL's declarative nature combined with powerful reasoning tools has effectively supported the development of very large and complex anatomy, disease, and clinical ontologies. OWL, however, is not a programming language, so using these models in applications necessitates both a *technical* means of integrating OWL models with programs and considerable *methodological* sophistication in knowing how to integrate them. In this paper, we present an analytical framework for evaluating various OWL-Java combination approaches. We have developed a software framework for what we call *hybrid modelling*, that is, building models in which part of the model exists and is developed directly in Java and part of the model exists and is developed directly in OWL. We analyse the advantages and disadvantages of hybrid modelling both in comparison to other approaches and by means of a case study of a large medical records system.

1 Introduction

A popular trend in software development is model driven engineering (MDE). In MDE, the primary artefact is not a program *per se*, but a model (which a program may instantiate). These models are typically expressed in a UML variant. Of course, programming languages, especially object oriented ones such as Java, themselves have modelling features and are often used to express (executable) models. The Web Ontology Language (OWL) [11] provides a modelling paradigm that is especially well suited for developing models of large, structurally complex domains such as those found in Health Care and the Life Sciences. OWL's declarative nature combined with powerful reasoning tools has effectively supported the development of very large and complex anatomy, disease, and clinical ontologies.

OWL, however, is not a programming language, so using OWL models in applications necessitates both a *technical* means of integrating OWL models with programs and considerable *methodological* sophistication in knowing how to integrate them. In this paper, we present an analytical framework for and evaluation of various OWL-Java combination approaches. We outline three distinct approaches to using ontologies to drive software architectures. The *direct* approach centres round the use of the ontological entities as *templates* for program classes. In this approach, the OWL based model is converted, statically, into a corresponding approximation in Java. The *indirect* approach represents the opposite extreme. In this case, the Java classes do not

A. Sheth et al. (Eds.): ISWC 2008, LNCS 5318, pp. 130–145, 2008.

model the domain concepts directly, but merely access an external model encoded in OWL. The third approach, presented in full in this paper, is a hybrid of the two. Here, the Java classes directly model a limited number of high level entities which are refined, dynamically, by aspects of the OWL ontology. The model is partly expressed by program classes and partly expressed by OWL classes but the two halves are integrated in a transparent way. The result is a model that can exploit the strengths of each side to compensate for weaknesses of the other, or to accommodate different skill sets and preferences of the modellers.

We developed the notion of a hybrid model, the supporting software, and the associated methodology in the course of the *Clinical E-Science Framework (CLEF)* project [11, 22, 33]. Part of the aims of CLEF were, broadly, to develop an architecture for representing series of *Electronic Patient Records* as coherent entities that capture a given patient's medical history in a unified form. We use the CLEF software [12] as the basis of a detailed case study of hybrid modelling.

Some key characteristics required of such a system drive us to hybrid modelling. First medical applications typically require a large knowledge base about medicine, e.g., disease, anatomy, treatment, *etc*. Given the specialised knowledge involved, the development and maintenance of this ontology needs to be performed by knowledge engineers with the requisite background and skills in medicine but who are not skilled software developers. The representation of this knowledge, and its use within the system, needs to be dynamic, in the sense that it can be modified or supplemented without any modifications being made to the software architecture that draws on it. Thus a good portion of the application's information is naturally modelled using logic based ontology languages (like OWL) common in the health informatics community.

However, capturing the types of complex temporally varying relationships that constitute a patient history, and, critically, doing this in a way that supports typical entry and searching patterns of the user base, is not a task for which OWL is particularly well suited. For this, complex data structures and procedures are needed within the model architecture. These requirements naturally suggested a combination of OWL, for its knowledge representation and reasoning functionalities, and an object oriented language, such as Java, for the procedural portions of the application.

In this paper, we examine and compare the three sorts of model especially for their distinct effects on software development. We show via a case study how a specific type of hybrid approach is well suited to a certain class of complex, information rich, dynamic applications.

2   Software Models

We now introduce our notion of what a software model is, and present a general framework to categorise the different varieties of software model.

Models: We use the term *model* specifically to mean a class based schema of some type that can be accessed via an API represented in some standard *Object-Oriented Programming Language (OOPL)*, such as Java or C++. The core of such a model is a hierarchically structured set of *classes* (not necessarily corresponding directly to specific classes in the host OOPL – see below), for each class an associated set of *fields*,

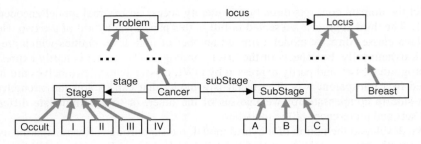

Fig. 1. Fragment of a simple software model - shows basic model entities (classes ▭, fields ➤, sub-class relationships ➤) whilst making no assumptions regarding mode of representation

and for each field a *type-constraint* defining the set of valid values. Figure 1 shows a fragment of such a model, concerned with the representation of patient problems, and specifically focusing on cancer and cancer-staging. Particular model formats may extend this structure in various ways, such as by providing cardinality constraints on fields, or providing data-type fields. Critically, the sorts of models we discuss are representations of a domain of interest, not (primarily) of the program itself. That is, the primary task is to represent the domain, not to structure the program.

Direct and Indirect Models (and Backing Models): One means of categorising such models is by the type of interface offered to the client code. There are two broad possibilities. A *direct* model is one in which the object model of the host programming language embodies the model directly, so that each OOPL class or field *is* a model entity. An *indirect* model is one in which the API presents the objects of a *backing*-model (BM) indirectly. Thus, instead of having a Java class called 'Cancer', an indirect model would use a generic Java class, say, ModelClass, a specific instance of which would then be used to represent a BM concept called 'Cancer'.

Figure 2 shows an instantiation of a fragment of an indirect model. A generic ModelInstance object represents an instance of a particular BM class, with the relevant class being specified via a ModelClass object (the current value of the instanceType field). Associated with the ModelInstance object are a set of model-fields that have been dynamically created, based on information derived from the relevant BM class. Each such dynamically-derived field will actually be represented via an object of an appropriate type (say, ModelInstanceField), with the current set of such objects providing the values for a single multi-valued OOPL field. We refer to such BM-derived, indirectly-represented model fields as *indirect* fields. We refer to model fields that are directly represented in the host OOPL as *direct* fields (such fields are found in both direct and hybrid models, but not in fully-indirect models).

An obvious difference between direct and indirect approaches is in their effect on the application development process. A direct model is tailor-made for a programmer writing domain-aware code, whereas an indirect model is more suitable for driving domain-neutral software. The converse usages are possible but more problematic. To drive generic software from a direct model requires the use of a reflection facility (as provided by the OOPL) in combination with appropriate coding conventions. Writing

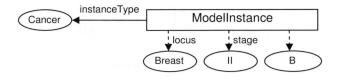

Fig. 2. Instantiation of a fragment of an indirect model – software entities represented both directly (object of named type ⌷, object field ⟶) and indirectly (reference to named model-class ⬭, model-derived field - - ▶)

domain-aware software to operate over an indirect model is awkward and unnatural for the programmer, and hence inefficient (see section 3 for further discussion).

There are other issues that arise from the contrasting approaches, with pros and cons on either side. With indirect models the BM will generally be represented in a standard format for which sophisticated tools are available. For instance with OWL, several editors are available, such as Protégé 4 [13] and Swoop [14], and a range of reasoners [66] and other services. This is important when the model must incorporate a large amount of domain knowledge, and particularly, as is often the case, when the encoding is to be performed by a domain expert. On the other hand, direct representations provide a more natural means of implementing processing beyond the modelling formalism, to either contribute to the dynamic aspects of the model itself, or to operate over its individual instantiations.

An additional advantage of indirect models is in the possibility of BM encapsulation, which in addition to facilitating the seamless mixing and matching of disparate BM formats, also enables the filtering of BM constructs not relevant to the application. For example model classes may be generated only for certain types of concept (excluding for instance compositional concepts that play a role in reasoning but are not relevant to the application) and model fields generated only for certain properties (possibly identified via appropriate super-properties).

Dynamic Models: A *dynamic* model is one in which the details of the model can vary depending on the current state of the specific instantiation. The variability can be in the composition of the field-sets associated with specific instances, in the constraints on specific fields, or even in the types of specific instances. In general, the more dynamic the model, the more natural it is to use an indirect representation.

Figure 3 depicts the dynamic interaction involved in representing cancer staging, where the set of potential stages is dependent on the type of the cancer, and the set of potential sub-stages (if any), on a combination of type and stage. The modification of field constraints manifests itself in the re-setting of the default fillers. It is desirable that such automated updating be fully dynamic, with any assertions, retractions or replacements producing appropriate responses. For example, if the type of the disease is now specialised to Leukemia, the locus should update to Blood and the default stage value to Leukemia+stage, and the sub-stage field should disappear.

There are two basic ways of achieving dynamics in a model. Firstly, one can require the client code, after setting specific field values, to explicitly make any resulting updates to other parts of the instantiation, in line with a set of stipulations provided as part of the model. Alternatively one can create an update mechanism that

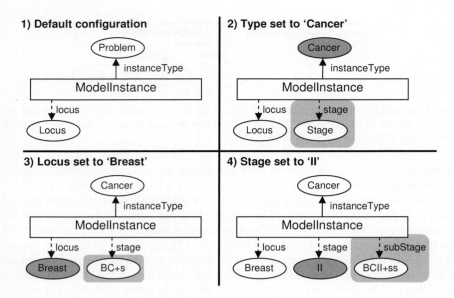

Fig. 3. Creation of an instantiation of a dynamic indirect model - basic key as for figure 2 - updates represented via shading (newly asserted field-value ⬤, area of automatic response by model ▬) - includes automatically-generated class names: BC+s = BreastCancer+stage = (I or II or II' or IV), *BCII+ss =BreastCancerStageII+subStage = (A or B)*

reacts appropriately to changes in model instantiations. The first alternative is the more flexible, imposing no restrictions on how the model is used. Hence, in addition to basic data-creation, the model could also be used to drive query-formulation, possibly with a variety of query schemas of varying expressivity. However, this flexibility comes at the price of additional complexity on the client side. Furthermore, the manner in which the updates are stipulated will be dependent on the BM format, ruling-out the possibility of BM encapsulation. The second alternative simplifies things on the client side, but does not necessarily provide the same flexibility. However, a suitable architecture can achieve the best of both worlds. For example, our framework provides an automatic update mechanism as part of an *instantiation building facility*. An associated *model-realisation plug-in facility* comes with alternative back-ends for data-creation and query-formulation. Additional back-ends (for e.g. alternative types of query-formulation) could be plugged-in if required.

Ontology-Backed Models: The types of dynamic model in which we are specifically interested are indirect models in which the BM is provided by an OWL ontology plus a suitable reasoner. In order for such a logic-based system to be used as the basis for a dynamic software model, some form of *sanctioning* scheme [88] must be used. Sanctioning provides a bridge between the constraint-based world of the ontology, and the field-based world of the software model. Specifically, a sanctioning scheme provides some means of associating a relevant set of fields with each OWL class. Exactly how this is achieved is not important here. Possible approaches include the use of heuristics to derive the field-sets directly from class

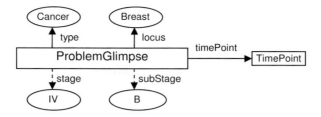

Fig. 4. Instantiation of a simple fragment of a hybrid model - software entities represented both directly (object of named type ▭, object field ⟶) and indirectly (reference to named ontology-concept ⬭, ontology-derived field ⁻ ⁻ ▶)

restrictions, or the explicit specification of the field-sets via some form of internal or external meta-data.

Hybrid Models: We define hybrid models as software models that integrate both direct and indirect sections into a coherent whole. The intention is to benefit from the strengths of the respective approaches whilst mitigating their weaknesses. The hybrid models in which we are specifically interested are exemplified by the *Patient Chronicle Model (PCM)*, described in detail in section 4. Such models are divided into a direct core section, in which a relatively small number of core entities provide the main structure of the domain, and an indirect peripheral section, in which a far larger number of entities provide the detailed domain knowledge. The BM for the indirect section of the current version of the PCM is provided by an OWL ontology, though this is not a defining feature of such hybrid models.

To illustrate the basics of such models we look at an example from the PCM. Figure 4 shows the representation of a single disconnected "glimpse" of a patient's cancer at a specific point-in-time. It can be seen that this is a very similar set-up to that shown in Figure 2. Differences to note are: (1) the main entity is a domain-specific ProblemGlimpse object rather than a domain-neutral ModelInstance, (2) locus is a direct field on the ProblemGlimpse class (although stage and subStage are still dynamically-derived indirect fields), and (3) an additional timePoint field has been added (although the representation of time is a central feature of the PCM, and provides additional motivation for the use of a hybrid model, for the purposes of the current discussion we can consider timePoint as just another field). An additional difference (not depicted) is that the fillers for the concept-valued direct fields (type and locus) are actually of domain-specific types, designed to provide a type-safe means of representing references to concepts from the relevant section of the ontology. Hence, the type field has value-type ProblemType, a class that represents references to concepts from the Problem section of the ontology. (Note that in the PCM the mappings between the concept-referencing classes and the relevant root-concepts in the ontology are provided via a configuration file and are not hard-coded in any way.)

Figure 5 shows how collections of domain objects, such as ProblemGlimpse, can be aggregated together to form larger networks. The core structure of such a network is provided by the domain objects and their interconnections, or in other words, is the

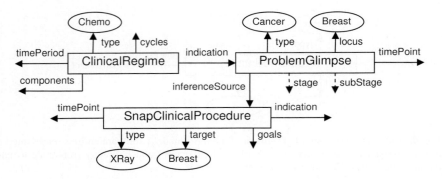

Fig. 5. Instantiation of a larger fragment of a hybrid model - key as for figure 4

instantiation of a direct model. It is only on the periphery of the model that the indirect elements intrude. On the other hand, the domain objects comprise only a small fraction of the model-entities – the vast majority residing within the ontology. For instance, the set-up in Figure 4 involves only three domain-specific classes, ProblemGlimpse, ProblemType and Locus (not to be confused with the Locus concept that provides the root of the hierarchy to which it maps), whilst the number of ontological concepts that can act as fillers for the type and locus fields, may number well into the thousands.

From the point-of-view of a programmer implementing a domain-aware application based on such a hybrid model, the direct nature of the core structure is a distinct advantage (as noted above in connection with fully direct models, and further discussed in section 3). However, the need for indirect model access has not been entirely eradicated. Providing fully direct access to the type of dynamic model with which we are dealing is simply not a practicable proposition. What the hybrid approach does do, however, is to greatly mitigate the problem by pushing the indirect representation to the edges of the model (in the case of the PCM, further mitigation is achieved by the provision of a dynamic model browser, which allows the programmer to explore the dynamic interaction in those areas of the model where it does need to be handled).

To provide a rough comparison of PCM-style hybrid models with both direct and ontology-backed indirect models, we have identified a number of potentially desirable features that the models may provide – see table 1. Although this set was derived directly from the requirements for the PCM, we feel that it is fairly comprehensive, though not necessarily exhaustive. Features include types of dynamic modification, as classified by modification-type (*field-constraint* or *model-shape - i.e.* the addition and removal of fields), and means of specification (*ontological* or *extra-ontological*). Also covered are type of API (*domain-neutral, domain-specific*), potential for attachment of processing mechanisms to operate over individual instantiations, knowledge maintenance by domain-experts, knowledge encapsulation, and potential for use in query-formulation. (*Note:* section 4 provides further discussion of some of the listed features.)

Table 1. Comparison of features offered by different types of software model

Feature	Direct	Ontology-Backed Indirect	Hybrid (PCM-style)
Dynamic modification (model-shape / ontological)	NO	YES	YES
Dynamic modification (model-shape / extra-ontological)	NO	NO	YES
Dynamic modification (field-constraints / ontological)	NO	YES	YES
Dynamic modification (field-constraints / extra-ontological)	YES	NO	YES
Domain-neutral API	YES[di1]	YES	YES
Domain-specific API	YES	NO	YES[hy1]
Model-instantiation processing	YES	NO	YES[hy1]
Knowledge maintenance by domain experts	NO	YES	YES[hy2]
Knowledge encapsulation	NO	YES	YES[hy2]
Query formulation	YES[di1]	YES[ob1]	YES

di1 = Given appropriate reflection based architecture
ob1 = Given appropriate architecture
hy1 = For core structure - not for detailed knowledge
hy2 = For detailed knowledge - not for core structure

Whilst both the fully-direct and fully-indirect approaches offer a subset of the listed features, PCM-style hybrids can, subject to certain trade-offs (see above discussion), be said to offer all of them. Obviously, when developing a software model one should select the approach that most closely meets the requirements of the particular domain, and where this implies a choice of options, one should probably go with the simplest. Hence, given the complexity overhead of the hybrid option, it should only be used if neither of the other options fits the bill. However, we feel that it is likely that for a large class of application areas this will indeed be the case.

3   Methodological Considerations

With a basic taxonomy of models in hand, we now turn to how various properties of the different sorts of model affect software development methodology via cognitive walkthroughs [10]. We consider how each sort of model handles the sequence of events shown in Figure 3, i.e. we (1) instantiate an instance of 'Cancer', (2) set the locus of the cancer to 'Breast', (3) set the stage of the breast cancer to 'II', and (4) set the sub-stage of the stage II breast cancer to some value. A key point about this sequence is that each setting of a field *alters* what other fields are available and the constraints upon those fields (and, thus, perhaps, the behaviour of the object in the application). Furthermore, the sequence of inputs and the particular values set (and thus the shape of the object) vary enormously. As described above, there are many different types of cancer, each with a different set of potential stages, and each combination of cancer and stage having a different set of sub-stages (or possibly none).

The "final" shape of the object (that is, when all its fields are set) is defined by a combination of (1) the model class of which the object is an instance, and (2) additional information provided by the model, specifying dynamic modifications specific to the evolving object. It is not always the case that there is a named model class that corresponds directly to the fully determined object (that is, the system can require runtime inference).

We take this sequence as an exemplar of how applications interact with entities in a model. It is easy to see that interactive applications (such as an electronic hospital chart) will need to modify entities in its model in this way. In all these cases, we presume that a sizable portion of the model will be expressed by a domain expert (who is probably not a programmer) in a suitable modelling language, such as OWL (*e.g.*, the specifics of cancer). Given this scenario, we can examine what costs and benefits model type offers. In order to keep things concrete, we confine the rest of the discussion to models expressed in OWL, Java, or a combination of the two, although the issues involved are potentially applicable to a range of modelling formalisms, and most general purpose OOPLs.

Consider direct models (*e.g.* as generated by an OWL2Java mapper [99]). One advantage of direct models is that the model is entirely captured in Java and the application programmer need never consider the ontology except as the input to the OWL2Java processor. Thus, the programmer can consider OWL to just be a funny kind of UML (and they may even view it as UML diagrams) and get on with the business of *using* the Java classes. If the ontology is small and unlikely to change, this is feasible. However, in typical medical applications, neither of these facts are the case. A recent version of the NCI Thesaurus contains 65,228 classes (up from 27,652 in 2003) and is updated monthly[1]. With a direct model, not only do we get a proliferation of Java classes that obscure the actual structure of the ontology, but the natural path to keeping the application in synch with the model is to regenerate and recompile[2]. Aside from the tedium of this procedure, it makes it practically impossible to modify the generated classes to introduce special behaviours, thereby eliminating a major benefit of the direct approach.

Furthermore, this sort of model is very difficult to work with given the sequence of operations in our example. In essence, to get the behaviour we want we need to determine in advance which specific class we are going to instantiate in step one (i.e., not just cancer, but *breast* cancer; and not just breast cancer, but *stage II* breast cancer; etc.). If we later want to change from stage II to stage I, or to correct the locus, we must discard our instance and create an instance of the relevant new sort.

Finally, since we do not have a reasoner available, we cannot query for aspects of the *ontology* that were not explicitly reflected into Java. Workarounds include trying to capture aspects of the semantics of OWL class expressions in Java (see [99]) or modifying the ontology to ensure that specific needed entailments get names, and are thus, reflected out to the application. In the first case, since the mapping is, at best, very partial and approximate, we still miss many possible entailments but now also get spurious ones. In the second case, we contaminate our model with various application

---

[1] http://nciterms.nci.nih.gov/NCIBrowser/Dictionary.do
[2] The Thesaurus is used in e.g.: http://cancerimages.nci.nih.gov/caIMAGE/index.jsp

specific classes and still cannot cover every case. Either way we would have great difficulty in replicating the type of behaviour illustrated by figure 3.

Of course all of this discussion concerning direct models assumes that the programmer is developing a domain-aware application. If alternatively the application is to operate in a domain-neutral fashion, the advantages described do not apply, whilst the difficulties in modelling dynamic behaviour are multiplied by the need to access that behaviour via some kind of reflection based mechanism.

In contrast, using an indirect model backed by an OWL ontology avoids many of these problems: The ontology is a separately modifiable component of the application. We have the full power of an OWL reasoner available and can even update the model in response to application events. Furthermore, the program does not have to incorporate thousands of classes, but only the small number of classes that provide the indirect model. As the API is domain independent, programmers can become expert in using that API and amortise the effort of learning it over many programs. Such APIs, as with SQL, provide a well defined interface for interacting with the ontology based model, so it is easy to analyse exactly where and how the application works with the model.

This flexibility can be accessed very nicely by a programmer developing a domain-neutral application, but for those developing domain-aware applications it comes at a considerable price. The indirect nature of the API becomes reflected in an unnatural indirect coding process, whereby the programmer must operate with API documentation in one hand, and some representation of the ontology in the other, to create code without type-safety or any other kind of API imposed constraints. Moreover, if parts of the behaviour required in the domain cannot be expressed in OWL (e.g., certain types of temporal relations, complex calculations, optimisations, *etc.*), the programmers must handle those aspects entirely on the Java side.

Of course, programmers could set up their own framework for mapping ontology classes into their Java based model on a case by case basis, but this is precisely building an *ad hoc* hybrid model where the details of the hybridization have to be managed explicitly.

With hybrid models we effectively split the difference. Consider the prototype scenario: the Java programmers developing the model can start with a high level model of the ontological aspects of the domain (say, a Problem class) as well as of other domain entities which do not appear in the ontology (e.g., a ProblemGlimpse). When the application is working with this abstract model the programmers can hook up a handful of key entities to corresponding entities in the ontology, in order to exploit the modelling done by the ontologist. Both the ontology and the program can dynamically modify the shape and constraints of their respective model elements without interfering with each other. Furthermore, programmers can naturally move the boundary between the part of the model which is in Java and the part which is OWL (or some other modelling formalism) as is appropriate. The fluidity of the boundary encourages programmers to use the formalism that is best for the job *given their tastes, experience, and skills*. Instead of having to jump directly into OWL (for example) they can defer that exploration until it is truly necessary. That is, *modelling* considerations, rather than limitations of their integration technique, drive shifts in the boundary. Obviously, hybrid models require *some* restrictions in the modelling on both sides.

All three approaches have sweet spots, though we believe that hybrid models are more generically useful. We suspect many domain applications will be better served by a hybrid model.

4 Case Study

We now look in detail at the *Patient Chronicle Model (PCM)*, as introduced in section 2, and at the generic framework with which it was built. The PCM provides the central component of an architecture designed for the representation of large bodies of patient record data in a richly-structured *chronicle* format, and their subsequent exploitation as a research resource over which interesting clinical queries can be formulated and executed. The framework comprises a fully generic *Core Model-Builder* and a temporally-focused but domain-neutral *Chronicle Model-Builder*[3], specifically for building chronicle-style models, such as the PCM. Although the framework was initially created with the PCM in mind, it is a generic entity that should have much wider applicability. Hence, our discussion here is concerned with the general class of hybrid model that the framework enables us to build, using the PCM for purposes of illustration. We concentrate on the external behaviour of the models. See [11] for a fuller description of their internal architecture. All of the software described here, including the PCM, the framework and the GUI-based tools, is available on the web (see [12]).

4.1 Design of Hybrid-Model Framework

An analysis of the requirements for the PCM resulted in the identification of the following set of elements to be incorporated into the design of the framework:

Ontological representation: An obvious requirement for a model representing medical data is that it in some way incorporates a large structured medical terminology. The fact that this terminology was required to support the type of representations and dynamic interactions illustrated in Figure 3 strongly implied some form of ontology with associated reasoning mechanisms. Furthermore, due to the size of the terminology, and the specialist knowledge that it was to embody, it was also necessary that the format facilitate maintenance by domain experts rather than software developers.

Temporal representation: Patient record data tends to come as a set of *snapshots*, representing such things as the current state of a patient's illness or the results of an x-ray procedure, at a particular point-in-time. However, to ask meaningful questions concerning a patient's *history*, we often need to aggregate together individual items of snapshot data into coherent entities representing, for instance, the entire history of a patient's condition. This implied a *SNAP/SPAN* representation [44] of some type, wherein the representation of temporal events is split between point-like *SNAP* events and temporally-protracted *SPAN* events. An associated requirement, to facilitate effective querying, was for the representation of temporal summarisations, or *temporal abstractions* [55] as they are known in the field of medical informatics. For instance,

---

[3] These components of the framework each comprise a set of classes and support utilities.

a set of measurements of the size of a tumour at various points-in-time can, with suitable interpolations, give rise to abstractions over selected time-periods, such as minimum-size, maximum-size, size-at-start-of-period, *etc*.

Temporal processing: Associated with the requirement for temporal abstraction structures, was a requirement for procedures to perform the relevant calculations. Also required was a *temporal-slicing* facility, for slicing SPAN objects up into sections representing arbitrary sub-periods. Such a facility is required for answering queries involving temporal-abstractions over dynamically-defined time-periods.

Ontological/temporal interaction: An additional requirement was for the orchestration of the higher-level interaction between the ontological representation, the SNAP/SPAN representation and the temporal abstraction structures (see below for details of such interaction).

Domain-specific API: The patient chronicle data is created programmatically by two data-creation applications, a heuristic-based *Chronicliser* that generates the richly-formatted patient chronicles from 'raw' patient record data, and a *Patient Chronicle Simulator* that generates realistic patient histories as an aid to system development. Both of these applications operate in a highly domain-aware manner and hence require a suitable domain-specific API.

Domain-neutral API: The model was also required to drive domain-neutral software, including an RDF-based *repository system*, with an associated *query-engine* (combining basic RDF querying with query expansion and dynamic temporal abstraction), and a set of GUIs for *model-browsing*, *record-browsing* and *query-formulation*. Hence, there was a strong requirement for a domain-neutral API (the alternative would be to let each application implement its own reflection-based interpretation of the model – obviously not a sensible option).

Query-formulation capability: The model was required to drive query formulation by domain-neutral applications, which, due to the dynamic nature of the model, implied a requirement for a flexible instantiation-builder with a model-realisation plug-in facility, to allow the incorporation of query-specific constructs into the instantiation (see discussion of dynamic models in section 2).

Fully-dynamic interaction: Since some of the model-driven software, such as the query-formulation system, needed to be highly interactive, it was required that instantiation-building be fully-dynamic in that the system respond appropriately to any assertions, retractions or replacements (see discussion of dynamic models in section 2).

Our hybrid model architecture, which was designed to incorporate this (partially conflicting) set of elements, is composed of (1) a central Java component incorporating both direct and indirect sections of the model, and (2) a set of one or more knowledge sources that collectively comprise the backing model (BM) for the indirect section. All BM access is via a clean API, which entirely encapsulates the underlying formalisms and associated reasoning mechanisms, allowing a range of formalisms to be mixed and matched. However, since the BM for the PCM currently consists of a single OWL ontology in combination with a FaCT++ reasoner [77] and suitable sanctioning mechanisms, we refer simply to the "ontology" throughout the following discussion.

4.2 Nature of Hybrid Models

We now look in more detail at the nature of the PCM-style models that can be built using our framework, building on the brief introduction provided in section 2. Specifically we look at (1) the type of complex dynamic interactions that the models can embody and the way in which they can exhibit behaviours not easily specifiable within a standard ontological representation, (2) the various distinct roles that procedural processing plays within the models, and (3) the 'network' representation, which enables both the models themselves and their individual instantiations to be accessed in a domain-neutral fashion.

Complex Dynamic Interaction: In section 2 we described how the ProblemGlimpse class can be used to represent a single disconnected 'glimpse' of a patient's cancer at a specific point-in-time (what we refer to as a *GLIMPSE* view). However, in practice ProblemGlimpse is used in only to represent transient conditions such as pain or headaches. For major conditions such as cancer, where we wish to track the progress of the condition through time, the more complex SNAP/SPAN-based representational pattern depicted in figure 6 is used. In this pattern, a series of ProblemSnapshot objects of the condition at specific points-in-time (the SNAPs) are aggregated together by a ProblemHistory object (the SPAN).

The first thing to note in this pattern is that the temporally-invariant type and locus fields are attached to the SPAN object, whereas copies of the temporally-variant stage and subStage fields are attached to each of the SNAP objects. This means that whereas the simpler GLIMPSE pattern could be mapped in a one-to-one fashion to a single ontological instance, the SNAP/SPAN version requires a collection of such instances, with each being mapped to a combination of the temporally-invariant fields on the SPAN object and the temporally-variant fields on a specific SNAP object.

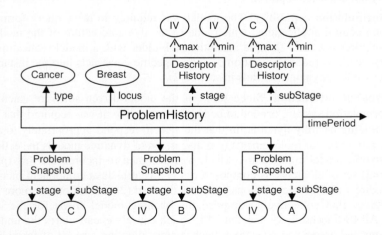

Fig. 6. Instantiation of a more complex fragment of a hybrid model (with problem description distributed between SPAN and SNAP entities) - basic key as for figures 4 and 5 - also show are fields derived from Temporal Abstraction System (– – ≫).(NOTE: the timePoint fields for the SNAP entities have been omitted.)

An additional element not present in the GLIMPSE version is the representation of temporal abstractions. For each current problem-descriptor (stage, subStage, etc.), there will be an abstraction field associated with the SPAN object, which provides an *abstraction-set* for that descriptor (via a DescriptorHistory object). Hence, the abstraction-sets for stage and subStage, both of which are defined (via extra-ontological meta-data) as ordinals, include attributes such as max and min. These abstraction-sets, as well as the methods for calculating the abstraction values, are ultimately provided by a *Temporal Abstraction System*, to which the PCM interfaces. This all adds additional complexity, involving (1) yet another ontological-instance, this one being mapped to a combination of the temporally-invariant fields on the SPAN object, and the set of abstraction fields, and (2) dynamic generation of the individual abstraction-sets, via interaction with the Temporal Abstraction System.

An important aspect of PCM-style hybrid models is the orchestration of interaction by the major domain classes within a representational pattern. For instance, locus has been specifically plucked out from the set of problem-descriptors to become a direct field on ProblemGlimpse (one reason being that it is an entity that the programmer will often wish to explicitly reference). However, even though the locus is explicitly represented in the direct model, it is still represented in the ontology (along with the more run-of-the-mill fields, such as stage and subStage, not represented in the direct model). Furthermore, we have seen that the locus value plays a part in the ontological reasoning behind the dynamic interaction illustrated in figure 3. Therefore classes such as ProblemGlimpse have to orchestrate the dynamic updating in a manner that maintains consistency between the model and the ontology.

With ProblemGlimpse, where there is a one-to-one correspondence between the entities in the direct model and those in the ontology, such orchestration adds nothing to the underlying ontological reasoning. It is merely the performance of a chore made necessary by the hybrid nature of the model. However, in the case of ProblemHistory where the corresponding interaction involves both multiple ontological-instances and the Temporal Abstraction System, its orchestration adds additional levels of complexity over and above that provided directly by the ontological reasoning.

Procedural Processing: Procedural processing plays three distinct roles within PCM-style hybrid models: *model-shape modification, field-constraint modification* and *model-instantiation processing*. Of these, the first two can be considered as providing an intrinsic part of the model, whereas the third acts on individual instantiations, but does not contribute to the model itself.

In the current PCM, the model-shape modification is always handled by fully generic mechanisms, although this is not something that is intrinsic to the task, and we could envisage a situation where shape modification of a more domain-specific nature was required. On the other hand, field-constraint modification and model-instantiation processing, as exemplified respectively by temporal abstraction and temporal slicing, are each, at different points in the PCM, handled by both generic and domain-specific mechanisms. The generic case can be seen from the ProblemHistory-centred pattern described above, where both the abstraction and slicing come as part of a configurable generic pattern. An example of the domain-specific case is provided by the DosagePattern class, used in representing sequences of drug administrations. This is an abstract base-class that provides both its own temporal abstraction fields (totalIntake, averageDailyIntake, *etc.*), and its own temporal slicing facility. The actual processing is farmed out to appropriate sub-classes (RegularDosagePattern, CyclicDosagePattern, *etc.*), each of which provides a distinct way of either summarising, or directly representing the individual administrations. It should be apparent that the flexibility offered by the object-oriented core of the hybrid models is very useful here, whereas associating any sort of procedural processing with a fully indirect model is less straightforward, and the greater the required flexibility, the less appealing such an option becomes.

Fully-Indirect Representation: In order to provide the required domain-neutral API the framework embodies a mechanism for automatically translating the *source* version of the hybrid representation into a fully-indirect *network* version (and back again). The translation process depends on the Java reflection facility and the conformance of the source version of the model to certain coding conventions (the necessary ingredients for obtaining generic access to the direct sections of the model - as discussed above).

The basic translation operation takes a source domain class, such as ProblemHistory, and uses it to generate a set of generic network objects, consisting of a ModelNode plus a set of ModelFields. The generated objects will collectively represent an instance of the class. A wider process takes a model instantiation and converts it into an entire network. An additional mechanism is provided to enable the specification of the dynamic behaviour required from the network. This specification is handled by the individual domain classes, each of which, upon being loaded at run-time, can register a set of *factory* objects, which as the relevant sections of network are generated, are used to create sets of *listener* objects that will implement the required interaction.

The network representation can be used in two distinct ways. Firstly, to provide a static representation of an existing model-instantiation, which can be used in the storage, retrieval and browsing of records. Secondly, with appropriate extensions to represent the required query-specific constructs, as a dynamic query-formulation system. In this case the network representation is acting as an instantiation-builder with a model-realisation plug-in facility, in the manner described above. (The network representation could in principle also be used as an instantiation-builder for dynamic record-creation, but since this has not been a requirement of the PCM, our framework does not currently provide this facility).

5   Conclusions

OWL ontologies offer a number of modelling advantages that have been fruitfully exploited by domain experts working in Health Care and the Life Sciences and other areas. However, to reap the benefit of those advantages requires that the resulting artefacts (i.e., the models expressed as ontologies) are effectively exploited by programmers building applications. In this paper, we have presented three mechanisms for integrating OWL ontologies with programs written in a statically typed OOPL (specifically Java), including a new approach based on hybrid-models. Hybrid-models allow for a smooth integration between Java based modelling and OWL based modelling wherein each modelling paradigm's strengths can be mobilised as needed to produce a model that, on the one hand, is a reasonable representation of the subject domain and, on the other, is a natural part of a program. We have shown that for a significant class of application this approach is especially effective.

Future work includes refining our hybrid-model supporting framework to better enable model refactoring and refinement. Right now, model development tools are entirely Java oriented or entirely OWL oriented, and thus do not allow for a unified view of the whole hybrid model. While one strong advantage of hybrid-models is that they allow different members of the development team to use the type of modelling technique that is most appropriate for the task or their own skill set, we believe that a holistic view of hybrid models has its own advantages.

References

1. Puleston, C., Cunningham, J., Rector, A.: A Generic Software Framework for Building Hybrid Ontology-Backed Models for Driving Applications. In: OWL Experiences and Directions Workshop (2008)
2. Rogers, J., Puleston, C., Rector, A.: The CLEF Chronicle: Patient Histories Derived from Electronic Health Records. In: IEEE Workshop on Electronic Chronicles, eChronicle 2006 (2006)
3. Taweel, A., Rector, A.L., Rogers, J., Ingram, D., Kalra, D., Gaizauskas, R., Hepple, M., Milan, J., Power, R., Scott, D., Singleton, P.: CLEF – Joining up Healthcare with Clinical and Post-Genomic Research. Current Perspectives in Healthcare Computing, 203–211 (2004)
4. Grenon, P., Smith, B.: SNAP and SPAN: Towards Dynamic Spatial Ontology. Spatial Cognition and Computation 4(1), 69–104 (2004)
5. Shahar, Y., Combi, C.: Temporal Reasoning and Temporal Data Maintenance: Issues and Challenges. Computers in Biology and Medicine 27(5), 353–368 (1997)
6. Baader, F., Calvanese, D., McGuinness, D., Nardi, D., Patel-Schneider, P.: The Description Logic Handbook. Cambridge University Press, Cambridge (2003)
7. Tsarkov, D., Horrocks, I.: FaCT++ Description Logic Reasoner: System Description. In: Furbach, U., Shankar, N. (eds.) IJCAR 2006. LNCS (LNAI), vol. 4130. Springer, Heidelberg (2006)
8. Bechhofer, A., Goble, C.: Using Description Logics to Drive Query Interfaces. In: DL 1997, International Workshop on Description Logics (1997)
9. Kalyanpur, A., Pastor, D., Battle, S., Padget, J.: Automatic Mapping of OWL Ontologies into Java. In: Software Engineering and Knowledge Engineering (SEKE), Banff, Canada, pp. 98–103 (2004)
10. Wharton, C.W., Reiman, J., Lewis, C., Polson, P.: The Cognitive Walkthrough Method: A Practitioner's Guide. In: Nielson, J.K., Mack, R.L. (eds.) Usability Inspection Methods. Wiley, New York (1994)
11. W3C: Web Ontology Language (OWL), http://www.w3.org/2004/OWL/
12. CLEF Chronicle Software: http://intranet.cs.man.ac.uk/bhig/clef_misc/chronicle/
13. Protégé 4 download page: http://www.co-ode.org/downloads/protege-x/
14. Swoop download page: http://code.google.com/p/swoop/

Extracting Semantic Constraint from Description Text for Semantic Web Service Discovery[*]

Dengping Wei[1], Ting Wang[1], Ji Wang[2], and Yaodong Chen[1]

[1] Department of Computer Science and Technology, School of Computer, National University of Defense Technology, Changsha,Hunan, 410073, P.R. China
{dpwei,tingwang,yaodongchen}@nudt.edu.cn
[2] National Laboratory for Parallel and Distributed Processing, Changsha,Hunan, 410073, P.R. China
jiwang@mail.edu.cn

Abstract. Various semantic web service discovery techniques have been proposed, many of which perform the profile based service signature (I/O) matching. However, the service I/O concepts are not sufficient to discover web services accurately. This paper presents a new method to enhance the semantic description of semantic web service by using the semantic constraints of service I/O concepts in specific context. The semantic constraints described in a constraint graph are extracted automatically from the parsing results of the service description text by a set of heuristic rules. The corresponding semantic web service matchmaker performs not only the profile's semantic matching but also the matching of their semantic constraints with the help of a constraint graph based matchmaking algorithm. The experiment results are encouraging when applying the semantic constraint to discover semantic web services on the service retrieval test collection OWLS-TC v2.

1 Introduction

Semantic web services (SWS) have attracted a significant amount of attention in recent years. The aggregation, including description and discovery of services plays an important role in various internet-based virtual computing environments [1]. SWS discovery is the process of locating existing web services based on the description of their functional and non-functional semantics [2]. Most SWS matchmakers perform the matching of service profile rather than service process model. SWS profile describes the services capabilities in terms of several elements, including its inputs(I), outputs(O), preconditions/assumptions(P) and effects/postconditions(E) [3].

[*] This research is supported by the National Grand Fundamental Research Program of China under Grant No. 2005CB321802, the Program for New Century Excellent Talents in University(NCET-06-0926), and the National Natural Science Foundation of China (60403050, 90612009).

A. Sheth et al. (Eds.): ISWC 2008, LNCS 5318, pp. 146–161, 2008.

Various SWS description languages such as OWL-S [3], WSMO [4], WSDL-S [5], SAWSDL [6], provide different frameworks to describe SWS. There are also various SWS matchmakers based on the respective profile elements: some perform logic based semantic IOPE matching [7] [8], and some others perform logic based semantic service signature (Input/Output) matching [9, 10, 11, 12, 13, 14, 15, 16].

A common characteristic of most current SWS matchmakers is that the semantic matching between a pair of SWS concepts annotated to the input and output parameters almost depends on the subsumption relations in the domain taxonomy. Most current SWS matchmakers treat the SWS signature as a set of concepts which are however not sufficient to discover SWS effectively when using logic based reasoning. Two services with similar real world semantics may fail to match, and even two services with the same input and output concepts may have essential differences in semantics which cannot be detected by logic based reasoning.

In order to overcome this problem, many recent researches have explored various information to complement service I/O concepts for SWS matchmaking. The ranked matching algorithm [9] explores the service category and service quality together with its I/O concepts to compute the combined degree of match between the request and the advertisement. Klusch et al. [7] [10] have proposed a hybrid method for SWS discovery which utilizes both the logic based reasoning and the content (unfolded concept expressions) based Information Retrieval(IR) techniques to remedy this limitation. Kiefer et al. [17] have proposed a new approach to perform SWS matchmaking based on iSPARQL strategies which combines structured and imprecise querying together on a diverse set of syntactic description information (service name, service description text, etc.). The work in [18] describes the relationships between inputs and outputs explicitly and uses OWL ontologies to fix the meaning of the terms used in a service description.

Most SWS discovery approaches consider each SWS signature as a bag of concepts and ignore the relationships between the concepts in the SWS profile. The relationships between the service I/O concepts, called semantic constraints in this paper, can be helpful for expressing the semantics of services and improving the existing SWS discovery methods in practice if they can be generated automatically. Motivated by this idea, we add some restriction relationships to the interface concepts to enhance the semantic description of services. These restriction relationships which may not be defined in the domain ontology can be extracted from the service description text automatically. A novel SWS discovery mechanism has been proposed to perform the matching on both the service I/O concepts and their semantic constraints which are represented by a constraint graph.

The rest of this paper is organized as follows. Section 2 gives the definition of semantic constraint for SWS and the constraint graph. Section 3 describes the semantic constraint extracting method. A Constraint Graph-based Matching (CGM) algorithm is proposed in section 4. Section 5 evaluates the proposed approach on the OWL-S service retrieval test collection TC v2[1]. Section 6 discusses the related work. The conclusion and future research are given in section 7 finally.

2   Semantic Constraint for Semantic Web Service Discovery

2.1   Motivation

Many SWS matchmakers like OWLM [19], OWLS-MX [10], OWLS-UDDI[2], Lumina[3] perform logic based semantic matchmaking on service I/O concepts. As discussed in section 1, representing SWS by the service I/O concepts is not sufficient enough for service discovery. Some important facts can be observed from the existing SWS collections.

(1)**The domain of concept is not specified**. The service I/O concepts usually describe abstract things like "price", "distance". Normally their meanings are heavily related to the context, that is, it is difficult to get its exact semantics unless they are set in a certain context, such as "the price of a book", "the price of a flight ticket", "the distance between two cities", "the distance between two stars". So if a service is only annotated with the concept *Price* as its output, it is still not clarified whose price is returned by this service.

(2) **The property of concept is not specified**. Every concept defined in domain ontology may have several properties to restrict its semantics. All the individuals of each concept can be divided into several categories according to different property restrictions and values. For example, the concept *Bag* has a property *hasColor*, so the individuals of *Bag* can be classified into different sets according to their values of the property *hasColor*, such as the red bags, the blue bags, the green bags and so on. Therefore, in specific context, the semantics of I/O concepts can be better clarified if they are associated with their property values. Meanwhile, the concepts with different property restrictions may correspond to various individuals in respective context. For example, service A returns a kind of food information with the maximum price, while service B returns the food information with brand "Coca Cola". We cannot assert that these two kinds of food information are similar. Also the functions of the two services cannot to be asserted to be similar.

(3) **The relationship between concepts is not specified**. Two web services annotated with the same input and output concepts may have essentially different semantics which cannot be detected by logic based reasoning. The difference may be caused by the diverse relationships between the input and output concepts. For example, both service A and service B have been annotated with concept *GroceryStore* for their input parameters and concept *Food* for their

---

[1] http://www-ags.dfki.uni-sb.de/~klusch/owls-mx/
[2] http://www.daml.ri.cmu.edu/matchmaker/
[3] http://lsdis.cs.uga.edu/projects/meteor-s/downloads/Lumina/

output parameters. But service A returns the food information contained in a certain grocery store while service B returns the food information sold by a certain grocery store. These two web services have the same interfaces and the same annotated concepts which the interface refers to, but their functions are totally different.

2.2 Constraint Types Definition

According to the above facts, the semantics of SWS will be better clarified if the constraint relationships of the concepts have been annotated. Each concept used to annotate web service can take some kinds of constraints. A concept and one of its constraints can be represented by a statement $\langle SC, CT, OC \rangle$ using RDF[4] terms:

- SC (Subject Concept), is the subject of the statement and usually corresponds to the service I/O concepts. It specifies the thing the statement is about.
- OC (Object Concept), is the object of statement. It can be described as another concept or a literal.
- CT (Constraint Type), is the predicate of the statement which identifies the property or characteristic of the subject concept that the statement specifies. There are many constraint relationships between two entities in the real world, which are also true in web service domain. We choose three important abstract constraint types, considering that not all the realistic constraints can be extracted accurately and automatically from description text.

  - *isPropertyObjectOf* Constraint: triple $\langle A, isPropertyObjectOf, B \rangle$ means that concept A is a property object of concept B. It specifies the domain which concept A belongs to, that is, the individuals of concept A in specific service are the property values of individuals of concept B. For example, if a service has been annotated with an output concept **Price** and the price has a constraint triple $\langle \boldsymbol{Price}, isPropertyObjectOf, \boldsymbol{Car} \rangle$, it means that this service returns the price of a certain car.
  - *hasPropertyObject* Constraint: this constraint relation is the inverse of *isPropertyObjectOf*. Triple $\langle A, hasPropertyObject, B \rangle$ means that concept A is the one which has an inherent property object concept B. For example, if a service has been annotated with an output concept **Car** that has a constraint triple $\langle \boldsymbol{Car}, hasPropertyObject, \boldsymbol{Price} \rangle$, it means that this service returns the information of a car associated with a value of price.
  - *Operation* Constraint: triple $\langle A, Operation, B \rangle$ means that the two concepts entities have a certain association between them. The word "Operation" is an abstract word representing all kinds of properties. For example, if a service has been annotated with an output concept **Book** and the concept **Book** in this web service has an "*Operation*" constraint triple $\langle Book, "published\ by", "Springer" \rangle$, it means that this service returns the books that are published by Springer.

---

[4] http://www.w3.org/TR/rdf-primer/

2.3   Constraint Graph Definition

After adding constraints defined above to service I/O concepts, the service semantics has been enriched and SWS is described by both service I/O concepts
and their constraints. In this paper, a concept together with its constraints is
described in a constraint graph. Let C be a set of concepts, a directed constraint
graph can be described as $ConstraintGraph(C) = \{\langle SC, CT, OC \rangle | SC \in C\}$.

The snippet of a food querying service profile is described in Fig.1(a). This
service returns the food information in a certain store together with their quantity, annotated with one input concept **Store** and two output concepts **Food**
and **Quantity**. After adding *isPropertyObjectOf* constraints to concepts **Food**
and **Quantity**, the connected constraint graph which is depicted in Fig.1(b)
indicates the relationships between **Food** and **Store**, **Quantity** and **Food**. The
description of service's semantics in Fig.1(b) is clearer than the set of three
concepts described in Fig.1(a).

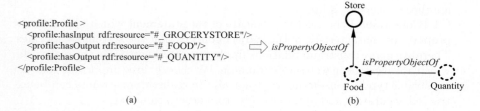

(a)   (b)

Fig. 1. (a) The snippet of a food service profile; (b) The constraint graph representation
of the service (the dashed circle denotes output concept and the solid circle denotes
input concept)

3   Extracting Semantic Constraint

The description texts in web services are important knowledge sources for service
discovery. The constraints of a certain concept can be extracted from the description texts according to the definition of concept in domain ontology, especially
the definition of property whose subject is the concept. However, in practice,
few domain ontologies are comprehensive enough to provide plenty constraints
information, as every concept in the domain ontology has only limited kinds
of properties. In this section, an extraction method based on the parse trees of
description text is proposed to obtain the semantic constraints of service I/O
concepts.

3.1   Overview

The semantic constraints for input/output concepts of a web service can be extracted from the description text of this web service. Each concept corresponds
to a sequence of words (called key-word) in the description text, and the syntactic relations between the key-word and other words may be the semantic

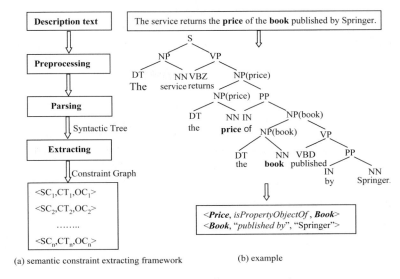

Fig. 2. Semantic constraint extraction

constraints of the corresponding concept. The syntactic structures of the service description text give rich information about the constraint types for service I/O concepts. Thus, the semantic constraints of I/O concepts could be derived from the syntactic tree. Unlike the ontology-based information extraction or relation extraction, we detect the semantic constraints for the service I/O concepts rather than their instances. In ontology-based information extraction, the relation extraction focuses on the relationship between two specified entities.

The semantic constraint extraction which is based on the parsing tree of the sentence consists of preprocessing, syntactic parsing and heuristic-based extracting (shown in Fig.2(a)). During preprocessing, some pre-selected key-words representing the service I/O concepts are tagged in the description text. And then, the text is parsed syntactically to identify the constituents modifying the key-words. Finally, several heuristics rules are used to extract constraint triples about the key-words from the syntactic constituents.

3.2   Preprocessing

The aim of preprocessing is to detect the key-words and process the text in order to improve the precision of syntactic parsing in the next step. Some more details are as follows.

- **Key-words detection.** The concepts in service I/O are annotated to a sequence of words in the description text, that is, each concept is instantiated by a word sequence through scanning all these fragments in text.
- **Name Entity Recognition.** ANNIE Gazetteer in GATE [20] is used to recognize the name entities in the text which are useful for the matching.

For example, the word "Japan" is annotated as a country, so it can match the concept ***Country*** in a geographical ontology.

- **Tokenization**. Corresponding to each service I/O concept, a key-word may include several tokens. A key-word should be a terminal node in the parsing tree in order to extract its constraints correctly. Therefore, each key-word is considered as one token.
- **Part Of Speech tagging for special words**. Several rules are designed to assign Part Of Speech (POS) tags to some important ambiguous words which can help improving the precision of parsing. Based on the observation that concepts in domain ontologies are usually nouns, we specify each key-word to noun and some important words to verb, e.g., "return", "provide".

3.3   Extracting Semantic Constraint

We firstly obtain the syntactic trees by parsing the sentences which contain the key-words in the description text. The semantic constraints of the key-words are identified according to syntactic relationships in parsing trees such as modification and represented in triples each of which includes a key-word, a constraint type and a constraint constituent. The extraction includes three phases as follows.

Candidate Constituent Detection. From the observation, the constraints of a key-word are probably contained in the phrase whose head word is the key-word. All such phrases can be detected by propagating the key-word from the bottom to the top of the syntactic tree. The propagation path is expressed as a sequence of interior nodes labeled by nonterminal categories in the parsing tree, e.g. a node sequence "NP NP" in Fig.2(b) is the propagation path of key-word "price".

The constituents contained in each phrase which is in the propagation path are called candidate constituents which may contain modification information of the key-word. For example, Fig.2(b) describes the parsing tree of the sentence "The service returns the price of the book published by Springer.", in which the key-word is "price". The propagation path of the key-word "price" is "NP NP". The candidate constituents contained in the propagation path are "DET" and "PP".

Constraint Constituents Filtering. In candidate constituents, some function words, such as "DET" in Fig.2(b), are not valuable modifiers for the key-word. Only the constituents that contain useful modifiers of the key-word are considered as constraint constituents. Thus, the constraint constituent of the key-word "price" is the constituent tagged with "PP".

Various constraint types defined in section 2 often have different syntactic characteristics and are expressed in diverse constituents, so respective rules are designed to filter the useful constraint constituents from all the candidate constituents. The second column in Table 1 presents the rules that can filter the useful constituents from the candidate constituents for the constraint types shown in the first column. As the *hasPropertyObject* constraint relationship is the inverse

Table 1. The semantic constraint extraction rules

Constraint Type	Candidate Constituents Filtering Rules	Modifiers Extraction Rules
isProperty ObjectOf	Rule 1: The candidate constituents tagged with "JJ", "PP", and "Pronoun" are indicator of constituents in which *isPropertyObjectOf* constraint locates.	Rule 2: Extract the noun string from the adjective if it's suffix is "'s".
		Rule 3: Extract the key phrase from the PP phrase whose head word is "of".
		Rule 4: Identify the reference word in possessive pronoun phrase.
Operation	Rule 5: The candidate constituents tagged with "VP" and "SBAR" are indicator of constituents in which *Operation* constraint locates.	Rule 6: Extract the verb, preposition, noun from the VP phrase.
		Rule 7: Extract the verb, preposition, noun from the SBAR phrase.

of *isPropertyObjectOf* constraint, we can get the *hasPropertyObject* constraints from the *isPropertyObjectOf* ones.

By analyzing the structure of the parsing tree, the *isPropertyObjectOf* constraint of the key-word can be extracted from either its sibling nodes or its parent's sibling nodes which are tagged with "JJ", "PP" or "Pronoun". The *Operation* constraint often locates in the constituents tagged with "VP" or "SBAR". Rule 1 and Rule 5 are designed to extract the candidate constituents for *isPropertyObjectOf* constraint and Operation constraint respectively. As depicted in Fig.2(b), the *isPropertyObjectOf* constraint of "price" locates in the constituent tagged with "PP".

Extracting Modifier. After the above two steps, the modifiers now can be extracted from the constraint constituents identified by the Rule 1 and Rule 5 in Table 1. Not all the constraint constituents provide useful constraints for the key-word. Some rules listed in the third column in Table 1 are also used to extract the modifiers of the key-word for specific type.

Rules 2-4 for *isPropertyObjectOf* constraint are motivated by the observation that the *isPropertyObjectOf* modifiers to nouns are usually locates in a PP phrase, an adjective or a possessive pronoun. Rule 2 states that only the adjectives like "book's" are the constraint constituents of the key-word. Taking the noun phrase "book's price" for example, the constraint constituent of the key-word "price" is the adjective "book's". Only the word "book" which can be extracted from the adjective is the *isPropertyObjectOf* modifier of the key-word "price". Rule 3 indicates that the *isPropertyObjectOf* modifier often locates in the PP phrase whose head word is "of". Rule 4 is supported by the observation that nouns are usually modified by a possessive pronoun like "its". The noun that refers to the pronoun is identified as the *isPropertyObjectOf* modifier.

Rules 6-7 for *Operation* constraint are supported by the observation that the candidate constituents tagged with "VP" and "SBAR" are good indicator of constituents where *Operation* constraint locates. The verbs in verb phrase and SBAR clause usually describe the relationship between two entities and the following preposition is a good indicator of the voice of the sub-sentence in which the VP phrase locates. If the verb is a transitive verb and the following word is a noun, then the *Operation* constraint is composed by the verb and the noun which is the object of the verb. A preposition "by" indicates that the noun following it would be extracted as the subject of the operation which the verb represents.

Finally, the constraint triples can be represented by the key-words and the extracted modifiers. In Fig.2(b), the modifier of "price" is "book" and the constraint triple is $\langle$***Price***, *isPropertyObjectOf*, ***Book***$\rangle$. For the snippet of sentence "the book published by Springer", the constraint triple $\langle$***Book***, *"published by"*, *"Springer"*$\rangle$ indicates that the subject of the verb "publish" is "Springer" and it's object is the noun "book".

In this paper, we only consider these three constraint relations of the concept at a coarse granularity level. The extraction rules are very specific in order to achieve the high precision of extracting. If there need to extract more information, more rules should be added accordingly.

4   Matching Algorithm

According to the definition of constraint graph introduced in section 2, we have designed a three levels' matching algorithm to measure the match between two constraint graphs.

4.1   Constraint Graph Matching(CGM)

The degree of matching between the $ConstraintGraph(C_r)$ of the request and the $ConstraintGraph(C_s)$ of a service is computed by the following formula:

$$ConstraintGraphMatch(C_r, C_s) = \sum_{i=1}^{P} \max_{j \in P'}(TripleMatch(RT_i, ST_i))/P$$

where P is the number of triples contained in the constraint graph $ConstraintGraph(C_r)$, P' is the number of the triples contained in the constraint graph $ConstraintGraph(C_s)$ and the function $TripleMatch(RT_i, ST_i)$ is used to estimate the match between two triples $RT_i \in ConstraintGraph(C_r)$ and $ST_j \in ConstraintGraph(C_s)$.

4.2   Triples Matching

If there is a triple T_r in the constraint graph of the request and a triple T_s in the constraint graph of web service, the matching between two triples is computed as following:

- *Step1*: compute the degree of match between two subjects.
- *Step2*: compute the degree of match between two objects if the constraint types of the two triples are similar and their subjects are matched.
- *Step3*: compute the weighted sum of match value obtained from the above operations as the match value if all the elements in the triple are matched.

This matching algorithm states that the subsumption relation between two subjects only means that the triples are possibly matching. Only when all the three elements in each triple are relative, can we say that the two triples are matched and the degree of match can be measured.

4.3   Concept Matching

The matching between two concepts is based on the subsumption relationship reasoning on the taxonomies of domain ontologies. The logic based semantic matchmaker we use here is much more relaxed compared to the algorithm described in [7]. Five different levels for the degree of semantic matching are defined, that is, Exact, Plug-in, Subsumed-by, Intersect, and Fail. Let c be a concept, then $Parents(c)$ returns the set of the generic concepts that are the parents of concept c and $Children(c)$ returns the set of the specific concepts that are the children of concept c. The details of the logic based semantic matching are as follows.

- **Exact match**: concept r of the request exactly matches the concept s of service if and only if $r = s$. The concept r of the request perfectly matches the concept s with respect to logic based equivalence of their formal semantic.
- **Plug-in match**: concept s of service plug-in matches the concept r of request if and only if $r \in Parents(s) \vee s \in Children(r)$.
- **Subsumed-by match**: concept r of the request subsumed-by matches the concept s of service if and only if $s \in Parents(r) \vee r \in Children(s)$.
- **Intersect match**: concept r of request intersect matches the concept s of service if and only if

$$\frac{\|Parents(r) \wedge Parents(s)\|}{\max\left(\|Parents(r)\|, \|Parents(s)\|\right)} \geq 0.5$$

- **Fails**: concept r of the request does not match any concept of service according to any of the above match levels.

The definitions of the plug-in match and the subsumed-by match seem reduplicate. A is the parent of B logically equals that B is the children of A, but it is not always true in ontology reasoning. In ontology library, there exist some kinds of reference relationships between two ontologies, such as "import" relation. If two concepts come from different ontologies and there is no reference relationship between these ontologies, then the two concepts match fails. Let ontology ont_1 be imported into another ontology ont_2, that the concept B in ont_2 is the child of concept A in ont_1 cannot infer that concept A is the parent of concept B, because the two concepts belong to different ontologies and the reasoning is based on different knowledge bases.

When the constraint graph based matchmaker coordinate with other matchmaker, the logic based semantic matchmaker used in it can be adapted accordingly. The semantic similarity between two literals is also measured by the distance in the lexical hierarchy which defined in WordNet dictionary. The degree of match is 1 if the two literals belong to the same synonym set.

5   Experiment Results and Analysis

The proposed method has been evaluated on the service retrieval test collection OWL-S TC v2, which consists of 576 web services from 7 domains, 28 queries (each query represents one request) with their relevance sets. Each web service in this dataset has only one operation. In this experiment, 27 queries are evaluated, except the one without output parameter. Two sets of web services using in this experiment have been transferred from OWL-S TC v2 by annotating output concepts with constraints: Dataset1 and Dataset2. In Dataset1, the semantic constraints of the output concepts in request and web service are manually annotated by two people and mainly described by service I/O concepts; while the semantic constraints of concepts in Dataset2 are automatically extracted using the method represented in section 3. The constraint graph based SWS matchmaker (described in section 4) called CGM is implemented in JAVA using Jena[5].

In the experiments, we measured the constraint graph based service retrieval performance. The Macro Averaged Recall Precision curves are shown in Fig. 3. OWLS-M0 is a pure logic based matchmaker on the service I/O concepts [10]. OWLS- M4 is reported to be the best-performing matchmaker variant of the OWLS-MX matchmaker [10] which uses Jensen-Shannon information divergence to compare the request and service based on the unfolded concept expressions. We also use the nearest-neighbor as the minimum degree of match and a value of 0.7 as syntactic similarity threshold for OWLS-M4. These values were suggested by the authors of OWLS-MX to obtain better results for OWLS-TC v2. InOut-Constraint matchmaker and AutoConstraint matchmaker use CGM to compare two services based on their semantic constraint graph, which have been evaluated on Dataset1 and Dataset2 respectively. M0+InOutConstraint matchmaker uses CGM to filter the results of OWLS-M0 on Dataset1. M0+AutoConstraint matchmaker uses CGM to filter the results of OWLS-M0 on Dataset2. M4+InOutConstraint matchmaker uses CGM to filter the results of OWLS-M4 on Dataset1. M4+AutoConstraint matchmaker uses CGM to filter the results of OWLS-M4 on Dataset2. For running OWLS-MX variants, we use OWLS-iMather[6]. All these datasets and matchmakers are available for download[7]. From

---

[5] http://jena.sourceforge.net/

[6] http://www.ifi.uzh.ch/ddis/research/semweb/imatcher/

[7] http://nlp.nudt.edu.cn/~dpwei/

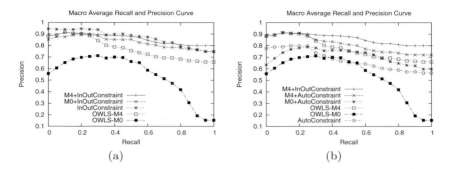

Fig. 3. Macro Average Recall-Precision Curves. (a). The performance on Dataset1; (b). The performance on Dataset2.

the preliminary experimental results depicted in Fig.3, the following facts can be observed.

(1) Both InOutConstraint and OWLS-M4 outperform OWLS-M0 as shown in Fig.3(a). This fact indicates that pure service I/O concepts based matchmaker is not effective to discovery web services. OWLS-M0 often returns some irrelevant services rather than relevant services according to the logic-based semantic filter criteria. While OWLS-M4 can use the syntactical similarity filter (matching of the unfolded concept expressions) to find relevant services that OWLS-M0 would fail to retrieve. InOutConstraint can perform the semantic matching more accurately by enriching the description of SWS. The logic based matchmaker for two concepts which differs from that of OWLS-MX variants uses more relaxed notions for matching in order to retrieve more services, and the constraints of I/O concepts can help filter the irrelevant ones. It also indicates that we should use more relaxed semantic based matchmaker when using the constraint graph based matchmaker.

(2) Fig.3(a) also shows that both M0+InOutConstraint and M4+InOutConstraint outperform their corresponding OWLS-MX variants in OWLS-TC v2. InOutConstraint can filter the irrelative services which are returned by OWLS-MX variants because of the constraints matching. In OWLS-TC v2, there are about 10% web services that have no input parameter and also some web services with matching I/O concepts but different functionalities, which increases the probability of returning irrelevant services by OWLS-MX variants. For example, the "_food_Exportservice.owls" service only has an output concept **Food** and returns the exported food information. The query "grocerystore_food_service.owls" which requires services that return the food owned by a certain grocery store has an input parameter **GroceryStore** and an output parameter **Food**. The query and the service implement different functions but OWLS-MX variants judge that the service matches the query because of the same output concept.

(3) InOutConstraint outperforms both the composed matchmaker M4+ InOutConstraint and OWLS-M4 as illustrated in Fig.3(a). The relaxed logic based semantic matchmaker defined in section 4.3 can retrieve more web services to

improve the recall, while the constraints which make the matching of concepts more accurately are benefit for filtering the irrelative services, because the constraints that InOutConstraint use are annotated by people carefully.

(4) M0+AutoConstraint outperforms OWLS-M0 and M4+AutoConstraint also outperforms OWLS-M4 as shown in Fig.3(b). This fact illustrates that the semantic constraint extracting method for service I/O concepts is effective to filter the irrelative services that OWLS-MX variants return, although the improved performance is lower than that in Dataset1. With the improvement of the extracting performance, the curve of M0+AutoConstraint would approach the curve of M0+InOutConstraint. AutoConstraint depicted in Fig.3(b) has lower performance than M4+AutoConstraint, even OWLS-M4, which suggests that the constraints extracted from the description text automatically are not good enough to filter all the irrelative services that the relaxed logic based semantic matchmaker returns.

From the above experiments, we can see that the semantic constraints of service input/output concepts enrich the description of services capabilities and alleviate the unclear problem of semantic web services that described by only I/O concepts. Semantic constraint of service I/O concepts can distinguish the similar web services to improve the precision of discovery task. The semantic constraint based matchmaker could combine to other SWS matchmakers to improve their performance. The matchmaker CGM explores the constraints information to filter the irrelevant services which probably match the request by pure logical reasoning, while the hybrid matchmaker emphasizes particularly on retrieve those services that fail to retrieve according to pure logic based matching. So they can work with each other without conflict. However, the performance of our matchmaker depends on the performance of constraints extraction. The best performance displayed in Fig.3(a) is InOutConstraint which indicates that semantic constraints of service I/O concepts are excellent in SWS discovery task if supported by a good constraints extracting result.

6   Related Work

Semantic web service discovery is a hot topic in the fields of both semantic web and web service. An abundance of different approaches for service matchmaking focuses on different aspects of service description including functional and non-functional ones. The work presented in this paper concerns only the function based matchmaking. The functional properties of SWS mainly include service inputs, service outputs, preconditions and effects [3]. Several studies concentrate on describing web service with richer semantics for discovery [7] [18], in which the matching has high complexity. Hull et al. [18] describe the relationships between inputs and outputs explicitly and use OWL ontologies to fix the meaning of the terms used in a service description. The description capability depends on the domain ontologies, and the service descriptions are mainly established by domain experts. The work presented in [21] [22] focus on annotating services I/O parameters with ontology concepts (semi-) automatically, while our work is to add the semantic constraints to these concepts.

The majority of the research work in the literature [9, 10, 12] focus on the matchmaking of service I/O, i.e. the data semantics of a web service. Different methods extend service I/O matching from different aspects of web service.

Paolucii et al. [12] consider the service profile and its inputs and outputs for determining the match between the request and advertisement. In [9], the ranked matching algorithm is based on service description, including service inputs, service outputs, service quality and service category. The matching of service inputs and outputs depends on the subsumption relations in domain ontology. The proposed method in this paper extends the matching of service I/O concepts by not only the taxonomy hierarchy in domain ontology but also the semantic constraints of each concept extracted from specific service context.

OWLS-MX [10] uses a hybrid SWS matching that complements logic based reasoning with approximate matching based on syntactic similarity computations. The matchmaker in this paper is similar to OWLS-MX but differs from it in several aspects. Firstly, they explore different information for matchmaking except service I/O concepts. The information of web service used in OWLS-MX is the unfolded concept expressions and the corresponding matching depends on the structures of the domain ontologies. In our method, the constraint information is extracted from the description text of web service and the matchmaker is less dependent on the domain ontologies. Secondly, they have different ways to improve the recall and precision. The reason that the hybrid variants in OWLS-MX outperforms the OWLS-M0 is that the matching of unfolded concept expressions can return those syntactically similar but logically disjoint services as the answer set. The reason that our method outperforms the OWLS-M0 is that it can remove those false web services that match the request on service I/O concepts by their constraints matching. Finally, our method mainly focuses on semantic constraint of service I/O concepts, it is expected to be easily utilized in other kinds of SWS matchmakers.

7   Conclusions and Future Work

Web service discovery is a significant challenge. Because of the low precision of current key word based discovery mechanism, most work has focused on logic based discovery of semantic web service recently. The majority of the work performs profile based service signature (I/O) matching. However, the service I/O is not sufficient to describe the function of web service clearly. This paper mainly works on enhancing the semantics of web service through introducing semantic constraints to service I/O concepts. Constraint graph is designed to describe the semantic constraints of the service I/O concepts. The semantic constraints of concepts can be extracted automatically from the parsing trees of the description text. Meanwhile, a matching algorithm for the constraint graph is proposed. The semantic similarity between the request and service is measured by the degree of matching between their corresponding constraint graphs.

Preliminary results of our comparative experiments show that building constraint graph based SWS matchmakers is more sufficient than purely service

I/O based matchmaker. Semantic constraints of service I/O concepts can improve the precision of semantic matching and easily be plugged into any other SWS matchmakers as long as they have the same logic based semantic filters.

The performance of our method depends on the performance of constraints extraction, finding more efficient extraction method to get better results of extraction is the work in the future. We are planning to extract more constraint relationships for the concepts that appear in the web service description, then web service can be represented by a more complicated graph and the matching algorithm will be more sophisticate.

References

1. Lu, X., Wang, H., Wang, J.: Internet-based virtual computing environment(ivce): Concepts and architecture. Science in China Series F: Information Sciences 49(6), 681–701 (2006)
2. Klusch, M.: Semantic services coordination. In: Schumacher, M., Helin, H., Schuldt, H. (eds.) CASCOM - Intelligent Service Coordination in the Semantic Web, ch. 4. Springer, Heidelberg (2008)
3. Martin, D., Burstein, M., Hobbs, J., Lassila, O., McDermott, D., Mcllraith, S., Narayanan, S., Paolucci, M., Parsia, B., Terry, P., Sirin, E., Srinivasan, N., Sycara, K.: Owl-s: Semantic markup for web services (2004), http://www.w3.org/Submission/OWL-S/
4. de Bruijn, J., Bussler, C., John, D., Fensel, D., Hepp, M., Kifer, M., König-Ries, B., Kopecky, J., Lara, R., Oren, E., Polleres, A., Scicluna, J., Stollberg, M.: D2v1.3. web service modeling ontology (wsmo) (2006), http://www.wsmo.org/TR/d2/v1.3/
5. Akkiraju, R., Farrell, J., Miller, J., Nagarajan, M., Schmidt, M.-T., Sheth, A., Verma, K.: Web service semantics - WSDL-S (2005), http://www.w3.org/Submission/WSDL-S/
6. Farrell, J., Lausen, H.: Semantic annotations for WSDL and XML schema (2007), http://www.w3.org/TR/sawsdl/
7. Kaufer, F., Klusch, M.: Wsmo-mx:a logic programming based hybrid service matchmaker. In: 4th European Conference on Web Service, Zurich, Switzerland, pp. 161–170. IEEE CS Press, Los Alamitos (2006)
8. Stollberg, M., Keller, U., Lausen, H., Heymans, S.: Two-phase web service discovery based on rich functional descriptions. In: Franconi, E., Kifer, M., May, W. (eds.) ESWC 2007. LNCS, vol. 4519, pp. 99–113. Springer, Heidelberg (2007)
9. Jaeger, M.C., Rojec-Goldmann, G., Liebetruth, C., Mühl, G., Geihs, K.: Ranked matching for service descriptions using owl-s. In: KiVS 2005, Informatik Aktuell, pp. 42–113 (2005)
10. Klusch, M., Fries, B., Sycara, K.: Automated semantic web service discovery with owls-mx. In: 5th International Joint Conference on Autonomous Agents and Multi-Agent Systems(AAMAS), Hakodate, Japan, pp. 915–922. ACM, New York (2006)
11. Verma, K., Sivashanmugam, K., Sheth, A., Patil, A., Oundhakar, S., Miller, J.: Meteor-s wsdi: A scalable p2p infrastructure of registries for semantic publication and discovery of web services. Information Technology and Management 6, 17–39 (2005)

12. Paolucci, M., Kawamura, T., Payne, T.R., Sycara, K.: Semantic matching of web services capabilities. In: Horrocks, I., Hendler, J. (eds.) ISWC 2002. LNCS, vol. 2342, pp. 333–347. Springer, Heidelberg (2002)
13. Constantinescu, I., Faltings, B.: Efficient matchmaking and directory services. In: IEEE/WIC International Conference on Web Intelligence, Washington, DC, USA. IEEE Computer Society, Los Alamitos (2003)
14. Srinivasan, N., Paolucci, M., Sycara, K.: Semantic web service discovery in the owl-s ide. In: 39th Hawaii International Conference on System Sciences, Washington, DC, USA, vol. 6. IEEE Computer Society, Los Alamitos (2005)
15. Klusch, M., Fries, B., Khalid, M., Sycara, K.: Owls-mx:hybrid semantic web service retrieval. In: 1st International AAAI Fall Symposium on Agents and the Semantic Web, Arlington VA, USA (2005)
16. Srinivasan, N., Paolucci, M., Sycara, K.: An efficient algorithm for owl-s based semantic search in uddi. In: Cardoso, J., Sheth, A.P. (eds.) SWSWPC 2004. LNCS, vol. 3387, pp. 96–110. Springer, Heidelberg (2005)
17. Kiefer, C., Abraham, B.: The creation and evaluation of isparql strategies for matchmaking. In: Bechhofer, S., Hauswirth, M., Hoffmann, J., Koubarakis, M. (eds.) ESWC 2008. LNCS, vol. 5021, pp. 463–477. Springer, Heidelberg (2008)
18. Hull, D., Zolin, E., Bovykin, A., Horrocks, I., Sattler, U., Stevens, R.: Deciding semantic matching of stateless services. In: 21st National Conference on Artificial Intelligence (AAAI 2006), pp. 1319–1324 (2006)
19. Jaeger, M.C., Tang, S.: Ranked matching for service descriptions using daml-s. In: CAiSE Workshops, Riga, Latvia, pp. 217–228 (2004)
20. Cunningham, H., Maynard, D., Bontcheva, K., Tablan, V.: Gate: A framework and graphical development environment for robust nlp tools and applications. In: 40th Anniversary Meeting of the Association for Computational Linguistics (ACL 2002) (2002)
21. Heß, A., Kushmerick, N.: Assam:a tool for semi-automatically annotating semantic web services. In: 3rd International Semantic Web Conference, pp. 320–334 (2004)
22. Oldham, N., Thomas, C., Sheth, A., Verma, K.: Meteor-s web service annotation framework with machine learning classification. In: Cardoso, J., Sheth, A.P. (eds.) SWSWPC 2004. LNCS, vol. 3387, pp. 137–146. Springer, Heidelberg (2005)

Enhancing Semantic Web Services with Inheritance

Simon Ferndriger[1], Abraham Bernstein[1], Jin Song Dong[2],
Yuzhang Feng[2], Yuan-Fang Li[3,*], and Jane Hunter[3]

[1] Department of Informatics
University of Zurich
Zurich, Switzerland
ferndriger@gmail.com, bernstein@ifi.uzh.ch
[2] School of Computing
National University of Singapore, Singapore
{dongjs, fengyz}@comp.nus.edu.sg
[3] School of ITEE
University of Queensland
Brisbane, Australia
{liyf, jane}@itee.uq.edu.au

Abstract. Currently proposed Semantic Web Services technologies allow the creation of ontology-based semantic annotations of Web services so that software agents are able to discover, invoke, compose and monitor these services with a high degree of automation. The OWL Services (OWL-S) ontology is an upper ontology in OWL language, providing essential vocabularies to semantically describe Web services. Currently OWL-S services can only be developed independently; if one service is unavailable then finding a suitable alternative would require an expensive and difficult global search/match. It is desirable to have a new OWL-S construct that can systematically support substitution tracing as well as incremental development and reuse of services. Introducing inheritance relationship (IR) into OWL-S is a natural solution. However, OWL-S, as well as most of the other currently discussed formalisms for Semantic Web Services such as WSMO or SAWSDL, has yet to define a concrete and self-contained mechanism of establishing inheritance relationships among services, which we believe is very important for the automated annotation and discovery of Web services as well as human organization of services into a taxonomy-like structure. In this paper, we extend OWL-S with the ability to define and maintain inheritance relationships between services. Through the definition of an additional "inheritance profile", inheritance relationships can be stated and reasoned about. Two types of IRs are allowed to grant service developers the choice to respect the "contract" between services or not. The proposed inheritance framework has also been implemented and the prototype will be briefly evaluated as well.

1   Introduction

Current Web Services technology such as WSDL, UDDI, and SOAP provide the means to describe the "syntax" of the code running in a distributed fashion over the Internet.

---

* Corresponding author.

A. Sheth et al. (Eds.): ISWC 2008, LNCS 5318, pp. 162–177, 2008.

They lack, however, the capabilities to describe the semantics of these code fragments, which is one of the major prerequisites for service recognition, service configuration and composition (i.e., realizing complex workflows and business logics with Web services), service comparison as well as automated negotiation.

To that end a number of languages such as OWL-S[1], SAWSDL[2], WSMO[3], and SWSF[4] have been proposed. Each of these languages allows connecting Web services with an ontology-based semantic description of what the service actually does. The OWL Services (OWL-S) ontology is an OWL ontology defining a set of essentia vocabularies to describe the "semantics" of Web services, defining its capabilities, requirements, internal structure and details about the interactions with the service. Other efforts provide similar vocabularies with different focus and coverage.

Based on the de-facto ontology language, OWL DL, OWL-S seems to be a promising candidate as an open standard. Currently OWL-S services can only be developed independently. Moreover, if one service is unavailable then finding a suitable alternative would require an expensive and difficult global search/match. It is desirable to have a new OWL-S construct that can support the systematic substitution tracing as well as the incremental development and reuse of services. Hence, in order for OWL-S to enjoy wider adoption, a more systematic, automated and effective mechanism of annotation and discovery of services is required.

The `owl:imports` construct of OWL can be seen as a rudimentary form of establishing links between OWL-S services to support easy service annotation. However, it does not provide the necessary flexibility since once a particular construct from a service ontology, say, a composite process in a service model, is imported, it can only be augmented by adding more triples describing it. Basically, the importing service cannot *revoke* any RDF statement already made in the imported ontology. Hence, only reusing constructs at very detailed level is possible for the importing approach, which we deem is neither desirable nor practical. An approach more flexible and powerful is needed.

Inspired by the object-oriented programming paradigm, we propose to extend OWL-S with service inheritance, which we believe improves the level of automation and effectiveness for carrying out the above tasks. So far, however, only the SWSF framework briefly discusses establishing connections between different Web services in order to reuse similar underlying elements and add additional relationship information. Furthermore, none of these standards defines a concrete and self-contained way of sharing specific elements among Web Services or a way of interpreting the relationship among these services.

1.1 Motivation

We believe adding inheritance relationships between services can help to automate and ease a number of tasks. In this subsection, we present some scenarios in which inheritance of services facilitates the completion of tasks.

---

[1] http://www.daml.org/services/owl-s/
[2] http://www.w3.org/TR/sawsdl/
[3] http://www.wsmo.org/wsml/wrl/wrl.html
[4] http://www.daml.org/services/swsf/

Semantic Service Annotation. The number of Semantic Web services (SWS) needs to reach a critical mass in order for SWS to gain wider acceptance and adoption. Hence, the creation of semantic annotation of Web services is an important first task. Currently, with tool support, annotation of services are still mostly created from scratch. Inheritance mechanism can greatly speed up the annotation of Web services by selectively reusing components from existing services.

Service Discovery. Automated Web service discovery is stated as a motivating task for OWL-S. Service discovery, however, depends heavily on (potentially large) service registries because there is yet no other way to discover those services otherwise.

An alternative to discovering relevant services without the need of a registry is to make use of inheritance relationships between services in order to find service substitutes more efficiently. Analogous to object-oriented concepts, when certain constraints are satisfied, a sub service may be used to substitute its super service for automated, dynamic service discovery and composition.

It may seem that existing language constructs such as `rdfs:subClassOf` can handle inheritance, by subclassing existing service annotations. However, as RDF Schema and OWL are based on monotonic logic, subclassing only represents a restricted form of inheritance.

Inspired by the MIT Process Handbook [1,2] we believe that service ontologies are central to the organization of business knowledge. As shown by Malone and colleagues, process repositories that build on the inheritance of process properties can be effectively used to (1) invent new business processes, (2) systematically explore the design space of possible service alternatives through recombination [3], (3) design robust services through the advanced usage of exceptions, (4) support knowledge management about services by improving their management process, ability to handle conflicts, support for communicative genres, (5) as well as improve software design and generation by increasing the coordination alternatives between pieces of code and achieve the flexible execution of workflows [4].

Based on the above motivating tasks, we propose to extend OWL-S with Inheritance Relationships (IRs) between services for more automated annotation and discovery. We draw inspirations from the Semantic Web Services Framework (SWSF) and expand the brief discussion in SWSF on inheriting and overriding processes among services.

In this paper, we present an inheritance framework for OWL-S ontology. Two versions of Inheritance Relationships are supported: normal and strict inheritance for OWL-S in the form of additional, independent service profiles. The *normal inheritance* does not impose additional restrictions on the inheritance relationship in order to allow for more flexible reuse of existing service components. As normal inheritance inevitably allows the alteration of existing services, a form of default inheritance advocated by SWSL [5] is employed. The *strict inheritance*, by imposing certain restrictions on IOPEs of the inherited process, dictate that the "contracts" of processes of a super service must be maintained by the inheriting service. This guarantees a proper refinement relationship between the super service and the sub service. Hence, a strictly inheriting sub service can substitute its super service whenever the super service is invoked, whereas this substitutability is not guaranteed with normal inheritance. Moreover, it enables a sub service to be more easily discovered.

The rest of the paper is organized as follows. Section 2 briefly presents background knowledge about OWL-S and SWSF. Section 3 discusses the two forms of inheritance relationships in detail. In Section 4, we extend the well-known CongoBuy example from OWL-S specification to illustrate the benefits of IRs. Finally, Section 5 concludes the paper and discusses future work directions.

2   Background Knowledge

In this section, we introduce the background knowledge necessary for the discussion of the following sections.

2.1   OWL-S

The OWL-S ontology has been developed to enrich Web Services with semantics. The semantic markup of OWL-S enables the automated discovery, invocation, composition, interoperation and monitoring of Web services. This automation is achieved by providing a standard ontology (OWL-S) for declaring and describing Web Services.

Being an OWL ontology, OWL-S defines a set of essential vocabularies to describe the three components of a service: profile, model and grounding. A service can have several profiles and one service model. The service model, in turn, may have one or more service groundings. In summary, a service profile describes what the service does; the service model describes how the service works and the grounding provides a concrete specification of how the service can be accessed.

The `ServiceProfile` class provides a bridge between service requesters and service providers. The instances are mainly meant to advertise an existing service by describing it in a general way that can be understood both by humans and computer agents. It is also possible to use a service profile to advertise a needed service request.

OWL-S provides a subclass of `ServiceProfile`, `Profile`. This default class should include provider information, a functional description and host properties of the described service. It is possible to define other profile classes that specify the service characteristics more precisely.

The `ServiceModel` class uses the subclass `Process` to provide a process view on the service. This view can be thought of as a specification of the ways a client may interact with a service. The service model defines the inputs, outputs, preconditions and effects (IOPEs) and the control flow of composite processes.

One useful language construct in the service model is the definition of `Expression`, which is used to express preconditions and effects in the logic language of choice by the service developer. Basically, an expression is characterized by an expression language, such as SWRL [6], KIF [7], etc., and an expression body, containing the logic expression in that language.

The `ServiceGrounding` class provides a concrete specification of how the service can be accessed. Of main interest here are subjects like protocol, message formats, serialization, transport and addressing. The grounding can be thought of the concrete part of the Semantic Web service description, compared to the service profile and service model which both describe the service on an abstract level.

2.2   SWSF

The Semantic Web Services Framework (SWSF) [8] includes two major components, The Semantic Web Services Language (SWSL) [5] and the Semantic Web Services Ontology (SWSO) [9]. SWSL is a generic language, used in the SWSF framework to formally specify Web service concepts and descriptions. It includes two sublanguages: SWSL-FOL (based on first-order logic) and SWSL-Rules (based on logic programming). SWSO serves essentially the same purposes as OWL-S: providing semantic specifications of Web services; namely (similarly to OWL-S) a comparable service profile, model and grounding.

However, SWSO also has a number of significant differences from OWL-S.

– Higher expressivity: the SWSF service ontology (called FLOWS) is expressed in first-order logic. OWL-S, in contrast, is expressed in OWL-DL, a variant of description logic language $\mathcal{SHOIN}(D)$ [10].
– Enhanced process model: SWSF claims to provide an enhanced process model as compared to OWL-S. It is based on the Process Specification Language [11], hence, it provides Web Services specific process concepts that include not only inputs and outputs, but also messages and channels.
– Non-monotonic language: In addition to the service language, SWSO makes use of SWSL-Rules, a non-monotonic language based on the logic-programming paradigm which is meant to support the use of the service ontology in reasoning and execution environments.
– Interoperability: an important final distinction between OWL-S and FLOWS is with respect to the role they play. Whereas both endeavor to provide an ontology for Web services, FLOWS had the additional objective of acting as a focal point for interoperability, enabling other business process modeling languages to be expressed or related to FLOWS.

Molecules are a language construct in the Frames layer of the SWSL-Rules language. We will use molecules to present some of the inheritance-related concepts in later sections. Here, we give a brief overview of molecules. A molecules can be viewed as an atomic term: a constant, a variable or an function application. Of the seven forms of molecules, we present the two forms that will be used in this work: value molecules and boolean molecules. In this paper, the molecules will be presented in `teletext` font.

Value molecules are of the form `t[m -> v]` where `t`, `m` and `v` are all terms where `t` denotes an object, `m` denotes a function invocation in the scope of `t` and `v` denotes a value returned by the invocation. The molecule `t[m *-> v]` denotes that this method is inheritable.

Boolean molecules are of the form `t[m]` where `t` and `m` are both terms. A Boolean molecules can be interpreted as `t[m -> true]`, meaning that the property `m` of object `t` is true.

Complex molecules can be formed from other molecules by grouping and nesting. For example, the molecules `t[m -> v] and t[p]`, which describe the same object `t`, can be grouped together to form the complex molecule `t[m -> v and p]`. Similarly, `t[m -> v] or t[p]` can be grouped to form `t[m -> v or p]`.

2.3   Related Works

The concept of inheritance is not new. It has been an active research area in programming languages and software engineering for over decades. In particular, the works on behavioral subtyping [12,13] of object-oriented languages and object-oriented specification languages are particularly related.

OWL-S defines an process hierarchy ontology[5] that describe a profile-based approach of creating service hierarchies. However, this approach, as the authors put it, "provides a useful means of constructing a 'yellow pages' style of service categorization". It does not support the extension/modification of services at the level of granularity presented in this paper.

3   IR Framework for OWL-S

Although inheritance has been widely used in computer science as a tool to encapsulate and manage program complexity and to improve code reuse and reliability, it has not been widely applied to the Web Services domain.

In this section, we present in detail our proposed inheritance relationships (IRs) framework for OWL-S services. The language constructs used to extend/modify inherited entities and conditions that must be satisfied by these constructs are presented.

We start this section with a discussion on the distinction between different types of inheritance relationships: normal vs strict inheritance and single vs multiple inheritance.

3.1   The Perspectives of Inheritance

Normal IR, closely related to default inheritance [14,15], allows for flexible alteration of inherited service components. It primarily facilitates the easy annotation of Web services.

In complete inheritance, information that is used by more than one element has to be stored in a more general element. This means that no redundant information is allowed and information has to be inherited down the inheritance chain: the generalization must be complete. Therefore, inherited information can neither be altered nor arbitrarily extended.

On the contrary, default inheritance is defined such that elements get inherited by default which can be modified and extended afterwards. Hence, new features/functionalities are allowed to be added in default inheritance. In the Web environment, it is often the case that an inheriting service intends to extend the functionality of the inherited service. Compared to the OWL `imports` approach, default inheritance allows for the possibility of freely modifying an inherited entity. Furthermore, default inheritance has been shown to be easier to understand by non-specialists [16] making it more suitable for the wide variety of users on the Semantic Web. For those reasons, default inheritance is adopted for this approach.

Strict IR aims at enabling more automated and accurate service discovery by following the inheritance chain between services. Seen as a form of normal IR, strict IR imposes

---

[5] http://www.daml.org/services/owl-s/1.1/ProfileHierarchy.owl

certain restrictions such that a strictly inheriting service can automatically be used as a faithful substitute for the inherited service.

The faithful substitute is achieved by following the principle of operational refinement [17] on the IOPEs of the inherited processes. Briefly, let the IOPEs of an OWL-S service process SP_{sb} and its ancestor process SP_{sp} be I, O, P, E and I', O', P', E', respectively. In order to establish a strict IR between S_{sb} and S_{sp}, the following conditions must hold.

$$P' \Rightarrow P \qquad\qquad E \Rightarrow E'$$

e.g., the preconditions P of the inheriting service SP_{sb} must be weaker than that of the inherited service, and vice versa for the effects. These follow from the well-established data refinement principles and covariance.

Besides preconditions and effects, the inputs and output must also satisfy similar constraints: (a) the number and all names of input/output parameters must match (up to permutation) and (b) for each matching input parameter in I and I', the type of parameter of the inheriting service must be a subtype of that of the inheriting service and (c) vice versa for the output parameters.

Single inheritance, from an orthogonal point of view, only allows an element to inherit from a single more general class.

Multiple inheritance has the advantage over single inheritance of providing the ability to inherit functionalities from several super services. It does, however, add additional complexity which might lead to inconsistencies such as catcalls [18].

Like normal inheritance, multiple inheritance enable service developers to reuse multiple services conveniently. Therefore, it is also incorporated. Hence, in our framework, default inheritance and multiple inheritance are allowed.

3.2   IR Syntax Extension

A number of OWL classes and properties are used to construc IRs and connect services to them. IRs are modeled in additional, independent service inheritance profiles, which are modeled as a subclass of the OWL-S class `ServiceProfile`. The inheritance can be modeled in two directions, meaning that a service can point to its super services, as well as its sub services. The `InheritanceProfile` is connected to the specific super/sub service through the *abstract* class `Relationship`, which is defined to be the disjoint union of classes `SuperService` and `SubService`, and the property `contains`. Definition 1 below defines inheritance profile and how it is linked to OWL-S services. The OWL code fragment is presented in the familiar "DL syntax" and in *math* font.

$$
\begin{array}{ll}
InheritanceProfile \sqsubseteq ServiceProfile & \geq 1\, contains \sqsubseteq InheritanceProfile \qquad (1) \\
Relationship = & \top \sqsubseteq \forall\, contains.Relationship \\
\qquad SuperService \sqcup SubService & SuperService \sqcap SubService = \bot
\end{array}
$$

Processes in service models are inheritable. Note that since the grounding of a service specifies the concrete physical Web service, groundings are not inheritable as an inherited service is a new service.

Note that in an inheritance profile, a service can refer to either its super service or its sub service. A service can have multiple super or sub services. A particularly interesting scenario arises when an inheritance relationship is stated in both the super service and sub service. In this case, the inheriting service may be interpreted as "endorsed" by the super service. Hence, if the super service is trusted, the sub service can also be trusted.

We distinguish two types of IRs in OWL-S: normal and strict (details are given in the subsection below). As introduced previously, the two types of inheritance tackles different problems an service developers have the freedom to choose the appropriate form. An OWL object property fromType and an enumerated class Type (with two instances normal and strict) are used to state whether a particular IR is normal or strict. An IR relationship has exactly one inheritance type[6].

The modification of default inheritance is modeled by an OWL property specifiedBy, with SuperService as its domain and the "abstract" class Specification as its range. The class Specification is a class with three disjoint subclasses: Customization, Extension and Manipulation. These three different types of modification are used to modify an inherited service and will be presented in detail later in this section.

As stated in Section 1, default inheritance alteration can be specified by SWSL-Rules language. In modeling IR, we define an OWL class SWSL-Expression and a property externallySpecifiedBy to link an normal IR to the SWSL-Rules expression that modifies it. An instance SWSL of the class LogicalLanguage (defined as part of the OWL-S framework) is also defined to represent the logic language SWSL.

Therefore, a super service can be further modified by at most one Specification or at most one SWSL expression, as given in Definition 2 below.

$$SuperService \sqsubseteq (\leq 1\ specifiedBy \sqcup \leq 1\ externallySpecifiedBy) \tag{2}$$

Finally, in order to link the current service to its super/sub service. An OWL object property hasSource is defined, with Relationship as its domain and Service as its range.

Modification of Inheritance In this subsection, we describe the language extensions used to modify the inheritance relationships between OWL-S services.

In Table 1 below, we briefly introduce the main differences among of the three types of modification of inherited services, which will be presented in more detail in the following subsections.

The language constructs in inheritance modification can be divided into three scenarios: customization, extension and manipulation, based on their intended usage. For better readability, the scenarios are presented in SWSL molecule syntax [5]. The same modeling can also be represented in OWL, which is more verbose.

Service customization allows one to choose whether to inherit service model; rename and replace an inherited entities (processes and parameters).

---

[6] For brevity reasons, the formal OWL definitions are not shown when not necessary. Full details of the modeling can be found online at http://www.fo-ss.ch/simon/ DiplomaThesis/InheritanceProfileInheritanceProfile./owl.

Table 1. Main differences between service customization, extension & manipulation

Type	IR type	Modification
Customization	strict	process & IDs
Extension	normal	process
Manipulation	normal & strict	IOPEs

$$\text{Inherit} [\text{AdoptServiceModel} (\text{PID}_{\text{INHERITED}})] \tag{3}$$
$$\text{Renaming} [\text{ID}_{\text{INHERITED}} \; *-> \; \text{ID}_{\text{REPLACEMENT}}]$$
$$\text{ProcessReplacement} [\text{PID}_{\text{INHERITED}} \; *-> \; \text{PID}_{\text{REPLACEMENT}}]$$

where $\text{PID}_{\text{INHERITED}}$ stands for the inherited process ID, which is to be replaced. $\text{PID}_{\text{REPLACEMENT}}$ represents the replacing process. The SWSL method AdoptService Model enables one to use one of the inherited processes from the super service as the new service model.

In the Inherit molecule, the term Processes must always be included. Therefore, by default, a copy of the ontology which contains the service model of the super service gets integrated into the service ontology of the sub service. Note that since an OWL-S service can only have one service model, in case of multiple inheritance, only the service model of one of the super services can be inherited.

For the molecules Renaming and ProcessReplacement it must be ensured that the new IDs do not conflict with existing ones. Moreover, in ProcessReplacement, the IOPEs of the replacement process must match those of the inherited process. In other words, the refinement relationship must be maintained.

Service extension allows one to extend an inherited process model by inserting into, detaching from or deleting an inherited process or the perform of the process.

$$\text{ProcessInsertion} [\{\text{after/before}\} \; \text{PPID}_{\text{INHERITED}} \; *-> \; \text{CCID}_{\text{NEW}}] \tag{4}$$
$$\text{ProcessDeletion} [\text{PPID}_{\text{INHERITED}}]$$

where the expression {after/before} means that the new process can be inserted either before or after the process perform. This allows on to directly use a process perform which is connected with a process, or one can wrap the process perform into another control construct (e.g. if-then-else or sequence, etc). The Boolean molecule ProcessDeletion models the fact that a process can be deleted from the inherited model.

Service manipulation allows one to modify the preconditions and effects of an inherited process. Moreover, in normal inheritance mode, service modification also allows one to add/remove the input/output of the inherited process.

$$\text{ExpressionReplacement} [\{ \tag{5}$$
$$\qquad \text{replaceCondition} (\text{CID}_1) \; *-> \; \text{CID}_2 \; \text{or}$$
$$\qquad \text{replaceResult} (\text{CID}_1) \; *-> \; \text{CID}_2$$
$$\}]$$
$$\text{AddInputsAndOutput} [\text{addIO} (\text{PID, OID}) \; ->* \; \text{NID}]$$
$$\text{DeleteInputsAndOutput} [\text{deleteIO} (\text{PID, OID}) \; ->* \; \text{NID}]$$

where CID_1 represents the ID of the replaced precondition/result and CID_2 represents the ID of the new one. NID and OID represent the RDF ID of the new and old input (or output) parameter, respectively. `PID` represents the ID of the process where the parameter is added/deleted.

When used in strict inheritance, the replacing precondition/effect must still satisfy the refinement relationship between the inherited and the current process, e.g., the replacing precondition must be weaker than the inherited precondition and vice versa for effects.

3.3   Satisfaction Conditions of IR

In this subsection, we present some conditions that the modifications of default inheritance presented above must satisfy. These conditions guarantee that, for example, the modifications allow proper process flow in case of strict inheritance. The following conditions are presented in first-order logic syntax with some notations/elements taken from the OWL abstract syntax and semantics [19].

Generally, these conditions need to be checked by software agents making use of the IRs. How the conditions are checked may be application-specific.

Service customization. Since an OWL-S service can only have one service model. in case of multiple-inheritance (i.e., multiple IRs), the service model can only be adopted once by a sub service, respectively it can only be inherited from one of all of its super services. This condition is formally captured in the following first-order predicate. Note that O, EC, ER and LV in the conditions below are entities of the abstract interpretation defined in OWL semantics [19].

$$\forall IHP, SS_1, SS_2, SP_1, SP_2, SM_1, SM_2 : O \bullet \tag{6}$$

$$SS_1 \in EC(SuperService) \land SS_2 \in EC(SuperService) \land$$
$$SP_1 \in EC(Inherit) \land SP_2 \in EC(Inherit) \land SM_1 \in EC(ServiceModel) \land$$
$$IHP \in EC(InheritanceProfile) \land SM_2 \in EC(ServiceModel) \land$$
$$\langle IHP, SS_1 \rangle \in ER(contains) \land \langle SS_1, SP_1 \rangle \in ER(specifiedBy) \land$$
$$\langle IHP, SS_2 \rangle \in ER(contains) \land \langle SS_2, SP_2 \rangle \in ER(specifiedBy) \land$$
$$\langle SP_1, SM_1 \rangle \in ER(adoptServiceModel) \land \langle SP_2, SM_2 \rangle \in ER(adoptServiceModel)$$
$$\Rightarrow$$
$$SS_1 = SS_2 \land SM_1 = SM_2$$

Formally, Definition 6 above specifies that for an arbitrary inheritance profile *IHP* (for an OWL-S service), if it *contains* two super services SS_1 and SS_2, and adopts the service model of each of these two services, then the two services are actually one service ($SS_1 = SS_2$) and the two service models are one model ($SM_1 = SM_2$) as well.

Similarly for renaming of IDs, it must be ensured the the original ID must be present in the inherited service and the new ID must not conflict with existing ones.

$$\forall SS, SP : O; \ XID : LV \bullet \exists X : O; \ OID : LV \bullet \tag{7}$$

$$SS \in EC(SuperService) \land SP \in EC(Renaming) \land \langle SS, SP \rangle \in ER(specifiedBy) \land$$
$$\langle SP, XID \rangle \in ER(oldID) \land (X, OID) \in ER(ID) \land X \in SS$$
$$\Rightarrow$$
$$XID = OID$$

The slight "abuse" of syntax in predicate $X \in SS$ above means that X is bound in SS. The above condition guarantees that the replaced ID is always present in the inherited service. The condition for new ID can be similarly defined.

The conditions for a process replacement are more complicated. First of all, the input and output IDs of the process replacement must match (Definitions 8 below specifies that for the inputs) and their types must be compatible. When input types of the replacement process are OWL classes, the input types must either be from the same OWL class or from an OWL sub class of the original ones, similarly for data types. It is specified in the second condition below.

Note that the formula of syntax $\{decl \mid pred \bullet proj\}$ is a set comprehension expression, meaning that for variables declared in *decl* part, the set contains elements specified in *proj* that satisfy the conditions specified in *pred*.

$$
\begin{aligned}
&\forall PR, OP, RP: O; \; OIDs, RIDs: \mathbb{P}\,LV \bullet \qquad\qquad \forall PR, OP, RP, RI: O; \hspace{2cm} (8) \\
&\quad PR \in EC(ProcessReplacement) \wedge \quad RT: V_r; \; RID: LV \bullet \\
&\quad \langle PR, OP \rangle \in ER(replaceProcess) \wedge \hspace{1.2cm} \exists\,OI: O; \; OT: V_o \bullet \\
&\quad OIDs = \{OID: LV, OI: O \mid \hspace{2cm} PR \in EC(ProcessReplacement) \wedge \\
&\qquad \langle OP, OI \rangle \in ER(hasInput) \wedge \hspace{1.4cm} \langle PR, OP \rangle \in ER(replaceProcess) \wedge \\
&\qquad \langle OI, OID \rangle \in ER(ID) \bullet OID\} \wedge \hspace{0.9cm} \langle OP, OI \rangle \in ER(hasInput) \wedge \\
&\quad \langle PR, RP \rangle \in ER(withProcess) \wedge \hspace{1.5cm} \langle PR, RP \rangle \in ER(withProcess) \wedge \\
&\quad RIDs = \{RID: LV, RI: O \mid \hspace{2.2cm} \langle RP, RI \rangle \in ER(hasInput) \wedge \\
&\qquad \langle RP, RI \rangle \in ER(hasInput) \wedge \hspace{1.2cm} \langle OI, OID \rangle \in ER(ID) \wedge \langle RI, OID \rangle \in ER(ID) \wedge \\
&\qquad \langle RI, RID \rangle \in ER(ID) \bullet RID\} \wedge \hspace{1cm} OI \in EC(V_o) \wedge RI \in EC(V_r) \\
&\qquad \Rightarrow \hspace{5.5cm} \Rightarrow \\
&\quad OIDs = RIDs \hspace{4cm} EC(V_r) \subseteq EC(V_o)
\end{aligned}
$$

Secondly, the preconditions and effects of the two processes must comply with the refinement concept, e.g., the preconditions of the modified process must be weaker than those of the original process. In case of multiple preconditions (each for a different scenario), their conjunction is taken into consideration.

$$
\begin{aligned}
&\forall PR, OP, RP: O; \; OPCs, RPCs: \mathbb{B} \bullet \hspace{4cm} (9) \\
&\quad PR \in EC(ProcessReplacement) \wedge \langle PR, OP \rangle \in ER(replaceProcess) \wedge \\
&\quad OPCs = \bigcap (OPC: \mathbb{B} \bullet \langle OP, OPC \rangle \in ER(hasPrecondition)) \wedge \\
&\quad \langle PR, RP \rangle \in ER(withProcess) \wedge \\
&\quad RPCs = \bigcap (RPC: \mathbb{B} \bullet \langle RP, RPC \rangle \in ER(hasPrecondition)) \\
&\qquad \Rightarrow \\
&\quad RPCs \Rightarrow OPCs
\end{aligned}
$$

Note that the symbols $\bigcap$ and $\bigcup$ represent distributed set intersection/union, respectively. For brevity reasons, those conditions regarding outputs and effects are omitted but they can be similarly defined.

Service manipulation. When service manipulation is used in strict mode, only the expression replacement statement can be made as the rest two (adding and removing inputs/outputs) would violate that relationship. For the first statement, it needs to be ensured that the altered precondition is logically weaker and the effect is stronger. The following two conditions model the case for preconditions and effects, respectively.

$$\forall EP, RP, CP \colon O; \ OPC, RPC \colon \mathbb{B} \bullet \qquad\qquad \forall EP, RP, CP \colon O; \ OPR, RPR \colon \mathbb{B} \bullet \qquad (10)$$

$$EP \in EC(ExpressionReplacement) \wedge \qquad EP \in EC(ExpressionReplacement) \wedge$$

$$\langle EP, OPC \rangle \in ER(replaceCondition) \wedge \qquad \langle EP, OPR \rangle \in ER(replaceResult) \wedge$$

$$\langle EP, RPC \rangle \in ER(withCondition) \qquad\qquad \langle EP, RPR \rangle \in ER(withResult)$$

$$\Rightarrow \qquad\qquad\qquad\qquad\qquad\qquad \Rightarrow$$

$$RPC \Rightarrow OPC \qquad\qquad\qquad\qquad\qquad OPR \Rightarrow RPR$$

3.4 Some Discussion on IR

During service discovery, the relationship may need to be interpreted and validated to ensure that it is valid. When there is a chain of inheritance, the validation/interpretation should be performed top down in order for inheritance information to propagate properly.

One interesting scenario may arise when a super service is modified, after an IR is established between it and a sub service. In this situation, it might be necessary in certain situations to revalidate the inheritance relationship.

Another reason to revalidate an IR is to benefit from possible side effects of such a revalidation. In case a super service changes not its service groundings, but its process composition, such a change would not affect the corresponding sub service as long as the IR stays valid.

For example, the super service could have a "Search Flight" process, which changes from being atomic to being composite in order to make it more efficient. Since this change happens in the service model which gets copied to the sub service during the validation of the IR, the service model of the sub service needs first to be updated by revalidating the IR in order to adopt this change.

Theoretically, service substitutes can not only be found via strict IR, but also via ordinary on-the-fly reasoning in a service registry using the refinement relationship as it is used for defining strict IRs. Without the IR relationship, however, this reasoning is likely to be very expensive in time, since every service has to be considered as a candidate substitute. Hence, IR helps to reduce service discovery time by providing guided exploration of service space and hence eliminating most necessary comparison.

4 The Benefits of IR – A Case Study

We have developed a prototype service repository[7] that implements the inheritance relationships framework presented in this paper. Through a Web interface, the prototype has the following four main functionalities: service annotation creation, service visualization, service discovery and inheritance validation.

In this section, we present one example in the service repository, the Congo book store web service from the OWL-S specification extended with inheritance. Congo is a fictitious online bookstore that uses OWL-S ontologies to semantically markup their services. We use both normal and strict IR in service annotation creation scenarios, demonstrating their benefits and differences.

---

[7] The prototype is accessible at http://www.fo-ss.ch/simon/DiplomaThesis/ IR_prototype/.

4.1   Normal Inheritance

FullCongoBuy is an example service published with the OWL-S specification. It provides a complete book buying service for physical books. In this example, we extend it with the capability of buying digital books also.

Given the existing FullCongoBuy service, it would be convenient to benefit from the work already done when creating the other book selling service E-BookBuy instead of starting from scratch. Without inheritance, the modifications needed to create E-BookBuy from FullCongoBuy is not possible.

The proposed IR, however, makes it possible to reuse the service model of FullCongoBuy within the context of service customization and extension. The IR allows not only the reuse but also the necessary altering, i.e. deletion and replacement of inherited properties. More concretely, the new service annotation E-BookBuy can be created by inheriting the service model from FullCongoBuy (11), replacing the process LocateBook with a new one (12), deleting the process SpecifyDeliveryDetails-Perform (13) while adding SpecifyDownloadDetailsPerform as an alternative, adding a new process ProvideDownloadOptions (14), making a new service profile and creating the grounding for the new processes. For better readability, the main process can be renamed (15).

```
Inherit[AdoptServiceModel(FullCongoBuy), Processes].          (11)
ProcessesReplacement[LocateBook *-> Locate_eBook].            (12)
ProcessDeletion[SpecifyDeliveryDetailsPerform].              (13)
ProcessInsertion[                                            (14)
    after(BuySequence) *-> SpecifyDownloadDetailsPerform,
    after(SpecifyDownloadDetails) *->
        ProvideDownloadOptionsPerform ].
Renaming[ FullCongoBuy *-> Full_eBookBuy ].                  (15)
```

4.2   Strict Inheritance

ExpressCongoBuy is an example service published with the OWL-S specification. It provides a one-step book buying service for the Congo shop with a standard delivery setting. In real life, however, there might be different delivery settings. Since a concrete delivery is not yet defined in the example, this use case defines a one-day delivery for ExpressCongoBuy and creates a new service annotation EconomyCongoBuy with a slower three-day delivery.

Given the existing ExpressCongoBuy service, it would be convenient to benefit from the work already done when creating the other book selling service EconomyCongoBuy instead of starting from scratch. Without inheritance, there is no way to benefit from existing atomic services in creating a new one other than using cut and paste.

The proposed IR, however, makes it possible to reuse the service model of ExpressCongoBuy within the context of service customization and manipulation. More concretely, the new service annotation EconomyCongoBuy can be created by inheriting the service model from ExpressCongoBuy, replacing the positive result and adding a new service profile and grounding. The necessary statements for this strict IR are described in SWSL below.

```
Inherit[AdoptServiceModel(ExpressCongoBuy), Processes ].
Renaming[ExpressCongoBuy *-> EconomyCongoBuy ].
ExpressionReplacement[
    Effect(ExpressCongoBuy, ExpressCongoOrderShippedEffect) *->
    Effect(EconomyCongoBuy, EconomyCongoOrderShippedEffect)
].
```

The benefits of using IRs in service creation can be summarized as follows.

- **Efficient service annotation creation**: First, the reuse of information facilitates
  the creation of the service annotations since the service model and groundings of
  existing services can be largely reused. Therefore, using IR can improve the efficiency of the creation of similar annotations. When normal IR is used, the process
  flow has to be taken care of by the service creator, and therefore, the creator has to
  be familiar with the original service model.
- **Additional relationship information**. The explicit statement of the strict IR provides additional information about the two services, which may be used later on to
  facilitate more smooth service discovery and substitute.

5   Conclusion

Semantic Web Services languages aim at providing semantic markups for Web Services
description in order to facilitate automated service discovery, composition, monitoring
and composition.

The OWL-S ontology provides a set of semantic service descriptions that describes a
service's functionality, internal structure and interfacing information for software agents
to automate the above task. However, the current OWL-S specification does not specify
a systematic way of creating and discovering services, limiting its wider adoption.

In this paper, we attempt to tackle the above problem by proposing a (default) inheritance relationship (IR) between OWL-S services. The IR is modeled in inheritance
profiles and a set of language constructs are provided to link a service to its super/sub
services. In addition, two modes of IRs, normal and strict, and their respective applications are presented and compared.

For service annotation creation, this approach provides service customization, extension and manipulation for sharing and modifying specific elements inherited from super
services. This ability is expected to substantially reduce the amount of work necessary
for creating and maintaining services. For service discovery, the approach provides a
solution to find service substitutes for the developed strict IRs, based on the concept of
refinement. These substitutes increase the choice of a service user or the availability of
services as a whole.

Additionally, the proposed IRs allow a service to point to both super and sub services.
The benefits are twofold. Firstly, if a particular IR is specified in both the super and sub
services, the inheritance relationship between the two can be seen as "endorsed" by the
services. Therefore, a stronger sense of trust can be established. Secondly, when used
extensively, the IRs may connect a potentially large amount of services and thereby
build a strong service graph without the need of a central registry. This facilitates the
distributed development and discovery of services.

The well-known frame problem [20] identified by Borgida, Mylopoulos and Reiter applies to most pre/post-condition style object-oriented and procedure specification languages. The problem is concerned about unwanted effects of a precedure/operation resulted from under specification of pre/post conditions. Essentially, the problem is caused by the inability of a specification language to express that an operation changes "only" those things that it intends to, and nothing else. This problem is particularly serious for specification languages with inheritance, where a sub class is usually constructed by conjoining specifications of super classes with additional predicates, where conjoining predicates may result in inconsistent pre/post-conditions.

As our work presented in this paper adds inheritance to OWL-S services and processes, it inherently has the above problem. It is an important future research task to investigate the impact and possible solutions of frame problems on both normal and strict IRs.

In Section 3 we discussed the conditions that the various types of modifications of default inheritance must satisfy. The validation of these conditions may be a computationally intensive process. In practical environments, a service may be modified after it is inherited by some other process. In this case, the conditions may need to be re-validated to ensure that they still hold. One future work direction would be to investigate under which circumstances these conditions need to be re-validated.

Another future work direction is to further develop the prototype into a more robust and usable form. We are currently investigating the possibility of developing a Protégé [21] plugin based on the current prototype. This plugin could directly communicate with the existing OWL-S plugin and would, therefore, be better accessible for future Web services developers.

Given that the OWL-S service ontology serves similar purposes as other languages such as SWSO, WSMO and WSDL-S, the investigation of the transfer of the proposed inheritance relationships is an interesting and important direction to pursue.

References

1. Malone, T.W., Crowston, K., Lee, J., Pentland, B., Dellarocas, C., Wyner, G., Quimby, J., Osborn, C.S., Bernstein, A., Herman, G., Klein, M., O'Donnell, E.: Tools for Inventing Organizations: Toward a Handbook of Organizational Processes. Manage. Sci. 45(3), 425–443 (1999)
2. Malone, T.W., Crowston, K., Herman, G.A.: Organizing Business Knowledge: The MIT Process Handbook. MIT Press, Cambridge (2003)
3. Bernstein, A., Klein, M., Malone, T.W.: The process recombinator: a tool for generating new business process ideas. In: ICIS 1999. Proceeding of the 20th international conference on Information Systems, Atlanta, GA, USA, pp. 178–192. Association for Information Systems (1999)
4. Bernstein, A.: How can cooperative work tools support dynamic group process? bridging the specificity frontier. In: CSCW 2000. Proceedings of the 2000 ACM conference on Computer supported cooperative work, pp. 279–288. ACM, New York (2000)
5. Battle, S., Bernstein, A., Boley, H., Grosof, B., Gruninger, M., Hull, R., Kifer, M., Martin, D., McIlraith, S., McGuinness, D., Su, J., Tabet, S.: Semantic Web Services Language (SWSL) (2005), http://www.daml.org/services/swsf/1.0/swsl/
6. Horrocks, I., Patel-Schneider, P.F., Boley, H., Tabet, S., Grosof, B., Dean, M.: SWRL: A Semantic Web Rule Language Combining OWL and RuleML (May 2004), http://www.w3.org/Submission/2004/SUBM-SWRL-20040521/

7. Ginsberg, M.L.: Knowledge Interchange Format – draft proposed American National Standard (dpANS). Technical Report 2/98-004, Stanford University (1998)
8. Battle, S., Bernstein, A., Boley, H., Grosof, B., Gruniger, M., Hull, R., Kifer, M., Martin, D., McIlraith, S., McGuinness, D., Su, J., Tabet, S.: Semantic Web Services Framework (SWSF). Technical report, Semantic Web Services Initiative (SWSI) (April 2005)
9. Battle, S., Bernstein, A., Boley, H., Grosof, B., Gruninger, M., Hull, R., Kifer, M., Martin, D., McIlraith, S., McGuinness, D., Su, J., Tabet, S.: Semantic Web Services Ontology (SWSO). Technical report, Semantic Web Services Initiative (May 2005)
10. Horrocks, I., Patel-Schneider, P.F.: Reducing OWL entailment to description logic satisfiability. In: Fensel, D., Sycara, K., Mylopoulos, J. (eds.) ISWC 2003. LNCS, vol. 2870, pp. 17–29. Springer, Heidelberg (2003)
11. Schlenoff, C., Gruninger, M., Tissot, F., Valois, J., Lubell, J., Lee, J.: The Process Specification Language (PSL): Overview and Version 1.0 Specification. Technical Report NISTIR 6459, National Institute of Standards and Technology (2000)
12. America, P.: Designing an object-oriented programming language with behavioural subtyping. In: de Bakker, J.W., de Roever, W.P., Rozenberg, G. (eds.) REX 1990. LNCS, vol. 489, pp. 60–90. Springer, Heidelberg (1991)
13. Findler, R.B., Latendresse, M., Felleisen, M.: Behavioral contracts and behavioral subtyping. In: ACM SIGSOFT 2001. Proceedings of the 9th Symposium on the Foundations of Software Engineering (FSE 2001). (2001)
14. Briscoe, T., Copestake, A., de Paiva, V. (eds.): Inheritance, Defaults and the Lexicon. Cambridge University Press, New York (1993)
15. Daelemans, W., De Smedt, K.: Default Inheritance in an Object-oriented Representation of Linguistic Categories. International Journal Human-Computer Studies 41, 149–177 (1994)
16. MacLean, A., Carter, K., Lövstrand, L., Moran, T.: User-tailorable systems: pressing the issues with buttons. In: CHI 1990. Proceedings of the SIGCHI conference on Human factors in computing systems, pp. 175–182. ACM, New York (1990)
17. Morgan, C.: Programming from Specifications. International Series in Computer Science. Prentice-Hall, Englewood Cliffs (1990)
18. Meyer, B.: Static Typing. In: OOPSLA 1995. Addendum to the proceedings of the 10th annual conference on Object-oriented programming systems, languages, and applications (Addendum), pp. 20–29. ACM, New York (1995)
19. Patel-Schneider, P.F., Hayes, P., Horrocks, I. (eds.): OWL Web Ontology Semantics and Abstract Syntax (2004),
http://www.w3.org/TR/2004/REC-owl-semantics-20040210/
20. Borgida, A., Mylopoulos, J., Reiter, R.: On the Frame Problem in Procedure Specification. IEEE Transactions on Software Engineering (1995)
21. Gennari, J., Musen, M.A., Fergerson, R.W., Grosso, W.E., Crubézy, M., Eriksson, H., Noy, N.F., Tu, S.W.: The evolution of protégé: An environment for knowledge-based systems development. Technical Report SMI-2002-0943, Stanford Medical Informatics, Stanford University (2002)

Using Semantic Distances for Reasoning with Inconsistent Ontologies

Zhisheng Huang and Frank van Harmelen

Computer Science Department, Vrije Universiteit Amsterdam, The Netherlands
{huang,Frank.van.Harmelen}@cs.vu.nl

Abstract. Re-using and combining multiple ontologies on the Web is bound to lead to inconsistencies between the combined vocabularies. Even many of the ontologies that are in use today turn out to be inconsistent once some of their implicit knowledge is made explicit. However, robust and efficient methods to deal with inconsistencies are lacking from current Semantic Web reasoning systems, which are typically based on classical logic. In earlier papers, we have proposed the use of *syntactic relevance functions* as a method for reasoning with inconsistent ontologies. In this paper, we extend that work to the use of semantic distances. We show how Google distances can be used to develop *semantic relevance functions* to reason with inconsistent ontologies. In essence we are using the implicit knowledge hidden in the Web for explicit reasoning purposes. We have implemented this approach as part of the PION reasoning system. We report on experiments with several realistic ontologies. The test results show that a mixed syntactic/semantic approach can significantly improve reasoning performance over the purely syntactic approach. Furthermore, our methods allow to trade-off computational cost for inferential completeness. Our experiment shows that we only have to give up a little quality to obtain a high performance gain.

> There is nothing constant in this world but inconsistency.
> -*Jonathan Swift (1667-1745)*

1 Introduction

A key ingredient of the Semantic Web vision is avoiding to impose a single ontology. Hence, merging ontologies is a key step. Earlier experiments (e.g. [10]) have shown that merging multiple ontologies can quickly lead to inconsistencies. Other studies have shown how migration [18] and evolution [9] also lead to inconsistencies. This suggests the importance and omnipresence of inconsistencies in ontologies in a truly web-based world.

At first sight, it might seem that many ontologies are semantically so lightweight (e.g. expressible in RDF Schema only[6]) that the inconsistency problem doesn't arise, since RDF Schema is too weak to even express an inconsistency[1]. However,

---

[1] Besides the rather limited case of disjoint datatypes.

A. Sheth et al. (Eds.): ISWC 2008, LNCS 5318, pp. 178–194, 2008.
© Springer-Verlag Berlin Heidelberg 2008

[17] has shown that on a closer look, many of these semantically lightweight ontologies make implicit assumptions such as the Unique Name Assumption, or assuming that sibling classes are disjoint. Such implicit assumptions, although not stated, are in fact used in the applications that deploy these ontologies. Not making these disjointness assumptions explicit harms the re-usability of these ontologies. However, if such assumptions are made explicit, many ontologies turn out to be in fact inconsistent.

One way to deal with inconsistencies is to first diagnose and then repair them. [18] proposes a nonstandard reasoning service for debugging inconsistent terminologies. This is a possible approach, if we are dealing with one ontology and we would like to improve this ontology. Another approach to deal with inconsistent ontologies is to simply avoid the inconsistency and to apply a non-standard reasoning method to obtain answers that are still meaningful, even though they have been obtained from an inconsistent ontology. The first approach could be dubbed "removing inconsistencies", while the second could be called "living with inconsistencies". This latter approach is more suitable for an open Web setting. where one would be importing ontologies from other sources, making it impossible to repair them, and where the scale of the combined ontologies would be too large to make repair effective. Therefore, this paper investigates the latter approach, namely, the approach of reasoning with inconsistent ontologies.

The classical entailment in logics is *explosive*: any formula is a logical consequence of a contradiction. Therefore, conclusions drawn from an inconsistent knowledge base by classical inference may be completely meaningless. The general task of a system of reasoning with inconsistent ontologies is: given an inconsistent ontology, return *meaningful* answers to queries. In [12] we developed a general framework for reasoning with inconsistent ontologies, in which an answer is "meaningful" if it is supported by a selected consistent sub-ontology of the inconsistent ontology, while its negation is not supported. In that work, we used relevance based selection functions to obtain meaningful answers. The main idea of the framework is: (1) a relevance function is used to select some consistent sub-theory from an inconsistent ontology; (2) then we apply standard reasoning on the selected sub-theory to try and find meaningful answers; (3) if a satisfying answer cannot be found, the relevance degree of the selection function is made less restrictive, thereby extending the consistent sub-theory for further reasoning. In this way the system searches for increasingly large sub-theories of an inconsistent ontology *until the selected sub-theory is large enough to provide an answer, but not yet so large so as to become itself inconsistent.*

In [13,11], several syntactic relevance based selection functions were developed. However, these approaches suffer several limitations and disadvantages. As we will show with a simple example later in this paper, such syntactic relevance functions are very sensitive to the accidental syntactic form of an ontology, which can easily lead to undesired conclusions on one syntactic form. A simple semantics preserving syntactic reformulation would have lead to the appropriate conclusion, but such careful design is unrealistic to require from knowledge engineers.

In this paper, we investigate the approach of *semantic relevance* selection functions as an improvement over the syntactic relevance based approach. We will examine the use of co-occurrence in web-pages, provided by a search engine like Google, as a measure of semantic relevance, assuming that when two concepts appear more frequently in the same web page, they are semantically more relevant. We will show that under this intuitive assumption, information provided by a search engine can be used for semantic relevance based selection functions for reasoning with inconsistent ontologies.

The main contributions of this paper are (1) to define some general formal properties of semantic relevance selection functions, (2) to propose the Google Distance as a particular semantic relevance function, (3) to provide an implementation of semantic relevance functions for reasoning with inconsistent ontologies in the PION system, (4) to run experiments with PION to investigate the quality of the obtained results, and (5) to highlight the cost/performance trade-off that can be obtained using our approach.

This paper is organised as follows. Section 2 briefly summarises the framework for reasoning with inconsistent ontologies. Section 3 introduces the notion of semantic relevance functions. Setion 4 presents a mixed approach which combine the advantages of both the syntactic approach and the semantic approach. Section 5 reports on our experiments of running PION on a realistic ontology before concluding the paper.

2   Reasoning with Inconsistent Ontologies

2.1   General Framework

Selection functions are central to the framework of reasoning with inconsistent ontologies. Such a selection function is used to determine which consistent subsets of an inconsistent ontology should be considered during its reasoning process. The selection function can either be based on a syntactic approach, like syntactic relevance [3], or based on semantic relevance. Examples of such semantic relevance are for example Wordnet distance [2], or (as we will propose in this paper) based on the co-occurrence of concepts in search engines like Google.

Given an ontology (i.e., a formula set) Σ and a query ϕ, a selection function s returns at each step $k > 0$ a subset of Σ. Let $\mathbf{L}$ be the ontology language, which is denoted as a formula set. A selection function s is then a mapping $s : \mathcal{P}(\mathbf{L}) \times \mathbf{L} \times N \to \mathcal{P}(\mathbf{L})$ such that $s(\Sigma, \phi, k) \subseteq \Sigma$.

In the following, we use $\Sigma \models \phi$ to denote that ϕ is a consequence of Σ in the standard reasoning, and we will use $\Sigma \approx\!\!\!| \phi$ to denote that ϕ is a consequence of Σ in the non-standard reasoning. The values of non-standard inference are defined as $\{Accepted, Rejected, Overdetermined, Undetermined\}$, following the 4-valued schema by [1].

Figure 1 shows a strategy to compute $\Sigma \approx\!\!\!| \phi$. This procedure is called a *linear extension* strategy because only one candidate Σ'' is chosen, and alternatives are not considered. This is attractive because the reasoner doesn't need to keep track of the extension chain. The disadvantage of the linear strategy is that it

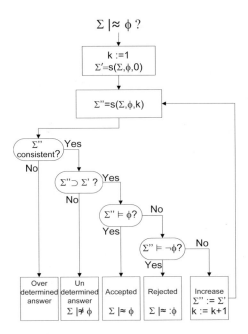

Fig. 1. Linear Extension Strategy

may result in too many 'undetermined' or 'overdetermined' answers when the selection function picks the wrong sequence of monotonically increasing subsets.

In the case of $s(\Sigma, \phi, k)$ being inconsistent, we can refine the procedure from figure 1 with a backtracking step, which tries to reduce $s(\Sigma, \phi, k)$ to a set that still extends $s(\Sigma, \phi, k-1)$, but that is still consistent. This would reduce the number of overdetermined answers, and hence improve the linear extension strategy. We call this procedure *overdetermined processing*(ODP). ODP introduces a degree of non-determinism: selecting different maximal consistent subsets of $s(\Sigma, \phi, k)$ may yield different answers to the query $\Sigma \not\approx \phi$. An easy solution to overdetermined processing is to return the first maximal consistent subset (FMC) of $s(\Sigma, \phi, k)$, based on certain search procedure. Query answers which are obtained by this procedure are still meaningful, because they are supported by a consistent subset of the ontology. However, it does not always provide intuitive answers because it depends on the search procedure of maximal consistent subset in overdetermined processing. A natural search procedure is to perform breadth-first search among the subsets of $s(\Sigma, \phi, k)$ in decreasing cardinality, until we find the first (and hence maximal) consistent subset.

2.2   Syntactic Selection Functions

Direct relevance between two *formulas* is defined as a binary relation on formulas: $\mathcal{R} \subseteq \mathbf{L} \times \mathbf{L}$. Given any direct relevance relation $\mathcal{R}$, we can extend it to a relation

$\mathcal{R}^+$ on a formula and a *formula set*, i.e. $\mathcal{R}^+ \subseteq \mathbf{L} \times \mathcal{P}(\mathbf{L})$, as follows:

$$\langle \phi, \Sigma \rangle \in \mathcal{R}^+ \text{ iff } \exists \psi \in \Sigma \text{ such that } \langle \phi, \psi \rangle \in \mathcal{R}.$$

In other words, a formula ϕ is relevant to a formula set Σ iff there exists a formula $\psi \in \Sigma$ such that ϕ and ψ are relevant. Two formulas ϕ, ϕ' are k-relevant with respect to a formula set Σ iff there exist formulas $\psi_0, \dots \psi_{k+1} \in \Sigma$ such that $\phi = \psi_0$ $\psi_{k+1} = \phi'$ and all ψ_i and ψ_{i+1} are directly relevant.

We can use such a relevance relation to define a selection function s as follows:

$s(\Sigma, \phi, 0) = \emptyset$
$s(\Sigma, \phi, 1) = \{\psi \in \Sigma | \phi \text{ and } \psi \text{ are directly relevant}\}$
$s(\Sigma, \phi, k) = \{\psi \in \Sigma | \psi \text{ is directly relevant to } s(\Sigma, \phi, k-1)\}$ *for* $k > 1$

There are various ways to define a syntactic relevance $\mathcal{R}$ between two formulas in an ontology. Given a formula ϕ, we use $I(\phi), C(\phi), R(\phi)$ to denote the sets of individual names, concept names, and relation names that appear in ϕ respectively. In [12], we proposed a direct relevance which considers the presence of a common concept/role/individual name in two formulas: two formulas ϕ and ψ are directly syntactically relevant, written $\mathcal{R}_{SynRel}(\phi, \psi)$, iff there is a common name which appears in both formulas.

In [11,12], we provided a detailed evaluation of the syntactic relevance approach by applying it to several inconsistent ontologies. The tests show that the syntactic relevance approach can obtain intuitive results in most cases for reasoning with inconsistent ontologies. The reason for this is that syntactic relevance mimics our intuition that real-world truth is (generally) preserved best by the argument with the shortest number of steps; and whatever process our intuitive reasoning uses, it is very likely that it would somehow privilege just these shortest path arguments[2]. However, as we will see, the problem is that the syntactic relevance approach requires that the syntactic encoding of the ontology by knowledge engineers correctly represents their intuitive understandings of the knowledge.

Example: A simple example where syntactic relevance works very well is the traditional penguin example in which birds are specified as flying animals and penguins are specified as birds which cannot fly. In this example, the reasoning path from *penguin* to $\neg fly$ is shorter than that from *penguin* to *fly*:
    penguin $\sqsubseteq \neg fly$,   *penguin* $\sqsubseteq bird \sqsubseteq fly$.

Example: However, the syntactic relevance approach does not work very well on the MadCow example[3], in which Cows are specified as vegetarians whereas MadCows are specified as Cows which eats brains of sheep (and hence are not vegetarians). Under the syntactic relevance selection functions, the reasoner returns the 'accepted' answer to the query 'is the_mad_cow a vegetarian?'. This counter-intuitive answer results from the weakness of the syntactic relevance approach, because it always prefers a shorter relevance path when a conflict occurs. In the MadCow example, the path *'mad cow - cow - vegetarian'* is shorter than

---

[2] Thanks to Norman Gray, for pointing this out in personal communication.
[3] The Mad Cow ontology is used in OilEd tutorials

the path *'mad cow - eat brain - eat bodypart - sheep are animals - eat animal - not vegetarian'*. Therefore, the syntactic relevance-based selection function finds a consistent sub-theory by simply ignoring the fact 'sheep are animals'.

2.3   Pro's and Cons of Syntactic Relevance

Empirically good results. In [11,12], we provided a detailed evaluation of the syntactic relevance approach by applying it to several inconsistent ontologies. The tests show that the syntactic relevance approach can obtain intuitive results in most cases for reasoning with inconsistent ontologies.

Sensitive to syntactic encoding. As shown above, the syntactic relevance approach is very dependent on the particular syntactic encoding that was chosen for the knowledge, since it selects short reasoning paths over longer ones. This works apparently works well in many cases (as shown in [11,12]), but it is not hard to think of natural examples where the shortest reasoning chain is not the correct one to follow.

Often needs a backtracking step. Because of the "fan out" behaviour of the syntactic selection function, the relevance set will grow very quickly, and will become very large after a small number of iterations. A very large relevance set is in danger of becoming inconsistent itself, causing the system to need the backtracking step that we called "overdetermined processing".

Backtracking is blind. To make matters worse, the backtracking step of the syntactic approach is essentially blind. It is hard to think of ways to make this backtracking more involved, based only on syntactic features.

3   Semantic Selection Functions

A wide space of semantic relevance measures exist, varying from Wordnet distance [2], to the co-occurrence of concepts in search engines like Google [5,4]. In this paper, we will use the latter, since we want to take advantage of the vast knowledge on the Web that is implicitly encoded in search engines. In this way, we can obtain light-weight semantics for selection functions.

The basic assumption here is that the more frequently two concepts appear in the same web page, the more semantically close they are, because most web pages are meaningful texts. Therefore, information provided by a search engine can be used to measure semantic relevance among concepts.

3.1   General Properties of Semantic Relevance

Semantic relevance is considered as the reverse relation of semantic dissimilarity: the more semantically relevant two concepts are, the smaller the distance between them. Assuming that both relevance and distance are taken from the [0,1] interval, this boils down to $Similarity(x, y) = 1 - Distance(x, y)$[4].

---

[4] In the following we use the terminologies *semantic dissimilarity* and *semantic distance* interchangeably.

To use semantic dissimilarity for reasoning with inconsistent ontologies, we define the dissimilarity measure between two formulas in terms of the dissimilarity measure between two concepts/roles/individuals from the two formulas. Moreover, in the following we consider only concept names $C(\phi)$ as the symbol set of a formula ϕ to simplify the formal definitions. However, note that the definitions can be easily generalised into ones in which the symbol sets contain also roles and individuals. We use $SD(\phi, \psi)$ to denote the semantic distance between two formulas. We expect the semantic distance between two formulas $SD(\phi, \psi)$ to satisfy the following intuitive properties:

Range. The semantic distance is a real number between 0 and 1: $0 \leq SD(\phi, \psi) \leq 1$ for any ϕ and ψ.

Reflexivity. Any formula is always semantically closest to itself: $SD(\phi, \phi) = 0$ for any ϕ.

Symmetry. The semantic distance between two formulas is symmetric: $SD(\phi, \psi) = SD(\psi, \phi)$ for any ϕ and ψ.

Maximum distance. If all symbols in a formula are semantically most-dissimilar from any symbol of another formula, then the two formulas are totally dissimilar: if $SD(C_i, C_j) = 1$ for all $C_i \in C(\phi)$ and $C_j \in C(\psi)$, then $SD(\phi, \psi) = 1$.

Intermediate values. If some symbols are shared between two formulas, and some symbols are semantically dissimilar, the semantic distance between the two formulas is neither minimal nor maximal: If $C(\phi) \cap C(\psi) \neq \emptyset$ and $C(\phi) \not\subseteq C(\psi)$ and $C(\psi) \not\subseteq C(\phi)$ then $0 < SD(\phi, \psi) < 1$.

3.2   Google Distance as Semantic Relevance

In [5,4], the Google Distance is introduced to measure the co-occurrence of two keywords on the Web. Normalised Google Distance (NGD) is introduced to measure the semantic distance between two concepts by the following definition:

Definition 1 (Normalised Google Distance [4]).

$$NGD(x, y) = \frac{max\{\log f(x), \log f(y)\} - \log f(x, y)}{\log M - min\{\log f(x), \log f(y)\}}$$

where $f(x)$ is the number of Google hits for the search term x, $f(y)$ is the number of Google hits for the search term y, $f(x, y)$ is the number of Google hits for the tuple of search terms x and y, and M is the number of web pages indexed by Google.

$NGD(x, y)$ can be understood intuitively as the symmetric conditional probability of co-occurrence of the search terms x and y. $NGD(x, y)$ is defined between two search items x and y. Simple ways to extend this to measure the semantic distance between two formulas are to take either the minimal, the maximal or the average NGD values between two concepts (or roles, or individuals) which appear in two formulas as follows:

$$SD_{min}(\phi, \psi) = min\{NGD(C_i, C_j)|C_i \in C(\phi) \text{ and } C_j \in C(\psi)\}$$
$$SD_{max}(\phi, \psi) = max\{NGD(C_i, C_j)|C_i \in C(\phi) \text{ and } C_j \in C(\psi)\}$$
$$SD_{ave}(\phi, \psi) = \frac{sum\{NGD(C_i, C_j)|C_i \in C(\phi) \text{ and } C_j \in C(\psi)\}}{(|C(\phi)| * |C(\psi)|)}$$

where $|C(\phi)|$ means the cardinality of $C(\phi)$. However, it is easy to see that SD_{min} and SD_{max} do not satisfy the Intermediate Values property, and SD_{ave} does not satisfy Reflexivity.

We therefore propose a semantic distance which is measured by the ratio of the summed distance of the difference between two formulas to the maximal distance between two formulas:

Definition 2 (Semantic Distance).

$$SD(\phi, \psi) = \frac{sum\{NGD(C_i, C_j)|C_i \in C(\phi) \backslash C(\psi), C_j \in C(\psi) \backslash C(\phi)\}}{(|C(\phi)| * |C(\psi)|)}$$

The intuition behind this definition is to sum the semantic distances between all terms that are not shared between the two formulae, but these must be normalised (divided by the maximum distance possible) to bring the value back to the [0,1] interval. It is easy to prove the following:

Proposition 1. *The semantic distance $SD(\phi, \psi)$ satisfies the properties Range, Reflexivity, Symmetry, Maximum Distance, and Intermediate Values.*

Using the semantic distance defined above, the obvious way to define a relevance relation for selection functions in reasoning with inconsistent ontologies is to take the semantically closest formulas as directly relevant:

$$\langle \phi, \psi \rangle \in \mathcal{R}_{sd} \text{ iff} \neg \exists \psi' \in \Sigma : SD(\phi, \psi') < SD(\phi, \psi).$$

(i.e. there exist no other formulas in the ontology that is semantically closer)

Given this semantic relevance relation, we now need to define a selection function. In the syntactic approach of the previous section, we used the query formula as the starting point for the selection function. We can define a similar selection function s in terms of the semantic relevance relation $\mathcal{R}_{sd}$. Namely, the newly defined selection function will track along the concept hierarchy in an ontology and always add to the selected set the closest formulas which have not yet been selected[5].

Example: Figure 2 shows how the semantic distance is used to obtain intuitive answers on the MadCow ontology (where the syntactic distance failed). By calculation of the Normalised Google Distance, we know that

$$NGD(MadCow, Grass) = 0.722911, \ NGD(MadCow, Sheep) = 0.612001.$$

---

[5] It is easy to see the definition about $SD(\phi, \psi)$ is easily extended into a definition about $SD(\phi, C)$, where ϕ, ψ are formulas, and C is a concept. Moreover, it is easy to see that $SD(\phi_1, C) < SD(\phi_2, C)$ iff $NGD(D_1, C) < NGD(D_2, C)$ for any ϕ_1 is of the form $C_1 \sqsubseteq D_1$ and any ϕ_2 is of the form $C_1 \sqsubseteq D_2$ where C, C_1, D_1 and D_2 are different concepts.

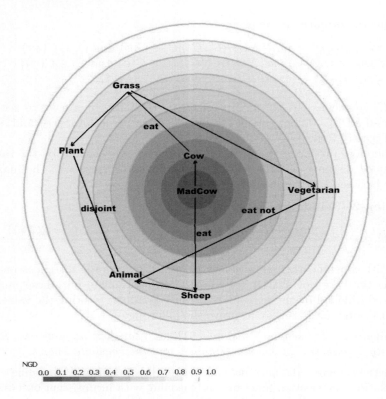

Fig. 2. NGD and MadCow Queries

Hence, the semantic distance between MadCow and Sheep is shorter than the semantic distance between MadCow and Grass (even though their syntactic distance is larger). Because of this, the reasoning path between *MadCow* and *Sheep* is preferred to the reasoning path between *MadCow* and *Grass*. Thus, we obtain the intuitive answer that *MadCow are not Vegetarians* instead of the previously obtained counter-intuitive answer that *MadCow are Vegetarians*. The intuition here is that although syntactically, the *MadCow - Sheep* path is the longer of the two, the accumulated semantic distance on this syntactically longer path is still shorter than the semantic distance on the syntactically short *MadCow - Grass* path.

3.3   Pro's and Cons of Semantic Relevance

Although empirical findings will only be discussed in section 5, we can already establish some of the advantages and disadvantages of the semantic approach to relevance.

Slower fan out behaviour. As is clear from the definition the growth of a relevance based on semantic distance is much slower than one based on syntactic relevance. In fact, at each step the semantic relevance set grows by a single formula (barring the exceptional case when some formulas share the same distance to the query).

Almost never needs a backtracking step. This slower growth of semantic relevance means that it will also hardly ever need a backtracking step, since the relevance set is unlikely to become "too large" and inconsistent.

Expensive to compute. Again by inspecting the definition, it is clear that computing the semantic relevance is expensive: it requires to know the semantic distance between the query and *every* formula ψ in the theory Σ. Furthermore, this must be done again for every new query concept C_1. With realistic modern ontologies often at a size of $O(10^5)$ concepts, and a computation time in the order of 0.2 secs for a single NGD-value, this would add a prohibitive cost to each query[6].

4   Mixed Approach

The picture that emerges from the pro's and cons in sections 2.3 and 3.3 is syntactic relevance is cheap to compute, but grows too quickly and then has to rely on a blind backtracking step, while semantic relevance has controlled growth, with no need for backtracking, but is expensive to compute.

In this section, we will propose a *mixed* approach which combine the advantages of both: we will use a syntactic-relevance selection function to grow the selection set cheaply, but we will use semantic relevance to improve the backtracking step. Instead of picking the first maximal consistent subset through a blind breadth-first descent, we can prune semantically less relevant paths to obtain a consistent set. This is done by removing the most dissimilar formulas from the set $s(\Sigma, \phi, k) - s(\Sigma, \phi, k - 1)$ first, until we find a consistent set such that the query ϕ can be proved or disproved.

Example: Taking the same example of the MadCow ontology above, we can see from Figure 2 that the path between *MadCow* and *Grass* can be pruned first, rather than pruning the path between *MadCow* and *Sheep*, because the NGD between *MadCow* and *Sheep* is smaller than the NGD between *MadCow* and *Grass*. Thus, the path *MadCow - Grass* (which lead to the counter-intuitive conclusion that *MadCow are vegetarians*) is pruned first.

We call this *overdetermined processing (ODP) using path pruning with Google distance*. While syntactic overdetermined processing (from section 2.1) can be seen as a blind breadth-first search, semantic-relevance ODP can be seen as a hill-climbing procedure, with the semantic distance as the heuristic.

Possible loss of completeness and soundness. Notice that semantic backtracking is not guaranteed to yield a *maximal* consistent subset. Consequently, the completeness of the algorithm may be affected, since we might have removed too many formulas from the relevance set in our attempt to restore consistency,

---

[6] Although some of this can be amortised over multiple queries by caching parts of the values that make up the NGD (definition 1).

thereby loosing the required implication to obtain the intuitive answer. Furthermore, it is possible that the semantic backtracking might lead to the *wrong* consistent subset, one supporting ϕ where $\neg\phi$ would have been the intuitive answer, or vice versa. In our experiment in section 5 we will find that indeed the completeness drops (as expected), but not by very much, while the unsoundness does not increase at all (making us belief that SD is a good heuristic for the hill-climbing search towards a consistent subset).

Cutting levels in ODP. Finally, the semantic distance provides the possibility for adjustable behaviour of the backtracking increments that are taken in the overdetermined processing phase. We introduce a cutting level α ($0 \leq \alpha \leq 1$), and instead of only pruning the semantically least relevant paths one by one until we obtain a consistent subset, we now prune in one step all formulas whose distance to the query is higher than α. In this way, α plays the role of a threshold, so that the processing can be sped up by pruning in a single step all those formulas which do not meet the relevance threshold. This might of course increase the amount of undetermined answers (since we may have overpruned), but it allows us to make a tradeoff between the amount of undetermined answers and the time performance. In Section 5 we will report an experiment in which this tradeoff obtains a 500% efficiency gain in exchange for only a 15.7% increase in undetermined answers.

5   Implementation and Experiments

We have implemented these definitions and algorithms in PION (Processing Inconsistent ONtologies)[7]. In this section, we report several experiments on reasoning with inconsistent ontologies using the selection functions introduced above.

5.1   Data

As already observed before, many ontologies on the Semantic Web (e.g. those indexed by Swoogle[8]) do not contain explicit inconsistencies. This would make it hard to obtain test-data for running our experiments and indeed, it would question the need for our inconsistency reasoning methods in the first place. The following brief analysis shows that under the surface, the situation is different.

Disjointness constraints between classes in an ontology are necessary for a suitable formalisation of a conceptualisation, and are required to draw the required inferences in tasks such as search, navigation, visualisation, service matching, etc [19]. However, as shown by [17] and [16], knowledge engineers often neglect to add disjointness statements to their ontologies, simply because they are not aware of the fact that classes which are not explicitly declared to be disjoint will be considered as potentially overlapping. Furthermore, an experiment by Völker

---

[7] Available for download at `wasp.cs.vu.nl/sekt/pion`

[8] `swoogle.umbc.edu`

and her colleagues in [19] showed that when prompted to add disjointness statements, human experts are very prone to introducing inadvertent inconsistencies.
Since we will take the ontologies that resulted from that experiment as our
dataset, we will describe that experiment in some detail.

The experiment in [19], takes as its starting point a subset of the PROTON
Ontology[9]. The selected subset of PROTON contains 266 classes, 77 object properties, 34 data-type properties and 1388 siblings. Each concept pair was randomly
assigned to 6 different people - 3 from a group of professional "ontologists", and
3 from a group of students without profound knowledge in ontological engineering. Each of the annotators was given between 385 and 406 concept pairs along
with a natural language descriptions of the classes whenever those were available, and were asked to annotate each concept pair as "disjoint", "overlapping"
or "unknown". Two enriched versions of the ontology were then constructed by
adding those disjointness statements that were agreed upon by 100% of the experts and of the students respectively.   We will call these the `experts` and the
`students` ontologies respectively. These two ontologies were both inconsistent.
For example, the `students` ontology alone already contained some 24 unsatisfiable concepts. Even more telling is the following example:

Example: 100 percent of students and experts (!) agree on the following axioms,
which are, however inconsistent:

$$Reservoir \sqsubseteq Lake$$
$$Lake \sqsubseteq WaterRegion$$
$$Reservoir \sqsubseteq HydrographicStructure$$
$$HydrographicStructure \sqsubseteq Facility$$
$$Disjoint(WaterRegion, Facility)$$

This case shows that inconsistency and incoherence occurs much more easily
than what is often expected. Interestingly enough, this problem would be handled by our semantic relevance approach. Normalised Google Distance tells us
that $Lake$ and $WaterRegion$ are more semantically relevant than $Facility$ and
$HydrographicStructure$ to $Reservoir$. Thus, using the semantic relevance based
selection function, we would conclude that $Reservoir$ is a $WaterRegion$.

The essence of all this is that although the original ontology did not contain
inconsistencies (due to lack of disjointness statements), the inconsistencies arise
as soon as human knowledge engineers are asked to add explicit disjointness
statements to the best of their capabilities. Thus, the resulting ontologies contain
"natural" inconsistencies. This makes the resulting set of inconsistent ontologies
a realistic data-set for our experiments.

5.2   Tests

Goal. Given that the mixed approach (using syntactic relevance for growing
the relevant set, and using semantic relevance for backtracking, possibly using
α-cuts) seems to be the best alternative to the purely syntactic approach of

---

[9] `proton.semanticweb.org/`

our earlier work, our experiment is aimed at (1) finding out the quality of the answers generated by the mixed approach, and (2) finding out the quality/cost trade-offs that can be obtained by varying the α-levels.

Test Queries and Answers. We created 529 subsumption queries randomly, and obtained PION's answers of these queries with backtracking done either blindly (First Maximal Consistent Subset, FMC), or via the semantic distance (SD). We compared these answers against a hand-crafted Gold Standard that contained the humanly-judged correct answer for all of these 529 queries. For each query, the answer given by PION can be classified in one of the following categories, based on the difference with the intuitive answer in the Gold Standard:

Intended Answer: PION's answer is the same as the intuitive answer from the Gold Standard.

Counter-intuitive Answer: PION's answer is opposite to the intuitive answer, i.e. the intuitive answer is "accepted" whereas PION's answer is "rejected", or vice versa.

Cautious Answer:The intuitive answer is "accepted" or "rejected", but PION's answer is "undetermined".

Reckless Answer: PION's answer is "accepted" or "rejected" while the intuitive answer is "undetermined".

Obviously, one would like to maximise the Intended Answers, and minimise the Reckless and Counter-intuitive Answers. Furthermore, we introduced different α-thresholds in the overdetermined processing to see how the tradeoff between the quality of query-answers and the time performance is effected by different cutting levels.

5.3   Results

Our results obtained by running PION with the data and the tests described above are shown in Figure 3. The first 4 rows show experiments on the `experts` ontology, the final 2 rows on the `students` ontology. In all cases, we use syntactic relevance for growing the relevance set until an answer can be found, but they differ on what happens when the relevance set becomes inconsistent, and backtracking is required. On the first line (labelled FMC, for First Maximal Consistent subset), the backtracking is done blindly, on the other lines, backtracking is guided by the semantic distance function, at different α-levels (i.e. with different sizes of the backtracking steps; smaller values for α, i.e. lower thresholds, means that more formulas are removed during backtracking). Not listed in the table is the fact that among the 529 queries, 414 (i.e. 78%) resulted in relevance sets that became inconsistent before the query could be answered meaningfully, hence they needed a backtracking phase.

Answer quality. The tables shows that when switching from syntactic backtracking (labelled FMC) to semantic backtracking (labelled SD) the intended

Ontology	Method	α	Query	IA(IA rate)	CA	RA	CIA	ICRate(%)	Time	TRatio
experts	FMC	n/a	529	266 (50%)	219	32	12	91.68	114.63	n/a
experts	SD	0.85	529	246 (47%)	238	32	13	91.49	54.28	2.11
experts	SD	0.80	529	239 (45%)	246	32	12	91.68	39.96	2.87
experts	SD	0.75	529	225 (43%)	260	32	12	91.68	22.37	5.12
students	FMC	n/a	529	234 (44%)	249	33	13	91.30	45.05	n/a
students	SD	1.00	529	189 (36%)	309	22	9	94.14	28.11	1.62

IA = Intended Answers, CA = Cautious Answers, RA = Reckless Answers, CIA = Counter-Intuitive Answers, IA Rate = Intended Answers(%), IC Rate = IA+CA(%), FMC = First Maximal Consistent subset, SD = Semantic Distance, α=Threshold, Time = Time Cost per Query (seconds), TRatio = TimeCost(FMC)/TimeCost(SD)

Fig. 3. PION test results by using FMC or SD for overdetermined processing

answer (IA) rate does indeed drop, as predicted in section 4. Furthermore, the IA-rate declines slowly with decreasing α-levels. Similarly, the cautious answer rate increases slowly with decreasing α-levels. This is again as expected: larger backtracking steps are more likely to remove too many formulas from the relevance set, hence potentially making the relevance set too small. Or put another way: the hill-climbing search performed in the ODP phase is aiming to get close to a maximal consistent subset, but larger hill-climbing steps make it harder to end up close to such a set, because of possible overpruning.

The combined IC-rate (combining intended and cautious answers, i.e. those answers that are not incorrect, but possibly incomplete), stays constant across between FMC and SD, and across all α-levels. It is important to note that the numbers of reckless and counter-intuitive answers remains constant. This means that although the semantically guided large-step reductions (at low α-levels) do of course remove formulas, they do not remove the wrong formulas, which could have potentially lead to reckless or counter-intuitive answers.

Summarising, *when switching from FMC to SD, and with decreasing α-levels, the completeness of the algorithm (IA Rate) gradually declines, while the soundness of the algorithm (IC rate) stays constant.*

Cost/Quality trade-offs. Although these findings on the answer quality are reassuring (the semantic backtracking doesn't damage the quality), they are not by themselves a reason to prefer semantic backtracking over syntactic backtracking. The strong point of the semantic backtracking becomes clear when we look at the computational costs of syntactic and semantic backtracking, particularly in the light of the answer quality.

Above, we have seen that the answer quality only degrades very gradually with decreasing α-levels. The final two columns of table 3 however show that the answer *costs* reduce dramatically when switching from syntactic to semantic backtracking, and that they drop further with decreasing α-levels. The absolute computation time is more than halved when switching from FMC to SD ($\alpha = 0.85$), and is again more than halved when dropping α from 0.85 to 0.75, leading to an overall efficiency gain of a factor of 5. Of course, this efficiency is gained at the cost of some loss of quality, but this loss of quality (the drop in completeness, the IA rate) is very modest: the twofold efficiency gain at $\alpha = 0.85$ is gained at a price of a drop of only 3 percentage points in completeness, and the fivefold efficiency gain at $\alpha = 0.75$ is gained at a price of a drop of only 7 percentage points in completeness. Summarising, *semantic backtracking with cut-off levels yields a very attractive cost/quality trade-off between costs in terms of run-time, and the quality in terms of soundness and completeness of the answers.*

6   Discussion and Conclusions

Research from a number of different areas is relevant to the current work. Semantic distances and similarity measures have been widely used in computational linguistics [2,14] and ontology engineering [8,15]. [7] proposes the use of a Google-based similarity measure to weigh approximate ontology matches. Our research in this paper is the first attempt to introduce the Google Distance for reasoning with inconsistent ontologies. In essence we are using the implicit knowledge hidden in the Web for explicit reasoning purposes.

The main contributions of this paper are: a) we investigated how a semantic relevance-based selection function can be developed by using information provided by a search engine, in particular, by using the Normalized Google Distance; b) we provided variants of backtracking strategies for reasoning with inconsistent ontologies, and c) we showed that semantic distances can be used for handling large scale ontologies through a tradeoff between run-time and the degree of incompleteness of the algorithm.

In our experiment we applied our PION implementation to realistic test data. The experiment used a high-quality ontology that became inconsistent after adding disjointness statements that had the full support of a group of experts. The test showed that the run-time of informed semantic backtracking is much better than that of blind syntactic backtracking, while the quality remains comparable. Furthermore the semantic approach can be parametrised so as to step-wise further improve the run-time with only a very small drop in quality.

Clearly, our experiments should be repeated on many different ontologies in order to see how generalisable our results are. We are now developing a benchmark system for reasoning with inconsistent ontologies, by which various approaches and selection functions can be tested with different application scenarios on much larger ontology data. This is an inherently difficult task, because existing ontologies will often need to be enriched by making disjointness statements explicit before they can be used as test data. Furthermore, a Gold Standard of intuitive answers can often only be created by hand. These high costs experiment-construction costs also justify why we did not run more experiments in the scope of this paper.

One of the future tasks is to make the NGD (Normalized Google Distance) component well integrated with the architecture of PION, so that the NGD values can be dynamically obtained at run time, rather than as the pre-loaded libraries, as it is done in the present implementation.

Acknowledgements. The work reported in this paper was partially supported by the EU-funded projects SEKT, KnowledgeWeb, and LarKC. We thank Annette ten Teije for useful clarifying discussions and Johanna Völker for providing the inconsistent PROTON ontologies.

References

1. Belnap, N.: A useful four-valued logic. In: Modern Uses of Multiple-Valued Logic, pp. 8–37. Reidel, Dordrecht (1977)
2. Budanitsky, A., Hirst, G.: Semantic distance in wordnet: An experimental, application-oriented evaluation of five measures. In: Workshop on WordNet, Pittsburgh, PA (2001)
3. Chopra, S., Parikh, R., Wassermann, R.: Approximate belief revision- prelimininary report. Journal of IGPL (2000)
4. Cilibrasi, R., Vitany, P.: The Google similarity distance. IEEE/ACM Transactions on Knowledge and Data Engineering 19(3), 370–383 (2007)
5. Cilibrasi, R., Vitanyi, P.: Automatic meaning discovery using Google. Technical report, Centre for Mathematics and Computer Science, CWI (2004)
6. d'Aquin, M., Baldassarre, C., Gridinoc, L., Angeletou, S., Sabou, M., Motta, E.: Characterizing knowledge on the semantic web with watson. In: 5th International EON Workshop at ISWC 2007 (2007)
7. Gligorov, R., Aleksovski, Z., ten Kate, W., van Harmelen, F.: Using Google distance to weight approximate ontology matches. In: Proceedings of WWW 2007 (2007)
8. Haase, P.: Semantic Technologies for Distributed Information Systems. PhD thesis at the Universität Karlsruhe (2006)
9. Haase, P., van Harmelen, F., Huang, Z., Stuckenschmidt, H., Sure, Y.: A framework for handling inconsistency in changing ontologies. In: Gil, Y., Motta, E., Benjamins, V.R., Musen, M.A. (eds.) ISWC 2005. LNCS, vol. 3729. Springer, Heidelberg (2005)
10. Hameed, A., Preece, A., Sleeman, D.: Ontology reconciliation. In: Staab, S., Studer, R. (eds.) Handbook on Ontologies in Information Systems, pp. 231–250. Springer, Heidelberg (2003)
11. Huang, Z., van Harmelen, F.: Reasoning with inconsistent ontologies: Evaluation. Project Report D3.4.2, SEKT (2006)
12. Huang, Z., van Harmelen, F., ten Teije, A.: Reasoning with inconsistent ontologies. In: Proceedings of IJCAI 2005 (2005)
13. Huang, Z., van Harmelen, F., ten Teije, A., Groot, P., Visser, C.: Reasoning with inconsistent ontologies: a general framework. Project Report D3.4.1, SEKT (2004)
14. Lin, D.: An information-theoretic definition of similarity. In: Proceedings of International Conference on Machine Learning, Madison, Wisconsin (July 1998)
15. Maedche, A., Staab, S.: Measuring similarity between ontologies. In: Gómez-Pérez, A., Benjamins, V.R. (eds.) EKAW 2002. LNCS (LNAI), vol. 2473, pp. 251–263. Springer, Heidelberg (2002)
16. Rector, A., Drummond, N., Horridge, M., Rogers, J., Knublauch, H., Stevens, R., Wang, H., Wroe, C.: Owl pizzas: Practical experience of teaching owl-dl - common errors and common patterns. In: Motta, E., Shadbolt, N.R., Stutt, A., Gibbins, N. (eds.) EKAW 2004. LNCS (LNAI), vol. 3257, pp. 63–81. Springer, Heidelberg (2004)

17. Schlobach, S.: Debugging and semantic clarification by pinpointing. In: Gómez-Pérez, A., Euzenat, J. (eds.) ESWC 2005. LNCS, vol. 3532, pp. 226–240. Springer, Heidelberg (2005)
18. Schlobach, S., Cornet, R.: Non-standard reasoning services for the debugging of description logic terminologies. In: Proceedings of IJCAI 2003 (2003)
19. Volker, J., Vrandecic, D., Sure, Y., Hotho, A.: Learning disjointness. In: Franconi, E., Kifer, M., May, W. (eds.) ESWC 2007. LNCS, vol. 4519. Springer, Heidelberg (2007)

Statistical Learning for Inductive Query Answering on OWL Ontologies

Nicola Fanizzi, Claudia d'Amato, and Floriana Esposito

Dipartimento di Informatica, Università degli Studi di Bari
Campus Universitario, Via Orabona 4, 70125 Bari, Italy
{fanizzi,claudia.damato,esposito}@di.uniba.it

Abstract. A novel family of parametric language-independent kernel functions defined for individuals within ontologies is presented. They are easily integrated with efficient statistical learning methods for inducing linear classifiers that offer an alternative way to perform classification w.r.t. deductive reasoning. A method for adapting the parameters of the kernel to the knowledge base through stochastic optimization is also proposed. This enables the exploitation of statistical learning in a variety of tasks where an inductive approach may bridge the gaps of the standard methods due the inherent incompleteness of the knowledge bases. In this work, a system integrating the kernels has been tested in experiments on approximate query answering with real ontologies collected from standard repositories.

1 Ontology Mining: Learning from Metadata

In the context of the Semantic Web (henceforth SW) many applications require the accomplishment of data-intensive tasks that can effectively exploit machine learning methods [1]. However, while a growing amount of metadata is being produced, most of the research effort addresses the problem of learning *for* the SW (mostly from structured or unstructured text [2]). Less attention was devoted to the advantages (and problems) of learning *from* SW data and metadata expressed in *Description Logics* (DLs) [3].

Classification is a central task for many applications. However, classifying through logic reasoning may be both too demanding because of its complexity and also too weak because of inconsistency or (inherent) incompleteness in the knowledge bases [4]. So far, for the sake of tractability, only simple DL languages have been considered in the development of logic-based learning methods [5, 6]. On the other end, efficient machine learning methods, that were originally developed for simple data, can be effectively upgraded to work with richer structured representations [7]. These methods have been shown to effectively solve *unsupervised* and *supervised* learning problems in DLs [8, 9], particularly those based on classification, clustering and ranking of individuals.

A. Sheth et al. (Eds.): ISWC 2008, LNCS 5318, pp. 195–212, 2008.
© Springer-Verlag Berlin Heidelberg 2008

Although the inductive methods that will be presented are general and could in principle be exploited in various scenarios, we will focus on methods for inducing efficient classifiers from examples and use them to carry out forms of approximate query answering (and concept retrieval). This task is normally performed by recurring to standard deductive reasoning procedures [3]. Hence it may turn out to be ineffective when (inconsistent or) incomplete knowledge is available, which is not infrequent with heterogeneous and distributed data sources.

As discussed in previous works [9], besides of approximated retrieval and query answering, alternative classification methods can be as effective as deductive reasoning, even suggesting new knowledge (membership assertions) that was not previously logically derivable. As an example, considering the well-known WINE ontology, a statistical classifier induced by machine learning methods presented in the following, is able to infer assertions that cannot be logically derived by a reasoner such as that KathrynKennedyLateral, which is known as a Meritage, is a CaliforniaWine and an AmericanWine as well as that CotturiZinfandel, which is only known as a Zinfandel, is not a CabernetSauvignon (a non-disjoint sibling class). This feature of inductive classifiers can be exploited during the time-consuming *ontology completion* task [10] since the knowledge engineer has only to validate such assertions.

Among the other learning methods, *kernel methods* [11] represent a family of very efficient algorithms, that ultimately solve linear separation problems (finding an optimal hyperplane in between positive and negative instances) in high-dimensional feature spaces whereto a kernel function implicitly maps the original feature space of the considered dataset (*kernel trick*). Ad hoc kernel functions allow for learning classifiers even when the instances are represented in rich languages.

In this work, we demonstrate the exploitation of a kernel method for inducing classifiers for individuals in OWL ontologies. Indeed, kernel functions have been recently proposed for languages of average expressiveness, such as the family of kernels for $\mathcal{ALC}$ [12, 13]. However, the scope of their applicability was limited because of two factors: the definition in terms of a normal form for concept descriptions and the employment of the notion of (approximations of) *most specific concepts* [3] in order to lift instances to the concept-level where the kernels actually work.

In order to overcome such limitations, we propose a novel parametric family of kernel functions for DL representations which is inspired to a semantic pseudo-metric for DLs [8]. These functions encode a notion of similarity between individuals, by exploiting only semantic aspects of the reference representation. Their definition is also related to other simple kernels that were recently proposed [14]. Yet, while each of these kernels acts separately on a different level of similarity [15], based on the concepts and properties of the ontology, ours may integrate these aspects being parametrized on a set of features (concept descriptions). Furthermore, these features are not fixed but may be induced enforcing the discernibility of different instances. Similarly to metric-learning procedures based on stochastic search [8], a method for optimizing the choice of the feature sets is also proposed. This procedure, based on genetic programming, can be exploited in case the concepts in the ontologies would turn out to be weak for *discriminative* purposes.

The basics of kernel methods are presented in the next section jointly with related works about kernels for complex representations. In Sect. 3 the new family of kernels is proposed together with an algorithm for optimizing the choice of its parameters. Then, the query answering problem and its solution through our inductive method are formally defined in Sect. 4 and experimentally evaluated in Sect. 5. Conclusions and further applications of ontology mining methods are finally outlined in Sect. 6.

2 Inducing Classifiers with Kernel Methods

Given the learning task of inducing classifiers from examples, kernel methods are particularly well suited from an engineering point of view because the learning algorithm (*inductive bias*) and the choice of the kernel function (*language bias*) are almost completely independent [1]. While the former encapsulates the learning task and the way in which a solution is sought, the latter encodes the hypothesis language, i.e. the representation for the target classes. Different kernel functions implement different hypothesis spaces (representations). Hence, the same kernel machine can be applied to different representations, provided that suitable kernel functions are available. Thus, an efficient algorithm may be adapted to work on structured spaces [7] (e.g. trees, graphs) by merely replacing the kernel function with a suitable one. Positive and negative examples of the target concept are to be provided to the machine that processes them, through a specific kernel function, in order to produce a definition for the target concept in the form of a decision function based on weights.

2.1 Learning Linear Classifiers with Kernel Methods

Most machine learning algorithms work on simple representations where a training example is a vector of boolean features x extended with an additional one y indicating the membership w.r.t. a target class: $(x, y) \in \{0, 1\}^n \times \{-1, +1\}$. Essentially these algorithms aim at finding a vector of coefficients $w \in \mathbb{R}^n$ which is employed by a linear function (i.e. a *hyperplane* equation) to make a decision on the y label for an unclassified instance $(x, \cdot)$:

$$\text{class}(x) = \text{sign}(w \cdot x)$$

if $w \cdot x \geq 0$ then predict x to be positive $(+1)$ else it is classified as negative (-1).

As an example, the PERCEPTRON is a well-known simple algorithm to learn such weights [1]. In the training phase, for each incoming training instance, the algorithm predicts a label according to mentioned decision function and compares the outcome with the correct label. On erroneous predictions, the weights w are revised depending on the set of examples that provoked the mistake (denoted by

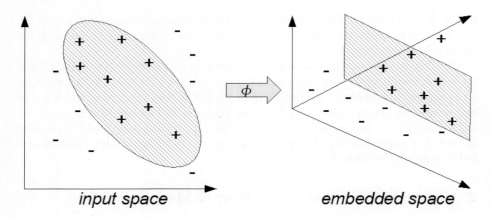

Fig. 1. The idea of the kernel trick

M): $\boldsymbol{w} = \sum_{\boldsymbol{v} \in M} l(\boldsymbol{v})\boldsymbol{v}$, where function l returns the label of the input example. Then, the resulting decision function can be written $\boldsymbol{w} \cdot \boldsymbol{x} = \sum_{\boldsymbol{v} \in M} l(\boldsymbol{v})(\boldsymbol{v} \cdot \boldsymbol{x})$. The dot product in these linear functions is the common feature of these methods.

Separating positive from negative instances with a linear boundary may be infeasible as it depends on the complexity of the target concept [1]. The *kernel trick* consists in mapping the examples onto a suitable different space (likely one with many more dimensions), allowing for the linear separation between positive and negative examples (*embedding space*, see Fig. 1). For example, the decision function for the perceptron becomes: $\sum_{\boldsymbol{v} \in M} l(\boldsymbol{v})(\phi(\boldsymbol{v}) \cdot \phi(\boldsymbol{x}))$, where ϕ denotes the transformation. Actually, such a mapping is never explicitly performed; a *valid* (i.e. definite positive) kernel function, corresponding to the inner product of the transformed vectors in the new space, ensures that an embedding exists [11]: $k(\boldsymbol{v}, \boldsymbol{x}) = \phi(\boldsymbol{v}) \cdot \phi(\boldsymbol{x})$. For instance, the decision function above becomes (*kernel perceptron*): $\sum_{\boldsymbol{v} \in M} l(\boldsymbol{v})k(\boldsymbol{v}, \boldsymbol{x})$.

Any algorithm for learning linear classifiers which is ultimately based on a decision function that involves an inner product could in principle be adapted to work on non-linearly separable cases by resorting to valid kernel functions which implicitly encode the transformation into the embedding space. Even more so, often many hyperplanes can separate the examples. Among the other kernel methods, the *support vector machines* (SVMs) aim at finding the hyperplane that maximizes the *margin*, that is the distance from the areas containing positive and negative training examples. The classifier is computed according to the closest instances w.r.t. the boundary (support vectors).

These algorithms are very efficient (polynomial complexity) since they solve the problem through quadratic programming techniques once the kernel matrix is produced [11]. The choice of kernel functions is very important as their computation should be efficient enough for controlling the complexity of the overall learning process.

2.2 Kernels for Structured Representations

When examples and background knowledge are expressed through structured (logical) representations, a further level of complexity is added. One way to solve the problem may involve the transformation of statistical classifiers into logical ones. However, while the opposite mapping has been shown as possible (e.g. from DL knowledge bases to artificial neural networks [16]), direct solutions to the learning problem are still to be investigated.

An appealing quality of the class of valid kernel functions is its closure w.r.t. many operations. In particular this class is closed w.r.t. the *convolution* [17]:

$$k_{\text{conv}}(x, y) = \sum_{\substack{\overline{x} \,\in\, R^{-1}(x) \\ \overline{y} \,\in\, R^{-1}(y)}} \prod_{i=1}^{D} k_i(\overline{x}_i, \overline{y}_i)$$

where R is a composition relationship building a single compound out of D simpler objects, each from a space that is already endowed with a valid kernel. Note that the choice of R is a non-trivial task which may depend on the particular application.

Then new kernels can be defined for complex structures based on simpler kernels defined for their parts using the closure property w.r.t. this operation. Many definitions have exploited this property, introducing kernels for strings, trees, graphs and other discrete structures. In particular, [7] provide a principled framework for defining new kernels based on type construction where types are defined in a declarative way.

While these kernels were defined as depending on specific structures, a more flexible method is building kernels as parametrized on concepts described with another representation. Such kernel functions allow for the employment of algorithms, such as the SVMs, that can simulate feature generation. These functions transform the initial representation of the instances into the related active features, thus allowing for learning the classifier directly from structured data. As an example, Cumby & Roth propose kernels based on a simple DL representation, the *Feature Description Language* [18].

Kernels for richer DL representations have been proposed in [12]. Such functions are actually defined for comparing $\mathcal{ALC}$ concepts based on the structural similarity of the AND-OR trees corresponding to a normal form of the input concept descriptions. However these kernels are not only structural since they ultimately rely on the semantic similarity of the primitive concepts (on the leaves) assessed by comparing their extensions through a set kernel. Although this proposal was criticized for possible counterintuitive outcomes, as it might seem that semantic similarity between two input concepts were not fully coped with, the kernels are actually applied to couples of individuals, after having lifted them to the concept level by means of (approximations of) their *most specific concept* [3]. Since these concepts are constructed on the grounds of the same ABox and TBox, it is likely that structural and semantic similarity tend to coincide.

A more recent definition of kernel functions for individuals in the context of the standard SW representations is reported in [14]. The authors define a set of kernels for individuals and for the various types of assertions in the ABox (on concepts, datatype properties, object properties). However, it is not clear how to integrate such functions which cope with different aspects of the individuals; the preliminary evaluation on specific classification problems regarded single kernels or simple additive combinations.

3   A Family of Kernels for Individuals in DLs

In the following we report the basic DL terminology utilized for this paper (see [3] for a thorough and precise reference).

Ontologies are built on a triple $\langle N_C, N_R, N_I \rangle$ made up by a set of concept names N_C, a set of role names N_R and a set of individual names N_I, respectively. An interpretation $\mathcal{I} = (\Delta^{\mathcal{I}}, \cdot^{\mathcal{I}})$ maps (via $\cdot^{\mathcal{I}}$) such names to the corresponding element subsets, binary relations, and objects of the domain $\Delta^{\mathcal{I}}$. A DL language provides specific constructors and rules for building complex concept descriptions based on these building blocks and for deriving their interpretation. The *Open World Assumption* (OWA) is made in the underlying semantics, which is convenient for the SW context.

A *knowledge base* $\mathcal{K} = \langle \mathcal{T}, \mathcal{A} \rangle$ contains a *TBox* $\mathcal{T}$ and an *ABox* $\mathcal{A}$. $\mathcal{T}$ is the set of terminological axioms of concept descriptions $C \sqsubseteq D$, meaning $C^{\mathcal{I}} \subseteq D^{\mathcal{I}}$, where C is the concept name and D is its description. $\mathcal{A}$ contains assertions on the world state, e.g. $C(a)$ and $R(a, b)$, meaning that $a^{\mathcal{I}} \in C^{\mathcal{I}}$ and $(a^{\mathcal{I}}, b^{\mathcal{I}}) \in R^{\mathcal{I}}$.

Subsumption w.r.t. the models of the knowledge base is the most important inference service. Yet in our case we will exploit *instance checking*, that amounts to decide whether an individual is an instance of a concept [3].

The inherent incompleteness of the knowledge base under open-world semantics may cause reasoners not to be able to assess the target class-membership. Moreover this can be a computationally expensive reasoning service. Hence we aim at learning efficient alternative classifiers that can help answering these queries effectively.

3.1   Kernel Definition

The main limitations of the kernels proposed in [12] for the space of $\mathcal{ALC}$ descriptions are represented by the dependency on the DL language and by the approximation of the most specific concept which may be computationally expensive. The use of a normal form has been also criticized since this is more a structural (syntactic) criterion that contrasts notion of semantic similarity.

In order to overcome these limitations, we propose a different set of kernels, based on ideas that inspired a family of inductive distance measures [8, 9], which can be applied directly to individuals:

Definition 3.1 (DL-kernels). *Let* $\mathcal{K} = \langle \mathcal{T}, \mathcal{A} \rangle$ *be a knowledge base. Given a set of concept descriptions* $\mathsf{F} = \{F_1, F_2, \ldots, F_m\}$, *a family of kernel functions* $k_p^{\mathsf{F}} : \mathsf{Ind}(\mathcal{A}) \times \mathsf{Ind}(\mathcal{A}) \mapsto [0, 1]$ *is defined as follows:*

$$\forall a, b \in \mathsf{Ind}(\mathcal{A}) \qquad k_p^{\mathsf{F}}(a, b) := \left[\sum_{i=1}^{m} \left| \frac{\kappa_i(a, b)}{m} \right|^p \right]^{1/p}$$

where $p > 0$ and $\forall i \in \{1, \ldots, m\}$ the simple concept kernel function κ_i is defined:
$\forall a, b \in \mathsf{Ind}(\mathcal{A})$

$$\kappa_i(a, b) = \begin{cases} 1 & (F_i(a) \in \mathcal{A} \wedge F_i(b) \in \mathcal{A}) \vee (\neg F_i(a) \in \mathcal{A} \wedge \neg F_i(b) \in \mathcal{A}) \\ 0 & (F_i(a) \in \mathcal{A} \wedge \neg F_i(b) \in \mathcal{A}) \vee (\neg F_i(a) \in \mathcal{A} \wedge F_i(b) \in \mathcal{A}) \\ \frac{1}{2} & otherwise \end{cases}$$

or, model-theoretically:

$$\kappa_i(a, b) = \begin{cases} 1 & (\mathcal{K} \models F_i(a) \wedge \mathcal{K} \models F_i(b)) \vee (\mathcal{K} \models \neg F_i(a) \wedge \mathcal{K} \models \neg F_i(b)) \\ 0 & (\mathcal{K} \models \neg F_i(a) \wedge \mathcal{K} \models F_i(b)) \vee (\mathcal{K} \models F_i(a) \wedge \mathcal{K} \models \neg F_i(b)) \\ \frac{1}{2} & otherwise \end{cases}$$

The rationale for these kernels is that similarity between individuals is determined by their similarity w.r.t. each concept in a given committee of features. Two individuals are maximally similar w.r.t. a given concept F_i if they exhibit the same behavior, i.e. both are instances of the concept or of its negation. Conversely, the minimal similarity holds when they belong to opposite concepts. Because of the OWA, sometimes a reasoner cannot assess the concept-membership, hence, since both possibilities are open, we assign an intermediate value to reflect such uncertainty.

As mentioned, instance-checking is to be employed for assessing the value of the simple similarity functions. Yet this is known to be computationally expensive (also depending on the specific DL language of choice). Alternatively, especially for ontologies that are rich of explicit class-membership information (assertions), a simple look-up may be sufficient, as suggested by the first definition of the κ_i functions.

The parameter p was borrowed from the form of the Minkowski's measures [19]. Once the feature set is fixed, the possible values for the kernel function are determined, hence p has an impact on the granularity of the measure.

3.2   Discussion

The most important property of a kernel function is its validity (it must correspond to a dot product in a certain embedding space).

Proposition 3.1 (validity). *Given an integer $p > 0$ and a committee of features F, the function k_p^{F} is a valid kernel.*

This result can be assessed by proving the function k_p^{F} definite-positive. Alternatively it is easier to prove the property by showing that the function can be obtained by composing simpler valid kernels through operations that guarantee the closure w.r.t. this property [17]. Specifically, since the simple kernel functions κ_i $(i = 1, \ldots, n)$ actually correspond to *matching kernels* [7], the property follows from the closure w.r.t. sum, multiplication by a constant and kernel multiplication [17].

One may note that such functions extend (and integrate) the kernels defined in [14]. For instance, the *common class* kernels may constitute a simplified version of the DL-kernels. They are essentially based on the intersection of the sets of common classes, considering only those occurring in the ontology. The new kernels, in principle, can be parametrized on any set of complex concept descriptions, including negated concepts. Moreover, they take also into account uncertain membership cases. As regards the data-property and object-property kernels, again the similarity is assessed by comparing (restrictions of) domains and ranges of defined relations related to the assertions on the input individuals. These may be encoded by further concepts to be added to the committee, especially when they can determine the separation of different individuals.

Furthermore, the uniform choice of the weights assigned to the various features in the sum $(1/m^p)$ may be replaced by assigning different weights reflecting the importance of a certain feature in discerning the various instances. A good choice may be based on the amount of *entropy* related to each feature (then the weight vector has only to be normalized) [9].

It is worthwhile to note that this is indeed a family of kernels parametrized on the choice of features. Preliminary experiments regarding instance-based classification, demonstrated the effectiveness of the kernel using the very set of both primitive and defined concepts found in the knowledge bases. However, the choice of the concepts to be included in the committee F is crucial and may be the object of a preliminary learning problem to be solved (*feature selection*).

3.3   Optimizing the Feature Set

As for the pseudo-metric that inspired the kernel definition [8], a preliminary phase may concern finding an optimal choice of features. This may be carried out by means of randomized optimization procedures, similar to the one developed for the pseudo-distance. However, the integration of the algorithm in suitable kernel machines guarantees that the feature construction job is performed automatically by the learning algorithm (the features correspond to the dimensions of the embedding space).

The underlying idea in the kernel definition is that similar individuals should exhibit the same behavior w.r.t. the concepts in F. Here, one may make the assumption that the feature-set F represents a sufficient number of (possibly redundant) features that are able to discriminate different individuals (in terms of a discernibility measure).

Namely, since the function is strictly dependent on the committee of features F, two immediate heuristics arise:

- The *number* of concepts of the committee,
- Their discriminating power in terms of a *discernibility factor*, i.e. a measure of the amount of difference between individuals.

```
GPOPTIMIZATION(K, maxGenerations, fitnessThr, FeatureSet)
input    K: current knowledge base
         maxGenerations: maximal number of generations
         fitnessThr: minimal required fitness threshold
output  FeatureSet: set of concept descriptions
static   currentFSs, formerFSs: arrays of feature sets
         currentBestFitness, formerBestFitness = 0: arrays of fitness values
         offsprings: array of generated feature sets
         fitnessImproved: improvement flag
         generationNo = 0: number of current generation
begin
currentFSs = MAKEINITIALFS(K,INIT_CARD)
formerFSs = currentFSs
repeat
     fitnessImproved = false
     currentBestFitness = BESTFITNESS(currentFSs)
     while (currentBestFitness < fitnessThr) and (generationNo < maxGenerations) do
         begin
         offsprings = GENERATEOFFSPRINGS(currentFSs)
         currentFSs = SELECTFROMPOPULATION(offsprings)
         currentBestFitness = BESTFITNESS(currentFSs)
         ++generationNo
         end
     if (currentBestFitness > formerBestFitness) and (currentBestFitness < fitnessThr) then
         begin
         formerFSs = currentFSs
         formerBestFitness = currentBestFitness
         currentFSs = EXTENDFS(currentFSs)
         end
     else
         fitnessImproved = true
     end
until not fitnessImproved
return SELECTBEST(formerFSs)
end
```

Fig. 2. Feature set optimization algorithm based on genetic programming

Finding optimal sets of discriminating features, should also profit by their composition, employing the specific constructors made available by the representation language.

These objectives can be accomplished by means of randomized optimization techniques, especially when knowledge bases with large sets of individuals are available. For instance in [8] we have proposed a metric optimization procedures based on stochastic search. Namely, part of the entire data can be drawn in order to learn optimal feature sets, in advance with respect to the successive usage for all other purposes.

A specific optimization algorithm founded in *genetic programming* has been devised to find optimal choices of discriminating concept committees. The resulting algorithm is shown in Fig. 2. Essentially, it searches the space of all possible feature committees, starting from an initial guess (determined by the call to the MAKEINITIALFS() procedure) based on the concepts (both primitive and defined) currently referenced in the knowledge base K, starting with a committee of a given cardinality (INIT_CARD). This initial cardinality may be determined as a function of $\lceil \log_3(N) \rceil$, where $N = |\mathsf{Ind}(\mathcal{A})|$, as each feature projection can categorize the individuals in three sets.

The outer loop gradually augments the cardinality of the candidate committees until the threshold fitness is reached or the algorithm detects some fixpoint: employing larger feature committees would not yield a better feature set w.r.t. the best fitness recorded in the previous iteration (with fewer features). Otherwise, the EXTENDFS() procedure extends the current committee by including a newly generated random concept.

The inner while-loop is repeated for a number of generations until a stop criterion is met, based on the maximal number of generations maxGenerations or, alternatively, when a minimal fitness threshold fitnessThr is crossed by some feature set in the population, which can be returned.

As regards the BESTFITNESS() routine, it computes the best fitness of the feature sets in the input vector. Fitness can be determined as the *discernibility factor* yielded by the feature set, as computed on the whole set of individuals or on a smaller sample. For instance, given the fixed set of individuals $IS \subseteq \mathsf{Ind}(\mathcal{A})$ the fitness function may be:

$$\mathrm{DISCERNIBILITY}(\mathsf{F}) := \nu \sum_{(a,b) \in IS^2} \sum_{i=1}^{|\mathsf{F}|} \mid 1 - \kappa_i(a,b) \mid$$

where ν is a normalizing factor that depends on the overall number of couples involved.

As concerns finding candidate sets of concepts to replace the current committee (the GENERATEOFFSPRINGS() routine), the function was implemented by recurring to some transformations of the current best feature sets:

− Choose $\mathsf{F} \in \mathsf{currentFSs}$;
− Randomly select $F_i \in \mathsf{F}$;
  • Replace F_i with $F_i' \in$ RANDOMMUTATION(F_i) randomly generated, or
  • Replace F_i with one of its refinements $F_i' \in$ REF(F_i)

The possible refinements of concept description are language-specific. E.g. for the case of $\mathcal{ALC}$ logic, refinement operators have been proposed in [6, 5].

This is iterated till a number of offsprings is generated (another parameter which determines the speed of the search process). Then these offspring feature sets are evaluated and the best ones are included in the new version of the currentFSs array; the best fitness value for these feature sets is also computed.

When the while-loop is over, the current best fitness is compared with the best one recorded for the former feature set length; if an improvement is detected then the outer repeat-loop is continued, otherwise (one of) the former best feature set(s) is selected and returned as the result of the algorithm.

4   Approximate Classification and Retrieval

SVMs based on kernel functions can efficiently induce classifiers that work by mapping the instances into an embedding feature space, where they can be discriminated by means of a linear classifier.

Given the kernel function for DLs defined in the previous section, we intend to use an SVM to induce a linear classifier which can be efficiently employed to solve the following problem:

Definition 4.1 (classification problem). *Let* $\mathcal{K} = \langle \mathcal{T}, \mathcal{A} \rangle$ *be a knowledge base, let* $\mathsf{Ind}(\mathcal{A})$ *be the set of all individuals occurring in* $\mathcal{A}$ *and let* $\mathcal{C} = \{C_1, \ldots, C_s\}$ *be the set of (both primitive and defined) concepts in* $\mathcal{K}$.
The classification problem *can be defined as follows:*
given *an individual* $a \in \mathsf{Ind}(\mathcal{A})$,
determine $\{C_1, \ldots, C_t\} \subseteq \mathcal{C}$ *such that:* $\mathcal{K} \models C_i(a)\ \forall i \in \{1, \ldots, t\}$.

In the general setting of the kernel algorithms, the target classes for the classification problem are normally considered as disjoint. This is unlikely to hold in the SW context, where an individual can be an instance of more than one concept. Then, a different setting has to be considered. The multi-class classification problem is decomposed into smaller binary classification problems (one per class). Therefore, a simple binary value set ($V = \{-1, +1\}$) may be employed, where $+1$ indicates that an individual x_i is instance of the considered concept C_j and -1 indicates that x_i is not instance of C_j.

This multi-class learning setting is valid when an implicit *Closed World Assumption* (CWA) is made. Conversely, in a SW context, where the OWA is adopted, this is not sufficient because of the uncertainty brought by the different semantics. To deal with this peculiarity, the absence of information on whether a certain instance x_i belongs to the extension of the concept C_j should not be interpreted negatively; rather, it should count as neutral information. Thus, a larger valued set has to be considered, namely $V = \{+1, -1, 0\}$, where the three values denote, respectively, class-membership, non-membership and uncertain assignment. Hence, given a query instance x_q, for every concept $C_j \in C$, the classifier will return $+1$ if x_q is an instance of C_j, -1 if x_q is an instance of $\neg C_j$, and 0 otherwise.

The classification is performed on the grounds of the linear models built from a set of training examples whose correct labels are provided by an expert (or a reasoner). For each concept, classifiers for membership and non-membership have to be learned.

Dually, statistical classifiers can be used to perform an approximate retrieval service. Considered a knowledge base $\mathcal{K}$ and a query concept Q, a learning problem can be solved providing a limited set of individuals that are (examples) and are not (counterexamples) in the concept extension. The learning algorithm will produce a classifier for deciding the class-membership of other individuals; then all other individuals in $\mathcal{A}$ can be classified w.r.t. Q, thus solving the concept retrieval problem inductively.

The classifier is generally very efficient (simple mathematical computation is carried out). As regards the effectiveness (see also the next section), its performance on query answering or retrieval tasks may be compared to that of a logic reasoner. Moreover, the classifier may be able, in some cases, to answer queries when the reasoner cannot; that is the classifier may be able to induce knowledge that is likely to hold but that is not logically derivable. One may also consider using binary classifiers only, in order to force the answer to belong to $\{+1, -1\}$, or provide a measure of likelihood for this answer [9], yet this goes beyond the scope of this work.

5   Experimental Evaluation

The new kernel functions were implemented and integrated with the support vector machines in the LIBSVM library[1]. They can be easily integrated also in the SVM^{light} extension[2] proposed in [14]. The experimental session was designed in order to evaluate the learning method on a series of query answering problems.

5.1   Setup

A number of different OWL ontologies were selected from the Protégé library[3]: NEWSPAPER, WINES, SURFACE-WATER-MODEL (S.W.M.), SCIENCE, and NEW TESTAMENT NAMES (N.T.N.). Details about them are reported in Tab. 1 (upper part).

Table 1. Facts about the ontologies employed in the experiments

ONTOLOGY	DL lang.	#concepts	#obj. prop.	#data prop.	#individuals
NEWSPAPER	$\mathcal{ALCF(D)}$	29	28	25	72
S.W.M.	$\mathcal{ALCOF(D)}$	19	9	1	115
WINES	$\mathcal{ALCIO(D)}$	112	9	10	149
SCIENCE	$\mathcal{ALCIF(D)}$	74	70	40	331
N.T.N.	$\mathcal{SHIF(D)}$	47	27	8	676
BIOPAX	$\mathcal{ALCIF(D)}$	74	70	40	323
LUBM	$\mathcal{ALR_+HI(D)}$	43	7	25	118
SWSD	$\mathcal{SHIF(D)}$	47	27	8	732
FINANCIAL	$\mathcal{ALCIO(D)}$	112	9	10	1000

For each ontology, all concepts in their turn were considered as queries. A ten-fold cross validation design[4] was adopted in order to overcome the variability in the composition of the training and test sets of examples. Examples were labeled according to the reasoner response; the classifier was then induced by the SVM exploiting the kernel matrix computed by the use of the DL-Kernel[5], for the subset of the training examples selected in each run of the experiment. The classifier was then tested on the remaining individuals assessing its performance with respect to the correct theoretical classification provided by the reasoner.

---

[1] http://www.csie.ntu.edu.tw/~cjlin/libsvm
[2] http://www.aifb.uni-karlsruhe.de/WBS/sbl/software/jnikernel/
[3] http://protege.stanford.edu/plugins/owl/owl-library
[4] The set of examples is randomly divided into ten parts then, in each fold, one part is used to validate the classifier induced using the instances in the other parts as training examples [1].
[5] The feature set for the DL-kernel was made by all concepts in the ontology and parameter p was set to 1 for simplicity and efficiency purposes.

A different setting has been considered in [14] with a simplified version of the GALEN ontology. There, the ontology was randomly populated and only seven selected concepts have been considered while no roles have been taken into account. We considered only populated ontologies with their genuine composition, with no change on their population. Differently from the mentioned experiments, the population was not randomly generated in order to avoid that the classifier resulting from the learning process were influenced by the specific generating algorithm.

The performance of the classifier was evaluated by comparing its responses on test instances to those returned by a standard reasoner[6] used as baseline. As mentioned, the experiment has been performed by adopting the ten-fold cross validation procedure. The results (percentages) presented in the following tables are averaged over the folds and over all the concepts occurring in each ontology. Particularly, for each concept in the ontology, the following parameters have been measured for the evaluation [9]:

- *match rate*: number of cases of individuals that got exactly the same classification by both classifiers with respect to the overall number of individuals;
- *omission error rate*: amount of unlabeled individuals while they actually were to be classified as instances or as counterexamples for the concept;
- *commission error rate*: amount of individuals labeled as instances of a concept, while they (logically) belong to the negation of that concept or viceversa;
- *induction rate*: amount of individuals that were found to belong to a concept or its negation, while this information is not logically derivable by the reasoner.

The experiment is aimed at showing that statistical classification is comparably effective w.r.t. logic classification. Meanwhile it is very efficient (because of the simple linear function it is based on) and is also able to suggest (by analogy) assertions that are not logically derivable from the ontologies.

5.2   Outcomes

The outcomes of the experiments regarding the classification of all the concepts occurring in each ontology are reported in Tab. 2. By looking at the table, it is important to note that, for every ontology, the commission error was null. This means that the classifier did not make critical mistakes, i.e. cases when an individual is deemed to be an instance of a concept while it really is an instance of another disjoint concept. At the same time it is important to note that very high match rates were registered for each ontology. Particularly, it is interesting to observe that the match rate increases with the increase of the number of individuals in the considered ontology. This is because the performance of statistical methods is likely to improve with the availability of large numbers of training examples, which means that there is more information for better separating the example space.

---

[6] PELLET 1.5.1: http://pellet.owldl.com

Table 2. Results (average rates ± standard deviation) of the experiments on classification using the SVM with the DL-kernel

ONTOLOGY	match	induction	omission	commission
NEWSPAPER	90.3 ± 8.3	0.0 ± 0.0	9.7 ± 8.3	0.0 ± 0.0
S.W.M.	95.9 ± 4.1	0.0 ± 0.0	4.1 ± 4.1	0.0 ± 0.0
WINES	95.2 ± 8.8	0.6 ± 5.2	4.2 ± 7.5	0.0 ± 0.0
SCIENCE	97.1 ± 2.0	1.8 ± 2.5	1.1 ± 1.6	0.0 ± 0.0
N.T.N.	98.2 ± 1.7	0.2 ± 0.9	1.6 ± 1.6	0.0 ± 0.0

Table 3. Results (average rates ± standard deviations) of the experiments on classification using the SVM with the $\mathcal{ALC}$ kernel ($\lambda = 1$)

ONTOLOGY	match	induction	omission	commission
NEWSPAPER	90.3 ± 8.3	0.0 ± 0.0	9.7 ± 8.3	0.0 ± 0.0
S.W.M.	87.1 ± 15.8	6.7 ± 16.0	6.2 ± 9.1	0.0 ± 0.0
WINES	95.6 ± 7.8	0.4 ± 3.4	4.0 ± 7.3	0.0 ± 0.0
SCIENCE	94.2 ± 7.8	0.7 ± 7.8	5.1 ± 7.8	0.0 ± 7.8
N.T.N.	92.5 ± 24.7	2.6 ± 8.4	0.1 ± 3.9	4.7 ± 11.3

A conservative behavior has been also observed, indeed the omission error rate was not null (although it was very low). This was probably due to a high number of training examples classified as unknown w.r.t. certain concepts. To decrease the tendency to a conservative behavior of the method, a threshold could be introduced for the consideration of the training examples with an *unknown* classification.

In almost all cases, the classifier was able to induce class-membership assertions that were not logically derivable. For example, in the NTN ontology JesusChrist was found to be an instance of the concepts Man and $\overline{\text{Woman}}$, while this could not be determined by deductive reasoning (it is known to be an instance of SonOfGod). However, the assessment of the quality of the induced knowledge is not possible because the correct answer to the inferred membership assertions is known by the experts that built and populated the ontologies.

The experiment has been repeated on the same ontologies, applying classifier induced using the SVM jointly with the $\mathcal{ALC}$ kernel [12, 13]. Since the languages of the ontologies are generally more complex than $\mathcal{ALC}$, we considered the individuals to be represented by approximations of the most specific concepts of such individuals w.r.t. the ABox [3]. Note that a separate random new ten-fold experiment was generated, hence the training / test subsets were different w.r.t. the previous run.

The outcomes of the experiments are reported in Tab. 3. By comparing the outcomes reported in the two tables, it is possible to note that the classifiers induced by the SVM with the new DL-kernel generally improve both match rate and omission rate with respect to the $\mathcal{ALC}$ kernel (in the cases where they do not improve the difference is not large). The observed induction rates are generally in favor of the classifiers induced with the $\mathcal{ALC}$ kernel. This can be explained with the higher precision of the classifiers induced by the DL-kernels, which increased the match rate in many cases when the reasoner was not able to give a certain classification. The commission rate for the experiments with the $\mathcal{ALC}$ kernel is null like in the experiments with the DL-kernel (but for one case). Finally, one may also observe that the outcomes of the classifiers induced by adopting the new kernel showed a more stable behavior as testified by the limited deviations reported in the tables (with some exceptions where the difference is limited).

5.3   Experiments on Query Answering

Another experimental session has been designed for evaluating the performance of the classifiers induced with the new kernels on solving query answering problems with randomly generated concepts.

Further larger ontologies were selected (see Tab. 1, lower part): the BioPax glycolysis ontology[7] (BioPax), an ontology generated by the Lehigh University Benchmark (LUBM), the Semantic Web Service Discovery dataset[8]

(SWSD) and FINANCIAL ontology[9] employed as a testbed for PELLET. This was to increase the diversity of the domain (as well as source and population) of the ontologies and to provide learning problems with many classified training instances (yet this also depends on the generality of query concepts).

Table 4. Results (average rates ± standard deviation) of the experiments on random query answering

ONTOLOGY	match	induction	omission	commission
S.W.M.	82.31 ± 21.47	9.11 ± 16.49	8.57 ± 8.47	0.00 ± 0.00
SCIENCE	99.16 ± 4.35	0.44 ± 3.42	0.39 ± 2.76	0.00 ± 0.00
N.T.N.	80.38 ± 17.04	8.22 ± 16.87	9.98 ± 10.08	1.42 ± 2.91
BIOPAX	84.04 ± 14.55	0.00 ± 0.00	0.00 ± 0.00	15.96 ± 14.55
LUBM	76.75 ± 19.69	5.75 ± 5.91	0.00 ± 0.00	17.50 ± 20.87
FINANCIAL	97.85 ± 3.41	0.42 ± 0.23	0.02 ± 0.07	1.73 ± 3.43
SWSD	97.92 ± 3.79	0.00 ± 0.00	2.09 ± 3.79	0.00 ± 0.00

Preliminarily, a number of individuals (30% of the entire number) was uniformly sampled; then the method for generating optimal feature sets was run for each ontology to better define the final kernel function (p was set again to 1). Random queries were also preliminarily generated for each ontology combining (2 through 8) atomic concepts or universal and existential restrictions (maximal depth 3), using the union and intersection operators. In order to be able to induce the classifier, the generated queries were required also to represent satisfiable concepts and that some individuals could be recognized as their examples and counterexamples.

The outcomes are reported in Tab. 4, from which it is possible to observe that the behavior of the classifier on these concepts is not very dissimilar with respect to the outcomes of the previous experiments. These queries were expected

---

[7] http://www.biopax.org/Downloads/Level1v1.4/
[8] https://www.uni-koblenz.de/FB4/Institutes/IFI/AGStaab/Projects/xmedia/dl-tree.htm
[9] http://www.cs.put.poznan.pl/alawrynowicz/financial.owl

to be harder than the previous ones which correspond to the very primitive or defined concepts for the various ontologies. Specifically, the commission error rate was low for all but two ontologies (BioPax and LUBM) for which some very difficult queries were randomly generated which raised this rate beyond 10% and consequently also the standard deviation values. The difficulty arose from the very limited number of training classified instances available for the target random concept (many unclassified training instances).

As for all methods that learn from examples, the number of positive and negative instances has an impact on the quality of the classifier, which is likely shown when their quality is assessed against the test set.

6   Conclusions and Future Work

Inspired from previous works on dissimilarity measures in DLs, a novel family of semantic kernel functions for individuals has been defined based on their behavior w.r.t. a number of features (concepts). The kernels are language-independent being based on instance-checking (or ABox look-up) and can be easily integrated with a kernel machine (a SVM in our case) for performing a broad spectrum of activities related to ontologies.

In this paper we focused on the application of statistical methods for inducing classifiers based on the individuals in an ontology. The resulting classifiers can be used to perform alternative classification and query answering in a more efficient yet effective way, compared with the standard deductive procedures. It has been experimentally shown that its performance is not only comparable to the one of a standard reasoner, but the classifier is also able to induce new knowledge, which is not logically derivable (e.g. by using a DL reasoner). Particularly, an increase in predictive accuracy was observed when the instances are homogeneously spread, as expected from statistical methods. The induced classifiers can be exploited for predicting / suggesting missing information about individuals, thus completing large ontologies. Specifically, it can be used to semi-automatize the population of an ABox. Indeed, the new assertions can be suggested to the knowledge engineer that has only to validate their acquisition.

This constitutes a new approach in the SW context, since the efficiency of the statistical-numerical approaches and the effectiveness of a symbolic representation have been combined [16]. As a next step, a more extensive experimentation of the proposed method has to be performed besides of a comparison with similar existing methods [9].

Further ontology mining methods can be based on kernels such as conceptual clustering which allows the discovery of interesting subgroups of individuals which may require the definition of a new concept or to track the drift of existing concepts over time (with the acquisition of new individuals) or even to detect new emerging concepts [8].

References

[1] Witten, I.H., Frank, E.: Data Mining: Practical Machine Learning Tools and Techniques, 2nd edn. Morgan Kaufmann, San Francisco (2005)

[2] Buitelaar, P., Cimiano, P., Magnini, B. (eds.): Ontology Learning from Text: Methods, Evaluation and Applications. IOS Press, Amsterdam (2005)

[3] Baader, F., Calvanese, D., McGuinness, D., Nardi, D., Patel-Schneider, P. (eds.): The Description Logic Handbook. Cambridge University Press, Cambridge (2003)

[4] Hitzler, P., Vrandečić, D.: Resolution-based approximate reasoning for OWL DL. In: Gil, Y., Motta, E., Benjamins, V., Musen, M.A. (eds.) Proceedings of the 4th International Semantic Web Conference, ISWC 2005, Galway, Ireland. LNCS, vol. 3279, pp. 383–397. Springer, Heidelberg (2005)

[5] Iannone, L., Palmisano, I., Fanizzi, N.: An algorithm based on counterfactuals for concept learning in the Semantic Web. Applied Intelligence 26, 139–159 (2007)

[6] Lehmann, J., Hitzler, P.: Foundations of refinement operators for description logics. In: Blockeel, H., Ramon, J., Shavlik, J., Tadepalli, P. (eds.) ILP 2007. LNCS (LNAI), vol. 4894, pp. 161–174. Springer, Heidelberg (2008)

[7] Gärtner, T., Lloyd, J., Flach, P.: Kernels and distances for structured data. Machine Learning 57, 205–232 (2004)

[8] Fanizzi, N., d'Amato, C., Esposito, F.: Randomized metric induction and evolutionary conceptual clustering for semantic knowledge bases. In: Silva, M., Laender, A., Baeza-Yates, R., McGuinness, D., Olsen, O., Olstad, B. (eds.) Proceedings of the ACM International Conference on Knowledge Management, CIKM 2007, Lisbon, Portugal, pp. 51–60. ACM, New York (2007)

[9] d'Amato, C., Fanizzi, N., Esposito, F.: Query answering and ontology population: An inductive approach. In: Bechhofer, S., Hauswirth, M., Hoffmann, J., Koubarakis, M. (eds.) ESWC 2008. LNCS, vol. 5021, pp. 288–302. Springer, Heidelberg (2008)

[10] Baader, F., Ganter, B., Sertkaya, B., Sattler, U.: Completing description logic knowledge bases using formal concept analysis. In: Veloso, M. (ed.) Proceedings of the 20th International Joint Conference on Artificial Intelligence, Hyderabad, India, pp. 230–235 (2007)

[11] Schölkopf, B., Smola, A.: Learning with Kernels. The MIT Press, Cambridge (2002)

[12] Fanizzi, N., d'Amato, C.: A declarative kernel for $\mathcal{ALC}$ concept descriptions. In: Esposito, F., Raś, Z.W., Malerba, D., Semeraro, G. (eds.) ISMIS 2006. LNCS (LNAI), vol. 4203, pp. 322–331. Springer, Heidelberg (2006)

[13] Fanizzi, N., d'Amato, C.: Inductive concept retrieval and query answering with semantic knowledge bases through kernel methods. In: Apolloni, B., Howlett, R.J., Jain, L. (eds.) KES 2007, Part I. LNCS (LNAI), vol. 4692, pp. 148–155. Springer, Heidelberg (2007)

[14] Bloehdorn, S., Sure, Y.: Kernel methods for mining instance data in ontologies. In: Aberer, K., Choi, K.-S., Noy, N., Allemang, D., Lee, K.-I., Nixon, L., Golbeck, J., Mika, P., Maynard, D., Mizoguchi, R., Schreiber, G., Cudré-Mauroux, P. (eds.) ASWC 2007 and ISWC 2007. LNCS, vol. 4825, pp. 58–71. Springer, Heidelberg (2007)

[15] Ehrig, M., Haase, P., Hefke, M., Stojanovic, N.: Similarity for ontologies - a comprehensive framework. In: Proceedings of the 13th European Conference on Information Systems, ECIS 2005 (2005)

[16] Hammer, B., Hitzler, P. (eds.): Perspectives of Neural-Symbolic Integration. Studies in Computational Intelligence, vol. 77. Springer, Heidelberg (2007)

[17] Haussler, D.: Convolution kernels on discrete structures. Technical Report UCSC-CRL-99-10, Department of Computer Science, University of California – Santa Cruz (1999)

[18] Cumby, C., Roth, D.: On kernel methods for relational learning. In: Fawcett, T., Mishra, N. (eds.) Proceedings of the 20th International Conference on Machine Learning, ICML 2003, pp. 107–114. AAAI Press, Menlo Park (2003)

[19] Zezula, P., Amato, G., Dohnal, V., Batko, M.: Similarity Search – The Metric Space Approach. In: Advances in database Systems. Springer, Heidelberg (2007)

Optimization and Evaluation of Reasoning in Probabilistic Description Logic: Towards a Systematic Approach

Pavel Klinov and Bijan Parsia

The University of Manchester
Manchester M13 9PL, UK

Abstract. This paper describes the first steps towards developing a methodology for testing and evaluating the performance of reasoners for the probabilistic description logic P-$\mathcal{SHIQ}$(D). Since it is a new formalism for handling uncertainty in DL ontologies, no such methodology has been proposed. There are no sufficiently large probabilistic ontologies to be used as test suites. In addition, since the reasoning services in P-$\mathcal{SHIQ}$(D) are mostly query oriented, there is no single problem (like classification or realization in classical DL) that could be an obvious candidate for benchmarking. All these issues make it hard to evaluate the performance of reasoners, reveal the complexity bottlenecks and assess the value of optimization strategies. This paper addresses these important problems by making the following contributions: First, it describes a probabilistic ontology that has been developed for the real-life domain of breast cancer which poses significant challenges for the state-of-art P-$\mathcal{SHIQ}$(D) reasoners. Second, it explains a systematic approach to generating a series of probabilistic reasoning problems that enable evaluation of the reasoning performance and shed light on what makes reasoning in P-$\mathcal{SHIQ}$(D) hard in practice. Finally, the paper presents an optimized algorithm for the non-monotonic entailment. Its positive impact on performance is demonstrated using our evaluation methodology.

1 Introduction

Probabilistic description logic P-$\mathcal{SHIQ}$(D) has been proposed to handle uncertainty in OWL ontologies [1]. Such formalisms have received significant research attention over the latest years, strongly driven by BioHealth and Semantic Web applications. In general, the capability of representing uncertain knowledge does not come for free: some extra reasoning complexity is usually incurred (not to mention various modeling difficulties) [2]. This problem is complicated because even classical DL reasoning is known to be worst case intractable for expressive languages, e.g., $\mathcal{SHIQ}$(D). Thus, optimization strategies are required to make the reasoning practical in real-life applications.

Optimization research can hardly be fruitful without a systematic evaluation methodolody and reasonably characteristic test data. Unfortunately, there were few, if any, tools for developing or using P-$\mathcal{SHIQ}$(D) ontologies, thus no modelers have used it, and thus there are no applications using such ontologies and, indeed, no such ontologies at all. This makes the optimization research unguided and the principled comparison of different reasoning algorithms, implementations and approaches nearly impossible.

A. Sheth et al. (Eds.): ISWC 2008, LNCS 5318, pp. 213–228, 2008.

Another difficulty is the lack of reasoning problems that can be easily used for benchmarking, like, for example, classification problem in classical DL. That is, we can treat classification time as a reasonable proxy for the efficacy of reasoner optimizations (at least, as a first approximation). Conversely, P-$\mathcal{SHIQ}$(D) reasoning services are mostly query-oriented and focused on individual, antecedently given, entailments. We address this problem by the generation of queries against a bespoke ontology such that both the ontology and the queries are sensible from an application perspective.

This paper presents the first steps towards a *systematic* evaluation methodology for P-$\mathcal{SHIQ}$(D) by making the following contributions:

1. It describes a custom P-$\mathcal{SHIQ}$(D) ontology about breast cancer which we believe is a solid starting point for evaluating P-$\mathcal{SHIQ}$(D) implementations. Breast cancer risk assessment (BRCA) is a rich field with several general models, e.g., Gail model [3], and a wealth of online information and risk calculators. Thus, there are both clear statements to be formalized and deployed applications that can be used for determining characteristic queries. The ontology we developed, though not large, is very challenging to reason with. We believe that reasoners that can handle this ontology will work for an interesting range of applications.
2. It proposes a methodology for generating P-$\mathcal{SHIQ}$(D) reasoning problems including fragments of probabilistic ontologies with a series of probabilistic queries for each. The methodology has been implemented in the library PREVAL-DL[1] and applied to the BRCA ontology. The results are presented and discussed.
3. It demonstrates the utility of the methodology by evaluating the optimization strategy of lexicographic entailment in P-$\mathcal{SHIQ}$(D) that is now implemented in the new version of Pronto[2] [4]. The results clearly show both positive impacts of the strategy and the remaining issues.

The remainder of the paper is organized as follows: Section 2 briefly provides preliminaries on P-$\mathcal{SHIQ}$(D) as a representation and reasoning formalism. Section 3 describes the modeling of the BRCA ontology, the approach to generating the reasoning problems including probabilistic models and queries. It also presents the results of evaluating Pronto that help to understand the complexity of P-$\mathcal{SHIQ}$(D) in general. Section 4 sketches the developed optimization strategy and discusses the results of its evaluation using the new approach. Finally, the future work in this line is delineated in Section 5.

2   Technical Preliminaries on P-$\mathcal{SHIQ}$(D)

2.1   Syntax and Semantics of P-$\mathcal{SHIQ}$(D)

The syntactic constructs of P-$\mathcal{SHIQ}$(D) include those of $\mathcal{SHIQ}$(D) together with *conditional constraints*. Constraints are expressions of the form $(D|C)[l, u]$ where D, C are $\mathcal{SHIQ}$(D) concept expressions (called *conclusion* and *evidence* respectively) and $[l, u] \subseteq [0, 1]$ is a closed interval. Constraints can be *default* or *strict* corresponding to

---

[1] PREVAL-DL is an open source framework for testing and evaluating P-$\mathcal{SHIQ}$(D) reasoners: http://www2.cs.man.ac.uk/~klinovp/projects/prevaldl/index.html

[2] Pronto 0.2: http://pellet.owldl.com/pronto

statements that are *generally* or *always* true respectively. Informally, default statements represent (probabilistic) knowledge that is true most of the time but might not apply in specific cases since details about the specific cases alters the probabilities. For example, we might have a general sense of the probability of the flu in the general population (say, low), whereas a subpopulation (say, old people and children) are more vulnerable thus have a higher probability of having the flu. There also could be a subsubpopulation (say, immunized old people and children) which has a very low probability of flu infection. P-$\mathcal{SHIQ}$(D) allows us to represent this situation using default statements.

A probabilistic TBox (PTBox) is a 2-tuple $PT = (T, P)$ where T is a classical DL TBox and P is a finite set of *default* conditional constraints (or just *defaults*). Informally, a PTBox axiom $(D|C)[l, u]$ means that "*generally*, if a randomly chosen individual belongs to C, its probability of belonging to D is in $[l, u]$". A probabilistic ABox (PABox) is a finite set of *strict* conditional constraints pertaining to a single probabilistic individual o [1]. All constraints in a PABox are of the restricted form $(D|\top)[l, u]$. Informally, they mean that "the individual o is a member of D with probability between $[l, u]$" [1]. A probabilistic knowledge base PKB is a combination of one PTBox and a set of PABoxes, one for each probabilistic individual.

The semantics of P-$\mathcal{SHIQ}$(D) is standardly explained in terms of the notion of a *possible world* which is a somewhat non-standard to DL and is defined with respect to a DL vocabulary (set of basic concepts) Φ [5]. A possible world I is a set of DL concepts from Φ such that $\{a : C|C \in I\} \cup \{a : \neg C|C \notin I\}$ is satisfiable for a fresh individual a. The set of all possible worlds with respect to Φ is denoted as $\mathcal{I}_\Phi$. A world I satisfies a concept C denoted as $I \models C$ if $C \in I$. Satisfiability of basic concepts is inductively extended to complex concepts as usual.

A world I is said to be a model of a DL axiom Ax denoted as $I \models Ax$ if $Ax \cup \{a : C|C \in I\} \cup \{a : \neg C|C \notin I\}$ is satisfiable for a fresh individual a. A world I is a model of a classical DL knowledge base KB denoted as $I \models KB$ if it is a model of all axioms of KB. Existence of a world that satisfies KB is equivalent to the satisfiability in the classical model-theoretic DL semantics [5].

We define probabilistic models in terms of the possible world semantics. A probabilistic interpretation Pr is a function $Pr : \mathcal{I}_\Phi \rightarrow [0, 1]$ such that $\sum_{I \in \mathcal{I}_\Phi} Pr(I) = 1$. Pr is said to *satisfy* a DL knowledge base KB denoted as $Pr \models KB$ iff $\forall I \in \mathcal{I}_\Phi, Pr(I) > 0 \Rightarrow I \models KB$. Next, the probability of a concept $C \in \Phi$, denoted as $Pr(C)$, is defined as $\sum_{I \models C} Pr(I)$. $Pr(D|C)$ is used as an abbreviation for $Pr(C \cap D)/Pr(C)$ given $Pr(C) > 0$. A probabilistic interpretation Pr satisfies a conditional constraint $(D|C)[l, u]$, denoted as $Pr \models (D|C)[l, u]$, iff $Pr(C) = 0$ or $Pr(D|C) \in [l, u]$. Finally, Pr satisfies a set of conditional constraints F iff it satisfies each of the constraints. A PTBox $PT = (T, P)$ is called *satisfiable* iff there exists a probabilistic interpretation that satisfies $T \cup P$.

A conditional constraint $(D|C)[l, u]$ is a *logical consequence* of a TBox T and a set of conditional constraints P, denoted as $T \cup P \models (D|C)[l, u]$, if $\forall Pr : Pr \models T \cup P \Rightarrow Pr(D|C) \in [l, u]$. It is a *tight logical consequence* of $T \cup P$ denoted as $T \cup P \models_{tight} (D|C)[l, u]$ if $l = inf_{Pr(C)>0 \wedge Pr \models T \cup P}(Pr(D|C))$ and $u = sup_{Pr(C)>0 \wedge Pr \models T \cup P}(Pr(D|C))$.

2.2   Reasoning in P-$\mathcal{SHIQ}$(D)

Lehmann's lexicographic entailment has been suggested as a non-monotonic conse-quence relation for P-$\mathcal{SHIQ}$(D) because of satisfying certain properties that are desir-able for default reasoning [6] [7]. A few definitions are required to formulate it:

- A probabilistic interpretation Pr *verifies* a default $(D|C)[l, u]$ iff $Pr(C) = 1$ and $Pr(D|C) \in [l, u]$.
- Pr *falsifies* a default $(D|C)[l, u]$ iff $Pr(C) = 1$ and $Pr(D|C) \notin [l, u]$.
- A default d is *tolerated* by a set of defaults P under a classical TBox T iff $\exists Pr :$ $Pr \models T \cup P$ and Pr verifies d.
- d *is in conflict* with P under T iff it is not tolerated by P under T.
- A default ranking σ is *admissible* for PTBox $PT = (T, P)$ iff $\forall P' \subseteq P, \forall d \in P$, d is in conflict with P' under $T \Rightarrow \exists d' \in P'$ s.t. $\sigma(d') < \sigma(d)$.
- A PTBox is called *consistent* iff an admissible default ranking exists [7].

An admissible default ranking, if one exists, can be computed in the form of an or-dered partition $\{P_i\}_{i=1}^k$ known as a *z-partition*. When using lexicographic entailment, those models that satisfy more defaults with higher ranks are considered *lexicograph-ically preferable*. Models such that no other model is lexicographically preferable to them are called *lexicographically minimal*. A conditional constraint $(D|C)[l, u]$ is a *lexicographic consequence* of a PTBox $PT = (P, T)$ and a set of conditional con-straints F if it is satisfied by every lexicographically minimal model of $F \cup PT$. It is a *tight* lexicographic consequence iff l (resp. u) is a minimum (resp. maximum) subject to all lexicographically minimal models [7].

It has been shown that lexicographically minimal models can be characterized via *lexicographically minimal sets* of conditional constraints [5]:

Definition 1 (Lexicographically minimal sets). *Given a consistent PTBox $PT = (T, P)$ with a z-partition $\{P_i\}_{i=1}^k$ and a set of conditional constraints $\mathcal{F}$, a set $P' \subseteq P$ is lexicographically preferable to $P'' \subseteq P$ given $\mathcal{F}$ iff:*

$$(T, P' \cup \mathcal{F}) \text{ and } (T, P'' \cup \mathcal{F}) \text{ are satisfiable.} \tag{1}$$

$$\textit{For some } i = \{1..k\}, |P' \cap P_i| > |P'' \cap P_i| \tag{2}$$

$$\textit{For all } j = \{i + 1..k\}, P' \cap P_i = P'' \cap P_i. \tag{3}$$

The set $P' \subseteq P$ given $\mathcal{F}$ is lexicographically minimal iff no $P'' \subseteq P$ is lexicographi-cally preferable to P' given $\mathcal{F}$.

The set of all lexicographically minimal sets of PTBox PT given $\mathcal{F}$ is denoted $LMS(PT, \mathcal{F})$

Informally, lexicographic entailment corresponds to standard logical entailment from lexicographically minimal sets. Computing $LMS(PT, F)$ is the first phase of comput-ing the entailment. Section 4 will explain how that step can be optimized and will also present the evaluation of the proposed optimization.

The following are the core reasoning problems of P-$\mathcal{SHIQ}$(D) [7]:

- *Probabilistic Satisfiability (PSAT)*. PSAT is the problem of deciding whether exists a probabilistic interpretation that satisfies given PTBox.
- *Probabilistic Generic Consistency (PGCon)*. PGCon is the problem of deciding whether an admissible default ranking exists for the given PTBox.
- *Tight Logical Entailment (TLogEnt)*. TLogEnt is the problem of computing the tightest probability intervals for logical consequences.
- *Tight Lexicographic Entailment (TLexEnt)*. TLexEnt is the problem of computing the tightest probability intervals for lexicographic consequences.

3   Performance Evaluation Methodology

Probabilistic deduction in general and lexicographic entailment in P-$\mathcal{SHIQ}$(D) in particular are known to be computationally hard [2] [5]. Both PSAT and TLexEnt problems in P-$\mathcal{SHIQ}$(D) are EXPTIME-Complete where hardness follows from the complexity of $\mathcal{SHIQ}$(D) [8] and completeness from the small model theorem for satisfiability problem in probabilistic first-order logic [2].

These theoretical results do not necessarily say much about the practicality of reasoning in P-$\mathcal{SHIQ}$(D). It is known that even harder tableau-based algorithms for classical DL can be successfully used in applications. However, the picture is much less clear with respect to P-$\mathcal{SHIQ}$(D). It has been recently shown that reasoning tasks in P-$\mathcal{SHIQ}$(D) require a massive amount of classical DL reasoning, namely, classical SAT instances to be solved [5] [4]. At the same time the number of SATs varies greatly over probabilistic inputs so that the distribution required deeper investigation.

In this paper we use present a systematic approach to performance evaluation that is based on random sampling. Both, fragments of probabilistic ontology (samples) and probabilistic queries will be randomly generated. The main dataset for sampling will be a probabilistic ontology for breast cancer risk assessment (BRCA).

3.1   The BRCA Ontology

The BRCA ontology [3] was created as an attempt to model the problem of breast cancer risk assessment in a clear, ontological manner. The central idea behind the design the ontology was to reduce risk assessment to probabilistic entailment in P-$\mathcal{SHIQ}$(D).

The ontology consists of two major parts: a classical OWL ontology and a probabilistic part that represents domain uncertainty. It is anticipated that extensive medical vocabularies will be used as classical parts of such models. To emphasize this possibility in our experiments, we used the NCI thesaurus[4] augmented with a collection of classes to represent the risk factors used by the NCI risk calculator. The thesaurus is a large medical ontology of more than 27,500 classes.

The ontology aims at modeling two types of risk of developing breast cancer. First, it models *absolute* risk, i.e., the risk that can be measured without reference to other categories of women. Statements like "*an average woman has up to 12.3% of developing*

---

[3] Available at: http://www2.cs.man.ac.uk/klinovp/pronto/brc/cancer_cc.owl
[4] http://www.mindswap.org/2003/CancerOntology/nciOncology.owl

breast cancer in her lifetime" are examples of absolute risk [9]. Such risk is modeled using subclasses of $WomanUnderAbsoluteBRCRisk$. Subclasses distinguish between the risk of developing cancer over a lifetime vs. in the short term (e.g., ten years).

Second, the ontology models relative breast cancer risk. This is useful for representing the impact of various risk factors by describing how they increase or decrease the risk compared to an average woman. Statements like "*having BRCA1 gene mutation increases the risk of developing breast cancer by a factor of four*" express relative risk [9]. The ontology provides classes for different categories of relative risk, e.g., for increased risk or decreased risk.

The ontology defines risk factors that are relevant to breast cancer using subclasses of $RiskFactor$. It makes the distinction between the factors that should be known to a woman, e.g., age, family cancer history, breastfeeding and those that can only be inferred on the basis of other factors or by examination, e.g., BRCA gene mutation, breast and bone densities, etc. It also defines different categories of women: first, those that have certain risk factors (subclasses of $WomanWithRiskFactors$); and, second, those distinct in terms of the risk of developing cancer (subclasses of $WomanUnderBRCRisk$).

With this classical ontology, it is possible to define the task of assessing the risk in terms of probabilistic entailment. The problem is to compute the conditional probability that a certain woman is an instance of some subclass of $WomanUnderBRCRisk$ given probabilities that she is an instance of some subclasses of $WomanWithRiskFactors$. This requires probabilistic entailment of PABox axioms. In addition, it might also be useful to infer the generic probabilistic relationships between classes under $WomanUnderBRCRisk$ and under $WomanWithRiskFactors$. This can be done by computing TLexEnt for the corresponding PTBox axioms.

Following the assumption that the subjective probabilities representing risk factors for a certain individual can be combined with objective probabilities representing the statistical knowledge, the model contains a set of PABox and PTBox axioms. The PABox axioms define risk factors that are relevant to a particular individual. The PTBox axioms model generic probabilistic relationships between classes in the ontology, i.e., those that are assumed to hold for a randomly chosen individual.

The model represents absolute risk using the subclasses of $WomanUnderAbsolute$ $BRCRisk$ as conclusions in conditional constraints. For example, the above statement that an average woman has risk up to 13.2% can be expressed as the following TBox axiom:

$$(WomanUnderAbsoluteBRCRisk|Woman)[0, 0.132].$$

Similarly, the model represents the impact of various risk factors by PTBox constraints with subclasses $WomanWithRiskFactors$ as evidence. For example, the influence of age can be represented by the following constraint:

$$(WomanWithBRCInShortTerm|Woman50Plus)[0.027, 0.041]$$

which expresses that a woman after the age of fifty has a certain risk of developing breast cancer in short term. Relative risk can be captured analogously by using the subclasses of $WomanUnderRelativeBRCRisk$ as conclusions. For example, the impact of BRCA gene mutation can be described as:

$$(WomanUnderStrongBRCRisk|WomanWithBRCAMutation)[0.9, 1]$$

which means that a woman having BRCA (BRCA1 or BRCA2) gene mutation is almost certainly in the highest risk category.

The model also allows one to express various inter-relationships between risk factors. One possibility is to represent how the presence of one risk factor allows one to guess on the presence of others. This is the principal way to use *inferred* risk factors, i.e., those unknown to a woman. For example, it is statistically true that Ashkenazi Jews are more likely to develop the BRCA gene mutation [9]. Although the person being questioned may not be aware of her chances of having a gene mutation, they can be estimated based on her ethnicity or other factors. Such relationships are captured using the PTBox constraints with evidence and conclusions being subclasses of $Woman$ or $WomanUnderBRCRisk$, such as:

$$(WomanWithBRCAMutation|AshkenaziJewishWoman)[0.025, 0.025]$$

In addition, the model allows to represent how different risk factors strengthen or weaken each other. The classical part of the ontology provides classes that are combinations of multiple risk factors. For example, $Woman50PlusMotherBRCA$ is a subclass of both $WomanAged50Plus$ and $WomanWithMotherBRCA$, i.e., it represents women after the age of 50 whose mothers developed breast cancer in the past. The model can define the risk for such women to be much higher than if they had just one of the factors. This is possible using the previously described *overriding* feature. Informally, PTBox axioms for the combination of factors, such as:

$$(WomanUnderStrongBRCRisk|Woman50PlusMotherBRCA)[0.9, 1]$$

overrides the axioms for each individual factor, thus allowing the system to make a more relevant and objective inference. It is theoretically possible to define an exponential number of such risk factor combinations but in practice only some of them require special attention.

Finally, the ontology contains a number of PABoxes that represent risk factors for specific individuals. The motivation is that while the generic probabilistic model that provides all the necessary statistics that can be developed and maintained by a central cancer research institute, individual women can supply the knowledge about the risk factors that are known to them, e.g., age. It is also possible to express uncertainty in having some particular risk factor. This is particularly important for inferred risk factors, for example, breast or bone density.

3.2   Random Sampling

Given the test data (BRCA ontology) the next step is to generate instances of reasoning problems to evaluate the performance. We chose to generate instances of PSAT and TLexEnt where TLexEnt also includes PGCon as a sub-problem.

Currently the full version of BRCA ontology cannot be handled by P-$\mathcal{SHIQ}$(D) reasoners mainly because linear system even for a single PSAT becomes too large (i.e., exponential in the number of conditional constraints). Therefore we decided to evaluate the performance on selected fragments of the ontology. As mentioned above, the performance varies significantly over fragments and it was originally unclear which fragments are "hard" and which are "easy". Thus it was natural to begin with the random sampling method.

In all the following experiments, the performance (or hardness) is measured in the number of classical DL SAT instances that need to be solved during probabilistic reasoning. This helps to abstract from platform-dependent metrics such as time.

Instances of PSAT have been generated using *simple random sampling*. Each sample was an independent probabilistic KB with the full classical part of the BRCA ontology and a subset of the PTBox constraints. The number of conditional constraints varied from 10 to 15 to maintain the balance between the size of each sample and the number of trials for each size. The latter was 200.

Instances of TLexEnt are less straighforward to generate. First note that entailments of PABox constraints are usually harder than PTBox because of interactions between default PTBox knowledge and strict PABox knowledge during non-monotonic reasoning. Simple random samples of PKB are insufficient for generating PABox queries. It is also required to have a probabilistic individual with PABox costraints. For example, in the case of BRCA ontology, such individual would be a woman with her personal probabilistic facts (risk factors that apply to her).

Such individual can be selected from the collection of predefined PABoxes (analogously to selecting a fragment of PTBox). But in this case it is hard to ensure the interaction between a randomly selected fragment of PTBox and a independently selected probabilistic individual. Intuitively, it is desirable to generate realistic problem instances so that the strict knowledge about the individual can be usefully combined with the statistical knowledge in the PTBox. Again, in the case of BRCA ontology, there should be PTBox constraints that represent statistics about the risk factors that are relevant to some probabilistic individual. Otherwise the latter are useless for assessing the breast cancer risk.

Our approach to generating such reaslistic TLexEnt instances is summarized by the following steps:

- Generate fragments of the PTBox using simple random sampling.
- Generate a probabilistic individual and the corresponding PABox. Each PABox constraint $(C|\top)[l, u]$ is generated such that C is a class appearing in some of the previously selected PTBox constraints and $[l, u]$ is a random interval.
- Generate a PABox query of the form $(C|\top)[?, ?]$ where C is selected from a domain-specific set of classes. In BRCA that set includes classes that represent women under absolute or relative breast cancer risk.

The effect of the first two steps is that the reasoner has to consider both PTBox and PABox constraints during reasoning, instead of eliminating some as irrelevant (which might have been the case if they had been generated completely independently). The third step ensures that the queries will be meaningful in that particular domain.

3.3   Results

We have applied the methodology to the latest version of Pronto. As expected it was observed that the hardness of PSAT grows exponentially with the number of conditional constraints. The interesting fact was that the exponential blowup did not happen in all cases. Moreover, some samples of size k (*k-samples*) happened to be easier that some samples of size $k - 1$. For example, for $k = 10$, the number of SATs varied from 699 to $14,200$ whereas for $k = 11$ it varied from $1,091$ to $38,522$. Such variation is important to investigate in order to understand what exactly makes probabilistic KBs hard or easy for reasoning. This might lead to developing reasoning algorithms that can exploit such characteristics of PKBs.

The lower bound on the number of needed SATs is the number of variables in the linear system generated during PSAT. Each variable corresponds to some world and a SAT should be solved in order to show that the world is *possible*, i.e., satisfies the classical part of the KB. Thus it is natural to investigate how the number of variables (or size of the *index set* [5]) varies over the random samples and what factors have an impact on it.

The variation of the index set size is similar to the variation of the number of SAT as expected: for 10-samples the minimal size was 447 and maximal was $8,064$. The more interesting problem is to identify what factors determine the size of the index set. Then it would be possible to assess hardness of samples *in advance* and potentially exploit this information during reasoning.

With this aim in mind we attempted to develop a metric for estimating hardness of a PTBox. It can be conjectured that the size of the index set should depend on the number of relations (e.g., subsumption, disjointness, etc.) that can be proven for classes appearing in conditional constraints [1]. In the extreme case, if no such relation exists, the size would be 3^N where N is the number of constraints [1]. In practice, however, many index set items can correspond to classical models that do not satisfy classical part of KB and should be pruned. As an example, consider TBox $T = \{Penguin \sqsubseteq Bird\}$ and the world $\{Penguin, \neg Bird\}$. Clearly this world is not possible. Thus the metric should reflect the number of such relations between classes in constraints which we call the *connectivity* of PTBox). We have implemented and experimented with this metric by computing, for each pair of constraints, the number of subsumptions between classes and their negations (i.e. 9 SAT tests for each pair of constraints). The results for 200 samples were compared with the actual hardness of PTBox, i.e., the number of SATs solved during PSAT, in Figure 1.

The results show the anticipated correspondence between the connectivity of PTBox and its actual hardness which means that some prediction of reasoning complexity can be done in advance. It is an interesting question whether the *phase transition* phenomenon [10] can be observed for PSAT. Phase transition is a property of many known NP-hard problems which says that the hardest instances are grouped in a relatively small region of the problem space which is characterized by a critical value of some order parameter. For example, for SAT in propositional logic such parameter would be the average number of literals in clauses. So around the critical value there is a transition from the set of underconstrained problems to overconstrained ones. Reasonable algorithms are often capable of solving most of the problems that do not fall into the

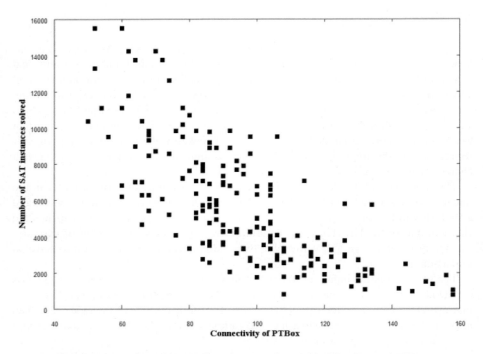

Fig. 1. Performance of PSAT plotted against the predicted hardness of PTBox

hard region efficiently, for example, such problems as 3-SAT, graph coloring, etc. are often tractable in practice.

The diagrams above do not show typical phase transition pattern although the connectivity metric is related to the extent to which a given instance is constrained. Less connectivity means that the classes in the conditional constraints are weakly related to each other so that the chance of conflicts is small. This is similar for underconstrained instances of SAT in propositional logic. Similarly, highly connected instances of PSAT are overconstrained. Thus it is reasonable to expect that some sort of phase transition phenomenon would occur. Why does it not happen?

The answer is that the PSAT algorithm [7] does not exploit the heuristic estimation of hardness in any way. Differently from many known algorithms for NP-complete problems it does not try to quickly find a solution for an underconstrained problem or quickly prove inexistence of solutions for an overconstrained problem. This might be one possible reason why PSAT is intractable for P-$\mathcal{SHIQ}$(D), and thus can be a promising direction for the optimization research on P-$\mathcal{SHIQ}$(D). More sophisticated evaluation techniques may need to be developed to support or falsify this conjecture.

The results for TLexEnt look similar to the results for single PSAT. The same metric proved to be predictive for a different problem. This is natural to expect because complexity of TLexEnt strongly depends on the complexity of PSAT which is its subproblem. The results plotted on the Figure 2 (again 200 samples were taken).

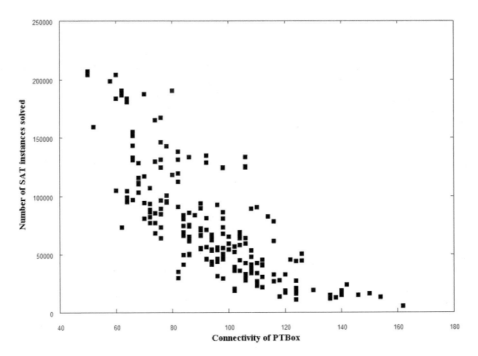

Fig. 2. Performance of TLexEnt plotted against the predicted hardness of PTBox

The important outcome is that for the latest TLexEnt algorithm (see Section 4 for details) there do not seem to be other factors except PSAT that affect its complexity. Interestingly this is not the case for the original algorithms that are due to Lukasiewicz [1] [5]. The same evaluation methodology can show that the original algorithm performs on some PKBs much worse than predicted by the metric. The reason is that the naive computation of lexicographically minimal sets during the non-monotonic phase of reasoning causes too many PSATs to be solved.

4   Evaluating Optimization Strategies

This section will demonstrate how new optimization strategies can be evaluated and compared to th existing algorithms using the proposed evaluation methodology. We start by briefly describing the optimization technique for computing lexicographically minimal models during TLexEnt.

4.1   Optimized TLexEnt Algorithm

The original TLexEnt algorithm computes the tightest interval for probabilistic query $(D|C)[?,?]$ in two phases [7] [1]:

1. *Model selection.* Conclusions in P-$\mathcal{SHIQ}$(D) are drawn from the set of lexicographically minimal models that are selected by computing lexicographically minimal sets (*LM-sets*) of constraints (see Definition 1).
2. *Entailment from preferred models.* Once models have been selected, the tightest interval can be computed by performing linear optimizations.

The complexity of the first phase determines the overall complexity of TLexEnt. Models selection can be done by solving $O(e^N)$ instances of PSAT each of which requires $O(e^N)$ instances SAT. Such a high complexity is caused by an uninformed search for LM-sets that runs over the powerset of constraints [5]. This is avoided in the improved algorithm that proceeds by eliminating the *minimal conflicting subsets*.

Definition 2 (Minimal conflict sets). *For a set of conditional constraints $\mathcal{F}$ and PTBox $PT = (T, P)$, a conflict set of PT given $\mathcal{F}$ is a set of conditional constraints Q s.t. $Q \subseteq P$ and $(T, Q \cup \mathcal{F})$ is unsatisfiable.*

A conflict set Q of $PT = (T, P)$ given $\mathcal{F}$ is minimal if $\forall Q' \subset Q$, $(T, Q' \cup \mathcal{F})$ is satisfiable.

The set of all minimal conflict sets of PT given $\mathcal{F}$ is denoted as $MCS(PT, \mathcal{F})$.

Informally, conflict sets identify those fragments of a probabilistic ontology that require conflict resolution during default reasoning. See the example below:

Example 1. Consider the following PTBox

$$PT = (\{Penguin \sqsubseteq Bird\},$$
$$\{(Fly|Bird)[0.9, 0.95], \tag{1}$$
$$(Fly|Penguin)[0, 0.05], \tag{2}$$
$$(Wings|Bird)[0.95, 1]\}) \tag{3}$$

$MCS(PT, \{(Penguin|\top)[1, 1]\}) = \{1, 2\}$, but $MCS(PT, \{(Bird|\top)[1, 1]\}) = \{\}$

As it will be shown below, conflict sets can be very useful for computing LM-sets.

Computing Minimal Conflict Sets. Finding *all* MCS is an NP-complete problem, so it may seem that an exponential number of PSAT instances will need to be generated and solved. However, it turns out that it is necessary to generate only a *single* PSAT instance to find all MCS thus avoiding a double exponential number of classical SAT tests. At the same time, it may be required to check an exponential number of linear systems for solvability. Fortunately, that step is computationally easier as it does not involve any classical DL reasoning.

The idea is as follows: First, *some initial* MCS is found by repeatedly removing linear inequalities from the linear system corresponding to $(T, P \cup \mathcal{F})$. The resulting system contains only those inequalities that correspond to conflicting constraints in MCS. Then it is possible to employ a standard technique for computing all explanation sets in classical DLs [11]. Each next MCS can be found by eliminating some constraints from all the previous MCS from the original PTBox and repeating the process of removing inequalities. The entire process terminates when no further MCS can be found.

It can be seen that there is only a *single* PSAT instance is generated during the computation of the first MCS. All other MCS are discovered by performing operations on linear systems and do not require any SAT tests at all.

Computing Lexicographically Minimal Sets. As mentioned before, the main goal of the optimization is to avoid solving an exponential number of PSATs during the search for lexicographically minimal sets while solving TLexEnt. It appears that it can be done by using the idea of conflict sets to compute maximal satisfiable subsets of PTBox by generating only a linear number of PSATs.

Definition 3 (Maximal satisfiable subsets). *Given a PTBox $PT = (T, P)$ and a set of conditional constraints $\mathcal{F}$, set $R \subseteq P$ is the maximal satisfiable subset of PT given $\mathcal{F}$ iff $(T, R \cup \mathcal{F})$ is satisfiable but $(T, S \cup \mathcal{F})$ is not for every $S \subseteq P$ s.t. $R \subset S$.*
The set of all maximal satisfiable subsets of PT given $\mathcal{F}$ is denoted as $MSS(PT, \mathcal{F})$.

The crucial observation is that lexicographically minimal sets can be computed by iterating over the z-partition and computing MSS at each subset. More formally:

Lemma 1. *Given a consistent PTBox $PT = (T, P)$ with z-partition $\{P_0, ..., P_k\}$ and a set of constraints $\mathcal{F}$, $LMS(PT, \mathcal{F})$ is equivalent to the set of all unions $\bigcup_{i=0}^{k} M_i$ where $M_k \in MSS((T, P_k), \mathcal{F})$ and $M_i \in \{MSS((T, P_i), M_{i+1}) | MSS((T, P_i), M_{i+1})$ has subsets of maximal cardinality subject to all $M_{i+1}\}$*

Lemma 1 essentially describes the algorithm for computing $LMS(PT, \mathcal{F})$. It is sufficient to iterate over all subsets of the z-partition in the order of decreasing specificity and compute MSS at each subset of the partition. It only remains to show how to compute $MSS(PT, \mathcal{F})$. It is well known that maximal satisfiable subsets are related with minimal unsatisfiable subsets in the following sense [12]:

Lemma 2. *Given a PTBox $PT = (T, P)$ and a set of constraints $\mathcal{F}$, $MSS(PT, \mathcal{F})$ is the set of all $M \subseteq P$ s.t. for every M there exists a set H s.t. $M = P \setminus H$, $H \cap Q \neq \emptyset$ for all $Q \in MCS(PT, \mathcal{F})$ and for any $H' \subset H$ there exists $Q \in MCS(PT, \mathcal{F})$ s.t. $H' \cap Q = \emptyset$*

Such sets H are called *minimal hitting sets* in the literature. Lemma 2 states a known approach to computing MSS that is based on computing all minimal hitting sets for all minimal conflict sets and then removing them from the initial set [12].

Using this technique the optimized algorithm computes a set of MSS that is linear in the number of subsets in the z-partition. Each MSS can be reduced to the computation of MCS and minimal hitting sets over the MCS. The latter is a known NP-complete problem but fortunately it is limited in its size and does not involve any classical DL reasoning. So, this algorithm computes lexicographically minimal sets by generating only a linear number of PSATs as opposed to the exponential number required by the Lukasiewicz algorithm. A simple example illustrates the advantage:

Example 2. Consider the following PTBox:

$$PT = (\{Penguin \sqsubseteq Bird\},$$
$$\{(Fly|Bird)[0.9, 0.95], \qquad\qquad (1)$$
$$(Fly|Penguin)[0, 0.05], \qquad\qquad (2)$$
$$(Wings|Bird)[0.95, 1]\}) \qquad\qquad (3)$$
$$\mathcal{F} = \{(Penguin|\top)[1, 1]\}$$

The z-partition is $\{\{1,3\},\{2\}\}$. Lukasiewicz's algorithm would compute $LMS(PT,$ $\mathcal{F})$ in the following steps (* means that a PSAT instance has to be generated):

1. Check satisfiability of $(T,\mathcal{F})^*$. Result: true.
2. Check satisfiability of $(T,\mathcal{F}\cup\{2\})^*$. Result: true.
3. Check satisfiability of $(T,\mathcal{F}\cup\{2,1,3\})^*$. Result: false.
4. Check satisfiability of $(T,\mathcal{F}\cup\{2,1\})^*$. Result: false.
5. Check satisfiability of $(T,\mathcal{F}\cup\{2,3\})^*$. Result: true.
6. $LMS(PT,\mathcal{F}):=\mathcal{F}\cup\{2,3\}$

There are two negative PSAT tests that are avoided in the new algorithm:

1. Check satisfiability of $(T,\mathcal{F})^*$. Result: true.
2. Compute $MSS((T,\{2\}),\mathcal{F})^*$. Result: $\{2\}$
3. Compute $MSS((T,\{1,3\}),\mathcal{F})^*$. Result: $\{3\}$
4. $LMS(PT,\mathcal{F}):=\mathcal{F}\cup\{2,3\}$

4.2   Evaluation of the Optimized Algorithm

The developed methodology enables us to systematically evaluate the performance of the optimized algorithm. The methodology can be applied to both algorithms and the results are easily comparable. This has been done by running both algorithms on random instances of TLexEnt generated as explained in the Section 3.2. We performed 200 runs where each PKB had 10 PTBox and 3 PABox constraints. The results are plotted on the Figure 3.

Simple visual comparison yields a few important observations. First, the new algorithm performs better as expected. Second, its behavior is more amenable to predictions using our connectivity metric. In other words, the relationship between the metric values and the actual hardness is apparent and resembles the same graph for PSAT. The naive algorithm, in contrast, produced a lot more outliers. There are some "hard" outliers — instances of TLexEnt that involve much more classical SATs than expected.

Finally, it can be noted that the fraction of such hard outliers is not large. This is a direct consequence of simple random sampling method which selects subsets of PTBox constraints with equal probability. Therefore, the chance that there will be conflicts similar to those shown in Example 2 is relatively small.

There is a question, however, whether such conflicts would be frequent in practice. At this point it is not fully clear because no P-$\mathcal{SHIQ}$(D) ontologies are employed in real applications. The BRCA ontology is the first attempt we know of to provide such model. In this ontology, conflicts can be expected because strict knowledge about particular women or their categories can often override general statistical knowledge. One example is African American and Ashkenazi Jew women for whom the statistical relationships from the Gail model are known to be imprecise or even incorrect.

In any case the evaluation methodology is useful because, first, it can systematically generate and run many random samples and therefore help to find "interesting cases", i.e., hard or easy outliers. Second, it can be used to compare different reasoning techniques and find inputs on which the techniques demonstrate similar or drastically different performance. At the same time it may be required to have a more intelligent problem generation method rather than random sampling. For example, a possible next step in the development of a benchmarking suite might be generation of only hard instances analogously to how it was done for other logics [13].

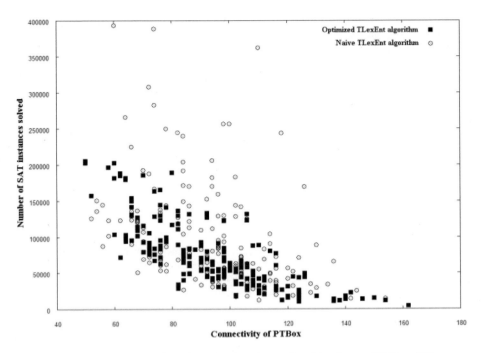

Fig. 3. Performance of TLexEnt plotted against the predicted hardness of PTBox. 200 runs.

5   Summary

The paper described first steps towards a systematic performance evaluation methodology for P-$\mathcal{SHIQ}$(D) reasoners. We have developed an approach to generating instances of the most important reasoning problems in P-$\mathcal{SHIQ}$(D) and provided a probabilistic ontology to serve as a basis for the generation. The methodology has been used to illustrate benefits of our optimizations for computing entailments.

Even though our approach is methodologically straighforward, to our knowledge, it has not been applied in this area before. Our experimental results show that being systematic in the evaluation of performances validates our analytical understanding of the reasoning tasks and algorithms but also yields important insights, such as the notion of connectivity for a set of conditional constraints and its impact on reasoning complexity.

The approach is flexible and extensible in the sense that one can contribute problem generators for their specific reasoning tasks. For example, as learned from the evaluation of the improved TLexEnt algorithm, a bias towards "hard" problem instances might be desirable. Also, there might be domain-specific evaluation. For instance, in the BRCA domain, it would be natural to generate PABoxes that only have constraints describing individual risk factors as opposed to randomly generated constraints. All such extensions can be smoothly plugged into the framework.

It is our expectation that the approach will also stimulate further reasoning optimization research for P-$\mathcal{SHIQ}$(D). The most important reasoning task to be optimized is

PSAT because it is currently responsible for the limited scalability of reasoners, e.g., Pronto. The evaluation strategy can highlight the problem instances on which the algorithm performs poorly so that specific optimization techniques might be developed to alleviate it. In this respect, current results can be considered as an important step towards practical reasoning in P-$\mathcal{SHIQ}$(D).

References

1. Giugno, R., Lukasiewicz, T.: $P-\mathcal{SHOQ}(D)$: A probabilistic extension of $\mathcal{SHOQ}$(D) for probabilistic ontologies in the semantic web. Technical Report Nr. 1843-02-06, Institut fur Informationssysteme, Technische Universitat Wien (2002)
2. Lukasiewicz, T.: Probabilistic logic programming with conditional constraints. ACM Transactions on Computational Logic 2(3), 289–339 (2001)
3. Gail, M.H., Brinton, L.A., Byar, D.P., Corle, D.K., Green, S.B., Shairer, C., Mulvihill, J.J.: Projecting individualized probabilities of developing breast cancer for white females who are being examined annually. Journal of the National Cancer Institute 81(25), 1879–1886 (1989)
4. Klinov, P.: Pronto: a non-monotonic probabilistic description logic reasoner. In: Proceeding of the European Semantic Web Conference (2008)
5. Lukasiewicz, T.: Expressive probabilistic description logics. Artificial Intelligence 172(6-7), 852–883 (2008)
6. Lehmann, D.: Another perspective on default reasoning. Annals of Mathematics and Artificial Intelligence 15(1), 61–82 (1995)
7. Lukasiewicz, T.: Probabilistic default reasoning with conditional constraints. Annals of Mathematics and Artificial Intelligence 34(1-3), 35–88 (2002)
8. Horrocks, I., Sattler, U., Tobies, S.: Practical reasoning for very expressive description logics. Journal of the IGPL 8(3) (2000)
9. Komen, S.G.: Breast cancer risk factors table (2007), Retrieved from:
   http://cms.komen.org/Komen/AboutBreastCancer/
10. Cheeseman, P., Kanefsky, B., Taylor, W.M.: Computational complexity and phase transitions. In: Proceedings of IJCAI, pp. 331–337 (1991)
11. Kalyanpur, A., Parsia, B., Horridge, M., Sirin, E.: Finding all justifications of OWL DL entailments. In: Proceedings of IJCAI, pp. 267–280 (2007)
12. Bailey, J., Stuckey, P.: Discovery of Minimal Unsatisfiable Subsets of Constraints Using Hitting Set Dualization. In: Hermenegildo, M.V., Cabeza, D. (eds.) PADL 2004. LNCS, vol. 3350, pp. 174–186. Springer, Heidelberg (2005)
13. Horrocks, I., Patel-Schneider, P.F.: Generating hard modal problems for modal decision procedures. In: First Methods for Modalities Workshop (1999)

Modeling Documents by Combining Semantic Concepts with Unsupervised Statistical Learning

Chaitanya Chemudugunta[1], America Holloway[1], Padhraic Smyth[1], and Mark Steyvers[2]

[1] Department of Computer Science
University of California, Irvine, Irvine, CA
{chandra,ahollowa,smyth}@ics.uci.edu
[2] Department of Cognitive Science
University of California, Irvine, Irvine, CA
msteyver@uci.edu

Abstract. Human-defined concepts are fundamental building-blocks in constructing knowledge bases such as ontologies. Statistical learning techniques provide an alternative automated approach to concept definition, driven by data rather than prior knowledge. In this paper we propose a probabilistic modeling framework that combines both human-defined concepts and data-driven topics in a principled manner. The methodology we propose is based on applications of statistical topic models (also known as latent Dirichlet allocation models). We demonstrate the utility of this general framework in two ways. We first illustrate how the methodology can be used to automatically tag Web pages with concepts from a known set of concepts without any need for labeled documents. We then perform a series of experiments that quantify how combining human-defined semantic knowledge with data-driven techniques leads to better language models than can be obtained with either alone.

Keywords: ontologies, tagging, unsupervised learning, topic models.

1 Introduction

An important step towards a semantic Web is automated and robust annotation of Web pages and online documents. In this paper we consider a specific version of this problem, namely, mapping of an entire document or Web page to concepts in a given ontology. To address this problem we propose a probabilistic framework for combining ontological concepts with unsupervised statistical text modeling. Here, and throughout, we use the term ontology to refer to *simple ontologies* [1] which are collections of human-defined concepts usually with a hierarchical structure. In this paper we focus on the simplest aspect of these ontologies, namely the ontological concepts and associated vocabulary (and to a lesser extent the hierarchical relations between concepts). We focus our investigation on the overall feasibility of the proposed approach—given the promise of the results obtained in this paper, the next step will be to develop models that can leverage the richer aspects of ontological knowledge representation.

We use statistical topic models (also known as latent Dirichlet allocation models [2,3]) as the underlying quantitative modeling framework. *Topics* from statistical models and *concepts* from ontologies both represent "focused" sets of words that relate to

A. Sheth et al. (Eds.): ISWC 2008, LNCS 5318, pp. 229–244, 2008.

Table 1. CIDE FAMILY concept and learned FAMILY topic

FAMILY Concept	FAMILY Topic	
beget	family	(0.208)
birthright	child	(0.171)
brood	parent	(0.073)
brother	young	(0.040)
children	boy	(0.028)
distantly	mother	(0.027)
dynastic	father	(0.021)
elder	school	(0.020)

some abstract notion—this similarity is the key idea we exploit in this paper. As an example, Table 1 lists some of the 204 words that have been manually defined as part of the concept FAMILY in the Cambridge International Dictionary of English (CIDE: more details on this ontology are provided later in the paper). The second column is a topic, also about families, that was learned automatically from a text corpus using a statistical topic model.

The numbers in parentheses are the probabilities that a word will be generated conditioned on the learned topic—these probabilities sum to 1 over the entire vocabulary of words, specifying a multinomial distribution. The concept FAMILY in effect puts probability mass 1 on the set of 204 words within the concept, and probability 0 on all other words. The topic multinomial on the other hand could be viewed as a "soft" version of this idea, with non-zero probabilities for all words in the vocabulary—but significantly skewed, with most of the probability mass focused on a relatively small set of words.

Many of the existing methods for semantic annotation of Web pages are focused on specific entity-tagging tasks, using a variety of natural language processing (NLP), information extraction (IE), and statistical language modeling techniques (e.g., [4,5,6]). A well-known semantic annotation system of this type is SemTag [7] which was built to annotate entity-rich web pages on a large scale. The main differences between this past work and our approach are that we map all words in a document, not just entities, onto a set of ontological concepts, we learn a probabilistic model over words and concepts, and we use an entirely unsupervised approach without any need for supervised labeling.

There has also been prior work that combines ontological concepts and data-driven learning within a single framework, such as using concepts as pre-processing for text modeling [8,9], using word-concept distributions as a form of background knowledge to improve text-classification [10], and combining human-derived linguistic knowledge with topic-based learning for word-sense disambiguation [11]. There has also been work on developing quantitative methods for evaluating how well ontologies fit specific text corpora [12,13] as well as a significant amount of research on ontology learning from data. Our work is different from all of this prior work in that we propose probabilistic models that combine concepts and data-driven topics within a single general framework, allowing (for example) the data to enable inferences about the concepts.

We begin the paper by reviewing the general ideas underlying statistical topic modeling and then show how these techniques can be directly adapted for the purposes of combining semantic concepts with text corpora. In the remainder of the paper we illustrate how the resulting models can be used to automatically tag words in Web pages and map each word into an ontological concept taking into account the context of the document. Additionally, we describe a set of quantitative experiments that evaluate the quality of the models when viewed as language models. We conclude that combining semantic concepts and data-driven topic learning opens up new opportunities and applications that would not be possible using either technique alone.

2   A Review of Statistical Topic Models

The latent Dirichlet allocation (LDA) model, also referred to as the topic model, is a state-of-the-art unsupervised learning technique for extracting thematic information from large document sets [2,3]. In this section we briefly review the fundamental ideas behind this model since it provides the basis for our approach later in the paper.

Let $\{w_1, \ldots, w_V\}$ be the set of unique words in a corpus, where V is the size of the vocabulary. Each document in the corpus is represented as a "bag of words", namely a sparse vector of length V where component i contains the number of times word i occurs in the document.

Table 2. Two example topics learned from a large corpus

HEALTH CARE		FARMING	
health	(0.064)	farm	(0.081)
care	(0.058)	crop	(0.027)
plan	(0.047)	cow	(0.018)
cost	(0.043)	field	(0.015)
insurance	(0.042)	corn	(0.015)
benefit	(0.032)	food	(0.012)
converage	(0.023)	bean	(0.010)
pay	(0.020)	cattle	(0.010)
program	(0.013)	market	(0.010)

A topic $z_j, 1 \leq j \leq T$ is represented as a multinomial probability distribution over the V words, $p(w_i|z_j), \sum_i^V p(w_i|z_j) = 1$. Simulating n words from a topic is analogous to throwing a die n times except that instead of 6 equiprobable outcomes on each throw we have V possible outcomes (where V can be on the order of 100,000 in practice) and the probabilities of individual outcomes (the words) may be significantly non-uniform. Table 2 shows two example topics that were learned from a large corpus (more details on learning below). The topic names are generally assigned manually. If we simulate data from one of these topics, the high probability words (shown in the figure) will occur with high frequency. A topic, in the form of a multinomial distribution over a vocabulary of words, can in a loose sense be viewed as a probabilistic representation of a semantic concept.

The topic model assumes that words in a document arise via a two-stage process: words are generated from topics and topics are generated by documents. More formally the distribution of words given a document, $p(w_i|d)$, is modeled as a mixture over topics:

$$p(w_i|d) = \sum_{j=1}^{T} p(w_i|z_j)p(z_j|d). \tag{1}$$

The topic variable z plays the role of a low-dimensional representation of the semantic content of a document.

Intuitively we can imagine simulating n words in a document by repeating the following steps n times: first, sample a topic z_j from the topic-document distribution $p(z|d)$, and then, given a topic z_j, sample a word from the corresponding word-topic distribution $p(w|z_j)$. For example, imagine that we have the following 5 topics with corresponding probability distributions over words: *earthquake, disaster response, international politics, China,* and *Olympic Games.* We could then represent individual documents as weighted combinations of this "basis set" of topics, e.g., one document could be a mixture of words from the topics *earthquake, disaster response,* and *China,* while another document could be a mixture of words from *China, international politics,* and *Olympic Games.*

By allowing documents to be composed of different combinations of topics, a topic model provides a more flexible representation of document content than clustering where each document is assumed to have been generated by a single cluster. Topics can also be considered a more natural representation for document content than the technique of latent semantic analysis (LSA) [14] since the multinomial basis of the topic model is better suited to predicting word counts than the inherently real-valued/least-squares framework that underlies LSA. A number of studies have shown that topic models provide systematically better results in document modeling and prediction compared to LSA ([15], [16]).

In the standard topic-modeling framework the word-topic distribution $p(w|z)$ and topic-document distributions $p(z|d)$ are learned in a completely unsupervised manner, without any prior knowledge of what words are associated with topics or what topics are associated with individual documents. The statistical estimation technique of Gibbs sampling is widely used [3]: starting with random assignments of words to topics, the algorithm repeatedly cycles through the words in the training corpus and samples a topic assignment for each word using the conditional distribution for that word given all other current word-topic assignments (see Appendix 1 for more details). After a number of such iterations through all words in the corpus (typically on the order of 100) the algorithm reaches a steady-state. The word-topic probability distributions can be estimated from the word-topic assignments. It is worth noting that topic model learning results in assignments of topics to each word in the corpus. This in turn directly enables "topic-tagging" of words, sentences, sections, documents, groups of documents, etc., a feature we will leverage later in this paper.

3   Semantic Concepts and Statistical Topic Modeling

We now return to the topic of concepts within ontologies and show how the statistical topic modeling techniques of the previous section can leverage text corpora to "overlay" probabilities on such concepts. As mentioned in the introduction, in this paper we focus

Table 3. Two example concepts from the CIDE thesaurus

Farming & Forestry		Earth & Outer Space	
crops	(0.135)	earth	(0.226)
plant	(0.076)	sky	(0.107)
grow	(0.050)	space	(0.082)
land	(0.040)	sun	(0.066)
fertilizers	(0.038)	scientists	(0.046)
soil	(0.037)	planets	(0.033)
earth	(0.034)	universe	(0.033)
farming	(0.034)	stars	(0.032)

on a simple aspect of ontological knowledge, namely sets of words associated with concepts.

Assume that we have been given a set of C human-defined concepts, where each concept c_j consists of a finite set of N_j unique words, $1 \leq j \leq C$. We also have available a corpus of documents such as Web pages. We propose to merge these two sources of information (concepts and documents) using a framework based on topic modeling. For example, we might be interested in "tagging" documents with concepts from the ontology, but with little or no supervised labeled data available (note that the approach we describe below can be easily adapted to include labeled documents if available). One way to approach this problem would be to assume a model in the form of a topic model, i.e.,

$$p(w_i|d) = \sum_{j=1}^{C} p(w_i|c_j)p(c_j|d). \qquad (2)$$

which is the same as Equation 1 but where we have replaced topics z with concepts c. We will refer to this type of model as the *concept model* throughout the paper. In the concept model the words that belong to a concept are defined by a human a priori (e.g., as part of an ontology) and are limited (typically) to a small subset of the overall vocabulary. In contrast, in a topic model, all words in the vocabulary can be associated with any particular topic but with different probabilities.

In Equation 2 above, the unknown parameters of the concept model are the word-concept probabilities $p(w_i|c_j)$ and the concept-document probabilities $p(c_j|d)$. Our goal (as in the topic model) is to estimate these from an appropriate corpus. Note for example that the probabilities $p(c_j|d)$ would address the afore-mentioned tagging problem, since each such distribution tells us the mix of concepts c_j that a document d is represented by.

We can use a modified version of statistical topic model learning algorithm to infer both $p(w_i|c_j)$ and $p(c_j|d)$. The process is to simply treat concepts as "topics with constraints," where the constraints consist of setting words that are not a priori mentioned in a concept to have probability 0, i.e., $p(w_i|c_j) = 0, w_i \notin c_j$. We can use Gibbs sampling to assign concepts to words in documents, using the same sampling equations as used for assigning topics to words in the topic model, but with the additional constraint that a word can only be assigned to a concept that it is associated with in the ontology[1]. Other than the constraint restriction, the learning algorithm is exactly the same as in standard learning of topic models, and the end result is that each word in the corpus is assigned to a concept in the ontology. In turn, these assignments allow us to directly estimate the terms of interest in Equation 2 above. To estimate $p(w_i|c_j)$ for a particular concept c_j we count how many words in the corpus were assigned by the sampling algorithm to concept c_j and normalize these counts (and typically also smooth them) to arrive at the probability distribution $p(w_i|c_j)$. To estimate $p(c_j|d)$ for a particular document d, we count how many times each concept is assigned to a word in document d and again normalize and smooth the counts to obtain $p(c_j|d)$. Table 3 shows an example of a set of learned probabilities for words (ranked highest by probability) for two different concepts from the CIDE ontolgy, after training on the TASA corpus (more details on ontologies and data sets are provided later).

The important point to note here is that we have defined a straightforward way to "marry" the qualitative information in sets of words in human-defined concepts with quantitative data-driven topics. The learning algorithm itself is not innovative, but the application is innovative in that it combines two sources of information (concepts from ontologies and statistical learning) that to our knowledge have not been combined in any general framework in prior work. We can use the learned probabilistic representation of concepts to map new documents into concepts within an ontology, and we can use the semantic concepts to improve the quality of data-driven topic models. We will explore both of these ideas in more detail in later sections of the paper.

There are numerous variations of the concept model framework that can be explored—we investigate some of the more obvious extensions below. For example, a baseline model is one where the word-concept probabilities $p(w_i|c_j)$ are defined to be uniform for all words within a concept. A related model is one where the word-concept probabilities are available a priori as part of the concept definition, e.g., where documents are provided with each concept allowing for empirical word-concept probabilities to be estimated. For both of these models, Gibbs sampling is still used as before to infer the word-concept assignments and the concept-document probabilities, but the $p(w|c)$ probabilities are held fixed and not learned. We will refer to these two models as ConceptU (concept-uniform) and ConceptF (concept-fixed) and use ConceptL (concept-learned) to refer to the more general concept model described earlier where the $p(w|c)$ probabilities are learned from the corpus.

Human-generated concepts not only come with words associated with concepts but are also often arranged in a hierarchical structure such as a concept tree, where each node is a concept with a set of associated words. A simple way to incorporate this hierarchical information is to propagate the words upwards in the concept tree, so that an internal concept node is associated with its own words and all the words associated with its children. When we use this propagation technique for representing the word-concept associations, we will refer to this by adding an "H" to the name of the learned model, e.g., ConceptLH, ConceptFH, etc.

---

[1] An alternative approach, not explored in this paper, would be to use the concept words to build an informative prior on topics rather than using them as a hard constraint. Under such an approach, each concept could be associated with any word in the corpus leading to significant computational demands since large ontologies could have tens of thousands of concepts.

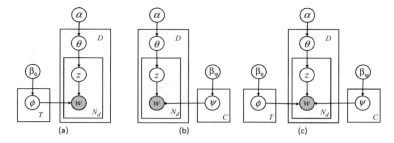

Fig. 1. Graphical models for (a) Topic model, (b) Concept model, and (c) Concept-topic model

Finally, a natural further extension of the model is to allow for incorporation of unconstrained data-driven topics alongside the concepts. This can be achieved by simply allowing the Gibbs sampling procedure to either assign a word to a constrained concept or to one of the unconstrained topics (see Appendix 1). In such a model a document is represented by a mixture over C concepts and T topics, allowing the model to use additional data-driven topics to represent themes that are not well-represented in the set of concepts in the ontology. We will in general refer to such models as concept-topic models and specific variations by ConceptL+Topics, ConceptLH+Topics etc.

Figure 1 shows a graphical model representation of the various models, including the standard topic model, the concept model, and the concept-topic model. Here, ϕ, ψ and θ represent word-topic, word-concept and topic-document/concept-document multinomial distributions respectively. β_ϕ, β_ψ and α represent the Dirichlet priors on ϕ, ψ and θ respectively. Further details on sampling equations for all of the model variants are provided in Appendix 1.

4   Concept Sets and Text Data

The experiments in this paper are based on one large text corpus and two different knowledge bases. For the text corpus, we used the Touchstone Applied Science Associates (TASA) dataset [14]. This corpus consists of $D = 37,651$ documents with passages excerpted from educational texts used in curricula from the first year of school to the first year of college. The documents are divided into 9 different educational topics. In this paper, we focus on the documents classified as SCIENCE and SOCIAL STUDIES, consisting of $D = 5356$ and $D = 10,501$ documents and 1.7M and 3.4M word tokens respectively.

The first set of concepts we used was the Open Directory Project (ODP), a human-edited hierarchical directory of the web (available at http://www.dmoz.org). The ODP database contains descriptions and urls on a large number of hierarchically organized topics. We extracted all the topics in the SCIENCE subtree, which consists of $C = 10,817$ nodes after preprocessing. The top concept in this hierarchy starts with SCIENCE and divides into concepts such as ASTRONOMY, MATH, PHYSICS, etc. Each of these topics divides again into more specific concepts with a maximum number of 11 levels. Each node in the hierarchy is associated with a set of urls related to the concept plus a set of human-edited descriptions of the site content. To create a bag of words

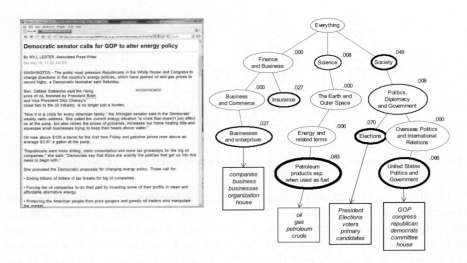

Fig. 2. Example of using the ConceptU model to automatically tag a Web page with CIDE concepts

representation for each node, we collected all the words in the textual descriptions and also crawled the urls associated with the node (a total of 78K sites). This led to a vector of word counts for each node.

The second source of concepts in our experiments was a thesaurus from the Cambridge International Dictionary of English (CIDE; www.cambridge.org/elt/cide). CIDE consists of $C = 1923$ hierarchically organized semantic categories. In contrast to other taxonomies such as WordNet [17], CIDE groups words primarily according to semantic concepts with the concepts hierarchically organized. The hierarchy starts with the concept EVERYTHING which splits into 17 concepts at the second level (e.g. SCIENCE, SOCIETY, GENERAL/ABSTRACT, COMMUNICATION, etc). The hierarchy has up to 7 levels. The concepts vary in the number of the words with a median of 54 words and a maximum of 3074. Each word can be a member of multiple concepts, especially if the word has multiple senses.

5   Tagging Documents with Concepts

One application of concept models is to tag documents such as Web pages with concepts from the ontology. The tagging process involves assigning likely concepts to each word in a document, depending on the context of the document. The document content can then be summarized by the probability distribution over concepts that reveal the dominant semantic themes. Because the concept models assign concepts at the word level, the results can be aggregated in many ways, allowing for document summaries at multiple levels of granularity. For example, tagging can be performed on snippets of text, individual sections of a Web page, whole Web pages or even collections of Web pages. Figure 2 illustrates the effect of tagging a Web page with CIDE concepts

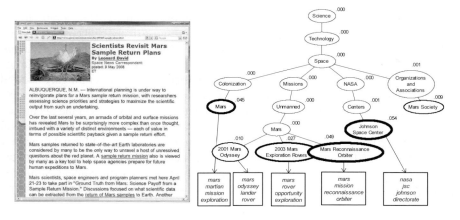

Fig. 3. Example of using the ConceptU model to automatically tag a Web page with ODP concepts

using the ConceptU model. For the purpose of illustration, the six highest probability concepts along will their parents and ancestors are shown. The thickness of the ellipse encapsulating a concept node is proportional to the probability of the concept in the Web page. The rectangular boxes contain words from the Web page that were assigned to the corresponding concept in decreasing order of frequency. Figure 3 shows an example of tagging another Web page using the ConceptU model with concepts from the ODP ontology, with "Johnson Space Center" and "Mars Reconnaissance Orbiter" among the high probability concepts. For these tagging illustrations, we ran 1500 Gibbs sampling chains and each chain was run for 50 iterations after which a single sample was taken.

Figure 4 illustrates concept assignments to individual words in a TASA document with CIDE concepts. The four most likely concepts are listed for this document. For

tag	$P(c\|d)$	Concept	$P(w\|c)$
a	0.1702	PHYSICS	electrons (0.2767) electron (0.1367) radiation (0.0899) protons (0.0723) ions (0.0532) radioactive (0.0476) proton (0.0282)
b	0.1325	CHEMICAL ELEMENTS	oxygen (0.3023) hydrogen (0.1871) carbon (0.0710) nitrogen (0.0670) sodium (0.0562) sulfur (0.0414) chlorine (0.0398)
c	0.0959	ATOMS, MOLECULES, AND SUB-ATOMIC PARTICLES	atoms (0.3009) molecules (0.2965) atom (0.2291) molecule (0.1085) ions (0.0262) isotopes (0.0135) ion (0.0105) isotope (0.0069)
d	0.0924	ELECTRICITY AND ELECTRONICS	electricity (0.2464) electric (0.2291) electrical (0.1082) current (0.0882) flow (0.0448) magnetism (0.0329)
o	0.5091	OTHER	

The hydrogen[b] ions[a] immediately[o] attach[o] themselves to water[o] molecules[c] to form[o] combinations[o] called[o] hydronium ions[a]. The chlorine[b] ions[a] also associate[o] with water[o] molecules[c] and become hydrated. Ordinarily[o], the positive[o] hydronium ions[a] and the negative[o] chlorine[b] ions[a] wander[o] about freely[o] in the solution[o] in all directions[o]. However, when the electrolytic cell[o] is connected[o] to a battery[o], the anode[d] becomes positively[o] charged[d] and the cathode[d] becomes negatively[o] charged[d]. The positively[o] charged[d] hydronium ions[a] are then attracted[o] toward the cathode[d] and the negatively[o] charged[d] chlorine[b] ions[a] are attracted[o] toward the anode[d]. The flow[d] of current[d] inside[o] the cell[o] therefore consists of positive[o] hydronium ions[a] flowing[d] in one direction[o] and negative[o] chlorine[b] ions[a] flowing[d] in the opposite[o] direction[o]. When the hydronium ions[a] reach[o] the cathode[d], which has an excess[o] of electrons[a], each takes[o] one electron[a] from it and thus neutralizes[o] the positively[o] charged[d] hydrogen[b] ion[a] attached[o] to it. The hydrogen[b] ions[a] thus become hydrogen[b] atoms[c] and are released[o] into the solution[o]. Here they pair[o] up to form[o] hydrogen[b] molecules[c] which gradually[o] come out of the solution[o] as bubbles[o] of hydrogen[b] gas[o]. When the chlorine[b] ions[a] reach[o] the anode[d], which has a shortage[o] of electrons[a], they give[o] up their extra[o] electrons[a] and become neutral[a] chlorine[b] atoms[c]. These pair[o] up to form[o] chlorine[b] molecules[c] which gradually[o] come out of the solution[o] as bubbles[o] of chlorine[b] gas[o]. The behavior[o] of hydrochloric acid[o] solution[o] is typical[o] of all electrolytes[o]. In general[o], when acids[b], bases[o], and salts[o] are dissolved[o] in water[o], many of their molecules[c] break[o] up into positively[o] and negatively[o] charged[d] ions[a] which are free[o] to move[o] in the solution[o].

Fig. 4. Example of tagging at the word level using the ConceptL model

each concept, the estimated probability distribution over words is shown next to the concept. In the document, words assigned to the four most likely concepts are tagged with letters a-d (and color coded if viewing in color). The words assigned to any other concept are tagged with "o" and words outside the vocabulary are not tagged. In the concept model, the distributions over concepts within a document are highly skewed such that most probability goes to only a small number of concepts. In the example document, the four most likely concepts cover about 50% of all words in the document.

The figure illustrates that the model correctly disambiguates words that have several conceptual interpretations. For example, the word *charged* has many different meanings and appears in 20 CIDE concepts. In the example document, this word is assigned to the physics concept which is a reasonable interpretation in this document context. Similarly, the ambiguous words *current* and *flow* are correctly assigned to the electricity concept.

6   Language Modeling Experiments

To quantitatively measure the quality of the concept models described in the earlier parts of the paper, we perform a set of systematic experiments that compare the quality of concept models and baselines. To do this we use standard techniques from language modeling that measure the predictive power of a model in terms of its ability to predict words in unseen documents.

6.1   Perplexity

Perplexity is widely used as a quantitative measure for comparing language models, e.g. [18]. It can be interpreted as being proportional to the distance (formally, the cross-entropy) between the word distribution learned by the model and the distribution of words in an unseen test document. Thus, lower scores are better since they indicate that the model's distribution is closer to that of the actual text. The perplexity of a test data set is defined as:

$$\text{Perp}(\mathbf{w}_{test}|\mathcal{D}^{\text{train}}) = \exp\left(-\frac{\sum_{d=1}^{D_{test}} \log p(\mathbf{w}_d|\mathcal{D}^{\text{train}})}{\sum_{d=1}^{D_{test}} N_d}\right)$$

where $\mathbf{w}_{test}$ is the words in test documents, $\mathbf{w}_d$ are words in document d of the test set, $\mathcal{D}^{\text{train}}$ is the training set, and N_d is the number of words in document d.

In the experiments that follow we partition the text corpus into disjoint training and test sets, with 90% of the documents being used for training and the remaining 10% for computing test perplexity. For each test document d, a randomly selected subset of 50% of the words in the document are assumed to be observed and used to estimate the document-specific parameters $p(c|d)$ and/or $p(z|d)$ via Gibbs sampling. Perplexity is then computed on the remaining 50% of the words in the document (a form of perplexity known as predictive-perplexity).

In our experiments below we use perplexity to evaluate the relative quality of different concept and concept-topic models. Although no single quantitative measure will necessarily provide an ideal measure of how well human concepts and a corpus are matched, we argue that perplexity scores have the appropriate behavior. In particular,

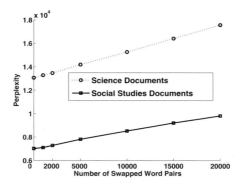

Fig. 5. Perplexity as a function of precision

perplexity will be sensitive to both the precision and recall of a knowledge-base in relation to a corpus. Precision in this context should measure the semantic coherence of words within a concept and recall should be sensitive to how well the concepts cover a body of knowledge (e.g., as represented by a corpus) [12]. Therefore, as precision or recall increase we expect perplexity to decrease. We illustrate this (for precision) with a simulated experiment where we swap words randomly between CIDE concepts (to intentionally "corrupt" the concepts) and then measure the quality of the resulting concept model on the TASA corpus using the ConceptU model. As the number of words swapped increases (x-axis in Figure 5) the precision decreases, and the resulting perplexity very clearly reflects the deterioration in the quality of the concepts. Thus, perplexity appears to be a reasonable surrogate measure for more ontology-specific notions of quality such as precision.

6.2   General Perplexity Results Across Models

We created a single $W = 33,635$ word vocabulary based on the 3-way intersection between the vocabularies of TASA, CIDE, and ODP. This vocabulary covers 89.9% of all of the word tokens in the TASA corpus and is the vocabulary that is used in all of the experiments reported in this paper. We also generated the same set of experimental results below using the union of words in TASA and CIDE and TASA and ODP, and found the same general behavior as with the intersection vocabulary. We report the intersection results below and omit the union results as they are essentially identical to the intersection results. A useful feature of using the intersection is that it allows us to evaluate two different sets of concepts (TASA and CIDE) on a common data set (TASA) and vocabulary, e.g., to evaluate which set of human-defined concepts better predicts a given set of text data. Note that selecting a predefined vocabulary (whether the intersection or the union) bypasses the important practical problem of modeling "out of vocabulary" words that may be seen in new documents. Although this is an important aspect of language modeling in general, in this paper our primary focus is on combining human defined concepts and data-derived topics.

Table 4. Perplexity scores for various models

Model	Science		SocialStudies	
	CIDE	ODP	CIDE	ODP
ConceptU	7019	5787	13071	9476
ConceptF	n/a	3651	n/a	7244
ConceptL	1461	1060	3479	2432
ConceptLH	1234	1014	2768	2298
ConceptLH+Topics (T=100)	1100	1014	2362	2297

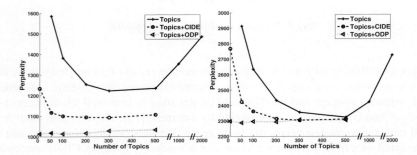

Fig. 6. Comparing perplexity for the Topics model with the ConceptsLH + Topics model on science (left) and social studies (right)

Table 4 shows predictive perplexity scores for a variety of models using the TASA corpus with the CIDE or ODP concepts. In terms of general trends, there is a systematic reduction in perplexity scores as more corpus-specific information is combined with the concepts. The concept models with uniform distributions (ConceptU) have relatively high perplexity scores, indicating that a uniform distribution over concept terms are a poor fit to the data as one would expect. Using the Web-derived distributions for ODP (ConceptF) leads to a significant reduction over uniform distributions.

Learning the word-concept distributions (ConceptL) yields a further significant decrease in perplexity scores compared to the fixed concept distributions as the concepts can now adapt to the corpus. Additionally, accounting for the hierarchy of the concepts (ConceptLH), by propagating words from child concepts to their parents as mentioned before, reduces perplexity even further. If we then add 100 topics to the ConceptLH model (ConceptLH+Topics (T=100) in Table 4), for the CIDE concepts we see another significant reduction in perplexity for both corpora, but no change for the ODP concepts. ODP concepts on their own (ConceptLH models) have lower perplexities than CIDE concepts, so there seems to be more room for improvement with CIDE when topics are added. In addition, ODP has far more concepts (over 10,000) than CIDE (1923), with the result that in the Topics+ODP model less than 1% of the words are assigned to Topics. Overall the ODP concepts produce lower perplexities than CIDE— probably because of the larger number of concepts in ODP, although in general it need not be the case that more concepts lead to better predictions.

6.3   Varying the Number of Unconstrained Topics

Natural next questions to ask are how would topic models on their own perform and how do the results vary as a function of the number of topics? We address these questions in Figure 6. In this and later experiments in the paper we are using the hierarchical (H) versions of the concept models. The curves in each graph represent topics on their own and topics combined with CIDE and ODP concepts. The x-axis represents the number of topics T used in each model. For example, the point $T = 0$ represents the conceptL models. The results clearly indicate that for any topic model with a fixed number of topics T (a particular point on the x-axis), the performance of the topic model is always improved when concepts are added. The performance improvement is particularly significant on the Science documents, which can be explained by the fact that both CIDE and ODP have well-defined science concepts. It is important to note that the performance difference between topic and concept-topic models is not because of a high number of effective topics $(T + C)$ in the concept-topic models. In fact, when we increase the number of topics to $T = 2,000$ for the topic model its perplexity increases significantly possibly due to overfitting. In contrast, the ODP model (for example) is using over 10,000 effective topics $(T + C)$ and achieving a lower perplexity score than topics alone. This is a direct illustration of the power of prior knowledge: the constraints represented by human-defined concepts lead to a better language model than what can be obtained with data-driven learning alone.

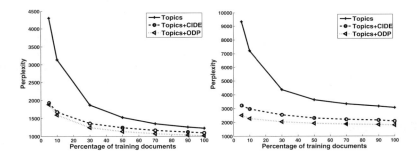

Fig. 7. Perplexity as a function of the amount of training data, testing on science documents, using training data from science (left) and social studies (right)

6.4   The Effect of Training Data Size

Finally we look at the effect of varying the amount of training data. The number of topics T used for each model was set to value that produced the lowest perplexity with all of the training data (based on results in Figure 6). Figures 7 and 8 show the perplexity results using science and social studies documents respectively as a test data set. The left plot in each figure shows the results when the training data set and test data set come from the same source and the right plot using different training and test data source.

When there is relatively little training data the concept-topic models have significantly lower perplexity than the topic model. This is a quantitative verification of the

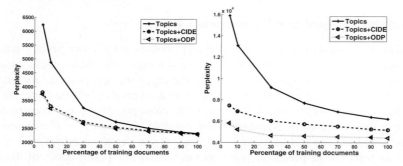

Fig. 8. Perplexity as a function of the amount of training data, testing on social studies documents, using training data from social studies (left) and science (right)

oft-quoted idea that "prior knowledge is particularly useful in learning when there is little data." The concept models are helped by the restricted word associations that are manually selected on the basis of their semantic similarity, providing an effective "prior" on words that are expected to co-occur together. The restricted word associations can also help in estimating more accurate word distributions with less data. While it may not be apparent from the figures due the scale used, even at the 100% training data point the concept-topic models have lower perplexity than the topic model (e.g. in Figure 7 at the 100% point on the left, the perplexities of the topic model and the concept-topic model using ODP are 1223.0 and 1013.9 respectively).

As expected, the perplexities are in general higher when a model is trained on one class and predictions are made on a different class (right plots in both the figures). What is notable is that the gap in perplexities between topics and topics+concepts is greater in such cases, i.e., prior knowledge in the form of concepts is even more useful when a model is used on new data that it is different to what it was trained on.

7   Conclusions

We have proposed a general probabilistic text modeling framework that can use both human-defined concepts and data-driven topics. The resulting models allow us to combine the advantages of prior knowledge from the form of ontological concepts and data-driven learning in a systematic manner—for example, the model can automatically place words and documents in a text corpus into a set of human-defined concepts. We also illustrated how concepts can be "tuned" to a corpus to obtain a probabilistic language model leading to improved language models compared with either concepts or topics on their own.

We view the framework presented in this paper as a starting point for exploring a much richer set of models that combine ontological knowledge bases with statistical learning techniques. In Chemudugunta, Smyth and Steyvers [19], we extend the model proposed in this paper to include explicit representation of concept hierarchies. Obvious next steps for exploration are treating concepts and topics differently in the generative model, integrating multiple ontologies and corpora within a single framework, and so forth.

Acknowledgments

The work of the authors was supported in part by the National Science Foundation under Award Number IIS-0083489 as part of the Knowledge Discovery and Dissemination program. In addition, the work of author PS was supported in part by a Google Research Award.

References

1. McGuinness, D.L.: Ontologies come of age. In: Fensel, D., Hendler, J.A., Lieberman, H., Wahlster, W. (eds.) Spinning the Semantic Web, pp. 171–194. MIT Press, Cambridge (2003)
2. Blei, D.M., Ng, A.Y., Jordan, M.I.: Latent Dirichlet allocation. J. Mach. Learn. Res. 3, 993–1022 (2003)
3. Griffiths, T.L., Steyvers, M.: Finding scientific topics. Proc. of Nat'l. Academy of Science 101, 5228–5235 (2004)
4. Handschuh, S., Staab, S., Ciravegna, F.: Scream — semi-automatic creation of metadata. In: International Conference on Knowledge Engineering and Knowledge Management (2002)
5. Popov, B., Kiryakov, A., Kirilov, A., Manov, D., Ognyanoff, D., Goranov, M.: Kim - semantic annotation platform. In: International Semantic Web Conference, pp. 834–849 (2003)
6. Tang, J., Hong, M., Li, J.Z., Liang, B.: Tree-structured conditional random fields for semantic annotation. In: International Semantic Web Conference, pp. 640–653 (2006)
7. Dill, S., Eiron, N., Gibson, D., Gruhl, D., Guha, R., Jhingran, A., Kanungo, T., Rajagopalan, S., Tomkins, A., Tomlin, J.A., Zien, J.Y.: Semtag and seeker: bootstrapping the semantic web via automated semantic annotation. In: WWW 2003, pp. 178–186. ACM, New York (2003)
8. Hotho, A., Staab, S., Stumme, G.: Text clustering based on background knowledge (technical report 425). Technical report, University of Karlsruhe, Institute AIFB (2003)
9. Gabrilovich, E., Markovitch, S.: Harnessing the expertise of 70,000 human editors: Knowledge-based feature generation for text categorization. J. Mach. Learn. Res. 8, 2297–2345 (2007)
10. Ifrim, G., Theobald, M., Weikum, G.: Learning word-to-concept mappings for automatic text classification. In: Proceedings of the 22nd ICML-LWS, pp. 18–26 (2005)
11. Boyd-Graber, D., Blei, D., Zhu, X.: A topic model for word sense disambiguation. In: Proc. 2007 Joint Conf. Empirical Methods in Nat'l. Lang. Processing and Compt'l. Nat'l. Lang. Learning, pp. 1024–1033 (2007)
12. Brewster, C., Alani, H., Dasmahapatra, S., Wilks, Y.: Data driven ontology evaluation. In: Int'l. Conf. Language Resources and Evaluation (2004)
13. Alani, H., Brewster, C.: Metrics for ranking ontologies. In: 4th Int'l. EON Workshop, 15th Int'l World Wide Web Conf. (2006)
14. Landauer, T.K., Dumais, S.T.: A solution to Plato's problem: The latent semantic analysis theory of the acquisition, induction and representation of knowledge. Psychological Review 104, 211–240 (1997)
15. Griffiths, T.L., Steyvers, M., Tenenbaum, J.B.: Topics in semantic representation. In: Psychological Review, vol. 114, pp. 211–244 (2007)
16. Chemudugunta, C., Smyth, P., Steyvers, M.: Modeling general and specific aspects of documents with a probabilistic topic model. In: NIPS, vol. 19, pp. 241–248 (2007)
17. Fellbaum, C. (ed.): WordNet: An Electronic Lexical Database (Language, Speech and Communication). MIT Press, Cambridge (1998)
18. Brown, P.F., de Souza, P.V., Mercer, R.L., Pietra, V.J.D., Lai, J.C.: Class-based n-gram models of natural language. Compt'l. Linguistics, 467–479 (1992)
19. Chemudugunta, C., Smyth, P., Steyvers, M.: Combining concept hierarchies and statistical topic models. In: 17th ACM Conference on Information and Knowledge Management (2008)

Appendix 1: Inference Using Collapsed Gibbs Sampling

Here, we briefly describe the sampling process for the concept-topic model and then describe how sampling for the other models can be viewed as special-cases of this model.

In the concept-topic model, ϕ, ψ and θ correspond to $p(w|t)$ word-topic distributions, $p(w|c)$ word-concept distributions and $p(z|d)$ document level mixtures of topics+concepts respectively. β_ϕ, β_ψ and α correspond to Dirichlet priors on ϕ, ψ and θ multinomial distributions respectively.

In the collapsed Gibbs sampling procedure, the topic assignment variables z_i can be efficiently sampled (after marginalizing the multinomial distributions θ, ϕ and ψ). Point estimates for the marginalized distributions θ, ϕ and ψ can be computed given the assignment labels z_i and predictive distributions are computed by averaging over multiple samples. The sampling equations for the concept-topic model are given by, case (i): $1 \leq \mathbf{z}_i \leq T$

$$P(\mathbf{z}_i = t | \mathbf{w}_i = w, \mathbf{w}_{-i}, \mathbf{z}_{-i}, \alpha, \beta_\phi) \propto \frac{C^{WT}_{wt,-i} + \beta_\phi}{\sum_{w'} C^{WT}_{w't,-i} + W\beta_\phi} (C^{(T+C)D}_{td,-i} + \alpha)$$

case (ii): $\mathbf{z}_i > T$

$$P(\mathbf{z}_i = t | \mathbf{w}_i = w, \mathbf{w}_{-i}, \mathbf{z}_{-i}, \alpha, \beta_\psi) \propto \frac{C^{WC}_{wc,-i} + \beta_\psi}{\sum_{w'} C^{WC}_{w'c,-i} + N_c\beta_\psi} (C^{(T+C)D}_{td,-i} + \alpha)$$

where C^{WT}_{wt}, C^{WC}_{wc} are the number of times word w is associated with topic t and concept c respectively, $C^{(T+C)D}_{td}$ is the number of times topic (or concept) t is associated with document d, $c = t - T$ and is only defined for case (ii) and N_c is the number of words associated with concept c. Subscript $-i$ denotes that the word w_i is removed from the counts.

When the concept distributions are fixed (e.g. for the ConceptU model), the inference becomes even simpler as we can just use the fixed distributions in the above equations. Also, note that the topic model and the concept models are special cases of the concept-topic model when $C = 0$ and $T = 0$ respectively. Therefore, we can easily adapt the sampling scheme described above to do inference for both these models. It is important to note that the inference for a concept model with N concepts is much faster than the inference of a topic model with N topics. This is because in the case of the concept model we can exploit the sparsity in the word-concept associations — for any word, only the probabilities over concepts that the word is a member of need to be calculated.

We use the standard setup from well-known publications and set $\alpha=50/(T+C)$, $\beta_\phi=\beta_\psi=0.01$ for models where they are defined. For all our models, we compute the predictive distributions by averaging over 10 different Gibbs chains that are run for 500 iterations and take the last sample to compute the point estimates for the various multinomial distributions.

Comparison between Ontology Distances
(Preliminary Results)

Jérôme David and Jérôme Euzenat

INRIA Grenoble Rhône-Alpes & LIG
Grenoble, France
Jerome.{David,Euzenat}@inrialpes.fr

Abstract. There are many reasons for measuring a distance between ontologies. In particular, it is useful to know quickly if two ontologies are close or remote before deciding to match them. To that extent, a distance between ontologies must be quickly computable. We present constraints applying to such measures and several possible ontology distances. Then we evaluate experimentally some of them in order to assess their accuracy and speed.

1 Motivations

The semantic web aims at exploiting formal knowledge at the world scale. It is, in particular, based on ontologies: a structure defining concepts used to represent knowledge and their relationships. These concepts are used for specifying semantic web services, annotating web resources (pictures, web pages, music) or for describing data flows.

It is however likely that different information sources will use different ontologies. It is thus necessary to find correspondences between ontologies in order to communicate from one ontology to another. Finding correspondences is called matching ontologies and the resulting set of correspondences is called an alignment [Euzenat and Shvaiko, 2007].

Together with matching ontologies, there are many occasions where it is useful to know if two ontologies are close to each others or not, or what is the closest ontology to another one. In particular,

- when one wants to find the community of people with whom she will be more likely to communicate easily, finding if they use similar ontologies can be useful information [Jung and Euzenat, 2007]; This can also help identifying communities in social networks [Jung et al., 2007];
- in semantic peer-to-peer systems, it will be easier to find information if queries can be sent to nodes using similar ontologies because query transformation will miss less information [Ehrig et al., 2005];
- in ontology engineering, it is useful to find similar ontologies that can be easily used in conjunction with other ones. For example, when developing an ontology for radiological diagnoses, it would be useful to find anatomy and pathology ontologies that can be used with each other;

A. Sheth et al. (Eds.): ISWC 2008, LNCS 5318, pp. 245–260, 2008.

- when modularising large ontologies into smaller parts [Stuckenschmidt and Klein, 2004], it is useful to consider the module candidates as sub-ontologies which will be more prone to be separated as they are distant to each others;
- in semantic search engines which return ontologies corresponding to a query [d'Aquin *et al.*, 2007], it would be useful to introduce a "Find similar ontologies" button. Distances can also be used in this case for ordering answers to such a query (ontology ranking, [Alani and Brewster, 2005]) with regard to ontology proximity;
- in some ontology matching algorithms [Gracia *et al.*, 2007] when one wants to use an intermediate ontology between two ontologies, it may be useful to select the closest ontology.

In these various applications, there are different requirements for an ontology distance measure. In particular, there is always a trade-off between speed and accuracy. We will review some of these possible measures and propose a first evaluation of their qualities.

The remainder of this paper is as follows: we first present and discuss previous work. Then, after recalling general definitions about distance measures, we introduce constraints applying to distances between ontologies. The next section introduces some ontology distances. Finally, we evaluate these measures and with regard to the criteria.

2   Related Works

Most of the work dealing with ontology distance [Mädche and Staab, 2002; Hu *et al.*, 2006; Vrandečić and Sure, 2007] is in reality concerned with concept distances. Such measures are widely used in ontology matching algorithms [Euzenat and Shvaiko, 2007]. They are quickly extended to ontologies without discussing the different ways to achieve this.

[Mädche and Staab, 2002] introduced a concept similarity based on terminological and structural aspects of ontologies. This very precise proposal combines an edit distance on strings and a structural distance on hierarchies (the cotopic distance). The ontology similarity strongly relies on the terminological similarity. This paper evaluates the ontology design process, but not ontology similarity.

The framework presented in [Ehrig *et al.*, 2005] aims at comparing concepts across ontologies instead of ontologies themselves. It provides a similarity combining string similarity, concept similarity – considered as sets – and similarity across usage traces.

There is also a quite elaborate framework in [Hu *et al.*, 2006]. This paper is mostly dedicated to the comparison of concepts but can be extended to ontologies. First, concepts are expanded so they are expressed in term of primitive concepts. Each concept is expressed as a disjunction of compound but conjunctive concepts. This works as long as no cycle occurs in the ontology. Then primitive concepts are considered as dimensions in a vector space and each concept is represented in this space. The weights used in this vector space are computed with TF·IDF. The distance between two concepts is the smallest cosine distance between vectors associated with disjuncts describing concepts. The way this is extended to ontology concepts is not clearly explained but the methods that will be explained in §4.3 would work.

Finally, [Vrandečić and Sure, 2007] more directly considered metrics evaluating ontology quality. This is nevertheless one step towards semantic measures since they introduce normal forms for ontologies which could be used for developing syntactically neutral measures.

A general comment about these works is that they rely of elaborate distance or similarity measures between concepts and they extend these measures to distance between ontologies. This extension is often considered as straightforward. However, they have barely been evaluated. This is what we attempt to do here.

3   Distances Properties

In this section, we first introduce the ontology model which we used and then review the general properties that distances between ontologies must satisfy.

3.1   Ontology Model

All measure will be based on a set of ontologies O which we refer to as the ontology space. For simplification purposes, an OWL ontology $o \in O$ is represented as a set of named entities E_o. These entities can be classes (C), properties (P) or individuals (I): $E = C \cup P \cup I$. Each entity is identified by a URI thanks to the function $uri : E \longrightarrow URI$. The function $l_{ln} : E \longrightarrow String$ returns the local name of the entity which is the specific part of the entity URI in the ontology. Each entity can be also described by annotations, i.e., labels, comments. The function $l_{annot} : E \longrightarrow \mathcal{P}(String)$ assigns a set of annotations to each entity.

3.2   Algebraic Distance Properties

A dissimilarity is a real positive function d of two ontologies which is as large as ontologies differ.

Definition 1 (Dissimilarity). *Given a set O of ontologies, a dissimilarity $\delta : O \times O \to \mathbb{R}$ is a function from a pair of ontologies to a real number such that:*

$$\forall o, o' \in O, \delta(o, o') \geq 0 \qquad \textit{(non-negativeness)}$$
$$\forall o \in O, \delta(o, o) = 0 \qquad \textit{(minimality)}$$
$$\forall o, o' \in O, \delta(o, o') = \delta(o', o) \qquad \textit{(symmetry)}$$

Some authors consider a 'non symmetric (dis)similarity', [Tverski, 1977]; we then use the term non symmetric measure or pre-similarity. There are more constraining notions of dissimilarity, such as distances and ultrametrics.

Definition 2 (Distance). *A distance (or metric) $\delta : O \times O \to \mathbb{R}$ is a dissimilarity function satisfying the definiteness and triangular inequality:*

$$\forall o, o' \in O, \delta(o, o') = 0 \text{ if and only if } o = o' \qquad \textit{(definiteness)}$$
$$\forall o, o', o'' \in O, \delta(o, o') + \delta(o', o'') \geq \delta(o, o'') \qquad \textit{(triangular inequality)}$$

There are in fact many reasons why an ontology measure may not be a distance. In particular, if we want to consider the semantics of ontologies, a sheer semantic measure should be 0 when the two arguments are semantically equivalent, even if they are not the same. For the sake of finding a distance, we must work in the quotient space in which the congruence relation is semantic equivalence. However, given the cost of computing semantic equivalence we will try to avoid that.

We will see below that there are good reasons to avoid symmetry as well.

Very often, the measures are normalised, especially if the dissimilarity of different kinds of entities must be compared. Reducing each value to the same scale in proportion to the size of the considered space is the common way to normalise.

Definition 3 (Normalised measure). *A measure is said to be* normalised *if it ranges over the unit interval of real numbers* [0 1]. *A normalised version of a measure* δ *is denoted as* $\bar{\delta}$.

In the remainder, we will consider mostly normalised measures and assume that a dissimilarity function between two entities returns a real number between 0 and 1.

3.3   Application-Specific Distance Properties

One could imagine some properties which are unrelated to the general notion of distance but are specific to its use. In addition to algebraic properties, we would like to express purpose-oriented constraints on the measure. Such constraints must ask that the smaller the distance,

- the faster it is to provide an alignment;
- the more entities correspond to entities of the other ontology;
- the more entities of the other ontology correspond to entities of this ontology;
- the closest are corresponding entities;
- the easier (the faster) it is to answer queries;
- ...

For example, we could take into account a property stating that the addition of specific information in one ontology implies an increase of the distance value:

$$\forall o, o', o'' \in O, o'' \cap o = \emptyset \Rightarrow \delta(o, o') \leq \delta(o, o' \cup o'')$$

Contrarily, the addition of information issued from the other ontology implies a decrease of the distance value:

$$\forall o, o', o'' \in O, o'' \subseteq o - o', \Rightarrow \delta(o, o' \cup o'') \leq \delta(o, o')$$

These first properties show that more ontologies share concepts, lesser is their distance. Nevertheless, they are useful only if we consider ontologies having entities which match perfectly. In concrete cases, the ontologies are sufficiently heterogeneous and consequently, this property cannot be satisfied.

4   Ontology Distances

When only two ontologies are available, ontology distances have to be computed by comparing them. On the basis of such measures, systems will decide between which ontologies to run a matching algorithm. They can measure the ease of producing an alignment (expected speed, expected quality). So naturally, one constraint is that the distance be computed faster than the actual alignment.

There are many possible ways to define a distance between ontologies. First of all, an ontology can be just viewed as a bag of terms. This approach is similar to those used in information retrieval based on the vector space model. These techniques relies on vector representations of ontologies and use distances measures between these vectors.

Another approach is to consider an ontology as a set of entities. These entities will depend on the techniques used for establishing the distance: they will generally be the classes or properties to be found within the ontologies. In this case, defining a distance between the ontologies will very often rely on:

- a distance (δ) or similarity (sim) measure between entities;
- a collection distance (Δ) which will use the distance between entities for computing a distance between ontologies.

We first present ontology distances based on the vector space model. Then, we consider some distances between entities before presenting various kinds of collection distances.

4.1   Ontology Distances Based on the Vector Space Model

A distance can be computed by comparing the sets of labels appearing in both ontologies and using a measure such as the Hamming distance, i.e., the complement to 1 of the ratio of common terms over the whole set of terms used by any of the ontologies. This distance would certainly run faster than any serious matching algorithm but does not tell a lot about the matching process. However, more elaborate measures based on the vector space model (VSM) have been designed.

In the VSM, each ontology is represented by a vector of terms. These terms are extracted from the annotations of the ontology entities. The set of terms T_o contained in an ontology o is build with the help of a term extraction function l_{te}:
$T_o = \bigcup_{e \in E_o} l_{te}(l_{annot}(e)) \cup l_{ln}(e)$.

Let O be, a set of ontologies and T be the set of terms contained in these ontologies. The vector of terms representing an ontology o is $\overrightarrow{D_O} = (w_1, ..., w_n)$ where each w_i represents the weight of term $t_i \in T$ for the ontology o. We have selected three types of weights :

- boolean weights: $w_i = 1$ if t_i occurs in o, $w_i = 0$ otherwise.
- frequency weights : $w_i = TF(t_i, o)$
- TF·IDF [Robertson and Spärck Jones, 1976] : $w_i = TF(t_i, o) . \ln \frac{|O|}{|\{o|t_i \in l_{te}(o)\}|}$

Then for comparing ontologies, it is possible to apply various similarity measures:

- Jaccard index with boolean weights;

- cosine index with frequency weights;
- cosine index with TF·IDF weights.

The first measure (Jaccard index with boolean weights) is close to the complement to 1 of the Hamming distance on class names.

4.2   Distances between Ontology Entities

The main way to measure a distance between ontologies is to compare their entities, e.g., their classes, properties, individuals. So any sort of distance that has been developed for matching ontologies can be extended as a distance between ontologies. There are such entity distances mentioned in [Euzenat and Shvaiko, 2007] since they are the most common basis of ontology matching.

Label-Based Distance

Lexical aggregation-based similarity measure. This first local measure only relies on the lexical information coming from annotations and the local name. Given an entity e, $T(e) = \{l_{ln}(uri(e))\} \cup l_{annot}(e)$ represents the set containing the annotations and the local name of e. The lexical similarity between two entities $e \in o$ and $e' \in o'$ is given by:

$$sim_l(e, e') = \frac{\sum_{(a,b) \in M(e,e')} sim_{jw}(a, b)}{min(|T(e)|, |T(e')|)}$$

where $M(e, e')$ is a maximum weight matching of $T(e) \times T(e')$, and sim_{jw} is the Jaro Winkler similarity.

Structural distances. There has been many proposals for distances between ontology concepts. Indeed most of the proposed distances in the literature are based on concept distances [Mädche and Staab, 2002; Euzenat and Valtchev, 2004; Hu *et al.*, 2006; Vrandečić and Sure, 2007].

OLA similarity. One good candidate as structural similarity is the those defined for OLA [Euzenat and Valtchev, 2004] because it relies on every feature of ontologies. OLA first encodes the ontologies into a labelled graph called OL-graph. Then, given an OL-Graph node, the similarity between two OL-graph nodes depends on:

- the similarity of the terms used to designate them, i.e., URIs, labels, names, etc.,
- the similarity of the pairs of neighbor nodes in the respective OL-Graphs that are linked by edges expressing the same relationships, e.g., class node similarity depends on similarity of superclasses, of property restrictions and of member objects,
- the similarity of other local descriptive features depending on the specific category, e.g., cardinality intervals, property types

Datatype and datavalue similarities are external and therefore they are either user-provided or measured by a standard function, e.g., string identity of values and datatype names/URIs.

Formally, given a category X together with the set of relationships it is involved in, $\mathcal{N}(X)$, the similarity measure $Sim_X : X^2 \to [0, 1]$ is defined as follows:

Table 1. Similarity function decomposition (`card` = `cardinality` and `all` = `all ValuesFrom`)

Funct.	Node	Factor	Measure
Sim_O	$o \in O$	$\lambda(o)$	sim_L
		$a \in A, (o,a) \in \mathcal{A}$	$MSim_A$
Sim_A	$a \in A$	$r \in R, (a,r) \in \mathcal{R}$	Sim_R
		$o \in O, (a,o) \in \mathcal{U}$	$MSim_O$
		$v \in V, (a,v) \in \mathcal{U}$	$MSim_V$
Sim_V	$v \in V$	value literal	type dependent
Sim_C	$c \in C$	$\lambda(c)$	sim_L
		$p \in P, (c,p) \in \mathcal{A}$	$MSim_P$
		$c' \in C, (c,c') \in \mathcal{S}$	$MSim_C$
sim_D	$d \in D$	$\lambda(r)$	XML-Schema
Sim_R	$r \in R$	$\lambda(r)$	sim_L
		$c \in C, (r, \mathtt{domain}, c) \in \mathcal{R}$	$MSim_C$
		$c \in C, (r, \mathtt{range}, c) \in \mathcal{R}$	$MSim_C$
		$d \in D, (r, \mathtt{range}, d) \in \mathcal{R}$	Sim_D
		$r' \in R, (r, r') \in \mathcal{S}$	$MSim_R$
Sim_P	$p \in P$	$r \in R, (p, r') \in \mathcal{S}$	Sim_R
		$c \in C, (p, \mathtt{all}, c) \in \mathcal{R}$	$MSim_C$
		$n \in \{0, 1, \infty\}, (p, \mathtt{card}, n) \in \mathcal{R}$	equality

$$Sim_X(x, x') = \sum_{\mathcal{F} \in \mathcal{N}(X)} \pi_{\mathcal{F}}^X MSim_Y(\mathcal{F}(x), \mathcal{F}(x')).$$

The function is normalized, i.e., the weights $\pi_{\mathcal{F}}^X$ sum to one, $\sum_{\mathcal{F} \in \mathcal{N}(X)} \pi_{\mathcal{F}}^X = 1$. The set functions $MSim_Y$ compare two sets of nodes of the same category. Table 1 illustrates the set of similarities used by OLA.

The value of these similarities is computed as a fix-point of the set of equations defining the similarity. This process always converges towards a solution. Since this similarity is already computed as an optimization problem, it generates a match of maximal weight.

Triple-based iterative similarity measure. We have also defined a new measure based on RDF triple similarity. This similarity is defined as a convergent sequence. Initially, some similarity values between nodes in the RDF graph are initialized via string similarity.

In a triple-based representation of an ontology, we consider 4 types of nodes:

- blank node: node having no URI,
- local node: named node defined in the ontology, i.e., node having the same namespace as the ontology,
- external node: named node not defined in the ontology : node imported from other ontologies, or node from the language,
- literal node.

$$simN_0(n_1, n_2) = \begin{cases} 1 \text{ if } n_1 = n_2, \\ simS(n_1, n_2) \text{ if } n_1 \text{ and } n_2 \text{ are literals,} \\ simS(n_1, n_2) \text{ if } n_1 \text{ and } n_2 \text{ are local nodes,} \\ 0 \text{ otherwise.} \end{cases}$$

where $simS$ is a syntactic similarity such as JaroWinkler or Levenstein.

Then, this measure is iteratively refined until the amount of change reaches a user-defined threshold. Given two nodes n_1 and n_2, the node similarity $simN_{i+1}$ between n_1 and n_2 is defined, for the stage $i + 1$, by:

$$simN_{i+1}(n_1, n_2) = \frac{1}{N_{max}} \sum_{k \in \{sub,pred,obj\}} \sum_{\substack{t_x \in T_x^k \\ t_y \in T_y^k}} \Delta \left(simT_i(t_x, t_y)\right) \times \max(|T_x^k|, |T_y^k|)$$

where $T_x^k = \{t_x | t_x.k = n_1\}$, $T_y^k = \{t_y | t_y.k = n_2\}$, and

$$N_{max} = \sum_{k \in \{sub,pred,obj\}} \max(|T_x^k|, |T_y^k|)$$

Δ is a collection similarity such as those introduced in Section 4.3. $simT_i$ is a similarity between two triples defined as the average values of similarities between nodes of two triples at stage i:

$$simT_i(t_1, t_2) = \frac{simN_i(t_1.pred, t_2.pred)}{3} \sum_{k \in \{sub,pred,obj\}} simN_i(t_1.k, t_2.k)$$

In the evaluations, we instantiate this measure with JaroWinkler similarity as $simS$ and MWMGM similarity (presented Section 4.3) as Δ. We choose to stop iterations when $\sum_{n_1,n_2 \in Nodes^2} |simN_i(n_1, n_2) - simN_{i+1}(n_1, n_2)| \leq 1$.

4.3   Collection Distances

Once one has a distance δ (or similarity sim) among concepts available, turning it into an ontology distance is not straightforward. There are different choices for extending measures at the concept level to the ontology level. This is achieved with the help of a collection measure Δ which computes the ontology measure value from the concept measures values. We present some of these below. These collection measures are defined as distance measures, but they can be turned into similarity measures easily.

Definition 4 (Average linkage). *Given a set of entities E and a dissimilarity function $\delta : E \times E \rightarrow [0\ 1]$, the average linkage measure between two ontologies is a dissimilarity function $\Delta_{alo} : 2^E \times 2^E \rightarrow [0\ 1]$ such that $\forall o, o' \subseteq E$,*

$$\Delta_{alo}(o, o') = \frac{\sum_{(e,e') \in o \times o'} \delta(e, e')}{|o| \times |o'|}.$$

Definition 5 (Hausdorff distance). *Given a set of entities E and a dissimilarity function $\delta : E \times E \rightarrow [0\ 1]$, the Hausdorff distance between two sets is a dissimilarity function $\Delta_{Hausdorff} : 2^E \times 2^E \rightarrow [0\ 1]$ such that $\forall o, o' \subseteq E$,*

$$\Delta_{Hausdorff}(o, o') = \max(\max_{e \in o} \min_{e' \in o'} \delta_K(e, e'), \max_{e' \in o'} \min_{e \in o} \delta_K(e, e'))$$

The problem with the Hausdorff distance, as with other linkage measures, is that its value in function of the distance between one pair of members of the sets. The average linkage, on the other hand, has its value function of the distance between all the possible comparisons. None of these are satisfactory. Matching-based dissimilarities [Valtchev, 1999] measure the dissimilarity between two ontologies by taking into account an alignment (matching) between these two ontologies. It can be defined independently of any alignment by using the minimum weight maximum matching. The quality of such a measure is thus that closeness depends on the actual correspondences between two ontologies (not an average). It will thus be possible to translate the knowledge of one ontology into another. However, these measures will be more difficult to compute.

Definition 6 (Minimum weight maximum graph matching distance). *Given a set of entities E and a dissimilarity function $\delta : E \times E \to [0\ 1]$, for any ontologies $o, o' \subseteq E$, a minimum weight maximum graph matching is a one-to-one matching $M \subseteq o \times o'$, such that for any one-to-one alignment $M' \subseteq o \times o'$,*

$$\sum_{\langle p,q \rangle \in M} \delta(p,q) \le \sum_{\langle p,q \rangle \in M'} \delta(p,q)$$

Then, one can define the distance between these two ontologies as:

$$\Delta_{mwmgm}(o, o') = \frac{\sum_{\langle p,q \rangle \in M} \delta(p,q) + \max(|o|, |o'|) - |M|}{\max(|o|, |o'|)}$$

Computing the minimum weight maximum graph matching distance (MWMGM) from a similarity, involves two related steps: extracting an alignment between the ontology and computing the distance value. The value depends on the extracted alignment and usual algorithms extract a matching, i.e., a one-to-one alignment. While this may be a reasonable choice mathematically, this may not be the needed alignment on which to ground such a distance. Hence, MWMGM leaves space for improvement.

5   Experimental Setting

The presented measures have to our knowledge, not been evaluated on ontology distances. We have emitted opinion on their relevance only grounded on their mathematical form. It is necessary to enhance this judgement through evaluation. We want to evaluate both the speed of distance computation and the accuracy with regard to asserted similarity.

The ideal experimental setting comprises a corpus of ontologies with clear expectations about the distances that should be found between them. We do not have such a corpus annotated with distances values between ontologies. However, the most important thing is to know the proximity order between ontologies.

Finding a relevant corpus is not an easy task. For this reason we provide only preliminary results here. We first describe the tested methods, the test set and various tests performed with this test set.

5.1   Selected Measures

In order to be representative, we selected both terminological and structural measures usually used in ontology matching. For each kind of measures, we chose to evaluate basic measures and more elaborated ones as shows Table 2. The JaccardVM(TF) measure

Table 2. Selected measures

	basic	elaborate
terminological	CosineVM(TF & TFIDF), JaccardVM(TF)	EntityLexicalMeasure
structural	TripleBasedEntitySim	OLAEntitySim

is the simplest one since it only represents the proportion of shared terms in two ontologies. CosineVM has been used with two types of weights: TF and TF·IDF. All other measures are entity based measures and then, they have been tested with the three collection measures presented Section 4.3: the Average Linkage, the Hausdorff distance, and the MWMGM distance. For normalization purpose, all measures evaluated here are similarity measures.

5.2 Evaluation on the OAEI Benchmark Suite

We have considered the Ontology Alignment Evaluation Initiative[1] benchmark test set because it offers a set of ontologies that are systematically altered from one particular ontology which will play the role of o. There are here 6 categories of alterations:

Name Name of entities that can be replaced by (R/N) random strings, (S)ynonyms, Name with different (C)onventions, (F) strings in another language than English.
Comments Comments can be (N) suppressed or (F) translated in another language.
Specialization hierarchy can be (N) suppressed, (E)xpansed or (F)lattened.
Instances can be (N) suppressed
Properties can be (N) suppressed or (R) having the restrictions on classes discarded.
Classes can be (E)xpanded, i.e., replaced by several classes or (F)latened.

Since, these ontologies are generated by applying successive transformations to o we know that the ontologies resulting from applying less transformations (for inclusion) should be closer to o. This is this property that we have exploited in this first test set.

Order between ontologies of benchmark. We can to build a partial order relation $\leq$ representing the alteration relation over all generated ontologies. $o \leq o'$ seems that the ontology o is an alteration of o' (o can be obtained by altering some features of o').
   For each category of alteration $c \in \{Name, Comments, Specialization, Instances, Properties, Classes\}$ and each ontology o, $c(o)$ represents the type of alteration made on the reference ontology for the category c. For each category, these alterations are ordered in the following way:

Name $\{R, N\} \leq \{S, C, F\} \leq \emptyset$ **Instance** $N \leq \emptyset$
Comments $N \leq F \leq \emptyset$ **Property** $N \leq R \leq \emptyset$
Specialization hierarchy $\{N, E, F\} \leq \emptyset$ **Classes** $\{E, F\} \leq \emptyset$

From these rules, an ontology o is an alteration of o', noted $o \leq o'$, if for each category of alterations c, we have $c(o) \leq c(o')$. Figure 1 displays a transitive reduction of this partial order.

---

[1] http://oaei.ontologymatching.org

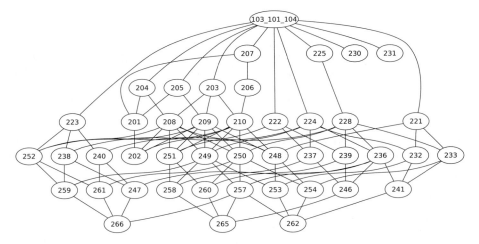

Fig. 1. Order lattice on benchmark ontologies set

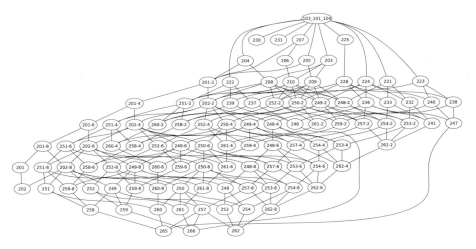

Fig. 2. Order lattice on the enhanced benchmark ontologies set

Initially we were considering all triples of ontologies related by a transformation path. But we observed that this procedure was biased towards lexical measures since it compares only labels and thus only changes its value between equal labels and non equal ones. Since our test is based on $\leq$, very often the distances are the same and thus the property was satisfied. Given the huge proportion of such tests in our test set we restricted ourselves comparing two ontologies with the initial ontology.

Another bias given by this test set is that the transformations were of the all-or-nothing kind: either all labels are changed, or they are preserved. For countering this bias, we produced a larger altered test set in which the label scrambling transformation is applied in (always the same) 20, 40, 60 and 80% of the labels. The new lattice is given in Figure 2.

6   Results

According to the experimental settings introduced in the previous section, we have collected and analyzed 3 kinds of results. The first results concern the comparison of the orders induced by measures and those really observed on the benchmark suites. The second results show how measures behave when we introduce unrelated ontologies. The last results are about the time consumption of evaluated measures.

6.1   Order Comparison on Benchmark

This first experiment aims at checking if the tested similarities are compatible with the order induced by the alterations. It checks if the assertion $sim(o, o'') \leq sim(o, o')$ (or $\delta(o, o') \leq \delta(o, o'')$ for distances) is verified for any triple o, o', o'' such as $o'' < o' < o$. 1372 triples can be formed with the original benchmark suite and 15780 triples with the enhanced benchmark suite.

Table 3 shows results obtained respectively on the original benchmark and the enhanced one. This table presents, for each measure, the proportions of triples o, o', o'' (such as $o'' < o' < o$) verifying $sim(o, o'') \leq sim(o, o')$.

On the original benchmark, the triple-based similarity performs the best with all structural measures. Cosine measure with TF weights also obtains good results and outperforms measures based on OLA similarity. Then, Jaccard similarity on TF weights gives satisfactory results. The measures based on lexical similarity and cosine with TF·IDF weights are the worst measures since they successfully passed only the half of the tests. Concerning, structural measures, MWMGM seems to be the best collection measure with the lexical and triple-based similarities. Hausdorff is only relevant with triple-based similarity because only 10% of tests have a reference similarity, $sim(o, o'')$, greater than 0 with OLA and none with lexical similarity (which explain the NaN value). Surprisingly, OLA similarity obtains its best result with AverageLinkage.

Measures tend to perform better on the enhanced benchmark than on the original benchmark. This improvement is especially noteworthy with lexical measures such as lexical entity, vector-based similarities. Triple-based similarity improves it results with all collection measures. OLA similarity almost obtain the same results. This owes to

Table 3. Results on the original and enhanced benchmark test sets

Measure	Tests Passed (ratio)	
	Original	Enhanced
MWMGM (EntityLexicalMeasure)	0.53	0.72
Hausdorff (EntityLexicalMeasure)	NaN	NaN
AverageLinkage (EntityLexicalMeasure)	0.44	0.31
MWMGM (OLAEntitySim)	0.75	0.78
Hausdorff (OLAEntitySim)	0.75	0.65
AverageLinkage (OLAEntitySim)	0.79	0.74
MWMGM (TripleBasedEntitySim)	0.86	0.92
Hausdorff (TripleBasedEntitySim)	0.86	0.89
AverageLinkage (TripleBasedEntitySim)	0.82	0.91
CosineVM (TF)	0.82	0.92
CosineVM (TFIDF)	0.57	0.81
JaccardVM (TF)	0.71	0.87

the fact that ontologies are, on average, more lexically similar in the enhanced benchmark. Results of MWMGM are better with all tested measures, but average linkage obtains worse results with lexical and OLA similarities. Results of Hausdorff are still not relevant with OLA and lexical similarities.

These results show that structural measures are more robust to the lexical alterations than lexical based measures. In cases where ontologies are lexically close, the use of vectorial measures seems to be relevant. The entity lexical measure is more sensitive to the collection measure used than structural measures. This can be explained by the fact that structural entity distances values are more dependent each other than lexical entity distances values. The advantage of TF against TF.IDF is probably due to the nature of the benchmark: since the altered ontologies are generated from one reference, a term tends to appear in a lot of ontologies and then its IDF weight is around 0.

6.2 Tests with Different Cardinality Matching

Finally, the benchmark test set is still biased towards 1-1 matching, so algorithms which are enforcing them (and in particular maximal matching algorithms) should be favored by our tests. In order to counter this problem, we used two imperfect tests:

- we compared with similar but different ontologies: 301, 303, 304 to be compared with the higher level of the hierarchy where what has changed is added/suppressed classes (248, 251, 252, 221, 222, 223, 228, 250) and suppressed properties (228, 250);
- we compared with irrelevant ontologies (confious, iasted and paperdyne from the conference test).

The expected result here is that the slightly altered ontologies are closer than the 30x which are still better than the conference ontologies (2xx > 3xx > CONFERENCE).

Table 4 presents, for each measure, the ontologies which have not been correctly ordered and the observed order. For example, 250 < 304 means that the measure finds that 304 is closer to 101 than 250 ($sim(101, 250) < sim(101, 304)$). In this experiment MWMGM similarities perform the best. Results given by Hausdorff measure combined with lexical and OLA similarities are not relevant since a lot of values are equals to 0. Nevertheless, triple-based similarity with Hausdorff gives good results. Results obtained by average linkage with lexical and triple-based similarities are not satisfactory since a lot of ontologies are not correctly ordered. Vectorial measures fail on the same tests: they do not work very well when the names of classes have been removed (ontologies 248, 250, 251, 252).

In this experiment, ontologies having different cardinality do not penalize MWMGM in comparison with other measures. These results also show the limits of lexical measures on tests where structure is preserved but not the lexical data.

6.3 Time Consumption

We compared the CPU time used by each measure. The testing platform is powered by a quad-core 3GHz Xeon processor with a Linux OS. All evaluated measures (but OLA) have been implemented and evaluated using the same framework. For each measure, we

Table 4. Ordering error between ontologies on a selection of ontologies

Measure	Observed disorders
MWMGM(EntityLexicalMeasure)	250 < 304 250 < {CONFERENCE} {301, 303} < iasted < 304 < {confious ,paperdyne}
MWMGM(OLAEntitySim)	250 < 304 303 < {confious, paperdyne}
MWMGM(TripleBasedEntitySim)	250 < {301,303} < 228 < 304 250 < {CONFERENCE}
Hausdorff(EntityLexicalMeasure)	All values equals to 0
Hausdorff(OLAEntitySim)	{228, 248, 250, 251, 252} < 304 {228, 248, 250, 251, 252} < {CONFERENCE} {301, 302 }< iasted < 303 < {confious, paperdyne}
Hausdorff(TripleBasedEntitySim)	250 < {3xx} 250 < confious
AverageLinkage(EntityLexicalMeasure)	{2xx} < {301,304} {228, 248, 250, 251, 252} < {confious, iasted} < {221,222, 223} < paperdyne 303 < {CONFERENCE}
AverageLinkage(OLAEntitySim)	{223, 248, 250, 251, 252} <301 {303} < {confious, paperdyne}
AverageLinkage(TripleBasedEntitySim)	{ 248, 250, 251, 252} < 303 < {221, 222, 223, 228}< {301,304} 250 < {confious, iasted}
CosineVM(TF)	{248, 250, 251, 252}<{3xx}
CosineVM(TFIDF)	{248, 250, 251, 252}<{3xx}
JaccardVM(TF)	{248, 250, 251, 252}<{3xx}

Table 5. CPU time consumption on the original benchmark

Measure	Total time (s)	Average time per similarity value (s)
MWMGM(EntityLexicalMeasure)	558	0.46
MWMGM (OLAEntitySim)	39 074	31.9
MWMGM (TripleBasedEntitySim)	7 950	6.49
Hausdorff (EntityLexicalMeasure)	451	0.37
Hausdorff (OLAEntitySim)	38 912	31.76
Hausdorff (TripleBasedEntitySim)	7 410	6.05
AverageLinkage (EntityLexicalMeasure)	444	0.36
AverageLinkage (OLAEntitySim)	38 995	31.83
AverageLinkage (TripleBasedEntitySim)	7 671	6.26
CosineVM (TF)	101	0.08
CosineVM (TFIDF)	102	0.08
JaccardVM (TF)	101	0.08

computed all similarity values between ontologies of the original benchmark. This test was performed two times with no significant differences in taken times. These results are those of the second round. Table 5 shows the CPU time spent to compute these 1225 similarities and the average time spent to compute one similarity value.

These results clearly shows that measures based on OLA are runtime intensive. Measures using triple-based entity similarity are 5 times less expensive than the first family. Lexical Entity based similarities and those based on the vector space model measures are computed largely faster.

Globally, these results confirm that entity-based measures are more time intensive than VSM measures. Among the entity-based measures, structural measures (OLA and triple-based entity similarities) are more extensive than lexical ones since they rely on an iterative refinement process. The observed runtime is consistent with theoretical complexity of measures (and the computation of all but the more complex is deterministic), so we do not observe a significant effect from coding.

This experimentation shows that only lexical measures are useable at large scale.

7   Conclusion

Measuring distances between ontologies can be useful in various tasks for different purposes (finding an ontology to replace another, finding an ontology in which queries can be translated, finding people using similar ontologies). Hence there is no universal criterion for deciding if an ontology is close or far from another.

There exist many measures using different aspects of ontologies. In order to know the behaviour of these methods, we have evaluated them on specific test benches. Results shows that structural similarities tends to more reliable and robust than lexical similarities. This is especially true when the ontologies to compare do not share a lot of common vocabulary. Nevertheless, due their complexity, structural measures are not adapted for real-time applications or for measuring similarities between large ontologies. We can also notice that some basic measures such cosine on TF vector give quite accurate results. Such kind of measures can be useful for quickly select a subset of close ontologies and thus allowing the use of structural measures in order to refine the proximity relation between the selected ontologies. Hence, more work must be developed for finding trade-offs between accuracy and efficiency.

This paper only restricts the study to measures used for matching ontologies. It could be interesting to test others measures relying on some global ontology characteristics (size, graphs densities, etc.).

References

[Alani and Brewster, 2005]Alani, H., Brewster, C.: Ontology ranking based on the analysis of concept structures. In: Proc. 3rd International conference on Knowledge Capture (K-Cap), Banff. (CA), pp. 51–58 (2005)

[d'Aquin et al., 2007]d'Aquin, M., Baldassarre, C., Gridinoc, L., Angeletou, S., Sabou, M., Motta, E.: Watson: a gateway for next generation semantic web applications. In: Proc. Poster session of the International Semantic Web Conference (ISWC), Busan (KR) (2007)

[Ehrig et al., 2005]Ehrig, M., Haase, P., Hefke, M., Stojanovic, N.: Similarity for ontologies – a comprehensive framework. In: Proc. 13th European Conference on Information Systems, Information Systems in a Rapidly Changing Economy (ECIS), Regensburg (DE) (2005)

[Euzenat and Shvaiko, 2007]Euzenat, J., Shvaiko, P.: Ontology matching. Springer, Heidelberg (DE) (2007)

[Euzenat and Valtchev, 2004]Euzenat, J., Valtchev, P.: Similarity-based ontology alignment in OWL-lite. In: Proc. 16th European Conference on Artificial Intelligence (ECAI), Valencia (ES), pp. 333–337 (2004)

[Gracia et al., 2007]Gracia, J., Lopez, V., d'Aquin, M., Sabou, M., Motta, E., Mena, E.: Solving semantic ambiguity to improve semantic web based ontology matching. In: Proc. 2nd ISWC Ontology matching workshop (OM), Busan (KR), pp. 1–12 (2007)

[Hu et al., 2006]Hu, B., Kalfoglou, Y., Alani, H., Dupplaw, D., Lewis, P., Shadbolt, N.: Semantic metrics. In: Staab, S., Svátek, V. (eds.) EKAW 2006. LNCS (LNAI), vol. 4248, pp. 166–181. Springer, Heidelberg (2006)

[Jung and Euzenat, 2007]Jung, J., Euzenat, J.: Towards semantic social networks. In: Franconi, E., Kifer, M., May, W. (eds.) ESWC 2007. LNCS, vol. 4519, pp. 267–280. Springer, Heidelberg (2007)

[Jung *et al.*, 2007]Jung, J., Zimmermann, A., Euzenat, J.: Concept-based query transformation based on semantic centrality in semantic peer-to-peer environment. In: Dong, G., Lin, X., Wang, W., Yang, Y., Yu, J.X. (eds.) APWeb/WAIM 2007. LNCS, vol. 4505, pp. 622–629. Springer, Heidelberg (2007)

[Mädche and Staab, 2002]Mädche, A., Staab, S.: Measuring similarity between ontologies. In: Gómez-Pérez, A., Benjamins, V.R. (eds.) EKAW 2002. LNCS (LNAI), vol. 2473, pp. 251–263. Springer, Heidelberg (2002)

[Robertson and Spärck Jones, 1976]Robertson, S., Spärck Jones, K.: Relevance weighting of search terms. Journal of the American Society for Information Science 27(3), 129–146 (1976)

[Stuckenschmidt and Klein, 2004]Stuckenschmidt, H., Klein, M.: Structure-based partitioning of large concept hierarchies. In: McIlraith, S.A., Plexousakis, D., van Harmelen, F. (eds.) ISWC 2004. LNCS, vol. 3298, pp. 289–303. Springer, Heidelberg (2004)

[Tverski, 1977]Tverski, A.: Features of similarity. Psychological Review 84(2), 327–352 (1977)

[Valtchev, 1999]Valtchev, P.: Construction automatique de taxonomies pour l'aide à la représentation de connaissances par objets. Thèse d'informatique, Université Grenoble 1, Grenoble (FR) (1999)

[Vrandečić and Sure, 2007]Vrandečić, D., Sure, Y.: How to design better ontology metrics. In: Franconi, E., Kifer, M., May, W. (eds.) ESWC 2007. LNCS, vol. 4519, pp. 311–325. Springer, Heidelberg (2007)

Folksonomy-Based Collabulary Learning

Leandro Balby Marinho, Krisztian Buza, and Lars Schmidt-Thieme

Information Systems and Machine Learning Lab (ISMLL) Samelsonplatz 1,
University of Hildesheim, D-31141 Hildesheim, Germany
{marinho,buza,schmidt-thieme}@ismll.uni-hildesheim.de

Abstract. The growing popularity of social tagging systems promises to alleviate the knowledge bottleneck that slows down the full materialization of the
Semantic Web since these systems allow ordinary users to create and share knowledge in a simple, cheap, and scalable representation, usually known as folksonomy. However, for the sake of knowledge workflow, one needs to find a
compromise between the uncontrolled nature of folksonomies and the controlled
and more systematic vocabulary of domain experts. In this paper we propose to
address this concern by devising a method that automatically enriches a folksonomy with domain expert knowledge and by introducing a novel algorithm based
on frequent itemset mining techniques to efficiently learn an ontology over the
enriched folksonomy. In order to quantitatively assess our method, we propose a
new benchmark for task-based ontology evaluation where the quality of the ontologies is measured based on how helpful they are for the task of personalized
information finding. We conduct experiments on real data and empirically show
the effectiveness of our approach.

1   Introduction

Due to the concrete advances towards the Semantic Web vision [4], ontologies are growing in use, specially in areas concerning information finding and organization. However,
their massive adoption is severely shortened because of the effort one needs to take to
assemble them, task which is usually assigned to domain experts and knowledge engineers. Although ontology learning can help to some extent, the participation of the
expert is still usually required since the learned representations are not free of incon-
sistences (in a semantic level at least) and therefore require manual validation and fine
tuning. A more promising solution to this problem lies in the rapid spread of the Web
2.0 paradigm as it has the potential to educate ordinary users towards voluntary semantic annotation, thereby decentralizing and cheapening knowledge acquisition. The
increasing popularity of Web 2.0 applications can be partly explained by the fact that
no specific skills are needed for participating, where anyone is free to add and categorize resources in the form of free keywords called *tags*. Tags do not need to conform
to a closed vocabulary and therefore reflect the latest terminology in the domain under
which the system operates. Furthermore, the exposure to each other tags and resources
creates a fundamental trigger for communication and sharing, thus lowering the barriers
to cooperation and contributing to the creation of collaborative lightweight knowledge
structures known as *folksonomies*. Despite the compelling idea of folksonomies, its uncontrolled nature can bring problems, such as: synonymy, homonymy, and polysemy,

A. Sheth et al. (Eds.): ISWC 2008, LNCS 5318, pp. 261–276, 2008.
© Springer-Verlag Berlin Heidelberg 2008

which lowers the efficiency of content indexing and searching. Another problem is that folksonomies usually disregard relations between their tags, what restricts the support for content retrieval. If tags are informally defined and continually changing, then it becomes difficult to automate knowledge workflow. In this sense, it is necessary to find a compromise between the flexibility and dynamics of folksonomies and the more systematic structure of controlled vocabularies. This compromise is usually known as *collabulary* [1], which corresponds to a portmanteau of the words *collaborative* and *vocabulary*. For our purposes we define a collabulary in terms of a special ontology that represents the knowledge of both users and experts in an integrated fashion.

In this paper we propose a method for collabulary learning. To this end, we first take a folksonomy and a domain-expert ontology as input and project them into an enriched folksonomy through semantic mapping; we then apply a fast and flexible algorithm based on frequent itemset techniques to learn an ontology over the enriched folksonomy. The main contributions of this paper are: (i) a definition for the new problem of *collabulary learning*, (ii) a method for automatically enriching folksonomies with domain-expert knowledge, (iii) a fast and flexible algorithm based on efficient frequent itemset mining techniques for ontology learning from folksonomies, and (iv) a new benchmark for task-based ontology evaluation in folksonomies.

An obvious question one could ask is to which extent this so called *collabulary* really helps. Looking at the literature on ontology learning from folksonomies (e. g., [26,15,9,23,22]) we see that most of the proposed approaches are motivated by facilitating navigation and information finding, even though they do not quantify to which extent the learned ontologies really help on this task. Instead, the quality of the learned ontologies is measured based on how good they match people's common sense or how similar they are to a reference ontology. We argue that in this context, an ontology is as good as it helps users finding useful information. Therefore, we propose, as contribution (iv), to plug the investigated knowledge structures in collaborative filtering algorithms for recommender systems and evaluate the outcome as an indicator of the ontologies' usefulness, given that collaborative filtering [21] is one of the most successful and prominent approaches for personalized information finding. To the best of our knowledge this is the first effort towards thorough empirical investigation of the trade-off between folksonomies and controlled vocabularies. We conduct experiments on a real-life dataset and demonstrate the effectiveness of our approach.

The paper is organized as follows. Section 2 presents a definition for the collabulary learning problem and our approach for enriching a folksonomy with domain-expert vocabulary. Section 3 introduces an algorithm based on fast frequent itemset mining techniques for ontology learning from folksonomies. Section 4 presents a new benchmark for task-based ontology evaluation and Section 5 discusses the conducted experiments and their results, followed by related work and conclusions (Sections 6 and 7).

2   Folksonomy Enrichment

Our approach for folksonomy enrichment is based on providing a semantic mapping between an ontology designed by domain experts and a folksonomy, assuming that both describe the same domain over the same set of instances.

2.1   Problem Definition

Before presenting our approach, we provide a simplified definition for some of the concepts used in this paper, namely, folksonomy, ontology, knowledge base and collabulary. Similarly to [16], we define a folksonomy as follows:

Definition 1. *A folksonomy[1] is a tuple* $\mathbb{F} := (U, T, R, Y)$ *where* U, T, *and* R *are finite sets, whose elements are called* users, tags *and* resources, *and* Y *is a ternary relation between them, i. e.,* $Y \subseteq U \times T \times R$, *whose elements are called tag assignments.*

Since taxonomies are central components of ontologies, we are going to focus on them first. Similarly to [12] we define an *ontology* as follows:

Definition 2. *An ontology is a tuple* $\mathcal{O} := (\mathcal{C}, root, \leq_{\mathcal{C}})$ *where* $\mathcal{C}$ *is a set of concept identifiers and* $\leq_{\mathcal{C}}$ *is partial order on* $\mathcal{C}$ *with one unique top element* root, *called taxonomy or concept hierarchy.*

A *knowledge base* in turn, is defined as follows:

Definition 3. *A knowledge base for an ontology* $\mathcal{O}$ *is a structure* $KB := (I, \iota_{\mathcal{C}})$ *where* I *is a set whose elements are instance identifiers and* $\iota_{\mathcal{C}} : \mathcal{C} \to 2^{I}$ *is a function associating concepts to instances called concept instantiation.*

To simplify our discussion, we assume that the relation between lexical terms and their associated concepts or instances is a bijection i. e., each lexical term is a identifier of a concept or an instance[2]. Finally we define the problem we want to address in this paper as follows:

Definition 4. *Given a folksonomy* $\mathbb{F}$, *an ontology* $\mathcal{O}$ *and a knowledge base* $KB_{\mathcal{O}} = (I, \iota_{\mathcal{C}})$ *for* $\mathcal{O}$, *an ontology* $\mathcal{P}$ *with concepts* $\mathcal{C}_{\mathcal{P}} = T_{\mathbb{F}} \cup \mathcal{C}_{\mathcal{O}}$ *and a knowledge base* $KB_{\mathcal{P}} = (I', \iota'_{\mathcal{C}})$ *with* $I' = I \cup R_{\mathbb{F}}$ *is called a* collabulary *over* $\mathbb{F}$ *and* $\mathcal{O}$. *The* **collabulary learning problem** *is here defined as finding a collabulary over* $\mathbb{F}$ *and* $\mathcal{O}$, *that best[3] represents the common knowledge between folksonomy users and domain experts.*

2.2   Semantic Mapping

Users of social tagging systems are heterogenous and thus have different levels of knowledge about a domain. Moreover, they can express very personal opinions about their resources, what lowers the potential for knowledge sharing. Tags like *stuff to chill*, *awesome artists*, or *makes me happy*[4] are very subjective and hence hard to make sense of, nevertheless they appear relatively often in real life folksonomies. In order to ensure interoperability, one needs to find a clear meaning for these tags such that we know, for example, that actually *stuff to chill* is related to *alternative*, *awesome artists* to *emo*

---

[1] In the original definition [16], it is additionally introduced a subtag/supertag relation, which we omit here since most of the real life folksonomies disregard such tag relations.

[2] For our purposes instances will correspond to resources.

[3] In this paper "best" is defined in terms of an ontology-based application scenario.

[4] Tags present in the online social radio station last.fm (http://last.fm)

and *makes me happy* to *rockabilly*, where *alternative*, *emo*, and *rockabilly* are concepts coming from a controlled and well agreed vocabulary[5]. We propose to address this issue by providing a semantic mapping between a folksonomy and a domain-expert ontology. As these two knowledge representations are structurally different (i. e., folksonomies do not have a partial order) we first turn the folksonomy, without loss of generality, into a *trivial ontology* $\mathcal{O}_{\mathbb{F}\downarrow T}$ over $\mathbb{F}$, i. e., a projection of the folksonomy to its tag space where all the concepts are leaf sibling nodes having root as father, more formally, $\forall t \in T : \leq_t := \emptyset$ and $\leq_{root} := T$ (e. g., Fig. 1.a). Now we can cast the semantic mapping, for our case, as an ontology matching problem, for which there is a well covered literature (e. g., [11,13,18]). Notice, however, that since the trivial ontology is structurally limited we can not rely on methods that heavily consider structural information.

Here we are interested in methods that depend only on the semantic content of the concepts involved since relying on syntactic descriptions of tags is not suitable, given the uncontrolled vocabulary of users. In [13] joint probability distributions are used as a framework for well-defined similarity measures that do not depend on the lexical layer of ontologies. Denoting the probability of two concepts A and B being identical by $P(A, B)$, in [13] it is shown that under certain general assumptions $P(A, B)$ for two concepts A and B coming from different ontologies, can be *approximated* as the fraction of instances that belong to both A and B[6], therefore reducing the problem to checking, for each instance, if it belongs to $A \cap B$. In our experiments we use the well known Jaccard coefficient

$$JS(A, B) := P(A \cap B)/P(A \cup B) := \frac{P(A, B)}{P(A, B) + P(A, \bar{B}) + P(\bar{A}, B)} \quad (1)$$

as a representative of this family of similarity measures, where $P(A, \bar{B})$ is the probability that a randomly chosen instance belongs to A but not to B and $P(\bar{A}, B)$ the other way around. Having defined the similarity measure to use, we build the enriched folksonomy as follows:

1. Let $\mathcal{O}_D$ denote the domain-expert ontology and $\mathcal{O}_\mathbb{F}$ a trivial ontology, first we need to define a function for mapping the concept identifiers of both ontologies, i. e., $\hat{T} : \mathcal{C}_{\mathcal{O}_\mathbb{F}} \rightarrow \mathcal{C}_{\mathcal{O}_D}$, where $\mathcal{C}_{\mathcal{O}_\mathbb{F}}$ denotes the set of concept identifiers for $\mathcal{O}_\mathbb{F}$ and $\mathcal{C}_{\mathcal{O}_D}$ for $\mathcal{O}_D$. In our case, for each concept A in the *trivial ontology* representing the folksonomy, the most similar concepts from a *domain-expert ontology* is found as follows: $\hat{T}(A) := \underset{x \in \mathcal{C}_{\mathcal{O}_D}}{\operatorname{argmax}} JS(A, x)$ (see Example 1).

2. After that, we add the best mappings in $Y_\mathbb{F}$ as additional triples, i. e., $Y := Y \cup \{(u, \hat{T}(t), r) | (u, t, r) \in Y\}$.

3. Finally, we create a "dummy" user u_θ representing the expert and integrate it to the folksonomy. In other words, we build additional triples reflecting the concept instantiation of the expert and add them in $Y_\mathbb{F}$. The *enriched folksonomy* is now composed by the triples $Y := Y \cup \{(u_\theta, c, r) | c \in \mathcal{C}_{\mathcal{O}_D}, r \in \iota(c)\}$.

---

[5] With respect to music genres.

[6] $P(A, B) \approx \frac{\text{count of instances belonging to both } A \text{ and } B}{\text{count of all instances}} = \frac{|\{\iota^*(A)\} \cup \{\iota^*(B)\}|}{\text{count of all instances}}$ where $\iota^*(C)$ denotes the extension of concept C, i. e., the set of resources belonging to C and its descendants.

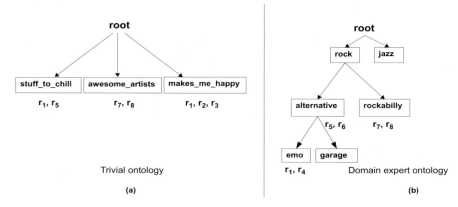

Fig. 1. Example of two knowledge bases associated with a trivial core ontology representing a folksonomy (a) and a domain-expert one (b) over the music domain

Example 1. Consider the concept *stuff to chill* in the ontology at Fig. 1.a. Among the concepts in the ontology at Fig. 1.b, we have to find the one with the highest similarity. Computing a term like $P(\textit{stuff to chill}, \textit{alternative})$ is very simple as we just need to find the instances belonging to both concepts. Looking at Fig. 1, we observe that $\iota_C(\textit{stuff to chill}) := \{r1, r5\}$ and $\iota_C^*(\textit{alternative}) := \{r1, r4, r5, r6\}$ and therefore *stuff to chill* $\cap$ *alternative* $:= \{r1, r5\}$. Now $P(\textit{stuff to chill}, \textit{alternative})$ is just the number of elements in this intersection divided by the total number of distinct instances, i. e., $\frac{2}{8} := 0.25$. The same procedure can be repeated to find the terms $P(\overline{\textit{stuff to chill}}, \textit{alternative})$ and $P(\textit{stuff to chill}, \overline{\textit{alternative}})$. Now we just need to plug all these terms in Eq. 1. Given that *alternative* is indeed the best mapping, for the triples containing *stuff to chill*, i. e., $\{(u, r1, \textit{stuff to chill}), (u, r5, \textit{stuff to chill})\} \in Y$, we add new triples having *alternative*, i. e., $Y := Y \cup \{(u, r1, \textit{alternative}), (u, r5, \textit{alternative})\}$. Finally the triples of the "expert user" are included, i. e., $Y := Y \cup \{(u_\theta, r5, \textit{alternative}), (u_\theta, r6, \textit{alternative}), (u_\theta, r7, \textit{rockabilly}), (u_\theta, r8, \textit{rockabilly}), (u_\theta, r1, \textit{emo}), (u_\theta, r4, \textit{emo})\}$.

Therefore, the enriched folksonomy has two major components: (i) a consistent vocabulary that properly matches the user vocabulary, and (ii) an additional user representing the expert point of view about the resources. The next step is to learn an ontology over this enriched folksonomy, which is the topic of the next section.

3   Frequent Itemsets for Learning Ontologies from Folksonomies

Most of the approaches concerning ontology learning from folksonomies rely on co-occurrence models (e. g., [26,17,9,23,22]). This is in line with the assumption that in sparse structures, such as folksonomies, positive correlations carry most of the essential information about the data (see [14] for a theoretical justification).

The idea of using frequent itemset mining for ontology learning in folksonomies is not new, in [22] for example, the authors conceptually proposed the exploitation of different projections of the folksonomy onto a two-dimensional formal context, where they applied association rule mining techniques. Our work is in line with this idea, however, we project the folksonomy to a transactional database[7], which is usual in the frequent itemset mining community and facilitates the direct application of highly efficient state-of-the-art methods. Furthermore, we explore additional assumptions on the users' resource tagging behaviour and how itemsets reflect relations between tags.

In this section, we propose a new algorithm for learning ontologies from folksonomies, which is based on frequent itemset mining on the one hand, and on extraction of taxonomic relationships from frequent itemsets on the other.

3.1 Frequent Itemset Mining

Searching for frequently co-occurring items is a well studied subject in data mining and is usually referred to as frequent itemset mining, for which there is a broad literature (e. g., [2,3,25,19]). A frequent itemset is a set of frequently co-occurring items, like products often purchased together in a supermarket. The task can formally be defined as follows.

Definition 5. *Let* $\Sigma = \{c_1, c_2, \ldots, c_s\}$ *be a set of items, a* transactional database D *is a subset of the power set of* Σ, *i. e.,* $D = \{C_1, C_2, \ldots, C_n\}, \forall k \in \{1, 2, \ldots, n\} : C_k \subseteq \Sigma$ *where the sets* $C_1, C_2, \ldots, C_n$ *denote co-occurring items and are called* transactions.

Definition 6. *The* support *for an* itemset $I \subseteq \Sigma$, *henceforth denoted as* $sup(I)$, *is defined as the number of supersets[8] of* I *in the transactional database* $sup(I) = |\{C_k : C_k \in D, I \subseteq C_k, k \in \{1, 2, \ldots, n\}\}|$

Definition 7. *Given* Σ, D *and a minimum support threshold* m, *the* frequent itemset mining problem *is defined as the task of finding all sets* $I \subseteq \Sigma$ *with* $sup(I) \geq m$.

We project a folksonomy to a transactional dataset as follows. Given that Σ now corresponds to the set of tags, the transactions are identified by user-resource pairs where a transaction is composed by the tags used by the same user to the same resource (e. g., Fig. 2). Considering the transactional database illustrated in Fig. 2 and given that the minimum support threshold is set to 2, a frequent itemset would be {*musical, modern* } for example.

3.2 Learning Ontologies from Folksonomies

Before introducing our method we define some intuitive **assumptions** used by the algorithm in the learning process:

1. **High Level Tag Assumption** – Users often associate resources with tags of different levels of an (eventually unknown) hierarchy. The more popular a tag is, the more general it is and therefore should occupy a higher level in the taxonomy to be learned. Notice that this in line with the assumption that users usually want to

---

[7] See Definiton 5.
[8] Note that for two subsets $C', I \subseteq \Sigma$, if $C' = I$, C' is a (trivial) superset of I.

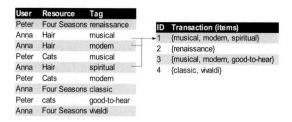

Fig. 2. Projection of a folksonomy to a transactional database

alleviate the cognitive effort by selecting tags representing broader concepts. Consider, for example, the well known singer *Elvis Presley* who is usually associated with the music genres *rock'n roll* and *rockabilly*, where *rock'n roll* is regarded as more general than *rockabilly*. If some users annotate this artist with *rock'n roll*, we expect that many of these users also use the tag *rockabilly*. However, as there are many other *rock'n roll* artists that are not necessarily *rockabilly*, we expect the tag *rock'n roll* to be used more often than *rockabilly*.

2. **Frequency Assumption** – If a frequent itemset F has significantly higher support than another frequent itemset F', we say that the items occuring in F are closer related to each other than the items in F', i.e. F should have more influence on the learned structure than F'.

3. **Large Itemset Assumption** – Suppose there are two frequent itemsets F_1 and F_2 and there is another frequent itemset F, $F \supseteq F_1 \cup F_2$. Suppose they all have approximatelly the same support. In this case (i. e., F, F_1, F_2 are frequent) we assume closer relation between the items included in $F_1 \cup F_2$, as if only F_1 and F_2 were frequent (but F not).

Our method is depicted in Algorithm 1. We apply an iterative process where the most frequent itemsets are mined first resulting in the learning of some pieces of the ontology.

Algorithm 1. Algorithm for Taxonomy Learning based on Frequent Itemsets

---

Require: Folksonomy $data$, Array of Min. Support Tresholds[] m,
    Goodness Treshold g, Array of Edge Tresholds[] e,
    Resource-Tag-Correlation Treshold c.
Ensure: Learned Ontology $\mathcal{O}$ and its Knowledge Base $KB_{\mathcal{O}}$.

1. Transactional Database $D = projectData(data)$;
2. S=empty taxonomy;
3. **for** $i = 0; i < m.length; ++i$ **do**
4. $fqSets = mineFrequentItemsets(D, m[i])$
5. $S' = buildTaxonomyPieces(fqSets, e)$
6. $S'' = prunePieces(S')$
7. $S = addPrunedPiecesToTaxonomy(S, S'')$
8. **end for**
9. $\mathcal{O}, KB_{\mathcal{O}} = addResourcesToOntology(S, data, c)$
10. **return** $\mathcal{O}, KB_{\mathcal{O}}$

This corresponds to our *Frequency Assumption*, i. e., we first learn the relations contained in the most frequent itemsets since this corresponds to stronger evidences. Then in each subsequent iteration we relax the minimum support threshold in order to mine less frequent itemsets leading to the learning of new pieces. These pieces are iteratively put together converging to the final ontology. The main steps of the algorithm are detailed below.

For the frequent itemset mining (line 1), we use a highly efficient implementation of the algorithm Apriori [3,2], which is based on a doubly recursive scheme to count how often a set of items co-occur (see [6,7] for details). Since frequent itemsets do not always reflect the true relation between the items, we use the following *goodness condition* to which an itemset $I = \{t_1, t_2, \ldots, t_k\}$ needs to fulfill

$$\forall i, i \in \{1, 2, \ldots, k\} : \frac{\frac{sup(I)}{|D|}}{\frac{sup(\{t_i\})}{|D|} \cdot \frac{sup(I \setminus \{t_i\})}{|D|}} > g \tag{2}$$

where g is a *goodness threshold*.

After the mining step, we turn to the building of the taxonomy. We then start building what we call for convenience *taxonomy-pieces* (line 1), i. e., the taxonomic relations to be learned in the current iteration. In this graph, two nodes t_x and t_y are connected, with t_x being a superconcept of t_y, if there are frequent itemsets I containing both t_x and t_y such that $sup(\{t_x\}) \geq e[j] \cdot sup(\{t_y\})$, where $j = |I|$ and $e[j]$ is a given edge threshold for itemsets of size j. In other words, according to the *High Level Tag Assumption*, $sup(\{t_x\})$ has to be "significantly larger" than $sup(\{t_y\})$. Moreover, due to the *Large Itemset Assumption* the co-occurrence of t_x and t_y in a large itemset means higher correlation between t_x and t_y than their co-occurrence in a smaller itemset, thus the meaning of "significantly larger" depends on the size of the itemset I, i. e., the edge threshold $e[j]$ is the largest for 2-itemsets ($j = 2$), and the larger the itemset, the smaller the threshold.

To avoid multiple-inheritance relations, we prune the graph restricting it to a tree[9] (line 1). In this step we give preference to long paths since they are usually more informative. Given the edges $t_x \rightarrow t_y$, $t_y \rightarrow t_z$ and $t_x \rightarrow t_z$ (note that t_z has two father concepts, t_x and t_y) for example, the edge $t_x \rightarrow t_z$ would be considered redundant and thus would be removed, since the path going through t_y is longer[10]. This step is done by a depth-first-search like traversal through the *taxonomy-pieces*, which also guarantees that one concept has only one father-concept in the pruned graph.

As pointed out before, the relations learned in each iteration are merged with relations learned in previous iterations (line 1), hence converging to the final ontology. According to the *Frequency Assumption*, we ensure that itemsets with higher support have higher priority than the itemsets with lower support at the merging step, i. e., the

---

[9] The decision of using a tree instead of an arbitrary directed acyclic graph is due to the fact, that the evaluation procedure (see section 4) operates only on trees.

[10] Note that the path $t_x \rightarrow t_y \rightarrow t_z$ leads to more specialized classification of the concept t_z, i. e., it is **not** classified as a subconcept of the general concept t_x, but it classified as a subconcept of a more specific concept t_y.

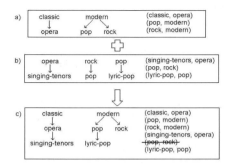

Fig. 3. Addition of taxonomical relations during the learning process. a) Relations already learned in previous iterations. b) Possible relations extracted in the current iteration. c) The learned taxonomy at the end of the current iteration. Note that only the new relations conforming to the old ones will be learned. In this example *pop* will not be defined as a subconcept of *rock*, because it is already known to be a subconcept of *modern* and a concept is only allowed to have one direct super-concept.

taxonomy-pieces learned in previous iterations with higher support have higher priority than the ones learned in the current iteration (e. g., Fig. 3).

Remarks on complexity. The most expensive steps of the algorithm are: (i) the projection of the folksonomy to a transactional database, (ii) the extraction of frequent itemsets and (iii) the population of the ontology (line 1), as these are the only steps operating on the original folksonomy[11]. The other steps operate on the extracted frequent itemsets, which have sizes of lower orders of magnitude. The step (i) requires a linear scan on the folksonomy and therefore has linear complexity. As for the step (ii), the algorithm described in [3] can be implemented as l linear scans on the transactional database, where l is the size of the largest itemsets to be found. However, we implemented this step using a sophisticated trie representation of the transactional database and the frequent itemsets, which further improves efficiency [6,7,5]. The last step, i. e., the creation of the knowledge base (line 1, outside the loop), is based on the counting of the co-occurrence between resources and tags, which means a linear scan on the folksonomy.

4   Recommender Systems for Ontology Evaluation

The non-hierarchical property of folksonomies can somewhat restrict the capabilities of the users for finding information, as the browsing is constrained to a flat structure where tag relations are disregarded. Most of the literature concerning ontology learning from folksonomies use this observation as a main motivation for providing users with a taxonomy of tags, even though the authors do not quantify how good it is for the task it was designed for. Instead, the quality of the ontologies is measured based on whether

---

[11] Or its projection, which have roughly the same size.

they match people's common sense or a reference ontology. Taking this into account, and given that collaborative filtering [21] is one of the most successful and prominent approaches for helping users finding useful information, we propose to plug the investigated knowledge structures in recommender systems and evaluate the outcome as an indicator of their usefulness.

According to Porzel et al. [20], the minimal elements necessary for a task-based evaluation of an ontology are: *a task, one or more ontologies, an application* and a *gold standard*, which are specified for our benchmark as follows.

The Task: Recommending useful resources (e. g., music tracks, videos, websites, etc.) to the users, based on their implicit feedback on resources.

One (or more) Ontologies: A *trivial ontology* representing a folksonomy, a *domain-expert* ontology and a *collabulary*.

The Application: A collaborative filtering algorithm that uses taxonomies for finding the resources of interest. The idea of collaborative filtering is to suggest new resources based on the opinion of like minded users usually called *neighborhood*.

A Gold Standard: In this case the gold standard is the set of resources that the user prefers the most. In a typical recommender systems evaluation scenario we know these resources in advance. We split this set into training and test sets and then use the ones in the training set to recommend other resources the user would eventually like. To measure the quality of the recommender, we compare the predicted resources with the ones in the test set.

The concrete application we have chosen is a taxonomy-driven approach for recommender systems proposed by Ziegler et al. [27], where taxonomies are not only regarded as background knowledge, but also the main cornerstone for efficient and personalized information discovery. The algorithm heavily relies on the taxonomy to make the recommendations and therefore we assume that the better the taxonomy is, the better are the recommendations. In their method a user profile is not composed by vectors such as $u_i \in \mathbb{R}^{|R|}$, where u_{i_k} indicates the user's rating for resource $r_k \in R$, common in traditional collaborative filtering techniques, but by vectors of interest scores assigned to concepts taken from an ontology $\mathcal{O}$ over resource's concepts. The main idea in representing user profiles in this way is the possibility to exploit the hierarchy of the ontology to generate more overlap and thus allow more meaningful similarity computation. The core idea lies in the users' profile assembly, which is briefly described as follows (see Example 2). For every concept $c \in \mathcal{C}_{\mathcal{O}}$ having the resources r_k that user u_i has implicitly rated as instances, it is also inferred an interest score for all super-concepts of c, where scores assigned to super-concepts decay with increasing distance from the concept c. There are other steps concerning score propagation and normalization (see [27] for more details) that will not be covered here due to space reasons.

Example 2. Consider the ontology depicted in Fig. 1.b. Let u_i have implicitly rated resource r_1. In this case this resource is assigned to just one concept, namely $\{emo\}$. Now let $s := 100$ denote the overall accorded interest score. After score propagation and normalization, the profile vector for user u_i is composed as $u_i := (emo := 53.3,$

Table 1. Characteristics of the knowledge bases. For convenience, we also let $|T|$ represent the count of concepts in musicmoz.

dataset	$|U|$	$|T|$	$|R|$	$|Y|$
last.fm	3532	7081	982	130899
musicmoz	-	555	982	-

Table 2. Examples of the three best semantic mappings between tags from *last.fm* and concepts from *musicmoz*

electro	*hip hop*	*chillout*	*old skool dance*	*anything else but death*	*depeche mode*
electronica	hip hop	alternative	house	heavy metal	experimental rock
dance	rhythm and blues	rock	hip hop	metal	synthpop
alternative	rap	electronica	electronica	rock	indie

alternative := 26.6, *rock* := 13.3, *root* := 6.6)^T. Note that the interest score (100) is unevenly distributed among the ancestors of *emo*, where higher concepts receive lower scores, thus reflecting the loss of specificity upwards to the root. After the taxonomy-based profiles are assembled for all users, traditional user-based collaborative filtering can be directly applied.

5   Experiments and Discussion

To evaluate our approach we used two different knowledge bases defined over the music domain, where musical resources are assigned to concepts either by the users of a folksonomy or by the experts in the domain. As the folksonomy representative we have chosen *Last.fm*[12], a social tagging system that provides personalized radio stations where users can tag artists and tracks they listen to. Representing the domain-expert we have chosen the *Open Music Project*[13](*musicmoz*), which is based on the *Open Directory*[14] philosophy and aims to be a comprehensive knowledge base about music. We extracted the *style* hierarchy representing a taxonomy of music genres[15] from *musicmoz* to constitute the domain-expert ontology. Since we consider that the aforementioned knowledge bases are defined over the same set of instances, we eliminated all the resources that are not present in both knowledge bases. Table 1 gives a brief overview on the datasets after this pre-processing.

Folksonomy Enrichment. In the semantic mapping step for the folksonomy enrichment described in Section 2, we also handled concept duplications by eliminating tags that are very similar to the concepts that are to be included in the enriched folksonomy. We

---

[12] This data can be easily gathered by the Web Services provided by last.fm (http://www.last.fm/api)

[13] http://musicmoz.org

[14] http://www.dmoz.org

[15] The hierarchy is composed by cross references which were disregarded in order to guarantee the tree structure.

do this by using the Levenshtein distance metric with a high threshold. Note that this also provides a lexical correction for misspellings, since if we detect that *altertative* and *alternative* are duplicates for example, we just include the syntactically correct one, in this case *alternative*. Table 5 illustrates some of the tags and their corresponding three best semantic mappings. While some mappings are trivial, others are very interesting. Consider the tag *old skool dance* for example. The best mapping refers to *house*, which according to *Wikipedia*[16]: *"House music is a style of electronic dance music that was developed by dance club DJs in Chicago in the early to mid-1980s"*, thus indeed "old skool dance". This can help both experts and users to evolve and specialize their vocabulary in order to further improve the knowledge sharing. Moreover, looking to the three best mappings of the tag *anything else but death* e. g., it is easy to infer that this tag refers to people who likes all styles of heavy music except the subgenre *death metal*, even though it is not explicit just by looking at the tag. Another interesting example is the tag *depeche mode*. Depeche Mode actually refers to an English band from the 80s, which according to *Wikipedia*[17] belongs to the genres *New Wave*, *Synth Pop*, *Post Punk* and *Alternative Dance*. Note that *synthpop* is the second best map in this case and therefore conforms to other authority sources such as *Wikipedia*.

Collabulary Learning. After some experiments for calibration, we used the following setting for the ontology learning process. The count of iterations was set to 7, the minimum support threshold in the i-th iteration was $\frac{0.025 \cdot |D|}{2^i}$ where $|D|$ is the count of transactions. This led in our case to a minimum support of about 500 in the first iteration, and 8 in the last one. This is reasonable, as we assume that tags occurring more than 500-times together certainly correlate no matter whether they co-occur for example 1100-times or just 700-times. We also assume that in our database of around 40000 transactions, two tags co-occur at least 8-times, if they are correlated. Edge thresholds were chosen to be 1.5 (for itemsets of size 2), 1.4 (for itemsets of size 3), 1.3 (for itemsets of size 4), 1.2 (for itemsets of size 5) and 1.1(for itemsets of size 6). The assumption behind this choice is that at least 50% of the users tagging a resource with a subconcept (subtag), also tags the same resource with the corresponding super-concept. It is also assumed, that a super-concept has at least 3 subconcepts (subtags). This results in the super-concept (supertag) being used at least $3 \cdot 0.5 = 1.5$-times more often than its subconcept (subtag). According to the *Large Itemset Assumption*, relaxed thresholds can be applied for itemsets of size more than 2. To filter "misleading" itemsets, we used a goodness threshold of $g = 2$.

Due to space reasons in Fig. 4 we just show an extract of the learned ontology [18] with root in *emo* and maximum depth set to 2. It is interesting to note that while *emo* is a leaf node in the domain expert ontology, it spans several other subtrees in our case. Looking at the definition of the genre in Wikipedia[19] we see that originally *emo* refers to *punk hardcore* and *indie rock* but starting in the mid-1990s, the term evolved and began to refer to a more melodic and less chaotic kind of *indie rock* style. Notice that as children of *emo* we have both concepts associated to the original definition, which

---

[16] http://en.wikipedia.org/wiki/House_music

[17] http://en.wikipedia.org/wiki/Depeche_Mode

[18] Generated with Pajek (http://pajek.imfm.si/doku.php)

[19] http://en.wikipedia.org/wiki/Emo

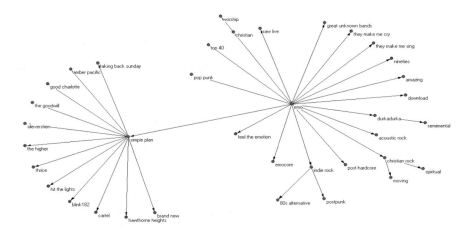

Fig. 4. Extract of the learned ontology

comes from the domain expert-ontology (e. g., *punk hardcore*, *indie rock*) and the ones associated with the term evolution (*nineties* and *post hardcore*). Also notice, that in the definition given in *musicmoz*[20] for the genre, the evolution of the term is not mentioned, denoting that controlled vocabularies evolve somewhat slower than folksonomies, as already intuitively expected.

Evaluation. As mentioned in Section 5, we have chosen three ontologies to evaluate according to our benchmark, namely, a trivial one representing the folksonomy (see Section 2) as the baseline, a domain-expert ontology represented by *musicmoz* and the collabulary. We have used a *AllBut1*[8] protocol to evaluate the obtained recommendations where the test set was obtained by randomly selecting one resource from every user. Note that this also means eliminating all the tag assignments of the corresponding user for the respective test resource. The rest of the data is used for the enrichment of the folksonomy and learning the collabulary that will be used by the recommender algorithm. We have repeated this procedure 5 times and averaged the outcomes in order to be more confident about the results[21]. Our evaluations considered any resource in the recommendation set that matches the resources in the testing set as a "hit", which is equivalent to the *Recall* metric, typical in such scenarios. The number of recommended resources was set to 10 and the size of the neighborhood to 20.

Fig. 5 shows the results of the experiments. Note that while the domain-expert ontology provides a significant improvement over the baseline, the colabullary largely outperforms both. Intuitively this is explained by the fact that we are not restricting the vocabulary of the user, on the contrary, we provide an enriched ontology composed by the best of both worlds. Based on this empirical evidences we can then conclude that while domain-expert ontologies indeed improve the information finding in comparison to pure flat folksonomies, the learned collabulary helps even further.

---

[20] http://musicmoz.org/Styles/Rock/Alternative/Emo/
[21] Standard deviation on the top of the bars in Fig. 5.

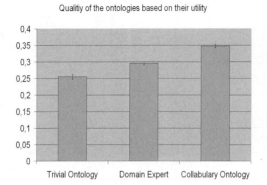

Fig. 5. Ontology evaluation based on recommender systems

6   Related Work

Given the novelty of the problem, there are still very few related works. In [24] e. g., the authors rely on external authority sources or on Semantic Web ontologies to make sense of tag semantics. Even though this can help finding more interesting relations than co-occurrence models, it can somewhat restrict the relation discovery, since if a relation is not defined in these external sources, it is assumed that the tags are not related, even if they frequently co-occur in the dataset. We instead, infer the relations directly from the data and thus are not dependant on external sources.

Other related areas are ontology learning and evaluation. As pointed out before, most of the literature concerning ontology learning from folksonomies base their approaches upon co-occurrence models. Mika [17] e. g., use co-occurrence of tags with resources and users to build graphs relating tags and users and also tags and resources. In [23] conditional probabilities are used to find subsumption relations and [26] use probabilistic unsupervised methods to derive a hierarchy of tags. Given that co-occurrence models are the core subject of frequent itemset mining, in [22] it is proposed the application of association rules between projections of pairs of elements from the triadic context model of folksonomies, although in a conceptual level only. Even though all these works qualitatively contribute for making folksonomies more useful, performance issues are rarely mentioned. Furthermore, all these works are only subjectively evaluated by checking whether the derived ontologies match a reference ontology for example.

7   Conclusions and Future Work

In order to take full advantage of folksonomies' potential to alleviate the knowledge bottleneck that slows down the Semantic Web realization, one needs to educate social tagging systems' users towards clear annotation, without however, taking out their freedom to tag. In this paper we proposed an approach to address this issue by combining the user vocabulary with the expert vocabulary in an integrated fashion. Various studies have shown (e. g., [10]) that the vocabulary of users in social tagging systems stabilize over time due to exposure to each other tags and resources. Taking this into account, we argue that exposing users to a collabulary where meaningful concepts are matched and put together with tags, have the potential to make the whole vocabulary converge to a more meaningful, shareable and useful knowledge representation. Furthermore, we have empirically evaluated the extent to which folksonomies, domain-expert ontologies and collabularies help recommender systems to deliver useful information to the users. Looking at the results, the main lessons learned were: (i) indeed hierarchies provide advantages over flat folksonomies, and (ii) collabularies provide clear benefits over pure domain-expert ontologies.

The main contributions can be summarized as follows: 1. The proposal of a new approach to address the trade-off between folksonomies and domain-expert ontologies. 2. The proposal of a new algorithm for ontology learning from folksonomies based on frequent itemset mining techniques. 3. The proposal of a new benchmark for ontology evaluation. 4. The evaluation of the proposed model on real-life data, namely, from *Last.fm* and *musicmoz*. In future work we plan to extend the method here proposed with the identification of non-taxonomical relations between the tags of the enriched folksonomy.

Acknowledgement. This work is supported by CNPq, an institution of Brazilian Government for scientific and technologic development, and the X-Media[22] (IST-FP6-026978) project.

References

1. Wikipedia article (accessed on May 2008), http://en.wikipedia.org/wiki/Folksonomy
2. Agrawal, R., Imielinski, T., Swami, A.: Mining association rules between sets of items in large databases. In: Proc. of SIGMOD 1993, pp. 207–216. ACM Press, New York (1993)
3. Agrawal, R., Srikant, R.: Fast algorithms for mining association rules in large databases. In: Proc. of the 20th international conference on Very Large Data Bases (VLDB 1994), pp. 478–499. Morgan Kaufmann, San Francisco (1994)
4. Berners-Lee, T., Hendler, J., Lassila, O.: The semantic web. Scientific American (May 2001)
5. Bodon, F.: A fast apriori implementation. In: Proc. 1st IEEE ICDM Workshop on Frequent Item Set Mining Implementations. CEUR Workshop Proc. CEUR-WS.org., vol. 90 (2003)
6. Borgelt, C.: Efficient implementations of apriori and eclat. In: FIMI, CEUR Workshop Proc. CEUR-WS.org., vol. 90 (2003)
7. Borgelt, C.: Recursion pruning for the apriori algorithm. In: FIMI, CEUR Workshop Proc. CEUR-WS.org., vol. 126 (2004)
8. Breese, J.S., Heckerman, D., Kadie, C.: Empirical analysis of predictive algorithms for collaborative filtering. In: Proceedings of the Fourteenth Conference on Uncertainty in Artificial Intelligence (UAI 1998), pp. 43–52. Morgan Kaufmann, San Francisco (1998)
9. Brooks, C.H., Montanez, N.: Improved annotation of the blogosphere via autotagging and hierarchical clustering. In: WWW 2006. Proc. of the 15th international conference on World Wide Web, pp. 625–632. ACM, New York (2006)
10. Cattuto, C., Loreto, V., Pietronero, L.: Collaborative tagging and semiotic dynamics (May 2006), http://arxiv.org/abs/cs/0605015

---

[22] http://www.x-media-project.org

11. Chalupsky, H.: Ontomorph: A translation system for symbolic knowledge. In: Proc. of the 17th International Conference on Knowledge Representation and Reasoning (2000)
12. Cimiano, P., Hotho, A., Staab, S.: Learning concept hierarchies from text corpora using formal concept analysis. Journal of Artificial Intelligence Research (JAIR) 24, 305–339 (2005)
13. Doan, A., Madhavan, J., Domingos, P., Halevy, A.Y.: Ontology matching: A machine learning approach. In: Handbook on Ontologies, International Handbooks on Information Systems, pp. 385–404. Springer, Heidelberg (2004)
14. Goldenberg, A., Moore, A.: Tractable learning of large bayes net structures from sparse data. In: Proc. of the 21st International Conference on Machine Learning (2004)
15. Heymann, P., Garcia-Molina, H.: Collaborative creation of communal hierarchical taxonomies in social tagging systems. Technical Report 2006-10, Stanford University (April 2006)
16. Hotho, A., Jaeschke, R., Schmitz, C., Stumme, G.: Information retrieval in folksonomies: Search and ranking. In: Sure, Y., Domingue, J. (eds.) ESWC 2006. LNCS, vol. 4011, pp. 411–426. Springer, Heidelberg (2006)
17. Mika, P.: Ontologies are us: A unified model of social networks and semantics. In: Gil, Y., Motta, E., Benjamins, V.R., Musen, M.A. (eds.) ISWC 2005. LNCS, vol. 3729, pp. 522–536. Springer, Heidelberg (2005)
18. Noy, N.F., Musen, M.A.: Prompt: Algorithm and tool for automated ontology merging and alignment. In: AAAI/IAAI, pp. 450–455 (2000)
19. Pei, J., Liu, J., Wang, K.: Discovering frequent closed partial orders from strings. IEEE Transactions on Knowledge and Data Engineering 18(11), 1467–1481 (2006)
20. Porzel, R., Malaka, R.: A task-based approach for ontology evaluation. In: Proc. of ECAI 2004 Workshop on Ontology Learning and Population, Valencia, Spain (August 2004)
21. Resnick, P., Iacovou, N., Suchak, M., Bergstorm, P., Riedl, J.: Grouplens: An open architecture for collaborative filtering of netnews. In: Proc. of ACM 1994 Conference on Computer Supported Cooperative Work, Chapel Hill, North Carolina, pp. 175–186. ACM, New York (1994)
22. Schmitz, C., Hotho, A., Jaeschke, R., Stumme, G.: Mining association rules in folksonomies. In: Data Science and Classification: Proc. of the 10th IFCS Conf., Studies in Classification, Data Analysis, and Knowledge Organization, pp. 261–270. Springer, Heidelberg (2006)
23. Schmitz, P.: Inducing ontology from flickr tags. In: Proc. of the Workshop on Collaborative Tagging at WWW 2006, Edinburgh, Scotland (May 2006)
24. Specia, L., Motta, E.: Integrating folksonomies with the semantic web. In: Franconi, E., Kifer, M., May, W. (eds.) ESWC 2007. LNCS, vol. 4519, pp. 624–639. Springer, Heidelberg (2007)
25. Sriphaew, K., Theeramunkong, T.: A new method for finding generalized frequent itemsets in generalized association rule mining. In: ISCC 2002. Proc. of the Seventh International Symposium on Computers and Communications (ISCC 2002), p. 1040 (2002)
26. Zhou, M., Bao, S., Wu, X., Yu, Y.: An unsupervised model for exploring hierarchical semantics from social annotations. In: Aberer, K., Choi, K.-S., Noy, N., Allemang, D., Lee, K.-I., Nixon, L., Golbeck, J., Mika, P., Maynard, D., Mizoguchi, R., Schreiber, G., Cudré-Mauroux, P. (eds.) ISWC 2007. LNCS, vol. 4825, pp. 673–686. Springer, Heidelberg (2007)
27. Ziegler, C., Schmidt-Thieme, L., Lausen, G.: Exploiting semantic product descriptions for recommender systems. In: Proc. of the 2nd ACM SIGIR Semantic Web and Information Retrieval Workshop (SWIR 2004), Sheffield, UK (2004)

Combining a DL Reasoner and a Rule Engine for Improving Entailment-Based OWL Reasoning

Georgios Meditskos and Nick Bassiliades

Department of Informatics
Aristotle University of Thessaloniki
{gmeditsk,nbassili}@csd.auth.gr

Abstract. We introduce the notion of the mixed DL and entailment-based (DLE) OWL reasoning, defining a framework inspired from the hybrid and homogeneous paradigms for integration of rules and ontologies. The idea is to combine the TBox inferencing capabilities of the DL algorithms and the scalability of the rule paradigm over large ABoxes. Towards this end, we define a framework that uses a DL reasoner to reason over the TBox of the ontology (hybrid-like) and a rule engine to apply a domain-specific version of ABox-related entailments (homogeneous-like) that are generated by TBox queries to the DL reasoner. The DLE framework enhances the entailment-based OWL reasoning paradigm in two directions. Firstly, it disengages the manipulation of the TBox semantics from any incomplete entailment-based approach, using the efficient DL algorithms. Secondly, it achieves faster application of the ABox-related entailments and efficient memory usage, comparing it to the conventional entailment-based approaches, due to the low complexity and the domain-specific nature of the entailments.

Keywords: Hybrid and Homogeneous Systems, Rule-based OWL Reasoning, Entailment Rules, Rule Engines, DL Reasoning.

1 Introduction

The Web Ontology Language (OWL) [29] is the W3C recommendation for creating and sharing ontologies on the Web and its theoretical background is based on the Description Logic (DL) [2] knowledge representation formalism, a subset of predicate logic. Existing sound and complete DL reasoners [12][44][45] implement tableaux algorithms [3]. However, although these systems perform well on complex TBox reasoning, they have a high ABox reasoning complexity on medium and large complexity TBoxes that constitutes a serious limitation regarding the efficient query answering capabilities needed in domains with large ABoxes [13][15][35].

Rules play an important role in the Semantic Web and, although there is not an unrestricted translation of DLs into the rule paradigm, they can be used in many directions, such as reasoning, querying, non-monotonicity, integrity constraints [4][11] [34]. Regarding rule-based OWL reasoning, the idea is to map OWL on a rule formalism that applies (a subset of) the OWL semantics in the KB of a rule engine. Practical examples of rule-based OWL reasoners are [21][23][27][28][33] that follow the entailment-based

A. Sheth et al. (Eds.): ISWC 2008, LNCS 5318, pp. 277–292, 2008.
© Springer-Verlag Berlin Heidelberg 2008

OWL reasoning (EOR) approach [17], and [19] that reduces OWL into disjunctive Datalog [18].

We present our effort to combine the strong points of the DL and EOR paradigms; the former performs efficient TBox reasoning while the latter is characterized by ABox reasoning scalability (comparing it to the DL paradigm), and the simplicity of implementation. We define a mixed DL and rule-based (DLE) framework that embeds the TBox inferencing results of the DL paradigm, and it is able to handle larger ABoxes than the conventional EOR systems. The latter is achieved by substituting the TBox-related entailments with TBox queries to the DL reasoner, generating the ABox-related entailments. The DLE framework is inspired from the hybrid and homogeneous approaches for integration of rules and ontologies [1], using the DL reasoner to answer TBox queries and the rule engine for instance queries.

The rest of the paper is structured as follows. In section 2 we present related background and our motivation for the DLE framework. In section 3 we describe the classification of entailments on which our methodology is based. In section 4 we give the general reasoning principles of the DLE framework. In section 5 we present experimental results about the reasoning activity of DLE-based implementations. Finally, in sections 6 and 7, we present related work and we conclude, respectively.

2   Background and Motivation

The RDF and RDFS semantics can be captured using *entailments* [14] that are rules that denote the RDF triples [10] that should be inferred (rule head) based on existing ones (rule body). A triple has a *subject*, a *predicate* and an *object*, represented as ⟨s p o⟩, where s is an RDF URI reference or a blank node, p is an RDF URI reference (in [17] p is allowed to take blank nodes) and o is an RDF URI reference, a blank node or a literal. Examples of entailments can be found in Table 1.

Horst [16] defines the pD^* semantics as a weakened variant of OWL Full and in [17] they are extended to apply to a larger subset of the OWL vocabulary, using 23 entailments and 2 inconsistency rules. Many practical OWL reasoners are based on the implementation of entailments in a rule engine, such as [21][23][27][28][31][33]. These systems apply a larger set of entailments than the pD^* semantics define. For example, class intersection or the eRDFS [14] entailments (RDF entailment regimes [5]) are not considered in pD^* semantics.

In general, the approaches towards the combination of rules and ontologies are either *hybrid* or *homogeneous* [1]. In homogeneous approaches, the rule and ontology predicates are treated homogeneously, as a new single logic language. Practically, ontologies are mapped on a rule-based formalism that coexists in the KB with rule predicates [32][40][41]. The EOR is a type of a homogeneous approach, since any rule predicate of a rule program coexists with the entailment rules [1].

In hybrid approaches, the rule and ontology predicates are separated and the ontology predicates can be used as constraints in rules. This is achieved by following a modular architecture, combining a DL reasoner for OWL reasoning and a rule engine for rule execution [6][7][9][24][39]. Therefore, while in homogeneous approaches OWL reasoning is performed only by rules, in the hybrid paradigm the OWL reasoning is performed only by the DL reasoner.

While the DL reasoners have poor ABox reasoning performance [13][15][35], the EOR paradigm has limited TBox reasoning completeness. For example, if p and g are both the inverse properties of q, then p and g should be inferred as equivalent properties. We observed that the [23][28][33] EOR systems do not deduce such a TBox relationship, in contrast, for example, to Pellet [44] which supports it. Notice that [23][28][33] treat the properties p and g as equivalent at the instance level through the implementation of the inverse entailment (*rdfp8ax*, Table 1). Thus, if $\langle x\ p\ y \rangle$ then $\langle y\ q\ x \rangle$, and therefore, $\langle x\ g\ y \rangle$. However, they do not infer that p and g are equivalent because they do not implement the corresponding entailment. In the EOR paradigm, some TBox entailments are either ignored, in order to speed up the TBox reasoning procedure, or they have not been considered during implementation. For both reasons, the result is an incomplete TBox reasoning procedure in the EOR paradigm.

The use of entailments makes the EOR paradigm also incomplete on ABox reasoning, for the same reasons we mentioned previously. Thus, the choice between an EOR system and a DL reasoner depends on the domain and the needs of the application (see also [4][38]). With DLE we tackle the TBox reasoning incompleteness of EOR. Our motivation can be summarized in the following observation: *"Why do we need to struggle to define the entailments for OWL TBox reasoning in the EOR paradigm, if we can make it effortless and more complete using the efficient DL algorithms?"*.

In [37] the RDFS(FA) sublanguage of RDFS and the RDF Model Theory are discussed. In brief, RDFS(FA) eliminates the dual roles of RDFS, stratifying built-in RDF primitives in different layers. On the other hand, the RDF MT handles dual roles by treating classes and properties as objects. Practically, DL reasoners treat the OWL extension of the bottom two layers of RDFS(FA), while the entailments are defined in RDF MT. For example, RDFS(FA) distinguishes built-in from user-defined properties, while in RDFS there is no such restriction and thus it is more expressive (although it is argued that this expressivity is too confusing). The existing EOR systems treat OWL as an extension of RDF MT, allowing *anyone to say anything about anything*. This was the domain of interest in [30] and [31], where only a rule engine is considered for TBox and ABox entailments in the RDF MT. On the other hand, the DLE framework is an RDFS(FA)-oriented approach, since it uses a DL component to reason on OWL ontologies, *following though the EOR paradigm*.

Our DLE framework considers the EOR paradigm as a *rule program* (ABox entailments) over the KB of a DL reasoner after TBox reasoning, following the architecture of the hybrid paradigm. Therefore, any subsequent user-defined rule program would then coexist in the rule base of the rule engine with the ABox rule program, in the same manner as in the homogeneous paradigm. Thus, in contrast to the existing EOR implementations, the OWL reasoning in the DLE framework is performed both by a DL component and a rule engine. More specifically, it combines the TBox inference capabilities of the DL component to compute the subsumption hierarchy and the related semantics to the ontology roles, e.g. domain constraints, property types, etc., with the ability of a rule engine to process a large number of instances, applying domain-specific entailment rules that are generated based on the DL reasoner.

Apart from scalability issues, one of the attracting features of the EOR paradigm is that it can be easily implemented in any rule engine, e.g. Jess [22], or a Prolog engine [42], enabling the use of ontological information into rule programs, exploiting the research on efficient rule engines with different capabilities. The DLE framework is based on this practicality and actually enhances the EOR paradigm in two directions:

- It simplifies the development of an EOR system, disengaging the TBox reasoning procedure from any (incomplete) TBox entailment implementation that should take into account all the possible OWL TBox semantic derivations.
- The rule engine applies faster the domain-specific ABox entailments than the corresponding generic entailments of the traditional EOR systems, enhancing their scalability in terms of ABox reasoning time and memory utilization.

3   Entailment Classification

The DLE framework is based on the classification of entailments into *terminological*, *hybrid* and *exceptional*. In this section we present the necessary and sufficient conditions for performing such a classification. Notice that, since we are based on a DL reasoner, the DLE framework is not compatible with RDF ontologies and handles only the OWL vocabulary [29]. Thus, we do not capture relationships such as that `owl:Class` is equivalent or subclass to `rdfs:Class`, `owl:Thing` is equivalent or subclass to `rdfs:Resource` or `owl:ObjectProperty` is equivalent or subclass to `rdfs:Property`. Furthermore, `owl:ObjectProperty` and `owl:DatatypeProperty` are disjoint sets [29]. Since we approach the entailment-based reasoning from the OWL perspective, we substitute any reference to `rdfs:Resource` and `rdfs:Class` in entailment rules with `owl:Thing` and `owl:Class`, respectively (Table 1).

We present a definition of an entailment rule that we follow in the rest of the paper.

Definition 1. *An entailment rule for an RDF graph G is of the form*

$$\langle s_1\, p_1\, o_1 \rangle \langle s_2\, p_2\, o_2 \rangle \, ... \, \langle s_n\, p_n\, o_n \rangle \rightarrow \langle s'_1\, p'_1\, o'_1 \rangle \langle s'_2\, p'_2\, o'_2 \rangle \, ... \, \langle s'_m\, p'_m\, o'_m \rangle,$$

where $n \geq 1$, $m \geq 1$, s_i, s'_i, p_i and p'_i are RDF URI references or blank nodes, and o_i and o'_i are RDF URI references, blank nodes or literals. The $\langle s_n\, p_n\, o_n \rangle$ triples denote the condition of the entailment and the $\langle s'_m\, p'_m\, o'_m \rangle$ triples the conclusion. The condition of the rule denotes the RDF triples that should exist in G, and the conclusion the RDF triples that should be added in G.

If $n = 0$, then all the conclusion triples should always exist in G (axiomatic triples [14][17]). If $m = 0$, then the entailment denotes that the triple pattern of the body should be viewed as inconsistent (inconsistency entailment).

3.1   Terminological and Assertional Triples

We present a classification of triples into *terminological* (*T-triples*) and *assertional* (*A-triples*), according to the OWL DL vocabulary *V* [29] of their components (prefixes have been omitted).

Definition 2. *A triple t = ⟨s p o⟩ is a terminological triple, denoted as t_T = ⟨s p o⟩$_T$, iff*

- *p* ∈ {domain, range, subClassOf, subPropertyOf, inverseOf, equivalentProperty, equivalentClass, intersectionOf, unionOf, complementOf, onProperty, hasValue, someValuesFrom, allValuesFrom, maxCardinality, minCardinality, cardinality, disjointWith}, *or*
- *p* = type ∧ *o* ∈ {ObjectProperty, DatatypeProperty, FunctionalProperty, InverseFunctionalProperty, SymmetricProperty, TransitiveProperty, Class, Restriction}.

A T-triple denotes information about the TBox of the ontology (class and property axioms), such as subclass relationships, class equivalence, property types, etc. The A-triples are defined as the complement of the terminological.

Definition 3. *A triple t = ⟨s p o⟩ is an assertional triple, denoted as t_A = ⟨s p o⟩$_A$, iff it is not a terminological, that is t_A ⟺ ¬t_T.*

Intuitively, an A-triple denotes information about the ABox of the ontology, such as instance class membership or instance equality/inequality (sameAs /different From) (we consider the oneOf construct as simple *i* : *C* assertions). Thus, ⟨*p* domain *c*⟩$_T$ and ⟨*p* type FunctionalProperty⟩$_T$, whereas ⟨*x* sameAs *y*⟩$_A$. Therefore, a triple component can either be bound to a term of the OWL vocabulary or not be bound. In the former case, we refer to the component as *constant*, whereas in the latter as *variable* (blank node). For example, the triple ⟨*p* domain *c*⟩ has a *variable* subject and object, and a *constant* predicate. We use the notation *var(c)* to denote that the *c* component of the triple is a *variable*.

3.2   Terminological, Hybrid and Exceptional Entailments

Based on the triple classification of section 3.1, we define the *terminological, hybrid* and *exceptional* entailments.

Definition 4. *An entailment is considered as a terminological (T-entailment), if and only if it contains only T-triples in its conclusion.*

Definition 5. *An entailment is considered as a hybrid (H-entailment), if and only if it contains both T- and A-triples in its condition and only A-triples in its conclusion.*

Definition 6. *An entailment is considered as an exceptional (E-entailment), if and only if it contains only A-triples in its condition and conclusion.*

Table 1 depicts some indicative examples of entailment classification, as well as some eRDFS entailments needed for OWL TBox reasoning [14] (denoted as *extX*).

4   Reasoning on the DLE Framework

The reasoning on the DLE framework is based on two reasoning paradigms over two distinct KBs that cooperate.

Table 1. Classification examples of some common entailment rules

Terminological Entailment Rules (T-entailments)	
rdfs8	$\langle c$ type Class$\rangle_T \rightarrow \langle c$ subClassOf Thing$\rangle_T$
rdfs11	$\langle c$ subClassOf $d\rangle_T \langle d$ subClassOf $k\rangle_T \rightarrow \langle c$ subClassOf $k\rangle_T$
rdfp12c	$\langle c$ subClassOf $d\rangle_T \langle d$ subClassOf $c\rangle_T \rightarrow \langle c$ equivalentClass $d\rangle_T$
rdfp13c	$\langle p$ subPropertyOf $q\rangle_T \langle q$ subPropertyOf $p\rangle_T \rightarrow \langle p$ equivalentProperty $q\rangle_T$
ext1	$\langle p$ domain $c\rangle_T \langle c$ subClassOf $d\rangle_T \rightarrow \langle p$ domain $d\rangle_T$
ext2	$\langle p$ domain $c\rangle_T \langle b$ subPropertyOf $p\rangle_T \rightarrow \langle b$ domain $c\rangle_T$
Hybrid Entailment Rules (H-entailments)	
rdfs2	$\langle p$ domain $c\rangle_T \langle x\ p\ y\rangle_A \rightarrow \langle x$ type $c\rangle_A$
rdfp8ax	$\langle p$ inverseOf $q\rangle_T \langle x\ p\ y\rangle_A \rightarrow \langle y\ q\ x\rangle_A$
rdfs9	$\langle c$ subClassOf $d\rangle_T \langle x$ type $c\rangle_A \rightarrow \langle x$ type $d\rangle_A$
rdfp1	$\langle p$ type FunctionalProperty$\rangle_T \langle x\ p\ y\rangle_A \langle x\ p\ z\rangle_A \rightarrow \langle y$ sameAs $z\rangle_A$
rdfp4	$\langle p$ type TransitiveProperty$\rangle_T \langle x\ p\ y\rangle_A \langle y\ p\ z\rangle_A \rightarrow \langle x\ p\ z\rangle_A$
rdfp14a	$\langle r$ hasValue $y\rangle_T \langle r$ onProperty $p\rangle_T \langle x\ p\ y\rangle_A \rightarrow \langle x$ type $r\rangle_A$
Exceptional Entailment Rules (E-entailments)	
rdfs4a	$\langle x\ p\ y\rangle_A \rightarrow \langle x$ type Thing$\rangle_A$
rdfs4b	$\langle x\ p\ y\rangle_A \rightarrow \langle y$ type Thing$\rangle_A$
rdfp6	$\langle x$ sameAs $y\rangle_A \rightarrow \langle y$ sameAs $x\rangle_A$
rdfp7	$\langle x$ sameAs $y\rangle_A \langle y$ sameAs $z\rangle_A \rightarrow \langle x$ sameAs $z\rangle_A$
rdfp11	$\langle x\ p\ y\rangle_A \langle x$ sameAs $x'\rangle_A \langle y$ sameAs $y'\rangle_A \rightarrow \langle x'\ p\ y'\rangle_A$

Definition 7. *The DLE framework consists of two distinct knowledge bases DLE =* $(\mathcal{DLKB}, \mathcal{RKB})$ *where:*

- $\mathcal{DLKB}$ *is the DL component's KB, with* $\mathcal{DLKB} = \langle \mathcal{T} \rangle$*, where* $\mathcal{T}$ *is the ontology TBox (concept and role axioms), and*
- $\mathcal{RKB}$ *is the rule engine's KB, with* $\mathcal{RKB} = \langle \mathcal{RB}, \mathcal{A} \rangle$*, where* $\mathcal{RB}$ *is the rule base of the rule engine that contains the entailment rules and* $\mathcal{A}$ *is the ABox of the ontology (instance and role assertions).*

The two KBs are distinct in the sense that the information flows only from the $\mathcal{DLKB}$ to the $\mathcal{RKB}$ (unidirectional) in order to populate the $\mathcal{RB}$ with entailments.

4.1 Reasoning on the DL Component

Basic DL reasoning problems include *class equivalence*, *concept subsumption*, *satisfiability* and *realization*. Since the $\mathcal{DLKB}$ does not consider ABoxes, the DL component of the DLE framework is not used for instance realization.

The use of DL TBox reasoning in the DLE framework makes redundant the T-entailments. These entailments generate T-triples (*Definition* 4), such as subclass and class equivalence relationships, domain and range restrictions, property equivalence, etc. Since these TBox semantics are handled by the DL reasoner, the T-entailments are ignored, decoupling the TBox inference procedure from any entailment-based approach. As we explain in the next section, each T-triple of an H-entailment is substituted with a query to the $\mathcal{DLKB}$. In that way, *we are not concerned about how to implement the TBox semantics in the DLE framework, but only how to use them at the instance level via ABox entailments.*

To exemplify, consider the *rdfs11* entailment for subclass transitivity. For every three concepts $C, D, E \in \mathcal{T}$ of the $\mathcal{DLKB}$:

$$\text{if } \mathcal{T} \models C \sqsubseteq D \text{ and } \mathcal{T} \models D \sqsubseteq E, \text{ then } \mathcal{DLKB} \models C \sqsubseteq E.$$

Practically, by querying the $\mathcal{DLKB}$ for the indirect superclasses of a concept, all the concepts that belong to the subclass transitive closure are returned (as well as their equivalents), thus the *rdfs11* entailment is natively supported.

Consider also the `intersectionOf` construct. A class C is contained in the intersection of the classes $C_1,\dots, C_n$ by saying that C is a subclass of each class C_n. This is also the inference result of a DL reasoner:

$$\text{if } \mathcal{T} \models C \equiv C_1 \sqcap \dots \sqcap C_n, \text{ then } \mathcal{DLKB} \models C \sqsubseteq C_n.$$

Thus, by querying the DL reasoner for the superclasses of C, the intersection classes C_n are also returned (a similar approach is followed for the semantics of the `unionOf` construct). Notice that the iff semantics of class intersection (and union) require ABox reasoning and thus, the DLE framework handles them at the instance level by entailments. The same holds for class restrictions, e.g. $\forall R.C$. The DL reasoner is used also to facilitate entailment-free TBox consistency check, e.g. inconsistent subclass relationships of disjoint classes.

4.2 Transforming H-Entailments

Since we do not consider T-entailments, the $\mathcal{RKB}$ does not contain any T-triple and thus, the H-entailments would never be activated. In order to cope with the missing TBox triple information, we substitute each T-triple of the condition of an H-entailment with a query to the DL component in order to ground the TBox-related variables (*vars*) of the remaining A-triples. We call this procedure *entailment reduction*, because the H-entailments are *reduced* to domain-specific E-entailments, which we call *dse-entailments*. Therefore, for a specific H-entailment, it is possible to generate more than one dse-entailment. There are two advantages behind the reduction:

- Since the dse-entailments are generated based on the ontology TBox axioms, the $\mathcal{RB}$ will contain only the entailments that are needed, in contrast to the traditional EOR paradigm where all the entailments are preloaded. For example, a transitive role-free ontology will result in a transitive entailment-free $\mathcal{RB}$, reducing the number of the rules need to be checked in each cycle.
- The dse-entailments contain less conditional elements than the initial H-entailments, since the T-triples are removed. Thus, the dse-entailments have lower complexity and thus, they are activated faster by the rule engine.

DL queries. A T-triple t_i can be transformed into a TBox query for the DL component, denoted as $t_T \twoheadrightarrow \mathcal{DLQ}(t_T)$, in order to retrieve the ontology values that correspond to the variables of the T-triple. Similarly, the set T of the conditional T-triples of an H-entailment H can be transformed into a conjunction query, denoted as $\mathcal{DLQ}_H$, that corresponds to each T-triple, that is $T \twoheadrightarrow \mathcal{DLQ}(t1_T) \wedge \dots \wedge \mathcal{DLQ}(tn_T), \forall tn_T \in T$.

To exemplify, using a predicate-like syntax for the DL queries, the T-triple $\langle p$ domain $c\rangle_T$ of the *rdfs2* entailment can be transformed into the query:

$$\mathcal{DLQ}_{rdfs2} : \langle p \text{ domain } c\rangle_T \rightarrow \text{domain}(var(p), var(c)),$$

retrieving all the properties of the $\mathcal{DLKB}$ with their domain constraints. Similarly, the T-triples of the *rdfp14a* entailment can be transformed into:

$$\mathcal{DLQ}_{rdfp14a} : \{\langle r \text{ hasValue } y\rangle_T, \langle r \text{ onProperty } p\rangle_T\} \rightarrow$$
$$\text{hasValue}(var(r), var(y)) \wedge \text{onProperty}(var(r), var(p)),$$

retrieving the properties and their restriction values for every hasValue restriction.

T-dependency. We introduce the notion of *T-dependency* between an A-triple (t_A) and a T-triple (t_T), according to whether t_A has a variable component that exists in t_T.

Definition 8. *An A-triple t_A is T-dependent to a T-triple t_T, denoted as $t_A \curvearrowright t_T$, iff $\exists var(c) \in t_A : var(c) \in t_T$. Each such c variable of a T-dependent triple is called T-dependent variable and it is denoted as [c] in the entailment rule.*

For example, both the $t2_A = \langle x \; p \; y\rangle_A$ and $t3_A = \langle x \text{ type } c\rangle_A$ A-triples of the *rdfs2* h-entailment are T-dependent to $t1_T = \langle p \text{ domain } c\rangle_T$, that is $t2_A \curvearrowright t1_T$ and $t3_A \curvearrowright t1_T$, since $var(p) \in t2_A$, $var(c) \in t3_A$ and $var(p), var(c) \in t1_T$. On the other hand, in the *rdfp1* H-entailment only the A-triples $t2_A = \langle x \; p \; y\rangle_A$ and $t3_A = \langle x \; p \; z\rangle_A$ are T-dependent to $t1_T = \langle p \text{ type FunctionalProperty}\rangle_T$, since $\nexists var(c) \in t1_T : var(c) \in t4_A = \langle y \text{ sameAs } z\rangle_A$. The T-dependency denotes the A-triples whose T-dependent variables should be grounded, due to the removal of the T-triple on which they depend.

Generating the dse-entailments. An H-entailment H is reduced by removing the T-triples of its condition and applying a $\mathcal{DLQ}_H$ query. The results of the query are used to ground the T-dependent variables of the A-triples, generating domain-specific versions of H (dse-entailments). The H-entailment reduction can be considered as the procedure of grounding the T-dependent variables of a *pseudo-rule*.

Definition 9. *The pseudo-rule PR_H for the H-entailment H, is the entailment-like rule we obtain after the removal of any T-triple of the H's condition, and it is of the form*

$$\langle [s_i] \; p_i \; o_i\rangle_A \ldots \langle s_k \; [p_k] \; o_k\rangle_A \ldots \langle s_n \; p_n \; [o_n]\rangle_A \rightarrow$$
$$\langle [s_l] \; p_l \; o_l\rangle_A \ldots \langle s_m \; [p_m] \; o_m\rangle_A \ldots \langle s_o \; p_o \; [o_o]\rangle_A \ldots \langle s_u \; p_u \; o_u\rangle_A,$$

where [x] denotes the T-dependent variables of the entailment. The pseudo-rule is actually a template rule which can be loaded in the $\mathcal{RB}$ as a valid rule (dse-entailment) after the grounding of its T-dependent variables.

For example, the pseudo-rule PR_{rdfs2} for the *rdfs2* entailment that we obtain after the removal of its T-triples is the following:

$$\langle x \; [p] \; y\rangle_A \rightarrow \langle x \text{ type } [c]\rangle_A,$$

where $[p]$ and $[c]$ are the two T-dependent variables of the entailment. Thus, based on the $\mathcal{DLQ}_{rdfs2}$ query, we can ground $(G[PR_{rdfs2}(p, c)])$ the T-dependent variables as:

$$\forall (p, c) \in \mathcal{DLQ}_{rdfs2} : G[\langle x \, [p] \, y \rangle_A \rightarrow \langle x \text{ type } [c] \rangle_A],$$

generating as many rules as the pairs (*property*, *domain*) are in the ontology. For example, for two properties p_k and p_m with the classes c_k and c_m as domain constraints, respectively, we will obtain the following dse-entailments:

$$\langle x \, p_k \, y \rangle_A \rightarrow \langle x \text{ type } c_k \rangle_A,$$
$$\langle x \, p_m \, y \rangle_A \rightarrow \langle x \text{ type } c_m \rangle_A.$$

One of the advantages of the entailment reduction is that the complexity of the dse-entailments is lower than the corresponding H-entailments. The time needed for the *rdfs2* H-entailment to be activated is $O(n^2)$, where n is the size of the partial closure graph under construction [17], whereas the reduced one can be handled in $O(pn)$, where p is the number of the grounded entailments generated from an H-entailment. Similarly, the *rdfp14a* H-entailment requires $O(n^3)$ time, while the reduced entailment runs in $O(pn)$. Generally, if $O(n^t)$ is the complexity of an H-entailment, where t is the number of triples of the condition, the reduced entailments have $O(pn^{t-k})$ complexity, where k is the number of T-triples (that are removed). We should mention that:

- The reduction results in a $\mathcal{RB}$ that contains more entailments than the initial H-entailments, since for each H-entailment more than one rule might be generated (p rules). However, in section 5 we show that such an $\mathcal{RB}$ terminates the ABox reasoning procedure faster than the corresponding generic H-entailment $\mathcal{RB}$.
- The $\mathcal{RB}$ contains only ABox-related entailments. Thus, only updates related to instances can be handled. We elaborate further on this in section 7.

4.3 Basic Reasoning Steps in a Forward Chaining DLE Framework

The E-entailments are the only entailments that are predefined in the DLE framework, since they cannot be reduced, following the approach of the convectional EOR paradigm. Assuming that E_A and PR are the sets of the A-entailments and the pseudo-rules (reduced H-entailments) that will be supported by the DLE-based implementation, the algorithm of Fig. 1 depicts the reasoning methodology using a forward chaining rule engine. Initially, the TBox of the ontology is loaded into the DL reasoner in order to classify the ontology (lines 1 and 2). Then, the ABox is loaded into the rule engine (line 3) in order to create the internal rule engine representation, for example triple facts. Moreover, the predefined E-entailments are loaded into the $\mathcal{RB}$ (line 4). In order to generate the dse-entailments, i.e. the grounded pseudo-rules, we conduct the necessary $\mathcal{DLQ}_H$ queries to the DL component in order to retrieve the values for the T-dependent variables of the H entailments. The resulting rules are loaded into the $\mathcal{RB}$, populating it with the domain specific dse-entailments (lines 5, 6 and 7). Finally, the rule engine runs and materializes the semantics in the form of derived triples.

BEGIN
1. $\mathcal{T} \leftarrow load(TBox)$
2. $classify(\mathcal{DLKB})$
3. $\mathcal{A} \leftarrow load(ABox)$
4. **for each** $e_A \in E_A$ **do** $\mathcal{RB} \leftarrow load(e_A)$
5. **for each** $pr_H \in PR$ **do**
6.   **for each** $(x_1,...,x_n) \in \mathcal{DLQ}_H$ **do**
7. $\mathcal{RB} \leftarrow load(G[pr_H(x_1,...,x_n)])$
8. $\mathcal{RKB}.run()$
END

Fig. 1. The reasoning steps involved in a production rule-based DLE system

5   Testing the ABox Reasoning Performance

We conducted experiments to test the scalability of the dse-entailments against the conventional implementation of the same set of entailments in the same rule engine. We used three rule engines (Bossam [33], the *forwardRETE* rule engine of Jena [28] and Jess [22]) and developed six prototype implementations: three DLE-based, using the Pellet reasoner as the DL component, and three generic, i.e. direct implementation of the entailments following the conventional EOR paradigm. Each prototype was tested on the UOBM [26], Vicodi and wine ontologies[1]. Table 2 depicts the number of entailment rules that each implementation involves. Notice that:

- Our intention is to test the behaviour of a rule engine following the DLE and the generic EOR paradigms and not to compare these two paradigms on different rule engines, since each rule engine has a different performance.
- We are not concerned about the completeness of reasoning, since it depends on the number of the implemented entailments[2]. We want only to test the scalability of the prototypes in terms of rule application time and memory utilization.
- The response time of queries over the ABoxes between a DLE-based and the corresponding generic implementation in the same rule engine is the same, since both approaches result in the same KB (inferred triples).

Moreover we should mention that a fair comparison of the DLE prototypes with existing EOR systems that use the same rule engines is not feasible since this requires the implementation of the same set of entailments that the reasoners support, as well as to follow the same implementation principles or potential optimizations. However, the Bossam OWL reasoner does not provide the full set of the supported entailments, and the Jena OWL reasoner and OWLJessKB implement some semantics internally without entailments, such as the class intersection (Jena reasoner) or using `defque-ries` and `deffunctions` (OWLJessKB). In order to conduct a fair comparison, we re-implemented directly the same set of entailments in the three rule engines. The experiments ran on Windows XP with 3.2 GHz processor, 2 GB RAM and 1.2 GB maximum Java heap space.

---

[1] We obtained Vicodi and wine from kaon2.semanticweb.org/download/test_ontologies.zip
[2] A set of OWL entailment rules can be found in http://www.agfa.com/w3c/euler/owl-rules

Table 2. The number of the entailments involved in the DLE and Generic implementations

	DLE Implementations (dse-entailments + exceptional)	Generic Implementations (generic entailments)
UOBM	323	34 (16 terminological + 13 hybrid + 5 exceptional)
Vicodi	1,164	
wine	592	

Fig. 2 depicts the ABox reasoning performance of each prototype in each ontology dataset. Each graph displays also the memory requirements of each implementation.
UOBM. We used a dataset of almost 810,000 triples (Fig. 2 (a)). DLE Bossam reasoned considerably faster than the Generic Bossam. In particular, it reasoned on five times more triples until it reached the memory limit. DLE Jena displayed a notably better performance than the Generic Jena, processing almost 200,000 more triples without reaching the memory limit. Finally, DLE Jess managed to process faster a dataset twice the size of the one processed by the Generic Jess.

Vicodi. These experiments were performed on three datasets (Fig. 2 (b)). DLE Jess demonstrated a better performance than the Generic Jess both in terms of reasoning time and memory utilization. DLE Jena processed the first two datasets in almost

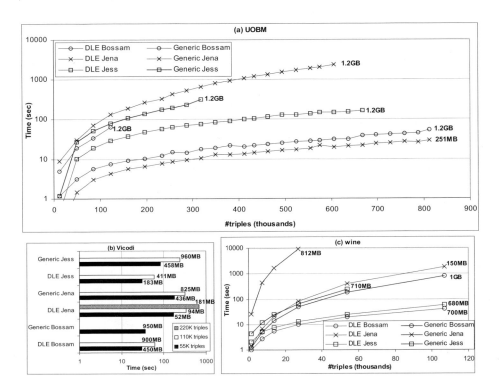

Fig. 2. Results on ABox reasoning

the same time to the Generic Jena but with better memory utilization, enabling the processing also of the third dataset without reaching the memory limit. The same behaviour observed in DLE Bossam that managed to process the first two datasets without reaching the memory limit. In contrast to DLE Jess, DLE Bossam and DLE Jena seem to be affected by the number of the dse-entailments (1,159) of their rule base. However, the memory utilization remains still in lower levels than the generic prototypes.

Wine. The wine experiments used a dataset of about 110,000 triples (Fig. 2 (c)). DLE Bossam processed the dataset significantly faster than the Generic Bossam, using half of the available memory. Generic Jena displayed a poor reasoning performance, while the DLE Jena managed to load the dataset in a reasonable time limit. Finally, Generic Jess processed only half triples than DLE Jess before reaching the memory limit.

Fig. 3 (a) depicts the TBox reasoning times of the prototypes. Since the DLE implementations are based on Pellet for TBox reasoning, they have the same TBox reasoning performance. Except for the wine ontology, Pellet achieves faster TBox inferencing than the generic entailment-based approaches. Bear in mind that the generic prototypes had been implemented with a limited number of T-entailments (16 entailments), while Pellet performs full TBox reasoning. The average dse-entailments generation time of the three DLE-prototypes is 250 ms (lines 5, 6 and 7 in Fig. 1).

In order to give a gist about the ABox reasoning performance of Pellet and KAON2, we present in Fig. 3 (b) the time needed by Pellet, KAON2 and DLE Jena to retrieve the instances on some datasets (since KAON2 performs reasoning on demand and thus, a query is required). We forced Pellet to completely realize the ABoxes which is close to the total materialization approach of our six prototypes, since we used forward chaining rule engines. For TBoxes with medium and large complexity, i.e. many class restriction, intersection or equivalence axioms, such as the UOBM and wine ontologies, Pellet does not perform well compared to the DLE approach, emphasizing the need for scalable implementations. On simple TBoxes, such as the Vicodi ontology which contains only simple subclass axioms, Pellet performs better, but KAON2 depicts the best performance, exploiting the ability to reason on demand. However, on the other two ontologies of medium and large TBox complexity, the DLE implementation performs better, especially on UOBM, even if it follows the

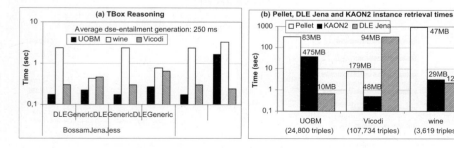

Fig. 3. (a) TBox reasoning times, and (b) Pellet, KAON2 and Jena DLE instance retrieval times

complete materialization approach. Notice that the results of Fig. 3 (b) cannot be considered as a fair comparison, due to the limited semantics that the DLE prototype supports and the different rule paradigm that it follows. A comparison of KAON2 with a backward chaining DLE implementation would be more meaningful. However, Fig. 3 (b) gives a gist about the weak and strong points of each reasoning paradigm.

6 Related Work

To the best of our knowledge, the existing EOR systems, such as OWLJessKB [23], Bossam [33], BaseVISor [27], Jena [28] and OWLIM [21], follow the same approach: the asserted ontological knowledge is transformed into facts and TBox and ABox entailment rules are applied. Although some systems offer the possibility to attach a DL reasoner for both TBox and ABox OWL reasoning, e.g. Jena, none of them has considered the possibility of using a DL reasoner in parallel with the rule engine for OWL reasoning. The DLE framework works towards this idea, combining the TBox inferencing capabilities of DL reasoners with the scalability of the EOR paradigm.

In [43] and [25], the entailments were enhanced with a dependency information, denoting the rules that should be checked after the firing of each entailment. Although this improves the performance, it is very difficult to manage such rule bases, since any modification needs the reconfiguration of the correlations. Furthermore, our experiments have shown that it is not the number of the rules that matters most, but the complexity of their condition. Although the $\mathcal{RB}$ of the DLE approach contains more rules than the conventional EOR paradigm, the inferencing procedure terminates faster. However, we believe that a combination of the DLE with the approach of [43] and [25] would increase even more the performance.

KAON2 [19] reduces OWL into disjunctive Datalog [18]. Its reasoning procedure is based totally on rules and it is focused mainly on query answering. As it was mentioned in [35], DL reasoners have better classification performance on complex TBoxes than KAON2. The DLE framework tries to embed this DL TBox efficiency into the EOR paradigm. The ABox performance of KAON2 depends on the TBox, having an increased performance on simple TBoxes.

Instance Store [15] combines a DL reasoner with a relational database. The idea is to use the DL reasoner for TBox reasoning and the database to store the ABox. The limitation of this approach is that it deals only with role-free ontologies, i.e. ontologies that contain only axioms of the form $i : C$. However, the use of a database is the only solution when the ABox exceeds the size of main memory (see also [43] and Owlgres [36] for DL-Lite). DLE has been only tested in main memory.

In [5] the embedding of different RDF entailments (including eRDFS) in F-Logic is presented that can be used to extend RDF or to align RDF and OWL DL. The DLE approach is focused only on the OWL language, defining an OWL reasoning framework based both on a DL reasoner and a rule engine.

The notion of generating ABox rules in the conventional EOR paradigm was briefly introduced in [30] and [31], where both the TBox and ABox reasoning are performed only by a rule engine (RDF MT). DLE extends and enhances our previous research, defining a new EOR reasoning paradigm (dedicated to OWL ontologies) using DL reasoning for TBox inferencing and ABox entailment generation.

7   Conclusions and Future Work

In this paper we presented an approach for embedding the TBox inferencing capabilities of DL reasoners into the EOR paradigm, resulting in an OWL-oriented reasoning framework. In that way we are able to capture OWL TBox semantics without applying TBox entailments, as well as to enhance the EOR scalability in terms of reasoning time and memory utilization. This is achieved by generating "engine-friendly" ABox entailments, with less conditional elements in their body (thus less complex) than the corresponding generic entailments of the traditional EOR paradigm.

We tested three DLE-based implementations against the three traditional EOR implementations, using three well known rule engines. The experiments have shown that although the DLE prototypes need to apply more rules than the corresponding EOR, they achieve better reasoning performance (at least on the tested ontologies) in terms of rule application time and memory utilization. We conclude that a DLE approach can considerably enhance the performance of existing EOR systems (regarding OWL reasoning), such as the Jena, OWLJessKB and Bossam OWL reasoners. More experiments, however, need to be conducted in order to investigate the impact of the number of the dse-entailments on the reasoning performance.

Although we define an entailment-free TBox framework, the DLE still depends on entailments for ABox reasoning. Thus, it is still a rule-based approach with all the modelling strengths and weaknesses comparing it to the DL paradigm, such as the limited modelling capabilities with incomplete information, the closed-world reasoning or the Unique Name Assumption (UNA) (see [4][34][38] for details).

In section 4 we mentioned that DLE is a unidirectional framework able to handle updates only on instances. However, in the hybrid paradigm, there is the notion of the *bidirectional* combination, allowing the rule program to alter the ontological information of the DL component [8][20][46]. This would be an interesting extension of the DLE framework and we plan to implemented it by indexing appropriately the dse-entailments in order to modify the $\mathcal{RB}$ according to TBox updates.

We are working on releasing a stable, DIG-compliant and more complete, in terms of supported ABox entailments, Jena-based DLE system, since with the Jena rule engine we achieved the best combination of memory utilization and reasoning performance. We plan also to implement a DLE system using the backward-chaining rule engine of Jena. As a practical application of the DLE framework, we consider the domain of OWL-S Semantic Web Service discovery and composition, where there is the need of efficient TBox reasoning on complex TBoxes, and scalable ABox reasoning on service advertisements that point to TBox concepts (inputs/outputs).

Acknowledgments. This work was partially supported by a PENED program (No. 03EΔ73), jointly funded by the European Union and the Greek government (GSRT).

References

1. Antoniou, G., Damasio, C.V., Grosof, B., Horrocks, I., Kifer, M., Maluszynski, J., Patel-Schneider, P.F.: Combining Rules and Ontologies, REWERSE Deliverables, REWERSE-DEL-2005-I3-D3 (2005)
2. Baader, F.: The Description Logic Handbook: Theory, Implementation and Applications. Cambridge University Press, Cambridge

3. Baader, F., Sattler, U.: An overview of tableau algorithms for description logics. Studia Logica 69(1), 5–40
4. de Bruijn, J., Lara, R., Polleres, A., Fensel, D.: OWL DL vs. OWL Flight: Conceptual Modeling and Reasoning for the Semantic Web. In: International Conference on World Wide Web, pp. 623–632. ACM Press, New York (2005)
5. de Bruijn, J., Heymans, S.: Logical foundations of (e)RDF(S): Complexity and reasoning. In: International Semantic Web Conference (+ 2ndASWC), pp. 86–99. Springer, Heidelberg (2007)
6. Donini, F.M., Lenzerini, M., Nardi, D., Schaerf, A.: AL-log: Integrating Datalog and Description Logics. Intelligent and Cooperative Information Systems, 227–252 (1998)
7. Drabent, W., Henriksson, J., Maluszynski, J.: HD-rules: A Hybrid System Interfacing Prolog with DL-reasoners. In: Applications of Logic Programming to the Web, Semantic Web and Semantic Web Services, CEUR-WS, vol. 287, pp. 76–90 (2007)
8. Eiter, T., Lukasiewicz, T., Schindlauer, R., Tompits, H.: Combining Answer Set Programming with Description Logics for the Semantic Web. Knowledge Representation and Reasoning, 141–151 (2004)
9. Eiter, T., Ianni, G., Schindlauer, R., Tompits, H.: NLP-DL: A Knowledge-Representation System for Coupling Nonmonotonic Logic Programs with Description Logics. In: International Semantic Web Conference, Galway, Ireland (2005)
10. Grant, J., Beckett, D.: RDF Test Cases (2004), http://www.w3.org/TR/rdf-testcases/
11. Grosof, B.N., Horrocks, I., Volz, R., Decker, S.: Description Logic Programs: Combining Logic Programs with Description Logic. In: WWW, pp. 48–57. ACM Press, New York (2003)
12. Haarslev, V., Moller, R.: Racer: A Core Inference Engine for the Semantic Web. In: International Workshop on Evaluation of Ontology-based Tools, pp. 27–36 (2003)
13. Haarslev, V., Moller, R.: An Empirical Evaluation of Optimization Strategies for ABox Reasoning in Expressive Description Logics. Description Logics, 22 (1999)
14. Hayes, P.: RDF Semantics (2004), http://www.w3.org/TR/rdf-mt/
15. Horrocks, I., Li, L., Turi, D., Bechhofer, S.: The Instance Store: Description Logic Reasoning with Large Numbers of Individuals. Description Logics, 104, 31–40 (2004)
16. Horst, H.J.: Extending the RDFS Entailment Lemma. In: Proceedings of the International Semantic Web Conference, pp. 77–91. Springer, Heidelberg (2004)
17. Horst, H.J.: Completeness, Decidability and Complexity of Entailment for RDF Schema and a Semantic Extension Involving the OWL Vocabulary. Journal of Web Semantics 3(2-3), 79–115 (2005)
18. Hustadt, U., Motik, B., Sattler, U.: Reducing SHIQ- Description Logic to Disjunctive Datalog Programs. Knowledge Representation and Reasoning, Canada, pp. 152–162 (2004)
19. KAON2, http://kaon2.semanticweb.org/
20. Kattenstroth, H., May, W., Schenk, F.: Combining OWL with F-Logic Rules and Defaults, Applications of Logic Programming to the Web. In: Semantic Web and Semantic Web Services. CEUR-WS, vol. 287, pp. 60–75 (2007)
21. Kiryakov, A., Ognyanov, D., Manov, F.: OWLIM - A Pragmatic Semantic Repository for OWL. In: Scalable Semantic Web Knowledge Base Systems, pp. 182–192. Springer, Heidelberg (2005)
22. Jess, http://herzberg.ca.sandia.gov/
23. Kopena, J.: OWLJessKB, http://edge.cs.drexel.edu/assemblies/software/owljesskb/
24. Levy, A.Y., Rousset, M.: Combining Horn Rules and Description Logics in CARIN. Artificial Intelligence 104(1-2), 165–209 (1998)

25. Li, H., Wang, Y., Qu, Y., Pan, J.Z.: A Reasoning Algorithm for pD*. In: 1st Asian Semantic Web Conference, pp. 293–299. Springer, Heidelberg (2006)
26. Ma, L., Yang, Y., Qiu, Z., Xie, G., Pan, Y., Liu, S.: Towards a Complete OWL Ontology Benchmark. In: Sure, Y., Domingue, J. (eds.) ESWC 2006. LNCS, vol. 4011, pp. 125–139. Springer, Heidelberg (2006)
27. Matheus, C., Dionne, B., Parent, D., Baclawski, K., Kokar, M.: BaseVISor: A Forward-Chaining Inference Engine Optimized for RDF/OWL Triples. In: Cruz, I., Decker, S., Allemang, D., Preist, C., Schwabe, D., Mika, P., Uschold, M., Aroyo, L.M. (eds.) ISWC 2006. LNCS, vol. 4273, Springer, Heidelberg (2006)
28. McBride, B.: Jena, Implementing the RDF Model and Syntax Specification. In: International Workshop on the Semantic Web, CEUR-WS, vol. 40 (2001)
29. McGuinness, D.L., Harmelen, F.: OWL Web Ontology Language Overview, W3C Recommendation, http://www.w3.org/TR/owl-features/
30. Meditskos, G., Bassiliades, N.: Rule-based OWL Ontology Reasoning Using Dynamic ABOX Entailments. In: 18th European Conference on Artificial Intelligence (ECAI), pp. 731–732. IOS Press, Patras (2008)
31. Meditskos, G., Bassiliades, N.: A Rule-Based Object-Oriented OWL Reasoner. IEEE Transactions on Knowledge and Data Engineering 20(3), 397–410 (2008)
32. Mei, J., Lin, Z., Boley, H.: ALC: An Integration of Description Logic and General Rules. In: Proceedings of the Web Reasoning and Rule Systems, pp. 163–177. Springer, Heidelberg (2007)
33. Minsu, J., Sohn, J.C.: Bossam: An Extended Rule Engine for OWL Inferencing. In: Rules and Rule Markup Languages for the Semantic Web, pp. 128–138 (2004)
34. Motik, B., Horrocks, I., Rosati, R., Sattler, U.: Can OWL and Logic Programming Live Together Happily Ever After? In: International Semantic Web Conference, pp. 501–514 (2006)
35. Motik, B., Sattler, U.: A Comparison of Reasoning Techniques for Querying Large Description Logic ABoxes. In: Logic for Programming Artificial Intelligence and Reasoning, pp. 227–241 (2006)
36. Owlgres, http://pellet.owldl.com/owlgres
37. Pan, J.Z., Horrocks, I.: RDFS(FA) and RDF MT: Two Semantics for RDFS. In: International Semantic Web Conference, pp. 30–46. Springer, Heidelberg (2003)
38. Patel-Schneider, P.F., Horrocks, I.: A Comparison of Two Modelling Paradigms in the Semantic Web. In: International Conference on World Wide Web, pp. 3–12. ACM Press, New York (2006)
39. Rosati, R.: Towards expressive KR systems integrating datalog and description logics. In: Workshop on Description Logics, CEUR-WS, vol. 22, pp. 160–164 (1999)
40. Rosati, R.: Semantic and Computational Advantages of the Safe Integration of Ontologies and Rules. In: Proc. Principles and Practice of Semantic Web Reasoning, pp. 50–64 (2005)
41. Rosati, R.: DL+log: Tight Integration of Description Logics and Disjunctive Datalog. In: Principles of Knowledge Representation and Reasoning, pp. 68–78. AAAI Press, Menlo Park (2006)
42. Sagonas, K., Swift, T., Warren, D.S.: XSB as an Efficient Deductive Database Engine. ACM SIGMOD Record 23(2), 442–453 (1994)
43. Sesame, http://openrdf.org/
44. Sirin, E., Parsia, B., Grau, B.C., Kalyanpur, A., Katz, Y.: Pellet: A Practical OWL-DL Reasoner. Journal of Web Semantics 5(2), 51–53 (2007)
45. Tsarkov, D., Horrocks, I.: Fact++ description logic reasoner: System description. In: Proceedings of Automated Reasoning, pp. 292–297. Springer, Heidelberg (2006)
46. Wang, K., Billington, D., Blee, J., Antoniou, G.: Combining Description Logic and Defeasible Logic for the Semantic Web. In: Rules and Rule Markup Languages for the Semantic Web, pp. 170–181. Springer, Heidelberg (2004)

Improving an RCC-Derived Geospatial Approximation by OWL Axioms

Rolf Grütter, Thomas Scharrenbach, and Bettina Bauer-Messmer

Swiss Federal Research Institute WSL, An Institute of the ETH Board, Zürcherstrasse 111,
CH-8903 Birmensdorf, Switzerland
{rolf.gruetter,thomas.scharrenbach,bettina.bauer}@wsl.ch

Abstract. An approach to improve an RCC-derived geospatial approximation is presented which makes use of concept inclusion axioms in OWL. The algorithm used to control the approximation combines hypothesis testing with consistency checking provided by a knowledge representation system based on description logics. Propositions about the consistency of the refined ABox w.r.t. the associated TBox when compared to baseline ABox and TBox are made. Formal proves of the divergent consistency results when checking either of both are provided. The application of the approach to a geospatial setting results in a roughly tenfold improved approximation when using the refined ABox and TBox. Ways to further improve the approximation and to automate the detection of falsely calculated relations are discussed.

Keywords: Geospatial approximation, Region Connection Calculus, Web Ontology Language, hypothesis testing, consistency checking.

1 Introduction

Topological relations play an important role for the description of geospatial phenomena. Accordingly, the Open GIS (OGIS) standard defines topological set operators for the retrieval of data in terms of spatial relations [1].[1] It is implemented by today's geographical information systems. However, there is currently no means to couple geometrical computations with symbolic reasoning services provided by a knowledge representation system. Such a coupling (or an alternative procedure with a similar effect) is necessary if users should be supported in constructing spatio-thematic queries which are consistent with their conceptualization of a given domain of discourse.

The coupling of geometrical computations with symbolic reasoning can be anticipated if the thematic (i.e., terminological) representation in the semantic layer of a system is complemented by a spatial representation. Ideally, the spatial representation uses topological relations which are compliant with the OGIS standard. It should also be based on a formalism which allows inferring implicit knowledge from the knowledge explicitly represented. Both requirements are fulfilled by the Region Connection Calculus (RCC) [2, 3].

---

[1] The OGIS consortium is formed by major software vendors to formulate an industry-wide standard related to GIS interoperability.

A. Sheth et al. (Eds.): ISWC 2008, LNCS 5318, pp. 293–306, 2008.
© Springer-Verlag Berlin Heidelberg 2008

A popular method for representing the terminology of a domain together with thematic descriptions in terms of the terminology is description logic, in the context of the Semantic Web particularly the Web Ontology Language (OWL) [4]. Therefore, in order to complement the thematic representation with a spatial representation, ways to combining OWL with RCC must be explored.

The herein presented work explores how an RCC-derived geospatial approximation can be improved by OWL axioms. It builds on an approach published in [5]. The idea is to calculate or approximate geospatial settings based on attributes which can easily be queried from spatial databases such that the process can be automated. For instance, non-administrative regions, such as biotopes, may be stored in data tables together with the administrative regions (e.g., cantons), they overlap, or administrative regions, such as communes, may be stored together with the administrative regions they are externally connected to. This information about the connectedness of regions can be used as a starting point for the calculation of more complex relations such as partOf. The calculated relations allow constructing and querying complex concepts with both thematic and spatial references such as public_park_containing_a_lake $\equiv$ park $\sqcap$ public $\sqcap$ $\exists$contains.lake which is taken from [6]. The contribution of the work, however, is not limited to the geographical domain. It rather adds to the knowledge about the combination of quantitative numerical approaches with qualitative symbolic (i.e., logic) approaches in general. The presented approach is related to previous work on combining RCC with OWL [7, 8] by addressing the calculation of a geospatial world description for assertion in the ABox of a knowledge base. Reasoning with both RCC and OWL will make use of the asserted world description.

The paper is organized as follows: In section 2, related work is discussed. An introduction to RCC is provided in section 3. In section 4, a geospatial approximation is presented which is derived from RCC. In section 5, the theoretical results are applied to a geospatial setting. The approach is discussed in section 6. Section 7 concludes with an outlook on future work.

2 Related Work

Spatio-thematic reasoning with the description logic $\mathcal{ALCRP(D)}$ has been introduced in [9]. The authors define an appropriate concrete domain $\mathcal{D}_P$ for polygons. $\mathcal{RP}$ stands for role definitions based on predicates. More specifically, $\mathcal{ALCRP(D)}$ extends $\mathcal{ALC(D)}$ [10] by a *role-forming operator* which is based on concrete domain predicates. The new operator allows the definition of roles with very complex properties and provides a close coupling of roles with concrete domains. A detailed account of $\mathcal{ALCRP(D)}$ is provided in [11]. In order to ensure termination of the satisfiability algorithm for the concrete domain $\mathcal{D}_P$, the authors impose restrictions on the syntactic form of the set of terminological axioms which impose tight constraints on modeling spatio-thematic structures [9, 11].

With the intention to augment a description logic like $\mathcal{ALC}$ with some kind of qualitative spatial reasoning capabilities, a rich variety of extensions to $\mathcal{ALC}$ is investigated in [12]. As a basic extension the author introduces role inclusion axioms of the form $S \circ T \sqsubseteq R_1 \sqcup \ldots \sqcup R_n$ which constrain the models $\mathcal{I}$ to $S^{\mathcal{I}} \circ T^{\mathcal{I}} \sqsubseteq R_1^{\mathcal{I}} \cup \ldots \cup R_n^{\mathcal{I}}$ ($\circ$ stands for the composition of roles). A set of these role inclusion axioms is referred to by the author as a *role box*. In previous work it has been shown that concept satisfiability in a related logic called $\mathcal{ALC}_{\mathcal{RA}}$, enforcing role disjointness on all roles R and S ($R^{\mathcal{I}} \cap S^{\mathcal{I}} = \emptyset$), is undecidable. In [12] specializations of $\mathcal{ALC}_{\mathcal{RA}}$ which specifically consider the family of RCC related calculi are investigated. Using role axioms of the above introduced form, the author shows for both $\mathcal{ALCI}_{RCC_5}$ and $\mathcal{ALCI}_{RCC_8}$ that satisfiability of concepts quantifying over roles ($\forall R.C$) can be undecidable in a practical application.

A general property of concrete domains that is sufficient for proving decidability of DLs equipped with them and General Concept Inclusions (GCIs) is identified in [13]. The authors further present a tableau algorithm for reasoning in DLs equipped with such concrete domains. In order to obtain their first result, they concentrate on a particular kind of concrete domains which they call *constraint systems*. According to the authors, a constraint system is a concrete domain that only has binary predicates, and these predicates are interpreted as jointly exhaustive and pairwise disjoint relations. The authors show that the spatial constraint system which is based on the real plane and the RCC-8 relations has the required property and that the description logic which allows defining concepts with reference to this constraint system is decidable. As a description logic they introduce $\mathcal{ALC}(\mathcal{C})$ which is $\mathcal{ALC}$ extended with two *constraint constructors*.

In [14] the authors aim at representing qualitative spatial information in OWL DL. On the basis of the (assumed) close relationship between the RCC-8 calculus and OWL DL they extend the latter with the ability to define reflexive roles. The extension of OWL DL with a reflexive property is motivated by the requirement that such a property, in addition to the transitive one, is needed in order to describe the accessibility relation. In order to represent RCC-8 knowledge bases the authors use a translation in which regions are expressed as non-empty regular closed sets. The RCC-8 relations are then translated into (sets of) concept axioms in OWL DL. The classes denoted by the introduced concepts are instantiated by asserting for each concept an individual in the ABox in order to ensure that the classes cannot be empty.

It seems to be more intuitive to define the RCC relations in terms of role descriptions than to translate them into concept axioms. In [15] it is shown that the extension of $\mathcal{SHIQ}$ with complex role inclusion axioms of the form $S \circ T \sqsubseteq R$ is undecidable, even when these axioms are restricted to the forms $S \circ T \sqsubseteq S$ or $T \circ S \sqsubseteq S$, but that decidability can be regained by further restricting them to be acyclic. Complex role inclusion axioms of the unrestricted form are supported by the description logic $\mathcal{SROIQ}$ which serves as a logical basis for OWL 1.1 [16]. However, in order to axiomatize the composition of RCC relations, a language must support an extension of the unrestricted form of role inclusion axioms, namely $S \circ T \sqsubseteq R_1 \sqcup \ldots \sqcup R_n$. If decidability should be preserved, complex role inclusion axioms are, therefore, not a solution to the translation problem of RCC. Axioms defining the basic RCC relations require additional role constructors such as intersection and complement. Extensions of $\mathcal{SHIQ}$ with these kinds of role constructors have, to our knowledge, not been investigated so far. $\mathcal{SROIQ}$ supports negation of roles (i.e. complement) but not intersection.

To summarize, the existing approaches show that the combination of formalisms for thematic and spatial reasoning is not straightforward. In order to uphold decidability, approaches based on $\mathcal{ALC}$ require that the resulting language is constrained. This bears on its expressivity for modeling spatio-thematic structures. $\mathcal{SHIQ}$ and $\mathcal{SROIQ}$ do not provide for the expression of role inclusion axioms of the form $S \circ T \sqsubseteq R_1 \sqcup \ldots \sqcup R_n$ which is a requirement for spatial reasoning with RCC. The approach based on $\mathcal{SHOIN(D)}$ (OWL DL) requires only a minimal extension of the language which has been considered in the draft to OWL 1.1 [17]. However, the notion of regions as sets in the abstract object domain prevents RCC from effectively combining with domain ontologies. The reason for this is that OWL DL requires type separation: a class cannot be an individual (or a property) at the same time [18]. Yet, in order to classify regions in a domain ontology they must be represented as individuals, and not as concepts.

Furthermore, in [19] it is proposed to encode spatial inferences in the Semantic Web Rule Language (SWRL) [20]. Even though not explicitly mentioned, the examples are provided in a RCC-like style. SWRL uses Horn-like rules which are combined with OWL DL (and OWL Lite). Horn rules do not allow complex heads (which refer to the expressions on the right hand side of the implication connective). However, complex heads in terms of disjunctions are required in order to formalize the RCC composition axioms.

In [6] a generic architectural framework for building ontology-based information systems is presented which covers regions in the system design space instead of isolated points. The framework introduces a graph-based *substrate data model* and a *substrate query language*. An ABox can be seen as a substrate, an ABox with an associated TBox as a substrate with a background theory. A substrate can also encode geometric or spatial structures in a geometric substrate which is called an *SBox* (Space Box). The authors propose four options to solve the spatial representation problem: (1) Use an ABox, (2) use a map substrate, (3) use a spatial ABox, (4) use an ABox + RCC substrate. The herein presented work addresses the fourth representation option.

3   The Region Connection Calculus

The Region Connection Calculus (RCC) is an axiomatization of certain spatial concepts and relations in first order logic [2, 3]. The basic theory assumes just one primitive dyadic relation: $C(x, y)$ read as "x connects with y". Individuals (x, y) can be interpreted as denoting spatial regions. The relation $C(x, y)$ is reflexive and symmetric.

Using the primitive relation $C(x, y)$ a number of intuitively significant relations can be defined. The most common of these are illustrated in figure 1 and their definitions together with those of additional relations are given in table 1. The asymmetrical relations P, PP, TPP and NTPP have inverses which we write, in accordance with [3], as Ri, where $R \in \{P, PP, TPP, NTPP\}$. These relations are defined by definitions of the form $Ri(x, y) \equiv_{def} R(y, x)$.

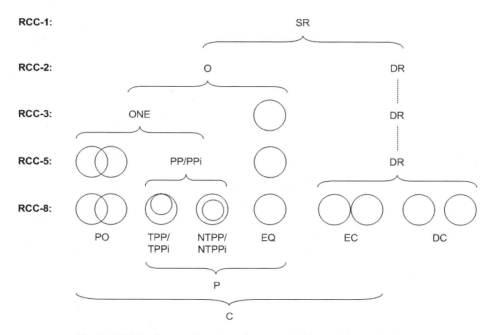

Fig. 1. RCC family tree (for the entire names of the relations cf. table 1)

Of the defined relations, DC, EC, PO, EQ, TPP, NTPP, TPPi and NTPPi have been proven to form a jointly exhaustive and pairwise disjoint set, which is known as RCC-8. Similar sets of one, two, three and five relations are known as RCC-1, RCC-2, RCC-3 and RCC-5, respectively: RCC-1 = {SR}, RCC-2 = {O, DR}, RCC-3 = {ONE, EQ, DR}, RCC-5 = {PP, PPi, PO, EQ, DR}. RCC also incorporates a constant denoting the universal region, a sum function and partial functions giving the product of any two overlapping regions and the complement of every region except the universe [3].

Table 1. RCC relations

$SR(x, y)$	$\equiv_{def} \top(x, y)$	(Spatially Related)
$C(x, y)$	(primitive relation)	(Connects with)
$DC(x, y)$	$\equiv_{def} \neg C(x, y)$	(DisConnected from)
$P(x, y)$	$\equiv_{def} \forall z[C(z, x) \rightarrow C(z, y)]$	(Part of)
$O(x, y)$	$\equiv_{def} \exists z[P(z, x) \land P(z, y)]$	(Overlaps)
$DR(x, y)$	$\equiv_{def} \neg O(x, y)$	(DiscRete from)
$EC(x, y)$	$\equiv_{def} C(x, y) \land \neg O(x, y)$	(Externally Connected to)
$EQ(x, y)$	$\equiv_{def} P(x, y) \land P(y, x)$	(EQual to)
$ONE(x, y)$	$\equiv_{def} O(x, y) \land \neg EQ(x, y)$	(Overlaps Not Equal)
$PP(x, y)$	$\equiv_{def} P(x, y) \land \neg P(y, x)$	(Proper Part of)
$PO(x, y)$	$\equiv_{def} O(x, y) \land \neg P(x, y) \land \neg P(y, x)$	(Partially Overlaps)
$TPP(x, y)$	$\equiv_{def} PP(x, y) \land \exists z[EC(z, x) \land EC(z, y)]$	(Tangential Proper Part of)
$NTTP(x, y)$	$\equiv_{def} PP(x, y) \land \neg \exists z[EC(z, x) \land EC(z, y)]$	(Non-Tangential Proper Part of)

According to [2], regions support either spatial or temporal interpretation. In case of spatial interpretation, there is a variety of models among which to choose. The authors provide some examples such as interpreting the relation C ("connects with") in terms of two regions whose closures share a common point or stating that two regions connect when the distance between them is zero.

In order to check consistency of a knowledge base holding spatial relations, so-called composition tables are used (cf. the composition table for RCC-5 in table 2). The entries in these tables share a uniform inference pattern which can be formalized as composition axioms of the general form $\forall x, y, z. \, S(x, y) \wedge T(y, z) \rightarrow R_1(x, z) \vee \ldots \vee R_n(x, z)$ where S, T, and R_i are variables for relation symbols.

A similar approach which is based on the description of topological relations between two spatial regions was introduced as the 9-intersection model in [21]. In this model, eight out of nine relations can be interpreted in the same way as we interpret the RCC-8 relations, namely as spatial relations between polygons in the integral plane [7]. Only the ninth relation is specific for the model. Since it is based on a topological framework – and not on a logical one – the 9-intersection model is harder to combine with OWL DL than RCC.

Table 2. RCC-5 composition table ($T(x, z) \equiv_{def} \{DR(x, z), PO(x, z), EQ(x, z), PP(x, z), PPi(x, z)\}$)

∘	DR(x, y)	PO(x, y)	EQ(x, y)	PPi(x, y)	PP(x, y)
DR(y, z)	T(x, z)	DR(x, z) PO(x, z) PPi(x, z)	DR(x, z)	DR(x, z) PO(x, z) PPi(x, z)	DR(x, z)
PO(y, z)	DR(x, z) PO(x, z) PP(x, z)	T(x, z)	PO(x, z)	PO(x, z) PPi(x, z)	DR(x, z) PO(x, z) PP(x, z)
EQ(y, z)	DR(x, z)	PO(x, z)	EQ(x, z)	PPi(x, z)	PP(x, z)
PP(y, z)	DR(x, z) PO(x, z) PP(x, z)	PO(x, z) PP(x, z)	PP(x, z)	PO(x, z) EQ(x, z) PP(x, z) PPi(x, z)	PP(x, z)
PPi(y, z)	DR(x, z)	DR(x, z) PO(x, z) PPi(x, z)	PPi(x, z)	PPi(x, z)	T(x, z)

4   An RCC-Derived Geospatial Approximation

In order to approximate a geospatial setting we derive a hypothesis for the RCC relation $P(x, y)$ from its definition in table 1 (section 4.1). For each pair of connecting regions, this hypothesis is tested against the role assertions in the ABox of a knowledge base (section 4.2). If the hypothesis is not falsified it is checked whether its assertion causes an inconsistency of the knowledge base or not. If the knowledge base remains consistent the relation $P(x, y)$ is asserted in the ABox.

We compare two procedures which use different levels of knowledge. The *baseline approximation* uses baseline knowledge in terms of the concept Region which is used for asserting individual regions in the ABox of the knowledge base. The *improved approximation* uses additional knowledge from cartographic analysis. In particular, it distinguishes communes as a special kind of regions and introduces a property restriction which constrains the values of the property to some communes. Since the approach is generic and can be applied to similar problems in other domains as well, the algorithm and the axioms of TBoxes $\mathcal{T}_1$ (baseline) and $\mathcal{T}_2$ (constrained) are introduced in an abstract way. The notation used for the DL axioms is taken from [22].

4.1 Deriving a Hypothesis from RCC

The relation $P(x, y)$ ("*x* is a part of *y*") plays a key role in the definitions of the RCC relations (cf. table 1): It is directly defined by the primitive relation and, conversely, a number of relations are defined in terms of $P(x, y)$. For this reason, we use the definition of the relation $P(x, y)$ as a starting point for our geospatial approximation. Since $P(x, y)$ is directly defined by the primitive relation it will be sufficient to assert the primitive relation in an input representation. The theory defines the relation $P(x, y)$ as follows:

$$P(x, y) \equiv_{def} \forall z[C(z, x) \rightarrow C(z, y)].$$

In accordance with the Semantic Web philosophy, this definition assumes an open world: *x* is a part of *y* if and only if for any imaginable region *z* the following holds: If *z* connects with *x* then *z* also connects with *y*. In order to adapt the definition to the closed world of a practical application we replace the universal quantifier by a conjunction ranging over all regions represented. In a closed world the condition on the right hand side of the expression is no longer sufficient – but still necessary (cf. below) – for the relation $P(x, y)$. Accordingly, we replace the equality sign by an inclusion sign.

$$P(x, y) \sqsubseteq \wedge_{z_i}[C(z_i, x) \rightarrow C(z_i, y)] \tag{1}$$

with $1 \leq i \leq n$, n the number of regions represented.

In the minimum case (i = 1) the region *x* connects only with itself (remember that the relation C is by definition reflexive) and it holds that $x = y = z$. In the maximum case (i = n) all regions, including *x* (*y*, respectively) connect with *x* (*y*, respectively). Intuitively, the calculation of $P(x, y)$ is expected to be more precise with a high number of regions z_i represented.

Note that from an epistemic viewpoint the shift from an open world to a closed world limits the range of the proposition – which is originally formalized as a definition – from the partially unknown universe to a known subset thereof. Both the universal proposition and the middle range proposition cannot be empirically verified but only falsified.[2] The first cannot be verified as a matter of principle: It is not possible to test for the infinite number of all imaginable regions *z* connecting with *x* whether they also connect with *y*. The second cannot be verified because the condition on the right hand side of the expression is not sufficient. However, both can be falsified: A single observation of a region *z* connecting with *x* but not with *y* is sufficient to falsify the hypothesis that *x* is a part of *y*. Following this line of argumentation a calculus for $P(x, y)$ is not expected to be sound. Instead, the question is whether it is complete or

---

[2] The expression *middle range* is adopted from the Middle Range Theory [23].

not. In a practical application it is further of interest how good a calculus approximates the geospatial setting and how the approximation can be controlled.

The question whether a calculus using formula (1) is complete or not can be answered by referring to the reflexive and symmetric properties of the primitive relation $C(x, y)$. If $z = y$ and x connects with y the formula $C(y, x) \rightarrow C(y, y) \equiv C(x, y) \rightarrow C(y, y)$ evaluates to true for $P(x, y)$. Thus, the condition that x connects with y is sufficient for hypothesizing that x is a part of y. This means that a calculus using formula (1) is expected to be complete in a practical application. $P(x, y)$ is generalized to the hypothesis $T(a, b)$ in the next section.

4.2   Combining Hypothesis Testing with Consistency Checking

The algorithm used to control the RCC-derived geospatial approximation combines hypothesis testing with consistency checking provided by a knowledge representation system based on description logics. Generically, it proceeds as follows:

FUNCTION   *geospatialApproximation*
Input: hypothesis $T(., .)$, knowledge base $KB = \{ABox, TBox\}$
Output: *counter*

```
0     counter ← 0
1     M ← {(a, b)} ⊆ ABox      /* the set of pairs of individuals in ABox */
2     WHILE M is NOT empty
3           SELECT (a, b) ∈ M
4           IF T(a, b) is NOT falsified in ABox THEN
5                 ABox ← ABox ∪ T(a, b)
6                 IF KB is consistent THEN
7                       counter ← counter + 1
8                 ELSE
9                       ABox ← ABox \ T(a, b)
10                ENDIF
11          ENDIF
12          M ← M \ (a, b)
13    ENDWHILE
```

For the consistency check in step 6 two different TBoxes are alternatively applied: TBox $\mathcal{T}_1$ consisting of axioms 0–1 and 4–7 or TBox $\mathcal{T}_2$ consisting of axioms 0–7:

0	$C \sqsubseteq \top$	$C^{\mathcal{I}} \subseteq \Delta^{\mathcal{I}}$
1	R	$R^{\mathcal{I}} \subseteq \Delta^{\mathcal{I}} \times \Delta^{\mathcal{I}}$
2	$D \sqsubseteq C$	$D^{\mathcal{I}} \subseteq C^{\mathcal{I}}$
3	$D \sqsubseteq \neg(\exists S.D)$	$D^{\mathcal{I}} \subseteq \Delta^{\mathcal{I}} \setminus \{a \in \Delta^{\mathcal{I}} \mid \exists b. (a, b) \in S^{\mathcal{I}} \wedge b \in D^{\mathcal{I}}\}$
4	$\exists R.\top \sqsubseteq C$	$\{a \in \Delta^{\mathcal{I}} \mid \exists b. (a, b) \in R^{\mathcal{I}}\} \subseteq C^{\mathcal{I}}$
5	$\top \sqsubseteq \forall R.C$	$\Delta^{\mathcal{I}} \subseteq \{a \in \Delta^{\mathcal{I}} \mid \forall b. (a, b) \in R^{\mathcal{I}} \rightarrow b \in C^{\mathcal{I}}\}$
6	$S \sqsubseteq R$	$S^{\mathcal{I}} \subseteq R^{\mathcal{I}}$
7	$T \sqsubseteq S$	$T^{\mathcal{I}} \subseteq S^{\mathcal{I}}$

The concept description $\exists S.D$ introduced by axiom 3 is interpreted as the set of those individuals each of which is in relationship $S^\mathcal{I}$ to some individuals of the set $D^\mathcal{I}$. The concept D is interpreted as a possibly improper subset of the complement of this set. Together with a refinement of the model in the ABox – a subset of the individuals in $C^\mathcal{I}$ are also members of the interpretation of the included concept D – axiom 3 is responsible for the divergent result of the consistency check when using two different TBoxes $\mathcal{T}_1$ or $\mathcal{T}_2$.

Proposition 1. Asserting the role $T(a, b)$ in the ABox $\mathcal{A}_1 = \{C(a), C(b)\}$ results in an ABox $\mathcal{A}_1' = \{C(a), C(b), T(a, b)\}$ which is consistent w.r.t. $\mathcal{T}_1$.

Proof

$(T \sqsubseteq S) \wedge T(a, b) \rightarrow S(a, b)$	Axiom 7
$(S \sqsubseteq R) \wedge S(a, b) \rightarrow R(a, b)$	Axiom 6
$(\exists R.\top \sqsubseteq C) \wedge R(a, b) \rightarrow C(a)$	Axiom 4
$(C \sqsubseteq \top) \wedge C(a) \rightarrow \top(a)$	Axiom 0
$(\top \sqsubseteq \forall R.C) \wedge R(a, b) \rightarrow C(b)$	Axiom 5
$(C \sqsubseteq \top) \wedge C(b) \rightarrow \top(b)$	Axiom 0

Proposition 2. Asserting the role $T(a, b)$ in the ABox $\mathcal{A}_2 = \{D(a), D(b)\}$ results in an ABox $\mathcal{A}_2' = \{D(a), D(b), T(a, b)\}$ which is *inconsistent* w.r.t. $\mathcal{T}_2$.

Proof

$(T \sqsubseteq S) \wedge T(a, b) \rightarrow S(a, b)$	Axiom 7
$(D \sqsubseteq \neg(\exists S.D)) \wedge S(a, b) \rightarrow (\neg D)(a)$	Axiom 3
$(\neg D)(a) \wedge D(a) \rightarrow \bot(a)$	

5 Applying the Approach to a Geospatial Setting

In order to demonstrate the approach, we use a sample of 44 two-dimensional spatial regions (polygons) from four layers of a productive GIS. The spatial regions refer to districts, communes, biotopes and are located in the canton of Zurich, Switzerland (cf. figure 2). The regions are asserted as individuals in the ABox of the knowledge base. For inference with TBox $\mathcal{T}_2$ 38 out of 44 regions are asserted as communes. The connections between regions – which were identified by cartographic analysis – are asserted as role assertions of type connectsWith. Overall, there are 262 relations asserted in our sample. The knowledge base is created from an OWL ontology using the reasoner Pellet (version 1.4). Most descriptions in the ontology are written in OWL DL. Only for the description of the role partOf the OWL 1.1 feature irreflexiveProperty is used. The description logic expressivity of the ontology is $\mathcal{ALHI}$ for $\mathcal{T}_1$ and $\mathcal{ALCHI}$ for $\mathcal{T}_2$. The algorithm used to compute formula (1) and to control the approximation as described in section 4.2 (steps 0–13) is programmed in Java. It accesses the knowledge base at the WonderWeb OWL API.[3]

---

[3] http://wonderweb.semanticweb.org/

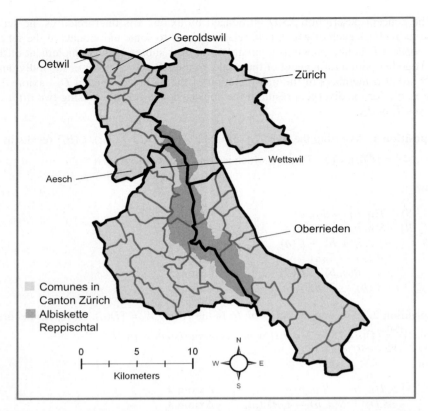

Fig. 2. Regions in the canton of Zurich. The dark grey shaded region depicts Albiskette-Reppischtal, a biotope of national interest. Regions with bold borderlines depict districts. Regions with regular borderlines depict communes. Note that the district of Zurich and the commune of Zurich share the same geometry, in terms of RCC: EQ (Bezirk_Zürich, Zürich).

The TBoxes $\mathcal{T}_1$ (axioms 0–1 and 4–7) and $\mathcal{T}_2$ (axioms 0–7) are instantiated as follows. Intuitively, axiom 3 says that communes must not overlap each other. The intuition behind axioms 4 and 5 is that regions are spatially related to each other.

0	Region $\sqsubseteq \top$	Region$^{\mathcal{I}} \subseteq \Delta^{\mathcal{I}}$
1	spatiallyRelated	spatiallyRelated$^{\mathcal{I}} \subseteq \Delta^{\mathcal{I}} \times \Delta^{\mathcal{I}}$
2	Commune $\sqsubseteq$ Region	Commune$^{\mathcal{I}} \subseteq$ Region$^{\mathcal{I}}$
3	Commune $\sqsubseteq \neg(\exists$overlaps.Commune$)$	Commune$^{\mathcal{I}} \subseteq \setminus \{a \in \Delta^{\mathcal{I}} \mid \exists b.\ (a, b) \in$ overlaps$^{\mathcal{I}} \wedge b \in$ Commune$^{\mathcal{I}}\}$
4	$\exists$spatiallyRelated.$\top \sqsubseteq$ Region	$\{a \in \Delta^{\mathcal{I}} \mid \exists b.\ (a,\ b) \in$ spatiallyRelated$^{\mathcal{I}}\} \subseteq$ Region$^{\mathcal{I}}$
5	$\top \sqsubseteq \forall$spatiallyRelated.Region	$\Delta^{\mathcal{I}} \subseteq \{a \in \Delta^{\mathcal{I}} \mid \forall b.\ (a,\ b) \in$ spatiallyRelated$^{\mathcal{I}} \rightarrow b \in$ Region$^{\mathcal{I}}\}$
6	overlaps $\sqsubseteq$ spatiallyRelated	overlaps$^{\mathcal{I}} \subseteq$ spatiallyRelated$^{\mathcal{I}}$
7	partOf $\sqsubseteq$ overlaps	partOf$^{\mathcal{I}} \subseteq$ overlaps$^{\mathcal{I}}$

Using TBox T_1 the algorithm calculates 109 relations of type P(x, y). The cartographic evaluation results in 27 relations being falsely calculated as P(x, y) whereas they are relations of type EC(x, y). Using TBox T_2 instead of T_1 the algorithm calculates 85 relations of type P(x, y). In this case only three relations of type EC(x, y) are falsely calculated as P(x, y). In both cases all relations of type P(x, y) verified by cartography are identified as such. As expected, the calculations are complete but not sound in our sample. The approximation with T_2 is roughly ten times better than that with T_1.

To give an example, one of the relations of type EC(x, y) which is falsely calculated as P(x, y) using TBox T_1 refers to the relation between Geroldswil and Oetwil (cf. figure 2). Since all regions connecting with Geroldswil also connect with Oetwil the relation between them is (falsely!) assumed to be of type P(x, y). Since TBox T_2 introduces the restriction that Geroldswil and Oetwil – which are both communes – must not overlap, partOf(Geroldswil, Oetwil) is removed from the ABox of the knowledge base in step 9 of the algorithm (cf. section 4).

At http://webgis.wsl.ch/rcc-webclient/faces/rcc-client.jspx the presented approach can be tested against the constraint (TBox T_2) and the unconstraint (TBox T_1) ontology, respectively.[4] The reader is encouraged to perform tests on their own by uploading a custom ontology. For being processible a custom ontology has to be consistent either with T_1 or with T_2.

6   Discussion

The three relations which are falsely calculated as P(x, y) in our sample – even with the improved algorithm – are connectsWith(Bezirk_Affoltern, Aesch), connectsWith (Bezirk_Dietikon, Wettswil), connectsWith(Albiskette-Reppischtal, Oberrieden). As can be seen from figure 2 the communes Aesch, Oberrieden and Wettswil share the property of being adjacent to the surrounding area. The question, therefore, is whether the assertion of additional connections between the border areas and the surrounding area in the ABox of the knowledge base would further improve the approximation. In our sample also the districts Bezirk_Affoltern and Bezirk_Dietikon and the biotope of national interest Albiskette-Reppischtal connect with the surrounding area. For this reason we do not expect an improvement in this case. However, an improvement can be expected in other cases.

Another question is whether the detection of the falsely calculated relations can be automated. This would substantially reduce the effort of a manual cartographic verification. To a certain degree this is indeed possible by introducing further axioms in the TBox of the knowledge base and by further refining the model in the ABox. With respect to the first, *qualified number restrictions* can be introduced, for instance in order to encode the restriction that communes may be part only of a single district. With respect to the last, those regions which are districts can be asserted as such. Qualified number restrictions are not supported by OWL DL. However, they are considered in the draft to OWL 1.1 (there a qualified number restriction is called an objectMinCardinality) [17].

---

[4] Please contact the authors if you wish to access the service and it is no longer maintained.

| 8 | District ⊑ Region | District$^{\mathcal{I}}$ ⊆ Region$^{\mathcal{I}}$ |
| 9 | Commune ⊑ ≤ 1 partOf.District | Commune$^{\mathcal{I}}$ ⊆ $\{a \in \Delta^{\mathcal{I}} \mid \mid\{b \mid (a, b) \in$ partOf$^{\mathcal{I}} \wedge b$ $\in$ District$^{\mathcal{I}}\}\mid \leq 1\}$ |

In our sample axioms 8 an 9 (together with a further refined model) are expected to cause an inconsistency of the knowledge base when asserting the relations part-Of(Aesch, Bezirk_Affoltern) and partOf(Wettswil, Bezirk_Dietikon) in the ABox (cf. step 5 of the algorithm introduced in section 4). This because the relations partOf(Aesch, Bezirk_Dietikon) and partOf(Wettswil, Bezirk_Affoltern) are (truly!) asserted in the ABox. Conversely, the falsely calculated partOf-relation between Oberrieden and Albiskette-Reppischtal would not be detected.

7   Conclusion and Outlook

We presented an approach to improve an RCC-derived geospatial approximation which makes use of concept inclusion axioms in OWL. The algorithm used to control the approximation combines hypothesis testing with consistency checking provided by a knowledge representation system based on description logics. We made propositions about the consistency of the refined ABox w.r.t. the associated TBox when compared to baseline ABox and TBox and provided formal proves of the divergent consistency results when checking either of both. The application of the approach to a sample of 44 two-dimensional regions which are related to each other through 262 spatial relations resulted in a roughly tenfold improved approximation when using the refined ABox and TBox.

Since we expect the approximation to be more precise with a high number of regions represented a next step will be to evaluate the approach on different scales. In addition to precision of the approximation the impact of scalability on performance will also be of interest.

For productive use it would be desirable to assess the degree of imprecision and to provide a measure of confidence along with the approximation. This measure could then be put into relationship with the inherent imprecision of the data. Real data are error-prone. For instance, the shape of the biotope of national interest Albiskette-Reppischtal which is geometrically computed in figure 2 is different from the shape which is reconstructed based on the textual description of the biotope (which is held in the same database!): According to the textual description, Albiskette-Reppischtal should also overlap the commune Wettswil which it does not geometrically.

Provided the relations of type P(x, y) can be sufficiently well approximated for a given spatial setting, a next step is to calculate the RCC-5 relations by using their definitions in table 1 and the relationships between RCC species in figure 1. The relations between individual regions can then be asserted in terms of RCC-5 in the ABox of a knowledge base. Based on this, a full-fledged spatio-terminological reasoning service as outlined in [7, 8] can be developed.

Since the presented approach is generic in nature it can be applied to other domains as well. It will be interesting to identify relevant problems in these domains.

Acknowledgments. The authors sincerely thank Jürg Schenker and Martin Hägeli for the fruitful discussions and leadership that made this research possible. They also acknowledge the stimulating discussions and exchange of ideas with Prof. Dr. rer. nat. habil. Ralf Möller from Hamburg University of Technology. This research has been funded and conducted in cooperation with the Swiss Federal Office for the Environment (FOEN).

References

1. Shekhar, S., Chawla, S.: Spatial Databases: A Tour. Pearson Education. Upper Saddle River, New Jersey (2003)
2. Randell, D.A., Cui, Z., Cohn, A.G.: A Spatial Logic based on Regions and Connections. In: Nebel, B., Rich, C., Swartout, W. (eds.) Principles of Knowledge Representation and Reasoning, pp. 165–176. Morgan Kaufmann, San Mateo (1992)
3. Bennett, B.: Logics for Topological Reasoning. In: 12th European Summer School in Logic, Language and Information (ESSLLI-MM), Birmingham, UK, August 2000, pp. 6–18 (2000)
4. Patel-Schneider, P.F., Hayes, P., Horrocks, I.: OWL Web Ontology Language: Semantics and Abstract Syntax. W3C Recommendation, February 10, 2004. World Wide Web Consortium (2004)
5. Grütter, R., Bauer-Messmer, B., Hägeli, M.: Extending an Ontology-based Search with a Formalism for Spatial Reasoning. In: Proceedings of the 23rd Annual ACM Symposium on Applied Computing (ACM SAC 2008), pp. 2266–2270. ACM, New York (2008)
6. Wessel, M., Möller, R.: Flexible Software Architectures for Ontology-Based Information Systems. Journal of Applied Logic, Special Issue on Emperically Sucessful Systems (2007)
7. Grütter, R., Bauer-Messmer, B.: Towards Spatial Reasoning in the Semantic Web: A Hybrid Knowledge Representation System Architecture. In: Fabrikant, S.I., Wachowicz, M. (eds.) The European Information Society: Leading the Way with Geo-information. LNGC, pp. 349–364. Springer, Heidelberg (2007)
8. Grütter, R., Bauer-Messmer, B.: Combining OWL with RCC for Spatioterminological Reasoning on Environmental Data. In: Golbreich, C., Kalyanpur, A., Parsia, B. (eds.) OWL: Experiences and Directions (OWLED). CEUR Workshop Proceedings, vol. 258 (2007)
9. Haarslev, V., Lutz, C., Möller, R.: Foundations of Spatioterminological Reasoning with Description Logics. In: Cohn, A.G., Schubert, L.K., Shapiro, S.C. (eds.) Principles of Knowledge Representation and Reasoning, pp. 112–123. Morgan-Kaufmann, San Mateo (1998)
10. Baader, F., Hanschke, P.: A scheme for integrating concrete domains into concept languages. In: Twelfth International Joint Conference on Artificial Intelligence, August 1991, pp. 452–457 (1991)
11. Haarslev, V., Lutz, C., Möller, R.A.: Description Logic with Concrete Domains and a Role-Forming Predicate Operator. Journal of Logic and Computation 9(3), 351–384 (1999)
12. Wessel, M.: On Spatial Reasoning with Description Logics. Position Paper. In: Horrocks, I., Tessaris, S. (eds.) Proceedings of the International Workshop on Description Logics (DL 2002), CEUR Workshop Proceedings, Toulouse, France, April 19–21, 2002, vol. 53, pp. 156–163 (2002)

13. Lutz, C., Miličić, M.: A Tableau Algorithm for Description Logics with Concrete Domains and General TBoxes. Journal of Automated Reasoning 38(1-3), 227–259 (2007)
14. Katz, Y., Grau, B.C.: Representing Qualitative Spatial Information in OWL-DL. In: Proceedings of OWL: Experiences and Directions, CEUR Workshop Proceedings, Galway, Ireland, vol. 188 (2005)
15. Horrocks, I., Sattler, U.: Decidability of SHIQ with Complex Role Inclusion Axioms. In: Proceedings of the 18th International Joint Conference on Artificial Intelligence (IJCAI 2003), pp. 343–348. Morgan Kaufmann, Los Altos (2003)
16. Horrocks, I., Kutz, O., Sattler, U.: The Even More Irresistible SROIQ. In: Proceedings of the 10th International Conference of Knowledge Representation and Reasoning (KR 2006), Lake District, United Kingdom (2006)
17. Grau, B.C., Motik, B.: OWL 1.1 Web Ontology Language: Model-Theoretic Semantics. Editor's Draft of (April 6, 2007), http://www.webont.org/owl/1.1/semantics.html
18. Patel-Schneider, P.F., Hayes, P., Horrocks, I.: OWL Web Ontology Language: Semantics and Abstract Syntax. W3C Recommendation February 10, 2004. World Wide Web Consortium (2004)
19. Bishr, Y.: Geospatial Semantic Web. In: Rana, S., Sharma, J. (eds.) Frontiers of Geographic Information Technology, pp. 139–154. Springer, Heidelberg (2006)
20. Horrocks, I., Patel-Schneider, P.F., Boley, H., Tabet, S., Grosof, B., Dean, M.: SWRL: A Semantic Web Rule Language. Combining OWL and RuleML. W3C Member Submission May 21, 2004. World Wide Web Consortium (2004)
21. Egenhofer, M., Franzosa, R.: Point-Set Topological Spatial Relations. International Journal of Geographical Information Systems 5(2), 161–174 (1991)
22. Baader, F., Nutt, W.: Basic Description Logics. In: Baader, F., Calvanese, D., McGuinness, D.L., Nardi, D., Patel-Schneider, P.F. (eds.) The Description Logic Handbook, pp. 47–100. Cambridge University Press, Cambridge (2003)
23. Merton, R.K.: Social Theory and Social Structure (Enlarged Edition). The Free Press, New York (1968)

OWL Datatypes: Design and Implementation

Boris Motik and Ian Horrocks

University of Oxford, UK

Abstract. We analyze the datatype system of OWL and OWL 2, and discuss certain nontrivial consequences of its definition, such as the extensibility of the set of supported datatypes and complexity of reasoning. We also argue that certain datatypes from the list of normative datatypes in the current OWL 2 Working Draft are inappropriate and should be replaced with different ones. Finally, we present an algorithm for datatype reasoning. Our algorithm is modular in the sense that it can handle any datatype that supports certain basic operations. We show how to implement these operations for number and string datatypes.

1 Introduction

The Web Ontology Language (OWL) has been phenomenally successful, and the OWL DL version of OWL is nowadays routinely used for conceptual modeling in fields as diverse as biomedicine, clinical sciences, astronomy, geography, aerospace and defence. OWL DL is grounded in *description logics* (DLs) [1]—a family of theoretically well-understood knowledge representation formalisms. The popularity of OWL DL is largely due to the availability of practically effective reasoners[1] that can be used in applications.

Applications of OWL often use properties with values such as strings and integers. OWL therefore supports *datatypes*, a simplified version of the concrete domain approach [2] that can be combined with most DLs in a decidable way [5,7]. A *datatype* can be seen as a unary predicate with a built-in interpretation; for example, the *xsd:integer* datatype is interpreted as the set of all integer values. Particular data values can be denoted at the syntax level using *constants*. Properties in OWL DL are separated into *object properties*, interpreted as relations between pairs of individuals, and *data properties*, interpreted as relations between individuals and data values. Data properties can be used in axioms such as "the range of the *a:name* property is *xsd:string*," or "each instance of the *a:Person* class must have an *xsd:integer* data value for the *a:age* property."

Practical experience with OWL DL has revealed several shortcomings of its datatype system. The datatypes in OWL DL are modeled after XML Schema [3], which provides a rich set of datatypes; however, only *xsd:string* and *xsd:integer* are normative in OWL DL, which is often not sufficient for applications. Furthermore, OWL DL provides no portable means for restricting datatypes, as

---

[1] http://www.cs.man.ac.uk/~sattler/reasoners.html

A. Sheth et al. (Eds.): ISWC 2008, LNCS 5318, pp. 307–322, 2008.

in "a person who is 70 or older" or "the values of the *a:name* property should be a string not containing a whitespace." Various OWL DL tools, such as the RACER reasoner [4], have provided proprietary solutions to these problems; however, there is currently little or no compatibility between extensions provided by different tools.

In order to address these shortcomings, as well as some shortcomings unrelated to datatypes, a W3C Working Group[2] has recently been established with the goal of developing a major revision of the language called OWL 2. The current OWL 2 Working Draft[3] lists most XML Schema datatypes as normative, and it supports XML Schema *facets* for restricting the range of built-in datatypes. For example, the minExclusive facet can be applied to *xsd:integer* to obtain a subset of integers larger than a particular value.

Extensions of DLs with concrete domains and datatypes have received in-depth theoretical treatment [2,9,10,5,7]. Furthermore, datatype groups [11] provide an architecture for integrating different datatypes, and the OWL-Eu approach [12] provides a way to restrict datatypes using expressions. These works assume that datatype reasoning can be performed using an external *datatype checking* procedure. Standard DL tableau calculi can then be extended to handle datatypes by invoking the datatype checker as an oracle.

Less attention has so far been paid to actual datatype checking algorithms. Datatype checking is a constraint satisfaction problem; however, general CSP algorithms [14] are typically too general and complex than what is necessary for datatype reasoning. Furthermore, the list of normative datatypes in the current OWL 2 Working Draft has been selected without paying attention to their suitability and implementability in a datatype checker.

In this paper we therefore formally define the OWL 2 datatype system, review its design, and investigate the problem of implementing a suitable datatype checker. Our analysis reveals several nonobvious consequences of the current design. In particular, we show that datatype checking in OWL 2 is NP-hard in the general case, but may become trivial in many (hopefully typical) cases. We also argue that certain datatypes listed as normative in the current OWL 2 Working Draft may be unsuitable from both a modeling and implementation perspective, and we suggest several changes to the Working Draft that address the problems identified. We then present a modular datatype checking algorithm that can support any datatype for which it is possible to implement a small set of basic operations that we call a *datatype handler*. Finally, we discuss how to implement datatype handlers for number and string datatypes.

The results of this paper thus provide important guidance for the designers of OWL 2, by pointing out potential design mistakes that could make implementation difficult, and for implementors of OWL 2 reasoners, by showing how to implement a suitable datatype checker.

---

[2] http://www.w3.org/2007/OWL/wiki/OWL_Working_Group
[3] http://www.w3.org/TR/owl2-syntax/

2   Preliminary Definitions

In this section, we present the formal definitions underlying the datatype system of OWL 2. For simplicity, we focus here only on unary datatypes; our definitions can be extended to n-ary datatypes as in [11]. We assume the reader to be familiar with the basics of DL syntax and semantics [1].

The central notion in the datatype system of OWL is the *datatype map*. In OWL 2, datatype maps can additionally support *facets*—expressions that can be applied to a datatype to restrict its interpretation.

Definition 1. *A* datatype map *is a 4-tuple* $\mathcal{D} = (N_D, N_C, N_F, \cdot^{\mathcal{D}})$, *where*

- N_D *is a set of* datatypes,
- N_C *is a function assigning a set of* constants $N_C(d)$ *to each* $d \in N_D$,
- N_F *is a function assigning a set of* facets $N_F(d)$ *to each* $d \in N_D$,
- $\cdot^{\mathcal{D}}$ *is a function assigning a* datatype interpretation $d^{\mathcal{D}}$ *to each datatype* $d \in N_D$, *a* facet interpretation $f^{\mathcal{D}} \subseteq d^{\mathcal{D}}$ *to each facet* $f \in N_F(d)$, *and a* data value $v^{\mathcal{D}} \in d^{\mathcal{D}}$ *to each constant* $v \in N_C(d)$.

By a slight abuse of notation, let $N_C = \bigcup_{d \in N_D} N_C(d)$; *the intended usage of* N_C *should be clear from the context.*

A facet expression *for a datatype* $d \in N_D$ *is a formula* φ *built using propositional connectives over the elements from* $N_F(d) \cup \{\top_d, \bot_d\}$. *The function* $\cdot^{\mathcal{D}}$ *is extended to facet expressions for* d *by setting, for* $f_{(i)} \in N_F(d)$, $\top_d^{\mathcal{D}} = d^{\mathcal{D}}$, $\bot_d^{\mathcal{D}} = \emptyset$, $(\neg f)^{\mathcal{D}} = d^{\mathcal{D}} \setminus f^{\mathcal{D}}$, $(f_1 \wedge f_2)^{\mathcal{D}} = f_1^{\mathcal{D}} \cap f_2^{\mathcal{D}}$, *and* $(f_1 \vee f_2)^{\mathcal{D}} = f_1^{\mathcal{D}} \cup f_2^{\mathcal{D}}$.

In the rest of this paper we additionally assume that the datatypes in N_D are pairwise disjoint—that is, that $d_1, d_2 \in N_D$ and $d_1 \neq d_2$ imply $d_1^{\mathcal{D}} \cap d_2^{\mathcal{D}} = \emptyset$. As we show in Sections 3 and 5, this leads to no loss of generality, simplifies our reasoning algorithm, and allows for a modular treatment of different datatypes. Our datatypes are, in this respect, comparable to datatype groups [11].

For example, $\mathcal{D}$ might be a datatype map with $N_D = \{str, real\}$, where $str^{\mathcal{D}}$ and $real^{\mathcal{D}}$ are the set of all strings and real numbers, respectively. The sets $N_C(str)$ and $N_C(real)$ would then contain all string constants and all decimal representations of real numbers. Finally, the set $N_F(real)$ might contain the facet int, interpreted as the set of all integers, and facets of the form $<_w$, $>_w$, $\leq_w$, and $\geq_w$ for each decimal number w. Thus, the facet expression $int \wedge >_{12} \wedge <_{15}$ would represent the integers 13 and 14.

We next show how to extend a description logic $\mathcal{DL}$ with a datatype map. We omit the definition of the concepts and axioms of $\mathcal{DL}$, and present only the concepts and axioms of the combined language that involve datatypes. We first introduce *data ranges*—expressions over the predicates in $\mathcal{D}$—and then show how to integrate data ranges into $\mathcal{DL}$ concepts and axioms.

Definition 2. *Let* $\mathcal{D} = (N_D, N_C, N_F, \cdot^{\mathcal{D}})$ *be a datatype map. The set of* data ranges *for* $\mathcal{D}$ *is the smallest set that contains* $\top_{\mathcal{D}}$, d, $d[\varphi]$, $\{v_1, \ldots v_n\}$, $\overline{dr}$, *for* $d \in N_D$, φ *a facet expression for* d, $v_i \in N_C$, *and* dr *a data range.*

Table 1. Model-Theoretic Semantics of $\mathcal{DL}+\mathcal{D}$

Semantics of Data Ranges	
$(\top_{\mathcal{D}})^{\mathcal{D}} = \triangle^{\mathcal{D}}$	$(d[\varphi])^{\mathcal{D}} = \varphi^{\mathcal{D}}$
$(\{v_1, \ldots, v_n\})^{\mathcal{D}} = \{v_1^{\mathcal{D}}, \ldots, v_n^{\mathcal{D}}\}$	$\overline{dr}^{\mathcal{D}} = \triangle^{\mathcal{D}} \setminus dr^{\mathcal{D}}$

Semantics of Concepts	Semantics of Axioms
$(\forall U.dr)^{\mathcal{I}} = \{x \mid \forall y : \langle x, y \rangle \in U^{\mathcal{I}} \to y \in dr^{\mathcal{D}}\}$	
$(\exists U.dr)^{\mathcal{I}} = \{x \mid \exists y : \langle x, y \rangle \in U^{\mathcal{I}} \wedge y \in dr^{\mathcal{D}}\}$	$\mathsf{Dis}(U_1, U_2) \Rightarrow U_1^{\mathcal{I}} \cap U_2^{\mathcal{I}} = \emptyset$
$(\leq n\, U.dr)^{\mathcal{I}} = \{x \mid \sharp\{y \mid \langle x, y \rangle \in U^{\mathcal{I}} \wedge y \in dr^{\mathcal{D}}\} \leq n\}$	$U_1 \sqsubseteq U_2 \quad \Rightarrow U_1^{\mathcal{I}} \subseteq U_2^{\mathcal{I}}$
$(\geq n\, U.dr)^{\mathcal{I}} = \{x \mid \sharp\{y \mid \langle x, y \rangle \in U^{\mathcal{I}} \wedge y \in dr^{\mathcal{D}}\} \geq n\}$	$U(a, v) \quad \Rightarrow \langle a^{\mathcal{I}}, v^{\mathcal{D}} \rangle \in U^{\mathcal{I}}$

Note: $\sharp N$ is the number of elements in a set N.

Let $\mathcal{DL}$ be a description logic, defined over a set of individuals N_I, and let N_{DP} be a set of data properties disjoint from each of the sets of symbols used in $\mathcal{DL}$. The logic $\mathcal{DL}+\mathcal{D}$, obtained by extending $\mathcal{DL}$ with $\mathcal{D}$, is defined as follows. The set of concepts of $\mathcal{DL}+\mathcal{D}$ extends the set of concepts of $\mathcal{DL}$ with datatype concepts of the form $\exists U.dr$, $\forall U.dr$, $\geq n\, U.dr$, and $\leq n\, U.dr$, for $U \in N_{DP}$, n a nonnegative integer, and dr a data range for $\mathcal{D}$. The set of axioms of $\mathcal{DL}+\mathcal{D}$ extends the set of axioms of $\mathcal{DL}$ with data property disjointness axioms $\mathsf{Dis}(U_1, U_2)$, data property inclusion axioms $U_1 \sqsubseteq U_2$, and data property assertions $U(a, v)$, for $U_{(i)} \in N_{DP}$, $a \in N_i$, and $v \in N_C$.

An interpretation for $\mathcal{DL}+\mathcal{D}$ is a triple $\mathcal{I} = (\triangle^{\mathcal{I}}, \triangle^{\mathcal{D}}, \cdot^{\mathcal{I}})$, where $\triangle^{\mathcal{I}}$ and $\triangle^{\mathcal{D}}$ are nonempty disjoint sets such that $d^{\mathcal{D}} \subseteq \triangle^{\mathcal{D}}$ for each $d \in N_D$, and $\cdot^{\mathcal{I}}$ is a function assigning to each concept C, property R, and individual a of $\mathcal{DL}$, and to each data property $U \in N_{DP}$, interpretations $C^{\mathcal{I}} \subseteq \triangle^{\mathcal{I}}$, $R^{\mathcal{I}} \subseteq \triangle^{\mathcal{I}} \times \triangle^{\mathcal{I}}$, $a^{\mathcal{I}} \in \triangle^{\mathcal{I}}$, and $U^{\mathcal{I}} \subseteq \triangle^{\mathcal{I}} \times \triangle^{\mathcal{D}}$ respectively. The functions $\cdot^{\mathcal{D}}$ and $\cdot^{\mathcal{I}}$ are extended to data ranges and datatype concepts as shown in Table 1. For $\mathcal{DL}+\mathcal{D}$ knowledge bases $\mathcal{K}$ and $\mathcal{K}'$, an interpretation $\mathcal{I}$ is a $\mathcal{D}$-model of $\mathcal{K}$, written $I \models_{\mathcal{D}} \mathcal{K}$, if all axioms of $\mathcal{DL}$ are satisfied in $\mathcal{I}$ as specified by $\mathcal{DL}$, and the additional axioms are satisfied in $\mathcal{I}$ and $\mathcal{D}$ as specified in Table 1; furthermore, $\mathcal{K}$ $\mathcal{D}$-entails $\mathcal{K}'$, written $\mathcal{K} \models_{\mathcal{D}} \mathcal{K}'$, if $I \models_{\mathcal{D}} \mathcal{K}'$ whenever $I \models_{\mathcal{D}} \mathcal{K}$.

3   The Architecture of the Datatype System of OWL 2

We now explain the rationale behind and some nontrivial consequences of the definitions presented in Section 2.

3.1   Openness of the Domain in a Datatype Map

A number of approaches for adding datatypes to DLs are based on the framework of concrete domains [2], in which the set $\triangle^{\mathcal{D}}$ is fixed in advance; for example, $\triangle^{\mathcal{D}}$ can be fixed to the set of all integers. Datatype groups [11] are similar to datatype maps, but they also fix the set $\triangle^{\mathcal{D}}$ as the union of the interpretations of all datatypes in the group. The semantics of OWL DL [13] is somewhat ambiguous regarding this point: it says that $\triangle^{\mathcal{D}}$ *contains* the interpretation of all datatypes, but it is not clear whether $\triangle^{\mathcal{D}}$ must be *exactly* that set (the intention was, however, that $\triangle^{\mathcal{D}}$ should *not* be fixed[4]).

The set of datatypes in OWL 2 should be extensible: future versions of OWL might want to add support datatypes that will not be normative in OWL 2, and implementations might also want to support custom datatypes. Extensibility, however, is impossible if the set $\triangle^{\mathcal{D}}$ is fixed in advance. For example, the axiom $\alpha = \top \sqsubseteq \forall U. <_5 \sqcup \exists U.real$ is a tautology w.r.t. a datatype map $\mathcal{D}$ where $\triangle^{\mathcal{D}}$ is fixed to the set of all real numbers: if some individual a is connected by U to a data value, this value must be a real number, so $\exists U.real(a)$ is satisfied; if a has no U-connections, then $\forall U. <_5 (a)$ is satisfied. The axiom α, however, is not a tautology w.r.t. a datatype map $\mathcal{D}'$ in which $\triangle_{\mathcal{D}'}$ is fixed to contain all real numbers and all strings: an individual a might be connected only to a string, which makes neither disjunct in α satisfied. This example shows that, in general, the consequences of a knowledge base $\mathcal{K}$ can depend on datatypes that are not mentioned in $\mathcal{K}$ at all—clearly an undesirable situation.

Therefore, in $\mathcal{DL}+\mathcal{D}$ (and in OWL 2) the set $\triangle^{\mathcal{D}}$ can be *any* set that at least contains the interpretations of all datatypes in $\mathcal{D}$. The following proposition shows that $\mathcal{K}$ can in fact be interpreted by considering only those datatypes explicitly mentioned in $\mathcal{K}$. Therefore, in the rest of this paper we simply talk of *models* and *entailment* instead of $\mathcal{D}$-models and $\mathcal{D}$-entailment.

Proposition 1. *Let* $\mathcal{D}_1 = (N_{D_1}, N_{V_1}, N_{F_1}, \cdot^{\mathcal{D}_1})$ *be a datatype map and* $\mathcal{K}$ *and* $\mathcal{K}'$ $\mathcal{DL}+\mathcal{D}_1$ *knowledge bases. For each datatype map* $\mathcal{D}_2 = (N_{D_2}, N_{V_2}, N_{F_2}, \cdot^{\mathcal{D}_2})$ *such that* $N_{D_1} \subseteq N_{D_2}$, $N_{V_1}(d) \subseteq N_{V_2}(d)$ *and* $N_{F_1}(d) \subseteq N_{F_2}(d)$ *for each* $d \in N_{D_1}$, *and* $\cdot^{\mathcal{D}_2}$ *coincides with* $\cdot^{\mathcal{D}_1}$ *on the elements from* $\mathcal{D}_1$, *we have* $\mathcal{K} \models_{\mathcal{D}_1} \mathcal{K}'$ *iff* $\mathcal{K} \models_{\mathcal{D}_2} \mathcal{K}'$.

Proof. We show the contrapositive: $\mathcal{K} \not\models_{\mathcal{D}_2} \mathcal{K}'$ iff $\mathcal{K} \not\models_{\mathcal{D}_1} \mathcal{K}'$. The $(\Rightarrow)$ direction is trivial. For the $(\Leftarrow)$ direction, let $\mathcal{I} = (\triangle^{\mathcal{I}}, \triangle^{\mathcal{D}}, \cdot^{\mathcal{I}})$ be an interpretation such that $\mathcal{I} \models_{\mathcal{D}_1} \mathcal{K}$ and $\mathcal{I} \not\models_{\mathcal{D}_1} \mathcal{K}'$. We construct $\mathcal{I}' = (\triangle^{\mathcal{I}'}, \triangle^{\mathcal{D}'}, \cdot^{\mathcal{I}'})$ such that $\triangle^{\mathcal{I}'} := \triangle^{\mathcal{I}}$ and $\cdot^{\mathcal{I}'} := \cdot^{\mathcal{I}}$, and $\triangle^{\mathcal{D}'}$ is obtained from $\triangle^{\mathcal{D}}$ by adding the interpretations of all datatypes from $N_{\mathcal{D}_2} \setminus N_{\mathcal{D}_1}$. Extending $\triangle^{\mathcal{D}}$ to $\triangle^{\mathcal{D}'}$ can change only the interpretation of complemented data ranges; hence, for each data range dr over $\mathcal{D}_1$, we have $dr^{\mathcal{D}} \subseteq dr^{\mathcal{D}'}$. Since $U^{\mathcal{D}'} = U^{\mathcal{D}}$ for each $U \in N_{DP}$, for each concept C of $\mathcal{DL}+\mathcal{D}_1$, we have $C^{\mathcal{I}'} = C^{\mathcal{I}}$; but then, $\mathcal{I}' \models_{\mathcal{D}_2} \mathcal{K}$ and $\mathcal{I}' \not\models_{\mathcal{D}_2} \mathcal{K}'$. $\qquad\square$

3.2 Giving Names to Data Ranges

The OWL 2 Working Group is currently considering whether to extend the language with the ability to explicitly name commonly used data ranges. For example, the axiom *Teens* $\equiv$ *real*$[int \wedge >_{12} \wedge <_{20}]$ would give a name *Teens* to the set of integers between 12 and 20, which could then be used in concept definitions such as *Teenager* $\equiv \exists hasAge.Teens$. The syntax and the semantics of such axioms can be formalized as shown in the following definition.

---

[4] Personal communication with Peter F. Patel-Schneider.

Definition 3. *Let* $\mathcal{D} = (N_D, N_C, N_F, \cdot^{\mathcal{D}})$ *be a datatype map, and let* N_N *be a set of* datatype names, *disjoint with* N_D, N_C, *and* N_F. *The set of data ranges of* $\mathcal{D}$ *is extended such that each* $dn \in N_N$ *is a data range. A* datatype naming axiom *then has the form* $dn \equiv dr$, *where* $dn \in N_N$ *and* dr *is a data range. An interpretation* $\mathcal{I} = (\triangle^{\mathcal{I}}, \triangle^{\mathcal{D}}, \cdot^{\mathcal{I}})$ *interprets each* $dn \in N_N$ *as* $dn^{\mathcal{D}} \subseteq \triangle^{\mathcal{D}}$. *Furthermore,* $\mathcal{I}$ *satisfies a datatype naming axiom* $dn \equiv dr$ *iff* $dn^{\mathcal{D}} = dr^{\mathcal{D}}$.

Although seemingly innocuous, datatype naming axioms can easily invalidate Proposition 1. For example, let $\mathcal{K} = \{A \equiv real, \ A \equiv \top_{\mathcal{D}}\}$. The datatype naming axioms in $\mathcal{K}$ behave similarly to general concept inclusions, allowing us to fix the set $\triangle^{\mathcal{D}}$ to the set of real numbers. Thus, $\mathcal{K}$ is satisfiable w.r.t. a datatype map that contains only real numbers, but it becomes unsatisfiable as soon as we extend the datatype map with a datatype disjoint with *real*.

Datatype naming axioms thus seem to be too expressive in general: datatypes are fully described by the datatype map, so allowing for additional axioms about the datatypes is likely to be undesirable. This problem can be solved by requiring datatype naming axioms $dn \equiv dr$ to be acyclic [1]: each name dn may occur in at most one such axiom, and it may neither directly nor indirectly be used in dr. With such restrictions, each datatype name dn can be unfolded—that is, it can be (recursively) replaced with its definition. Thus, datatype names can always be eliminated from a knowledge base, so Proposition 1 still holds.

3.3   Disjointness of Datatypes in a Map

In Definition 1 and in OWL 2, the datatypes in a datatype map need not be disjoint. Without losing generality, however, we can assume the contrary: two nondisjoint datatypes d_1 and d_2 can be replaced with a datatype d_{1+2} that is interpreted as a union of d_1 and d_2 and that provides d_1 and d_2 as facets. For example, a datatype system with strings, real numbers, and integers can be formalized as a datatype map in which strings and real numbers are datatypes, and integers are modeled as a facet for the real numbers.

Assuming that datatypes in a map are disjoint allows us to obtain a modular algorithm for datatype reasoning. In particular, only four basic operations (see Definition 5) are needed to support a datatype d in our datatype reasoning algorithm; if d is disjoint from any other datatype in the map, these operations need not consider any of the other supported datatypes. For example, in Sections 5.2 and 5.3 we present datatype handlers for numbers and strings, respectively. Although the handler for numbers needs to know about all kinds of numbers, it does not need to know about strings and vice versa. Therefore, the notion of a datatype provides us with a natural modularization boundary for reasoning.

3.4   The Semantics of Complemented Data Ranges

Complemented data ranges have been added to $\mathcal{DL+D}$ mainly to support the representation of axioms in negation-normal form. In (1), for example, the concept $\exists hasAge.real[<_{18}]$ is implicity negated, which becomes visible if the axiom is brought into negation-normal form, cf. (2).

(1) $\exists hasAge.real[<_{18}] \sqsubseteq YoungPerson$

(2) $\top \sqsubseteq \forall hasAge.\overline{real[<_{18}]} \sqcup YoungPerson$

Note that the semantics of complemented data ranges is defined w.r.t. the entire set $\triangle^{\mathcal{D}}$; hence, the interpretation of $\overline{real[<_{18}]}$ includes all real numbers greater than or equal to 18, as well all data values that are not numbers—that is, $\overline{real[<_{18}]}^{\mathcal{D}} = \overline{real}^{\mathcal{D}} \cup real[\geq_{18}]^{\mathcal{D}}$. This may seem counterintuitive, but it is necessary if $\mathcal{DL}+\mathcal{D}$ is to have a standard first-order semantics. For example, if $\overline{real[<_{18}]}$ were interpreted as "the set of real numbers greater than or equal to 18," then $\forall hasAge.\overline{real[<_{18}]}$ would not be the complement of $\exists hasAge.real[<_{18}]$, which would invalidate the basic assumptions of first-order logic.

3.5   Reasoning with a Datatype Map

Several tableau algorithms for DLs extended with datatypes have been proposed [2,10,5,7,11]; we illustrate them using the ABox $\mathcal{A}$, shown in (3). First, standard tableau expansion rules are used to expand $\mathcal{A}$ to $\mathcal{A}'$, shown in (4). To obtain a model from $\mathcal{A}'$, one must check whether individuals t_1 and t_2 can be assigned values from $\triangle^{\mathcal{D}}$ in a consistent way. For this purpose, algorithms such as [7] invoke an external *datatype checker*—an oracle that decides satisfiability of conjunctions of the form $d_1(x_1) \wedge \ldots \wedge d_n(x_n)$, where d_i are datatypes. In our example, $\varphi = \{5\}(x_1) \wedge int[>_4 \wedge <_6](x_2)$ is satisfied by an assignment $x_1 = x_2 = 5$, so we can conclude that $\mathcal{A}'$ and $\mathcal{A}$ are satisfiable.

(3) $\mathcal{A} = \{\quad \exists U_1.\{5\}(a),\quad \exists U_2.int[>_4 \wedge <_6](a)\quad \}$

(4) $\mathcal{A}' = \mathcal{A} \cup \{\quad U_1(a,t_1),\quad \{5\}(t_1),\quad U_2(a,t_2),\quad int[>_4 \wedge <_6](t_2)\quad \}$

Although the set $\triangle^{\mathcal{D}}$ in a datatype map is not fixed (see Section 3.1), a tableau algorithm for $\mathcal{DL}$ can be combined with a datatype checker in much the same way as in [2,10,5,7,11]. A minor problem in $\mathcal{DL}+\mathcal{D}$ arises due to disjointness of data properties. Assume that the knowledge base shown in (3) also contains the axiom $\mathsf{Dis}(U_1, U_2)$. The assignment $x_1 = x_2 = 5$ is the only one satisfying φ; however, setting t_1 and t_2 equal to 5 clearly invalidates $\mathsf{Dis}(U_1, U_2)$. A similar problem has been observed and solved in a slightly different context [10]. Roughly speaking, the solution is to derive the inequality $t_1 \not\approx t_2$ whenever $U_1(s, t_1)$ and $U_2(s, t_2)$ are derived and U_1 and U_2 are disjoint. In our example, deriving $t_1 \not\approx t_2$ gives rise to a conjunction $\varphi' = \varphi \wedge x_1 \not\approx x_2$, which is clearly unsatisfiable. The model construction therefore fails, and we can correctly conclude that $\mathcal{A} \cup \{\mathsf{Dis}(U_1, U_2)\}$ is unsatisfiable.

The following definition formalizes the datatype checking problem, as applicable in our setting, and introduces some useful notation. For convenience, we treat conjunctions of datatype assertions as sets.

Definition 4. *Let* $\mathcal{D} = (N_D, N_C, N_F, \cdot^{\mathcal{D}})$ *be a datatype map and* N_V *a set of variables disjoint with* N_C. *A* $\mathcal{D}$-conjunction *is a finite set of assertions of the* *form $dr(t)$ and $t_1 \not\approx t_2$, for dr a data range over $\mathcal{D}$ and $t_{(i)} \in N_V \cup N_C$. A $\mathcal{D}$-conjunction Γ is $\mathcal{D}$-satisfiable if a set $\triangle^{\mathcal{D}}$ and a mapping $\delta : N_V \cup N_C \to \triangle^{\mathcal{D}}$ exist such that (i) $d^{\mathcal{D}} \subseteq \triangle^{\mathcal{D}}$ for each $d \in N_D$, (ii) $\delta(c) = c^{\mathcal{D}}$ for each $c \in N_C$, (iii) $dr(t) \in \Gamma$ implies $\delta(t) \in (dr)^{\mathcal{D}}$, and (iv) $t_1 \not\approx t_2 \in \Gamma$ implies $\delta(t_1) \neq \delta(t_2)$.*

Let Γ be a $\mathcal{D}$-conjunction. Each assertion $t_1 \not\approx t_2$ in Γ should also be read as $t_2 \not\approx t_1$ —that is, the $\not\approx$ predicate has built-in symmetry. For x a variable, $\Gamma/_{-x}$ is the result of deleting all assertions in Γ that contain x; for t a variable or a constant, $\Gamma/_{x \mapsto t}$ is the result of replacing x with t in all assertions in Γ; finally, let $\mathsf{cv}(\Gamma, x) = \{x' \mid x' \not\approx x \in \Gamma\}$ and $\mathsf{cc}(\Gamma, x) = \{c \mid c \not\approx x \in \Gamma\}$.

Boolean combinations of facets are thus dealt with in the datatype checker rather than the tableau algorithm, since knowledge about datatypes and facets can be used to optimize the handling of common cases (see Sections 5.2 and 5.3).

3.6 Complexity of Datatype Checking

We now turn our attention to the complexity of datatype reasoning, and show that datatype checking is intractable in general (Proposition 2), but that an important case exists in which the problem becomes trivial (Proposition 3).

Proposition 2. *Checking $\mathcal{D}$-satisfiability of a $\mathcal{D}$-conjunction is NP-hard.*

Proof. The proof is by reduction from the NP-hard GRAPH 3-COLORABILITY problem: for a finite undirected graph $G = (V, E)$, decide whether it is possible to label each vertex in V with a number from the set $\{1, 2, 3\}$ such that adjacent vertices are not labeled with the same number.

For $G = (V, E)$ a finite graph, let x_i be a variable uniquely assigned to each vertex $i \in V$, and let Γ_G be the following $\mathcal{D}$-conjunction, where $dr = \{1, 2, 3\}$:

$$\Gamma_G = \bigcup_{i \in V} \{ \ dr(x_i) \ \} \cup \bigcup_{\langle i,j \rangle \in E} \{ \ x_i \not\approx x_j \ \}$$

It is easy to see that G is 3-colorable if and only if Γ_G is $\mathcal{D}$-satisfiable.   □

The proof of Proposition 2 requires inequality predicates, which are already necessary for the proper handling of number restriction datatype concepts. Such concepts, however, generate sets of pairwise unequal variables, which may be easier to handle; for example, it is not trivial to see if they can encode GRAPH 3-COLORABILITY. In contrast, $\mathcal{D}$-conjunctions of the form used in the proof of Proposition 2 can be obtained using only axioms of the form $\exists U.dr(a)$ and $\mathsf{Dis}(U_1, U_2)$. Proposition 2 thus suggests that data property disjointness axioms might make datatype checking harder in practice.

Proposition 3. *Let Γ be a $\mathcal{D}$-conjunction and x a variable such that (i) x occurs in Γ in exactly one assertion of the form $dr(x)$,[5] (ii) $x \not\approx x \notin \Gamma$, and (iii) $\sharp dr^{\mathcal{D}} \geq \sharp \mathsf{cv}(\Gamma, x) + \sharp \mathsf{cc}(\Gamma, x) + 1$.[6] Then, Γ is $\mathcal{D}$-satisfiable if and only if $\Gamma/_{-x}$ is $\mathcal{D}$-satisfiable.*

---

[5] If Γ does not contain such assertion, we can always take $dr = \top_{\mathcal{D}}$.

[6] $\sharp S$ denotes the cardinality of the set S.

Proof. The ($\Rightarrow$) direction is obvious. For the ($\Leftarrow$) direction, assume that a set $\triangle^{\mathcal{D}}$ and a mapping δ of variables in $\Gamma/_{-x}$ to the elements of $\triangle^{\mathcal{D}}$ exist such that $\Gamma/_{-x}$ is satisfied. By (iii), the set $dr^{\mathcal{D}} \setminus \{c^{\mathcal{D}} \mid c \not\approx x \in \Gamma\} \setminus \{\delta(x') \mid x' \in \mathsf{cv}(\Gamma, x)\}$ contains at least one element that is not mapped to a variable in $\mathsf{cv}(\Gamma, x)$ or a constant in $\mathsf{cc}(\Gamma, x)$. Let $\delta(x)$ be an arbitrarily selected element from this set. By (i) and (ii), all assertions in $\Gamma \setminus \Gamma/_{-x}$ are of the form $dr(x)$, $x \not\approx x'$ with $x' \in \mathsf{cv}(\Gamma, x)$, or $x \not\approx c$ with $c \in \mathsf{cc}(\Gamma, x)$. Clearly, δ satisfies Γ. $\square$

4   Selecting the Set of Datatypes for OWL 2

The current OWL 2 Working Draft[7] contains a normative list of datatypes and facets, most of which are taken from XML Schema [3]. Although these datatypes may be quite useful in XML applications, some of them do not seem appropriate for a logic-based language such as OWL 2.

4.1   String-Based Datatypes

The base string datatype in OWL 2 is *xsd:string*, and it is equipped with facets length n, minLength n, maxLength n, which restrict the length of a string, and pattern re, which restricts the form of a string to the regular expression re. We present algorithms for handling strings and all these facets in Section 5.3. XML Schema includes a number of string-derived datatypes [3], which can be seen as shortcuts for the pattern facet with particular values. The *xsd:anyURI* datatype represents Uniform Resource Locators (URIs). The OWL 2 specification needs to clarify whether this datatype is a subset of *xsd:string*.

4.2   Numbers

XML Schema provides a multitude of datatypes for numbers: *xsd:decimal* represents arbitrarily long numbers in decimal notation, *xsd:integer* represents unbounded integers, and *xsd:double* and *xsd:float* represent floating-point numbers in double and single precision, respectively. Other numeric datatypes are derived from these base ones by imposing various restrictions; for example, *xsd:nonNegativeInteger* represents all nonnegative integers. The supported facets are minInclusive x, minExclusive x, maxInclusive x, and maxExclusive x, which restrict the range of numbers, and pattern re, which restricts numbers to those whose string representation matches the regular expression re.

These datatypes exhibit a number of different problems. First, *xsd:decimal* is not closed under division, so it does not provide a suitable basis for possible extensions of OWL 2 with arithmetic. Second, the floating-point datatypes have a very large but finite number of values, and can also exhibit complex behavior due to rounding of values that cannot be exactly represented; these features could lead to unexpected inferences, and they might be a source of inefficiency in implementations. Third, the pattern facet seems to be of limited utility for

---

[7] http://www.w3.org/TR/2008/WD-owl2-syntax-20080411/

number types, and it might place an unreasonable burden on implementers as it allows for data ranges such as "all decimal numbers greater than 5 that match a particular regular expression."

In view of these problems, we propose that the OWL 2 datatypes *xsd:decimal*, *xsd:double*, and *xsd:float* be replaced with a new datatype *owl:real*, interpreted as the set of all real numbers. Clearly, not all data values in the interpretation of *owl:real* can be represented using a constant (i.e., a finite string over a finite alphabet). This should, however, not pose a problem in practice: one could define constants for all rational numbers, and possibly also for "important" irrational numbers such as π or e. The *xsd:integer* datatype can be supported as a facet of *owl:real*, as can the other facets apart from pattern. In Section 5.2 we present a reasoning algorithm for this datatype.

4.3   Date and Time

XML Schema provides the *xsd:dateTime* datatype, interpreted as a set of time points in the Gregorian calendar. A number of other datatypes represent possibly recurring intervals and time points. For example, *xsd:date* represents intervals of length one day; *xsd:time* represents an instance in time that recurs each day; and *xsd:gMonthDay* represents a Gregorian date that recurs every year.

Reasoning about recurring time points and intervals is difficult due to their complex and ill defined semantics: the recurrences are irregular due to exceptions such as leap years; furthermore, the occurrence of future time points cannot be determined in advance due to leap seconds, which are introduced into the calendar by the International Earth Rotation and Reference Systems Service as necessary. Therefore, only the *xsd:dateTime* datatype seems amenable to implementation, and it can be handled by techniques similar to the ones for numbers.

5   Reasoning with Datatypes in OWL 2

While the principles of integrating a datatype checker with a tableau algorithm are well understood, little attention has been paid in the literature to the details of actual datatype checking algorithms. We next present such an algorithm that is extensible w.r.t. the set of supported datatypes and show its correctness.

5.1   An Extensible Datatype Checking Algorithm

We first identify the basic operations that are needed to support a particular datatype in a datatype map. In Sections 5.2 and 5.3, we discuss how to implement these operations for the real and string datatypes, respectively.

Definition 5. *Let* $\mathcal{D} = (N_D, N_C, N_F, \cdot^{\mathcal{D}})$ *be a datatype map where the interpretations of different datatypes are pairwise disjoint. A datatype handler for a datatype* $d \in N_D$ *is a 4-tuple* $(\mathsf{minc}_d, \mathsf{enu}_d, \mathsf{in}_d, \mathsf{eq}_d)$ *of functions where, for each data range dr of the form* $d[\varphi]$,

- $\mathsf{minc}_d(dr, n) = \mathsf{true}$ *for an integer n iff $dr^{\mathcal{D}}$ contains at least n elements,*
- $\mathsf{enu}_d(dr)$ *is defined only if $dr^{\mathcal{D}}$ is finite, and it is a set $\{c_1, \ldots, c_n\}$ such that $dr^{\mathcal{D}} = \{c_1^{\mathcal{D}}, \ldots, c_n^{\mathcal{D}}\}$—that is, $\mathsf{enu}_d(dr)$ returns a finite set of constants that enumerate the interpretation of dr,*
- $\mathsf{in}_d(c, dr) = \mathsf{true}$ *for $c \in N_C(d)$ iff $c^{\mathcal{D}} \in dr^{\mathcal{D}}$, and*
- $\mathsf{eq}_d(c_1, c_2) = \mathsf{true}$ *for $c_1, c_2 \in N_C(d)$ iff $c_1^{\mathcal{D}} = c_2^{\mathcal{D}}$.*

Our algorithm for checking $\mathcal{D}$-satisfiability of a $\mathcal{D}$-conjunction Γ consists of Procedures 1 and 2. For convenience, we assume that all data ranges in Γ are of the form $d[\varphi]$ (i.e., data ranges without facet expressions are represented as $d[\top_d]$). The letter c (possibly with subscripts or superscripts) denotes a constant. We additionally use the following two auxiliary functions: for c_1 and c_2 constants, $\mathsf{eq}(c_1, c_2) = \mathsf{true}$ iff c_1 and c_2 are constants of the same datatype d and $\mathsf{eq}_d(c_1, c_2) = \mathsf{true}$; furthermore, for c a constant and $d[\varphi]$ a data range, $\mathsf{in}(c, d[\varphi]) = \mathsf{true}$ iff c is a constant of the datatype d and $\mathsf{in}_d(c, d[\varphi]) = \mathsf{true}$.

We next explain the intuition behind Procedure 1. First, Γ is checked for trivial unsatisfiability (lines 1–3), after which all complemented enumerations are rewritten using inequalities (lines 4–6) in order to simplify the rest of the algorithm. Lines 7–22 form the core part of the algorithm. For each variable x in Γ, the set of data ranges in which x occurs is normalized (line 8)—that is, it is reduced to $*$ (meaning that x is trivially satisfiable), a data range of the form $d[\varphi]$, or a finite enumeration $\{c_1, \ldots, c_n\}$. In the second case, a call is made to the datatype handler to see whether Proposition 3 is applicable (line 9); if so, the variable x is removed from Γ (line 11). If $D = d[\varphi]$ and the test in line 9 fails, then the interpretation of D is finite, so it is enumerated (line 13). Thus, by line 15, D is either $*$ or a finite enumeration. If D is empty, then Γ unsatisfiable (lines 15–16). If D is not empty, the normalized data range D is reintroduced into Γ (lines 17–21): if D is a singleton, then x must be assigned the only value in D (line 18); otherwise, all original data range assertions involving x are replaced with $D(x)$ (line 20). By line 23, therefore, in all assertions of the form $dr(x) \in \Gamma$, for dr a nonempty enumerated data range.

Lines 23–33 try to further simplify Γ, first by considering all assertions not containing variables (lines 23–30), and then by applying Proposition 3 to the remaining assertions (lines 31–33). All that now remains is to check if Γ is satisfied for at least one assignment of values to variables and constants. This part of the algorithm is nondeterministic, and can be implemented using search. To reduce the search space, Γ is first decomposed into variable-disjoint subsets (line 34), each of which is tested for $\mathcal{D}$-satisfiability independently (lines 35–43).

Procedure 2 normalizes a set of data ranges S to a finite enumeration of constants of the form $\{c_1, \ldots, c_n\}$, a single data range of the form $d[\varphi]$, or $*$ (meaning that the corresponding variable is trivial to satisfy). Lines 1–13 handle the case where S contains at least one enumerated data range: the result is then an enumeration containing only those constants c_i that are contained in all data ranges in S. If S does not contain at least one positive data range (line 14), then the corresponding variable can be assigned a fresh distinct data value not contained in the interpretation of any of the datatypes (note that $\triangle^{\mathcal{D}}$ can be

---

Procedure 1. $\mathcal{D}$-SATISFIABLE(Γ)

Require: a $\mathcal{D}$-conjunction Γ containing assertions of the form $dr(t)$ and $t_1 \not\approx t_2$

1: **if** $\top_{\mathcal{D}}(t) \in \Gamma$ for t a variable or a constant, or $x \not\approx x \in \Gamma$ **then**
2:     **return** false
3: **end if**
4: **for each** $\overline{\{c_1, \ldots, c_n\}}(t) \in \Gamma$ for t a variable or a constant **do**
5:     $\Gamma := (\Gamma \setminus \{\ \overline{\{c_1, \ldots, c_n\}}(t)\ \}) \cup \{t \not\approx c_1, \ldots, t \not\approx c_n\}$
6: **end for**
7: **for each** variable x occurring in Γ **do**
8:     $D := \text{NORMALIZE}(\{dr \mid dr(x) \in \Gamma\})$
9:     **if** $D = *$, or $D = d[\varphi]$ and $\text{minc}_d(D, \sharp\text{cv}(\Gamma, x) + \sharp\text{cc}(\Gamma, x) + 1) = $ true **then**
10:         $\Gamma := \Gamma/_{-x}$ $\triangleright$ apply Proposition 3 to x
11:         $D := *$
12:     **else if** $D = d[\varphi]$ **then**
13:         $D := \text{enu}_d(D)$
14:     **end if**
15:     **if** $D = \emptyset$ **then**
16:         **return** false
17:     **else if** $D = \{c\}$ **then**
18:         $\Gamma := (\Gamma \setminus \{dr(x) \mid dr(x) \in \Gamma\})/_{x \mapsto c}$
19:     **else if** $D \neq *$ **then**
20:         $\Gamma := (\Gamma \setminus \{dr(x) \mid dr(x) \in \Gamma\}) \cup \{D(x)\}$
21:     **end if**
22: **end for**
23: **for each** $\alpha \in \Gamma$ that does not contain a variable **do**
24:     **if** $\alpha = c_1 \not\approx c_2$ and $\text{eq}(c_1, c_2) = $ true, or
25:         $\alpha = d[\varphi](c)$ and $\text{in}(c, d[\varphi]) = $ false, or $\alpha = \overline{d[\varphi]}(c)$ and $\text{in}(c, d[\varphi]) = $ true, or
26:         $\alpha = \overline{\{c_1, \ldots, c_n\}}(c)$ and $\text{eq}(c, c_i) = $ false for each $1 \leq i \leq n$ **then**
27:         **return** false
28:     **end if**
29:     $\Gamma := \Gamma \setminus \{\ \alpha\ \}$
30: **end for**
31: **while** Γ contains some $\overline{\{c_1, \ldots, c_n\}}(x)$ such that $n \geq \sharp\text{cv}(\Gamma, x) + \sharp\text{cc}(\Gamma, x) + 1$ **do**
32:     $\Gamma := \Gamma/_{-x}$ $\triangleright$ apply Proposition 3 to x
33: **end while**
34: **decompose** Γ into nonempty mutually disjoint subsets $\Gamma_1, \ldots, \Gamma_n$ such that no Γ_i and Γ_j, $i \neq j$, have variables in common
35: **for each** $1 \leq i \leq n$ **do**
36:     **if** an assignment δ to variables and constants in Γ_i such that
37:         $\delta(c) = c$ for each constant c, and
38:         $\overline{\{c_1, \ldots, c_m\}}(x) \in \Gamma$ implies $\delta(x) = c_i$ for some $1 \leq i \leq m$, and
39:         $t_1 \not\approx t_2 \in \Gamma$ implies $\text{eq}(\delta(t_1), \delta(t_2)) = $ false
40:     does not exist **then**
41:         **return** false
42:     **end if**
43: **end for**
44: **return** true

Procedure 2. NORMALIZE(S)

Require: a nonempty set of data ranges S of the form $d[\varphi]$, $\overline{d[\varphi]}$, or $\{c_1, \ldots, c_n\}$
1: **if** S contains a data range of the form $\{c_1, \ldots, c_n\}$ **then**
2:        $R := \{c_1, \ldots, c_n\}$
3:        **for each** $c \in R$ **do**
4:                **for each** $dr \in S$ **do**
5:                        **if** $dr = \{c'_1, \ldots, c'_m\}$ and $\mathsf{eq}(c, c'_i) = \mathsf{false}$ for each $1 \le i \le m$, or
6:                                $dr = d[\varphi]$ and $\mathsf{in}(c, d[\varphi]) = \mathsf{false}$, or
7:                                $dr = \overline{d[\varphi]}$ and $\mathsf{in}(c, d[\varphi]) = \mathsf{true}$ **then**
8: $R := R \setminus \{c\}$
9:                        **end if**
10:                **end for**
11:        **end for**
12:        **return** R
13: **end if**
14: **if** S contains no data range of the form $d[\varphi]$ **then**
15:        **return** $*$
16: **else if** S contains data ranges $d_1[\varphi_1]$ and $d_2[\varphi_2]$ such that $d_1 \ne d_2$ **then**
17:        **return** $\emptyset$
18: **end if**
19: **let** d be the datatype of all the data ranges in S of the form $d[\varphi]$
20: $\psi := \top_d$
21: **for each** $dr \in S$ **do**
22:        **if** $dr = d[\varphi]$ **then**
23:                $\psi := \psi \wedge \varphi$
24:        **else if** $dr = \overline{d[\varphi]}$ **then**
25:                $\psi := \psi \wedge \neg\varphi$
26:        **end if**
27: **end for**
28: **return** $d[\psi]$

---

any set that contains the interpretations of all the datatypes in $\mathcal{D}$). If S contains two positive data ranges with different datatypes (line 16), then S is clearly unsatisfiable. Lines 21–27 then combine the facet expressions in all the data ranges in S. Note that the simplification of complemented data ranges in line 25 is possible because $\overline{d[\varphi]}^{\mathcal{D}} = \overline{d}^{\mathcal{D}} \cup \overline{d[\neg\varphi]}^{\mathcal{D}}$ and, since S contains at least one data range of the form $d[\varphi]$, no data value can be in both $d[\varphi]^{\mathcal{D}}$ and $\overline{d}^{\mathcal{D}}$.

The correctness of our algorithm follows easily from Proposition 3. Therefore, we only sketch the proof of the following theorem.

Theorem 1. $\mathcal{D}$-SATISFIABLE(Γ) *returns* true *if and only if* Γ *is $\mathcal{D}$-satisfiable.*

Proof (Sketch). It is easy to see that, for each $D = $ NORMALIZE(S), the following holds (†): $D = *$ if and only of all data ranges in S are of the form $\overline{d[\varphi]}$; otherwise,

$$D^{\mathcal{D}} = \bigcap\nolimits_{dr \in S_p} dr^{\mathcal{D}} \cap \bigcap\nolimits_{\overline{d[\varphi]} \in S \setminus S_p} (d[\neg\varphi])^{\mathcal{D}}, \text{ where}$$
$$S_p = \{dr \in S \mid dr \text{ is not of the form } \overline{d[\varphi]}\}.$$

We now prove the claim of this theorem. If Γ is $\mathcal{D}$-satisfiable, then the condition in line 1 cannot be satisfied. Furthermore, the transformation in line 5 clearly preserves $\mathcal{D}$-satisfiability of Γ. Consider now the iteration over the variables occurring in Γ in lines 7–22. In lines 9–11, assertions containing x can be deleted from Γ without affecting its $\mathcal{D}$-satisfiability if either $D = *$, or if $D = d[\varphi]$ and Proposition 3 is applicable; in the former case, this is because (†) tells us that we can interpret x as an arbitrary unique element not contained in any datatype in $\mathcal{D}$. If $D = d[\varphi]$ and Proposition 3 is not applicable, then $d[\varphi]$ must be finite and equal to the enumeration obtained in line 13; hence, in lines 15–22 we can either return false (if the enumeration is empty), replace x with c in Γ (if $D(x)$ is a singleton enumeration $\{c\}$), or replace all assertions of the form $dr(x)$ in Γ with a single assertion $D(x)$, all of which preserve the $\mathcal{D}$-satisfiability of Γ. Lines 23–30 detect obvious inconsistencies involving assertions in Γ that do not contain variables; clearly, this does not affect the $\mathcal{D}$-satisfiability of Γ. Lines 31–33 apply Proposition 3 again, which by definition preserves the $\mathcal{D}$-satisfiability of Γ. Since all data ranges in Γ are now finite enumerations, only a finite number of assignments need to be considered, and lines 34–44 will detect if one of these satisfies Γ.   □

In practice, the number of variables in a $\mathcal{D}$-conjunction Γ is likely to be of the same order of magnitude as the numbers occurring in number restrictions, which are usually quite small. Furthermore, data ranges are rarely constrained to small interpretations in practice, so test (9) is likely to succeed. The satisfiability of such a Γ can thus be decided without the need to enumerate data ranges and perform combinatorial reasoning. The performance of our algorithm in practice is thus mainly limited by the efficiency of minc_d, which, as we discuss next, can be efficiently implemented for numbers and strings.

5.2   A Datatype Handler for Numbers

To implement a datatype handler for the *owl:real* datatype of OWL 2, here abbreviated as *real*, we devise an efficient representation of the interpretation of a facet expression ψ. In particular, we represent fragments of this interpretation as *intervals* of the form $t[l, u]$, $t[l, u)$, $t(l, u]$, and $t(l, u)$, where $t \in \{real, int, \overline{int}\}$ determines the *type* of the interval, l is either $-\infty$ or a real number, and u is either $+\infty$ or a real number, such that $l \leq u$. Such an interval represents the set of all real numbers of type t between l and u; the round parenthesis means that the end-point is not included, and the square parenthesis means that the end-point is included in the set. By taking into account that $int \cap \overline{int} = \emptyset$, it is straightforward to define the intersection $\alpha \cap \beta$ of intervals α and β.

We can now represent the interpretation of each facet expression ψ using a set of intervals S_ψ, inductively defined as follows:

$$
\begin{aligned}
S_{<w} &= \{real(-\infty, w)\} & S_{>w} &= \{real(w, +\infty)\} \\
S_{\leq w} &= \{real(-\infty, w]\} & S_{\geq w} &= \{real[w, +\infty)\} \\
S_{int} &= \{int(-\infty, +\infty)\} & S_{\neg\psi} &= \{real(-\infty, +\infty) \cap \alpha \mid \alpha \in S_\psi\} \\
S_{\psi_1 \vee \psi_2} &= S_{\psi_1} \cup S_{\psi_2} & S_{\psi_1 \wedge \psi_2} &= \{\alpha \cap \beta \mid \alpha \in S_{\psi_1} \text{ and } \beta \in S_{\psi_2}\}
\end{aligned}
$$

In practice, it is beneficial to ensure that each S_ψ does not contain overlapping, empty, or adjoining intervals; the sets S_ψ can then be efficiently implemented by storing interval end-points in a sorted array.

The function $\mathsf{minc}_{real}(real[\varphi], n)$ can then be implemented by computing S_φ, checking whether it consists only of finite integer-restricted intervals, and if so, comparing their total length to n. Similarly, the function $\mathsf{enu}_{real}(real[\varphi])$ can be implemented by computing S_φ, checking whether it consists of only finite integer-restricted intervals, and if so, enumerating all the relevant integers. The function $\mathsf{in}_{real}(c, real[\varphi])$ can be implemented in a straightforward manner if φ is a facet, and it can be computed recursively for φ a general expression by taking into account the standard semantics of propositional connectives. Finally, the function $\mathsf{eq}_{real}(c_1, c_2)$ can be implemented by normalizing the lexical representation of c_1 and c_2 and then comparing the result.

5.3   A Datatype Handler for Strings

We now discuss the implementation of the datatype handler for the *xsd:string* datatype of OWL 2, here abbreviated as *str*. The function $\mathsf{eq}_{str}(c_1, c_2)$ can be implemented as an identity. The function $\mathsf{in}_{str}(d[\varphi], c)$ can be implemented in a straightforward manner if φ is a facet (if φ is a regular expression, membership of a string in a regular language can be checked as in [6]), and it can be computed recursively for a general expression φ in the obvious way.

The implementations of $\mathsf{minc}_{str}(d[\varphi], n)$ and $\mathsf{enu}_{str}(d[\varphi])$ differ based on the facets used in φ. If φ contains no regular expressions—that is, if the only restrictions are on the length of the string—then the intervals of allowed string lengths can be computed as in Section 5.2. If φ contains regular expressions, then each of the facets in φ can be represented using a finite state automaton [6]. (Note that the languages of all strings longer or shorter than some integer are regular.) Regular languages are closed under Boolean connectives so using, say, the results from [6], one can compute a finite state automaton $\mathcal{A}_\varphi$ accepting the language $\mathcal{L}_\varphi$ of φ. The next step is to test whether $\mathcal{L}_\varphi$ is finite. This can be done by identifying states that can occur on a path between the starting and the accepting states of $\mathcal{A}_\varphi$ and checking whether these states can occur in a loop. If the language is finite, it can be enumerated by identifying all finite paths between the starting and the accepting states of $\mathcal{A}_\varphi$.

We note that checking emptiness for intersections of regular languages is known to be PSPACE-complete [8]. This source of complexity, however, has been well studied in the literature and several optimization techniques are available.

6   Conclusion

In this paper, we have formalized the datatype system of OWL 2 and have discussed some nontrivial consequences of our definitions. Furthermore, we have discussed the normative datatypes listed in the current OWL 2 Working Draft and have proposed some modifications to the list. Finally, we have presented a general algorithm for datatype checking—the basic reasoning problem involving datatypes. Our algorithm is applicable to any datatype for which a small set of basic operations can be implemented. We have also discussed how to realize these operations for strings and numbers. The main challenge for our future work is to implement these algorithms in our reasoner and test them in practice.

References

1. Baader, F., Calvanese, D., McGuinness, D., Nardi, D., Patel-Schneider, P.F. (eds.): The Description Logic Handbook: Theory, Implementation and Applications, 2nd edn. Cambridge University Press, Cambridge (2007)
2. Baader, F., Hanschke, P.: A Scheme for Integrating Concrete Domains into Concept Languages. In: Proc. IJCAI 1991, Sydney, Australia, pp. 452–457 (1991)
3. Biron, P.V., Malhotra, A.: XML Schema Part 2: Datatypes, 2nd edn. (October 28, 2004), http://www.w3.org/TR/xmlschema-2/
4. Haarslev, V., Möller, R.: RACER System Description. In: Goré, R.P., Leitsch, A., Nipkow, T. (eds.) IJCAR 2001. LNCS (LNAI), vol. 2083, pp. 701–706. Springer, Heidelberg (2001)
5. Haarslev, V., Möller, R., Wessel, M.: The Description Logic $ALCNH_{R+}$ Extended with Concrete Domains: A Practically Motivated Approach. In: Goré, R.P., Leitsch, A., Nipkow, T. (eds.) IJCAR 2001. LNCS (LNAI), vol. 2083, pp. 29–44. Springer, Heidelberg (2001)
6. Hopcroft, J.E., Motwani, R., Ullman, J.D.: Introduction to Automata Theory, Languages, and Computation, 2nd edn. Addison-Wesley, Reading (2000)
7. Horrocks, I., Sattler, U.: Ontology Reasoning in the $\mathcal{SHOQ}(\mathbf{D})$ Description Logic. In: Proc. IJCAI 2001, Seattle, WA, USA, pp. 199–204 (2001)
8. Kozen, D.: Lower Bounds for Natural Proof Systems. In: FOCS 1977, pp. 254–266. Providence, RI, USA (1977)
9. Lutz, C.: The Complexity of Reasoning with Concrete Domains. PhD thesis, Teaching and Research Area for Theoretical Computer Science, RWTH Aachen, Germany (2002)
10. Lutz, C., Areces, C., Horrocks, I., Sattler, U.: Keys, Nominals, and Concrete Domains. Journal of Artificial Intelligence Research 23, 667–726 (2005)
11. Pan, J.Z., Horrocks, I.: Web Ontology Reasoning with Datatype Groups. In: Fensel, D., Sycara, K.P., Mylopoulos, J. (eds.) ISWC 2003. LNCS, vol. 2870, pp. 47–63. Springer, Heidelberg (2003)
12. Pan, J.Z., Horrocks, I.: OWL-Eu: Adding customised datatypes into OWL. Journal of Web Semantics 4(1), 29–39 (2006)
13. Patel-Schneider, P.F., Hayes, P., Horrocks, I.: OWL Web Ontology Language: Semantics and Abstract Syntax, W3C Recommendation, (February 10, 2004), http://www.w3.org/TR/owl-semantics/
14. Rossi, F., van Beek, P., Walsh, T.: Handbook of Constraint Programming. Elsevier Science Inc., New York (2006)

Laconic and Precise Justifications in OWL

Matthew Horridge, Bijan Parsia, and Ulrike Sattler

School of Computer Science
The University of Manchester
Oxford Road
Manchester
M13 9PL

Abstract. A justification for an entailment in an OWL ontology is a minimal subset of the ontology that is sufficient for that entailment to hold. Since justifications respect the syntactic form of axioms in an ontology, they are usually neither *syntactically* nor *semantically* minimal. This paper presents two new subclasses of justifications—*laconic justifications* and *precise justifications*. Laconic justifications only consist of axioms that do not contain any superfluous "parts". Precise justifications can be derived from laconic justifications and are characterised by the fact that they consist of flat, small axioms, which facilitate the generation of semantically minimal repairs. Formal definitions for both types of justification are presented. In contrast to previous work in this area, these definitions make it clear as to what exactly "parts of axioms" are. In order to demonstrate the practicability of computing laconic, and hence precise justifications, an algorithm is provided and results from an empirical evaluation carried out on several published ontologies are presented. The evaluation showed that laconic/precise justifications can be computed in a reasonable time for entailments in a range of ontologies that vary in size and complexity. It was found that in half of the ontologies sampled there were entailments that had more laconic/precise justifications than regular justifications. More surprisingly it was observed that for some ontologies there were fewer laconic justifications than regular justifications.

1 Introduction

Since the Web Ontology Language, OWL, became a W3C standard, there has been a notable increase in the number of people building, extending and using ontologies. As a result of this, a large number of people have been enticed into using some kind of description logic reasoning service as an aid during the construction and deployment of ontologies. As people have gained in confidence, they have begun to move from creating or using modestly expressive ontologies, through to using richly axiomatised ontologies that exercise the full expressivity of OWL-DL. Experience of delivering a variety of tools to a wide range of users has made it evident that there is a significant demand for editing environments that provide sophisticated editing and browsing support services. In particular, the generation of *justifications* [1] for entailments is now recognised as highly desirable functionality for both ontology development and ontology reuse. A clear

A. Sheth et al. (Eds.): ISWC 2008, LNCS 5318, pp. 323–338, 2008.

demonstration of the need for *practical* explanation services that provide justifications was demonstrated by the fact that many users switched from Protégé 3.2 to Swoop purely for the benefits of automatic explanation facilities.

The ability to obtain justifications for entailments was originally exposed to the masses in the ontology editing and browsing tool Swoop [2]. Since then, other OWL tools, such as Protégé-4 [3], OWLSight,[1] and Top Braid Composer[2] have adopted the ability to generate these justifications, showing the importance of explanations to end users.

Intuitively, a justification is a set of axioms from an ontology that is sufficient for an entailment to hold. A key aspect of justifications is that they operate on the level of *asserted* axioms. That is, the axioms in a justification directly correspond to axioms that have been asserted in the ontology in which the entailment holds. Therefore, if an ontology contains "long" axioms, for example, ones containing many complex class expressions, then there may be *parts* of axioms in a justification that are not required for the entailment that is supported by the justification. For example, consider the set of axioms $\mathcal{J} = \{A \sqsubseteq B \sqcap C, C \sqsubseteq D, A \sqsubseteq \neg D\}$ which entails $A \sqsubseteq \bot$. The right hand side of first axiom in the set contains the conjunct B. However, this conjunct does not play any part in causing A to be unsatisfiable. In this sense, if this set of axioms is a justification for $A \sqsubseteq \bot$ then it could be more "fine-grained"—it should be somehow possible to indicate that only "part" of the first axiom is required for the entailment $A \sqsubseteq \bot$ to hold. Justifications that only contain parts of axioms that are relevant for the entailment to hold, have been referred to as "fine-grained" justifications [4], and also "precise justifications" [5].

While there is a general feeling that fine-grained justifications should only consist of the *parts of axioms* that are relevant to the entailment in question, there have not been any attempts to produce a rigorous formalisation of these kinds of justifications. This means that while it is cut and dried as to what exactly a justification is, the same cannot be said for fine-grained justifications. In particular, it is not clear what it means to talk about *parts of axioms*. Not only does this make it difficult for implementers to be sure they can generate fine-grained justifications in a sound and complete manner, it also makes it difficult to compare the approach taken by one system in generating fine-grained justifications to the approach taken by other systems.

The purpose of this paper is to identify the desirable characteristics of fine-grained justifications, propose a formal definition for these types of justifications, and show how they can be computed for OWL-DL. In order to demonstrate that computing these justifications according to this definition is feasible, an algorithm is provided, which is evaluated on a sample of published ontologies.

2   Preliminaries

Throughout this paper, the following nomenclature is used.

---

[1] http://pellet.owldl.com/ontology-browser/
[2] http://www.topbraid.org

α	an axiom; subscripts and primes are used to denote different axioms
$|\alpha|$	refers to the length of α
$\mathcal{O}$	an ontology
$\delta(\mathcal{O})$	a set of axioms that is the result of a structural transformation on $\mathcal{O}$
η	an arbitrary entailment that is assumed to hold in some ontology or set of axioms
$\mathcal{O}^*$	the deductive closure of an ontology $\mathcal{O}$
$(\delta(\mathcal{O}))^*$	the deductive closure of the structural transformation of $\mathcal{O}$
$\mathcal{O}^+$	a subset of the deductive closure of an ontology $\mathcal{O}$
$\mathcal{S}$	a set of sets of axioms
$\mathcal{J}$	a justification
$\mathcal{L}$	a description logic, e.g. $\mathcal{ALC}$, $\mathcal{SHOIQ}$

A, B, C, D, E are used as concept names, R and S as role names, n and n' are place holders for positive integers. $\mathcal{T}$ refers to a T-Box, $\mathcal{R}$ a Role-Box and $\mathcal{A}$ an A-Box.

Given an ontology, $\mathcal{O}$, and a description logic $\mathcal{L}$ the *deductive closure* of $\mathcal{O}$, is written as $\mathcal{O}_{\mathcal{L}}^{\star}$, where $\mathcal{O}_{\mathcal{L}}^{\star} = \{\alpha \in \mathcal{L} \mid \mathcal{O} \models \alpha\}$. In other words the deductive closure contains *all* well formed $\mathcal{L}$-axioms that are entailed by the ontology $\mathcal{O}$. When it is clear from the context, the subscript $\mathcal{L}$ is dropped.

An axiom α' is said to be *weaker* than another axiom, α iff $\alpha \models \alpha'$ and $\alpha' \not\models \alpha$.

OWL and Description Logics. This paper focuses on OWL-DL or its rough syntactic variant $\mathcal{SHOIN}(\mathcal{D})$ [6], but the approach can be applied to other description logics such as $\mathcal{SROIQ}$, which will underpin the next version of OWL.

For the purposes of this paper, an *ontology* is regarded as a finite set of $\mathcal{SHOIN}$ axioms $\{\alpha_0, \dots, \alpha_n\}$. An axiom is of the form of $C \sqsubseteq D$ or $C \equiv D$, where C and D are (possibly complex) concept descriptions, or $S \sqsubseteq R$ or $S \equiv R$ where S and R are (possibly inverse) roles. OWL contains a significant amount of syntactic sugar, such as $DisjointClasses(C, D)$, $FunctionalObjectProperty(R)$ or $Domain(R, C)$, however, these kinds of axioms can be represented using subclass axioms [6].

Justifications. A justification [1,7,8] for an entailment in an ontology is a *minimal* set of axioms from the ontology that is sufficient for the entailment to hold. The set is minimal in that the entailment does not follow from any proper subset of the justification. It should be noted that there may be several, potentially overlapping, justifications for a given entailment.

Definition 1 (Justification). *For an ontology $\mathcal{O}$ and an entailment η where $\mathcal{O} \models \eta$, a set of axioms $\mathcal{J}$ is a justification for η in $\mathcal{O}$ if $\mathcal{J} \subseteq \mathcal{O}$, $\mathcal{J} \models \eta$ and if $\mathcal{J}' \subsetneq \mathcal{J}$ then $\mathcal{J}' \not\models \eta$.*

3   Motivation for Fine-Grained Justifications

Justifications have proved to be very useful in general. However, there are at least four reasons that motivate the investigation of fine-grained justifications:

1. An axiom in a justification can contain irrelevant parts. Consider $\mathcal{O} = \{B \sqsubseteq C \sqcap D, D \sqsubseteq E\} \models B \sqsubseteq E$. Clearly, $\mathcal{O}$ is a justification for $B \sqsubseteq E$, but the first conjunct in the first axiom is irrelevant for this entailment and might *distract* a user from identifying the relevant parts. It is arguable that focusing a user's attention on the relevant parts of an axiom can make it easier for them to understand a justification.

2. A justification can conceal relevant information. Consider $\mathcal{O} = \{B \sqsubseteq \neg C \sqcap D, B \sqsubseteq C \sqcap \neg D\} \models B \sqsubseteq \bot$. B is unsatisfiable for two distinct reasons, but there is only a single justification for it (containing both axioms from $\mathcal{O}$). If this entailment is broken by deleting one of the axioms then modelling errors could be introduced—it may be that the repaired ontology should contain $B \sqsubseteq D \sqcap \neg C$. We refer to this condition as *internal masking*.

3. Justifications can mask relevant axioms. Consider $\mathcal{O} = \{B \sqsubseteq D \sqcap \neg D \sqcap C, B \sqsubseteq \neg C\} \models B \sqsubseteq \bot$. There is no justification in $\mathcal{O}$ for $B \sqsubseteq \bot$ that includes $\{B \sqsubseteq \neg C\}$, yet, it clearly plays a role in entailing the unsatisfiability. Users working with justifications would most likely be unaware of this. We refer to this condition as *external masking*

4. Multiple justifications can conceal a fine-grained core. In certain cases there may be multiple justifications for an entailment but fewer fine-grained justifications for the same entailment. Consider $\{A \sqsubseteq B \sqcap C, B \sqsubseteq D\}$ and $\{A \sqsubseteq B \sqcap F, B \sqsubseteq D\}$ as two justifications for $A \sqsubseteq B$. There is just one fine-grained justification: $\{A \sqsubseteq B, B \sqsubseteq D\}$. Besides making the entailment easier to understand, this scenario might also indicate modelling errors or redundancies.

A common point running through all of the above is the issue of repair. Since a justification is a subset of an ontology, and consists of *asserted* axioms, it is relatively straightforward, and intuitive, to devise a repair for the ontology that breaks the entailment in question: Given some undesired entailment η that holds in an ontology $\mathcal{O}$, and a set of justifications $\mathfrak{J}$ for η, a simple method of breaking the entailment is to choose one axiom from each justification $\mathcal{J} \in \mathfrak{J}$ and remove these chosen axioms from $\mathcal{O}$. However, from the above examples, it should be fairly clear that when working with "regular" justifications there is a potential to "over repair" an ontology so that more entailments are lost than is necessary. In this sense, it is desirable that any definition of fine-grained justifications should result in justifications that make it easy to devise and enact a minimal and consistent repair of an ontology. Ideally, the underlying repair process should mimic the intuitive process of repair when working with regular justifications— one axiom from each fine-grained justification should be identified as a candidate for removal from the ontology or some suitable variant.

4   Related Work

In [5], Kalyanpur et al. propose an algorithm for computing "precise" justifications. The algorithm rewrites axioms into smaller axioms in order to obtain the relevant "parts".

The ontology editor Swoop [2] features the ability to "strike out" irrelevant parts of axioms in a justification. However, this is based on a heuristic approach, and while it is very efficient, and is strongly expected to be sound, it is incomplete.

In [4], Lam presents "fine grained" justifications for $\mathcal{ALC}$ with general TBoxes. A tableaux reasoning technique, which is an extension of the technique from Meyer et al. [9] and Baader and Hollunder [10] is used.

In [8], Schlobach and Cornet focus on computing explanations for unsatisfiable classes. They define the concepts of *MUPS* (Minimal Unsatisfiability Preserving Sub-TBoxes) and MIPS (Minimal Incoherence Preserving Sub-TBoxes), which are special cases of justifications. Schlobach and Cornet describe a procedure that *syntactically generalises* all of the axioms in each MIPS to produce a generalised TBox that contains smaller axioms which are responsible for any unsatisfiable classes.

Deng et al. [11] take a novel approach to "measuring inconsistencies in description logic ontologies" by using results obtained from Game Theory. Although no details are provided, Deng claims that the technique could easily be extended to pinpoint the "proportions" of axioms that are responsible for an unsatisfiable class, via the use of clause pinpointing.

Finally, in [12] Baader et al. pinpoint axioms for entailments in the description logic $\mathcal{EL}$. Although the work is not concerned with fine-grained justifications, the $\mathcal{EL}$ subsumption algorithm uses a normalisation procedure that flattens axioms and makes them smaller. These smaller axioms could be used to indicate the parts of axioms responsible for an entailment.

A common aim of all previous approaches for computing fine-grained justifications is to determine the *parts* of axioms that are responsible for a particular entailment. However, none of these approaches define exactly what they mean by *parts* of axioms. Moreover, each approach is specific to a particular implementation technique and is defined in an *operational* sense. This means that it is generally unclear as to what exactly constitutes a fine-grained justification. As a consequence, it is unclear as to whether any one approach for computing fine-grained justifications would result in the obtaining the same set of fine-grained justifications for an entailment when compared to another approach.

In summary, a *general definition of fine-grained justifications* is needed. Ideally, such a definition would not be tied to a particular DL. This definition would then permit the evaluation and comparison of algorithms for computing fine-grained or "precise" justifications and, it would make it possible to investigate the underlying problem in a thorough way.

5   Laconic and Precise Justifications Defined

There appear to several desirable properties that a definition for fine-grained justifications should satisfy. In particular:

- **Minimality** Each axiom in a fine-grained justification should, in some sense, be as *small* as possible—each axiom should only capture the parts of the asserted form that are required for the entailment in question to hold.

- **Repair** As with regular justifications, fine-grained justifications should suggest as simple a repair as possible. Ideally, removing an axiom from a fine-grained justification should generate a repair that is minimal in terms of lost entailments.

In what follows a definition of fine-grained justifications is proposed. This definition consists of two parts: 1) a definition of what we term *laconic justifications*, which informally are justifications that do not contain any superfluous parts; 2) a definition of what we term *precise* justifications, that can be derived from laconic justifications, and are such that their axioms are flat, small and semantically minimal. Precise justifications are primarily geared towards repair.

5.1 δ–The Structural Transformation

The definition of laconic justifications below uses $\delta(\mathcal{J})$, where δ is a satisfiability preserving *structural transformation* on $\mathcal{J}$ that removes all nested descriptions and hence produces axioms that are as small and flat as possible. An appropriate transformation, a version of which is shown below, is the well known structural transformation described in Plaisted and Greenbaum [13] and used in [14].

$$\delta(\mathcal{O}) := \bigcup_{\alpha \in \mathcal{R} \cup \mathcal{A}} \delta(\alpha) \cup \bigcup_{C_1 \sqsubseteq C_2 \in \mathcal{T}} \delta(\top \sqsubseteq \mathsf{nnf}(\neg C_1 \sqcup C_2))$$

$$\delta(D(a)) := \delta(\top \sqsubseteq \neg\{a\} \sqcup \mathsf{nnf}(D))$$

$$\delta(\top \sqsubseteq \mathbf{C} \sqcup D) := \delta(\top \sqsubseteq A'_D \sqcup \mathbf{C}) \cup \bigcup_{i=1}^{n} \delta(A'_D \sqsubseteq D_i) \text{ for } D = \prod_{i=1}^{n} D_i$$

$$\delta(\top \sqsubseteq \mathbf{C} \sqcup \exists R.D) := \delta(\top \sqsubseteq A_D \sqcup \mathbf{C}) \cup \{A_D \sqsubseteq \exists R.A'_D\} \cup \delta(A'_D \sqsubseteq D)$$

$$\delta(\top \sqsubseteq \mathbf{C} \sqcup \forall R.D) := \delta(\top \sqsubseteq A_D \sqcup \mathbf{C}) \cup \{A_D \sqsubseteq \forall R.A'_D\} \cup \delta(A'_D \sqsubseteq D)$$

$$\delta(\top \sqsubseteq \mathbf{C} \sqcup {\geq} nR.D) := \delta(\top \sqsubseteq A_D \sqcup \mathbf{C}) \cup \{A_D \sqsubseteq {\geq} nR.A'_D\} \cup \delta(A'_D \sqsubseteq D)$$

$$\delta(\top \sqsubseteq \mathbf{C} \sqcup {\leq} R.D) := \delta(\top \sqsubseteq A_D \sqcup \mathbf{C}) \cup \{A_D \sqsubseteq {\leq} nR.A'_D\} \cup \delta(A'_D \sqsubseteq D)$$

$$\delta(A'_D \sqsubseteq D) := \delta(A'_D \sqsubseteq D) \text{ (If } D \text{ is of the form } A \text{ or } \neg A)$$

$$\delta(A'_D \sqsubseteq D) := \delta(\top \sqsubseteq \neg A'_D \sqcup D) \text{ (If } D \text{ is } \textit{not} \text{ of the form } A \text{ or } \neg A)$$

$$\delta(\beta) := \beta \text{ for any other axiom}$$

Note. A is an atomic concept in the signature of $\mathcal{O}$, A_D and A'_D are fresh concept names that are not in the signature of $\mathcal{O}$. C_i and D are arbitrary concepts, excluding $\top$, $\bot$ and literals of the form X or $\neg X$ where X is not in the signature of $\mathcal{O}$, $\mathbf{C}$ is a possibly empty disjunction of arbitrary concepts. $C \equiv D$ is syntactic sugar for $C \sqsubseteq D$ and $D \sqsubseteq C$, as is $=nR.D$ for ${\geq} nR.D \sqcap {\leq} nR.D$. Domain and range axioms are GCIs so that $Domain(R, C)$ means $\exists R.\top \sqsubseteq C$, and $Range(R, C)$ means $\top \sqsubseteq \forall R.C$. The negation normal form of D is $\mathsf{nnf}(D)$.

The transformation ensures that concept names that are in the signature of $\mathcal{O}$ only appear in axioms of the form $X \sqsubseteq A$ or $X \sqsubseteq \neg A$, where X is some concept name *not* occurring in the signature of $\mathcal{O}$. Note that the structural transformation does not use structure sharing[3].

---

[3] For example, given $\top \sqsubseteq C \sqcup \exists R.C$, two new names should be introduced, one for each use of C, to give $\{\top \sqsubseteq X_0 \sqcup X_1, X_1 \sqsubseteq \exists R.X_2, X_2 \sqsubseteq C\}$. The preclusion of structure sharing ensures that the different positions of C are captured.

5.2   Axiom Length

The definition of laconic justifications uses the notion of the length of an axiom. Length is defined as follows: For X, Y a pair of concepts or roles, A a concept name, and R a role, the length of an axiom is defined as follows:

$$|X \sqsubseteq Y| := |X| + |Y|, \quad |X \equiv Y| := 2(|X| + |Y|), \quad |Sym(R)| = |Trans(R)| := 1,$$

where

$$|\top| = |\bot| := 0,$$
$$|A| = |\{i\}| = |R| := 1,$$
$$|\neg C| := |C|$$
$$|C \sqcap D| = |C \sqcup D| := |C| + |D|$$
$$|\exists R.C| = |\forall R.C| = | \geq nR.C| = | \leq nR.C| := 1 + |C|$$

Note. This definition is slightly different from the usual definition, but it allows cardinality axioms such as $A \sqsubseteq \, \leq 2R.C$ to be weakened to $A \sqsubseteq \, \leq 3R.C$ without increasing the length of the axiom.

5.3   Laconic Justifications

With a suitable structural transformation, δ, and the notion of axiom length in hand, laconic justifications can be defined. (Recall that $\mathcal{O}^{\star}$ is the deductive closure of $\mathcal{O}$)

Definition 2. *(**Laconic Justification**) Let $\mathcal{O}$ be an ontology such that $\mathcal{O} \models \eta$. $\mathcal{J}$ is a laconic justification for η over $\mathcal{O}$:*

1. *$\mathcal{J}$ is a justification for η in $\mathcal{O}^{\star}$*
2. *$\delta(\mathcal{J})$ is a justification for η in $(\delta(\mathcal{O}))^{\star}$*
3. *For each $\alpha \in \delta(\mathcal{J})$ there is no α' such that*
   (a) *$\alpha \models \alpha'$ and $\alpha' \not\models \alpha$*
   (b) *$|\alpha'| \leq |\alpha|$*
   (c) *$\delta(\mathcal{J}) \setminus \{\alpha\} \cup \delta(\{\alpha'\})$ is a justification for η in $(\delta(\mathcal{O}))^{\star}$*

Intuitively, a laconic justification is a justification where all axioms only contain sub-concepts (parts) that are relevant for the entailment in question, and moreover, these sub-concepts (parts) are *as weak as possible*.

5.4   Notes on Definition 2

$\mathcal{O}^{\star}$—**the deductive closure of $\mathcal{O}$.** It is apparent from Definition 2(1) that laconic justifications for an entailment in an ontology may be drawn from the deductive closure of that ontology. Therefore, unlike regular justifications, laconic justifications are not specific to the asserted axioms in an ontology. This ensures that it is possible to capture the internal and external masking cases highlighted in Section 3.

δ—a structural transformation. The primary use of a structural transformation in Definition 2 is to transform a justification into an equi-satisfiable set of axioms, $\delta(\mathcal{J})$, where each axiom does not have any nested complex descriptions (each axiom is "flattened out") and each axiom is as small as possible. These axioms can be thought of as a representation of all of the "parts" of the axioms in $\mathcal{J}$. Thus, ensuring that the axioms in $\delta(\mathcal{J})$ are as weak as possible ensures that all "parts" of axioms in a laconic justification are as weak as possible.

Applying the transformation to a justification results in two possibilities:

1. $\delta(\mathcal{J})$ is *not* a justification for η—it is a superset of a justification for η because $\mathcal{J}$ consists of axioms that contains parts or strong parts that are not required for $\mathcal{J} \models \eta$. Hence condition 2 of Definition 2 is violated and $\mathcal{J}$ is not laconic.
2. $\delta(\mathcal{J})$ *is* a justification for η, which implies that all sub-concepts of all axioms (in their existing or in a weakened form) are required for $\mathcal{J} \models \eta$. Hence, if each axiom in $\delta(\mathcal{J})$ is as weak as possible, as dictated by condition 3 of Definition 2, then $\mathcal{J}$ is laconic.

Example 1. Consider the following ontology $\mathcal{O} = \{\alpha_1 \colon A \sqsubseteq B, \alpha_2 \colon B \sqsubseteq D, \alpha_3 \colon A \sqsubseteq B \sqcap C\} \models A \sqsubseteq D$. There are two justifications for $\mathcal{O} \models A \sqsubseteq D$, $J_1 = \{\alpha_1, \alpha_2\}$ and $J_2 = \{\alpha_2, \alpha_3\}$. By Definition 2, J_1 *is* a laconic justification since $\delta(J_1) = \{\top \sqsubseteq X_0 \sqcup X_1, X_0 \sqsubseteq \neg A, X_1 \sqsubseteq B, \top \sqsubseteq X_2 \sqcup X_3, X_2 \sqsubseteq \neg B, X_3 \sqsubseteq D\}$ neither of these axioms can be weakened further without lengthening them or without resulting in $\mathcal{J}_1 \not\models A \sqsubseteq D$. Conversely, J_2 *is not* a laconic justification since after performing the structural transformation to obtain $\delta(\mathcal{J})$ a superset of a justification for η is obtained.

Definition 3. *(Precise Justification)*
Let $\mathcal{O}$ be an ontology such that $\mathcal{O} \models \eta$. Let $\mathcal{J}$ be a justification for $\mathcal{O} \models \eta$ and let $\mathcal{J}' = \delta(\mathcal{J})$. $\mathcal{J}'$ is precise with respect to $\mathcal{J}$ if $\mathcal{J}$ is a laconic justification for $\mathcal{O} \models \eta$.

Intuitively, a precise justification is a version of a laconic justification where the axioms contained in the precise justification are as flat, small and as weak as possible. In essence, a precise justification is a "repair friendly" version of a laconic justification. Note that *a precise justification is precise with respect to a laconic justification*—a justification cannot in itself be precise.

Lemma 1. *All laconic justifications can be converted to precise justifications by means of the structural transformation, δ.*

5.5   Repair

Although "repair" is not the primary subject of this paper, it should be noted that the motivation behind Definition 3 is based on the notion that it should be possible to generate a repair that makes semantically minimal changes to an ontology. Because, by definition, a precise justification contains axioms that are maximally flat, small and weak, it is only necessary to remove one of these axioms in order to generate a minimal repair. A semantically minimal repair will be generated if an axiom of the form $X \sqsubseteq A$ or $X \sqsubseteq \neg A$, where X is any named introduced in the structural transformation and A is a concept name occurring in the signature of $\mathcal{O}$, is removed (hence the requirement that δ produces axioms where concept names from the signature of $\mathcal{O}$ only occur in the aforementioned pattern).

In order to generate a semantically minimal repair for an entailment, laconic justifications for the entailment should be first generated, then precise justifications for these laconic justifications should be computed. The precise justifications can then be used to guide the process of axiom selection to indicate the parts of asserted axioms that should be removed.

6   Filtering Laconic Justifications

Since laconic justifications are defined with respect to the deductive closure of an ontology, it is not difficult to see that there could be many laconic justifications for an entailment. For example, given $\mathcal{O} = \{C \sqsubseteq D \sqcap \neg D \sqcap E, A \sqsubseteq B\}$, laconic justifications for $\mathcal{O} \models C \sqsubseteq \bot$ include $\{C \sqsubseteq D \sqcap \neg D\}$ and $\{C \sqsubseteq D, C \sqsubseteq \neg D\}$. It is noticeable that both of these justifications are somewhat structurally or syntactically related to the asserted axioms in $\mathcal{O}$. However, Definition 2, also admits other laconic justifications such as $\{C \sqsubseteq B \sqcap \neg B\}$ (since $C \sqsubseteq B \sqcap \neg B$ is also in the deductive closure of $\mathcal{O}$). Despite the fact that this is a valid laconic justification, it is arguable that justifications of this ilk, which could be considered to be syntactically irrelevant or "incidental", are not of general interest to an ontology modeller who is trying to understand the reasons for an entailment.

In order to focus on syntactically relevant laconic justifications, a filter on the deductive closure, called $\mathcal{O}^+$, is introduced. $\mathcal{O}^+$ is essentially a *representative* of the deductive closure of an ontology, which gives rise to *preferred* laconic justifications.

Definition 4. *($\mathcal{O}^+$ completeness) Let $\mathcal{O}$ be an ontology, $\mathcal{O}^+$ a set of axioms such that $\mathcal{O} \subseteq \mathcal{O}^+ \subseteq \mathcal{O}^\star$, and η such that $\mathcal{O} \models \eta$. $\mathcal{O}^+$ is complete for η and $\mathcal{O}$ if for $\mathfrak{J}$ the set of all laconic justifications for η w.r.t. $\mathcal{O}^+$, for any $\mathcal{O}'$ such that $\mathcal{O}' \subseteq \mathcal{O}^+$, if $\mathcal{O}' \not\models \mathcal{J}$ for all $\mathcal{J} \in \mathfrak{J}$, then $\mathcal{O}' \not\models \eta$.*

Completeness ensures that the laconic justifications can be used for a simple repair: if an ontology is weakened so that it entails none of the laconic justifications, then it no longer entails η.

The exact details of how to construct a suitable $\mathcal{O}^+$ depend somewhat on how laconic justifications will be used. For example, an application that presents laconic justifications to end users may well prefer justifications that maintain conjunctions or disjunctions. For example given $\mathcal{O} = \{C \sqsubseteq D \sqcap \neg D \sqcap E\}$, it might be preferable to generate $\{C \sqsubseteq D \sqcap \neg D\}$ as opposed to $\{C \sqsubseteq D, C \sqsubseteq \neg D\}$ as a laconic justification for $C \sqsubseteq \bot$. The reverse may be true if generating laconic justifications to display in a repair tool, where smaller axioms might suggest a more appropriate repair.

In what follows an $\mathcal{O}^+$ is specified for the description logic $\mathcal{SHOIQ}$ $(\mathcal{O}^+_{\mathcal{SHOIQ}})$ so as to capture the sort of laconic (fine-grained) OWL-DL justifications in play in the literature. Importantly, $\mathcal{O}^+_{\mathcal{SHOIQ}}$ produces laconic justifications that are syntactically relevant—there is a direct correspondence between asserted axioms in $\mathcal{O}$ and axioms that appear in laconic justifications, which is essential for usability in tools such as browsers and editors, and directly corresponds with the strikeout feature that is available in Swoop. It should be noted that $\mathcal{O}^+_{\mathcal{SHOIQ}}$ is finite, which means that the set of precise justifications with respect to $\mathcal{O}^+_{\mathcal{SHOIQ}}$ is also finite. Even though it is possibly exponentially larger than $\mathcal{O}$, it will later be seen that it is not necessary to compute $\mathcal{O}^+_{\mathcal{SHOIQ}}$ in its entirety.

Definition 5. *Let* $\mathcal{O}^+_{\mathcal{SHOIQ}} = \{\alpha' \mid \alpha' \in \sigma(\alpha) \text{ where } \alpha \in \mathcal{O}\}$. *We define the mappings* $\sigma(\alpha)$, $\tau(C)$ *and* $\beta(C)$ *inductively as follows, where* $X \in \{\tau, \beta\}$ *is used as a meta-variable with* $\overline{\tau} = \beta$, $\overline{\beta} = \tau$, $\max_\tau = \top$, $\max_\beta = \bot$, *and* n' *be the maximum number in number restrictions in* $\mathcal{O}$:

$$\sigma(C_1 \sqcup \cdots \sqcup C_n \sqsubseteq D_1 \sqcap \cdots \sqcap D_m) := \{C_i' \sqsubseteq D_j' \mid C_i' \in \beta(C_i), D_j' \in \tau(D_j)\}$$
$$\sigma(C \equiv D) := \sigma(C \sqsubseteq D) \cup \sigma(D \sqsubseteq C) \cup \{C \equiv D\}$$
$$\sigma(R \sqsubseteq S) := \{R \sqsubseteq S\}$$
$$\sigma(R \equiv S) := \{R \sqsubseteq S\} \cup \{S \sqsubseteq R\}$$
$$\sigma(Trans(R)) := \{Trans(R)\}$$
$$X(A) := \{\max_X, A\} \text{ for } A \text{ a concept name or } \{i\}$$
$$X(C_1 \sqcap \cdots \sqcap C_n) := \{C_1' \sqcap \cdots \sqcap C_n' \mid C_i' \in X(C_i)\}$$
$$X(C_1 \sqcup \cdots \sqcup C_n) := \{C_1' \sqcup \cdots \sqcup C_n' \mid C_i' \in X(C_i)\}$$

$$X(\neg C) := \{\neg C' \mid C' \in \overline{X}(C)\}$$
$$X(\exists R.C) := \{\exists R.C' \mid C' \in X(C)\} \cup \{\top\}$$
$$X(\forall R.C) := \{\forall R.C' \mid C' \in X(C)\} \cup \{\top\}$$
$$X(\geq nR.C) := \{\geq mR.C' \mid C' \in X(C), n \leq m \leq n'\} \cup \{\max_X\}$$
$$\tau(\leq nR.C) := \{\leq mR.C' \mid C' \in \beta(C), 0 \leq m \leq n\} \cup \{\top\}$$
$$\beta(\leq nR.C) := \{\leq mR.C' \mid C' \in \tau(C), n \leq m \leq n'\} \cup \{\bot\}$$
$$X(\{j_1 \ldots j_n\}) := X(\{j_1\} \sqcup \cdots \sqcup \{j_n\})$$

In essence $\mathcal{O}^+_{\mathcal{SHOIQ}}$ generates weaker, shorter axioms from asserted axioms in $\mathcal{O}$ by, for example, stepwise replacement of sub-concepts with either $\top$ or $\bot$. In fact, $\mathcal{O}^+_{\mathcal{SHOIQ}}$ parallels the well know structural transformation without transforming axioms into negation normal form or clausal form. The benefit of this being a close syntactic correspondence of axioms in $\mathcal{O}^+_{\mathcal{SHOIQ}}$ with axioms in $\mathcal{O}$. From now on the subscript $\mathcal{SHOIQ}$ is dropped so that $\mathcal{O}^+$ refers to $\mathcal{O}^+_{\mathcal{SHOIQ}}$.

7 Computing Laconic Justifications

In order to compute laconic justifications for display in editors such as Swoop and Protégé-4, or for use in other tools such as automatic repair tools, it is necessary to compute the *preferred laconic justifications*. Recall that these particular kinds of justifications are computed with respect to a representative of the deductive closure of an ontology, namely $\mathcal{O}^+$. Therefore, one of the conceptually simplest methods of computing precise justifications for an entailment η in an ontology $\mathcal{O}$ would be to first compute $\mathcal{O}^+$ directly from $\mathcal{O}$ and then compute justifications with respect to $\mathcal{O}^+$. This would yield a set of justifications that is the *superset* of the set of *preferred* laconic justifications for $\mathcal{O} \models \eta$. The actual set of laconic justifications could then be obtained by some post processing of the justifications that were computed from $\mathcal{O}^+$. However, the size of $\mathcal{O}^+$ is exponential in the size of axioms in $\mathcal{O}$. Since for a given entailment not all axioms and their weakenings will participate in the laconic justifications that are obtained from $\mathcal{O}^+$, computing $\mathcal{O}^+$ for a whole ontology $\mathcal{O}$ could be regarded as being wasteful both in terms of space and time.

Algorithm. *ComputeLaconicJustifications*
Input: $\mathcal{O}$ an ontology, η and entailment such that $\mathcal{O} \models \eta$
Output: $\mathcal{S}$, a set of precise justifications for $\mathcal{O} \models \eta$
1. $\mathcal{S} \leftarrow ComputeOPlusJustifications(\mathcal{O}, \eta)$
2.   **for** $S \in \mathcal{S}$
3.   **if** $IsLaconic(S, \eta) = $ **false**
4. $\mathcal{S} \leftarrow \mathcal{S} \setminus S$
5.   **return** $\mathcal{S}$

Algorithm. *ComputeOPlusJustifications*
Input: $\mathcal{O}$ an ontology, η and entailment such that $\mathcal{O} \models \eta$
Output: $\mathcal{S}$, a set of justifications for $\mathcal{O}^+ \models \eta$
1. $\mathcal{O}' \leftarrow \mathcal{O}$
2. $\mathcal{S} \leftarrow \emptyset$
3. $\mathcal{S}' \leftarrow Justifications(\mathcal{O}', \eta)$
4.   **repeat**
5. $\mathcal{S} \leftarrow \mathcal{S}'$
6.   **for** $S \in \mathcal{S}$
7. $\mathcal{O}' \leftarrow (\mathcal{O}' \setminus S) \cup ComputeOPlus(S)$
8. $\mathcal{S}' \leftarrow Justifications(\mathcal{O}', \eta)$
9.   **until** $\mathcal{S} = \mathcal{S}'$
10. **return** $\mathcal{S}$

Algorithm. *IsLaconic*
Input: J, a justification, η and entailment such that $J \models \eta$
Output: **true** if J is laconic, otherwise **false**
1. $\mathcal{S} \leftarrow Justifications(ComputeOPlus(\delta(J)), \eta)$
2.   **return** $\mathcal{S} = \{\delta(\mathcal{J})\}$

It is also tempting to assume that the laconic justifications can efficiently be computed directly from regular justifications without reference to rest of the ontology. While it may be sufficient to utilise this strategy when implementing a strikeout feature similar to that found in Swoop, this approach would not capture all laconic justifications with respect to $\mathcal{O}^+$. As described in Section 3 point 3, the *external masking* condition means that there could be laconic justifications that would not be found using this technique.

With these points in mind, an optimised algorithm for computing precise justifications, *ComputeLaconicJustifications*, is presented below. Since the algorithm does not require a specific reasoner, or indeed a particular reasoning procedure such as tableau, it is a *Black-Box* algorithm. The algorithm *incrementally* computes the set of all laconic justifications for a given entailment by incrementally computing $\mathcal{O}^+$ from previously found justifications. This yields a set of justifications that is a superset of the laconic justifications for the entailment in question. The algorithm then processes each justification in this set to extract the justifications that are laconic justifications using the *IsLaconic* subroutine. This subroutine essentially tests whether a justification J is laconic by computing $ComputeOPlus(\delta(\mathcal{J}))$ and then computing laconic justifications from this set of axioms. If the justification is laconic then its singleton set will be equal to the justifications computed from $ComputeOPlus(\delta(\mathcal{J}))$.

The algorithm requires two main subroutines that are not defined below. *Justifications*, which computes the (regular) justifications for an entailment η that holds in some set of axioms (ontology). This subroutine can be implemented using any strategy that computes justifications in accordance with Definition 1. Additionally, the *ComputeOPlus* subroutine takes a set of axioms and returns a set of axioms that represents $\mathcal{O}^+$ computed from this set of axioms in accordance with Definition 5.

7.1   Performance

In order to evaluate the practicability of computing laconic justifications, the above algorithm and subroutines were implemented using the latest version of the OWL API[4] backed with the Pellet reasoner [15]. This API has clean and efficient support for manipulating an ontology at the level of axioms, and has a relatively efficient and direct wrapper for Pellet. A selection of publicly available ontologies, shown in Table 1 were chosen for number of entailments that hold in them and to provide a range of expressivity.[5]

Each ontology was classified in order to determine the unsatisfiable classes and atomic subsumptions. These kinds of entailments were selected as input to the compute laconic justifications algorithm because they are usually exposed by tools such as Protégé-4 or Swoop and are therefore the kinds of entailments that users typically seek justifications for. For each entailment the time to compute all regular justifications and all laconic justifications was recorded.

---

[4] http://owlapi.sourceforge.net
[5] All of the ontologies used may be found in the TONES ontology repository at http://owl.cs.manchester.ac.uk/repository. Entailments include atomic subsumptions and unsatisfiable classes.

Table 1. Ontologies used in experiment

ID	Ontology	Expressivity	Axioms No.	Entailments
1	Generations	$\mathcal{ALCOIF}$	38	24
2	Economy	$\mathcal{ALCH(D)}$	1625	51
3	People+Pets	$\mathcal{ALCHOIN}$	108	33
4	MiniTambis	$\mathcal{ALCN}$	173	66
5	Nautilus	$\mathcal{ALCHF}$	38	10
6	Transport	$\mathcal{ALCH}$	1157	62
7	University	$\mathcal{SOIN}$	52	10
8	PeriodicTableComplex	$\mathcal{ALU}$	58	366
9	EarthRealm	$\mathcal{ALCHO}$	931	543
10	Chemical	$\mathcal{ALCHF}$	114	44
11	DOLCE	$\mathcal{SHIF}$	351	2

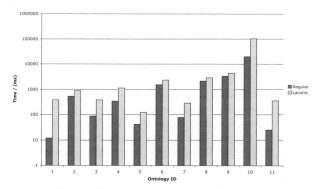

Fig. 1. Times to compute justifications

Figure 1 displays the times for computing regular justifications and laconic justifications. It is clear to see that computing laconic justifications takes longer than computing regular justifications. This is to be expected since the computation of regular justifications is used as a first step for computing laconic justifications. Nevertheless, the mean time for computing laconic justifications is acceptable for the purposes of computing laconic justifications on demand in an ontology development environment. It can be seen from Figure 1 that Ontology 10, which is the Chemical ontology, required the most time computationally—the average time was 100000 milliseconds (1 min 40 seconds) per entailment. Indeed, for the Chemical ontology, the average time per entailment to compute regular justifications is 20 seconds. The reason for this longer computation time, is that on average, each entailment in the chemical ontology has 9 justifications, with one entailment topping 26 justifications.

8   Observations on Computed Laconic Justifications

The laconic justifications that were computed in the experiment, and the relationship that these have with their corresponding regular justifications, exhibited several properties that verify the motivational reasons for laconic justifications (Section 3). Examples of masking and larger numbers of regular justifications than laconic justifications for a given entailment were present. This section discusses these examples.

8.1   Masking

As described in Section 3 one of the main issues with regular justifications is that for a given entailment they can *mask* other justifications. An example of such masking occurs in the DOLCE ontology. The entailment `quale ⊑ region` has a single justification: {`quale ≡ region⊓∃ atomic-part-of.region`}. However, computing laconic justifications for this entailment reveals that there are further justifications that are masked by this regular justification. There are three laconic justifications, the first being {`quale ⊑ region`}, which is directly obtained as a weaker form of the regular justification. This first laconic justification could be identified in Swoop using the strike out feature (The conjunct ∃ `atomic-part-of.region` would be struck out). More interestingly, there are two additional laconic justifications: {`quale ⊑ ∃atomicPartOf.region, atomicPartOf ⊑ partOf, partOf ⊑ part⁻, region ⊑ ∀part.region`} and also {`quale ⊑ atomicPartOf.region, atomicPartOf ⊑ atomicPart⁻, atomicPart ⊑ part, region ⊑ ∀part.region`}

Masking is surprising in general, and the above example is a nice illustration of how this information would not be revealed with regular justifications. In such cases the user or developer of an ontology would be completely unaware of these further justifications when attempting to formulate a repair strategy or when simply trying to gain a deeper understanding of the ontology.

8.2   Number of Justifications Versus Number of Laconic Justifications

Figure 2 displays the mean number of justifications per entailment. For ontologies 1, 2, 5, 6 and 9 (five out of eleven ontologies) the number of laconic justifications coincides with the number of regular justifications. However, for ontologies 3, 4, 7, 8 and 11, the mean number of laconic justifications per entailment is greater than that of regular justifications. This is an indication that internal or external masking is occurring for a significant number of ontologies, corresponding to the second and third motivations in Section 3. Again, these extra justifications would not be salient to a user who only works with regular justifications, and would mean that it might be impossible to gain a full understanding of an ontology when devising a repair plan.

In ontology 10, the Chemical ontology, it is evident that the mean number of laconic justifications per entailment is *less* than the number of regular justifications. In fact, in this particular ontology, there is an entailment with *six* regular justifications and only *two* laconic justifications. This situation also occurs in places in the PeriodicTableComplex ontology, where there are a large number of entailments that have two regular justifications and one laconic justification. In

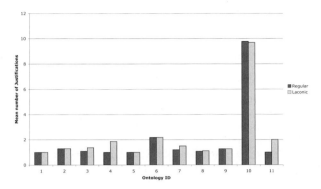

Fig. 2. Number of regular justifications versus the number of laconic justifications

the case where there are fewer laconic justifications than regular justifications, the laconic justifications highlight a common core amongst the regular justifications that is responsible for the entailment. Working with this information when repairing an ontology potentially minimises the possibility of applying an "over-repair" that could occur by quasi-independently examining each regular justification.

9 Exploiting Laconic and Precise Justifications

It should be noted that the work presented in this paper has not covered how laconic, or in particular how precise justifications, should be presented to users, incorporated into workflows or how they might be exploited in various reasoning services. While it is easy to imagine how they can be used to provide an enhanced and complete service for striking out irrelevant parts of axioms, they could also be used as a basis for measuring incoherence, measuring complexity of understanding, as metrics for repair services, and in ontology refactoring and simplification services. The issue of presenting laconic and precise justifications and incorporating them into various services is the topic of future work.

10 Conclusions and Future Work

This paper has presented a formal definition of fine-grained justifications in the form of laconic justifications and precise justifications. The definition of laconic justifications captures the intuitive notion of fine-grained justifications from previous related work, while the definition of precise justifications captures the notion of being able to generate a semantically minimal repair. An optimised algorithm to computed laconic justifications in accordance with this definition has been given and it has been shown that it is feasible to compute laconic justifications (and hence precise justifications) in practice. The definition and evaluation has provided a deeper insight into the properties of laconic justifications and how laconic justifications and their precise counterparts might eventually be exploited. Finally, it should be noted that the definition that has been presented in this paper assumes a consistent ontology. Dealing with inconsistent ontologies is the subject of future work.

References

1. Kalyanpur, A.: Debugging and Repair of OWL Ontologies. PhD thesis, The Graduate School of the University of Maryland (2006)
2. Kalyanpur, A., Parsia, B., Hendler, J.: A tool for working with web ontologies. International Journal on Semantic Web and Information Systems 1 (2005)
3. Horridge, M., Tsarkov, D., Redmond, T.: Supporting early adoption of owl 1.1 with protégé-owl and fact++. In: OWL: Experiences and Directions (2006)
4. Lam, S.C.J.: Methods for Resolving Inconsistencie. In Ontologies. PhD thesis, Department of Computer Science, Aberdeen (2007)
5. Kalyanpur, A., Parsia, B., Grau, B.C.: Beyond asserted axioms: Fine-grain justifications for OWL-DL entailments. In: Proc. of DL (2006)
6. Horrocks, I., Patel-Schneider, P.F., van Harmelen, F.: From $\mathcal{SHIQ}$ and RDF to OWL: The making of a web ontology language. J. of Web Semantics 1(1), 7–26 (2003)
7. Baader, F., Hollunder, B.: Embedding defaults into terminological representation systems. J. of Automated Reasoning 14, 149–180 (1995)
8. Schlobach, S., Cornet, R.: Non-standard reasoning services for the debugging of description logic terminologies. In: Proc. of IJCAI (2003)
9. Meyer, T., Lee, K., Booth, R., Pan, J.Z.: Finding maximally satisfiable terminologies for the description logic $\mathcal{ALC}$. In: Proc. of AAAI (2006)
10. Baader, F., Hollunder, B.: Embedding defaults into terminological knowledge representation formalisms. In: Proc. of KR 1992, pp. 306–317. Morgan Kaufmann, San Francisco (1992)
11. Deng, X., Haarslev, V., Shiri, N.: Measuring inconsistencies in ontologies. In: Franconi, E., Kifer, M., May, W. (eds.) ESWC 2007. LNCS, vol. 4519, Springer, Heidelberg (2007)
12. Baader, F., Peñaloza, R., Suntisrivaraporn, B.: Pinpointing in the description logic el. In: Hertzberg, J., Beetz, M., Englert, R. (eds.) KI 2007. LNCS (LNAI), vol. 4667, pp. 52–67. Springer, Heidelberg (2007)
13. Plaisted, D.A., Greenbaum, S.: A structure-preserving clause form translation. Journal of Symbolic Computation (1986)
14. Motik, B., Shearer, R., Horrocks, I.: Optimized reasoning in description logics using hypertableaux. In: Pfenning, F. (ed.) CADE 2007. LNCS (LNAI), vol. 4603, pp. 67–83. Springer, Heidelberg (2007)
15. Sirin, E., Parsia, B., Grau, B.C., Kalyanpur, A., Katz, Y.: Pellet: A practical OWL-DL reasoner. Journal of Web Semantics 5(2) (2007)

Learning Concept Mappings from Instance Similarity

Shenghui Wang[1], Gwenn Englebienne[2], and Stefan Schlobach[1]

[1] Vrije Universiteit Amsterdam
[2] Universiteit van Amsterdam

Abstract. Finding mappings between compatible ontologies is an important but difficult open problem. Instance-based methods for solving this problem have the advantage of focusing on the most active parts of the ontologies and reflect concept semantics as they are actually being used. However such methods have not at present been widely investigated in ontology mapping, compared to linguistic and structural techniques. Furthermore, previous instance-based mapping techniques were only applicable to cases where a substantial set of instances was available that was doubly annotated with both vocabularies. In this paper we approach the mapping problem as a classification problem based on the similarity between instances of concepts. This has the advantage that no doubly annotated instances are required, so that the method can be applied to any two corpora annotated with their own vocabularies. We evaluate the resulting classifiers on two real-world use cases, one with homogeneous and one with heterogeneous instances. The results illustrate the efficiency and generality of this method.

1 Introduction

Motivation. The problem of semantic heterogeneity and the resulting problems of interoperability and information integration have been studied for over 40 years now. It is at present an important hurdle to the realisation of the Semantic Web. Solving matching problems is one step to the solution of the interoperability problem. To address it, the Database and Semantic Web communities have invested significant efforts over the past few years [1,2,3]. More directly, the current work was motivated by our work in the Cultural Heritage domain, in which we address interoperability problems within the Dutch National Library, and across collections with the Dutch Institute for Sound and Vision.

Previous work. A common way of judging whether two concepts from different ontologies are semantically linked is to observe their *extensional* information [4,5], that is, the instance data they classify. The idea behind such instance-based matching techniques is that similarity between the extensions of two concepts reflects the semantic similarity of these concepts. A first and straightforward way is to measure the *common extension* of the concepts — the set of objects that are simultaneously classified by both concepts [6,7]. This method has a number of important benefits. Contrarily to lexical methods, it does not depend on the concept labels, which is particularly important when the ontologies or thesauri where written in a multi-lingual setting. Moreover, as opposed to structure-based methods, it does not depend on a rich ontology structure; this is important in the case of thesauri, which often have a very weak, and sometimes even almost flat structure.

A. Sheth et al. (Eds.): ISWC 2008, LNCS 5318, pp. 339–355, 2008.

However, measuring the common extension of concepts requires the existence of sufficient amounts of shared instances, something which is often not the case. Furthermore, it only uses part of the available information, *i.e.*, ignores similarity between instances that have not been doubly annotated. Similarity on the instance-level is often ignored. In this paper we apply a more general *similarity-based extension comparison*, deriving concept mappings from similarity of their instances.

Method. In this paper we formulate matching problems as classification problems and develop a machine learning technique to learn the relationship between the similarity of instances and the validity of mappings between concepts. In many application contexts, information exists about the instances that are annotated. It is therefore possible to compute a measure of similarity between the instances. The main idea of our method is to use this similarity between instances to determine similarity (mappings) between concepts. Unlike previous methods, our method does not rely on the presence of doubly annotated instances.

We extend our previous work [7] in several ways: we apply a more fine-grained measure of instance-similarity by taking the meta-data description of instances into account. This allows us to go even further in two steps: firstly, to apply our method to collections for which no joint instances exist, and secondly, to collections in which the instances are described in different ways (heterogeneous collections).

Research questions. The above described method is based on a number of implicit assumptions, and the purpose of this paper is to evaluate their influence on the quality of the resulting mappings. The most important research questions are:

1. **RQ1:** Are the benefits from feature-similarity of instances in extensional mapping significant when compared to existing methods, such as based on simple co-occurrence information?

More specifically:

2. **RQ2 Joint instances:** Can our approach be applied to corpora for which there are no doubly annotated instances, i.e. for which there are no joint instances.
3. **RQ3 Heterogeneous collections:** Can our approach be applied to corpora in which instances are described in a heterogeneous way? To answer this question we have to answer a more technical question:

   - **RQ3.1 Feature selection:** Can we maintain high mapping quality when features are selected (semi)-automatically?
   - **RQ3.2 Training data:** can we maintain high mapping quality when there is no initial training set available in the first place?

and finally, from a domain perspective:

4. **RQ4 Feature weightings:** Can we make qualitative use of the learned model, more concretely the weightings of importance of similarities?

Experiments. To answer the above research questions we evaluate our method in the context of two real-world cases: (1) collections of books which have been annotated with two thesauri that are to be matched and (2) a collection of books and a "multimedia" collection, both of which have been annotated with their own thesauri, between which the mapping will improve the interoperability across collections.

The first application scenario stems from the Dutch National Library (Koninklijke Bibliotheek, or KB) which requires mappings between two thesauri both used to annotate two *homogeneous* book collections. The second scenario is related to supporting integrated online access of parts of the collections of the Dutch institute of Sound and Vision (Beeld en Geluid, or BG) and the KB, *i.e.*, a mapping between two thesauri, each used for describing a *heterogeneous* collection.

Findings. We show that our method is effective to map both thesauri which are used for homogeneous and thesauri used for heterogeneous collections. It improves significantly over the simple lexical and co-occurrence based method and over one state-of-the-art tool. Moreover, it also works with a disjoint instance space, when no common instances exist. We demonstrate how to use our method when initially no training data is available, *i.e.*, when there are no pairs of concepts for which we know that they should be mapped. This makes this methods generalisable to many other applications. A qualitative analysis on the learning results shows this method can also contribute to achieve a metadata-level interoperability.

Relevance for the Semantic Web. The paper is relevant for the Semantic Web in two aspects: first, as an application of Semantic Web technology (of course, all data and ontologies are represented in SW standards (RDF(S) and SKOS). Secondly, we contribute to the problem of ontology mappings, as our methods can be extended to any ontology mapping problem where information about concepts can be expressed as sets of similarity features.

What to expect from this paper. Beside contributing a novel formulation of the mapping problem and the definition of a mapping method, **instance-similarity mapping**, we provide a thorough empirical evaluation showing that our proposed method improves on the state of the art, even when no initial training data is available, and investigate how it can be generalised when no joint instances are available and the collections are heterogeneous. Finally, as a nice by-product, our method can be used for meta-data schema mapping.

Section 2 introduces our application context and matching problem statement. Section 3 presents the mapping method employed. In Section 4 we describe our experimental setup to validate our research questions before concluding.

2   Application Problems

Our research has been motivated by practical problems in the Cultural Heritage domain, an interoperability problem within Dutch National Library (KB), and the problem of unified access to two heterogeneous collections, one from the KB, one from the BG, Dutch archive for Sound and Vision.

2.1 Homogeneous Collections with Multiple Thesauri

Our first task is to match the GTT and Brinkman thesauri, which contain 35K and 5K concepts respectively. The average concept depths are 0.689606 and 1.03272 respectively.[1] These two thesauri are individually used to annotate two book collections in KB. Both thesauri have similar coverage but differ in granularity.

In order to improve the interoperability between these two collections, for example, using GTT concepts to search books annotated only with Brinkman concepts, we need to find mappings between these two thesauri. Among nearly 1M books whose subjects are annotated by concepts from these two thesauri, 307K books are annotated with GTT concepts only, 490K with Brinkman concepts only and 222K with both. The instances in both collections are books annotated with the same metadata structure, more specially, using an extension of the Dublin Core metadata standard.[2]

2.2 Heterogeneous Collections with Multiple Thesauri

Our second task is to match the two thesauri (Brinkman and GTT) from the KB to the thesaurus GTAA, which contains 160K concepts and has an average concept depth of 1.30817. The GTAA thesaurus is used to annotate the multimedia collection in the BG. The BG serves as the archive of the Dutch national broadcasting corporations, radio and television programmes that have been broadcast come into the archive continuously. Besides over 700,000 hours of material, the BG also houses 2,000,000 still images and the largest music library of the Netherlands. For our experiments, we used nearly 60K instances from this archive. Each object in the BG collection is annotated by several concepts from the GTAA thesaurus.

Mapping GTAA to one or both of the KB thesauri is very interesting from a Cultural Heritage perspective. For example, one could be interested to search for some broadcasts from the BG about the author of the book he is reading in the KB. Different from the KB case, now the instance meta-data differs significantly.

In both cases, objects (books or multimedia objects) which are annotated by a thesaurus concept are considered as the instances of this concepts. In the next section, we will introduce in details the instance-similarity based mapping technique.

3   Mapping Method: Classification Based on Instance Similarity

Our concept mapping method is based on the similarity between instances, and automatic classification based on some training or seeding mappings. More concretely, we apply the following steps:

1. ontology **concepts are represented as feature vectors** (mostly information about instances, *e.g.* the content of their meta-data fields), as shown in Figure 1.
2. **similarity between two concepts** is represented as a vector of similarities between these features.

---

[1] Nearly 20K GTT concepts have no parents.
[2] http://dublincore.org/documents/dces/

3. a **classifier** learns the relation between instance similarity and concept mappings based on some training data. This classifier then estimates the probability whether an unseen pair of concepts should be mapped or not.

The trained classifier can then be used to determine whether new pairs of concepts should be mapped or not. Let us discuss each of the steps in a bit more detail.

3.1   Representing Concepts as Feature Vectors

It is common to most ontology mapping approaches that properties of concepts are collected, and compared in order to calculate a similarity score between pairs of concepts. In our use-cases, the most prominent knowledge about our concepts are from the books and multimedia objects annotated with the concepts. In [7] we showed already that the information about co-occurrence of instances can provide good mappings. In this paper, we extend this approach by including further information into the mappings process, in our case the metadata about the book and multimedia objects. This has the advantage that we can ignore whether the books are dually annotated, because similarity between the meta-data of the instances also reflects the relatedness of their annotation concepts.

We take the KB case as an example. As Figure 1 shows, books are normally described by their title, creator, publisher, *etc.* These features together represent an individual book instance. For each concept, all its instances are grouped into an integrated representation of the concept, feature by feature. For example, all titles of these books are put together as a "bag of words." Term frequencies are measured within these bags. Thus, a concept is represented by a set of high-dimensional vectors of term frequencies, one per feature of the instances, which we consider as the features of the concept. When the instances share the same features, the similarity between the corresponding concepts is calculated with respect to each feature, using the cosine similarity between the term frequency vectors of corresponding features. This is the "homogeneous collection" case that we mentioned in our introduction.

Notice that, although in our case the whole corpus is in Dutch, these similarity features can be chosen to be language-independent, *e.g.* ISBN numbers, proper names of creators or actors, publishers, dates, *etc.* This method is therefore usable in a multi-language context.

When no straightforward relationship exists between the metadata of instances, as is the case with heterogeneous collections, we compute the similarity between all pairs of the metadata fields and evaluate which of those are informative; see Section 4.2 for more details. In the end we obtain set of similarity measures, encoded in one vector per pair of concepts, which reflects the similarity of their instances. The classifier deals with this vector only, and sees it as the feature vector of the pair of concepts.

In fact, the feature vector needs not be limited to the instance similarity used here. It is trivial to extend it with the lexical similarity between the concept labels or with structure-based measures of similarity. The classifier would then learn to weigh those appropriately, creating a powerful, integrated solution. However, these measures are not always available, *e.g.* in a multilingual setting or when dealing with ontologies of little structure. We do not include such features in this paper and focus on instance-based features only.

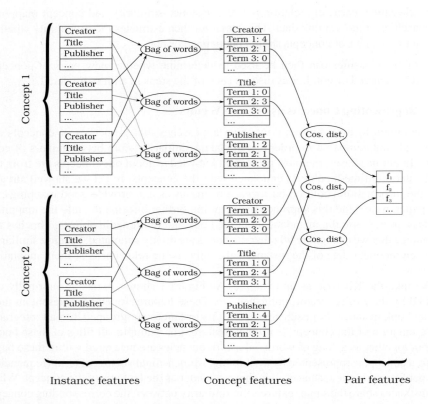

Fig. 1. Feature extraction for a single concept pair. Term frequencies are calculated from the combined features of the instances of the concepts. The cosine distances of these frequency vectors are used as the features of the pair. See text for details.

3.2 Representing Similarity of Concepts

The similarity between a pair i of concepts is measured and represented by a vector $\mathbf{x}^{(i)}$, where each element j of $\mathbf{x}^{(i)}$, denoted $\mathbf{x}_j^{(i)}$, represents the similarity between the features j of the concepts. These similarity vectors can be treated as points in a space. In this "similarity space" of concept pairs, each dimension corresponds to the similarity between features of the concepts. As we know, some points (*i.e.*, some pairs of concepts) correspond to real mappings but some are not. Our hypothesis is that the *label* of a point — which represents whether the pair is a *positive* mapping or *negative* one — is correlated with the position of this point in this space.

Given some known mappings, *e.g.* from a manual selection by experts, our goal is to *learn* this correlation. Therefore the mapping problem is transformed into a classification problem. With already labelled points and the actual similarity values of concepts involved, it is possible to classify a point, *i.e.*, to give it a right label, based on its location given by the actual similarity values.

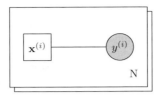

Fig. 2. Graphical representation of the MRF used in this work. The shaded, circular node denotes a hidden, discrete variable and the clear square node denotes the observed, multidimensional variables. This model is repeated N times, once for each data element.

3.3   The Classifier Used: Markov Random Field

We use a Markov Random Field (MRF, [8]) to model the classification-based mapping problem. Let $T = \{ (\mathbf{x}^{(i)}, y^{(i)}) \}_{i=1}^{N}$ be the training set containing N mappings, with, for each given pair of concepts i, a feature vector $\mathbf{x}^{(i)} \in \mathbb{R}^{K}$ and an associated label $y^{(i)} \in \mathcal{Y}$, where K is the number of features of a pair of instances and $\mathcal{Y}$ is the set of possible values the label can take. Here, the label $y^{(i)}$ is either *positive* or *negative*, although this can be extended to a set of possible mapping relations, such as *exactMatch*, *broadMatch*, *narrowMatch*, *relatedMatch* or *noLink*.

We consider a simple graphical model, consisting of an observed multivariate input $\mathbf{x}$ and a single random variable y (see Figure 2). The input is a vector of the similarity features, the random variable represents the possible values of the label and associated probabilities. We assume the mappings are conditionally independent and identically distributed (*i.i.d.*), conditionally on the observations, and model the conditional probability of a mapping given the input, $p(y^{(i)}|\mathbf{x}^{(i)})$, using a probability distribution from the exponential family. That is:

$$p(y^{(i)}|\mathbf{x}_i, \theta) = \frac{1}{Z(\mathbf{x}_i, \theta)} \exp \big(\sum_{j=1}^{K} \lambda_j \phi_j(y^{(i)}, \mathbf{x}^{(i)}) \big), \tag{1}$$

where $\theta = \{ \lambda_j \}_{j=1}^{K}$ are the weights associated to the potential function and $Z(\mathbf{x}_i, \theta)$, called the partition function, is a normalisation constant ensuring that the probabilities of the mutually exclusive labels sum to 1. It is given by

$$Z(\mathbf{x}_i, \theta) = \sum_{y \in \mathcal{Y}} \exp \big(\sum_{j=1}^{K} \lambda_j \phi_j(y^{(i)}, \mathbf{x}^{(i)}) \big). \tag{2}$$

The resulting model can be seen as a Conditional Random Field (CRF, [9]) of length zero. Since we assume that the mappings of different concept pairs are independent of each other, the likelihood of the data set for given model parameters $p(T|\theta)$ is given by:

$$p(T|\theta) = \prod_{i=1}^{N} p(y^{(i)}|\mathbf{x}^{(i)}) \tag{3}$$

During learning, our objective is to find the most likely values for θ for the given training data. We can obtain this using Bayes' rule if we assume some *prior probability* distribution $p(\theta)$ on the parameters. We here chose a prior which favours small values, that we model with a normal distribution with zero mean and covariance σ^2 for each λ_i, as this penalises overly flexible models and this reduces over-fitting. The *posterior* probability of θ is then given by

$$p(\theta|T) = \frac{p(T|\theta)p(\theta)}{p(T)}, \tag{4}$$

where $p(T)$ is a normalisation term which does not depend on θ and can therefore be ignored during optimisation. Moreover, since the logarithm is a monotonically increasing function, we can optimise $\log p(\theta|T)$ rather than $p(\theta|T)$. This is simpler, as it involves taking the derivative of a sum rather than of a product over all data points. Ignoring additive constants, which do not affect the derivative, the function we optimise is then:

$$\log p(\theta|T) = \sum_{i=1}^{N} \left[\sum_{j=1}^{K} \lambda_j \phi_j(y^{(i)}, \mathbf{x}^{(i)}) - \log Z(\mathbf{x}^{(i)}) \right] - \sum_{j=1}^{K} \frac{\lambda_j^2}{2\sigma^2}. \tag{5}$$

This function cannot be optimised in closed form because of the logarithm of a sum in the partition function. However it is a convex function which can easily be optimised numerically using any variation of gradient ascent, although (quasi-)Newton methods have proven best suited [10]. We used the limited memory BFGS method to obtain the results presented here [11]. The first derivative of eq. 5 is given by

$$\frac{\partial p(\theta|T)}{\partial \lambda_j} = \sum_{i=1}^{N} \left[\phi_j(y^{(i)}, \mathbf{x}^{(i)}) - \sum_{L \in \{0,1\}} \phi_j(y^{(i)}, \mathbf{x}^{(i)}) p(y^{(i)}|\mathbf{x}^{(i)}, \theta) \right] - \frac{\lambda_j}{\sigma^2} \tag{6}$$

The variance of the prior, σ, is a parameter that has to be set by hand and can be seen as a regularisation parameter which prevents over-fitting of the training data.

Once the model is trained, we use the resulting parameters to compute the probability of a label for a pair of concepts. The decision criterion for assigning a label $y^{(i)}$ to a new pair of concepts i is then simply given by:

$$y^{(i)} = \underset{y}{\operatorname{argmax}}\, p(y|\mathbf{x}^{(i)}) \tag{7}$$

That is, for a given pair of concepts, the label with the highest probability given the pair's feature vector is assigned. Note that in settings where a higher cost is associated with one type of error than the other, another threshold could be set. For example, if the system were used to propose candidates for mapping, missed mappings would be worse than erroneously proposed candidate pairs. In such a case, we could set the system to propose the pair if, say, $p(y = \text{positive}|\mathbf{x}^{(i)}) > 0.3$.

4   Description of the Experiments

The goal of our experiments is to show the effectiveness of our approach in general, and to evaluate the influence of three factors: the existence of joint instances, the use of heterogeneous rather than homogeneous collections for mapping, and how to build a representative training data set when no hand-made initial training set is available.

Table 1. Numbers of positive examples in the training sets

Thesauri	lexical equivalent mapping	non-lexical mapping
GTAA vs. GTT	2720	116
GTAA vs. Brinkman	1372	323

Experimental setup. All of our experiments are set up in the same way. We map two thesauri which are used to annotate instances either from the homogeneous, or from heterogeneous collections.

Training data. Ideally, the training set should be representative enough to model the relation between instance similarity and concept mappings. For the first KB case, we used the manually-built golden standard from [7] as the training set. This set contains a balanced number of positive and negative examples; where all positive mappings are non-lexically but semantically equivalent pairs. For the second BG case, no such hand-crafted golden standard is available. Of course, we can build a training set manually, as we did in the GTT and Brinkman case. But this is a very time consuming task, especially when two thesauri to be mapped are very big. To overcome this problem we used two ways to construct a training set automatically.

- *(lexical seeding)* One assumption we take is that concepts with the same label form a valid mapping. Therefore, we applied a simple lexical mapper[3] to select the positive examples. The same number of pairs of concepts are selected at random as negative data from the set of pairs that were not lexically matched. Some true mappings may therefore conceivably be present among the negative training examples, but the probability of this occurring is negligible. A more serious problem, however, is the strong bias towards lexical similarity of this data set.
- *(background seeding)* A way to find non-lexically equivalent concepts which are semantically equivalent is to use "background knowledge" [12]. More concretely, for mapping Brinkman concepts to those of GTAA, we use GTT as a background knowledge. In our previous work, many Brinkman concepts are mapped to GTT concepts using the co-occurrence based techniques [7]; some of these mappings are non-lexical ones. For each Brinkman with no lexical mapping to any GTAA concept but which is mapped to a GTT concept, we check whether the corresponding GTT concept is lexically mapped to a GTAA concept. If there is such a link, then the Brinkman and the GTAA concept are considered as a mapping.

Table 1 lists the size of the training sets built by the above two ways.

Evaluation. We apply two different types of evaluation: standard 10-fold cross-validation, and testing on a specific test set. The former is applied whenever possible as it provides us with an estimate of the reliability of the results. However, as we will see, when evaluating

---

[3] This Dutch language-specific lexical mapper makes use of the CELEX (http://www.ru.nl/celex/) morphology database, which allows to recognise lexicographic variants of a word-form, as well as its morphological components.

how the selection of the training data affects the results, we need to keep the training sets separate and cross validation is not possible.

The quality of our methods is measured quantitatively using the *misclassification rate* or *error rate*, *i.e.*, the number of wrongly classified pairs over the total number of pairs. This is an appropriate measure when the training and test data sets have balanced numbers of positive and negative examples. At the end of Section 4.3, we will briefly discuss the case of more skewed distributions of positive and negative examples.

From a more qualitative point of view, we also analyse the respective importance of the different features, based on the explicit (learned) weights of these features.

4.1   Experiment I: RQ1: Feature-Similarity Based Mapping Versus Existing Methods

The purpose of the following experiment is to compare our new method with existing methods. The task is to map GTT and Brinkman concepts given books from the two KB collections. We compare our approach with the co-occurrence based method detailed in [7], a simple lexical approach, and the state-of-the-art Falcon ontology mapper.[4] An existing golden standard that was built manually by experts, including 747 positive and negative examples, is used for the evaluation.

We compare the following methods:

Falcon: we apply the Falcon mapper to map GTT and Brinkman. We then calculate the error rate by considering all mappings returned by Falcon as classified positive and all the rest pairs as classified negative.

S_{jacc}, the Jaccard similarity between concepts is measured based on 222K dually annotated books. It is defined as the number of books that have been annotated by both concepts over the total number of books that have been annotated by those concepts. As described in [7] we apply a simple adaption to exclude concepts with too few instances.

S_{lex}, the relative edit distance between the labels (including "prefLabel" and "altLabel") of the two concepts is measured and the minimum distance is kept as the lexical similarity of these two concepts.

S_{bag}, where all information was put into a single bag, and similarity is calculated based on one bag of words.

$\{f_1, \ldots, f_{28}\}$ is an instance of our mapping method, where the similarity between each field of the instances was computed separately as depicted in Figure 1.

We train four classifiers on S_{jacc}, S_{lex}, S_{bag} and $\{f_1, \ldots, f_{28}\}$, respectively, and estimate the error-rate using 10-fold cross-validation. When trained on a single feature (*e.g.*, S_{jacc}, S_{lex} and S_{bag}), the classifier simply learns a threshold that separates mapped from non-mapped concepts.

The results are summarised in Table 2: The classifiers based solely on lexical or Jaccard similarity perform slightly better than chance level. Just calculating similarity between the complete information of all instances in S_{bag} is even worse. It is obvious

---
[4] http://iws.seu.edu.cn/projects/matching/

Table 2. Comparison between existing methods and similarities-based mapping, in KB case

Mapping method	Error rate
Falcon	0.28895
S_{lex}	0.42620 ± 0.049685
S_{jacc80}	0.44643 ± 0.059524
S_{bag}	0.57380 ± 0.049685
$\{f_1, \ldots f_{28}\}$ (our new approach)	0.20491 ± 0.026158

Table 3. Comparison between classifiers using joint and disjoint instances, in KB case

Collections	Testing set	Error rate
Joint instances	golden standard (representative)	0.20491 ± 0.026158
(original KB corpus)	lexical only	0.137871
No joint instances	golden standard (representative)	0.28378 ± 0.026265
(double instances removed)	lexical only	0.161867

that based on this information alone, the classification does not work. Falcon outperforms all three of these methods.

Using the similarity of different fields separately reduces the error rate to a more acceptable level of around 20%, and significantly outperforms all other methods.

4.2   Experiment II: RQ2: Extending to Corpora without Joint Instances

As mentioned in Section 2, there is an overlap between the two collections in KB, so that there are books which are dually annotated, by both GTT and Brinkman concepts. This allows us to apply methods based on co-occurrence to find mappings [7]. However, it is not always the case that two thesauri have joint instances. In this section we evaluate whether our approach can be applied to the case where there are no joint instances, *i.e.*, where there are no doubly annotated instances.

To determine the influence of joint instances on our mapping approach we trained two classifiers: one on the complete KB corpus, and one on the same corpus minus the joint instances, *i.e.* keeping books which are annotated by either GTT or Brinkman concepts, but not by both. For this purpose, we used the same golden standard used above, containing 747 concept pairs for all of which joint instances exist.

We applied two evaluation methods: first, 10-fold cross-validation on the golden standard, and second, testing on a set of purely lexically equivalent concepts. We did this to check whether instance-similarity-based methods could recover the mappings found by lexical techniques.

The results in Table 3 show that inferring the mapping from disjoint instances results in a higher error rate. This is not surprising, as instance co-occurrence information is implicitly distributed among the features, and the existence of joint instances is indeed a more direct indicator of the similarity between concepts. Yet even without joint instances, the classifier performs reasonably well, surely good enough for many applications. This indicates that our method can be extended to situations where joint instances are not available.

On lexically equivalent pairs of concepts the error rate is actually significantly lower than the average error rate tested on the golden standard. This means that by using instances alone, the classifier trained on semantically equivalent concepts works well enough on classifying lexically equivalent mappings. Finding mappings between lexically equivalent concepts tends to be easier than between concepts which do not have lexically equivalent labels: some instance features, such as title or subject, tend to contain words that are closely related to the concept labels. As we will see below, classifiers trained on concept pairs that are lexically matched therefore tend to perform worse when tested on pairs that should be mappings but have no lexically equivalent labels, than the other way around.

4.3   Experiment III: RQ3: Extending to Heterogeneous Collections

The instances used in the previous experiments are books from two collections of KB. These books are described using the same metadata structure, which allows a straightforward measure of similarity. Such a shared structure is however not available when finding mappings between thesauri which are used for different collections with heterogeneous metadata structures. In this section, we evaluate the effect of applying our method to corpora for which instances are described in a heterogeneous way.

In order to work with heterogeneous collections we need to select features to model the similarity space.

RQ3.1: *Feature selection.* The method to construct the pair features outlined in Figure 1 requires corresponding metadata fields. In heterogeneous collections these are most likely not really available. We have some choices to construct a vector of pair features. We will first discuss our options before evaluation the effect on the mapping quality.

1. *(exhaustive combination)* As instances in different collections have different metadata structures, a naïve approach is to ignore the meaning of the fields and to calculate the similarity between all possible pairs of fields exhaustively. In our case, this similarity vector then has 28×38 (1064) dimensions. Theoretically, as the dimensionality of the data grows, an exponentially large number of training examples is required, which is unrealistic in practice. However, the similarity vector in practice can be very sparse — as in our case, fields such as "date" and "creator" do not have any similarity — so that over-fitting might not be too serious.

In order to avoid the potential over-fitting problem, we need to reduce the dimensionality of the data. We have two options:

2. *(manual selection)* To manually select corresponding metadata fields and calculate the similarity between the selected fields. In our case, among 28 fields in KB collection and 38 fields in the BG collection, we found eight shared fields which were therefore chosen for the similarity measuring.
3. *(mutual information)* To select the most informative fields automatically. We computed the mutual information between the label (*positive* or *negative*) and each of

Table 4. Comparison of the performance with different methods of feature selection, using non-lexical dataset

Thesaurus	Feature selection	Error rate	Thesaurus	Error rate
	manual selection	0.11290 ± 0.025217		0.10000 ± 0.050413
GTAA *vs.*	mutual information	**0.09355 ± 0.044204**	GTAA vs. GTT	**0.07826 ± 0.044904**
Brinkman	exhaustive	0.10323 ± 0.031533		0.11304 ± 0.046738

the exhaustively computed similarities. This indicates how informative the similarity between each combination of instance features is to predict the label. We then select a number of the most informative field combinations: In our case, we arbitrarily selected the 30 most informative fields.

We now investigate how this choice influences the performance of classifiers. We map GTAA to GTT/Brinkman and test the performance using three selections of similarity features. Results are given in Table 4. We see that, due to the nature of our data, the differences in error rate are not statistically significant here. However, the exhaustive enumeration of all feature pairs results in a very high-dimensional feature space. The corresponding classifier therefore much more prone to over-fitting than the other feature selections, not to mention the computational overhead in time and space. Yet, it should also be noted that training with the exhaustive features still only takes around 8 minutes for a set of 4896 examples, on a 2.3GHz Core II processor, as compared to around 15 seconds for the features selected by mutual information. Classification of new pairs is essentially instantaneous.

Automatic selection of features according to mutual information gives the best results, thus answering our third research question: automatic feature selection does result in good mappings. This is also of practical value, as the lower dimensionality of the resulting data leads to reduced computational costs.

RQ3.2: *Training set.* We mentioned above that the construction of a training set by hand is expensive and often impractical, and that we devised two automatic seeding methods: lexical and background seeding. Background seeding can provide us with semantically valid mappings, but the background knowledge is often hard to obtain. Lexical seeding has the advantage that it is easily applied and that the training-sets can easily be quite large, compared to manually created sets. However, we will have to evaluate how this *biases* the training set and thus the classifier.

In the following experiment, we train two classifiers based on two training sets which were automatically generated by these two methods. Table 5 compares the results, using automatically selected features.

As we mentioned in Section 4.2, when tested on lexical mappings, both classifiers, whether trained on lexical mappings or on non-lexical mappings perform well. However when tested on non-lexical mappings, the classifier trained on lexical mappings performs much worse than the one trained on non-lexical mappings. This indicates that when trained on lexical mappings only, the resulting classifier is biased towards pairs of concepts with lexically similar labels and exhibits a degraded performance on

Table 5. Comparison using different datasets (feature selected using mutual information)

Thesauri	Training set	Test set	Error rate	Thesauri	Error rate
	non-lexical	non-lexical	0.09355 ± 0.044204		0.07826 ± 0.044904
GTAA vs.	lexical	non-lexical	0.11501	GTAA vs.	0.098712
Brinkman	non-lexical	lexical	0.07124	GTT	0.088603
	lexical	lexical	0.04871 ± 0.029911		0.06195 ± 0.008038

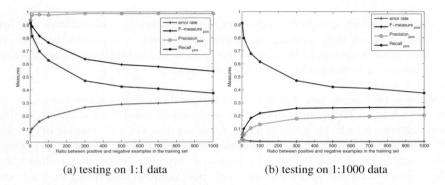

(a) testing on 1:1 data   (b) testing on 1:1000 data

Fig. 3. The influence of positive-negative ratios in Brinkman-GTAA mapping

non-lexical mappings. This is an interesting finding, as it suggest that the use of a non-lexical training set is more generic.

We further investigated the influence of the ratio of positive-negative examples during training and testing. We trained on different datasets with the positive-negative ratio varying from 1:1 to 1:1000 and tested on two datasets with 1:1 and 1:1000 positive-negative examples. When training and testing on the data with the same or similar positive-negative ratio, our method performs well, see the left of Fig. 3 (a) and the right of (b). Note, when training on very few positive examples but many negative examples, the error rate could stay very low due to the correct classification of negative examples while the the precision for positive mappings could be low, as the classifier is focused on classifying negative examples and the predictive power on positive examples is therefore not optimised. In practice, the training data should be chosen so as to contain a representative ratio of positive and negative examples, while still providing enough material for the classifier to have good predictive capacity on both types of examples.

4.4   Experiment IV: RQ4: Qualitative Use of Feature Weights

The final set of experiments is a qualitative analysis of the explicit knowledge extracted from the classifiers: the weighting of the features. The training process results in a set of weightings for the features used, *i.e.*, $\theta = \{\lambda_j\}_{j=1}^{K}$, as introduced in Section 3. The value of λ_j reflects the importance of the feature f_j in the process of determining similarity (mappings) between concepts.

When mapping between the GTT and Brinkman thesauri with disjoint instances from the two KB collection, we indeed found fields with descriptive information, such as

Table 6. Examples of informative pairs of metadata fields from the exhaustive feature list

KB fields	BG fields
kb:title	bg:subject
kb:abstract	bg:subject
kb:annotation	bg:LOCATIES
kb:annotation	bg:SUBSIDIE
kb:creator	bg:contributor
kb:creator	bg:PERSOONSNAMEN
kb:Date	bg:OPNAMEDATUM
kb:dateCopyrighted	bg:date
kb:description	bg:subject
kb:publisher	bg:NAMEN
kb:temporal	bg:date

kb:abstract, kb:subject, kb:title, are informative for the classification process. We also find some language-independent fields, such as kb:ISBN and kb:contributor, are similarly important. This indicates this method can be applied in a multi-lingual setting, as long as the features are chosen so as to be language-independent.

When mapping from GTT and Brinkman to GTAA in the heterogeneous BG case, the features are similarities calculated from the exhaustive combination of all metadata fields. By observing the features with large λ values, we can find interesting links between those fields. Basically, we expect a feature in the exhaustive set (which is the Cartesian product between the two instance-feature sets) that corresponds to a high value of λ to indicate that the meta-data fields are related.

As introduced above, eight pairs of fields, such as kb:subject–bg:subject, kb:issued–bg:date, *etc.*, were manually selected. Among the top 30 features with the highest mutual information, about half of these manually selected pairs were present. Investigating the λ values of the features, from the classifier trained on the exhaustive feature set, shows that other pairs of fields, listed in Table 6, are also informative. That is, the similarity between these fields is important to determine the similarity between concepts. This in itself provides useful information for mapping metadata fields, and can also help to achieve interoperability at the meta-data level across different collections.[5]

5   Conclusion

In this paper, we have proposed a machine learning method to automatically use the similarity between instances to determine mappings between concepts from different thesauri/ontologies. This method has the advantage that it does not rely on concept labels or ontology structure. It can therefore be used when other methods fail, *e.g.* in multi-lingual settings or when the ontologies have very little structure.

---

[5] Similar (and sometimes complementary) results can be obtained using the mutual information as done for feature selection, however this is out of the scope of this paper.

A major improvement of this method, compared to previous instance-based methods, is that it does not require dually annotated instances, *i.e.*, common instances. Instead it uses a more fine-grained similarity at the instance level and uses a classifier to learn the relationship between instance similarity and concept mappings. We have shown that using feature-similarity of instances provides significant improvements when compared to existing methods, such as methods based on the co-occurrence of instances, or on the lexical similarity of concept labels. Moreover, we have demonstrated that our method can be applied when instances are heterogeneous. In our experiments, we have obtained good results when mapping thesauri annotating book and multimedia collections for which the instances are strongly dissimilar and only heterogeneous metadata was available. Finally, the method also works when no initial hand-crafted training mappings are available, as we have shown that using training sets of lexical mappings, or mappings generated using some background knowledge, can still provide high quality results.

A qualitative analysis on the resulting parameters λ allowed us (1) in the homogeneous case, to observe which metadata fields play important roles for mapping decisions; (2), more interestingly, in heterogeneous case, to find some links between metadata fields from different collections. Though a by-product rather than the core of our research, we consider this to be a nice contribution to the field of meta-data mapping.

In the future, we intend to apply our method to other collections, *e.g.* multilingual collections and to investigate integration with other techniques, *e.g.* based on lexical similarity or structure-based methods.

References

1. Rahm, E., Bernstein, P.A.: A survey of approaches to automatic schema matching. VLDB J. 10(4) (2001)
2. Doan, A., Halevy, A.Y.: Semantic integration research in the database community: A brief survey. AI Magazine 26(1) (2005)
3. Euzenat, J., Shvaiko, P.: Ontology Matching. Springer, Heidelberg (2007)
4. Li, W.S., Clifton, C., Liu, S.Y.: Database integration using neural networks: Implementation and experiences. Knowledge and Information Systems 2, 73–96 (2000)
5. Doan, A.H., Madhavan, J., Domingos, P., Halevy, A.: Learning to map between ontologies on the semantic web. In: Proceedings of the 11th international conference on World Wide Web, pp. 662–673 (2002)
6. Ichise, R., Takeda, H., Honiden, S.: Integrating multiple internet directories by instance-based learning. In: Proceedings of the eighteenth International Joint Conference on Artificial Intelligence (2003)
7. Isaac, A., van der Meij, L., Schlobach, S., Wang, S.: An empirical study of instance-based ontology matching. In: Proceedings of the 6th International Semantic Web Conference, Busan, Korea (2007)
8. Kindermann, R., Snell, J.L.: Markov Random Fields and their applications. American Mathematical Society (1980)
9. Lafferty, J., McCallum, A., Pereira, F.: Conditional random fields: Probabilistic models for segmenting and labeling sequence data. In: Proc. 18th International Conf. on Machine Learning, pp. 282–289 (2001)

10. Sha, F., Pereira, F.: Shallow parsing with conditional random fields. In: Proceedings of the 2003 Conference of the North American Chapter of the Association for Computational Linguistics on Human Language Technology, Edmonton, Canada, vol. 1, pp. 134–141. Association for Computational Linguistics (2003)
11. Nocedal, J.: Updating quasi-newton matrices with limited storage. Mathematics of Computation 35, 773–782 (1980)
12. Aleksovski, Z., ten Kate, W., van Harmelen, F.: Ontology matching using comprehensive ontology as background knowledge. In: Shvaiko, P., et al. (eds.) Proceedings of the International Workshop on Ontology Matching at ISWC 2006, CEUR, pp. 13–24 (2006)

Instanced-Based Mapping between Thesauri and Folksonomies

Christian Wartena and Rogier Brussee

Telematica Instituut
P.O. Box 589
7500 AN Enschede, The Netherlands
{Christian.Wartena,Rogier.Brussee}@telin.nl

Abstract. The emergence of web based systems in which users can annotate items, raises the question of the semantic interoperability between vocabularies originating from collaborative annotation processes, often called folksonomies, and keywords assigned in a more traditional way. If collections are annotated according to two systems, e.g. with tags and keywords, the annotated data can be used for instance based mapping between the vocabularies. The basis for this kind of matching is an appropriate similarity measure between concepts, based on their distribution as annotations. In this paper we propose a new similarity measure that can take advantage of some special properties of user generated metadata. We have evaluated this measure with a set of articles from Wikipedia which are both classified according to the topic structure of Wikipedia and annotated by users of the bookmarking service del.icio.us. The results using the new measure are significantly better than those obtained using standard similarity measures proposed for this task in the literature, i.e., it correlates better with human judgments. We argue that the measure also has benefits for instance based mapping of more traditionally developed vocabularies.

1   Introduction

Describing collections of books, articles, pictures or movies by assigning keywords to the objects in the collection has a long tradition. Traditionally this has been done by authors, publishers and librarians. Recently, keyword-like metadata are also provided by readers through collaborative tagging systems (1). The nature of these reader provided metadata, usually called tags, differs from the traditional keywords (see e.g. (2)). In particular, keywords are often taken from a restricted vocabulary, e.g. a thesaurus or ontology, while the vocabulary for tagging is always unrestricted. However, only a small part of all tags for a given collection is used frequently (2; 3; 4). The system of terms used in a tagging system, resources (e.g. documents), users and the relations between them is often called a folksonomy (5). More precisely, we will understand a folksonomy as a set of assignments of tags to resources by distinguishable users.

The fact that different collections are described with different vocabularies gives rise to interoperability problems. These problems have been acknowledged

A. Sheth et al. (Eds.): ISWC 2008, LNCS 5318, pp. 356–370, 2008.

as one of the most important obstacles for realizing a large scale semantic web and has led to a large research area on ontology matching (6). The emergence of folksonomies adds the problem of finding mappings between these vocabularies and traditional thesauri and ontologies as a new and interesting issue to this field. One of the main differences between folksonomies and ontologies is the fact that ontologies are usually designed carefully and subsequently might be used to annotate data, whereas folksonomy terms are in the first place used for annotation and the resulting system is only subsidiary. Together with the absence of structure and relations between the terms in a folksonomy this makes instance based mapping a natural choice for finding relations with concepts from a folksonomy.

This paper proposes a new method to map tags, to terms from thesauri or taxonomies (and vice versa), and gives an information theoretic measure for the quality of that mapping. We evaluate our method by mapping Wikipedia categories onto del.icio.us tags and comparing the found mappings to correspondences established by existing methods.

The organization of this paper is as follows: After an overview of related work, we introduce some of the basic concepts used in this paper (section 3). In section 4 we give an overview of dissimilarity measures and introduce a new measure that is especially suited for mapping terms of folksonomies. Section 5 describes an experiment carried out for evaluation and presents its results. We conclude the paper with a discussion for further applications of the mapping method proposed in this paper.

2   Related Work

Euzenar and Shvaiko (6) give an overview of ontology matching systems based on similarity of instances. Isaac e.a. (7) focus more specifically on instance based mapping between ontologies of keywords that uses annotated data to compute similarities between terms. As pointed out by (7) one of the crucial factors for this kind of mapping is the dissimilarity measure used to compare terms. They compare the effects of choosing different dissimilarity measures and find that in their case a slightly modified variant of the well known Jaccard coefficient gives best results. Our focus is also on the dissimilarity measure. We define a new dissimilarity measure that takes advantage of the property of tagged data, that we know the number of people that assigned a tag to an item. The results obtained using this dissimilarity measure are much better than using the other measures we tested.

The FCA-Merge algorithm (8), an approach to ontology merging based on formal concept analysis (FCA), is in fact also a good example of an instance based mapping technique. In FCA concepts are characterized by their instances. Concepts from two ontologies that are characterized by a similar set of instances are likely to be related. This observation is exploited in the FCA-Merge approach. In order to get enough data for merging Stumme e.a. (8) consider occurrences of concepts in documents instead of common instances. Our approach can be regarded as a statistical version of FCA-merge, in that we do not consider a (binary) occurrence relation of concepts in documents but a probability that a concept occurs in a document. Another difference is that we consider collections of terms and neither use ontological relations between terms nor produce them.

3   Annotated Data

Tags are terms that users give to items, like photos, movies or articles, usually on the internet. Users have different motivations to tag items, the most important being (1) organizing and finding back their favorite items and (2) describing non-textual items. A typical example of the first usage is provided by the book-marking service del.icio.us. Examples for the latter usage are given by websites for sharing photos or videos. On these websites people tag the items they add to the site to make them findable for other people. In both cases the tags are very similar to keywords in that they provide one word descriptions for (part of) the content of the tagged object. Keywords assigned in a more traditional way differ from tags in that they are often taken from a predetermined list of terms and that they are chosen carefully to reflect the content of an item. Thus, tags contain more noise. Moreover, not all tags describe the content, e.g. opinionating tags ('interesting'), tags like 'to_read' or tags describing a personal context ('thesis') are found (see (1) for an overview of tag types). However, many tagging systems keep track of the number of times a tag was assigned to an item. It is likely that only the relevant descriptional tags reach high frequencies. Halpin e.a. (4) found that the distribution of tags for frequently tagged items tends to become stable over time.

In Wikipedia articles are classified according to categories by the article's authors. These categories are organized hierarchically. Since moreover the category system of Wikipedia is rather stable and the result of many debates on the correct structure, this system and its usage is more similar to a classical taxonomy and its typical usage than to a folksonomy (9), (10).

3.1   Formal Setup

For the following we consider a collection of tagged items (or documents) $C = \{d_1, \ldots d_M\}$. Furthermore, we consider two collections of n, respectively n' annotations or tag occurrences W and W'. Each tag occurrence is an instance of a tag t in $T = \{t_1, \ldots t_m\}$ and $T' = \{t'_1, \ldots t'_{m'}\}$, respectively. In the following we will assume that T and T' (and hence W and W') are disjoint. Each occurrence occurs on a tagged item (e.g. document) d in C. Let $n(d, t)$ be the number of occurrences of tag t on d, $n(t) = \sum_d n(d, t)$ be the number of occurrences of tag t, $N(d) = \sum_t n(d, t)$ the number of tag occurrences in d and $D(t) = \{d \mid n(d, t) > 0\}$ the set of documents tagged by T. The size of this set $df(t) = |D(t)|$ is called the document frequency of t.

4   Similarity of Terms

Instance based ontology mapping relies on the presence of a similarity concept for terms based on their instances, or in our case, on their usage as annotations. One of the most obvious things to do is to look at the co-occurrence of annotations from different vocabularies on items in a collection that is annotated according to both systems. In the discussion (section 6) we will also sketch another possibility.

4.1   Co-occurrence Coefficients

A well known family of measures for the degree in which terms co-occur is provided by the co-occurrence coefficients, like the Dice coefficient, the overlap coefficient or the Jaccard coefficient (see e.g. (11) for an overview). In (7) the Jaccard coefficient was used for instance based mapping. We will also use this coefficient to make results comparable. The Jaccard coefficient is given by:

$$JC(t, t') = \frac{|D(t) \cap D(t')|}{|D(t) \cup D(t')|} \tag{1}$$

Isaac e.a. give a slight variation of the Jaccard coefficient that gives smaller scores to low frequency co-occurring annotations (7). They got slightly better results using this coefficient that is defined by

$$JC_{corr}(t, t') = \frac{\sqrt{|D(t) \cap D(t')| \cdot (|D(t) \cap D(t')| - 0.8)}}{|D(t) \cup D(t')|} \tag{2}$$

Both coefficients give values between 0 and 1, where 1 indicates perfect similarity. As a measure for dissimilarity we therefore use $1 - JC(t, t')$ and $1 - JC_{corr}(t, t')$.

4.2   Co-occurrence Distributions

There are two important types of information on the annotations that are not used by the co-occurrence coefficients discussed above. In the first place the number of occurrences of an annotation for an object is not taken into account. This type of information is usually not available for collections annotated with keywords, but is a very important source of information for user tagged data collections, since it allows to suppress "noise" that is always present in these data. In the second place the co-occurrence coefficients look only at the co-occurrence of two annotations but not at other annotations that co-occur with the annotations that are compared: if two terms co-occur often with the same terms, they are likely to be similar, even if their mutual co-occurrence is not very high.

The first type of information could be used by considering annotations as vectors in a document space and computing some geometrical distance between the vectors or by taking the angle between two vectors as a dissimilarity measure. In our experiments it turned out that almost all annotations are completely orthogonal to each other and the mapping based on these dissimilarity measures does not produce any useful results. Nevertheless, in other experiments useful results were obtained using the cosine similarity (4). For other tasks, like clustering of keywords this measure also gives decent results (12). Taking into account the co-occurrence of other annotations is typical for latent semantic indexing (13). In the following we will introduce a more direct approach that takes both types of information into account.

For a term (tag or keyword) t we compute the co-occurrence probabilities with all other terms. More precisely, for each term t' we compute the probability that an annotation for an item annotated with t is an instance of t', weighted with the importance of t for that item. Arranged in the right way, this gives us for each term a probability distribution over all terms. This approach is very similar to the setup in (14) (section 3). The difference is that we keep track of the density of a term in an item rather than just the mere occurrence or non occurrence of a term. Finally, we can take a standard information theoretic dissimilarity measure between probability distributions in order to compare terms.

To make things more precise we consider (conditional) probability distributions Q on $\mathcal{C}$ and q on $\mathcal{T}$.

$$Q_t(d) = n(d,t)/n(t) \text{ on } \mathcal{C}$$
$$q_d(t) = n(d,t)/N(d) \text{ on } \mathcal{T}$$

The distribution $Q_t(d)$ is called the *source distribution of* t and can be interpreted as the probability that a randomly selected occurrence of term t has source d. Similarly, $q_d(t)$, the *term distribution of* d is the probability that a randomly selected term occurrence from item d is an instance of term t. Now we define the average co-occurrence distribution as

$$\bar{p}_z(t) = \sum_d q_d(t) Q_z(d). \tag{3}$$

We use the notation $\bar{p}_z$ since this distribution is just the weighted average (hence the bar) of the tag distributions of documents containing z where the weight is the probability to find (an instance of) z on item d. We can also interpret this distribution as the transformation of the simple distribution p_z, that is defined by

$$p_z(t) = \begin{cases} 1 & \text{if } t = z, \\ 0 & \text{otherwise.} \end{cases}$$

The transformation is given by

$$\sum_{d,t'} q_d(t) Q_{t'}(d) p_z(t') = \sum_d q_d(t) Q_z(d) = \bar{p}_z(t) \tag{4}$$

which is a two step evolution in a Markov chain that connects terms to documents and document to terms.

4.3   Similarity of Distributions

We will use the distributions of co-occurring terms as a base for the definition of the dissimilarity between terms. A standard measure for this is the Jensen-Shannon divergence. The Jensen-Shannon divergence or information radius (11; 15) between two distributions p and q is defined as

$$\mathrm{JSD}(p||q) = \frac{1}{2}D(p||m) + \frac{1}{2}D(q||m)$$

where $m = 1/2(p+q)$ is the mean distribution and $D(p||q)$ is the relative entropy or Kullback-Leibler divergence between p and q which is defined by

$$D(p||q) = \sum_{i=1}^{n} p_i \log\left(\frac{p_i}{q_i}\right)$$

The dissimilarity between two terms based on the average co-occurrence distributions defined above, is thus given by

$$\mathrm{JSD\,dis}(s,t) = \mathrm{JSD}(\bar{p}_s, \bar{p}_t).$$

This distribution provides a way to express the similarity of the contexts in which two terms occur.

5   Evaluation

To evaluate the quality of instance based ontology mapping using the tag similarity defined in the previous section we have performed two experiments. In the first small scale experiment we have mapped tags assigned to a small set of video fragments by high-school students onto the thesaurus based keywords provided by archive of the Dutch public broadcasting companies, and vice versa. Since the results from this experiment were very encouraging, we performed a second experiment with a much larger data set. In this larger scale experiment we compared the categories of English Wikipedia articles with the tags assigned by del.icio.us users, and evaluated the relation between the dissimilarity of the term mapping and the quality of the mapping. We also evaluated the influence of tag frequency on the quality of the mapping, and compared the dissimilarity measure proposed here with other measures proposed for this purpose.

5.1   The Data Sets

For the first experiment we used tags that were assigned to a set of 115 video fragments by high-school students from different schools in an experiment on tagging (16). 244 students participated in this experiment. They assigned 4,359 different tags to the fragments with a total of 12,414 assignments (tag occurrences). The video fragments were also provided with keywords by the Dutch

Institute of Sound en Vision, the archive of the Dutch public broadcasting companies. The keywords are taken from the Gemeenschappelijke Thesaurus voor Audiovisuele Archieven (GTAA, Common Thesaurus for Audiovisual Archives), containing about 9,000 subject terms and extensive lists of person names, company names and geographical names (17). For the annotation of the selected 115 fragments 269 different keywords were used, with a total of 638 assignments.

For the second experiment we used articles from the English Wikipedia that were also bookmarked by users in a sample of del.icio.us data. To access the category information for the Wikipedia pages we used an SQL dump of Wikipedia from January 3th, 2008 (http://download.wikimedia.org/enwiki/20080103/). Besides a large number of categories that are used to classify the content of an article, Wikipedia also has a small number of categories that keep track of the status of an article, e.g. that it needs references, violates copyrights etc. Most of these categories can easily be identified by unique prefixes. We have left out these categories from our data by filtering on the following prefixes: *Wikipedia, All_ , cleanup, Unprintworthy, Articles, Redirects*. Moreover, we have restricted the dataset to article pages, and did not consider previous versions, discussion or history pages etc. From the cleaned up set of pages we selected the subset for which we have at least one tag from a sample of del.icio.us bookmarks, obtained by continuous aggregation at Klagenfurt University and kindly provided to us by Mathias Lux. This gives us 58,345 pages (i.e. about a quarter of all English Wikipedia articles), 42,445 different Wikipedia categories for these pages and 222,640 category assignments together with 49,603 different tags for the selected articles and 278,693 tag assignments.

5.2   Experimental Setup

In the experiments we computed for each tag from a vocabulary T the nearest tag from vocabulary T' and vice versa. Thus we have produced two mappings for each experiment and each dissimilarity measure. Since we cannot expect to find useful results for very low frequency terms we only computed the mapping for terms t for which $df(t) > 3$ in the first experiment and $df(t) > 10$ in the second experiment. In order to reduce computation time we also restricted the set of possible candidates to tags with document frequencies higher than 3 and 10, respectively. This restriction has an influence on a very small part of the results only, since these very low frequency tags are unlikely to match the more frequent ones. Thus, we have computed 33 mappings from user tags to GTAA terms and 97 mappings the other way around in the first experiment. In the second experiment 2355 tags were mapped onto a Wikipedia category and 1827 categories onto tags for each evaluated dissimilarity measure.

5.3   Evaluation Criteria

Since there exist no reference mappings for the vocabularies we used, any evaluation will always be somewhat subjective. Moreover, rather than classifying mappings as good or bad, we wanted to have a more fine grained evaluation. We have therefore defined a number of categories for the quality of a mapping and manually classified a sample of the mappings. We used the following classes:

i Identical. Since the same term might have different meanings in different ontologies or folksonomies, mapping of a term to a literally identical term might not be correct per se. Nevertheless, in the absence of more detailed knowledge of the vocabularies we will consider these mappings as good. Terms with variations in capitalization, usage of blanks, underscores and hyphens and singular/plural variations are classified also classified as identical. Note that we have assumed that the vocabularies T and T' are disjoint. Since we keep track of the source of the annotations this is satisfied, even if both vocabularies contain terms with identical string values.

s Synonym. This categories contains synonyms and abbreviations. Examples are pairs like *vista – Windows Vista* or *Human-Computer interaction – hci*.

b Broader term. A mapping is classified as 'broader' if the source term is mapped onto a broader term. Broader term has to be understood in an informal and intuitive way, and not according to some formal ontology. Examples are pairs like *Windows software – Windows* or *War correspondents – journalist*.

n Narrower term. The opposite from the previous category.

r Related term. The term is clearly related but does not fall in any of the previous categories. Examples are pairs like *Pharmacology – drug* or *Digital typography – font*. Note that related terms are not necessarily worse than broader or narrower terms. E.g. *presidential elections* is only a related term of *presidential candidates*, while *people* is a broader term.

u Unrelated. Mappings between terms that are not, or only very loosely related. In this category we find many pairs the relatedness of which terms can only be understood in a specific collection, like *Vermont culture – poetry* or *People from Texas – presidents*

x The source term does not classify the content of an article. Thus it cannot be expected that a meaningful mapping can be found. Examples are tags like *important, to_read* or *Wikipedia*. Also some Wikipedia categories that escaped from our filtering, fall into this class.

q We did not know the exact meaning of one of the terms.

In the second experiment we did not evaluate all mappings but evaluated every tenth mapping, the mappings being sorted by the frequency of the source term. Since we are interested in the relation between the quality of the mapping and the frequency of the source terms and between the quality and the dissimilarity of the mapped terms we also evaluated the mappings for the 100 most and 100 least frequent terms, the 100 mappings with the largest dissimilarity between the found terms and the 100 with the smallest dissimilarity. This resulted in the numbers of evaluated mappings as given in Table 1.

5.4   Results

Results for mappings using divergence of average co-occurrence distributions. Fig. 1 shows the fraction of mappings that can be classified according

Table 1. Number of evaluated mappings for two different mapping directions and three different dissimilarity measures

	JSD dis	Jaccard	Jaccard corr.
Categories to tags	522	498	511
Tags to categories	584	568	587

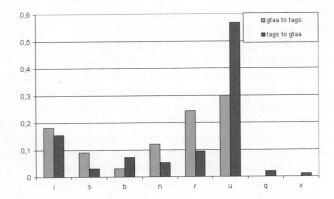

Fig. 1. Fraction of mappings from GTAA terms onto tags and vice versa using JSD dis for each evaluation category

to each of the evaluation classes discussed above[1]. For the thesaurus terms, in about 70% of the cases a synonym or related term could be found. In the opposite direction, for more than half of the tags no related thesaurus term was found. These results are largely due to the very small data set. Recall that we computed matching terms for terms with only more than tree occurrences.

The corresponding results for the experiment with del.icio.us tags and Wikipedia categories is given in Fig. 2, again using JSD dis of average co-occurrence distributions to compute similarities. Again, we see that the mapping from keywords onto tags is much better than the mapping the other way around. However, the overall quality is clearly better. Furthermore, we observe a strong tendency to map the Wikipedia categories to more general tags, whereas the tags tend to be mapped to more specific categories. This suggests that the Wikipedia categories in general are more specific than the user tags. This can also be observed by inspecting the data more closely. The category names are often rather long and specific, whereas the corresponding tags tend to be short and hence in many cases more general, e.g. *20th century classical composers* is mapped onto the tag *composers* or *Software development process* onto *softwaredevelopment*. We should also note that del.icio.us does not support tags consisting of more than one word, but uses a blank as a tag separator. Many tags suggest moreover

---

[1] The complete set of data from the experiment is available at
https://doc.telin.nl/dsweb/View/Collection-19536

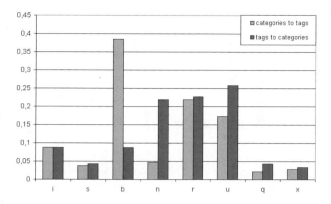

Fig. 2. Fraction of mappings from Wikipedia categories onto tags and vice versa using JSD dis for each evaluation category

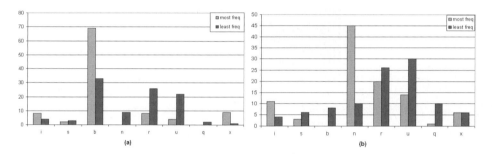

Fig. 3. Evaluation of mappings from the 100 most and least frequent Wikipedia categories onto tags (a) and tags onto categories (b) using JSD dis for each evaluation category

that many users are not aware of this feature. On the other hand in the Wikipedia category system more general terms are available. However, the most specific terms are used to annotate articles. E.g. the term *20th century classical composers* is used to annotate 1,706 articles, the more general terms *classical composers* and *composers* only for 14 and 313 articles, respectively.

Next we inspect the influence of the frequency of terms (tags or categories) on the quality of the found mappings. The results are presented in Fig. 3. Clearly, the results for the high frequency terms are much better than for the least frequent ones. Nevertheless, for both directions the results for the low frequency terms still show substantially more mappings to related terms (including synonyms and broader and narrower terms) than to unrelated terms.

We also investigated whether, for a mapping from t onto t', the divergence of the average co-occurrence distributions, JSD dis(t, t'), can serve as an indication for the quality of the mapping. This is an important feature in practical applications, since this gives the possibility to automatically decide whether a mapping

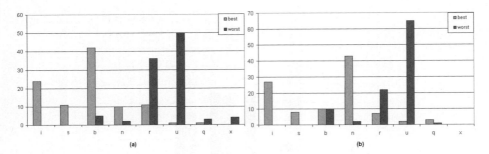

Fig. 4. Evaluation of mappings of the 100 mappings from Wikipedia categories onto tags (a) and tags onto categories (b) with smallest (best) and largest (worst) dissimilarity using JSD dis for each evaluation category

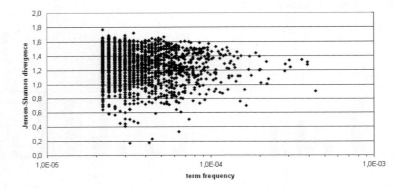

Fig. 5. Frequency of a Wikipedia category (logarithmic scale) vs. the Jensen-Shannon divergence of the category and the best fitting tag

is good enough to be used or not. The results for the evaluation of the mappings with the smallest and largest dissimilarity are given in Fig. 4. The tendency is rather clear and suggest that we can indeed use the dissimilarity as an indication of mapping quality.

Finally, we did not find a strong correlation between the frequency of a Wikipedia category and the dissimilarity of the category and the best fitting tag (see Fig. 5 for the direction from Wikipedia categories to del.icio.us tags).

Comparison between divergence of average co-occurrence distributions and Jaccard coefficient. For a quite similar task (7) found that the Jaccard coefficient and a modification introduced by them (above repeated as 2) gave best results. We compared the results using these two coefficients with the results already discussed above. The fraction of identical mappings is given in Table 2. We see that the Jaccard and the modified Jaccard coefficient give almost the same results, whereas these are rather different from the mapping produced using JSD dis.

Table 2. Fraction of identical assignments by using three different dissimilarity measures for mapping of Wikipedia categories onto tags (first number) and vice versa (second number)

	JSD dis	Jaccard	Jaccard corr.
JSD dis	1 / 1		
Jaccard	0.42 / 0.50	1 / 1	
Jaccard corr.	0.47 / 0.50	0.86 / 0.97	1 / 1

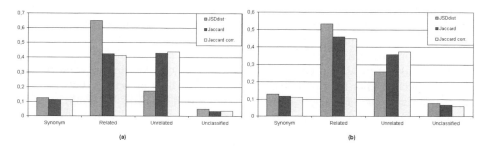

Fig. 6. Fraction of mappings from Wikipedia categories onto tags (a) and vice versa (b) using different dissimilarity measures for each evaluation category

Moreover, the mappings found with JSD dis are substantially better, as can be seen from Fig. 6. We did not find better results using the modified Jaccard coefficient as compared to the simple Jaccard coefficient.

Theoretically, the contexts of two terms t and t' can be very similar and the terms might be mapped onto each other using JSD dis even if t and t' never co-occur. In order to see whether this happens indeed, we computed the Jaccard coefficient for the pairs produced by the mapping from Wikipedia categories onto tags using JSD dis. We found that for 11 mappings the Jaccard coefficient was 0, indicating that there is no overlap. Moreover, there were many mappings with a Jaccard coefficient that was almost 0. Most of the 11 mappings were onto weakly related terms, e.g. *History of science* was mapped onto *philosophy-of-science*. These annotations never co-occur, but it is no surprise that they have similar contexts. In this case the tag found using the Jaccard coefficient was *sci*. Other examples are *1990 deaths – people* (*art-deco* using the Jaccard coefficient) or *science fiction critics – science-fiction* (*batman* using Jaccard coefficient).

6   Discussion

One of the main contributions of this paper is the introduction and usage of a novel similarity measure for terms, the Jensen-Shannon divergence of average co-occurrence distributions. In the experiments we found that this similarity measure gives better results that the Jaccard coefficient for finding corresponding terms in a taxonomy and a folksonomy. It is likely that this is, to a large extent, the consequence of taking into account the frequency of tag assignments, while the Jaccard coefficient only uses the information whether an article is tagged with a term or not. However, we expect that our measure also gives better results in domains in which such frequency information is not available, since in contrast to simple co-occurrence coefficients like the Jaccard coefficient, we also make use of the context in which tags appear.

In another paper (12) we also obtained good results for clustering keywords using this measure. Together with the relative simplicity of this measure and its natural information theoretical interpretation, Jensen-Shannon divergence of co-occurrence distributions seems to be an interesting new way to compare terms. The theoretical time complexity of computing the underlying distributions $\bar{p}_z$ is a disadvantage for this approach. However, by coding distributions as efficient sparse vectors, the necessary computations are still practicable.

As we have seen above, contexts for two terms can be similar even if they never co-occur. This feature makes it possible to find mappings between the annotation systems of collections with only a small overlap. It should even be possible to find similarities between annotations in collections without overlap, if there is a number of annotations that is common (or for which the mappings are known from another source) to both collections. The context of terms can then be expressed in terms of distributions over these common terms. The divergence of these distributions again serves as a dissimilarity measure of terms. Whether this gives satisfying results, and how many common terms are needed, are questions that are subject for future work.

Finally, we want to remark that in this paper we have focused on the choice for a dissimilarity measure, not on the design of a terminology matching system. In such a system, substantially better results could be obtained, e.g. by reducing the noise arising from different spellings and variations of tags (especially hyphens, underscores, etc.) and by also using lexical similarity as a matching cue.

7   Conclusions

In this paper we introduced a novel measure for the similarity of terms that are used for annotation of items in large collections, like books in libraries, movies in archives, URLs on the internet, etc. This measure takes into account the contexts in which annotations occur and is based on the distribution of co-occurring annotations. We used this measure for instance based mapping between Wikipedia categories and tags from the bookmarking service del.icio.us. We compared the results with mappings produced using the Jaccard coefficient, that is reported to give best results in similar experiments in the literature. In a human evaluation we found that our similarity measure gives substantially better results than the Jaccard coefficient.

A second contribution of this paper is that we have investigated the correspondence of terms from a folksonomy and a more traditionally structured thesaurus. For most frequently used terms a correspondence could be found between categories assigned by the authors of Wikipedia articles and tags used by readers to bookmark these articles on del.icio.us. However, some advanced statistical methods are needed to detect these correspondences and distinguish them from noise present in folksonomies.

Acknowledgements

We would like to thank Mathias Lux (Klagenfurt University) for kindly making available his collection of tags from del.icio.us. This research was funded by MultimediaN (http://www.multimedian.nl), sponsored by the Dutch government under contract BSIK 03031 and by the European Commission FP7 project MyMedia (http://www.mymediaproject.org) under the grant agreement no. 215006.

References

1. Golder, S.A., Huberman, B.A.: The structure of collaborative tagging systems. CoRR abs/cs/0508082 (2005)
2. Noll, M.G., Meinel, C.: Authors vs. readers: a comparative study of document metadata and content in the www. In: King, P.R., Simske, S.J. (eds.) ACM Symposium on Document Engineering, pp. 177–186. ACM, New York (2007)
3. Lux, M., Granitzer, M., Kern, R.: Aspects of broad folksonomies. In: DEXA Workshops, pp. 283–287. IEEE Computer Society, Los Alamitos (2007)
4. Halpin, H., Robu, V., Shepherd, H.: The complex dynamics of collaborative tagging. In: WWW, pp. 211–220 (2007)
5. Hotho, A., Jäschke, R., Schmitz, C., Stumme, G.: BibSonomy: A Social Bookmark and Publication Sharing System. In: Proceedings of the Conceptual Structures Tool Interoperability Workshop at the 14th International Conference on Conceptual Structures, pp. 87–102 (2006)
6. Euzenat, J., Shvaiko, P.: Ontology matching. Springer, Heidelberg (DE) (2007)
7. Isaac, A., van der Meij, L., Schlobach, S., Wang, S.: An empirical study of instance-based ontology matching. In: Aberer, K., Choi, K.-S., Noy, N., Allemang, D., Lee, K.-I., Nixon, L., Golbeck, J., Mika, P., Maynard, D., Mizoguchi, R., Schreiber, G., Cudré-Mauroux, P. (eds.) ISWC 2007. LNCS, vol. 4825, pp. 253–266. Springer, Heidelberg (2007)
8. Stumme, G., Maedche, A.: FCA-Merge: Bottom-up merging of ontologies. In: 7th Intl. Conf. on Artificial Intelligence (IJCAI 2001), pp. 225–230 (2001)
9. Ponzetto, S.P., Strube, M.: Deriving a large-scale taxonomy from Wikipedia. In: AAAI, pp. 1440–1445. AAAI Press, Menlo Park (2007)
10. Huijsen, W.O., Wartena, C., Brussee, R.: Learning ontologies from wikipedia for semantic annotation of texts. In: Proceedings of the 13th Knowledge Management Forum, Milano (November 2008) (to appear)
11. Manning, C.D., Schütze, H.: Foundations of Statistical Natural Language Processing. The MIT Press, Cambridge, Massachusetts (1999)
12. Wartena, C., Brussee, R.: Topic detection by clustering keywords. In: DEXA Workshops. IEEE Computer Society, Los Alamitos (to appear, 2008)
13. Landauer, T., Foltz, P., Laham, D.: Introduction to latent semantic analysis. Discourse Processes 25, 259–284 (1998)

14. Li, H., Yamanishi, K.: Topic analysis using a finite mixture model. Inf. Process. Manage. 39(4), 521–541 (2003)
15. Fuglede, B., Topsoe, F.: Jensen-shannon divergence and hilbert space embedding. In: Proc. of the Internat. Symposium on Information Theory, p. 31 (2004)
16. Melenhorst, M., Grootveld, M., Veenstra, M.: Tag-based information retrieval of educational videos. EBU Technical Review Q2 (2008), http://www.ebu.ch/en/technical/trev/trev_2008-Q2_social-tagging.pdf
17. Malaisé, V., Gazendam, L., Brugman, H.: Disambiguating automatic semantic annotation based on a thesaurus structure. In: Actes de la 14e conférence sur le Traitement Automatique des Langues Naturelles, pp. 197–206 (2007)

Collecting Community-Based Mappings in an Ontology Repository

Natalya F. Noy, Nicholas Griffith, and Mark A. Musen

Stanford University, Stanford, CA 94305, US
{noy,ngriff,musen}@stanford.edu

Abstract. Several ontology repositories provide access to the growing collection of ontologies on the Semantic Web. Some repositories collect ontologies automatically by crawling the Web; in other repositories, users submit ontologies themselves. In addition to providing search across multiple ontologies, the added value of ontology repositories lies in the metadata that they may contain. This metadata may include information provided by ontology authors, such as ontologies' scope and intended use; feedback provided by users such as their experiences in using the ontologies or reviews of the content; and *mapping metadata* that relates concepts from different ontologies. In this paper, we focus on the ontology-mapping metadata and on community-based method to collect ontology mappings. More specifically, we develop a model for representing mappings collected from the user community and the metadata associated with the mapping. We use the model to bring together more than 30,000 mappings from 7 sources. We also validate the model by extending BioPortal—a repository of biomedical ontologies that we have developed—to enable users to create single concept-to-concept mappings in its graphical user interface, to upload and download mappings created with other tools, to comment on the mappings and to discuss them, and to visualize the mappings and the corresponding metadata.

1   Ontology Mapping and the Wisdom of the Crowds

As the number of ontologies available for Semantic Web applications grows, so does the number of ontology repositories that index and organize the ontologies. Some repositories crawl the Web to collect ontologies (e.g., Swoogle [4], Watson [3] and OntoSelect [2]). In other repositories, users submit their ontologies themselves (e.g., the DAML ontology library[1] and SchemaWeb[2]). These repositories provide a gateway for users and application developers who need to find ontologies to use in their work. In our laboratory, we have developed BioPortal[3]—an open repository of biomedical ontologies. Researchers in biomedical informatics submit their ontologies to BioPortal and others can access the ontologies through the BioPortal user interface or through web services. The BioPortal users can browse and search the ontologies, update the ontologies in the repository by uploading new versions, comment on any ontology (or portion of an ontology) in the repository, evaluate it, describe their experience in using the ontology,

---

[1] http://www.daml.org/ontologies/
[2] http://www.schemaweb.info/
[3] http://alpha.bioontology.org

A. Sheth et al. (Eds.): ISWC 2008, LNCS 5318, pp. 371–386, 2008.

or make suggestions to ontology developers. At the time of this writing, BioPortal has 72 biomedical ontologies with more than 300,000 classes. While the BioPortal content focuses on the biomedical domain, the BioPortal technology is domain-independent.

Ontologies in BioPortal, as in almost any ontology repository, overlap in coverage. Thus, **mappings** among ontologies in a repository constitute a key component that enables the use of the ontologies for data and information integration. For example, researchers can use the mappings to relate their data, which had been annotated with concepts from one ontology, to concepts in another ontology. We view ontology mappings as an essential part of the ontology repository: Mappings between ontology concepts are first-class objects in the BioPortal repository. Users can browse the mappings, create new mappings, upload the mappings created with other tools, download mappings that BioPortal has, or comment on the mappings and discuss them.

The mapping repository in BioPortal address two key problems in ontology mapping. First, our implementation enables and encourages **community participation in mapping creation**. We enable users to add as many or as few mappings as they like or feel qualified to do. Users can use the discussion facilities that we integrated in the repository to reach consensus on controversial mappings or to understand the differences between their points of view. Most researchers agree that, even though there has been steady progress in the performance of the automatic alignment tools [6], experts will need to be involved in the mapping task for the foreseeable future. Enabling community participation in mapping creation, we hope to have more people contributing mappings and, hence, to get closer to the critical mass of users that we need to create and verify the mappings. Second, the integration of an ontology repository with a mapping repository provides users with a **one-stop shopping for ontology resources**. The BioPortal system integrates ontologies, ontology metadata, peer reviews of ontologies, resources annotated with ontology terms, and ontology mappings, adding value to each of the individual components. The services that use one of the resources can rely on the other resources in the system. For instance, we can use mappings when searching through OBR. Alternatively, we can use the OBR data to suggest new mappings. The BioPortal mapping repository contains not only the mappings created by the BioPortal users, but also (and, at the time of this writing, mostly) mappings created elsewhere and by other tools, and uploaded in bulk to BioPortal.

In recent years, Semantic Web researchers explored community-based approaches to creating various ontology-based resources [18]. For example, SOBOLEO [28] uses an approach that is similar to collaborative tagging to have users create a simple ontology. Collaborative Protégé [17] enables users to create OWL ontologies collaboratively, discussing their design decisions, putting forward proposals, and reaching consensus. BioPortal harnesses collective intelligence to provide peer reviews of ontologies and to have users comment on ontologies and ontology components [23].

Researchers have also proposed using community-based approaches to create mappings [16]. For example, McCann and colleagues[14] asked users to identify mappings between database schemas as a "payment" for accessing some services on their web site. The authors then used these mappings to improve the performance of their mapping algorithms. They analyzed different characteristics of the user community and their contributions in terms of how the mappings produced by the community affect the accuracy of their mapping algorithm.

Zhdanova and Shvaiko [31] proposed collecting mappings as one of the services provided by an ontology repository. The authors focused on enabling users to run one of the automatic mapping algorithms and then to validate the mappings produced by the algorithm, rejecting some mappings and creating new ones. In many aspects, this work is a precursor for the implementation that we describe here. However, Zhdanova and Shvaiko did not address the issues of scalability, visualization, mapping metadata, maintainability over different versions, and mechanisms for reaching consensus.

In this paper, we make the following contributions:

- We analyze use cases and requirements for supporting community-based mappings in the context of an ontology repository (Section 2).
- We define an extensible annotation model to represent community-based mappings that focuses on mappings between individual concepts rather than ontologies and that contains a detailed metadata model for describing mappings (Section 3).
- We validate the flexibility and coverage of our annotation model by representing more than 30,000 mappings from 7 sources created by biomedical researchers in different contexts (Section 4).
- We validate the practical application of our annotation model by using it to extend BioPortal with a web-based user interface to create mappings, to visualize mappings that are already in the BioPortal, and to download mappings (Section 5). These features are also accessible to developers through a web-service interface.

2   Use Cases and Requirements for Community-Based Mappings

We now identify several scenarios and the corresponding requirements that a mapping repository can support. We have collected this list through our informal interactions and through formal surveys and discussions with the biomedical-informatics researchers who participate in the BioPortal user group.[4] The scenarios include the following:

Defining new mappings interactively. As a user browses an ontology in a repository, he may come across a concept for which he knows there is a similar concept in another ontology. The user can create the mapping on-the-fly, linking the two concepts.

Uploading mappings to a repository. We do not envision that BioPortal will be the primary environment for creating large volumes of mappings. We expect that users will use custom-tailored ontology-mapping tools (e.g., PROMPT [19]) to create many of the mappings. Users can then upload the mappings to BioPortal.

Adding metadata to mappings. In the repository where mappings can come from many different sources, mapping metadata is a critical component. We must know what the source of each mapping is, how the mapping was created and in which application context, who uploaded it to BioPortal and when.

Maintaining mappings across ontology versions. The BioPortal repository maintains successive versions of the ontologies.[5] When users define a mapping, they define this

---

[4] http://www.bioontology.org/usergroups.html
[5] This feature will be available in the July 2008 BioPortal release.

mapping for a particular version. If necessary, the users must be able to see the context of a particular version for the mapping. At the same time, we do not want to discard all mappings for an ontology O once a developer submits a new version of the ontology O to the repository.

Using mappings for ontology navigation. As users browse ontologies in BioPortal, mappings can serve as navigation mechanisms, enabling users to "enter" a new ontology through the concept that is familiar to them in another ontology.

Reaching consensus on mappings. Researchers have found that mappings can be subject of discussion themselves, just as ontology components are [26]. BioPortal enables users to comments on mappings, and to have discussions about each mapping.

Visualizing mappings. With more than 30,000 mappings already in BioPortal, visualization of mappings becomes a critical issue. Users must be able to see where the mappings are, where are the contradictory or controversial mappings, where the disagreements are or where discussions are taking place.

Searching, filtering, and downloading. As the number of mappings in the repository grows larger, the users may want to focus only on specific mappings. For instance, a user may ask to show only the mappings that have been supported by more than one source, or for mappings supported by the users in his or her web of trust, or mappings created buy a particular algorithm, and so on. The user may then browse the filtered mappings or download them to use in his own applications.

Accessing algorithms for creating alignments automatically. Researchers have developed a wide variety of algorithm for identifying correspondences between ontologies automatically or semi-automatically [7]. For users, it is desirable to be able to invoke these algorithms on ontologies in BioPortal or parts of the ontologies, to specify parameters for the algorithms, and to accept or reject candidate mappings.

BioPortal currently supports all but two of the requirements (versioning and accessing automatic algorithms).[6] We expect to support all requirements by the time the final version of this paper is due.

3   Representing Community-Based Mappings in BioPortal

For the purposes of the discussion in this paper, a **concept mapping**, or simply a **mapping**, is a relationship between two concepts in different ontologies. Each mapping has a *source concept*, a *target concept*, and a mapping *relationship*. The most common mapping in BioPortal is a *similarity* mapping: For instance, there are several ontologies in BioPortal that represent some aspects of human anatomy, such as the Foundational Model of Anatomy (FMA) [22] and the National Cancer Institute's (NCI) Thesaurus [24]. We can create a similarity mapping between the class `Body_Tissue` in the NCI Thesaurus and the class `Body tissue` in the FMA. A collection of all mappings from ontology O_1 to another ontology O_2 is a **mapping between** O_1 **and** O_2.

---

[6] As the July 2008 BioPortal release includes maintenance of multiple ontology versions, we will implement the maintenance of mappings across different versions.

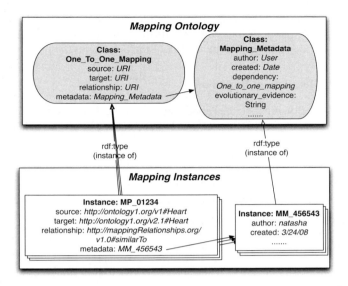

Fig. 1. Mapping ontology and its instances. Each mapping is an instance of the class One_to_One_Mapping, which refers to the source and target concepts of the mapping, and to the metadata associated with the mapping.

Formally, we define a mapping as a four-tuple: $\langle C_s, C_t, R, M \rangle$, where C_s is the source concept of the mapping, C_t is the target concept, R is the mapping relationship, and M is a set of metadata fields and values describing the mapping. C_s, C_t, and R are fully-qualified references to the definition of the corresponding concept or property in an ontology in BioPortal or elsewhere. Here, a fully-qualified name of a concept includes a reference to the ontology, the version, and the concept itself.

We represent mappings in BioPortal as instances in the **mapping ontology** (Figure 1). Each instance corresponds to a single mapping between concepts (not ontologies). Each mapping instance points to the two concepts being mapped (the *source* concept and the *target* concept), the mapping relationship, and to the metadata about this mapping. All mappings are directional: they connect source to target. Thus, for symmetric mappings (such as similarity), there are two instances, each corresponding to a different direction of the mapping.

Note that our model is different from the model defined by the Ontology-Alignment API [5] that is commonly used to represent the mappings in the OAEI [6]. In our model, we focus on mappings between individual concepts rather than sets of mappings between ontologies; we identify each concept by a fully qualified URI that includes the ontology and its specific version. By contrast, the mappings in the Ontology-Alignment API focus on mappings between ontologies, rather than individual concepts, grouping several mappings between two ontologies in a single collection. Representing individual fully-qualified mappings as first-class objects is more consonant with our model where users can create single concept-to-concept mappings, where metadata pertains to each individual mapping, where alternative mappings can exist for the same concept, and where we need to maintain mappings as ontology versions evolve. Thus, BioPortal has a single knowledge base that contains all mappings among all ontologies in the repository. We store mappings independently of the ontologies themselves.

3.1   Mapping Metadata in BioPortal

We represent the following metadata about each mapping. Note that not all the metadata values are required and, in practice, we rarely have values for all the fields.

General comment: General comment about the mapping is usually added by the person who created the mapping. For example, there is a set of mappings in BioPortal that is based on the information in the Unified Medical Language System (UMLS) [12]. UMLS integrates a large number of biomedical ontologies and terminologies, mapping concepts from these resources to a Concept Unique Identifier (CUI) in UMLS Metathesaurus. A general comment for a UMLS-based mapping in BioPortal may contain the CUI that served as the basis for the mapping.

Discussions and user comments: There can be a discussion thread associated with a mapping; mappings are first-class objects that others can comment on and discuss. Discussion messages themselves are instances of an `Annotation` ontology in BioPortal, which are linked to mappings or ontology components, or other messages that they annotate.

Application context: Researchers have demonstrated that "correct" mapping between ontologies may depend on the specific application scenario for the use of the mappings [10]. Therefore, we store a (free-text) description of the intended application of the mapping as a metadata field.

Mapping dependency: One mapping can depend on another: *"If X is Y, then A is B"*. Because mappings are first-class objects (individuals), we can refer to them easily. Thus, the value for the dependency field is another mapping individual (or individuals).

Mapping algorithm: Information about the algorithm that was used to create the mappings, if the mappings where created outside of BioPortal and uploaded. This property is a string, but can contain a link to a web page describing the algorithm. When we display the mappings to the user (cf. Figure 3), we can link directly to that web page. It is also important to record the specific version of the algorithm that was used to create the mappings and any parameters that were used to tune the algorithm in case users want to reproduce the results: algorithms change over time and may produce different results.

The date the mapping was created: This property is a simple `date` field that records when the mapping was created.

The user who performed the mapping: This property contains the name of the registered user who created or uploaded the mapping.

External references: If the mapping is based on some references to external sources (e.g., publications), this information can be part of the metadata.

3.2   What Relationship Does a Mapping Represent?

It is customary to think about mappings as *equivalence* mapping, and many researchers suggested using a logical equivalence relationship (`owl:equivalentClass`) to link concepts from different ontologies. In most cases of inter-ontology mapping, however, the mapping is not a true logical equivalence; the concepts are similar in their intended meaning but do not share all their instances or defining characteristics. In our experience, many mappings between ontologies that the users create can be described more accurately as similarity, rather than equivalence, mappings. In our framework, we store the exact mapping relationship, as specified by the user, as part of the mapping.

Note that for many reasoning and querying tasks, we can treat similarity mappings in the same way as equivalence. For instance, when we look for data annotated with a concept C_s, we may also bring in the data annotated with a concept C_t that is similar to C_s.

Many researchers think of ontology mapping as a *bridge* between ontologies: each ontology stands on its own, is used on its own, but the mapping indicates the point of overlap between the two ontologies. For example, when we create a mapping between the anatomy part of the NCI Thesaurus and the FMA, our goal is not to merge the two ontologies, but rather to help applications integrate the data that was annotated with terms from either ontology. We expect, however, that many applications will use only one or the other ontology. In the field of biomedical ontologies, researchers often think of ontology mapping not as a bridge between two ontologies, but rather as a *glue* that brings the two ontologies together to create a single whole, with clearly identifiable components. In this case, the ontologies that are mapped are intended to be used together, as a single unit. For example, consider the following mapping (from C. Mungall [15]): `ZFA:heart` **is_a** `CARO:cavitated_compound_organ`. There is no intention in the zebrafish anatomy ontology (ZFA) to define organs at the general level, as the Common Anatomy Reference Ontology (CARO) does. Thus, we use the mapping to make the definition of `ZFA:heart` to be more precise.

The line between the two settings can be fuzzy, and sometimes it is discernible only through the intention of those who created a mapping. The distinction, however, is an important one in the biomedical community.

Pragmatically, with mappings of the first kind (a bridge), equivalence, similarity, or generalization and specialization mappings are the more common mapping relationships. In the second case, any mappings are possible: for instance, a class in one ontology could be a range for a property for another (e.g. `CL:nucleate_erythrocyte` **has_part** `GO:nucleus` [15]). This last type of mapping is hardly present in the bridge setting.

4   Using the Annotation Model to Represent Mappings Among Biomedical Ontologies

We have extracted mappings from different sources to populate the mapping repository in BioPortal. We currently have more than 30,000 mappings, involving 20 ontologies (Figure 2). Many of these mappings were created manually by developers of biomedical

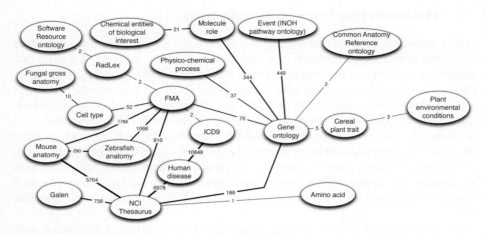

Fig. 2. Ontologies in BioPortal and mappings between them. The diagram shows the ontologies that have any mappings defined for their concepts at the time of this writing. Each edge between a node representing ontology O_1 and a node representing ontology O_2 represents the mappings between concepts in O_1 and O_2 (in both directions). The number on the edge indicates the number of mappings.

ontologies. Several of these sets of mappings were provided by members of our user community. We accept any mappings for BioPortal ontologies that our users submit. The current set of BioPortal mappings comes from the following sources:

Unified Medical Language System (UMLS). As mentioned earlier, UMLS integrates a large number of biomedical ontologies and terminologies, mappings concepts from these resources to concepts in its Metathesaurus. For BioPortal ontologies that are part of UMLS (Gene Ontology, ICD-9, FMA, the NCI Thesaurus), we created correspondences for classes that are mapped to the same concept in UMLS Metathesaurus.

Lexical mappings of names and synonyms to UMLS. For ontologies that are not in UMLS, our colleagues used exact matching of preferred names and synonyms for concepts in the domain ontologies that represent anatomy to concepts in UMLS as the basis for mappings.[7]

OBO xref property. Developers of biomedical ontologies that use the OBO format[8] frequently use the property obo:xref to relate concepts from their ontology to concepts in other OBO ontologies. The property obo:xref is similar to rdfs:seeAlso. We have used the values of this property to establish links between concepts in different OBO ontologies that are represented in BioPortal.

Mappings produced by the NCI Center for Bioinformatics (NCICB). The NCICB researchers are developing the NCI Thesaurus. They have also manually established mappings between concepts in the NCI Thesaurus and the Mouse anatomy ontology. We uploaded these mappings to BioPortal.

---

[7] Nigam Shah, personal communication
[8] http://oboedit.org

Mappings from participants in OAEI-07. The mapping between the NCI Thesaurus and the Mouse anatomy ontology was one of the tasks in OAEI 2007 [6]. We have included the results from the system that performed the best on that task [30].

Results from the PROMPT *algorithm.* We included the results of using the simple mapping based on lexical comparison in the PROMPT mapping algorithm to create mappings between the NCI Thesaurus and the Galen ontology [21].

Lexical mappings between ontologies representing anatomy. Our colleagues have used simple string-matching techniques to match class names and synonyms for ontologies that represent anatomy (FMA, adult mouse anatomy, zebrafish anatomy).[9]

There are also a small number of mappings that users have created directly in BioPortal, using the interface shown in Figure 5 (see Section 5). We expect that the set of mappings in the repository will continue to grow significantly over the next few months. We also plan to run one of the more advanced automatic mapping algorithms on all pairs of ontologies to add mappings to BioPortal.

5   Web-Based User Interface for Mappings in BioPortal

BioPortal is a java application that uses Protégé[10] and the Mayo Clinic's Lexgrid[11] system to store ontologies and uses RESTful web services to serve those ontologies. The web front-end is a Ruby On Rails application that consumes the java RESTful services to display the ontologies, their concepts, and the metadata associated with them. Ontologies may be reprented in OWL, RDF, OBO Format, or the Protégé frame language.

Figure 3 shows a snapshot of the mapping user interface for BioPortal. More specifically, it shows the summary of mappings between the NCI Thesaurus and the Mouse anatomy. BioPortal has two sets of mappings between these ontologies (see Section 4): one set was created manually, in an effort at NCI Center for Bioinformatics (NCICB) that was led by Terry Hayamizu. The second set was produced automatically by an algorithm developed by Songmao Zhang and Olivier Bodenreider [29]. The listing starts with the concepts from the NCI Thesaurus that have multiple mappings to concepts in the Mouse anatomy ontology (e.g., `Pelvic bone`, `Sural Artery`). The display shows the target concepts for the mappings, and the source of each mapping. The mappings that are supported by more than one source (e.g., the mapping between `Sural Artery` in the NCI Thesaurus and `external sural artery` in Mouse anatomy), are presented in larger font (similar to tag-cloud displays).

The user can filter the mappings, by choosing, for example, only mappings from a particular user or a particular source. The user can also download the filtered mappings as an RDF file. Applications can access the filter and the resulting mappings as a web service.

The user can click on any concept to bring up the definition of that class and to explore the mappings in more detail (Figure 4). From there, registered user can comment

---

[9] Chris Mungall, personal communication.
[10] http://protege.stanford.edu
[11] http://informatics.mayo.edu/LexGrid/index.php

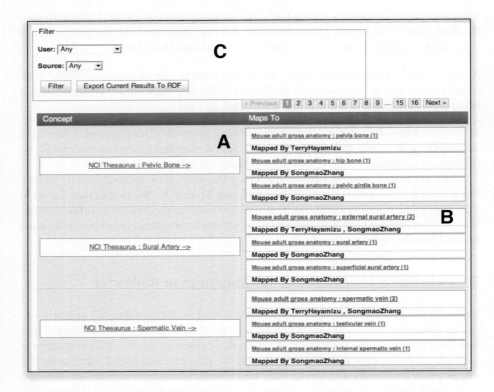

Fig. 3. The user interface for mappings in BioPortal. The snapshot shows part of the summary of mappings between the NCI Thesaurus and the Mouse anatomy (A). The list starts with the concepts that have the largest number of mappings. The mappings that are agreed by more than one user are shown in a larger font (B). The user can filter the mappings and download the filtered mappings as an RDF file (C).

on a mapping, follow the link by displaying the mapped concept, or create a new mapping. Figure 5 shows the interface to create new mappings. The user starts with the page for the source concept of a mapping. He then gets a dialog with the list of ontologies in the repository, can select the ontology for the target concept and search for the concept of interest. Before creating the mapping, the user can view its definition and visualize its neighborhood in the ontology.

6   Using a Mapping Repository

Many of the existing ontology repositories are simply collections of ontologies that can be searched, with either ontologies collected by crawling the web, or ontologies submitted by their developers. We believe, however, that much of the power of a repository comes from the metadata and additional resources that it makes available. The metadata includes information that the authors may provide about their ontologies such as their

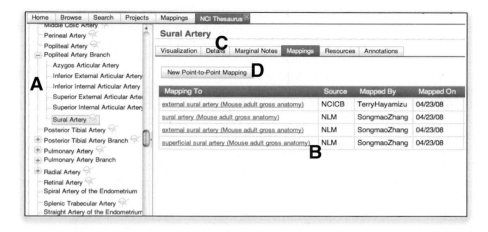

Fig. 4. Details of mappings for a selected class in the ontology. When a user clicks on a class name, as shown in Figure 3, BioPortal opens the corresponding ontology and shows that class in the class hierarchy (A), and the details of its definition and details of the mappings (B) (Definition is available under the "Details" tab (C)). From this screen, registered users can also create new mappings (D) (See Figure 5).

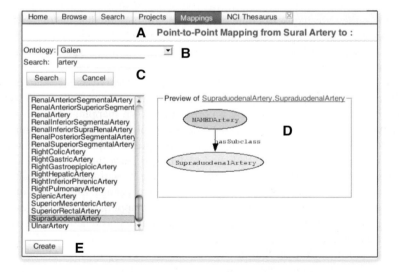

Fig. 5. An interface to create a new mapping in BioPortal. The user has chosen to create a new mapping for the class `Sural artery` (A) (Figure 4). The user wants to create a mapping to a class in the Galen ontology (B), searches for all classes with the string "artery" in them (C), selects a class of interest to see its details (D). Once he finds the class that he needs, he can create the mapping (E). At the moment, this interface allows the creation of only similarity mappings. The back-end store supports arbitrary mapping relations, as we described in Section 3.

intent in designing the ontology and its intended scope [20]; the metadata that the users provide about their use of ontologies in their repository, their reviews and ranking of the ontologies [16]; the reviews and quality control that selected experts can provide, similar to editorial boards in journals [25]; the metadata that analytical tools can provide, such as results of running an inference engine over the ontology data or analysis of connections in the ontology; finally, the resources and data that are annotated with the ontology data, with the repository providing access to these resources and data [11]. Mappings between ontologies in a repository is one such metadata. We envision a variety of uses for the kind of mapping repository that we described in this paper.

First, we plan to use the mappings to *augment many services provided by the Bio-Portal repository*. For instance, one of the core functionalities of the BioPortal is to enable access to biomedical resources, such as papers, experiment results, standard terminologies, and so on. The Open Biomedical Resources (OBR) component automatically indexes important biomedical data sets available online (e.g., entries in PubMed, the Gene Expression Omnibus, ClinicalTrials.gov) on the basis of metadata annotations, and links the underlying data sets to the terms in the ontologies in BioPortal [11]. As the users browse or query the ontologies they can access the resources annotated with a specific ontology term. If the repository contains mappings, the users can access resources annotated not only with the term that they specified, but also with the related terms from other ontologies.

Second, as biomedical researchers explore the ontologies in BioPortal—for instance, to understand whether or not a specific ontology would be useful in their own application—they are likely to come across some ontologies that they already know. If they can view the new ontologies through the "prism" of these familiar ontologies, by accessing new ontologies through mappings to the familiar one, they may find it *easier to understand* the new ontologies.

Third, the corpus of mappings that we provide could serve as a resource for *new algorithms for mapping discovery*. For instance, Madhavan and colleagues suggested machine-learning algorithms that use a corpus of known mappings to discover mappings between a pair of new database schemas or ontologies [13]. The techniques that they propose are similar to machine-learning techniques in information extraction: the authors use the evidence from established matches to learn the rules for finding new ones. They also learn statistics about elements and their relationships and use them to infer constraints that they later use to prune candidate mappings. Developers of such algorithms can use web-service access to the BioPortal mappings to get the latest corpus.

Fourth, as researchers in many application domains try to reach consensus on one or a small number of ontologies that they use, they must *reconcile* the larger number of ontologies in that domain that already exists. One group of our collaborators that faces this task, plans to use the community-based mapping facilities in BioPortal, to help them agree on how the existing ontologies relate to one another, which concepts could be merged, which concepts from each ontology should be brought into the consensus ontology, and so on. BioPortal provides a forum for discussions that can occur in context, as an integral part of ontology exploration. The discussion gets stored along with the ontologies and therefore provides an auditing track for the process.

Finally, as our repository becomes available, we expect that users will find other ways to exploit the large and diverse collection of mappings that we provide.

7 Future Work

As we gain more experience with mappings in BioPortal and as more users start contributing the mappings, we hope that the data that we collect will help us understand the **dynamics of ontology mapping as a collaborative and open process** and will help us understand how users reach consensus on mappings. We would like to answer the following questions: How much disagreement do users in biomedical domain have about the mappings? Biomedical researchers have been using ontologies probably more actively and for a longer period of time than researchers in other fields. We would like to understand if such experience leads to faster and easier consensus on mappings (as measured by the volume of the discussion threads and time to reach consensus) or do they take longer? How much do ontology mappings differ based on application context [10]? Researchers have long noted the *cumulative-advantage* phenomenon [27]: the users who put in the data first have disproportionate effect on the community, with other users often reluctant to override their suggestions. In other words, the mappings that get in first tend may have an implicit priority over the mappings that are added later. And in fact, users might not even consider re-evaluating the mappings that are already there and that come from an authoritative source. We would like to understand how strong the phenomenon of cumulative advantage is in community-based ontology mappings. If developers deploy our repository in a different domain (recall that our technology is completely domain-independent), it would be interesting to compare the dynamics of these process among researchers in different domains.

We can also use the infrastructure to evaluate different ways of composing mappings [1]. Consider the mappings between ontologies in Figure 2. There are independently created mappings between Mouse anatomy and the NCI Thesaurus; between Mouse anatomy and FMA; and between the NCI Thesaurus and FMA. We can also use transitivity of mappings to infer, for instance, mappings between Mouse anatomy and the NCI Thesaurus based on the other two sets of mappings. We can then compare these inferred mappings with the ones that were created independently. We can use such data to test different mapping-composition approaches and investigate the effect of different mapping relationships on the composition results. We can also present users with the mappings that were inferred by composing existing mappings and evaluate the users level of agreement with these mappings.

There are a number of **technical challenges** in implementing the mapping repository that we are only now starting to address.

First, we have not yet implemented a strategy to maintain mappings as developers submit *new versions of their ontologies*. We can envision two "extreme" approaches to this maintenance problem. On the one extreme, any time an ontology author submits a new version of the ontology, we discard all the mappings and other metadata that were associated with the old version. This approach is clearly not practical as most of the metadata are still valid for the new version. At the other extreme, we can associate mappings with a *name* of a concept, rather than with a concept in a specific version.

Thus, a mapping added to a concept C in a version V_1 of an ontology O will be indistinguishable from a mapping added to a concept C in a newer version V_2 of the same ontology O. This solution also creates problems, because occasionally mappings will no longer be valid in a new version, because concept definitions evolve. Thus, a "wholesale" migration of mappings is not necessarily a practical approach either. We are currently implementing a middle-ground approach: each mapping is associated with a concept in the specific ontology version that was considered when the mapping was created. However, when we access the mappings for a concept in the latest version, we retrieve the mappings for that concept for all the previous version as well. The user gets the context for the mappings and knows whether the mapping was created for the current version of the ontology or for some earlier one (and if it is the latter, which earlier version).

Second, we need to develop a strategy for *invalidating or deleting* mappings. Our infrastructure supports multiple mappings from the same concepts, and even mappings that might contradict one another, and we do not impose any quality control on the mappings that the users submit. However, occasionally, the users may want to delete a mapping, either because a concept definition has changed in a new version of the ontology and the mapping is no longer valid, or because the discussion with other users convinced the author of the mapping that it was incorrect. There are several possible options in handling mapping removal or invalidation. For example, we need to decide who can delete mappings (e.g., only the original author, or an administrator, or the ontology author)? Should the mappings be deleted permanently or simply marked as deprecated and still available for viewing if necessary? Should the deleted mappings be archived and available only under certain conditions? We are currently consulting with our users to develop the best strategy.

Third, mapping *visualization* becomes a more pertinent issue as the number of mappings in BioPortal grows. Users must be able to find the mappings, understand relationships between ontologies in BioPortal as defined by the mappings, determine where controversial mappings are. If the number of mappings becomes large enough, we are considering such navigation mechanisms as tag clouds, where the size of the node reflects the number of mappings a concept or an ontology has or the intensity level of discussion at that item. In general, to date, researchers have not studied mapping visualization as actively as mapping discovery or representation [8]. With large sets of mappings in the repository, we must deal with visualization as well.

Fourth, if our vision is realized, BioPortal will have large sets of mappings that overlap with one another and contradict one another. We built our model explicitly supporting the idea that *alternative mappings can co-exist*. However, users and application developers must be able to filter the mappings based on different criteria. These criteria may be based on mapping metadata such as mappings created for a specific application context, or mappings created for specific versions, or mappings coming from specific sources or specific users. The criteria may use social metrics, such as filtering all mappings that were corroborated by more than one source. Another example is filtering all mappings that came from a specific authoritative source. In fact, members of the BioPortal user group told us that simply being able to download all UMLS-based mappings in RDF (something the UMLS Knowledge Services do not provide) would be extremely valuable to them. In the future, we plan also to use web of trust [9] to help users focus on the mappings from those whom they trust to be the experts in the field. For example, a user can ask to see only the mappings on which others in this user's web of trust agree. All these filters will also be available to applications through a web service interface.

In general, we believe that the infrastructure and the application that we described in this paper not only provides a valuable and evolving resource to the biomedical community, but also enables the Semantic Web researchers to understand ontology mapping better and to improve the technologies in mapping discovery and maintenance. Our technology and implementation is open-source and domain independent. Readers can access the BioPortal at `http://alpha.bioontology.org`.

Acknowledgments

This work was supported by the National Center for Biomedical Ontology, under roadmap-initiative grant U54 HG004028 from the National Institutes of Health. Chris Mungall, Nigam Shah, Terry Hayamizu, and Songmao Zhang provided some of the mappings. We are very grateful to Harith Alani, Sean Falconer, Chris Mungall, and Daniel Rubin for their comments on an earlier version of this paper.

References

1. Bernstein, P., Green, T., Melnik, S., Nash, A.: Implementing mapping composition. The VLDB Journal The International Journal on Very Large Data Bases 17(2), 333–353 (2008)
2. Buitelaar, P., Eigner, T., Declerck, T.: OntoSelect: A dynamic ontology library with support for ontology selection. In: Demo Session at the International Semantic Web Conference, ISWC 2004 (2004)
3. d'Aquin, M., Baldassarre, C., Gridinoc, L., Angeletou, S., Sabou, M., Motta, E.: Watson: A gateway for next generation semantic web applications. In: Poster session at the International Semantic Web Conference, ISWC 2007 (2007)
4. Ding, L., Finin, T., Joshi, A., Pan, R., Cost, R.S., Sachs, J., Doshi, V., Reddivari, P., Peng, Y.: Swoogle: A search and metadata engine for the semantic web. In: Thirteenth ACM Conference on Information and Knowledge Management (CIKM 2004), Washington DC (2004)
5. Euzenat, J.: An api for ontology alignment. In: McIlraith, S.A., Plexousakis, D., van Harmelen, F. (eds.) ISWC 2004. LNCS, vol. 3298, pp. 698–712. Springer, Heidelberg (2004)
6. Euzenat, J., Isaac, A., Meilicke, C., Shvaiko, P., Stuckenschmidt, H., Šváb, O., Svátek, V., Hage, W.R.V., Yatskevich, M.: Results of the Ontology Alignment Evaluation Initiative 2007. In: 2nd International Workshop on Ontology Matching (OM 2007) at ISWC (2007)
7. Euzenat, J., Shvaiko, P.: Ontology matching. Springer, Berlin (2007)
8. Falconer, S.M., Storey, M.-A.: A cognitive support framework for ontology mapping. In: Aberer, K., Choi, K.-S., Noy, N., Allemang, D., Lee, K.-I., Nixon, L., Golbeck, J., Mika, P., Maynard, D., Mizoguchi, R., Schreiber, G., Cudré-Mauroux, P. (eds.) ASWC 2007 and ISWC 2007. LNCS, vol. 4825. Springer, Heidelberg (2007)
9. Guha, R., Kumar, R., Raghavan, P., Tomkins, A.: Propagation of trust and distrust. In: 13th International Conference on World Wide Web (WWW 2004), New York, USA (2004)
10. Isaac, A., Matthezing, H., van der Meij, L., Schlobach, S., Wang, S., Zinn, C.: Putting ontology alignment in context: usage scenarios, deployment and evaluation in a library case. In: Bechhofer, S., Hauswirth, M., Hoffmann, J., Koubarakis, M. (eds.) ESWC 2008. LNCS, vol. 5021. Springer, Heidelberg (2008)

11. Jonquet, C., Musen, M.A., Shah, N.: A System for Ontology-Based Annotation of Biomedical Data. In: Bairoch, A., Cohen-Boulakia, S., Froidevaux, C. (eds.) DILS 2008. LNCS (LNBI), vol. 5109, pp. 144–152. Springer, Heidelberg (2008)
12. Lindberg, D., Humphreys, B., McCray, A.: The unified medical language system. Methods of Information in Medicine 32(4), 281 (1993)
13. Madhavan, J., Bernstein, P.A., Doan, A., Halevy, A.: Corpus-based schema matching. In: International Conference on Data Engineering (ICDE), Tokyo, Japan (2005)
14. McCann, R., Shen, W., Doan, A.: Matching schemas in online communities: A Web 2.0 approach. In: The 24th Int. Conf. on Data Engineering (ICDE 2008), Cancun, Mexico (2008)
15. Mungall, C.: Mappings in OBO Foundry (2008),
    http://obofoundry.org/wiki/index.php/Mappings
16. Noy, N., Guha, R., Musen, M.A.: User ratings of ontologies: Who will rate the raters? In: AAAI Symp. on Knowledge Collection from Volunteer Contributors, Stanford, CA (2005)
17. Noy, N., Tudorache, T.: Collaborative ontology development on the (semantic) web. In: AAAI Spring Symposium on Semantic Web and Knowledge Engineering (SWKE), Stanford, CA (2008)
18. Noy, N.F., Chugh, A., Alani, H.: The CKC Challenge: Exploring tools for collaborative knowledge construction. IEEE Intelligent Systems 23(1), 64–68 (2008)
19. Noy, N.F., Musen, M.A.: The PROMPT suite: Interactive tools for ontology merging and mapping. International Journal of Human-Computer Studies 59(6), 983–1024 (2003)
20. Palma, R., Hartmann, J., Haase, P.: OMV: Ontology Metadata Vocabulary for the Semantic Web. Technical report (2008), http://ontoware.org/projects/omv/
21. Rector, A., Rogers, J., Pole, P.: The GALEN High Level Ontology. In: 14th International Congress of the European Federation for Medical Informatics, MIE 1996, Denmark, (1996)
22. Rosse, C., Mejino, J.L.V.: A reference ontology for bioinformatics: The Foundational Model of Anatomy. Journal of Biomedical Informatics (2004)
23. Rubin, D.L., et al.: The National Center for Biomedical Ontology: Advancing biomedicine through structured organization of scientific knowledge. OMICS: A Journal of Integrative Biology 10(2) (2006)
24. Sioutos, N., de Coronado, S., Haber, M., Hartel, F., Shaiu, W., Wright, L.: NCI Thesaurus: A semantic model integrating cancer-related clinical and molecular information. Journal of Biomedical Informatics 40(1), 30–43 (2007)
25. Smith, B., et al.: The OBO Foundry: coordinated evolution of ontologies to support biomedical data integration. Nature Biotechnology 25(11), 1251–1255 (2007)
26. Šváb, O., Svátek, V., Stuckenschmidt, H.: A study in empirical and "casuistic" analysis of ontology mapping results. In: Franconi, E., Kifer, M., May, W. (eds.) ESWC 2007. LNCS, vol. 4519, Springer, Heidelberg (2007)
27. Watts, D.: Six Degrees: The New Science of Networks, Vintage (2004)
28. Zacharias, V., Braun, S.: SOBOLEO - social bookmarking and lightweight ontology engineering. In: Workshop on Social and Collaborative Construction of Structured Knowledge at WWW 2007 (2007)
29. Zhang, S., Bodenreider, O.: Alignment of multiple ontologies of anatomy: Deriving indirect mappings from direct mappings to a reference. In: AMIA Annual Symposium (2005)
30. Zhang, S., Bodenreider, O.: Hybrid alignment strategy for anatomical ontologies: results of the 2007 ontology alignment contest. In: 2nd International Workshop on Ontology Matching (OM 2007) at ISWC 2007 (2007)
31. Zhdanova, A., Shvaiko, P.: Community-driven ontology matching. In: 3rd European Semantic Web Conference, Budva, Montenegro, p. 3449 (2006)

Algebras of Ontology Alignment Relations

Jérôme Euzenat

INRIA & LIG
Grenoble, France
Jerome.Euzenat@inrialpes.fr

Abstract. Correspondences in ontology alignments relate two ontology entities with a relation. Typical relations are equivalence or subsumption. However, different systems may need different kinds of relations. We propose to use the concepts of algebra of relations in order to express the relations between ontology entities in a general way. We show the benefits in doing so in expressing disjunctive relations, merging alignments in different ways, amalgamating alignments with relations of different granularity, and composing alignments.

1 Motivations

The heterogeneity of ontologies on the semantic web requires finding the correspondences between them in order to interoperate. The operation of finding correspondences is called ontology matching and its result is a set of correspondences called an alignment [8]. Alignments are used for importing data from one ontology to another or for translating queries.

In general, a correspondence relates an entity, e.g., a class, a property, an instance, of a first ontology to an entity of the second ontology by a specific relation. This relation can be the equivalence or subsumption between these entities or more complex relations, e.g., mereologic relations such as partOf.

Within an alignment, correspondences are interpreted conjunctively, but it may sometimes be necessary to express disjunctions of relations, e.g., when one is only able to establish a subset of the possibly holding relations. Moreover, in the wider context of sharing alignments on the web and composing ontology matchers, it is necessary to manipulate alignments: combining alignments either conjunctively or disjunctively, composing alignments when a direct alignment between two ontologies does not exist or converting alignments using a different set of relations. Current support for alignment is not adapted to this: correspondences are usually expressed with respect to simple relations and the connection between relations is not explicit.

Example 1 (Background). We consider the example of three geographic ontologies designed for statistical purposes. They are loosely built on the Eurostat Nomenclature for Territorial Units for Statistics (or NUTS). We deal with three ontologies adapted to the German (o), English (o') and French (o'') territory. In order to be able to aggregate information from the German and the British sources, engineers need an alignment between o and o'. They will take advantage of alignments provided by various sources.

A. Sheth et al. (Eds.): ISWC 2008, LNCS 5318, pp. 387–402, 2008.
© Springer-Verlag Berlin Heidelberg 2008

We propose to solve this problem by expressing alignment relations within the formalism of algebras of relations. At first sight, algebras of relations may seem like just one possible solution to express disjunctions. However, we show that, in addition to allowing disjunctive relations, this formalism provides many advantages in the manipulation of alignments.

We first present in more detail the notion of ontology alignment and relations between ontology entities (§2), as well as algebra of relations (§3). We then show how, in addition to expressing disjunction, algebras of relations can support several types of relation aggregation (§4), alignment composition (§5), algebraic reasoning on alignments (§6), and weakening of representations (§7). We finally show that this kind of relations can still be manipulated coherently with confidence measures (§8).

2   Alignments and Relations

Alignments express the correspondences between entities belonging to different ontologies (we restrict ourselves to two ontologies here). We provide the definition of the alignment following the work in [6,3].

Definition 1 (Correspondence). *Given two ontologies o and o' with associated entity languages Q_L and $Q_{L'}$, a set of alignment relations Θ, and a confidence structure over Ξ, a correspondence is a quadruple:*

$$\langle e, e', r, n \rangle,$$

such that

- $e \in Q_L(o)$ and $e' \in Q_{L'}(o')$;
- $r \in \Theta$;
- $n \in \Xi$.

The correspondence $\langle e, e', r, n \rangle$ asserts that the relation r holds between the ontology entities e and e' with confidence n.

The entities can be simply made of all the formulas of the ontology language based on the ontology vocabulary. They can be restricted to particular kinds of formulas from the language, such as atomic formulas, or even to terms of the language, like class expressions. It can also restrict the entities to be only named entities. The entity language can be an extension of the ontology language. For instance, it can be a query language, such as SPARQL [14], adding operations for manipulating ontology entities that are not available in the ontology language itself, like concatenating strings or joining relations.

In some tradition, e.g., schema matching [16,15], some authors tend to consider that a correspondence like:

address = street + " " + number

is some kind of ternary complex relation ($\cdot = \cdot + $ " " $ + \cdot$) between three entities address, street and number. In our setting, this is simply a normal correspondence in which the binary relation is equivalence ($=$) and the ontology entities are address and street+" "+number. This is the main reason why we consider ontology entities, the latter entity is a term built on strings and operations on strings (here concatenation +).

The next important component of the alignment is the relation that holds between the entities. We identify a set of relations Θ that is used for expressing the relations between the entities. Matching algorithms primarily use the equivalence relation (=) meaning that the matched objects are the same or are equivalent if these are formulas. It is possible to use relations from the ontology language within Θ. For instance, using OWL, it is possible to take advantage of the `owl:equivalentClass`, `owl:disjointWith` or `rdfs:subClassOf` relations in order to relate classes of two ontologies. These relations correspond to set-theoretic relations between classes: *equivalence* (=), *disjointness* ($\bot$), *less general* ($\sqsubseteq$). They can be used without reference to any ontology language.

For pragmatic reasons, the relationship between two entities is assigned a degree of confidence which can be viewed as a measure of trust in the fact that the correspondence holds – 'I trust 70% the fact that the correspondence is correct or reliable' – and can be compared with the certainty measures provided with meteorological forecasts. These values are taken from a bounded ordered set Ξ that we call a confidence structure. We will come back on this in Section 8 and ignore it until then.

Finally, an alignment is defined as a set of correspondences.

Definition 2 (Alignment). *Given two ontologies o and o', an* alignment *is made up of a set of correspondences between pairs of entities belonging to $Q_L(o)$ and $Q_{L'}(o')$ respectively.*

Example 2 (Alignment). Consider two alignments A_1 and A_2, relating respectively the German to the French ontology and the French ontology to the British one, containing the following correspondences (A_1 is on the left, A_2 on the right):

Konstruktion$\bot$Commune	Commune $\geq$ Municipality
Stadtgebiet $>$ Ville	Ville $\between$ Municipality

This means that A_1 considers that a Konstruction, i.e., a Building, is disjoint from a Commune, i.e., a Ward, and a Stadtgebiet, i.e., a Urban area, is more general than a Ville, i.e., a Town. A_2 expresses that a Commune is more general or equivalent to a Municipality and Ville overlaps withMunicipality, i.e., that both concepts have common instances but none is more general than the other.

This definition does not tell how to interpret this set of correspondences. However, it is clear from usage that it has to be interpreted in a conjunctive manner: all the correspondences are asserted to hold when asserting an alignment.

Hence, the problem of expressing disjunctions of correspondences can be raised. This can be because it is necessary to aggregate the result of methods which address the ontology matching problem from different dimensions, this can be because the person or the program generating the alignment is unsure about the exact relation but knows that this relation is constrained to a specific set of alternative relations.

Example 3 (Disjunctive relations). For instance, an engineer may know that a Stadt, i.e., Town, and a Town are similar things but may not know exactly the nature of the overlaps. She can express that they are not disjoint by the disjunction of relations $<, >$, $\lozenge$ and $=$, thus prohibiting $\perp$. This can also be because the alignment has been generated by composing two alignments. This operation does not usually return a simple relation but a disjunction of such relations, e.g., if Stadtgebiet, i.e., Urban area, is more general than Ville, i.e., Town, and Ville overlaps Municipality, then Stadtgebiet either is more general or overlaps Municipality, it cannot be disjoint with it. Hence, the result is a disjunction of relations.

This is also the case of the $\geq$ (more-general-or-equal) and $\leq$ (more-specific-or-equal) relations used by some systems. These are typically the disjunction of $<$ and $=$ or $>$ and $=$. In fact, practice which considers that if both $\leq$ and $\geq$ hold (conjunction), then $=$ holds, only reflects the set operation: $\{<, =\} \cap \{>, =\} = \{=\}$ or the logical interpretation that:

$$\forall a, b, (a < b \vee a = b) \wedge (a > b \vee a = b) \models a = b \text{ if } <, > \text{ and } = \text{ are exclusive}$$

The first goal of this paper is to consider a systematic treatment for disjunctive alignment relations. For that purpose, we use algebra of relations and we show that this has many advantages.

3   Relation Algebra

An algebra of binary relations (hereafter referred to as relation algebra) [19] is a structure $\langle \Theta, \wedge, \vee, *, 1, 0, 1', \neg \rangle$ such that $\langle \Theta, \wedge, \vee, 1, 0 \rangle$ is a Boolean algebra; $*$ is an associative internal composition law with (left and right) unity element $1'$, that distributes over $\vee$; $\neg$ is an internal involutive unary operator, that distributes over $\vee, \wedge$ and $*$.

We consider a particular type of relation algebras[1] in which Θ is the powerset of a generating set Γ closed under $\neg$ (hereafter $^{-1}$) and $\wedge/\vee$ are set intersection/union $(\cap/\cup)$. Such an algebra of (binary) relations is defined by $\langle 2^\Gamma, \cap, \cup, \cdot, \Gamma, \varnothing, \{=\}, ^{-1} \rangle$ such that:

- Γ is a set of jointly exhaustive and pairwise disjoint (JEPD) relations between two entities. This means that, in any situation, the actual relation between two objects is one and only one of these relations. Sets of relations allow to express uncertainty: the full Γ set is the "I do not know" relation since it is satisfied by any pair of entities;
- $\cap$ and $\cup$ are set operations used to meet and join two sets of base relations, hence if xry or $xr'y$, then $xr \cup r'y$;
- $\cdot$ is the composition operator such that if xry and $y\, r'\, z$, then $x\, r \cdot r'\, z$; "$=$" is such that $\forall r \in \Gamma, (r \cdot =) = (= \cdot r) = r$;
- $^{-1}$ is the converse operator, i.e., such that $\forall e, e' \in \Gamma, ere' \Leftrightarrow e'r^{-1}e$.

These operations are applied to sets of base relations by distributing them on each element, e.g., $R \cdot R' = \bigcup_{r \in R, r' \in R'} r \cdot r'$.

---

[1] [12] shows that a weaker structure than algebras of relations, non associative, can be used in most of the purposes of qualitative calculi. However, we will need associativity later.

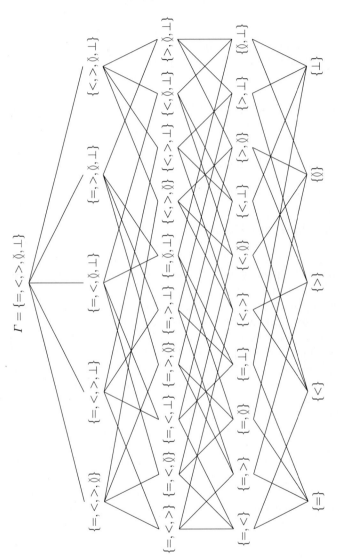

Fig. 1. The lattice of 31 disjunctive relations in $A5$

A typical example of such an algebra is the Allen algebra of temporal interval relations [1]. Here, we will consider, as an example, a simpler algebra, called $A5$, isomorphic to that applying to sets in which the base (JEPD) relations between two sets are equivalent ($=$), includes ($>$), is-included-in ($<$), overlaps ($\lozenge$) and disjoint ($\perp$). In this algebra, all base relations but $<$ and $>$ are there own converse while $>^{-1}=<$ and $<^{-1}=>$.

The complete set of $2^5 - 1 = 31$ valid relations that can be made out of these 5 base relations is depicted in Figure 1. Among these relations, Γ means "I do not know" as it contains all the base relations. $\neq$ is equivalent to $\{<, >, \lozenge, \perp\}$, $\leq$ is equivalent to $\{=, <\}$, $\geq$ to $\{=, >\}$, $\gtrsim$ to $\{<, >, \lozenge\}$ and $\not\perp$ to $\{=, <, >, \lozenge\}$. The composition table is given in Table 1.

Relation algebras can still be used when the ontology entities are formulas (or queries) and the base relations are logical connectives between formulas ($\Rightarrow, \equiv$). Indeed, it is sufficient to split the disjunctive relations into a disjunction of formulas:

$$\phi\{\Rightarrow, \equiv\}\psi \text{ would be equivalent to } \phi \equiv \psi \vee \phi \Rightarrow \psi$$

(these relations are not exclusive anymore).

4   Aggregating Matcher Results

The set Θ is closed for $\vee$, $\wedge$ and $\neg$. This means that any combination of these operations yields an element of Θ. This is very powerful if one wants to combine relations, e.g., for combining correspondences and alignments.

When matching methods bring new evidences for a correspondence from a different perspective, they are though of as bringing new arguments in favor of a correspondence. Hence, its results must be aggregated with union ($R \cup R'$). When the matching methods, instead, are competing algorithms providing all the possible base relations, i.e., providing arguments against the non selected correspondences, then intersection ($R \cap R'$) should be used. Because, we now have two distinguished operations, they can be used together in the same application.

These operations can be used for describing two cases of matching process: an expanding matching process which starts with the empty relation ($\varnothing$) between each pair of entities and which finds evidences for more base relations between these entities aggregating them with $\cup$ and a contracting process which starts with Γ between each pair of entities and which discards support for some base relations between entities aggregating the result with $\cap$. In the first case, the more matching methods are used, the less precise the alignment becomes: this can be balanced by confidence measures as we will see below. In the second case, the more methods are used, the more precise the alignment becomes.

These operations on Θ are used in correspondence aggregation: an alignment is interpreted as all the correspondences it contains hold, and a distributed system is interpreted as all alignments hold. Hence, the disjunctive aggregation of alignments is based on the combination of their set of correspondences with the union of relations; the conjunctive aggregation of alignments is based on the combination of their set of correspondences with the intersection of relations. Hence, we define a normalisation operation $\bar{A}$, which implements the conjunctive interpretation of alignments. It provides exactly one cor-respondence per pair of entities and makes explicit all the relations between entities in A:

$$A^0 = \{\langle e, e', \Gamma \rangle | e \in Q_L(o), e' \in Q_{L'}(o')\}$$
$$\bar{A} = \{\langle e, e', \cap_{\langle e, e', r \rangle \in A \cup A^0} r \rangle\}$$

It is then easy to define intersection:

$$A \wedge A' = \{\langle e, e', r \cap r' \rangle | \langle e, e', r \rangle \in \bar{A}, \langle e, e', r' \rangle \in \bar{A}'\}$$

as well as additional operators such as disjunction and converse of alignments:

$$A \vee A' = \{\langle e, e', r \cup r' \rangle | \langle e, e', r \rangle \in \bar{A}, \langle e, e', r' \rangle \in \bar{A}'\}$$
$$A^{-1} = \{\langle e', e, r^{-1} \rangle | \langle e, e', r \rangle \in \bar{A}\}$$

Within this paper, alignments are always presented in a reduced way, i.e., without trivial $\langle e, e', \Gamma \rangle$ correspondences added by normalisation.

Example 4 (Alignment aggregation). Consider two alignments A_3 and A_5, resulting from two different matchers which match ontologies using different features for ruling out correspondences (A_3 is on the left, A_5 on the right):

Konstruktion$\{\perp\}$Municipality   Stadt$\{<\}$Town
Stadtgebiet$\{>, \emptyset\}$Municipality   Stadtgebiet$\{\perp, \emptyset\}$Municipality

Since these matchers provide competing alignments between ontology o and o', their result can be aggregated conjunctively. The result $A_6 = A_3 \wedge A_5$, is given below as the left-hand side alignment:

Konstruktion$\{\perp\}$Municipality
Stadt$\{<\}$Town   Stadt$\{=, <, >, \emptyset\}$Town
Stadtgebiet$\{\emptyset\}$Municipality   Stadtgebiet$\{\perp\}$Municipality

This alignment is aggregated with the right-hand side alignment A_4. Since they provide evidence for alignments from different perspective, they are aggregated disjunctively, yielding $A_8 = A_4 \vee A_6$:

Stadt$\{=, <, >, \emptyset\}$Town
Stadtgebiet$\{\emptyset, \perp\}$Municipality

Algebras of relations are useful because they can account for these two behaviours. However, there are other benefits brought by algebra of relations.

5   Composing Alignments

Another way of reusing alignments is to deduce new alignments from existing ones. One way to do so, is to compose alignments. If there exists an alignment between ontology o and ontology o'', and another alignment between o'' and a third ontology o', we would like to find which correspondences hold between o and o'. The operation that returns this set of correspondences is called composition.

Alignment composition has already considered [21] with the idea that, in an open system like the Alignment API, the rules for composing alignment relations, e.g., instanceOf · subClassOf = instanceOf, should be given by a composition table. Composition tables come directly from algebra of relations and they naturaly extend from base relations to disjunctions of base relations.

Alignment composition can thus be reduced to combining correspondences with regard to their relations and the structure of related entities and computing the confidence degree of the result. The composition table between the base relations of $A5$ is given in Table 1.

Table 1. Composition table for the $A5$ relation algebra

	$=$	$>$	$<$	$\lozenge$	$\perp$
$=$	$=$	$>$	$<$	$\lozenge$	$\perp$
$>$	$>$	$>$	$><=\lozenge$	$>\lozenge$	$>\lozenge\perp$
$<$	$<$	Γ	$<$	$<\lozenge\perp$	$\perp$
$\lozenge$	$\lozenge$	$>\lozenge\perp$	$<\lozenge$	Γ	$>\lozenge\perp$
$\perp$	$\perp$	$\perp$	$<\lozenge\perp$	$<\lozenge\perp$	Γ

The composition of two alignments A and A' is defined by:

$$A \cdot A' = \overline{\{\langle e, e'', r \cdot r' \rangle | \langle e, e', r \rangle \in A, \langle e', e'', r' \rangle \in A'\}}$$

One can compose an alignment with itself (self-composition) through: $A^2 = A \cdot A^{-1} \cdot A$. This operation may provide new correspondences.

Example 5 (Composing alignments). The alignment A_3 of Example 4, is the result of the composition of alignments A_1 and A_2 of Example 2: $A_3 = A_1 \cdot A_2$. The first simple application of Table 1 occurs when composing Konstruktion $\{\perp\}$ Commune and Commune $\{>,=\}$ Municipality, then it can be deduced that Konstruktion $\{\perp\}$ Municipality because $\{\perp\} \cdot \{>,=\} = (\perp \cdot >) \cup (\perp \cdot =) = \{\perp\}$. Things can be more complex, when composing Stadtgebiet $\{>\}$ Ville and Ville $\{\lozenge\}$ Municipality, then Table 1 allows to deduce that Stadtgebiet $\{>, \lozenge\}$ Municipality because $\{>\} \cdot \{\lozenge\} => \cdot \lozenge = \{<, \lozenge\}$. The result provided by the table in this case is a disjunction of relations because it is not possible to obtain more precise information from the alignments alone.

Very often the composition of two base relations is not a base relation but a disjunction of relations. Hence, if we were not dealing with sets of base relations, it would not be possible to represent the composition of two alignments by an alignment.

Moreover, defining composition by an algebra of relations automatically satisfies all the constraints on the categorical characterisation of alignments defined in [21]: it must be associative and have an identity element. This is true from the definition of algebra of relations.

6   Algebraic Reasoning with Alignments

α-consequences are correspondences which are entailed by two aligned ontologies [7]; they can be extended as the correspondences entailed by a system of many ontologies and many alignments between them. [20] introduced the notion of quasi-consequences as the set of formulas entailed by a set of alignments alone (without considering ontologies). This notion can be straightforwardly extended to correspondences as quasi-α-consequences: the correspondences which are entailed by the set of alignments, when considering the ontologies as void of axioms. Quasi-α-consequences are also α-consequences.

Reasoning on alignments aims at using existing alignments in order to deduce more and more complete alignments. Such a reasoning procedure can be considered, for soundness and completeness, with respect to α-consequences.

Algebraic reasoning (using combination of composition, converse, and intersection) can be used as a practical and efficient way to reason with alignments. The algebraic closure of a set of alignments S is the set of normalised alignments, containing $\bar{S}$, closed under composition, converse and intersection.

This procedure is correct (the algebraic operations can be transformed into their logical equivalent). However, since it does not consider ontologies, it can only deduce quasi-α-consequences. We have no guarantee that it is complete even for finding quasi-α-consequences.

However, this can already be used for two purposes: (1) improving the existing alignments by deducing new correspondences coming either from the alignment itself or from other alignments, and (2) checking the consistency of a set of alignments (or one alignment). Indeed, if we can deduce $x\{\}y$, e.g., because $x\{<\}y$ and $x\{>\}y$ for two competing matchers, since the intersection is empty we know that the alignment itself is inconsistent. This kind of reasoning can be more complex, involving several alignments as well as composition operations, i.e., checking a whole distributed system. Then, the set of alignments as a whole would be inconsistent, hence the distributed system would have no model.

Example 6 (Algebraic reasoning). The simplest instance of an inconsistent alignment is to have two contradictory statements like Konstruktion$\{\perp\}$Town and Konstruktion$\{<\}$Town in the same alignment. The conjunction of these two relations, obtained by normalisation, is empty. Such an inconsistent alignment can also be obtained by combining consistent alignments. For instance, aggregating conjunctively alignments A_3 and A_4 of Example 4, will generate the inconsistent Stadtgebiet $\{\}$ Municipality correspondence.

Algebraic reasoning allows to expand alignments. For instance, Stadt$\{>\}$Town, Stadtgebiet$\{<\}$Town, Stadtgebiet$\{\perp\}$Municipality entails Stadt$\{>, \emptyset, \perp\}$Municipality. Computing the compositional closure of this alignment will find this correspondence.

Moreover, if the initial alignment also contain Stadt$\{=, <\}$Municipality, the compositional closure will bring the inconsistence to light.

Once again, it is possible to use disjunctive relations to better evaluate alignments. Indeed, the problem is that if a matcher returns a correspondence between two entities with relation $\leq$ while the expected (and exact) relation was $<$, then the use of syntactic precision and recall measures would count this relation as incorrect. Hence, if the expected alignment was made of this correspondence alone, both precision and recall would be 0. This is unfair because it cannot be said that this correspondence is both incorrect and incomplete. In fact, it is incomplete, because it does not provide the exact relation, but not incorrect, because the relation is more general than the correct one.

This is indeed what happens with semantic precision and recall [7]: since the relation $\{<, =\}$, which is the disjunction of $<$ and $=$, can be deduced from $\{<\}$ alone, it would count as correct for semantic precision and still as incorrect for semantic recall because $\{<, =\}$ does not entail $\{<\}$.

We could introduce an algebraic precision and recall for evaluating ontology alignments as an intermediary step between classic precision and recall and semantic precision and recall. It would simply use the inclusion between the relations as suggested above instead of the entailment between correspondences of [7] and would be far easier to compute. The resulting measure would be a relaxation of precision and recall in the sense of [4].

7   Algebra Granularity

In order to investigate granularity within algebras of relations, we introduced the notion of weakening [5]. Weakening an algebra of relations simply consists of grouping together several base relations and taking the result as the base relations of the, less precise, weaker algebra.

In fact, taking any maximal antichain[2] that preserves converse in the lattice of Figure 1 yields a base for an algebra of relations. Other constraints can be put on weakening, such as requiring that they preserve a neighbourhood structure. Neighbourhood structures for algebras of relations have been introduced in [10]. They are based on a connectivity relation between relations that is used for defining neighbourhood. This connectivity relation can be based on different properties of the domain the relations apply to. We have shown that granularity operators, at least in time and space algebras, can be automatically built on such neighbourhoods [5].

In terms of alignment, the interesting aspect of this weakening operation is that it helps considering that alignments using different sets of relations are compatible and can still be used together. Coming back to the example of set-relations, there can be systems that provides only the $=$ base relation leaving implicitly all the others as Γ, there are other systems like the one considered previously which consider $=, <, >, \emptyset$ and $\perp$. The set of base relations is different and thus it is not easy to combine two such alignments. However, if we consider that the first one is a weakening of the second one (grouping $<, >, \emptyset$ and $\perp$ into $\neq$), then it is possible to import one alignment into

---

[2] An antichain is a set of relations such that no one is comparable to the other.

another formalism and vice-versa (at the expense of completeness when we export to the weaker algebra).

Example 7 (Algebra granularity). Different sources of alignments may provide alignments with different kinds of relations between objects. For instance, the following A_7 alignment (left-hand side) is expressed in the simple $\{\bot, \cancel{\bot}\}$ algebra, i.e., identifying only incompatible elements.

Stadt$\{\cancel{\bot}\}$Town   Stadt$\{\cancel{\bot}\}$Town

Stadtgebiet$\{\bot\}$Municipality   Stadtgebiet$\{\bot, \cancel{\bot}\}$Municipality

Thanks to the use of compatible algebras, A_7 can be expressed in the more expressive algebra. In fact, the alignment A_4 of Example 4 is the transcription of A_7 in the $A5$ algebra. On the other hand, it is possible to degrade an alignment into the coarser algebra at the expense of precision. The alignment on the right-hand side is the result of converting the alignment A_5 of Example 4 to the $\{\bot, \cancel{\bot}\}$ algebra.

Figure 2 shows the "interesting" weakened algebras of relations from the initial algebra. It features the $\{=, \neq\}$ algebra but also shows that the usually considered $\{=, \leq, \geq, \bot\}$ is not a correct base for such an algebra because it is neither jointly exhaustive ($\leq$ and $=$ can occur at the same time), nor pairwise disjoint ($\lozenge$ is missing).

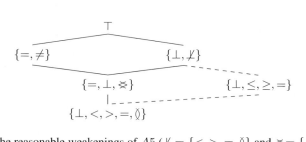

Fig. 2. The reasonable weakenings of $A5$ ($\cancel{\bot} = \{<, >, =, \lozenge\}$ and $\lessgtr = \{<, >, \lozenge\}$)

8   Compatibility with Confidence Measures

Most matchers assign confidences to the correspondences they produce. They express to what extent they trust the correspondence. This confidence is expressed in a confidence structure:

Definition 3 (Confidence structure). *A confidence structure is an ordered set of degrees $\langle \Xi, \leq \rangle$ for which there exists a greatest element $\top$ and a smallest element $\bot$.*

The usage of confidence degrees is that the higher the degree with regard to $\leq$, the most likely the relation holds. This means that a particular confidence degree entails all the inferior confidence degrees.

 With algebras of relations, the confidence assigned to a relation applies to the disjunction as a whole. Hence, when a confidence is assigned of some relation, this confidence can be assigned to all its supersets, but not to its subsets. The larger the relation, the stronger the confidence can be: typically, Γ should be given full confidence ($\top$) and $\varnothing$ the lowest one ($\bot$).

A correspondence $\langle e, e', r, n \rangle$ entails another correspondence $\langle e, e', r', n' \rangle$ if and only if $r \subseteq r'$ and $n \geq n'$. The normalisation operation will then only retain maximal elements for the induced order:

$$A^0 = \{\langle e, e', \Gamma, \top \rangle | e \in Q_L(o), e' \in Q_{L'}(o')\}$$
$$\bar{A} = max^{\subseteq, \geq} A \cup A^0$$
$$A^{-1} = \{\langle e', e, r^{-1}, n \rangle | \langle e, e', r, n \rangle \in \bar{A}\}$$

The previous definition of $\bar{A}$ satisfies this definition, either with $\subseteq$ as an order or with $\top$ as the only confidence grade.

The most widely used structure is based on the real number unit interval $[0\ 1]$, but some systems simply use the boolean lattice. It is convenient to interpret the greatest element as the boolean true and the smallest element as the boolean false. Some other possible structures are fuzzy degrees, probabilities or other lattices. [11] has investigated the structure of fuzzy confidence relations. Below, we simply consider the unit interval with the usual arithmetic operations.

Example 8 (Normalisation with confidences). For instance,

$$\textsf{Stadt}\{<\}_{.8}\textsf{Town}, \textsf{Stadt}\{<, =\}_{.9}\textsf{Town}$$

is a normalised alignment (if it is added $\textsf{Stadt}\Gamma_\top\textsf{Town}$), which is interpreted (still conjunctively) as: I am confident with .8 that Stadt is strictly subsumed by Town, and confident with .9 that Stadt subsumed or equivalent to Town. If the confidences were stripped down, this conjunctive statement would be reduced to $\textsf{Stadt}\{<\}\textsf{Town}$. This corresponds to having the confidences set to 1. Similarly, the alignment:

$$\textsf{Stadt}\{<\}_{.9}\textsf{Town}, \textsf{Stadt}\{<, =\}_{.8}\textsf{Town}$$

is not minimal and can be reduced to $\textsf{Stadt}\{<\}_{.9}\textsf{Town}$. This is because $\{<\} \subseteq \{<, =\}$ and $.9 \geq .8$, hence $\textsf{Stadt}\{<\}_{.9}\textsf{Town}$ entails $\textsf{Stadt}\{<, =\}_{.8}\textsf{Town}$.

Designing operators for merging two alignments with disjunctive relations and confidence, is more open than considering the disjunctive relations alone. The implementation of the aggregation operators can differ as well as the measure for aggregating confidence. Instead of applying a union or intersection, each pair of entities will be assigned for each disjunction of relation (returned by a matcher) a confidence measure which will result in the aggregation of the confidence of all other matchers. There are many different ways to aggregate matcher results depending on confidence [8]:

- Triangular norms (min, weighted products) are useful for selecting only the best results in case of competing alignments;
- Multidimensional distances (Euclidean distance, weighted sum) are useful for taking into account all dimensions in case of complementary alignments;
- Fuzzy aggregation (min, weighted average) is useful for aggregating competing algorithms and averaging their results;
- Other specific measures, e.g., ordered weighted average, may also be used.

The natural aggregation measure is the conjunctive one induced by the normalisation procedure:

$$A \wedge A' = \overline{\bar{A} \cup \bar{A}'}$$

However, other aggregation operations can be designed from a confidence aggregation function f and a relation combination $\times$ as:

$$A \times^f A' = \overline{\{\langle e, e', r \times r', f(n, n')\rangle | \langle e, e', r, n\rangle \in \bar{A}, \langle e, e', r', n'\rangle \in \bar{A}'\}}$$

The $\times$ operation can be, for instance, $\cap$, $\cup$ or id (which applies only if both sets of relations are equal). We have presented it with two alignments, but this extends straightforwardly to n alignments.

Example 9 (Aggregation with confidence). Consider two alignments A_9 and A_{10} containing the following correspondences (A_9 is on the left, A_{10} on the right):

Stadt$\{=, <\}_{.5}$Town	Stadt$\{=\}_{1.0}$Town
Stadt$\{=, \perp\}_{.2}$Town	
Stadtgebiet$\{\perp\}_{.1}$Municipality	Stadtgebiet$\{\perp, \emptyset\}_{.2}$Municipality

Applying $\cap^{\min}(A_9, A_{10})$ yields, once normalised, the following correspondences:

$$\text{Stadt}\{=\}_{1.}\text{Town}$$
$$\text{Stadtgebiet}\{\perp, \emptyset\}_{.2}\text{Municipality}$$
$$\text{Stadtgebiet}\{\perp\}_{.1}\text{Municipality}$$

The same operation with $\cup$ and weighted sum (weight being $1/3$ and $2/3$), would yield:

$$\text{Stadt}\{<, =\}_{.83}\text{Town}$$
$$\text{Stadt}\{\perp, =\}_{.4}\text{Town}$$
$$\text{Stadtgebiet}\{\perp, \emptyset\}_{.16}\text{Municipality}$$

Which cannot be reduced.

Ideally, we would like that these aggregation functions preserve the opportunity given in Section 4 to have different operations usable in different situations.

This works well for min and weighted product functions together with intersection.

Property 1 (Reduction to intersection). The (normalised) weighted products and min functions, if applied to correspondences with confidences in $\{0, 1\}$ with $\cap$ computes $\cap$.

This works well for intersection because the methods used have 0 as an absorbing element. Since the other operations do not have a (upper) absorbing element, the confidence that they return are not necessary equal to 1 (or $\top$).

Property 2 (Reduction to union). The weighted sum, Euclidean distance and other instance of the Minlowski distance, if applied to correspondences with confidences in $\{0, 1\}$ and $\cup$ computes $\cup$. However, the confidence value assigned to the resulting correspondences may not be $\top$.

9   Related Work

There are very few papers about alignment relations in the literature. [2] uses a spatial relation algebra for the purpose of expressing correspondences between spatio-temporal ontologies. However, the relations are only used in the ontology language and the alignment relations are still the classical $=$ and $\leq$ relations. [18] went one step further by explicitly considering a set of base relations similar to $A5$, but they do not consider using disjunction of relations as alignment relations (hence most of what is in this paper does not apply).

In database schema matching, the notion of mapping composition is prominent and has been thoroughly investigated [13]. The problem there is to design a composition operator that guarantee that the successive application of two mappings yields the same results as the application of their composition [9]. The approach is relatively different since relations in this context are always subsumption and their applications involve manipulating the ontology language instead of the alignment language (here the alignment relations). It is even shown that in general the result of composition may require a stronger language than the alignment language (hence the actual result of the composition is not an alignment). The approach taken with composition in algebra of relations is weaker – results are always in the alignment language – but is not complete.

[17] suggests to transform the XML schema matching problem into a constraint optimisation problem. Constraint optimisation problems are constraint satisfaction problems whose solutions maximise a quantity. Though the authors do not consider algebras of relations, this work suggests to use them, since Allen's constraint propagation algorithm [1] is an instance of a constraint propagation algorithm (which can be generalised as arc-consistency).

10   Conclusions

Starting from the need to express disjunctions of relations between ontology entities, we have introduced algebras of relations as a tool for expressing alignment relations. We have shown that this tool can easily express the most common relations used in ontology matching. However, algebras of relations are very flexible tools and new ones can be created for specific applications, e.g., mixing set and mereological relations.

What makes algebras of relations particularly attractive, besides expressing disjunctions is their ability to support genuinely other needed operations, and, in particular:

- conjunction and disjunction operators can be used as a more flexible means of combining alignments;
- composition had already been identified as the perfect tool for composing alignments: disjunction in alignments enables the expression of composition results as alignments;
- it can also be used for reasoning directly at the algebraic level and detecting valid consequences and constraints on alignments, before even considering the deeper (onto)logical level;
- weakening algebras can be used for combining alignments expressed in different (but compatible) algebras of relations.

This paper is a theoretical paper. We have no experimental setting to demonstrate the superiority of the approach over an eventual previous one. However, we claim that it conveniently demonstrate the benefits brought by the use of algebra of relations within ontology alignments. Algebra of relations is a well-studied domain and the fact that it can apply straight away to ontology alignment is very precious in a context when we want to freely share alignments and combine matching methods on the web.

No implementation is available yet: the full support for algebra of relations remains to be implemented in our Alignment API [6]. Implemented operations would effectively compute composition so as to be able to provide new alignments to clients.

Acknowledgements

This work started from a discussion with José Ángel Ramos and Asunción Gómez Pérez at Universidad Politécnica de Madrid. José Ángel provided further comments on the previous version of this paper. It also received comments from Jérôme David and Pavel Shvaiko.

References

1. Allen, J.: Maintaining knowledge about temporal intervals. Communication of the ACM 26(11), 832–843 (1983)
2. Bennacer, N.: Formalizing mappings for OWL spatiotemporal ontologies. In: Bressan, S., Küng, J., Wagner, R. (eds.) DEXA 2006. LNCS, vol. 4080, pp. 368–378. Springer, Heidelberg (2006)
3. Bouquet, P., Ehrig, M., Euzenat, J., Franconi, E., Hitzler, P., Krötzsch, M., Serafini, L., Stamou, G., Sure, Y., Tessaris, S.: Specification of a common framework for characterizing alignment. Deliverable D2.2.1, Knowledge web NoE (2004)
4. Ehrig, M., Euzenat, J.: Relaxed precision and recall for ontology matching. In: Proc. K-CAP Workshop on Integrating Ontologies, Banff (CA), pp. 25–32 (2005)
5. Euzenat, J.: Granularity in relational formalisms with application to time and space representation. Computational intelligence 17(4), 703–737 (2001)
6. Euzenat, J.: An API for ontology alignment. In: McIlraith, S.A., Plexousakis, D., van Harmelen, F. (eds.) ISWC 2004. LNCS, vol. 3298, pp. 698–712. Springer, Heidelberg (2004)
7. Euzenat, J.: Semantic precision and recall for ontology alignment evaluation. In: Proc. 20th International Joint Conference on Artificial Intelligence (IJCAI), Hyderabad (IN), pp. 348–353 (2007)
8. Euzenat, J., Shvaiko, P.: Ontology matching. Springer, Heidelberg (2007)
9. Fagin, R., Kolaitis, P., Popa, L., Tan, W.-C.: Composing schema mappings: Second-order dependencies to the rescue. ACM Transactions on Database Systems 30(4), 994–1005 (2005)
10. Freksa, C.: Temporal reasoning based on semi-intervals. Artificial intelligence 54(1), 199–227 (1992)
11. Gal, A., Anaby-Tavor, A., Trombetta, A., Montesi, D.: A framework for modeling and evaluating automatic semantic reconciliation. The VLDB Journal 14(1), 50–67 (2005)
12. Ligozat, G., Renz, J.: What is a qualitative calculus? a general framework. In: Zhang, C., W. Guesgen, H., Yeap, W.-K. (eds.) PRICAI 2004. LNCS (LNAI), vol. 3157, pp. 53–64. Springer, Heidelberg (2004)

13. Madhavan, J., Halevy, A.: Composing mappings among data sources. In: Proc. 29th International Conference on Very Large Data Bases, Berlin (DE), pp. 572–583 (2003)
14. Prud'hommeaux, E., Seaborne, A. (eds.): SPARQL query language for RDF. Working draft, W3C (2007)
15. Rahm, E., Bernstein, P.: A survey of approaches to automatic schema matching. The VLDB Journal 10(4), 334–350 (2001)
16. Sheth, A., Larson, J.: Federated database systems for managing distributed, heterogeneous, and autonomous databases. ACM Computing Surveys 22(3), 183–236 (1990)
17. Smiljanić, M., van Keulen, M., Jonker, W.: Formalizing the XML schema matching problem as a constraint optimization problem. In: Andersen, K.V., Debenham, J., Wagner, R. (eds.) DEXA 2005. LNCS, vol. 3588, pp. 333–342. Springer, Heidelberg (2005)
18. Sotnykova, A., Vangenot, C., Cullot, N., Bennacer, N., Aufaure, M.-A.: Semantic mappings in description logics for spatio-temporal database schema integration. Journal on Data Semantics III, 143–167 (2005)
19. Tarski, A.: On the calculus of relations. Journal of symbolic logic 6(3), 73–89 (1941)
20. Zimmermann, A.: Sémantique des connaissances distribuées. PhD thesis, Université Joseph-Fourier, Grenoble (FR) (2008)
21. Zimmermann, A., Krötzsch, M., Euzenat, J., Hitzler, P.: Formalizing ontology alignment and its operations with category theory. In: Proc. 4th International Conference on Formal Ontology in Information Systems (FOIS), Baltimore (MD US), pp. 277–288 (2006)

Scalable Grounded Conjunctive Query Evaluation over Large and Expressive Knowledge Bases

Julian Dolby[1], Achille Fokoue[1], Aditya Kalyanpur[1], Li Ma[2], Edith Schonberg[1], Kavitha Srinivas[1], and Xingzhi Sun[2]

[1] IBM Watson Research Center, P.O. Box 704, Yorktown Heights, NY 10598, USA
{dolby,achille,adityakal,ediths,ksrinivs}@us.ibm.com
[2] IBM China Research Lab, Beijing 100094, China
{malli,sunxingz}@cn.ibm.com

Abstract. Grounded conjunctive query answering over OWL-DL ontologies is intractable in the worst case, but we present novel techniques which allow for efficient querying of large expressive knowledge bases in secondary storage. In particular, we show that we can effectively answer grounded conjunctive queries without building a full completion forest for a large Abox (unlike state of the art tableau reasoners). Instead we rely on the completion forest of a dramatically reduced summary of the Abox. We demonstrate the effectiveness of this approach in Aboxes with up to 45 million assertions.

1 Introduction

Scalable conjunctive query answering is an important requirement for many large-scale Semantic Web applications. In this paper, we present a tableau-based reasoning solution for answering grounded conjunctive queries over large and expressive Aboxes. Grounded conjunctive queries, which use distinguished variables only, are more realistic in practice and can also be answered more efficiently than the more general case. Our approach is sound and complete for DL $\mathcal{SHIN}$ (OWL-DL minus nominals and datatypes).

The naive tableau-based algorithm for grounded conjunctive query is to split the query into its component membership atoms and relationship atoms, solve each atom separately, and join the respective bindings at the end. For example, consider a conjunctive query $C(x) \wedge R(x,y) \wedge D(y)$ where x and y are distinguished variables, C and D are concepts, and R and S are roles. Naively, the membership atoms $C(x)$ and $D(x)$ and the relationship atom $R(x,y)$ are solved as three separate queries, and then the result bindings are joined. Without any optimization, solving each membership atom requires testing every individual in the Abox, and solving each relationship query requires testing every pair of individuals in the Abox. For the membership atom $C(x)$, for each individual a, the assertion $a : \neg C$ is added to the Abox, and the new Abox is tested for

A. Sheth et al. (Eds.): ISWC 2008, LNCS 5318, pp. 403–418, 2008.
© Springer-Verlag Berlin Heidelberg 2008

consistency. For the relationship atom $R(x, y)$, for each pair of individuals a and b, the assertions $a : \neg \exists R N_b$ and $b : N_b$ are added to the Abox, where N_b is a new concept. This abox is then tested for consistency. Optimizations reduce the number of tests that need to be performed. However, this approach remains fundamentally impractical for large Aboxes.

This paper builds on our previous technique for solving membership queries over $\mathcal{SHIN}$ KBs containing millions of assertions [1]. The technique applies a standard tableau algorithm to a summary Abox $\mathcal{A}'$ rather than the original Abox $\mathcal{A}$. The summary $\mathcal{A}'$ is created by aggregating individuals with the same concept sets (i.e., the same set of explicit types) into a single summary individual (of that same type). For a given membership query, its negation is added to the concept set of each individual a in $\mathcal{A}'$. If the summary is consistent, then all individuals mapped to a can be ruled out as solutions to the query. If inconsistent, it is possible that either (i) a subset of individuals mapped to a are instances of the query or (ii) the inconsistency is a spurious effect of the summarization. We determine the answer through *refinement*, a process which selectively expands the summary Abox by focusing on inconsistency *justifications* (minimal assertion sets implying the inconsistency), and making them more precise w.r.t the original Abox. Precise justifications are then used to find query solutions. A key point here is that the individuals in the summary Abox after each refinement step, even individuals in precise justifications, still represent many individuals in the original Abox. The scalability of the approach comes from the fact that it makes decisions on groups of individuals as a whole in the summary.

We extend our summarization-based technique to solve grounded conjunctive queries. Like the naive tableau-based conjunctive algorithm, we split a conjunctive query into its atomic parts, solve each atom separately, and join the respective bindings at the end. To find bindings for membership atoms, we apply our membership algorithm on the summary Abox (Section 3). The relationship atoms provide additional optimization opportunities: potential membership query bindings that do not satisfy relationship constraints are filtered out as candidates. We use the completion forest of the summary Abox to filter candidates, and reduce the search space significantly. We prove that this optimization is correct for summary Aboxes in Section 3.1.

Solving relationship atoms efficiently using the summary Abox is not as straightforward. For a relationship atom $R(x, y)$ and $\mathcal{A}'$ individuals a and b, we cannot simply add $a : \neg \exists R N_b$ to $\mathcal{A}'$ and apply our summary Abox membership algorithm. This is because both a and b in $\mathcal{A}'$ each represent many individuals in $\mathcal{A}$. If a and b satisfy this relationship atom, we need to find all pairs a_i, b_i in $\mathcal{A}$ that map to a and b respectively, such that $R(a_i, b_i)$ holds in $\mathcal{A}$. This requires testing some individual pairs in the original Abox for completeness, which defeats the advantage of summarization[1]. Instead, we apply datalog reasoning over the original Abox to conservatively estimate candidates, and apply a modified summary graph algorithm to determine which of these candidates are real solutions. We present details in Section 4.

---

[1] For details, see [2].

Our contributions in this paper are as follows: (a) we present a technique to perform scalable grounded conjunctive query answering over large and expressive Aboxes which relies on an important new property – using the completion forest of the summary Abox for various optimizations; (b) we demonstrate the effectiveness of this technique with very large Aboxes on the UOBM benchmark; (c) we demonstrate graceful degradation of our algorithm's performance, such that queries whose solutions do not exploit non-determinism in the KB (e.g., do not require non-deterministic mergers between individuals) are performed very efficiently.

2 Background

2.1 Definition of Conjunctive Query

Given a knowledge base (KB) $\mathcal{K}$ and a set of variables V disjoint with the set Ind of named individuals in $\mathcal{K}$, a conjunctive query Q is of the form $(x_1, ..., x_n)$ $\leftarrow q_1 \wedge ... \wedge q_m$ where, for $1 \leq i \leq n$, $x_i \in V$ and, for $1 \leq j \leq m$, q_j is a query term. A query term q is of the form $C(x)$ or $R(x, y)$ where x and y are either variables or named individuals in $\mathcal{K}$, C is a concept expression and R is a role. $Var(Q)$ refers to the set of variables occurring in query Q. Let $\pi : Var(Q) \rightarrow Ind$ be a total function from variables in Q to named individual in $\mathcal{K}$. For a query term q, $\pi.q$ denotes the query term obtained by substituting in q all occurrences of a variable x by $\pi(x)$.

$(a_1, ..., a_n)$ is a solution in the KB $\mathcal{K}$ of the conjunctive query Q of the form $(x_1, ..., x_n) \leftarrow q_1 \wedge ... \wedge q_m$ iff. there is a total function $\pi : Var(Q) \rightarrow Ind$ such that the following hold : (1), for $1 \leq i \leq n$, $\pi(x_i) = a_i$, and (2), for $1 \leq j \leq m$, $\mathcal{K} \models \pi.q_j$ (i.e. $\mathcal{K}$ entails $\pi.q_j$).

2.2 Summarization and Refinement

In our earlier work, we presented an algorithm based on summarization and refinement to scale consistency checking and membership query answering to large Aboxes in secondary storage. A key feature of our algorithm is that we perform consistency detection on a summarized version of the Abox rather than the Abox in secondary storage [3]. A summary Abox $\mathcal{A}'$ can be constructed by mapping all individuals in the original Abox $\mathcal{A}$ with the same concept set to a single individual in the summary $\mathcal{A}'$. Formally, an Abox $\mathcal{A}'$ is a summary Abox of a $\mathcal{SHIN}^2$ Abox $\mathcal{A}'$ if there is a mapping function $\mathbf{f}$ that satisfies the following constraints:

(1) if $a : C \in \mathcal{A}$ then $\mathbf{f}(a) : C \in \mathcal{A}'$
(2) if $R(a, b) \in \mathcal{A}$ then $R(\mathbf{f}(a), \mathbf{f}(b)) \in \mathcal{A}'$
(3) if $a \neq b \in \mathcal{A}$ then $\mathbf{f}(a) \neq \mathbf{f}(b) \in \mathcal{A}'$

---

[2] We assume without loss of generality that $\mathcal{A}$ does not contain an assertion of the form $a \doteq b$

If the summary Abox $\mathcal{A}'$ obtained by applying the mapping function $\mathbf{f}$ to $\mathcal{A}$ is consistent w.r.t. a given Tbox $\mathcal{T}$ and a Rbox $\mathcal{R}$, then $\mathcal{A}$ is consistent w.r.t. $\mathcal{T}$ and $\mathcal{R}$. However, the converse does not hold. In general, an inconsistency in the summary may reflect either a real inconsistency in the original Abox, or could simply be an artifact of the summarization process.

In the case of an inconsistent summary, we use a process of iterative refinement described in [1] to make the summary more precise, to the point where we can conclude that an inconsistent summary $\mathcal{A}'$ reflects a real inconsistency in the actual Abox $\mathcal{A}$. Refinement is a process by which only the part of the summary that gives rise to the inconsistency is made more precise, while preserving the summary Abox properties(1)-(3). To pinpoint the portion of the summary that gives rise to the inconsistency, we focus on the *justification* for the inconsistency, where a justification is a minimal set of assertions which, when taken together, imply a logical contradiction.

2.3   Tableau Completion Forest

As described in [4], the tableau algorithm operates on a completion forest $F = (G, \mathcal{L}, \neq, \doteq)$ where G is a graph, with nodes corresponding to individuals and edges corresponding to relations; $\mathcal{L}$ is a mapping from a node x in G to a set of concepts, $\mathcal{L}(x)$, and from an edge $< x, y >$ in G to a set of roles, $\mathcal{L}(< x, y >)$, in $\mathcal{R}$; $\doteq$ is an equivalence relation corresponding to the equality between nodes of G; and $\neq$ is the binary relation *distinct from* on nodes of G. At the beginning of the execution of the tableaux algorithm on an Abox $\mathcal{A}$, the completion forest is initialized as follows: There is a node x in G iff there is an individual x in $\mathcal{A}$. $< x, y >$ is an edge in G with $R \in \mathcal{L}(< x, y >)$ iff $R(x, y) \in \mathcal{A}$. For x and y in G, $x \neq y$ iff $x \neq y \in \mathcal{A}$. Initially, there are no x and y in G such that $x \doteq y$. The tableaux algorithm consists of executing a set of non-deterministic rules to satisfy constraints in $\mathcal{A}$. As soon as an obvious inconsistency, a clash, is detected, the algorithm either backtracks and selects a different non-deterministic choice or stops if all non-deterministic choices have already been made. A *root* node a is a node present in the initial completion forest (it corresponds to the named individual with the same name in $\mathcal{A}$).

For a root node c in the completion forest F, the root node $\alpha(c)$ is defined as follows (informally, $\alpha(c)$ corresponds to the node in which c has been directly or indirectly merged):

$$\alpha(c) = \begin{cases} c \text{ if } \mathcal{L}(c) \neq \emptyset \\ d \text{ if } \mathcal{L}(c) = \emptyset, d \text{ is the unique root node in } F \\ \quad \text{with } \mathcal{L}(d) \neq \emptyset \text{ and } d \doteq c \end{cases}$$

3   Solving the Membership Query Part

The algorithm for solving membership queries on the summary Abox consists of two phases: finding candidate individuals in the summary Abox, and then applying the membership query algorithm as described in [1] to all of the candidates. The technique for finding candidates is described in Section 3.1. The complete algorithm is described in Section 3.2.

3.1 Optimizing Conjunctive Querying with the Summary Completion Forest

To evaluate all the membership query atoms in the conjunctive query efficiently, we restrict our tests to candidate individuals that conservatively satisfy all the relationship atoms in the conjunctive query by making use of the completion forest of the summary Abox. Note that in general (as described in [5]), the completion forest of an Abox can be used to rule out candidates a, b to test for a relationship query $R(x, y)$. The intuition here is that a completion forest F represents an abstraction of a model of the Abox, and thus if b is not an R-neighbor[3] of a in F (and R is not transitive), the relation $R(a, b)$ cannot be entailed by the KB. We apply the same principle to the completion forest of the summary Abox, which is possible due to theorem 1 below.

By theorem 1, if F' denotes the clash-free completion forest resulting from the consistency check on the summary $\mathcal{A}'$ of $\mathcal{A}$, then there exists a complete and clash-free completion forest F resulting from a direct application of tableau rules on $\mathcal{A}$, such that for two named individuals in $\mathcal{A}$, a and b, if $\alpha(\mathbf{f}(b))$ is not a R-neighbor of $\alpha(\mathbf{f}(a))$ then b is not a R-neighbor of a. In other words, we can rule out the existence of R-neighbors of a in F based on the non-existence of R-neighbors of $\alpha(\mathbf{f}(a))$ in F'. Therefore, candidate solutions for a query of the form $R(x, y)$ can be pruned based on completion forest checking on F' instead of F.

Theorem 1. *Let $K = (\mathcal{A}, \mathcal{T}, \mathcal{R})$ be a consistent knowledge base. Let $\mathbf{f}$ be a summary mapping function that maps $\mathcal{A}$ to a consistent summary Abox $\mathcal{A}'$. Let F' be the complete and clash-free completion forest resulting from a consistency check on $\mathcal{A}'$, $\mathcal{T}$ and $\mathcal{R}$. There exists a complete and clash-free completion forest F resulting from an application of tableau rules directly on $\mathcal{A}$ such that, for named individuals a and b in F originally present in $\mathcal{A}$ and a role S in $\mathcal{R}$,*

(1) *$\mathcal{L}(a) \subseteq \mathcal{L}'(\alpha(\mathbf{f}(a)))$ (where $\mathcal{L}(a)$ denotes the concept set of a in F, and $\mathcal{L}'(\alpha(\mathbf{f}(a)))$ is the concept set of the $\alpha(\mathbf{f}(a))$ in F'*
(2) *if b is S-neighbor of a in F, then, in F', $\alpha(\mathbf{f}(b))$ is a S-neighbor of $\alpha(\mathbf{f}(a))$.*

Proof. The proof relies on the following main ideas:

– First, to make sure that properties (1) and (2) of Theorem 1 hold, we use F' to guide the execution of non-deterministic rules on $\mathcal{A}$ (i.e. we make the same choices as in F').

---

[3] By definition, y is a R-neighbor of x iff. $S(x, y) \in \mathcal{A}$ or $P(y, x) \in \mathcal{A}$ where S and P^- are subroles of R

– Second, we maintain, during the execution of the tableau algorithm on $\mathcal{A}$, a mapping σ that maps nodes x in the completion forest F obtained from $\mathcal{A}$ to nodes in F', regardless of whether x refers to a root node, or to a generated node. Furthermore, the relationship between a node x in F and $\sigma(x)$ should be compatible with properties (1) and (2) of Theorem 1. This mapping of x in F to nodes in F' is not straightforward in the presence of blocking, because there is no guarantee that an unblocked generated node x in F always maps to a node in F' that is also not blocked.

We therefore formally define the function σ as mapping a node x in F to a pair (u, u') of nodes in F', to handle the case when x is related to a blocked node u'. The node u in the pair is the node that blocks u' if u' is blocked; if u' is not blocked, then u and u' are the same $(u = u')$.

Let F be the completion forest initialized from $\mathcal{A}$ in the standard way. Before the start of the execution of tableau rules on F, the function σ maps a root node in F to a pair of nodes in F' as follows:

– For a root node a in F, we define $\sigma(a) = (\alpha(\mathbf{f}(a)), \alpha(\mathbf{f}(a)))$

As new generated nodes are introduced during the execution of the tableau rules on F, the mapping σ is extended to these new nodes as explained in the treatment of the $\exists$-rule and $\geq$-rule. $\sigma(a)[1]$ denotes the first element of the pair $\sigma(a)$, and $\sigma(a)[2]$ is its second element.

We show by induction that at any given step k of a particular execution[4] of tableau rules on F the following holds: for all nodes x and y in F

(A') $\mathcal{L}_k(x) \subseteq \mathcal{L}'(\sigma(x)[1])$ (where $\mathcal{L}_k(x)$ denotes the concept set of a at step k of the execution of the standard tableau algorithm on $\mathcal{A}$, and $\mathcal{L}'(\sigma(x)[1])$ is the concept set of the $\sigma(x)[1]$ in F')

(B') if y is a S-neighbor of x and y is either a root node or a generated child of x, then, in F', $\sigma(y)[2]$ is a S-neighbor of $\sigma(x)[1]$.

(C') for $\sigma(x) = (u, u')$, $u = u'$ iff. u is not blocked

(D') for $\sigma(x) = (u, u')$, $u \neq u'$ iff. u' is blocked by u.

(E') if $x \dot{\neq} y$ holds in F, then $\sigma(x)[2] \dot{\neq} \sigma(y)[2]$ holds in F'

It is very important to note that, since F' is clash-free, if, at any step k, (A'), (B') and (E') hold, then, at any step k, F is clash-free.

The details of the induction proof is given in [2].

3.2 Membership Query Algorithm

The algorithm SELECT-CANDIDATES-MQ to select test candidates is shown below. Basically, the algorithm transforms the relationship atoms in the original conjunctive query into a SPARQL query Q_r and issues it over the completion forest of the summary F'. Solutions to Q_r give us candidates to test for the membership constraints.

---

[4] An execution in which non-deterministic choices are made based on choices made in F' as explained in the treatment of non-deterministic rules

SELECT-CANDIDATES-MQ(F', $\mathcal{R}$, f, $R_j(x_k, x_l)$ $(1 \leq j \leq n)$)
Input: F' Completion forest of Summary Abox, $\mathcal{R}$ Rbox of the original KB, **f** Abox Summary mapping function, $R_j(x_k, x_l)$ Set of role atoms in original conjunctive query
Output: $\tau(x \mapsto i)$ mapping from variables to summary individuals

(1) $V_s \leftarrow$ set of all variables in role atoms R_j $(1 \leq j \leq n)$
(2) $R_s \leftarrow$ set of all role atoms $R_j(x_k, x_l)(1 \leq j \leq n)$
(3)   For any constant c in any of the role atoms in R_s, obtain the summary individual $s \leftarrow f(c)$, and replace c by s
(4)   Create a SPARQL query Q_r whose SELECT clause is V_s and whose WHERE clause is $\bigwedge R_s$.
(5)   Issue Q_r over F' with only Rbox inferencing using $\mathcal{R}$ to obtain solution mapping $\tau(x \mapsto i)$
(6)   Remove individual solutions from τ which are considered 'anonymous' in F'
(7)   Since F' may contain mergers between individuals in $\mathcal{A}'$, expand any individual binding i in τ by its equivalence set ($sameAs(i)$)
(8)   **return** $\tau(x \mapsto i)$

During the transformation, special care is taken for constants appearing in role atoms. Since Q_r is evaluated on the summary, constants are replaced by the corresponding summary individuals that they are mapped to. Since we assume that all variables in the original conjunctive query are distinguished, we need to consider the variables in role atoms in the select clause of Q_r. The query is evaluated considering the Rbox $\mathcal{R}$ of the original KB, as we would like to capture relationships that can be inferred due to sub-property, inverse or transitive axioms in it (Note that the Tbox need not be considered since we do not care about concepts and concept-related axioms at this point). Since F' is small, evaluating this query is straightforward.

The result of executing Q_r is a mapping τ from variable to summary individuals, the latter becoming test candidates for the membership query constraints on the former. Note that the completion forest of the summary Abox may contain 'anonymous' individuals that are generated due to the presence of existential quantifiers in the KB. Obviously, these anonymous summary individuals are not present in the original Abox either and so we do not need to test them. Therefore, we discard any anonymous individuals from τ.

Having identified suitable test candidates, we now proceed to test them for their respective membership query atoms, using our summarization and refinement algorithm [1]. While the previous work focused on testing a single membership query on the summary, it can be easily extended to test multiple membership queries on the summary at the same time. The main difference is that we now start by adding the negation of all the membership types to their respective summary individual candidates, before testing for inconsistency (for details of other optimizations to membership querying, see [2]).

SOLVE-MQ, sketched below, captures the essence of the evaluation of membership queries.

> SOLVE-MQ(Q, $\mathcal{A}$,$\mathcal{T}$,$\mathcal{R}$)
> **Input:** Q the conjunctive query, $\mathcal{A}$ Abox, $\mathcal{T}$ Tbox, $\mathcal{R}$ Rbox
> **Output:** $\mathcal{A}'_c$ consistent version of summary Abox, $\mathbf{f}_c$ summary mapping function for $\mathcal{A}'_c$, F'_c completion forest of $\mathcal{A}'_c$, β mapping from a variable to summary individuals satisfying its type constraints
> (1) $(\mathcal{A}', \mathbf{f}) \leftarrow$ compute summary abox of $\mathcal{A}$ and its mapping function $\mathbf{f}$
> (2) $(\mathcal{A}'_c, \mathbf{f}_c) \leftarrow$ consistent version of $\mathcal{A}'$ and its mapping function obtained through refinement
> (3) $F'_c \leftarrow$ complete and clash-free completion forest of $\mathcal{A}'_c$
> (4) $\tau \leftarrow$ SELECT-CANDIDATES-MQ($F'_c, \mathcal{R}, \mathbf{f}_c, R_j(x_k, x_l) \in Q$)
> (5)   **foreach** variable x_k in Q
> (6)   **if** variable x_k has type constraints in Q
> (7) $\beta(x_k) \leftarrow$ compute, through refinement, summary individuals in $\tau(x_k)$ instances of concept $\bigcap_{C_p(x_k) \in Q} C_p$
> (8)   **else**
> (9) $\beta(x_k) \leftarrow \tau(x_k)$
> (10)   **return** $(\mathcal{A}'_c$, $\mathbf{f}_c, F'_c, \beta)$

4   Solving the Relationship Query Part

In this section, we discuss how we evaluate each of the role atoms $R(x,y)$ in the conjunctive query. We solve an atomic role query in three steps:

1. Section 4.1: We estimate an upper bound on potential relationship solutions for $R(x,y)$ in the Abox by capturing all possible ways in which relationships can be inferred in $\mathcal{SHIN}$. We do this efficiently using the completion forest of the summary Abox and a set of Datalog rules. The rules are restricted to the membership query solutions that are output in the previous step.
2. Section 4.2: After estimating potential role assertion solutions in the Abox, we identify *definite* or deterministically-derived role assertions, since we do not have to test for them.
3. Section 4.3: Finally, we test and solve the remaining potential relationship solutions in the summary Abox.

4.1   Estimating Potential Solutions for an Atomic Role Query $R(x, y)$

Our approach to estimate potential solutions to role queries consists in first understanding how, in the completion forest F of an Abox $\mathcal{A}$, a root node can acquire new root node R-neighbors (i.e. root node R-neighbors that were not

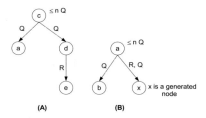

Fig. 1. Acquisition of named individual R-neighbors

(Init)	*InfTriple(X, R, Y)*	:- $R(X,Y) \in \mathcal{A}$
(SameSym)	*same(X,Y)*	:- *same(Y, X)*
(SameTrans)	*same(X, Y)*	:- *same(X,Z) and same(Z,Y)*
(NamedMerge)	*same(X, Y)*	:- $\mathbf{f}(Z) = A$ *and* $\leq nR \in \mathcal{L}'(\alpha(A))$ *and*
		$X \neq Y$ *and InfTriple(Z, R, X)*
		and InfTriple(Z, R, Y)
(SameRel1)	*InfTriple(X, R, Y)*	:- *same(X,Z) and InfTriple(Z, R, Y)*
(SameRel2)	*InfTriple(X, R, Y)*	:- *same(Y,Z) and InfTriple(X, R, Z)*
(UnnamedMerge)	*InfTriple(X, R, Y)*	:- $\mathbf{f}(X) = A$ *and* $\mathbf{f}(Y) = B$ *and*
		$\leq nT \in \mathcal{L}'(\alpha(A))$ *and*
		($\exists S.C \in \mathcal{L}'(\alpha(A))$ *or* $\geq mS \in \mathcal{L}'(\alpha(A))$)
		and $\alpha(B)$ *is a R-neigbhor of* $\alpha(A)$ *in* F'
		and $\{R,T\} \subseteq \widehat{\phi_A}$ *and InfTriple(X, T, Y)*
(SubRole)	*InfTriple(X, R, Y)*	:- $S \sqsubseteq^* R$ *and InfTriple(X, S, Y) and* $S \neq R$
(InvRole)	*InfTriple(X, R, Y)*	:- $S^- = R$ *and InfTriple(Y, S, X)*
(Relevance)	*RelInfTriple(X, R, Y)*	:- *InfTriple(X, R, Y) and* $\mathbf{f}(X) = A$
		and $\mathbf{f}(Y) = B$ *and* $R(x_1, x_2) \in Q$
		and $A \in \beta(x_1)$ *and* $B \in \beta(x_2)$

Fig. 2. PotentialRuleSet: Rules to compute potential new named individual neighbors that are relevant to conjunctive query Q. Main output: $RelInfTriple$.

present before the beginning of rule execution). Then, we devise a set of simple rules (see Figure 2) to conservatively estimate potential R-neighbors. These rules are simple enough to be efficiently evaluated using a datalog engine. Figure 1 illustrates the two ways a root node a in F can acquire new R-neighbors that are root nodes during the execution of the tableaux algorithm on F:

(A) The root node a is merged with another root node d and acquires root node R-neighbors of d. The merger is performed to satisfy the maximum cardinality restriction $\leq nQ$ in the concept set of c. This case also captures acquisition of R-neighbors through mergers involving root neighbors of a.

(B) The root node b is merged with a generated node x to satisfy the maximum cardinality restriction $\leq nQ$ in the concept set of a. As a result of this merger, b becomes a R-neighbor of a since x was a R-neighbor of a.

Let us assume that F' is a complete and clash-free completion forest of the summary $\mathcal{A}'$ of the Abox $\mathcal{A}$, and F is the complete and clash-free completion forest of $\mathcal{A}$ given by Theorem 1.

We can conservatively account for acquisition of named R-neighbors of a through mergers with named individuals by applying rules (see $NamedMerge$, $SameRel1$, and $SameRel2$ in Figure 2) on the Abox that trigger a merger between a and d if (1) a is a Q-neighbor of c in $\mathcal{A}$ (explicitly or as a result of evaluation of our simple rules), (2) d is a Q-neighbor of c in $\mathcal{A}$ (explicitly or as a result of evaluation of our simple rules), and (3), in the completion forest F', $\leq nQ \in \mathcal{L}'(\alpha(\mathbf{f}(c)))$. If the last condition is not satisfied, Theorem 1 guarantees that a merger between a and d is not possible in F since $\leq nQ$ cannot be the concept set of c in F.

One way to account for mergers between root nodes and generated nodes is to have rules that create these generated nodes. However, this is not practical because too many nodes might be generated, complex blocking mechanism will be required to ensure termination, and the resulting rules will not be simple enough to be efficiently evaluated by a datalog engine.

Our approach to conservatively account for mergers illustrated in Figure 1 (B) is to first observe that in order for them to occur in F, the following conditions must be satisfied:

- a role generator ($\exists S.C$ or $\geq mS$, where S is a role in the Rbox) must be in the concept set of a (otherwise, a cannot have a generated node as its neighbor), and
- a maximum cardinality restriction $\leq nQ$ must be in the concept set of a and, the following must hold:
  - b must be a Q-neigbhor of a, and
  - x must be a Q-neigbhor of a.

For a named individual a in the abox $\mathcal{A}$, Theorem 1 allows us to check whether a maximum cardinality $\leq nQ$ and a role generator concept ($\exists S.C$ or $\geq mS$) can be present in the concept set of a in the completion forest F of $\mathcal{A}$ simply by checking whether they are in concept set of $\alpha(\mathbf{f}(a))$ in F'. This reduces the number of potential individuals a and b such that b can become a R-neighbor of a through mergers of type (B). To further reduce this number, we need a good upper bound on the set ϕ_a of roles P such that there is a generated node x P-neighbor of a in F, since R has to be in ϕ_a. A direct consequence of property (B') in the proof of Theorem 1 is that the following set is such an upper bound: $\{P|$ there is a P-neigbhor of $\alpha(\mathbf{f}(a))$ in $F'\}$.

Let $\widehat{\phi_{\mathbf{f}(a)}}$ be an upper bound of the set ϕ_a that depends only on information in F'. We can now express all the necessary conditions for b to possibly become a R-neighbor of a in F through a merger of type (B) in terms of information present in F':

- an existential restriction $\exists S.C$ or a minimum cardinality restriction $\geq mS$ must be in the concept set of $\alpha(\mathbf{f}(a))$ in F'.
- a maximum cardinality restriction $\leq nQ$ must be in the concept set of $\alpha(\mathbf{f}(a))$ and, the following must hold:
  - b must be a Q-neigbhor of a (either explicitly in $\mathcal{A}$ or through the application of rules to estimate potential new mergers)
  - $\{Q, R\} \subseteq \widehat{\phi_{\mathbf{f}(a)}}$ (because there must be a generated node x which is both a Q-neighbor of a and a R-neighbor of a in F).
- finally, $\alpha(\mathbf{f}(b))$ must be a R-neighbor of $\alpha(\mathbf{f}(a))$ (direct consequence of Theorem 1 and the fact that b has become R-neighbor of a in F)

Based on the previous necessary conditions, rule $UnnamedMerge$ in Figure 2 accounts for potential acquisition of new R-neighbors in F through merger of type (B).

For transitive roles, we perform the transitive closure over the estimated inferred neigbhors (computed by rules in Figure 2). It is important to note that new relations found after the application of the transitive closure cannot cause merger rules to trigger because, in $\mathcal{SHIN}$, maximum cardinality restrictions can only be defined on simple roles (i.e. roles which are not transitive and do not have transitive subrole).

Finally, the rule $Relevance$ in Figure 2 forces the rule engine to focus only on relationships appearing in the conjunctive query Q, and on the individual solutions which satisfy the membership constraints in Q, specified by the mapping β in the output of algorithm SOLVE-MQ.

4.2 Finding Definite Role Assertions

After estimating potential role assertion solutions in the Abox, we identify *definite* or deterministically-derived role assertions, since we do not have to test for them.

In particular, consider the rule $NamedMerge$ in Figure 2 which conservatively estimates potential mergers between named Abox individuals. We can be more precise here for deterministic mergers if we somehow identify which Abox individuals mapped to summary individual A are entailed to be of type $\leq 1.R$. Conceptually, this amounts to solving the membership query $\leq 1.R(x)$, which we evaluate efficiently using our membership query answering solution. Similar analysis is done for the rule $UnnamedMerge$ to identify Abox individuals that have role-generators ($\geq m.S$ or $\exists S.C$) as an entailed type. This gives us two new rules $- DefnNamedMerge, DefnUnnamedMerge -$ shown in Figure 3, which replace the rules $NamedMerge, UnnamedMerge$ in the $PotentialRuleSet$ (Figure 2)

(SummaryKB Defn)	$\mathcal{K}$	$= (\mathcal{A},\ \mathcal{T},\ \mathcal{R})$
(DefnNamedMerge)	*same(X, Y)*	:- $\mathcal{K} \models_{\leq} 1R(Z)$ *and* $X \neq Y$
		and InfTriple(Z, R, X)
		and InfTriple(Z, R, Y)
(DefnUnnamedMerge)	*InfTriple(X, R, Y)* :-	$\mathcal{K} \models_{\leq} 1T(X)$
		and ($\mathcal{K} \models \exists S.C(X)$ *or* $\mathcal{K} \models_{\geq} mS(X)$)
		and $S \sqsubseteq^{*} R$ *and* $S \sqsubseteq^{*} T$
		and InfTriple(X, T, Y)

Fig. 3. DefnRuleSet: Obtained by replacing *NamedMerge* and *UnnamedMerge* in the PotentialRuleSet with the rules shown

to produce the rule set *DefnRuleSet* that computes definite Abox relationship solutions.

4.3 Solving Remaining Potential Role Assertions

Having found potential role assertions solutions for $R(x, y)$ in the Abox and identifying the definite ones, we are left with testing the remaining potential solutions.

Suppose the remaining potential tuples to test are $\{R(u_1, v_1), ... R(u_n, v_n)\}$, where u_k, v_k, $(1 \leq k \leq n)$ are Abox individuals. Instead of testing these tuples in the Abox, we test them in the summary, i.e., for a given tuple $R(u_k, v_k)$ we identify the summary individuals to which u_k, v_k are mapped, say a_i, b_j respectively, and test whether $R(a_i, b_j)$ is entailed in the summary KB. This test is done by reducing the problem to membership query answering as described in the introduction. However, the limitation here is that when we find a tuple solution $R(b_i, b_j)$ in the summary (where b_i, b_j are summary individuals), we cannot compute all Abox relationship solutions from it – all we know is that every individual mapped to b_i is entailed to have an R-relation to some individual in b_j (and vice-versa, every individual mapped to b_j has an R^- relation to some individual mapped to b_i)[5].

In this case, for the sake of completeness, we are left with no choice other than to split one of the summary individuals down to the level of the Abox individuals mapped to it and test for relationships subsequently. Obviously, we choose to split the summary individual which has less Abox individuals mapped to it, to restrict the size of our summary Abox. Even in this worst case scenario, the performance of the algorithm is not severely affected as only one end of the tuple is split and the grouping of individuals is still preserved at the other end. Also, other than the tested tuples, the rest of the summary remains unchanged (so typically a large part of the Abox is still summarized).

We combine the three steps discussed in this section into an algorithm SOLVE-RQ that finds all solutions to a relationship query.

---

[5] From a precise summary justification for $R(b_i, b_j)$, we can issue an SQL query based on the justification pattern to get some relationship pair solutions in the Abox, but this would not be complete. For details, see [2]

SOLVE-RQ($R(x_i, x_j)$, $\mathcal{A}$, $\mathcal{T}$, $\mathcal{R}$, $\mathcal{A}'_c$, $\mathbf{f}_c$, F'_c, β)

Input: $R(x_i, x_j)$ Relationship query, $\mathcal{A}$ Abox, $\mathcal{T}$ Tbox, $\mathcal{R}$ Rbox, $\mathcal{A}'_c$
Consistent Summary of $\mathcal{A}$, $\mathbf{f}_c$ Abox $\mapsto$ Summary mapping function,
F'_c Completion forest of $\mathcal{A}'_c$, β output mapping from SOLVE-MQ(..)

Output: S set of pairs (a, b) s.t. $(\mathcal{A}, \mathcal{T}, \mathcal{R}) \models R(a, b)$

(1) $DefnInfTriple \leftarrow RelInfTriple$ computed after evaluation
        of $DefnRuleSet$ using $\mathcal{T}$, $\mathcal{R}$, β, $\mathcal{A}'_c$ (for $\mathcal{A}'$), $\mathbf{f}_c$ (for $\mathbf{f}$)

(2) $S \leftarrow DefnInfTriple$

(3) $PotentialInfTriple \leftarrow RelInfTriple$ computed after evalua-
        tion of $PotentialRuleSet$ using $\mathcal{R}$, β, $\mathcal{A}'_c$ (for $\mathcal{A}'$), $\mathbf{f}_c$ (for $\mathbf{f}$)
        and F'_c (for F') (Note: *Init* rule here initializes $InfTriple$
        as a union of role assertions in $\mathcal{A}$ and $DefnInfTriple$)

(4) $PotentialInfTriple \leftarrow PotentialInfTriple - DefnInfTriple$ (re-
        maining potential Abox role assertion solutions)

(5)   Test and Solve $PotentialInfTriple$ as described in Section
        4.3 to get solution pairs S'

(6) $S \leftarrow S \cup S'$

(7)   **return** S

5   Computational Experience

5.1   Correctness and Scalability Tests

We evaluated our approach on the UOBM benchmark [6], which was modified
to $\mathcal{SHIN}$ expressivity. We used 14 of the 15 queries defined in the benchmark
(query Q2, which is a pure membership query, was not included in our evaluation). The results are reported for 1, 5, 10, 30, 100 and 150 universities. We
compared our results against KAON2 [7]. (Pellet [8] did not scale to even one
university). For KAON2, we set all maximum cardinality restrictions to one
because of KAON2 limitations. Our experiments were conducted on a 2-way
2.4GHz AMD Dual Core Opteron system with 16GB of memory running Linux,
and a maximum heap size of 2G. The Abox was stored in a IBM DB2 V9.1 for
SHER and MySQL V5.0 for KAON2.

The size of the datasets are given in Table 1 (a). Table 1 (b) summarizes the
times taken (in seconds) by KAON2 and SHER solely for query answering, i.e.,
in both cases, the times do not include the knowledge base pre-processing and
setup costs. KAON2 ran out of memory on UOBM-30. In 13 out of 14 queries
SHER and KAON2 had 100% agreement. The difference on query Q15 was due
to differences in the constraints used. As can be seen, the average runtimes for
SHER are significantly lower, usually by an order of magnitude, than those for
KAON2. [2] presents more detailed data on the evaluation performance for each
query on each KB. On all queries, except query 9, SHER scales almost linearly
from UOBM-1 to UOBM-150. Query 9, which has 3 role atoms, is an example
of a query where we can improve our performance by using a cost model based
approach to control the order of evaluation of query atoms as explained in [5].

Table 1. Evaluation data

(a) Dataset Statistics

Dataset	type assertions	role assertions
1	25K	214K
5	120K	928K
10	224K	1,816K
30	709K	6.5M
100	7.8M	22.4M
150	11.7M	33.5M

(b) Runtimes in sec

Reasoner	Dataset	Avg. Time	St.Dev	Range
KAON2	1	18	5	14
KAON2	5	166	102	376
KAON2	10	667	508	1872
SHER	1	12	2	7
SHER	5	25	6	19
SHER	10	46	14	44
SHER	30	150	50	140
SHER	100	531	322	1222
SHER	150	1066	706	2818

5.2   Handling Non-deterministic Mergers

In experiments described in the previous subsection, UOBM queries did not exploit non-deterministic mergers between individuals in the Abox to produce new inferred results. Therefore, we modified the UOBM dataset to generate new relationships from non-deterministic mergers between named individuals, and considered a new query whose solutions required this.

We added disjoint relations between the four UOBM concepts FineArts, Science, HumanitiesAndSocial, Engineering representing course subjects, and a set of Abox assertions each resembling the pattern shown in Figure 4. The newly added individual LS_1 had type LeisureStudent, which is defined as (≤ 3. takesCourse) in the UOBM Tbox. LS_1 was assigned four takesCourse relations to individuals $C_1..C_4$ respectively. In general, we randomly added any one of the four course subjects mentioned above as a type to C_i, $1 \leq i \leq 3$ (C_4 is always assigned the type Course). In the case shown, C_1, C_2, C_3 are mutually disjoint concepts and hence the maxCardinality restriction in the type of LS_1 causes a non-deterministic merger between C_4 and any one C_j ($1 \leq j \leq 3$), which in turn causes C_4 to acquire a new isTaughtBy relation to the Lecturer individual L_1. To exploit this behavior, we considered the query: Q_{ND}: *(x, y, z)* $\leftarrow$ *LeisureStudent(x)* $\wedge$ *takesCourse(x, y)* $\wedge$ *Course(y)* $\wedge$ *isTaughtBy(y, z)* $\wedge$ *Lecturer(z)*. In the example shown, there are 4 tuple solutions to Q_{ND}, three of which are explicit $(LS_1, C_1/C_2/C_3, L_1)$, and one is inferred (LS_1, C_4, L_1) .

We modified UOBM-1, UOBM-5 and UOBM-10 by adding 100, 200 and 300 instances of LeisureStudent respectively. These numbers and datasets were chosen since as the pattern in Figure 4 shows, generation of new relationships due to non-deterministic mergers is non-trivial and seldom seen in large quantities in practice. We then evaluated Q_{ND} on the modified datasets. KAON2 is unable to handle this query since it cannot deal with non-deterministic mergers. Results of this query evaluation using SHER are shown in Figure 5. In the table, the column E (resp. I) stands for the number of explicit (resp. inferred) solutions for the query introduced by our script, $P_{\mathcal{A}}$, computed in step (4) of SOLVE-RQ, is the number of potential relationship pairs in the Abox that need to be

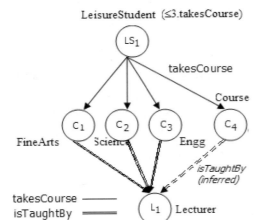

LeisureStudent (≤ 3.takesCourse)

Dataset	Time (in s)	E	I	P_A	$P_{A'}$	$S_{A'}$
1	38	210	70	182	32	1
5	76	480	160	335	75	1
10	165	786	152	319	100	15

Fig. 4. Abox pattern creating new `isTaughtBy` relations from non-deterministic mergers

Fig. 5. Evaluating Q_{ND}

tested, $P_{A'}$ is the number of summary pairs corresponding to the Abox pairs counted in P_A, and $S_{A'}$ is the number of summary solution tuples found using the procedure described in Section 4.3, which are eventually split down to the individual level.

As the results show, the algorithm demonstrates a graceful degradation for this query. For example, in UOBM-10, there are a total of $786+152 = 938$ entailed `isTaughtBy` relationships (152 due to non-deterministic mergers[6]), however our algorithm finds, in step(4) of SOLVE-RQ, that only 319 need to be tested. Moreover, they are first tested through their corresponding summary pairs as explained in Section 4.3. As result, only 15 out of 100 summary pairs are found to be solutions[7], and only one end of these 15 pairs are split down to the individual level. We feel that the times shown are acceptable for realistic use-cases.

6   Related Work and Conclusions

We have focused on answering grounded conjunctive queries instead of general conjunctive queries because, to our knowledge, there is currently no practical algorithm for answering general conjunctive queries with respect to SHIN ontologies [9]. It is for this reason that OWL-DL reasoner implementations which support conjunctive query answering, such as Pellet [8], RACER [10] and KAON2 [7], do so with the grounded conjunctive query semantics.At the same time, even after using the grounded conjunctive query semantics, tableau-based

---

[6] Only `isTaughtBy` relations can be inferred due to non-deterministic mergers.
[7] Not all potential relationships are solutions since the script may not necessarily add disjoint subject types to individuals C_1, C_2, C_3.

reasoners such as Pellet and RACER do not scale to several millions of assertions as SHER does. The fundamental limitation is that they work with the complete Abox, and the complexity of the tableau reasoning algorithm makes it infeasible to build a completion forest for a large and expressive Abox, which affects both solution pruning and testing.

On the other hand, KAON2, which we included in our evaluation, is a non-tableau based approach that relies on translating Description Logic to disjunctive datalog and is able to scale to an Abox with a million assertions. However, KAON2 has problems dealing with max-cardinality restrictions (for cardinality greater than 1) and even excluding such restrictions, is unable to scale to an Abox with 7 million assertions.

In our experiments, our technique appears to scale almost linearly for conjunctive queries of large, expressive Aboxes composed of 30-45 million Abox assertions, and conceptually, nothing in our approach prevents it from scaling to much larger datasets. As future work, we plan to integrate a cost-model to determine an efficient join order for the query atoms.

References

1. Dolby, J., Fokoue, A., Kalyanpur, A., Kershenbaum, A., Ma, L., Schonberg, E., Srinivas, K.: Scalable semantic retrieval through summarization and refinement. In: Proc. of the 22nd Conf. on Artificial Intelligence, AAAI 2007 (2007)
2. Dolby, J., Fokoue, A., Kalyanpur, A., Ma, L., Schonberg, E., Srinivas, K., Sun, X.: Scalable conjunctive query evaluation: Technical report (2008), http://domino.research.ibm.com/comm/research_projects.nsf/pages/iaa.index.html/$FILE/techReportCQ.pdf
3. Fokoue, A., Kershenbaum, A., Ma, L., Schonberg, E., Srinivas, K.: The summary abox: Cutting ontologies down to size. In: Cruz, I., Decker, S., Allemang, D., Preist, C., Schwabe, D., Mika, P., Uschold, M., Aroyo, L.M. (eds.) ISWC 2006. LNCS, vol. 4273, pp. 136–145. Springer, Heidelberg (2006)
4. Horrocks, I., Sattler, U., Tobies, S.: Reasoning with individuals for the description logic SHIQ*. In: Proc. of 17th Int.Conf. on Automated Deduction, pp. 482–496 (2000)
5. Sirin, E., Parsia, B.: Optimizations for answering conjunctive abox queries: First results. In: Proc. of the Description Logics Workshop, DL 2006 (2006)
6. Ma, L., Yang, Y., Qiu, Z., Xie, G., Pan, Y.: Towards a complete owl ontology benchmark. In: Sure, Y., Domingue, J. (eds.) ESWC 2006. LNCS, vol. 4011, pp. 124–139. Springer, Heidelberg (2006)
7. Hustadt, U., Motik, B., Sattler, U.: Reducing shiq description logic to disjunctive datalog programs. In: Proc. of 9th Intl. Conf. on Knowledge Representation and Reasoning (KR 2004), pp. 152–162 (2004)
8. Sirin, E., Parsia, B.: Pellet: An OWL DL reasoner. In: Description Logics (2004)
9. Glimm, B., Horrocks, I., Lutz, C., Sattler, U.: Conjunctive query answering for the description logic shiq. In: IJCAI, pp. 399–404 (2007)
10. Haarslev, V., Moller, R.: Racer system description. In: Goré, R.P., Leitsch, A., Nipkow, T. (eds.) IJCAR 2001. LNCS (LNAI), vol. 2083, pp. 701–705. Springer, Heidelberg (2001)

A Kernel Revision Operator for Terminologies — Algorithms and Evaluation*

Guilin Qi[1], Peter Haase[1], Zhisheng Huang[2], Qiu Ji[1], Jeff Z. Pan[3],
and Johanna Völker[1]

[1] Institute AIFB, University of Karlsruhe, Germany
{gqi,pha,qiji,jvo}@aifb.uni-karlsruhe.de
[2] Department of Mathematics and Computer Science, Vrije University Amsterdam
huang@cs.vu.nl
[3] Department of Computing Science, The University of Aberdeen
jpan@csd.abdn.ac.uk

Abstract. Revision of a description logic-based ontology deals with the problem of incorporating newly received information consistently. In this paper, we propose a general operator for revising terminologies in description logic-based ontologies. Our revision operator relies on a reformulation of the kernel contraction operator in belief revision. We first define our revision operator for terminologies and show that it satisfies some desirable logical properties. Second, two algorithms are developed to instantiate the revision operator. Since in general, these two algorithms are computationally too hard, we propose a third algorithm as a more efficient alternative. We implemented the algorithms and provide evaluation results on their efficiency, effectiveness and meaningfulness in the context of two application scenarios: Incremental ontology learning and mapping revision.

1 Introduction

Ontologies are typically not static entities, but they evolve over time and need to revised. Changes to an ontology may be caused, e.g., by modifications in the application domain, the reorganization of existing information, or the incorporation of additional knowledge according to changes in the users' needs.

An important problem in revising ontologies is maintaining the consistency of the ontology, i.e. the accommodation of new knowledge in an ontology without introducing logical contradictions. Due to the variety of sources and consequences of changes, such a revision is not a trivial process and thus cannot be left as manual work to the ontology engineer. Especially in the context of semi-automated ontology engineering, in which the ontology engineer is supported by agents (e.g. in the form of ontology learning tools) that suggest ontology changes, an automated revision is desired.

---

* Guilin Qi, Peter Haase, Qiu Ji and Johanna Völker are partially supported by the EU in the IST project NeOn (http://www.neon-project.org/). Jeff Z. Pan is partially supported by the EU MOST project (http://www.most-project.eu/). Zhisheng Huang is partially supported by EU-funded Projects OpenKnowledge and LarKC. We thank the reviewers for very helpful comments to improve the quality of our work.

A. Sheth et al. (Eds.): ISWC 2008, LNCS 5318, pp. 419–434, 2008.

Generally, we can distinguish two kinds of logical contradictions: inconsistency and incoherence. An ontology is inconsistent iff it has no model, i.e., it is inconsistent in the first-order sense. An ontology is incoherent iff there exists some unsatisfiable concept (i.e, an unsatisfiable concept stands for the empty set). There is a close relationship between inconsistency and incoherence [5], i.e., inconsistency is often caused by adding instances of concepts and relations to an incoherent ontology. However, an ontology can be incoherent but consistent. Incoherence is a problem that occurs in terminologies of ontologies. Resolving incoherence in a single terminology has been widely discussed (for example, see [19, 20]). However, there is very little work on resolving incoherence between terminologies of different ontologies.

There exists a number of prior work on revision in DLs, such as those reported in [5, 6, 7, 16]. Most of them focus on postulates for revision operators. For example, an important principle is that one should delete information in the original ontology as little as possible to accommodate the new knowledge consistently. Theoretically, it is important to know how to characterize a revision operator in terms of postulates. However, for practical applications, we require concrete revision operators that can be used. There are concrete revision operators defined to deal with inconsistency [7, 16]. But to the best of our knowledge, there is no revision operator dealing specifically with incoherence (as opposed to inconsistency) in the context of revision.

In this paper, we propose a kernel revision operator in Description Logic-based ontologies based on MIPS (minimal incoherence-preserving sub-terminologies) and an incision function. The notion of MIPS is originally developed for non-standard reasoning service in debugging incoherent terminologies [19, 20]. It is similar to the notion of a *kernel set* in belief base change defined in [10]. In order to resolve the logical contradiction, the incision function is used to select from each MIPS the axioms to be removed from the original ontology. Our revision operator focuses on revising terminologies, i.e. the TBox-part of ontologies. Two algorithms are developed to define specific kernel revision operators. The first algorithm is based on the reformulation of Reiter's Hitting Set Tree (HST) algorithm given in [20] and a scoring function. In this algorithm, we first compute all the MIPSs of the original ontology w.r.t. the new ontology. Then we calculate for each axiom in the MIPS a score corresponding to the number of MIPS which contain this axiom. Finally, we take subsets of those MIPSs that contain axioms with maximal scores and apply the reformulated HST algorithm to get a set of axioms to be deleted. The second algorithm is applied to ontologies where each axiom is attached a confidence value which indicates the reliability of the axiom. Such confidence values, as well as other kinds of provenance information, are typically generated by automated agents such as ontology learning or matching tools. The motivation for exploiting confidence information in this algorithm is to delete only axioms that are least reliable from each MIPS of the original ontology w.r.t. the new ontology. In this algorithm, we need to compute all the MIPSs, which is computationally hard in general. Therefore, we propose a third, alternative algorithm, which utilizes confidence values attached to axioms in the ontology to resolve unsatisfiable concepts without computing all the MIPSs. Compared to the second algorithm, this algorithm is computationally easier, but it does not necessarily remove more axioms from the original ontology after revision. Although it does not produce a kernel revision operator, it can be viewed as a good variant of the second algorithm.

We implemented the three algorithms and provide evaluation results on their efficiency and effectiveness in the context of two application scenarios: Incremental ontology learning and mapping revision. To evaluate the scalability of our algorithms, we iteratively add a set of new terminology axioms to an ontology. We also evaluate the effectiveness of the algorithms by counting the number of axioms deleted from the old ontology by our algorithms in each iteration. Finally, we evaluate the meaningfulness of the revision results by means of a user study, where the meaningfulness is measured by the ratio of correct removals.

The rest of this paper is organized as follows: Section 2 provides a preliminary introduction to Description Logics and various notions in ontology debugging. Section 3 overviews related work of revision in DLs. Section 4 presents our revision operator for terminologies. Section 5 proposes some algorithms to instantiate the revision operator. In Section 6, we report on evaluation results with real life data. We conclude in Section 7.

2   Preliminaries

This section introduces some basic notions of Description Logics (DLs) as well as the essential notions of debugging terminologies. Since our revision operator is independent of a specific DL language, and thus can be applied to any DL, we only give a general overview of description logics.

In our work, we focus on DL-based terminological ontologies: A terminology (TBox) $\mathcal{T}$ consists of concept axioms and role axioms. A subset of a TBox is called a sub-TBox. Concept axioms have the form $C \sqsubseteq D$ where C and D are (possibly complex) concept descriptions[1], and role axioms are expressions of the form $R \sqsubseteq S$, where R and S are role descriptions. We will refer to both concept axioms and role axioms as terminology axioms.

The semantics of DLs is defined via a model-theoretic semantics, which explicates the relationship between the language syntax and the model of a domain: An interpretation $\mathcal{I} = (\triangle^{\mathcal{I}}, \cdot^{\mathcal{I}})$ consists of a non-empty domain set $\triangle^{\mathcal{I}}$ and an interpretation function $\cdot^{\mathcal{I}}$, which maps from concepts and roles to elements of the domain, subsets of the domain and binary relations on the domain, respectively.

Given an interpretation $\mathcal{I}$, we say that $\mathcal{I}$ satisfies a concept axiom $C \sqsubseteq D$ (respectively, a role inclusion axiom $R \sqsubseteq S$) if $C^{\mathcal{I}} \subseteq D^{\mathcal{I}}$ ($R^{\mathcal{I}} \subseteq S^{\mathcal{I}}$, respectively). An interpretation $\mathcal{I}$ is called a *model* of a TBox $\mathcal{T}$, iff it satisfies each axiom in $\mathcal{T}$. We use $Mod(\mathcal{T})$ to denote all the models of a TBox $\mathcal{T}$. A named concept C in a terminology $\mathcal{T}$ is unsatisfiable iff, for each model $\mathcal{I}$ of $\mathcal{T}$, $C^{\mathcal{I}} = \emptyset$. A terminology $\mathcal{T}$ is incoherent iff there exists an unsatisfiable named concept in $\mathcal{T}$. Two TBoxes $\mathcal{T}$ and $\mathcal{T}'$ are equivalent, denoted by $\mathcal{T} \equiv \mathcal{T}'$, iff $Mod(\mathcal{T}) = Mod(\mathcal{T}')$.

We now introduce the notions of MIPS and MUPS which will be used to define our revision operator. Both of these terms have originally been defined in [19] and are used to pinpoint errors in an ontology.

---

[1] A complex concept is a concept that is formed by some atomic concepts and constructors such as conjunction $\sqcap$ and disjunction $\sqcup$.

Definition 1. *Let A be a named concept which is unsatisfiable in a TBox $\mathcal{T}$. A set $\mathcal{T}' \subseteq \mathcal{T}$ is a* minimal unsatisfiability-preserving sub-TBox (MUPS) *of $\mathcal{T}$ w.r.t. A if A is unsatisfiable in $\mathcal{T}'$, and A is satisfiable in every sub-TBox $\mathcal{T}'' \subset \mathcal{T}'$.*

A MUPS of $\mathcal{T}$ *w.r.t.* A is a minimal sub-TBox of $\mathcal{T}$ in which A is unsatisfiable.

Example 1. Let $\mathcal{T} = \{A \sqsubseteq B, A \sqsubseteq \neg B, C \sqsubseteq A, C \sqsubseteq D, C \sqsubseteq \neg D\}$. There are two unsatisfiable concepts in $\mathcal{T}$: A and C. It is easy to check that there are two MUPSs of $\mathcal{T}$ w.r.t. C: $\{A \sqsubseteq B, A \sqsubseteq \neg B, C \sqsubseteq A\}$ and $\{C \sqsubseteq D, C \sqsubseteq \neg D\}$, and there is one MUPS of $\mathcal{T}$ w.r.t. A: $\{A \sqsubseteq B, A \sqsubseteq \neg B\}$.

MUPSs are useful for relating sets of axioms to the unsatisfiability of specific concepts, but they can also be used to calculate minimal incoherence preserving sub-TBoxes, which relate sets of axioms to the incoherence of a TBox in general and are defined as follows.

Definition 2. *Let $\mathcal{T}$ be an incoherent TBox. A TBox $\mathcal{T}' \subseteq \mathcal{T}$ is a* minimal incoherence-preserving sub-TBox (MIPS) *of $\mathcal{T}$ if $\mathcal{T}'$ is incoherent, and every sub-TBox $\mathcal{T}'' \subset \mathcal{T}'$ is coherent.*

A MIPS of $\mathcal{T}$ is the minimal sub-TBox of $\mathcal{T}$ which is incoherent. For $\mathcal{T}$ in Example 1, we get the following MIPSs: $\{A \sqsubseteq B, A \sqsubseteq \neg B\}$ and $\{C \sqsubseteq D, C \sqsubseteq \neg D\}$.

3   Related Work and Motivation

This work is related to belief revision which has been widely discussed in the literature. The theory of belief revision in propositional and first-order logic deals with logical inconsistency resulting from revising a knowledge base by newly received information. Alchourrón, Gardenfors and Markinson (AGM for short) [1] propose a set of postulates to characterize a revision operator. In AGM's work, beliefs of an agent are represented by a set of formulas closed under logical consequence, called a belief set. A revision operator is an operation that maps a belief set and a formula to a belief set. This representation is afflicted by a number problems. For example, there is potentially an infinite number of formulas in a belief set. Therefore, several researchers have proposed to use a belief base which is a set of formulas that is not closed under logical consequence to represent the beliefs of an agent [11, 13]. In the scenario of ontology change, this later representation seems to be more natural because we do not require that an ontology should be closed under logical consequence.

The problem of revision in DLs has been extensively studied in the literature. In [6], Flouris, Plexousakis and Antoniou generalize the AGM framework in order to apply the rationales behind the AGM framework to a wider class of logics, i.e. a larger class of logics which are AGM-compliant. In [5], a framework for the distinction between incoherence and inconsistency of an ontology is proposed. A set of rational postulates for a revision operator in DLs is proposed based on the distinction between coherent negation and consistent negation. However, in [5] no concrete revision operator is proposed. In [16], reformulated AGM postulates for revision are adapted to DLs. Two revision operators that satisfy the adapted postulates are given, but no algorithm to implement any of the operators is introduced.

Similar to our revision operator, the revision operator defined in [18] also utilizes an *incision function* to select axioms to be removed from the original ontology. Our work differs from theirs in several aspects. First, our revision operator deals with incoherence instead of inconsistency. Second, we provide algorithms for computation of specific revision operators and discuss evaluation results on their implementation. This work is also related to the work presented in [8], in which an algorithm is given to determine consistent sub-ontologies by adding an axiom to an ontology. The algorithm is based on a selection function by assuming that all axioms in the ontology are connected. Recently, a revision operator has been defined to repair erroneous mappings derived by automated ontology alignment systems [14]. Their revision operator, however, calculates neither MUPS nor MIPS and may remove too much information.

According to the discussion of related work, although there is no revision operator dealing with incoherence, it is possible to define such a revision operator based on the result of debugging and diagnosis. However, there are several problems to be solved. First, the notions of MUPS and MIPS are defined on a single ontology, whilst we need to consider two ontologies such that one of them is more important than the other. Therefore, we need to generalize the notions of MUPS and MIPS. Second, it has been shown in [20] that finding all the MUPS and MIPS in $\mathcal{ALC}$ is time-consuming and efficiency is a problem that prevents us from calculating all the MUPS and MIPS. This problem is even more serious for more expressive DLs (thus computationally harder). Third, even if we can find an efficient algorithm for calculating all the MIPS, we must find an efficient way to remove as few axioms as possible to restore coherence. We tackle these problems by first defining a generalized MIPS and a general revision operator based on it. We then give three algorithms to instantiate the general revision operator.

4   Kernel Revision Operator for Terminologies

In this section, we define our revision operator based on the notion of MIPS. Originally, the notion of a MIPS is defined on a single TBox, whereas a revision operator deals with conflicts between two TBoxes. We therefore generalize MIPS by considering two TBoxes: the TBox $\mathcal{T}$ to be revised, and the newly received TBox $\mathcal{T}_0$. In the following, we further assume that both $\mathcal{T}$ and $\mathcal{T}_0$ are coherent.

Definition 3. *Let $\mathcal{T}$ and $\mathcal{T}_0$ be two TBoxes. A minimal incoherence-preserving sub-TBox (MIPS) $\mathcal{T}'$ of $\mathcal{T}$ w.r.t. $\mathcal{T}_0$ is a sub-TBox of $\mathcal{T}$ which satisfies (1) $\mathcal{T}'\cup\mathcal{T}_0$ is incoherent; (2) $\forall \mathcal{T}''\subset\mathcal{T}'$, $\mathcal{T}''\cup\mathcal{T}_0$ is coherent. We denote the set of all MIPSs of $\mathcal{T}$ w.r.t $\mathcal{T}_0$ by $MIPS_{\mathcal{T}_0}(\mathcal{T})$.*

A MIPS of TBox $\mathcal{T}$ *w.r.t.* TBox $\mathcal{T}_0$ is a minimal sub-TBox of $\mathcal{T}$ that is incoherent with $\mathcal{T}_0$. This definition of MIPS is similar to the notion of a minimal axiom set given in [2] where an ontology is split into a *static part* and a *rebuttal part*. It can be considered as the *kernel* defined by Hansson in [10]. Similar to Definition 3, we can define a MUPS of $\mathcal{T}$ w.r.t. $\mathcal{T}_0$ and an unsatisfiable concept of $\mathcal{T} \cup \mathcal{T}_0$. When $\mathcal{T}_0$ is an empty set, then Definition 3 is reduced to Definition 2. In classical logic, given a knowledge base A which is a set of classical formulas and a formula ϕ, a ϕ-kernel of A is the minimal subbase of A that implies ϕ. To define a contraction function for removing knowledge from a knowledge base, called kernel contraction, Hansson defines an incision function which selects formulas to be discarded in each ϕ-kernel of A. We adapt the incision function to define our revision operator.

Definition 4. *Let $\mathcal{T}$ be a TBox. An* incision function *for $\mathcal{T}$, denoted as σ, is a function $(\sigma : 2^{2^{\mathcal{T}}} \to 2^{\mathcal{T}})$ such that for each TBox $\mathcal{T}_0$*

(i) $\sigma(MIPS_{\mathcal{T}_0}(\mathcal{T})) \subseteq \bigcup_{\mathcal{T}_i \in MIPS_{\mathcal{T}_0}(\mathcal{T})} \mathcal{T}_i$;
(ii) if $\mathcal{T}' \in MIPS_{\mathcal{T}_0}(\mathcal{T})$, then $\mathcal{T}' \cap \sigma(MIPS_{\mathcal{T}_0}(\mathcal{T})) \neq \emptyset$.

An incision function for a TBox $\mathcal{T}$ is a function such that for each TBox $\mathcal{T}_0$, it selects formulas from every MIPS of $\mathcal{T}$ w.r.t. $\mathcal{T}_0$ if this MIPS is not empty. Condition (i) says the axioms selected by an incision function must belong to some MIPSs of $\mathcal{T}$ w.r.t. $\mathcal{T}_0$. Condition (ii) says each MIPS of $\mathcal{T}$ w.r.t. $\mathcal{T}_0$ must have at least one axiom selected. The incision function plays a similar role as concept pinpointing in [19]. However, the latter is only applied to a single ontology.

An important incision function is the one which is called *minimal incision function* [4]. The idea of this incision function is to select a minimal subset of elements from the set of kernel sets. We adapt this incision function as follows.

Definition 5. *Let $\mathcal{T}$ be a TBox. An incision function σ for $\mathcal{T}$ is* minimal *if there is no other incision function σ' for $\mathcal{T}$ such that there is a TBox $\mathcal{T}_0$, $\sigma'(MIPS_{\mathcal{T}_0}(\mathcal{T})) \subset \sigma(MIPS_{\mathcal{T}_0}(\mathcal{T}))$.*

A minimal incision function selects a minimal subset of $\mathcal{T}$ w.r.t. the set inclusion. However, among all the minimal incision functions, some of them select more axioms than others. To make the number of selected axioms minimal, we define a cardinality-minimal incision function.

Definition 6. *Let $\mathcal{T}$ be a TBox. An incision function σ for $\mathcal{T}$ is* cardinality-minimal *if there is no other incision function σ' such that there is a TBox $\mathcal{T}_0$, $|\sigma'(MIPS_{\mathcal{T}_0}(\mathcal{T}))| < |\sigma(MIPS_{\mathcal{T}_0}(\mathcal{T}))|$.*

It is clear that a cardinality-minimal incision function is always a minimal incision function.

Proposition 1. *Let $\mathcal{T}$ be a TBox. Suppose σ is a cardinality-minimal incision function for $\mathcal{T}$, then it is a minimal incision function.*

From each incision function, we can define an operator for revising a TBox $\mathcal{T}$ by a newly received TBox $\mathcal{T}_0$. The idea is that we first calculate the MIPS of TBox $\mathcal{T}$ *w.r.t* TBox $\mathcal{T}_0$, then delete axioms in $\mathcal{T}$ selected by the incision function. After that, we take the union of the modified TBox and $\mathcal{T}_0$ as the result of the revision.

Definition 7. *Let $\mathcal{T}$ be a TBox, and σ be an incision function for $\mathcal{T}$. The* kernel revision operator $\circ_\sigma$ *for $\mathcal{T}$ is defined as follows: for each TBox $\mathcal{T}_0$,*

$$\mathcal{T} \circ_\sigma \mathcal{T}_0 = (\mathcal{T} \setminus \sigma(MIPS_{\mathcal{T}_0}(\mathcal{T}))) \cup \mathcal{T}_0.$$

The result of a revision by the kernel revision operator only contains a single TBox. According to the definition of an incision function, the resulting TBox of the kernel revision operator is always a unique coherent TBox.

Proposition 2. *Let T be a TBox, and σ be an incision function for T. The operator $\circ_\sigma$ satisfies the following properties: for any TBoxes T_0, T_0'*
(R1) $T_0 \subseteq T \circ_\sigma T_0$.
(R2) If $T \cup T_0$ is coherent, then $T \circ_\sigma T_0 = T \cup T_0$.
(R3) If T_0 is coherent, then $T \circ_\sigma T_0$ is coherent.
(R4) If $T_0 \equiv T_0'$, then $T \circ_\sigma T_0 \equiv T \circ_\sigma T_0'$.
(R5) If $\phi \in T$ and $\phi \notin T \circ_\sigma T_0$, then there is a subset S of T and a subset S_0 of T_0 such that $S \cup S_0$ is coherent, but $S \cup S_0 \cup \{\phi\}$ is not.

Proof. (sketch) It is clear that (R1)-(R3) hold. We show that (R4) holds. Suppose $T_0 \equiv T_0'$. According to Definition 3, we must have $MIPS_{T_0}(T) = MIPS_{T_0'}(T)$. Therefore, we have $\sigma(MIPS_{T_0}(T)) = \sigma(MIPS_{T_0'}(T))$. It follows that $T \circ_\sigma T_0 \equiv T \circ_\sigma T_0'$. (R5) holds because the incision function that is used to define a kernel revision operator only selects axioms from MIPSs of T w.r.t. T_0. Therefore, these axioms must be in a subset of T that is in conflict with some axioms in T_0.

Properties (R1)-(R4) are adapted from postulates (O+1), (O+2*), (O+3*) and (O+4) in [5]. (R1) says that every axiom in the new TBox should be accepted after revision. (R2) says, if two TBoxes have no contradiction, then we do not need to change anything. (R3) means that if the new TBox is coherent, then the result of revision should also be coherent. (R4) is a weakened form of syntax-independence. That is, the revision operator is independent of the syntactic form of axioms in the new TBox. (R5) is a new property which is adapted from the core-retainment postulate in [10]. It states that if an axiom is deleted after revision, then it must be responsible for the conflict.

5   Algorithms

The kernel revision operator is defined by an incision function. However, we have not given any incision function up to now. In the following, inspired by the work reported in [21], we propose some algorithms for computing an incision function based on Reiter's Hitting Set Tree (HST) algorithm [17] which is reformulated in [20]. We briefly introduce Reiter's theory. Given a *universal set U*, and a set $S = \{s_1, ..., s_n\}$ of subsets of U which are *conflict sets*, i.e. subsets of the system components responsible for the error. A *hitting set T* for S is a subset of U such that $s_i \cap T \neq \emptyset$ for all $1 \leq i \leq n$. A *minimal hitting set T* for S is a hitting set such that no $T' \subset T$ is a hitting set for S. Reiter's algorithm is used to calculate minimal hitting sets for a collection $S = \{s_1, ..., s_n\}$ of sets by constructing a labeled tree, called a Hitting Set Tree (HST). We select one arbitrary minimal hitting set of $MIPS_{T_0}(T)$ given by HST algorithm in [20]. We denote the revised HST algorithm as $HSTree$. We do not apply the revised HST algorithm to $MIPS_{T_0}(T)$ because there may have a large number of hitting sets if we use all the MIPSs and the algorithm will be very slow. Instead, we apply the revised HST algorithm to the set of subsets of the MIPSs in $MIPS_{T_0}(T)$.

The first algorithm is based on the *scoring function on axioms*[2] which is defined as follows.

---

[2] A scoring function has been used in [15] to measuring inconsistency in a single ontology and is defined by MIPS, whilst ours is not defined by MIPS.

Algorithm 1. Algorithm for Repair based on scoring function

Data: Two TBoxes $\mathcal{T}$ and $\mathcal{T}_0$, where $\mathcal{T}$ is the TBox to be revised
Result: A repaired coherent TBox $\mathcal{T} \circ_\sigma \mathcal{T}_0$
begin

    $\mathcal{C} = \emptyset$

    calculate $MIPS_{\mathcal{T}_0}(\mathcal{T})$

    for $ax \in \bigcup_{\mathcal{T}_i \in MIPS_{\mathcal{T}_0}(\mathcal{T})} \mathcal{T}_i$ **do**

        $w_{ax} := S_{MIPS_{\mathcal{T}_0}(\mathcal{T})}(\{ax\})$

    for $\mathcal{T}_i \in MIPS_{\mathcal{T}_0}(\mathcal{T})$ **do**

        $A_i := \{ax \in \mathcal{T}_i : \nexists ax' \in \mathcal{T}_i, w_{ax'} > w_{ax}\}$

        $\mathcal{C} := \mathcal{C} \cup \{A_i\}$

    $\sigma(MIPS_{\mathcal{T}_0}(\mathcal{T})) := HSTree(\mathcal{C})$

    $\mathcal{T} \circ_\sigma \mathcal{T}_0 := \mathcal{T} \setminus \sigma(MIPS_{\mathcal{T}_0}(\mathcal{T})) \cup \mathcal{T}_0$

    return $\mathcal{T} \circ_\sigma \mathcal{T}_0$

end

Definition 8. *Let $\mathcal{T}$ be a TBox and $\mathcal{M}$ be a set of sub-TBoxes of $\mathcal{T}$. The scoring function for $\mathcal{T}$ w.r.t. $\mathcal{M}$, is a function $S_{\mathcal{M}} : \mathcal{P}(\mathcal{T}) \mapsto N$ such that for all $\mathcal{T}' \in \mathcal{P}(\mathcal{T})$*

$$S_{\mathcal{M}}(\mathcal{T}') = |\{\mathcal{T}_i \in \mathcal{M} : \mathcal{T}_i \cap \mathcal{T}' \neq \emptyset\}|.$$

The scoring function $S_{\mathcal{M}}$ for $\mathcal{T}$ returns for each subset $\mathcal{T}'$ of $\mathcal{T}$ the number of elements of $\mathcal{M}$ that have an overlap with $\mathcal{T}'$. If we apply the scoring function to each singleton $\{ax_i\}$, where ax_i is an axiom in $\mathcal{T}$, we can attach a degree to each axiom in $\mathcal{T}$.

In Algorithm 1, we first calculate all the MIPSs of $\mathcal{T}$ w.r.t. $\mathcal{T}_0$ (MIPSs for short). The approach for calculating all the MIPSs is based on a black-box algorithm for finding all justifications proposed in [12]. We then compute the score of each axiom in the union of the MIPSs (see the first "for" loop) by applying the scoring function $S_{MIPS_{\mathcal{T}_0}}$. For each MIPS, we select a subset of it containing those axioms whose scores are maximal among all the axioms in the MIPS, and we apply the modified HST algorithm to these axioms (see the second "for" loop and the line after it). The result of the modified HST algorithm is the set of axioms to be deleted, i.e., $\sigma(MIPS_{\mathcal{T}_0}(\mathcal{T}))$. After removing them, we restore coherence of the TBox $\mathcal{T}$ w.r.t. $\mathcal{T}_0$. In our algorithm, we use subsets of MIPSs consisting of those axioms with highest scores as an input to the HST algorithm, instead of using all the MIPSs. So, the number of removed axioms may not be minimal.

Example 2. Suppose that we have two TBoxes:

$\mathcal{T} = \{E \sqsubseteq B, D \sqsubseteq \neg B, F \sqsubseteq B, F \sqsubseteq C\}$, and $\mathcal{T}_0 = \{D \sqsubseteq E, G \sqsubseteq D, F \sqsubseteq D, H \sqsubseteq A\}$. The MIPSs of $\mathcal{T}$ w.r.t. $\mathcal{T}_0$ are $\mathcal{T}' = \{E \sqsubseteq B, D \sqsubseteq \neg B\}$ and $\mathcal{T}'' = \{D \sqsubseteq \neg B, F \sqsubseteq B\}$.

The score of the disjointness axiom $D \sqsubseteq \neg B$ is 2, because it belongs to both MIPSs. The scores of the other axioms are 1. Therefore, $\mathcal{C} = \{\{D \sqsubseteq \neg B\}, \{D \sqsubseteq \neg B\}\}$ and $\sigma(MIPS_{\mathcal{T}_0}(\mathcal{T})) = \{D \sqsubseteq \neg B\}$. So we delete $D \sqsubseteq \neg B$ and the result of revision is $\mathcal{T} \circ_\sigma \mathcal{T}_0 = \{E \sqsubseteq B, F \sqsubseteq B, F \sqsubseteq C, D \sqsubseteq E, G \sqsubseteq D, F \sqsubseteq D, H \sqsubseteq A\}$.

In some cases, there are confidence values attached to axioms in an ontology. These confidence values can be generated during an ontology learning process (see [9]) or

---

Algorithm 2. Algorithm for Repair based on confidence values

Data: Two TBoxes $\mathcal{T}$ and $\mathcal{T}_0$, where $\mathcal{T}$ is the TBox to be revised, each axiom ax in $\mathcal{T}$ is
    attached a confidence value w_{ax}
Result: A repaired coherent TBox $\mathcal{T} \circ_c \mathcal{T}_0$
begin
$\quad$ $\mathcal{C} = \emptyset$
$\quad$ **calculate** $MIPS_{\mathcal{T}_0}(\mathcal{T})$
$\quad$ **for** $\mathcal{T}_i \in MIPS_{\mathcal{T}_0}(\mathcal{T})$ **do**
$\qquad$ $A_i := \{ax \in \mathcal{T}_i : \nexists ax' \in \mathcal{T}_i, w_{ax'} < w_{ax}\}$
$\qquad$ $\mathcal{C} := \mathcal{C} \cup \{A_i\}$
$\quad$ $\sigma(MIPS_{\mathcal{T}_0}(\mathcal{T})) := HSTree(\mathcal{C})$
$\quad$ $\mathcal{T} \circ_c \mathcal{T}_0 := \mathcal{T} \setminus \sigma(MIPS_{\mathcal{T}_0}(\mathcal{T})) \cup \mathcal{T}_0$
$\quad$ **return** $\mathcal{T} \circ_c \mathcal{T}_0$
end

---

given by human experts. When confidence values are attached to axioms in the TBox, we can choose an axiom with least confidence from a MIPS and delete it. Note that we do not need to know the exact values attached to the axioms. What matters is the total ordering between axioms. A natural idea is to replace the score w_{ax} of each axiom ax in Algorithm 1 by its confidence value if applicable. This leads us to Algorithm 2.

In Algorithms 1 and 2, we extract a subset of each MIPS which consists of either those axioms with the maximal score or those with least confidence values. We then apply the modified HST algorithm to these subsets to find a hitting set. Our algorithms clearly compute an incision function, which is not the minimal incision function. However, according to our experiment on real life ontologies, our algorithms delete only a small number of axioms in order to restore consistency and have acceptable evaluation of meaningfulness.

Example 3. (Example 2 Continued) Suppose axioms in the TBox $\mathcal{T}$ are attached with confidence values as follows:
$w_{E \sqsubseteq B} = 0.4$, $w_{D \sqsubseteq \neg B} = 0.5$, $w_{F \sqsubseteq B} = 0.6$, $w_{F \sqsubseteq C} = 0.9$.
The axioms in $\mathcal{T}_0$ are assigned weight 1, i.e., they are firmly believed.

It is clear that $\mathcal{A}' = \{E \sqsubseteq B\}$ and $\mathcal{A}'' = \{D \sqsubseteq \neg B\}$. Therefore, $\mathcal{C} = \{\{E \sqsubseteq B\}, \{D \sqsubseteq \neg B\}\}$ and $\sigma(MIPS_{\mathcal{T}_0}(\mathcal{T})) = \{E \sqsubseteq B, D \sqsubseteq \neg B\}$. Therefore, $\mathcal{T} \circ_c \mathcal{T}_0 = \{F \sqsubseteq B, G \sqsubseteq D, D \sqsubseteq E, H \sqsubseteq A, F \sqsubseteq C, F \sqsubseteq D\}$.

In Algorithm 3, when resolving incoherence of a TBox, we do not compute all the MIPSs. Instead, we resolve incoherence by iteratively dealing with unsatisfiable concepts. That is, we remove axioms in the MUPSs of an unsatisfiable concept and make it satisfiable before dealing with another unsatisfiable concept, and so on. The function which computes all the MUPSs of $\mathcal{T}$ w.r.t. $\mathcal{T}_0$ and C is similar to the algorithm to compute MUPS in [12], and it is denoted by $GETMUPS_{\mathcal{T}_0}(C, \mathcal{T})$. The only difference is that after computing a single MUPS of $\mathcal{T} \cup \mathcal{T}_0$ w.r.t. C, we only take the intersection of the MUPS and $\mathcal{T}$ as the node in the Hitting Set Tree. For each unsatisfiable concept, we take the subset of every MUPS which contains axioms with minimal confidence values and then apply the HST algorithm to select the axioms to be deleted. In this sense, this algorithm still achieves some kind of minimal change when resolving unsatisfiability of

---

Algorithm 3. Adapted algorithm for Repair based on confidence values

---

Data: Two TBoxes $\mathcal{T}$ and $\mathcal{T}_0$, where $\mathcal{T}$ is the TBox to be revised, axioms in $\mathcal{T}$ are attached
with confidence values

Result: A repaired coherent TBox $\mathcal{T} \circ_w \mathcal{T}_0$

begin

$\quad \mathcal{C} := \emptyset$

$\quad$ **for** $C \in GET ALL CONCEPTS(\mathcal{T} \cup \mathcal{T}_0)$ **do**

$\quad\quad$ **while** $\mathcal{T} \cup \mathcal{T}_0 \models C \sqsubseteq \bot$ **do**

$\quad\quad\quad \mathcal{M}_{C,\mathcal{T},\mathcal{T}_0} := GETMUPS_{\mathcal{T}_0}(C, \mathcal{T})$

$\quad\quad\quad$ **for** $\mathcal{T}_i \in \mathcal{M}_{C,\mathcal{T},\mathcal{T}_0}$ **do**

$\quad\quad\quad\quad \underline{\mathcal{T}_i} := \{ax \in \mathcal{T}_i : \nexists ax' \in \mathcal{T}_i, w_{ax'} < w_{ax}\}$

$\quad\quad\quad\quad \mathcal{C} := \mathcal{C} \cup \{\underline{\mathcal{T}_i}\}$

$\quad\quad\quad \mathcal{T}_C := HSTree(\mathcal{C})$

$\quad\quad\quad \mathcal{T} := \mathcal{T} \setminus \mathcal{T}_C$

$\quad\quad\quad \mathcal{C} := \emptyset$

$\quad$ **return** $\mathcal{T} \cup \mathcal{T}_0$

end

---

a concept, even if the revision operator implemented by this algorithm is not a kernel revision operator. As we we do not need to calculate all the MIPSs, the algorithm is much more efficient than Algorithm 2 as long as all the MIPS in Algorithms 1 and 2 are calculated from all the MUPS as suggested by Schlobach and Cornet in [19].

Example 4. (Example 3 Continued) There are three unsatisfiable concepts in $\mathcal{T} \cup \mathcal{T}_0$: G, D and F. Suppose our algorithm chooses F first. The MUPS of F in $\mathcal{T}$ w.r.t. $\mathcal{T}_0$ is $\mathcal{T}' = \{D \sqsubseteq \neg B, F \sqsubseteq B\}$. So $\mathcal{M}_{F,\mathcal{T},\mathcal{T}_0} = \{\mathcal{T}'\}$. Since $w_{D \sqsubseteq \neg B} < w_{F \sqsubseteq B}$, we have $\mathcal{C} = \{\{D \sqsubseteq \neg B\}\}$. So $\mathcal{T}_C = \{D \sqsubseteq \neg B\}$.

We replace $\mathcal{T}$ by $\mathcal{T} \setminus \{D \sqsubseteq \neg B\}$. It is easy to check that $\mathcal{T} \cup \mathcal{T}_0$ is coherent now. So, the algorithm terminates and the result of the revision is $\mathcal{T} \circ_w \mathcal{T}_0 = \{F \sqsubseteq B, E \sqsubseteq B, G \sqsubseteq D, D \sqsubseteq E, H \sqsubseteq A, F \sqsubseteq C, F \sqsubseteq D\}$.

6   Experimental Evaluation

Our algorithms have been implemented in Java as part of the RaDON plugin[3] for the NeOn Toolkit.[4] In this section, we provide an evaluation and comparison of the algorithms with respect to efficiency, effectiveness and meaningfulness. The experiments have been performed on a Linux server running Sun's Java 1.5.0 with a maximum heap space 2048 MB. For each revision operation, the maximal time limit is 1 hour.

6.1   Application Scenarios and Data Sets

We performed the evaluation in an ontology learning scenario and an ontology mapping scenario. All data sets can be downloaded from RaDON website[5]. In the ontology

---

[3] http://radon.ontoware.org/

[4] http://www.neon-toolkit.org/

[5] http://radon.ontoware.org/downloads/data-revision-iswc08.zip

learning scenario, an ontology is automatically and incrementally generated using ontology learning algorithms. Dealing with incoherence is especially important in ontology learning: Due to the nature of ontology learning algorithms, the acquired ontologies inherently represent uncertain and possibly contradicting knowledge. In the ontology mapping scenario, we start with two heterogeneous source ontologies, which are then extended and revised by adding mappings relating elements of the two ontologies. The mappings are created by an ontology matching system. As in the case of ontology learning, also the matching systems produce uncertain and potentially erroneous mappings. As a result, the integrated ontologies become incoherent in many cases. Resolving the incoherence caused by the mappings is a critical task to improve the quality of ontology mapping results.

Ontology learning scenario: We applied the ontology learning framework Text2Onto[6] on a text corpus consisting of abstracts from the "knowledge management" information space of the BT Digital Library. We extracted concepts, taxonomic and non-taxonomic relationships, as well as disjointness axioms from the documents in the information space. The generated axioms are annotated with confidence values based, e.g., on lexical context similarity or the frequency of lexico-syntactic patterns matched in the text. The generated ontology bt_km comprises $4,000$ terminological axioms in total.

Starting with an initially empty ontology, in every revision step we incrementally[7] add an ontology $\mathcal{T}_0$ of 100 randomly generated axioms to the ontology $\mathcal{T}$. For each iteration, if $\mathcal{T}$ w.r.t. $\mathcal{T}_0$ turns incoherent, we apply our algorithms to obtain a coherent revised ontology. Otherwise, we simply add $\mathcal{T}_0$ to $\mathcal{T}$. Then the revised ontology (i.e. the modified $\mathcal{T}$) serves as input for the next iteration.

Ontology mapping scenario: Here we address the scenario of integrating two heterogeneous source ontologies via mappings. While the individual source ontologies are locally coherent, relating them with mapping axioms may turn the integrated ontologies globally incoherent. In this scenario, we assume the two source ontologies to be fixed and the generated mappings to be revised in the case of logical contradictions. Therefore, we apply our revision algorithm to remove only mapping axioms and treat the source ontology axioms as stable.

For this scenario, we use the ontology mapping data sets provided by the University of Mannheim.[8] The data sets include some source ontologies and mappings used in the ontology alignment evaluation initiative[9], which provides a platform to evaluate ontology matching systems. For our test, we use as source ontologies different ontologies about the domain of scientific conferences: *CONFTOOL* (a $\mathcal{SIF}(\mathcal{D})$ ontology), *CMT* (a $\mathcal{ALCIF}(\mathcal{D})$ ontology), *EKAW* (a $\mathcal{SHIN}$ ontology), *CRS* (a DL-Lite ontology) and *SIGKDD* (a $\mathcal{ALI}(\mathcal{D})$ ontology) with 197, 246, 248, 69 and 122 axioms respectively. The pairwise mappings were generated automatically by the HMatch system [3]. They

---

[6] http://ontoware.org/projects/text2onto/

[7] By adding set of axioms incrementally, we are actually doing iterated revision. The purpose of doing this is to evaluate the scalability of our algorithms. Discussions on iterated revision using our revision operators are out of the scope of this paper and will be left as future work.

[8] http://webrum.uni-mannheim.de/math/lski/ontdebug/index.html

[9] http://om2006.ontologymatching.org/

include *CONFTOOL-CMT* with 14 mapping axioms, *EKAW-CMT* with 46 mapping axioms and *CRS-SIGKDD* with 22 mapping axioms. We selected these ontologies and mappings for our experiments because they exhibit inconsistencies when integrated, and are thus interesting for revision experiments.

6.2 Evaluation Measures

In the following, we evaluate our algorithms with respect to aspects of efficiency, effectiveness and meaningfulness.

For measuring its efficiency, we provide the revision time t including the time to check whether the ontology is incoherent as well as the time to debug and resolve the incoherence. We further measure the effectiveness of our algorithms in terms of the number R of axioms which need to be removed from T to restore the coherence. The fewer axioms are removed, the better the algorithm complies with the principle of minimal change.

In order to measure the meaningfulness of our algorithms, four users are asked to assess whether the removal of an axiom in a particular revision was correct from their point of view. Specifically, we provide several axioms which are selected for removal by our algorithms as well as the MIPSs and MUPSs containing them (and scores of the axioms or confidence degrees attached to the axioms if applicable). For each removed axiom, we ask the users to decide whether the removal: (1) was correct, (2) was incorrect, or (3) whether they are unsure. For the evaluated results returned by each user, the meaningfulness is then measured by the ratio of correct removals:

$$\text{Correctness} = \frac{\#Correct\_Removals}{\#Total\_Removals}$$

Similarly we can define an "Error_Rate" based on the incorrect removals and an "Unknown_Rate" based on the removals where the users were unsure. We combine the obtained Correctness (respectively Error_Rate and Unknown_Rate) values from different users by averaging them.

6.3 Evaluation Results

Analysis of Efficiency and Effectiveness

Results for the ontology learning scenario: The runtime performance of our algorithms over ontology bt_km is depicted by Figure 1. Additional details for the entire ontology (4, 000 axioms) are shown in Table 1. It can be seen from the figure that the accumulative revision time does not linearly increase with the number of TBox axioms. This is because the revision time is related not only to the size of the input TBox, but also to the number of MUPSs. Take the iteration when the size of the current TBox T reaches about 2, 900 as an example. In this iteration, Algorithm 1 computes 124 unsatisfiable concepts and 154 MUPSs based on its previous revision results. The MUPSs found in this iteration are much more than those obtained in previous iterations, and thus the accumulative time for Algorithm 1 increases sharply in such case. This explanation can be also applied to Algorithm 2, while for Algorithm 3 in this iteration, only 3 unsatisfiable

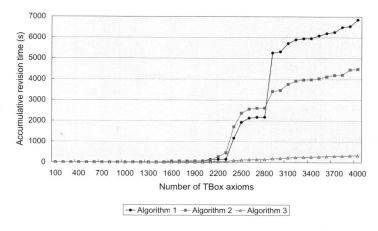

Fig. 1. The runtime performance of ontology revision for ontology bt_km

Table 1. Accumulative results for the entire ontology bt_km

Algorithm	# of Unsatisf. Concepts	# of MUPS	# of Removals	Revision time t (Seconds)
Algorithm 1	581	790	27	6, 856
Algorithm 2	506	717	34	4, 486
Algorithm 3	30	39	33	338

concepts and 3 MUPSs are computed, similarly to the previous iterations. Thus in this iteration, the accumulative time for this algorithm increases smoothly.

Let us now compare Algorithm 1 with Algorithm 2, since they share the same procedure to debug incoherence[10], but apply different strategies to resolve incoherence: Table 1 shows that Algorithm 1 removes fewer axioms than Algorithm 2, while taking more time to revise. On the one hand, the strategy using a scoring function better follows the principle of minimal change. On the other hand, as Algorithm 2 removes more axioms, more potential incoherence is resolved which may exist when new information is added. Thus, less MUPSs are computed in the end.

Second, we compare Algorithm 2 with Algorithm 3, because they use the same strategy to restore coherence while relying on different debug procedures. Algorithm 3 is considerably faster than Algorithm 2, since less MUPSs need to be computed. From the MUPSs obtained by the two algorithms, we observe that most of the unsatisfiable concepts can be derived from others. In such case, if we resolve the unsatisfiability of some concepts, others will be resolved automatically. Therefore, Algorithm 3 is much more efficient than Algorithm 2.

Results for the ontology mapping scenario: Table 2 presents the evaluation results of our algorithms based on the mapping data set described above. According to table 2, Algorithm 3 outperforms the other two w.r.t. efficiency. The reason is that Algorithm 3

---

[10] When we say debug incoherence, we mean finding all the MUPSs of an unsatisfiable concept or finding all MIPS.

Table 2. Evaluation results to revise mappings

Mappings	Strategy	# of Unsatisf. Concepts	# of MUPS All	# of MUPS Avg	MUPS_Size Avg	# of Removed Axioms	Time seconds
	Algorithm 1	26	351	14	6	4	331
CONFTOOL-CMT	Algorithm 2	26	351	14	6	8	322
	Algorithm 3	4	15	4	5	8	12
	Algorithm 1	18	372	21	5	16	867
EKAW-CMT	Algorithm 2	18	372	21	5	15	863
	Algorithm 3	5	62	12	5	14	51
	Algorithm 1	19	64	3	5	5	18
CRS-SIGKDD	Algorithm 2	19	64	3	5	10	18
	Algorithm 3	5	13	3	4	7	6

does not need to handle all the unsatisfiable concepts, for example, 4 by Algorithm 3 versus 26 by other algorithms for *CONFTOOL-CMT*. Algorithm 1 has similar efficiency as Algorithm 2. This shows the efficiency to resolve incoherence using confidence values is similar to that using scoring function, since both algorithms share the same procedure to debug incoherence, but apply different strategies to resolve it.

Regarding to the effectiveness, Algorithm 2 removes more axioms than Algorithm 1 to restore the coherence in most cases. The reason is that for each found MIPS, there is always one axiom with the lowest confidence value. In such case, we have no other choice but removing this axiom when using confidence values to resolve incoherence. But for Algorithm 1, usually we have several choices for each MIPS. Therefore, by applying the Hitting Set Tree algorithm, Algorithm 1 can find a hitting set which is cardinality-smaller than that of Algorithm 2. For example, Algorithm 2 removes 8 axioms when repairing mappings in *CONFTOOL-CMT*, while Algorithm 1 removes only 4 axioms. But for *EKAW-CMT*, Algorithm 1 removes a few more axioms than Algorithm 2, because in most cases there are at least two axioms with the lowest confidence values for each MIPS. For all the test ontologies, Algorithm 3 removes less axioms than Algorithm 2. The reason is that Algorithm 3 may remove an axiom in a MUPS which belongs to several MIPS and Algorithm 2 always removes one axiom with the lowest confidence value in each MIPS.

To sum up, Algorithm 1 removes the least number of axioms in most cases, best complying with the requirement of minimal change. Algorithm 3 has excellent runtime performance compared with other two algorithms. At the same time, it sometimes removes fewer axioms than Algorithm 2. Thus it is the preferable option to deal with incoherence for large data sets when we have information about confidence values (or other ranking information) for axioms in the ontology.

Analysis of Meaningfulness. Table 3 shows the results for the meaningfulness of the repair based on the expert users' assessment whether the removal was correct. That is, if the definition of a removed axiom does not make sense according to the expert users' experience, we consider the removal as correct.

From Table 3 we can see that for all data sets and algorithms the rate of correct removals is considerable higher than that of the erroneous removals. This shows that

Table 3. Evaluation results for meaningfulness

Data set	Algorithm	# of Removed Axioms	Correctness	Error_Rate	Unknown_Rate
bt_km	Algorithm 1	27	0.41	0.28	0.31
	Algorithm 2	34	0.53	0.19	0.28
	Algorithm 3	33	0.65	0.13	0.22
CONFTOOL-CMT	Algorithm 1	4	0.56	0.31	0.13
	Algorithm 2	8	0.97	0.03	0
	Algorithm 3	8	0.97	0.03	0
EKAW-CMT	Algorithm 1	16	0.68	0.11	0.21
	Algorithm 2	15	0.64	0.05	0.31
	Algorithm 3	14	0.84	0.07	0.09
CRS-SIGKDD	Algorithm 1	5	0.60	0.40	0
	Algorithm 2	10	0.50	0.25	0.25
	Algorithm 3	7	0.79	0.07	0.14

generally that the ranking of axioms in our approach works well for resolving incoherence. The exact ratios largely depend on the data set. Especially the Unknown_Rate varies considerably for the different data sets; this is due to the nature of the data sets: For *bt_km*, there are many cases in which the users do not know whether the removal make sense or not, as the concepts in this data set are quite abstract like "model", "knowledge" and "order", it is hard to decide the relationships among those concepts. Comparing the meaningfulness results obtained by different algorithms, Algorithm 1 using scoring function is designed to comply with the principle of minimal change, and thus it typically removes fewer axioms. Yet, as it does not take any information about the confidence into account, Algorithm 2 and 3 using confidence values to resolve incoherence outperform the Algorithm 1 in terms of meaningfulness in most cases. For data set *CRS-SIGKDD*, the correctness for Algorithm 2 is higher than that for Algorithm 1, but the Error_Rate is much lower. Algorithm 3 consistently yields the most meaningful results. This shows that relying on confidence values, as provided by ontology learning tools applied to *bt_km*, or generated by ontology matching systems, leads to considerably more meaningful results when applying them for resolving incoherence.

7   Conclusions

In this paper, we have proposed a kernel revision operator for terminologies using an incision function. We have shown that our operator satisfies desirable logical properties. Further, we have provided two algorithms to instantiate our revision operator, one based on a scoring function and another one based on confidence values. Since these two algorithms need to compute all the MIPSs of the original ontology w.r.t. the new ontology, they are computationally very hard. Therefore, we have proposed an alternative algorithm which repairs the ontology by calculating MUPSs of the original ontology w.r.t. the new ontology and an unsatisfiable concept. According to our experimental results with real life ontologies, this last algorithm shows good scalability, although it may potentially remove slightly more axioms than the first one. An interesting future work is to explore efficient algorithms for generating minimal (or cardinality minimal) incision functions.

References

1. Alchourrón, C.E., Gärdenfors, P., Makinson, D.: On the logic of theory change: Partial meet contraction and revision functions. J. Symb. Log. 50(2), 510–530 (1985)
2. Baader, F., Peñaloza, R., Suntisrivaraporn, B.: Pinpointing in the description logic EL+. In: Hertzberg, J., Beetz, M., Englert, R. (eds.) KI 2007. LNCS (LNAI), vol. 4667, pp. 52–67. Springer, Heidelberg (2007)
3. Castano, S., Ferrara, A., Montanelli, S.: Matching ontologies in open networked systems: Techniques and applications. pp. 25–63 (2006)
4. Falappa, M.A., Fermé, E.L., Kern-Isberner, G.: On the logic of theory change: Relations between incision and selection functions. In: Proc. of ECAI 2006, pp. 402–406 (2006)
5. Flouris, G., Huang, Z., Pan, J.Z., Plexousakis, D., Wache, H.: Inconsistencies, negations and changes in ontologies. In: Proc. of AAAI 2006, pp. 1295–1300 (2006)
6. Flouris, G., Plexousakis, D., Antoniou, G.: On applying the AGM theory to DLs and OWL. In: Gil, Y., Motta, E., Benjamins, V.R., Musen, M.A. (eds.) ISWC 2005. LNCS, vol. 3729, pp. 216–231. Springer, Heidelberg (2005)
7. Haase, P., Stojanovic, L.: Consistent evolution of OWL ontologies. In: Gómez-Pérez, A., Euzenat, J. (eds.) ESWC 2005. LNCS, vol. 3532, pp. 182–197. Springer, Heidelberg (2005)
8. Haase, P., van Harmelen, F., Huang, Z., Stuckenschmidt, H., Sure, Y.: A framework for handling inconsistency in changing ontologies. In: Gil, Y., Motta, E., Benjamins, V.R., Musen, M.A. (eds.) ISWC 2005. LNCS, vol. 3729, pp. 353–367. Springer, Heidelberg (2005)
9. Haase, P., Völker, J.: Ontology learning and reasoning - dealing with uncertainty and inconsistency. In: Proc. of URSW 2005, pp. 45–55 (2005)
10. Hansson, S.O.: Kernel contraction. Journal of Symbolic Logic 59(3), 845–859 (1994)
11. Hansson, S.O.: A Textbook of Belief Dynamics: Theory Change and Database Updating. Kluwer Academic Publishers, Dordrecht (1999)
12. Kalyanpur, A., Parsia, B., Horridge, M., Sirin, E.: Finding all justifications of OWL DL entailments. In: Aberer, K., Choi, K.-S., Noy, N., Allemang, D., Lee, K.-I., Nixon, L., Golbeck, J., Mika, P., Maynard, D., Mizoguchi, R., Schreiber, G., Cudré-Mauroux, P. (eds.) ASWC 2007 and ISWC 2007. LNCS, vol. 4825, pp. 267–280. Springer, Heidelberg (2007)
13. Katsuno, H., Mendelzon, A.O.: Propositional knowledge base revision and minimal change. Artificial Intelligence 52(3), 263–294 (1992)
14. Meilicke, C., Stuckenschmidt, H.: Applying logical constraints to ontology matching. In: Hertzberg, J., Beetz, M., Englert, R. (eds.) KI 2007. LNCS (LNAI), vol. 4667, pp. 99–113. Springer, Heidelberg (2007)
15. Qi, G., Hunter, A.: Measuring incoherence in description logic-based ontologies. In: Aberer, K., Choi, K.-S., Noy, N., Allemang, D., Lee, K.-I., Nixon, L., Golbeck, J., Mika, P., Maynard, D., Mizoguchi, R., Schreiber, G., Cudré-Mauroux, P. (eds.) ASWC 2007 and ISWC 2007. LNCS, vol. 4825, pp. 381–394. Springer, Heidelberg (2007)
16. Qi, G., Liu, W., Bell, D.A.: Knowledge base revision in description logics. In: Fisher, M., van der Hoek, W., Konev, B., Lisitsa, A. (eds.) JELIA 2006. LNCS (LNAI), vol. 4160, pp. 386–398. Springer, Heidelberg (2006)
17. Reiter, R.: A theory of diagnosis from first principles. Artificial Intelligence 32(1), 57–95 (1987)
18. Ribeiro, M.M., Wassermann, R.: Base revision in description logics - preliminary results. In: Proc. of IWOD 2007, pp. 69–82 (2007)
19. Schlobach, S., Cornet, R.: Non-standard reasoning services for the debugging of description logic terminologies. In: Proc. of IJCAI 2003, pp. 355–362 (2003)
20. Schlobach, S., Huang, Z., Cornet, R., van Harmelen, F.: Debugging incoherent terminologies. Journal of Automated Reasoning 39(3), 317–349 (2007)
21. Wassermann, R.: An algorithm for belief revision. In: Proc. of KR 2000, pp. 345–352 (2000)

Description Logic Reasoning with Decision Diagrams

Compiling $\mathcal{SHIQ}$ to Disjunctive Datalog*

Sebastian Rudolph, Markus Krötzsch, and Pascal Hitzler

Institut AIFB, Universität Karlsruhe, Germany
{sru,mak,phi}@aifb.uni-karlsruhe.de

Abstract. We propose a novel method for reasoning in the description logic $\mathcal{SHIQ}$. After a satisfiability preserving transformation from $\mathcal{SHIQ}$ to the description logic $\mathcal{ALCIb}$, the obtained $\mathcal{ALCIb}$ Tbox $\mathcal{T}$ is converted into an ordered binary decision diagram (OBDD) which represents a canonical model for $\mathcal{T}$. This OBDD is turned into a disjunctive datalog program that can be used for Abox reasoning. The algorithm is worst-case optimal w.r.t. data complexity, and admits easy extensions with DL-safe rules and ground conjunctive queries.

1 Introduction

In order to leverage intelligent applications for the Semantic Web, scalable reasoning systems for the standardised Web Ontology Language OWL[1] are required. OWL is essentially based on description logics (DLs), with the DL known as $\mathcal{SHIQ}$ currently being among its most prominent fragments. State-of-the art OWL reasoners, such as Pellet, FaCT++, or RacerPro use tableau methods with good performance results, but even those successful systems are not applicable in all practical cases. This motivates the search for alternative reasoning approaches that build upon different methods in order to address cases where tableau algorithms turn out to have certain weaknesses. Successful examples are recent works based on resolution and hyper-tableau calculi, as realised by the systems KAON2 and HermiT.

In this paper, we pursue a new DL reasoning paradigm based on the use of ordered binary decision diagrams (OBDD). These reasoning tools have been successfully applied in the domain of large-scale model checking and verification, but have hitherto seen only little investigation in DLs [1]. Our work bases on a recent adoption of OBDDs for terminological reasoning in $\mathcal{SHIQ}$ [2]. This approach, however, is inherently inapt of dealing with assertional knowledge directly. We therefore adopt the existing OBDD method for terminological reasoning, but use its output for generating a disjunctive datalog program that can in turn be combined with Abox data to obtain a correct reasoning procedure. The main technical contribution of the paper is to show this adoption to be sound and complete based on suitable model constructions. Considering possible applications, the work establishes the basis for applying OBDD-based methods for $\mathcal{SHIQ}$ reasoning, including natural support for DL-safe rules and ground queries.

---

* Supported by the European Commission under contracts 027595 NeOn and 215040 ACTIVE, and by the Deutsche Forschungsgemeinschaft (DFG) under the ReaSem project.
[1] http://www.w3.org/2004/OWL/

A. Sheth et al. (Eds.): ISWC 2008, LNCS 5318, pp. 435–450, 2008.

The structure of the paper is as follows. In Section 2, we recall some essential definitions and results on which we base our approach. Section 3 then discusses the decomposition of models into sets of *dominoes*, which are then computed with OBDDs in Section 4. The resulting OBDD presentation is transformed to disjunctive datalog in Section 5, where we also show the correctness of the approach. Section 6 concludes.

2   The Description Logics $\mathcal{SHIQ}$ and $\mathcal{ALCIb}$

We first recall some basic definitions of DLs (see [3] for a comprehensive treatment of DLs) and introduce our notation. Next we define a rather expressive description logic $\mathcal{SHIQb}$ that extends $\mathcal{SHIQ}$ with restricted Boolean role expressions [4]. We will not consider $\mathcal{SHIQb}$ knowledge bases, but the DL serves as a convenient umbrella logic for the DLs used in this paper.

Definition 1. *A $\mathcal{SHIQb}$ knowledge base is based on three disjoint sets of* concept names N_C, role names N_R, *and* individual names N_I. *A set of* atomic roles **R** *is defined as* $\mathbf{R} := N_R \cup \{R^- \mid R \in N_R\}$. *In addition, we set* $\mathrm{Inv}(R) := R^-$ *and* $\mathrm{Inv}(R^-) := R$, *and we will extend this notation also to sets of atomic roles. In the sequel, we will use the symbols R, S to denote atomic roles, if not specified otherwise.*

   The set of Boolean role expressions **B** *is defined as*

$$\mathbf{B} ::= \mathbf{R} \mid \neg \mathbf{B} \mid \mathbf{B} \sqcap \mathbf{B} \mid \mathbf{B} \sqcup \mathbf{B}.$$

We use $\vdash$ to denote standard Boolean entailment between sets of atomic roles and role expressions. Given a set $\mathcal{R}$ of atomic roles, we inductively define:

- *For atomic roles R, $\mathcal{R} \vdash R$ if $R \in \mathcal{R}$, and $\mathcal{R} \nvdash R$ otherwise,*
- *$\mathcal{R} \vdash \neg U$ if $\mathcal{R} \nvdash U$, and $\mathcal{R} \nvdash \neg U$ otherwise,*
- *$\mathcal{R} \vdash U \sqcap V$ if $\mathcal{R} \vdash U$ and $\mathcal{R} \vdash V$, and $\mathcal{R} \nvdash U \sqcap V$ otherwise,*
- *$\mathcal{R} \vdash U \sqcup V$ if $\mathcal{R} \vdash U$ or $\mathcal{R} \vdash V$, and $\mathcal{R} \nvdash U \sqcup V$ otherwise.*

*A Boolean role expression U is restricted if $\emptyset \nvdash U$. The set of all restricted role expressions is denoted **T**, and the symbols U and V will be used throughout this paper to denote restricted role expressions. A $\mathcal{SHIQb}$ Rbox is a set of axioms of the form $U \sqsubseteq V$ (role inclusion axiom) or $\mathrm{Tra}(R)$ (transitivity axiom). The set of non-simple roles (for a given Rbox) is inductively defined as follows:*

- *If there is an axiom $\mathrm{Tra}(R)$, then R is non-simple.*
- *If there is an axiom $R \sqsubseteq S$ with R non-simple, then S is non-simple.*
- *If R is non-simple, then $\mathrm{Inv}(R)$ is non-simple.*

*A role is simple if it is atomic (simplicity of Boolean role expressions is not relevant in this paper) and not non-simple. Based on a $\mathcal{SHIQb}$ Rbox, the set of concept expressions **C** is the smallest set containing N_C, and all concept expressions given in Table 1, where $C, D \in \mathbf{C}$, $U \in \mathbf{T}$, and $R \in \mathbf{R}$ is a simple role. Throughout this paper, the symbols C, D will be used to denote concept expressions. A $\mathcal{SHIQb}$ Tbox (or terminology) is a set of general concept inclusion axioms (GCIs) of the form $C \sqsubseteq D$. A $\mathcal{SHIQb}$ Abox (containing assertional knowledge) is a set of statements of the form $C(a)$ or $R(a, b)$, where $a, b \in N_I$. We assume throughout that all roles and concepts occurring in the*

Table 1. Semantics of constructors in $\mathcal{SHIQ}b$ for an interpretation $\mathcal{I}$ with domain $\Delta^{\mathcal{I}}$

Name	Syntax	Semantics
inverse role	R^-	$\{\langle x,y\rangle \in \Delta^{\mathcal{I}} \times \Delta^{\mathcal{I}} \mid \langle y,x\rangle \in R^{\mathcal{I}}\}$
role negation	$\neg U$	$\{\langle x,y\rangle \in \Delta^{\mathcal{I}} \times \Delta^{\mathcal{I}} \mid \langle x,y\rangle \notin U^{\mathcal{I}}\}$
role conj.	$U \sqcap V$	$U^{\mathcal{I}} \cap V^{\mathcal{I}}$
role disj.	$U \sqcup V$	$U^{\mathcal{I}} \cup V^{\mathcal{I}}$
top	$\top$	$\Delta^{\mathcal{I}}$
bottom	$\bot$	$\emptyset$
negation	$\neg C$	$\Delta^{\mathcal{I}} \setminus C^{\mathcal{I}}$
conjunction	$C \sqcap D$	$C^{\mathcal{I}} \cap D^{\mathcal{I}}$
disjunction	$C \sqcup D$	$C^{\mathcal{I}} \cup D^{\mathcal{I}}$
univ. rest.	$\forall U.C$	$\{x \in \Delta^{\mathcal{I}} \mid \langle x,y\rangle \in U^{\mathcal{I}} \text{ implies } y \in C^{\mathcal{I}}\}$
exist. rest.	$\exists U.C$	$\{x \in \Delta^{\mathcal{I}} \mid y \in \Delta^{\mathcal{I}}: \langle x,y\rangle \in U^{\mathcal{I}}, y \in C^{\mathcal{I}}\}$
qualified	$\leq n\,R.C$	$\{x \in \Delta^{\mathcal{I}} \mid \#\{y \in \Delta^{\mathcal{I}} \mid \langle x,y\rangle \in R^{\mathcal{I}}, y \in C^{\mathcal{I}}\} \leq n\}$
number rest.	$\geq n\,R.C$	$\{x \in \Delta^{\mathcal{I}} \mid \#\{y \in \Delta^{\mathcal{I}} \mid \langle x,y\rangle \in R^{\mathcal{I}}, y \in C^{\mathcal{I}}\} \geq n\}$

Abox are atomic (which can be done without loss of generality). A $\mathcal{SHIQ}b$ knowledge base KB is a triple $\langle \mathcal{A}, \mathcal{R}, \mathcal{T} \rangle$, where $\mathcal{A}$ is an Abox, $\mathcal{R}$ is an Rbox, and $\mathcal{T}$ is a Tbox.

As mentioned above, we will consider only fragments of $\mathcal{SHIQ}b$. In particular, a $\mathcal{SHIQ}$ knowledge base is a $\mathcal{SHIQ}b$ knowledge base without Boolean role expressions, and an $\mathcal{ALCI}b$ knowledge base is a $\mathcal{SHIQ}b$ knowledge base that contains no Rbox axioms and no number restrictions (i.e. axioms $\leq n\,R.C$ or $\geq n\,R.C$). Consequently, an $\mathcal{ALCI}b$ knowledge base only consists of a pair $\langle \mathcal{A}, \mathcal{T} \rangle$, where $\mathcal{A}$ is an Abox and $\mathcal{T}$ is a Tbox. The related DL $\mathcal{ALCQI}b$ has been studied in [4].

An interpretation $\mathcal{I}$ consists of a set $\Delta^{\mathcal{I}}$ called *domain* (the elements of it being called *individuals*) together with a function $\cdot^{\mathcal{I}}$ mapping individual names to elements of $\Delta^{\mathcal{I}}$, concept names to subsets of $\Delta^{\mathcal{I}}$, and role names to subsets of $\Delta^{\mathcal{I}} \times \Delta^{\mathcal{I}}$. The function $\cdot^{\mathcal{I}}$ is extended to role and concept expressions as shown in Table 1. An interpretation $\mathcal{I}$ *satisfies* an axiom φ if we find that $\mathcal{I} \models \varphi$, where

- $\mathcal{I} \models U \sqsubseteq V$ if $U^{\mathcal{I}} \subseteq V^{\mathcal{I}}$,
- $\mathcal{I} \models \mathsf{Tra}(R)$ if $R^{\mathcal{I}}$ is a transitive relation,
- $\mathcal{I} \models C \sqsubseteq D$ if $C^{\mathcal{I}} \subseteq D^{\mathcal{I}}$,
- $\mathcal{I} \models C(a)$ if $a^{\mathcal{I}} \in C^{\mathcal{I}}$,
- $\mathcal{I} \models R(a,b)$ if $(a^{\mathcal{I}}, b^{\mathcal{I}}) \in R^{\mathcal{I}}$.

$\mathcal{I}$ satisfies a knowledge base KB, $\mathcal{I} \models$ KB, if it satisfies all axioms of KB. *Satisfiability*, *equivalence*, and *equisatisfiability* of knowledge bases are defined as usual.

For convenience of notation, we abbreviate Tbox axioms of the form $\top \sqsubseteq C$ by writing just C. Statements such as $\mathcal{I} \models C$ and $C \in$ KB are interpreted accordingly. Note that $C \sqsubseteq D$ can thus be written as $\neg C \sqcup D$.

Finally, we will often need to access a particular set of quantified and atomic subformulae of a DL concept. These specific parts are provided by the function $P : \mathbf{C} \to 2^{\mathbf{C}}$:

$$P(C) := \begin{cases} P(D) & \text{if } C = \neg D \\ P(D) \cup P(E) & \text{if } C = D \sqcap E \text{ or } C = D \sqcup E \\ \{C\} \cup P(D) & \text{if } C = \mathsf{Q}U.D \text{ with } \mathsf{Q} \in \{\exists, \forall, \geq n, \leq n\} \\ \{C\} & \text{otherwise} \end{cases}$$

We generalise P to DL knowledge bases KB by defining $P(\text{KB})$ to be the union of the sets $P(C)$ for all Tbox axioms C in KB.

We will usually express all Tbox axioms as simple concept expressions as explained above. Given a knowledge base KB we obtain its negation normal form NNF(KB) by converting every Tbox concept into its negation normal form as usual. It is well-known that KB and NNF(KB) are equivalent.

For $\mathcal{ALCIb}$ knowledge bases KB, we will usually require another normalisation step that simplifies the structure of KB by *flattening* it to a knowledge base FLAT(KB). This is achieved by transforming KB into negation normal form and exhaustively applying the following transformation rules:

- Select an outermost occurrence of $QU.D$ in KB, such that $Q \in \{\exists, \forall\}$ and D is a non-atomic concept.
- Substitute this occurrence with $QU.F$ where F is a fresh concept name (i.e. one not occurring in the knowledge base).
- Add $\neg F \sqcup D$ to the knowledge base.

Obviously, this procedure terminates yielding a flat knowledge base FLAT(KB) all Tbox axioms of which are Boolean expressions over formulae of the form A, $\neg A$, or $QU.A$ with A an atomic concept name. As shown in [2], any $\mathcal{ALCIb}$ knowledge base KB is equisatisfiable to FLAT(KB). This work also detailed a reduction of $\mathcal{SHIQ}$ knowledge bases to $\mathcal{ALCIb}$ that we summarise as follows:

Theorem 2. *Any $\mathcal{SHIQ}$ knowledge base* KB *can be transformed in polynomial time into an equisatisfiable $\mathcal{ALCIb}$ knowledge base* KB$'$.

It is easy to see that the algorithm from [2] is still applicable in the presence of Aboxes, and that ground Abox conclusions are preserved – with the exception of entailments of the form $R(a, b)$ for non-simple roles R which fall victim to the standard elimination of transitivity axioms.

3   Building Models from Domino Sets

Our approach towards terminological reasoning in $\mathcal{ALCIb}$ exploits the fact that models for this DL can be decomposed into small parts, which we call *dominoes*. Intuitively, each domino abstractly represents two individuals in an $\mathcal{ALCIb}$ interpretation, based on their concept properties and role relationships. We will see that suitable sets of such two-element pieces suffice to reconstruct models of $\mathcal{ALCIb}$ Tboxes, and satisfiability of $\mathcal{ALCIb}$ terminologies can thus be reduced to the existence of suitable sets.

We first introduce the basic notion of a domino set, and its relationship to interpretations. Given a DL language with concepts **C** and roles **R**, a *domino* is an arbitrary triple $\langle \mathcal{A}, \mathcal{R}, \mathcal{B} \rangle$, where $\mathcal{A}, \mathcal{B} \subseteq \mathbf{C}$ and $\mathcal{R} \subseteq \mathbf{R}$. We will generally assume a fixed language and refer to dominoes over that language only. Interpretations can be deconstructed into sets of dominoes as follows:

Definition 3. *Given an interpretation* $\mathcal{I} = \langle \Delta^{\mathcal{I}}, \cdot^{\mathcal{I}} \rangle$, *and a set* $C \subseteq \mathbf{C}$ *of concept expressions, the* domino projection *of* $\mathcal{I}$ *w.r.t.* C, *denoted by* $\pi_C(\mathcal{I})$ *is the set that contains for all* $\delta, \delta' \in \Delta^{\mathcal{I}}$ *the triple* $\langle \mathcal{A}, \mathcal{R}, \mathcal{B} \rangle$ *with*

- $\mathcal{A} = \{C \in C \mid \delta \in C^{\mathcal{I}}\}$,
- $\mathcal{R} = \{R \in \mathbf{R} \mid \langle \delta, \delta' \rangle \in R^{\mathcal{I}}\}$,
- $\mathcal{B} = \{C \in C \mid \delta' \in C^{\mathcal{I}}\}$.

An inverse construction of interpretations from arbitrary domino sets is as follows:

Definition 4. *Given a set $\mathbb{D}$ of dominoes, the induced domino interpretation $\mathcal{I}(\mathbb{D}) = \langle \Delta^{\mathcal{I}}, \cdot^{\mathcal{I}} \rangle$ is defined as follows:*

1. $\Delta^{\mathcal{I}}$ *consists of all finite nonempty words over $\mathbb{D}$ where, for each pair of subsequent letters $\langle \mathcal{A}, \mathcal{R}, \mathcal{B} \rangle$ and $\langle \mathcal{A}', \mathcal{R}', \mathcal{B}' \rangle$ in a word, we have $\mathcal{B} = \mathcal{A}'$.*
2. *For $\delta = \langle \mathcal{A}_1, \mathcal{R}_1, \mathcal{A}_2 \rangle \langle \mathcal{A}_2, \mathcal{R}_2, \mathcal{A}_3 \rangle \ldots \langle \mathcal{A}_{i-1}, \mathcal{R}_{i-1}, \mathcal{A}_i \rangle$ a word and $A \in \mathsf{N}_C$ a concept name, we define $\mathrm{tail}(\delta) := \mathcal{A}_i$, and set $\delta \in A^{\mathcal{I}}$ iff $A \in \mathrm{tail}(\delta)$,*
3. *For each $R \in \mathsf{N}_R$, we set $\langle \delta_1, \delta_2 \rangle \in R^{\mathcal{I}}$ if either $\delta_2 = \delta_1 \langle \mathcal{A}, \mathcal{R}, \mathcal{B} \rangle$ with $R \in \mathcal{R}$ or $\delta_1 = \delta_2 \langle \mathcal{A}, \mathcal{R}, \mathcal{B} \rangle$ with $\mathrm{Inv}(R) \in \mathcal{R}$.*

Mark that – following the intuition – the domino interpretation is constructed by conjoining matching dominoes. This process is also similar to the related method of "unravelling" models in order to obtain tree-like interpretations.

Domino projections do not faithfully represent the structure of the interpretation that they were constructed from, yet they capture enough information to reconstruct models of a Tbox $\mathcal{T}$, as long as C is chosen to contain at least $P(\mathcal{T})$. Indeed, it was shown in [2] that, for any $\mathcal{ALCI}b$ terminology $\mathcal{T}$, $\mathcal{J} \models \mathcal{T}$ iff $\mathcal{I}(\pi_{P(\mathcal{T})}(\mathcal{J})) \models \mathcal{T}$. This observation allows us to devise an algorithm that directly constructs a suitable domino set from which one could obtain a model that witnesses the satisfiability of some knowledge base. The following algorithm therefore considers all possible dominoes, and iteratively eliminates those that cannot occur in the domino projection of any model:

Definition 5. *Consider an $\mathcal{ALCI}b$ terminology $\mathcal{T}$, and define $C = P(\mathsf{FLAT}(\mathcal{T}))$. Sets $\mathbb{D}_i$ of dominoes based on concepts from C are constructed as follows:*
   $\mathbb{D}_0$ consists of all dominoes $\langle \mathcal{A}, \mathcal{R}, \mathcal{B} \rangle$ which satisfy:

kb: *for every concept $C \in \mathsf{FLAT}(\mathcal{T})$, we have that $\sqcap_{D \in \mathcal{A}} D \sqsubseteq C$ is a tautology[2],*
ex: *for all $\exists U.A \in C$, if $A \in \mathcal{B}$ and $\mathcal{R} \vdash U$ then $\exists U.A \in \mathcal{A}$,*
uni: *for all $\forall U.A \in C$, if $\forall U.A \in \mathcal{A}$ and $\mathcal{R} \vdash U$ then $A \in \mathcal{B}$.*

   Given a domino set $\mathbb{D}_i$, the set $\mathbb{D}_{i+1}$ consists of all dominoes $\langle \mathcal{A}, \mathcal{R}, \mathcal{B} \rangle \in \mathbb{D}_i$ satisfying the following conditions:

delex: *for every $\exists U.A \in C$ with $\exists U.A \in \mathcal{A}$, there is some $\langle \mathcal{A}, \mathcal{R}', \mathcal{B}' \rangle \in \mathbb{D}_i$ such that $\mathcal{R}' \vdash U$ and $A \in \mathcal{B}'$,*
deluni: *for every $\forall U.A \in C$ with $\forall U.A \notin \mathcal{A}$, there is some $\langle \mathcal{A}, \mathcal{R}', \mathcal{B}' \rangle \in \mathbb{D}_i$ such that $\mathcal{R}' \vdash U$ but $A \notin \mathcal{B}'$,*
sym: $\langle \mathcal{B}, \mathrm{Inv}(\mathcal{R}), \mathcal{A} \rangle \in \mathbb{D}_i$.

   The construction of domino sets $\mathbb{D}_{i+1}$ is continued until $\mathbb{D}_{i+1} = \mathbb{D}_i$. The final result $\mathbb{D}_{\mathcal{T}} := \mathbb{D}_{i+1}$ defines the canonical domino set of $\mathcal{T}$.

---

[2] Note that formulae in $\mathsf{FLAT}(\mathcal{T})$ and in $\mathcal{A} \subseteq C$ are such that this can easily be checked by evaluating the Boolean operators in C as if $\mathcal{A}$ was a set of true propositional variables.

Note that the algorithm must terminate, since it starts from a finite initial set $\mathbb{D}_0$ that is reduced in each computation step. Intuitively, the algorithm implements a kind of greatest fixed point construction that yields the domino projection of the largest possible model of the terminological part of an $\mathcal{ALCI}b$ knowledge base. The following result makes this intuition more explicitly:

Lemma 6. *Consider an $\mathcal{ALCI}b$ terminology $\mathcal{T}$ and an arbitrary model $\mathcal{I}$ of $\mathcal{T}$. Then the domino projection $\pi_{P(\text{FLAT}(\mathcal{T}))}(\mathcal{I})$ is contained in $\mathbb{D}_{\mathcal{T}}$.*

Proof. The claim is shown by a simple induction. In the following, we use $\langle \mathcal{A}, \mathcal{R}, \mathcal{B} \rangle$ to denote an arbitrary domino of $\pi_{P(\text{FLAT}(\mathcal{T}))}(\mathcal{I})$. For the base case, we must show that $\pi_{P(\text{FLAT}(\mathcal{T}))}(\mathcal{I}) \subseteq \mathbb{D}_0$. Let $\langle \mathcal{A}, \mathcal{R}, \mathcal{B} \rangle$ to denote an arbitrary domino of $\pi_{P(\text{FLAT}(\mathcal{T}))}(\mathcal{I})$ which was generated from elements $\langle \delta, \delta' \rangle$. Then $\langle \mathcal{A}, \mathcal{R}, \mathcal{B} \rangle$ satisfies condition **kb**, since $\delta \in C^{\mathcal{I}}$ for any $C \in \text{FLAT}(\mathcal{T})$. The conditions **ex** and **uni** are obviously satisfied.

For the induction step, assume that $\pi_{P(\text{FLAT}(\mathcal{T}))}(\mathcal{I}) \subseteq \mathbb{D}_i$, and let $\langle \mathcal{A}, \mathcal{R}, \mathcal{B} \rangle$ again denote an arbitrary domino of $\pi_{P(\text{FLAT}(\mathcal{T}))}(\mathcal{I})$ which was generated from elements $\langle \delta, \delta' \rangle$.

- For **delex**, note that $\exists U.A \in \mathcal{A}$ implies $\delta \in (\exists U.A)^{\mathcal{I}}$. Thus there is an individual δ'' such that $\langle \delta, \delta'' \rangle \in U^{\mathcal{I}}$ and $\delta'' \in A^{\mathcal{I}}$. Clearly, the domino generated by $\langle \delta, \delta'' \rangle$ satisfies the conditions of **delex**.
- For **deluni**, note that $\forall U.A \notin \mathcal{A}$ implies $\delta \notin (\forall U.A)^{\mathcal{I}}$. Thus there is an individual δ'' such that $\langle \delta, \delta'' \rangle \in U^{\mathcal{I}}$ and $\delta'' \notin A^{\mathcal{I}}$. Clearly, the domino generated by $\langle \delta, \delta'' \rangle$ satisfies the conditions of **deluni**.
- The condition of **sym** for $\langle \mathcal{A}, \mathcal{R}, \mathcal{B} \rangle$ is clearly satisfied by the domino generated from $\langle \delta', \delta \rangle$. $\qquad\square$

We will also exploit this observation in the later construction of models for knowledge bases with individual assertions. The following was again shown in [2]:

Theorem 7. *An $\mathcal{ALCI}b$ terminology $\mathcal{T}$ is satisfiable iff its canonical domino set $\mathbb{D}_{\mathcal{T}}$ is non-empty. Definition 5 thus defines a decision procedure for satisfiability of $\mathcal{ALCI}b$ terminologies.*

4   Sets as Boolean Functions

The algorithm of the previous section may seem to be of little practical use, since it requires the computations on an exponentially large set of dominoes. The required computation steps, however, can also be accomplished with a more indirect representation of the possible dominoes based on Boolean functions. Indeed, any propositional logic formula represents a set of interpretations for which the function evaluates to *true*. Using a suitable encoding, each interpretation can be understood as a domino, and a propositional formula can represent a domino set.

In order for this approach to be more feasible than the naive algorithm given above, an efficient representation of propositional formulae is needed. For this we use binary decision diagrams (BDDs), that have been applied to represent complex Boolean functions in model-checking (see, e.g., [5]). A particular optimisation of these structures are ordered BDDs (OBDDs) that use a dynamic precedence order of propositional variables to obtain compressed representations. We provide a first introduction to OBDDs below. A more detailed exposition and pointers to the literature are given in [6].

Boolean Functions and Operations. We first explain how sets can be represented by means of Boolean functions. This will enable us, given a fixed finite base set S, to represent every family of sets $\mathbb{S} \subseteq 2^S$ by a single Boolean function.

A *Boolean function* on a set Var of variables is a function $\varphi : 2^{\text{Var}} \rightarrow \{true, false\}$. The underlying intuition is that $\varphi(V)$ computes the truth value of a Boolean formula based on the assumption that exactly the variables of V are evaluated to *true*. A simple example are so-called *characteristic functions* of the form $[\![v]\!]_\chi$ for some $v \in$ Var, which are defined as $[\![v]\!]_\chi(V) := true$ iff $v \in V$, or the functions $[\![true]\!]$ and $[\![false]\!]$ mapping any input to *true* or *false*, respectively.

Boolean functions over the same set of variables can be combined and modified in several ways. Firstly, there are the obvious Boolean operators for negation, conjunction, disjunction, and implication. By slight abuse of notation, we will use the common (syntactic) operator symbols $\neg$, $\wedge$, $\vee$, and $\rightarrow$ to also represent such (semantic) operators on Boolean functions. Given, e.g., Boolean functions φ and ψ, we find that $(\varphi \wedge \psi)(V) = true$ iff $\varphi(V) = true$ and $\psi(V) = true$. Note that the result of the application of $\wedge$ results in another Boolean function, and is not to be understood as a syntactic formula. Another operation on Boolean functions is existential quantification over a set of variables $V \subseteq$ Var, written as $\exists V.\varphi$ for some function φ. Given an input set $W \subseteq$ Var of variables, we define $(\exists V.\varphi)(W) = true$ iff *there is some* $V' \subseteq V$ such that $\varphi(V' \cup (W \setminus V)) = true$. In other words, there must be a way to set truth values of variables in V such that φ evaluates to *true*. Universal quantification is defined analogously, and we thus have $\forall V.\varphi := \neg\exists V.\neg\varphi$ as usual. Mark that our use of $\exists$ and $\forall$ overloads notation, and should not be confused with role restrictions in DL expressions.

Ordered Binary Decision Diagrams. Binary Decision Diagrams (BDDs), intuitively, are a generalisation of decision trees which allow the reuse of nodes. Structurally, BDDs are directed acyclic graphs whose nodes are labelled by variables from some set Var. The only exception are two *terminal* nodes that are labelled by *true* and *false*, respectively. Every non-terminal node has two outgoing edges, corresponding to the two possible truth values of the variable.

Definition 8. *A BDD is a tuple* $\mathbb{O} = (N, n_{\text{root}}, n_{\text{true}}, n_{\text{false}}, \text{low}, \text{high}, \text{Var}, \lambda)$ *where*

- N *is a finite set called* nodes,
- $n_{\text{root}} \in N$ *is called the* root *node,*
- $n_{\text{true}}, n_{\text{false}} \in N$ *are called the* terminal *nodes,*
- $\text{low}, \text{high} : N \setminus \{n_{\text{true}}, n_{\text{false}}\} \rightarrow N$ *are two* child functions *assigning to every non-terminal node a* low *and a* high *child node. Furthermore the graph obtained by iterated application has to be acyclic, i.e. for no node n exists a sequence of applications of* low *and* high *resulting in n again.*
- Var *is a finite set of* variables.
- $\lambda : N \setminus \{n_{\text{true}}, n_{\text{false}}\} \rightarrow$ Var *is the labelling function assigning to every non-terminal node a variable from* Var.

OBBDs are a particular realisation of BDDs where a certain ordering is imposed on variables to achieve more efficient representations. We will not require to consider the background of this optimisation in here. Now every BDD based on a variable set Var = $\{x_1, \ldots, x_n\}$ represents an n-ary Boolean function $\varphi : 2^{\text{Var}} \rightarrow \{true, false\}$.

Definition 9. *Given a BDD* $\mathbb{O} = (N, n_{\text{root}}, n_{\text{true}}, n_{\text{false}}, \text{low}, \text{high}, \text{Var}, \lambda)$ *the Boolean function* $\varphi_{\mathbb{O}} : 2^{\text{Var}} \to \{true, false\}$ *is defined recursively as follows:*

$$\varphi_{\mathbb{O}} := \varphi_{n_{\text{root}}} \qquad \varphi_{n_{\text{true}}} = [\![true]\!] \qquad \varphi_{n_{\text{false}}} = [\![false]\!]$$

$$\varphi_n = \left(\neg [\![\lambda(n)]\!]_\chi \wedge \varphi_{\text{low}(n)} \right) \vee \left([\![\lambda(n)]\!]_\chi \wedge \varphi_{\text{high}(n)} \right) \text{ for } n \in N \setminus \{n_{\text{true}}, n_{\text{false}}\}$$

In other words, the value $\varphi(V)$ for some $V \subseteq \text{Var}$ is determined by traversing the BDD, beginning from the root node: at a node labelled with $v \in \text{Var}$, the evaluation proceeds with the node connected by the high-edge if $v \in V$, and with the node connected by the low-edge otherwise. If a terminal node is reached, its label is returned as a result.

BDDs for some Boolean formula might be exponentially large in general, but often there is a representation which allows for BDDs of manageable size. Finding the optimal representation is NP-complete, but heuristics have shown to yield good approximate solutions. Hence (O)BDDs are often conceived as efficiently compressed representations of Boolean functions. In addition, many operations on Boolean functions – such as the aforementioned "point-wise" negation, conjunction, disjunction, implication as well as propositional quantification – can be performed directly on the corresponding OBDDs by fast algorithms.

Translating Dominos into Boolean Functions. To apply the above machinery to DL reasoning, consider a flattened $\mathcal{ALCIb}$ terminology $\mathcal{T} = \text{FLAT}(\mathcal{T})$. A set of propositional variables Var is defined as $\text{Var} := \mathbf{R} \cup (P(\mathcal{T}) \times \{1, 2\})$. We thus obtain an obvious bijection between sets $V \subseteq \text{Var}$ and dominoes over the set $P(\mathcal{T})$ given as $\langle \mathcal{A}, \mathcal{R}, \mathcal{B} \rangle \mapsto (\mathcal{A} \times \{1\}) \cup \mathcal{R} \cup (\mathcal{B} \times \{2\})$. Hence, any Boolean function over Var represents a domino set as the collection of all variable sets for which it evaluates to *true*. We can use this observation to rephrase the construction of $\mathbb{D}_{\mathcal{T}}$ in Definition 5 into an equivalent construction of a function $[\![\mathcal{T}]\!]$.

We first represent DL concepts C and role expressions U by characteristic Boolean functions over Var as follows. Note that the application of $\wedge$ results in another Boolean function, and is not to be understood as a syntactic formula.

$$[\![C]\!] := \begin{cases} \neg[\![D]\!] & \text{if } C = \neg D \\ [\![D]\!] \wedge [\![E]\!] & \text{if } C = D \sqcap E \\ [\![D]\!] \vee [\![E]\!] & \text{if } C = D \sqcup E \\ [\![\langle C, 1 \rangle]\!]_\chi & \text{if } C \in P(\mathcal{T}) \end{cases} \qquad [\![U]\!] := \begin{cases} \neg[\![V]\!] & \text{if } U = \neg V \\ [\![V]\!] \wedge [\![W]\!] & \text{if } U = V \sqcap W \\ [\![V]\!] \vee [\![W]\!] & \text{if } U = V \sqcup W \\ [\![U]\!]_\chi & \text{if } U \in \mathbf{R} \end{cases}$$

We can now define an inferencing algorithm based on Boolean functions.

Definition 10. *Given a flattened $\mathcal{ALCIb}$ terminology $\mathcal{T}$ and a variable set* Var *defined as above, Boolean functions $[\![\mathcal{T}]\!]_i$ are constructed based on the definitions in Fig. 1:*

- $[\![\mathcal{T}]\!]_0 := \varphi^{\text{kb}} \wedge \varphi^{\text{uni}} \wedge \varphi^{\text{ex}}$,
- $[\![\mathcal{T}]\!]_{i+1} := [\![\mathcal{T}]\!]_i \wedge \varphi_i^{\text{delex}} \wedge \varphi_i^{\text{deluni}} \wedge \varphi_i^{\text{sym}}$

The construction terminates as soon as $[\![\mathcal{T}]\!]_{i+1} = [\![\mathcal{T}]\!]_i$, and the result of the construction is then defined as $[\![\mathcal{T}]\!] := [\![\mathcal{T}]\!]_i$. The algorithm returns "unsatisfiable" if $[\![\mathcal{T}]\!](V) = false$ for all $V \subseteq \text{Var}$, and "satisfiable" otherwise.

As shown in [2], the above algorithm is a correct procedure for checking consistency of terminological $\mathcal{ALCIb}$ knowledge bases. Moreover, all required operations and checks

$$\varphi^{\mathbf{kb}} := \bigwedge_{C \in \mathcal{T}} [\![C]\!]$$

$$\varphi^{\mathbf{uni}} := \bigwedge_{\forall U.C \in P(\mathcal{T})} [\![\langle \forall U.C, 1\rangle]\!]_\chi \wedge [\![U]\!] \rightarrow [\![\langle C, 2\rangle]\!]_\chi \qquad \varphi^{\mathbf{ex}} := \bigwedge_{\exists U.C \in P(\mathcal{T})} [\![\langle C, 2\rangle]\!]_\chi \wedge [\![U]\!] \rightarrow [\![\langle \exists U.C, 1\rangle]\!]_\chi$$

$$\varphi_i^{\mathbf{delex}} := \bigwedge_{\exists U.C \in P(\mathcal{T})} [\![\langle \exists U.C, 1\rangle]\!]_\chi \rightarrow \exists (\mathbf{R} \cup C\times\{2\}).([\![\mathcal{T}]\!]_i \wedge [\![U]\!] \wedge [\![\langle C, 2\rangle]\!]_\chi)$$

$$\varphi_i^{\mathbf{deluni}} := \bigwedge_{\forall U.C \in P(\mathcal{T})} [\![\langle \forall U.C, 1\rangle]\!]_\chi \rightarrow \neg\exists (\mathbf{R} \cup C\times\{2\}).([\![\mathcal{T}]\!]_i \wedge [\![U]\!] \wedge \neg[\![\langle C, 2\rangle]\!]_\chi)$$

$$\varphi_i^{\mathbf{sym}}(V) := [\![\mathcal{T}]\!]_i\big(\{\langle D, 1\rangle \mid \langle D, 2\rangle \in V\} \cup \{\mathrm{Inv}(R) \mid R \in V\} \cup \{\langle D, 2\rangle \mid \langle D, 1\rangle \in V\}\big)$$

Fig. 1. Boolean functions for defining the canonical domino set in Definition 10

$$PhDStudent \sqsubseteq \exists has.Diploma$$
$$Diploma \sqsubseteq \forall has^-.Graduate$$
$$Diploma \sqcap Graduate \sqsubseteq \top$$
$$Diploma(laureus) \qquad PhDStudent(laureus)$$

Fig. 2. An example $\mathcal{ALCI}b$ knowledge base

are provided by standard OBDD implementations, and thus can be realised in practice. Correctness follows from the next observation, which is also relevant for extending reasoning to Aboxes below:

Proposition 11. *For any $\mathcal{ALCI}b$ terminology $\mathcal{T}$ and variable set $V \in \mathsf{Var}$ as above, we find that $[\![\mathcal{T}]\!](V) = true$ iff V represents a domino in $\mathbb{D}_{\mathcal{T}}$ as defined in Definition 5.*

In the remainder of this section, we illustrate the above algorithm by an extended example, to which we will also come back to explain the later extensions of the inference algorithm. Therefore, consider the $\mathcal{ALCI}b$ knowledge base given in Fig. 2. For now, we are only interested in the terminological axioms, the consistency of which we would like to establish. As a first transformation step, all Tbox axioms are transformed into the following universally valid concepts in negation normal form:

$$\neg PhDStudent \sqcup \exists has.Diploma \qquad \neg Diploma \sqcup \forall has^-.Graduate \qquad \neg Diploma \sqcup \neg Graduate$$

The flattening step can be skipped since all concepts are already flat. Now the relevant concept expressions for describing dominoes are as follows given by the set $P(\mathcal{T}) = \{\exists has.Diploma, \forall has^-.Graduate, Diploma, Graduate, PhDStudent\}$. We thus obtain the following set Var of Boolean variables (though Var is just a set, our presentation follows the domino intuition):

$\langle \exists has.Diploma, 1\rangle$	has	$\langle \exists has.Diploma, 2\rangle$
$\langle \forall has^-.Graduate, 1\rangle$	has^-	$\langle \forall has^-.Graduate, 2\rangle$
$\langle Diploma, 1\rangle$		$\langle Diploma, 2\rangle$
$\langle Graduate, 1\rangle$		$\langle Graduate, 2\rangle$
$\langle PhDStudent, 1\rangle$		$\langle PhDStudent, 2\rangle$

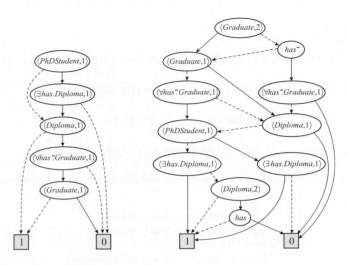

Fig. 3. OBDDs arising when processing the terminology of Fig. 2. Following traditional BDD notation, solid arrows indicate high successors, and dashed arrows indicate low successors.

We are now ready to construct the OBDDs as described. Figure 3 (left) displays an OBDD corresponding to the following Boolean function:

$$\varphi^{kb} := (\neg[\![\langle PhDStudent, 1 \rangle]\!] \vee [\![\langle \exists has.Diploma, 1 \rangle]\!])$$
$$\wedge (\neg[\![\langle Diploma, 1 \rangle]\!] \vee [\![\langle \forall has^-.Graduate, 1 \rangle]\!])$$
$$\wedge (\neg[\![\langle Diploma, 1 \rangle]\!] \vee \neg[\![\langle Graduate, 1 \rangle]\!])$$

and in Fig. 3 (right) shows the OBDD representing the function $[\![\mathcal{T}]\!]_0$ obtained from φ^{kb} by conjunctively adding

$$\varphi^{ex} = \neg[\![\langle Diploma, 2 \rangle]\!] \vee \neg[\![has]\!] \vee [\![\langle \exists has.Diploma, 1 \rangle]\!] \text{ and}$$
$$\varphi^{uni} = \neg[\![\langle \forall has^-.Graduate, 1 \rangle]\!] \vee \neg[\![has^-]\!] \vee [\![\langle Graduate, 2 \rangle]\!].$$

Then, after the first iteration of the algorithm, we arrive at an OBDD representing $[\![\mathcal{T}]\!]_1$ which is displayed in Fig. 4. This OBDD turns out to be the final result $[\![\mathcal{T}]\!]$.

5   Abox Reasoning with Disjunctive Datalog

The above algorithm does not yet take any assertional information about individuals into account. Now the proof of Theorem 7 given in [2] hinges upon the fact that the constructed domino set $\mathbb{D}_{\mathcal{T}}$ induces a model of the terminology $\mathcal{T}$, and Lemma 6 states that this is indeed the *greatest* model in a certain sense. This provides some first intuition of the problems arising when Aboxes are to be added to the knowledge base: $\mathcal{ALCIb}$ knowledge bases with Aboxes do generally not have a greatest model.

We thus employ *disjunctive datalog* as a paradigm that allows us to incorporate Aboxes into the reasoning process. The basic idea is to forge a datalog program that – depending on two given individuals a and b – describes possible dominoes that may

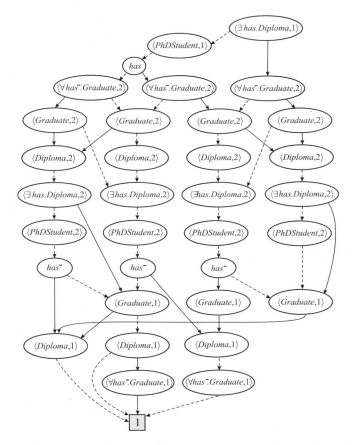

Fig. 4. Final OBDD obtained when processing Fig. 2, using notation as in Fig. 3. Arrows to the 0 node have been omitted for better readability.

connect a and b in models of the knowledge base. There might be various, irreconcilable such dominoes in different models, but disjunctive datalog supports such choice since it admits multiple minimal models. As long as the knowledge base has some model, there is at least one possible domino for every pair of individuals (possibly without connecting roles) – only if this is not the case, the datalog program will infer a contradiction.

In earlier sections, we have already reduced terminological reasoning in $\mathcal{ALCIb}$ to iterative constructions of Boolean formulae, and one might be tempted to directly cast these constructions into datalog. However, the terminological reasoning must take into account *all* possible individuals occurring in the constructed greatest model. If we want to represent individuals by constants in datalog, this would require us to declare exponentially many individuals in datalog. This would give up on the possible optimisation of using OBDDs, and basically just mirror the naive domino set construction in datalog.

So we use the OBDD computed from the terminology as a kind of pre-compiled version of the relevant terminological information. Abox information is then considered as a kind of incomplete specification of dominoes that must be accepted by the OBDD, and the datalog program simulates the OBDD's evaluation for each of those.

Definition 12. *Consider an $\mathcal{ALCIb}$ knowledge base* KB $= \langle \mathcal{A}, \mathcal{T} \rangle$ *such that $\mathcal{A}$ contains only atomic concepts, and let* $\mathbb{O} = (N, n_{root}, n_{true}, n_{false}, \text{low}, \text{high}, \text{Var}, \lambda)$ *denote an OBDD obtained as a representation of* $[\![\text{FLAT}(\mathcal{T})]\!]$ *as in Definition 10. A disjunctive datalog program* DD(KB) *is defined as follows.* DD(KB) *uses the following predicate symbols:*

- *a unary predicate S_C for every concept expression $C \in P(\text{FLAT}(\mathcal{T}))$,*
- *a binary predicate S_R for every atomic role $R \in N_R$,*
- *a binary predicate A_n for every OBDD node $n \in N$.*

The constants in DD(KB) *are just the individual names used in $\mathcal{A}$. The disjunctive datalog rules of* DD(KB) *are defined as follows:*

(1) DD(KB) *contains rules* $\rightarrow A_{n_{root}}(x, y)$ *and* $A_{n_{false}}(x, y) \rightarrow$.
(2) If $C(a) \in \mathcal{A}$ then DD(KB) *contains* $\rightarrow S_C(a)$.
(3) If $R(a, b) \in \mathcal{A}$ then DD(KB) *contains* $\rightarrow S_R(a, b)$
(4) If $n \in N$ with $\lambda(n) = \langle C, 1 \rangle$ then DD(KB) *contains rules*
$\quad S_C(x) \wedge A_n(x, y) \rightarrow A_{high(n)}(x, y)$ *and* $A_n(x, y) \rightarrow A_{low(n)}(x, y) \vee S_C(x)$.
(5) If $n \in N$ with $\lambda(n) = \langle C, 2 \rangle$ then DD(KB) *contains rules*
$\quad S_C(y) \wedge A_n(x, y) \rightarrow A_{high(n)}(x, y)$ *and* $A_n(x, y) \rightarrow A_{low(n)}(x, y) \vee S_C(y)$.
(6) If $n \in N$ with $\lambda(n) = R$ for some $R \in N_R$ then DD(KB) *contains rules*
$\quad S_R(x, y) \wedge A_n(x, y) \rightarrow A_{high(n)}(x, y)$ *and* $A_n(x, y) \rightarrow A_{low(n)}(x, y) \vee S_R(x, y)$.
(7) If $n \in N$ with $\lambda(n) = R^-$ for some $R \in N_R$ then DD(KB) *contains rules*
$\quad S_R(y, x) \wedge A_n(x, y) \rightarrow A_{high(n)}(x, y)$ *and* $A_n(x, y) \rightarrow A_{low(n)}(x, y) \vee S_R(y, x)$.

Note that the number of variables per rule in DD(KB) is bounded by 2. The semantically equivalent grounding of DD(KB) thus is a propositional program of quadratic size, and the worst-case complexity for satisfiability checking is NP, as opposed to the NExpTime complexity of disjunctive datalog in general. Note that, of course, DD(KB) may still be exponential in the size of KB in the worst case. It remains to show the correctness of the datalog translation.

Lemma 13. *Given an $\mathcal{ALCIb}$ knowledge base* KB *such that $\mathcal{I}$ is a model of* KB, *there is a model $\mathcal{J}$ of* DD(KB) *such that $\mathcal{I} \models C(a)$ iff $\mathcal{J} \models S_C(a)$, and $\mathcal{I} \models R(a, b)$ iff $\mathcal{J} \models S_R(a, b)$, for any $a, b \in N_I$, $C \in N_C$, and $R \in N_R$.*

Proof. Let KB $= (\mathcal{A}, \mathcal{T})$. We define an interpretation $\mathcal{J}$ of DD(KB). The domain of $\mathcal{J}$ is the domain of $\mathcal{I}$, i.e. $\Delta^{\mathcal{I}} = \Delta^{\mathcal{J}}$. For individuals a, we set $a^{\mathcal{J}} := a^{\mathcal{I}}$. The interpretation of predicate symbols is now defined as follows (note that A_n is defined inductively):

- $\delta \in S_C^{\mathcal{J}}$ iff $\delta \in C^{\mathcal{I}}$,
- $\langle \delta_1, \delta_2 \rangle \in S_R^{\mathcal{J}}$ iff $\langle \delta_1, \delta_2 \rangle \in R^{\mathcal{I}}$,
- $\langle \delta_1, \delta_2 \rangle \in A_{n_{root}}^{\mathcal{J}}$ for all $\delta_1, \delta_2 \in \Delta^{\mathcal{J}}$,
- $\langle \delta_1, \delta_2 \rangle \in A_n$ for $n \neq n_{root}$ if there is a node n' such that $\langle \delta_1, \delta_2 \rangle \in A_{n'}$, and one of the following is the case:

- $\lambda(n') = \langle C, i \rangle$, for some $i \in \{1, 2\}$, and $n = \mathsf{low}(n')$ and $\delta_i \notin C^{\mathcal{I}}$
- $\lambda(n') = \langle C, i \rangle$, for some $i \in \{1, 2\}$, and $n = \mathsf{high}(n')$ and $\delta_i \in C^{\mathcal{I}}$
- $\lambda(n') = R$ and $n = \mathsf{low}(n')$ and $\langle \delta_1, \delta_2 \rangle \notin R^{\mathcal{I}}$
- $\lambda(n') = R$ and $n = \mathsf{high}(n')$ and $\langle \delta_1, \delta_2 \rangle \in R^{\mathcal{I}}$

Mark that, in the last two items, R is any role expression from Var, and hence is a role name or its inverse. Also note that due to the acyclicity of $\mathbb{O}$, the interpretation of the A-predicates is indeed well-defined. We now show that $\mathcal{J}$ is a model of DD(KB). To this end, first note that the extensions of predicates S_C and S_R in $\mathcal{J}$ were defined to coincide with the extensions of C and R in $\mathcal{I}$. Since $\mathcal{I}$ satisfies $\mathcal{A}$, all ground facts of DD(KB) are satisfied by $\mathcal{J}$. This settles cases (2) and (3) of Definition 12.

Similarly, we find that the rules of cases (4)–(7) are satisfied by $\mathcal{J}$. Consider the first rule of (4), $S_C(x) \wedge A_n(x, y) \rightarrow A_{\mathsf{high}(n)}(x, y)$, and assume that $\delta_1 \in S_C^{\mathcal{J}}$ and $\langle \delta_1, \delta_2 \rangle \in A_n^{\mathcal{J}}$. Thus $\delta_1 \in C^{\mathcal{I}}$, and, using the preconditions of (4), we conclude that $\langle \delta_1, \delta_2 \rangle \in A_{\mathsf{high}(n)}^{\mathcal{J}}$ follows from the definition of $\mathcal{J}$. The second rule of case (4) covers the analogous negative case, and all other cases can be treated similarly.

Finally, for case (1), we need to show that $A_{n_{\mathsf{false}}}^{\mathcal{J}} = \emptyset$. For that, we first explicate the correspondence between domain elements of $\mathcal{I}$ and sets of variables of $\mathbb{O}$: Given elements $\delta_1, \delta_2 \in \Delta^{\mathcal{I}}$ we define $V_{\delta_1, \delta_2} := \{\langle C, n \rangle \mid C \in P(\mathsf{FLAT}(\mathcal{T})), \delta_n \in C^{\mathcal{I}}\} \cup \{R \mid \langle \delta_1, \delta_2 \rangle \in R^{\mathcal{I}}\}$, the set of variables corresponding to the $\mathcal{I}$-domino between δ_1 and δ_2.

Now $A_{n_{\mathsf{false}}}^{\mathcal{J}} = \emptyset$ clearly is a consequence of the following claim: for all $\delta_1, \delta_2 \in \Delta^{\mathcal{I}}$ and all $n \in N$, we find that $\langle \delta_1, \delta_2 \rangle \in A_n$ implies $\varphi_n(V_{\delta_1, \delta_2}) = true$ (using the notation of Definition 9). The proof proceeds by induction. For the case $n = n_{\mathsf{root}}$, we find that $\varphi_{n_{\mathsf{root}}} = [\![\mathcal{T}]\!]$. Since V_{δ_1, δ_2} represents a domino of $\mathcal{I}$, the claim thus follows by combining Proposition 11 and Lemma 6.

For the induction step, let n be a node such that $\langle \delta_1, \delta_2 \rangle \in A_n$ follows from the inductive definition of $\mathcal{J}$ based on some predecessor node n' for which the claim has already been established. Note that n' may not be unique. The cases in the definition of $\mathcal{J}$ must be considered individually. Thus assume n', n, and δ_1 satisfy the first case, and that $\langle \delta_1, \delta_2 \rangle \in A_n$. By induction hypothesis, $\varphi_{n'}(V_{\delta_1, \delta_2}) = true$, and by Definition 9 the given case yields $\varphi_n(V_{\delta_1, \delta_2}) = true$ as well. The other cases are similar. □

Lemma 14. *Given an $\mathcal{ALCIb}$ knowledge base* KB *such that $\mathcal{J}$ is a model of* DD(KB), *there is a model $\mathcal{I}$ of* DD(KB) *such that $\mathcal{I} \models C(a)$ iff $\mathcal{J} \models S_C(a)$, and $\mathcal{I} \models R(a, b)$ iff $\mathcal{J} \models S_R(a, b)$, for any $a, b \in \mathsf{N}_I$, $C \in \mathsf{N}_C$, and $R \in \mathsf{N}_R$.*

Proof. Let KB $= (\mathcal{A}, \mathcal{T})$. We construct an interpretation $\mathcal{I}$ whose domain $\Delta^{\mathcal{I}}$ consists of all sequences starting with an individual name followed by a (possibly empty) sequence of dominoes from $\mathbb{D}_{\mathcal{T}}$ such that, for every $\delta \in \Delta^{\mathcal{I}}$,

- if δ begins with $a\langle \mathcal{A}, \mathcal{R}, \mathcal{B} \rangle$, then $\{C \mid C \in P(\mathsf{FLAT}(\mathcal{T})), a^{\mathcal{J}} \in S_C^{\mathcal{J}}\} = \mathcal{A}$, and
- if δ contains subsequent letters $\langle \mathcal{A}, \mathcal{R}, \mathcal{B} \rangle$ and $\langle \mathcal{A}', \mathcal{R}', \mathcal{B}' \rangle$, then $\mathcal{B} = \mathcal{A}'$.

For a sequence $\delta = a\langle \mathcal{A}_1, \mathcal{R}_1, \mathcal{A}_2 \rangle \langle \mathcal{A}_2, \mathcal{R}_2, \mathcal{A}_3 \rangle \ldots \langle \mathcal{A}_{i-1}, \mathcal{R}_{i-1}, \mathcal{A}_i \rangle$, we define $\mathsf{tail}(\delta) := \mathcal{A}_i$, whereas for a $\delta = a$ we define $\mathsf{tail}(\delta) := \{C \mid C \in P(\mathsf{FLAT}(\mathcal{T})), a^{\mathcal{J}} \in S_C^{\mathcal{J}}\}$. Now the mappings of $\mathcal{I}$ are defined as follows:

- for $a \in \mathsf{N}_I$, we have $a^{\mathcal{I}} := a$,
- for $A \in \mathsf{N}_C$, we have $\delta \in A^{\mathcal{I}}$ iff $A \in \mathsf{tail}(\delta)$,

- for $R \in N_R$, we have $\langle \delta_1, \delta_2 \rangle \in R^I$ if one of the following holds
  - $\delta_1 = a \in N_I$ and $\delta_2 = b \in N_I$ and $\langle a, b \rangle \in S_R^J$, or
  - $\delta_2 = \delta_1 \langle \mathcal{A}, \mathcal{R}, \mathcal{B} \rangle$ with $R \in \mathcal{R}$, or
  - $\delta_1 = \delta_2 \langle \mathcal{A}, \mathcal{R}, \mathcal{B} \rangle$ with $\mathrm{Inv}(R) \in \mathcal{R}$.

Thus, intuitively, I is constructed by extracting the named individuals as well their concept (and mutual role) memberships from J, and appending an appropriate domino-constructed tree model to each of those named individuals. We proceed by showing that I is indeed a model of KB.

We begin with the following auxiliary observation: For every two individual names $a, b \in N_I$, and $\mathcal{R}_{ab} := \{R \mid \langle a^J, b^J \rangle \in S_R^J\} \cup \{\mathrm{Inv}(R) \mid \langle b^J, a^J \rangle \in S_R^J\}$, the domino $\langle \mathrm{tail}(a), \mathcal{R}_{ab}, \mathrm{tail}(b) \rangle$ is contained in $\mathbb{D}_{\mathcal{T}}$ (Claim †). Using Proposition 11, it suffices to show that the Boolean function $[\![\mathcal{T}]\!]$ if applied to $V_{a,b} := \{\mathrm{tail}(a) \times \{1\} \cup \mathcal{R}_{ab} \cup \mathrm{tail}(b) \times \{2\}\}$ yields *true*. Since $[\![\mathcal{T}]\!] = \varphi_{n_{\mathrm{root}}}$, this is obtained by showing the following: For any $a, b \in N_I$, we find that $\langle a^J, b^J \rangle \in A_n^J$ implies $\varphi_n(V_{a,b}) = true$. Indeed, the intended claim follows since we have $\langle a^J, b^J \rangle \in A_{n_{\mathrm{root}}}^J$ due to the first rule of (1) in Definition 12. We proceed by induction, starting at the leaves of the OBDD. The case $\langle a, b \rangle \in A_{n_{\mathrm{true}}}^I$ is immediate, and $\langle a, b \rangle \in A_{n_{\mathrm{false}}}^I$ is excluded by the second rule of (1). For the induction step, consider nodes $n, n' \in N$ such that either $\lambda(n) \in V_{a,b}$ and $n' = \mathrm{high}(n)$, or $\lambda(n) \notin V_{a,b}$ and $n' = \mathrm{low}(n)$. We assume that $\langle a^J, b^J \rangle \in A_n^J$, and, by induction, that the claim holds for n'. If $\lambda_n = \langle C, 1 \rangle$, then one of the rules of case (4) applies to a^J and b^J. In both cases, we can infer $\langle a^J, b^J \rangle \in A_{n'}^J$, and hence $\varphi_{n'}(V_{a,b}) = true$. Together with the assumptions for this case, Definition 9 implies that $\varphi_n(V_{a,b}) = true$ as required. The other cases are analogous.

It is easy to see that I satisfies all Abox axioms from KB by definition, due to the ground facts in $\mathrm{DD}(\mathrm{KB})$ (case (2) and (3) in Definition 12). To show that the Tbox is also satisfied, we need to show that all individuals of I are contained in the extension of each concept expression of $\mathrm{FLAT}(\mathcal{T})$. To this end, we first show that $\delta \in C^I$ iff $C \in \mathrm{tail}(\delta)$ for all $C \in P(\mathrm{FLAT}(\mathcal{T}))$. If $C \in N_C$ is atomic, this follows directly from the definition of I. The remaining cases that may occur in $P(\mathrm{FLAT}(\mathcal{T}))$ are $C = \exists U.A$ and $C = \forall U.A$.

First consider the case $C = \exists U.A$, and assume that $\delta \in C^I$. Thus there is $\delta' \in \Delta^I$ with $\langle \delta, \delta' \rangle \in U^I$ and $\delta' \in A^I$. The construction of the domino model admits three possible cases:

- $\delta, \delta' \in N_I$ and $\mathcal{R}_{\delta\delta'} \vdash U$ and $A \in \mathrm{tail}(\delta')$. Now by †, the domino $\langle \mathrm{tail}(\delta), \mathcal{R}_{\delta\delta'}, \mathrm{tail}(\delta') \rangle$ satisfies condition **ex** of Definition 5, and thus $C \in \mathrm{tail}(\delta)$ as required.
- $\delta' = \delta \langle \mathrm{tail}(\delta), \mathcal{R}, \mathrm{tail}(\delta') \rangle$ with $\mathcal{R} \vdash U$ and $A \in \mathrm{tail}(\delta')$. Since $\mathbb{D}_{\mathcal{T}} \subseteq \mathbb{D}_0$, we find that $\langle \mathrm{tail}(\delta), \mathcal{R}, \mathrm{tail}(\delta') \rangle$ satisfies condition **ex**, and thus $C \in \mathrm{tail}(\delta)$ as required.
- $\delta = \delta' \langle \mathrm{tail}(\delta'), \mathcal{R}, \mathrm{tail}(\delta) \rangle$ with $\mathrm{Inv}(\mathcal{R}) \vdash U$ and $A \in \mathrm{tail}(\delta')$. By condition **sym**, $\mathbb{D}_{\mathcal{T}}$ contains the domino $\langle \mathrm{tail}(\delta), \mathrm{Inv}(\mathcal{R}), \mathrm{tail}(\delta') \rangle$, and we can again invoke **ex** to conclude $C \in \mathrm{tail}(\delta)$.

For the converse, assume that $\exists U.A \in \mathrm{tail}(\delta)$. So $\mathbb{D}_{\mathcal{T}}$ contains a domino $\langle \mathcal{A}, \mathcal{R}, \mathrm{tail}(\delta) \rangle$. This is obvious if the sequence δ ends with a domino. If $\delta = a \in N_I$, then it follows by applying † to a with the first individual being arbitrary. By **sym** $\mathbb{D}_{\mathcal{T}}$ also contains the domino $\langle \mathrm{tail}(\delta), \mathcal{R}, \mathcal{A} \rangle$. By condition **delex**, the latter implies that $\mathbb{D}_{\mathcal{T}}$ contains a domino $\langle \text{tail}(\delta), \mathcal{R}', \mathcal{A}' \rangle$ such that $\mathcal{R}' \vdash U$ and $A \in \mathcal{A}'$. Thus $\delta' = \delta\langle \text{tail}(\delta), \mathcal{R}', \mathcal{A}' \rangle$ is an $\mathcal{I}$-individual such that $\langle \delta, \delta' \rangle \in U^{\mathcal{I}}$ and $\delta' \in A^{\mathcal{I}}$, and we obtain $\delta \in (\exists U.A)^{\mathcal{I}}$ as claimed.

For the second case, consider $C = \forall U.A$ and assume that $\delta \in C^{\mathcal{I}}$. As above, we find that $\mathbb{D}_{\mathcal{T}}$ contains some domino $\langle \mathcal{A}, \mathcal{R}, \text{tail}(\delta) \rangle$, where † is needed if $\delta \in \mathsf{N}_I$. By **sym** we find a domino $\langle \text{tail}(\delta), \mathcal{R}, \mathcal{A} \rangle$. For a contradiction, suppose that $\forall U.A \notin \text{tail}(\delta)$. By condition **deluni**, the latter implies that $\mathbb{D}_{\mathcal{T}}$ contains a domino $\langle \text{tail}(\delta), \mathcal{R}', \mathcal{A}' \rangle$ such that $\mathcal{R}' \vdash U$ and $A \notin \mathcal{A}'$. Thus $\delta' = \delta\langle \text{tail}(\delta), \mathcal{R}', \mathcal{A}' \rangle$ is an $\mathcal{I}$-individual such that $\langle \delta, \delta' \rangle \in U^{\mathcal{I}}$ and $\delta' \notin A^{\mathcal{I}}$. But then $\delta \notin (\forall U.A)^{\mathcal{I}}$, which is the required contradiction.

For the other direction, assume that $\forall U.A \in \text{tail}(\delta)$. According to the construction of $\mathcal{I}$, for all elements δ' with $\langle \delta, \delta' \rangle \in U^{\mathcal{I}}$, there are three possible cases:

- $\delta, \delta' \in \mathsf{N}_I$ and $\mathcal{R}_{\delta\delta'} \vdash U$. Now by †, the domino $\langle \text{tail}(\delta), \mathcal{R}_{\delta\delta'}, \text{tail}(\delta') \rangle$ satisfies condition **uni**, whence $A \in \text{tail}(\delta')$.
- $\delta' = \delta\langle \text{tail}(\delta), \mathcal{R}, \text{tail}(\delta') \rangle$ with $\mathcal{R} \vdash U$. Since $\mathbb{D}_{\mathcal{T}} \subseteq \mathbb{D}_0$, $\langle \text{tail}(\delta), \mathcal{R}, \text{tail}(\delta') \rangle$ must satisfy condition **uni**, and thus $A \in \text{tail}(\delta')$.
- $\delta = \delta'\langle \text{tail}(\delta'), \mathcal{R}, \text{tail}(\delta) \rangle$ with $\text{Inv}(\mathcal{R}) \vdash U$. By condition **sym**, $\mathbb{D}_{\mathcal{T}}$ also contains the domino $\langle \text{tail}(\delta), \text{Inv}(\mathcal{R}), \text{tail}(\delta') \rangle$, and we can again use **uni** to conclude $A \in \text{tail}(\delta')$.

Thus, $A \in \text{tail}(\delta')$ for all U-successors δ' of δ, and hence $\delta \in (\forall U.A)^{\mathcal{I}}$ as claimed.

To finish the proof, note that any domino $\langle \mathcal{A}, \mathcal{R}, \mathcal{B} \rangle \in \mathbb{D}_{\mathcal{T}}$ satisfies condition **kb**. Using **sym**, we have that for any $\delta \in \Delta^{\mathcal{I}}$, the axiom $\bigsqcap_{D \in \text{tail}(\delta)} D \sqsubseteq C$ is a tautology for all $C \in \mathsf{FLAT}(\mathcal{T})$. As shown above, $\delta \in D^{\mathcal{I}}$ for all $D \in \text{tail}(\delta)$, and thus $\delta \in C^{\mathcal{I}}$. Hence every individual of $\mathcal{I}$ is an instance of each concept of $\mathsf{FLAT}(\mathcal{T})$ as required.   □

Lemma 13 and 14 show that DD(KB) faithfully captures both positive and negative ground conclusions of KB, and in particular that DD(KB) and KB are equisatisfiable. As discussed in Section 2, $\mathcal{SHIQ}$ knowledge bases can be transformed into equisatisfiable $\mathcal{ALCIb}$ knowledge bases, and hence the above algorithm can also be used to decide satisfiability in the case of $\mathcal{SHIQ}$. The transformations used to convert $\mathcal{SHIQ}$ to $\mathcal{ALCIb}$, however, do not preserve all ground consequences. In particular, $\mathcal{SHIQ}$ consequences of the form $R(a, b)$ with R being non-simple may not be entailed by DD(KB). Such positive non-simple role atoms are the only case where entailments are lost, and thus DD(KB) behaves similar to the disjunctive datalog program created by the KAON2 approach [7].

The above observation immediately allow us to add reasoning support for *DL-safe rules* [8], simply by adding the respective rules to DD(KB) after replacing C and R by S_C and S_R. A special case of this are *DL-safe* conjunctive queries, i.e. conjunctive queries that assume all variables to range only over named individuals. It is easy to see that, as a minor extension, one could generally allow for concept expressions $\forall R.A$ and $\exists R.A$ in queries and rules, simply because DD(KB) represents these elements of $P(\mathsf{FLAT}(\mathcal{T}))$ as atomic symbols in disjunctive datalog.

6   Discussion

We have presented a new reasoning algorithm for $\mathcal{SHIQ}$ knowledge bases that compiles $\mathcal{SHIQ}$ terminologies into disjunctive datalog programs, which are then combined with assertional information for satisfiability checking and (ground) query answering. The approach is based on our earlier work on terminological $\mathcal{SHIQ}$ reasoning with ordered binary decision diagrams (OBDDs), which fails when introducing Aboxes as it hinges upon a form of greatest model property [2]. OBDDs now are still used to process terminologies, but are subsequently transformed into disjunctive datalog programs that can incorporate Abox data. The generation of disjunctive datalog may require exponentially many computation steps, the complexity of which depends on the concrete OBDD implementation at hand – finding *optimal encodings* is NP-complete but heuristic approximations are often used in practice. Querying the disjunctive datalog program then is co-NP-complete w.r.t. the size of the Abox, so that the data complexity of the algorithm is worst-case optimal [7].

The presented method exhibits similarities to the algorithm underlying the KAON2 reasoner [7]. In particular, pre-transformations are first applied to $\mathcal{SHIQ}$ knowledge bases, so that the resulting datalog program is not complete for querying instances of non-simple roles. Besides this restriction, extensions with DL-safe rules and ground conjunctive queries are straightforward. The presented processing, however, is very different from KAON2. Besides using OBDDs, it also employs Boolean role constructors that admit an efficient binary encoding of number restrictions [2].

For future work, the algorithm needs to be evaluated in practice. A prototype implementation was used to generate the examples within this paper, but this software is not fully functional yet. It is also evident that redundancy elimination techniques are required to reduce the number of generated datalog rules, which is also an important aspect of the KAON2 implementation. Another strand for future development is the extension of the approach to take nominals into account – significant revisions of the model-theoretic considerations are needed for that case.

References

1. Pan, G., Sattler, U., Vardi, M.Y.: BDD-based decision procedures for the modal logic K. Journal of Applied Non-Classical Logics 16(1-2), 169–208 (2006)
2. Rudolph, S., Krötzsch, M., Hitzler, P.: Terminological reasoning in SHIQ with ordered binary decision diagrams. In: Proc. 23rd National Conference on Artificial Intelligence (AAAI 2008), pp. 529–534. AAAI Press, Menlo Park (2008)
3. Baader, F., Calvanese, D., McGuinness, D., Nardi, D., Patel-Schneider, P. (eds.): The Description Logic Handbook: Theory, Implementation and Applications. Cambridge University Press, Cambridge (2007)
4. Tobies, S.: Complexity Results and Practical Algorithms for Logics in Knowledge Representation. PhD thesis, RWTH Aachen, Germany (2001)
5. Burch, J., Clarke, E., McMillan, K., Dill, D., Hwang, L.: Symbolic model checking: 10^{20} states and beyond. In: Proc. 5th Annual IEEE Symposium on Logic in Computer Science, Washington, D.C., pp. 1–33. IEEE Computer Society Press, Los Alamitos (1990)
6. Huth, M.R.A., Ryan, M.D.: Logic in Computer Science: Modelling and reasoning about systems. Cambridge University Press, Cambridge (2000)
7. Motik, B.: Reasoning in Description Logics using Resolution and Deductive Databases. PhD thesis, Universität Karlsruhe (TH), Germany (2006)
8. Motik, B., Sattler, U., Studer, R.: Query answering for OWL-DL with rules. Journal of Web Semantics 3(1), 41–60 (2005)

RDF123: From Spreadsheets to RDF[*]

Lushan Han[1], Tim Finin[1], Cynthia Parr[2], Joel Sachs[1], and Anupam Joshi[1]

[1] University of Maryland, Baltimore County, USA
{lushan1,finin,jsachs,joshi}@cs.umbc.edu
[2] University of Maryland, College Park, USA
csparr@umd.edu

Abstract. We describe RDF123, a highly flexible open-source tool for translating spreadsheet data to RDF. Existing spreadsheet-to-rdf tools typically map only to star-shaped RDF graphs, i.e. each spreadsheet row is an instance, with each column representing a property. RDF123, on the other hand, allows users to define mappings to arbitrary graphs, thus allowing much richer spreadsheet semantics to be expressed. Further, each row in the spreadsheet can be mapped with a fairly different RDF scheme. Two interfaces are available. The first is a graphical application that allows users to create their mapping in an intuitive manner. The second is a Web service that takes as input a URL to a Google spreadsheet or CSV file and an RDF123 map, and provides RDF as output.

Keywords: RDF, Spreadsheets, Web services, Data interoperability.

1 Introduction

A significant amount of the world's data is maintained in spreadsheets. In this paper we present RDF123, a highly flexible open-source tool for translating spreadsheet data to RDF. Our work is motivated by the fact that spreadsheets are easy to understand and use, offer intuitive interfaces, and have representational power adequate for many common purposes. Moreover, online spreadsheets are increasingly popular and have the potential to boost the growth of the Semantic Web by providing well-formed and publicly shared data sources that can be directly maintained by users and automatically translated into RDF.

A drawback of spreadsheets is that their simplicity often results in data tables that do not follow the best practices of database design, such as attention to keys and normalization, let alone the richer features enabled by knowledge bases. Moreover, the liberty that people take with spreadsheets will sometimes require different rows to be translated with different schemata. RDF123 addresses both of these issues. RDF123's translation from a spreadsheet to an RDF graph is driven by a map which permits a rich schema to apply to a row, rather than just creating a single instance of a RDF/OWL class. We also adopt a general

---

[*] Apologies to Jonathan Sachs and Mitch Kapor. Partial support for this research was provided by the National Science Foundation through awards ITR-IIS-0325172 and NSF-ITR-IDM-0219649.

A. Sheth et al. (Eds.): ISWC 2008, LNCS 5318, pp. 451–466, 2008.

approach that allows different rows to use fairly different schemata. For example, depending on the value in the spreadsheet's column labeled 'sex', we generate a 'Man' instance or a 'Woman' instance.

In our approach, we borrow the idea from GRDDL [1] of placing a link in an online spreadsheet referencing the RDF123 translation map, which is itself an RDF document, specifying the desired translation. When an agent comes to the spreadsheet, it follows the link, reads the map file, applies it to the spreadsheet and thus generates RDF data. Moreover, RDF123's Web service also allows users to apply map files to other users' online spreadsheets and generate their customized RDF data.

The remainder of the paper proceeds as follows. Section 2 briefly contrasts our approach to other systems that map spreadsheets or database tables into RDF graphs and also to GRDDL. Section 3 describes in detail the workings of the RDF123 translation. Section 4 describes the approach to representing a *map graph* as an RDF document. Section 5 explains how to specify metadata in RDF123. Section 6 provides an architectural overview of the system. We conclude the paper with some brief remarks and identify issues for future research.

2   Related Work

Several programs have been developed to convert or export data from spreadsheets to RDF. The Maryland Mindswap Lab developed two early systems: Excel2RDF [2] and the more flexible ConvertToRDF [3,4]. Both these application assumed that an instance of a given class should be created for each row in the spreadsheet. The row's cells are used to populate the instance with property values and, typically, one provides an RDF node id for the instance itself.

The Babel system that is part of the MIT Simile suite of tools [5] can extract data from excel spreadsheets and from tab-delimited tabular data and render it in JSON and eventually RDF.

The TopBraid Composer [6] Semantic Web development system can extract class and instances information from spreadsheets and these can be further manipulated and transformed using additional tools in the suite.

One limitation of the approaches described above is that the RDF schema used for one row or a group of rows is quite simple, usually having the shape of a star in which all property edges come out from a single center – the ID resource. This works well for normalized database tables, but is not flexible enough for general purpose spreadsheets. Another limitation in the above approaches is that one fixed RDF schema is applied to all rows of a table.

A problem that is very similar is generating RDF data from a relational database. A more sophisticated translation system, such as like D2R [7], can specify mappings from rows in the result set of a SQL query to instances of a RDF/OWL class. The approach uses D2R MAP [8], which is a declarative language to define mappings between relational database schemata and OWL/RDFS ontologies. Using the SQL query language to define mappings yields a system with considerable representational power and flexibility, but requires that its users have considerable familiarity with relational databases and SQL. Moreover, its applicability is limited to databases, which are, for the most part, developed and used by IT professionals rather than IT users.

An alternative approach is to create an XML representation of the spread-sheet, and then to use GRDDL, which has high flexibility, for the translation. However, the need to generate intermediate XML documents is a barrier. Consider an example. Suppose a restaurant maintains a published online spreadsheet showing the up-to-date menu items the restaurant is providing. The manager of the restaurant may want the menu items to not only be read by humans but also be read by software agents and therefore available for semantic web queries. He also wants the machine-readable menu item data to be the most recent available. Since the data to be translated by GRDDL must be in XML (XHTML) format, the online spreadsheet translation has to include an additional step, that is, transforming the data in the spreadsheet to data in XML. This means that every time the restaurant manager modifies the menu items, he must take extra steps to push the spreadsheet data to XML and publish it; otherwise, an agent who reads the online XML will get stale data. Another significant drawback of GRDDL translation is that the XSLT transform, which GRDDL relies on, is hard to create for users who are not XSLT specialists. The mapping from tabular data to RDF should and can be done more intuitively than XSLT transformation.

3   Translation Design

3.1   Mapping Design

In order to define a more general mapping, we treat an RDF graph as a directed labeled graph, disregarding for the moment RDF schema concepts like classes and instances. Each vertex is either a resource or a literal and each edge is an RDF triple. A resource with exactly the same label is treated as the same resource. A triple is also unique in a RDF graph. Every row of a spreadsheet will generate a row graph, and the RDF graph produced for the whole spreadsheet is the result of merging all of the row graphs, eliminating duplicated resources and triples as necessary. We would like to define simple mappings that allow the row graphs to take any shape, and also to vary significantly from one another.

We formally define the mapping from a spreadsheet to a RDF graph as the following, where G_i is the row graph for the ith row and G_{final} is the ultimate graph.

$$G_{final} = \bigcup_{i=1}^{row\ count} G_i \ . \tag{1}$$

$$G_i = map\,(rowCells[\ ], i) \ . \tag{2}$$

The *map* function produces a row graph for the ith row given an array of its cell values and the row number i. The computation of the function *map* only relies on the inputs of current row and not on the previous computation or future computation of the *map* on other rows. The row number i is a required input,

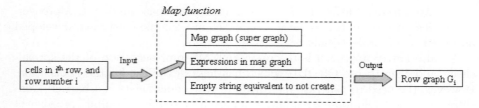

Fig. 1. Three elements in our implementation of the *map* function

used to generate unique IDs or labels spanning the whole RDF graph. Two row graphs may differ in their number of vertices and/or edges, but they will typically have a similar pattern. For example, some edges in different row graphs will have the same label, or the labels of vertices in different row graphs come from the same column. If we overlap these row graphs by unifying vertices and edges, and then we look from the top, we end up with a graph that is a super graph of every row graph, with similar vertices/edges in different graphs converging on a single vertex/edge. (There can be other ways to merge the similar vertices/edges so that the super graph may not be unique.) This super graph is the basis of our mapping design, which we call the *map graph* or template.

When the *map graph* should produce different labels for a converged vertex or edge in different row graphs, an expression is used for the vertex or edge rather than a static label. The expression is evaluated for each row and the result used as the ultimate label of vertices/edges in each row graph. The inputs of an expression are the same as the inputs of the *map* function: the array rowCells [ ] and the row number i. Expressions can use if-then-else sub-expressions and string manipulation operators to compute a string as the final label for a vertex or edge. Since the *map graph* is a super graph of every row graph, for those vertices and edges which are in the *map graph* but absent from a row graph, the expressions will output empty strings, which signal that no vertex or edge should be created. Note that if a vertex is not created, no incident edges are created as well.

The three elements, the *map graph*, *map graph* expressions, and the convention that empty strings generate no vertices or edges, characterize the *map* function and render it able to generate all row graphs, as shown in Figure 1.

The *map* function has high expressiveness, as we don't impose any constraints on every row graph; it can be arbitrary RDF graph. Because spreadsheets used by end users may not be normalized tables, arbitrary row graphs can maintain the expressiveness of spreadsheets. On the other hand, a RDF123 mapping is more intuitive than an XSLT transformation because it is expressed as a graph and can be visualized and authored with RDF123 graphical application. Typically this *map graph* resembles a diagram of entities and their relationships that captures what users have interpreted from a spreadsheet. As you would expect, the *map graph* can be serialized in an RDF document with RDF/XML, N3 and other common RDF serializations.

$$Ex: @If(\$2! = \$3; foaf+'knows';'')$$

Fig. 2. The terms \$2 and \$3 in this RDF123 expression denote the cell values in columns 2 and 3. The expression computes a dynamic edge: if the two cell values are not equal we generate 'foaf:knows' for the converged edge; otherwise nothing is generated. \$[i] denotes cell values in column i for current row when $i > 0$ while \$[0] gives the current row number.

3.2   RDF123 Expression Design

The role of RDF123 expression is to compute the final label for a converged vertex/edge in a *map graph*, depending on the input of cell values of a row and the row number. The expressiveness and simplicity of RDF123 expression are both important because they determine the complexity of the *map graph*. RDF123 expression is defined by a context-free grammar and is able to do branch, arithmetic and string processing operations. All these operations, including branch, are themselves expressions that can be recursively embedded in other expressions. Expressions strings used as input to a parent expressions, with the value of the outermost expression serving as a final label for a vertex or edge, provided it is not the null string. While string concatenation and equality use an infix notation, other operations employ a functional notation. For example, branch expression is defined as $@If(arg1; arg2; arg3)$ and addition as $@Add(arg1, arg2)$.

Since RDF123 is implemented in Java, string manipulating methods in the java.lang.String class are easily exposed as RDF123 expressions and common $@Length$, $@IndexOf$ and $@Substr$ methods are available. To maintain a conceptually simple model, there are no other data types such as number or boolean in RDF123 expressions. However, strings are coerced to the appropriate data type when the semantics of the operation and the operand require other types. Exceptions may happen during conversion, which leads to the two running modes of RDF123 program. One mode is to produce as many triples as possible. In this mode, any exception will result in an empty output string which means not creating the vertex or edge but the whole program will continue running for other vertexes and edges. The other mode simply terminates the whole program and returns an error message. In order to have a neat display, RDF123 expression also allows using prefix instead of writing whole namespace.

Not every converged vertex/edge has a label that must be computed or transformed; some are simple static labels. To distinguish dynamic and static labels for converged vertexes/edges, we introduce a pseudo namespace 'Ex'. If a label begins with 'Ex:', it simply means that the following string is a RDF123 expression; otherwise a static label. The use of this pseudo namespace makes the *map graph* have the form of an RDF graph and enables it to be serialized in many forms, such as RDF/XML and N3. (See Figure 2 for an example.) The normal RDF semantics does not apply to RDF123 map graphs, of course. Luckily, they are easily recognizable via their metadata annotations and should ultimately pose no more problems that document templates and samples do for human readers.

3.3   Determining the Type of a Converged Vertex

The role of an RDF123 expression is to produce a final label for a converged vertex or edge. It is more like a process of data extraction and transformation. However, we also need know the RDF element type for the converged vertex or edge before we can output the data as RDF. For edges, it is very simple because they are always *rdf:Property*. But for vertices, it is a little bit tricky because the potential type could be one of several data types (e.g., *rdf:Resource*, *rdf:Literal*, XML data types) or even composite data types like RDF container, collection or object group. We can divide the possibilities into two general cases. For those vertices which have outgoing edges, we can conclude that they should be of type *rdf:Resource*. When it comes to those leaf vertices, we allow users to explicitly append a vertex type at the end of the static label or RDF123 expression. For example, *Ex:$1^^integer*. When lacking an explicit data type, we take the following heuristic: if the final label is a valid URI, we make it a *rdf:Resource* otherwise a *rdf:Literal*.

When the specified vertex type is a composite data type, like *rdf:Bag*, we require that the vertex label must follow a certain syntax, such as Prolog list, so that a parser can understand it and put it to the corresponding RDF data. The atomic elements in the list can also have vertex type appended so that it is possible to generate a bag of *rdf:Resource* or *rdf:Literal* dependently. RDF123 expressions can provide some basic functions to help users format their data to syntactically correct list. Composite vertex types are not supported in current version of RDF123.

An object group is also a composite data type, but is different from RDF container or collection with respect to how the data is transformed to RDF. For each element in the object group, we create a separate assertion instead of one assertion to the whole set. For example, consider a spreadsheet for school classes. A class can have one instructor and multiple students, which are stored in only two columns 'instructor' and 'students'. We would like to generate a foaf:knows assertion from the instructor to every students respectively. In this case, we can use an object group vertex type for 'students'.

RDF123 also supports blank nodes. To create a blank node, just leave the label of a vertex completely empty. Be careful that 'Ex:' has a completely different semantic because it is interpreted as not creating the vertex. Actually all blank nodes have internal IDs in a physical RDF storage model. In RDF123, the row number i is used to generate a unique internal ID for a blank node.

3.4   A Simple Translation Example

People like spreadsheets because they provide a convenient way to capture the similarity of data, group and store similar data together in a succinct, informal

Table 1. A simple spreadsheet for the members of a research club

Name	Email	Office	Faculty	Coffee Due	Advisor
Tim Finin	finin@umbc.edu	ITE329	Yes	$10	
Lushan Han	lushan@umbc.edu	ITE377	No		Tim Finin
Wenjia Li	wenjia@umbc.edu	ITE377	No		Anupam Joshi

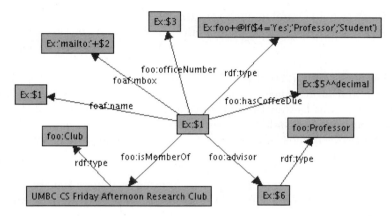

Fig. 3. The corresponding *map graph* made with RDF123 graphical application

schema. This schema may be easily criticized by database specialists because it is hard to store and query. However, it has the advantage of being intuitive. RDF123 *map graph* is a template that copies the intuitive schema from a spreadsheet and allows subtleties and dissimilarities within similarity to be expressed with RDF123 expressions. Generally speaking, a vertex in a *map graph* can often find its corresponding column in a spreadsheet and an edge simply comes from an interpreted semantic relation between two columns. RDF123 expressions and vertex type play a role of refining data and transform data to RDF, a machine-understandable schema, from an intuitive but informal schema. Let's see one example.

Suppose that the UMBC CS department has a research club that includes faculty and students. Faculty are required to pay a small amount of money for monthly coffee dues, but students are not. Table. 3.4 shows a spreadsheet and Figure 3 the corresponding *map graph*.

The *map graph* follows the intuitive schema of the spreadsheet but expresses some subtleties using RDF123 expression. The expression *Ex:foo+@If($4='Yes'; 'Professor'; 'Student')* will produce a resource 'foo:Professor' for the rows having 'Yes' in FACULTY column and 'foo:Student' for the others (where 'foo' is a hypothetical namespace). The expression *Ex:$5^^decimal* specifies a vertex type 'decimal' because the coffee due has monetary value. For the rows representing students, *Ex:$5^^decimal* will output an empty string and therefore the corresponding vertex along with the incident edge 'foo:hasCoffeeDue' will not be created for students. Besides information in columns, we add a general assertion that all generated instances are members of a club instance named 'UMBC CS Friday Afternoon Research Club'. Because no namespace is specified for the local name 'UMBC CS Friday Afternoon Research Club', the club instance will, by default, have the online document base URI as its namespace. Although the club instance is created for each row graph, they share the same resource URI. Therefore, after doing a union of all row graphs, only one club instance remains.

This spreadsheet example implicitly uses the 'unique name assumption' because a person name's is used as node id of instances, such as *Ex:$1* and *Ex:$6*. Typically, we would like to have a unique resource URI for referencing the same instance appearing in different places of a spreadsheet. Doing so allows different assertions about an instance to come together. For example, the second row of the spreadsheet tells us that Lushan Han's adviser is Tim Finin. And the person Tim Finin has an email address 'finin@umbc.edu', which is actually obtained from the first row. In most spreadsheets, the unique name assumption is implied because their authors would certainly hope to see that potential readers can disambiguate person names. If two people share the same name, the author might introduce the middle name or use another notation to differentiate them. When such a unique name assumption is not appropriate, we can use a map graph that generates blank nodes or use unique numeric IDs.

4   Serializing a Map Graph as RDF

Since a *map graph* is a template for producing row RDF graphs, it shares many characteristics with an ordinary RDF graph. It is beneficial to serialize *map graph* with standard RDF serializations like RDF/XML or N3 because it enables people who are familiar with RDF to edit the *map graph* manually or with some existing popular RDF tools. After we serialize the *map graph* to a file, we can publish the file online to encourage reuse.

There are two subtleties about serializing a *map graph* with RDF123 expressions. First, we have to forge a namespace for the expressions, as every resource is required to have a namespace. In RDF123, the forged namespace is 'Ex:' which is not a prefix but a full namespace. The second involves the W3C namespaces recommendation. Because it is quite likely that the ending character of an RDF expression is not in the required *NCNameStartChar* class (i.e., a letter or underscore), this will result in an empty local name when splitting the URI consisting of a RDF expression. A property with empty local name is not permitted in RDF/XML serialization, but we can work around this by appending a character '_', which is, of course, in the *NCNameStartChar* class, to the end of a RDF expression. The optional ending character '_' has no effect on the interpretation of a RDF expression. It is not necessary for a *map graph* to exactly follow

RDF syntax because it is just a template rather than a true RDF document. A serialized file of the *map graph* in Figure 3 is shown below.

```
<rdf:RDF
    xmlns:foaf="http://xmlns.com/foaf/0.1/"
    xmlns:foo="http://www.foo.org/"
    xmlns:rdf="http://www.w3.org/1999/02/22-rdf-syntax-ns#">
  <foo:Club rdf:about="#UMBC CS Friday Afternoon Research Club"/>
  <rdf:Description rdf:about="Ex:$1">
    <rdf:type rdf:resource="Ex:foo+@If($4='Yes';'Professor';'Student')"/>
    <foaf:mbox>Ex:'mailto:'+$2</foaf:mbox>
    <foaf:name>Ex:$1</foaf:name>
    <foo:officeNumber>Ex:$3</foo:officeNumber>
    <foo:hasCoffeeDue>Ex:$5^^decimal</foo:hasCoffeeDue>
    <foo:advisor>
      <foo:Professor rdf:about="Ex:$6"/>
    </foo:advisor>
    <foo:isMemberOf rdf:resource="#UMBC CS Friday Afternoon Research Club"/>
  </rdf:Description>
</rdf:RDF>
```

5   Incorporating Metadata

RDF123 allows people to specify metadata both in map file and spreadsheets. The metadata serves two functions. One is to provide parameters to the translation procedure, such as specifying the spreadsheet region containing the table to be translated, whether the table has a header, and the map file's URI. The other is to add RDF descriptions to the produced RDF graph, such as title, author, and comment. Besides functioning as annotations, the descriptions also provide an identifier via a map file or spreadsheet template to facilitate discovering and collecting a certain type of RDF documents on the Web using a search engine like Swoogle [9] or Sindice [10]. It is possible that metadata specified in the map file can conflict with the one specified in the spreadsheet. When this occurs, if the map file exists as an embedded link in an online spreadsheet, the metadata of the spreadsheet will override the one in the map file because the transformation is controlled by the spreadsheet owner. If a map file is applied to other people's online spreadsheets, the metadata of the map file will override the one in the spreadsheets because the transformation is invoked by the map file owner.

5.1   Metadata in a Spreadsheet

RDF123 allow users to specify metadata in a spreadsheet. In this case, users should be owners or co-authors of the spreadsheet. Unlike the case where metadata is specified in a map file, an embedded URL to the online map file is required.

Spreadsheet metadata is embedded into a contiguous and isolated tabular area with two columns and a header 'rdf123:metadata'. When the RDF123

File Edit Table

	A	B	C	D
1		employee	email	phone
2		mary smith	msmith@foo.com	410-455-1000
3		Bob Jones	bjones@foo.com	410-455-1001
4		Bill Gates	dtrump@foo.com	410-455-1002
5				
6				
7		rdf123:metad...		
8		title	employee table	
9		creator	Lushan Han	
10		start row		1
11		end row		4
12		start column		2
13		row head	yes	
14		comment	this is a test example	
15		type	abc:EmployeeSheet	
16		abc	http://abc#	
17		map file	http://userpages.umbc.edu/~lushan1/employeeMap2.rdf	
18				
19				

Fig. 4. RDF123 uses a simple convention for embedding metadata for the translation using RDF123. This metadata can define properties of the RDF document produced (e.g., title), the range of the spreadsheet to be transformed, and the location of the RDF123 map.

application or service processes a spreadsheet, it first scans all cells for a recognizable metadata block. If one is found, the RDF123 metadata is extracted, used in the translation process and stored in the resulting RDF graph. If no block is found, the entire spreadsheet is considered as a regular table with the first row being the header row.

In the RDF123 metadata area, people are allowed to tag the spreadsheet in a manner reminiscent to machine tags [11]. The value of a tag can be a literal string or a RDF resource. Some common tags are recognized without defining namespaces using a predefined mapping to 'machine tags'. For example, the 'comment' tag is interpreted as 'rdfs:comment'. For additional convenience, RDF123 also predefines the prefixes of popular namespaces, such as 'rdf', 'rdfs', 'owl', 'dc', 'foaf', 'sioc', 'vcard', and 'swrc'. Figure 4 shows an example of embedded metadata.

5.2   Metadata in the *Map Graph*

The following is an example for specifying metadata in a map file. The RDF123 expression '*Ex:?*' stands for the base URI of the online RDF document to be translated to. The properties defined in the namespace 'rdf123', such as 'rdf123:startRow' and 'rdf123:endRow' are used to specify the translation metadata. But you can also create annotation metadata or identification metadata by making RDF descriptions about '*Ex:?*'.

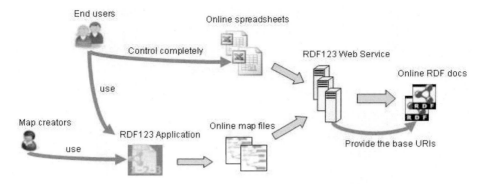

Fig. 5. RDF123 provides an application that allows users to create and edit maps and to generate RDF documents from spreadsheets as well as a Web service that generates RDF documents on demand from online spreadsheets

```
<rdf:RDF
    xmlns:emp="http://emp.example.org/"
    xmlns:foaf="http://xmlns.com/foaf/0.1/"
    xmlns:rdf="http://www.w3.org/1999/02/22-rdf-syntax-ns#"
    xmlns:rdfs="http://www.w3.org/2000/01/rdf-schema#"
    xmlns:rdf123="http://rdf123.umbc.edu/ns/">
  <rdf:Description rdf:about="Ex:?">
    <rdf123:startCol>3</rdf123:startCol>
    <rdf123:startRow>6</rdf123:startRow>
    <rdf123:endRow>9</rdf123:endRow>
    <rdfs:comment>use metadata in a map file</rdfs:comment>
  </rdf:Description>
  <foaf:Person>
    <foaf:name>Ex:$1^^string</foaf:name>
    <emp:supervisor>
      <foaf:Person>
        <foaf:name>Ex:$4^^string</foaf:name>
      </foaf:Person>
    </emp:supervisor>
  </foaf:Person>
</rdf:RDF>
```

6   RDF123 Architecture

As shown in Figure 5, RDF123 consists of two components: the RDF123 application and Web service. The application allows users to create and edit RDF123 maps as well as to generate RDF documents from local spreadsheet files. The Web service is designed to automatically generate RDF documents from online spreadsheets in any of several forms using RDF123 maps specified in the service or the spreadsheet itself.

6.1   RDF123 Application

The main purpose of the RDF123 application is to give users an interactive and easy-to-use graphical interface for creating the *map graph* and outputting the

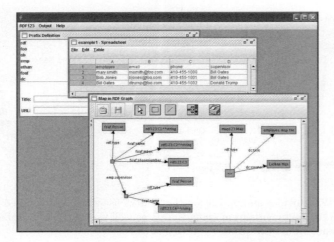

Fig. 6. An RDF123 application (downloadable from [12]) is available for Windows and Unix that provides a graphical interface for creating, inspecting and editing RDF123 maps and using them to generate RDF documents from local spreadsheets

map graph in RDF syntax like RDF/XML and N3. As an RDF document, the serialized *map graph* can be manually edited by people and published online to encourage reuse and extensibility. The application also supports a full work cycle of translating a spreadsheet to RDF, including importing CSV file into a graphical spreadsheet editor and translating the spreadsheet to RDF by applying a *map graph*.

The RDF123 application shown in Figure 6 is composed of three internal frames. The first, 'prefix definition', works as a prefix library in which users store (namespace, prefix) pairs. Namespaces are typically long and forgettable URLs that are hard to manage for many users. By using the 'prefix definition' list, users need not write the full namespace again but use prefix wherever a namespace is required. The second frame is a spreadsheet editor, which enable users to open a CSV file, edit, and save the file. The third frame is an interactive graphical editor that allows user to create and remove a vertex/edge, drag a vertex, and change properties of a vertex/edge. With this graph editor, users create their *map graph*s and saved them to local files, along with the positions of the vertices/edges, for the purpose of future modification. They can also be serialized as RDF documents in RDF/XML or N3.

6.2   RDF123 Web Service

The RDF123 Web service is a public service that translates online spreadsheets to RDF and also works as the host of the URIs of the produced RDF documents. The service is built on the HTTP Get protocol. The service URL is `http://rdf123.umbc.edu/server/` and it takes three basic parameters: 'src', 'map' and 'out'. If a spreadsheet has an embedded link to its online map file, we just need to specify the URL of the spreadsheet with the 'src' parameter; otherwise, we also need give the location of the map file with the 'map' parameter. The parameter 'out' is used to specify the output syntax. An additional parameter 'gid' is used to specify the sheet id within a spreadsheet that has multiple sheets. Examples are available at `http://rdf123.umbc.edu/examples/`. Note the the RDF123 Web service need not be a centralized service and should, in fact, be replicated by different individuals and organizations.

6.3   RDF123 Map Layer

Adding a map layer between the original data in spreadsheets and converted data in RDF can smooth data reusability and maintenance. People may have different aspects and interests in interpreting spreadsheet data. By using different RDF123 maps, the same data can be available in different domains just by associating it with different map files. Figure 7 gives an example. Moreover, when the domain ontology evolves, the map file can be modified, rather than the physical RDF documents, in order to have the data adapt to the change Thus, data maintenance is eased, since data is directly maintained by spreadsheet owners and the RDF data is always rendered current.

In other cases, the map layer can also play a role in integrating data from heterogeneous spreadsheets created by different organizations, and making them available in a unique domain. For example, researchers who do statistics sometimes need to collect data from different sources, many of which are in the form of spreadsheets. However, these spreadsheets usually have different formats and duplicated data. In order to conduct statistical analyses, researchers must do considerable pre-processing to ensure the data have the same format. With the help of RDF123, researchers can accomplish this merging task easily by defining a map for each spreadsheet and translating all of them to RDF using a unique ontology. Then, they can load converted RDF data to a triple store and use a SPARQL query to output data in tabular format which can be accepted by a statistical analysis program. Figure 8 gives an illustration.

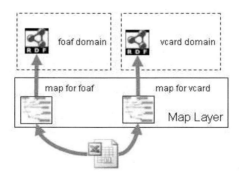

Fig. 7. Separating the spreadsheet data and the maps used to convert them to RDF makes it easy to generate different RDF encodings from the same spreadsheet, encouraging data reuse

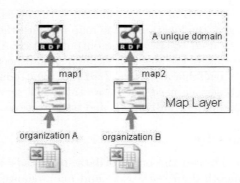

Fig. 8. Separating the data and maps also enables different organizations to have spreadsheet data in their own unique formats but mapped to the same RDF ontology

6.4   A Case Study

We evaluated an early version of RDF123 during the first annual Blogger Bioblitz [13,14]. A Bioblitz [15] is a 24 hour long inventory of the living organisms in a given location, typically by a team of scientists leading a larger group of students and hobbyists with the goals of raising interest in and awareness about biological diversity.

We found that this application demonstrated the strength of RDF123 as a means of publishing and collating distributed data maintained in public spreadsheets. Participants were bloggers who spent one day observing as many different organisms in their chosen location(s). Last year, a common spreadsheet of observations was completed by each individual blogger and then sent to a central location for collation. Idiosyncrasies in the way the spreadsheets were completed made manual integration into a single spreadsheet difficult, but we did successfully use RDF123 to make over 1500 observations available in RDF. However, the data could not easily be corrected or updated. In the 2008 Blogger Blitz we will urge participants to maintain their data publicly using Google spreadsheets. Some map files may need to be created to capture idiosyncrasy, but this should be easier and more dynamic than reformatting to a common template.

6.5   Publishing and Harvesting RDF Data from Spreadsheets

How can we publish the RDF data converted from spreadsheets? One way is to publish the URI provided by the RDF123 web service in the same way we publish a physical RDF document: submitting the URI to a semantic web search engine such as Swoogle [9].

Could we use traditional search engines like Google to help us find possible spreadsheets that could be converted to RDF? Google already supports searching on CSV and Excel files on the whole web and has indexed over 1,350,000 CSV files and 14,700,000 XLS files. If we use a search engine to query for spreadsheet files using keywords that are particular to RDF123 metadata like 'rdf123:metadata', 'map file' and tag values, we are able to harvest all spreadsheets of a particular type that can be converted by RDF123. Thus, there could be a very simple way for end users to publish their own data. Many RDF123 spreadsheet templates about different subjects can be distributed among end users. End users can fill in their own data and publish the instantiated spreadsheet. Once Google indexes these documents, a semantic web search engine can find them via Google API and convert them to RDF without the involvement of end users.

7   Conclusions and Future Work

Spreadsheets are widely used to store and maintain simple data collections. The structural simplicity of the data stored in many spreadsheets makes it relatively easy to export the data into an RDF format. We have described RDF123 as an application designed to make it easy for end users to develop a map between their spreadsheet data and RDF and to use this map to generate RDF data serialized as either XML or N3. The RDF123 web service allows agents to translate spreadsheets as they are encountered, ensuring that data is always current, and obviating the need for maintaining a separate RDF repository online.

Our experience in using RDF123 in the 2007 Blogger Bioblitz convinced us that RDF123 did a good job in meeting our design goals. Users who were familiar with spreadsheets and general computer applications found the application and its tools for modeling data both easy to understand and use. Bioblitz participants found their familiar spreadsheet systems convenient for entering and editing data. The flexibility of the RDF123 mapping language allowed more sophisticated users (i.e., the authors) to refine and publish the maps to produce the desired target encoding in RDF.

There is still one barrier left for common users to use RDF123 to contribute their data to the Semantic Web. Although drawing a *map graph* in the RDF123 application is not hard, choosing proper Semantic Web terms (classes and properties) requires familiarity with appropriate ontologies and the terms they define. We are developing a system that suggests appropriate RDF terms given semantically related English words and general domain and context information [16]. The Swoogle Semantic Web search engine is used to provide RDF term and namespace statistics, the WordNet lexical ontology to find semantically related words, and a naive Bayes classifier to suggest terms. Our initial results show good performance in predicting appropriate RDF terms as measured by precision and recall and we are optimistic that it will be a useful extension to the RDF123 application.

References

1. Hazael-Massieux, D., Connolly, D.: Gleaning Resource Descriptions from Dialects of Languages (GRDDL), W3C Team Submission. Technical report, World Wide Web Consortium (May 2005)

2. Reck, R.P.: Excel2rdf for microsoft windows (January 2003),
   http://www.mindswap.org/rreck/~excel2rdf.shtml
3. Grove, M.: Mindswap Convert To RDF Tool,
   http://www.mindswap.org/~mhgrove/convert/
4. Golbeck, J., Grove, M., Parsia, B., Kalyanpur, A., Hendler, J.: New Tools for the Semantic Web. In: Proceedings of the European Knowledge Acquisition Workshop, pp. 392–400 (2002)
5. Huynh, D., Karger, D., Miller, R.: Exhibit: lightweight structured data publishing. In: Proceedings of the 16th international conference on World Wide Web, pp. 737–746. ACM Press, New York (2007)
6. TopQuadrant: TopBraid Composer Web site, http://www.topbraidcomposer.com/
7. Bizer, C., Seaborne, A.: D2RQ-Treating Non-RDF Databases as Virtual RDF Graphs. In: McIlraith, S.A., Plexousakis, D., van Harmelen, F. (eds.) ISWC 2004. LNCS, vol. 3298. Springer, Heidelberg (2004)
8. Bizer, C.: D2R MAP-A Database to RDF Mapping Language. In: Proceedings of the 12th International World Wide Web Conference (May 2003), pp. 20–24 (2003)
9. Ding, L., Finin, T., Joshi, A., Pan, R., Cost, R.S., Peng, Y., Reddivari, P., Doshi, V.C., Sachs, J.: Swoogle: A search and metadata engine for the semantic web. In: Proceedings of the Thirteenth ACM Conference on Information and Knowledge Management (2004)
10. Tummarello, G., Delbru, R., Oren, E.: Sindice. com: Weaving the open linked data. In: Proceedings of the Sixth International Semantic Web Conference, Springer, Heidelberg (2007)
11. Schmitz, P.: Inducing ontology from Flickr tags. In: Proceedings of the WWW 2006 Collaborative Web Tagging Workshop (May 2006)
12. RDF123 Web site, http://rdf123.umbc.edu/
13. Scienceblogs: Blogger bioblitz, http://scienceblogs.com/voltagegate/2007/03/announcing_the_first_annual_bl.php
14. Parr, C., Sachs, J., Han, L., Wang, T., Finin, T.: RDF123 and Spotter: Tools for generating OWL and RDF for biodiversity data in spreadsheets and unstructured text. In: Proceedings of the Biodiversity Information Standards (TDWG) Annual Conference 2008 (October 2007) (poster abstract)
15. Lundmark, C.: BioBlitz: Getting into Backyard Biodiversity. BioScience 53(4), 329–329 (2003)
16. Han, L., Finin, T.: Predicting Semantic Web Terms from Words. In: Proceedings of the Twenty-Third AAAI Conference on Artificial Intelligence. AAAI Press, Menlo Park (2008) (student abstract)

Evaluating Long-Term Use of the Gnowsis Semantic Desktop for PIM

Leo Sauermann and Dominik Heim

Knowledge Management Department
German Research Center for Artificial Intelligence (DFKI) GmbH,
Kaiserslautern, Germany
{firstname.lastname}@dfki.de

Abstract. The Semantic Desktop is a means to support users in Personal Information Management (PIM). Using the open source software prototype Gnowsis, we evaluated the approach in a two month case study in 2006 with eight participants. Two participants continued using the prototype and were interviewed after two years in 2008 to show their long-term usage patterns. This allows us to analyse how the system was used for PIM. Contextual interviews gave insights on behaviour, while questionnaires and event logging did not. We discovered that in the personal environment, simple has-Part and is-related relations are sufficient for users to file and re-find information, and that the personal semantic wiki was used creatively to note information.

1 Introduction

People gather information on their desktop computers, but current systems lack the ability to integrate this information based on concepts or across applications. The vision of the *Semantic Desktop* [16] is to use Semantic Web technology on the desktop to support *Personal Information Management* (PIM). In addition to providing an interface for managing your personal data it also provides interfaces for other applications to access this, acting as a central hub for semantic information on the desktop. *"Here I have the possibility to gather things"* is a quote from a user of our prototype. Previous work published about Semantic Desktop applications [3,15,17] did show that this approach is promising to support users in filing and finding information, and to work with information in new ways. The *challenge* of our field is that evaluations with real end-users are scarce, and especially there exist no long-term study of Semantic Desktop usage. In this paper, we present two long-term evaluations of the *gnowsis* system.

The rest of the paper is structured as follows: First, the underpinning ideas of the Semantic Desktop and the *gnowsis* implementation are described. The *research question* is to see if our *PIMO ontology* and the *software prototype* were used by the participants for PIM tasks, namely filing and finding information. A short introduction to the methods of HCI evaluations and the problems faced when evaluating PIM system is given, which influenced our selection of tools. In Section 2 a two-month case study with eight participants is presented. The

A. Sheth et al. (Eds.): ISWC 2008, LNCS 5318, pp. 467–482, 2008.

second evaluation is a contextual inquiry after 22 months with two users who continued using the system, in Section 3. A discussion follows in Section 4. The paper concludes with an outlook on future work. Readers should be familiar with the approach of the Semantic Desktop which is described in the related work.

1.1 Semantic Desktop and Gnowsis

In [17] we presented the basic architecture of the evaluated system. The core services are: a service to *crawl data* on the desktop and convert it to RDF (the Aperture[1] framework), store data in an *RDF database*, and *infer new knowledge* from the data. A *search* service provides fulltext and semantic search, a *semantic wiki*[2] provides means to store text, and an *ontology service* provides a programmatic API to *tag* documents or manipulate classes and instances. These services have been described in [17] and were further improved and extended in the NEPOMUK project [7].

The underlying ontologies consist of generic ontologies for files, e-mails, and other document types, and higher level ontologies to represent the mental model of the user, our application domain. The *Personal Information Model* (*PIMO*) [18][3] is a model to represent a single users' concepts, such as projects, tasks, contacts, organizations, allowing files, e-mails, and other resources of interest to the user to be categorized, independent of application and with multiple relations [4]. It is based on RDF/S, it was used as main information representation ontology for the end-user. The generic upper-class is Thing and we will refer to it throughout this paper. In Section 3.1 the predefined sub-classes of Thing are listed. They were selected based on their generality and applicability independent of domain. Latif and Tjoa [12] came to a similar selection that influenced us.

In the rest of this paper RDFS classes, instances, and properties defined in a user's PIMO are distinguished from text by underlining. It is a reference to hyperlinks and emphasizes the fact that each element points to an RDF resource and can be browsed by users.

Based on the core services and the PIMO ontologies, various applications exist to help the user filing and finding information. A short description of each component is given now. *Miniquire*[4] is a sidebar application showing an overview of the user's PIMO. As shown in Figure 1 the sidebar contains a search interface on the top and below an overview on the user's PIMO. It allows users to quickly find things inside their PIMO or manipulate them. It contains functionality for adding and deleting (sub-) classes or things as well as more sophisticated options like hiding or highlighting things.

The *Thing Editor* shown in figure 2 enables the user to focus on a specific Thing to see all relations as well as providing the possibility to edit the relations

---

[1] http://aperture.sourceforge.net

[2] The open source Kaukolu wiki: http://kaukoluwiki.opendfki.de

[3] An improved an more current draft version is available at
http://dev.nepomuk.semanticdesktop.org/wiki/PimoOntology

[4] The name is a pun referencing the work that was before, Tim Berners-Lee's Enquire.

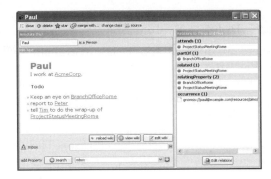

Fig. 1. The sidebar user interface "Miniquire"

Fig. 2. The "ThingEditor" browser and editor

and meta-data. To the left the semantic wiki *kaukolu* is embedded. On the right side, the relations are shown as a list. Clicking on the related things navigates to them. When editing, the core relations easily editable are (defined in the PIMO ontology): <u>related</u>, <u>part Of</u>, <u>has Part</u>, <u>has Topic</u>, <u>is Topic Of</u>. More properties can be added. The *Personal Semantic Wiki* is realized using the *Kaukolu* [11] open source software. The main idea is that a wiki page can be created for every <u>Thing</u> in the user's PIMO. Using a semantic wiki syntax, it is possible to annotate concepts.

As identified by Barreau and Nardi [1], support for filing information is a crucial task in personal information management. The *Drop-Box*[5] is an application to (semi-) automatically move and classify a file. It consists of a normal file folder called "dropbox" that is observed by the gnowsis system. It usually resides on the desktop. If a file is dropped, a "Drop-Box" window appears showing possible tags to classify the file and possible folders where to move it. New tags and folders can be created ad hoc. The system *suggests possible tags* based on text analysis and document similarity to previously tagged documents[6]. Two *Plugins* were developed for Mozilla Thunderbird and Microsoft Outlook to tag items in these applications. These plugins don't support recommendations.

At this point it is important to know that all interfaces work on the same level of abstraction — the <u>Thing</u>. The name of a *wiki page* equals the name of the <u>Thing</u> equals the name of a *tag*, and can relate (if assigned) to a file folder on disk. This makes a user's PIMO a data structure used across applications and data domains. Another application is a *Semantic Search* interface combining full-text search and semantic search. It was described in [17], the latest version is [19]. More details about the software and its features can be found in [17], or by watching the tutorial

---

[5] The name is derived from the Mac OS drop-box.
[6] For a detailed description, refer to our previous publications or `http://www.gnowsis.org/statisch/0.9/doc/use_cases/_dropbox.html`

videos[7]. The software used for the evaluations is *gnowsis* 0.9[8]. It is written in Java and is available under the open source BSD license.

1.2   Related Evaluations

There are different approaches to evaluate software in Human Computer Interaction (HCI). One approach is to set up an experiment in a lab-environment, reducing factors that may influence the experiment and invite test users to use the system in a supervised way. Such evaluation is typically done in a short time span (one day, or a few hours).

In a well-noted paper, Greenberg and Buxton [6] questioned if this practice of HCI evaluations is good for all cases. Their key argument is *the choice of evaluation methodology—if any—must arise from and be appropriate for the actual problem or research question under consideration. Eat-your-own-dogfood* is one of these possible methods.

In the field of PIM, the main problem is the long-term nature of any activity. Storing and retrieving tasks cannot be observed in laboratory settings in a realistic way. For PIM, privacy concerns are a problem and also the stability of the software prototypes, as people will depend on the usefulness of the system when evaluating it. Kelly [10] showed that in PIM, an evaluation has a longer duration and aims at providing information about the usefulness[9] of the software itself, and not only the usability of the GUI. She also recommends not to evaluate with fictional tasks, as PIM is highly associated to real user activities. Furthermore Kelly showed that it is important to let the user work his or her own personal space of information. In related work on PIM [15,2] we find that evaluations based on interviews with users, accompanied by implementing a prototype are common.

In terms of quality, Nielsen [14] argues that five users are sufficient to discover 85% of the usability problems in an interface. Many insights about Microsoft's "MyLifeBits" [5] prototype were found with one very dedicated user, Gordon Bell, who is also co-author of the publications.

1.3   Decisions for the Gnowsis Evaluation

We agree with the view of Greenberg and Buxton and consider *eating your own dogfood* and long-term contextual inquiries as methods that help us to learn about Semantic Web technologies. Our research question is to see how our *PIMO ontology* and the *software prototype* were used by the participants for PIM tasks, namely filing and finding information[10]. Especially it is important to know how users file information, find and re-find information, and maintain their structures using the Semantic Desktop. Also the relation between mental models and the explicit PIMO structures is interesting. Secondary it is interesting to get feedback about the usability of the software.

---

[7] http://gnowsis.opendfki.de/wiki/GnowsisUsage
[8] http://www.gnowsis.org/Download/
[9] Distinguishing between usability and usefulness is stressed by [6].
[10] These tasks are considered essential PIM activities in [9].

2   Usability Evalution July 2006

To find the usage patterns of the users, a two-month evaluation was planned with participants using the software on their own computers in their daily work.

2.1   Participants

In the beginning of the evaluation, we asked the participants at a scientific workshop to participate in the experiment, 15 agreed. When the actual evaluation period started, only a small fraction of the initial volunteers was still available. To get useful results, 8 volunteers on site where then asked to participate. Compared to remote users, these participants were available for the contextual inquiry interviews at the end and for the usability test at the beginning. Also, it was possible to fix software problems in-situ. More background information about the selection of participants is given in [8, p81]. The participants were not financially compensated for their effort.

All of the participants worked within DFKI, our company, two were from departments that are not related to Semantic Web, six were related to Semantic Web and our Knowledge Management Lab. This biased them to rate the prototype better than it actually was, but also let them be more forgiving when bugs and problems occurred. Their ages ranged from 25 to 40, one participant was female. All participants were familiar with desktop computers and general PIM activities. All participants but one were native German speakers, their feedback was translated to English by the authors of this paper.

Participant A was a male senior researcher who was also occupied with software project management and consulting. He installed gnowsis in early 2006 and was still using it in May 2008. He has experience in Semantic Web technologies and semantic modelling. He was using *gnowsis* often each day. His work documents include 8300 files within 1160 folders. He created 1196 elements in his PIMO. B was a female junior researcher engaged in writing a PhD thesis and research project work. She has been using *gnowsis* since July 2006 and was still using it at the time of writing. She has experience in Semantic Web technologies and semantic modelling. Her work documents include 75900 files and 11700 folders. She created 465 elements in her PIMO. Participant C and D were student workers of other departments who had programming experience but did not know about Semantic Web. The four remaining participants were researchers from the Knowledge Management department.

2.2   Procedure

The evaluation started with the installation of the software on the user's desktop in July 2006. Then first interactions with the user interface were explained by the interviewer and done by the user. We used a software tutorial to guide users through the first steps with the system and make them acquainted with the features, which we called *use cases*. This was done for each participant individually at their office. The interviewer took written notes of the user's feedback and encouraged users to *think aloud*. This proper user training took about one hour for each participant, but turned out to be an important and time-saving step considering the following two month long-term evaluation.

The *team approach* introduced by Morse [13] is an idea to accompany an evaluation with sessions during which the users exchange their experiences with the software. Users explained each other how to use the software and shared their experiences, how to solve practical problems with it, which turned out to be a good support of the users.

In parallel, an *activity logger* collected statistical data of which actions the user did with the system. The last part of the evaluation was a *final contextual inquiry*. Contextual Inquiry (also known as proactive field studies) is a structured field interviewing method that aims to fully understand the users working-environment. The evaluator can ask each user individual questions to collect information about the processes themselves as well as the consequence of the behaviour [10].

In former evaluations (in 2005 and January 2006) we realized the importance of direct conversation with the participants. The interviews were conducted in the office of the participant, while the user was working with the system. The inquiry consisted of two parts. The first one mirrored the *training use cases*, asking when and how the participant used each feature. The user had to rate each use case by importance and frequency of usage as well as the most often used features. In the second part we asked for information about the frequency of using the software as well as features the participant missed during his interaction.

One of the most important question to us was for what tasks and goals the software was used, and how they relate to PIM. Given such a generic tool as the Semantic Desktop, what problems will users solve with it, and what creative ways did they invent to reach their goals?

2.3   Results of the Usability Evaluation 2006

The results of this evaluation can be divided into three main parts: *Expectation Questionnaire, Usability and GUI*, and *PIM use cases*. The *activity-log* file was used to cross-validate answers.

The *Expectation Questionnaire* consisted of ten questions that were answered using a six-point *Likert-scale* about expectations and ten yes-no questions about previous experience. Users emphasized that the system should be easy to use, help structure their documents, and provide a good search function. Interestingly, more than half of the participants used desktop search engines and nearly all have used wikis before. The rest of the results can be found in [8, pp85-87].

Results from the *Usability and GUI* inquiry were divided into positive and negative feedback. The spoken feedback from the contextual inquiry at the end and during the evaluation was gathered, categorized and grouped. This results in a "most mentioned features" list. The autocomplete functionality of the wiki was mentioned first. It suggests what things can be inserted in the wiki after the user has typed the first letters and pressed a key combination. Second was miniquire and Thing Editor.This was followed by the drag-drop functionalities, "starring" things, and the easy installer.

Negative feedback was about the slow search and the inability to stop a search once started. The need to switch from the browser-based wiki to the Java-Swing based Thing Editor was described as a "loss of focus". Users wanted to filter the ontology tree in miniquire. This was added later and turned out to be a key feature for the two long-term users. An in-application help-system (tooltips, help buttons, manual) was also requested and found missing. All users noted problems with parts of the GUI.

Results from the *activity logger* showed that one user had the system running permanently, one user 3 hours per week, two user 2 hours, one user 1 hour, and one user 30 minutes per week. The rest of the activity log was used to see what features the participants used, but did not help us much to find out if the system supports PIM or not. The *Semantic Wiki* was used by 75% of the participants for note-taking and one third noted that semantic relations can be created faster using the semantic wiki syntax than using the graphical Thing Editor. Half of the participants did not use the *e-mail tagging plugin*, because either they use another email client or no local client at all. 25% of the participants used the plugin but stopped because they did not get any advantage of it (e.g. retrieval of all tagged emails is not possible).

The *most frequently used* components were (multiple denomination was allowed): 87.5% used Miniquire, 75% the wiki, 25% the Thing editor, 12.5% the DropBox. This was also backed in the activity log-files. The rest of the results of the first contextual interview is presented in [8, pp87-93]. About the usefulness of the features, participants were very positive, but cross-validation with the activity log showed that some participants did not use the rated feature more than 10 times and therefore these results are not relevant.

The results of the evaluation logger show, that users did *extend the default PIMO* ontology with custom classes, 11 classes in the mean with a mean derivation of 9. Thus, some users did create many classes while other none. Only half of the users created custom properties, and then not in a significant amount (less than five). Users did create instances though, altogether 371 instances, but with a mean derivation of 81 between users, so some were not active at all.

Main Purpose Of Usage. The purpose of usage divides users. 37.5% of the participants stated to use gnowsis for project management, but 25% also stated explicitly *not* to use gnowsis for project managemant, as it did not provide sufficient features for them. The majority of the participants used *gnowsis* primarily for managing purposes (e.g. events, conferences). This is not surprising, as *gnowsis* is a tool for the management of personal information. This is approved by 37.5% of the users, who used gnowsis to organize their knowledge (including their ideas). The fact that relations within PIMO can help while creating new knowledge is documented by 12.5% of the participants that used gnowsis as support during writing papers.

2.4   Discussion of the First Evaluation

A discussion of the results of this evaluation is given now, more results from the second evaluation are shown later. The most feedback was given about the wiki, miniquire, and the Thing Editor, so we focus on them.

Semantic Wiki. The expectation questionnaire already showed that the majority contribute to wikis as well as use them to organize themselves. This was also reflected during the evaluation. The majority of the participants used the wiki as personal notepad. Some of them used it for entering text as a notepad and some for entering text in a quick manner that was remodelled as PIMO structures later on. The auto-completion feature helped to add relations with a minimal effort. The participants found manifold fields of application for the wiki: documentation, comments on files, contact information of persons, to-do lists, notepad.

Miniquire and Thing Editor. When browsing and annotating in the Thing Editor, the basic default relations of PIMO seemed to be enough for the users to structure their knowledge in a way that allows them to retrieve it. Here the simple graphical view of related items in a list to the right did help users to navigate. Users did not create custom relations, but did create custom classes.

The question is now, how users fulfil their PIM tasks using the system. Based on the two months of use and the questionnaires, we can say that the system assists in PIM, but not how exactly and what the key semantic features are.

3   Contextual Interviews April 2008

Nearly two years after the first evaluation, three users were still available and kept using the system. They were interviewed in a contextual inquiry in April 2008. One of them did only sporadically use gnowsis for daily work and is excluded from the results. The participants A and B (described above in 2.1) remain, to keep their anonymity we further only speak of "one participant" in the male form. Both participants described themselves humorously as *nitpicking information keepers*, one used the German term *"Strukturierungszwang"*. Also, both are colleagues of the main author of this paper. They should be seen as *"eat your own dogfood"* users, that are biased positively towards the system and behave enthusiastic about it. Other users do not share this enthusiasm, only 2 of 8 have continued using the system voluntarily. Nevertheless, the behaviour the users show and the structures they created in their PIMO tell us something about the use of ontologies for PIM.

As *procedure* we concentrated on a contextual inquiry, as this method brought the most interesting results in the last study. Instead of taking written notes we used a video of the interview. The contextual interviews started with setting up a video-camera for audio and screen recording and an introduction to the process. Basic questions about the participant were asked for warm-up. These were questions about name, gender, occupation, since when and how often they use gnowsis. Going from the warm-up into the contextual part was by the question *For what did you use gnowsis last?*. Then participants began showing their structures and telling about experiences. If needed, specific questions were asked about the PIMO classes and instances they created, whether they associated files with instances, their web bookmark keeping behavior, usage of the wiki, and how they created associations and instances. Then, the interviewer asked participants to continue working on a task that they need to do at that moment anyway as part of their normal work. At the end of the inquiry, feedback about the interview process could be given and the interviewer asked the participants to provide a copy of the *Evaluation Logger* logfile that contains the activities the users have been doing. Altogether, each interview lasted for about two hours including a coffee break.

3.1   Results

The interviews were screened and transcribed to text, some open questions were asked two weeks after the interview to clarify facts. In the following observed usage patterns are presented and classified into filing, finding, and maintenance activities. A similar classification was used by [1] and is proposed in [9]. First observations about the PIMO structures created by the users are given.

Classes and Instances. The predefined classes of PIMO are [18]: Group, Location, Building, City, Country, Room, Document, BlogNote, Contract, Organization, Person, Event, SocialEvent, Meeting, Task, Project, Topic.

Users extended them with the following classes: Application, Domain, Hardware, Book, Notes, Paper Collection, Presentation, Proceedings, Project Documents, Project Plan, Project Proposal, Survey, Paper, Story (war stories, usage stories), Tips Tricks, Diploma Thesis, Thesis, PhD Thesis (or Dissertation), Project Work (a document), Pro-Seminar, Department, Research Institute, External Project, DFKI Project, Private Project, Conference (an Event), Conference Series, Phone Call, Workshop, Work Package, Student.

There may be more custom-created classes (but they were not shown to us). In the activity log we find that one user created 10 classes, the other 31. The question is, why and when users create classes. One user said about the possible subclass of Organisation, Ministry: *For example, I could have created "Ministry". But the effort to create it without having a benefit for it was not worth it. ...I model when I think that I can use it. Like Research Company. When I had 2-3 research institutes amongst the companies the pressure was big enough — I created research institute and changed the class (of the existing instances).* The same case was with External Project. The participant started with the predefined Project assuming the semantics of "my own project" until faced with projects run by others. Then, another class was created for those. Both participants did not use the system for *private data*, although one created a class Private Project but did not create instances.

The class Location was scarcely used. Participants articulated no need for locations as they do not classify information by location. Upon further inquiry

Table 1. Interpretation of relations

Domain	Predicate	Range	Interpretation
project	has part	person	Person is member of the project.
organization	has part	person	Person works for organization.
project	has part	workpackage	The workpackage is part of the project plan.
meeting	has part	person	Person attends the meeting.
topic	has part	document	Topic isTopicOf document.
project	has topic	topic	Project is about this topic.

they both have used the Google-Maps integration of locations at the beginning, but the lack of support to automatically geo-code and create locations made it unattractive.

Also the possibility to create explicit Groups to collect similar things was not used, instead participants used to collect things by making them part Of another thing. One participant created 43 Meetings, the other 2, with the explanation *if the Outlook integration worked better, I would use it more.* Both participants used automated features of gnowsis to create Persons from address book entries. But at a later stage, this feature turned out to be buggy and was not used but maintained persons by hand. One participant entered telephone numbers, and e-mail addresses into the system. 101 and 154 persons were created. Both participants used Topics to classify things (59 and 201 instances). Especially people and documents were annotated with topics.

In total the first participant created 288 instances with 92 of them having wiki-pages, the other 959 with 148 wiki-pages. Most instances were created in the classes Topic, Person (and subclasses of person), Document, and Organization.

The system provides *basic relations* which were used extensively: part-of (used 566 and 63 times) with the inverse is-part-of, related (122, 193) has-topic (401, 78), and the inverse is-topic-of. Besides these generic relations, more precise relations were available but the user interface supports the basic relations better. Hence the basic relations were used more and interpreted depending on context, see Table 1. Over the longer period of two years, these structures are getting blurred and unprecise. Although the participant had no problem finding elements and navigating, he noticed that the structures are "wrong". An example was a final meeting *FinalMeeting* about a report *Report* for company *Acme*. A kickoff meeting *Kickoff* started the process. The structures were (in simplified N3):

```
Kickoff a Meeting;
  hasPart PersonA, PersonB, PersonC, PersonD, FinalMeeting;
  partOf  Report;
  related TopicA, TopicB, ReportX;
  occurrence fileA, another_report_about_topicA, interviewnotes.
FinalMeeting a Meeting;
  hasTopic Report;
  partOf Report.
```

When looking at these structures during the inquiry, the participant noted that the has-topic relation between *FinalMeeting* and *Report* is redundant and removed it. Later the participant said he created the relations over time, to navigate faster.

Filing Information. Following is behaviour specific to filing information, in the tasks the participants have chosen to work on during the inquiry. One participant chose to create a Person representing a new student worker. He created an instance of <u>Student</u>, added firstname, lastname, and fullname ("for the search"). A relation to a project was created and a folder on the hard-disk associated. The participant wished to enter the skype-id of the student, but was not able as the property was missing, and the user interface made creating properties complicated.

Another task was to create a new task for this student. An instance of <u>Task</u> was created, then opened and the previously created student selected as <u>related</u>. To express that the task is about a certain topic, two topics were assigned.

Participants used their filesystem and e-mail system in parallel with gnowsis. As e-mail was not integrated well (the plugins had many bugs and were de-activated by the users), participants did not annotate e-mails, but expressed the dire need for annotating e-mails. One participant used *gnowsis'* web-tagging tool to file websites. With files, the drop-box application saves time and helps decision-making when filing, therefore it was used much. The quality of automated tag suggestions was described as very bad, participants complained that they often had to do the tag assignment by hand. The rate of files annotated with gnowsis varies from folder to folder and application. For example, one folder with scientific papers was annotated heavily, whereas others weren't. Compared to file keeping using folders and folder hierarchies, the possibility of multiple classification was both heavily used and expressed as very positive both for filing and for retrieval. Both used the Drop-Box extensively, 386 and 149 times.

Noting text in the *personal semantic wiki* proved to be a key feature. Each thing in the ontology can have a correlated wiki page. Participants used the wiki page for different purposes: to write short notes defining what the concept is, longer notes with copy-pasted quotation and text snippets or to write down meeting notes. The wiki was also used to create web link collections by copy-pasting URIs into the wiki text (by the user who did not use the *gnowsis* web-tagging tool). Both participants discovered many hidden features of the Semantic Wiki by reading the documentation and used them creatively.

One participant created an instance of <u>Task</u> called <u>Todo</u>. On its wiki-page, he used the option of Kaukoluwiki to include other pages, and sections of other pages. The participant then created todo-sections in various other pages and included them all in the master <u>Todo</u> page. For example, the section *Things to do* in the wiki page of <u>ProjectX</u> was integrated in the master todo page. The inclusion was never removed or maintained from the master page, included pages do not show up once they are empty. This system allowed the user to *gather all todos in one place.*

Finding and Reminding. Both participants used the miniquire visualization as main entry point to the system. It is possible to filter the tree using text-search. This feature keeps the spatial arrangement of information of miniqure but filters it. At the end of the inquiry, one participant noted that the miniquire search box is most important for him.

During the interviews, both participants did always use miniquire when opening a specific thing. Once a thing was open, the linked relations to other things were used to navigate. One participant described a certain PersonX as *entry point* to more data. It was an external project partner being responsible for a certain part of the project. The participant navigated from PersonX to find telephone notes, workpackages, and documents related to the project. It seems that once a certain path to information elements is followed, it gets trained and revisited many times later. The preference to follow paths and navigate amongst items was already noticed by Barreau and Nardi [1], *"users prefer to be able to go to the correct location on the first try"*. Using fulltext search was often used when navigation fails.

One participant mentioned to gather information about a person X before doing an important business phone call. The relations allowed to step from X to previously entered phone notes, and to the project. When asked about *what would you do if gnowsis would be taken away from you*, the participant said that the missing text-search functionality would not be such a problem, but the structures and relations. He would not have *confidence* in himself when searching files, because he relies on the structures.

Maintenance activities. In PIM research, maintenance activities are tasks to reorganize or think about information [9]. One effective example use of the system for maintenance was the preparation for a survey. The participant did come back from an interview, having taken notes as a text. Later, he had to deconstruct the interview into parts. This is described as a "creative step": to read the transcribed text and create relations from the interview to other things, such as the project or the customer. For example, the topic S was brought up in a meeting by a customer. In the notes, the topic was mentioned but the participant did not know about it. As the customer will likely mention it again, he decided to create S as a Topic and add some text to it. An internet search brought up the homepage of organization SU which works primarily in S. The participant changed the type of S to Company and attached the homepage. Two documents describing S were related. The participant mentioned that S will be important for other projects also, that was the main reason to create it in his PIMO as an instance. If not for the possibility to relate it to future projects, he *would have just added some text about S in the wiki*.

Generally, participants needed some time to learn how to model effectively. One participant decided to start a completely new PIMO after four months of use, to clean up.

From the larger pool of available features, the two long-term users both only used a very limited set. We learn that these features help them to fulfill their day-to-day PIM. Namely, the miniquire view as an entry point, the wiki to keep notes (and search them), the relations between things, and the relations between things and files and folders are used.

3.2   Perceived Structures Versus Real Structures

Both participants knew their PIMO very well. They verbally used the terms as written in the PIMO. Upon asked what a certain thing in the PIMO represents, participants gave verbal explanations that are very similar to the wiki texts they have written. This confirms the nature of "supplement to memory" we envisioned in the first definition of the Semantic Desktop [16]. For both final participants, it was apparent that they can use the PIMO structures effectively, without always being modelled precise, correct, or exhaustive. They were able to find things effectively using the miniquire sidebar. But their *perceived mental model* of the PIMO structures differed from the *stored ontology*. Especially the relations written in PIMO are different from the mental models.

One participant said: *If I have a topic, I always know whom to ask.* Upon inquiry if this was in the past used to contact people, the participant did first only remember the first name of a person and showed a wrong person, and later remembered the right success story. Another example: *This person, Donald X, when I was looking at his homepage, I added the topics he works on.* Looking into the data, the topics were not annotated but the papers published by Donald X.

4   Discussion

For PIM we can conclude that the combination of wiki, tagging system, and ontology is a good basis to the Semantic Desktop. The wiki was the second most used component and all of the users did use it during the evaluation.

Looking at the small number of custom-created classes and the even smaller amount of created properties shows that the granularity of semantic modeling is not so important when used for **personal** information management. Users did remember where things can be found and how to navigate to them, and followed paths along "entry point" things. For the navigation to work, the nature of the relation (part-of, is-related, has-topic) is not relevant. A daunting hypothesis is, that for PIM, the only needed relation is <u>has Tag</u> expressing that two things are related. This remains to be evaluated.

In general, the approach of the users is to only model when it is necessary and needed later. Participants repeatedly said *I do not want to model the whole world.* Rather, the model is used to explicitly remember important things or facts. As a side-effect this also kept the system usable. A technical limitation of the user interface is that the performance gets worse when many thousand instances are modelled, and the miniquire tree-visualization would then be crowded too much.

From the created classes, we learn that they refine a specific PIMO class and not the generic Thing. Also, the low amount of created classes shows that classes as such are not such an important modeling tool. Removing the ability to create classes at all may remove one burden of the user to decide when to create a class. Also, the selection of classes identified by [12] as useful for PIM were affirmed by the structures created by our participants. Only geographic location was not used as much as expected.

A problem is, that *classes* cannot be annotated like instances can. A class cannot have relations to things, nor can they be annotated in the wiki or be used as Tags for occurrences. So classes are excluded from most editing functionalities. Users wanted to annotate classes to add meta-information about the class. Given the class Requirements Specification (a subclass of Document, instances are concrete requirements specification reports from customers), the user wanted to relate this class with a document that has instructions how to write a Requirements Specification. The same happened with Survey where there are documents about "doing surveys" in parallel to the instances of the class Survey as such.

The contextual inquiry also influenced the participant and the data as such. At the beginning of the interview, one participant noted that *"this should be part of that"* and modified his PIMO. The behaviour of users doing organization work during interviews was also noticed by Barreau and Nardi [1]. Key results of the 2006 case-study were:

− The *drop-box* component increased productivity as it allows to file items faster than without the assistance.
− The possibility to add multiple categories to a document was used, in the mean 2.5 categories were attached to a file, which is significantly more than the single category a hierarchical file system provides.
− The participants agreed that the PIMO reflects their personal mental models.
− The *gnowsis desktop search* was used very frequently and users found unexpected information.

It is interesting to note that the two long-term users who were interviewed in 2008 did not use the desktop search frequently. The key patterns learned from the interview in 2008 are:

− The ontology tree-view presented in the miniquire user interface is the major entry point to the data. This confirms the results of [1] where users also browsed first and only when not finding elements used search.
− The semantic wiki feature is crucial. The text notes were used daily and for various tasks. Users found creative ways to realize task management and note-taking.
− Applications which are not yet integrated are described as a problem, and a hindrance to use the Semantic Desktop. Plugins for all desktop applications are needed to further support users.
− The PIMO structures enable the users to replicate their mental models using associations and multi-criterial classification [4]. The ability to structure information helped users to creatively file their information and remember elements. They used the structures as entry point to their files.

Both *methods* used for evaluation, the long term case study and the contextual inquiry brought insight about how users cope with Semantic Technology. Clearly, more experience is needed to improve the quality of the studies, here a cooperation with usability experts would be fruitful. Besides that, the technical effort is not to be underestimated. For example, participants had the possibility to integrate their MS-Outlook e-mails, contacts, and appointments with gnowsis, but didn't use this option because installation was too complicated and they feared that bugs could damage their data. Hence, software problems will always influence your evaluation.

5   Future Work

Looking at the *limited* results that were given by the questionnaire and the evaluation logging, and the *rich* information about practical experiences retrieved by video-recording contextual interviews, more evaluations with interviews could be done. Another evaluation with the NEPOMUK prototype is planned for mid-2008. The question is how the Semantic Desktop and PIMO compare to existing file systems and software. With the deployment of KDE 4.0 in July 2008, the Semantic Desktop and the idea of the PIMO will be delivered to more than a million users, which is partly a success of our work and the Semantic Web, but also opens a challenging field for research. For KDE, we reported the issues found in the long-term contextual inquiry to the main developer Sebastian Trüg, who is addressing them there.

Acknowledgements

This work has been supported by the European Union IST fund (Grant FP6-027705, Project NEPOMUK[11]). Special thanks to Rósa Gudjónsdóttir and Kristina Groth from the Royal Institute of Technology, KTH, Sweden for supporting us with guidelines how to do end-user evaluations. We also want to thank the various developers of *gnowsis*, Aperture and NEPOMUK. As a special credit, we wish Gunnar Grimnes to rock as hard as 300 Spartans. He did a lot of the implementation, good science, and proof-read this paper. Finally, we want to thank the participants of the study for their time and investment.

References

1. Barreau, D., Nardi, B.A.: Finding and reminding: File organization from the desktop (1995)
2. Boardman, R.: Improving Tool Support for Personal InformationManagement. PhD thesis (July 13, 2004)
3. Cheyer, A., Park, J., Giuli, R.: IRIS: Integrate. Relate. Infer. Share. In: Proc. of Semantic Desktop WS at ISWC (2005)
4. Dengel, A.R.: Six Thousand Words about Multi-Perspective Personal Document Management. In: Proc. EDM IEEE Workshop. IEEE, Los Alamitos (2006)
5. Gemmell, J., Bell, G., Lueder, R., Drucker, S., Wong, C.: Mylifebits: Fulfilling the memex vision. In: ACM Multimedia, France (December 2002)

---

[11] http://nepomuk.semanticdesktop.org/

6. Greenberg, S., Buxton, B.: Usability evaluation considered harmful (some of the time). In: Proc. CHI 2008, pp. 111–120. ACM, New York (2008)
7. Groza, T., Handschuh, S., Moeller, K., Grimnes, G., Sauermann, L., Minack, E., Mesnage, C., Jazayeri, M., Reif, G., Gudjonsdottir, R.: The NEPOMUK Project — On the way to the Social Semantic Desktop. In: Pellegrini, T., Schaffert, S. (eds.) Proceedings of I-Semantics 2007, JUCS, pp. 201–211 (2007)
8. Heim, D.: Semantic Wikis in knowledge management - Evaluating the Gnowsis approach. Master's thesis, Fachhochschule Kaiserslautern (August 2006)
9. Jones, W., Teevan, J. (eds.): Personal Information Management, October 2007. University of Washington Press (2007)
10. Kelly, D.: Evaluating personal information management behaviors and tools. Communications of the ACM 49(1), 84–86 (2006)
11. Kiesel, M.: Kaukolu: Hub of the semantic corporate intranet. In: Proc. of the Workshop SemWiki 2006 - From Wiki to Semantics at the ESWC Conference (2006)
12. Latif, K., Tjoa, A.M.: Combining context ontology and landmarks for personal information management. In: Proc. ICOCI, Kuala Lumpur, Malaysia (June 2006)
13. Morse, E.L.: Evaluation methodologies for information management systems. D-Lib Magazine 8(9) (September 2002)
14. Nielsen, J.: Why you only need to test five users (March 2000), http://www.useit.com/alertbox/20000319.html
15. Quan, D.: Designing End User Information Environments Built on Semistructured Data Models. PhD thesis, MIT (2003)
16. Sauermann, L., Bernardi, A., Dengel, A.: Overview and Outlook on the Semantic Desktop. In: Decker, S., Park, J., Quan, D., Sauermann, L. (eds.) Proceedings of the 1st Workshop on The Semantic Desktop at the ISWC 2005 Conference (2005)
17. Sauermann, L., Grimnes, G.A., Kiesel, M., Fluit, C., Maus, H., Heim, D., Nadeem, D., Horak, B., Dengel, A.: Semantic Desktop 2.0: The Gnowsis Experience. In: Cruz, I., Decker, S., Allemang, D., Preist, C., Schwabe, D., Mika, P., Uschold, M., Aroyo, L.M. (eds.) ISWC 2006. LNCS, vol. 4273, pp. 887–900. Springer, Heidelberg (2006)
18. Sauermann, L., van Elst, L., Dengel, A.: PIMO - a Framework for Representing Personal Information Models. In: Proc. I-Semantics 2007. JUCS (2007)
19. Schuhmacher, K., Sintek, M., Sauermann, L.: Combining Metadata and Document Search with Spreading Activation for Semantic Desktop Search. In: Proc. of ESWC (to appear, 2008)

Bringing the IPTC News Architecture into the Semantic Web

Raphaël Troncy

CWI Amsterdam, P.O. Box 94079, 1090 GB Amsterdam, The Netherlands
raphael.troncy@cwi.nl

Abstract. For easing the exchange of news, the International Press Telecommunication Council (IPTC) has developed the NewsML Architecture (NAR), an XML-based model that is specialized into a number of languages such as NewsML G2 and EventsML G2. As part of this architecture, specific controlled vocabularies, such as the IPTC News Codes, are used to categorize news items together with other industry-standard thesauri. While news is still mainly in the form of text-based stories, these are often illustrated with graphics, images and videos. Media-specific metadata formats, such as EXIF, DIG35 and XMP, are used to describe the media. The use of different metadata formats in a single production process leads to interoperability problems within the news production chain itself. It also excludes linking to existing web knowledge resources and impedes the construction of uniform end-user interfaces for searching and browsing news content.

In order to allow these different metadata standards to interoperate within a single information environment, we design an OWL ontology for the IPTC News Architecture, linked with other multimedia metadata standards. We convert the IPTC NewsCodes into a SKOS thesaurus and we demonstrate how the news metadata can then be enriched using natural language processing and multimedia analysis and integrated with existing knowledge already formalized on the Semantic Web. We discuss the method we used for developing the ontology and give rationale for our design decisions. We provide guidelines for re-engineering schemas into ontologies and formalize their implicit semantics. In order to demonstrate the appropriateness of our ontology infrastructure, we present an exploratory environment for searching and browsing news items.

1 Introduction

Nearly every European citizen reads, watches or listens to the news, at home, while commuting to and from work, at work and even as part of their work. As voting citizens, we need to understand local, national and international politics to allow us to cast our vote. As company employees, we need to understand the state and development of local, national and international economies to enable us to understand our markets. As part of our leisure time, we want to know about our favorite sports teams, the lives of our soap idols or the most recent

A. Sheth et al. (Eds.): ISWC 2008, LNCS 5318, pp. 483–498, 2008.

books available. Nowadays, this information is online, and hence easily accessible from anywhere.

In existing news workflow processes, news items are typically *i)* produced by news agencies, independent journalists or citizen media, *ii)* consumed and enhanced by newspapers, magazines or broadcasters then finally *iii)* delivered to end users. News items are typically accompanied by a set of metadata and descriptions that facilitate their storage and retrieval. However, much of the metadata is lost because of interoperability problems occurring along the workflow. In addition, at the end user interface, opportunities for making use of the available metadata are often lost. Consequently, users are forced to explore news information in environments that contain large amounts of irrelevant, unreliable and repeated information, with insufficient access to background knowledge.

Our ultimate goal is to create an environment that facilitates end-users in seeing meaningful connections among individual news items (stories, photos, graphics, videos) through underlying knowledge of the descriptions of the items, their relationships and related background knowledge. This requires semantic metadata models to improve metadata interoperability along the entire news production chain. The underlying research problem tackled in this paper covers the two ends of the news workflow spectrum: how to model and represent semantic multimedia metadata along the news workflow and the consequences of this modeling at the user interface.

The contribution of this paper is twofold. On one hand, we report on the modeling of the ontologies for the IPTC family of languages and we convert the IPTC NewsCodes into SKOS thesaurus for demonstrating how the news metadata can be automatically enriched and further integrated with the knowledge already formalized on the Web. We generalize our approach and provide guidelines for re-engineering schemas into ontologies and formalize their implicit semantics. On the other hand, we discuss these modeling decisions with respect to their consequences on the end-user interfaces. We present exploratory interfaces for searching and browsing news that require rich semantic descriptions of the data.

This paper is organized as follows. In the next section, we briefly introduce the main news and multimedia standards used by the media industry. Readers who are already familiar with these formats can skip this section. In Section 3, we discuss the existing methods for engineering ontologies from schemas and porting thesauri to the Semantic Web. We also present the existing attempts for integrating multimedia and news ontologies. We detail in Section 4 the steps for building an ontology-based news infrastructure. We discuss the design decisions and we provide guidelines for re-engineering schemas into ontologies. In order to demonstrate the appropriateness of our ontology infrastructure, we present a semantic search system for multimedia news in the Section 5. Finally, we give our conclusions and outline future work in Section 6.

2   News and Multimedia Standards

2.1   News Standards

Historically, the International Press Telecommunication Council (IPTC) has developed NITF[1] and NewsML, two XML-based languages for describing the structure and the content of news articles. These languages proved, however, to be inadequate to describe all kind of multimedia news and were often judged too verbose. Recently, IPCT has released the News Architecture framework (NAR[2]) which provides the framework for the second generation of IPTC G2 standards.

NAR is a generic model that defines four main objects (`newsItem`, `packageItem`, `conceptItem` and `knowledgeItem`) and the processing model associated with these structures. Specific languages such as NewsML G2 or EventsML G2 are then built on top of this architecture. For example, the generic `newsItem` is specialized into media objects (textual stories, images or audio clips) in NewsML G2.

Finally, IPTC maintains a number of controlled vocabularies called the IPTC NewsCodes, that are used as values while annotating news items. Among others, the Subject Codes is a thesaurus of 1300 terms used for categorizing the main topics (*subjects*) of each news items.

2.2   Multimedia Standards

Although the NAR architecture defines the basic concepts for representing the various media (text, photo, audio, video, graphics), a multitude of other standards are used in the media industry [12].

Pictures taken by a journalist come with EXIF[3] metadata related to the image data structure (height, width, orientation), the capturing information (focal length, exposure time, flash) and the image data characteristics (transfer function, color space transformation). Both Kanzaki[4] and Norm Walsh[5] have proposed an RDFS ontology of EXIF and services for extracting and converting the metadata stored in the header of the images.

These technical metadata are generally completed with other standards aiming at describing the subject matter. DIG35[6] is a specification of the International Imaging Association (I3A). It defines, within an XML Schema, metadata related to image parameters, creation information, content description (who, what, when and where), history and intellectual property rights. In collaboration with Ghent University, we have recently modeled these metadata blocks into

---

[1] News Industry Text Format: http://www.nitf.org/
[2] http://www.iptc.org/NAR/
[3] Exchangeable Image File Format:
   http://www.digicamsoft.com/exif22/exif22/html/exif22_1.htm
[4] http://www.kanzaki.com/ns/exif
[5] http://sourceforge.net/projects/jpegrdf
[6] http://www.i3a.org/resources/dig35/

a DIG35 ontology[7], following the same guidelines detailed in Section 4. XMP[8] provides a native RDF data model and predefined sets of metadata property definitions such as Dublin Core, basic rights and media management schemas for describing still images. IPTC has itself integrated XMP in its Image Metadata specifications[9]. PhotoRDF[10] is also an attempt to standardize a set of categories for personal photo management using Dublin Core and a minimal RDF schema defining 10 terms for the `dc:subject` property.

Video can be decomposed and described using MPEG-7, the *Multimedia Content Description* ISO Standard [15]. This language provides a large and comprehensive set of descriptors including multimedia decomposition descriptors, management metadata properties, audio and visual low-level features and more abstract semantic concepts. The ambiguity and lack of formal semantics of MPEG-7 have been largely pointed out, and several OWL ontologies modeling this standard have been proposed and recently compared [18]. Among them, the Core Ontology for Multimedia Annotation (COMM) proposes to re-engineer completely MPEG-7 using DOLCE as upper ontology and multimedia design patterns [1]. From the broadcast world, the European Broadcaster Union[11] (EBU) has recently adopted the NAR architecture for describing videos, providing some extensions in order to be able to associate metadata to arbitrary parts of videos and to have a vocabulary for rights management.

In conclusion, we end up with an environment that uses numerous languages and formats, often XML-based, that leads to interoperability problems within the news production chain itself and that excludes linking to other vocabularies and existing web knowledge resources. We propose to use Semantic Web languages for leveraging all these standards and ease their integration. This requires a proper ontology infrastructure. Based on the related literature detailed in the next section and our own experience, we discuss the rationale of the design decisions and we formulate guidelines for modeling ontologies from existing schemas.

3   Related Work

Many approaches have been reported to build ontologies [11]. For example, Uschold and Grüninger methodology [20] provides the general steps for the whole process of ontology engineering while METHONTOLOGY [6] proposes to build the ontology at the knowledge level using a set of *intermediate representations*. Specific methods focus on the conceptualization of the ontology, that is, how to structure the taxonomy of concepts [4]. These methodologies, however, do not consider the (supposedly easier) case where a schema (UML diagrams, XML Schema, thesaurus) formalizing already the domain pre-exists but still needs to be ported to the Semantic Web.

---

[7] http://multimedialab.elis.ugent.be/users/chpoppe/Ontologies/index.html
[8] Adobe's Extensible Metadata Platform: http://www.adobe.com/products/xmp/
[9] http://www.iptc.org/IPTC4XMP/
[10] http://www.w3.org/TR/photo-rdf/
[11] http://www.ebu.ch

3.1   Porting Schemas and Thesauri to the Semantic Web

Semantic Web and object-oriented languages are compared in [14] which further explains how to develop ontology-driven software. The "SKOSification" of thesauri in the cultural heritage domain has lead to a general method for porting thesauri to the Semantic Web [22,3,2]. This method advocates four steps (preparation, syntactic conversion, semantic conversion, standardization) and provides a number of guidelines for each step. Our method follows the same recommendations and add more guidelines regarding the modeling of existing UML diagrams in OWL ontologies.

The alignment of the resulting thesaurus with existing semantic web resources is particularly addressed in [17], that leads to the AnnoCultor[12] conversion tool. We have used this tool for converting the IPTC NewsCodes into SKOS thesauri.

3.2   NewsML and Multimedia Ontologies

Various attempts for building a news ontology have been reported. NEWS[13] is a completed EU project that aims to combine Semantic Web technologies and web services for improving the news agencies workflow. The project has developed a lightweight RDFS news ontology (in English, Spanish and Italian) based on the IPTC Subject Codes for categorizing the news items and on NITF and NewsML for the metadata management [7,9]. The Neptuno[14] research project has also modeled a lightweight RDFS news ontology representing a newspaper archive. The ontology is again a mix between news management metadata based on the NewsML standard and on the IPTC Subject Codes aligned with a news agency thesaurus for categorizing the news items [5]. Finally, MESH[15] is an ongoing EU project that focuses on multimedia analysis for enriching automatically news metadata and deliver personalized news summary. A news ontology seems to have been developed but it is not available.

In contrast to these projects, our approach is to decouple the thesauri used in the metadata values from the ontology that describes the management of the news items according to the journalist point of view. This separation of concern provides a more flexible infrastructure where the Subject Codes can be aligned to other thesauri. We expose these aligned thesauri on the Semantic Web, providing dereferencable URIs for every terms. Furthermore, we conform to the latest standard for the news metadata (NAR) and we design the ontology to be linked with other media ontologies.

The XML Semantics Reuse methodology consists in converting automatically XML Schemas into OWL ontologies[16]. This methodology is used in the journalism domain for converting the NewsML and NITF document formats, the IPTC Subject Codes taxonomy and the MPEG-7 multimedia format into

---

[12] http://sourceforge.net/projects/annocultor
[13] http://www.news-project.com/
[14] http://seweb.ii.uam.es/neptuno/
[15] http://www.mesh-ip.eu/
[16] See the ReDeFer project: http://rhizomik.net/redefer

OWL/RDF [10]. The resulting ontology, however, fails to capture the intended semantics of these standards that cannot be represented in XML Schema [18]. It recreates the complex nested structures used in the original schema (e.g. the definition of intermediate containers defining the XML Schema types and elements) that should generally not be modeled in the ontology. We advocate, on the contrary, to re-engineer the ontology following some good practices that we detail in the next section.

4   Building a Semantic Web Infrastructure for News

As we have described in the Section 2, NAR is a generic model for describing the news items as well as their management, packaging, and the way they are exchanged. Interestingly, this model shares the principles underlying the Semantic Web:

– News items are distributed resources that need to be uniquely identified like the Semantic Web resources;
– News items are described with shared and controlled vocabularies.

NAR is however defined in XML Schema and has thus no formal representation of its intended semantics (e.g. a `NewsItem` can be a `TextNewsItem`, a `PhotoNewsItem` or a `VideoNewsItem`). Extension to other standards is cumbersome since it is hard to state the equivalence between two XML elements.

By modeling a NAR ontology, we do expect the following benefits:

– Better control of NewsML G2 descriptions enabled by logical consistency check;
– Enhanced search of news items enabled by logical inferences from the thesaurus and the knowledge formalized on the web;
– Unified semantic interfaces for searching and browsing seamlessly news content and background knowledge.

In the following, we describe the necessary steps for modeling such an ontology infrastructure. The various interconnected ontologies (NAR, NewsML-G2, EventsML-G2) are available at http://newsml.cwi.nl/ontology/.

4.1   Step 1: Modeling the NAR Ontology

The first step aims to capture formally the intended semantics of NAR and the family of IPTC G2 standards. Even though these models exist in UML diagrams, their "ontologisation" is not trivial. We discuss below the rationale of our modeling decisions.

Flattening the XML structure. XML Schema provides the means to have very rich structure but is rather limited when expressing the meaning of this structure as the language is (only) concerned with providing typing and structuring information for isolated chunk of data. Consequently, the NAR model defines intermediate structures and *containers* whose only goal are to group a number of properties without particular semantics. These structures should not be represented in the ontology, as they will generate blank nodes in the RDF graph at the instance level, complexifying its visualization in any Semantic Web browser. While modeling the ontology, we therefore advocate to flatten the XML structure keeping only the properties that will be instantiated.

Reification. Statements about news items need often to be reified. For example, an editor registered as `team:md` can classify a news item as `diplomacy` at `2005-11-11T08:00:00Z`. Using the RDF reification and the N3 syntax, this yields the following statements:

```
{<> nar:subject cat:11002000} dc:creator   team:md ;
                              dc:modified  ''2005-11-11T08:00:00Z''.
```

The RDF reification having no model theory semantics, we advocate the use of Networked Graphs where the relationships between graphs are described declaratively using SPARQL queries and an extension of the SPARQL semantics [16].

Modeling unique identifiers. News items metadata make use of numerous thesauri that implements a coding scheme for identifying the terms in order to be language agnostic. For example:

```
<pubStatus code="stat:usable"/>
<locCreated code="city:Paris"/>
<creator code="team:DOM"/>
<subject code="cat:04000000"/>
<subject code="isin:NL0000361939"/>
<subject code="pers:021147"/>
```

IPTC has therefore defined the notion of QCODES (by analogy to the XML QNAMES) with the following properties:

- Each coding scheme is associated with a URI. That URI must resolve to a resource (or resources) containing information about the scheme.
- The prefix represents the URI of the scheme within which the local part is allocated.
- There are almost no constraints on the values of the local part. For example, the local part (the code) is allowed to start with a digit.
- The two taken together must form a legal URI.
- This URI should provide access to a definition of the concept represented by that code within that scheme, i.e. it is dereferencable.

The tuple `prefix:localname` is however not identical to a CURIE[17] since the two parts (scheme and code) have each a meaning. For solving this issue, we advocate the "slash" rule, i.e. the concatenation of the scheme URI, a slash and the code, for the construction of a valid and dereferencable code URI.

---

[17] http://www.w3.org/TR/curie/

4.2   Step 2: Linking with Media Ontologies

As we have discussed in the Section 2.2, other multimedia standards such as
EXIF, Dublin Core, XMP, DIG35 or MPEG-7 are used in the media industry.
These standards have generally been converted into OWL ontologies and can
thus be integrated within our ontology infrastructure. Therefore, this step consists in adding OWL axioms stating the relationship between resources defined
in different but strongly overlapping ontologies. For example, the NAR ontology
contains the following axioms:

```
nar:subject owl:equivalentProperty dc:subject
nar:Person owl:equivalentClass foaf:Person
```

Semantic Web search engines such as Sindice[18], Watson[19] or Falcon[20] are
useful tools for discovering concepts and properties defined in other ontologies
that share the same semantics as the ones defined in our news infrastructure and
could be linked to them.

4.3   Step 3: Converting IPTC News Codes into SKOS Thesaurus

The IPTC NewsCodes define 36 thesauri used as metadata values in the NAR
architecture. Although the terms are sometimes organized in a taxonomy, the
subsumption relationship is not explicit but instead encoded into the coding
scheme identifying the terms. For example, "cancer" (cat:07001004) is narrower
than "disease" (cat:07001000) which is narrower than "health" (cat:07000000)
because they share a number of digits. We have converted these thesauri into
SKOS, an application of RDF, making the subsumption relationships explicit
(skos:narrower, skos:broader).

This RDF compatibility allows us to define some concepts in the NAR ontology
in terms of owl:Restriction: the value of a property can be a skos:Concept or
must come from a given skos:ConceptScheme. For example, the nar:subject
object property is defined as having all its values from the IPTC Subject Codes
skos:ConceptScheme.

Finally, we have exposed all these thesauri at http://newsml.cwi.nl/
NewsCodes/ following the Best Practice Recipes for Publishing RDF Vocabularies[21] and Cool URIs for the Semantic Web[22] notes. Each term is thus identified by
a dereferencable URI. Consequently, sending an http request with the requested
type Accept:text/html will deliver the original XML human readable version
from IPTC of the thesauri, while the requested type Accept:application/
rdf+xml will return the SKOS/RDF machine processable version of the
thesaurus.

---

[18] http://sindice.com/
[19] http://watson.kmi.open.ac.uk/WatsonWUI/
[20] http://www.falcons.com.cn/
[21] http://www.w3.org/TR/swbp-vocab-pub/
[22] http://www.w3.org/TR/cooluris/

4.4   Step 4: Enriching the News Metadata

Once the NAR ontology has been modeled, linked to other media ontologies, and the thesauri converted into SKOS, the conversion of the metadata of individual news items into RDF according to this ontology infrastructure is straightforward. However, we advocate a further step aiming at enriching semantically the news metadata following the linked data principle[23]. In our case, we apply linguistic processing of textual news items and visual analysis of photo and video news items in order to extract more semantic metadata (Figure 1).

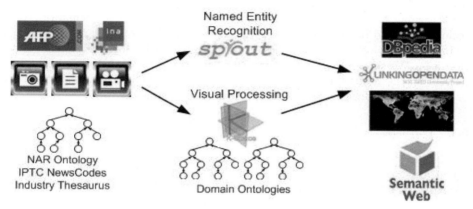

Fig. 1. Metadata enrichment of NewsItems

The linguistic processing consists in extracting named entities such as persons, organisations, locations, brands, etc. from the textual stories. Named Entity Recognizers such as GATE[24], SPROUT[25] or the most recent OpenCalais infrastructure[26] can be used. Once the named entities have been extracted, we map them to formalized knowledge on the web available in Geonames for the locations, or in DBPedia for the persons and organisations. Visual analysis provides additional metadata useful for organizing the results of a semantic news search engine. For example, an unsupervised clustering of photo news items can be obtained using texture and color histograms, allowing to distinguish the photos of a football player such as Zinedine Zidane on the field, versus in a suit while he receives some award.

5   Semantic Search of Multimedia News

In order to demonstrate the appropriateness of our ontology infrastructure, we present an exploratory environment for searching and browsing news items. We

---

[23] http://linkeddata.org/

[24] http://gate.ac.uk/

[25] http://sprout.dfki.de/

[26] http://www.opencalais.com/

use the Cliopatria[27] semantic search web-server [13,21]. ClioPatria is a SWI-Prolog based platform for Semantic Web Applications that provides a scalable in-core RDF triple store and joins the SWI-Prolog RDF and HTTP infrastructure with a SeRQL/SPARQL query engine, interfacing to the The Yahoo! User Interface Library (YUI) and libraries that support semantic search. In contrast to client-only architectures such as Simile's Exhibit, ClioPatria has a client-server architecture. The core functionality is provided as HTTP APIs by the server. The results are served as presentation neutral data objects and can thus be combined with different presentation and interaction strategies. We have decided to use this open source software as it allows us to create customized presentations for searching and exploring news items, while being based on Semantic Web technologies and benefit from inference reasoning and SPARQL querying.

In the following, we present first the dataset used in our experiment (Section 5.1). We show then how we use the semantic metadata as dimensions for presenting the results of semantic search (Section 5.2) and for guiding a faceted browser like interface (Section 5.3) in the news domain.

5.1   Dataset

The dataset used in our experiment consists of the ontology infrastructure detailed previously, 60000 news stories in English, 40000 news stories in French, 2557 photos and 8 hours of broadcasted video (Table 1).

Following the four steps detailed in the Section 4, we have processed the news items in order to enrich the metadata. We have used SPROUT together with a specific football gazetteer in order to extract named entities from the caption of the 2557 photos contained in our dataset. The use of a domain specific ontology allows us to extract more semantic information such as the role of a football player (goalkeeper, middlefielder), the name of a team, etc. The Figure 2 (resp. 3) shows the algorithm for linking the entities of type `Person` (resp. `Location`) with DBPedia (resp. Geonames).

This processing step provided 217 DBPedia persons and 426 Geonames locations. The assessment of the results shows that the Geonames web service tend to return primarily a US city when a single string is passed as an argument. Fortunately, news items contain always information about the city and the country yielding accurate recognition of the location mentioned in the story. The few errors we noticed come from an incorrect typing of the named entity from the SPROUT system, e.g. `Australia` as been typed as a `Person`. More sophisticated disambiguation heuristics such as IdentityRank [8] can be further employed to minimize these errors.

5.2   Semantic Search of News Items

The Figure 4 shows the result for the query "Lyon" in our semantic search system. The news items are grouped according to the path in the RDF graph

---

[27] http://e-culture.multimedian.nl/software/ClioPatria.shtml

Table 1. Number of RDF triples loaded in our semantic search web-server

Description	Number of RDF triples
General ontologies: NAR, NewsML-G2, DC, VRA, FOAF	7,336
Domain specific ontologies: Football ontology	104,358
Thesauri: IPTC NewsCodes, INA Thesaurus	34,903
External resources: Geonames, DBPedia	53,468
AFP News Feed in English from June and July 2006	804,446
AFP Photos from the 2006 World Cup	61,311
INA Broadcast Video from June and July 2006	1,932
Total	1,067.754

For each named entity of type `Person` recognized, do:

1. Construct a SPARQL query for DBPedia using the `rdfs:label` property and all supported languages and return the first resource
2. Construct a SPARQL query for the Football ontology using the `dolce:firstName` and `dolce:lastName` properties and return the first resource
3. If a resource is found both in DBPedia and in the Football ontology, then add a `owl:sameAs` statement between these two resources
4. If no resource is found in DBPedia and in the Football ontology, then create a new instance of `Person` in the knowledge base

Fig. 2. Pseudo algorithm for linking the extracted named entities with DBPedia

that leads to the property for which the value has matched the query. In our case, the system returns the news items where "Lyon" occurs in the `title`, the `headline`, the `slugline`, etc. Each group can be collapsed or expanded. Furthermore, the information about the type of news item allows us to customize the visual rendering of each group: text news items have a snippet view displaying the first three lines of the stories, while photo news items are displayed in a thumbnail carousel.

Interestingly, the last group of result contain a single news photo item depicting three football players. The metadata of this photo does not contain the string "Lyon". Instead, the caption of this photo mentions Juninho Pernambucano, recognized by SPROUT as a football player, later on linked with the DBPedia resource identifying this person. DBPedia contains information about this person such as his birthdate and all the past teams where he played. Among them, "Lyon" is his current club and this is why this image has been retrieved, even though at the end of the list because of the length in the RDF graph necessary to reach this term.

Similarly, the query for "Saksamaa" returns a single group of 679 photos while none of them contain this term in the metadata. Again, the explanation is that all these photos have been captured in Germany (during the World Cup), a named

For each named entity of type `Location` recognized, do:

1. Get the location and if available the broader location
2. Construct a textual query for Geonames with either the exact location name or with both terms and return the first resource, for example:
   http://ws.geonames.org/search?maxRows=1&type=rdf&name_equals=Germany
   or http://ws.geonames.org/search?maxRows=1&type=rdf&q=Berlin,Germany

Fig. 3. Pseudo algorithm for linking the extracted named entities with Geonames

Fig. 4. Search for "Lyon" in the semantic search engine

entity recognized by SPROUT as a location, later on linked with Geonames. The Geonames resource contains information about all the alternative names of Germany in all languages, Saksamaa being the Ethiopian name of Germany.

5.3 Semantic Browsing of Multimedia News

Additionally to the semantic search interface, we provide a faceted browser like interface for better exploring the news dataset. Facets correspond to properties of interest in the metadata, and can be selected by the end-user. We have defined a soccer view gathering the properties `subject`, `slugline`, `locCreated`, `location` and `person`. Using the information provided by Geonames, we are able to propose more views for presenting the information. The Figure 5 shows such a view, where the football player `Zinedine Zidane` has been selected as

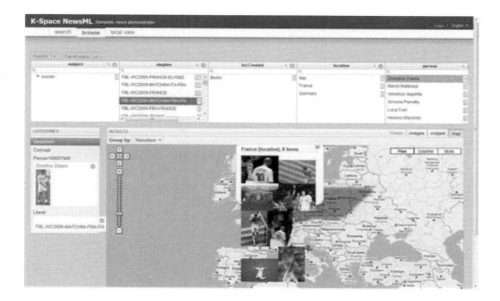

Fig. 5. Browsing the photos captured in France during the World Cup finale

Fig. 6. Video local view: arbitrary temporal segment can be played

a filter: the red flags correspond then to the countries mentioned in the news stories, while the blue flags correspond to the cities where the news stories have been produced.

Interestingly, this view allows the end-user to immediately distinguish between the two sluglines `FBL-WC2006-MATCH64-ITA-FRA` and `FBL-WC2006-MATCH64-FRA-ITA` that look really similar but actually correspond to the news stories produced in Italy (Italian point of view) and in France (French point of view). Such a subtlety is hard to see in the metadata while the map view gives an immediate insight of the data.

Finally, the Figure 6 shows the local view of a video resource. The metadata corresponds to a particular sequence in a TV news broadcast program. Arbitrary sequences of a video can be played in the semantic browser using the `tcin` and `tcout` buttons. An auto-play has been considered but the non-ability of the current web infrastructure to address temporal fragments of a video file prevent such a functionality [19].

6   Conclusion and Future Work

In this paper, we have described a method composed of four steps for building an ontology-based news infrastructure. These guidelines are complementary to the development of ontology design patterns and best practices for publishing semantic web vocabularies that is central in the web of data. We have discussed the design decisions regarding the modeling of the NAR ontology from existing XML Schemas. At the ontology level, we advocate to flatten the XML structure, to identify properly the resources and to reuse as much as possible existing ontologies. At the instance level, we recommend to enrich and link the metadata with existing SKOS thesauri and formalized knowledge existing on the web such as DBPedia and Geonames. The NAR ontology is currently reviewed by the IPTC and could be endorsed by the standardisation body. We presented a semantic search system and various exploratory interfaces for searching and browsing news items. These interfaces use the richness of the semantic metadata for grouping, ranking and presenting the results of a given query. The system is publicly available at `http://newsml.cwi.nl/explore/search`.

Time is an essential dimension in the news domain and our system provides also a timeline view. Nevertheless, reasoning on time information is a complex task. From the representation point of view, we plan to include the Time Ontology[28] and the temporal relations module from the DOLCE upper ontology in order to propose histogram views aggregating the stories per topic and per day, week or month. Our current system works on static, pre-processed and staged data. A natural evolution is to create a dynamic environment where incoming news feed is processed in live and immediately available to the end-user. Finally, an evaluation of the system by AFP journalists is planned.

---

[28] `http://www.w3.org/TR/owl-time/`

Acknowledgments

The dataset used has been kindly provided by AFP for the news stories and the photos, and by INA for the videos. The author would like to particularly thank Laurent Le Meur from AFP for fruitful discussions on the design of the NAR ontology. The author would also like to thank the following colleagues at CWI, Amsterdam (Lynda Hardman, Michiel Hildebrand, Michiel Kauw-A-Tjoe, Zeljko Obrenovic and Jacco van Ossenbruggen) and at IBBT Multimedia Lab, Ghent University (Erik Mannens, Gaëtan Martens and Chris Poppe) for their feedback on the prototype and earlier versions of this paper. The research leading to this paper was supported by the European Commission under contract FP6-027026, Knowledge Space of semantic inference for automatic annotation and retrieval of multimedia content - K-Space.

References

1. Arndt, R., Troncy, R., Staab, S., Hardman, L., Vacura, M.: COMM: Designing a Well-Founded Multimedia Ontology for the Web. In: Aberer, K., Choi, K.-S., Noy, N., Allemang, D., Lee, K.-I., Nixon, L., Golbeck, J., Mika, P., Maynard, D., Mizoguchi, R., Schreiber, G., Cudré-Mauroux, P. (eds.) ASWC 2007 and ISWC 2007. LNCS, vol. 4825, pp. 30–43. Springer, Heidelberg (2007)
2. van Assem, M., Malaisé, V., Miles, A., Schreiber, G.: A Method to Convert Thesauri to SKOS. In: Sure, Y., Domingue, J. (eds.) ESWC 2006. LNCS, vol. 4011, pp. 95–109. Springer, Heidelberg (2006)
3. van Assem, M., Menken, M.R., Schreiber, G., Wielemaker, J., Wielinga, B.: A Method for Converting Thesauri to RDF/OWL. In: McIlraith, S.A., Plexousakis, D., van Harmelen, F. (eds.) ISWC 2004. LNCS, vol. 3298, pp. 17–31. Springer, Heidelberg (2004)
4. Bachimont, B., Isaac, A., Troncy, R.: Semantic Commitment for Designing Ontologies: A Proposal. In: Gómez-Pérez, A., Benjamins, V.R. (eds.) EKAW 2002. LNCS (LNAI), vol. 2473, pp. 114–121. Springer, Heidelberg (2002)
5. Castells, P., Perdrix, F., Pulido, E., Rico, M., Benjamins, R., Contreras, J., Lorés, J.: Neptuno: Semantic Web Technologies for a Digital Newspaper Archive. In: Bussler, C.J., Davies, J., Fensel, D., Studer, R. (eds.) ESWS 2004. LNCS, vol. 3053, pp. 445–458. Springer, Heidelberg (2004)
6. Fernández, M., Gómez-Pérez, A., Juristo, N.: METHONTOLOGY: From Ontological Art Towards Ontological Engineering. In: AAAI 1997 Spring Symposium Series on Ontological Engineering, Stanford, California, USA, pp. 33–40 (1997)
7. Fernández, N., Blázquez, J.M., Arias, J., Sánchez, L., Sintek, M., Bernardi, A., Fuentes, M., Marrara, A., Ben-Asher, Z.: NEWS: Bringing Semantic Web Technologies into News Agencies. In: Cruz, I., Decker, S., Allemang, D., Preist, C., Schwabe, D., Mika, P., Uschold, M., Aroyo, L.M. (eds.) ISWC 2006. LNCS, vol. 4273, pp. 778–791. Springer, Heidelberg (2006)
8. Fernández, N., Blázquez, J.M., Sánchez, L., Bernardi, A.: IdentityRank: Named Entity Disambiguation in the Context of the NEWS Project. In: Franconi, E., Kifer, M., May, W. (eds.) ESWC 2007. LNCS, vol. 4519, pp. 640–657. Springer, Heidelberg (2007)

9. Fernández, N., Sánchez, L., Blázquez, J.M., Villamor, J.: The NEWS Ontology for Professional Journalism Applications. In: A Handbook of Principles, Concepts and Applications in Information Systems. Integrated Series in Information Systems, vol. 14. Springer, Heidelberg (2007)
10. Garcia, R., Perdrix, F., Gil, R., Oliva, M.: The semantic web as a newspaper media convergence facilitator. Journal of Web Semantics 6(2), 151–161 (2008)
11. Gómez-Pérez, A., Fernandez-Lopez, M., Corcho, O.: Ontological Engineering with examples from the areas of Knowledge Management, e-Commerce and the Semantic Web, 1st edn. Advanced Information and Knowledge Processing. Springer, Heidelberg (2004)
12. Hausenblas, M., Boll, S., Bürger, T., Celma, O., Halaschek-Wiener, C., Mannens, E., Troncy, R.: Multimedia Vocabularies on the Semantic Web. W3C Multimedia Semantics Incubator Group Report (2007), http://www.w3.org/2005/Incubator/mmsem/XGR-vocabularies/
13. van Ossenbruggen, J., Hardman, L., Hildebrand, M.: /facet: A Browser for Heterogeneous Semantic Web Repositories. In: Cruz, I., Decker, S., Allemang, D., Preist, C., Schwabe, D., Mika, P., Uschold, M., Aroyo, L.M. (eds.) ISWC 2006. LNCS, vol. 4273, pp. 272–285. Springer, Heidelberg (2006)
14. Knublauch, H., Oberle, D., Tetlow, P., Wallace, E.: A Semantic Web Primer for Object-Oriented Software Developers. W3C Note (2006), http://www.w3.org/TR/sw-oosd-primer/
15. MPEG-7. Multimedia Content Description Interface. ISO/IEC 15938 (2001)
16. Schenk, S., Staab, S.: Networked Graphs: A Declarative Mechanism for SPARQL Rules, SPARQL Views and RDF Data Integration on the Web. In: 17th International World Wide Web Conference (WWW 2008), Beijing, China (2008)
17. Tordai, A., Omelayenko, B., Schreiber, G.: Semantic Excavation of the City of Books. In: Semantic Authoring, Annotation and Knowledge Markup Workshop (SAAKM 2007), pp. 39–46 (2007)
18. Troncy, R., Celma, Ó., Little, S., García, R., Tsinaraki, C.: MPEG-7 based Multimedia Ontologies: Interoperability Support or Interoperability Issue? In: 1st International Workshop on Multimedia Annotation and Retrieval enabled by Shared Ontologies (MAReSO), Genova, Italy (2007)
19. Troncy, R., Hardman, L., van Ossenbruggen, J., Hausenblas, M.: Identifying Spatial and Temporal Media Fragments on the Web. In: W3C Video on the Web Workshop (2007)
20. Uschold, M., Grüninger, M.: Ontologies: Principles, Methods and Applications. Knowledge Engineering Review 2, 93–155 (1996)
21. Wielemaker, J., Hildebrand, M., van Ossenbruggen, J., Schreiber, G.: Thesaurus-based search in large heterogeneous collections. In: 7th International Semantic Web Conference (ISWC 2008), Karlsruhe, Germany (2008)
22. Wielinga, B., Wielemaker, J., Schreiber, G., van Assem, M.: Methods for Porting Resources to the Semantic Web. In: Bussler, C.J., Davies, J., Fensel, D., Studer, R. (eds.) ESWS 2004. LNCS, vol. 3053, pp. 299–311. Springer, Heidelberg (2004)

RDFS Reasoning and Query Answering on Top of DHTs

Zoi Kaoudi*, Iris Miliaraki**, and Manolis Koubarakis

Dept. of Informatics and Telecommunications
National and Kapodistrian University of Athens, Greece

Abstract. We study the problem of distributed RDFS reasoning and query answering on top of distributed hash tables. Scalable, distributed RDFS reasoning is an essential functionality for providing the scalability and performance that large-scale Semantic Web applications require. Our goal in this paper is to compare and evaluate two well-known approaches to RDFS reasoning, namely backward and forward chaining, on top of distributed hash tables. We show how to implement both algorithms on top of the distributed hash table Bamboo and prove their correctness. We also study the time-space trade-off exhibited by the algorithms analytically, and experimentally by evaluating our algorithms on PlanetLab.

1 Introduction

As the Semantic Web has become a reality, there is an emerging need not only for dealing with a huge amount of distributed metadata, but also for being able to reason with it. Previous work on *centralized* RDF stores has considered forward chaining, backward chaining and hybrid approaches to implement RDFS reasoning and query processing [6,27,2,12]. In the forward chaining approach, new statements are exhaustively generated from the asserted ones. In contrast, a backward chaining approach only evaluates RDFS entailments on demand, i.e., at query processing time. Intuitively, we expect that a forward chaining approach has minimal requirements during query answering, but needs a significant amount of storage. In contrast, the backward chaining approach has minimal storage requirements, at the cost of an increase in query response time. There is a time-space trade-off between these two approaches [24], and only by knowing the query and update workload of an application, we can determine which approach would suit it better. This trade-off has never been studied in detail in a *distributed* Internet-scale scenario, and this is one of the challenges that we undertake in this paper.

* Supported by project "Peer-to-Peer Techniques for Semantic Web Services" (funded from the Greek General Secretariat for Research and Technology).
** Supported by Microsoft Research through its European PhD Scholarship Programme.

A. Sheth et al. (Eds.): ISWC 2008, LNCS 5318, pp. 499–516, 2008.

P2P networks and especially distributed hash tables (DHTs) [3] have gained much attention recently, given the scalability, fault-tolerance and robustness features they can provide to Internet-scale applications. Since current centralized RDF repositories lack the required scalability and fault tolerance for such applications [10], DHTs have been proposed for the storage and querying of RDF data at Internet scale [7,19,14,1]. However, these works are solely concerned with query processing for RDF data, and pay no attention to RDF Schema (RDFS) reasoning and query processing. The only DHT-based RDF store that has dealt with RDFS reasoning in the past is BabelPeers [5,14]. It is implemented on top of Pastry [23] and supports a subclass of the SPARQL query language [21]. BabelPeers uses a forward chaining approach in order to provide the RDFS inference capability required to answer the supported class of SPARQL queries.

In this paper we design and implement both forward and backward chaining algorithms for RDFS reasoning and query answering on top of the Bamboo DHT [22]. Our algorithms have been integrated in the RDF query processing system Atlas (http://atlas.di.uoa.gr) [15] and have been used to enable the processing of RQL [17] schema queries. To the best of our knowledge, our backward chaining algorithm is the first distributed backward chaining algorithm proposed on top of DHT-based RDF stores. Another contribution of this work is proving the correctness of our algorithms and providing a comparative study of forward and backward chaining algorithms both analytically and experimentally. In addition, we propose an optimization technique for backward chaining which decreases query response time and allow us to minimize the response time difference between the two approaches. For the experimental part of our study, we deploy our system on PlanetLab to obtain measurements in a realistic large-scale distributed environment. The results obtained in our experiments agree with the predictions of our analytical model. An important result of our experiments is that forward chaining is a very expensive algorithm in terms of storage load, bandwidth and time and does not scale for a large number of triples.

2   Data and Query Model

In the rest of the paper, we will constantly use the notion of an RDF triple. RDF data as well as RDFS descriptions (we will further use the term RDF(S) to refer to both) can be written as triples and constitute an *RDF(S) database*.

To support RDFS reasoning, the RDFS entailment rules of RDF Semantics [13] constitute a vital element of our data model. Following a datalog-like notation with extensional database relation (edb) *triple* and intensional database relations (idb) *subClass*, *subProperty* and *type*, the RDFS entailment rules can be written as shown in Table 1. Each rule is indexed by a number that we will use to refer to it. Rules are also indexed with their symbolic name from [13] for co-reference reasons. Certain rules are deliberately omitted (such as the ones with the axiomatic triples) since we are more interested in rules needed for the computation of the transitive closure. However, our algorithms work with all the

Table 1. RDFS Entailment Rules

Rule	Head	Body
1	$type(X, Y)$	$triple(X, rdf{:}type, Y)$
2 (rdfs2)	$type(X, Y)$	$triple(X, P, Z), triple(P, rdfs{:}domain, Y)$
3 (rdfs3)	$type(X, Y)$	$triple(Z, P, X), triple(P, rdfs{:}range, Y)$
4 (rdfs9)	$type(X, Y)$	$type(X, Z), subClass(Z, Y)$
5	$subProperty(X, Y)$	$triple(X, rdfs{:}subPropertyOf, Y)$
6 (rdfs5)	$subProperty(X, Y)$	$triple(X, rdfs{:}subPropertyOf, Z), subProperty(Z, Y)$
7	$subClass(X, Y)$	$triple(X, rdfs{:}subClassOf, Y)$
8 (rdfs11)	$subClass(X, Y)$	$triple(X, rdfs{:}subClassOf, Z), subClass(Z, Y)$

RDFS entailment rules except the ones with blank nodes. We leave it as future work to consider various implications that these rules might have.

In our notation, arguments beginning with a capital letter (such as X and Y) denote variables, and arguments starting with a lowercase letter denote constants. Predicate names always start with a lowercase letter. Namespaces *rdf* and *rdfs* are the namespaces of the core RDF and RDFS vocabulary and will be used throughout the paper. To avoid confusion with the double meaning of the word *predicate*, from now on we will refer to the predicate of a triple with the word *property* and to a term of a datalog rule with the word *predicate*.

Rules 1-4 compute all possible instances of a class, rules 5 and 6 compute the transitive closure of an RDFS property hierarchy, and rules 7 and 8 compute the transitive closure of an RDFS class hierarchy. We note that all recursive rules above are *linear* (with at most one recursive predicate in their body) and *safe* (all variables appear as an argument in the rule bodies).

In the rest of the paper, we consider queries consisting of a *single* edb or idb predicate with arguments that are constants or variables. A formula of the form (s, p, o) where s, p and o can be URIs, literals or variables is called a *triple pattern*. We will use the equivalence of triple patterns (s, p, o) with p equal to *rdf:type*, *rdfs:subClassOf*, or *rdfs:subPropertyOf*, and idb predicates *type*, *subClass* and *subProperty*, to navigate freely among these two representations.

Let DB be an RDF(S) database. The *answer* to a query is defined as the answer to the same query posed over the logic program formed by the union of the triples in DB and the RDFS entailment rules of Table 1. We omit detailed formal definitions since they are very well understood [20,9].

3   Distributed RDFS Reasoning

Firstly, let us briefly describe the functionality of DHTs. DHTs are *structured* P2P systems which try to solve the *lookup problem*: given a data item x, find the node which holds x. Each node and each data item is assigned a unique m-bit identifier by using a hashing function such as SHA-1. The identifier of a node can be computed by hashing its IP address. For data items, we first have to compute a *key* and then hash this key to obtain an identifier *id*. The lookup problem is then solved by providing a simple interface of two requests; PUT(id, x) and GET(id). In Bamboo [22], when a node receives a PUT request, it efficiently routes the request to a node with an identifier that is numerically closest to *id* using a technique called prefix routing. This node is responsible for storing the data item *x*. In the same way, when a node receives a GET request, it routes it to the responsible node to fetch data item *x*. Such requests can be done in $O(log n)$ hops, where n is the number of nodes in the network.

Although for the implementation of our algorithms we used Bamboo [22], our algorithms are DHT-agnostic; they can be implemented on top of any DHT.

3.1   Storing Protocol

In both approaches, the same protocol is followed for storing RDF(S) triples. We have adopted the triple indexing algorithm originally presented in [7] where each triple is indexed in the DHT *three times*, once for its subject, once for its property and once for its object. Whenever a node receives a request to store a triple, it sends three DHT PUT requests using as key the subject, property and object respectively, and the triple itself as the item. The key is hashed to create the identifier that leads to the responsible node where the triple is stored. We call that node the *responsible node* for this key or identifier. Notice that a node responsible for a key which is a class name *C* (*responsible node for class C*), will have in its local database all triples that contain class *C* either as a subject or as an object (class *C* cannot be a property).

Since an RDF(S) database is actually a graph, we can exploit the fact that many of the triples share a common key (i.e., they have the same subject, property or object) and end up to be stored in the same node. So, instead of sending different PUT messages for each triple, we group them in a list *triples* based on the distinguished keys that exist, hash these keys to obtain identifiers and send a MULTIPUT(*id, triples*) message for each identifier. The node responsible for the identifier *id*, which receives this message, stores in its local database all triples included in the list *triples*.

3.2   Forward Chaining

The general idea of forward chaining (FC) is that all inferred triples are precomputed and stored in the network a priori. Each time a node receives a triple to be stored in its local database, it computes all inferred triples and sends them to the network to be stored.

Let us now introduce the notation that will be used in the algorithms description. Keyword **event** precedes every event handler for handling messages, while keyword **procedure** declares a procedure. In both cases, the name is prefixed by the node identifier in which the handler or the procedure is executed. Keywords **sendto** and **receive** declare the message that we want to send to a node with known either its identifier (thus DHT routing will be used) or the IP address, and the message we receive from a node respectively.

Figure 1 shows in pseudocode how the FC algorithm works. Suppose a PUT(*id, triple*) request arrives at node n and a new triple should be stored in the local database of n. First, node n retrieves from the local database all triples that have been stored under the identifier *id* and puts them in list *triples*. Then, it computes the inferred triples from this list according to the RDFS entailment rules using local function INFER(*triples*). For all newly inferred triples, three identifiers are created based on the subject, property and object of each triple, and three PUT requests are sent to the network. Each node holds a list *infTriples* with all inferred triples that it has computed, so that it can check when it reaches a fixpoint where no new triples are generated. The algorithm terminates when all nodes have reached a fixpoint.

The following proposition states that the FC algorithm is sound and complete. By *sound* we mean that if t is a new triple produced by FC and stored in the network, then t is entailed by the set of logical formulas formed by the RDF(S) database before the algorithm is executed, the input triple that fires the algorithm and the RDFS entailment rules of Table 1. By *complete* we mean that if a triple t is entailed by the set of logical formulas formed by the triples stored in the network before FC execution, the input triple that fires the algorithm and the RDFS entailment rules of Table 1, then t will eventually be generated by FC and will be stored in the network.

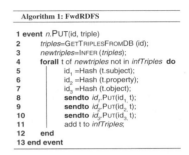

Fig. 1. FC algorithm

Proposition 1. *The FC algorithm is sound and complete.*

Proof (Sketch). The algorithm is sound since it is based on the RDFS entailment rules of Table 1. To prove that the algorithm is complete, we need to show that triples which are used to satisfy the body of a rule and generate a new triple will meet at the same node. If we check the rules of Table 1, we will see that predicates of rule bodies always have a common argument. Triples are indexed three times based on three identifiers, namely the hash values of their subject, property and object. Therefore, triples with a common part will meet at the node responsible for the identifier of this common part.   □

Note that the proof of the proposition depends on the assumption that no messages are lost due to network churn, i.e., nodes joining, leaving voluntarily or failing. Although the Bamboo DHT has many recovery mechanisms and can handle churn using various methods [22], we leave it as a subject of future work how our algorithms can be extended to deal with dynamic networks.

In order to evaluate a triple pattern after FC has taken place, we choose a *key* from the triple pattern and then hash it to create the identifier that will lead to the appropriate node. The key is the constant part of the triple pattern. When there is more than one constant parts, we select keys in the order "subject, object, property" based on the fact that we prefer keys with lower selectivity and the reasonable assumption that subjects or objects have more distinct values

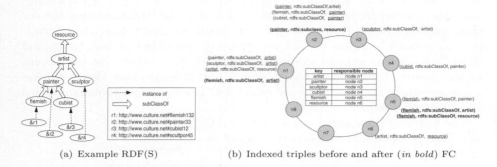

(a) Example RDF(S)   (b) Indexed triples before and after (*in bold*) FC

Fig. 2. Example for forward chaining

than properties. At the destination node, the triple pattern is checked against the local database and matching triples are found.

Since the RDFS entailment rules are highly-redundant [18], even a centralized forward chaining approach can be very expensive. As we will also show in the experimental evaluation of FC, the distributed version of FC results in generating more redundancies than expected in a centralized environment and leads to a significant increase in network traffic and load. Let us demonstrate that with an example. Figure 2(a) depicts a small RDFS class hierarchy of the cultural domain [17] with some sample instances populated underneath. Figure 2(b) shows the indexed triples that concern the subclass relation. Triples that are not in bold are initially inserted in the network. The key of each triple that led to a specific node is underlined. After two iterations of FC, both nodes n_1 and n_2 will result in generating the triple ($flemish, rdfs{:}subClassOf, resource$). This triple will be sent twice in the network to be stored. This could have been avoided in a centralized environment where all triples are stored in one local database.

3.3   Backward Chaining

In contrast to the data driven nature of FC, backward chaining (BC) starts from the given query and tries to find rules that can be used to derive answers. Thus, each time a node receives a request for evaluating a query, it should also use the RDFS entailment rules to compute the complete answer.

The challenge here is to construct an algorithm that can process recursive rules in a distributed environment such as DHTs. To achieve that, considering the RDFS entailment rules, it is helpful to transform the rules presented in Section 2 to a set of *adorned rules* that indicate which variables are bound and which are free. This is useful for finding the optimal order in which predicates should be evaluated.

We will extend the concept of rule adornment from recursive query processing [26] in order to exploit the distributed philosophy of DHTs. As already mentioned, in order to evaluate a triple pattern, a *key* has to be computed and then hashed to create the identifier that will lead to the responsible node. Therefore,

Table 2. Adorned RDFS Entailment Rules

	Head	Body
1a	$type^{kf}(X,Y)$	$triple^{kbf}(X, rdf{:}type, Y)$
1b	$type^{fk}(X,Y)$	$triple^{fbk}(X, rdf{:}type, Y)$
2a	$type^{kf}(X,Y)$	$triple^{kff}(X, P, Z),\ triple^{fbf}(P, rdfs{:}domain, Y)$
2b	$type^{fk}(X,Y)$	$triple^{fff}(X, P, Z),\ triple^{fbk}(P, rdfs{:}domain, Y)$
3a	$type^{kf}(X,Y)$	$triple^{ffk}(Z, P, X),\ triple^{fbf}(P, rdfs{:}range, Y)$
3b	$type^{fk}(X,Y)$	$triple^{fff}(Z, P, X),\ triple^{fbk}(P, rdfs{:}range, Y)$
4a	$type^{kf}(X,Y)$	$triple^{kbf}(X, rdf{:}type, Z),\ subClass^{ff}(Z, Y)$
4b	$type^{fk}(X,Y)$	$type^{ff}(X, Z),\ triple^{fbk}(Z, rdfs{:}subClassOf, Y)$
5a	$subProperty^{kf}(X,Y)$	$triple^{kbf}(X, rdfs{:}subPropertyOf, Y)$
5b	$subProperty^{fk}(X,Y)$	$triple^{fbk}(X, rdfs{:}subPropertyOf, Y)$
6a	$subProperty^{kf}(X,Y)$	$triple^{kbf}(X, rdfs{:}subPropertyOf, Z),\ subProperty^{ff}(Z, Y)$
6b	$subProperty^{fk}(X,Y)$	$subProperty^{ff}(X, Z),\ triple^{fbk}(Z, rdfs{:}subPropertyOf, Y)$
7a	$subClass^{kf}(X,Y)$	$triple^{kbf}(X, rdfs{:}subClassOf, Y)$
7b	$subClass^{fk}(X,Y)$	$triple^{fbk}(X, rdfs{:}subClassOf, Y)$
8a	$subClass^{kf}(X,Y)$	$triple^{kbf}(X, rdfs{:}subClassOf, Z),\ subClass^{ff}(Z, Y)$
8b	$subClass^{fk}(X,Y)$	$subClass^{ff}(X, Z),\ triple^{fbk}(Z, rdfs{:}subClassOf, Y)$

the corresponding predicate of the triple pattern has an argument that not only is bound, but it is also the *key* that led to the responsible node.

Definition 1. *An* adornment *of a predicate p with n arguments is an ordered string a of k's, b's and f's of length n, where k indicates an argument which is the key, b indicates a bound argument which is not the key, and f a free argument.*

Following this definition, a predicate p^a indicates which argument of p is the key, which ones are bound and which are free. Table 2 shows all possible adornments of the rules presented in Table 1.

Let us now describe the BC algorithm which is shown in Fig. 3. Suppose that a GETREQ request with unique identifier *rid* arrives at node n and a query should be evaluated. Node n firstly checks if the predicate corresponding to the triple pattern *tp* matches the head of any of the adorned rules. If no rule can be found, the triple pattern is simply checked against the node's local database and the bindings of the triple pattern's variables are returned to the node that made the request. In this case no backward chaining takes place. If there are rules that can be applied, local procedure BwdRDFS is called, which takes as an input the adorned predicate p^a and the request identifier *rid* and outputs a relation R which contains the bindings of the free arguments (i.e., the variables) of the predicate. These bindings form the answer to the query.

When BwdRDFS is called, the input predicate p^a is checked against the rules. Rules that can be applied to the predicate are added to the list *adornedRules*. Each rule can have one or two predicates in its body. Rules that have one predicate in their body (e.g., rule 1a) can always be evaluated locally since this predicate is an edb relation. In this case, node n calls local procedure MATCH-PREDICATE(p^a) and assigns to relation R the bindings of the predicate's variables that match the triples locally stored in its database.

Algorithm 2: BwdRDFS

```
1 event  n.GETReq (key, tp, rid) from  m          8 procedure  n.BwdRDFS (pᵃ, rid)
2    if no rule can be applied to the predicate of tp then    9    if rid in processedRequests return {};
              R=MATCH (tp);                       10      add rid to processedRequests ;
3    else                                         11      R=MATCHPREDICATE (pᵃ);
4      Let pᵃ be the adorned predicate of tp;     12      adornedRules =APPLYRULE(pᵃ);
5      R=BwdRDFS (pᵃ, rid);                        13      forall rules in adornedRules do
6        sendto  m.GetResp(R);                     14        r <-- REMOVEFIRST(adornedRules);
7 end event                                        15        if r has one predicate then break;
                                                   16        else
                                                   17          Let pₖ be the adorned predicate of r with a
31 event  n.BwdRDFSReq (pᵃ, rid) from  m                        k element in its adornment and pₗ the free
32    R=BwdRDFS (pᵃ);                                            predicate;
33      sendto  m.BwdRDFSResp (R)                  18          R' = MATCHPREDICATE (pₖ);
34 end event                                       19          if R' = {} then return R;
                                                   20          foreach value vi of the common variable Z in R' do
                                                   21            idᵢ =Hash (vᵢ);
                                                   22            rewrite pₗ to p'ᵢ ;
                                                   23            sendto  idᵢ.BwdRDFSReq (p'ᵢ)
                                                   24            receive  BwdRDFSResp (Rᵢ) from idᵢ
                                                   25            R = R U Rᵢ;
                                                   26          end
                                                   27        end
                                                   28      end
                                                   29      return R;
                                                   30 end procedure
```

Fig. 3. BC algorithm

For rules with two predicates in their body, we have to decide which predicate should be evaluated first. We select to evaluate first the predicate that can be processed locally. There always will be one such predicate since one of the arguments of the head predicate will be the key that led to the specific node. Therefore, there will be a body predicate (p_k) which has an adornment containing the letter k (always possible as seen in Table 2) and can be processed locally. Then, p_k is checked against the local database to find matching triples and the variable bindings are returned in a relation R'. By evaluating the first predicate locally, we have values that can be passed to the second predicate which is sent to be evaluated remotely at a different node. Notice in Table 2 that all rule bodies with two predicates have a single common variable (Z). Therefore, each tuple in relation R' will include a binding for this common variable Z. For each of these bindings (Z/v_i), node n rewrites the second predicate p_f to a new predicate p'_f where it has substituted the variable Z with its value v_i and made the corresponding letter of the adornment equal to k. Then, it sends a BwdRDFSReq message to the node responsible for the hashed value of key v_i. This part of the procedure is executed *in parallel* for each value v_i since the messages are sent to different nodes. Node n sends $|R'|$ number of messages (equal to the number of bindings found) and receives the responses asynchronously. When node n has collected all responses BwdRDFSResp(R_i), it adds the tuples of each R_i to relation R and returns R.

This procedure is recursive and terminates when the node that received the initial query has collected all responses. A recursion path ends when the predicate which is evaluated locally returns no bindings and therefore there are no values to pass to the second predicate. Cyclic hierarchies are handled by keeping a list of all processed requests (lines 9-10) so that an infinite loop is avoided.

The following proposition states that the BC algorithm is sound and complete. In this case, by *sound* we mean that if R is the relation with the bindings of a query q produced by BC and triple t is obtained from replacing the variables of the query q by their value in R, then t is entailed by the set of logical formulas formed by the RDF(S) database and the RDFS entailment rules. By *complete* we mean that if t is entailed by the set of logical formulas formed by the triples stored in the network and the RDFS entailment rules, then a relation R will be eventually produced, such that by replacing the variables of q by their values in R, triple t will be obtained. Similarly with FC, we make the same assumption regarding the stability of the network.

Proposition 2. *The BC algorithm is sound and complete.*

Proof (Sketch). The algorithm is sound since the answers are computed using the edb relations and the RDFS entailment rules. To prove that the algorithm is complete, it is important to stress the following. The local database of each node is part of the edb relation *triple*. The adorned predicate of a rule body that has the letter k is always the edb predicate *triple*. Therefore, these predicates will be evaluated locally according to the algorithm. Now, it is sufficient to show that each time an adorned predicate *triple* is checked against a node's local database *all* triples needed are found at this node. The adornment letter k indicates an argument of the predicate *triple* which is the key that was hashed and led to the specific node. This node is the responsible node for this key. Based on our indexing scheme all triples that contain this key either as a subject, property or object were sent to be stored to this node. As a result the edb relation of this node will include all triples that contain the specific key.   □

Figure 4 shows how the proof tree of backward chaining is distributed in the various nodes of the network using the example RDF(S) hierarchy of Fig. 2(a). We want to pose the query "Find all the instances of class artist", which is expressed as $(X, rdf{:}type, artist)$. Node n_1 responsible for key *artist* receives a request for evaluating this triple pattern. Now BC should take place by starting from the adorned predicate $type^{fk}(X, artist)$.

4   An Analytical Cost Model

In this section, we present an analytical cost model for both FC and BC. We will show in the experimental evaluation that our implementation follow the predictions of this cost model. We focus on a frequently used type of queries which asks for all the instances of a certain class in an RDFS hierarchy. Intuitively, this type of queries is the most expensive and most frequently used in RDFS reasoning, so we regard it as the most representative one for comparing our algorithms. However, the algorithms are able to answer any type of queries considered in the paper. Our results for class hierarchies trivially hold for property hierarchies too. We consider as future work the evaluation of more complex queries such as the ones of the LUBM benchmark [11].

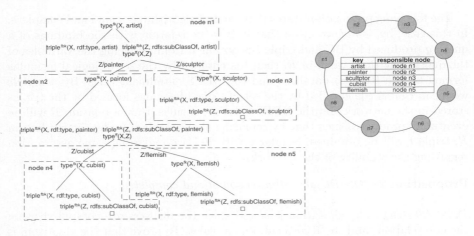

Fig. 4. Distribution of proof tree for backward chaining

We assume a complete tree-shaped RDFS class hierarchy of depth d and branching factor b as our RDF Schema (i.e., $(b^{d+1} - 1)/(b - 1)$ classes) with instances distributed under classes following either a uniform or a Zipfian distribution. RDF Schema triples are of the form $(C_k, rdfs:subClassOf, C_l)$ while RDF data triples are of the form $(r_j, rdf:type, C_k)$. In the analytical calculations presented below we start with an RDF(S) database. Then, we apply the FC and BC algorithms to answer a query of the type mentioned above, and estimate its cost. We will denote the number of RDF data triples (i.e., also total number of instances) in our initial database by T_d. Similarly, we use T_s to denote the number of RDF Schema triples in the initial database. For a uniform distribution, given the total number of instances (i.e., T_d), the number of instances under *each* class is $I_u = (T_d * (b - 1))/(b^{d+1} - 1)$. Considering a Zipfian distribution of instances with a skew parameter of 1, a class with rank r has $I_r = T_d/(r * h)$ instances where $h = \sum_{j=1}^{N} 1/j$ for N classes ($N = (b^{d+1} - 1)/(b - 1)$). Leaf classes are given a lower rank.

In the following, we constantly use the fact that the total number of subclasses of class C including itself is $(b^{d-\ell+1} - 1)/(b - 1)$ where ℓ is the level of the class. The proof is omitted due to space limitations. Furthermore, we utilize the fact that the reasoning and query answering algorithms for the type of queries we consider are essentially transitive closure computations.

4.1   Storage Cost Model

We first estimate analytically the costs associated with the storage of triples by both algorithms.

Storage load. We define as *storage load* the total number of triples that are stored in the network. In BC, the storage load (SL_b) is three times the number of RDF(S) triples that were inserted in the network based on our indexing scheme.

In FC, it is sufficient to compute the total number of triples that result from the transitive closure computations of the hierarchy (triples initially in the database plus inferred ones). Then, the storage load incurred in FC (SL_f) is three times this total number of triples.

Lemma 1. *The total number of triples generated after the computation of the transitive closure of the RDFS class hierarchy is* $T_s + \sum_{i=1}^{d} b^i(i-1) + [(b^{d+1})/(b-1)] - 2$.

Proof. For each level i of the tree, we have b^i classes, and for each class we infer $i - 1$ triples of the form $(C_k, rdfs:subClassOf, C_l)$ for the upper levels of the tree. Furthermore, we have the inferred triples (which are $[(b^{d+1} - 1)/(b-1)] - 1$) that state that all classes are subclasses of the $rdfs:Resource$ class.   □

Lemma 2. *The total number of triples generated after the computation of the transitive instances of the RDFS class hierarchy is* $T_d + I_u * \sum_{i=0}^{d} b^i(i+1)$.

Proof. Each class has I_u direct instances. For each level i of the tree, we have b^i classes, and for each class we infer i triples of the form $(r_j, rdf:type, C_k)$ for its superclasses plus the triple $(r_j, rdf:type, rdfs:Resource)$.   □

Based on these two lemmas, the sum of the formulas computed above is the total number of triples that result from the transitive closure computations of FC. Depending on the distribution of instances, the storage load of FC changes based on the number of instances per class (i.e., I_u and I_r). Table 3 summarizes the storage load of FC and BC for both kinds of instance distributions. For the Zipfian distribution we also made use of the following proposition. The proof is omitted due to space limitations.

Proposition 3. *Given a class with rank r in a Zipfian distribution of instances, the level of the class in the hierarchy is* $\ell_r = \lfloor \log_b((b - 1) * [(b^{d+1} - 1)/(b-1) - r + 1]) \rfloor$.

Store messages. We define as *store messages* the number of DHT messages sent for storing triples. In BC, the number of store messages sent (SM_b) is three times the number of triples stored and therefore it is equal to the storage load incurred. It is also independent from the instance distribution.

However, the total number of messages sent by FC is much larger than the storage load incurred as it was already illustrated in Section 3.2. Each node responsible for a certain class C that is in the i level of the class hierarchy generates for each instance i triples of the form $(r_j, rdf:type, C_k)$, where C_k is a superclass of class C, and one triple of the form $(r_j, rdf:type, Resource)$, and $i - 1$ triples of the form $(C, rdfs:subClassOf, C_l)$, where C_l is an ancestor class of class C. Note that messages are sent not only for each direct instance of each class but also for the instances of its subclasses. The number of messages sent in FC is depicted in Table 3 for both kinds of instance distribution.

4.2   Querying Cost Model

In this section, we calculate the cost of answering the query $(X, rdf:type, C)$ where class C is at level ℓ of the hierarchy.

Table 3. Storage cost summary table

Storage cost	Uniform	Zipfian
SL_b	$3 * (T_s + T_d)$	$3 * (T_s + T_d)$
SL_f	$3 * [T_s + T_d + \sum_{i=1}^{d} b^i(i-1) + [(b^{d+1})/(b-1)] - 2 + I_u * \sum_{i=0}^{d} b^i(i+1)]$	$3 * [T_s + T_d + \sum_{i=1}^{d} b^i(i-1) + \sum_{r=1}^{N} I_r * \ell_r]$
SM_b	$3 * (T_s + T_d)$	$3 * (T_s + T_d)$
SM_f	$3 * [T_s + T_d + \sum_{i=1}^{d} b^i(i-1) * ((b^{d-i+1} - 2)/(b-1)) + I_u * \sum_{i=1}^{d} b^i * (i+1) * ((b^{d-i+1} - 1)/(b-1))]$	$3 * [T_s + T_d + \sum_{r=1}^{N} I_r * \ell_r * (\ell_r + 1)/2]$

Query processing load. We define as *query processing load* of a node the number of triples that this node retrieves from its local database in order to answer a query.

The FC algorithm generates query processing load (QL_f) only at the node that is responsible for class C. This load is equal to the total number of instances of class C: $QL_f = ((b^{d-\ell+1} - 1) * I_u)/(b-1)$. On the other hand, BC will generate load at $(b^{d-\ell+1} - 1)/(b-1)$ nodes, namely the nodes that are responsible for the subclasses of class C including C. The load of each of these nodes (QL_b) is simply I_u. Note that the total query load is the same for both approaches.

For a Zipfian distribution, the number of instances of the class would be $I_r = T_d/(h * r)$, where the rank of the class is in the interval $[(b^d - 1)/(b-1) + 1, (b^{d+1} - 1)/(b-1)]$.

Query messages. We define as *query messages* the messages sent while answering a query. The case of FC is straightforward since just one message is sent to the node responsible for class C. In BC, the number of messages sent (QM_b) is as many as the number of the subclasses of class C. Therefore, we have: $QM_b = [(b^{d-\ell+1} - 1)/(b-1)] - 1$. The distribution of the instances does not affect the number of messages sent for the query answering.

5   Experimental Evaluation

In this section, we present an experimental evaluation of both FC and BC as we have implemented them on top of the Bamboo DHT [22]. Our goal is to evaluate the performance of FC and BC in a real distributed system and compare it with the analytical cost model of the previous section. We used as a testbed the PlanetLab network (http://www.planet-lab.org/) with 123 nodes that were available and lightly loaded at the time of the experiments.

The RDF(S) data we used was produced synthetically from the RBench generator[25]. The generator produces binary-tree-shaped RDFS class hierarchies parameterized on three different aspects: the depth of the tree, the total number of instances under the tree, and the distribution of the instances under the nodes of the tree. The queries we measure are queries that ask for all the transitive instances of the root class of the RDFS hierarchy. We used both uniform and Zipfian distribution of instances under the RDFS class hierarchy. In the Zipfian distribution, we used a skew parameter of value 1. Leaf classes were given a lower rank and therefore more instances of the lower level classes were generated.

Storing RDF(S) data. In this section, we measure the performance of both algorithms while storing RDF(S) data. Unless otherwise specified, in the experiments below we have generated and inserted in the network 10^4 instances uniformly distributed under an RDFS class hierarchy of varying depth.

Initially, we present results concerning the network traffic that is generated by the system measured in terms of both total number of messages sent in the network and bandwidth usage. As shown in Fig. 5(a), the number of messages sent by FC increases exponentially with the tree-depth while it remains constant in BC. In this experiment, we used the MULTIPUT functionality described in Section 3.1 for both approaches and thus the number of messages measured in the experiment is less than the number of messages computed by the analytical cost model. By using this functionality, we decrease significantly the number of messages sent by FC. Figure 5(b) shows how bandwidth is increasing exponentially with the tree-depth in FC while remaining constant in BC. When we increase the number of initially inserted triples from 10^3 to 10^4, we observe a huge increase in FC's bandwidth consumption which is also affected by the tree depth. Finally, in Fig. 5(c) the difference between a uniform and a Zipfian distribution of instances is depicted for both bandwidth (left y-axis) and messages (right y-axis). BC is not affected by the distribution and as a consequence the number of messages sent as well as the bandwidth consumption remain unchanged. However, FC's bandwidth usage deteriorates when a Zipfian distribution is used. The more skewed the instances are to the lower level classes, the more triples are needed to be sent to the upper level classes and thus the more bandwidth is spent.

Figure 5(d) shows the total storage load incurred in the network. BC's storage load is significantly lower than FC and is independent of the tree-depth. However, FC's storage load is increasing linearly with the tree-depth. In the same graph, we have included the total number of triples that are sent to be stored at various nodes during FC including the redundant ones generated as we have discussed in Section 3.2. The difference between these two measurements, named *fwd actual load* and *fwd generated load*, shows that the redundant information generated by FC is remarkable and increases significantly with the tree-depth. Figure 5(d) also shows that our cost model (*bwd load model* and *fwd load model* in the graph) precisely predicts the results obtained in the experiments.

In Fig. 5(e) and 5(f), we show the time needed by each approach to complete the insertion of 10^3 and 10^4 triples that either follow a uniform or a Zipfian distribution. In BC, this time represents the time needed until all inserted triples are stored at the respective nodes. In FC, we also take into account the time spent for the inferred triples to be stored in the network. We observe here that when we go from 10^3 to 10^4 initial triples, there is a blow up in FC's storage

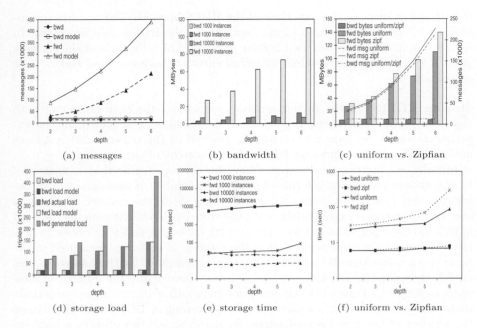

(a) messages   (b) bandwidth   (c) uniform vs. Zipfian

(d) storage load   (e) storage time   (f) uniform vs. Zipfian

Fig. 5. Storing RDF(S) data

time (y-axis is in logarithmic scale). Generally, FC needs an enormous amount of time to reach a fixpoint and makes the measurement of inserting 10^5 initial triples infeasible. Our experiment for inserting this number of triples using FC was active for *16 hours* and still a fixpoint had *not* been reached.

Finally, we conducted some measurements concerning the storage load distribution in both algorithms. The results showed that as the depth of the hierarchy increases, the number of classes increases exponentially and more nodes share the load resulting in a more balanced distribution. Furthermore, BC distributes the load slightly better than FC for larger tree depths. This is a result of a characteristic property of FC, namely that classes of higher levels of the hierarchy have more instances than classes from lower levels (since each class keeps all the instances of its subclasses). Therefore nodes that are responsible for classes of higher levels are more loaded with triples than nodes responsible for classes of lower levels. Due to space limitations we are not presenting these graphs and we consider as future work further improvements [16,4] in both approaches that will distribute the storage load more evenly among nodes.

Querying RDFS class hierarchies. In this experiment, we present results while evaluating queries of the form "give me all instances of the root class". We generated and stored 10^4 instances for both uniform and Zipfian distributions under an RDFS class hierarchy of varying depth. We run 100 queries of the above form and averaged the measurements taken.

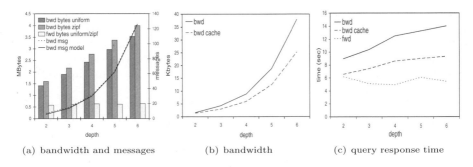

(a) bandwidth and messages   (b) bandwidth   (c) query response time

Fig. 6. Querying 10^4 instances

Figure 6(a) shows two metrics concerning the network traffic; bandwidth (left y-axis) and messages sent (right y-axis). Since FC sends a single message regardless of the tree-depth, we do not depict it in the graph. In BC, the number of messages sent in the network is analogous to the number of classes in the RDFS hierarchy since it sends one message for each class. Therefore, as the hierarchy becomes deeper, the number of classes increases and the total number of messages sent also increases. This was also shown by the analytical cost model of Section 4.2 and is depicted in the graph as well. Figure 6(a) also shows the bandwidth used in the network for both approaches and for both types of instance distribution. FC bandwidth consumption is limited to the number of bytes sent for the delivery of the results of the query and is independent of both the tree-depth and the instance distribution. In comparison, the bandwidth of BC increases with the tree-depth as a result of the increasing number of messages that are sent in the network. Furthermore, a skewed instance distribution slightly affects the bandwidth used since more instances belong to lower classes and need to be sent towards the root class.

We also experimented with the query load. As already shown in the analytical evaluation the total query load occurred is similar for both approaches and only the distribution differs. While FC involves a single node in the query processing, in BC the load is shared among the nodes that are responsible for the subclasses of the root class. Due to space limitations we have omitted these graphs. As ongoing work, we are exploring various load balancing techniques [16,4] for both approaches.

Optimizations. In this section, we measure the effect of a caching optimization technique for the BC algorithm. Since there is a need to traverse the subclass hierarchy quite often, we could take advantage of the first time it is traversed and cache useful routing information. Assuming that different nodes are responsible for different classes in a hierarchy, we need to make $d*O(logn)$ hops to reach from a node responsible for the root class of the hierarchy to the nodes responsible for the leaf classes, where d is the depth of the hierarchy tree and n the number of the nodes in the network. We can minimize this by adding extra routing information to each node x which is responsible for a certain class C of the hierarchy. The first time a node x contacts a node y which is responsible for a subclass of C, it keeps the IP address of y and uses it for further communication. In this way, a direct subclass is found in just 1 hop and the whole hierarchy is traversed in d hops. This technique minimizes network traffic and decreases query response time, while the overhead of maintaining such a table is not significant and it is only kept in memory.

In Fig. 6(b), we show how the bandwidth of BC is decreased when using this caching technique. Finally, Figure 6(c) shows the query response time. A query needs an almost constant time to be evaluated when in FC. This is reasonable, considering that only one node participates in the query processing, although the query processing it needs to handle is quite heavy. On the contrary, BC response time increases with the tree-depth. However, using the caching technique we manage to improve query response time at a satisfying degree.

6   Related Work

A representative centralized RDF store that supports the forward chaining approach is Sesame [6]. Each time an RDF Schema is uploaded in Sesame, an inference module computes the schema closure of the RDFS and asserts the inferred RDF statements. Jena [27] can support both approaches depending on the underlying rule engine. RSSDB [2] follows a totally different approach in which the taxonomies are stored using the underlying DBMS inheritance capabilities so that retrieval is more efficient. Nevertheless, this approach is still an *on demand* approach and resembles the backward chaining evaluation algorithm. 3store [12] follows a hybrid approach in order to gain from the advantages of both approaches. Finally, in Oracle RDBMS [8], RDFS inference is done at query execution time using appropriate SQL queries.

From the distributed point of view, RDFPeers [7], GridVine [1], and [19] consider RDF query processing on top of structured overlay networks, such as DHTs. [7] and [19] consider no RDFS reasoning while [1] provides semantic interoperability through schema mappings. The only DHT-based RDF store that is closely related with our work is BabelPeers [5] which enables RDFS reasoning. It is implemented on top of Pastry [23] and only a forward chaining approach is supported. The reasoning process runs in regular intervals on each node and checks for new triples that have arrived to the node. Then, it exhaustively generates new inferred triples based on the RDFS entailment rules and sends them to be stored in the network. [5] presents no experimental evaluation of the forward chaining algorithm. However, the results of our experiments show how expensive FC is in terms of storage load, time and bandwidth.

7   Conclusions and Future Work

We presented and evaluated both analytically and experimentally two algorithms, namely forward and backward chaining, that enable RDFS reasoning on top of DHTs. We proposed the first distributed backward chaining algorithm on top of DHT-based RDF stores together with an optimization which improves significantly query response time. A main result of our experiments is that a simple forward chaining implementation as it is also supported in BabelPeers [5] cannot scale and thus related optimizations should be considered.

In future work, we plan to evaluate exhaustively the query processing algorithms of Atlas [15,19] taking into account RDFS reasoning as presented in this paper and examine how various load balancing techniques [4,16] can be applied to our algorithms. We also plan to investigate how forward chaining can be made more efficient as well as hybrid approaches that combine forward and backward chaining.

References

1. Aberer, K., Cudre-Mauroux, P., Hauswirth, M., Pelt, T.V.: GridVine: Building Internet-Scale Semantic Overlay Networks. In: WWW 2004 (2004)
2. Alexaki, S., Christophides, V., Karvounarakis, G., Plexousakis, D.: On Storing Voluminous RDF Descriptions: The case of Web Portal Catalogs. In: WebDB 2001 (2001)
3. Balakrishnan, H., Kaashoek, M.F., Karger, D.R., Morris, R., Stoica, I.: Looking up data in P2P systems. Communications of the ACM (2003)
4. Battre, D., Heine, F., Hoing, A., Kao, O.: Load-balancing in P2P based RDF stores. In: SSWS 2006 (2006)
5. Battre, D., Hoing, A., Heine, F., Kao, O.: On Triple Dissemination, Forward-Chaining, and Load Balancing in DHT based RDF stores. In: Moro, G., Bergamaschi, S., Joseph, S., Morin, J.-H., Ouksel, A.M. (eds.) DBISP2P 2005 and DBISP2P 2006. LNCS, vol. 4125. Springer, Heidelberg (2007)
6. Broekstra, J., Kampman, A.: Sesame: A Generic Architecture for Storing and Querying RDF and RDF Schema. In: Horrocks, I., Hendler, J. (eds.) ISWC 2002. LNCS, vol. 2342. Springer, Heidelberg (2002)
7. Cai, M., Frank, M.: RDFPeers: a scalable distributed RDF repository based on a structured peer-to-peer network. In: WWW 2004 (2004)
8. Chong, E.I., Das, S., Eadon, G., Srinivasan, J.: An Efficient SQL-based RDF Querying Scheme. In: VLDB 2005 (2005)
9. Cyganiak, R.: A relational algebra for SPARQL, Technical Report (2005)
10. Guo, Y., Pan, Z., Heflin, J.: An Evaluation of Knowledge Base Systems for Large OWL Datasets. In: McIlraith, S.A., Plexousakis, D., van Harmelen, F. (eds.) ISWC 2004. LNCS, vol. 3298, pp. 274–288. Springer, Heidelberg (2004)
11. Guo, Y., Pan, Z., Heflin, J.: LUBM: A Benchmark for OWL Knowledge Base Systems. J. Web Sem. 3(2-3), 158–182 (2005)
12. Harris, S., Gibbins, N.: 3Store: Efficient Bulk RDF Storage. In: PSSS 2003 (2003)
13. Hayes, P.: RDF Semantics. W3C Recommendation (February 2004)
14. Heine, F., Hovestadt, M., Kao, O.: Processing Complex RDF Queries over P2P Networks. In: P2PIR 2005 (November 2005)
15. Kaoudi, Z., Miliaraki, I., Magiridou, M., Liarou, E., Idreos, S., Koubarakis, M.: Semantic Grid Resource Discovery in Atlas. Knowledge and Data Management in Grids (2006)
16. Karger, D.R., Ruhl, M.: Simple efficient load balancing algorithms for peer-to-peer systems. In: SPAA 2004 (2004)
17. Karvounarakis, G., Alexaki, S., Christophides, V., Plexousakis, D., Scholl, M.: RQL: A Declarative Query Language for RDF. In: WWW 2002 (2002)

18. Lassila, O.: Taking the RDF Model Theory Out For a Spin. In: Horrocks, I., Hendler, J. (eds.) ISWC 2002. LNCS, vol. 2342. Springer, Heidelberg (2002)
19. Liarou, E., Idreos, S., Koubarakis, M.: Evaluating Conjunctive Triple Pattern Queries over Large Structured Overlay Networks. In: Cruz, I., Decker, S., Allemang, D., Preist, C., Schwabe, D., Mika, P., Uschold, M., Aroyo, L.M. (eds.) ISWC 2006. LNCS, vol. 4273, pp. 399–413. Springer, Heidelberg (2006)
20. Polleres, A.: From SPARQL to Rules (and back). In: WWW 2007 (2007)
21. Prud'hommeaux, E., Seaborn, A.: SPARQL Query Language for RDF
22. Rhea, S., Geels, D., Roscoe, T., Kubiatowicz, J.: Handling Churn in a DHT. In: USENIX Annual Technical Conference (2004)
23. Rowstron, A., Druschel, P.: Pastry: Scalable, Distributed Object Location and Routing for Large-Scale- Peer-to-Peer Storage Utility. In: Guerraoui, R. (ed.) Middleware 2001. LNCS, vol. 2218. Springer, Heidelberg (2001)
24. Stuckenschmidt, H., Broekstra, J.: Time-Space Trade-offs in Scaling up RDF Schema Reasoning. In: WISE Workshop 2005 (2005)
25. Theoharis, Y., Christophides, V., Karvounarakis, G.: Benchmarking Database Representations of RDF/S Stores. In: Gil, Y., Motta, E., Benjamins, V.R., Musen, M.A. (eds.) ISWC 2005. LNCS, vol. 3729. Springer, Heidelberg (2005), http://athena.ics.forth.gr:9090/RDF/RBench/
26. Ullman, J.D.: Principles of Database and Knowledge-Base Systems, vol. I. Computer Science Press (1988)
27. Wilkinson, K., Sayers, C., Kuno, H.A., Raynolds, D.: Efficient RDF Storage and Retrieval in Jena2. In: SWDB 2003 (2003)

An Interface-Based Ontology Modularization Framework for Knowledge Encapsulation

Faezeh Ensan and Weichang Du

Faculty of Computer Science,
University of New Brunswick, Fredericton, Canada
{faezeh.ensan,wdu}@unb.ca

Abstract. In this paper, we present a framework for developing ontologies in a modular manner, which is based on the notions of interfaces and knowledge encapsulation. Within the context of this framework, an ontology can be defined and developed as a set of ontology modules that can access the knowledge bases of the others through their well-defined interfaces. An important implication of the proposed framework is that ontology modules can be developed completely independent of each others' signature and language. Such modules are free to only utilize the required knowledge segments of the others. We describe the interface-based modular ontology formalism, which theoretically supports this framework and present its distinctive features compared to the exiting modular ontology formalisms. We also describe the real-world design and implementation of the framework for creating modular ontologies by extending OWL-DL and modifying the Swoop interfaces and reasoners.

1 Introduction

OWL-DL has been well established and widely used in the recent years as an expressive description logic based language for representing ontologies. Nonetheless, several challenges still exist in efficiently creating large-scale OWL-DL ontologies specially for complex domains. Developing a large monolithic ontology can lead to performance difficulties in reasoning, management challenges when some parts of the ontology changes based on new domain requirements, and also issues in ontology integration when several parts of an ontology have been developed by different groups of experts.

Recently, the development of ontologies in a modular manner has been proposed to address the above mentioned issues [19]. The idea of modularization can also be seen in the software engineering field and mainly in Object Oriented design where complex software systems are modeled as a set of self-contained components [18]. The behavior of these components are defined through their interfaces which are separated from their later detailed implementation. Consequently, components can utilize each others' functions without being consciously aware of each other detailed implementation. Then implementation changes can occur even after logical component inter-connections have been specified.

A. Sheth et al. (Eds.): ISWC 2008, LNCS 5318, pp. 517–532, 2008.

Due to their perceived advantages, a considerable amount of work has been dedicated to creating new formalisms for ontologies which support developing ontologies in a modular manner. Distributed Description Logics (DDL) [3], Package Based Description Logics P-DL) [1], E-connections [14] and Semantic Import [2] are among such formalisms. DDL defines a modular ontology as a set of ontology modules which are connected through 'bridge rules'. A bridge rule forms a mapping between two concepts of two different ontology modules. Further, Context OWL (C-OWL) [4] has been introduced as an extension to OWL to support the syntax and semantics of such bridge rules. E-connections is another formalism, which supports ontology modularization by introducing a new type of roles (called links) whose domain and range belong to different ontology modules. The authors in [10] describe their extension to OWL for supporting E-connections 'links' and also the modification of the Pellet reasoning engine to support its semantics. The Semantic Import and P-DL formalisms allow ontology modules to import the knowledge base elements of each other.

Despite these numerous efforts, it seems that ontology modularization is still far from the level of maturity needed to be accepted as an established method for developing ontologies. For instance, E-connections is based on a limiting precondition that the domain of the ontology modules need to be completely disjoint. DDL restricts mappings to ontology concepts and does not support role mappings or creating complex concepts using foreign terms. The decidability of consistency checking in P-DL in its current form can only be proven when it restricts importing terms to concepts and also when all of the component modules are in the same description logics language [1]. Moreover, as it is shown in [9] P-DL consists of some ambiguities in its introduced semantics such that for performing a reasoning task on a module, the union of the knowledge bases of all modules should be processed. Semantic Import provides reasonable expressiveness that allows a module to use the others' terms in its complex concepts, however, the properties of the formalization is being discussed under the assumption that a module imports all of the symbols of the others [2].

In addition to the technical issues discussed above, there is a missing feature in the existing formalisms, that is support for 'knowledge encapsulation', which would benefit ontology modularization. By knowledge encapsulation we mean providing support for ontology modules to define their main contents using well-defined interfaces, such that their knowledge bases can only be accessed by other modules through these interfaces. The advantages of knowledge encapsulation in ontology design and development can be enumerated as follows:

- Since ontology modules are connected indirectly through their interfaces, they can evolve independent of each others' signature and knowledge bases. While the interfaces of a module do not change, its entire knowledge base can change without requiring other connected modules to change their signatures.
- An ontology module can express its knowledge content through different interfaces with different levels of complexity and completeness. Hence, modules can access those parts of an ontology module they need without being required to go through the complicated knowledge base.

- Using interfaces, the specification of the knowledge of an ontology module and the exact meaning of these content can be separated. Consequently, an ontology module can provide new meaning for a concept which is used by other modules through its interfaces.

In this paper, we propose a new framework for developing ontologies in a modular manner based on an interfaces-based formalism. This framework supports a type of knowledge encapsulation that allows ontology modules to define various interfaces through which other ontology modules can access their knowledge base. Based on this interface-based formalism, a modular ontology is defined as a set of ontology modules, which can be developed independent of each others' language and signature. Furthermore, the formalism addresses the technical issues existing in the current ontology modularization proposals. The interface-based formalism enjoys a great expressiveness power, which allows a module to create its knowledge base from the other modules' knowledge expressed through their interfaces. At the same time, it allows for partial reuse, i.e., it lets a module use only the necessary parts of the knowledge base of the other modules. In addition, the consistency checking of a modular ontology is decidable even though the modules are in different description languages.

The rest of this paper is organized as follow: Section 2 provides preliminaries regarding basics of description logics and also epistemic queries. Section 3 introduces the syntax and semantics of the proposed interface-based formalism. Section 4 presents the important features of the formalism. Section 5 illustrate the implementation of the framework for OWL ontologies. Section 6 discusses related works and finally, section 7 concludes the paper.

2   Preliminaries

OWL-DL provides an expressiveness equivalent to the $\mathcal{SHOIN}(D)$ Description Logic (DL). A DL knowledge base is defined as $\Psi = \langle \mathcal{T}, \mathcal{A} \rangle$, where $\mathcal{T}$ denotes TBox and comprises of a set of general inclusion axioms and A stands for ABox and comprise of a set of instance assertions. The signature of an ontology is defined as a set of all concept names (C_N), role names (R_N) and individulas (I_N) which are included its knowledge base. The semantic of a DL is defined by an interpretation $\mathcal{I} = (\Delta^{\mathcal{I}}, \cdot^{\mathcal{I}})$ where $\Delta^{\mathcal{I}}$ is a non-empty set of individuals and $\cdot^{\mathcal{I}}$ is a function which maps each $C \in C_N$ to $C^{\mathcal{I}} \subseteq \Delta^{\mathcal{I}}$, each $R \in R_N$ to $R^{\mathcal{I}} \subseteq \Delta^{\mathcal{I}} \times \Delta^{\mathcal{I}}$ and each $a \in I_N$ to an $a^{\mathcal{I}} \in \Delta^{\mathcal{I}}$. An interpretation $\mathcal{I}$ satisfies a TBox axiom $C \sqsubseteq D$ iff $C^{\mathcal{I}} \subseteq D^{\mathcal{I}}$, satisfies an ABox assertion $C(a)$ iff $a^{\mathcal{I}} \in C^{\mathcal{I}}$ and an ABox assertion $R(x, y)$ iff $\langle x^{\mathcal{I}}, y^{\mathcal{I}} \rangle \in R^{\mathcal{I}}$. An interpretation $\mathcal{I}$ is a model of a knowledge base Ψ if it satisfies every TBox axiom and ABox assertion of Ψ. A knowledge base is consistent iff it has a model. A concept C is *satisfiable* if there is a model $\mathcal{I}$ for Ψ such that $C^{\mathcal{I}} \neq \emptyset$.

DL ontologies are based an open-world assumption, i.e., the knowledge base of an ontology is not considered to be complete and consequently if something cannot be proven, it cannot be assumed to be false based on the knowledge base.

Nevertheless, there have been some proposals in the literature that attempt to augment the semantics of DLs with close-world reasoning capabilities. Epistemic operator K is introduced in [7] and allows queries whose result can be captured by the closed-world assumption approach. K queries ask about the facts that are known to be true with the extent of information available in the current knowledge of a given knowledge base. [5] investigates mechanisms for posing K epistemic queries to expressive DL knowledge bases. [17] shows the capability of the Pellet reasoning engine for answering K queries for concepts and roles that are posed to simple knowledge bases.

To have a formal understanding of K queries, let C be a concept in a description logic knowledge base Ψ, KC reports a set of individuals which are known to belong to C in every model of Ψ. An *epistemic interpretation* for Ψ is defined as $\mathfrak{I} = (\mathcal{J}, \mathcal{M})$, where $\mathcal{J}$ is an interpretation of Ψ with the domain Δ, and $\mathcal{M}$ is a set of interpretations for Ψ over Δ. The epistemic interpretation for simple epistemic concepts and roles are defined as below:

$(KC)^{\mathfrak{I}} = \bigcap_{j \in \mathcal{M}} (C)^j$

$(KR)^{\mathfrak{I}} = \bigcap_{j \in \mathcal{M}} (R)^j$

An epistemic model for a knowledge base Ψ is a maximal non-empty set $\mathcal{M}$ such that for every $\mathcal{J} \in \mathcal{M}$, $(\mathcal{J}, \mathcal{M})$ satisfies all TBox inclusion axioms and ABox assertions of Ψ. Consider an epistemic query $KC(x)$ posed to a knowledge base Ψ, $\Psi \vDash KC(x)$ if for every epistemic model $\mathfrak{I} = (\mathcal{J}, \mathcal{M})$ for Ψ, $x \in KC^{\mathfrak{I}}$. An epistemic query $KR(x, y)$ is also defined in the same way as $\Psi \vDash KR(x, y)$ if for every epistemic model $\mathfrak{I} = (\mathcal{J}, \mathcal{M})$ for Ψ, $(x, y) \in KR^{\mathfrak{I}}$.

3   An Interface-Based Framework for Modular Ontologies

3.1   Formalization

In this section, we introduce the interface-based formalization of modular ontologies and highlight its main features using the following example.

Example 1. *Consider a case where we want to develop an ontology for the tourism domain. We have found an ontology describing different places in Canada and an ontology which covers North America. We desire to utilize these existing ontologies rather than gathering and categorizing geographical information regarding these places in the tourism ontology from scratch. In addition, we want to have a reliable way to understand and use the main features of the these ontologies without being required to go through their knowledge bases and figure out their taxonomies and axioms.*

Furthermore, suppose that we introduce the Sightseeing *concept as a notion to represent the places where tourists are interested to visit. Each sight needs to have a name and a specific address, so that a tourist can easily locate it. However, it is possible to specialize the definition of a sight from different perspectives. For example describing from a scientific perspective, a sight refers to a place that exhibits scientific value. This place may relate to one or more branches of science and can be visited by various scientists. On the other hand, from the natural perspective, a sight relates to a place with natural attractions, such as beaches, mountains, parks and jungles. Visiting the places in this category is most suitable in certain seasons of the year and the visitors may need to take specialized equipments with them to be able to enjoy their visit. In the tourism ontology, we desire to only know the common sense of sightseeing concept, while its different specializations can be later bound to it based on different requirements.*

Through the interface-based modular ontology formalism, a *'modular ontology'* is defined as a set of *'ontology modules'* (modules) and *'interfaces'*. An interface is a set of knowledge base expressions, which are used in a module but their exact meaning are provided by other ontology modules. An ontology module may utilize or realize a set of interfaces. Referring to Example 1 we can define a modular ontology as a set of ontology modules: Tourism and Canada Destination, where the Canada Destination ontology module 'realizes' the interface concept `Place` and provides its meaning, properties and instances. The interface concept `Place` would be 'utilized' by the Tourism ontology. Definitions 2 and 3 give the formal specifications of interfaces and modules in the proposed formalism.

Definition 2. *An interface I is defined as $I = \langle C_N, R_N, \mathcal{T} \rangle$ where $\mathcal{T}$ is the TBox of the interface and C_N and R_N are sets of concept and role names used in $\mathcal{T}$. I has no ABox assertions. We say an interface I' extends I if $C_N^{I'} \sqsupseteq C_N^I$ and $R_N^{I'} \sqsupseteq R_N^I$ or $\mathcal{T}^{I'} \equiv \mathcal{T}^I \sqcup \alpha$ where α is a set of general inclusion axioms defined using the signature of I'.*

It is easy to see that if I' extends I and I'' extends I', I'' also extends I. We use $Exd(I)$ to denote a set of all interfaces that extends I.

Definition 3. *An ontology module M is defined as $M = \langle \Psi, I_r, I_u \rangle$ where Ψ is the knowledge base and I_r is a set of all interfaces which is realized by M and I_u is a set of all interfaces which is utilized by M. M can be in any description logic language, but it must support nominals.*

We define a module M as consistent with regards to its interfaces iff $\Psi \cup (\bigcup_{i \in I_r} \mathcal{T}_i) \cup (\bigcup_{j \in I_u} \mathcal{T}_j)$ is consistent. A module which utilizes or realizes an interface must be consistent regarding to it. Let P be a concept or role name in an interface I, it is referred in the knowledge base of modules as $I : P$. A module M realizes an interface I, either if $I \in I_r$ or there is an $i \in Exd(I)$ such that $i \in I_r$.

Given an interface, we refer to the module which uses it as utilizer module and the module which gives semantics to its terms as realizer module.

A module which utilizes an interface needs to access the instances provided by the realizer modules. In the interface-based formalism, we follow a query-based approach to augment the semantic of a utilizer module with the individuals provided by the realizer modules. For instance, regarding Example 1, the Tourism ontology module may pose a query to the Canada Destination ontology on the location of an accommodation. Through the proposed formalism, the Tourism KB is augmented with the individuals that are provided by Canada Destination for the Place concept. This augmentation approach brings considerable advantages for the framework. First, after augmenting a module with appropriate individuals from other modules, reasoning engines are not required to take into account the other modules' knowledge bases anymore. Analogous to the idea of knowledge compilation which is employed in [19] for DDL, knowledge augmentation leads to local reasoning instead of reasoning on external modules, which entails lower time complexity. Secondly, The augmentation process does not pose any limitations on the semantics of the module which realizes an interface. This module's semantics can be changed independently of the semantics of those modules which utilize its interfaces. In other words, those modules which require an interface are dependent on the interface of the realizer modules but not vice versa.

In order to augment the domain of a utilizer module, the proposed formalism uses epistemic queries to retrieve the individuals of interfaces' concepts and roles from the realizer module. The hypothesis behind this approach is that a utilizer module looks at the realizers as black-boxes whose knowledge about the interface terms are compete enough for reasoning. As an explanatory example assume that in the case of Example 1, the Tourism ontology uses an interface concept `SingleLingualCity` which refers to those cities that have only one official language:

`SingleLingualCity` $\equiv$ `City` $\sqcap$ ≤ 1 `hasOfficialLang.Language`

North America Destination ontology realizes this interface concept with the following expressions in its knowledge base:

```
City(New York),
hasOfficialLang(New York, English),
Language(English)
```

From the point of view of the Tourism ontology, the knowledge base of North America is complete for reasoning about places and cities, so New York will be recognized as a `SingleLingualCity`. Observe that without using epistemic queries, `SingleLingualCity` would not match any instance from the realizer module.

Based on the definition of interfaces and modules, we now define a modular ontology as follows:

Definition 4. *A modular ontology is a triple* $O = \langle M, I, F \rangle$ *where* M *is a set of ontology modules,* I *is a set of interfaces whose description logic is less or equally expressive with regards to the description logic of the ontology modules (description constructors of any given interface is the subset of the description constructors of any ontology module) and* F *is a configuration function* $F :$ $M \times I \rightarrow M$ *which chooses one realizing module for every utilizer module-Interface pair.* $F(M, I) = M'$ *if:*

(c1) $I \in I_u^M$ and ($I \in I_r^{M'}$ or there is an $i \in Exd(I)$ such that $i \in I_r^{M'}$)
(c2) M and M' are consistent regarding I and i
(c3) Let C_i and R_j be the result sets of queries $K I : C_i$ and $K I : R_j$ posed to M', $\Psi_M \bigcup_{C_i \in I} (I : C_i \equiv C_i) \bigcup_{R_j \in I} (I : R_j \equiv R_j)$ is consistent.

A path PATH in an modular ontology is defined as a set of modules which are connected through the configuration function F. PATH(M) specifies the path which includes module M.

Based on Definition 4, the final form of a modular ontology is specified by the configuration function F. This function shows the connected modules thorough interfaces and its value can be set at configuration time. Introduction of the configuration function F in Definition 4 implies the development and configuration times of a modular ontology are distinguishable. The development time is when an ontology module is developed, its necessary interfaces as well as those interfaces that it realizes are specialized. The configuration time is the time when the required modules are selected to realize the interfaces of a particular module. (e.g. someone may develop the Tourism ontology through the proposed framework and specify that it needs the place interface concept. However, at configuration time it would be finalized whether the Canada ontology or the North America ontology will realize the place concept).

For being connected through the configuration function, two ontology modules should satisfy the three conditions mentioned in Definition 4. First of all a module should realize an interface or one of its extension in order to be selected by the configuration function and be connected to the utilizer module. Secondly, two modules should be consistent with regards to their interfaces. And finally, the third condition ensures that the integration of two modules does not entail inconsistencies. Since the domain of the utilizer module would be augmented by the individuals of the interface terms from the realizer modules through epistemic K queries, condition three ensures that this augmentation does not lead to an inconsistency in the utilizer module. Example 5 shows the formal representation of the situation which is described in Example 1.

Example 5. *Regarding Example 1, Let 'Tourism' be an ontology module which utilizes the interface 'Location'. Furthermore let the signature and TBox of them be as follows:*

$C_N^{Tourism}$ ={Accommodation}
$R_N^{Tourism}$ ={hasAddress}
$C_N^{Location}$ ={Place}

The ontology module Tourism can use the interface terms for creating complex concepts and for defining general inclusion axiom. For instance the TBox of the tourism ontology has the following axiom:

Accommodation⊑ ∃hasAddress.Location:Place.

Let 'Canada Destination' (CD) and 'North America Destination' (NAD) be two ontology modules which realize the interface Location. Two different values for the configuration F lead to two different modular ontologies O1 and O2 as follows:

$O1$={ { Tourism, CD, NDA},{Location}, $F1$ },
*where F1(*Tourism,Location*)=CD*
$O2$={ { Tourism, CD, NDA},{Location}, $F2$ },
*where F2(*Tourism,Location*)=NDA*

3.2   Augmented Semantics

To give a formal specification for the notion of augmentation for an ontology module, we define an augmentation function as follows:

Definition 6. *Let $PTBox(M)$ be a set of all axioms in the TBox of all interfaces in $PATH(M)$ for a given ontology module M, an augmentation function $Aug : M \rightarrow M$ is a function such that $\Psi^{Aug(M)}$ is defined as the union of the following elements:*

   (i) $\mathcal{T}^M$,
   (ii) $PTBox(M)$,
   (iii) $\{I : c_1\} \sqcup \ldots \sqcup \{I : c_n\}$ where c_i is a member of the result set of $KI : C$ posed to $F(M, I)$ for all concepts C in all $I \in I_u^M$,
   (iv) $\{I : x_1\} \sqcup \ldots \sqcup \{I : x_n\} \sqcup \{I : y_1\} \sqcup \ldots \sqcup \{I : y_m\}$ where $\langle x_i, y_j \rangle$ is a member of the result set of $KI : R$ posed to $F(M, I)$ for all roles R in all $I \in I_u^M$.

Based on Definition 6, the result set of the epistemic queries posed to the realizer module is being inserted to the knowledge base of the utilizer module as new nominals. The following definition gives the exact semantic of the augmented module.

Definition 7. *An augmented semantics for a module M_j in a modular ontology $O = \langle M, I, F \rangle$, is defined as $\mathcal{I}_j = (\triangle^{\mathcal{I}_j}, \cdot^{\mathcal{I}_j})$, where $\triangle^{\mathcal{I}_j}$ is a non-empty domain for $Aug(M_j)$ and a mapping function $\cdot^{\mathcal{I}_j}$ which maps each concept of $Aug(M_j)$ to a subset of $\triangle^{\mathcal{I}_j}$, each role of $Aug(M_j)$ to a subset $\triangle^{\mathcal{I}_j} \times \triangle^{\mathcal{I}_j}$ and each individual name from $Aug(M_j)$ to an element $a^{\mathcal{I}} \in \triangle^{\mathcal{I}_j}$. $\cdot^{\mathcal{I}_j}$ maps M_j concept expressions based on the semantic of concept constructors of M_j. For the concepts and roles of the utilized interfaces, the function develops mapping as follows:*

   (i) For every interface concept $I : C$, $x \in (I : C)^{\mathcal{I}_j}$ iff $\{x\}^{\mathcal{I}_j} \subseteq \triangle^{\mathcal{I}_j}$ and $\Psi_k \models KC(x)$, where KC is an epistemic query posed to $M_k = F(M_j, I)$,
   (ii) For every interface role $I : R$, $\langle x, y \rangle \in (I : R)^{\mathcal{I}_j}$ iff $\{x\}^{\mathcal{I}_j} \subseteq \triangle^{\mathcal{I}_j}$ and $\{y\}^{\mathcal{I}_j} \subseteq \triangle^{\mathcal{I}_j}$ and $\Psi_k \models KR(\langle x, y \rangle)$, where KR is an epistemic query posed to $M_k = F(M_j, I)$.

An ontology module is *augmentedly consistent* if there is an augmented interpretation $\mathcal{I}$ (augmented model), which satisfies all axioms and assertions in $\Psi_{Aug(M)}$. Let α be an inclusion axiom or an ABox assertion $\Psi_{Aug(M)} \models \alpha$ if for every augmented model $\mathcal{I}$, α is satisfied by $\mathcal{I}$. For a concept expression α, $\Psi_{Aug(M)} \models \alpha$ if for every augmented model $\mathcal{I}$, α is satisfiable.

In order to augment the knowledge base of a module, we exploit the notion of nominals. The conditions (i) and (ii) of the Definition 7 ensure that the interface concepts and roles in a utilizer module is interpreted the same as their interpretation in the realizer modules. For instance in the situation described in Example 5, suppose the realizer module expresses that {Toronto, Montreal, Vancouver} individuals are of the type of the Place concept. These places are inserted as nominal to the domain of the Tourism ontology module and the concept `Location:Place` is interpreted in such way to be equal to {`Toronto`, `Montreal`, `Vancouver`}.

Let us make two remarks about the proposed semantics. First, since the augmented semantic is defined for an ontology module, the $\top$ and $\neg$ symbols in a modular ontology are interpreted from the point of view of each augmented module. For example $\neg I : C$ in a utilizer module M_i is interpreted as $\triangle^{\mathcal{I}_i} \setminus (I : C)^{\mathcal{I}_i}$ when $I : C$ has been interpreted to be equal to the result set of $KI : C$ posed to $F(M_i, I)$. Second, we do not make a *unique name assumption* and hence two nominals may refer to the same individual; therefore, an inserted nominal to a module can be interpreted to be equivalent to an existing nominal in its knowledge base.

4   Properties of the Interface Based Formalism

In this section we describe the significant properties of the proposed modularization framework. Initially we point out two features, 'directed semantic' [1] and 'polymorphism' which are driven from the interface base nature of the formalism. Secondly we prove decidability of the formalism and its capability to propagate the logical consequences of the public parts of inter-connected modules to each others.

4.1   Directed Semantic

In the previous section, we pointed out that according to the proposed interface-based modular formalism, the realizer modules are semantically independent from those modules which utilize their interfaces . The importance of such independency is brightened when we observe that it leads to 'directed semantic' [1]. Directed semantic means that if a module which uses a set of interface terms, gives new semantics to these terms, this semantic does not affect their meaning in the original modules. For instance, in the case of Example 1, suppose that the Tourism ontology uses the interface concept `BeautifulPlaces` and `RuralPlaces` and the modular ontology has been configured in such way that the Canada Destination realizes these two interface concepts. The Tourism ontology may add an inclusion axiom that `BeautifulPlaces` $\sqsubseteq$ `RuralPlaces`, however, through our formalism, this subsumption does not necessarily hold in the Canadian Destination ontology.

4.2   Polymorphism

The proposed formalism supports *polymorphism* in the development of modular ontologies in the sense that the meaning of an interface term is subject to specialization based on the configuration of the ontology. When one develops an ontology, she only needs to work with general interfaces, and the specialized meaning of interface concepts would be bound to it in configuration time. For instance, in Example 1, we may define the Sightseeing as an interface concept while Natural Sightseeing and Scientific Sightseeing are modeled as two extensions for this concept. The Tourism ontology may configured to use the one of specialized meaning of Sightseeing. Furthermore, it will be possible to extend the meaning of Sightseeing in the future through the introduction of more perspectives without any changes on the syntax of the Tourism ontology. Example 8 shows this capability of the formalism:

Example 8. *Regarding Example 1, suppose that the ontology module 'Tourism' utilizes the interface Attractions which represents the notion of sightseeing and its related properties and concepts. Knowledge base of the Tourism module includes the following axioms:*

PlaceToGo ⊑ Attractions:Sightseeing

Let ScientificAttractions be an extension for Attractions with the following axiom:
Sightseeing ⊑ (∀ HasScienceBranch.ScienceBranch)
⊓ (∃ IsVisitedBy.Scientist)
Further, let NaturalAttractions be another extension for Attractions with the following axioms:
Sightseeing ⊑ (∃HaveBestSeason.Season)
⊓ (∃ HavePreCondition.Equipment),
Sightseeing ⊑ Beach ⊔ Jungle⊔Park ⊔ mounts

Here, ScientificAttractions and NaturalAttractions provide two morphs (forms) for the Sightseeing *concept, hence the tourism ontology can answer different type of queries related to the concept* Sightseeing *based on the type of configuration.*

4.3 Decidability

Consistency checking is a basic problem in description logic knowledge bases to which the other reasoning problems can be reduced. This issue is more vital in the proposed interface-based formalism because different modules can be defined using different description logics; hence their integration may lead to new inconsistency problems. In the following, we show that the consistency checking of augmented modules in a modular ontology is decidable in our proposed formalism.

Lemma 9 [1]

Let M be an ontology module in a modular ontology, such that it supports nominals and there exists an algorithm for creating completion graphs for deciding the consistency of its description logic[2]. The problem of consistency checking for Aug(M) is decidable.

Hence, while the formalism provides an expressiveness power which allows creating complex concepts using foreign terms, it ensures that the consistency checking of the integration of different ontology modules is decidable.

---

[1] For the proofs of all discussed lemmas in this paper see: http://falcon.unb.ca/ ∼ m4742/mo.pdf

[2] Such algorithms can be found in [12,11].

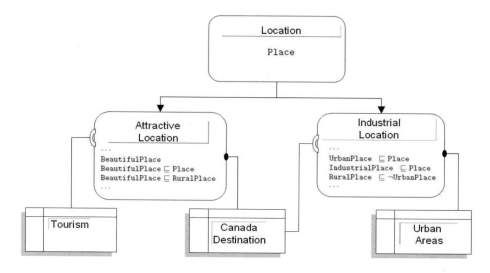

Fig. 1. An example of propagating logical consequences in the interface-based modularization formalism

4.4   Transitive Logical Consequences and Partial Reuse

The interface-based formalism supports propagating the logical consequences of the public part of an ontology module to all of the connected ontology modules, while its private parts do not have such consequence propagation. Hence, the formalism does not require the ontology modules to use or import all of the terms of the other connected modules and does not require reasoning engines to process the union of all inter-connected modules. The public section of each ontology module is the knowledge base of its interfaces. Lemma 10 shows that the logical consequences of a module's interfaces propagate to all other modules that are connected to it.

Lemma 10

(1) Let $M_1 = \langle \mathcal{T}, I_u, I_r \rangle$ be an augmentedly consistent module in a modular ontology, $I_1 = \bigcup_{I \in (I_u, I_r)} I$ and α be a concept expression (General inclusion axiom) whose signature is a subset of the signature I_1. If $I_1 \models \alpha$, for all augmentedly consistent module such as M on $PATH(M_1)$, $\Psi_{Aug(M)} \models \alpha$.

(2) Furthermore, consider an augmentedly consistent module M_2 on $PATH(M_1)$ and let I_2 be defined for M_2 similarly to I_1 for M_1. Let $I : A$ and $I : B$ be two concepts in I_1 and $I : C$ a concept in I_2. If $I_1 \models I : A \sqsubseteq I : B$ and $I_2 \models I : B \sqsubseteq I : C$, for all augmentedly consistent modules such as M on $PATH(M_1)$, we have $\Psi_{Aug}(M) \models I : A \sqsubseteq I : C$.

As an explainer example, consider the situation shown in Figure 1. According to the figure, there are three ontology modules: Tourism, Canada Destination and Urban Areas and three interfaces: Location, Attractive Location and Industrial Location where the last two interfaces are extensions of the Location interface. Tourism and Canada Destination utilize Attractive Location and Industrial Location, respectively. Moreover Canada Destination realizes Attractive Location and Urban Areas realizes Industrial Location. Since Industrial Location is the public part of the Urban Areas ontology modules, all of its axioms is propagated to the connected ontology modules. According to Industrial Location `RuralPlace` $\sqsubseteq$ `¬UrbanPlace`, and since Canada destination has `BeautifulPlace` $\sqsubseteq$ `¬RuralPlace` in its own interface, its augmented semantics entails `BeautifulPlace` $\sqsubseteq$ `¬UrbanPlace`.

The following proposition shows the unsatisfiability of a concept in a realizer module will be preserved in all of its utilizer modules.

Proposition 11. *Let M be a module which realizes interface I, for all modules M' such that $F(M', I) = M$, $\Psi_{Aug(M')} \models I : C \sqsubseteq \bot$, if $\Psi_{Aug(M)} \models I : C \sqsubseteq \bot$*

Proof Sketch: For every unsatisfiable concept $I : C$, the result set of the query $KI : C$ is the empty set.

In contrast to the public section, the private section of an ontology module does not necessarily propagate monotonically through connected modules. For example in the case of Figure 1, consider that Urban Areas module indicates that `IndustrialPlace(Toronto)`, `UrbanPlace(Toronto)` and `UrbanPlace(Montreal)` in its ABox. The Canada destination ontology module may conclude that `IndustrialPlace` is a subclass of `UrbanPlace` even though this axiom does not necessary hold in the realizer module Urban Areas.

5   Implementation

In this section, we present the implementation of the interface-based formalism introduced in the previous sections. The objective of this implementation is to allow ontology developers to define a modular ontology based on the definitions provided by the formalism as a set of ontology modules and interfaces, configure the modular ontology and select the connected ontologies, and also be able to perform reasoning on the developed modular ontology.

For the purpose of the implementation, we perform two tasks. First, we extend OWL-DL in order to allow ontology modules to use or realize a set of interfaces and second we extend the architecture of the Swoop ontology editor and browser in order to be able to work with interfaces and perform reasoning on the modular ontologies based on the semantics described earlier.

In order to extend OWL-DL, we define 'useInterface' and 'realizeInterface' as two new built-in ontology properties, analogous to the definition of 'owl:imports'. 'useInterface' and 'realizeInterface' are followed by the ID of interfaces they use and realize, respectively. We also modify OWL-API, a set of java interfaces for manipulating owl ontologies, such that an 'ontology' object has references to its used and realized interfaces.

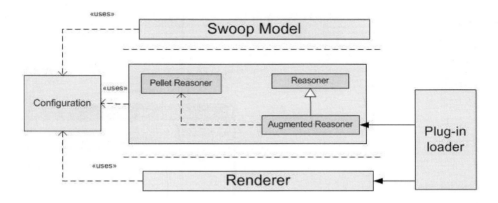

Fig. 2. A new extension to the architecture of Swoop for supporting interface-based modular ontologies

Swoop [13] is an ontology browser and editor which is tailored for OWL ontologies. It provides a convenient environment for manipulating multiple ontologies. The architecture of Swoop is comprised of three components: Model, Reasoning and Rendering. In addition, it consists of a plug-in loader which loads the appropriate reasoner or renderer in the environment. We modify the architecture of Swoop by introducing an augmented reasoner as well as a configuration object which can be shared among different layers of the architecture. Figure 2 depicts the modified architecture of the Swoop ontology editor.

As it is illustrated in Figure 2, the augmented reasoner can be defined as an extension to any existing reasoner available for Swoop. The augmented reasoner augments the knowledge base of the ontology module with the result set of the epistemic queries posed to the modules which realize its required interfaces before performing a reasoning task. The augmented reasoner uses the capability of performing epistemic queries from Pellet for doing its augmentation process. It uses the configuration component in order to recognize the appropriate realizer modules.

We also modify the Model component of the Swoop architecture such that it provides capabilities for loading and working with interfaces as well. Using the extended OWL-API, the new version of Swoop supports loading interfaces and configuring modular ontologies in such a way that for each ontology module the users can select a realizer module for each of the interfaces it uses. Figure 3 shows a snapshot of the modified Swoop environment for creating a modular ontology.

As it can be seen it this figure, a user can graphically see the configuration of the modular ontology through the newly introduced tab: "Configuration". Moreover, clicking on the "Configure Module" button, a pop-up menu is shown to the user that contains the list of all modules and for each selected module, the list of all interfaces it uses and for any selected interface the list of its realizer modules. The users can use this menu to change the configuration of a modular ontology.

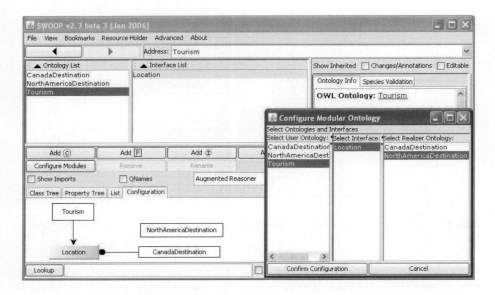

Fig. 3. Interface-based modular ontologies in Swoop

6   Related Work

We can categorize the accomplished efforts on modular ontologies into two classes: (*i*) those that attempt to decompose a large and comprehensive ontology into a set of smaller and self-contained modules and (*ii*) those that introduce new formalisms for developing modular ontologies.

With regards to the first category, [6] proposes an algorithm for extracting a module from an ontology which describes a given concept. The extracted module captures the meaning of that concept and should be 'locally correct' and 'locally complete' which means that any assertions which are provable in a module should also be provable in the ontology. Also any assertion which is provable in the ontology and asserted using the signature of a module should be provable in that module. The notion of local complete modules is close to the notion of 'conservative extensions' which is discussed in [8] and employed for extracting an approximation of the smallest meaningfull module related to a set of concept and role names from an ontology. [16] proposes an algorithm for segmentation of an ontology by traversing through the ontology graph starting from a given concept name. In [20], authors propose a method for partitioning a large ontology into disjoint sets of concepts. It has been assumed that the given ontology is mostly comprised of hierarchal relationships between concepts instead of more complex roles and binary relationships.

The efforts in the second category focus on proposing new formalisms for modular ontologies. These formalisms mostly provide new extensions to existing description logics syntax and semantics in order to make automated reasoning over ontology modules feasible. The interface-based modularity formalism introduced in this paper and also E-connections, DDL, P-DL and semantic import which are discussed in the first section of this paper can be categorized in this class of works. [15] describes a new formalism for ontology modularization. Based on [15] a module is described by its identifier, a set of the identifiers of other modules which are imported, a set of interfaces, a set of mapping assertions between different concept names of imported modules and its concepts and an export interface. It uses a mapping approach and defines a mapping function in order to let a module use the others' elements. Through this function, concepts of different modules are mapped to a global domain. The notion of interfaces in [15] are different from those that are defined in this paper. Contrary to our approach for defining interfaces as independent ontology units which ensures indirect connection of ontology modules, in [15] interfaces are a set of concept names of a specific module that can be imported by others. Furthermore, in [15] a module can only have one export interface, consequently it is not possible to provide different perspectives for describing an ontology module.

7   Concluding Remarks

In this paper, we have described a framework for developing ontologies in a modular manner. The core of this framework is the interface-based modular ontology formalism which supports knowledge encapsulation, i.e., it allows ontology modules to describe their content through well-defined interfaces. We showed that this formalism provides a considerable expressiveness power by allowing ontology modules to create complex roles and concepts using the interface terms. We have also proven the decidability of the formalism and its capability for propagating the logical consequences of the public parts of the ontology modules. In order to make the application of the frameworks feasible in real-world applications, we have extended the syntax and semantic of OWL-DL and the Swoop architecture.

References

1. Bao, J., Caragea, D., Honavar, V.: On the semantics of linking and importing in modular ontologies. In: International Semantic Web Conference, pp. 72–86 (2006)
2. Bao, J., Slutzki, G., Honavar, V.: A semantic importing approach to knowledge reuse from multiple ontologies. In: AAAI, pp. 1304–1309 (2007)
3. Borgida, A., Serafini, L.: Distributed description logics: Assimilating information from peer sources. J. Data Semantics 1, 153–184 (2003)
4. Bouquet, P., Giunchiglia, F., Harmelen, F., Serafini, L., Stuckenschmidt, H.: C-OWL: Contextualizing ontologies. In: Fensel, D., Sycara, K.P., Mylopoulos, J. (eds.) ISWC 2003. LNCS, vol. 2870. Springer, Heidelberg (2003)
5. Calvanese, D., Giacomo, G.D., Lembo, D., Lenzerini, M., Rosati, R.: Eql-lite: Effective first-order query processing in description logics. In: IJCAI, pp. 274–279 (2007)
6. Cuenca Grau, B., Parsia, B., Sirin, E., Kalyanpur, A.: Modularity and web ontologies. In: Proceedings of KR 2006, pp. 198–209. AAAI Press, Menlo Park (2006)

7. Donini, F.M., Lenzerini, M., Nardi, D., Nutt, W., Schaerf, A.: An epistemic operator for description logics. Artif. Intell. 100(1-2), 225–274 (1998)
8. Grau, B.C., Horrocks, I., Kazakov, Y., Sattler, U.: Just the right amount: extracting modules from ontologies. In: WWW 2007, pp. 717–726. ACM, New York (2007)
9. Grau, B.C., Kutz, O.: Modular ontology languages revisited. In: Proceedings of the IJCAI 2007 Workshop on Semantic Web for Collaborative Knowledge Acquisition (2007)
10. Grau, B.C., Parsia, B., Sirin, E.: Combining owl ontologies using e-connections. Journal of Web Semantics 4(1) (2005)
11. Horrocks, I., Sattler, U.: Ontology reasoning in the shoq(d) description logic. In: IJCAI, pp. 199–204 (2001)
12. Horrocks, I., Sattler, U., Tobies, S.: Practical reasoning for expressive description logics. In: Ganzinger, H., McAllester, D., Voronkov, A. (eds.) LPAR 1999. LNCS, vol. 1705, pp. 161–180. Springer, Heidelberg (1999)
13. Kalyanpur, A., Parsia, B., Sirin, E., Grau, B.C., Hendler, J.A.: Swoop: A web ontology editing browser. J. Web Sem. 4(2), 144–153 (2006)
14. Kutz, O., Lutz, C., Wolter, F., Zakharyaschev, M.: E-connections of abstract description systems. Artif. Intell. 156(1), 1–73 (2004)
15. Haase, P., Rudolph, S.: J. E. A. Z. M. D. M. I. Y. J. C. C. C. B. A. J. M. G. Deliverable d1.1.3 neon formalisms for modularization: Syntax, semantics, algebra, NEON EU-IST-2005-027595 (2008)
16. Seidenberg, J., Rector, A.: Web ontology segmentation: analysis, classification and use. In: WWW 2006, pp. 13–22. ACM, New York (2006)
17. Sirin, E., Parsia, B., Grau, B.C., Kalyanpur, A., Katz, Y.: Pellet: A practical owl-dl reasoner. Web Semant. 5(2), 51–53 (2007)
18. Snyder, A.: Encapsulation and inheritance in object-oriented programming languages. In: OOPLSA 1986, pp. 38–45. ACM, New York (1986)
19. Stuckenschmidt, H., Klein, M.C.A.: Integrity and change in modular ontologies. In: IJCAI, pp. 900–908 (2003)
20. Stuckenschmidt, H., Klein, M.C.A.: Structure-based partitioning of large concept hierarchies. In: International Semantic Web Conference, pp. 289–303 (2004)

On the Semantics of Trust and Caching in the Semantic Web

Simon Schenk

ISWeb, University of Koblenz-Landau, Germany
sschenk@uni-koblenz.de
http://isweb.uni-koblenz.de/

Abstract. The Semantic Web is a distributed environment for knowledge representation and reasoning. The distributed nature brings with it failing data sources and inconsistencies between autonomous knowledge bases. To reduce problems resulting from unavailable sources and to improve performance, caching can be used. Caches, however, raise new problems of imprecise or outdated information. We propose to distinguish between certain and cached information when reasoning on the semantic web, by extending the well known $\mathcal{FOUR}$ bilattice of truth and knowledge orders to $\mathcal{FOUR} - \mathcal{C}$, taking into account cached information. We discuss how users can be offered additional information about the *reliability* of inferred information, based on the availability of the corresponding information sources. We then extend the framework towards $\mathcal{FOUR} - \mathcal{T}$, allowing for multiple *levels of trust* on data sources. In this extended setting, knowledge about trust in information sources can be used to compute, how well an inferred statement can be trusted and to resolve inconsistencies arising from connecting multiple data sources. We redefine the stable model and well founded semantics on the basis of $\mathcal{FOUR} - \mathcal{T}$, and reformalize the Web Ontology Language OWL2 based on logical bilattices, to augment OWL knowledge bases with trust based reasoning.

The Semantic Web is envisioned to be a *Web of Data* [2]. As such, it integrates information from various sources, may it be through rules, data replication or similar mechanisms. Obviously, in a distributed scenario, information sources may become unavailable. In order to still be able to answer queries in such cases, mechanisms like caching can be used to reduce the negative implications of failure. Alternatively, some default truth value could be assumed for unavailable information. However, cached values may be inaccurate or outdated, default assumptions can be wrong. Moreover, also available information sources may be trusted to different extents. We propose a framework for reasoning with such trust levels, which allows to give additional information on the reliability of results to the user. In particular, we are able to tell whether a statement's truth value is inferred based on really accessible information, or whether it might change in the future, when cached or default values are updated. Consequently, we extend the framework towards multiple levels of trust, taking into account a *trust order* over information sources, which can possibly be *partial*. While assigning absolute trust values has little semantics, users are usually good at comparing the trustworthiness of *two* information sources.

A. Sheth et al. (Eds.): ISWC 2008, LNCS 5318, pp. 533–549, 2008.

This problem is highly relevant, because fault tolerance and reliable data integration is a main prerequisite for a distributed system like the semantic web. The level of reliability of a piece of information can strongly influence further usability of derived information. For this reason, our approach can be seen as a bridge between the rules, proof and trust layers of the semantic web layercake.

The availability of multiple integration mechanisms for distributed resources makes formulating a generic framework a non-trivial task. Moreover, classical two-valued logic fails to capture the 'unknown' truth value of unavailable information. In fact, many applications today rely on simple replication of all necessary data, instead of more flexible mechanisms.

Our approach extends a very flexible basis of most logical frameworks, namely bilattices, which allow to formalize many logics in a coherent way [9]. Hence, it is applicable to a broad range of logical languages. We propose extensions $\mathcal{FOUR} - \mathcal{C}$ and $\mathcal{FOUR} - \mathcal{T}$ to the the $\mathcal{FOUR}$ bilattice. These extensions add trust orders to truth values. As we allow possibly partial orders and also use multiple $\top$ and $\bot$ values, $\mathcal{FOUR} - \mathcal{T}$ is strictly more expressive than for example fuzzy logics.

We investigate support for connected and interlinked autonomous and distributed semantic repositories (for RDF or OWL) — a basic idea behind the semantic web effort. These repositories exchange RDF and OWL data statically (e.g. by copying whole RDF graphs, as in the caching scenario) or dynamically using views or rules. We model our example scenario as follows: Information sources consist of facts (à la RDF) or axioms (à la OWL). Information sources are connected through views, which are clauses of normal programs. This for example subsumes SPARQL queries and SPARQL based views [13], [16]. One possible syntactic and semantic realization of this scenario are Networked Graphs, which we describe in [16].

We show how $\mathcal{FOUR} - \mathcal{C}$ and $\mathcal{FOUR} - \mathcal{T}$ can be used for an extension of the stable model and well founded semantics, which are very popular for rule based mechanisms and also underlie our Networked Graphs mechanism. We extend the web ontology language OWL2 [8] to be based on logical bilattices and use trust levels for inconsistency resolution in OWL2. Finally, we review related work in section 8 before concluding the paper.

1   Use Case

Oscar is a project officer at a large funding agency and supervises several research projects. One of his regular tasks is to check the timely publication of project deliverables. Fortunately, the Semantic Web has made his life much easier. All his projects are advised to publish their deliverables on their websites using the FOAF vocabulary[1].

In the following, we annotate predicates $p(x)$ with the information *source*, where the corresponding data is expected to come from as follows: $p(x)|source$. Rules and DL axioms A are written in the usual syntaxes. Analogously, we

---

[1] http://xmlns.com/foaf/spec

annotate them with the repositories they are stored in as follows: $A||source$. We use functional style syntax for all facts.

Oscar sets up a view listing all deliverables of all projects and their due dates. The view makes use of his own data and the projects' websites to determine all timely deliverables.

(1) $deliverable(report1).||oscar$

(2) $due(report1, 20081005).||oscar$

(3) $hasDeliverable(project1, report1).||oscar$

(4) $TimelyDeliverable(X) \leftarrow deliverable(X)|oscar, due(X, Y)|oscar,$
$published(X, Z)|project1 \vee project2, Z \leq Y|oscar.||oscar$

Rule (4) uses information from Oscars knowledge base to find all deliverables and due dates, as well as information about publishing dates from the project websites, to infer whether deliverables are published on time. When it is time for Oscar's next check, he opens his knowledge base and lists all delayed deliverables. Unfortunately on this day, Project1's website is not working due to technical problems. However, Oscar's webserver still has partial results cached from last month. Now Oscar would like to be able to infer, which deliverables have been delivered, and which information about deliveries might be outdated. He does not want to send a formal reminder to Project1 by mistake. The next day, Project1's website works again, listing

(5) $published(report1, 20081001).||project1$

Oscar wants to easily produce reports, so he adds the following DL axioms to distinguish between good and bad projects, and some additional facts:

(5) $GoodProject = Project \sqcap \forall hasDeliverable.TimelyDeliverable.||oscar$

(6) $BadProject = Project \sqcap \neg GoodProject.||oscar$

(7) $hasDeliverable(project1, report2).||oscar$

(8) $\leq 2 project1.hasDeliverable.||oscar$

(9) $report1 \not\approx report2.||oscar$

(10) $timelyDeliverable(report2).||oscar$

(11) $Project(project1).||oscar$

When Oscar goes on vacation, he asks his secretary Susan to monitor deliverables, while he is away. Susan does her own bookkeeping, also using OWL. When Oscar returns, he imports Susan's data into his knowledge base, causing an inconsistency, as Susan has added an axiom

(12) $\neg timelyDeliverable(report2).||susan$

As Oscar assumes his own data to be more reliable, he would like to automatically resolve such inconsistencies in the future by discarding lowly trusted information. Additionally, he wants to make sure, that projects can not cheat. He trusts projects less than his secretary. As he does not prefer any project over others, however, there is no trust order among projects. So if we have

(13) $SuccessfulProject(project1).||project1$ and
$\neg SuccessfulProject(project1).||project2,$

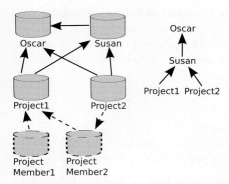

Fig. 1. Repositories in Use Case (left) and Oscar's Trust Order (right)

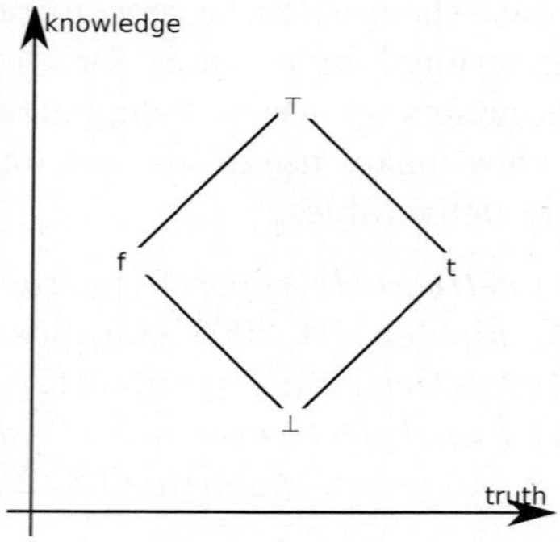

Fig. 2. The knowledge and truth order $\mathcal{FOUR}$

he wants to discard both axioms, while in the case of

(14) $SuccessfulProject(project1).\|project1$ and $\neg SuccessfulProject(project1).\|susan,$

only $SuccessfulProject(project1).\|project1$ would be discarded.

All repositories involved in this scenario are shown in the left part of fig 1. In the lower part we see that also additional repositories might be involved, which are not directly relevant and known to Oscar. In the right part, we see Oscar's trust order. Obviously, there is a need for multiple levels of trust in information sources in our scenario. However, not all sources need to be comparable. In this paper, we propose a very flexible mechanism for reasoning with such partial trust orders.

2 $\mathcal{FOUR}$

Most logic programming paradigms, including classical logic programming, stable model and well founded semantics, and fuzzy logics can be formalized based on bilattices of truth values and fixpoints of a direct consequence operator on such a bilattice. Therefore, if we build our extension into this foundational layer, it will directly be available in many different formalisms.

A logical bilattice [5] is a set of truth values, on which two partial orders are defined, which we call the truth order $\leq_t$ and the knowledge order $\leq_k$. Both $\leq_t$ and $\leq_k$ are complete lattices, i.e. they have a maximal and a minimal element and every two elements have exactly one supremum and infimum.

In logical bilattices, the operators $\vee$ and $\wedge$ are defined as supremum and infimum wrt. $\leq_t$. Analogously join ($\oplus$) and meet ($\otimes$) are defined as supremum and infimum wrt. $\leq_k$. As a result, we have multiple distributive and commutative laws, which all hold. Negation ($\neg$) simply is an inversion of the truth order. Hence, we can also define material implication ($a \rightarrow b = \neg a \vee b$) as usual.

The smallest non trivial logical bilattice is $\mathcal{FOUR}$, shown in figure 2. In addition to the truth values t and f, $\mathcal{FOUR}$ includes $\top$ and $\bot$. $\bot$ means "unknown", i.e. a fact is neither true or false. $\top$ means "overspecified" or "inconsistent", i.e. a fact is both true and false.

In traditional, two valued logic programming without negation, only t and f would be allowed as truth values. In contrast, e.g. the stable model semantics, allows to use $\top$ and $\bot$. In this case, multiple stable models are possible. For example, we might have a program with three clauses:

$man(bob) \leftarrow person(bob), \neg woman(bob).$
$woman(bob) \leftarrow person(bob), \neg man(bob).$
$person(bob).$

Using default f, we might infer both $man(bob) \land \neg woman(bob)$ and $woman(bob) \land \neg man(bob)$. While in two valued logics we would not be able find a model, in four values, we could assign truth values $t \oplus f = \top$ and $t \otimes f = \bot$. In fact, both would be allowed under the stable model semantics, resulting in multiple models for a single program.

The well founded semantics distinguishes one of these models — the minimal one, which is guaranteed to always exist and only uses t, f, and $\bot$. In a similar way, other formalisms can be expressed in this framework as well. Particularly, we can also formalize open world based reasoning, using $\bot$ instead of f as default value. In order to keep this paper short, we refer the reader to the very good overview in [3].

3 $\mathcal{FOUR} - \mathcal{C}$

To apply our work to a variety of different logical formalisms, we directly extend $\mathcal{FOUR}$ as the theoretical basis. We will give two examples of how the extensions can be used by applying them to the well founded semantics and to OWL in the remainder of this paper.

To distinguish between certain information, which is local or currently available online, and cached information (or information derived from cached information), we extend the set of possible truth values: For information, of which we know the actual truth value we use the truth values $\{t_k, f_k, \top_k, \bot_k\}$. For cached information, we use a different set: $\{t_c, f_c, \top_c, \bot_c\}$. The basic idea of the extension is that cached information is always potentially outdated. For example a cached *false* value might actually be *true*. Therefore, we assume cached information to be always a bit less false or true than certain information — as the truth value might have changed.

In our scenario, let us assume Project1's website is currently inaccessible. In a normal closed world setting, we would assume $published(report1, \_)$ to be f, hence also $timelyDeliverable(report1)$, by rule (4). Changing our default to $\bot$ – unknown – would also not help us determine, whether Project1's website is just updated slowly, or whether the available information might be inaccurate. In $\mathcal{FOUR} - \mathcal{C}$, we assign f_c (or $\bot_c$ in an open world setting) to $published(report1, \_)$. We can then conclude from (1-4) and the unavailability

of (5) that $timelyDeliverable(report1)$ is $t_k \wedge f_c = f_c$, and hence possibly outdated. Therefore, Oscar will simply update his report later, when all relevant
data sources are available again, instead of sending a reminder by mistake. Analogously, if we run into an inconsistency, we want to be aware, if this inconsistency
could potentially be resolved by updating the cache. Summarizing, our operators
should act as in $\mathcal{FOUR}$, if we only compare truth values on the same trust level.
If we compare values from multiple trust levels, we would like to come up with
analogous truth values as in the four valued case, but on the trust level, which
is the lowest of the compared values.

Ginsberg [5] describes how we can obtain a logical bilattice: Given two dis-
tributive lattices $\mathcal{L}_1$ and $\mathcal{L}_2$, create a bilattice $\mathcal{L}$, where the nodes have values
from $\mathcal{L}_1 \times \mathcal{L}_2$, such that the following orders hold:

- $\langle a, b \rangle \leq_k \langle x, y \rangle$ iff $a \leq_{\mathcal{L}_1} x \wedge b \leq_{\mathcal{L}_2} y$ and
- $\langle a, b \rangle \leq_t \langle x, y \rangle$ iff $a \leq_{\mathcal{L}_1} x \wedge y \leq_{\mathcal{L}_2} b$

If $\mathcal{L}_1$ and $\mathcal{L}_2$ are infinitely distributive — that means distributive and commutative laws hold for infinite combinations of the lattice based operators from
section 2 — then $\mathcal{L}$ will be as well.

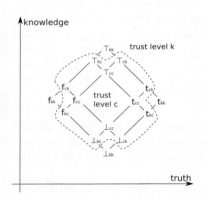

Fig. 3. $\mathcal{FOUR} - \mathcal{C}$

We use $\mathcal{L}_1 = \mathcal{L}_2 = t_k > t_c > f_c > f_k$
as input lattices, resulting from our basic
idea that cached values are a bit less true
and false. As $\mathcal{L}_1$ and $\mathcal{L}_2$ are totally ordered
sets, they are complete lattices and hence
infinitely distributive. The resulting
$\mathcal{FOUR} - \mathcal{C}$ bilattice shown in fig. 3. In fig.
3 we label nodes of the form $\langle f_x, t_y \rangle$ with
$\top_{xy}$, $\langle t_x, f_y \rangle$ with $\bot_{xy}$, $\langle f_x, f_y \rangle$ with f_{xy} and
$\langle t_x, t_y \rangle$ with t_{xy}.

The artificial truth values $f_{kc}, f_{ck}, t_{kc}, t_{ck},$
$\top_{kc}, \top_{ck}, \bot_{kc}$ and $\bot_{ck}$ are only used for reasoning purposes. Users will only be inter-
ested in trust levels, which are equivalence
classes of truth values: Given an order of
$k > c$, the trust level of a truth value is the
minimal element in its subscript. For example the trust level of t_c, $\bot_{kc}$ and $\top_{ck}$
is c. In fig. 3, these equivalence classes are separated by dotted lines. As we only
have two trust levels here, there are exactly two equivalence classes - one for currently accessible (truth values $t_{kk}, f_{kk}, \top_{kk}, \bot_{kk}$) and one for cached information
(truth values $t_{cc}, f_{cc}, \top_{cc}, \bot_{cc}$) and information derived from cached information
(truth values $t_{kc}, f_{kc}, \top_{kc}, \bot_{kc}, t_{ck}, f_{ck}, \top_{ck}, \bot_{ck}$).

Obviously, $\mathcal{FOUR} - \mathcal{C}$ meets our requirements from the beginning of this
section: We have two sub-bilattices isomorphic to $\mathcal{FOUR}$, one on each trustlevel.
Additionally, we always come up with truth values on the right trust level, e.g.
$f_{kk} \oplus t_{cc} = \top_{kc}$, which is on trust level c and $\top_{kk} \wedge \top_{cc} = \top_{ck}$, which is on trust
level c and correctly reflects the fact that the result may be inaccurate in case
$\top_{cc}$ needs to be corrected to some f_{xx}.

In the caching scenario, we can assume a default truth value of $\perp$ or f (depending on whether we do open or closed world reasoning) to all statements, where the actual truth value can not be determined at the moment. However, some more information may be available for example due to caching, statistics or similar. Using $\mathcal{FOUR} - \mathcal{C}$, we can still do inferencing in the presence of such unreliable sources. Moreover, a user or application can determine, whether a piece of information is completely reliable, or if more accurate information may become available.

4   Extension Towards Trust Levels

In the previous chapter we have focused on two levels of reliability of information. More generally, we would like to be able to infer multiple levels of trust in a distributed setting, as we have seen in the use case in Oscar's trust order.

Definition 1 (Trust Order)
A trust order $\mathcal{T}$ is a partial order over a finite set of information sources with a maximal element, called ∞.

∞ is the information source with the highest trust level, assigned to local data. For any two information sources a and b comparable wrt. $\mathcal{T}$, we have a $\mathcal{FOUR} - \mathcal{C}$ lattice as described above, with the less trusted information source corresponding to the inner part of the lattice. Extended to multiple information sources, this results in a situation as depicted in fig. 4, where the outermost bilattice corresponds to local, fully trusted information.

If a and b are not comparable we introduce virtual information sources inf_{ab} and sup_{ab}, such that

- $inf_{ab} < a < sup_{ab}$ and $inf_{ab} < b < sup_{ab}$;
- $\forall_{c<a,c<b} : c < inf_{ab}$ and
- $\forall_{d>a,d>b} : d > sup_{ab}$

To understand the importance of this last step, assume that $c > a > d$ and $c > b > d$ and a, b are incomparable. Then the truth value of $a \vee b$ would have a trust level of c, as c is the supremum in the trust order. Obviously this escaping to a higher trust level is not desirable. Instead, the virtual information sources represent that we need to trust both, a and b, if we believe in the computed truth value. We illustrate this situation in fig. 5 (We abbreviate inf_{ab} by $<$ and sup_{ab} by $>$).

In the general case (≥ 3 incomparable sources) such a trust order results in a non-distributive lattice. This can be fixed, however, by introducing additional virtual nodes. The basic idea here is to create a virtual node for each element in the powerset of the incomparable sources, with set inclusion as the order. We will call this modified trust order *completed*. We can again derive a complete lattice from the completed trust order. As it can become quite large, we only show for $\top$ how a fragment of the logical bilattice is derived from the trust order for the case of two incomparable information sources in fig. 5.

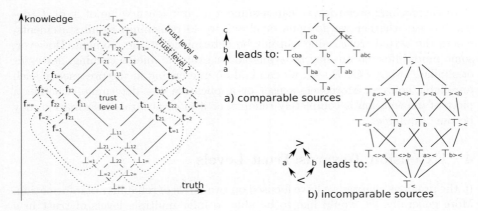

Fig. 4. $\mathcal{FOUR} - \mathcal{T}$ **Fig. 5.** Virtual Sources

Using the same method as in the previous chapter, we can construct the corresponding bilattice from a given completed truth order as follows: Given a trust order $\mathcal{T}$, generate a lattice $\mathcal{L}_1$, such that

- $f_a <_{\mathcal{L}_1} f_b$, iff $a >_{\mathcal{T}} b$;
- $t_b <_{\mathcal{L}_1} t_a$, iff $a >_{\mathcal{T}} b$ and
- $\forall a : f_a <_{\mathcal{L}_1} t_a$.

The result is a lattice with t_∞ and f_∞ as maximal and minimal elements. Now create the logical bilattice $\mathcal{L}$ from $\mathcal{L}_1 \times \mathcal{L}_1$ as described in the previous section.

In our scenario, (10) and (12) (as well as (13) and (14)) are inconsistent. We could obtain a consistent knowledge base by removing any of the axioms, however we need to choose which one. Making use of the trust order, we can choose to discard the axiom obtained from Susan. In chapter 6 we define how to do such choices for arbitrarily complex ontologies in an extension of $\mathcal{SROIQ}$. Please note that in spite of the rather simple example, we can have complex derived facts instead and also multiple ways to derive them, as shown in the dependency graph in the left part of in fig. 1.

5   Extended Stable and Well Founded Semantics

The stable model semantics is popular for assigning models to normal logic programs (which are used for modeling our scenario) without limitations on the use of negation. For every normal program there is a distinguished stable model, called the well founded model. We now define a well-founded semantics [19], which takes into account the extensions defined in the previous chapter. We follow the definitions in [3] and [14], which are based on $\mathcal{FOUR}$ and $\{f, \perp, t\}$ respectively. As we use complete, distributive bilattices, the results extend to $\mathcal{FOUR} - \mathcal{T}$ [3], [5]. In the following we refer to normal logic programs when using the term program. In fact, it would even be possible to use $\oplus$ and $\otimes$ as operators in programs.

Definition 2 (Valuations)
Let P be a program. ground(P) is the set of ground instantiated clauses of P. A valuation of P wrt. a bilattice $\mathcal{L}$ is a mapping from ground(P) to truth values of $\mathcal{L}$. For any ground atom $A \in$ ground(P) and valuation v, $v(A)$ denotes the truth value assigned to A by v.

Let v and w be valuations wrt. $\mathcal{L}$. We define $(v\Delta w)$ as a mapping from ground literals to truth values of $\mathcal{L}$ as follows: $(v\Delta w)(A) = v(A)$ and $(v\Delta w)(\neg A) = w(A)$. $(v\Delta w)$ extends to more complex terms in the natural way: $(v\Delta w)(B\vee C) = (v\Delta w)(B) \vee (v\Delta w)(C)$, $(v\Delta w)(B \oplus C) = (v\Delta w)(B) \oplus (v\Delta w)(C)$, analogous for $\wedge$ and $\otimes$.

As usual, we use a fixpoint operator to define the semantics of a program:

Definition 3 (Single Step Immediate Consequence Operator)
$$\psi_P(v, w) = \{\langle A, u\rangle | A \leftarrow B \in \text{ground}(P) \wedge u = (v\Delta w)(B)\}$$

ψ_P uses v to assign values to positive literals in rules and w to assign values to negative literals. Hence, we treat negation symmetrically. In our caching scenario we would use the local, cached and default information to initialize the valuations v and w. If no cached information is available, we start from the known true facts only. We have the following properties of ψ_P:

Proposition 1 (Properties of ψ_P [3])
  - $\psi_P(v, w)$ *is monotone wrt.* $\leq_k$ *in both v and w;*
  - $\psi_P(v, w)$ *is monotone wrt.* $\leq_t$ *in v and*
  - $\psi_P(v, w)$ *is anti-monotone wrt.* $\leq_t$ *in w.*

An operator $O(x)$ is anti-monotone, if $x_1 \leq x_2 \rightarrow O(x_2) \leq O(x_1)$. Obviously, $O^2(x)$ must be monotone. Since $\psi_P(v, w)$ is monotone in v, we can define

Definition 4 (Immediate Consequence Operator)
$\psi'_P(w)$ *is the least fixpoint of* $(\lambda v)\psi_P(v, w)$ *wrt.* $\leq_t$ *and $\mathcal{L}$.*

Similarly to ψ_P, we have the following properties of ψ'_P:

Proposition 2 (Properties of ψ'_P [3])
  - $\psi'_P(w)$ *is monotone wrt.* $\leq_k$ *and*
  - $\psi'_P(w)$ *is anti-monotone wrt.* $\leq_t$.

$\mathcal{L}$ is a complete lattice, hence the first item guarantees, that fixpoints of ψ'_P exist and that there is a minimal one wrt. $\leq_k$. Obviously, as we still have w as parameter, $\psi'_P(w)$ can have multiple fixpoints.

Definition 5 (Stable Model)
Let P be a program and $\mathcal{L}$ a complete bilattice. A fixpoint of ψ'_P wrt. $\mathcal{L}$ is called a stable model *of P. The minimal stable model is called the* well founded model *of P.*

In contrast to the usual three or four valued definitions of the well founded and stable models, we can have true and false values with trust levels, but also $\top$ and $\bot$ values with trust levels. While this may seem a bit odd at first, it is very useful in our caching and trust setting: A $\top$ value with a trust level represents the maximally trusted information source, which is responsible for an inconsistency. We will use this idea in section 7 to resolve inconsistencies by dropping lowly trusted information. $\bot$ values with trust levels lead to a propagation of lower trust levels through the reasoning process, also in the presence of $\bot$ defaults or cache entries. As a result, we may come up with lowly trusted t, f or $\top$ values later. Intuitively, this reflects the assumption that the inferred value may be wrong, if the initial $\bot$ was already wrong in the cache or untrusted information source.

Theorem 1 (Complexity)
The data complexity of the well founded semantics wrt. $\mathcal{FOUR} - \mathcal{T}$ is polynomial, the combined complexity is EXPTIME.

Proof (sketch) We start from the known complexities of the well founded semantics in $\mathcal{FOUR}$. Our semantics is defined analogously to that in $\mathcal{FOUR}$. The only difference is that we no longer have a fixed logical bilattice. Therefore, we need to consider possible additional complexity for finding the supremum or infimum of two truth values. As our bilattices are built based on orders over a finite set of information sources, they must be finite. A finite bilattice $\mathcal{L}$ has finitely many edges $n \leq \mathcal{L}^2$. Assume we use a very basic algorithm to find a supremum (infimum): starting from both values to be compared, we follow all possible $\geq$-edges in the relevant order, until we find the other value. To do so, we need to follow less than $2n$ edges. Hence, the additional effort is polynomial.

Based on existing work for the well founded semantics on $\mathcal{FOUR}$ we can also define an operational semantics: The alternating fixpoint procedure proposed by Gelder [19] computes the two $\leq_t$-extremal stable models m_t and M_t of a program P, using the anti-monotonicity of ψ_P wrt. $\leq_t$. The well founded model of P is then obtained as $m_t \otimes M_t$ [3]. The alternating fixpoint procedure in turn gives rise to an implementation.

6   Extension Towards OWL

In this section we extend $\mathcal{SROIQ}$, the description logic underlying the proposed OWL2 [8], to $\mathcal{SROIQ} - \mathcal{T}$ evaluated on a logical bilattice. The extension towards logical bilattices works analogously to the extension of $\mathcal{SHOIN}$ towards a fuzzy logic as proposed in [18]. Operators marked with a dot, e.g. $\dot{\geq}$ are the lattice operators described above, all other operators are the usual (two valued) boolean operators. For two valued operators and a logical bilattice $\mathcal{L}$ we map t to $max_t(\mathcal{L})$ and f to $min_t(\mathcal{L})$ to model that these truth values are absolutely trusted[2]. Please note that while we limit ourselves to $\mathcal{SROIQ}$ here, analogous

---

[2] In $\mathcal{FOUR} - \mathcal{T}$ these would be f_∞ and t_∞, but we start with the general case.

extensions are possible for $\mathcal{SROIQ}(\mathcal{D})$ to support datatypes. Please also note that we do not include language constructs, which can be expressed by a combination of other constructs defined below. In particular, $Sym(R) = R^- \sqsubseteq R$ and $Tra(R) = R \circ R \sqsubseteq R$.

Definition 6 (Vocabulary)
A vocabulary $V = (N_C, N_P, N_I)$ is a triple where
- *N_C is a set of OWL classes,*
- *N_P is a set of properties and*
- *N_I is a set of individuals.*
N_C, N_P, N_I need not be disjoint.

A first generalization is that interpretations assign truth values from any given bilattice. In contrast, $\mathcal{SROIQ}$ is defined via set membership of (tuples of) individuals in classes (properties) and uses two truth values only.

Definition 7 (Interpretation)
Given a vocabulary V an interpretation $\mathcal{I} = (\Delta^{\mathcal{I}}, \mathcal{L}, \cdot^{\mathcal{I}_C}, \cdot^{\mathcal{I}_P}, \cdot^{\mathcal{I}_i})$ is a 5-tuple where
- *$\Delta^{\mathcal{I}}$ is a nonempty set called the object domain;*
- *$\mathcal{L}$ is a logical bilattice and Λ is the set of truth values in $\mathcal{L}$*
- *$\cdot^{\mathcal{I}_C}$ is the class interpretation function, which assigns to each OWL class $A \in N_C$ a function: $A^{\mathcal{I}_C} : \Delta^{\mathcal{I}} \to \Lambda$;*
- *$\cdot^{\mathcal{I}_P}$ is the property interpretation function, which assigns to each property $R \in N_P$ a function $R^{\mathcal{I}_P} : \Delta^{\mathcal{I}} \times \Delta^{\mathcal{I}} \to \Lambda$;*
- *$\cdot^{\mathcal{I}_i}$ is the individual interpretation function, which assigns to each individual $a \in N_I$ an element $a^{\mathcal{I}_i}$ from $\Delta^{\mathcal{I}}$.*

$\mathcal{I}$ is called a complete *interpretation, if the domain of every class is $\Delta^{\mathcal{I}}$ and the domain of every property is $\Delta^{\mathcal{I}} \times \Delta^{\mathcal{I}}$.*

We extend the property interpretation function $\cdot^{\mathcal{I}_P}$ to property expressions:

$$(R^-)^{\mathcal{I}} = \{(\langle x, y \rangle, u) | (\langle y, x \rangle, u) \in R^{\mathcal{I}}\}$$

The second generalization over $\mathcal{SROIQ}$ is the replacement of all quantifiers over set memberships with conjunctions and disjunctions over Λ. We extend the class interpretation function $\cdot^{\mathcal{I}_C}$ to descriptions as shown in table 1.

Satisfaction of axioms in an interpretation $\mathcal{I}$ is defined in table 2. With $\circ$ we denote the composition of binary relations. For any function f, $dom(f)$ returns the domain of f. The generalization is analogous to that of $\cdot^{\mathcal{I}_C}$. Note that for equality of individuals, we only need two valued equality.

Satisfiability in $\mathcal{SROIQ} - \mathcal{T}$ is a bit unusual, because when using a logical bilattice we can always come up with interpretations satisfying all axioms by assigning $\top$ and $\bot$. Therefore, we define satisfiability wrt. a truth value:

Definition 8 (Satisfiability)
We say an axiom E is u-satisfiable in an ontology O wrt. a bilattice $\mathcal{L}$, if there exists a complete interpretation $\mathcal{I}$ of O wrt. $\mathcal{L}$, which assigns a truth value $val(E, \mathcal{I})$ to E, such that $val(E, \mathcal{I}) \geq_k u$.

Table 1. Extended Class Interpretation Function

$$\top^{\mathcal{I}}(x) = \top_{yy}, \text{where y is the information source, defining} \top^{\mathcal{I}}(x)$$
$$\bot^{\mathcal{I}}(x) = \bot_{yy}, \text{where y is the information source, defining} \bot^{\mathcal{I}}(x)$$
$$(C_1 \sqcap C_2)^{\mathcal{I}}(x) = C_1^{\mathcal{I}}(x) \dot\wedge C_2^{\mathcal{I}}(x)$$
$$(C_1 \sqcup C_2)^{\mathcal{I}}(x) = C_1^{\mathcal{I}}(x) \dot\vee C_2^{\mathcal{I}}(x)$$
$$(\neg C)^{\mathcal{I}}(x) = \dot\neg C^{\mathcal{I}}(x)$$
$$(S^-)^{\mathcal{I}}(x,y) = S^{\mathcal{I}}(y,x)$$
$$(\forall R.C)^{\mathcal{I}}(x) = \dot\bigwedge_{y \in \Delta^{\mathcal{I}}} R^{\mathcal{I}}(x,y) \dot\to C^{\mathcal{I}}(y)$$
$$(\exists R.C)^{\mathcal{I}}(x) = \dot\bigvee_{y \in \Delta^{\mathcal{I}}} R^{\mathcal{I}}(x,y) \dot\wedge C^{\mathcal{I}}(y)$$
$$(\exists R.\text{Self})^{\mathcal{I}}(x) = R^{\mathcal{I}}(x,x)$$
$$(\geq nS)^{\mathcal{I}}(x) = \dot\bigvee_{\{y_1,\dots,y_m\} \subseteq \Delta^{\mathcal{I}}, m \geq n} \dot\bigwedge_{i=1}^{n} S^{\mathcal{I}}(x,y_i)$$
$$(\leq nS)^{\mathcal{I}}(x) = \dot\neg \dot\bigvee_{\{y_1,\dots,y_{n+1}\} \subseteq \Delta^{\mathcal{I}}} \dot\bigwedge_{i=1}^{n+1} S^{\mathcal{I}}(x,y_i)$$
$$\{a_1,\dots,a_n\}^{\mathcal{I}}(x) = \dot\bigvee_{i=1}^{n} a_i^{\mathcal{I}} = x$$

Table 2. Satisfaction of Axioms

$$(R \sqsubseteq S)^{\mathcal{I}} = \dot\bigwedge_{x,y \in \Delta^{\mathcal{I}}} R^{\mathcal{I}}(x,y) \dot\to S^{\mathcal{I}}(x,y)$$
$$(R = S)^{\mathcal{I}} = \dot\bigwedge_{x,y \in \Delta^{\mathcal{I}}} R^{\mathcal{I}}(x,y) \dot\leftrightarrow S^{\mathcal{I}}(x,y)$$
$$(R_1 \circ \dots \circ R_n \sqsubseteq S)^{\mathcal{I}} = \dot\bigwedge_{\langle x_1, x_{n+1}\rangle \in dom(S^{\mathcal{I}})} \dot\bigvee_{\{x_2,\dots,x_n\}} \dot\bigwedge_{i=1}^{n} R_i^{\mathcal{I}}(x_i, x_{i+1})$$
$$(Asy(R))^{\mathcal{I}} = \dot\bigwedge_{x,y \in \Delta^{\mathcal{I}}} \dot\neg(R^{\mathcal{I}}(x,y) \dot\wedge R^{\mathcal{I}}(y,x))$$
$$(Ref(R))^{\mathcal{I}} = \dot\bigwedge_{x \in \Delta^{\mathcal{I}}} R^{\mathcal{I}}(x,x)$$
$$(Irr(R))^{\mathcal{I}} = \dot\bigwedge_{x \in \Delta^{\mathcal{I}}} \dot\neg R^{\mathcal{I}}(x,x)$$
$$(Dis(R,S))^{\mathcal{I}} = \dot\bigwedge_{x,y \in \Delta^{\mathcal{I}}} R^{\mathcal{I}}(x,y) \dot\to \dot\neg S^{\mathcal{I}}(x,y)$$
$$(C \sqsubseteq D)^{\mathcal{I}} = \dot\bigwedge_{x \in \Delta^{\mathcal{I}}} C^{\mathcal{I}}(x) \dot\to D^{\mathcal{I}}(x)$$
$$(a : C)^{\mathcal{I}} = C^{\mathcal{I}}(a^{\mathcal{I}})$$
$$((a,b) : R)^{\mathcal{I}} = R^{\mathcal{I}}(a^{\mathcal{I}}, b^{\mathcal{I}})$$
$$a \approx b = a^{\mathcal{I}} = b^{\mathcal{I}}$$
$$a \not\approx b = a^{\mathcal{I}} \neq b^{\mathcal{I}}.$$

We say an ontology O is u-satisfiable, if there exist a complete interpretation $\mathcal{I}$, which u-satisfies all axioms in O and for each class C we have $|\{a|\langle a, v\rangle \in C \wedge v \geq_t u\}| > 0$, that means no class is empty.

Now we define a special kind of satisfiability, which reflects our trust order:

Definition 9 (Trust Satisfiability)
Let $\mathcal{I}$ be a complete interpretation, O an ontology, which is composed from multiple data sources $\{S_1, ..., S_n\}$, and $\mathcal{T}$ a trust order over $\{S_1, ..., S_n\}$. Let $source(E)$ denote the $\mathcal{T}$-maximal datasource, which axiom E comes from. O is trust satisfiable, if there exists an $\mathcal{I}$, which satisfies O, such that for all axioms in table 2 and all $E \in O : val(E, \mathcal{I}) \geq_t t_{source(E)}$.

Analogously we define consistency wrt. the knowledge order:

Definition 10 (Consistency)
Let $\mathcal{I}$ be a complete interpretation, O an ontology, which is composed from multiple data sources $\{S_1, ..., S_n\}$ and $\mathcal{T}$ a trust order over $\{S_1, ..., S_n\}$. Let $source(E)$ denote the $\mathcal{T}$-maximal datasource axiom E comes from.

We say O is u-consistent, if there exists an $\mathcal{I}$, which assigns a truth value $val(E, \mathcal{I})$ to all axioms E in O, such that $u \leq_k val(E, \mathcal{I})$. $\mathcal{I}$ is called a u-model of O.

We say O is consistent, if there exist an $\mathcal{I}$, which assigns a truth value $val(E, \mathcal{I})$ to all axioms E in O, such that $\forall x : val(E, \mathcal{I}) \notin \{\top_x, \bot_x\}$. We say $\mathcal{I}$ is a model of O.

If $\mathcal{I}$ models O and trust satisfies O, $\mathcal{I}$ is called a trusted model of O.

Finally, we define entailment:

Definition 11 (Entailment)
O entails a $\mathcal{SROIQ} - \mathcal{T}$ ontology O' ($O \vDash O'$), if every model of O is also a model of O'. O and O' are equivalent if O entails O' and O' entails O.

The following theorem shows that we have indeed defined a strict extension of $\mathcal{SROIQ}$:

Theorem 2. *If $\mathcal{FOUR}$ is used as logical bilattice, $\mathcal{SROIQ} - \mathcal{T}$ is isomorphic to $\mathcal{SROIQ}$.*

Proof (sketch). From a model of a $\mathcal{SROIQ} - \mathcal{T}$ ontology O wrt. $\mathcal{FOUR}$, we can derive a model for the same ontology in $\mathcal{SROIQ}$ by doing the following steps:

- For each class C, replace $C(a) = t$ by $a \in C$ and $C(a) = f$ by $a \notin C$. Analogous for properties.
- As O is consistent, we do not have $\top$ and $\bot$ truth values in a model.
- The only connectives used in the ontology language are $\neg, \vee, \wedge$, these are equivalent to their boolean counterparts in $\mathcal{FOUR}$.
- As we only have one trust level, we can simply ignore it.

– replace "trust-consistent" in $\mathcal{SROIQ} - \mathcal{T}$ with "consistent" in $\mathcal{SROIQ}$. Analogous for satisfiability.

For a complete proof we need to show for every rule in tables 1 and 2 that we can transform it to the $\mathcal{SROIQ}$ form (wrt. $\mathcal{FOUR}$) by replacing conjunctions and disjunctions by quantifiers over set membership or inclusion.

7   Resolving Inconsistencies

Inconsistencies in ontologies often emerge, when ontologies are integrated from various sources using ontology modules, ontology mappings and similar mechanisms [12]. In this section we propose to use trust based reasoning for inconsistency resolution.

Definition 12 (Maximally Trust Consistent Interpretation)
We say an interpretation $\mathcal{I}$ is maximally trust consistent, *if it does not assign any artificial $\top$ values, i.e. $\top_{xy}$ with $x \neq y$. An ontology O is said to be maximally trust consistent, if it has an interpretation $\mathcal{I}$, which is maximally trust consistent.*

This means, a trust maximal interpretation can still be inconsistent, but such inconsistency then arises from information obtained from a single information source. We now define, how a maximally trust consistent ontology can be derived from any given ontology.

Various approaches for *repairing* inconsistent ontologies have been proposed. In most approaches, axioms are removed from the ontology until the rest is a consistent ontology (cf.[12]). There usually are multiple possible choices for axioms to remove. While this might not seem too bad in our example, consider a similar scenario involving a red traffic light and we accidentally remove *RedLightSituation $\sqsubseteq$ BetterBreakSituation*. Here trust based reasoning comes into play. We use the trust level as input to a selection function, which determines axioms to be removed, a point left open in [12].

Definition 13 (Minimal Inconsistent Subontology)
Let O be an inconsistent ontology. A minimal inconsistent subontology $O' \subseteq O$ *is an inconsistent ontology, such that every $O'' \subset O'$ is consistent.*

Now trust maximal consistency is reestablished by iteratively removing all axioms with the lowest trust level from O', until the resulting ontology is trust maximal consistent. This captures the idea, that in the case of an inconsistency, humans tend to ignore lowly trusted information first.

If a trust maximal consistent ontology still is inconsistent, we need a different selection function. However, in this case already the *local* knowledge is inconsistent. A similar situation arises, if we have an inconsistency resulting from two incomparable information source. In the latter case, however, we can choose to discard information from both. Of cause, depending on the actual application, we can also use a more sophisticated selection function, choosing among axioms on the lowest trust level.

8   Related Work

Relevant related work comes from the fields of semantic caching, multi-valued logics — particularly based on logical bi-lattices — from belief revision and trust. The following works are closely related:

The term Semantic Caching refers to the caching of semantic data. Examples are such diverse topics as caching results of semantic webservice discovery [17], caching of ontologies [1] and caching to improve the performance of query engines [10]. These approaches have in common, that they discuss how to best do caching of semantic data. In this paper, we describe which additional information about the reliability of knowledge we can infer, given a heterogeneous infrastructure containing semantic caches.

Katz and Golbeck propose to use trust levels obtained from the analysis of online social networks to prioritize default logics [11]. A trust level in a rule then is a global value. Here, we do not specify, how the trust order is determined, but assume it is supplied by the user. [11] is based on a two valued default logic, that means trust levels of inferred facts are not computed. In [15], rule based reasoning over annotated information sources is done to establish a trust relation between the provider and the requester of a resource. The focus, is on establishing a trust relation using fine grained negotiation, instead of determining trust in a statement.

Much work has been done about basing logical formalisms on bilattices (cf. [3], [9]). Most of these works, however propose a certain logic by manually designing a suitable bilattice, or discuss how a particular logic can be formalized using a bilattice. In contrast to these works, we do not propose a fixed bilattice or logic. Instead, we automatically derive logical bilattices for trust based reasoning. Hence, we propose a whole family of logics, which can automatically be tailored to the problem at hand.

Deschrijver et al. propose a bilattice based framework of handling graded truth [4]. They extend fuzzy logics, which has a single, continuous order $\leq_t$ towards bilattices which also have a "fuzzy $\leq_k$". While this is obviously closely related to ours, we propose to use possibly partial orders, instead of strict orders as in fuzzy logics. Additionally we show, how the logical framework an be used with rule based and description logics.

Using belief revision, a single, consistent world view is retained in the presence of contradictory information by discarding e.g. lowly trusted information. In contrast, paraconsistent reasoning, as applied here, limits the influence to inconsistencies to fragments of the knowledge base. An approach similar to ours, but based on belief revision is proposed in [7].

Other works can be considered orthogonal to our approach: Following Golbeck's categorization of trust [6], our approach deals with trust in content (vs. trust in people or services). Further, we provide means for computing trust. In contrast to existing systems (cf. [6], chap. 2), we allow to infer trust levels on the very fine level of axioms, instead of the usual level of documents. As our approach is agnostic to the actual trust order, e.g. social trust derived from social networks or P2P based algorithms for computing trust measures can be used to provide this order.

9   Conclusion

We have proposed an extension to the logical bilattice $\mathcal{FOUR}$, called $\mathcal{FOUR} - \mathcal{T}$, which allows to reason with trust levels. As bilattices are a basis for various logical formalisms, this allows to extend many languages with trust based reasoning.

We have started by applying the extension to the well founded semantics. We have re-formalized $\mathcal{SROIQ}$ to work on bilattices, so our extension is applicable in both, open and closed world reasoning. As applications of $\mathcal{FOUR} - \mathcal{T}$ we have investigated caching and inconsistency resolution. We are sure that our mechanism can be a good component of a future Semantic Web trust layer. As part of our future work we will investigate additional applications. We will investigate the complexity of trust based reasoning with description logics and plan an implementation, extending existing reasoning engines.

Acknowledgements. This work has been supported by the European project Life-cycle Support for Networked Ontologies (NeOn, IST-2006-027595).

References

1. Liang, B., et al.: Semantic Similarity Based Ontology Cache. In: Zhou, X., Li, J., Shen, H.T., Kitsuregawa, M., Zhang, Y. (eds.) APWeb 2006. LNCS, vol. 3841. Springer, Heidelberg (2006)
2. Berners-Lee, T.: Semantic Web Road Map [2008-05-12] (1998), http://www.w3.org/DesignIssues/Semantic.html
3. Fitting, M.: Fixpoint Semantics for Logic Programming - A Survey. Theoretical Computer Science 278(1-2) (2002)
4. Deschrijver, G., et al.: A Bilattice-based Framework for Handling Graded Truth and Imprecision. Uncertainty, Fuzziness and Knowledge-Based Systems 15(1) (2007)
5. Ginsberg, M.L.: Multivalued Logics: A Uniform Approach to Inference in Artificial Intelligence. Computational Intelligence 4(3) (1992)
6. Golbeck, J.: Trust on the World Wide Web: A Survey. Web Science 1(2), 131–197 (2006)
7. Golbeck, J., Halaschek-Wiener, C.: Trust-Based Revision for Expressive Web Syndication. Logic and Computation (to appear, 2008)
8. Grau, B.C., Motik, B.: OWL 2 Web Ontology Language: Model-Theoretic Semantics [2008-05] (2008), http://www.w3.org/TR/owl2-semantics/
9. Hitzler, P., Wendt, M.: A uniform approach to logic programming semantics. TPLP 5(1-2) (2005)
10. Kaplunova, A., Kaya, A., Möller, R.: Experiences with load balancing and caching for semantic web applications. In: Proc. of DL Workshop (2006)
11. Katz, Y., Golbeck, J.: Social Network-based Trust in Prioritized Default Logic. In: Proc. of AAAI (2006)
12. Haase, P., et al.: A Framework for Handling Inconsistency in Changing Ontologies. In: Gil, Y., Motta, E., Benjamins, V.R., Musen, M.A. (eds.) ISWC 2005. LNCS, vol. 3729. Springer, Heidelberg (2005)
13. Polleres, A.: From SPARQL to rules (and back). In: Proc. of WWW (2007)
14. Przymusinski, T.C.: The Well-Founded Semantics Coincides with the Three-Valued Stable Semantics. Fundamenta Informaticae 13(4) (1990)

15. Gavriloaie, R., et al.: No Registration Needed: How to use Declarative Policies and Negotiation to Access Sensitive Resources on the Semantic Web. In: Bussler, C.J., Davies, J., Fensel, D., Studer, R. (eds.) ESWS 2004. LNCS, vol. 3053. Springer, Heidelberg (2004)
16. Schenk, S., Staab, S.: Networked Graphs: A Declarative Mechanism for SPARQL Rules, SPARQL Views and RDF Data Integration on the Web. In: Proc. of WWW (2008)
17. Stollberg, M., Hepp, M., Hoffmann, J.: A Caching Mechanism for Semantic Web Service Discovery. In: Aberer, K., Choi, K.-S., Noy, N., Allemang, D., Lee, K.-I., Nixon, L., Golbeck, J., Mika, P., Maynard, D., Mizoguchi, R., Schreiber, G., Cudré-Mauroux, P. (eds.) ASWC 2007 and ISWC 2007. LNCS, vol. 4825. Springer, Heidelberg (2007)
18. Straccia, U.: A Fuzzy Description Logic for the Semantic Web. In: Fuzzy Logic and the Semantic Web. Elsevier, Amsterdam (2006)
19. van Gelder, A., Ross, K., Schlipf, J.S.: The Well-Founded Semantics for General Logic Programs. J. of the ACM 38(3) (1991)

Semantic Web Service Choreography: Contracting and Enactment

Dumitru Roman[1] and Michael Kifer[2]

[1] STI Innsbruck, University of Innsbruck, Austria
[2] State University of New York at Stony Brook, USA

Abstract. The emerging paradigm of service-oriented computing requires novel techniques for various service-related tasks. Along with automated support for service discovery, selection, negotiation, and composition, support for automated service contracting and enactment is crucial for any large scale service environment, where large numbers of clients and service providers interact. Many problems in this area involve reasoning, and a number of logic-based methods to handle these problems have emerged in the field of Semantic Web Services. In this paper, we build upon our previous work where we used Concurrent Transaction Logic (CTR) to model and reason about service contracts. We significantly extend the modeling power of the previous work by allowing iterative processes in the specification of service contracts, and we extend the proof theory of CTR to enable reasoning about such contracts. With this extension, our logic-based approach is capable of modeling general services represented using languages such as WS-BPEL.

1   Introduction

The area of Semantic Web Services is one of the most promising subareas of the Semantic Web. It is the main focus of large international projects such as OWL-S,[1] SWSL,[2] WSMO,[3] DIP,[4] and SUPER,[5] which deal with service discovery, service choreography, automated contracting for services, service enactment and monitoring. In this context, the focus of this paper is modeling of behavioral aspects of services, and service contracting and enactment.

In a service-oriented environment, interaction is expected among large numbers of clients and service providers, and contracts through human interaction is out of question. To enable automatic establishment of contracts, a formal contract description language is required and a reasoning mechanism is needed to be able to verify that the contract terms are fulfilled. A contract specification has to describe the functionality of the service, values to be exchanged, procedures,

---

[1] http://www.daml.org/services/owl-s/
[2] http://www.w3.org/Submission/SWSF-SWSL/
[3] http://www.wsmo.org/
[4] http://dip.semanticweb.org/
[5] http://ip-super.org/

A. Sheth et al. (Eds.): ISWC 2008, LNCS 5318, pp. 550–566, 2008.

and guarantees. This paper develops such a formal framework with a particular focus on service contracting.

Our approach is based on *Concurrent Transaction Logic* (CTR) [3] and continues the line of research that looks at CTR as a unifying formalism for modeling, discovering, choreographing, contracting, and enactment of Web services [5,12,6,9]. This previous work, however, was confined to straight-line services, which cannot do repeated interactions—a serious limitation in view of the fact that the emerging languages for specifying the behavior of Web services, such as the Web Services Choreography Description Language (WS-CDL)[6] or Web Services Business Process Execution Language (WS-BPEL),[7] identify iterative processes as core elements of their models. The present paper closes this gap by extending the previous work to cover choreography of iterative processes. CTR can thus be used to model and reason about general service choreographies, including the ones definable by very expressive languages such as [13], WS-CDL, and WS-BPEL. We obtain these results by significantly extending the language for service contracts and through a corresponding extension to the proof theory of CTR. In this way, we also contribute to the body of results about CTR itself.

This paper is organized as follows. Section 2 describes the basic techniques from process modeling, such as control flow and constraints. We then review the modeling framework of [9]. Section 3 gives a short introduction to CTR to help keep the paper self-contained. Section 4 shows how the framework outlined in Section 2 is formalized in CTR. Section 5 describes the reasoning procedure—the key component of service contracting in our framework. Section 6 presents related work, and Section 7 concludes this paper.

2   Modeling, Contracting, and Enactment of Services

In [9] we identified three core service behavior aspects of service contracting: (1) *Web service choreography*—a specification of how to invoke and interact with the service in order to get results; (2) *Service policies*—a set of additional constraints imposed on the choreography and on the input; and (3) *Client contract requirements*—the contractual requirements of the user, which go beyond the basic functions (such as selling books or handling purchase orders) of the service.

The choreography of a service is described with *control* and *data flow graphs*, and service policy and clients' contract requirements are described with *constraints*. We defined the problem of *service contracting* as: given a service choreography, a set of service policies, and a set of client contract requirements, decide whether an execution of the service choreography exists, that satisfies both the service policies and the client contract requirements. Furthermore, the problem of *service enactment* was defined as finding out whether enactment of a service is possible according to a contract and, if so, finding an actual sequence of interactions that fulfils the contract.

---

[6] http://www.w3.org/TR/ws-cdl-10/
[7] http://www.oasis-open.org/committees/wsbpel/

The present paper deals with the same type of problems, but the settings are significantly different: the service choreographies (control-flow graphs) are more complex since now we allow iterative interactions (whereas [9] could deal only with iteration-free interactions), and the set of allowed constraints is likewise different: it allows complex constraints on iterative interactions. We illustrate the new approach through an example of a virtual manufacturing service, which handles purchase orders (POs). Apart from the clients, the virtual manufacturing service may interact with service providers to purchase the various items requested in the purchase order. It may also contract with shippers to deliver the purchased items to the client. For each item in a PO, the virtual manufacturing service checks item availability with producers. Depending on the availability and for other reasons (e.g., the reputation of the item producers) the virtual service may choose a certain item producer-service, or it may inform the client that an item is unavailable. If an item is available, the client may choose to accept or to reject the offered item. To ship the items in the PO, the virtual service contacts shipping services. Depending on the availability of shipping services and taking a host of other considerations into account (e.g., shipper's reputation) the virtual service may book a specific shipper for delivering some or all of the items. Clients are required to provide payment guarantees. If a guarantee is provided, payment can be made for all items at once or separately for each item.

Service Choreographies. Figure 1 shows a *service choreography* depicted as a *control flow graph* of a fairly complex virtual manufacturing service. Control flow graphs are typically used to specify local execution dependencies among the interactions of the service, since they are a good way to visualize the overall flow of control. The nodes in such a graph represent *interaction tasks*, which can be thought of as functions that take inputs and produce outputs. In Figure 1, tasks are represented as labeled rectangles. The label of a rectangle is the name of the task represented by that rectangle, and a graph inside the rectangle is the *definition* or *decomposition* of that task. Such a task is called *composite*. Tasks that do not have associated graphs are *primitive*. A control flow graph can thus be viewed as a hierarchy of tasks. There is always one composite task at the root of the hierarchy. In our example, **handle_order** is the root; its subtasks include **handle_items** and **handle_shippers**. These subtasks are composite and their rectangles are shown separately (to avoid clutter). The task **place_order** is an example of a primitive task. To differentiation primitive and composite tasks, we use rectangles with gray background to depict primitive tasks.

Each composite task has an initial and the final interaction task, the successor task for each interaction task, and a sign that tells whether these successors must *all* be executed concurrently (indicated by **AND**-split nodes), or whether only one of the alternative branches needs to be executed non-deterministically (indicated by **OR**-nodes). For instance, in Figure 1(a), all successors of the initial interaction **place_order** must be executed, whereas in Figure 1(e) either **item_available** or **item_unavailable** is executed. The node **payment_guarantee** is also an **OR**-split, but with a twist. The lower branch going out of this node represent a situation where the client provides a payment.

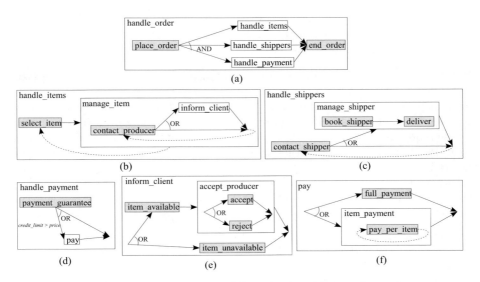

Fig. 1. A hierarchy of control-flow graphs with iterative tasks

The upper branch, has no interactions, and it joins the lower branch. This means that the lower branch is *optional*: the service may or may not accept a payment.

Control flows of some of the composite tasks in Figure 1 have dashed arrows pointing from the final task to the initial task in the respective graphs. Such arrows represent the fact that a composite task can execute iteratively multiple times. We call these *iterative* tasks and differentiate them from *non-iterative* tasks. For example, Figure 1(b) depicts **handle_items** as an iterative task where a sequence of two sub-tasks, **select_item** and **manage_item**, can be executed multiple times (for example for each item in the PO). Note also the condition *credit_limit > price* attached to the arc leaving the node **payment_guarantee**. Such a condition is called *transition condition*. It says that in order for the next interaction with the service to take place the condition must be satisfied. The parameters *credit_limit* and *price* may be obtained by querying the current state of the service or they may be passed as parameters from one interaction to another—the actual method depends on the concrete representation. In general, transition conditions are Boolean expressions attached to the arcs in control flow graphs. Only the arcs whose conditions are true can be followed at run time.

With these explanations, it should be clear how the control flow graph in Figure 1 represents the virtual manufacturer scenario described earlier. The top-level composite task, **handle_order** is used to process orders. Figure 1(a) depicts its decomposition: first the order is placed (**place_order**), then the items in the PO are processed (**handle_items**), delivery is arranged (**handle_shippers**), and a payment process is initiated (**handle_payment**). These three tasks are executed in parallel. Once all of them completes, the order handling is finished (**end_order**). The other parts of the figure show how each of the above subtasks are executed. The important things to observe here is that some tasks are complex and some primitive; some are to be executed in parallel (the **AND**-nodes) and some in sequence; some tasks have non-deterministic choice (the **OR**-nodes) and some are iterative (those that have dashed arcs in their definitions).

Service Policies and Client Contract Requirements. Apart from the local dependencies represented directly in control flow graphs, *global* constraints often arise as part of *policy* specification. Another case where global constraints arise is when a client has specific requirements to the interaction with the service. These requirements usually have little to do with the functionality of the service (e.g., handling orders), but rather with the particular guarantees that the client wants before entering into a contract with the service. We call such constrains *client contract requirements*. In (1) we give an example of global constraints that represent service policy and client contract requirements for our running example. Note that all the constraints in this example are on iterative tasks.

> **Service policy:**
>
> *1. A shipper is booked only if the user accepts at least 7 items.*
>
> *2. If pay per item is chosen by the user, then the payment must happen immediately before each item delivery.*
>
> *3. Payment guarantee must be given before the client is informed about the availability of items.*
>
> **Client contract requirements:**
>
> *4. All items purchased by the client must be shipped at once.*
>
> *5. If full payment is chosen by the client, then it must happen only after all purchased items are delivered.*
>
> *6. Before the client purchases items the service must have booked a shipper.*

$$(1)$$

Service Contracting and Enactment. With this modeling mechanism in place, we define service contracting and enactment as follows:

- *Service contracting*: given a service choreography (i.e. a hierarchical control-flow graph containing iterations), a set of service policies and client contract requirements (i.e. constraints), decide if there is an execution of the service choreography that complies both with the service policies and the client contract requirements.
- *Service enactment*: if service contracting is possible, find out an actual order of interactions that fulfils the contract, and execute it.

3   Overview of CTR

Concurrent Transaction Logic (CTR) [3] is an extension of the classical predicate logic, which allows programming and reasoning about state-changing processes. Here we summarize the relevant parts of CTR's syntax and give an informal explanation of its semantics. For details we refer the reader to [3].

Basic syntax. The atomic formulas of CTR are identical to those of the classical logic, *i.e.*, they are expressions of the form $p(t_1, \ldots, t_n)$, where p is a predicate symbol and the t_i's are function terms. More complex formulas are built with the help of connectives and quantifiers. Apart from the classical $\vee$, $\wedge$, $\neg$, $\forall$, and $\exists$, CTR has two additional connectives, $\otimes$ (*serial conjunction*) and $\mid$ (*concurrent conjunction*), and a modal operator $\odot$ (*isolated execution*). For instance, $\odot(p(X) \otimes q(X)) \mid (\forall Y (r(Y) \vee s(X, Y)))$ is a well-formed formula.

Informal semantics. Underlying the logic and its semantics is a set of database *states* and a collection of *paths*. For the purpose of this paper, the reader can think of the states as just a set of relational databases, but the logic does not rely on the exact nature of the states—it can deal with a wide variety of them.

A *path* is a finite sequence of states. For instance, if $s_1, s_2, ..., s_n$ are states, then $\langle s_1 \rangle$, $\langle s_1\ s_2 \rangle$, and $\langle s_1\ s_2\ ...\ s_n \rangle$ are paths of length 1, 2, and n, respectively.

Just as in classical logic, CTR formulas assume truth values. However, *unlike* classical logic, the truth of CTR formulas is determined over paths, *not* at states. If a formula, ϕ, is true over a path $\langle s_1, ..., s_n \rangle$, it means that ϕ can *execute* starting at state s_1. During the execution, the current state will change to s_2, s_3, ..., etc., and the execution terminates at state s_n.

With this in mind, the intended meaning of the CTR connectives can be summarized as follows:

- $\phi \otimes \psi$ *means*: execute ϕ then execute ψ. Or, model-theoretically, $\phi \otimes \psi$ is true over a path $\langle s_1, ..., s_n \rangle$ if ϕ is true over a prefix of that path, say $\langle s_1, ..., s_i \rangle$, and ψ is true over the suffix $\langle s_i, ..., s_n \rangle$.
- $\phi \mid \psi$ *means*: ϕ and ψ must both execute concurrently, in an interleaved fashion.
- $\phi \wedge \psi$ *means*: ϕ and ψ must both execute along the *same* path. In practical terms, this is best understood in terms of *constraints* on the execution. For instance, ϕ can be thought of as a transaction and ψ as a constraint on the execution of ϕ. It is this feature of the logic that lets us specify constraints as part of service contracts.
- $\phi \vee \psi$ *means*: execute ϕ *or* execute ψ non-deterministically.
- $\neg \phi$ *means*: execute in any way, provided that this will *not* be a valid execution of ϕ. Negation is an important ingredient in temporal constraint specifications.
- $\odot \phi$ *means*: execute ϕ in isolation, *i.e.*, without interleaving with other concurrently running activities. This operator enables us to specify the transactional parts of service contacts.

CTR contains a predefined propositional constant, `state`, which is true only on paths of length 1, that is, on database states. As we shall see in the next section, `state` is used in the definition of iterative tasks in service choreographies. Another propositional constant that we will use in the representation of constraints is `path`, defined as `state` $\vee \neg$`state`, which is true on every path.

Concurrent-Horn subset of CTR. Implication $p \longleftarrow q$ is defined as $p \vee \neg q$. The form and the purpose of the implication in CTR is similar to that of Datalog: p can be thought of as the name of a procedure and q as the definition of that procedure. However, unlike Datalog, both p and q assume truth values on execution paths, not at states.

More precisely, $p \longleftarrow q$ means: if q can execute along a path $\langle s_1, ..., s_n \rangle$, then so can p. If p is viewed as a subroutine name, then the meaning can be re-phrased as: one way to execute p is to execute its definition, q.

The control flow parts of service choreographies are formally represented as *concurrent-Horn goals* and *concurrent Horn rules*. A *concurrent Horn goal* is:

- any atomic formula is a concurrent-Horn goal;
- $\phi \otimes \psi$, $\phi \mid \psi$, and $\phi \vee \psi$ are concurrent-Horn goals, if so are ϕ and ψ;
- $\odot \phi$ is a concurrent-Horn goals, if so is ϕ.

A *concurrent-Horn rule* is a CTR formula of the form *head* $\longleftarrow$ *body*, where *head* is an atomic formula and *body* is a concurrent-Horn goal. The concurrent-Horn fragment of CTR is efficiently implementable, and there is an SLD-style proof procedure that proves concurrent-Horn formulas and *executes* them at the same time [3]. Observe that the definition of concurrent-Horn rules and goals *does not* include the connective $\wedge$. In general, $\wedge$ represents *constrained execution*. The present work deals with a much larger class of constraints than [5,9], which includes iterative processes, and formulas of the form $ConcurrentHornGoal \wedge Constraints$ are handled by the extended proof theory in Section 5.

Elementary updates. In CTR *elementary updates* are formulas that change the underlying database state. Semantically they are binary relations over states. For instance, if $\langle s_1 \ s_2 \rangle$ belongs to the relation corresponding to an elementary update u, it means that u can cause a transition from state s_1 to state s_2.

Constraints. Because transactions are defined on paths, CTR can express a wide variety of constraints on the way transactions execute. One can place conditions on the state of the database during transaction execution (constraints based on serial conjunction), or may forbid certain sequences of states (constraints based on serial implication). To express the former, the proposition constant `path`, introduced above, is used; for example, `path` $\otimes \psi \otimes$ `path` specifies a path on which ψ must be true. To express the latter, the binary connectives "$\Leftarrow$" and "$\Rightarrow$" are used. The formula $\psi \Leftarrow \phi$ means that whenever ψ occurs, then ϕ occurs right after it. The formula $\psi \Rightarrow \phi$ means that whenever ϕ occurs, then ϕ must have occurred just before it.

4   Formalizing Service Contracts

We begin our formalization by showing how service choreographies are represented in CTR (Section 4.1). Section 4.2 proceeds to formalize service policies and contract requirements as constraints expressed in CTR. Section 4.3 provides a discussion on the assumption we take when modeling service contracts.

4.1   Modeling Service Choreography with CTR

In CTR, *tasks* are represented as formulas of the form $p(X_1, \ldots, X_n)$, where p is a predicate symbol and the X_i's are variables. The predicate symbol is the name of the task, and the variables are placeholders for data items that the task manipulates (e.g. inputs, outputs, etc.). A *task instance* is a task whose variables are substituted with concrete values.

Definition 1. (Dependency between tasks) *A task p_1 depends on a task p_2 if p_2 appears in the body of a rule that has p_1 as its head.*

Definition 2. (Primitive task) *A task is* primitive *if it does not depend on any other task.*

We conceptualize primitive tasks as opaque actions that produce some external action. In CTR, such actions are represented as elementary updates and so ground instances of primitive tasks are treated as CTR's elementary updates.

Definition 3. (Non-iterative task) *A non-iterative* composite task, p, *is a task defined by a rule of the form $p \leftarrow q$, where q is a CTR goal none of whose tasks depends on p.*

Definition 4. (Iterative task) *An* iterative task, p, *is a task defined by a rule of the form $p \leftarrow (q \otimes p) \vee \mathsf{state}$, where q is a CTR goal none of whose tasks depends on p. This is equivalent to a pair of concurrent-Horn rules.*

Definition 5. (Service choreography) *A service choreography is an iterative or non-iterative composite task that represents the root of the task hierarchy, along with the rules defining it.*

A CTR representation of the service choreography from Figure 1 is shown in (2), below. The *handle_order* task is the root of task hierarchy; it is followed by the rules that define it.

$$
\begin{aligned}
handle\_order \leftarrow\ & place\_order \otimes \\
& (handle\_items \mid handle\_shippers \mid handle\_payment) \otimes end\_order \\
handle\_items \leftarrow\ & (select\_item \otimes manage\_item \otimes handle\_items) \vee \mathsf{state} \\
manage\_item \leftarrow\ & (contact\_producer \otimes (inform\_client \vee state) \otimes manage\_item) \vee \mathsf{state} \\
handle\_shippers \leftarrow\ & (contact\_shipper \otimes (manage\_shipper \vee \mathsf{state}) \otimes handle\_shippers) \vee \mathsf{state} \\
manage\_shipper \leftarrow\ & book\_shipper \otimes deliver \\
handle\_payment \leftarrow\ & payment\_guarantee \otimes ((credit\_limit > price \otimes pay) \vee \mathsf{state}) \\
inform\_client \leftarrow\ & (item\_available \otimes accept\_producer) \vee item\_unavailable \\
accept\_producer \leftarrow\ & accept \vee reject \\
pay \leftarrow\ & full\_payment \vee item\_payment \\
item\_payment \leftarrow\ & (pay\_per\_item \otimes item\_payment) \vee \mathsf{state}
\end{aligned}
\tag{2}
$$

4.2 Modeling Constraints Using CTR

We now define an algebra of constraints, $\mathcal{C}_{ONSTR}$, which we will use in this paper to model service policies.

Definition 6. (Constraints) *The following constraints form the algebra $\mathcal{C}_{ONSTR}$ (and nothing else):*

1. **Primitive constraints**: *If a is a task in a service choreography, then the following are **primitive** constraints:*
    - *existence(a, n)*—task a must execute at least n times (n $\geq$ 1): $\nabla_{\geq n} a \equiv \underbrace{\nabla_{\geq 1} a \otimes ... \otimes \nabla_{\geq 1} a}_{n \geq 1}$, where $\nabla_{\geq 1} a \equiv \mathsf{path} \otimes a \otimes \mathsf{path}$. [8]

---

[8] We will also use ∇a as an abbreviation for $\nabla_{\geq 1} a$ as this constraint occurs very frequently.

- $absence(a)$ - task a must not execute: $\neg \nabla a \equiv \neg(\text{path} \otimes a \otimes \text{path})$
- $exactly(a, n)$ - task a must execute exactly n times ($n \geq 1$): $\nabla_n a \equiv \underbrace{\nabla_1 a \otimes ... \otimes \nabla_1 a}_{n \geq 1}$, where $\nabla_1 a \equiv \neg \nabla a \otimes a \otimes \neg \nabla a$.

2. **Serial constraints**: *If a, b are tasks in a service choreography then the following are **serial** constraints:*

   - $after(a, b)$ - whenever a executes, b has to be executed after it. Task b does not have to execute immediately after a, and several other instances of a might execute before b does: $(\text{path} \otimes a) \Rightarrow \nabla b$
   - $before(a, b)$ - whenever b executes, it must be preceded by an execution of a. Task a does not have to execute immediately before b: $\nabla a \Leftarrow (b \otimes \text{path})$
   - $blocks(a, b)$ - if a executes, b can no longer be executed in the future: $(\text{path} \otimes a) \Rightarrow \neg \nabla b$
   - $between(a, b, a)$ - b must execute between any two executions of a, i.e. after an execution of a, any subsequent execution of a is blocked until b is executed: $(\text{path} \otimes a) \Rightarrow \nabla b \Leftarrow (a \otimes \text{path})$
   - $not\text{-}between(a, b, a)$ - b must not execute between any pair of executions of a. If b executes after a, no future execution of a is possible: $(\text{path} \otimes a) \Rightarrow \neg \nabla b \Leftarrow (a \otimes \text{path})$

3. **Immediate serial constraints**: *If a, b are tasks in a service choreography then the following are **immediate serial** constraints:*

   - $right\text{-}after(a, b)$ - whenever a executes, b has to execute immediately after it: $(\text{path} \otimes a) \Rightarrow (b \otimes \text{path})$
   - $right\text{-}before(a, b)$ - whenever b executes, a has to be executed immediately before it: $(\text{path} \otimes a) \Leftarrow (b \otimes \text{path})$
   - $not\text{-}right\text{-}after(a, b)$ - whenever a and b execute, b must not execute immediately after a, i.e. between the execution of a and b there must be an execution of a task other than a and b: $(\text{path} \otimes a) \Rightarrow (\neg \text{state} \land \neg \nabla a \land \neg \nabla b) \Leftarrow (b \otimes \text{path})$

   The negation of $right\text{-}before(a, b)$ is equivalent to $not\text{-}right\text{-}after(b, a)$, so we do not define it explicitly.

4. **Complex constraints**: *If C_1, $C_2 \in \mathcal{C}_{ONSTR}$ then so are $C_1 \land C_2$, and $C_1 \lor C_2$.*

The following examples illustrate the diverse set of constraints that can be expressed with the help of $\mathcal{C}_{ONSTR}$.

- $\neg \nabla a \lor \nabla_1 a \lor \nabla_2 a \lor ... \lor \nabla_n a$ — task a must execute at most n times. We denote this constraint by $at\text{-}most(a, n)$. This constraint together with the primitive constraints introduced earlier capture the set of *existence formulas* from [13].
- $\neg \nabla a \lor \nabla b$ — if a is executed, then b must also execute (before or after a).
- $(\neg \nabla a \lor \nabla b) \land (\neg \nabla b \lor \nabla a)$ — if a is executed, then b must also be executed, and vice versa.
- $after(a, b) \land before(a, b)$ — every occurrence of task a must be followed by an occurrence of task b and there must be an occurrence of a before every occurrence of b.
- $\neg \nabla a \lor between(a, b, a)$ — if task a is executed then b must execute after it, and before that b there can be no other a.
- $\neg \nabla b \lor (before(a, b) \land between(b, a, b))$ — if task b is executed, it has to be preceded by an occurrence of a. The next instance of b can executed only after another occurrence a.
- $between(a, b, a) \land between(b, a, b)$ — tasks a and b must alternate.

- $right\text{-}after(a, b) \wedge right\text{-}before(a, b)$ — executions of a and b must be next to each other with no intervening tasks in-between.
- $\neg \nabla a \vee \neg \nabla b$ — it is not possible for a and b to execute in the same choreography run.
- $not\text{-}between(a, b, a) \wedge not\text{-}between(b, a, b)$ — b must not execute between any two executions of a, and a must not execute between any two executions of b.

With the modeling mechanism in place, we can now represent the constraints from (1) formally:

1. $(at\text{-}most(accept, 6) \wedge absence(book\_shipper)) \vee$
   $(existence(accept, 7) \wedge after(accept, book\_shipper))$
2. $absence(item\_payment) \vee right\_before(pay\_per\_item, deliver)$
3. $before(payment\_guarantee, inform\_client)$ $\hspace{3cm}$ (2)
4. $exactly(deliver, 1)$
5. $absence(full\_payment) \vee (before(deliver, full\_payment) \wedge block(full\_payment, deliver))$
6. $before(book\_shipper, pay)$

4.3 Service Contracts Assumption

We now introduce the assumptions about the forms of the constraints and tasks involved in service choreography. These assumptions do not limit the modeling power of the language in the sense that any service choreography can be simulated by another choreography that satisfies these assumptions.

Primitive Tasks Independence Assumption. A service choreography, G, satisfies the independence assumption iff all of its primitive tasks are independent of each other; two primitive tasks are said to be *independent* iff they are represented by *disjoint* binary relations over database states.[9] This assumption means that a transition between two states is caused by precisely one primitive task, and no other task can cause the transition between those states. It is easy to see that the independence assumption does not limit the modeling power in the following sense: there is a 1-1 correspondence between executions of the original choreography and executions of the instrumented choreography.

Constraints Based on Primitive Tasks. Our service contracting reasoning technique (developed in Section 5) assumes that constraints are based on primitive tasks: a set of constraints, C, is said to be based on primitive tasks iff all tasks appearing in C are primitive tasks. As with the independence assumption, the above restriction on constraints does not limit the modeling power of the language. It is easy to instrument composite tasks in such a way that constraints that the resulting set of constraints will be based on primitive tasks only. More specifically, every composite task, p, can be changed as follows: $p_{start} \otimes p \otimes p_{end}$, where p_{start} and p_{end} are new *unique* primitive tasks. The effect is that now each composite task has a clearly identified begin- and end-subtask, which can be used in constraints. For instance, the constraint *between(a,b,a)* is now equivalent to *between(a,b_{start},a)* $\wedge$ *between(a,b_{end},a)*. We can also have constraints such as *before(a_{start},b_{end})* and *between(a_{start},b_{start},a_{end})*.

Unique Task Occurrence Assumption. Some of our results depend on the *unique task occurrence assumption*, which informally says that each task can

---

[9] Recall that primitive tasks are represented by elementary updates of CTR, and an elementary update is a binary relation over database states.

occur only once in the conjunctive part of the definition of any composite task. The unique task occurrence assumption does not limit the modeling power of our language, since the different occurrences of such tasks can be renamed apart.

Definition 7. (Service Contracts Assumption) *A service choreography G and a set of constraints C satisfy the* service contract assumption *iff the primitive tasks of G satisfy the independence assumption and G has the unique task occurrence property. In addition, the set of constraints C must be based on primitive tasks.*

5   Reasoning about Contracts

Let C be a constraint from $\mathcal{C}_{ONSTR}$, which includes the service policy and the client contract requirements. Let G be a a service choreography. Suppose G and C satisfy the service contracts assumption. Then

1. **Contracting**: The problem of determining if contracting for the service is possible is finding out if an execution of the CTR formula $G \wedge C$ exists.
2. **Enactment**: The problem of enactment is formally defined as finding a constructive *proof* for formulas of the form $G \wedge C$. A constructive proof is a sequence of inference rules of CTR that starts with an axiom and end with the formula $G \wedge C$. Each such proof gives us a way to execute the choreography so that all constraints are satisfied.

The rest of this section develops a proof theory for formulas of the form $G \wedge C$, where G is a service choreography and C is a constraint in $\mathcal{C}_{ONSTR}$. Section 5.1 presents a simplification operation used by the extended proof theory, and Section 5.2 presents the actual proof theory.

5.1   The Simplification Transformation

First, we define an auxiliary simplification transformation, $\mathcal{S}$. If G is a choreography and σ a primitive constraint, then $\mathcal{S}(G, \sigma)$ is also a service choreography (in particular, it does not contain the logical connective $\wedge$). If G has the unique task occurrence property then $\mathcal{S}(G, \sigma)$ is defined in such a way that the following is true: $\mathcal{S}(G, \sigma) \equiv G \wedge \sigma$. In other words, $\mathcal{S}$ is a transformation that eliminates the primitive constraint σ from $G \wedge \sigma$ by "compiling" it into the service choreography.[10] The following defines the simplification transformation $\mathcal{S}$.

Definition 8. (Simplification transformation) Let s be a primitive task from $\mathcal{T}_{ASKS}$. Let t be a another primitive task from $\mathcal{T}_{ASKS}$. Then:

$$\mathcal{S}(t, \triangledown t) = t; \qquad \mathcal{S}(t, \triangledown_{\geq n} t) = \neg\textbf{path}; \qquad \mathcal{S}(t, \neg\triangledown t) = \neg\textbf{path}; \qquad \mathcal{S}(t, \triangledown_1 t) = t;$$
$$\mathcal{S}(t, \triangledown_n t) = \neg\textbf{path}; \qquad \mathcal{S}(t, \triangledown s) = \neg\textbf{path}; \qquad \mathcal{S}(t, \triangledown_{\geq n} s) = \neg\textbf{path};$$
$$\mathcal{S}(t, \neg\triangledown s) = t; \qquad \mathcal{S}(t, \triangledown_1 s) = \neg\textbf{path}; \qquad \mathcal{S}(t, \triangledown_n s) = \neg\textbf{path};$$

---

[10] Note that the conjunction $G \wedge \sigma$ can be an inconsistency.

We remind that $\neg\texttt{path}$ means inconsistency so if a conjunct reduces to $\neg\texttt{path}$ then the whole conjunction is inconsistent and if a disjunct is found to be inconsistent then it can be eliminated.

Let $p \in \mathcal{T}_{ASKS}$ be an *iterative* task of the form (4) (i.e. $p \leftarrow (q \otimes p) \vee \texttt{state}$) that satisfies the unique task assumption. Then:

$$\mathcal{S}(p, \nabla_{\geq n}s) =_{n=k_1+\ldots+k_m} \overset{\vee}{} (p \otimes \mathcal{S}(q, \nabla_{\geq k_1}s) \otimes p \otimes \mathcal{S}(q, \nabla_{\geq k_2}s) \otimes p \otimes$$
$$\ldots \otimes \mathcal{S}(q, \nabla_{\geq k_m}s) \otimes p)$$

$$\mathcal{S}(p, \neg\nabla s) = p', \text{ where } p' \text{ is defined as: } p' \leftarrow (\mathcal{S}(q, \neg\nabla s) \otimes p') \vee \texttt{state}$$

$$\mathcal{S}(p, \nabla_n s) =_{n=k_1+\ldots+k_m} \overset{\vee}{} (\mathcal{S}(p, \neg\nabla s) \otimes \mathcal{S}(q, \nabla_{k_1}s) \otimes \mathcal{S}(p, \neg\nabla s) \otimes$$
$$\mathcal{S}(q, \nabla_{k_2}s) \otimes \mathcal{S}(p, \neg\nabla s) \otimes$$
$$\cdots \qquad \cdots$$
$$\mathcal{S}(q, \nabla_{k_m}s) \otimes \mathcal{S}(p, \neg\nabla s))$$

Let $r \in \mathcal{T}_{ASKS}$ be a composite *non-iterative* task of the form (3) (i.e. $r \leftarrow q$) that satisfies the unique task assumption. Let δ stand for $\neg\nabla s$, $\nabla_{\geq n}s$, or $\nabla_n s$, where $n \geq 1$. Then, $\mathcal{S}(r, \delta) = \mathcal{S}(q, \delta)$. Since q can have the forms $u \otimes v$, $u \mid v$, $\odot u$, or $u \vee v$, $\mathcal{S}(q, \delta)$ is obtained as follows:

$$\mathcal{S}(u \otimes v, \delta) = \begin{cases} (\mathcal{S}(u, \delta) \otimes v) \vee (u \otimes \mathcal{S}(v, \delta)), & \text{if } \delta \text{ is } \nabla_{\geq n}s \text{ or } \nabla_n s \\ \mathcal{S}(u, \delta) \otimes \mathcal{S}(u \otimes v, \delta), & \text{if } \delta \text{ is } \neg\nabla s \end{cases}$$

$$\mathcal{S}(u \mid v, \delta) = \begin{cases} (\mathcal{S}(u, \delta) \mid v) \vee (u \mid \mathcal{S}(v, \delta)), & \text{if } \delta \text{ is } \nabla_{\geq n}s \text{ or } \nabla_n s \\ \mathcal{S}(u, \delta) \mid \mathcal{S}(v, \delta), & \text{if } \delta \text{ is } \neg\nabla s \end{cases}$$

$$\mathcal{S}(\odot u, \delta) = \odot\mathcal{S}(u, \delta)$$
$$\mathcal{S}((u \vee v), \delta) = \mathcal{S}(u, \delta) \vee \mathcal{S}(v, \delta)$$

5.2   Extended Proof Theory

This section develops a proof theory for formulas of the form $G \wedge C$, where G is a service choreography and $C \in \mathcal{C}_{ONSTR}$.

It is easy to see that C is equivalent to $\vee_i(\wedge_j C_{ij})$, where each C_{ij} is either a primitive or serial constraint. To check if there is an execution of $\psi \wedge C$, we need to use the inference rules introduced below and apply them to each disjunct $\psi \wedge (\wedge_j C_{ij})$ separately. Therefore, we can assume that our constraint C is a set of primitive or serial constraints.

Hot Components. We remind the notion of *hot components* of a formula from [3]: $hot(\psi)$ is a set of subformulas of ψ which are "ready to be executed." This set is defined inductively as follows:

1. $hot(()) = \{\}$, where $()$ is the empty goal
2. $hot(\psi) = \psi$, if ψ is an atomic formula
3. $hot(\psi_1 \otimes \ldots \otimes \psi_n) = hot(\psi_1)$
4. $hot(\psi_1 \mid \ldots \mid \psi_n) = hot(\psi_1) \cup \ldots \cup hot(\psi_n)$
5. $hot(\odot\psi) = \{\odot\psi\}$
6. $hot(\psi_1 \vee \ldots \vee \psi_n) = hot(\psi_1)$ or $\ldots$ or $hot(\psi_n)$

Eligible Components. The set of *eligible* components is used in deciding which inference rules are applicable at any given moment in a proof. Let ψ be a service choreography and C a set of constraints. Let $tasks(C)$ denote the set of all tasks mentioned by the constraints in C. The set of eligible components of a CTR goal ψ with respect to a set of constraints C is initially defined as follows:

$$eligible(\psi, C) = \{t \mid t \in hot(\psi), \text{ and } C \text{ has no constraints of the form}$$
$$before(X, t) \text{ or } right\text{-}before(X, t),$$
$$\text{where } X \in tasks(C) \text{ or } X = ?\}$$

The *eligible* set keeps changing as the tasks in the choreography execute. The exact mechanism of these changes is described in the inference rule 4. Note the use of the "?" symbol in the definition of eligible: it appears in constraints of the form $before(?,t)$, which are added or deleted during the execution, by inference rule 4. The constraint $before(?, t)$ means that for t to execute, one task (which is different from t), denoted by "?", must execute prior to t. The symbol "?" also occurs as part of a new kind of constraints which are used internally by the proof procedure: $right\text{-}before^+(a, b) \stackrel{def}{=} ? \otimes right\text{-}before(a, b)$.

This constraint means that the first task can be anything (denoted by "?"), but beginning with the second action the constraint $right\text{-}before(a, b)$ must hold during the rest of the execution. Such constraints are not present initially, but they are introduced by the proof theory system.

Sequents. Let P be a set of composite task definitions. The proof theory manipulates expressions of the form $P, D \text{---} \vdash (\exists)\,\phi$, called *sequents*, where P is a set of task definitions and D is the underlying database state. Informally, a sequent is a statement that the transaction $(\exists)\,\phi$, which is defined by the rules in P, can execute starting at state D. Each inference rule has two sequents, one above the other, which is interpreted as: If the upper sequent is inferred, then the lower sequent should also be inferred. As in classical resolution, any instance of an answer-substitution is a valid answer to a query.

This inference system extends the system for Horn CTR given in [3] with one additional inference rule (rule 3). The other rules from [3] are also significantly modified. The new system reduces to the old one when the set C of constraints is empty. The new system also extends the proof theory developed in [9].

Axioms. P, D --- $\vdash ()\wedge C$, for any database state D, where C does not contain constraints of the form $\nabla_{\geq n}s$ or $\nabla_n s$, where $n \geq 1$.

Inference Rules. In rules 1-5 below, σ denotes a substitution, ψ and ψ' are service choreographies, C and C' are constraint sets, D, D_1, D_2 denote database states, and a is an atomic formula in $eligible(\psi)$.

1. *Applying transaction definitions*: Let $b \leftarrow \beta$ be a rule in P, and assume that its variables have been renamed so that none are shared with ψ. If a and b unify with the most general unifier σ then

$$\frac{P,\ D\ \text{---}\vdash\ (\exists)\ (\psi'\wedge C)\ \sigma}{P,\ D\ \text{---}\vdash\ (\exists)\ \psi\wedge C}$$

where ψ' is obtained from ψ by replacing an eligible occurrence of a by β.

2. *Querying the database*: If $(\exists)a\sigma$ is true in the current state D and $a\sigma$ and $\psi'\sigma$ share no variables then

$$\frac{P,\ D\ \text{---}\vdash\ (\exists)\ (\psi'\wedge C)\ \sigma}{P,\ D\ \text{---}\vdash\ (\exists)\ \psi\wedge C}$$

where ψ' is obtained from ψ by deleting an eligible occurrence of a.

3. *Simplification*: If δ is a primitive constraint, then

$$\frac{P,\ D\ \text{---}\vdash\ (\exists)\ (\mathcal{S}(\psi,\delta)\wedge C)}{P,\ D\ \text{---}\vdash\ (\exists)\ \psi\wedge (C\wedge\delta)}$$

4. *Execution of primitive tasks*: If $a\sigma$ is a primitive task that changes state D_1 to D_2 then

$$\frac{P,\ D_2\ \text{---}\vdash\ (\exists)\ (\psi'\wedge C')\ \sigma}{P,\ D_1\ \text{---}\vdash\ (\exists)\ \psi\wedge C}$$

where ψ' is obtained from ψ by deleting an eligible occurrence of a. C' is obtained from C as follows. Suppose $T,S\in tasks(C)$ are arbitrary task names, and that $T\neq a$. Then:

- Step 1: Initially C' is C.
- Step 2:
    (a) replace every constraint of the form $right\text{-}before^+(T,S)$ in C' with a constraint of the form $right\text{-}before(T,S)$
    (b) delete every constraint of the form $before(a,T)$ in C'
    (c) delete every constraint of the form $before(?,T)$ in C'
    (d) replace every constraint of the form $right\text{-}before(a,T)$ in C' with a constraint of the form $right\text{-}before^+(a,T)$
    (e) for every constraint of the form $not\text{-}between(a,T,a)$ in C', add a constraint of the form $blocks(T,a)$ to C'
    (f) for every constraint of the form $after(a,T)$ in C', add a constraint of the form $existence(T)$ to C'
    (g) for every constraint of the form $blocks(a,T)$ in C', add a constraint of the form $absence(T)$ to C'
    (h) for every constraint of the form $right\text{-}after(a,T)$ in C', for all S in $tasks(C)$, $S\neq T$, add $before(T,S)$ to C'
    (i) for every constraint of the form $not\text{-}right\text{-}after(a,T)$ in C', add a constraint of the form $before(?,T)$ to C'
    (j) for every constraint of the form $between(a,T,a)$ in C', add a constraint of the form $before(T,a)$ to C'

5. *Execution of atomic transactions*: If $\odot\alpha$ is a hot component in ψ then

$$\frac{P,\ D \ \text{---}\vdash (\exists)\ (\alpha\otimes\psi')\wedge C}{P,\ D \ \text{---}\vdash (\exists)\ \psi\wedge C}$$

where ψ' is obtained from ψ by deleting an eligible occurrence of $\odot\alpha$.

Theorem 1. *The above inference system is sound and complete for proving constraint service choreographies, if the service choreographies and constraints satisfy the service contracts assumption.*

Proof: The proof is given in the technical report [10].

6   Related Work

The closest to the present paper is our earlier work [9]. By allowing iterative processes in choreography descriptions, as well as new kind of constraints that can be applied to iterative processes, the current paper significantly extends the modeling power of the service contract framework developed in [9]. The extension of the reasoning mechanism developed in this paper is also quite different and more general than the approach taken in [9].

DecSerFlow [13] is a service flow language that is closely related to our service behavior modeling framework. It uses Linear Temporal Logic to formalize service flows and automata theory to enact service specifications. The relations between tasks are entirely described in terms of constraints. First, our constraint algebra $\mathcal{C}_{ONSTR}$ is more expressive than DecSerFlow. Second, by combining constraints with control-flow graphs, our framework appears closer to the current practices in workflow modeling. Third, data flow and conditional control flow are easily available in our framework [9], while, to the best of our knowledge, they have not been developed in the context of DecSerFlow.

An emerging area related to our work is that of compliance checking between business processes and business contracts. For example, in [7,8] both processes and contracts are represented in a formal contract language called FCL. FCL is based on a formalism for the representation of contrary-to-duty obligations, i.e., obligations that take place when other obligations are violated as typically applied to penalties in contracts. Based on this, the authors give a semantic definition for compliance, but no practical algorithms. In contrast, our work provides a proof theory for the logic we use for service contracting.

Several other works, although not directly related to our approach to service contracting and enactment, are relevant [11,1,2,4]. Most of them present logical languages for representing contracts in various contexts. However, they are mainly based on normative deontic notions of obligation, prohibition, and permission, and thus could be seen as complementary to our approach.

In process modeling, the main other tools are Petri nets, process algebras, and temporal logic. The advantage of CTR over these approaches is that it is a *unifying* formalism that integrates a number of process modeling paradigms ranging from conditional control flows to data flows to hierarchical modeling to constraints, and even to game-theoretic aspects of multiagent processes (see, for example, [6]). Moreover, CTR models the various aspects of processes in distinct ways, which enabled us to devise algorithms with better complexity than the previously known techniques from the area of model checking for temporal logic specifications [5].

7   Conclusions

We have extended the CTR-based logic language for specifying Web service choreography and contracts to include iterative processes. As mentioned in the introduction, many practical languages for describing service behavior include iterative processes in their models and enabling reasoning about iterative processes opens up new possibilities for automated service contracting and enactment on top of existing behavioral languages. In this way, we have closed most of the outstanding problems in logic-based process modeling, which were raised in [5]. We have also extended the proof theory of CTR and made it capable of handling complex practical problems in process modeling and enactment. Due to space limitation, we did not include such modeling aspects as dataflow and conditional control flow, but this can be handled similarly to [9].

Some problems still remain. For instance, reasoning about dynamically created multiple instances of subprocesses is largely future work.

Acknowledgments. Part of this work was done while Michael Kifer was visiting Free University of Bozen-Bolzano, Italy. His work was partially supported by the BIT Institute, NSF grant IIS-0534419, and US Army Research Office under a subcontract from BNL. Dumitru Roman was partly funded by the BIT Institute, and the projects SUPER (FP6-026850), Knowledge Web (FP6-507482), SWING (FP6-26514), SHAPE (ICT-2007-216408), and EASTWEB (Contract TH/Asia Link/010 (111084)).

References

1. Alberti, M., Chesani, F., Gavanelli, M., Lamma, E., Mello, P., Montali, M., Torroni, P.: Expressing and verifying business contracts with abductive logic programming. Number 07122 in Dagstuhl Seminar Proceedings (2007)
2. Andersen, J., Elsborg, E., Henglein, F., Simonsen, J.G., Stefansen, C.: Compositional specification of commercial contracts. Int. J. Softw. Tools Technol. Transf. 8(6), 485–516 (2006)
3. Bonner, A.J., Kifer, M.: Concurrency and Communication in Transaction Logic. In: Joint International Conference and Symposium on Logic Programming (1996)
4. Carpineti, S., Castagna, G., Laneve, C., Padovani, L.: A formal account of contracts for web services. In: Bravetti, M., Núñez, M., Zavattaro, G. (eds.) WS-FM 2006. LNCS, vol. 4184, pp. 148–162. Springer, Heidelberg (2006)
5. Davulcu, H., Kifer, M., Ramakrishnan, C.R., Ramakrishnan, I.V.: Logic Based Modeling and Analysis of Workflows. In: PODS, pp. 25–33 (1998)

6. Davulcu, H., Kifer, M., Ramakrishnan, I.: CTR–S: A Logic for Specifying Contracts in Semantic Web Services. In: WWW 2004, p. 144 (2004)
7. Governatori, G., Milosevic, Z., Sadiq, S.: Compliance checking between business processes and business contracts. In: EDOC 2006, pp. 221–232 (2006)
8. Governatori, G., Milosevic, Z., Sadiq, S., Orlowska, M.: On compliance of business processes with business contracts. Technical report, File System Repository (Australia) (2007), http://search.arrow.edu.au/apps/ArrowUI/OAIHandler
9. Roman, D., Kifer, M.: Reasoning about the behavior of semantic web services with concurrent transaction logic. In: VLDB, pp. 627–638 (2007)
10. Roman, D., Kifer, M.: Service contracting: A logic-based approach. Tech. report (2008), http://www.wsmo.org/TR/d14/ServiceContracting140508.pdf
11. Angelov, P.G.S.: B2B E-Contracting: A Survey of Existing Projects and Standards. Report I/RS/2003/119, Telematica Instituut (2003)
12. Senkul, P., Kifer, M., Toroslu, I.: A Logical Framework for Scheduling Workflows under Resource Allocation Constraints. In: VLDB 2002, pp. 694–705 (2002)
13. van der Aalst, W., Pesic, M.: DecSerFlow: Towards a Truly Declarative Service Flow Language. In: The Role of Business Processes in Service Oriented Architectures, number 06291 in Dagstuhl Seminar Proceedings (2006)

Formal Model for Semantic-Driven Service Execution

Tomas Vitvar[1], Adrian Mocan[1], and Maciej Zaremba[2]

[1] Semantic Technology Institute (STI2)
University of Innsbruck, Austria
`firstname.lastname@sti2.at`
[2] Digital Enterprise Research Institute (DERI)
National University of Ireland in Galway, Ireland
`maciej.zaremba@deri.org`

Abstract. Integration of heterogeneous services is often hard-wired in service or workflow implementations. In this paper we define an execution model operating on semantic descriptions of services allowing flexible integration of services with solving data and process conflicts where necessary. We implement the model using our WSMO technology and a case scenario from the B2B domain of the SWS Challenge.

1 Introduction

Existing technologies for service invocation and interoperation usually depend on ad-hoc or hard-wired solutions for interoperability. In particular, message level interoperability is often maintained in business processes using XSLT, and process level interoperability is often achieved through manual configuration of workflows. Such rigid solutions are a drawback to services' flexibility and adaptability: changes in service descriptions require changes in service implementation or workflows. One possible approach to improve the interoperability is to use semantics in service descriptions. With help of semantics and a logical reasoning, it is possible to automate the integration process and achieve an integration that is more adaptive to changes in business requirements.

In Semantic Web Services (SWS), there are two phases in the service integration process, namely *late-binding phase* and *execution phase* [15]. Late-binding phase allows binding a user request and a set of services "on-the-fly" through semi-automation of the *service lifecycle* by applying tasks for service discovery, composition, selection, mediation, etc. Execution phase allows for the invocation and conversation of bound services. While services may have heterogeneous descriptions in terms of data and protocols, it is important to achieve their interoperability within the both phases. In this paper we elaborate on the execution phase and show how semantic services can be decoupled in the integration process and how their interoperability can be achieved through combined data and process mediation. Particular contributions of our paper are as follows:

A. Sheth et al. (Eds.): ISWC 2008, LNCS 5318, pp. 567–582, 2008.

- We define a sound conceptual model for data and process mediation for SWS exectuion extending our previous, more technical and implementation-driven work in [7].
- We built on top of existing results from the area of ontology-based data mediation [11] by providing a formal algorithm that shows how a run-time mediation can be interlaced with other type of mediation, that is, process mediation.
- We show how the formal model for service exectution can provide a solution for a real-world case scenario. For this purpose we describe the implementation using the WSMO[13], WSML[13], WSMX[1] including a solution architecture for a case scenario from the SWS Challenge[2].

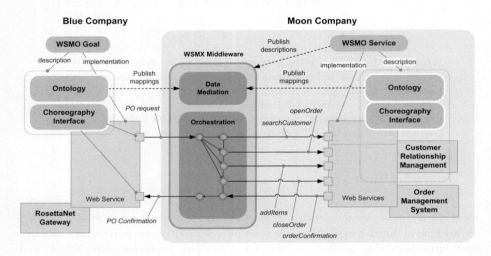

Fig. 1. Case Scenario and Solution Architecture

In order to demonstrate a problem we target in our work, Figure 1 depicts a case scenario of the SWS Challenge mediation problem. In the scenario, a trading company, called Moon, uses a Customer Relationship Management system (CRM) and an Order Management System (OMS) to manage its order processing. Moon has signed agreements to exchange Purchase Order (PO) messages with a company called Blue using the RosettaNet standard PIP3A4. There are two interoperability problems in the scenario: At the *data level*, the Blue uses PIP3A4 to define the PO request and confirmation messages while Moon uses a proprietary XML Schema for its OMS and CRM systems. At the *process level*, the Blue follows PIP3A4 Partner Interface Protocol (PIP), i.e. it sends out a PIP3A4 PO message, including all items to be ordered, and expects to receive a

---

[1] http://www.wsmx.org
[2] SWS Challenge, http://www.sws-challenge.org, defines a testbed together with a set of increasingly difficult problems on which various SWS solutions can be objectively evaluated.

PIP3A4 PO confirmation message. On the other hand, various interactions with the CRM and OMS systems must be performed in Moon in order to process the order, i.e. get the internal ID for the customer from the CRM system, create the order in the OMS system, add line items into the order, close the order, and send back the PO confirmation.

In section 4 we further describe a solutuion for the scenario building on our SWS technology implementing the service execution model. In Section 2 we provide some background definitions for this model and in Section 3 we formaly define this model.

2   Definitions

A service engineer or a client (depending on the level of automation) must decide whether to bind with services according to the descriptions of *service contracts*. In SWS, we represent service contracts at the *semantic level* and *non-semantic level*. For the non-semantic level we use Web Service Description Language (WSDL) and for the semantic level we use four types of descriptions: *information, functional, non-functional* and *behavioral*. In addition, grounding defines a link between semantic and non-semantic descriptions of services.

For purposes of the execution phase, we provide the background definitions for information and behavioral semantic descriptions of services together with grounding to WSDL. In this paper we do not use the other types of semantic descriptions, please refer to e.g. [14] for more information about these descriptions. In addition, we provide definitions for the two major stages of the execution phase, that is, data mediation and process mediation.

2.1   Information Semantics

Information semantics is the formal definition of some domain knowledge used by the service in its *input* and *output* messages. We define the information semantics as an ontology

$$O = (C, R, E, I) \tag{1}$$

with a set of classes (unary predicates) C, a set of relations (binary and higher-arity predicates) R, a set of explicit instances of C and R called E (extensional definition), and a set of axioms called I (intensional definition) that describe how new instances are inferred.

2.2   Behavioral Semantics

Behavioral semantics is a description of the public and the private service behaviors. For our work we only use the public behavior, called choreography, describing a protocol, that is, all messages sent to the service from the network and all messages sent from the service back to the network[3]. We do not use the private behavior of the service, i.e. the internal workflow, in our model. We define the choreography X using the Abstract State Machine (ASM) as

$$X = (\Sigma, L), \tag{2}$$

where $\Sigma \subseteq (\{x\} \cup C \cup R \cup E)$ is the signature of symbols, i.e., variable names $\{x\}$ or identifiers of elements from C, R, E of some ontology O; and L is a set of rules. Further, we denote by Σ_I and Σ_O the input and output symbols of the choreography (subsets of $C \cup R \cup E$), corresponding to the input data sent to the service and the returned output data. Each rule $r \in L$ is defined as $r^{cond} \rightarrow r^{eff}$ where r^{cond} is an expression in logic $\mathcal{L}(\Sigma)$ which must hold in a state before the rule is executed; r^{eff} is an expression in logic $\mathcal{L}(\Sigma)$ describing how the state changes when the rule is executed.

2.3   Grounding

Grounding defines a link between semantic descriptions of services and various components of WSDL. We denote the WSDL schema as S and the WSDL interface as N. Further, we denote $\{x\}_S$ as a set of all element declarations and type definitions of S, and $\{o\}_N$ as a set of all operations of N. Each operation $o \in \{o\}_N$ may have one input message element $m \in \{x\}_S$ and one output message element $n \in \{x\}_S$.

There are two types of grounding used for information and behavioral semantics. The first type of grounding specifies *references* between input/output symbols of a choreography $X = (\Sigma, L)$ and input/output messages of respective WSDL operations $\{o\}_N$ with schema S. We define this grounding as

$$ref(c, m) \tag{3}$$

where $m \in \{x\}_S$, $c \in \Sigma$ and *ref* is a binary relation between m and c. Further, m is the input message of operations in $\{o\}_N$ if $c \in \Sigma_I$ or m is the output message of operations in $\{o\}_N$ if $c \in \Sigma_O$.

The second type of grounding specifies *transformations* of data from schema S to ontology $O = (C, R, E, I)$ called *lifting* and vice-versa called *lowering*. We define this grounding as

$$lower(c_1) = m \qquad \text{and} \qquad lift(n) = c_2, \tag{4}$$

where $m, n \in \{x\}_S$, $c_1, c_2 \in (C \cup R)$, *lower* is a *lowering transformation function* transforming the semantic description c_1 to the message m, and *lift* is a *lifting transformation function* transforming the message n to the semantic description c_2.

---

[3] Please note, that our notion of the choreography is different from the one used by the Web Service Choreography Description Language (WS-CDL) defining the choreoraphy as a common behavior of collaborating parties (http://www.w3.org/TR/ws-cdl-10/)

Table 1. MEPs, Rules and WSDL operations

MEP and Rule	WSDL Operation

in-out:
if c_1 then $add(c_2)$
$c_1 \in \Sigma_I$, $ref(c_1, \text{msg1})$
$c_2 \in \Sigma_O$, $ref(c_2, \text{msg2})$
	`<operation name="oper1" pattern="w:in-out">`
	  `<input messageLabel="In" element="msg1"/>`
	  `<output messageLabel="Out" element="msg2"/>`
	`</operation>`

in-only:
if c_3 then *no action*
$c_3 \in \Sigma_I$, $ref(c_3, \text{msg3})$
	`<operation name="oper2" pattern="w:in-only">`
	  `<input messageLabel="In" element="msg3"/>`
	`</operation>`

out-only:
if *true* then $add(c_4)$
$c_4 \in \Sigma_O$, $ref(c_4, \text{msg4})$
	`<operation name="oper3" pattern="w:out-only">`
	  `<output messageLabel="Out" element="msg4"/>`
	`</operation>`

out-in:
if *true* then $add(c_5)$
if $c_5 \wedge c_6$ then *no action*
$c_5 \in \Sigma_O$, $ref(c_5, \text{msg5})$
$c_6 \in \Sigma_I$, $ref(c_6, \text{msg6})$
	`<operation name="oper4" pattern="w:out-in">`
	  `<output messageLabel="Out" element="msg5"/>`
	  `<inpput messageLabel="In" element="msg6"/>`
	`</operation>`

A client uses both types of grounding definitions when processing the choreography rules and performing the communication with the service while following the underlying definition of WSDL operations and their Message Exchange Patterns (MEPs). Table 1 shows basic choreography rules for four basic WSDL 2.0 MEPs[4], i.e. *in-out, in-only, out-only, out-in* (please note that we currently do not handle fault messages), and corresponding WSDL operations. Here, a rule $r^{cond} \rightarrow r^{eff}$ is represented as if r^{cond} then r^{eff}; the symbols msg1...msg6 refer to schema elements used for input/output messages of operations; the symbols $c_1 \ldots c_6$ refer to identifiers of semantic descriptions of these messages; $ref(m, c)$ denotes a reference grounding, and w: is a shortening for the URI http://www.w3.org/ns/wsdl/.

2.4   Data Mediation

Data mediation resolves interoperability conflicts between two services that use two different ontologies. In general, the data mediation has two stages: 1) creation of alignments between *source* and *target* ontologies during *design-time* and 2) applying the alignments to resolve interoperability conflicts during *run-time*. Since the interoperability problems can greatly vary in their nature and severity, fully automatic solution for the creation of alignments are not feasible in real-world case scenarios due to the lower than 100% precision and recall of existing methods[5]. From this reason, the design-time data mediation stage is still dependent on manual support of a service engineer.

---

[4] http://www.w3.org/TR/wsdl20-adjuncts/#meps
[5] The *"Ontology Alignment Evaluation Initiative 2006"* [5] shows that the best five systems' scores vary between 61% and 81% for precision and between 65% and 71% for recall.

An alignment consists of a set of mappings (rules) expressing the semantic relationships that exist between the two ontologies. In particular, a mapping can specify that classes from two ontologies are equivalent while corresponding rules use logical expressions to unambiguously define how the data encapsulated in an instance of one class can be encapsulated in instances of the second class. Formally, we define an alignment A between source and target ontologies $O_s = (C_s, R_s, E_s, I_s)$ and $O_t = (C_t, R_t, E_t, I_t)$ as

$$A_{s,t} = (O_s, O_t, \Phi_{s,t}) \tag{5}$$

where $\Phi_{s,t}$ is the set of mappings m in the form

$$m = < \varepsilon_s, \varepsilon_t, \gamma_{\varepsilon_s}, \gamma_{\varepsilon_t} > \tag{6}$$

where ε_s, ε_t represent the mapped entities from the two ontologies while γ_{ε_s}, γ_{ε_t} represent restrictions (i.e. conditions) on these entities such as $\varepsilon_s \in C_s \cup R_s$, $\varepsilon_t \in C_t \cup R_t$ while γ_{ε_s} and γ_{ε_t} are expressions in logic $\mathcal{L}(C_s \cup R_s \cup E_s)$ and $\mathcal{L}(C_t \cup R_t \cup E_t)$, respectively.

In order to execute the mappings during the execution phase, these mappings must be *grounded*[6] to rules expressed in some logical language for which a reasoning support is available (in Section 4 we use the WSML language for this grounding). We obtain the set of rules $\rho_{s,t} = \Phi_{s,t}^G$ by applying the grounding G to the set of mappings Φ. In the following definitions, $\{x\}$ stands for the set of variables used by the mapping rule and x' and x'' are two particular variables.

Every mapping rule $mr \in \rho_{s,t}$ has the following form:

$$mr : \bigwedge_{i=1..n}^{\{x\}} mr_i^{head} \rightarrow \bigwedge_{i=1..n}^{\{x\}} mr_i^{body} \tag{7}$$

where

$$
\begin{aligned}
mr^{head} \in \ & \{x' \textbf{ instanceOf } \varepsilon \mid \varepsilon \in C_t \wedge x' \in \{x\}\} \ \cup \\
& \{\varepsilon(x', x'') \mid \varepsilon \in R_t \ \wedge \varepsilon(x', x'') \in E_t \wedge x', x'' \in \{x\}\}
\end{aligned}
\tag{8}
$$

$$
\begin{aligned}
mr^{body} \in \ & \{x' \textbf{ instanceOf } \varepsilon \mid \varepsilon \in C_s \wedge x' \in \{x\}\} \ \cup \\
& \{\varepsilon(x', x'') \mid \varepsilon \in R_s \ \wedge \varepsilon(x', x'') \in E_s \wedge x', x'' \in \{x\}\} \ \cup \\
& \{\gamma_s \mid \gamma_s \in \mathcal{L}(C_s \cup R_s \cup E_s \cup \{x\})\} \ \cup \\
& \{\gamma_t \mid \gamma_t \in \mathcal{L}(C_t \cup R_t \cup E_t \cup \{x\})\}
\end{aligned}
\tag{9}
$$

A mapping rule is formed of a head and a body. The head is a conjunction of logical expressions over the target elements and describes the result of the mediation in terms of instances of the target ontology. The body is formed of a set of logical expressions over the source entities which represent the data to be mediated, plus a set of logical expressions representing conditions over both the source and the target data.

---

[6] Please note, that this grounding is different to the grounding defined in Section 2.3.

There are situations when there is no corresponding data in the source ontology as required by the target ontology such as when mapping prices with different currency units. These issues are, however, dependant on implementation of the data mediation and the reasoning engine. In our implementation, it is possible to specify an URI for a transformation function and its parameters as placeholders for the missing target values. It is the role of the reasoning engine to fill the parameters placeholders with data from the source ontology. The data mediation engine then executes the function and gets the data for the target ontology.

2.5   Process Mediation

Process Mediation handles interoperability issues which occur in descriptions of choreographies of the two services. In [3] Cimpian defines five process mediation patterns:

a. **Stopping an unexpected message** when one service sends a message which is not expected by the other service.
b. **Inverting the order of messages** when one service sends messages in a different order than the other service expects them to receive.
c. **Splitting a message** when a service sends a message which the other service expects to receive in multiple different messages.
d. **Combining messages** when a service expects to receive a message which is sent by the other service in multiple different messages.
e. **Generating a message** when one service expects to receive a message which is not supplied by the other service.

3   Execution Phase

Figure 2 depicts the main states of the execution phase In Section 3.1 we define the algorithm for the execution phase and in Section 3.2 we discuss some relevant aspects for the data and process mediation applied within the algorithm.

3.1   Algorithm

Input:

- Service W_1 and service W_2. Each such a service W contains the ontology (information semantics) $W.O$ (Eq. 1), the choreography $W.X$ (Eq. 2) with set of rules $W.X.L$, WSDL description and grounding (Eq. 3, 4). In addition, for a rule $r \in W.X.L$, the condition r^{cond} is a logical expression with set of semantic descriptions $\{c\}$, and the effect r^{eff} is a logical expression with set of actions $\{a\}$. For each element a we denote its action name as $a.action$ with values *delete* or *add* and a semantic description as $a.c$.
- Mappings Φ_{12} of $W_1.O$ to $W_2.O$ and mappings Φ_{21} of $W_2.O$ to $W_1.O$.

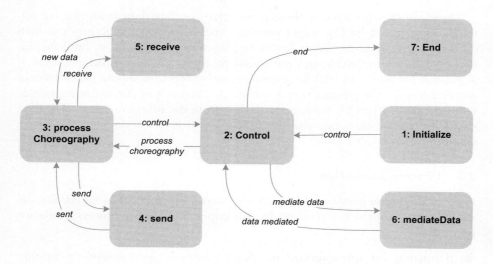

Fig. 2. Control State Diagram for the Execution Phase

Uses:
- Symbols M_1 and M_2 corresponding to the processing memory of the choreography $W_1.X$ and $W_2.X$ respectively (a memory M is a populated ontology $W.O$ with instance data). The content of each memory M determines at some point in time a state in which a choreography $W.X$ is. In addition, each memory has methods $M.add$ and $M.remove$ allowing to add or remove data to/from M and a flag $M.modified$ indicating whether the memory was modified. The flag $M.modified$ is set to *true* whenever the method $M.add$ or $M.remove$ is used.
- Symbols D_1 and D_2 corresponding to the set of data to be added to the memory M_1 and M_2 after one or more rules of a choreography are processed. Each D has a method $D.add$ for adding new data to the set.
- A symbol A corresponding to all actions to be executed while processing the choreography. Each element of A has the same definition as the element of the rule effect r^{eff}. A has methods $A.add$ and $A.remove$ for adding and removing actions to/from the set.
- A symbol o corresponding to a WSDL operation of a service and symbols m, n corresponding to some XML data of the message (input or output) of the operation o.

States 1, 2, 7: Initialize, Control, End
1: $M_1 \leftarrow \emptyset$; $M_2 \leftarrow \emptyset$
2: **repeat**
3:   $M_1.modified \leftarrow false$; $M_2.modified \leftarrow false$
4:   $D_1 \leftarrow processChoreography(W_1, M_1)$
5:   $D_2 \leftarrow processChoreography(W_2, M_2)$
6:   **if** $D_1 \neq \emptyset$ **then**

```
 7:        Dₘ ← mediateData(D₁, W₁.O, W₂.O, Φ₁₂)
 8:        M₁.add(D₁); M₂.add(Dₘ)
 9:     end if
10:     if D₂ ≠ ∅ then
11:        Dₘ ← mediateData(D₂, W₂.O, W₁.O, Φ₂₁)
12:        M₁.add(Dₘ); M₂.add(D₂)
13:     end if
14: until not M₁.modified and not M₂.modified
```

After the initialization of the processing memory M_1 and M_2 (line 1), the execution gets to the control state when the algorithm can process choreographies (state 3), mediate the data (state 6) or end the execution (state 7). The execution ends when no modifications of the processing memories M_1 or M_2 has occurred.

State 3: $D = processChoreography(W, M)$

```
 1: A ← ∅; D ← ∅
 2: {Performing rule's conditions and sending data}
 3: for all r in W.X.L : holds(rᶜᵒⁿᵈ, M) do
 4:    A.add(rᵉᶠᶠ)
 5:    for all c in rᶜᵒⁿᵈ : c ∈ W.X.Σ_I do
 6:       send(c, W)
 7:    end for
 8: end for

 9: {Performing delete actions}
10: for all a in A : a.action = delete do
11:    M.remove(a.c)
12:    A.remove(a)
13: end for

14: {Receiving data and performing add actions}
15: while A ≠ ∅ do
16:    c ← receive(W)
17:    if c ≠ null then
18:       for all a in A: (a.action = add and a.c = c) do
19:          D.add(c)
20:          A.remove(a)
21:       end for
22:    end if
23: end while

24: return D
```

The algorithm executes each rule of the choreography which condition holds in the memory by processing its condition and effect, i.e. the algorithm collects all data to be added to the memory or removes existing data from the memory in three major steps as follows.

- **Performing rule's conditions and sending data (lines 2-8):** the algorithm adds the effect of the rule which condition holds in the memory (line 3) to the set of effects A (line 4). Then, for each input symbol of the rule's condition (line 5), the algorithm sends the data to the service (line 6, see State 4).
- **Performing delete actions (lines 9-13):** the algorithm processes all effects with *delete* action, removes the data of the effect from the memory (line 11) as well as from A (line 12).
- **Receiving data and performing add actions (lines 14-24):** when there are effects to be processed in A and the new data is received from the service (line 16), the algorithm checks if the new data corresponds to some of the *add* effect from A. In this case, it adds the data to the set D (line 19) and removes the effect from A (line 20).

The result of the algorithm is the set D which contains all new data to be added to the memory M. The actual modification of the memory M with the new data is done in State 2. During the choreography processing, the algorithm relies on a consistent definition of the reference grounding (see Eq. 3), i.e. choreography rules are consistent with WSDL operations and their MEPs, as well as assumes no failures occur in services. In lines 14-23 the algorithm waits for every message from the service for every *add* action of the rule's effect. If the definition of the rules was not consistent with WSDL description, the algorithm would either ignore the received message which could in turn affect the correct processing of the choreography (in case of missing *add* action) or wait infinitely (in case of extra *add* action or a failure in a service). For the latter, the simplest solution would be to introduce a timeout in the loop (lines 14-23), however, we do not currently handle this situation.

State 4: $send(c, W)$

1: $m \leftarrow lower(c)$
2: **for all** o of which m is the input message **do**
3:    send m to W
4: **end for**

In order to send the data c the algorithm first retrieves a corresponding message definition according to the grounding and transforms c to the message m using the lowering transformation function (line 1). Then, through each operation of which the message m is the input message, the algorithm sends the m to the service W.

State 5: $c = receive(W)$

1: **if** receive m from W **then**
2:    $c \leftarrow lift(m)$
3:    **return** c
4: **else**
5:    **return** *null*
6: **end if**

When there is new data from the service W, the algorithm lifts the data (message m in XML) to the semantic representation using lifting transformation function associated with the message (line 2).

State 6: $D_m = mediateData(D, O_s, O_t, \Phi)$

1:   $\rho \leftarrow \emptyset$; $\xi_m \leftarrow \emptyset$
2: **for all** $c \in D$ **do**
3:    $\varepsilon \leftarrow getTypeOf(c)$;
4:    $\varepsilon_m \leftarrow null$
5:    **for all** $m = <\varepsilon_s, \varepsilon_t, \gamma_{\varepsilon_s}, \gamma_{\varepsilon_t}> \in \Phi$ **do**
6:     **if** $\varepsilon = \varepsilon_s$ **then**
7:      **if** $isBetterFit(\varepsilon_t, \varepsilon_m)$ **then**
8:       $\varepsilon_m \leftarrow \varepsilon_t$
9:      **end if**
10:      $m_G \leftarrow ground(m)$; $\rho \leftarrow \rho \cup \{m_G\}$
11:     **end if**
12:    **end for**
13:    $\xi_m \leftarrow \xi_m \cup \varepsilon_m$
14: **end for**
15: **if** $\xi_m = null$ **then**
16:   **return** $null$
17: **end if**
18: $D_m \leftarrow getDataForType(\xi_m, \rho)$
19: **return** D_m

The algorithm performs two steps during data mediation. Firstly, the algorithm processes mappings in order to determine the most suitable target concepts to mediate the source data to, and secondly, the algorithm transforms the mappings into an executable form and executes the mappings. Since current reasoning engines does not scale well in terms of processing time, keeping these steps separate enable high-performance in processing of alignments independent on the logical language and reasoning engine used. In other words, this approach minimizes the use of the reasoning during the data mediation.

- **Step 1.** The algorithm first determines a concept for an instance data to be mediated (line 3). After that, the algorithm traverses through a set of mappings in order to determine the type of the target data (mediated data) (lines 5-12). Since there could be more mappings from a given source entity to the several other target entities, the algorithm determines the most suitable concept (lines 7-9). In particular, if a concept ε_s is mapped to two target concepts ε_t^1 and ε_t^2, then ε_t^1 is more suitable if ε_t^1 is a sub-concept of ε_t^2 (the most specific) or if ε_t^2 can be reached via binary relationships (i.e. attributes) starting from ε_t^1 (maximal coverage).
- **Step 2:** While traversing the set of mappings, the algorithm grounds each mapping to a logical language by transforming them to a set of logical mapping rules (line 10). Finally, by using a reasoner engine, the algorithm queries

and retrieves all the data of the selected target type according to the source data and the set of mapping rules (line 18).

3.2 Discussion

Data mediation ensures that all new data coming from one service is translated to the other's service ontology. Thus, no matter from where the data originates the data is always ready to use for the both services. From the process mediation point view, the data mediation also handles the splitting of messages (pattern c) and combining messages (pattern d). Since the mediated data is always added to the both memories (see State 2, lines 8, 12 and the next paragraph for additional discussion) the patterns (a) and (b) are handled automatically through processing of the choreography rules. In particular, the fact that a message will be stopped (pattern a) means that the message will never be used by the choreography because no rule will use it. In addition, the order of messages will be inverted (pattern b) as defined by the choreography rules and the order of ASM states in which conditions of rules hold. This means that the algorithm automatically handles the process mediation with help of data mediation through rich description of choreographies when no central workflow is necessary for that purpose. In order to fulfill the pattern (e), the algorithm might need a third-party data for which an integration workflow might be neccessary. Although some of the third-party data can be gathered through transformation functions of the data mediation which can in turn facilitate some cases of pattern (e), we do not provide a general solution for this pattern. A special case of pattern (e) could be "generating an acknowledgement message" for which the algorithm should distinguish types of interactions. For example, if the algorithm is able to understand control interactions (such as acknowledgements) among all the interactions between services, it could generate an acknowledgment message (evaluation of successfull reception of the message by the other service is, however, another issue).

4   Implementation

In this section we describe the implementation of the execution model using the WSMO, WSML and WSMX technology on the use case from the SWS Challenge as Figure 1 depicts. We use WSMO to model ontologies, services, mediators and goals according to the scenario. WSMO uses WSML family of ontology languages to define concrete semantics of these elements. In addition, WSMX is the execution environment for WSMO allowing to run the execution of WSMO services. In the core of the solution, the WSMX middleware is located between Blue and Moon systems. WSMX functionality can be customized to conform to particular integration needs through choosing appropriate components and their configuration. For our solution, we use the *orchestration* which executes the conversation and the *data mediation* which resolves the heterogeneity issues, both operating according to the execution model. In addition, WSMX contains the base components such as *reasoning* which performs logical reasoning over semantic descriptions and *communication* or *persistence*. For brevity, we do not show them in the figure.

We use WSMO ontology to model the information semantics of services, and the *choreography interface* definition of WSMO Service/Goal to model the behavioral semantics of services. In addition, we need to create grounding to underlying WSDL and XML Schema (The SWS Challenge provides all services in WSDL together with endpoints accessible via SOAP over HTTP) and mapping rules between ontologies. Firstly, we create ontologies in WSML language as semantic representations of the PIP3A4, CMR and OMS XML schemata. Secondly, we create lifting and lowering transformations in XSLT between underlying XML and the ontologies. Finaly, we define mappings between the both ontologies. Although we could define one overarching ontology for the both XML schema together with lifting/lowering transformations to this ontology, we want to demonstrate the use of mappings and data mediation. Hence we define two heterogeneous ontologies for the two heterogeneous XML schema.

```
axiom aaMappingRule23
    definedBy
        mediated(?X21, SearchCustomerReq)[searchString hasValue ?Y22] memberOf o1#
            SearchCustomerReq
  :− ?X21[businessName hasValue ?Y22] memberOf o2#BusinessDescription.
```

Listing 1.1. Mapping Rules in WSML

Listing 1.1 shows a sample mapping rule between the *SearchCustomerReq* concept of the CMR ontology (denoted using *o1* prefix) and *BusinessDescription* concept of the PIP3A4 ontology (denoted using *o2* prefix). The construct *mediated* (X, C) represents the identifier of the newly created target instance, where X is the source instance that is transformed, and C is the target concept we map to.

In line with Eq. 2, the WSMO service choreography contains the definition of the input, output and shared symbols (called state signature or vocabulary) and a set of rules. Using these rules we model the choreography of the both PIP3A4 and CRM/OMS services separately and for each define the order in which the operations should be correctly invoked. Listing 1.2 shows a fragment of the choreography for the CRM/OMS service. There are two rules defined. The first rule (lines 17-22) defines that the *SearchCustomerReq* will be sent to the service and on result the *SearchCustomerResp* will be expected as the output message. The *SearchCustomerReq* message must be available in the memory (in our case the data for the message is provided by the Blue after the data mediation). The second rule (lines 24-30) defines that the *SearchCustomerResp* must be available in the memory while its *customerId* will be used for the *customerId* of the *CreateNewOrderReq* which will be sent to the service. On result, the *CreateNewOrderResp* will be expected to be received back. The data for the *CreateNewOrderReq* will be again supplied by the Blue after the data mediation. All the messages used in the choreography as the input or output symbols refer

to the definition of concepts in the ontology imported in line 3 while at the
same time the grounding of these concepts to the underlying WSDL messages is
defined in lines 5-14.

```
1    choreography MoonWSChoreography
2     stateSignature _"http://example.com/ontologies/MoonWS#statesignature"
3     importsOntology {_"http://example.com/wsml/Moon" }
4     // input symbols
5     in moon#SearchCustomerReq
6        withGrounding { _"http://example.com/MoonCRM#wsdl.interfaceMessageReference(search/in0)
                 " }
7        moon#CreateNewOrderReq
8        withGrounding { _"http://example.com/MoonOMS#wsdl.interfaceMessageReference(openorder/
                 in0)" }
9
10    // output symbols
11    out moon#SearchCustomerResp
12        withGrounding { _"http://example.com/MoonCRM#wsdl.interfaceMessageReference(search/
                 out0)" }
13        moon#CreateNewOrderResp
14        withGrounding { _"http://example.com/MoonOMS#wsdl.interfaceMessageReference(openorder
                 /out0)" }
15    ...
16    transitionRules _"http://example.com/ontologies/MoonWS#transitionRules"
17    // rule 1: search the customer in CRM
18    forall {?customerReq} with (
19    ?customerReq memberOf moon#SearchCustomerReq
20    ) do
21    add(_# memberOf moon#SearchCustomerResp)
22    endForall
23
24    // rule 2: open the order in OMS
25    forall {?orderReq, ?customerResp} with (
26    ?customerResp[customerId hasValue ?id] memberOf moon#SearchCustomerResp and
27    ?orderReq[customerId hasValue ?id] memberOf moon#CreateNewOrderReq
28    ) do
29    add(_# memberOf moon#CreateNewOrderResp)
30    endForall
```

Listing 1.2. Moon CRM/OMS Choreography

5   Related Work

The most relevant related work is among other submissions addressing the SWS-
Challenge mediation scenario, namely WebML [10] and dynamic process binding
for BPEL[8]. They are based on software engineering methods focusing on modelling of integration process as a central point of integration. They do not use
logical languages in their data model. In addition, Preist et al [12] presented a
solution covering all phases of a B2B integration life-cycle, starting from discovering potential partners to performing integrations including mediations. Their
solutions is rather conceptual with missing details about the actual components
and algorithms used. More general SWS related work include IRS-III[4] which
is an execution environment also based on WSMO. In addition, there are related works that apply ASM for various stages of service integration process.
Altenhofen et al [1] also address Process Mediation of services modelled using

ASM, however they focus only on Process Mediation aspect while we provide a complete conceptual model and implementation addressing both Data and Process Mediation. In [6] a composition algorithm based on Web service process ASMs is described where formal, mathematical model of Web services and orchestrations is provided. Their model of ASMs varies from our ontologized ASMs model for example with respect to modelling incoming and outgoing messages. In our model it is implicitly inferred from concept grounding in choreography in/out states whether ASM Knowledge Base modifications entail communication with the service as opposed to ASM used in [6] explicitly models communication constructs. The purpose of describing Web service public processes in [6] is different – it is primarily focused on composition of ASM services, while we utilize ASMs for achieving Process Mediation between communicating Web services. Lerner [9] focuses on analysis and verification of parameterized State Machines applied to reusable processes specified in Little-JIL language. Provided algorithm is able to detect deadlock and other anomalies of analyzed processes. Underlying language is based on State Machines similarly like in our case. We might consider in the future to focus more on the ASM process analysis and verification using similar methods as proposed by Lerner. Benatallah et al [2] present a conceptual model for Web service protocol specifications. They provide framework supporting analysis of commonalities and differences between protocols supported by different Web services. Similarly, like in case of Lemcke [6] most of the focus was given to public process analysis (called Web service protocol by Benatallah) while in our work we presented models, mediation algorithms and working implementation addressing both data and public process heterogeneity issues.

6   Conclusion and Future Work

One of the main advantages of our approach is the strong partner de-coupling. This means that when changes occur in back-end systems of one partner, consequent changes in service descriptions does not affect changes in the integration. The integration automatically adapts to the changes in service descriptions as there is no central integration workflow. On the other hand, changes in back-end system still require manual effort in making changes in semantic descriptions such as ontologies and mapping rules. Although our SWS technology allows for semi-automated approaches in modelling and mapping definitions, it is still a human user who must adjust and approve the results. It is important to note, however, that this type of integration where no central workflow is necessary is only usable in situations when two public processes (ASM choreographies) are compatible, that is, they may have different order/structure of messages but by adjusting the order/structure the integration is possible. In general, there could be cases where third-party data need to be obtained (e.g. from external databases) for some interactions. Although some of the third-party data can be gathered through the transformation functions of the mapping rules, in some cases, an external workflow could be required to accommodate the integration process. It is our open research work to further investigate such cases in detail.

Acknowledgments

This work is supported by the Science Foundation Ireland Grant No. SFI/02/ CE1/I131, and the EU projects SUPER (FP6-026850), and SemanticGov (FP-027517).

References

1. Altenhofen, M., Börger, E., Lemcke, J.: An abstract model for process mediation. In: Lau, K.-K., Banach, R. (eds.) ICFEM 2005. LNCS, vol. 3785, pp. 81–95. Springer, Heidelberg (2005)
2. Benatallah, B., Casati, F., Toumani, F.: Representing, analysing and managing web service protocols. Data Knowl. Eng. 58(3), 327–357 (2006)
3. Cimpian, E., Mocan, A.: Wsmx process mediation based on choreographies. In: Business Process Management Workshops, pp. 130–143 (2005)
4. Domingue, J., et al.: Irs-iii: A broker-based approach to semantic web services. J. Web Sem. 6(2), 109–132 (2008)
5. Euzenat, J., et al.: Results of the Ontology Alignment Evaluation Initiative 2006. In: Proceeding of International Workshop on Ontology Matching (OM 2006), CEUR Workshop Proceedings, Athens, Georgia, USA, November 2006, vol. 225, pp. 73–95 (2006)
6. Friesen, A., Lemcke, J.: Composing web-service-like abstract state machines (asm). In: Autonomous and Adaptive Web Services (2007)
7. Haselwanter, T., et al.: WSMX: A Semantic Service Oriented Middleware for B2B Integration. In: ICSOC, pp. 477–483 (2006)
8. Kuster, U., Konig-Ries, B.: Dynamic binding for bpel processes - a lightweight approach to integrate semantics into web services. In: Second International Workshop on Engineering Service-Oriented Applications: Design and Composition (WESOA 2006) at 4th International Conference on Service Oriented Computing (ICSOC 2006), Chicago, Illinois, USA (December 2006)
9. Lerner, B.S.: Verifying process models built using parameterized state machines. In: ISSTA, pp. 274–284 (2004)
10. Margaria, T., et al.: The sws mediator with webml/webratio and jabc/jeti: A comparison. In: ICEIS (4), pp. 422–429 (2007)
11. Mocan, A., Cimpian, E.: An Ontology-Based Data Mediation Framework for Semantic Environments. International Journal on Semantic Web and Information Systems (IJSWIS) 3(2) (April - June, 2007)
12. Preist, C., et al.: Automated Business-to-Business Integration of a Logistics Supply Chain using Semantic Web Services Technology. In: Proc. of 4th Int. Semantic Web Conference (2005)
13. Roman, D., et al.: Web Service Modeling Ontology. Applied Ontologies 1(1), 77–106 (2005)
14. Vitvar, T., Kopecky, J., Fensel, D.: WSMO-Lite: Lightweight Semantic Descriptions for Services on the Web. In: ECOWS (2007)
15. Vitvar, T., et al.: Semantically-enabled service oriented architecture: concepts, technology and application. Service Oriented Computing and Applications 2(2), 129–154 (2007)

Efficient Semantic Web Service Discovery in Centralized and P2P Environments

Dimitrios Skoutas[1,2], Dimitris Sacharidis[1], Verena Kantere[3],
and Timos Sellis[2,1]

[1] National Technical University of Athens
Athens, Greece
{dskoutas,dsachar}@dblab.ece.ntua.gr
[2] Institute for the Management of Information Systems (R.C. "Athena")
Athens, Greece
timos@imis.athena-innovation.gr
[3] Ecole Polytechnique Fédérale de Lausanne
Lausanne, Switzerland
verena.kantere@epfl.ch

Abstract. Efficient and scalable discovery mechanisms are critical for enabling service-oriented architectures on the Semantic Web. The majority of currently existing approaches focuses on centralized architectures, and deals with efficiency typically by pre-computing and storing the results of the semantic matcher for all possible query concepts. Such approaches, however, fail to scale with respect to the number of service advertisements and the size of the ontologies involved. On the other hand, this paper presents an efficient and scalable index-based method for Semantic Web service discovery that allows for fast selection of services at query time and is suitable for both centralized and P2P environments. We employ a novel encoding of the service descriptions, allowing the match between a request and an advertisement to be evaluated in constant time, and we index these representations to prune the search space, reducing the number of comparisons required. Given a desired ranking function, the search algorithm can retrieve the top-k matches progressively, i.e., better matches are computed and returned first, thereby further reducing the search engine's response time. We also show how this search can be performed efficiently in a suitable structured P2P overlay network. The benefits of the proposed method are demonstrated through experimental evaluation on both real and synthetic data.

1 Introduction

Web services enable interoperability and integration between heterogeneous systems and applications. Current industry standards for describing and locating Web services (WSDL, UDDI), describe the structure of the service interface and of the exchanged messages. Even though this provides interoperability at the syntactic level, it limits the discovery process to essentially keyword-based search. To increase the precision of the discovery process, appropriate services

A. Sheth et al. (Eds.): ISWC 2008, LNCS 5318, pp. 583–598, 2008.

should be identified and selected in terms of the semantics of the requested and offered capabilities. To that end, Semantic Web services combine the benefits of Semantic Web and Web services technology. Several approaches have been proposed for semantically enhancing the descriptions of Web services (OWL-S [1], WSDL-S [2], WSMO [3]), and automating the service discovery, composition, and execution. Service requests and advertisements are annotated by concepts from associated ontologies, and the matchmaking is based on subsumption reasoning between concepts corresponding to the requested and offered parameters.

As the number of services on the Web increases, the efficiency and the scalability of service discovery techniques become a critical issue. Moreover, several applications are inherently distributed. Consider, for example, a network of businesses or institutions, each providing its own services; creating and maintaining a centralized registry would not be desirable. However, the majority of current approaches focuses on centralized architectures, i.e., a single service registry or multiple service registries synchronizing periodically. This introduces bottlenecks and single points of failure, and fails to scale when the availability and demand for services grows significantly. On the other hand, P2P networks support large-scale, decentralized applications, offering scalability and reliability. In addition, structured P2P overlays provide guarantees for retrieving all search results in bounded time and distributing the load among peers. Hence, there has been recently a lot of research interest in issues overlapping the two fields, Semantic Web and P2P computing, primarily focusing on distributed RDF stores [4,5,6].

Regarding Semantic Web services, proposed approaches for service discovery in P2P environments typically rely on the use of ontologies to partition the network topology into concept clusters, and then forward requests to the appropriate cluster. However, constructing concept clusters in a fully automated way is not straightforward, as well as providing guarantees regarding search times and load balancing. In this paper we address the issue of Semantic Web service discovery, focusing on the aspects of efficiency and scalability. We present a method for fast search and selection of services at query time that is suitable for both centralized and P2P environments. In particular, our contributions are summarized in the following:

- We employ a novel encoding of the service descriptions, allowing the match between a service request and a service advertisement to be evaluated in constant time, avoiding the overhead of invoking the reasoner at query time.
- We index the service representations to prune the search space, minimzing the number of comparisons required to locate the matching services.
- We discuss the need for ranking the matched services, and present an algorithm that, given a desired ranking function, fetches the top-k matches progressively, thereby further reducing the search engine's response time.
- We extend our method to a structured P2P overlay network, showing that the search process can be done efficiently in a decentralized, dynamic environment.
- We demonstrate the efficiency and the scalability of our approach through experimental evaluation.

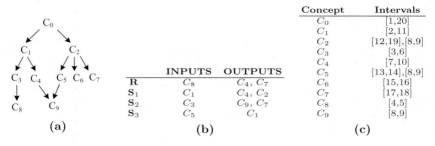

Fig. 1. (a) A sample ontology fragment, (b) A service request (R) and three service advertisements (S_1, S_2, S_3), (c) Intervals assigned to ontology concepts

The rest of the paper is organized as follows. Section 2 discusses Semantic Web services matchmaking and ranking, and presents our encoding for service descriptions. Section 3 presents the indexing and searching of services in a centralized registry. Section 4 shows how service descriptions can be distributed and searched efficiently in a structured P2P overlay network. Experimental evaluation of the proposed approach is presented in Section 5, while related work is reviewed in Section 6. Finally, Section 7 concludes the paper.

2   The Matchmaking Framework

In this section we present our framework for efficient Semantic Web service matchmaking in centralized and P2P environments. First, we describe the service selection and ranking process in Section 2.1. Our framework is based on the encoding and indexing of service descriptions discussed in Section 2.2.

2.1   Semantic Selection of Services

In the following, we consider an ontology as a set of hierarchically organized concepts. Since multiple inheritance is allowed, the concepts form a rooted directed acyclic graph. The nodes of the graph correspond to concepts, with the root corresponding to the top concept, e.g., `owl:Thing` in an OWL ontology, whereas the edges represent subsumption relationships between the concepts, directed from the father to the child. To allow for semantic search of services on the Web, the description of a service is enhanced by annotating its parameters (typically inputs and outputs) with concepts from an associated ontology [1,2,3]. A service request is the description of a desired service, also annotated with ontology concepts. Figure 1a illustrates a sample ontology fragment, while a sample set of a service request and 3 service advertisements is shown in Figure 1b. The underlying assumption is that if a service provides as output (resp., accepts as input) a concept C, then it is also expected to likely provide (resp., accept) the subconcepts of C. For instance, a service advertised as selling computers is expected to sell servers, desktops, laptops, PDAs, etc.; similarly, a service offering delivery in Europe is expected to provide delivery within all (or at least most) European countries.

Matchmaking of semantically annotated Web services is then based on subsumption reasoning between the semantic descriptions of the service request and the service advertisement. Along the lines of earlier works [7,8], we specify the match between a service request R and a service advertisement S based on the semantic match between the corresponding parameters in their descriptions. More specifically, for a service parameter C_S and a request parameter C_R, we consider the match as *exact*, if C_S is equivalent to C_R ($C_S \equiv C_R$); *plug-in*, if C_S subsumes C_R ($C_S \sqsupset C_R$); *subsumes*, if C_S is subsumed by C_R ($C_S \sqsubset C_R$); *fail*, otherwise. Exact matches are preferable to plug-in matches, which in turn are preferable to subsumes matches. In the example of Figure 1, service S_1 provides one exact and two plug-in, service S_2 provides one plug-in, one subsumes, and one exact, whereas S_3 provides two fail and one plug-in matches.

Given that a large number of services may provide a partial match to the request, differentiating between the results within the same type of match is also required. Further following the aforementioned assumption regarding the semantics of a service description, we use as a criterion for assessing the degree of match between two concepts C_1 and C_2 the portion of their common subconcepts, or in other words, the extend to which the subtrees (more generally, subgraphs) rooted at C_1 and C_2 overlap. Intuitively, the higher the overlap, the more likely it is for the service to match the request. Thus, in the following, we consider the degree of match between two concepts C_1 and C_2 as

$$degreeOfMatch(C_1, C_2) = \frac{|\{C \mid C \sqsubseteq C_1 \wedge C \sqsubseteq C_2\}|}{\max(|\{C \mid C \sqsubseteq C_1\}|, |\{C \mid C \sqsubseteq C_2\}|)} \qquad (1)$$

Returning to our example from Figure 1, notice that regarding the requested input, services S_1 and S_2 provide a plug-in match. However, using Equation (1), the degree of match for the service S_1 is $1/5$, whereas for the service S_2 is $1/2$. Notice that the proposed approach for service selection is not limited by this criterion. Different ranking criteria may be appropriate in different applications (for example, see [9] for a more elaborate similarity measure for ranking Semantic Web services). Our approach is generic and it can accommodate different ranking functions (see Section 3 for details). Retrieving services in descending order of their degree of match to the given request constitutes an important feature for a service discovery engine. In the case that the requester is a human user, it can be typically expected that he/she will navigate only the first few results. In fact, experiments conducted in a recent survey [10] showed that the users viewed the top-1 search result in about 80% of the queries, whereas results ranked below 3 were viewed in less than 50% of the queries. Even though this study refers to Web search, it is reasonable to assume a roughly similar behavior for users searching for services. On the other hand, Semantic Web service discovery plays an important role in fully automated scenarios, where a software agent, such as a travel planning agent, acting on behalf of a human user, selects and composes services to achieve a specific task. Typically, the agent will select the top-1 match, ignoring the rest of the results. Hence, computing only the best possible match would be sufficient in this case. In fact, this often makes sense for human users as well; Google's "I'm Feeling Lucky" feature is a characteristic example based on this assumption.

2.2   Encoding of Service Descriptions

Invoking the reasoner to check for subsumption relationships between the ontology concepts annotating the service parameters constitutes a significant overhead, which has to be circumvented in order to allow for fast service selection at query time. For this purpose, we employ an appropriate service encoding based on labeling schemes [11]. The main idea works as follows. In the case of a tree hierarchy, each concept is labeled with an interval of the form $[begin, end]$. This is achieved by performing a depth-first traversal of the tree, and maintaining a counter, which is initially set to 1 and is incremented by 1 at each step. Each concept is visited twice, once before visiting any of its subconcepts and once after all its subconcepts have been visited. The interval assigned to the concept is constructed by setting its lower (resp., upper) bound to the value of the counter when the concept is visited for the first (resp., second) time. Observe that due to the way intervals are assigned, a concept C_1 is subsumed by another concept C_2 if and only if its interval is contained in that of C_2, i.e., $I_{C_1} \subset I_{C_2}$. This scheme generalizes to the case of graphs, which is the typical case for ontologies on the Semantic Web, by first computing a spanning tree T and applying the aforementioned process. Then, for each non spanning tree edge, the interval of a node is propagated recursively upwards to its parents. Therefore, more than one intervals may be assigned to each concept. As before, subsumption relationships are checked through interval containment: C_1 is subsumed by C_2 if and only if every interval of C_1 is contained in some interval of C_2.

In our example, the intervals assigned to the ontology concepts are shown in Figure 1c, and have been computed considering the spanning tree formed by removing the edge (C_5, C_9). Notice how the interval assigned to the concept C_9 is then propagated to the concepts C_5, C_2, and C_0 (in the latter, it is subsumed by the initially assigned interval).

Consequently, a service request or advertisement can be represented by the set of intervals associated to its input and output concepts. With this encoding, determining the type of match between two service parameters is reduced to checking for containment relationship between the corresponding intervals; a constant time operation. In particular, we can rewrite the conditions determining the type of match between a request parameter C_R and a service parameter C_S, as shown in Table 1, where $\mathcal{I}_C$ denotes the set of intervals assigned to C.

Furthermore, the ranking criterion discussed in Section 2.2 can be expressed by means of the intervals based representation. For a concept C, the size of the subgraph rooted at C, G_C, is given by

Table 1. Types of match using the intervals based encoding

Type of match	Condition
exact	$\mathcal{I}_{C_R} = \mathcal{I}_{C_S}$
plug-in	$\mathcal{I}_{C_R} \subset \mathcal{I}_{C_S}$
subsumes	$\mathcal{I}_{C_R} \supset \mathcal{I}_{C_S}$

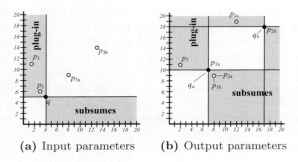

(a) Input parameters   (b) Output parameters

Fig. 2. Interval based search

$$|G_C| = \sum_{I \in \mathcal{I}_C} \left\lceil \frac{|I|}{2} \right\rceil \tag{2}$$

Hence, for two concepts C_1, C_2, where $C_1 \sqsubseteq C_2$ or $C_1 \sqsupseteq C_2$, Equation 1 becomes:

$$degreeOfMatch(C_1, C_2) = \frac{\min\{|G_{C1}|, |G_{C2}|\}}{\max\{|G_{C1}|, |G_{C2}|\}} \tag{3}$$

The above presented service representation allows the evaluation of the type and degree of match between a pair of requested and offered services in constant time. Still, the number of comparisons required is proportional to the number of available services. To further reduce the time required by the matcher, an index structure is employed for pruning the search space, keeping the number of comparisons required to a minimum. For this purpose, each interval is represented as a point in a 2-dimensional space, with the coordinates corresponding to the intervals' lower and upper bounds respectively, i.e., *begin* and *end*. Then, checking for containment between intervals is translated to a range query on this space. Figures 2a and 2b draw the input and output parameters, respectively, of the example in Figure 1. Points labeled as q_x, correspond to the parameters of the requested service, whereas p_{ix} correspond to parameters of the i-th offered service. For example, the output parameters of service S_2 is represented by points $p_{2a} = (8, 9)$ for class C_9 and $p_{2b} = (17, 18)$ for class C_7. For a given interval, the intervals contained by it are those located in its lower-right region, whereas those containing it are located in its upper-left region.

3   Centralized Service Discovery

In a centralized environment a single registry contains the information about all the advertised services and is responsible for performing the matchmaking and ranking process. Under our framework, this registry encodes all service descriptions and uses multi-dimensional indexes, such as the R-tree [12], to expedite service selection. The R-tree partitions points in hierarchically nested, possibly overlapping, *minimum bounding rectangles* (MBR). Each node in the tree stores a variable number of entries, up to some predefined maximum. Leaf nodes contain data points, whereas internal nodes contain the MBRs of their children.

We use two R-trees, T_{in}, T_{out} to index the services, where T_{in} (T_{out}) stores the intervals associated with the input (output) parameters. Consider as an example the 3 services discussed in Section 2.2. Figure 3 shows the MBRs and the structure of the two R-trees. An MBR is denoted by N_i and its corresponding entry as e_i. Notice that points that are close in the space (e.g., p_1, p_2 in Figure 3a) are grouped and stored in the same leaf node (N_2 in Figure 3b).

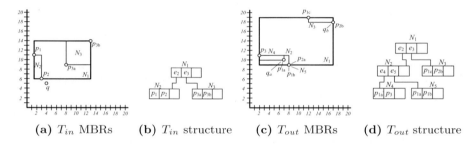

(a) T_{in} MBRs   **(b)** T_{in} structure   **(c)** T_{out} MBRs   **(d)** T_{out} structure

Fig. 3. R-trees example

In the following we describe the algorithm (shown in Figure 4) for finding the services matching a request using our running example. The algorithm examines all request parameters in turn (Line 2). Assume that the first examined parameter *par* is the input corresponding to concept C_8; thus, T_{in} is examined (Line 3). The intervals, in this case $[4, 5]$, associated with the ontology concept is inserted in $\mathcal{I}$ (Line 5). Subsequently, three queries are posed to T_{in} retrieving the exact matches under point $(4, 5)$ (Line 7), the plug-in matches inside the range extending from $(0, 5)$ up to $(4, \infty)$ (Line 8) and the subsumes matches inside $(4, 0)$ up to $(\infty, 5)$ (Line 9). A range query is processed traversing the R-tree starting from the root. At each node, only its children whose MBR overlaps with the requested range are visited. Similarly, for the case of a point query, only children whose MBR contains the requested point are visited. A small performance optimization is to perform the three queries in parallel minimizing, thus, node accesses. Subsequently, all matches to *par* are merged into m_{par} (Line 10). Once all parameters have been examined, the candidate services $\mathcal{S_R}$ are constructed by intersecting the parameter matching results (Line 11). This retains

---

Search Algorithm

Input: request R, available services S indexed in T_{in}, T_{out}
Output: services S_R matching R

1 **begin**
2 | **foreach** $par \in IN_R \cup OUT_R$ **do**
3 | **if** $par \in IN_R$ **then** $T \leftarrow T_{in}$
4 | **else** $T \leftarrow T_{out}$
5 | $\mathcal{I} \leftarrow$ the intervals associated with par
6 | **foreach** $interval$ $I = (i_s, i_e) \in \mathcal{I}$ **do**
7 | $m_{par}^{ex} \leftarrow$ point $[i_s, i_e]$ query in T
8 | $m_{par}^{pl} \leftarrow$ range $(0, i_e) \times (i_s, \infty)$ query in T
9 | $m_{par}^{sb} \leftarrow$ range $(i_s, 0) \times (\infty, i_e)$ query in T
10 | $m_{par} = m_{par}^{ex} \cup m_{par}^{pl} \cup m_{par}^{sb}$
11 | $S_R = \bigcap_{par} m_{par}$
12 | $S_R = S_R \setminus \{S : \exists IN_S$ not matched by any $IN_R\}$
13 | **return** S_R
14 **end**

Fig. 4. Algorithm for index-based service matchmaking

only the services which match all request parameters. Since some services in S_R can have additional input parameters that are not satisfied by the request, they are filtered out from the final result (Line 12).

As discussed in Section 2.1, in many cases a ranked list of the top-k best matching services is preferred as the result of the matchmaking process. Figure 5 illustrates the Progressive Search Algorithm to retrieve the top-k services given a request R using our framework. As before we present the algorithm using our running example. Initially, all intervals associated with the request parameters are inserted into $\mathcal{I}$ (Line 2). In particular, $\mathcal{I}$ contains $[4, 5]$ (represented by point q in Figure 3a) for the input parameter, and $[7, 10]$, $[17, 18]$ (represented by points q_a, q_b, respectively, in Figure 3c) for the two output parameters. A heap H_I is associated with each interval $I = [i_s, i_e] \in \mathcal{I}$ (Lines 3–7); in our case there are 3 heaps for q, q_a and q_b. Initially, these heaps contain the root node of T_{in} or T_{out}, depending on the interval's parameter type (Lines 5–6). Entries e_I in I's heap are R-tree nodes and are sorted increasingly by their *minimum distance* (MINDIST) to (i_s, i_e). The MINDIST of a leaf node, i.e., a point, is its distance from (i_s, i_e). The MINDIST of an internal node, i.e., an MBR, is the minimum distance of the MBR from (i_s, i_e).

The Progressive Search Algorithm proceeds examining heap entries until k services have been retrieved (Lines 8–19). The heap whose head entry has the minimum MINDIST is selected (Line 9). In our example both heaps for q_a and q_b have MINDIST 0 as their head entry (T_{out}'s root) contains both q_a and q_b; assume q_a's heap is selected. The entry (node N_1 in *out*) is popped from the heap (Line 10) and since it is an internal node all its children are inserted in the

```
Progressive Search Algorithm
    Input: request R, available services S indexed in T_in, T_out, k
    Output: services S_R matching R, in descending order of degree of match
 1  begin
 2  │   I ← the intervals associated with all parameters in IN_R ∪ OUT_R
 3  │   foreach I = [i_s, i_e] ∈ I do
 4  │   │   create a heap H_I
 5  │   │   if I corresponds to some par ∈ IN_R then  insert in H_I root of T_in
 6  │   │   else  insert in H_I root of T_out
 7  │   └   H_I entries are sorted increasingly by their MINDIST to (i_s, i_e)
 8  │   while k > 0 do
 9  │   │   find the heap H_I whose head entry has the minimum MINDIST
10  │   │   e_I ← pop(H_I)
11  │   │   if e_I is an internal node then  insert in H_I all children of e_I
12  │   │   else
13  │   │   │   let S be the service corresponding to e_I
14  │   │   │   let par_I be the parameter corresponding to interval I
15  │   │   │   mark that S has a match for par_I
16  │   │   │   if S has matches for all parameters in IN_R ∪ OUT_R then
17  │   │   │   │   insert S in S_R ;                        // S is a result
18  │   │   │   │   k ← k − 1
19  │   │   │   └   if k = 0 then  return S_R
20  end
```

Fig. 5. Algorithm for progressively returning matches

heap (Line 11). Then, the heaps are examined again and q_a's heap is selected, as node N_2 is in its head and has MINDIST 0. N_2 is popped and its children are inserted. Repeating the process once more, a leaf entry p_{1a} is popped, which corresponds to the first output parameter of service S_1 (Lines 13–14). We mark that S_1 has a match for a request parameter (Line 15). Then, S is checked if it has matches for all parameters, i.e., it is a result (Lines 16–19). The algorithm returns when k results have been found.

The output of the algorithm is the ranked service list S_2, S_1, S_3. Notice that S_2 has a subsumes match but it is ranked higher than S_1, having only exact and plug-in matches. Further, S_3 is included even though is has two fail matches. This is due to the fact that the MINDIST function described does not discriminate among points in different regions with respect to the point corresponding to a request parameter's interval. For example, in Figure 3c p_{2a} and p_{1b} are closer to q_a than p_3 and are regarded as better matches to parameter q_a, even though they are only subsumes matches (they lie in the lower right quadrant w.r.t. q_a). To obtain arbitrary rankings as described in Section 2.1, MINDIST can be trivially modified to be region aware. For example, it can evaluate heap entries that correspond to plug-in as closer compared to subsumes matches.

4   P2P Service Discovery

As the availability and demand for Web services grows, the issue of managing Semantic Web services in a distributed environment becomes vital. We describe a scalable and fault-tolerant solution that is adaptable and efficient in a distributed environment. From the variety of paradigms of distributed systems, we choose to focus on flat P2P overlays, as the latter represent the current trend for distributed data management. More specifically, we employ a structured P2P overlay, since it provides self-maintenance and robustness, as well as efficiency in data management. In the following we discuss the adaptation of our framework to the distributed setting.

4.1   The Underlying P2P Overlay

Before discussing distributed service discovery, we have to choose a suitable framework. To support the adaptation of the algorithms presented in Section 3, such a framework must support both point and range queries, so as to allow for the retrieval of both exact and plug-in/subsumes matches, respectively. Furthermore, since our proposed service encoding and service search algorithm are based on the 2-dimensional space, it is necessary to select a P2P framework that is efficient and scalable for 2-dimensional data. More specifically, the selected P2P framework should preserve locality and directionality, if possible.

SpatialP2P [13] is a recently proposed structured P2P framework, targeted to spatial data. It handles areas, which are either cells of a grid-partitioned space or sets of cells that form a rectangular. The basic assumption of the framework is that each area has knowledge of its own coordinates and the coordinates of some other areas to which it is directly linked. The goal of SpatialP2P is to guarantee that any stored area can be searched and reached from any other, solely by exploiting local area knowledge.

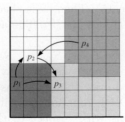

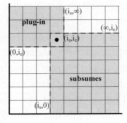

Fig. 6. Illustrative SpatialP2P overlay   **Fig. 7.** Search regions

Figure 6 shows an example of a SpatialP2P overlay with four peers. Each peer maintains links to others towards the four directions of the 2D space. The grid is hashed to the four peers, such that each cell is stored and managed by the closest peer. In the figure cells and their storing peers share the same color.

In SpatialP2P, search is routed according to locality and directionality. This means that search is propagated to the area that is closer to and towards the same direction with the sought area, choosing from the available areas that are linked to the one on which the search is currently iterated.

4.2   Managing Services in the P2P Overlay

The management of services in the P2P overlay consists of two basic operations: insertion of services and search for services in the system. Search can be either exhaustive, i.e., seeking for any possible results, or top-k, i.e., seeking the k best-matching results. In the following we discuss the details of these operations.

Service insertion. In order to use the P2P framework for the distributed management of Semantic Web services, we assume that the ID space of the overlay (i.e. the space of values for node and data IDs) corresponds to the space of values defined by the encoding of the service descriptions.

When a new service is published, its description is encoded using the intervals based representation presented in Section 2, and then it is inserted in the network. Specifically, each encoded service parameter is hashed to and eventually stored by the peer whose ID is closer to its value in the 2-dimensional space. Each inserted service parameter is accompanied by some meta-data about the respective type, (input or output), as well as the service it belongs to.

The locality-preserving property of the SpatialP2P overlay guarantees that similar services are stored by the same or neighboring peers. By similar, we mean services whose input and output parameters correspond to matching concepts. Moreover, the preservation of directionality means that following subsequent peers in a particular direction results, for example, in locating concepts subsuming or subsumed by the ones previously found. As described below, these properties are essential for minimizing the search time, and this applies to both exhaustive range and top-k queries.

Service search. Searching for services in the P2P overlay is performed by an adaptation of the search algorithm of Section 3 to the SpatialP2P API. For each requested service parameter, a point or a range query is performed, depending on the requirement of exact, plug-in or subsumes match with the available service parameters. An exact request for a service parameter corresponding to interval $I = [i_s, i_e]$ is performed by a point query asking the retrieval of the point (i_s, i_e), if such data exists in the overlay. For plug-in and subsumes requests for a parameter associated to the interval $I = [i_s, i_e]$, a pair of range queries is initiated. Since the data space is bounded (recall the intervals construction from Section 2.2), these requests are represented by range queries for rectangular areas. Specifically, for plug-in matches, a query requesting the range extending from $(0, i_e)$ up to (i_s, ∞) is issued, while for the subsumes request, the corresponding range is $(i_s, 0) \times (\infty, i_e)$ (see Figure 7). The results of these two queries are unified to provide the answer to the requested parameter. Parallel searches are conducted for each requested parameter, and the results are finally intersected to compute the final matches.

Finding the top-k matches. SpatialP2P supports top-k search by extending search for range queries to dynamically increase the respective range. In detail, a search for a service parameter represented by an interval $I = [i_s, i_e]$ is initiated as the minimum range query that includes (i_s, i_e); thus, the minimum range is extended only in the grid cell in which the point (i_s, i_e) resides. After the search is performed in this minimum range, if the number of retrieved results is lower than k, then the range is increased towards the desired direction of the 2D space by the minimum, i.e., by one grid cell. The process repeats iteratively, until k results have been retrieved (or the whole space has been searched).

5   Experimental Evaluation

Experimental setup. We have evaluated our approach on two data sets. For the first data set, to simulate a real-world scenario, we used the OWL-S service retrieval test collection OWLS-TC v2[1]. This collection contains services retrieved mainly from public IBM UDDI registries, and semi-automatically transformed from WSDL to OWL-S. More specifically, it comprises: (a) a set of ontologies, derived from 7 different domains (education, medical care, food, travel, communication, economy and weapons), comprising a total of 3500 concepts, used to semantically annotate the service parameters, (b) a set of 576 OWL-S services, (c) a set of 28 sample requests, and (d) the relevance set for each request (manually identified).

The second data set was synthetically generated, based on the first one, so as to maintain the properties of real-world service descriptions. In particular, we constructed a set of approximately 10K services, by creating variations of the 576 services of the original data set. For each original service, we selected randomly one or more input or output parameters, and created a new service description by replacing them with randomly chosen superconcepts or subconcepts from the corresponding domain ontology. A set of 100 requests was generated following the same process, based on the original 28 requests. All the experiments were conducted on a Pentium D 2.4GHz with 2GB of RAM, running Linux.

Ranking. In the first set of experiments we used the first data set to evaluate the effectiveness of the ranking approach. For each of the 28 queries we retrieved the ranked list of match results, and compared them against the provided relevance sets. We use well-established IR metrics[2] to evaluate the performance of the search and ranking process. In particular, Figure 8a depicts the micro-averaged recall-precision curves for all the 28 queries, i.e., the precision (averaged over all queries) for different recall levels. Observe that a 30% of the relevant services can be retrieved with precision higher than 80%, whereas for retrieving more than 70% of the relevant services the precision drops below 50%. Also, the following metrics are presented in Figure 8b: (a) precision at k, i.e., the (average) precision

---

[1] http://projects.semwebcentral.org/projects/owls-tc/
[2] http://trec.nist.gov/

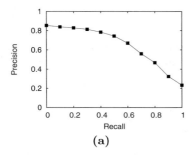

P@5	0.764
P@10	0.721
P@15	0.631
S@1	0.857
S@2	0.964
S@4	1

(a)   **(b)**

Fig. 8. (a) Recall-precision curve, (b) Precision and Success at k (P@k, S@k)

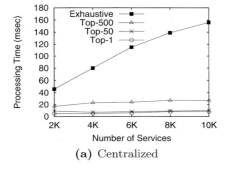

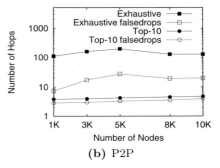

(a) Centralized   **(b)** P2P

Fig. 9. Search cost for: (a) centralized registry, (b) P2P registry

after k results have been retrieved; (b) success at k, i.e., whether a relevant result has been found after k results have been retrieved.

As we can see, the precision drops below 70% after the top-10 matches have been retrieved. Moreover, for the set of 28 queries, in 24 of them the top-1 match is a relevant one, in 27 queries there is a relevant result among the top-2 matches, and in all cases there is a relevant result among the top-4 matches. The above results opt for the emphasis on top-k queries and on fetching results progressively, as discussed in Section 2.

Experiments for centralized search. In this set of experiments we measured the time required by our search algorithm to discover and rank services in a centralized registry. In particular, we investigated the performance benefits, i.e., the reduction in response time, resulting from restricting the search to retrieving only the top-k matches. For this purpose, we used the synthetically generated data set described previously. We varied the number of services from 2K up to 10K, and we measured the processing time (averaged over 100 queries) for retrieving: (i) all matches, and (ii) the top-k matches for $k \in \{1, 50, 500\}$. The experimental results are illustrated in Figure 9a. Notice the significant savings in the processing time when restricting the search to top-k matches, as well as the fact that the processing time in the latter case is significantly less sensitive to the number of available services.

Experiments for search in P2P environment. In the last set of experiments, we evaluated our search method in a P2P environment, as described in Section 4. We varied the size of the P2P network, from 1K up to 10K peers, and we inserted a total of 10K services. We conducted two experiments. In the first experiment, we retrieved, for each request, all the identified matches, whereas in the second, we restricted the search range to obtain (approximately) the top-10 results for each request. For each of the experiments we report two measures: (i) total number of hops (i.e., number of peers processing the query), and (ii) number of falsedrops (i.e., number of peers on the search path not contributing to the result set). The results are shown in Figure 9b. Both measures are quite low and relatively stable w.r.t. the size of the network. As discussed in Section 4, this is due to the fact that SpatialP2P is particularly designed to preserve the locality and the directionality of the data space, thus queries are effectively routed towards peers containing relevant information. As in the centralized case, the search cost is significantly lower, when retrieving only the top-k matches.

6   Related Work

Service discovery is an important issue for the Semantic Web, hence several works have dealt with this problem. Matchmaking for Semantic Web services based on inputs and outputs has been studied in [7,8], and more recently in [14,15]. These form the basis of our matching approach, however they do not deal with the aspects of efficiency, ranking, and discovery in P2P networks, which are the main issues of our work. Implemented systems for matchmaking of OWL-S and WSMO services are described in [16,17]. In [9,18] similarity measures for ranking Semantic Web services are presented. These measures can be used by our top-k search algorithm, and hence are complementary to our work. The efficiency of the discovery process is considered in [19]. However, it is based on pre-computing and storing, for each concept in the ontology, the list of services matching this concept (together with the type of match). This imposes excessive storage requirements, and fails to scale as the number of available services (i.e., the size of the stored lists) and the size of the ontologies (i.e., the number of lists to store) grow significantly. Efficient matchmaking, together with ranked retrieval, is presented in [20]. Similar to our work, it uses intervals and indexing, which are however constructed in a different way. Moreover, search in P2P networks is not considered.

A P2P approach for Web service discovery is presented in [21]. However, the services are not semantically described; instead, the search is based on (possibly partial) keywords. Semantic Web service discovery in P2P networks has been studied in [22,23]. In contrast to our work, these approaches deal with unstructured networks. In [24] Web service descriptions are indexed by keywords taken from domain ontologies, and are then stored on a DHT network. In [25] the peers are organized in a hypercube and the ontology is used to partition the network into concept clusters, so that queries are forwarded to the appropriate cluster. However, the subset of concepts to be used as structuring concepts should be known in advance. The approach in [26] distributes semantic service advertisements among available registries, by categorizing concepts into different groups based on their semantic similarity, and assigning groups to peers. In [27] services are distributed to registries depending on their type, e.g., a registry related to the travel domain will only maintain Web services specific to this domain.

7   Conclusions

We have presented and evaluated an efficient and scalable approach to Semantic Web service discovery and ranking. Efficiency is achieved by employing a suitable encoding for the service descriptions, and by indexing these representations to effectively prune the search space, consequently reducing the search engine's response time. To allow for scalability, we describe how the service representations can be distributed in a suitable structured P2P overlay network, and we show how the search is performed in this setting.

In this work we have treated the matching of service requests and advertisements as a matching of their inputs and outputs. However, service descriptions may also contain preconditions and effects, as well as QoS parameters. The described ideas can be easily extended to consider these additional criteria. In the future we plan to incorporate such parameters in our search algorithm.

References

1. Burstein, M., et al.: OWL-S: Semantic Markup for Web Services. In: W3C Member Submission (2004)
2. Akkiraju, R., et al.: Web Service Semantics - WSDL-S. In: W3C Member Submission (2005)
3. Lausen, H., Polleres, A., Roman, D. (eds.): Web Service Modeling Ontology (WSMO). In: W3C Member Submission (2005)
4. Liarou, E., Idreos, S., Koubarakis, M.: Evaluating Conjunctive Triple Pattern Queries over Large Structured Overlay Networks. In: Cruz, I., Decker, S., Allemang, D., Preist, C., Schwabe, D., Mika, P., Uschold, M., Aroyo, L.M. (eds.) ISWC 2006. LNCS, vol. 4273, pp. 399–413. Springer, Heidelberg (2006)
5. Sidirourgos, L., Kokkinidis, G., Dalamagas, T., Christophides, V., Sellis, T.K.: Indexing Views to Route Queries in a PDMS. Distributed and Parallel Databases 23(1), 45–68 (2008)
6. Staab, S., Stuckenschmidt, H. (eds.): Semantic Web and Peer-to-Peer. Springer, Heidelberg (2006)
7. Paolucci, M., Kawamura, T., Payne, T.R., Sycara, K.P.: Semantic Matching of Web Services Capabilities. In: Horrocks, I., Hendler, J. (eds.) ISWC 2002. LNCS, vol. 2342, pp. 333–347. Springer, Heidelberg (2002)
8. Li, L., Horrocks, I.: A Software Framework for Matchmaking based on Semantic Web Technology. In: WWW, pp. 331–339 (2003)
9. Skoutas, D., Simitsis, A., Sellis, T.: A Ranking Mechanism for Semantic Web Service Discovery. In: IEEE SCW, pp. 41–48 (2007)

10. Joachims, T., Radlinski, F.: Search Engines that Learn from Implicit Feedback. IEEE Computer 40(8), 34–40 (2007)

11. Christophides, V., Karvounarakis, G., Plexousakis, D., Scholl, M., Tourtounis, S.: Optimizing Taxonomic Semantic Web Queries Using Labeling Schemes. J. Web Sem. 1(2), 207–228 (2004)

12. Guttman, A.: R-Trees: A Dynamic Index Structure for Spatial Searching. In: SIGMOD Conference, pp. 47–57 (1984)

13. Kantere, V., Skiadopoulos, S., Sellis, T.: Storing and Indexing Spatial Data in P2P Systems. IEEE Transactions on Knowledge and Data Engineering (to appear)

14. Bellur, U., Kulkarni, R.: Improved Matchmaking Algorithm for Semantic Web Services Based on Bipartite Graph Matching. In: ICWS, pp. 86–93 (2007)

15. Skoutas, D., Sacharidis, D., Simitsis, A., Sellis, T.: Serving the Sky: Discovering and Selecting Semantic Web Services through Dynamic Skyline Queries. In: ICSC (2008)

16. Klusch, M., Fries, B., Sycara, K.P.: Automated Semantic Web Service Discovery with OWLS-MX. In: AAMAS, pp. 915–922 (2006)

17. Kaufer, F., Klusch, M.: WSMO-MX: A Logic Programming Based Hybrid Service Matchmaker. In: ECOWS, pp. 161–170 (2006)

18. Cardoso, J.: Discovering Semantic Web Services with and without a Common Ontology Commitment. In: IEEE SCW, pp. 183–190 (2006)

19. Srinivasan, N., Paolucci, M., Sycara, K.P.: An Efficient Algorithm for OWL-S Based Semantic Search in UDDI. In: Cardoso, J., Sheth, A.P. (eds.) SWSWPC 2004. LNCS, vol. 3387, pp. 96–110. Springer, Heidelberg (2005)

20. Constantinescu, I., Binder, W., Faltings, B.: Flexible and Efficient Matchmaking and Ranking in Service Directories. In: ICWS, pp. 5–12 (2005)

21. Schmidt, C., Parashar, M.: A Peer-to-Peer Approach to Web Service Discovery. WWW 7(2), 211–229 (2004)

22. Paolucci, M., Sycara, K.P., Nishimura, T., Srinivasan, N.: Using DAML-S for P2P Discovery. In: ICWS, pp. 203–207 (2003)

23. Basters, U., Klusch, M.: RS2D: Fast Adaptive Search for Semantic Web Services in Unstructured P2P Networks. In: Cruz, I., Decker, S., Allemang, D., Preist, C., Schwabe, D., Mika, P., Uschold, M., Aroyo, L.M. (eds.) ISWC 2006. LNCS, vol. 4273, pp. 87–100. Springer, Heidelberg (2006)

24. Li, Y., Su, S., Yang, F.: A Peer-to-Peer Approach to Semantic Web Services Discovery. In: ICCS (4), pp. 73–80 (2006)

25. Schlosser, M.T., Sintek, M., Decker, S., Nejdl, W.: A Scalable and Ontology-Based P2P Infrastructure for Semantic Web Services. In: P2P Computing, pp. 104–111 (2002)

26. Vu, L.H., Hauswirth, M., Aberer, K.: Towards P2P-Based Semantic Web Service Discovery with QoS Support. In: BPM Workshops, pp. 18–31 (2005)

27. Verma, K., Sivashanmugam, K., Sheth, A., Patil, A., Oundhakar, S., Miller, J.: METEOR-S WSDI: A Scalable P2P Infrastructure of Registries for Semantic Publication and Discovery of Web Services. Inf. Tech. and Manag. 6(1), 17–39 (2005)

Exploring Semantic Social Networks Using Virtual Reality

Harry Halpin[1], David J. Zielinski[2], Rachael Brady[2], and Glenda Kelly[2]

[1] School of Informatics
University of Edinburgh
2 Buccleuch Place
EH8 9LW Edinburgh
Scotland, UK
H.Halpin@ed.ac.uk
[2] Visualization Technology Group
Duke University
130 Hudson Hall Box 90291
Durham, NC 27708, USA
djzielin@duke.edu, rbrady@duke.edu, glenda@ee.duke.edu

Abstract. We present Redgraph, the first generic virtual reality visualization program for Semantic Web data. Redgraph is capable of handling large data-sets, as we demonstrate on social network data from the U.S. Patent Trade Office. We develop a Semantic Web vocabulary of virtual reality terms compatible with GraphXML to map graph visualization into the Semantic Web itself. Our approach to visualizing Semantic Web data takes advantage of user-interaction in an immersive environment to bypass a number of difficult issues in 3-dimensional graph visualization layout by relying on users themselves to interactively extrude the nodes and links of a 2-dimensional graph into the third dimension. When users touch nodes in the virtual reality environment, they retrieve data formatted according to the data's schema or ontology. We applied Redgraph to social network data constructed from patents, inventors, and institutions from the United States Patent and Trademark Office in order to explore networks of innovation in computing. Using this data-set, results of a user study comparing extrusion (3-D) vs. no-extrusion (2-D) are presented. The study showed the use of a 3-D interface by subjects led to significant improvement on answering of fine-grained questions about the data-set, but no significant difference was found for broad questions about the overall structure of the data. Furthermore, inference can be used to improve the visualization, as demonstrated with a data-set of biotechnology patents and researchers.

Keywords: Visualization, Virtual Reality, RDF, Semantic Web, 3-D Interaction, User Interface, Network Analysis, CAVE.

1 Introduction

While researchers have become interested in the large amounts of network-structured data available on the Web, intuitive understanding of the structure

A. Sheth et al. (Eds.): ISWC 2008, LNCS 5318, pp. 599–614, 2008.

of networks remains more of a black art than science. Mathematical frameworks developed to analyze networks can be difficult to interpret, so visualizing networks has become a common tool for users to gain intuitive understanding of the data. While research into network visualization algorithms is extensive on the Semantic Web, much of this research has focused on two dimensional visualization which often produces dense and confusing "spaghetti-like" structures that elude visual analysis and comprehension.

Redgraph, our virtual reality-based network visualization program for Semantic Web data, relies on user-directed 3-D extrusion to transform the visualization from 2 into 3-dimensions. While this application builds on previous virtual reality work that allows users to reposition nodes [6], Redgraph differs from previous applications by allowing users to "extrude" nodes from 2-D into 3-D in a fully immersive and interactive CAVE^{TM}-like environment, as shown in Figure 1 [7]. Redgraph is not customized for any particular data-set or application, but is a generic toolset is capable of visualizing any network that can be described using the RDF (Resource Description Framework) data model, which naturally maps its subjects and objects to nodes in a network and properties to links [13].

2   Related Work

Visualization of Semantic Web data is nearly synonymous with graph visualization, which has a long history prior to the advent of the Semantic Web [15]. Visualization is necessary since "implicit information embedded in semantic web graphs, such as topography, clusters, and disconnected subgraphs, is difficult to extract from text files" [18]. Almost all graph visualizers for Semantic Web data, as exemplified by IsaViz, produce 2-D graphs [25]. These tools have in turn been built on top of tools developed for generic graph visualization such as GraphViz [10], so most researchers have simply applied pre-existing 2-D visualization algorithms from applications such as GraphViz to Semantic Web visualization [23]. Researchers who study user interfaces in the Semantic Web have begun criticizing graph visualization as the primary method of visualizing Semantic Web data. In particular, Karger and Schraefel have noted that the "Big Fat Graph" approach is popular because visualization researchers are "allowing the computer's internal representation of data to influence the way their tools present information to users, when instead they should be developing interfaces that are based on the users' needs, independent of the computer's particular information representation" [18]. In other words, just because RDF has a graph structure does not mean a graph structure should be used to interface with data. To the extent that Karger and Schraefel believe that more human-centric user interfaces are needed for Semantic Web, we agree wholeheartedly. However, work on alternative methodologies for visualizing Semantic Web data suffers from its fair share of limitations as well. If graph visualization can be characterized as the "Big Fat Graph" approach, the alternatives cited by Karger and Schraefel such as mSpace [18] and the Tabulator browser [3] can be characterized as the "Lots of Confusing Little Menus" approach. It is the careful choice of menus and other interface components that creates a functional user-interface. Therefore prior to designing any application-centric user interface, an overall intuitive understanding of the particular data-set and what users want out of this data is needed. For data sets that researchers are just beginning to explore, visualization can be vital for gaining an intuitive understanding of the data set so that these questions can even be asked. Graph visualization is then an important default mode of visualization for Semantic Web data, as it makes minimal assumptions about the data.

A pragmatic approach to visualization of Semantic Web data recognizes that there is no one perfect visualization technique, much less a fully generic user interface for the entire Semantic Web. Despite its critics, there are definitely cases where graph visualization is the most appropriate technique. First, graph visualization is appropriate when what is being graphed is an actual network, such as a social networks. Visualization is important in network analysis, as most networks are less amendable to analytic analysis than other types of data, primarily due to their violation of the Gaussian distribution. For identifying clusters, there are wide variety of statistics, each with their own drawbacks and advantages, which can lead to confusion for the user [21]. Typical networks in "the wild" (as opposed to Erdoś-Renyi random graph models [2]) obey a power-law distribution of connectivity between nodes and links [1] so that typical descriptive statistics such as means cannot be computed since these rely on properties that do not hold in power-law distributions [24]. Due to the above-mentioned factors, visual inspection can be crucial when understanding social networks in particular. While Karger and Schraefel complain that one problem with graph visualization is that it puts "next to each other the things that are connected by links," if the link represents something important, such as a relationship in a social network, it is crucial for user understanding that the node be placed in close proximity with the other node. While more traditional user interfaces may be developed once a data-set is understood thoroughly, for the initial exploration of the data-set it is actually useful for the visualization to follow the data quite closely, as to not give any a priori bias to the presentation of subsets of the data when creating interfaces using tools such as Exhibit [16]. Due to this request from researchers in the social networks of patents, we decided to use graph visualization as our method of exploring the data before developing a more traditional user interface.

In order to address some of the traditional problems with graph visualization, in particular the limitations of two-dimensions in visualizing complex networks, many researchers have studied graph visualization in virtual reality [15]. Studies in virtual reality have found that viewing a graph using three-dimensions allows "three times as much information can be perceived in the head coupled stereo view as in the 2-D view." [29]. Moving beyond mere 3-D visualization to fully immersive virtual reality, with head-tracking and stereoscopic vision, allows users to gain an intuitive understanding of the network by literally letting them walk "around" and "into" the data, facilitating kinesthetic comprehension [5]. This is not to underestimate the effect of the quality of the visual display for understanding complex data in the presence of information clutter. The high fidelity graphics provided by virtual reality systems like the CAVE have been shown to help performance, as immersion "provides many depth cues that other technologies do not; in particular, stereo images and head tracking let users exercise their built-in capacity for understanding stereopsis[1] and motion parallax. Thus a higher level of immersion can lead to greater spatial understanding, which can result in greater effectiveness for many applications such as scientific visualization" [5].

Despite these advantages, no previous graph visualization research in the Semantic Web has used immersive virtual reality technology in order to enable the employment of an interactive third dimension. This stands in contrast to research in Topic Maps, where immersive virtual reality technology has been demonstrated [19]. Researchers have developed OntoSphere3D, a Protege plugin that in 2-D visualizes an ontology projected on a 3-D sphere that allows "panning, rotation, and zoom" but does not allow interaction with the graph's 3-D layout itself, much less immersion and stereoscopic 3-D [4]. In particular, the usage of virtual reality technologies can confront one of the primary objections of Karger and Schraefel to using graphs as a way of visualizing Semantic Web data: the fact that graphs "are flat" [18].

To explore the advantages and disadvantages of immersive and interactive visualization, a particular data-set featuring a social network of people and institutions involved in filing patents was selected. It is social since researchers are interested in the co-authoring of patents and movements of researchers through institutions, and a network since these relationships can be represented as links between people and institutions, as well as patents. The literature on social networks, including their interaction with the Semantic Web, is immense, and those interested in exploring the interface of social networks with the Semantic Web are encouraged to reference the work of Mika and others [21]. Note that this data-set is a semantic social network since the connections between the various actors in the network have been given a semantic basis by being formulated using Semantic Web standards [13]. For our purposes, it is enough that this network is a social network where the data naturally takes a graph format, and so graph visualization is a sensible visualization for this data.

3   Virtual Reality Vocabulary

There is currently no standardized Semantic Web vocabulary for graph visualization, including virtual reality. Proposals for virtual reality markup-languages on the Web have proliferated since the beginning of the Web itself. The current ISO standard is the Web3D's consortium's X3D, the successor of VRML (Virtual Reality Markup Language) [30]. X3D is far too complex for our application, as it is meant to cover all possible cases of virtual reality modeling from humanoid movement to landscape scenes, and furthermore, it is defined only via XML and

---

[1] In other words, "depth perception."

XML Schema, although there are proposals to to introduce the use of Semantic Web standards to X3D [26].

One alternative is to use a Semantic Web version of GraphXML, an easy-to-use XML language meant to describe and annotate graphs in both 3-D and 2-D [14]. GraphXML is used in several open source projects and has parsers available in a number of languages. While merely creating a Semantic Web version of an existing XML vocabulary is not a contribution in of itself, creating a Semantic Web vocabulary for 3-D visualization allows already existing Semantic Web data to be easily annotated with the properties needed for 3-D visualization using the same format. Users can then store and query both visualization meta-data and the data itself using the same tools, and seamlessly merge formerly separate graphs of data into a single visualization, offering new and useful capabilities not provided by GraphXML itself. We developed an XSLT transformation capable of transforming GraphXML to a RDF(S) vocabulary we call Vis3D. This mapping is given in Table 1, illustrating the equivalence between GraphXML elements and attributes with Vis3D classes and properties, where a property's domain is the class it is listed with and its range is given in parenthesis with XML Schema data-typing assumed. If there is no class listed, the property gives both its domain and range as *property(domain, range)*. If multiple properties are listed with the same domain and range, only the final property explicitly lists these. The advantage of Vis3D over GraphXML is that while GraphXML is a "closed-world" for sharing graph visualizations, arbitrary data-sets can be visualized by just giving Vis3D visualization properties or Vis3D classes if the data is or can be converted to the Semantic Web, a byproduct of the "meta" nature of having the visualization vocabulary and data share the same underlying model. A `visualize` property turns visualization of that element on or off, allowing a user to visualize employee relationships and not co-author relationships. In this manner, users can customize exactly what subset of their data is to be visualized, and configure the precise details.

Table 1. Mapping of GraphXML to Vis3D

GraphXML Element: attribute(s)	Vis3D Class: Property(Range)
Graph: vendor, version	Graph: vendor(string), version(positiveInt)
node: isMetaNode	Node, isMeta(boolean)
edge: source, target	link(Node,Node)
position: x, y, z	x, y, z(Node,int)
size: width, height, depth	width, height, depth(Node,positiveInt)

4   Redgraph Capabilities

Redgraph is the first generic cross-platform Semantic Web virtual reality tool for visualization. It uses a custom parser to load the data file into a virtualized data structure, and has optimizations for speed over large data-sets. Vis3D annotations are used to determine which nodes (subjects and objects) and edges

(properties) are rendered, and which nodes and edges are considered meta-data
that is shown only as "details on demand" when a user "touches" the node. When
the user touches the node, data associated with the node is presented to the
user. Whatever schema is available can be retrieved, and the XML Schema data
type provides the information needed to format the data in a human-readable
form and provides captions for the data as well. The network visualization and
extrusion technique is implemented in the Syzygy virtual reality library [27].
GraphViz [15] or Boost (only available on Linux) can both be used for the initial 2-D network layout. Vis3D is used to save and load visualizations. Pictures
showing Redgraph in action are shown rightmost in Figure 1 and a movie may
be viewed on the Web at http://www.redgraph.org.

Fig. 1. DiVE: Left - Person in DiVE, Right - Exploring Social Networks in the DiVE

5   Network Layout

From the point of view of the user, the initial network layout is of primary
importance. A large number of algorithms have been developed for both 2-D
and 3-D network visualizations. Most 2-dimensional layout algorithms consider
links to be simulated "springs" and nodes to have some repulsive force, and
these simulated springs are used to model forces between each pair of nodes.
"Springer embedding" algorithms like the Kamada-Kawai algorithm minimize
a function which is the sum of the forces on all nodes in the network [17].
This algorithm has been shown to produce diagrams that accurately model the
structure of networks like social networks and also are aesthetically pleasing
[23]. However, Redgraph separates the initial 2-D layout of the network, which
can use in theory any algorithm to present data to the user. This is because
the algorithm used for network layout often takes much longer to layout. Even
using a relatively fast algorithm such as Fruchterman-Reingold, which has been
optimized for large data-sets, the network layout takes much more time than the
actual loading of the data and virtual reality materialization [11]. For example,
for a network of 15734 nodes, loading and visualizing the data took only 2.6

seconds, while layout took 54.7 seconds. By separating the two steps, a user can lay out large data-sets only once, and then easily and quickly load the data, manipulate node locations, and save any modifications made, and not have to repeat the network layout step. For the current user study, the Kamada-Kawai algorithm was utilized [17]. For the data-set used in our foray into using inference in Section 11 the Fruchterman-Reingold algorithm was used [11].

6   Retrieving Data through Extrusion

One common challenge in graph visualization is the complexity in understanding unfamiliar data-sets in order to conceptualize underlying patterns and points of interest. This problem is especially pronounced in data sets in which either the higher-dimensional structure does not map well onto 2 or even 3 dimensions. This becomes apparent when the number of connections is so high that visualizing the network as a graph leads to confusing "spaghetti-like" clusters that are so dense as to not be amendable to interpretation [6]. Furthermore, while algorithms like Kamada-Kawai can be extended to three dimensions, the problem of determining where precisely a user should be placed in the 3-D visualization remains an open area of research [8]. Placing the user's view point "in the middle" of a three-dimensional visualization may hinder rather than help the visualization process, since the viewpoint may hide parts of the network. One effective way to circumvent both these problems in one fell swoop is the use of extrusion. Extrusion allows the user to select nodes of interest from a 2-dimensional network and then pull these out into a 3-dimensional space, which literally extrudes the 2-dimensional data into 3 dimensions.

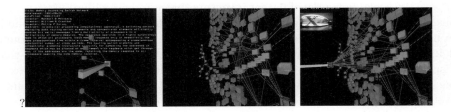

Fig. 2. Redgraph Snapshots: Left - Node Metadata, Center - 2-D Network Layout, Right - Interactively extruding a Node into 3-D

In detail, the user will first see the data in 2-D on a plane of the virtual environment, as shown in the center picture in Figure 2 above. The subject can then use the virtual wand, which they control through a handset, and their point of view that is monitored by head-tracking, to rapidly get an overview of the network and "zoom in" onto relevant details. Upon intersection between the wand and a node, the visualization program will display all information contained in the Semantic Web data model as shown in the leftmost picture of Figure 2. When the subject discovers a node that they are interested in, they can continue holding down a button on the wand, and pull or push the node into the third dimension, as shown in the rightmost picture in Figure 2 above. As a result of this movement of the node, the subject can stretch the links between the nodes into three dimensions. Furthermore, the user can also use a key combination on the virtual want to do "group pullout," where a node and its children nodes are pulled out into 3-D together.

This method has a number of advantages for data interpretation. First, many users can quickly identify clusters and other interesting phenomenon in 2-dimensional data-structures. By beginning the visualization in 2-D, this capitalizes on users ability to conceptualize the data quickly in 2-D and then use 3-D only as needed. Extrusion avoids the problem of the 3-D layout of data by using well-understood and optimized 2-D visualization algorithms for the initial layout. Most importantly, extrusion allows users to interact with the data using both their visual and proprioceptic-motor abilities, thus giving them potential to optimally position the data display according to their preferences. This interaction with network data via extrusion allows users to dynamically re-cluster data using parameters difficult or impossible for computers alone to detect, and thus offers the third dimension as a "sketch pad" for the placement of nodes according to their particular preferences and task at hand.

7   Exploring Social Networks in Patents

The data-set used was available at no cost from the United States Patent and Trademark Office (USPTO) and allows exploration of the social networks of innovation in computer science. The goal was to map the data available in the USPTO to the Semantic Web, and then use a visualization of this data-set to expose the social networks of patent inventors through their affiliation with various institutions and their co-authoring of patents. Researchers with backgrounds in history of science had been using non-interactive 2-D visualization but found it insufficient when confronted with these dense networks, and so thought that visualizing them in immersive 3-D would help. Although they did not have a background in the Semantic Web, they correctly thought the use of Semantic Web technologies would allow them to "mash-up" data from several sources in order to help their visualization. Using Semantic Web technologies, the U.S. Patent Data was integrated with pictures, video interviews, employee records, and other material of interest [12], and stored in the OWL ontology given briefly in Table 2. This table uses the same conventions as used in Table 1, with dates always given as month, day, and year. If the exact date is not known, the date will default to the 1st of January.

Since the USPTO patent database totaled over seven million patent records in 2006, a subset of patents were selected that relate to computing history linking work in personal computing at Xerox PARC by gathering a list of employees from Xerox PARC and then searching the entire patent database for any patents filed under their names. This resulted in a data-set of patents with 7667 RDF triples.

Table 2. Social Network Ontology for Patents

Class	Properties(range)
Person	imageURI, interviewURI, movieURI(URI)
	born, died, started, ending, worked(date)
	name, university, location, bio(string)
Institution	foundedYear, endedYear(date)
	location, name(string)
	employees, creditScore(positiveInt)
	NAIC, sales(float)
Patent	number(positiveInt)
	name, classification, title, abstract, fulltext(string)
	dateFiled, dateIssued(date)
	cited, citedBy(Patent)
	inventor(Person)
	assignee(Institution)

After the visualization was developed and explored using Redgraph, the researchers used their discoveries in 3-D to architect a 2-D exhibit using Exhibit [16].

8 User Study

A user study was conducted to assess the efficacy of the 3-D data extrusion technique when compared with the 2-D method of data presentation as applied to the social network data-set described in Section 7. In other words, did pulling out and interacting with the data in an immersive 3-D modality help any tasks? Comparison of the time taken by subjects to correctly answer 8 quantitatively-measured questions was made between these two methods of data presentation. While the 3-D method allowed full interaction, including extrusion, the 2-D method, while also taking place in the DiVE, but did not allow extrusion, allowing only re-positioning the nodes on a 2-D plane using the virtual wand. It was hypothesized that subjects would give correct answers faster when allowed to use the 3-D technique as compared to the 2-D technique. Subjects were also asked 4 additional qualitative questions to provide formative evaluative feedback to better understand the ways that subjects interacted with the interface. A total of 21 subjects completed the study, 15 male and 6 female, with a mean age of 19.6 years and ranging from 18 to 28 years. Only 3 subjects had used immersive interaction before, and of these, their average number of times in an immersive virtual reality environment was 1.6. The study was conducted in the DiVE (Duke Immersive Virtual Environment), a 6-sided CAVE-like system, as shown in Figure 1 [7]. The structure is 3m x 3m x 3m with a rigid ceiling and floor, flexible walls, and a sliding door. The DiVE uses a 7 PC workstation cluster running XP with NVidia Quadro FX 3000G graphics cards as the graphics-rendering engine. Christie Digital Mirage 2000 projectors connect with Stererographics active stereo glasses to provide the visual interface, while head-tracking is supported by the Intersense IS-900 system.

For this network visualization, patents, inventors, and institutions (colored green, blue, and red respectively) were all represented as cubes and the links connecting the nodes represented relationships between these elements. The effect of using different colors and shapes was not investigated, but standard results from visualization were assumed to hold, and hence basic shapes and primary colors with high contrast were used. Auxiliary information for each element extracted from the Semantic Web encoding was displayed when the user touched that node with the virtual wand. For example, when subjects touched a given patent node with their virtual wands, the patent abstract, patent filing date, any images associated with the abstract, and other associated information were displayed on the wall of the virtual chamber, as in the rightmost picture of Figure 2. Other nodes also displayed relevant information when touched. Before the experiment was run, subjects were given a tutorial to help familiarize them with the data display. Subjects were shown how the virtual 3-D extrusion, head-tracking, and virtual wand controls worked and allowed 3-5 minutes to experiment with the system. They were asked to find and name a company, an inventor and a patent to see if they understood the node representation system, and then received two simple basic warm-up questions. After their training, they were given instructions to answer each question aloud as soon as they could using the visualization. The experimenter recorded the length of time that was taken by the subject to correctly answer the question. After each question, the DiVE was re-set and the visualization re-loaded in 2-D without any 3-D extrusion left over from previous question-answering.

8.1   2-D and 3-D Experimental Conditions

For the quantitatively-measured questions, subjects were assigned to alternating 2-D versus 3-D conditions per question. Assignment was made so that there were equal numbers of subjects assigned to both the 2-D and 3-D conditions for each question. Many of these quantitatively-measured questions are essentially queries of the data, while the qualitative questions help measure the utility of the visualization for more free-form exploration and understanding of the network.

1. What patent had the most inventors?
2. What company has the 2nd largest number of patents?
3. How many patents did Charles Thacker have?
4. Name a patent by Robert Kahn?
5. Name the patent by Robert Metcalfe that has "collision detection" in the title.
6. Name the company that filed a patent by Vinton Cerf.
7. Name the company that Ivan Sutherland works for.
8. Find the name of the inventor who filed patents for both BBN and Xerox.

Prior to answering the qualitative questions, subjects were given the task of exploring the data network for a few minutes using whatever strategies they chose (including 2-D and 3-D) via these instructions: "See how you can use this network to get a better understanding of the history and flow of ideas in Computer Science." After exploration, subjects gave written responses to the following questions:

- What discoveries did you make using this virtual data display today?
- What did you discover about what was or is important in the area of computer science using this data display system today?
- What were the most helpful features of this display system for helping you learn and/or make discoveries today?
- What suggestions do you have for making this system more helpful as a tool?

9   Data Analysis

For each of the 8 quantitatively-measured questions, separate t-tests were computed comparing subjects' time to answer correctly the question posed by the experimenter when allowed to use 2-D versus 3-D pull-out strategies. F-tests for equality of variances were computed and found to be significant at $p < .05$ for questions 3, 4, and 5 so for these questions t-tests were computed using the unequal variance model. Subjects using the 3-D condition were found to correctly answer Questions 3 and 5 were significantly faster ($p < .05$) than subjects using only the 2-D condition, although for Question 4 2-D was faster. Examination of the mean answer time for these 8 questions in Table 3 below indicates that for all questions except Question 4, subjects were faster in giving correct answers using 3-D versus 2-D strategies, though these differences were not statistically significant for three-quarters of the questions due to high individual differences between users in their ability to exploit the third dimension. However, the mean time of subjects using the 3-D condition were quicker for all except one question, suggesting that this effect should be assessed further utilizing a larger sample size.

The answers to the qualitative questions listed in Section 8.1 about the discoveries subjects made using this display indicated that when allowed to experiment with both 2 and 3-D displays to explore the data, all subjects preferred the 3-D display. Particular aspects of the 3-D extrusion that subjects found most helpful were "being able to bring everything into three dimensions allowed the

Table 3. Mean Time to Solve Tasks in 2-D and 3-D

Question	2-D Mean	2-D S.D.	3-D Mean	3-D SD
1	190.6	230.54	95.18	55.31
2	25.18	22.17	37.13	36.85
3	85.40	39.77	45.90	20.21
4	38.90	14.00	54.20	23.89
5	44.20	24.68	22.00	10.35
6	31.55	17.38	29.70	13.74
7	42.90	37.97	33.55	17.21
8	100.45	91.26	70.70	64.83

connections between company, inventor and patent to be seen very clearly" and "crucial in sorting overlapping lines that connected patents and their creators." Furthermore, the use of 3-D extrusion "allows the user to separate and identify how nodes are related to each other-it was like one of those cognitive thought maps" and "since the entire 2-D workspace was filled with data it was difficult to do any sort of mass organizing with lack of open space - the 3-D did just that. The added dimension gave me a lot of free space for me to use in my organization."

10 Discussion

These findings suggests that in general questions users were faster using 3-D extrusion to answer directed timed questions than 2-D inspection, although more further studies with a greater number of subjects are needed to assess this effect. In particular, the effect is more pronounced (i.e. significant) on questions that focus on finding fine-grained structure, such as Question 5, which requires "tracing" a route in a dense cluster, and that require searching through and "picking-apart" a dense cluster either for particular information (Question 5) or for purposes of counting (Question 3). Question 1 and 2 were questions whose answer could be deduced by just looking at the overall structure of the network, and for these there was no significant difference. The one exception where subjects achieved a correct answer faster using 2-D than 3-D was on Question 4, which is a simple question in a non-dense section of the visualization that requires no fine-grained tracing of nodes or any inspection of multiple nodes, so in this case extrusion served significantly as a distraction. In conclusion, for tasks involving navigating dense networks for fine-grained results that involve tracing connections between nodes and information search, extrusion into the third-dimension is useful, while it may not be useful and can even be a distraction in making broad observations or finding information that is not hidden in dense clusters. The qualitative feedback from subjects suggests that the added value of the 3-D data extrusion technique lies in the area of being able to explore the organization of the data and the relationships in the underlying structure of the data.

11 Inference and Large Data-Sets

Once the utility of exploring social networks in patent data in 3-D were demonstrated, researchers wanted to use the same technique on larger data-sets. A substantially larger data-set, consisting of 47202 triples in comparison with the data-set used in Section 8 that consisted of 7667 triples, was created by querying the United States Patent Office with the names of all biotechnology companies either based or having research facilities in North Carolina, the "Silicon Valley" of biotechnology. This data-set was aggregated, using GRDDL [12], with large amounts of data about employees, net income of the institution per year, and other information stored in various traditional spreadsheets. By converting

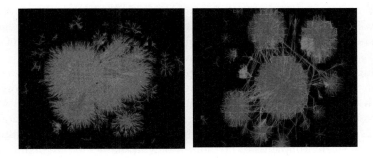

Fig. 3. DiVE: Left - Before inference, Right - After inference and parameter change

these diverse data-sets to RDF and combining them with visualization annotations given by Vis3D, the data was visualized, as shown in Figure 3. The users of the Redgraph were happy that it "quickly showed them the hubs and clusters" but felt that the data-set, even in an immersive environment, was "just overwhelming." In response to this, inference was used to add a new property that let the patent creator be directly linked to the company they filed a parent for. After the inference, the Vis3D annotations were changed to only visualize this inferred property as opposed to the earlier `assignee` and `inventor` properties. In this manner, the data-set was "filtered" for easier browsing and manipulation, as shown in the contrast between the leftmost and rightmost pictures in Figure 3. As one researcher noted "this makes everything easier." Ongoing work with these researchers is using Redgraph to understand value chains and platforms in biotechnology that without visualization would be difficult to extract from masses of diverse textual records [20]. Further work aims to elucidate the general principles of abstraction that can be used to help visualize inference.

12   Conclusion and Future Work

There are a number of improvements that can be made to Redgraph. The next visual component to be implemented is a hyperbolic browsing mode that would make nodes and links diminish more rapidly in size the farther away they are from where the user's point of view as determined by the head-tracker, thus bringing closely related items into higher resolution. This may increase the ability of users to utilize the space of virtual reality systems more effectively, since "the volume of hyperbolic 3-space increases exponentially, as opposed to the familiar geometric increase of Euclidean 3-space" [22]. Some work has begun using hyperbolic browsers for the Semantic Web such as Ontorama, but it allows only very limited types of data to be displayed (i.e. only class hierarchies in OWL ontologies, not properties or instances), and it is unsuitable for many tasks, including the social network visualization done in our study where a large amount of interaction is wanted by the users and when the network is very large [9]. Development of better filtering techniques via dynamic SPARQL querying and inference animation are also being investigated. Modifications in further user studies will incorporate the findings of the current study to apply Redgraph to understanding connections and underlying structures in other types of data as well as assessing user efficiency at discovering types of information.

Although Redgraph is available as open source,[2] the full advantages of using fully immersive techniques virtual reality programs like Redgraph are only available to those institutions with a CAVE-like virtual reality environment, although some advantages may be gained by using the program in "2-1/2 dimension" high-resolution environments as well [7]. Yet in order for the advantages of interactive and immersive environments to be more widely available, we are planning to develop a version of Redgraph for conventional desktop use, and ideally a version that would allow it to be incorporated in popular environments like Second Life[3], which while lacking immersion, provide popular and three-dimensional forums for the social and collaborative creation and manipulation of visualizations. The lack of open standards in these virtual environments makes development more difficult, but the recent progress of open source 3-D environments like Open-Croquet is encouraging [28]. Standards-based virtual reality integrated with the Semantic Web is the long-term goal for our project. Despite the long path ahead, the future of bringing the Semantic Web into 3-D and immersive environments is bright. The users of the patent innovation data-set, all of whom where ordinary people well-acquainted neither with the Semantic Web nor virtual reality, found using Redgraph to be an enjoyable and exciting way to discover relationships in Semantic Web data.

References

1. Barabasi, A., Albert, R.: Emergence of scaling in random networks. Science 286, 509–512 (1999)
2. Bollobas, B.: Random Graphs. Academic Press, London (1985)
3. Berners-Lee, T., Chen, Y., Chilton, L., Connoly, D., Dhanara, R., Hollenbach, J., Lerer, A., Sheets, D.: Tabulator: Exploring and Analyzing Linked Data on the Web. In: Proceedings of the Third International Semantic Web User Interaction Workshop, Athens, Georgia, USA (2006)
4. Bosca, A., Bonino, D.: OntoSphere3D: a multidimensional visualization tool for ontologies. In: The Proceedings of 5th International Workshop on Web Semantics, Krakow, Poland (2006)
5. Bowman, D., McMahan, R.: Virtual Reality: How Much Immersion Is Enough? Computer 40(7), 36–43 (2007)
6. Crutcher, L.A., Lazar, A.A., Feiner, S.K., Zhou, M.X.: Managing networks through a virtual world. IEEE Parallel Distributed Technology 3(2), 4–13 (1995)
7. Cruz-Neira, C., Sandin, D., Defanti, T.: Surround-screen Projection-Based Virtual Reality: the design and implementation of the CAVE. In: The Proceedings of SIGGRAPH 2000, Anaheim, CA (2000)

---

[2] See http://www.redgraph.org
[3] http://www.secondlife.com

8. Eades, P., Houlse, M., Webber, R.: Finding the best viewpoints for three-dimensonal graph drawings. In: The Proceedings of Symposium on Graph Drawing, Rome, Italy (1997)
9. Eklund, P., Roberts, N., Green, S.: OntoRama: Browsing RDF Ontologies using a Hyperbolic-style Browser. In: The Proceedings of First International Symposium on Cyber Worlds, Theory and Practices, Toyko, Japan (2002)
10. Ganser, E., North, S.: An open graph visualization system and its applications to software engineering. Software Practice and Experience 30(11) (1993)
11. Fruchterman, T., Reingold, E.: Graph drawing by force-directed placement. Software Practice and Experience 21(11), 1129–1164 (1991)
12. Halpin, H.: Social Semantic Mashups with GRDDL and Microformats. In: The Proceedings of XML 2006, Boston, Massachusetts, USA (2006)
13. Hayes, P.: RDF Semantics, W3C Recommendation (2004),
    http://www.w3.org/TR/2004/REC-rdf-mt-20040210/
14. Herman, I., Marshall, M.: GraphXML - An XML-Based Graph Description Format. In: The Proceedings of the 8th International Symposium on Graph Drawing, Williamsburg, Virginia, USA (2000)
15. Herman, I., Melancon, G., Marshall, M.: Graph visualization and navigation in information visualization: A survey. IEEE Transactions on Visualization and Computer Graphics 6(1), 24–43 (2000)
16. Huynh, D., Miller, R., Karger, D.: Exhibit: Lightweight Structured Data Publishing. In: The Proceedings of the World Wide Web Conference (WWW 2007), Banff, Canada (2007)
17. Kamada, T., Kawaii, S.: An algorithm for drawing general undirected graphs. Information Processing Letters 31(1), 7–15 (1989)
18. Karger, D., Schraefel, M.: Position Paper: The Pathetic Fallacy of RDF. In: The Proceedings of the International Semantic Web User Interaction Workshop, Athens, Georgia, USA (2006)
19. Le Grand, B., Soto, M.: Information management: Topic maps visualization. In: The Proceedings of XML Europe 2000, Paris, France (2000)
20. Lenoir, T., Giannella, E.: Networks of Innovation: The Emergence and Diffusion of DNA Microarray Technology. Journal of Biomedical Discovery 11(1) (2006)
21. Mika, P.: Social Networks and the Semantic Web. Springer, London, United Kingdom (2007)
22. Munzer, T.: H3: Laying out Large Directed Graphs in 3D Hyperbolic Space. In: The Proceedings of the IEEE Symposium on Information Visualization, Phoenix, Arizona, USA (1997)
23. Mutton, P., Golbeck, J.: Visualization of semantic metadata and ontologies. In: The Proceedings of Seventh International Conference on Information Visualization, Zurich, Switzerland (2003)
24. Newman, M.: Power laws, Pareto distributions and Zipf's law. Contemporary Physics 46, 323–351 (2005)
25. Pietriga, E.: IsaViz: a Visual Environment for Browsing and Authoring RDF Models. In: The Developer's Track of the World Wide Web Conference (WWW 2002), Honululu, Hawaii, USA (2002)
26. Pittarello, F., Faveri, D.: A Semantic description of 3D environments: a proposal based on web standards. In: The Proceedings of the Eleventh international Conference on 3D Web Technology, Columbia, Maryland (2006)
27. Schaeffer, B., Goudeseene, C.: Syzygy: native PC cluster VR. In: The Proceedings of Virtual Reality (VR 2003), Los Angeles, CA, USA (2003)

28. Smith, D., Raab, A., Reed, D., Kay, A.: Croquet: A Menagerie of New User Interfaces. In: The Proceedings of Conference on Creating, Connecting and Collaborating through Computing (C5), Kyoto, Japan (2004)
29. Ware, C., Franck, G.: Viewing a graph in a virtual reality display is three times as good as a 2D diagram. In: The Proceedings of IEEE Conference on Visual Languages and Human-centric Computing, Rome, Italy (1994)
30. Web3D-Consortium. X3D Specification (2005),
    http://www.web3d.org/x3d/specifications

Semantic Grounding of Tag Relatedness in Social Bookmarking Systems

Ciro Cattuto[1], Dominik Benz[2], Andreas Hotho[2], and Gerd Stumme[2]

[1] Complex Networks Lagrange Laboratory, Institute for Scientific Interchange (ISI) Foundation,
10133 Torino, Italy
ciro.cattuto@isi.it
[2] Knowledge & Data Engineering Group, University of Kassel,
34121 Kassel, Germany
{benz,hotho,stumme}@cs.uni-kassel.de

Abstract. Collaborative tagging systems have nowadays become important data sources for populating semantic web applications. For tasks like synonym detection and discovery of concept hierarchies, many researchers introduced measures of tag similarity. Even though most of these measures appear very natural, their design often seems to be rather ad hoc, and the underlying assumptions on the notion of similarity are not made explicit. A more systematic characterization and validation of tag similarity in terms of formal representations of knowledge is still lacking. Here we address this issue and analyze several measures of tag similarity: Each measure is computed on data from the social bookmarking system del.icio.us and a semantic grounding is provided by mapping pairs of similar tags in the folksonomy to pairs of synsets in Wordnet, where we use validated measures of semantic distance to characterize the semantic relation between the mapped tags. This exposes important features of the investigated similarity measures and indicates which ones are better suited in the context of a given semantic application.

1   Introduction

Social bookmarking systems have become extremely popular in recent years. Their underlying data structures, known as *folksonomies*, consist of a set of users, a set of free-form keywords (called *tags*), a set of resources, and a set of tag assignments, i. e., a set of user/tag/resource triples. As folksonomies are large-scale bodies of lightweight annotations provided by humans, they are becoming more and more interesting for research communities that focus on extracting machine-processable semantic structures from them. The structure of folksonomies, however, differs fundamentally from that of e.g., natural text or web resources, and sets new challenges for the fields of knowledge discovery and ontology learning. Central to these tasks are the concepts of similarity and relatedness. In this paper, we focus on similarity and relatedness of tags, because they carry the semantic information within a folksonomy, and provide thus the link to ontologies. Additionally, this focus allows for an evaluation with well-established measures of similarity in existing lexical databases.

A. Sheth et al. (Eds.): ISWC 2008, LNCS 5318, pp. 615–631, 2008.

Budanitsky and Hirst pointed out that similarity can be considered as a special case of relatedness [1]. As both similarity and relatedness are semantic notions, one way of defining them for a folksonomy is to map the tags to a thesaurus or lexicon like Roget's thesaurus[1] or WordNet [2], and to measure the relatedness there by means of well-known metrics. The other option is to define measures of relatedness directly on the network structure of the folksonomy. One important reason for using measures grounded in the folksonomy, instead of mapping tags to a thesaurus, is the observation that the vocabulary of folksonomies includes many community-specific terms which did not make it yet into any lexical resource. Measures of tag relatedness in a folksonomy can be defined in several ways. Most of these definitions use statistical information about different types of *co-occurrence* between tags, resources and users. Other approaches adopt the *distributional hypothesis* [3,4], which states that words found in similar contexts tend to be semantically similar. From a linguistic point of view, these two families of measures focus on orthogonal aspects of structural semiotics [5,6]. The co-occurrence measures address the so-called syntagmatic relation, where words are considered related if they occur in the same part of text. The contextual measures address the paradigmatic relation (originally called associative relation by Saussure), where words are considered related if they can replace one another without affecting the structure of the sentence.

In most studies, the selected measures of relatedness seem to have been chosen in a rather ad-hoc fashion. We believe that a deeper insight into the semantic properties of relatedness measures is an important prerequisite for the design of ontology learning procedures that are capable of harvesting the emergent semantics of a folksonomy.

In this paper we analyse five measures of tag relatedness: the *co-occurrence count*, *three distributional measures* which use the cosine similarity [7] in the vector spaces spanned by users, tags, and resources, respectively, and *FolkRank* [8], a graph-based measure that is an adaptation of PageRank [9] to folksonomies. Our analysis is based on data from a large-scale snapshot of the popular social bookmarking system del.icio.us.[2] To provide a semantic grounding of our folksonomy-based measures, we map the tags of del.icio.us to synsets of WordNet and use the semantic relations of WordNet to infer corresponding semantic relations in the folksonomy. In WordNet, we measure the similarity by using both the taxonomic path length and a similarity measure by Jiang and Conrath [10] that has been validated through user studies and applications [1]. The use of taxonomic path lengths, in particular, allows us to inspect the edge composition of paths leading from one tag to the corresponding related tags, and such a characterization proves to be especially insightful.

The paper is organized as follows: In Section 2 we discuss related work. In Section 3 we provide a formal definition of a folksonomy and describe the del.icio.us data on which our experiments are based. Section 4 describes the measures of relatedness that we will analyze. Section 5 provides examples and qualitative insights. The semantic grounding of the measures in WordNet is described in Section 6. We discuss our results in the context of ontology learning and related tasks in Section 7, where we also point to future work.

---

[1] http://www.gutenberg.org/etext/22
[2] http://del.icio.us/

2   Related Work

One of the first studies about folksonomies is Ref. [11], where several concepts of bottom-up social annotation are introduced. Ref. [12,13,11] provide overviews of the strengths and weaknesses of such systems. Ref. [14,15] introduce a tri-partite graph representation for folksonomies, where nodes are users, tags and resources. Ref. [16] provides a first quantitative analysis of del.icio.us. We investigated the distribution of tag co-occurrence frequencies in Ref. [17] and the network structure of folksonomies in Ref. [18]. Tag-based metrics for resource distance have been introduced in Ref. [19]. To the best of our knowledge, no systematic characterization of tag relatedness in folksonomies is available in the literature.

Ref. [20] generalizes standard tree-based measures of semantic similarity to the case where documents are classified in the nodes of an ontology with non-hierarchical components. The measures introduced there were validated by means of a user study. Ref. [21] analyses distributional measures of word relatedness and compares them with measures of semantic relatedness in thesauri like WordNet. They concluded that "even though ontological measures are likely to perform better as they rely on a much richer knowledge source, distributional measures have certain advantages. For example, they can easily provide domain-specific similarity measures for a large number of domains, their ability to determine similarity of contextually associated word pairs more appropriately [...]."

The distributional hypothesis is also at the basis of a number of approaches to synonym acquisition from text corpora [22]. As in other ontology learning scenarios, clustering techniques are often applied to group similar terms extracted from a corpus, and a core building block of such procedure is the metric used to judge term similarity. In order to adapt these approaches to folksonomies, several distributional measures of tag relatedness have been introduced in theoretical studies or implemented in applications [23,24]. However, the choice of a specific measure of relatedness is often made without justification and often it appears to be rather ad hoc.

A task which depends heavily on quantifying tag relatedness is that of tag recommendation in folksonomies. Scientific publications in this domain are still sparse. Existing work can be broadly divided in approaches that analyze the content of the tagged resources with information retrieval techniques [25,26] and approaches that use collaborative filtering methods based on the folksonomy structure [27]. An example of the latter class of approaches is Ref. [28], where we used our FolkRank algorithm [8] for tag recommendation. FolkRank-based measures will be also covered in this paper.

Relatedness measures also play a role in assisting users who browse the contents of a folksonomy. Ref. [29] shows that navigation in a folksonomy can be enhanced by suggesting tag relations grounded in content-based features.

A considerable number of investigations are motivated by the vision of "bridging the gap" between the Semantic Web and Web 2.0 by means of ontology-learning procedures based on folksonomy annotations. Ref. [15] provides a model of semantic-social networks for extracting lightweight ontologies from del.icio.us. Other approaches for learning taxonomic relations from tags are provided by Ref. [23,24]. Ref. [30] presents a generative model for folksonomies and also addresses the learning of taxonomic relations. Ref. [31] applies statistical methods to infer global semantics from a folksonomy.

The results of our paper are especially relevant to inform the design of such learning methods.

3   Folksonomy Definition and Data

In the following we will use the definition of folksonomy provided in Ref. [8]: [3]

Definition 1. *A folksonomy is a tuple* $\mathbb{F} := (U, T, R, Y)$ *where* U, T, *and* R *are finite sets, whose elements are called* users, tags *and* resources, *respectively., and* Y *is a ternary relation between them, i. e.,* $Y \subseteq U \times T \times R$. *A* post *is a triple* (u, T_{ur}, r) *with* $u \in U$, $r \in R$, *and a non-empty set* $T_{ur} := \{ t \in T \mid (u, t, r) \in Y \}$.

Users are typically represented by their user ID, tags may be arbitrary strings, and resources depend on the system and are usually represented by a unique ID. For instance, in del.icio.us the resources are URLs, while in YouTube the resources are videos.

For our experiments we used data from the social bookmarking system del.icio.us, collected in November 2006. In total, data from $667, 128$ users of the del.icio.us community were collected, comprising $2, 454, 546$ tags, $18, 782, 132$ resources, and $140, 333, 714$ tag assignments. As one main focus of this work is to characterize tags by their properties of co-occurrence with other tags, we restricted our dataset to the $10, 000$ most frequent tags of del.icio.us, and to the resources/users that have been associated with at least one of those tags. One could argue that tags with low frequency have a higher information content in principle — but their inherent sparseness makes them less useful for the study of both co-occurrence and distributional measures. The restricted folksonomy consists of $|U| = 476, 378$ users, $|T| = 10, 000$ tags, $|R| = 12, 660, 470$ resources, and $|Y| = 101, 491, 722$ tag assignments.

4   Measures of Relatedness

A folksonomy can be also regarded as an undirected tri-partite hyper-graph $G = (V, E)$, where $V = U \cup T \cup R$ is the set of nodes, and $E = \{\{u, t, r\} \mid (u, t, r) \in Y\}$ is the set of hyper-edges. Alternatively, the folksonomy hyper-graph can be represented as a three-dimensional (binary) adjacency matrix. In Formal Concept Analysis [32] this structure is known as a *triadic context* [33]. All these equivalent notions make explicit that folksonomies are special cases of three-mode data. Since measures of similarity and relatedness are not well developed for three-mode data yet, we will consider two- and one-mode views on the data. These views will be complemented by a graph-based approach for discovering related tags (FolkRank) which makes direct use of the three-mode structure.

4.1   Co-occurrence

Given a folksonomy (U, T, R, Y), we define the *tag-tag co-occurrence graph* as a weighted undirected graph whose set of vertices is the set T of tags. Two tags t_1 and t_2

---

[3] Ref. [8] additionally introduces a user-specific sub-tag/super-tag relation, which we will ignore here as it is not relevant for del.icio.us.

are connected by an edge, iff there is at least one post (u, T_{ur}, r) with $t_1, t_2 \in T_{ur}$. The *weight* of this edge is given by the number of posts that contain both t_1 and t_2, i. e.,

$$w(t_1, t_2) := \operatorname{card}\{(u, r) \in U \times R \mid t_1, t_2 \in T_{ur}\} \ . \tag{1}$$

Co-occurrence relatedness between tags is given directly by the edge weights. For a given tag $t \in T$, the tags that are most related to it are thus all the tags $t' \in T$ with $t' \neq t$ such that $w(t, t')$ is maximal. We will denote this co-occurrence relatedness by *co-occ*. For its computation, we first create a sorted list of all tag pairs which occur together in a post. The complexity of this can be estimated as $O(\frac{|Y|^2}{2|P|} \log(\frac{|Y|^2}{2|P|}))$. Then, we group this list by each tag and sort by count, which corresponds to an additional complexity of $O(|T|^2 \log(|T|^2))$. Y, P, T denote the set of tag assignments, posts and tags, respectively (see Section 3).

4.2   Distributional Measures

We introduce three distributional measures of tag relatedness that are based on three different vector space representations of tags. The difference between the representations – and thus between the measures – is the feature space used to describe the tags, which varies over the possible three dimensions of the folksonomy. Specifically, for $X \in \{U, T, R\}$ we consider the vector space $\mathbb{R}^X$, where each tag t is represented by a vector $\boldsymbol{v}_t \in \mathbb{R}^X$, as described below.

Tag Context Similarity. The Tag Context Similarity (TagCont) is computed in the vector space $\mathbb{R}^T$, where, for tag t, the entries of the vector $\boldsymbol{v}_t \in \mathbb{R}^T$ are defined by $v_{tt'} := w(t, t')$ for $t \neq t' \in T$, where w is the co-occurrence weight defined above, and $v_{tt} = 0$. The reason for giving weight zero between a node and itself is that we want two tags to be considered related when they occur in a similar context, and not when they occur together. The complexity of this measure comprises the cost of computing co-occurrence (see above), i.e., $O(\frac{|Y|^2}{2|P|} \log(\frac{|Y|^2}{2|P|}) + |T|^2 \log(|T|^2))$, plus the cost of comparing each tag pair, which is $O(|T|^2 2|X|), X \subseteq T$. In our case $|X| = 10,000$.

Resource Context Similarity. The Resource Context Similarity (ResCont) is computed in the vector space $\mathbb{R}^R$. For a tag t, the vector $\boldsymbol{v}_t \in \mathbb{R}^R$ is constructed by counting how often a tag t is used to annotate a certain resource $r \in R$: $v_{tr} := \operatorname{card}\{u \in U \mid (u, t, r) \in Y\}$. In terms of complexity, the tag-resource counts amount for $O(|Y| \log(|Y|))$, plus the pairwise comparison cost of $O(|T|^2 2|R|)$.

User Context Similarity. The User Context Similarity (UserCont) is built similarly to ResCont, by swapping the roles of the sets R and U: For a tag t, the vector $\boldsymbol{v}_t \in \mathbb{R}^U$ is defined as $v_{tu} := \operatorname{card}\{r \in R \mid (u, t, r) \in Y\}$. In this case, the complexity is $O(|Y| \log(|Y|) + |T|^2 2|U|))$.

In all three representations, we measure vector similarity by using the cosine measure, as is customary in Information Retrieval [7]: If two tags t_1 and t_2 are represented by $\boldsymbol{v}_1, \boldsymbol{v}_2 \in \mathbb{R}^X$, their cosine similarity is defined as: $\operatorname{cossim}(t_1, t_2) := \cos \angle(\boldsymbol{v}_1, \boldsymbol{v}_2) = \frac{\boldsymbol{v}_1 \cdot \boldsymbol{v}_2}{||\boldsymbol{v}_1||_2 \cdot ||\boldsymbol{v}_2||_2}$. The cosine similarity is thus independent of the length of the vectors. Its value ranges from 0 (for totally orthogonal vectors) to 1 (for vectors pointing into the same direction).

4.3 FolkRank

The PageRank algorithm [34] reflects the idea that a web page is important if there are many pages linking to it, and if those pages are important themselves. We employed the same principle for folksonomies [8]: a resource which is tagged with important tags by important users becomes important itself. The same holds, symmetrically, for tags and users. By modifying the weights for a given tag in the random surfer vector, FolkRank can compute a ranked list of relevant tags.

More specifically, FolkRank considers a folksonomy (U, T, R, Y) as an undirected graph $(U \cup T \cup R, E)$ with $E := \{\{u, t\}, \{u, r\}, \{t, r\} \mid (u, t, r) \in Y\}$. For a given tag t, it computes in this graph the usual PageRank [34] with a high weight for t in the random surfer vector.[4] Then, the resulting vector is compared to the case of PageRank without random surfer (which equals the simple edge count, as the graph is undirected). This way we compute the winners (and losers) that arise when giving preference to a specific tag in the random surfer vector. The tags that, for a given tag t, obtain the highest FolkRank are considered to be the most relevant in relation to t. Ref. [8] provides a detailed description of the algorithm. The complexity of FolkRank can be estimated as $O(i|Y|)$, where i is the number of iterations (the typical values used in this study were 30-35).

5   Qualitative Insights

Using each of the measures introduced above, we computed, for each of the $10,000$ most frequent tags of del.icio.us, its most closely related tags. As we used different (partially existing) implementations for the measures we investigated, runtimes do not provide meaningful information on the computational cost of the different measures. We refer the reader to the discussion of Section 4 on computational complexity.

Table 1 provides a few examples of the related tags returned by the measures under study. A first observation is that in many cases the tag and resource context similarity provide more synonyms than the other measures. For instance, for the tag *web2.0* they return some of its alternative spellings.[5] For the tag *games*, the tag and resource similarity also provide tags that could be regarded as semantically *similar*. For instance, the morphological variations *game* and *gaming*, or corresponding words in other languages, like *spiel* (German), *jeu* (French) and *juegos* (Spanish). This effect is not obvious for the other measures, which tend to provide rather *related* tags instead (*video*, *software*). The same observation holds for the "functional" tag *tobuy* (see Ref. [16]), for which the tag context similarity provides tags with equivalent functional value (*to_buy*, *buyme*), whereas the FolkRank and co-occurrence measures provide categories of items one could buy. The user context similarity also yields a remarkable amount of functional tags, but with different target actions (*toread*, *todownload*, *todo*).

---

[4] In this paper, we have set the weights in the random surfer vector as follows: Initially, each tag is assigned weight 1. Then, the weight of the given tag t is increased according to $w(t) = w(t) + |T|$. Afterwards, the vector is normalized. The random surfer has an influence of 15 % in each iteration.

[5] The tag *"web"* at the fourth position (tag context) is likely to stem from users who typed *"web 2.0"*, which the early del.icio.us interpreted as two separate tags, *"web* and *2.0"*.

Table 1. Examples of most related tags for each of the presented measures

rank	tag	measure	1	2	3	4	5
13	web2.0	*co-occurrence*	ajax	web	tools	blog	webdesign
		folkrank	web	ajax	tools	design	blog
		tag context	web2	web-2.0	webapp	"web	web_2.0
		resource context	web2	web20	2.0	web_2.0	web-2.0
		user context	ajax	aggregator	rss	google	collaboration
15	howto	*co-occurrence*	tutorial	reference	tips	linux	programming
		folkrank	reference	linux	tutorial	programming	software
		tag context	how-to	guide	tutorials	help	how_to
		resource context	how-to	tutorial	tutorials	tips	diy
		user context	reference	tutorial	tips	hacks	tools
28	games	*co-occurrence*	fun	flash	game	free	software
		folkrank	game	fun	flash	software	programming
		tag context	game	timewaster	spiel	jeu	bored
		resource context	game	gaming	juegos	videogames	fun
		user context	video	reference	fun	books	science
30	java	*co-occurrence*	programming	development	opensource	software	web
		folkrank	programming	development	software	ajax	web
		tag context	python	perl	code	c++	delphi
		resource context	j2ee	j2se	javadoc	development	programming
		user context	eclipse	j2ee	junit	spring	xml
39	opensource	*co-occurrence*	software	linux	programming	tools	free
		folkrank	software	linux	programming	tools	web
		tag context	open_source	open-source	open.source	oss	foss
		resource context	open-source	open	open_source	oss	software
		user context	programming	linux	framework	ajax	windows
1152	tobuy	*co-occurrence*	shopping	books	book	design	toread
		folkrank	toread	shopping	design	books	music
		tag context	wishlist	to_buy	buyme	wish-list	iwant
		resource context	wishlist	shopping	clothing	tshirts	t-shirts
		user context	toread	cdm	todownload	todo	magnet

An interesting observation about the tag *java* is that *python*, *perl* and *c++* (provided by tag context similarity) could all be considered as siblings in some suitable concept hierarchy, presumably under a common parent concept like *programming languages*. An approach to explain this behavior is that the tag context is measuring the frequency of co-occurrence with other tags *in the global context* of the folksonomy, whereas the co-occurrence measure and — to a lesser extent — FolkRank measure the frequency of co-occurrence with other tags *in the same posts*.

Another insight offered by this first visual inspection is that context similarities for tags and resources seem to yield equivalent results, especially in terms of synonym identification. The tag context measure, however, seems to be the only one capable of identifying sibling tags, as it is visible for the case of *java* in Table 1. This is also visible in Fig. 1, which displays the tag co-occurrence vectors of 5 selected tags. The vectors are restricted to co-occurrence with the 30 most frequent tags of the folksonomy, i.e., to only 30 dimensions of the vector space $\mathbb{R}^T$ introduced in Section 4.[6] The figures shows that both *java* and *python* appear frequently together with *programming*, and (to a lesser degree) with *development*. These two common peaks alone contribute approx. 0.68 to the total cosine similarity of the two tags *java* and *python* of 0.85. A similar behavior can be seen for *game* and *games* both displaying peaks at *fun* and (to a lesser degree) *free*. Here we also see the effect of imposing $v_{tt} = 0$ in the definition of the

---

[6] The length of all the vectors was normalized to 1 in the 2-norm.

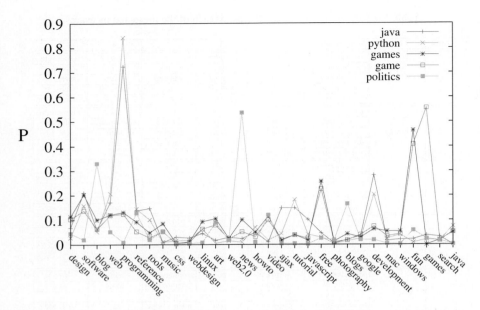

Fig. 1. Tag co-occurrence fingerprint of five selected tags in the first 30 dimensions of the tag vector space

Table 2. Overlap between the 10 most closely related tags

	co-occurrence	FolkRank	tag context	resource context
user context	1.77	1.81	1.35	1.55
resource context	3.35	2.65	2.66	
tag context	1.69	1.28		
FolkRank	**6.81**			

cosine measure: while the tag *game* has a very high peak at *games*, the tag *games* has by definition a zero component there.

The high value for tag *game* in the dimension *games* shows that these two tags are frequently assigned together to resources (probably because users anticipate that they will not remember a specific form at the time of retrieval).

In the case of *python*, on the other hand, we observe that it seldom co-occurs with *java* in the same posts (probably because few web pages deal with both java and python). Hence — even though *python* and *java* are "most related" according to the tag context similarity, they are less so according to the other measures. In fact, in the lists of tags most closely related to *java*, *python* is at position 21 according to FolkRank, 34 according to co-occurrence, 97 according to user context similarity, and 476 according to resource context similarity.

Our next step is to substantiate these first insights with a more systematic analysis. We start by using simple observables that provide qualitative insights into their behavior.

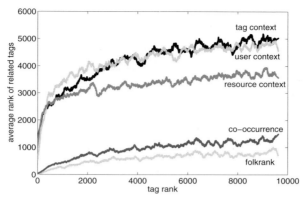

Fig. 2. Average rank of the related tags as a function of the rank of the original tag

The first natural aspect to investigate is whether the most closely related tags are shared across measures of relatedness. We consider the $10,000$ most popular tags in del.icio.us, and for each of them we compute the 10 most related tags according to each of the relatedness measures. Table 2 reports the average number of shared tags for the relatedness measures we investigate. We first observe that the user context measure does not exhibit a strong similarity to any of the other measures. The same holds for the tag context measure, with a slightly higher overlap of 2.65 tags with the resource context measure. Based on the visual inspection above, this can be attributed to shared synonym tags. A comparable overlap also exists between resource context and FolkRank / co-occurrence similarity, respectively.

Based on the current analysis, it is hard to learn much on the nature of these over-lapping tags. A remarkable fact, however, is that relatedness by co-occurrence and by FolkRank share a large fraction (6.81) of the 10 most closely related tags. That is, given a tag t, its related tags according to FolkRank are – to a large extent – tags with a high frequency of co-occurrence with t. In the case of the context relatedness measures, in-stead, the suggested tags seem to bear no special bias towards high-frequency tags. This is due to the normalization of the vectors that is implicit in the cosine similarity (see Section 4), which disregards information about global tag frequency.

To better investigate this point, for each of the $10,000$ most frequent tags in del.icious we computed the average rank (according to global frequency) of its 10 most closely related tags, according to each of the relatedness measures under study. Fig. 2 shows the average rank of the related tags as a function of the original tag's rank. The av-erage rank of the tags obtained by co-occurrence relatedness and by FolkRank is low and increases slowly with the rank of the original tag: this points out that most of the related tags are high-frequency tags, independently of the original tag. On the contrary, the context (distributional) measures display a different behavior: the rank of related tags increases much faster with that of the original tag. That is, the tags obtained from context relatedness span a broader range of ranks.[7]

---

[7] Notice that the curves for the tag and user context relatedness approach a value of $5\,000$ for high ranks: this is the value one would expect if the rank of the related tags was independent from the rank of the original tags.

Table 3. WordNet coverage of del.icio.us tags

# top-frequency tags	100	500	1,000	5,000	10,000
fraction in WordNet	82 %	80 %	79 %	69 %	61 %

6   Semantic Grounding

In this section we shift perspective and move from the qualitative discussion of Section 5 to a more formal validation. Our strategy is to ground the relations between the original and the related tags by looking up the tags in a formal representation of word meanings. As structured representations of knowledge afford the definition of well-defined metrics of semantic similarity, one can investigate the type of *semantic* relations that hold between the original tags and their related tags, defined according to any of the relatedness measures under study.

In the following we ground our measures of tag relatedness by using WordNet [2], a semantic lexicon of the English language. In WordNet words are grouped into *synsets*, sets of synonyms that represent one concept. Synsets are nodes in a network and links between synsets represent semantic relations. WordNet provides a distinct network structure for each syntactic category (nouns, verbs, adjectives and adverbs). For nouns and verbs it is possible to restrict the links in the network to (directed) *is-a* relationships only, therefore a subsumption hierarchy can be defined. The *is-a* relation connects a *hyponym* (more specific synset) to a *hypernym* (more general synset). A synset can have multiple hypernyms, so that the graph is not a tree, but a directed acyclic graph. Since the *is-a* WordNet network for nouns and verbs consists of several disconnected hierarchies, it is useful to add a fake top-level node subsuming all the roots of those hierarchies, making the graph fully connected and allowing the definition of several graph-based similarity metrics between pairs of nouns and pairs of verbs. We will use such metrics to ground and characterize our measures of tag relatedness in folksonomies.

In WordNet, we will measure the semantic similarity by using both the taxonomic shortest-path length and a measure of semantic distance introduced by Jiang and Conrath [10] that combines the taxonomic path length with an information-theoretic similarity measure by Resnik [35]. We use the implementation of those measures available in the WordNet::Similarity library [36]. It is important to remark that [1] provides a pragmatic grounding of the Jiang-Conrath measure by means of user studies and by its superior performance in the context of a spell-checking application. Thus, our semantic grounding in WordNet of the similarity measures is extended to the pragmatic grounding in the experiments of [1].

The program outlined above is only viable if a significant fraction of the popular tags in del.icio.us is also present in WordNet. Several factors limit the WordNet coverage of del.icio.us tags: WordNet only covers the English language and contains a static body of words, while del.icio.us contains tags from different languages, tags that are not words at all, and is an open-ended system. Another limiting factor is the structure of WordNet itself, where the measures described above can only be implemented for nouns and verbs, separately. Many tags are actually adjectives [16] and although their grounding is possible, no distance based on the subsumption hierarchy can be computed

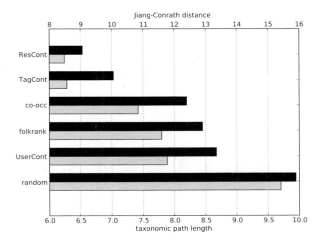

Fig. 3. Average semantic distance, measured in WordNet, from the original tag to the most closely related one. The distance is reported for each of the measures of tag similarity discussed in the main text (labels on the left). Grey bars (bottom) show the taxonomic path length in WordNet. Black bars (top) show the Jiang-Conrath measure of semantic distance.

in the adjective partition of WordNet. Nevertheless, the nominal form of the adjective is often covered by the noun partition. Despite this, if we consider the popular tags in del.icio.us, a significant fraction of them is actually covered by WordNet: as shown in Table 3, roughly 61% of the 10 000 most frequent tags in del.icio.us can be found in WordNet. In the following, to make contact with the previous sections, we will focus on these tags only.

A first assessment of the measures of relatedness can be carried out by measuring – in WordNet – the average semantic distance between a tag and the corresponding most closely related tag according to each one of the relatedness measures we consider. Given a measure of relatedness, we loop over the tags that are both in del.icio.us and WordNet, and for each of those tags we use the chosen measure to find the corresponding most related tag. If the most related tag is also in WordNet, we measure the semantic distance between the synset that contains the original tag and the synset that contains the most closely related tag. When measuring the shortest-path distance, if either of the two tags occurs in more than one synset, we use the pair of synsets which minimizes the path length.

Figure 3 reports the average semantic distance between the original tag and the most related one, computed in WordNet by using both the (edge) shortest-path length and the Jiang-Conrath distance. The tag and resource context relatedness point to tags that are semantically closer according to both distances. We remark once more that the Jiang-Conrath measure has been validated in user studies [1], and because of this the semantic distances reported in Fig. 3 correspond to distances cognitively perceived by human subjects.

The best performance is achieved by similarity according to resource context. This is not surprising as this measure makes use of a large amount of contextual information (the large vectors of resources associated with tags). While similarity by resource

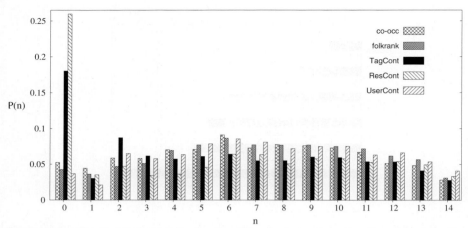

Fig. 4. Probability distribution for the lengths of the shortest path leading from the original tag to the most closely related one. Path lengths are computed using the subsumption hierarchy in WordNet.

context is computationally very expensive to compute, it can be used as a reference for comparing the performance of other measures. To this end, we also computed the distances for the worst case scenario of a measure (marked as *random* in Figure 3) that associates every tag with a randomly chosen one. All the other measures of relatedness fall between the above extreme cases. Overall, the taxonomic path length and the Jiang-Conrath distance appear strongly correlated, and they induce the same ranking by performance of the similarity measures. Remarkably, the notion of similarity by tag context (*TagCont*) has an almost optimal performance. This is interesting because is it computationally lighter that the similarity by resource context, as it involves tag co-occurrence with a fixed number (10 000) of popular tags, only. The closer semantic proximity of tags obtained by tag and resource context relatedness was intuitively apparent from direct inspection of Table 1, but now we are able to ground this statement through user-validated measures of semantic similarity based on the subsumption hierarchy of WordNet.

As already noted in Section 5, the related tags obtained via tag context or resource context appear to be "synonyms" or "siblings" of the original tag, while other measures of relatedness (co-occurrence and FolkRank) seem to provide "more general" tags. The possibility of looking up tags in the WordNet hierarchy allows us to be more precise about the nature of these relations. In the rest of this section we will focus on the shortest paths in WordNet that lead from an initial tag to its most closely related tag (according to the different measures of relatedness), and characterize the length and edge composition (hypernym/hyponym) of such paths.

Figure 4 displays the normalized distribution $P(n)$ of shortest-path lengths n (number of edges) connecting a tag to its closest related tag in WordNet. All similarity measures share the same overall behavior for $n > 3$, with a broad maximum around $n \simeq 6$, while significant differences are visible for small values of n. Specifically, similarity by tag context and resource context display a strong peak at $n = 0$. Tag context similarity

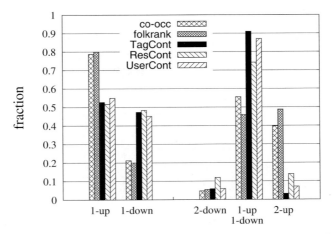

Fig. 5. Edge composition of the shortest paths of length 1 (left) and 2 (right). An "up" edge leads to a hypernym, while a "down" edge leads to a hyponym.

also displays a weaker peak at $n = 2$ and a comparatively depleted number of paths with $n = 1$. For higher values of n, the histograms for resource context and tag context have the same shape as the others, but are systematically lower due to the abundance of very short paths and the normalization of $P(n)$. The peak at $n = 0$ is due to the detection of actual synonyms in WordNet. As nodes in WordNet are synsets, a path to a synonym appears as an edge connecting a node to itself (i. e., a path of length 0). Similarity by tag context points to a synonym in about 18 % of the cases, while using resource context this figure raises to about 25 %. In the above cases, the most related tag is a tag belonging to the same synset of the original tag. In the case of tag context, the smaller number of paths with $n = 1$ (compared with $n = 0$ and $n = 2$) is consistent with the idea that the similarity of tag context favors siblings/synonymous tags: moving by a single edge, instead, leads to either a hypernym or a hyponym in the WordNet hierarchy, never to a sibling. The higher value at $n = 2$ (paths with two edges in WordNet) for tag context may be compatible with the sibling relation, but in order to ascertain this we have to characterize the typical edge composition of these paths.

Figure 5 displays the average edge type composition (hypernym/hyponym edges) for paths of length 1 and 2. The paths analyzed here correspond to $n = 1$ and $n = 2$ in Figure 4. For tag context, resource context and user context, we observe that the paths with $n = 2$ (right-hand side of Figure 5) consist almost entirely of one hypernym edge (up) and one hyponym edge (down), i. e., these paths do lead to siblings. This is especially marked for the notion of similarity based on tag context, where the fraction of paths leading to a sibling is about 90% of the total. Notice how the path composition is very different for the other non-contextual measures of relatedness (co-occurrence and FolkRank): in these cases roughly half of the paths consist of two hypernym edges in the WordNet hierarchy, and the other half consists mostly of paths to siblings. We observe a similar behavior for paths with $n = 1$, where the contextual notions of similarity have no statistically preferred direction, while the other measures point preferentially to hypernyms (i. e., 1-up in the WordNet taxonomy).

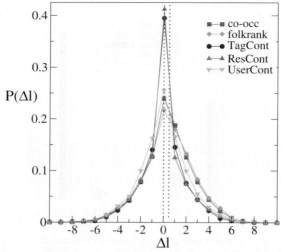

Fig. 6. Probability distribution of the level displacement Δl in the WordNet hierarchy

We now generalize the analysis of Figure 5 to paths of arbitrary length. Specifically, we measure for every path the *hierarchical displacement* Δl in WordNet, i. e., the difference in hierarchical depth between the synset where the path ends and the synset where the path begins. Δl is the difference between the number of edges towards a hypernym (up) and the number of edges towards a hyponym (down). Figure 6 displays the probability distribution $P(\Delta l)$ measured over all tags under study, for the five measures of relatedness. We observe that the distribution for the tag context and resource context is strongly peaked at $\Delta l = 0$ and highly symmetric around it. The fraction of paths with $\Delta l = 0$ is about 40%. The average value of Δl for all the contextual measures is $\overline{\Delta l} \simeq 0$ (dotted line at $\Delta l = 0$) . This reinforces, in a more general fashion, the conclusion that the contextual measures of similarity involve no hierarchical biases and the related tags obtained by them lie at the same level of the original one, in the Word-Net hierarchy. Tag context and resource context are more peaked, while the distribution for user context, which is still highly symmetric around $\Delta l = 0$, is broader. Conversely, the probability distributions $P(\Delta l)$ for the non-contextual measures (co-occurrence and FolkRank), look asymmetric and both have averages $\overline{\Delta l} \simeq 0.5$ (righ-hand dotted line). This means that those measures – as we have already observed – point to related tags that preferentially lie higher in the WordNet hierarchy.

7   Discussion and Perspectives

The contribution of this paper is twofold: First, it introduces a systematic methodology for characterizing measures of tag relatedness in a folksonomy. Several measures have been proposed and applied, but given the fluid and open-ended nature of social bookmarking systems, it is hard to characterize – from the semantic point of view – what kind of relations they establish. As these relations constitute an important building block for extracting formalized knowledge, a deeper understanding of tag relatedness is needed.

In this paper, we grounded several measures of tag relatedness by mapping the tags of the folksonomy to synsets in WordNet, where we used well-established measures of semantic distance to characterize the investigated measures of tag relatedness. As a result, we showed that distributional measures, which capture the context of a given tag in terms of resources, users, or other co-occurring tags, establish – in a statistical sense – *paradigmatic* relations between tags in a folksonomy. Strikingly, our analysis shows that the behavior of the most accurate measure of similarity (in terms of semantic distance of the indicated tags) can be matched by a computationally lighter measure (tag context similarity) which only uses co-occurrence with the popular tags of the folksonomy. In general, we showed that a semantic characterization of similarity measures computed on a folksonomy is possible and insightful in terms of the type of relations that can be extracted. We showed that despite a large degree of variability in the tags indicated by different similarity measures, it is possible to connotate *how* the indicated tags are related to the original one.

The second contribution addresses the question of emergent semantics: Our results indicate clearly that, given an appropriate measure, globally meaningful tag relations can be harvested from an aggregated and uncontrolled folksonomy vocabulary. Specifically, we showed that the measures based on tag and resource context are capable of identifying tags belonging to a common semantic concept. Admittedly, in their current status, none of the measures we studied can be seen as *the* way to instant ontology creation. However, we believe that further analysis of these and other measures, as well as research on how to combine them, will help to close the gap towards the Semantic Web.

In an application context, the semantic characterization we provided can be used to guide the choice of a relatedness measure as a function of the task at hand. We will close by briefly discussing which of the relatedness measures we investigated is best for ...

- ... *synonym discovery.* The tag or resource context similarities are clearly the first measures to choose when one would like to discover synonyms. As shown in this work, these measures deliver not only spelling variants, but also terms that belong to the same WordNet synset (see especially Fig. 4). This kind of information could be applied to suggest concepts in tagging system or to support users by cleaning up the tag cloud.
- ... *concept hierarchy.* Both FolkRank and co-occurrence relatedness seem to yield more general tags in our analyses. This is why we think that these measures provide valuable input for algorithms to extract taxonomic relationships between tags.
- ... *tag recommendations.* The applicability of both FolkRank and co-occurrence for tag recommendations was demonstrated in Ref. [28]. Both measures allow for recommendations by straightforward modifications. Our evaluation in Ref. [28] showed that FolkRank delivered superior and more personalized results than co-occurrence. On the other hand, similar tags and spelling variants as frequently provided by the context similarity are less accepted by the user in recommendations.
- ... *query expansion.* Our analysis suggests that resource or tag context similarity could be used to discover synonyms and – together with some string edit distance – spelling variants of the tags in a user query. The original tag query could be expanded by using the tags obtained by these measures.

Future work includes the application of different measures of relatedness in the context of the tasks listed above. In particular, we plan to adapt existing ontology learning techniques to the case of folksonomies, building upon the semantic characterization of tag relateness that we presented here.

Acknowledgment

This research was partly supported by the TAGora project (FP6-IST5-34721) funded by the Future and Emerging Technologies program (IST-FET) of the European Commission. We thank A. Baldassarri, V. Loreto, F. Menczer, V. D. P. Servedio and L. Steels for many stimulating discussions.

References

1. Budanitsky, A., Hirst, G.: Evaluating wordnet-based measures of lexical semantic relatedness. Computational Linguistics 32(1), 13–47 (2006)
2. Fellbaum, C. (ed.): WordNet: an electronic lexical database. MIT Press, Cambridge (1998)
3. Firth, J.R.: A synopsis of linguistic theory 1930-55. Studies in Linguistic Analysis (special volume of the Philological Society) 1952-59, 1–32 (1957)
4. Harris, Z.S.: Mathematical Structures of Language. Wiley, New York (1968)
5. de Saussure, F.: Course in General Linguistics. Duckworth, London (trans. Roy Harris) ([1916] 1983)
6. Chandler, D.: Semiotics: The Basics, 2nd edn. Taylor & Francis, Abington (2007)
7. Salton, G.: Automatic text processing: the transformation, analysis, and retrieval of information by computer. Addison-Wesley Longman Publishing Co., Inc., Boston (1989)
8. Hotho, A., Jäschke, R., Schmitz, C., Stumme, G.: Information retrieval in folksonomies: Search and ranking. In: Sure, Y., Domingue, J. (eds.) ESWC 2006. LNCS, vol. 4011, pp. 411–426. Springer, Heidelberg (2006)
9. Page, L., Brin, S., Motwani, R., Winograd, T.: The PageRank citation ranking: Bringing order to the web. In: WWW 1998, Brisbane, Australia, pp. 161–172 (1998)
10. Jiang, J.J., Conrath, D.W.: Semantic Similarity based on Corpus Statistics and Lexical Taxonomy. In: Proceedings of the International Conference on Research in Computational Linguistics (ROCLING), Taiwan (1997)
11. Mathes, A.: Folksonomies – Cooperative Classification and Communication Through Shared Metadata (December 2004), http://www.adammathes.com/academic/computer-mediated-communication/folksonomies.html
12. Hammond, T., Hannay, T., Lund, B., Scott, J.: Social Bookmarking Tools (I): A General Review. D-Lib Magazine 11(4) (April 2005)
13. Lund, B., Hammond, T., Flack, M., Hannay, T.: Social Bookmarking Tools (II): A Case Study - Connotea. D-Lib Magazine 11(4) (April 2005)
14. Lambiotte, R., Ausloos, M.: Collaborative tagging as a tripartite network. In: Alexandrov, V.N., van Albada, G.D., Sloot, P.M.A., Dongarra, J. (eds.) ICCS 2006. LNCS, vol. 3993, p. 1114. Springer, Heidelberg (2006)
15. Mika, P.: Ontologies are us: A unified model of social networks and semantics. In: International Semantic Web Conference. LNCS, pp. 522–536. Springer, Heidelberg (2005)
16. Golder, S., Huberman, B.A.: The structure of collaborative tagging systems. Journal of Information Science 32(2), 198–208 (2006)

17. Cattuto, C., Loreto, V., Pietronero, L.: Semiotic dynamics and collaborative tagging. Proc. Natl. Acad. Sci. USA 104, 1461–1464 (2007)
18. Cattuto, C., Schmitz, C., Baldassarri, A., Servedio, V.D.P., Loreto, V., Hotho, A., Grahl, M., Stumme, G.: Network properties of folksonomies. AI Communications Journal, Special Issue on Network Analysis in Natural Sciences and Engineering 20(4), 245–262 (2007)
19. Cattuto, C., Baldassarri, A., Servedio, V.D.P., Loreto, V.: Emergent community structure in social tagging systems. Advances in Complex Physics. In: Proceedings of the European Conference on Complex Systems ECCS 2007 (to appear)
20. Maguitman, A.G., Menczer, F., Erdinc, F., Roinestad, H., Vespignani, A.: Algorithmic computation and approximation of semantic similarity. World Wide Web 9(4), 431–456 (2006)
21. Mohammad, S., Hirst, G.: Distributional measures as proxies for semantic relatedness (Submitted for publication), http://ftp.cs.toronto.edu/pub/gh/Mohammad+Hirst-2005.pdf
22. Cimiano, P.: Ontology Learning and Population from Text — Algorithms, Evaluation and Applications. Springer, Berlin, Heidelberg, Germany, Originally published as PhD Thesis, 2006, Universität Karlsruhe (TH), Karlsruhe, Germany (2006)
23. Heymann, P., Garcia-Molina, H.: Collaborative creation of communal hierarchical taxonomies in social tagging systems. Technical Report 2006-10, Computer Science Department (April 2006)
24. Schmitz, P.: Inducing ontology from Flickr tags. In: Collaborative Web Tagging Workshop at WWW 2006, Edinburgh, Scotland (May 2006)
25. Mishne, G.: Autotag: a collaborative approach to automated tag assignment for weblog posts. In: WWW 2006. Proceedings of the 15th international conference on World Wide Web, pp. 953–954. ACM Press, New York (2006)
26. Brooks, C.H., Montanez, N.: Improved annotation of the blogosphere via autotagging and hierarchical clustering. In: WWW 2006. Proceedings of the 15th international conference on World Wide Web, pp. 625–632. ACM Press, New York (2006)
27. Xu, Z., Fu, Y., Mao, J., Su, D.: Towards the semantic web: Collaborative tag suggestions. In: Proceedings of the Collaborative Web Tagging Workshop at the WWW 2006, Edinburgh, Scotland (May 2006)
28. Jäschke, R., Marinho, L.B., Hotho, A., Schmidt-Thieme, L., Stumme, G.: Tag recommendations in folksonomies. In: Kok, J.N., Koronacki, J., López de Mántaras, R., Matwin, S., Mladenič, D., Skowron, A. (eds.) PKDD 2007. LNCS (LNAI), vol. 4702, pp. 506–514. Springer, Heidelberg (2007)
29. Aurnhammer, M., Hanappe, P., Steels, L.: Integrating collaborative tagging and emergent semantics for image retrieval. In: Proceedings WWW 2006, Collaborative Web Tagging Workshop (May 2006)
30. Halpin, H., Robu, V., Shepard, H.: The dynamics and semantics of collaborative tagging. In: Proceedings of the 1st Semantic Authoring and Annotation Workshop (SAAW 2006) (2006)
31. Zhang, L., Wu, X., Yu, Y.: Emergent semantics from folksonomies: A quantitative study. Journal on Data Semantics VI (2006)
32. Ganter, B., Wille, R.: Formal Concept Analysis: Mathematical Foundations. Springer, Heidelberg (1999)
33. Lehmann, F., Wille, R.: A triadic approach to formal concept analysis. In: Ellis, G., Rich, W., Levinson, R., Sowa, J.F. (eds.) ICCS 1995. LNCS, vol. 954. Springer, Heidelberg (1995)
34. Brin, S., Page, L.: The Anatomy of a Large-Scale Hypertextual Web Search Engine. Computer Networks and ISDN Systems 30(1-7), 107–117 (1998)
35. Resnik, P.: Using Information Content to Evaluate Semantic Similarity in a Taxonomy. In: Proceedings of the XI International Joint Conferences on Artificial, pp. 448–453 (1995)
36. Pedersen, T., Patwardhan, S., Michelizzi, J.: Wordnet:similarity - measuring the relatedness of concepts (2004), http://citeseer.ist.psu.edu/665035.html

Semantic Modelling of User Interests Based on Cross-Folksonomy Analysis

Martin Szomszor[1], Harith Alani[1], Ivan Cantador[2], Kieron O'Hara[1],
and Nigel Shadbolt[1]

[1] Intelligence, Agents, Multimedia
School of Electronics and Computer Science
University of Southampton, Southampton, UK
{mns03r,h.alani,kmo,nrs}@ecs.soton.ac.uk
[2] Escuela Politcnica Superior
Universidad Autnoma de Madrid
28049 Madrid, Spain
ivan.cantador@uam.es

Abstract. The continued increase in Web usage, in particular participation in folksonomies, reveals a trend towards a more dynamic and interactive Web where individuals can organise and share resources. Tagging has emerged as the de-facto
standard for the organisation of such resources, providing a versatile and reactive
knowledge management mechanism that users find easy to use and understand. It
is common nowadays for users to have multiple profiles in various folksonomies,
thus distributing their tagging activities. In this paper, we present a method for
the automatic consolidation of user profiles across two popular social networking
sites, and subsequent semantic modelling of their interests utilising Wikipedia as
a multi-domain model. We evaluate how much can be learned from such sites, and
in which domains the knowledge acquired is focussed. Results show that far richer
interest profiles can be generated for users when multiple tag-clouds are combined.

1 Introduction

With the growth of Web2.0, it is becoming increasingly common for users to maintain a presence in more than one site. For example, one could be bookmarking pages
in del.icio.us, uploading images in Flickr, listening to music in Last.fm, blogging in
Technorati, etc. The nature of these pursuits naturally leads users to express the relevant aspects of their interests, which are likely to be different across the sites. If such
multiple identities and distributed activities could be brought together independently of
the Web 2.0 sites, far richer user profiles could be generated.

There would be a number of potential gains for recommender systems from the
greater profile depth. Usually, such systems monitor in-house user activities over a
certain period of time to build up profiles that support recommendations. As a result,
they will be limited to the activities of users within those systems, and thus may fail
to capture other user interests, resulting in potential recommendations, and eventually
transactions, being lost. Furthermore, a fuller set of user activities can be captured when
expanding data gathering to multiple sites, thus ensuring dynamic updates. For example, if someone reduces their use of Last.fm for a few months, their opinions of the

A. Sheth et al. (Eds.): ISWC 2008, LNCS 5318, pp. 632–648, 2008.

latest music may not be properly captured, leading to a tailing-off of recommendation quality.

There is a strong push towards opening up social networking to support portability of data across various sites. Many popular sites are racing to develop tools to allow their users to port their personal profiles to other sites. Within days from each other in May 2008, Google, MySpace, and Facebook announced new initiatives for increasing social profile portability called Friend Connect[1], Data Availability [16], and Connect[18] respectively. Efficient cross-linking of user profiles should reduce tag-cloud maintenance, and facilitate search and retrieval of tagged resources from multiple sites. Tagging, a fast spreading activity where users assign terms to online resources, is an important discourse within the Web 2.0 phenomenon. Tags serve various purposes, such as for resource organisation, promotions, sharing with friends, with the public, etc. [14,1]. However, studies have shown that tags are generally chosen to reflect their user's interests. Golder and Huberman [8] analysed tags on del.icio.us, and found that (a) the overwhelming majority of tags identify the topic of the tagged resource, and (b) almost all tags are added for personal use, rather than for the benefit of the public. These findings lend support to the idea of using tags to derive user profiles. But tags are free text, and thus suffer from various vocabulary problems [15,8,10]. If it were possible to *clean* such tags and render them somewhat more standardised, this could be helpful to improve tag-cloud compatibility.

The issue of modelling user interests based on cross-folksonomy activity is likely to become increasingly significant. In a recent survey, Ofcom found that 39% of UK adults with at least one folksonomy profile have indeed two or more profiles [19]. It has even been predicted that by 2010, each of us will have between 12 and 24 online identities [21]. Users are often forced to create separate accounts to participate in different activities. There are signs that many of these users are keen to link up their separate accounts. For example, many Last.fm users provided their Flickr or del.icio.us account URL as their homepages.

In this paper we will explore an approach for unifying distributed user profiles, and building semantic profiles of interests using FOAF and Wikipedia ontologies. The next section will review related work. Section 3 provides a full description of the approach, followed by an experiment in section 4 and an evaluation of results in section 5. Discussion and future works are covered in sections 6 and 7 respectively.

2   Related Work

2.1   Analytic Studies

The spread of tagging and the derivation of folksonomies is providing valuable data sources and environments for studying various user-related issues, such as online behaviour, tagging patterns, incentives for sharing, social networking, and opinion formation. A number of studies have focused on analysing user incentives and motivations behind tagging: Marlow and colleagues studied the effect of system design on tagging style, and the various incentives behind tagging [14]. Similarly, Ames and Naaman [1] studied the reasons why people tag images in Flickr, and articulated a taxonomy

---

[1] http://www.google.com/friendconnect/

of social and functional motivations. They found that users tag for various reasons, such as for organising their resources, sharing them with others, or simply to promote their work. As noted above, the motivations behind tagging tend to be almost always personally-focused [8], and the connection between tagging practice, user preferences and maximally effective profiling.

In a study from Yahoo! on the del.icio.us data, Li and colleagues found that tags are better representatives of users' interests than the keywords of the tagged Web pages, because (a) they offer a higher level of content abstraction, and (b) they are better representations of the user's perception of that content [12]. The authors investigated matching users based on the similarity of tag clusters in del.icio.us. In our work however, we are interested in identifying the specific interest of the users as an independent attribute, and not only the similarity of his/her interests with others.

Investigations in related fields have shown that there are interesting correlations between social networking environments and the domains to which they relate. For instance, De Choudhury and colleagues found a correlation between certain blogs and the movement of the stock market [5], while Singla and Richardson analysed MSN Messenger chat logs and the search queries of the chatters, and found that those who exchanged short messages frequently were more likely to issue similar search queries [22]. In our work, we are investigating the correlation between user tagging activities across multiple folksonomies.

2.2 Normative Accounts of Tagging Practice

Tags are free text, and users can tag resources with any terms they wish to use. On the one hand, this total freedom simplifies the process and thus attracts users to contribute. It also avoids the problem of forcing users into using terms they do not feel apply, a situation that arrises when vocabularies are enforced. For these reasons, the lack of constraints seems essential. On the other hand, it generates various vocabulary problems: tags can be too personalised, made of compound words, mix plural and singular terms, meaningless, synonymous, etc. [15,8,10]. This total lack of control is resulting in some sort of tagging chaos, thus obstructing search [10] and analysis [12].

Guy and Tonkin [10] suggest that users should be educated about how to author better tags, and that systems should implement procedures to check for problematic tags and suggest alternatives. While such steps could be useful for improving tag quality, in our work we follow the approach of *cleaning* existing tags using a number of term filtering processes. In the same spirit of our tag filtering, Hayes and colleagues [11] in their work on tag clustering have performed a number of filtering operations, such as stemming, stop word removal, tokenisation, and removal of highly frequent tags. Clustering of tags has been used by Begelman and colleagues for tag disambiguation [2], where similar tags were grouped together to facilitate distinguishing between their different meaning when searching.

2.3 Collection and Semantic Representation of User Interests

This paper is mainly concerned with learning about user interests, and there is a strong tradition of work in this area. Mori and colleagues investigated extracting information from web pages using term co-occurrence analysis to build FOAF files [17]. Diederich and Iofciu [7] tried to identify user interests based on tag clustering. However, the tags used were in fact DBLP keywords, resulting in a system serving a different purpose to free tagging. Demartini suggests using the history of users' edits in Wikipedia to find out about their expertise [6]. Such an approach will obviously only work for users who actively edit Wikipedia pages. In contrast, the work reported in this paper exploits the resources of Wikipedia, but our primary interest is in identifying and semantically representing the general interests of users, based on what they tag and how they tag across several folksonomies.

Semantic representation should ideally involve associating user interests with appropriate URIs, thus moving folksonomy user profiles closer to the Semantic Web and moving the agenda of using Semantic Web technology to organise collectively assembled information characteristic of Web 2.0 [9]. Semantically-Interlinked Online Communities (SIOC) is an ontology that provides a foundation for semantically representing user activities in blogs and forums [3]. To facilitate representing tags with URIs, Meaning Of A Tag (MOAT) was developed as a framework to help users manually select appropriate URIs for their tags from existing ontologies [20], in contrast to the work reported in this paper, which explores the strategy of automating the selection of URIs to maintain the essential simplicity of tagging. Specia and Motta [23] investigated reusing existing ontologies to link tags automatically with pre-crafted concepts and relations. Here we are concerned with selecting URIs that represent *topics*, and not just any concept in any ontology.

The novelty of the work reported here is in the amalgamation of multiple Web 2.0 user-tagging histories to build up personal semantically-enriched models of interest. Correlating different folksonomies is a very new art, and has received surprisingly very little attention so far. In a previous study, we compared the tag clouds of users in both Flickr and del,icio.us and found that they tend to be more similar to each other than to other users tag clouds [24]. That indicated that users often carry some of their tagging selections and patterns across different folksonomies, and this insight is an important motivation for this work.

3   An Architecture for Building Semantic Profiles of Interest

The objective of this work is to supply an architecture that constructs a model of user interests by examining their interaction with various folksonomy sites. We are working under the assumption that the tags used most often by an individual correspond to the topics, places, events, and people they are most interested in. To maximise the utility of such profiles, semantic modeling is essential - tags themselves are only string literals and have no explicit semantics so there are no relationships between terms. For example, resources related to programming languages may be tagged in del.icio.us using the terms perl, c++, or python. While it is clear to the user that these tags are related, such a relationship is not modeled within the folksonomy. Hence, our approach relies not only on identifying the most important tags used, but also correlating them to a URI that has explicit references describing its semantics.

While previous semantic profiling work has concentrated on using well defined ontologies for this purpose, it is not practical for a general solution since information within folksonomy sites such as del.icio.us and Flickr is extremely diverse. Further-more, folksonomies are dynamic systems that constantly evolve to accommodate new terminology and trends. Therefore, we decided to use Wikipedia categories to model user interests because Wikipedia covers a wide range of topics and is constantly up-dated by the community. Referring to the example above, the Wikipedia categories for perl and c++ are both subcategories of "Programming language families".

Broadly, the architecture is split into four sections, as depicted in Figure 1:

1. **Account Correlation.** The first step is to identify the accounts held by a particular individual across a range of social networking sites. By using the Google Social Graph API, we are able to take a URL denoting the user (such as their homepage) and discover the various accounts they hold.
2. **Data Collection Module.** Once the user accounts have been identified, the Data Collection Module harvests a complete history of their tagging activity within each site.
3. **Tag Filtering.** After collecting an individual's raw tagging activity, we utilise a Tag Filtering architecture, developed in previous work [24], to filter and merge tags into a canonical representation. This stage allows us to resolve compound nouns (for example, the tags second_life and second-life are merged), cater for misspellings, identify acronyms, and identify synonyms.
4. **Profile Building.** The final stage in the process consumes an individual's filtered tag-clouds and attempts to match each term to a Wikipedia category. Once the list of categories has been generated, a FOAF file is generated to express their interests using references to Wikipedia category URLs.

The following Subsections give a more detailed account of each of these processes.

3.1   Account Correlation

Many users create multiple profiles across a range of folksonomy sites to meet different social and information requirements. Since many of these sites are provided by different

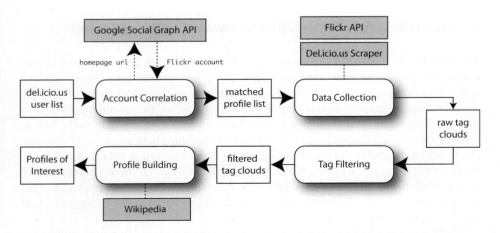

Fig. 1. An Architecture for building Semantic Profiles of Interest

vendors, there are no provisions made to explicitly link accounts that belong to the same individual. In previous work [24], we matched user accounts between del.icio.us and Flickr by examining the usernames chosen by individuals. If the same username was found in both systems, and the string given as their real name was identical in both profiles, the accounts were matched. While such an approach is not particularly robust, the accuracy can be increased by matching other profile information such as age, sex, and location.

Through closer examination, it was apparent that many social networking sites supplied users with a field in their profile page to link to another resource that described them, such as a homepage or blog URL. When we examined a number of Last.fm profiles, we found that many individuals linked to their del.icio.us or Flickr profile. This kind of approach is more robust than matching on strings only since it is unlikely that two accounts that point to the same URL are *not* owned by the same individual. Fortunately, Google recently released an implementation of this matching technique as part of their Social Graph API [2] providing our profile building architecture with a powerful account tracing facility.

3.2   Data Collection

The Data Collection module is responsible for harvesting information from a range of social networking sites. In the case of sites such as Flickr and Last.fm, public APIs are provided that allow us to download a complete history of user tagging activity. For other sites, such as del.icio.us, public APIs are very limited so custom screen-scraping scripts were developed.

3.3   Tag Filtering

When users choose to tag a resource, be it a web page, photo, or video, they are free to choose any tag(s) they please. While it has been shown that this uncontrolled behaviour does result in meaningful structures at the global level, the tag-clouds of particular individuals often contain misspellings, synonyms and morphologic variety. As a result, important correlations between resources and users are sometimes lost simply because of the syntactic mismatches in the tags they have used. To cater for this problem, we developed a tag filtering architecture that cleans and reduces user generated tag-clouds [24].

The filtering process is a sequential execution of different morphologic filtering modules: the output from one filtering step is used as input to the next. The output of the entire filtering process is a set of new tags and their frequencies. Figure 2 provides a visual representation of the filtering process where a set of raw tags is transformed into a set of filtered tags and a set of discarded tags. Each of the numbers in the diagram corresponds to a step outlined below:

Step 1: Syntactic Filtering. After the raw tags have been loaded, they are passed to the *Syntactic Filter*. First, tags that are too small (with length $= 1$) or too large (length > 25) are removed. Due to discrepancies regarding the use of *special* characters (such

---

[2] http://code.google.com/apis/socialgraph/

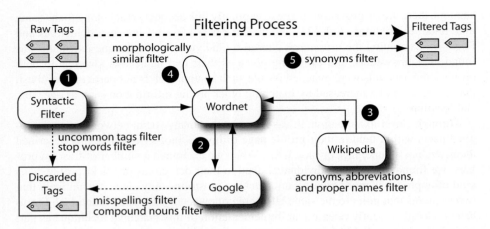

Fig. 2. The tag filtering process

as accents, dieresis and caret symbol), special characters are all converted to their base form. For example, the tag *Zürich* is converted to *Zurich*.

Tags containing numbers are also filtered according to a set of custom heuristics: To maintain salient numbers, such as dates (2006, 2007, etc), common references (911, 666, etc), or combinations of alphanumeric characters (7up, 4x4, 35mm), we consider the global tag frequency and discard any unpopular tags. Finally, common stop-words, such as pronouns, articles, prepositions, and conjunctions are discarded. After syntactic filtering, tags are verified against WordNet. If the tag has an exact match in WordNet, we pass it directly to the set of filtered tags to avoid unnecessary processing.

Step 2: Compound Nouns and Misspellings. If the tags were not found in WordNet, we consider possible misspellings and compound nouns. It is common for users to misspell tags, for example, the use of barclona instead of barcelona. To solve this problem, we make use the Google *did you mean* mechanism. When a search term is entered, Google will check to see if more relevant search results would be found using an alternative spelling. Because Google's spell check is based on occurrences of all words on the Internet, it is able to suggest common spellings for proper nouns (e.g. names and places) that would not appear in a standard dictionary.

The Google "did you mean" mechanism also provides an excellent way to resolve compound nouns. Since most tagging systems prevent users from entering white spaces into the tag name, users create compound nouns by concatenating two nouns together or delimiting them with a non-alphanumeric character such as a _ or -. This is an obvious source of complication when aligning folksonomy activity: users do not consistently use the same compound noun creation schema. By entering a compound terms into Google, we can resolve the tag into its constituent parts. For example, the tag sanfrancisco is corrected to san francisco. After using Google to check for compound nouns and misspellings, the results are validated against WordNet. Any unmatched or un-processed tags are passed to Step 3.

Step 3: Wikipedia Correlation. Many of the popular tags appearing in communal tagging systems do not appear in grammatical dictionaries, such as WordNet, because they correspond to nouns (such as famous people, places, or companies), contemporary terminology (such as `web2.0` and `podcast`), or are widely used acronyms (such as `tv` and `diy`). In order to provide an agreed representation for such tags, we correlate them to their appropriate Wikipedia page. For example, when searching Wikipedia using the tag `nyc`, the entry for New York City is returned. If the search term `ny` is used, the entry for New York state is returned. The advantage of using Wikipedia to agree on tags from folksonomies is that Wikipedia is a community-driven knowledge base, much like folksonomies are, so it will rapidly adapt to accommodate new terminology. For example, Wikipedia contains extensive entries for terms such as `web2.0`, `ajax`, and `blog`.

Step 4: Morphologically Similar. An additional issue to be considered during the tag filtering process is that users often use morphologically similar terms to refer to the same concept. One very common example of this is the discrepancy between singular and plural terms, such as `blog` and `blogs`. Using a custom singularisation algorithm, and the stemming functions provided by the *snowball* library[3], we reduce morphologically similar tags to a single tag. The shortest term in WordNet is used as the representative term.

Step 5: WordNet Synonyms. The final step in the filtering process is to identify tags that are non-ambiguous synonyms, and merge them. This process must be carefully executed because many terms have ambiguous meaning. The algorithm for this process is present in [24] and explained in full with pseudocode.

3.4   Building Profiles of User Interests

The final stage of our profile building architecture turns a set of filtered tag-clouds to a single FOAF file representing as many of the user's interests as possible. To accomplish this, a three-stage process is followed: **(Stage 1)** Each filtered tag-cloud is transformed to a weighted list of Wikipedia categories. For example, if a del.icio.us and Flickr account are discovered for a particular individual, a separate category list is generated for each. **(Stage 2)** then combines these category lists and filters out the uncommon terms to produce the final interest list. **(Stage 3)** turns this list into an RDF representation using the FOAF and Wikipedia ontologies.

The process of transforming a filtered tag-cloud to a Wikipedia category list (Stage 1) is explained below:

1. **Identify Wikipedia Page:** For every tag, we attempt to identify its corresponding Wikipedia page. For example the tag `perl` is matched to the Wikipedia page `http://en.wikipedia.org/wiki/Perl`.
2. **Extract Category List:** For some terms, such as `directory`, the page returned by Wikipedia is a disambiguation page - one that does not define the term itself, but simply references a list of other pages associated with the title. In these cases, no Wikipedia category is found and we move on to the next tag. In the cases where a page is found, the list of categories (found at the bottom of the page) is extracted.

---

[3] http://snowball.tartarus.org/

3. **Selection of Representative Categories:** Initially, we believed it would be useful to record *all* the categories associated with a particular page. For example, the Wikipedia page for Blogs is associated with the categories Blogs, Blogging, Digital Revolution, Internet terminology, Politics and technology, and Technology in society. However, due to the diversity of categories used in Wikipedia, the final category list was often be dominated by spurious terms such as "host cities of the summer olympic games" (from the entry for London), "Christmas nomenclature and language" (from the entry Christmas), as well as Wikipedia specific meta-categories such as "needs more sources". To compensate for this, we decided to only include a category if: a) there is only one category associated with the page, b) the category matches the page name exactly to maximise accuracy, or c) the category is a pluralisation of the page name (e.g. http://en.wikipedia.org/wiki/Blog has the category http://en.wikipedia.org/wiki/Category:Blogs). If an appropriate category is found, a weight is associated to it corresponding to the frequency of the tag. If more than one tag is matched to the same category, the category weight is the sum of all the tag frequencies.

For Stage 2, a global category list is generated by combining all the category lists generated from stage 1. If a category appears in more than one list (e.g. the user has used the same tag in del.icio.us and Flickr), its final weight is the sum of all weights. These final lists often have a characteristic long-tail: For most users, there are many categories that appear with a weight of only 1 or 2. Since these categories are the product of a tag with a low frequency, and therefore do not necessarily correspond to something which the user is particularly interested in, they are filtered out - the final category list contains only categories with a weight above the average for that user.

Finallly, Stage 3 constructs the RDF profile of interest using the FOAF interest property to link the person to each of the Wikipedia categories. Figure 3 presents a partial example FOAF file (for an anonymous user), emphasising how tags extracted from del.icio.us and Flickr tag-clouds are associated with Wikipedia categories. In this example, the popular tags Flickr, Youtube, C++, and Perl have been extracted from their del.icio.us tag-cloud and correlated with the appropriate Wikipedia categories. Such terms are often related by a common super category such as "Online Social Networking" and "Programming Languages". From their Flickr tag-cloud, the terms London, and Southampton have been correlated. Furthermore, the tag cloisters has been correlated to the category "Church Architectures", a match that would not be possible without semantic techniques.

4   Experiment

To build a suitable test-set to evaluate our semantic profiling architecture, we bootstrapped our system with a list of 667,141 del.icio.us users obtained in previous work [4]. For each user that specified an account url in their del.icio.us profile, we queried the Social Graph API to find all other accounts held. By filtering out those who also held a Flickr account, as well as those with low activity (i.e. less than 50 distinct tags in del.icio.us and Flickr), we obtained a final list of 1,392 users. For each individual, a complete history of their tagging activity was harvested using the Data Collection

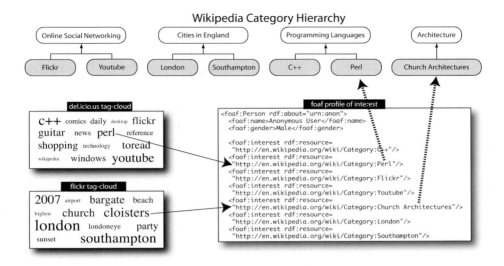

Fig. 3. An example FOAF file. Its contents relates to the users tag clouds, and the categories matched are represented in the Wikipedia hierarchy.

Module. The table below provides a summary of the data collected. In general, users had posted more information in Flickr, using a wider variety of tags.

Del.icio.us		Flickr	
Total Posts	1,134,527	Total Posts	2,215,913
Distinct Tags	138,028	Distinct Tags	307,182

After collecting the user tag-clouds, they were filtered and then passed to the Profile Builder which constructed a list of Wikipedia categories describing their interests (see section 3).

5   Evaluation of Results

We present and evaluate our semantic profiling architecture in four ways: (1) the performance of the Tag Filtering and mapping to Wikipedia entries, (2) the difference between the most common categories (or interests) in del.icio.us and Flickr, (3) the amount learnt from merging profiles from the two folksonomies, and (4) the accuracy of the matching of tags to Wikipedia categories (concepts).

5.1   Tag Filtering and Category Matching

The process of correlating a raw user tag to a Wikipedia category follows three steps: i) first the tag is filtered ii) it is matched to a Wikipedia page, and iii) a suitable category is selected. The table below provides as summary of how many terms were matched during the filtering step, the page matching step, and the category selection step.

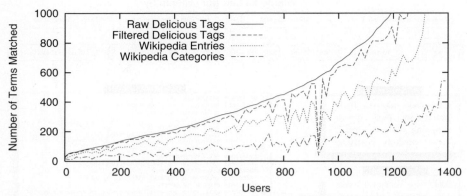

(a) Matching del.icio.us tags to Wikipedia ontology. Graph shows how many tags the user had in the raw tag cloud, how many tags were filtered, how many corresponded to a Wikipedia entry, and finally how many categories were selected to represent the given tag cloud.

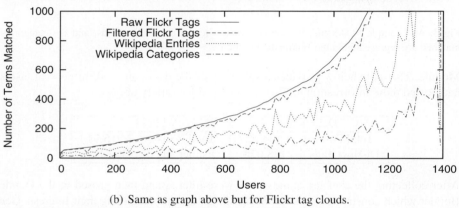

(b) Same as graph above but for Flickr tag clouds.

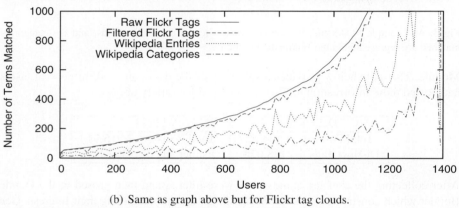

(c) Category increase from adding Flickr to del.icio.us. Graph shows that the higher the user activity is in del.icio.us or in Flickr, the more new Wikipedia categories are found.

	del.icio.us		Flickr	
	Average	Std. Dev.	Average	Std. Dev.
Tag Filtering	90.6%	6.8%	90.7%	8.0%
Wikipedia Page Matching	70.6%	9.6%	59.1%	12.2%
Wikipedia Category Matching	40.9%	7.2%	41.5%	7.5%

On average, 40.9% of del.icio.us tags and 41.5% of Flickr tags were correlated to a Wikipedia category (see 5.4). Figures 4(a) and 4(b) provide detailed plots representing each of these steps. For every user, four points are plotted corresponding to the number of terms matched at each stage (including the number of raw tags). For both del.icio.us and Flickr, the general trends are approximately the same, but there are some anomalies. For example, user 928 has 571 del.icio.us tags, yet only 7 were matched with a Wikipedia category. On closer examination of this user's activity, we discovered that many posts were made using a single tag that itself was a concatenation of many tags. For example, popular tags were `culture.humor`, `politics.party`, and `tech.computers`.

5.2 Global Category View

To understand the difference in what can be learnt from a user's del.icio.us and Flickr activity, we generated a global category frequency table. Each time a tag was matched to a category, we increment the global frequency by the number of times that tag was used. The following table shows the top 15 categories found in del.icio.us and Flickr.

del.icio.us				Flickr			
Wikipedia Categ.	Total Freq.	Wikipedia Categ.	Total Freq.	Wikipedia Categ.	Total Freq.	Wikipedia Categ.	Total Freq.
design	69,215	blogs	68,319	travel	51,674	australia	51,617
music	45,063	photography	41,356	london	46,623	festivals	42,504
tools	35,795	video	34,318	music	40,943	cats	38,230
arts	29,966	software	28,746	holidays	37,610	family	37,100
maps	26,912	teaching	22,120	japan	36,513	concerts	35,374
games	21,549	how-to	19,533	surnames	34,947	washington	33,924
technology	18,032	news	17,737	given names	32,843	dogs	32,206
humor	15,816			birthdays	22,290		

These results are a good indication of the types of interest one can learn from the two different domains. Del.icio.us tells us about the bookmarking habits of the user, and subsequently, the topics they are interested in reading about on the Web. For example, `design`, `software`, and `humor` account for many of the posts made. In Flickr, the tags tell us more about locations and events. This shows that it is very likely to learn about other user interests when such different folksonomies are correlated.

5.3 Learning More about Users

One central argument behind our approach is that different online profiles for an individual tell us different things about what that person is interested in. In Section 5.2 (above), we summarised the different types of categories we learnt from del.icio.us and

Flickr. To evaluate this at a user level, we consider the difference in profiles that would be generated if only their del.icio.us tag-cloud is used versus a profile generated from del.icio.us and Flickr. The underlying hypothesis is that one should increase the number of categories found by including their Flickr profile. On average, 94.8 new categories were found (15 were with above average frequency and thus added to the profile on-tology) for the individuals in our test-set, with a standard deviation of 100. This high variation is accounted for by the fact that many users have a high activity in del.icio.us, but a low activity in Flickr (or vice versa). To account for this, the plot given in Figure 4(c) shows the number of new categories added for a user as a function of the their flickr activity.

5.4 Evaluating Category Matching

To evaluate the approach in terms of how well it identifies the relevant Wikipedia cat-egories from tags, we generated a random sample of 100 users, and randomly selected from their tag clouds 1 tag from del.icio.us and 1 from Flickr. The following procedure was then followed:

1. Open the del.icio.us and Flickr pages for the user.
2. Open the list of resources that the user tagged with the randomly selected tag.
3. Establish from the content of these resources, as well as from the other tags in those postings, what the interest is likely to be.
4. Check if that interest is in the list of Wikipedia categories selected for that user.

The table below summarises the results of this evaluation.

	Represented with correct Wikipedia concepts	Unresolved	Ambiguous
del.icio.us	66	20	14
Flickr	63	25	12

As the table shows, about 13% of the tags lead to Wikipedia disambiguation pages, and thus were discarded (see Future Work section). On average, 64.5% of the tags were correctly represented with a Wikipedia Category. However, 22.5% of the tags were not mapped to any Wikipedia concepts, a situation that arrises when the tag is not well covered in Wikipedia (i.e. no Wikipedia entry was found), or when there is no unique category for representing it (i.e. multiple categories, non of which with the same label as the given tag - section 3). As noted earlier, our system currently ignores highly ambiguous terms in Wikipedia, hence the true accuracy of the results according to the results above is 76.7% for del.icio.us and 71.6% for Flickr (average is 74.2%).

During this evaluation, two false positives were found (i.e. tag mapped to the wrong category). Those were the tags "oracle", and "labrador". The first was represented with the Wikipedia category "Divinity", whereas the user was interested in Oracle the data-base. As for "labrador", which was used to tag photos of a dog in Flickr, was incorrectly mapped to a city in Canada. To avoid such cases, a disambiguation process is required as explained in section 7.

6   Discussion

There were some interesting properties of the strategy reported here of using Wikipedia. The use of a rich source like Wikipedia means that tags that cannot be matched can still gain from matching related tags. For example, one user tagged a resource with "KL". This we were unable to match against Wikipedia, but other tags for similar resources by the user matched to "Kuala Lumpur" and "Malaysia".

Some ambiguities are the result not of using ambiguous terms so much as the abstraction with which interests are represented, which may render the exact nature of the interest opaque. Again, other tags of the user may help resolve the abstraction down to a level which is meaningful. For instance, one person tagged a resource with "time". This is over-general and not too helpful. However, when we looked at other tags for this user on comparable posts, it became clear that the particular interest of the user was astronomy, in which the concept of time plays a specific role. Both "time" and "astronomy" were selected for this user as *interests*, but it is clear that the real interest lies in the combination of the two.

At the other end of the abstraction spectrum, specific tags do not map easily onto concepts when considered as an expression of interest. Examples of specific tags include names of individual (non-celebrity) people, including friends and family of the tagger. These by their nature were more likely to turn up in Flickr than in, say, del.icio.us. Similarly, low-frequency tags were found to be less likely to indicate users interests, especially if the frequency did not grow over time. Again, exploiting the richness of the tag-cloud aided investigation of such anomalous issues, so for example one person tagged some contact "visa", which is perhaps not best characterised as an 'interest', but a better expression of their interests – "travel" – emerged from their other tags.

In the current implementation, tags that lead to a Wikipedia disambiguation page (e.g. directory) are discarded and hence are not represented with an interest. As, noted, with the free text structure of tags, ambiguity and unclarity are endemic problems, and various methods for disambiguation have been proposed in the literature (see section 2). Some of these methods are based on clustering the whole collection of tags (e.g. [2]), or resources [13]), but such techniques are more suitable for static environments where the data do not change too often. In the world of tagging, this is hardly the case. Hence, we need less demanding methods to disambiguate tags, to cope with the highly dynamic nature of folksonomies, even if those methods could never be perfect (see next section).

An alternative approach to disambiguation is proposed in [23] where pairs of tags (e.g. "apple" and "computer") are searched for in a collection of ontologies, and the ontology that contains both concepts will define the disambiguated domain. This approach has a number of limitations, such as when multi-domain ontologies exist (WordNet, Cyc), or when the tags are not found in any single ontology.

Nevertheless, representing user interests with an ontology enables us to benefit from the hierarchical structure when dealing with the user's interests at different levels of granularity. For example, if someone is interested in "Visualbasic", "Perl", and "C++", then one can infer that this person is into "Programming languages". The hierarchy can show how general the user interest is, so one user may use the tag "music" very often, while another might tag with "jazz" or "Hip hop", which are more specific concepts than "music". People tag with different levels of specificity, and this usually reflects their level of expertise in the subject [8,25]. More specific concepts (i.e. interests in our case) can be found at lower levels in the ontology hierarchy.

7   Future Work

According to Wikipedia, the majority of terms can be disambiguated (e.g. "iTunes" could be the device or the store, "furniture" is the object or the UK band). We will investigate using the *distance* between Wikipedia categories to disambiguate the tags. For example, if the user tagged a resource with "apple" and "computer", then these will match to Wikipedia categories with the same labels, but there will be ambiguity to resolve. The path between these two concepts in the Wikipedia ontology is shorter if "apple" is matched to the technological concept, than if it was matched to the fruit. One approach is to select the category with the shortest path to represent this interest.

The use of an ontology will allow recommendation systems to find out how specific the user interest is, and use this information to fine tune recommendations. Inferring interests by analysing links or paths in the ontology can help uncover implicit interests. For example if the user is found to be interested in "Science fiction" and in "Books" then the system might assume that the user will be interested in science-fiction books. This can be refined further depending on, for example, places or authors of interest to the user. We plan to explore using the profiles of interest we produced for making cross-domain recommendations. In addition, we plan to expand this work to also include last.fm accounts, which we have already gathered for all the users in our current dataset.

Cross-domain interests could also be served, given the range of social networking sites and activities, by exploring the taggers social environment. Both Flickr and del.icio.us now allow users to form links with others (e.g. friends, groups). Such social links could be explored for further interest and recommendation analysis. We are currently collecting this information for the users in our dataset and will investigate whether there is a correlation between social links and user interests.

8   Conclusions

This paper investigated a novel idea of merging users' distributed tag clouds to build richer profile ontologies of interests, using FOAF interest properties and Wikipedia categories. We experimented with over 1300 users with high activities in both del.icio.us and Flickr, and the result showed that on average 15 new concepts of interest were learnt for each users when expanding tag analysis to their tag cloud in the other folksonomy. We have also introduced a process to "clean" the data to maximise tag-cloud matching. Our initial evaluation showed that on average, 72% of the filtered tags have been correctly represented with a Wikipedia category (i.e identification of interest).

Acknowledgement

This research has been supported by the TAGora project funded by the Future and Emerging Technologies program (IST-FET) of the European Commission under the contract IST-34721. The information provided is the sole responsibility of the authors and does not reflect the Commission's opinion. The Commission is not responsible for any use that may be made of data appearing in this publication.

References

1. Ames, M., Naaman, M.: Why we tag: Motivations for annotation in mobile and online media. In: Proc. of Computer and Human Interaction (CHI), San Jose, CA (2007)
2. Begelman, G., Keller, P., Smadja, F.: Automated Tag Clustering: Improving search and exploration in the tag space. In: Proc. 17th Int. World Wide Web Conf., Edinburgh, UK (2006)
3. Bojars, U., Breslin, J.G., Finn, A., Decker, S.: Using the semantic web for linking and reusing data across web2.0 communities. Web Semantics: Science, Services and Agents on the World Wide Web 6(1), 21–28 (2008)
4. Cattuto, C.: Semiotic dynamics in online social communities. The European Physical Journal C - Particles and Fields 46, 33–37 (2006)
5. Choudhury, M.D., Sundaram, H., John, A., Seligmann, D.: Can Blog Communication Dynamics be correlated with Stock Market Activity? In: Proc. Int. Conf. Hypertext (HT 2008), Pittsburgh, PA, USA (2008)
6. Demartini, G.: Finding experts using wikipedia. In: Proc. ExpertFinder Workshop, at ISWC, Busan, Korea (2007)
7. Diederich, J., Iofciu, T.: Finding Communities of Practice from User Profiles Based on Folksonomies. In: Proc. Workshop on Building Technology Enhanced Learning solutions for Communities of Practice, EC-TEL, Crete, Greece (2006)
8. Golder, S.A., Huberman, B.A.: Usage patterns of collaborative tagging systems. Journal of Information Science 32, 198–208 (2006)
9. Gruber, T.: Collective Knowledge Systems: Where the Social Web meets the Semantic Web. Journal of Web Semantics 6(1) (2008)
10. Guy, M., Tonkin, E.: Tidying up tags? D-Lib Magazine 12(1) (2006)
11. Hayes, C., Avesani, P., Veeramachaneni, S.: An analysis of the use of tags in a log recommender system. In: Int. Joint Conf. Artificial Intelligence (IJCAI), Hyderabad, India (2007)
12. Li, X., Guo, L., Zhao, Y.E.: Tag-based social interest discovery. In: Proc. 19th Int. World Wide Web Conf. (WWW), Beijing, China (2008)
13. Yeung, C.M.A., Gibbins, N., Shadbolt, N.: Tag Meaning Disambiguation through Analysis of Tripartite Structure of Folksonomies. In: Proc. Workshop on Collective Intelligence on Semantic Web (CISW), IEEE/WIC/ACM, Los Alamitos, CA, USA (2007)
14. Marlow, C., Naaman, M., Boyd, D., Davis, M.: Ht06, tagging paper, taxonomy, flickr, academic article, to read. In: Proc. Int. Conf. Hypertext (HT 2006), Odense, Denmark (2006)
15. Mathes, A.: Folksonomies - cooperative classification and communication through shared metadata. Computer Mediated Communication - LIS590CMC (December 2004)
16. McCarthy, C.: Myspace announces 'data availability' project with yahoo, ebay, photobucket, twitter. CNET (2008),
    http://www.news.com/8301-13577_3-9939286-36.html
17. Mori, J., Matsuo, Y., Ishizuka, M., Faltings, B.: Keyword Extraction from the Web for FOAF Metadata. In: Workshop on Friend of a Friend, Social Networking and the Semantic Web, Galway, Ireland (2004)
18. Morin, D.: Announcing facebook connect. facebook developers (2008),
    http://developers.facebook.com/news.php?blog=1&story=108
19. Ofcom. Social networking: A quantitative and qualitative research report into attitudes, behaviours, and use (2008),
    http://news.bbc.co.uk/1/shared/bsp/hi/pdfs/02_04_08_ofcom.pdf

20. Passant, A., Laublet, P.: Meaning of A Tag: A Collaborative Approach to Bridge the Gap Between Tagging and Linked Data. In: Workshop on Linked Data on the Web (LDOW), Int. Word Wide Web Conference, Beijing, China (2008)
21. Silver, D.: Smart Start-ups: How to Make a Fortune from Starting Online Communities. John Wiley and Sons, Inc., Chichester (2007)
22. Singla, P., Richardson, M.: Yes, There is a Correlation - From Social Networks to Personal Behavior on the Web. In: Proc. Int. World Wide Web Conf., Beijing, China (2008)
23. Specia, L., Motta, E.: Integrating folksonomies with the semantic web. In: Franconi, E., Kifer, M., May, W. (eds.) ESWC 2007. LNCS, vol. 4519. Springer, Heidelberg (2007)
24. Szomszor, M., Cantador, I., Alani, H.: Correlating user profiles from multiple folksonomies. In: Proc. Int. Conf. Hypertext (HT 2008), Pittsburgh, PA, USA (2008)
25. Tanaka, J.W., Taylor, M.: Object categories and expertise: Is the basic level in the eye of the beholder? Cognitve Psychology 23, 457–482 (1991)

ELP: Tractable Rules for OWL 2

Markus Krötzsch, Sebastian Rudolph, and Pascal Hitzler

Institut AIFB, Universität Karlsruhe, Germany

Abstract. We introduce ELP as a decidable fragment of the Semantic Web Rule Language (SWRL) that admits reasoning in polynomial time. ELP is based on the tractable description logic $\mathcal{EL}^{++}$, and encompasses an extended notion of the recently proposed *DL rules* for that logic. Thus ELP extends $\mathcal{EL}^{++}$ with a number of features introduced by the forthcoming OWL 2, such as disjoint roles, local reflexivity, certain range restrictions, and the universal role. We present a reasoning algorithm based on a translation of ELP to Datalog, and this translation also enables the seamless integration of DL-safe rules into ELP. While reasoning with DL-safe rules as such is already highly intractable, we show that DL-safe rules based on the Description Logic Programming (DLP) fragment of OWL 2 can be admitted in ELP without losing tractability.

1 Introduction

The description logic (DL) family of knowledge representation formalisms has been continuously developed for many years, leading to highly expressive (and complex), yet decidable languages. The most prominent such language is currently $\mathcal{SROIQ}$ [1], which is also the basis for the ongoing standardisation of the new *Web Ontology Language OWL 2*.[1] On the other hand, there has also been considerable interest in more light-weight languages that allow for polynomial time reasoning algorithms. DL-based formalisms that fall into that category are $\mathcal{EL}^{++}$ [2], DL Lite [3], and DLP [4]. While DL Lite strives for sub-polynomial reasoning, $\mathcal{EL}^{++}$ and DLP both are P-complete fragments of $\mathcal{SROIQ}$. In spite of this similarity, $\mathcal{EL}^{++}$ and DLP pursue different approaches towards tractability, and the combination of both is already highly intractable [5].

In this paper, we reconcile $\mathcal{EL}^{++}$ and DLP in a novel rule-based knowledge representation language ELP. While ELP can be viewed as an extension of both formalisms, however, it limits the interactions between the expressive features of either language and thus preserves polynomial time reasoning complexity. ELP also significantly extends $\mathcal{EL}^{++}$ by local reflexivity, concept products, conjunctions of simple roles, and limited range restrictions as in [6]. These features in part are already anticipated for the $\mathcal{EL}^{++}$ based language profile of OWL 2, but, to the best of our knowledge, this work is the first to establish their joint tractability.

The reasoning algorithms presented herein are based on a polynomial reduction of ELP knowledge bases to a specific kind of Datalog programs that can be evaluated in polynomial time. Since the Datalog reduction as such is comparatively simple, this

---

[1] OWL 2 is the forthcoming W3C recommendation for updating OWL, and is based on the OWL 1.1 member submission. See http://www.w3.org/2007/OWL

A. Sheth et al. (Eds.): ISWC 2008, LNCS 5318, pp. 649–664, 2008.

outlines an interesting new implementation strategy for the $\mathcal{EL}^{++}$ profile of OWL 2: Besides the possibility of reusing optimisation methods from deductive databases, the compilation of $\mathcal{EL}^{++}$ to Datalog also provides a practical approach for extending $\mathcal{EL}^{++}$ with DL-safe rules [7]. In these respects, the presented approach bears similarities with the KAON2 transformation of $\mathcal{SHIQ}$ knowledge bases into disjunctive Datalog programs [8], though the actual algorithms are very different due to the different DLs that are addressed. DL-safe rules add new expressivity but their entailments are specifically restricted for preserving decidability – an extended example will illustrate the effects.

For this paper, we chose a presentation of ELP based on *DL rules*, a decidable subset of the Semantic Web Rule Language SWRL that has been recently proposed in two independent works [9,10]. As shown in [9], it is possible to indirectly express such rules by means of the expressive features of $\mathcal{SROIQ}$, and large parts of ELP can still be regarded as a subset of $\mathcal{SROIQ}$. The following examples illustrate the correspondence between DLs and DL rules, and give some intuition for the expressivity of ELP:

Concept inclusions. DL *Tbox* axioms $C \sqsubseteq D$ for subconcept relationships correspond to rules of the form $C(x) \rightarrow D(x)$.

Role inclusions. DL *Rbox* axioms $R \circ S \sqsubseteq T$ express inclusions with role chains that correspond to rules of the form $R(x, y) \wedge S(y, z) \rightarrow T(x, z)$.

Local reflexivity. The DL concept $\exists R.\mathsf{Self}$ of all things that have an R relation to themselves is described by the expression $R(x, x)$. For example, the axiom $\exists\,\mathsf{loves.Self} \sqsubseteq$ Narcist corresponds to $\mathsf{loves}(x, x) \rightarrow \mathsf{Narcist}(x)$.

Role disjointness. Roles in $\mathcal{SROIQ}$ can be declared disjoint to state that individuals related by one role must not be related by the other. An according example rule is $\mathsf{HasSon}(x, y) \wedge \mathsf{HasHusband}(x, y) \rightarrow \bot(x)$ ($\bot$ denoting the empty concept).

Concept products and the universal role. Concept products have, e.g., been studied in [11]. The statement that all elephants are bigger than all mice corresponds to the axiom $\mathsf{Elephant} \times \mathsf{Mouse} \sqsubseteq \mathsf{biggerThan}$ and to the rule $\mathsf{Elephant}(x) \wedge \mathsf{Mouse}(y) \rightarrow \mathsf{biggerThan}(x, y)$. The universal role U that relates all pairs of individuals can be expressed by the rule $\rightarrow U(x, y)$ or as the product of the $\top$ concept with itself.

Qualified role inclusions. Rules can be used to restrict role inclusions to certain concepts, which is not directly possible in $\mathcal{SROIQ}$. An example is given by the rule $\mathsf{Woman}(x) \wedge \mathsf{hasChild}(x, y) \rightarrow \mathsf{motherOf}(x, y)$.

While this work is conceptually based on [9], it significantly differs from the latter by following a completely new reasoning approach instead of extending the known algorithm for $\mathcal{EL}^{++}$. While our use of Datalog may still appear similar in spirit, the model constructions in the proofs expose additional technical complications that arise due to the novel combination of concept products, role conjunctions, and local reflexivity. Moreover, the proposed integration of DL-safe rules is not trivial since, in the absence of inverse roles, it cannot be achieved by the usual approach for "rolling-up" nested expressions, and termination of the modified transformation is less obvious.

The paper proceeds by first recalling some minimal preliminaries regarding DLs, SWRL rules, and DL-safety. Thereafter, we introduce ELP based on DL Rules for the DL $\mathcal{EL}^{++}$, and continue by giving an extended example of an ELP rule base. The next section then presents the Datalog reduction as the basis of our reasoning algorithms, before we proceed to establish the overall reasoning complexity for ELP. We conclude the paper with a discussion of our results and some further pointers to related work. Many proofs were omitted or replaced by intuitive sketches due to space restrictions. The complete technical details can be found in the technical report [12].

2   DLs, Rules, and DL-Safety

This section gives some basic notions of description logics (DL) [13], and introduces rules that are logically similar to the Semantic Web Rule Language SWRL [14]. Such rules may include DL concept expressions, and thus generalise the common DL axiom types of Abox, Tbox, and Rbox. We thus restrict our presentation to rules, the general form of which we will later restrict to obtain favourable computational properties.

The logics considered in this paper are based on three disjoint sets of *individual names* N_I, *concept names* N_C, and *role names* N_R. Throughout this paper, we assume that these basic alphabets are finite, and consider them to be part of the given knowledge base when speaking about the "size of a knowledge base." We assume N_R to be the union of two disjoint sets of *simple roles* N_R^s and *non-simple roles* N_R^n. Later on, the use of simple roles in conclusions of logical axioms will be restricted to ensure, intuitively speaking, that relationships of these roles are not implied by *chains* of other role relationships. In exchange, simple roles might be used in the premises of logical axioms as part of role conjunctions and reflexivity statements where non-simple roles might lead to undecidability. Fixing sets of simple and non-simple role names simplifies our presentation – in practice one could of course also check, for a given knowledge base, whether each role name satisfies the requirements for belonging to either N_R^n or N_R^s.

Definition 1. *The set* **C** *of concept expressions of the DL* $\mathcal{SHOQ}$ *is defined as follows:*

- $N_C \subseteq \mathbf{C}$, $\top \in \mathbf{C}$, $\bot \in \mathbf{C}$,
- *if* $C, D \in \mathbf{C}$, $R \in N_R$, $S \in N_R^s$, $a \in N_I$, *and* n *a non-negative integer, then* $\neg C$, $C \sqcap D$, $C \sqcup D$, $\{a\}$, $\forall R.C$, $\exists R.C$, $\leq n\,S.C$, *and* $\geq n\,S.C$ *are also concept expressions.*

The semantics of these concepts is recalled below (see also Table 1). We present $\mathcal{SHOQ}$ as a well-known DL that contains all expressive means needed within this paper, but we will not consider $\mathcal{SHOQ}$ as such. Additional features of the yet more expressive DLs $\mathcal{SHOIQ}$ and $\mathcal{SROIQ}$ can be expressed by using $\mathcal{SHOQ}$ concepts in rules.

Definition 2. *Consider some DL* $\mathcal{L}$ *with concept expressions* **C**, *individual names* N_I, *and role names* N_R, *and let* **V** *be a countable set of first-order variables. A* term *is an element of* $\mathbf{V} \cup N_I$. *Given terms* t, u, *a* concept atom (role atom) *is a formula of the form* $C(t)\ (R(t, u))$ *with* $C \in \mathbf{C}\ (R \in N_R)$.

A rule *for* $\mathcal{L}$ *is a formula* $B \rightarrow H$, *where* B *and* H *are conjunctions of (role and concept) atoms of* $\mathcal{L}$. *To simplify notation, we will often use finite sets* S *of atoms for representing the conjunction* $\bigwedge S$.

Semantically, rules are interpreted as first-order formulae, assuming that all variables are universally quantified, and using the standard first-order logic interpretation of DL

Table 1. Semantics of concept constructors in $\mathcal{SHOQ}$ for an interpretation $\mathcal{I}$ with domain $\Delta^{\mathcal{I}}$

Name	Syntax	Semantics	Name	Syntax	Semantics
top	$\top$	$\Delta^{\mathcal{I}}$	nominal con.	$\{a\}$	$\{a^{\mathcal{I}}\}$
bottom	$\bot$	$\emptyset$	univ. rest.	$\forall U.C$	$\{x \in \Delta^{\mathcal{I}} \mid \langle x, y \rangle \in U^{\mathcal{I}} \text{ implies } y \in C^{\mathcal{I}}\}$
negation	$\neg C$	$\Delta^{\mathcal{I}} \setminus C^{\mathcal{I}}$	exist. rest.	$\exists U.C$	$\{x \in \Delta^{\mathcal{I}} \mid y \in \Delta^{\mathcal{I}} : \langle x, y \rangle \in U^{\mathcal{I}}, y \in C^{\mathcal{I}}\}$
conjunction	$C \sqcap D$	$C^{\mathcal{I}} \cap D^{\mathcal{I}}$	qualified	$\leq n\,R.C$	$\{x \in \Delta^{\mathcal{I}} \mid \#\{y \in \Delta^{\mathcal{I}} \mid \langle x, y \rangle \in R^{\mathcal{I}}, y \in C^{\mathcal{I}}\} \leq n\}$
disjunction	$C \sqcup D$	$C^{\mathcal{I}} \cup D^{\mathcal{I}}$	number rest.	$\geq n\,R.C$	$\{x \in \Delta^{\mathcal{I}} \mid \#\{y \in \Delta^{\mathcal{I}} \mid \langle x, y \rangle \in R^{\mathcal{I}}, y \in C^{\mathcal{I}}\} \geq n\}$

concepts (see Definition 3 below). In general, a DL knowledge base may entail the existence of *anonymous* domain elements that are not directly represented by some individual name, and it may even require models to be infinite. The fact that rules generally apply to all domain elements can therefore be problematic w.r.t. computability and complexity. It has thus been suggested to consider rules within which variables may only represent a finite amount of *named* individuals, i.e. individuals of the interpretation domain that are represented by some individual name in the knowledge base. Hence, effectively, these so-called *DL-safe* rules [7] apply to named individuals, but not to further anonymous individuals which have been inferred to exist.

Technically, this restriction can be achieved in various ways. The most common approach is to introduce a new concept expression HU that is asserted to contain the named individuals, and that is then used to restrict safe variables to that range. On the other hand, one can also dispense with this additional syntax by building the safety restriction directly into the semantics of variables – this is the intuition behind the use of *safe variables* in the following definition.

Definition 3. *An interpretation $\mathcal{I}$ consists of a set $\Delta^{\mathcal{I}}$ called* domain *(the elements of it being called* individuals*) together with a function $\cdot^{\mathcal{I}}$ mapping individual names to elements of $\Delta^{\mathcal{I}}$, concept names to subsets of $\Delta^{\mathcal{I}}$, and role names to subsets of $\Delta^{\mathcal{I}} \times \Delta^{\mathcal{I}}$. The function $\cdot^{\mathcal{I}}$ is inductively extended to role and concept expressions as shown in Table 1. An element $\delta \in \Delta^{\mathcal{I}}$ is a* named individual *if $\delta = a^{\mathcal{I}}$ for some $a \in \mathsf{N}_I$.*

Let $\mathbf{V}_s \subseteq \mathbf{V}$ be a fixed set of safe variables. *A variable assignment Z for an interpretation $\mathcal{I}$ is a mapping from the set of variables $\mathbf{V}$ to $\Delta^{\mathcal{I}}$ such that $Z(x)$ is named whenever $x \in \mathbf{V}_s$. Given a term $t \in \mathsf{N}_I \cup \mathbf{V}$, we set $t^{\mathcal{I},Z} := Z(t)$ if $t \in \mathbf{V}$, and $t^{\mathcal{I},Z} := t^{\mathcal{I}}$ otherwise. Given a concept atom $C(t)$ (role atom $R(t, u)$), we write $\mathcal{I}, Z \models C(t)$ ($\mathcal{I}, Z \models R(t, u)$) if $t^{\mathcal{I},Z} \in C^{\mathcal{I}}$ ($\langle t^{\mathcal{I},Z}, u^{\mathcal{I},Z} \rangle \in R^{\mathcal{I}}$), and we say that $\mathcal{I}$ and Z satisfy* the atom in this case.

An interpretation $\mathcal{I}$ satisfies *a rule $B \to H$ if, for all variable assignments Z for $\mathcal{I}$, either $\mathcal{I}$ and Z satisfy all atoms in H, or $\mathcal{I}$ and Z fail to satisfy some atom in B. In this case, we write $\mathcal{I} \models B \to H$ and say that $\mathcal{I}$ is a* model *for $B \to H$. An interpretation satisfies a set of rules (i.e. it is a* model *for this set) whenever it satisfies all elements of this set. A set of rules is* satisfiable *if it has a model, and* unsatisfiable *otherwise. Two sets of rules are* equivalent *if they have exactly the same models, and they are* equisatisfiable *if either both are unsatisfiable or both are satisfiable.*

Note that we have assumed earlier that N_I is always finite – typically it may comprise exactly the symbols that are actually used in the knowledge base –, and hence there are only a finite number of assignments for safe variables. Also note that empty rule bodies are considered to be vacuously satisfied, and expressions of the form $\rightarrow H$ encode (sets of) facts. It is well-known that the satisfiability of sets of rules for DLs that support $\exists$ is undecidable, and we will introduce various restrictions to recover decidability below. One simple option is to restrict to so-called *Datalog* programs which we will later use to simulate inferences of more expressive rule languages:

Definition 4. *A* rule *is a* Datalog rule *if all concept atoms contained in it are of the form $C(t)$ with $C \in N_C$, $\top(t)$, and $\bot(t)$. A Datalog program* is a set of Datalog rules.

3   DL Rules and ELP

In this section, we define the rule-based knowledge representation language ELP, and note that it subsumes several other existing languages in terms of expressivity. It is easy to see that unrestricted (SWRL) rules encompass even the very expressive DL $\mathcal{SROIQ}$ [1], since Tbox and Rbox axioms can readily be rewritten as rules. On the other hand, rules in their general form do not impose any of the restrictions on, e.g., *simple roles* or *regularity of Rboxes* that are crucial to retain decidability in $\mathcal{SROIQ}$. Recent works therefore have proposed *DL rules* as a decidable subset of SWRL that can be combined with various DLs without increasing the worst-case complexity of typical reasoning problems [9,10].

We first recall DL rules (with conjunctions of simple roles) and apply them to the tractable DL $\mathcal{EL}^{++}$. The resulting formalism is the core of ELP, and significantly extends the expressivity of $\mathcal{EL}^{++}$ rules as considered in [9].

Definition 5. *Consider a rule $B \rightarrow H$ and terms $t, u \in N_I \cup V$. A* direct connection *from t to u is a non-empty set of atoms of the form $R(t, u)$. If B contains a direct connection between t and u, then t is* directly connected *to u. The term t is* connected *to u (in B) if the following inductive conditions apply:*

- *t is directly connected to u in B, or*
- *u is connected to t in B, or*
- *there is a variable $x \in V$ such that t is connected to x and x is connected to u.*

An extended DL rule *is a rule $B \rightarrow H$ such that if variables $x \neq y$ in B are connected, then there is some direct connection $S \subseteq B$ such that x and y are not connected in $B \setminus S$.*

A path *from t to some variable x in B is a non-empty sequence of the form $R_1(x_1, x_2)$, $\ldots, R_n(x_n, x_{n+1}) \in B$ where $x_1 = t$, $x_2, \ldots, x_n \in V$, $x_{n+1} = x$, and $x_i \neq x_{i+1}$ for $1 \leq i \leq n$. A term t in B is* initial *if there is no path to t. An extended DL rule is a* DL rule *if the following hold, where we assume x, y to range over variables V, and t, t' to range over terms $N_I \cup V$:*

(1) for every variable x in B, there is a path from at most one initial term t to x,
(2) if $R(x, t) \in H$ or $C(x) \in H$, then x is initial in B,
(3) whenever $R(x, x) \in B$, we find that $R \in N_R^s$ is simple,
(4) whenever $R(t, x), R'(t, x) \in B$, we find that $R, R' \in N_R^s$ are simple,
(5) if $R(t, y) \in H$ with $R \in N_R^s$ simple, then all role atoms of the form $R'(t', y) \in B$ are such that $t' = t$ and $R' \in N_R^s$.

The above ensures that bodies of extended DL rules essentially correspond to sets of undirected trees, though reflexive "loops" $R(t, t)$ are also possible. Note that connections are essentially transitive but may not span over individual names. The notion of a connection turns out to be most convenient to establish the later decomposition of rules to accomplish the main tractability result in Theorem 14.

Bodies of DL rules are sets of directed trees due to item (1) in Definition 5. Two exceptions to that structure are admitted. Firstly, the definition of connections admits two elements of a path to be connected by multiple roles, corresponding to conjunctions of such roles. Secondly, atoms $R(x, x)$ are not taken into account for defining paths, such that local reflexivity conditions are admitted. Note that items (3) and (4) restricts both cases to simple roles.

Item (2) above ensures that the first variable in the rule head occurs in the rule body only as the root of some tree. Without this restriction, DL rules would be able to express inverse roles, even for DLs that deliberately exclude this feature to retain tractability. Extended DL rules waive requirements (1) and (2) to supply the expressivity of inverse roles, and indeed any extended DL rule that satisfies the additional requirements (3) to (5) on simplicity can be rewritten as a DL rule if inverse roles are available.

Item (5), finally, imposes the necessary restrictions on the use of simple roles, and, as an alternative presentation, one could also have *defined* the set of simple roles as the (unique) largest set of roles for which this requirement holds in a given rule base. In classical definitions of DLs, simple roles R are usually only admitted in role inclusion axioms of the form $S \sqsubseteq R$. Our definition relaxes this requirement to allow for further DL rules as long as these do not include certain role chains. For example, rules $C(x) \wedge D(y) \rightarrow R(x, y)$ and $R'(x, y) \wedge D(y) \rightarrow R(x, y)$ are possible even if R is simple.

We now apply DL rules to the description logic $\mathcal{EL}^{++}$ [2], for which many typical inference problems can be solved in polynomial time. We omit concrete domains in our presentation as they can basically be treated as shown in [2].

Definition 6. *An $\mathcal{EL}^{++}$ concept expression is a $\mathcal{SHOQ}$ concept expression that contains only the following concept constructors: $\sqcap$, $\exists$, $\top$, $\bot$, as well as nominal concepts $\{a\}$. An $\mathcal{EL}^{++}$ rule is a DL rule for $\mathcal{EL}^{++}$, and an $\mathcal{EL}^{++}$ rule base is a set of such rules.*

An $\mathcal{EL}^{++}$ knowledge base is a set of $\mathcal{EL}^{++}$ concept inclusions $C \sqsubseteq D$ and role inclusion axioms $R_1 \circ \ldots \circ R_n \sqsubseteq R$. See [2] for details. It is easy to see that any $\mathcal{EL}^{++}$ knowledge base can be written as an equivalent $\mathcal{EL}^{++}$ rule base. The above notion of $\mathcal{EL}^{++}$ rule bases extends [9] in two ways. Firstly, we now also allow conjunctions of simple roles, and secondly we allow atoms of the form $R(x, x)$ in rule bodies. Both extensions are non-trivial and require additional mechanisms during reasoning.

As we will see later, reasoning with $\mathcal{EL}^{++}$ rules is indeed possible in polynomial time. However, extending $\mathcal{EL}^{++}$ rules with further forms of rules, even if restricting to Datalog, readily leads to undecidability. This can be prevented if only *DL-safe* Datalog rules are permitted: a Datalog rule is DL-safe, if all of its variables are safe. Yet, this formalism can still capture all Datalog programs, and therefore satisfiability checking remains ExpTime hard [15].

Our strategy for extending $\mathcal{EL}^{++}$ rules into ELP therefore is to blend them with tractable fragments of DL-safe Datalog. As we will see below, one particular such Datalog fragment can again be characterised by the above notion of (extended) DL rule.

Another option is to allow only DL-safe Datalog rules of a particular form, namely those for which the number of variables per rule is bounded by some fixed finite number n. Indeed, it is easy to see that any DL-safe (Datalog) rule is equivalent to the set of rules obtained by replacing all safe variables by individual names in all possible ways. Since the replacements for each variable are independent, this leads to up to $|N_I|^n$ different rules – which is a polynomial bound if n is a constant. Note, however, that large n might render practical computation infeasible.

In addition to various forms of DL-safe rules, ELP also allows for special rules of the form $R(x,y) \to C(y)$ expressing *range restrictions* on the role R. Such restrictions are neither DL-safe Datalog nor DL rules, and in general they do indeed lead to undecidability of $\mathcal{EL}^{++}$. However, it has recently been observed that range restrictions can still be admitted under certain conditions [6]. Therefore, even though this special form of rules is somewhat orthogonal to the other types of rules considered herein, we will include range restrictions into our considerations to give credit to their practical relevance.

Definition 7. *A rule $B \to H$ is a* basic ELP *rule if:*

- *$B \to H$ is an extended $\mathcal{EL}^{++}$ rule, and*
- *the rule $B' \to H'$ obtained from $B \to H$ by replacing all safe variables by some individual name is a DL rule.*

An ELP *rule base RB is a set of basic* ELP *rules together with* range restriction *rules of the form $R(x,y) \to C(y)$, that satisfies the following condition:*

- *If* RB *contains rules of the form $R(x,y) \to C(y)$ and $B \to H$ with $R(t,z) \in H$, then $C(z) \in B$.*

Whenever a set of range restriction rules satisfies the above condition for some set of ELP *rules, we say that the range restrictions are* admissible *for this rule set.*

A rule $B \to H$ is an ELP_n *rule for some natural number $n > 2$ if it is either an* ELP *rule, or a DL-safe Datalog rule with at most n variables.*

We remark that the above condition on admissibility of range restrictions is not quite the same as in [6]. Both versions ensure that, whenever an axiom entails some role atom $R(x,y)$, domain restrictions of R have no effect on the classification of y. The interaction of rules implying role atoms and range restrictions thus is strongly limited. In the presence of DL rules, we can accomplish this by restricting the applicability of rules by additional concept atoms $C(z)$ as in Definition 7. In [6], in contrast, additional range restrictions are required, and these, if added to an existing knowledge base, may also lead to new consequences. Any set of axioms that meets the requirements of [6] can clearly be extended to a semantically equivalent set of admissible ELP axioms, so that the approach of Definition 7 does indeed subsume the cases described in [6].

Before providing an extended example in the next section, we show how ELP subsumes some other tractable languages. One interesting case is DLP, a formalism introduced as the intersection of the DL $\mathcal{SHOIQ}$ and Datalog [4]. DLP can also be generalised using DL rules [9]: A *DLP head concept* is any $\mathcal{SHOQ}$ concept expression that includes only concept names, nominals, $\sqcap$, $\top$, $\bot$, and expressions of the form $\leq 1\, R.C$

where C is an $\mathcal{EL}^{++}$ concept expression. A *DLP rule* $B \to H$ is an extended DL rule such that all concept expressions in B are $\mathcal{EL}^{++}$ concept expressions, and all concept expressions in H are DLP head concepts.

Even the combination of DLP and $\mathcal{EL}$ contains the DL Horn-$\mathcal{FLE}$ and is thus Exp-Time complete [5]. Yet, DLP and $\mathcal{EL}^{++}$ inferences can be recovered in ELP without losing tractability. In this sense, the following simple theorem substantiates our initial claim that ELP can be regarded as an extension both of DLP and $\mathcal{EL}^{++}$.

Theorem 8. *Consider any ground atom α of the form $C(a)$ or $R(a, b)$. Given a DLP rule base* RB *and an $\mathcal{EL}^{++}$ description logic knowledge base* KB, *one can compute an* ELP *rule base* RB' *in linear time, such that: If* RB $\models \alpha$ *or* KB $\models \alpha$ *then also* RB' $\models \alpha$, *and, if* RB' $\models \alpha$ *then* RB $\cup$ KB $\models \alpha$.

Proof. The proof in [12] is based on observing that replacing all variables in a DLP rule base with safe variables does not affect satisfiability, since DLP does not infer the existence of new individuals. Now rules of the form $B \to \forall R.C(t)$ can be rewritten to $B \wedge R(t, y) \to C(y)$, and the result is easily seen to be in ELP given that all variables are safe. Rules of the form $B \to \leq 1\, R.C(t)$ are expressed by rules $B \wedge R(t, y_1) \wedge C(y_1) \wedge R(t, y_2) \wedge C(y_2) \to \approx_S (y_1, y_2)$, where $\approx_S$ is a new role for which the standard equality axioms (using safe variables) are added. The $\mathcal{EL}^{++}$ knowledge base can be added using the basic transformations given in the introduction (with new unsafe variables).   □

Note that the resulting ELP rule base entails all individual consequences of RB and KB, and some but not all consequences of their (unsafe) union. ELP thus provides a means of combining $\mathcal{EL}^{++}$ and DLP in a way that prevents intractability, while still allowing for a controlled interaction between both languages. We argue that this is a meaningful way of combining both formalisms in practice since only some DLP axioms must be restricted to safe variables. Simple atomic concept and role inclusions, for example, can always be considered as $\mathcal{EL}^{++}$ axioms, and all concept subsumptions entailed from the $\mathcal{EL}^{++}$ part of a combined knowledge base do also affect classification of instances in the DLP part. DLP thus gains the terminological expressivity of $\mathcal{EL}^{++}$ while still having available specific constructs that may only affect the instance level.

4   Example

We now provide an extended example to illustrate the expressivity of ELP. The rules in Table 2 express a simplified conceptualisation of some preferences regarding food ordered in a restaurant: rule (1) states that all people that are allergic to nuts dislike all nut products, which is a kind of concept product. Rule (2) expresses the same for vegetarians and fish products. Rule (3) is a role conjunction, stating that anyone who ordered a dish he does not like will be unhappy. Rule (4) says that people generally dislike dishes that contain something that they dislike. Rule (5) is a range restriction for the role orderedDish. Rules (6) and (7) claim that any Thai curry contains peanut oil and some fish product, and the facts (8)–(12) assert various concept memberships.

We first verify that this is indeed a valid ELP rule base where all roles are simple. Indeed, the relaxed simplicity constraints on DL rules as given in Definition 5 are not

Table 2. A simple example rule base about food preferences. The variable v is assumed to be safe.

$$
\begin{aligned}
&(1) & \mathsf{NutAllergic}(x) \wedge \mathsf{NutProduct}(y) &\rightarrow \mathsf{dislikes}(x, y) \\
&(2) & \mathsf{Vegetarian}(x) \wedge \mathsf{FishProduct}(y) &\rightarrow \mathsf{dislikes}(x, y) \\
&(3) & \mathsf{orderedDish}(x, y) \wedge \mathsf{dislikes}(x, y) &\rightarrow \mathsf{Unhappy}(x) \\
&(4) & \mathsf{dislikes}(x, v) \wedge \mathsf{Dish}(y) \wedge \mathsf{contains}(y, v) &\rightarrow \mathsf{dislikes}(x, y) \\
&(5) & \mathsf{orderedDish}(x, y) &\rightarrow \mathsf{Dish}(y) \\
&(6) & \mathsf{ThaiCurry}(x) &\rightarrow \mathsf{contains}(x, \mathsf{peanutOil}) \\
&(7) & \mathsf{ThaiCurry}(x) &\rightarrow \exists\mathsf{contains}.\mathsf{FishProduct}(x) \\
&(8) & &\rightarrow \mathsf{NutProduct}(\mathsf{peanutOil}) \\
&(9) & &\rightarrow \mathsf{NutAllergic}(\mathsf{sebastian}) \\
&(10) & &\rightarrow \exists\mathsf{orderedDish}.\mathsf{ThaiCurry}(\mathsf{sebastian}) \\
&(11) & &\rightarrow \mathsf{Vegetarian}(\mathsf{markus}) \\
&(12) & &\rightarrow \exists\mathsf{orderedDish}.\mathsf{ThaiCurry}(\mathsf{markus})
\end{aligned}
$$

violated in any of the rules. All rules other than (4) and (5) are readily recognised as $\mathcal{EL}^{++}$ rules. By first considering the connections in the respective rule bodies of (1)–(3), (6), and (7), we find that only rule (3) actually has connected terms at all, connected only by a single direct connection {$\mathsf{orderedDish}(x, y), \mathsf{dislikes}(x, y)$}. Both roles occurring in that connection are indeed simple. Similarly, the variable x is initial for these rules, and expressions of the form $R(z, z)$ do not occur.

It remains to check that also rules (4) and (5) are legal ELP statements. For rule (5), this requires us to check whether this range restriction rule is admissible, which is easy since no rule head contains atoms of the form $\mathsf{orderedDish}(t, y)$. For rule (4), we first need to check that it qualifies as an extended DL rule for $\mathcal{EL}^{++}$. This is easy to see since the direct connections in (4) do indeed form an undirected tree. Next, we assume that v was replaced by some individual name, and consider the paths in the rule. By Definition 5, paths must not end with individual names, and hence the modified rule contains no paths, such that it satisfies all conditions of an $\mathcal{EL}^{++}$ rule.

We can now investigate the semantics of the example. An interesting inference that can be made is $\mathsf{Unhappy}(\mathsf{sebastian})$. Indeed, combining (1), (8), and (9), we find that Sebastian dislikes peanut oil. Rule (10) implies that any interpretation must contain some domain element that is a Thai curry ordered by Sebastian, where we note that there is no individual name that explicitly refers to that curry. By (5) this unnamed curry is a dish, and by (6) it contains peanut oil. At this point we can apply rule (4), where v is mapped to the individual denoted by $\mathsf{peanutOil}$, x is mapped to the individual denoted by $\mathsf{sebastian}$, and y is mapped to the unnamed Thai curry. Hence we find that Sebastian dislikes his curry, and thus by rule (3) he is unhappy.

It is instructive to point out the use of safe and unsafe variables in that case. In contrast to plain Datalog, the above example involves computations relating to some unnamed individual – the Thai curry – to which rules are applied. On the other hand, rule (4) could only be invoked since the individual represented by v is named.

The impact of safety restrictions becomes clear by checking the happiness of Markus. Using similar inferences as above, we find that Markus ordered some (unnamed) Thai curry (12) – note that this need not be the same that was ordered by Sebastian – and that this Thai curry contains some fish product (7) that Markus dislikes (2). However, this fish product is again unnamed, and hence we cannot apply rule (3), and we cannot conclude that Markus dislikes the dish he ordered. Thus, colloquially speaking, Markus is not unhappy since there is no information about some concrete (named) fish product in his curry.

5   Polytime ELP Reasoning with Datalog

We now introduce a polytime algorithm for compiling ELP rule bases into equisatisfiable Datalog programs. A useful feature of this transformation is that it does not only preserve satisfiability but also instance classification. Firstly, we observe that range restrictions in $\mathcal{EL}^{++}$ rule bases can be eliminated:

Proposition 9. *Consider an $\mathcal{EL}^{++}$ rule base* RB *and a set* RR *of range restrictions that are admissible for* RB. *Then there is a rule base* RB′ *that is equisatisfiable to* RB $\cup$ RR, *and which can be computed in polynomial time.*

The proof given in [12] extends the elimination strategy given in [6] to $\mathcal{EL}^{++}$ rules in a straightforward way. The main observation is that the formalisation of admissibility given above sufficiently generalises the conditions from [6] to encompass also concept-product-like rules that entail role relations without explicitly using roles in the antecedent. Next, we expand nested concept expressions in rules:

Definition 10. *An $\mathcal{EL}^{++}$ rule base* RB *is in* normal form *if all concept atoms in rule bodies are either concept names, $\top$, or nominals, all variables in a rule's head also occur in its body, and all rule heads contain only atoms of one of the following forms:*

$$A(t) \qquad \exists R.B(t) \qquad R(t, u)$$

where $A \in \mathsf{N}_C \cup \{\{a\} \mid a \in \mathsf{N}_I\} \cup \{\bot\}$, $B \in \mathsf{N}_C$, $R \in \mathsf{N}_R$, and $t, u \in \mathsf{N}_I \cup \mathbf{V}$.

Proposition 11. *Every $\mathcal{EL}^{++}$ rule base* RB *can be transformed in polynomial time into an equisatisfiable $\mathcal{EL}^{++}$ rule base* RB′ *in normal form.*

The following transformation of $\mathcal{EL}^{++}$ rules to Datalog is the core of our approach for reasoning in ELP:

Definition 12. *Given an $\mathcal{EL}^{++}$ rule base* RB *in normal form, the Datalog program $\bar{\mathsf{P}}$(RB) is defined as follows. The following new symbols are introduced:*

- *a role name $R_\approx$ (the* equality predicate*),*
- *concept names C_a for each $a \in \mathsf{N}_I$,*
- *concept names Self_R for each simple role $R \in \mathsf{N}_R^s$,*
- *individual names $d_{R,C}$ for each $R \in \mathsf{N}_R$ and $C \in \mathsf{N}_C$.*

In the following, we will always use N_I, N_C, N_R, N_R^n, N_R^s to refer to the original sets of symbols in RB, *not including the additional symbols added above. The program $\bar{\mathsf{P}}$(RB) is obtained from* RB *as follows:*

(a) *For each individual name a occurring in* RB, *the program $\bar{\mathsf{P}}$(RB) contains rules* $\rightarrow C_a(a)$ *and* $C_a(x) \rightarrow R_\approx(x, a)$.

(b) *For each concept name C and role name R occurring in $\bar{\mathsf{P}}(\mathrm{RB})$, the program $\bar{\mathsf{P}}(\mathrm{RB})$ contains the rules*

$$\rightarrow R_{\approx}(x,x) \qquad R(z,x) \wedge R_{\approx}(x,y) \rightarrow R(z,y)$$
$$R_{\approx}(x,y) \rightarrow R_{\approx}(y,x) \qquad R(x,z) \wedge R_{\approx}(x,y) \rightarrow R(y,z)$$
$$C(x) \wedge R_{\approx}(x,y) \rightarrow C(y) \quad R_{\approx}(x,y) \wedge R_{\approx}(y,z) \rightarrow R_{\approx}(x,z)$$

(c) *For all rules $B \rightarrow H \in \mathrm{RB}$, a rule $B' \rightarrow H' \in \bar{\mathsf{P}}(\mathrm{RB})$ is created by replacing all occurrences of $R(x,x)$ by $\mathsf{Self}_R(x)$, all occurrences of $\{a\}(t)$ by $C_a(t)$, and all occurrences of $\exists R.C(t)$ with $C \in \mathsf{N}_C$ by the conjunction $R(t,d_{R,C}) \wedge C(d_{R,C})$.*

(d) *For all rules $B \rightarrow H \in \mathrm{RB}$ with $R(x,y) \in H$ and $R \in \mathsf{N}_R^s$ simple, $\bar{\mathsf{P}}(\mathrm{RB})$ contains a rule $B' \rightarrow \mathsf{Self}_R(x) \in \bar{\mathsf{P}}(\mathrm{RB})$, where B' is obtained from B by replacing all occurrences of y with x, all occurrences of $\{a\}(t)$ by $C_a(t)$, and (finally) all expressions $S(x,x)$ with $\mathsf{Self}_S(x)$.*

(e) *For each $R \in \mathsf{N}_R^s$ and $a \in \mathsf{N}_I$, the rule $C_a(x) \wedge R(x,x) \rightarrow \mathsf{Self}_R(x)$ is in $\bar{\mathsf{P}}(\mathrm{RB})$.*

Theorem 13. *Given an $\mathcal{EL}^{++}$ rule base RB in normal form, RB is unsatisfiable iff $\bar{\mathsf{P}}(\mathrm{RB})$ is unsatisfiable.*

Proof sketch. The proof in [12] proceeds by constructing models of $\bar{\mathsf{P}}(\mathrm{RB})$ from models of RB, and vice versa. We omit the technical details here for space reasons, and merely sketch some of the relevant methods and insights.

It is well-known that, in the case of $\mathcal{EL}^{++}$, models can be generated by introducing only a single individual for each atomic concept [2]. For $\mathcal{EL}^{++}$ rules, however, the added features of role conjunction and local reflexivity change the situation: considering only one characteristic individual per atomic concept leads to undesired entailments in both cases. Our model constructions therefore deviate from the classical $\mathcal{EL}^{++}$ construction that worked for the simple $\mathcal{EL}$ rules in [9] with only minor modifications.

For instance, the rule base $\{a\}(x) \rightarrow \exists R.C(x), \{a\}(x) \rightarrow \exists S.C(x)$ does not entail any conjunction of the form $R(a,x) \wedge S(a,x)$. Yet, every interpretation in which the extension of C is a singleton set would necessarily entail this conjunction. This motivates the above use of $d_{R,C}$ in $\bar{\mathsf{P}}(\mathrm{RB})$, which, intuitively, represent individuals of C that have been "generated" by a rule head of the form $\exists R.C(x)$. Thus we admit $|\mathsf{N}_R|$ distinct characteristic individuals for each class, and this suffices for the proper model construction in the presence of role conjunctions.

The second problematic feature are expressions of the form $R(x,x)$, which again preclude the consideration of only one characteristic individual per class. The use of concept atoms $\mathsf{Self}_R(x)$ enables the translation of models for RB to models of $\bar{\mathsf{P}}(\mathrm{RB})$ (the soundness of the satisfiability checking algorithm). The latter may indeed entail additional statements of type $R(x,x)$ without impairing the validity of the Datalog rules that use $\mathsf{Self}_R(x)$.

In the other direction, models of RB are built from models of $\bar{\mathsf{P}}(\mathrm{RB})$ by creating infinitely many "parallel copies" of a basic model structure. These copies form an infinite sequence of levels in the model, and simple roles relate only to successors in higher levels. Exceptions to this construction principle, such as the concept product rules discussed earlier, make the exact formalisation technically involved. The proof in [12] for this case hinges upon the simplicity of roles in concepts Self_S, and it is not clear if a relaxation of this requirement would be possible.   □

We are now ready to show the tractability of ELP.

Theorem 14. *Satisfiability of any* ELP_n *rule base* RB *can be decided in time polynomial in the size of* RB *and exponential in n. More precisely,* RB *can be transformed into an equisatisfiable Datalog program* $\mathsf{P}(\mathrm{RB})$ *which contains at most* $\max(3, n)$ *variables per rule, and this transformation is possible in polynomial time in the size of* RB. *Moreover, for any* $C \in \mathsf{N}_C$, $R \in \mathsf{N}_R$, *and* $a, b \in \mathsf{N}_I$, *we find that*

- $\mathrm{RB} \models C(a)$ *iff* $\mathsf{P}(\mathrm{RB}) \models C(a)$
- $\mathrm{RB} \models \{a\}(b)$ *iff* $\mathsf{P}(\mathrm{RB}) \models C_a(b)$
- $\mathrm{RB} \models R(a, b)$ *iff* $\mathsf{P}(\mathrm{RB}) \models R(a, b)$

Proof. We present some core parts of the proof in [12]. Grounding all safe variables of a rule base in all possible ways is a feasible reasoning method, but may lead to exponential increases in the size of the rule base. This can be prevented, however, by ensuring that any rule contains only a limited number of variables. A similar method can be used to ensure that the Datalog program $\bar{\mathsf{P}}(\cdot)$ as obtained in Definition 12 can be evaluated in polynomial time. We thus provide a satisfiability preserving polytime reduction of ELP rule bases into ELP rule bases that contain only a bounded number of variables per rule. We consider only basic ELP rules for the reduction, since range restrictions do not require any transformation. One should, however, observe that the transformation does not violate the admissibility restrictions for range restrictions.

Let $\mathrm{RB}' \subseteq \mathrm{RB}$ denote the set of ELP rules in RB (i.e. excluding only additional DL-safe rules of n variables that might be available in ELP_n). We first transform the ELP rule base into a normal form by applying the algorithm from Proposition 11. It is easy to see that this transformation can also be applied to ELP rules by treating safe variables like individual names. Hence, this transformation preserves satisfiability, and yields a rule base RB_1 the size of which is polynomial in the size of RB'. The new rule base RB_1 is then of a normal form similar to the one of Definition 10 but with additional safe components per rule.

Next, we reduce conjunctions in rule heads in the standard way: any rule of the form $B \to H_1 \wedge H_2$ is replaced by two rules $B \to H_1$ and $B \to H_2$ until all conjunctions in rule heads are eliminated. Again, the resulting rule base RB_2 is clearly equisatisfiable to RB_1 and can be obtained in polynomial time.

As the next step, we transform the extended DL rules of RB_2 into extended DL rules with at most 3 variables per rule. Besides the notions defined in Definition 5, we use a number of auxiliary notions in describing the transformation. In the following, we assume that all direct connections (cf. Definition 5) between terms t and u in some set B are *maximal*, i.e. contain all role atoms of the form $R(t, u) \in B$. Consider some rule $B \to H$:

- A *connected component* of B is a non-empty subset $S \subseteq B$ such that, for all terms $t \neq u$ occurring in S, we find that t and u are connected in S. A *maximal connected component* (MCC) is a connected component that has no supersets that are connected components.
- A variable x is *initial for* H if H is of the form $C(x)$ or $R(x, t)$.
- A variable x is *final for* H if H is of the form $R(t, x)$. If H is not of this form but $B \to H$ contains some variable, then some arbitrary but fixed variable in $B \to H$ is selected to be final for H.

- Given a subset S of B, we say that S is *reducible* if it contains variables that are neither initial nor final in H.
- Let S be an MCC of B, and consider a direct connection T from a term t to a term u in S. Let $S_{T,t}$ be the set of all atoms in S that contain some term t' connected to t in $S \setminus T$. Similarly, let $S_{T,u}$ be the set of all atoms in S that contain some term u' connected to u in $S \setminus T$.

Intuitively, the sets $S_{T,t}$ and $S_{T,u}$ consist of all atoms to the "left" or to the "right" of the connection T that can be reached from t and u, respectively, without using the atoms of T.

We can now proceed to reduce the forest structure of rule bodies.

In each iteration step of the reduction, select some rule $B \to H$ in RB_2 that contains more than three variables and some reducible MCC S of B, and do one of the following:

(1) If S contains no variable that is final for H, then select an initial element t as follows: if S contains a variable x that is initial for H then $t = x$; otherwise set $t = a$ for an arbitrary individual name $a \in N_I$. The rule $B \to H$ is replaced by two new rules $(B \setminus S) \cup \{C(t)\} \to H$ and $S \to C(t)$, where C is a new concept name.

For all other cases, assume that the variable y in S is final for H.

(2) There is a direct connection T from y to some variable u such that $S_{T,u}$ is reducible but contains no variable initial for H. Then rule $B \to H$ is replaced by three new rules $B \cup \{C(y)\} \setminus (S_{T,u} \cup T) \to H$, $T \cup \{D(u)\} \to C(y)$, and $S_{T,u} \to D(u)$, where C, D are new concept names.

(3) There is a direct connection T from some variable t to y such that $S_{T,t}$ is reducible, and contains a variable x that is initial for H. Then rule $B \to H$ is replaced by three new rules $B \cup \{R(x, y)\} \setminus (S_{T,t} \cup T) \to H$, $\{R'(x, t)\} \cup T \to R(x, y)$, and $S_{T,t} \to R'(x, t)$, where R, R' are new non-simple role names.

(4) There is a direct connection T from some variable t to y such that $S_{T,t}$ is reducible but contains no variable that is initial for H. Then rule $B \to H$ is replaced by three new rules $B \cup \{R(a, y)\} \setminus (S_{T,t} \cup T) \to H$, $\{R'(a, t)\} \cup T \to R(a, y)$, and $S_{T,t} \to R'(a, t)$, where $a \in N_I$ is an arbitrary individual name, and R, R' are new non-simple role names.

(5) There is a direct connection T from y to some variable u such that $S_{T,u}$ is reducible, and contains a variable x that is initial for H, and some further variable z besides x and u. We distinguish various cases:

   (a) There is a direct connection from some term $t \neq y$ to u. Then rule $B \to H$ is replaced by two new rules $B \cup \{R(x, u)\} \setminus S_{T,u} \to H$ and $S_{T,u} \to R(x, u)$, where R is a new non-simple role name.

   (b) The above is not the case, and there is some direct connection T' from u to some variable u' such that $S_{T',u'}$ is reducible but does not contain x. Then rule $B \to H$ is replaced by two new rules $B \cup \{C(u)\} \setminus (S_{T',u'} \cup T') \to H$ and $S_{T',u'} \cup T' \to C(u)$, where C is a new concept name.

   (c) None of the above is the case, and u is involved in a direct connection T' besides T, which connects u to some variable u' such that $S_{T',u'}$ contains x. Let S_u denote the set $S_u := S \setminus (S_{T,y} \cup S_{T',u'})$. The rule $B \to H$ is replaced by two new rules $B \cup \{R(y, u')\} \setminus S_u \to H$ and $S_u \to R(y, u')$, with R a new non-simple role name.

This iteration is repeated until no further transformation is applicable. The proof in [12] proceeds by establishing various properties of the above reduction:

- All rules created in the above transformation are valid ELP rules.
- After the above translation, all rules in RB_2 have at most three variables in the body.
- The transformation terminates after a finite number of steps that is polynomially bounded in the size of RB_2.
- The above translation preserves satisfiability of RB_2.

Thus, the transformed rule base RB_2 is polynomial in the size of RB and contains at most three variables per rule. We can now compute the grounding of all safe variables in RB_2, i.e. the set of rules obtained by replacing safe variables in each rule of RB_2 with individual names in all possible ways. The obtained rule base is called RB_3 and its size clearly is polynomially bounded by $|RB_2|^3$. Moreover, RB_3 is clearly equivalent to RB_2 and, by Definition 7, contains only $\mathcal{EL}^{++}$ rules and range restrictions. We can now apply the elimination of range restrictions of Proposition 9, and then use the normalisation from Proposition 11 to again obtain a set RB_4 of normalised $\mathcal{EL}^{++}$ rules. Again, RB_4 is equivalent to RB_3, and the transformations are easily seen to preserve the bound on the number of variables per rule, especially since rule bodies had already been normalised when computing RB_1.

Now, finally, the Datalog program $\bar{P}(RB_4)$ is constructed. By inspecting the cases of Definition 12, we find that $\bar{P}(RB_4)$ still contains at most 3 (unsafe) variables per rule. Since $\bar{P}(RB_4)$ and the initial set of basic ELP rules RB′ are equisatisfiable, we can show that $\bar{P}(RB_4) \models C(a)$ iff RB′ $\models C(a)$ for all $C \in N_C$ and $a \in N_I$. The claim clearly holds if RB′ is unsatisfiable. Otherwise, consider RB″ = RB′ $\cup \{C(a) \to \bot(a)\}$, and again apply the above construction to obtain an according Datalog program $\bar{P}(RB''_4)$. Clearly, RB″ is unsatisfiable iff RB′ $\models C(a)$. But the former is equivalent to $\bar{P}(RB_4)$ being unsatisfiable. Since $\bar{P}(RB_4)$ is satisfiable, and since clearly $\bar{P}(RB''_4) = \bar{P}(RB_4) \cup \{C(a) \to \bot(a)\}$ (assuming that C and a occur in RB′, and were thus already considered for the rules (a), (b), and (e) of $\bar{P}(RB_4)$), this is in turn equivalent to $\bar{P}(RB''_4) \models C(a)$ as claimed. In a similar fashion, one can show the correspondence for entailments of the form $\{a\}(b)$ $(C_a(b))$ and $R(a, b)$, similar to the statement claimed for the theorem.

The last result enables us to safely combine $\bar{P}(RB_4)$ with any additional DL-safe rule with n variables that may be present in ELP_n. For that purpose, one merely needs to introduce a concept HU and add facts $\to HU(a)$ for all $a \in N_I$. For each n-variable Datalog rule $B \to H$, a rule $B' \to H'$ then is created by replacing any atom of the form $\{a\}(t)$ by $C_a(t)$, and by adding a body atom $HU(x)$ for any variable x occurring in $B \to H$. The resulting set of transformed Datalog rules is denoted LP, and we define $P(RB) := \bar{P}(RB_4) \cup LP$.

It is easy to see that $P(RB)$ is equisatisfiable to RB, since RB′ and $\bar{P}(RB_4)$ contain the corresponding ground facts, and since the rules of LP are applicable only to such ground facts, where the above construction of LP establishes the required syntactic transformations and explicit safety conditions. Similarly, we also find that $P(RB)$ entails the same ground facts as RB, as required in the theorem. Since $\bar{P}(RB)$ is a Datalog program with at most $\max(3, n)$ variables per rule, it can naively be evaluated by computing its grounding, which is again bounded in size by $|\bar{P}(RB)|^{\max(3,n)}$. Together with the polynomial size restrictions established for $\bar{P}(RB)$, this shows the claimed worst-case complexity of reasoning.   □

6   Discussion and Future Work

We have introduced ELP as a rule-based tractable knowledge representation language that generalises the known tractable description logics $\mathcal{EL}^{++}$ and DLP, where polynomial time reasoning was established using a novel reduction to Datalog. ELP in particular extends the DL $\mathcal{EL}^{++}$ with local reflexivity, concept products, conjunctions of simple roles, and limited range restrictions [6].

The notion of simple roles has been slightly extended as compared to the definition commonly used in DL, such that, e.g., the universal role can also be defined to be simple. A natural question is whether further extensions of ELP might be admissible. Regarding the simplicity restriction on role conjunctions, it is well-known that conjunctions of arbitrary roles in $\mathcal{EL}^{++}$ lead to undecidability. Querying for such conjunctions remains intractable [16] even when adopting regularity restrictions similar to the ones in $\mathcal{SROIQ}$. The complexity of using this feature in rules remains open, as does the question whether or not arbitrary roles could be used in reflexivity conditions of the form $R(x, x)$. The presented proofs, however, strongly depend on these restrictions.

The use of Datalog as an approach to solving DL reasoning tasks has been suggested in various works. KAON2 [8] provides an exponential reduction of $\mathcal{SHIQ}$ into disjunctive Datalog programs. The outcome of this reduction resembles our case since it admits for the easy extension with DL-safe rules and safe conjunctive queries. The model-theoretic relationships between knowledge base and Datalog program, however, are somewhat weaker than in our case. In particular, our approach admits queries for non-simple roles. Various other approaches used reductions to Datalog in order to establish mechanisms for conjunctive query answering [17,18,19]. These works differ from the presented approach in that they focus on general conjunctive query answering for $\mathcal{EL}$ and $\mathcal{EL}^{++}$, which is known to be more complex than satisfiability checking [16]. Another related approach is [20], where resolution-based reasoning methods for $\mathcal{EL}$ have been investigated (where we note that resolution is also the standard approach for evaluating Datalog). The methodology used there, however, is technically rather different from our presented approach.

Acknowledgements. The authors wish to thank Carsten Lutz, Boris Motik, and Uli Sattler for useful discussions, and the anonymous reviewers for helpful comments. Research reported herein is supported by the EU in projects ACTIVE (IST-2007-215040) and NeOn (IST-2006-027595), and by the German Research Foundation under the ReaSem project.

References

1. Horrocks, I., Kutz, O., Sattler, U.: The even more irresistible $\mathcal{SROIQ}$. In: Proc. of the 10th Int. Conf. on Principles of Knowledge Representation and Reasoning (KR 2006), pp. 57–67. AAAI Press, Menlo Park (2006)
2. Baader, F., Brandt, S., Lutz, C.: Pushing the EL envelope. In: Proc. 19th Int. Joint Conf. on Artificial Intelligence (IJCAI 2005), Edinburgh, UK. Morgan-Kaufmann Publishers, San Francisco (2005)

3. Calvanese, D., Giacomo, G.D., Lembo, D., Lenzerini, M., Rosati, R.: Tractable reasoning and efficient query answering in description logics: The DL-Lite family. J. of Automated Reasoning 9, 385–429 (2007)
4. Grosof, B., Horrocks, I., Volz, R., Decker, S.: Description logic programs: Combining logic programs with description logics. In: Proc. of WWW 2003, Budapest, Hungary, pp. 48–57. ACM, New York (2003)
5. Krötzsch, M., Rudolph, S., Hitzler, P.: Complexity of Horn description logics. In: Proc. 22nd AAAI Conf., AAAI 2007 (2007)
6. Baader, F., Lutz, C., Brandt, S.: Pushing the EL envelope further. In: Proc. 4th Int. Workshop on OWL: Experiences and Directions (OWLED 2008 DC), Washington, DC (2008)
7. Motik, B., Sattler, U., Studer, R.: Query answering for OWL DL with rules. J. of Web Semantics 3, 41–60 (2005)
8. Hustadt, U., Motik, B., Sattler, U.: Data complexity of reasoning in very expressive description logics. In: Proc. 18th Int. Joint Conf. on Artificial Intelligence (IJCAI 2005), Edinburgh, UK, pp. 466–471. Morgan-Kaufmann Publishers, San Francisco (2005)
9. Krötzsch, M., Rudolph, S., Hitzler, P.: Description logic rules. In: Proc. 18th European Conf. on Artificial Intelligence (ECAI 2008), pp. 80–84. IOS Press, Amsterdam (2008)
10. Gasse, F., Sattler, U., Haarslev, V.: Rewriting rules into $\mathcal{SROIQ}$ axioms. In: Poster at 21st Int. Workshop on Description Logics, DL 2008 (2008)
11. Rudolph, S., Krötzsch, M., Hitzler, P.: All elephants are bigger than all mice. In: Proc. 21st Int. Workshop on Description Logics, DL 2008 (2008)
12. Krötzsch, M., Rudolph, S., Hitzler, P.: ELP: Tractable rules for OWL 2. Technical report, Universität Karlsruhe, Germany (May 2008), http://korrekt.org/page/ELP
13. Baader, F., Calvanese, D., McGuinness, D., Nardi, D., Patel-Schneider, P. (eds.): The Description Logic Handbook: Theory, Implementation and Applications. Cambridge University Press, Cambridge (2007)
14. Horrocks, I., Patel-Schneider, P.F.: A proposal for an OWL rules language. In: Feldman, S.I., Uretsky, M., Najork, M., Wills, C.E. (eds.) Proc. 13th Int. Conf. on World Wide Web (WWW 2004), pp. 723–731. ACM, New York (2004)
15. Dantsin, E., Eiter, T., Gottlob, G., Voronkov, A.: Complexity and expressive power of logic programming. ACM Computing Surveys 33, 374–425 (2001)
16. Krötzsch, M., Rudolph, S., Hitzler, P.: Conjunctive queries for a tractable fragment of OWL 1.1. In: Aberer, K., Choi, K.-S., Noy, N., Allemang, D., Lee, K.-I., Nixon, L., Golbeck, J., Mika, P., Maynard, D., Mizoguchi, R., Schreiber, G., Cudré-Mauroux, P. (eds.) ASWC 2007 and ISWC 2007. LNCS, vol. 4825, pp. 310–323. Springer, Heidelberg (2007)
17. Pérez-Urbina, H., Motik, B., Horrocks, I.: Rewriting conjunctive queries under description logic constraints. In: Proc. Int. Workshop on Logic in Databases, LID 2008 (2008)
18. Pérez-Urbina, H., Motik, B., Horrocks, I.: Rewriting conjunctive queries over description logic knowledge bases. In: Proc. Int. Workshop on Semantics in Data and Knowledge Bases, SDKB 2008 (2008)
19. Rosati, R.: Conjunctive query answering in EL. In: Proc. 20th Int. Workshop on Description Logics, DL 2007 (2007)
20. Kazakov, Y.: Saturation-Based Decision Procedures for Extensions of the Guarded Fragment. PhD thesis, Universität des Saarlandes, Germany (2005)

Term Dependence on the Semantic Web

Gong Cheng and Yuzhong Qu

Institute of Web Science, School of Computer Science and Engineering
Southeast University, Nanjing 210096, P.R. China
{gcheng,yzqu}@seu.edu.cn

Abstract. A large amount of terms (classes and properties) have been published on the Semantic Web by various parties, to be shared for describing resources. Terms are defined based on other terms, and thus a directed dependence relation is formed. The study of term dependence is a foundation work and is important for many other tasks, such as ontology maintenance, integration, and distributed reasoning on the Web scale. In this paper, we analyze the complex network characteristics of the term dependence graph and the induced vocabulary dependence graph. The graphs analyzed in the experiments are constructed from a large data set that contains 1,278,233 terms in 3,039 vocabularies. The results characterize the current status of schemas on the Semantic Web in many aspects, including degree distributions, reachability, and connectivity.

1 Introduction

As with the decentralized linkage nature of the Semantic Web, terms (classes and properties) are usually defined based on other terms in various vocabularies. The meaning of a term is dependent on the meanings of those terms used in its definition. In other words, a change of the meaning of a term may affect the meanings of those terms that are dependent on it. Therefore, term dependence on the Semantic Web is a fundamental problem concerned with ontology maintenance on the Web scale and the evolution of the Semantic Web. Furthermore, the term dependence topology is an important factor that influences how applications access the meanings of semantically interlinked terms, as well as distributed reasoning on the Web scale.

Recently, graph analysis of the Semantic Web has been performed from various aspects [7,9,11,12,14,17]. However, the graph structure of schemas on the Semantic Web on a large scale has not yet been well studied. In this paper, we propose a notion of term dependence on the Semantic Web, and analyze the complex network characteristics of the term dependence graph constructed from a data set that contains 1,278,233 terms defined in 3,039 vocabularies, discovered by our Falcons search engine.[1] We analyze its degree distributions, reachability, and connectivity. We also generalize the dependence from the term level to the vocabulary level, and study its characteristics.

The remainder of this paper is structured as follows. Section 2 discusses related work. Section 3 defines basic terminology used in this paper. Section 4 provides

---

[1] http://iws.seu.edu.cn/services/falcons/.

A. Sheth et al. (Eds.): ISWC 2008, LNCS 5318, pp. 665–680, 2008.

an overview of the data set used in the experiments. In sections 5 and 6, complex network analysis techniques are applied to the term dependence graph and the vocabulary dependence graph, respectively. Section 7 concludes the paper and presents future work.

2   Related Work

Graph analysis has been successfully performed to measure the World Wide Web. Albert et al. [2] analyzed the distributions of incoming and outgoing links between HTML documents on the World Wide Web, and observed power law tails. Barabási et al. [1] found similar results at the site level. As an early work, Gil et al. [9] performed graph analysis on the Semantic Web. They combined ontologies from DAML Ontology Library into a single graph, which included 56,592 vertices and 131,130 arcs. They observed that the Semantic Web is a small world with an average path length 4.37, and the degree distribution follows a power law. Ding et al. [7] studied social networks induced by over 1.5 million of FOAF documents, in which power laws were also observed and interesting patterns of connected components were revealed. Ding and Finin [6] collected 1,448,504 RDF documents and focused on aspects such as the distribution of documents over hosts and the sizes of documents. They measured the complexity of terms by counting the number of RDF triples used to define them, and measured the instance space by counting the meta-usages of terms. Power laws were observed in both experiments. Tummarello et al. [15] found that the distribution (reuse) of URIs over documents follows a power law.

Recently, graph analysis techniques have also been applied to single ontology. Hoser et al. [11] illustrated the benefits of applying social network analysis to ontologies by measuring SWRC and SUMO ontologies. They discussed how different notions of centrality (degree, betweenness, eigenvector, etc.) describe the core content and structure of an ontology, and compared ontologies in size, scope, etc. Ma and Chen [12] surveyed the topology of two TCMLS sub-ontologies. They reported that the analyzed networks, composed of concepts and instances, are typical small-world and scale-free networks. Zhang [17] studied NCI-Ontology, Full-Galen, and other 5 ontologies, and discovered that the degree distributions of entity networks fit power laws well. Theoharis et al. [14] analyzed graph features of 250 ontologies. For each ontology, they constructed a property graph and a class subsumption graph. They found that the majority of ontologies with a significant number of properties approximate a power law for total-degree distribution, and each ontology has a few focal classes that have numerous properties and subclasses.

3   Preliminaries

3.1   Term and Vocabulary

Basically, vocabularies and related definitions in this paper are in accordance with [3]. A *vocabulary* on the Semantic Web is a non-empty set of URIs (called

Table 1. URI namespaces and corresponding prefixes

Prefix	URI Namespace
cyc	`http://www.cyc.com/2004/06/04/cyc#`
dc	`http://purl.org/dc/elements/1.1/`
dcterms	`http://purl.org/dc/terms/`
foaf	`http://xmlns.com/foaf/0.1/`
food	`http://www.w3.org/TR/2003/PR-owl-guide-20031209/food#`
owl	`http://www.w3.org/2002/07/owl#`
rdf	`http://www.w3.org/1999/02/22-rdf-syntax-ns#`
rdfs	`http://www.w3.org/2000/01/rdf-schema#`
skos	`http://www.w3.org/2004/02/skos/core#`
vcard	`http://www.w3.org/2001/vcard-rdf/3.0#`
vin	`http://www.w3.org/TR/2003/PR-owl-guide-20031209/wine#`

its *constituent terms*) that denote a class or a property with a common URI namespace. For example, the URIs `http://xmlns.com/foaf/0.1/Person` and `http://xmlns.com/foaf/0.1/knows`, both containing the URI namespace `http://xmlns.com/foaf/0.1/`, are two constituent terms of the FOAF vocabulary. For convenience, qualified names [4] are used to give URIs in this paper, e.g., `foaf:Person` for `http://xmlns.com/foaf/0.1/Person`. Well-known URI namespaces and corresponding prefixes used in the paper are listed in Table 1. The *authoritative description* of a vocabulary is an RDF graph (a set of RDF triples) encoded by its *namespace document* as well as those RDF documents retrieved by dereferencing the URIs of its constituent terms. Whereas anyone can say anything on the Semantic Web, the authoritative description of a vocabulary is considered to be the most trustable with regard to its constituent terms.

A vocabulary v on the Semantic Web is formulated as $\langle id, C, P, G \rangle$, where id is the URI namespace that identifies v; C and P are the sets of constituent classes and properties of v, respectively, s.t. $C \cup P \neq \emptyset$; G is the authoritative description of v. A URI t is a constituent class (property) of a vocabulary v iff two conditions are satisfied: (a) the URI namespace of t is $v.id$; (b) $v.G$ entails the RDF triple $\langle t, \texttt{rdf:type}, \texttt{rdfs:Class} \rangle$ ($\langle t, \texttt{rdf:type}, \texttt{rdf:Property} \rangle$). The entailment in the experiments is performed by an implemented reasoning engine, based on RDF(S) [10] and OWL DL [13] entailment rules. For example, t is a constituent class of v if $v.G$ contains an RDF triple whose subject is t and predicate is `rdfs:subClassOf`; t is a constituent property of v if $v.G$ contains an RDF triple whose predicate is `owl:onProperty` and object is t. All such rules are not listed in the paper due to space restrictions.

3.2   RDF Sentence

Two RDF triples are *b-connected* [18] if they contain common blank nodes. The b-connected relation is defined as transitive. In an RDF graph, an *RDF sentence* is a maximum subset of b-connected RDF triples. Formally, in an RDF graph G, an RDF sentence $\tilde{s}$ is a subset of RDF triples that satisfy the following conditions: (a) $\forall \tau_i, \tau_j \in \tilde{s}$, τ_i, τ_j are b-connected; (b) $\forall \tau_i \in \tilde{s}, \tau_j \in G \setminus \tilde{s}$, τ_i, τ_j are not b-connected. Figure 1 illustrates an RDF sentence. Let U be the set of all URIs. For an RDF sentence $\tilde{s}$, define $\text{Subj}(\tilde{s}) = \{s | s \in U \wedge \exists \langle s, p, o \rangle \in \tilde{s}\}$. Analogously define $\text{Pred}(\tilde{s}) = \{p | p \in U \wedge \exists \langle s, p, o \rangle \in \tilde{s}\}$ and $\text{Obj}(\tilde{s}) = \{o | o \in U \wedge \exists \langle s, p, o \rangle \in \tilde{s}\}$. For example, for the RDF sentence $\tilde{s}$ depicted in Fig. 1, $\text{Subj}(\tilde{s}) = \{\texttt{food:SeafoodCourse}\}$, $\text{Pred}(\tilde{s}) = \{\texttt{rdfs:subClassOf, rdf:type, owl:onProperty, owl:allValuesFrom, owl:hasValue}\}$, and $\text{Obj}(\tilde{s}) = \{\texttt{owl:Restriction, food:hasDrink, vin:hasColor, food:White}\}$.

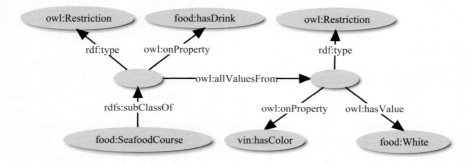

http://www.w3.org/TR/2003/PR-owl-guide-20031209/food

Fig. 1. An RDF sentence

Evidently, two distinct RDF sentences do not share blank nodes. RDF semantics [10] treats blank nodes as existential variables, which are not addressable from outside a graph and are usually created to connect URIs and literals. Besides, an RDF graph G can be decomposed into a unique set of RDF sentences, denoted by $\text{Sent}(G)$. For more details of RDF sentence, refer to [18] or [16]. In [16], the Minimum Self-contained Graph is an equivalent definition of RDF sentence.

3.3   Term Dependence

On the Semantic Web, terms are related to each other in various ways. Most previous work analyzed specific kinds of relations between terms, such as property graph [14], class subsumption graph [14], or a combination of several specific relations [11]. This paper generalizes from these specific relations to a single relation called term dependence, and analyzes its complex network characteristics.

For term $t_1 \in v_1.C \cup v_1.P$ and term t_2, t_1 *directly depends on* t_2, or t_2 *directly influences* t_1, iff $\exists \tilde{s} \in \text{Sent}(v_1.G), t_1 \in \text{Subj}(\tilde{s}), t_2 \in \text{Pred}(\tilde{s}) \cup \text{Obj}(\tilde{s})$. For example, in Fig. 1, $\texttt{food:SeafoodCourse}$ directly depends on all the other terms occurring in that RDF sentence. Using RDF sentences rather than RDF triples to induce dependence causes that dependence is always from URIs to URIs, and blank nodes are not involved in the dependence graph, whereas they still make contributions.

Direct dependence between terms is a very general directed relation, which covers many important specific relations. For example, in RDFS expressions, a class directly depends on its super-classes, and a property directly depends on its super-properties, domain and range; in OWL expressions (after translating OWL axioms to RDF graphs according to [13]), a class directly depends on the properties and classes in its property restrictions (as shown in Fig. 1), and a property directly depends on its inverse property. And naturally, terms often directly depend on those language-level terms in RDF(S) and OWL when using the expressions thereof. Characterizing relations between terms with dependence could greatly simplify the analysis since the relations become homogeneous. Compared with specific relations, term dependence gives a more comprehensive view. However, it is limited by its origin from RDF syntax, e.g., OWL equivalence axioms will only be transformed into unidirectional dependence, whereas bidirectional dependence may be better in some cases.

Generally, to understand the meaning of a term, it is necessary to understand the terms it depends on. In other words, a change of the meaning of some term may affect the meanings of the terms that depend on it, which explains why the word "influence" is used as the inverse relation of dependence.

Direct dependence/influence can be naturally extended to more general dependence/influence in a recursive way: for terms t_1 and t_2, t_1 *depends on* t_2, or t_2 *influences* t_1, iff t_1 directly depends on t_2 or there exists a term t_3 satisfying that t_1 directly depends on t_3 and t_3 depends on t_2.

4   Data Set

All the experiments described in this paper were run on a snapshot of the Semantic Web data collected by the Falcons search engine [5] until April 2008. This section introduces how the data set is constructed, including the seed set collection and the crawling strategy, and then characterizes the distributions of the data set.

4.1   Crawler

RDF document, each identified by a URI, is a basic unit of the data set. The construction of the data set was bootstrapped by submitting to the crawler a set of seed URIs of RDF documents, which were obtained in two ways. Firstly, a list of phrases were extracted from the category names at the top three levels of the Open Directory Project,[2] randomly combined as keyword queries, and sent to the Swoogle search engine[3] and Google search engine (for "filetype:rdf" and "filetype:owl") to retrieve URIs of potential RDF documents. Secondly, the

---

[2] http://www.dmoz.org/.
[3] http://swoogle.umbc.edu/.

URIs of RDF documents from several online repositories were manually added to the seed set, including Ping the Semantic Web.com,[4] SchemaWeb,[5] etc.

A multi-thread crawler was then implemented to dereference URIs with content negotiation and download RDF documents. For simplicity, the "Accept" field in the header of HTTP requests was always set to "application/rdf+xml", and only well-formed RDF/XML documents would be included in the data set. After parsing an RDF document by using Jena,[6] all the URIs mentioned in the document were submitted for further crawling. During a six-month running, 24 million URIs have been pinged, and 11 million documents have been downloaded, 9.8 million of them confirmed as well-formed RDF/XML documents.

4.2 Distributions and Statistics of the Data Set

The 9.8 million RDF documents analyzed in this paper come from 114,408 hosts, or 7,290 registered domain names.[7] The distribution of the number of RDF documents on registered domain names, shown in Fig. 2(a), approximates a power law. The long tail of the distribution is caused by several registered domain names that host large numbers of RDF documents, including bio2rdf.org, dbpedia.org, openlinksw.com, buzznet.com, bibsonomy.org, l3s.de, etc.

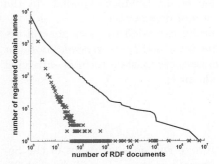 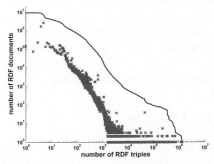

(a) Distribution (*crosses*) and cumulative distribution (*curve*) of the number of registered domain names versus the number of RDF documents per registered domain name.

(b) Distribution (*crosses*) and cumulative distribution (*curve*) of the number of RDF documents versus the number of RDF triples per document.

Fig. 2. Distributions of the data set

The data set contains 401 million RDF triples altogether. The distribution of sizes of RDF documents, shown in Fig. 2(b), also approximates a power law,

---

[4] http://pingthesemanticweb.com/.

[5] http://www.schemaweb.info/.

[6] http://jena.sourceforge.net/.

[7] A registered domain name is more general than the host part of a URI. For example, the host part of http://iswc2008.semanticweb.org/ is iswc2008.semanticweb. org, but its registered domain name is semanticweb.org.

except for the initial segment. Actually the distribution has a maximum at 5 RDF triples, and the cumulative distribution curve exhibits that about half of the RDF documents (51.6%) in the data set contain no more than 5 RDF triples. Generally, each of these small RDF documents encodes a snippet of RDF triples to describe only one specific entity, and such style has been widely adopted in the data sources from the Linking Open Data project.[8] There are also 237 thousand RDF documents (2.4%) that do not contain any RDF triples, but may declare some URI namespaces. It is partially because several servers do not return the HTTP response code 404 for unknown URIs but return such "skeleton" RDF documents. Besides, the only two RDF documents that contain more than 1 million RDF triples are the NCI Thesaurus[9] and WordNet.[10]

Based on the definitions introduced in Sect. 3.1, a total of 3,039 vocabularies and 1,278,233 constituent terms have been recognized from the data set, including 1,158,480 classes (90.6%), 118,808 properties (9.3%), and the other 945 (0.07%) that are both classes and properties. Although RDFS and OWL Full do not require disjointness of classes and properties, such "ambiguous" definitions may on one hand become an obstacle to attract inexperienced developers into the promotion of the Semantic Web, and may on the other hand increase the complexity of computation (e.g., reasoning), especially when some popular terms fall into this group, such as vcard:Orgname.

Actually, if the definition of a term is relaxed from the authoritative description of its vocabulary to any description discovered on the Semantic Web, the numbers of classes and properties in the data set will increase to 2,196,855 (+89.5%) and 195,812 (+63.5%), respectively. However, to best ensure the quality, the following analysis will only focus on the previous 1,278,233 terms, denoted by $\mathbb{T}$.

Figure 3(a) shows the distribution of constituent terms of vocabularies, which approximates a power law especially when the number of constituent terms is larger than 10. The largest vocabulary observed is EthanAnimals,[11] which contains 196,591 terms, followed by FMA,[12] containing 75,245 terms. Different vocabularies are created for different domains and purposes, and they may contain more classes or more properties. Figure 3(b) shows a scatter plot of such data. There are 2,385 vocabularies (78.5%) containing at least one class and one property, 557 vocabularies (18.3%) containing only classes, and 97 vocabularies (3.2%) containing only properties. EthanAnimals, as the vocabulary that contains the most classes, does not contain any properties. Actually, out of the 23 vocabularies that contain more than 10,000 terms, 19 contain less than 10 properties, and most of these large vocabularies describe the medical domain by

---

[8] http://esw.w3.org/topic/SweoIG/TaskForces/CommunityProjects/
LinkingOpenData.

[9] http://www.berkeleybop.org/ontologies/obo-all/ncithesaurus/
ncithesaurus.owl.

[10] http://www.w3.org/2006/03/wn/wn20/rdf/full/wordnet-wordsensesandwords.
rdf.

[11] http://spire.umbc.edu/ontologies/EthanAnimals.owl.

[12] http://onto.eva.mpg.de/fma/fma.owl.

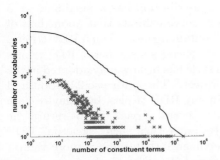

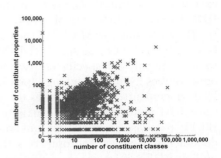

(a) Distribution (crosses) and cumulative distribution (curve) of the number of vocabularies versus the number of constituent terms of a vocabulary.

(b) Distribution of the number of constituent properties versus the number of constituent classes of a vocabulary.

Fig. 3. Distributions of terms

using only large class hierarchies. A vocabulary[13] used by DBpedia contains the most properties. Besides, there are 6 vocabularies that contain more than 1,000 classes and more than 1,000 properties, all of which are different versions of the OpenCyc ontology.[14]

5   Complex Network Analysis of Term Dependence

Dependence between terms on the Semantic Web can be characterized by a directed graph, called the *term dependence graph*, denoted by TDG = {$\mathbb{T}$, $\mathbb{TD}$}, where $\mathbb{T}$ is the vertex set, each vertex labeled with a term $t \in \mathbb{T}$; $\mathbb{TD}$ is the arc set, and an arc $\langle t_1, t_2 \rangle$ exists iff t_1 directly depends on t_2. The TDG analyzed in this paper includes 1,278,233 vertices and 7,312,657 arcs (after removing self-loops). The remainder of this section will analyze TDG to study its complex network characteristics and show how terms are defined and related to each other on the real Semantic Web.

5.1   Degree Analysis

Two basic measures of TDG are the distributions of in-degrees and out-degrees, which are called *direct influence degrees* and *direct dependence degrees* of terms, respectively. A term of a higher direct influence degree is referenced by more other terms in their definitions, and a term of a higher direct dependence degree references more other terms in its definition. It is worth noting that direct dependence is derived from explicitly specified information by data owners, which exhibits their original biases and customs of defining terms.

---

[13] http://dbpedia.org/property/.
[14] http://www.opencyc.org/.

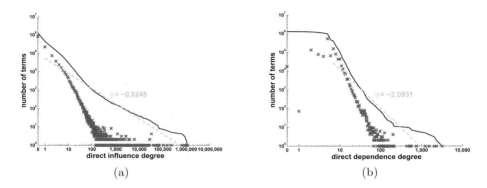

Fig. 4. Distribution (*crosses*) and cumulative distribution (*curve*) of the number of terms versus the (a) direct influence degree and (b) direct dependence degree

Figure 4 shows the distributions of direct influence degrees and direct dependence degrees on a log-log scale. The average in-degree/out-degree is 5.72. In Fig. 4(a), the cumulative distribution of direct influence degrees follows a power law with the exponent $\gamma = -0.8245$. There are 7 terms that are of a direct influence degree higher than 100,000, which are (in descending order) `rdf:type`, `rdfs:subClassOf`, `owl:Class`, `rdfs:label`, `rdfs:comment`, `rdfs:Class`, and `owl:equivalentClass`. It indicates that class hierarchy (including class equivalence) is the most observed structure when defining terms and publishing vocabularies, whereas developers are also inclined to attach human-readable information to terms by using annotation properties. Not surprisingly, all these terms are in language-level vocabularies. The most observed non-language-level terms are mainly those properties for generating unique identifiers in large vocabularies, such as `cyc:guid`. Besides, 829,101 terms (64.9%) do not directly influence any other terms, which covers 64.6% classes and 67.0% properties.

As shown in Fig. 4(b), the distribution of direct dependence degrees does not fit a power law quite well, especially for the initial segment, and has a maximum at 5 degrees, which covers 40.9% terms. It is interesting that 17,505 terms (1.4%) do not depend on any other terms. It is mainly because, some large vocabularies, such as the NEWT taxonomy,[15] do not encode all the term definitions in one RDF document but only returns information principally about just one term when its URI is dereferenced. Then, it is possible that some term definitions have not been crawled but they can still be confirmed as terms since they have been found in other term definitions in the same vocabulary. Besides, the cumulative distribution curve shows that 10 terms are of a direct dependence degree higher than 400. A case-by-case study reveals that all these terms are classes and are also of a direct influence degree higher than 1,400. Actually each of them, called a focal class in [14], is a central term in the vocabulary, depending on and being depended on by many other terms.

---

[15] `http://purl.uniprot.org/taxonomy/`.

The Pearson's correlation coefficient between direct influence degrees and direct dependence degrees is 0.006 (ranging from -1 to 1), which means there is almost no linear relationship between them.

5.2   Reachability Analysis

The previous subsection analyzed the direct dependence and influence between terms in graph view. According to the definitions in Sect. 3.3, the more general dependence and influence can also be clearly characterized in the view of graph theory: $\forall t_1, t_2 \in \mathbb{T}$, t_1 depends on t_2, or t_2 influences t_1, iff t_2 is reachable from t_1 in TDG. For each term, the number of its reachable terms is called its *dependence degree*, and the number of the terms that can reach it is called its *influence degree*.

When retrieving a term definition or understanding a term, it is often the case that those terms it directly depends on still need to be explored and their definitions will also be retrieved, and goes on. A term of a higher dependence degree requires more steps of such retrieval. Correspondingly, the influence degree indicates how important a term exhibits on the Semantic Web because a change of the meaning of a high-influence-degree term will affect the meanings of a large amount of other terms.

Figure 5 shows the distributions of term influence degrees and term dependence degrees on a log-log scale. In average, each term depends on 1,105 other terms. In Fig. 5(a), the initial segment of the distribution follows a power law, but the rest part is a mess. One reason is that many large strongly connected components (SCC) are observed in TDG, and all the terms in an SCC have exactly the same influence degree and dependence degree. Particularly, there are 13 terms, including rdf:type, rdfs:Resource, rdfs:Class, rdfs:subClassOf, rdf:Property, rdfs:subPropertyOf, rdfs:domain, rdfs:range, rdfs:label, rdfs:comment, rdfs:seeAlso, rdfs:isDefinedBy, and rdfs:Literal that compose an SCC, all of which influence almost all the terms on the Semantic Web. It also explains why few terms has a dependence degree between 1 and 12, as shown in Fig. 5(b). These results demonstrate that RDF and RDFS should be kept stable because a change of their meanings will almost change the whole Semantic Web. It is also a best practice for all the Semantic Web applications to be equipped with the ability to understand and use these terms.

In graph theory, the distance between two vertices is the length of a shortest path between them, and the eccentricity of a vertex is the maximum distance from the vertex to any other reachable vertices. In TDG, the eccentricity of a term is called its *dependence depth*. When retrieving a complete definition of a high-dependence-depth term, more rounds of breadth-first search (BFS) are required; and when understanding such terms, people are more likely to become lost in long-distance paths. Besides, it may take more steps to reflect a change of the meaning of a high-dependence-depth term caused by a change of the meaning of some term it depends on, due to the long distance. Figure 6(a) shows the distribution of dependence depths of terms on the Semantic Web. The average dependence depth is 10.05. About half of the terms (51.4%) have

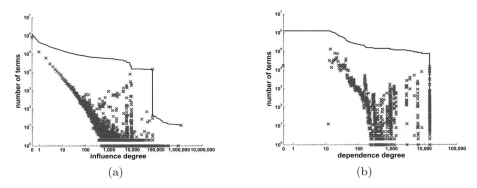

Fig. 5. Distribution (*crosses*) and cumulative distribution (*curve*) of the number of terms versus the (a) influence degree and (b) dependence degree

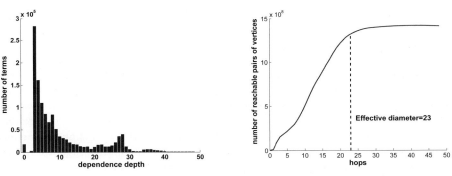

(a) Distribution of the number of terms versus the dependence depth.

(b) Hop plot and effective diameter.

Fig. 6. Eccentricities and hop plot of TDG

a dependence depth not higher than 6. However, there are still 11.5% terms that have a dependence depth higher than 25, which often occur at the bottom of class hierarchies. The highest dependence depth observed is 48, owned by 4 classes at the bottom of a deep class hierarchy in FMA.

In some cases, to process a term, it is not necessary to retrieve all the terms it depends on, but instead, a significantly large subset (e.g., 90%) is enough for specific applications. In graph theory, hop plot [8] is used to measure the rate of increase of reachable vertices with increasing the distance threshold (called hops). The effective diameter of a graph is the minimum number of hops in which 90% of all reachable pairs of vertices can reach from one to the other. Figure 6(b) shows the hop plot of TDG, which approximates a linear correlation when hops is less than 23, the effective diameter of TDG. It means that in average, when retrieving the definitions of a term and those terms it depends on in a BFS way, the number of newly found terms does not remarkably decrease until 23 rounds later. Evidently, 23 seems too large a value for human beings.

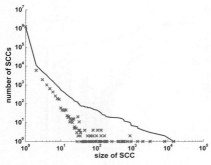
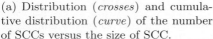

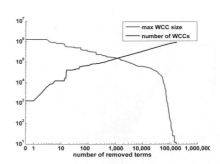

(a) Distribution (*crosses*) and cumulative distribution (*curve*) of the number of SCCs versus the size of SCC.

(b) Connectivity versus the number of removed terms.

Fig. 7. Connectivity of TDG

5.3   Connectivity Analysis

It is possible that a set of related terms are defined in respect of each other, i.e., they are reachable from each other in TDG and thus form an SCC. Figure 7(a) shows the distribution of sizes of SCCs. Most terms (93.4%) are within trivial SCCs (SCC with only one vertex), i.e., they are not involved in circular dependence. The largest SCC is with 14,883 terms in FMA, interlinked by class subsumption relations and property restriction structures. Although there are 10,994 non-trivial SCCs, only 23 of them are with terms in more than one vocabularies, out of which 22 are with terms in a "family" of vocabularies, i.e., a set of vocabularies that have a significantly long common prefix of URI namespaces, such as FOOD and VIN. The only real cross-vocabulary SCC is with 19 properties in DC, DCTERMS, and SKOS, including `dc:creator`, `dcterms:date`, `skos:note`, etc., each of which is used to describe some of the others.

To further examine the connectivity of TDG, the terms of the highest degree are removed one at each step. Figure 7(b) shows that TDG is rapidly broken into over 40 thousand weakly connected components (WCCs) after only 16 terms are removed, revealing that the connectivity of TDG heavily depends on a few popular terms. Specifically, if all the terms in language-level vocabularies (RDF, RDFS, OWL, and DAML) are removed, the average in-degree/out-degree will decrease from 5.72 to 1.92.

6   Vocabulary Dependence

Out of the 7,312,657 arcs in TDG, 5,315,615 (72.7%) are between terms in different vocabularies. Thus, it is interesting to generalize the dependence from term level to vocabulary level, and study its characteristics.

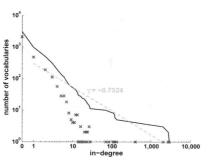

(a) Distribution (crosses) and cumulative distribution (curve) of in-degrees.

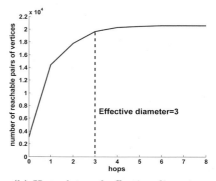

(b) Hop plot and effective diameter.

Fig. 8. In-degree distribution and hop plot of VDG

Dependence between vocabularies on the Semantic Web can also be characterized by a directed graph, called the *vocabulary dependence graph*, denoted by $VDG = \{\mathbb{V}, \mathbb{VD}\}$, where $\mathbb{V}$ is the vertex set, each vertex labeled with a vocabulary v; $\mathbb{VD}$ is the arc set, and an arc $\langle v_1, v_2 \rangle$ exists iff $\exists t_1 \in v_1.C \cup v_1.P, t_2 \in v_2.C \cup v_2.P$, t_1 directly depends on t_2. The VDG analyzed in this paper includes 3,039 vertices and 11,392 arcs (after removing self-loops). The average in-degree/out-degree is 3.75. Figure 8(a) shows the cumulative distribution of in-degrees of VDG, which approximates a power law with the exponent $\gamma = -0.7524$. The four vocabularies of the highest in-degree are RDF, RDFS, OWL, and DAML, all of which are language-level vocabularies. If these vocabularies are removed, the average in-degree/out-degree will decrease to 1.06. It exhibits that most dependence relations are attributed to the dependence to language-level vocabularies.

To measure the reachability and distance features of VDG, Fig. 8(b) shows its hop plot. Over half of all reachable pairs of vertices can reach from one to the other with no more than 1 hop, and the effective diameter is just 3, which is much smaller than 23, the effective diameter of TDG. It means that long-distance term dependence is principally within vocabularies.

To examine the connectivity of VDG, the vocabularies of the highest degree are removed one at each step. Figure 9(a) shows that VDG is totally fragmented to the isolation of single vertices only after 695 vocabularies (22.9%) are removed. Actually, VDG is rapidly broken into 1,320 WCCs just after four language-level vocabularies are removed, as depicted in Figure 9(b).[16] However, there is still a large WCC with 871 vocabularies (28.7%), mainly due to the use of annotation properties in DC and SKOS. Other small non-trivial WCCs are usually composed of families of vocabularies.

---

[16] This figure is generated by Pajek (`pajek.imfm.si`).

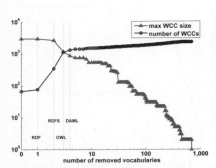

(a) Connectivity versus the number of removed vocabularies.

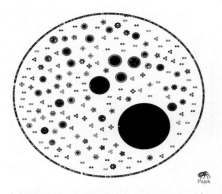

(b) WCCs of VDG after removing four language-level vocabularies.

Fig. 9. Connectivity of VDG

Table 2. Indicators of TDG and VDG before/after removing four language-level vocabularies

Indicator	TDG before Avg.	TDG before Max.	TDG after Avg.	TDG after Max.	VDG before Avg.	VDG before Max.	VDG after Avg.	VDG after Max.
In-degree	5.72	1,187,173	1.92	60,836	3.75	2,947	1.06	133
Out-degree	5.72	3,239	1.92	3,235	3.75	20	1.06	17
#Reachable_from	1,105	1,260,727	1,088	196,512	5.77	2,968	2.95	332
#Reachable	1,105	15,259	1,088	15,240	5.77	46	2.95	43
Eccentricity	10.05	48	9.55	48	1.77	8	1.25	8
Effective diameter	23		22		3		3	
γ (in-degree)	-0.8245				-0.7524			

7 Conclusion

This paper proposed term dependence on the Semantic Web, based on the RDF sentence structure extracted from authoritative description of terms, and analyzed the complex network characteristics of the term dependence graph as well as the induced vocabulary dependence graph. Experiments were performed on a real data set collected by our Falcons search engine, which is much larger than those in previous graph analysis of the Semantic Web. The main results are summarized in Table 2. The data set, analyzed graphs, and statistical results are available online.[17]

We observed that term dependence on the Semantic Web forms a scale-free network, i.e., with a power-law degree distribution. The graph structure is very

---

[17] http://iws.seu.edu.cn/projects/ontosearch/dependence_graph/.

complex, and a change of the meaning of a term may affect a large amount of other terms (in average) through long-distance paths. However, complex structures mainly exist within vocabularies. To define terms, developers establish most cross-vocabulary dependence to language-level terms or other popular annotation properties, and they rarely link their terms to other domain vocabularies even on overlapped topics. The schema-level of the Semantic Web is still far away from a Web of interlinked ontologies, which indicates that ontologies are rarely reused and it will lead to difficulties for data integration.

In future work, as with the growth of the numbers of terms and vocabularies on the Semantic Web, their evolution model deserves to be investigated in the future. Besides, exploring the macrostructure of the instance level of the Semantic Web is also an attractive research topic.

Acknowledgments. The work is supported in part by the NSFC under Grant 60773106, and in part by the 973 Program of China under Grant 2003CB317004. We would like to thank Jun Ye for his effort in the experiments. We are also grateful to Weiyi Ge for his work in implementing the crawler.

References

1. Adamic, L.A., Huberman, B.A., Barabási, A.-L., Albert, R., Jeong, H., Bianconi, G.: Power-Law Distribution of the World Wide Web. Science 287(5461), 2115a (2000)
2. Albert, R., Jeong, H., Barabási, A.-L.: Internet: Diameter of the World-Wide Web. Nature 401(6749), 130–131 (1999)
3. Berrueta, D., Phipps, J.: Best Practice Recipes for Publishing RDF Vocabularies. W3C Working Draft (2008)
4. Bray, T., Hollander, D., Layman, A., Tobin, R.: Namespaces in XML 1.0 (Second Edition). W3C Recommendation (2006)
5. Cheng, G., Ge, W., Qu, Y.: Falcons: Searching and Browsing Entities on the Semantic Web. In: 17th International Conference on World Wide Web, pp. 1101–1102. ACM Press, New York (2008)
6. Ding, L., Finin, T.: Characterizing the Semantic Web on the Web. In: Cruz, I., Decker, S., Allemang, D., Preist, C., Schwabe, D., Mika, P., Uschold, M., Aroyo, L.M. (eds.) ISWC 2006. LNCS, vol. 4273, pp. 242–257. Springer, Heidelberg (2006)
7. Ding, L., Zhou, L., Finin, T., Joshi, A.: How the Semantic Web is Being Used: An Analysis of FOAF Documents. In: 38th Annual Hawaii International Conference on System Sciences, pp. 113–113. IEEE Computer Society, Washington (2005)
8. Faloutsos, M., Faloutsos, P., Faloutsos, C.: On Power-Law Relationships of the Internet Topology. In: Annual Conference of the Special Interest Group on Data Communication, pp. 251–262. ACM Press, New York (1999)
9. Gil, R., García, R., Delgado, J.: Measuring the Semantic Web. AIS SIGSEMIS Bulletin 1(2), 69–72 (2004)
10. Hayes, P.: RDF Semantics. W3C Recommendation (2004)
11. Hoser, B., Hotho, A., Jäschke, R., Schmitz, C., Stumme, G.: Semantic Network Analysis of Ontologies. In: Sure, Y., Domingue, J. (eds.) ESWC 2006. LNCS, vol. 4011, pp. 514–529. Springer, Heidelberg (2006)

12. Ma, J., Chen, H.: Complex Network Analysis on TCMLS Sub-Ontologies. In: 3rd International Conference on Semantics, Knowledge and Grid, pp. 551–553. IEEE Computer Society, Washington, DC (2007)
13. Patel-Schneider, P.F., Hayes, P., Horrocks, I.: OWL Web Ontology Language Semantics and Abstract Syntax. W3C Recommendation (2004)
14. Theoharis, Y., Tzitzikas, Y., Kotzinos, D., Christophides, V.: On Graph Features of Semantic Web Schemas. IEEE Trans. Knowl. Data Eng. 20(5), 692–702 (2008)
15. Tummarello, G., Delbru, R., Oren, E.: Sindice.com: Weaving the Open Linked Data. In: Aberer, K., Choi, K.-S., Noy, N., Allemang, D., Lee, K.-I., Nixon, L., Golbeck, J., Mika, P., Maynard, D., Mizoguchi, R., Schreiber, G., Cudré-Mauroux, P. (eds.) ASWC 2007 and ISWC 2007. LNCS, vol. 4825, pp. 552–565. Springer, Heidelberg (2007)
16. Tummarello, G., Morbidoni, C., Bachmann-Gmür, R., Erling, O.: RDFSync: Efficient Remote Synchronization of RDF Models. In: Aberer, K., Choi, K.-S., Noy, N., Allemang, D., Lee, K.-I., Nixon, L., Golbeck, J., Mika, P., Maynard, D., Mizoguchi, R., Schreiber, G., Cudré-Mauroux, P. (eds.) ASWC 2007 and ISWC 2007. LNCS, vol. 4825, pp. 537–551. Springer, Heidelberg (2007)
17. Zhang, H.: The Scale-Free Nature of Semantic Web Ontology. In: 17th International Conference on World Wide Web, pp. 1047–1048. ACM Press, New York (2008)
18. Zhang, X., Cheng, G., Qu, Y.: Ontology Summarization Based on RDF Sentence Graph. In: 16th International Conference on World Wide Web, pp. 707–716. ACM Press, New York (2007)

Semantic Relatedness Measure Using Object Properties in an Ontology

Laurent Mazuel and Nicolas Sabouret

Laboratoire Informatique de Paris 6 - LIP6, 104 av du Président Kennedy,
75016, Paris, France
{laurent.mazuel,nicolas.sabouret}@lip6.fr

Abstract. This paper presents a new semantic relatedness measure on ontologies which considers especially the object properties between the concepts. Our approach relies on two hypotheses. Firstly, using only concept hierarchy and object properties, only a few paths can be considered as "semantically corrects" and these paths obey to a given set of rules. Secondly, following a given edge in a path has a cost (represented as a weight), which depends on its type (*is-a*, *part-of*, etc.), its context in the ontology and its position in this path. We propose an evaluation of our measure on the lexical base WordNet using *part-of* relation with two different benchmarks. We show that, in this context, our measure outperforms the classical semantic measures.

1 Introduction

Whereas semantic similarity focuses on common points in the concepts definitions, semantic relatedness permits to take into account functional relations between concepts. Thus, as stated by Resnik [1], semantic similarity "represents a special case of semantic relatedness", that only uses the subsumption relation of the knowledge representation. For example, "car" and "gasoline" have a low similarity degree but a high relatedness degree [1]. However, automatic computation of a relatedness degree is considered to be much more difficult than computing a similarity measure. For this reason, much work on semantic measures has focused on computing similarity degrees using well-known hierarchy of concepts (*e.g.* MeSH [2], WordNet [3]). Recent work on relatedness measure tries to adopt a different point of view with the use of glosses [4,5], since a gloss contains a functional description of the concept (by means of other concepts), or terms frequencies in Internet using search engine [6,7]. However, no work, since work of Hirst & St-Onge in 1998 [8], has focused on the issue of semantic relatedness using heterogeneous relations in a graph-based knowledge representation, as semantic network or ontology limited to hierarchy and object properties. In this paper, semantic relatedness between two concepts in a semantic net can be materialized by a path, starting from one concept, following different kinds of relations (subsumption (*is-a*), meronymy (*part-of*) or any other domain specific relation) to the other concept. However, the Hirst & St-Onge measure assumes that all edges have the same weight (*i.e.* the information-content of all edges is uniform in the ontology), which has been demonstrated to be an incorrect hypothesis [1].

A. Sheth et al. (Eds.): ISWC 2008, LNCS 5318, pp. 681–694, 2008.

Computing semantic relatedness in graph-based knowledge base rather that using glosses is however a challenging and important issue. Indeed, the great majority of existing Human-Machine interaction systems with Natural Language (questions/answering systems, dialogue systems...) makes use of ontologies for the semantic interpretation of requests [9,10,11]. Moreover, in web services composition, it is often interesting to compute the semantic similarity between two ontologies before searching for an alignment (to avoid useless attempts if the ontologies don't represent the same data) [12,13]. However in most work, this ontology is reduced to a hierarchy of concepts and the semantic interpretation relies only on similarity measure, although the literature underlines the need for semantic relatedness measures [14,9,15].

Some recent work has tried to define semantic relatedness using complex ontology languages, as OWL-Lite [16]. For instance, in [15], the authors take into account negation, intersection or disjunction of classes, to compute the relatedness between web services. However, these approaches are currently not evaluated nor implemented and remain at a theoretical point of view. Hence, define a semantic relatedness measure based on ontology limited to concept hierarchy and object properties which can be evaluated can be seen as a first step towards semantic relatedness in complex and real ontologies.

In this paper, we present and evaluate a new semantic relatedness measure on an ontology voluntarily limited to concept hierarchy and object properties with the following attributes:

- It considers a set of patterns to filter the paths which are not "semantically correct". Actually, when using more relations than only the subsumption, there exists a lot of paths between two concepts and only a sub-set of these paths are "semantically correct" (*i.e.* their structure correspond to a semantic meaning).
- It uses the information-theoretic definition of semantic similarity to weight the hierarchical edges in the graph,
- It computes the weight of non-hierarchical edges.
- It combines these three points in a unique measure.

The next section presents related work on semantic similarity and semantic relatedness. Moreover, it outlines the problem of finding a semantically correct path in a graph-based knowledge representation. Section 3 presents our approach for computing semantic relatedness using subsumption links and heterogeneous links. Section 4 presents the evaluation and discusses the results.

2   Related Work

The first part of this section presents work on similarity measures. The second part deals with semantic relatedness measures and discusses the problem of finding a semantically correct path in a given ontology.

2.1   Semantic Similarity Measures

This kind of measures is also called *taxonomic measures* or *attributional measures* [17]. An intuitive way to compute quickly semantic similarity between two nodes of a hierarchy is to count the number of edges in the shortest path between these two nodes.

The idea captured with this hypothesis is that the semantic distance of two concepts is correlated with the length of the shortest path to join these concepts. This measure was firstly defined by Rada [2]:

$$dist_{rad}(c_1,c_2) = len(c_1,c_2)$$

where *len* is the length of the shortest path (in terms of the number of edges) between c_1 and c_2 in the hierarchy. However, this measure relies upon the assumption that each edge carries the same amount of information, which is not true in most ontology [1]. Thus, many other formulae extend this measure by computing weights on edges using additional information, like the depth of each concept in the hierarchy [18,19] or the *closest common parent* of the two nodes [20,19]. All these methods are called *edge-based measures*.

On the contrary, *node-based measures* associate a weight to each node. This weight represents the *information content* (IC) of the concept[1]. The more specialised a concept is, the more the weight will be. The first node-based similarity measure, proposed by Resnik in [1], is defined by the information content of the *closest common parent* (*ccp*) of the two concept c_1 and c_2:

$$sim_{res}(c_1,c_2) = IC(ccp(c_1,c_2))$$

where $IC(c)$ is the information content of the concept c. Many other propositions have been made after Resnik to combine the IC of the two target nodes and their ccp (*e.g.* [22,23]).

An interesting attempt to mix node-based and edge-based methods is the Jiang & Conrath measure [23]. They define their distance measure as "derived from the edge-based notion" (by analogies with the Rada measure [2]) "by adding the information content as a decision factor". In this measure, each edge is linked to a weight and the semantic distance is computed by adding all the edge weight along the shortest path. The weight $LS(x,y)$ ("*LS* for link strength") of an edge $\{x,y\}$ between the node x and the node y is computing regarding to their information content:

$$LS(x,y) = |IC(x) - IC(y)|$$

Then, the Jiang & Conrath measure is defined as:

$$dist_{JC}(c_1,c_2) = \sum_{\{x,y\} \in sp(c_1,c_2)} LS(x,y)$$
$$= IC(c_1) + IC(c_2) - 2 \times IC(ccp(c_1,c_2))$$

where $sp(c_1,c_2)$ is the shortest taxonomic path between c_1 and c_2.

---

[1] We don't focus in this article on the way to compute the information content of a concept. The most classical ways are the Resnik's approach with a corpus [1] and the Seco's approach with taxonomy analysis [21].

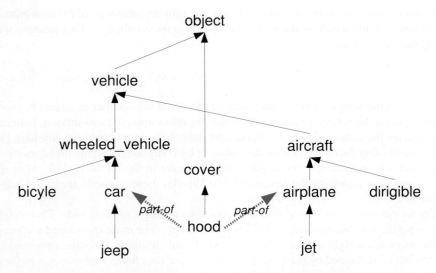

Fig. 1. Example of paths with relations. The black arrows represent *is-a* links.

2.2 Semantic Relatedness Measures and Notion of Semantically Correct Path

This kind of measures is also called *relational measure* [17]. For instance, these measures can capture the semantic link between couples like *gasoline-car* or *bee-honey* [1]. To do this, they have to consider several kinds of relations (and not only the taxonomic relation) as *partOf*, *madeWith*, etc. However, using different kinds of relations will give rise to the problem of multiple existing paths. Indeed, if the shortest path is unique in a hierarchy, many possible paths exist and are conceivable in a graph-based knowledge representation although most of them are not *correct semantically* [24,8]. For this reason, any relational measure must provide (implicitly or explicitly) a set of constraints to ensure that a path is semantically correct.

For instance, in Aleksovski's work [24], a path is considered to be semantically correct if and only if no hierarchical links appear after a non-hierarchical one. Hence, the path[2] {*cover, includes, hood, part-of, airplane, is-a, aircraft*} in figure 1 is incorrect for Aleksovski since (*airplane, is-a, aircraft*) follows the non-hierarchical relation (*hood, part-of, airplane*).

In [8], Hirst & St-Onge associate a direction in *Upward, Downward* and *Horizontal* for each relation type and give three rules to define a semantically correct path in terms of the three directions. It is interesting to note that the rule R1 of Hirst & St-Onge (*i.e.* "No other direction may precede an upward link") subsumes the conclusion of Aleksovski work (*is-a* being an Upward link). Finally, Hirst & St-Onge enumerate only 8 patterns of semantically-correct paths which match their three rules: *{U, UD, UH, UHD, D, DH, HD, H}*. The difficulty is to determine the direction of each relation. For instance, considering the WordNet [25] relations, the authors define in their paper

---

[2] In this article, we will represent a path in a graph using a set-notation as $\{x_1, e_{1,2}, x_2, e_{2,3}, ..., x_n\}$ where x_i are nodes and $e_{i,i+1}$ are links type.

the hypernymy and meronymy relations as Upward link, the hyponymy and holonymy relations as Downward links and the synonymy and antonymy relations as Horizontal links. As a consequence, for Hirst & St-Onge, the path from *jeep* to *airplane* in figure 1 is semantically correct (pattern *UD*) whereas the path from *bicycle* to *hood* is not (*Upward* link after a *Downward* link), even if these two paths have both the same length.

Once a semantically-correct path has been found, one must evaluate its cost so as to determine whether the concepts are semantically close or far (taking into account that several candidate paths can exist). In Hirst & St-Onge's work, the semantic relatedness value is computed by the following formula:

$$rel_{hso}(c_1, c_2) = C - len(c_1, c_2) - k \times turns(c_1, c_2)$$

where C and k are constant defined empirically ($C = 8$ and $k = 1$ in [8]), *len* the length of the shortest path between c_1 and c_2 *considering all relations and the patterns* and *turns* the number of changes of direction in this path. The main drawback of this measure is that it considers, as Rada measure does with similarity, that each edges of each types represents the same information content. However, apart Hirst & St-Onge measure, few work has been made on semantic relatedness measures using heterogeneous relations.

In the next section, we define our semantic relatedness measure, which mixes the information theoretic approach of non-uniformity of the hierarchical edge values, the Hirst & St-Onge patterns to find semantically correct paths and a new proposition to compute a non-hierarchical link weight.

3   Our Semantic Relatedness Measure

Preliminary note. The measure we present in this section is analogue to a distance (*i.e.* a non-relatedness measure): the lower is the score, the higher are the concepts related. However, we will show in section 3.3 a linear transformation from this distance measure to a bounded relatedness measure.

Our work is based on the assumption (used in information-theoretic measure for similarity) that two different hierarchical edges do not carry the same information content, and extends this assumption to non-hierarchical links. We first present the computation of a weight for a single-relation path. We then explain how it can be combined for mixed-relation paths. Finally, we present our measure on the set of semantically correct weighted paths.

3.1   Single-Relation Path

We call *single-relation path* a path whose edges are all of the same type X. For instance, in figure 1, the path $\{jeep, is\text{-}a, car, is\text{-}a, wheeled\_vehicle, is\text{-}a, vehicle\}$ is a single-relation path of type *is-a*. To compute the weight W of a single-relation path, we separate hierarchical relations (X is the *is-a* or the *includes* relation) and non-hierarchical relations.

Let us consider a path $path_X(x, y)$ between two concepts x and y in the ontology, following *only* the relation X:

- If X is a hierarchical relation, we chose to consider the Jiang & Conrath definition for the weight of an edge (see section 2.1):

$$W(path_{X \in \{is\text{-}a, include\}}(x,y)) = |IC(x) - IC(y)|$$

Note that even if *is-a* and *includes* are symmetric relation, the definition of single relation-path does not allow them to be in the same path. For instance, the path $\{jeep, is\text{-}a, car, is\text{-}a, wheeled\_vehicle, includes, bicycle\}$ is not a single-relation path but the mix of the two single relation-path $\{jeep, is\text{-}a, car, is\text{-}a, wheeled\_vehicle\}$ and $\{wheeled\_vehicle, includes, bicycle\}$. This definition is consistent with the similarity measure of Jiang & Conrath defined in section 2.1.
- If X is not a hierarchical relation, we cannot use the information content of nodes, because this value is computed regarding to the hierarchy structure ([21,1]). We then suggest a new proposition for computing this weight. This weight formula is based upon two parts:

  1. A static weight TC_X, which corresponds to the "strength" of a given relation type. This strength will represent the maximum information content that this kind of link can carried. For a given link of type X, if we have $TC_X < 1$, then this type of link is considered being informative and the cost of this edge must be low. For $TC_X = 1$, the cost will be equals the cost of a hierarchical link and for $TC_X > 1$ the edge will be costly. This allows us to study if different kinds of relations carry different types of information. For instance, in most system the meronymy will have a $TC_{hasPart}$ inferior to 1 and the antonymy relation will, in the contrary, have a $TC_{antonymy}$ superior to 1.
  2. A formula to compute the impact of the length of the path for the cost of the path. We want that the cost of a path with a single relation X must respect the following properties: 1) it increases with the length of the path, 2) it is bounded by TC_X which represents the worst possible value for an X-relation path (*i.e.* the value of an infinite-length path that uses only X relations) and 3) it follows a polynomial function to be compared with the *log* progression of IC value for hierarchical links.

Actually, information-theoretic measures [1,3] have outlined the adequacy of the *log* function to compute the weight of a node (and, by extension, an edge). However, to respect our three requirements and since the *log* function is not bounded, we use the $n/n+1$ function to simulate a logarithmic bounded function (figure 2).

As a consequence, the weight of $path_X(x,y)$ when X is not a hierarchical relation, is defined by:

$$W(path_X(x,y)) = TC_X \times \left(\frac{|path_X(c_1, c_2)|}{|path_X(c_1, c_2)| + 1} \right)$$

3.2   Mixed-Relation Path

Let us consider a path $path(x,y)$ between two concepts x and y in the ontology. It can be factorized as an ordered set of n single-relation sub-paths such as:

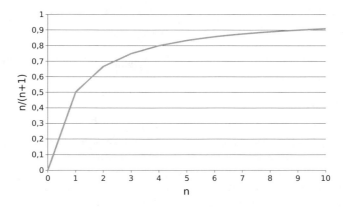

Fig. 2. The progression of the function $n/n+1$, which is bounded and close to the *log* function

$$path(x,y) = path_{X_1}(x,z_1) \oplus path_{X_2}(z_1,z_2) \oplus \dots \oplus path_{X_n}(z_{n-1},y)$$

Therefore, there exist several possible factorizations of a given path $path(x,y)$. We define the minimal factorization $T_{min}(path(x,y))$ as the factorization which minimizes the value n. As a consequence, in the minimal factorization $T_{min}(path(x,y))$ of n sub-paths, we have the property $\forall i \in [1,n-1], X_i \neq X_{i+1}$.[3] This property can be used to build the $T_{min}(path(x,y))$ factorization.

For instance, in the figure 1, $T_{min}(path(jeep,object)) =$

$$\left\{ \begin{array}{l} \{jeep,\ is\text{-}a,\ car\}, \\ \{car,\ has\text{-}part,\ hood\}, \\ \{hood,\ is\text{-}a,\ cover,\ is\text{-}a,\ object\} \end{array} \right\}$$

Note that $\{jeep,is\text{-}a,car\}$ and $\{hood,is\text{-}a,cover,is\text{-}a,object\}$ are two disjoint sub-paths in $T_{min}(path(x,y))$ even if they have the same link type, because they don't have any common nodes.

We finally suggest to define the weight of the path as the sum of the weights of all sub-paths of the minimal factorization. Hence, the weight of the mixed path $path(x,y)$ is then defined as the sum of all sub-paths of T_{min}:

$$W(path(x,y)) = \sum_{p \in T_{min}(path(x,y))} W(p)$$

For instance, let consider the preceding path $path(jeep,object)$ with[4] $IC(jeep) = 1.0$, $IC(car) = 0.68$, $IC(hood) = 1.0$, $IC(cover) = 0.55$, $IC(object) = 0.08$ and the relation factor $TC_{has\text{-}part} = 0.4$. The weights of the three single-relation paths are:

---

[3] Proof by contradiction: Assume that in $T_{min}(path(x,y))$ with n sub-paths, $\exists i \in [1,n-1], X_i = X_{i+1}$. This would mean that $path_{X_i}(z_i,z_{i+1}) \oplus path_{X_{i+1}}(z_{i+1},z_{i+2})$ is a single-relation sub-path. Thus, there exists a factorization of $path(x,y)$ with $n-1$ single-relation sub-paths, which is smaller than the minimal factorization $T_{min}(path(x,y))$. This is a contradiction. Therefore, in $T_{min}(path(x,y))$, we have the property $\forall i \in [1,n-1], X_i \neq X_{i+1}$.

[4] The five IC values are real values computes with Seco algorithm [21] and WordNet 3.0.

- $W(\{jeep, is\text{-}a, car\}) = |IC(jeep) - IC(car)| = 0.32$
- $W(\{car, has\text{-}part, hood\}) = 0.4 \times \frac{1}{1+1} = 0.2$
- $W(\{hood, is\text{-}a, cover, is\text{-}a, object\}) = |IC(hood) - IC(object)| = 0.92$

And finally, we sum these weights and we obtain that the weight of this path from *jeep* to *object* is $W(path(jeep, object)) = 0.32 + 0.2 + 0.92 = 1.44$.

3.3   Final Measure

To compute the semantic distance between two concepts, we consider only the semantically correct paths between these two concepts and we will select the best one.

We chose to use the Hirst & St-Onge rules (see section 2.2) to filter the semantically correct paths. Let us consider two concepts c_1 and c_2. We note $\pi(c_1, c_2)$ the set of acyclic paths between c_1 and c_2 and $HSO : \pi(c_1, c_2) \longrightarrow \mathbb{B}$ the function such that $HSO(p)$ is *true* if and only if p is a valid path *w.r.t.* Hirst & St-Onge patterns. Our semantic distance between c_1 and c_2 is then defined as follows:

$$dist(c_1, c_2) = \min_{\{p \in \pi(c_1, c_2) | HSO(p) = true\}} W(p)$$

Let consider again the example *path(jeep, object)*. The path using the *part-of* relation gives a weight of 1.44. The hierarchical path between *jeep* and *object* is (figure 1) $\{jeep, is\text{-}a, car, is\text{-}a, wheeled\_vehicle, is\text{-}a, vehicle, is\text{-}a, object\}$ and its weight is $|IC(jeep) - IC(object)| = 0.92$. Thus, in this example, the relational path does not give more information than the hierarchical path and the final semantic score obtained is: $dist(jeep, object) = 0.92$. It is an expected result, since *object* is a very general concept and does not contain relations with other concepts. On the contrary, if we consider the distance from *jeep* to *hood* the relational path $\{jeep, is\text{-}a, car, has\text{-}part, hood\}$ correspond to a weight of 0.52 whereas the hierarchical path (using the *ccp object*) correspond to a weight of 1.50. In this last example, the relation path gives the final result: $dist(jeep, hood) = 0.52$.

Note that, *by construction*, the hierarchical path between c_1 and c_2 (corresponding to the semantic similarity measure) is a semantically correct path and has value $dist_{JC}(c_1, c_2)$.[5] Thus, since $dist(c_1, c_2)$ is the minimum value considering all semantically correct paths, $dist(c_1, c_2) \leq dist_{JC}(c_1, c_2)$. As a consequence, if IC is bounded between $[0, IC_{max}]$ (which is the case with the Seco formula [21], with $IC_{max} = 1.0$), the weight of a path can be bounded in $[0, 2 \times IC_{max}]$ ($2 \times IC_{max}$ being the maximal value of the Jiang & Conrath distance). Thus, our distance measure can, if necessary, be linearly transformed into a relatedness measure [1,23]:

$$rel(c_1, c_2) = 2 \times IC_{max} - dist(c_1, c_2)$$

4   Evaluation

The purpose of our evaluation is to show the relevancy of our hypothesis on weight and path-validity for semantic relatedness measure. In this section, we present the evaluation on two different benchmarks using WordNet relations and several values for TC_X.

---

[5] Our measure can be seen as a relatedness generalisation of the Jiang & Conrath similarity measure.

4.1   Implementation

The Miller & Charles test [26] is well-know for semantic measure evaluation (*e.g.* [3,23,22,1]). This test is composed of 30 couples of words. For each couple of words, a significantly number of persons had associated a value between 0 and 4 of "synonymy judgment". This gives us a vector of 30 semantic rates which can be used as a benchmark for semantic computation. Testing a semantic measure simply consists of computing the correlation factor (usually the Pearson-product moment correlation factor) between this Miller & Charles vector and the vector generated by the computer using the semantic measure.

However, since this test was defined for "synonymy judgment", its accuracy to test a semantic relatedness measure is not evident. Most of the chosen couples do not have any functional relationship and the subjects were explicitly asked to evaluate the *synonymy* between the words. Only a few couple (such as "journey-car" or "furnace-stove") appears to have a possible non-hierarchical relation. It is interesting to note that in known similarity evaluations (as [3]), these couples are the ones which mostly differ from the human reference. However, since "semantic similarity represents a special case of semantic relatedness", a semantic relatedness measure should at least work on the Miller & Charles test, even if it does not allow to validate the relational aspects of the measure. This is the reason why we first evaluate our measure on the Miller & Charles benchmark. The human reference for the Miller & Charles test was taken in [3].

Table 1. Pearson product-moment correlation factor for Rada, Resnik, Lin, Jiang & Conrath, Hirst & St-Onge and our approach ($TC_X = 0.4$)

Measures	Correlation	
	Miller & Charles	WordSimilarity-353
Rada	0.638	0.249
Resnik	0.804	0.375
Lin	0.836	0.377
Jiang & Conrath	0.880	0.362
Hirst & St-Onge	0.847	0.380
Our measure, $TC_X = 0.4$	0.902	0.400

Then, we needed another benchmark with to test the relational part of our measure. We have found the WordSimilarity-353[6] test [27], that was essentially constructed with couples of words which are relationally connected (*e.g.* "computer-keyboard", "telephone-communication", etc.). Classical similarity measures logically tend to fail this test (*i.e.* their correlation factors are very low). We can expect that our measure will outperform similarity measure, but we will see that the choice of the ontology can limit this impact.

To compute our semantic score, we will consider as an ontology the *noun* sub-part of the lexical base WordNet 3.0 [25], since this knowledge base is easily accessible and

---

[6] http://www.cs.technion.ac.il/~gabr/resources/data/wordsim353/
wordsim353.html

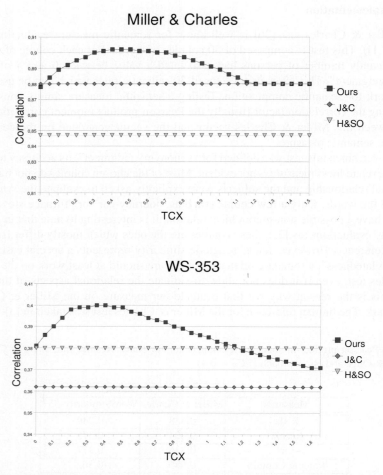

Fig. 3. Correlation factor evolution regarding to the TC_X factor in Miller & Charles and WS-353 test. The maximal value is obtained for $TC_X = 0.4$ in both test.

contains a large number of concepts. For the WS-353 test, we have removed 9 couples of words that did not exist in the *noun* part of WordNet (*e.g.* the couple "fighting-defeating"). Because of WordNet relation definition, we will consider only the *part-of* relations for non-hierarchical relations[7]. Moreover, we chose to consider only one fixed maximal weight TC_X for all the 6 *part-of* relations of WordNet.

Since we cannot anticipate the correct value for this $TC_{part-of}$, we evaluated all values from 0 (free non-hierarchical links) to 1.5 (very costly non-hierarchical links) by steps of 0.05.

---

[7] In WordNet, it is separated in three meronym relations (*meronym_member, meronym_part, meronym_substance*) and three holonym relations (*holonym_member, holonym_part, holonym_substance*).

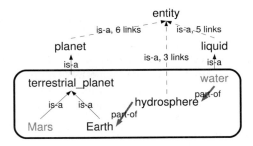

Fig. 4. Example of semantically bad path for Hirst & St-Onge patterns. From "water" to "Mars", the shortest path in a graph-based view is not semantically correct since the last link (*terrestrial_planet, includes, Mars*) is not authorised.

The information content *IC* of a given node is computed using the Seco formula [21], which does not require a corpus to give a good representation of *IC* and is bounded by [0, 1]. We consider for correlation comparison 4 classical similarity measure (Rada [2], Resnik [1], Lin [22] and Jiang & Conrath [23]) and one relatedness measure (Hirst & St-Onge, see section 2.2).[8]

The results are given in the table 1. The evolution of the correlation factor regarding to the value of TC_X is given in the figure 3.

4.2 Discussions

In both tests, our measure outperforms the correlation of the others measures. Moreover, to our knowledge, it is the first time that a semantic measure based on WordNet reaches a correlation of 0.4 for the WS-353 test (see Strube work [5] for the last known evaluation).

Since, the Miller & Charles test is based on similarity, the major part of couples use a hierarchy-only path and, thus, our result corresponds to the Jiang & Conrath result. However, some results are different with the use of non-hierarchical link. For instance, if we consider the "furnace-stove" couple, the common closest concept is "artifact", which makes a weak relation. Using relation, our measure was able to find the path "furnace *has-part* grate *part-of* stove" which uses the functional common property between the two concepts. This path is also identified by the Hirst & St-Onge measure, but the Hirst & St-Onge suffers from the uniformity hypothesis for the edge weight. Our measure takes advantages of the two paradigms: the semantically correct path for semantic relatedness and the information-theoretic approach to refine result.

Also note that no relevant link for the "journey-car" couple was found, neither by our measure nor by the Hirst & St-Onge one. This comes from the fact that the "common-sense relation" between these two concepts is not expressed in terms of meronym/holonym relations as defined in WordNet. This underlines the lack of relation types in WordNet, which is a limitation to computes semantic relatedness.

---

[8] To be able to compare correlation factors, we have recalculated them for all these measures with WordNet 3.0.

As expected, relatedness measures (Hirst & St-Onge and our) do better than similarity measure on the WS-353 test, since this test correspond to couples of words which are connected relationally. Relatedness measures can find relational functions between words (*e.g.* "keyboard-computer", etc.) whereas similarity measures only consider the hierarchy. In addition, the Hirst & St-Onge patterns permits to invalidate some path which are very shorter than the hierarchical, but incorrect in a semantic point of view (figure 4). However, some couples were not connected in WordNet and, thus, by our measure (*e.g.* "telephone-communication", etc.). This can be explained, again, by the lack of relations types in WordNet. Moreover, as stated by Strube [5], the WS-353 test contains many couples which are connected by common-sense link and that cannot be connected in WordNet (*e.g.* "popcorn-movie", etc.). Then, it explains why it is very difficult to obtain a good correlation using WordNet for this test. For this reason, we believe that it will be very difficult to go beyond the 0.35-0.4 limit on the WS-353 test using only WordNet as an ontology.

5   Conclusion and Future Work

In this paper, we have presented a new semantic measure to evaluate the semantic relatedness between two concepts. This measure makes use of the Hirst & St-Onge [8] patterns for semantically correct paths and the information-theoretic paradigm introduced by Resnik [1]. We have implemented our hypotheses on WordNet with the Miller & Charles [26] and the WS-353 test [27] and have shown that our measure outperforms known measures.

Note that by construction, our relatedness measure is always higher than the Jiang & Conrath similarity measure. Thus, we should always "fail" on couples with a J&C result higher than the human reference. However, one can hardly conceive an ontology that contains a relation between two concepts that are not associated from the human point of view. This would mean that this knowledge base would not be consistent with the domain. Moreover, if these links yet exists, we can study for a new weight allocation to invalidate it.

In addition, the evaluation underlines the lack of non-hierarchical relations in Word-Net, as first mentioned in [8]. For instance, in WordNet, there is no relational path between concepts like "journey-car" or "telephone-communication". This allows us to conclude that to use the capabilities of a semantic relatedness measure on ontologies, we need a *real* domain ontology. For this reason, our next aim is to test the impact of our measure on the performance rate of our semantic heterogeneity management system for multi-agent system [28]. Firstly, this measure will be used to enhance the alignment of the two agent's ontologies, and secondly will be used for semantic interpretation of requests exchanged. This kind of result will allow us to conclude for the scalability of our approach in different applications.

Another open issue is the allocation of the weight for each link type. In our evaluation, we used a single TC_X and the best results are obtained in both tests for $TC_X = 0.4$. However, there is no proof that this value will always lead to the best result, nor than having one single TC_X for all edges is appropriate. One idea could be to query a domain expert to fix an initial set of value the weights TC_X. The expert should know what kind of relation is important for semantic proximity. Since we consider the application of our semantic heterogeneity system in semantic interpretation of natural language commands (considering that an user is a special agent [29]), we think that the user feedback can be used as a background knowledge for a reinforcement learning algorithm [30,31] on the weight evolution. When the user confirms the system's interpretation of the command (*i.e.* the selected path in the ontology), the TC_X factor on the concerned edges will decrease. On the contrary, if the user denies the command, the TC_X factor will increase. This algorithm, currently in study, should be in charge to learn the optimal weight (or to accurate the expert weight) of the TC_X function, regarding to a given ontology.

Our final objective is to propose and to evaluate a measure of semantic similarity for more complex language of knowledge representation. For instance, Hau & al [15] has proposed a similarity measure based upon the Lin measure template [22] which considers OWL restriction, cardinality, intersection, etc. Alas, this measure was neither evaluated nor implemented and remains a theoretical proposition. We believe that, based on such work, it is however possible to extend our measure to model specific relations between concepts, such as intersection or disjunctive classes.

References

1. Resnik, P.: Using information content to evaluate semantic similarity in a taxonomy. In: 14th International Joint Conference on Artificial Intelligence (IJCAI 2005), pp. 448–453 (1995)
2. Rada, R., Mili, H., Bicknell, E., Blettner, M.: Development and Application of a Metric on Semantic Nets. IEEE Transactions on Systems, Man, and Cybernetics 19, 17–30 (1989)
3. Budanitsky, A., Hirst, G.: Evaluating wordnet-based measures of semantic distance. Computational Linguistics 32, 13–47 (2006)
4. Patwardhan, S., Pedersen, T.: Using WordNet Based Context Vectors to Estimate the Semantic Relatedness of Concepts. In: Proc. of the EACL 2006 Workshop Making Sense of Sense - Bringing Computational Linguistics and Psycholinguistics Together (2006)
5. Strube, M., Ponzetto, S.: WikiRelate! Computing semantic relatedness using Wikipedia. In: Proc. of AAAI, vol. 6, pp. 1419–1424 (2006)
6. Cilibrasi, R., Vitanyi, P.: Automatic Extraction of Meaning from the Web. In: Proc. IEEE International Symposium on Information Theory, pp. 2309–2313 (2006)
7. Iosif, E., Potamianos, A.: Unsupervised Semantic Similarity Computation usingWeb Search Engines. In: International Conference on Web Intelligence (WI), pp. 381–387. IEEE, Los Alamitos (2007)
8. Hirst, G., St-Onge, D.: Lexical chains as representation of context for the detection and correction malapropisms. In: Fellbaum, C. (ed.) WordNet: An Electronic Lexical Database, pp. 305–332. MIT Press, Cambridge (1998)
9. Corby, O., Dieng-Kuntz, R., Faron-Zucker, C.: Querying the Semantic Web with the CORESE search engine. In: Press, I. (ed.) Proc. of the ECAI 2004, Valencia, pp. 705–709 (2004)
10. Milward, D., Beveridge, M.: Ontology-based dialogue systems. In: Proc. 3rd Workshop on Knowledge and reasoning in practical dialogue systems (IJCAI 2003), pp. 9–18 (August 2003)
11. Dzikovska, M.O., Allen, J.F., Swift, M.D.: Integrating linguistic and domain knowledge for spoken dialogue systems in multiple domains. In: Proc. of IJCAI 2003 Workshop on Knowledge and Reasoning in Practical Dialogue Systems (2003)

12. Euzenat, J., Shvaiko, P.: Ontology matching. Springer, Heidelberg (DE) (2007)
13. Maedche, A., Staab, S.: Measuring similarity between ontologies. In: Gómez-Pérez, A., Benjamins, V.R. (eds.) EKAW 2002. LNCS (LNAI), vol. 2473, pp. 251–263. Springer, Heidelberg (2002)
14. Eliasson, K.: Case-Based Techniques Used for Dialogue Understanding and Planning in a Human-Robot Dialogue System. In: Proc. of IJCAI 2007, pp. 1600–1605 (2007)
15. Hau, J., Lee, W., Darlington, J.: A Semantic Similarity Measure for Semantic Web Services. In: Proc. Workshop on Web Service Semantics (2005)
16. Smith, M.K., Welty, C., McGuinness, D.L.: Owl web ontology language guide (February 2004), http://www.w3.org/TR/owl-guide/
17. Turney, P.: Similarity of Semantic Relations. Computational Linguistics 32(3), 379–416 (2006)
18. Sussna, M.: Word Sense Disambiguation for Free-text Indexing Using a Massive Semantic Network. In: Bhargava, B.K., Finin, T.W., Yesha, Y. (eds.) Proc. of the 2nd International Conference on Information and Knowledge Management (CIKM 1993), Washington, DC, USA, November, pp. 67–74. ACM, New York (1993)
19. Wu, Z., Palmer, M.: Verb semantics and lexical selection. In: 32nd. Annual Meeting of the Association for Computational Linguistics, New Mexico State University, Las Cruces, New Mexico, pp. 133–138 (1994)
20. Zhong, J., Zhu, H., Li, J., Yu, Y.: Conceptual graph matching for semantic search. In: ICCS 2002. Proceedings of the 10th International Conference on Conceptual Structures, London, UK, pp. 92–196. Springer, Heidelberg (2002)
21. Seco, N., Veale, T., Hayes, J.: An Intrinsic Information Content Metric for Semantic Similarity in WordNet. In: Proc. ECAI 2004, the 16th European Conference on Artificial Intelligence, pp. 1089–1090 (2004)
22. Lin, D.: An information-theoretic definition of similarity. In: Proc. 15th International Conf. on Machine Learning, pp. 296–304. Morgan Kaufmann, San Francisco, CA (1998)
23. Jiang, J., Conrath, D.: Semantic similarity based on corpus statistics and lexical taxonomy. In: Proc. on International Conference on Research in Computational Linguistics, Taiwan, pp. 19–33 (1997)
24. Aleksovski, Z., ten Kate, W., van Harmelen, F.: Exploiting the structure of background knowledge used in ontology matching. In: Proc. Workshop on Ontology Matching in ISWC 2006, CEUR Workshop Proceedings (2006)
25. Fellbaum, C. (ed.): WordNet, An Electronic Lexical Database. MIT Press, Cambridge (1998)
26. Miller, G., Charles, W.: Contextual correlates of semantic similarity. Language and Cognitive Processes 6(1), 1–28 (1991)
27. Finkelstein, L., Gabrilovich, E., Matias, Y., Rivlin, E., Solan, Z., Wolfman, G., Ruppin, E.: Placing search in context: the concept revisited. In: WWW 2001. Proceedings of the 10th international conference on World Wide Web, pp. 406–414. ACM Press, New York (2001)
28. Ferber, J.: Multi-Agent Systems: An Introduction to Distributed Artificial Intelligence. Addison-Wesley Longman Publishing Co., Inc., Boston, MA (1999)
29. Mazuel, L., Sabouret, N.: Generic command interpretation algorithms for conversational agents. Web Intelligence and Agent Systems 6(2) (April 2008)
30. Sutton, R.: Learning to predict by the methods of temporal differences. Machine Learning 3(1), 9–44 (1988)
31. Watkins, C., Dayan, P.: Q-learning. Machine Learning 8(3), 279–292 (1992)

Thesaurus-Based Search in Large Heterogeneous Collections

Jan Wielemaker[1], Michiel Hildebrand[2],
Jacco van Ossenbruggen[2], and Guus Schreiber[3]

[1] University of Amsterdam, Human Computer Studies (HCS), The Netherlands
[2] CWI Amsterdam, The Netherlands
[3] VU University Amsterdam, The Netherlands

Abstract. In cultural heritage, large virtual collections are coming into existence. Such collections contain heterogeneous sets of metadata and vocabulary concepts, originating from multiple sources. In the context of the E-Culture demonstrator we have shown earlier that such virtual collections can be effectively explored with keyword search and semantic clustering. In this paper we describe the design rationale of ClioPatria, an open-source system which provides APIs for scalable semantic graph search. The use of ClioPatria's search strategies is illustrated with a realistic use case: searching for "Picasso". We discuss details of scalable graph search, the required OWL reasoning functionalities and show why SPARQL queries are insufficient for solving the search problem.

1 Introduction

Traditionally, cultural heritage, image and video collections use proprietary database systems and often their own thesauri and controlled vocabularies to index their collection. Many institutions have made or are making (parts of) their collections available online. Once on the web, each institution, typically, provides access to their own collection. The cultural heritage community now has the ambition to integrate these isolated collections and create a potential source for many new inter-collection relationships. New relations may emerge between objects from different collections, through shared metadata or through relations between the thesauri.

The MultimediaN E-culture project[1] explores the usability of semantic web technology to integrate and access museum data in a way that is comparable to the MuseumFinland project [1]. We focus on providing two types of end-user functionality on top of heterogeneous data with weak domain semantics. First, keyword search, as it has become the de-facto standard to access data on the web. Secondly, thesaurus-based annotation for professionals as well as amateurs.

This document is organised as follows. In Sect. 2 we first take a closer look at our data and describe our requirements by means of a use case. In section Sect. 3 we take a closer look a search and what components are required to

---

[1] http://e-culture.multimedian.nl

A. Sheth et al. (Eds.): ISWC 2008, LNCS 5318, pp. 695–708, 2008.

realise keyword search in a large RDF graph. The ClioPatria infrastructure is described in section Sect. 4, together with some illustrations on how ClioPatria can be used. We conclude the paper with a discussion where we position our work in the Semantic Web community.

2   Materials and Use Cases

Metadata and vocabularies. In our case study we collected descriptions of 200,000 objects from six collections annotated with six established thesauri and several proprietary controlled keyword lists, which adds up to 20 million triples. We assume this material is representative for the described domain. Using semantic web technology, it is possible to unify the data while preserving its richness. The procedure is described elsewhere [2] and summarised here.[2]

The MultimediaN E-Culture demonstrator harvests metadata and vocabularies, but assumes the collection owner provides a link to the actual data object, typically an image of a work such as a painting, a sculpture or a book. When integrating a new collection into the demonstrator we typically receive one or more XML/database dumps containing the metadata and vocabularies of the collection. Thesauri are translated into RDF/OWL, where appropriate with the help of the W3C SKOS format for publishing vocabularies [3]. The metadata is transformed in a merely syntactic fashion to RDF/OWL triples, thus preserving the original structure and terminology. Next, the metadata schema is mapped to VRA[3], a specialisation of Dublin Core for visual resources. This mapping is realized using the 'dumb-down' principle by means of `rdfs:subPropertyOf` and `rdfs:subClassOf` relations. Subsequently, the metadata goes through an enrichment process in which we process plain-text metadata fields to find matching concepts from thesauri already in the demonstrator. For example, if the `dc:creator` field contains the string *Pablo Picasso*, than we will add the concept `ulan:500009666` from ULAN[4] to the metadata. Most enrichment concerns named entities (people, places) and materials. Finally, the thesauri are aligned using `owl:sameAs` and `skos:exactMatch` relations. For example, the art style *Edo* from a local ethnographic collection was mapped to the same art style in AAT[5] (see the use cases for an example why such mappings are useful). Our current database (April 2008) contains 38,508 `owl:sameAs` and 9,635 `skos:exactMatch` triples and these numbers are growing rapidly.

After this harvesting process we have a graph representing a connected network of works and thesaurus lemmas that provide background knowledge. VRA and SKOS provide —weak— structure and semantics. Underneath, the richness of the original data is still preserved. The data contains many relations that are not covered by VRA or SKOS, such as relations between artists (e.g. ULAN `teacherOf` relations) and between artists and art styles (e.g. relations between

---

[2] The software can be found at `http://sourceforge.net/projects/annocultor`
[3] Visual Resource Association, http://www.vraweb.org/projects/vracore4/
[4] Union List of Artist Names is a thesaurus of the Getty foundation.
[5] Art & Architecture Thesaurus, another Getty thesaurus.

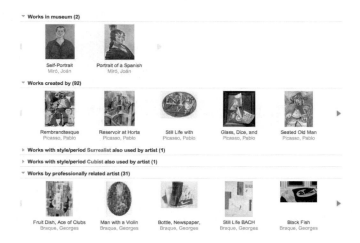

Fig. 1. Clustered result searching "picasso"

AAT art styles and ULAN artists [4]). These relations are covered by their original schema. Their diversity and lack of defined semantics make it hard to map them to existing ontologies and provide reasoning based on this mapping.

Use cases. Assume a user is typing in the query "picasso". Despite the name *Picasso* is a reasonably unique in the art world, the user may still have many different intentions with this simple query: a painting by Picasso, a painting of Picasso, the styles Picasso has worked in? Without an elaborate disambiguation process it is impossible to tell in advance.

Fig. 1 show part of the results of this query in the MultimediaN demonstrator. We see several clusters of search results. The first cluster contains works from the Picasso Museum, the second cluster contains works by Pablo Picasso (only first five hits shown; clicking on the arrow allows the user to inspect all results); clusters of surrealist and cubist paintings (styles that Picasso worked in; not shown for space reasons), and works by George Braque (a prominent fellow Cubist painter, but the works shown are not necessarily cubist). Other clusters include works made from *picasso marble* and works with *Picasso* in the title (includes two self portraits). The basic idea is that we are aiming to create clusters of related objects such that the user can afterwards choose herself what she is interested in. We have found that even in relatively small collections of 100K objects, users discover interesting results they did not expect. We have termed this type of search tentatively 'post-query disambiguation': in response to a simple keyword query the user gets (in contrast to, for example, Google image search) semantically-grouped results that enable further detailing of the query. It should be pointed out that the knowledge richness of the cultural heritage domain allows this approach to work. In less rich domains such an approach is less likely to provide added value. Notably typed resources and relations give meaning to the path linking a literal to a target object.

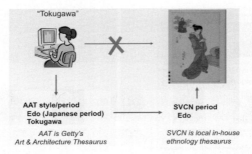

Fig. 2. Explore alignment to find Edo painting from "tokugawa"

Another typical use case for search concerns the exploitation of vocabulary alignments. The Holy Grail of the unified cultural-heritage thesaurus does not exist and many collection owners have their own home-grown variants. Consider the situation in Fig. 2, which is based on real-life data. A user is searching for "tokugawa". This Japanese term has actually two major meanings in the heritage domain: it is the name of a 19th century shogun and it is a synonym for the Edo style period. Assume for a moment that the user is interested in finding works of the latter type. The Dutch ethnographic museum in Leiden actually has works in this style in its digital collection, such as the work shown in the top-right corner. However, the Dutch ethnographic thesaurus SVCN, which is being used by the museum for indexing purposes, only contains the label "Edo" for this style. Fortunately, another thesaurus in our collection, the aforementioned AAT, does contain the same concept with the alternative label "Tokugawa". In the harvesting process we learned this equivalence link (quite straightforward: both are Japanese styles with matching preferred labels). The objective of our graph search is to enable to make such matches.

Although this is actually an almost trivial alignment, it is still extremely useful. The cultural-heritage world (like any knowledge rich domain) is full of such small local terminology differences. Multilingual differences should also be taken into consideration here. If semantic-web technologies can help making such matches, there is a definite added value for users.

3   Required Methods and Components

In this section we study the methods and components we need to realise the keyword search described above. Our experiments indicate that meaningful matches between keyword and target often involve chains up to about five relations. At this distance there is a potentially huge set of possible targets. The targets can be organised by rating based on semantics or statistics and by clustering based on the graph pattern linking a literal to the target. We discuss three possible approaches: querying using a fixed set of graph patterns, completely unconstrained graph search and best-first exploration of the graph.

Using a set of fixed queries. A cluster as shown in Fig. 1 is naturally represented as a graph pattern as found in many semantic web query languages. If we can enumerate all possible meaningful patterns of properties that link literals to targets, we reduce the search process to finding instances of all these graph patterns. This would be a typical approach in Semantic Web applications such as DBin [5]. This approach is, however, not feasible for highly heterogenous data sets. Our current data contains over 600 properties, most of which do not have a very well defined meaning (e.g. `detailOf`, `cooperatedWith`, `usesStyle`). If we combine this with our observation that is it quite common to find valuable results at 4 or even 5 steps from the initial keywords, we have to evaluate a very large number of possible patterns. To a domain expert, it is obvious that the combination of `cooperatedWith` and `hasStyle` can be meaningful while "*A* died in *P*, where *B* was born" is generally meaningless, but the set of possible combinations to consider is very large. Automatic rating of this type of relation pattern is, as far as we know, not feasible. Even if the above is possible, new collections and vocabularies often come with new properties, which must all be considered in combination to the already created patterns.

Using graph exploration. Another approach is to explore the graph, looking for targets that have, often indirectly, a property with matching literal. This implies we search the graph from *Object* to *Subject* over arbitrary properties, including triples entailed by `owl:inverseOf` and `owl:SymmetricProperty`. We examine the scalability issues using unconstrained graph patterns, after which we examine an iterative approach.

Considering a triple store that provides reasoning over `owl:inverseOf` and `owl:SymmetricProperty` it is easy to express an arbitrary path from a literal to a target object with a fixed length. The total result set can be expressed as a union of all patterns of fixed length up to (say) distance 5. Table 1 provides the statistics for some typical keywords at distances 3 and 5. The table shows total visited and unique results for both visited nodes and targets found which indicates that the graph contains a large number of alternative paths and the implementation must deal with these during the graph exploration to reduce the amount of work. Even without considering the required post-processing to rank and cluster the results it is clear that we cannot obtain interactive response times (of at most a few seconds) using this approach.

Fortunately, a query system that aims at human users only needs to produce the most promising results. This can be achieved by introducing a distance measure and doing *best-first* search until our resources are exhausted *(anytime algorithm)* or we have a sufficient number of results. The details of the distance measure are still subject of research [6], but not considered vital to the architectural arguments in this article. The complete search and clustering algorithm is given in Fig. 3. In our experience, the main loop requires about 1,000 iterations to obtain a reasonable set of results, which leads to acceptable performance when the loop is pushed down to the triple store layer.

Table 1. Statistics for exploring the search graph for exactly *Distance* steps (triples) from a set of literals matching *Keyword*. *Literals* is the number of literals holding a word with the same stem as *Keyword*; *Nodes* is the number of nodes explored and *Targets* is the number of target objects found. *Time* is on an Intel Core duo X6800.

Keyword	Distance	Literals	Nodes Visited	Unique	Targets Visited	Unique	Time (sec.)
tokugawa	3	21	1,346	1,228	913	898	0.02
steen	3	1,070	21,974	7,897	11,305	3,658	0.59
picasso	3	85	9,703	2,399	2,626	464	0.26
rembrandt	3	720	189,611	9,501	141,929	4,292	3.83
impressionism	3	45	7,142	2,573	3,003	1,047	0.13
amsterdam	3	6,853	1,327,797	421,304	681,055	142,723	39.77
tokugawa	5	21	11,382	2,432	7,407	995	0.42
steen	5	1,070	1,068,045	54,355	645,779	32,418	19.42
picasso	5	85	919,231	34,060	228,019	6,911	18.76
rembrandt	5	720	16,644,356	65,508	12,433,448	34,941	261.39
impressionism	5	45	868,941	50,208	256,587	11,668	18.50
amsterdam	5	6,853	37,578,731	512,027	23,817,630	164,763	620.82

Term search. The combination of best-first graph exploration with semantic clustering, as described above, works well for 'post-query' disambiguation of results in exploratory search tasks. It is, however, less suited for quickly selecting a known thesaurus term. The latter is often needed in semantic annotation and 'pre-query' disambiguation search tasks. For such tasks we rely on the proven *autocompletion* technique, which allows us to quickly find resources related to the prefix of a label or a word inside a label, organise the results (e.g. organise cities by country) and provide sufficient context (e.g. date of birth and death of a person). Often results can be limited to a sub-hierarchy of a thesaurus, expressed as an extra constraint using the transitive skos:broader property. Although the exact technique differs, the technical requirements to realise this type of search is similar to the keyword search described above.

Literal matching. Similar to document retrieval, we start our search from a rated list of literals that contain words with the same stem as the searched keyword. Unlike document retrieval systems such as Swoogle [7] or Sindice [8], we are not interested in which RDF documents contain the matching literals, but which semantically related target concepts are connected to them. Note that term search as described above requires finding literals from the prefix of a contained word that is sufficiently fast to be usable in autocompletion interfaces (see also [9]).

Using SPARQL. If possible, we would like our search software to connect to an arbitrary SPARQL endpoint. Considering the *fixed query* approach, each pattern is naturally mapped onto a SPARQL graph pattern. *Unconstrained graph search* is easily expressed too. Expressed as a CONSTRUCT query, the query engine can return a minimal graph without duplicate paths.

Unfortunately, both approaches proved to be infeasible implementation strategies. The best-first graph exploration requires one (trivial) SPARQL query to

1. Find literals that contain the same stem as the keywords, rate them on minimal edit distance (short literal) or frequency (long literal) and sort them on the rating to form the initial *agenda*
2. Until satisfied or empty *agenda*, do
   (a) Take highest ranked value from *agenda* as O. Find **rdf**(S,P,O) terms. Rank the found S on the ranking of O, depending on P. If P is a subProperty of owl:sameAs, the ranking of S is the same as O. If S is already in the result set, combine their values using $R = 1 - ((1 - R_1) \times (1 - R_2))$. If S is new, insert it into *agenda*, else reschedule it in the agenda.
   (b) If S is a target, add it to the *targets*. Note that we must consider **rdf**(O,IP,S) if there is an **inverseOf**(P,IP) or P is symmetric.
3. Prune resulting graph from branches that do not end in a target.
4. Smush resources linked by owl:sameAs, keeping the most linked resource.
5. Cluster the results
   (a) Abstract all properties to their VRA or SKOS root property (if possible).
   (b) Abstract resources to their class, except for instances of skos:Concept and the top-10 ranked instances.
   (c) Place all triples in the abstract graph. Form (RDF) Bags of resources that match to an abstracted resource and use the lowest common ancestor for multiple properties linking two bags of resources.
6. Complete the nodes in the graph with label information for proper presentation.

Fig. 3. Best first graph search and clustering algorithm

find the neighbours of the next node in the *agenda* for each iteration to update the agenda and to decide on the next node to explore. Latency and volume of data transfer make this infeasible when using a remote triple store.

The reasoning for clustering based on the property hierarchy cannot be expressed in SPARQL, but given the size and stability of the property hierarchy we can transfer the entire hierarchy to the client and use local reasoning. After obtaining the clustered results, the results need to be enriched with domain specific key information (title and creator) before they can be presented to the user. Requesting the same information from a large collection of resources can be realised using a rather inelegant query as illustrated below.

```
SELECT ?l1 ?l2 ...
WHERE { { ulan:artists1 rdfs:label ?l1 } UNION
        { ulan:artists2 rdfs:label ?l2 } UNION
        ...
```

Regular expression literal matching cannot support match on stem. Prefix and case insensitive search for contained word can be expressed. Ignoring diacritic marks during matching as generally needed for multi-script matching is not supported by the SPARQL regular expression syntax.

We conclude that remote access is inadequate for adaptive graph exploration and SPARQL is incapable of expressing lowest common parent problems and relevant literal operations and impractical for enriching computed result sets.

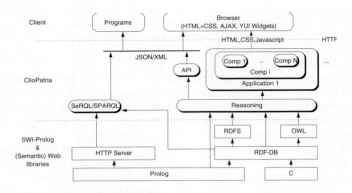

Fig. 4. Overall architecture of the ClioPatria server

Summary of requirements for search.

- Obtain rated list of literals from stem and prefix of contained words.
- Entailment over `owl:inverseOf` and `owl:SymmetricProperty`.
- Entailment over `owl:TransitiveProperty` to limit the domain of term search.
- Entailment over `owl:sameAs` for term search.
- Graph exploration requires tight connection to the triple store.
- Reasoning with types as well as the class, concept and property hierarchy. This includes finding the lowest common parent of a set of resources.

4   The ClioPatria Search and Annotation Toolkit

We have realised the functionality described in the previous section on top of the SWI-Prolog[6] web and semantic web libraries [10,11]. This platform provides a scalable in-core RDF triple store [12] and a multi-threaded HTTP server library. ClioPatria[7] is the name of the reusable core of the E-culture demonstrator, the architecture of which is illustrated in Fig. 4. First, we summarise some of the main features of ClioPatria.

- Running on a Intel core duo X6800@2.93GHz, 8GB, 64-bit Linux it takes 120 seconds elapsed time to load the 20 million triples. The server requires 4.3Gb memory for 20 million triples (2.3Gb in 32-bit mode). Time and space requirements grow practically linear in the amount of triples.
- The store provides safe persistency and maintenance of provenance and change history based on a (documented) proprietary file format.
- Deleting and modifying triples complicates maintenance of the pre-computed entailment. Therefore, reasoning is as much as possible based on backward chaining, which fits with Prolog's search resolution.

---

[6] http://www.swi-prolog.org
[7] Open source from http://e-culture.multimedian.nl/software.html

4.1 Client-Server Architecture

In contrast to client-only architectures such as Simile's Exhibit [13], ClioPatria has a client-server architecture. The core functionality is provided as HTTP APIs by the server. The results are served as presentation neutral data objects using the JSON[8] serialization and can thus be combined with different presentation and interaction strategies. Within ClioPatria, the APIs are used by its web applications. In addition, the APIs can be used by third party applications to create mashups. The ClioPatria toolkit contains web applications for search and annotation. The end-user applications are a combination of server side generated HTML and client side JavaScript interface widgets.

In the MultimediaN E-Culture demonstrator[9] ClioPatria's web application for search and annotation are used. The K-Space European Network of Excellence is using ClioPatria to search news[10]. At the time of writing Europeana[11] is setting up ClioPatria as a demonstrator to provide multilingual access to a large collection of very divers cultural heritage data. The ClioPatria API provided by the E-Culture Project is also used by the CATCH/CHIP project Tour Wizard that won the 3rd prize at the Semantic Web Challenge of 2007. For the semantic search functionality CHIP uses the web services provided by the ClioPatria API.

4.2 Output Formats

Server side we have two types of presentation oriented output routines. *Components* are Prolog grammar rules that define reusable parts of a page. Components can embed each other. *Applications* produce an entire HTML page that largely consists of configured components. Applications automatically add the required CSS and JavaScript based on dependency declarations.

Client side presentation and interaction is built on top of YUI JavaScript widget library.[12] ClioPatria contains widgets for autocompletion, a search result viewer, a detailed view on a single resource, and widgets for semantic annotation fields. The result viewer can visualise data in thumbnail clusters, a geographical map, Simile Exhibit, Simile Timeline and a Graphviz[13] graph visualisation.

4.3 APIs

ClioPatria provides programmatic access to the RDF data through HTTP. http://e-culture.multimedian.nl/demo/doc/ The query API provides standardized access to the data via the SeRQL and SPARQL. As we have shown in Sect. 3 such a standard query API is not sufficient to provide the intended keyword search functionality. Therefore, ClioPatria provides an additional search API for

---

[8] http://www.json.org
[9] http://e-culture.multimedian.nl/demo/search
[10] http://newsml.cwi.nl/explore/search
[11] http://www.europeana.eu/
[12] http://developer.yahoo.com/yui/
[13] http://www.graphviz.org/

keyword-based access to the RDF data. In addition, ClioPatria provides APIs to get resource-specific information, update the triple store and cache media items. This paper only discusses the query and search API in more detail.

Query API. The SeRQL/SPARQL library provides a semantic web query interface that is compatible to Sesame [14] and provides open and standardised access to the RDF data stored in ClioPatria. Both SeRQL and SPARQL are translated into a Prolog query that relies on the **rdf**(S,P,O) predicate provided by the RDF store and on auxiliary predicates that realise functions and filters defined by SeRQL and SPARQL. Conjunctions of **rdf/3** statements and filter expressions are optimised through reordering based on statistical information provided by the store [15].

Search API. The search API provides services for graph search (Fig. 3) and term search (Sect. 3). Both services return their result as a JSON object (using the serialisation for SPARQL SELECT queries [16]). Both services can be configured with several parameters. General search API parameters are:

- **query**(*string* | *URI*): the search query.
- **filter**(false | *Filter*): constrains the results to match a combination of *Filter* primitives, typically OWL class descriptions that limit the results to instances that satisfy these descriptions. Additional syntax restricts results to resources used as values of properties of instances of a specific class.
- **groupBy**(false | path | *Property*): if path, cluster results by the abstracted path linking query to target. If a property is given, group the result by the value on the given property.
- **sort**(path_length | score | *Property*): Sort the results on path-length, semantic distance or the value of *Property*.
- **info**(false | *PropertyList*): augment the result with the given properties and their values. Examples are skos:prefLabel, foaf:depicts and dc:creator.
- **view**(thumbnail | map | timeline | graph | exhibit): shorthands for specific property lists of info.
- **sameas**(*Boolean*): smushes equivalent resources, as defined by owl:sameAs or skos:exactMatch into a single resource.

Consider the use case discussed in Sect. 2. Clustered results that are semantically related to keyword "picasso" can be retrieved through the graph search API with this HTTP request:

```
/api/search?query=picasso&filter=vra:Work&groupBy=path&view=thumbnail
```

Parameters specific to the graph search API are:

- **abstract**(*Boolean*): enables the abstraction of the graph search paths over rdfs:subClassOf and rdfs:subPropertyOf, reducing the number of clusters.
- **bagify**(*Boolean*): puts (abstracted) resources of the same class with the same (abstracted) relations to the rest of the graph in an RDF bag. I.e. convert a set of triples linking a painter over various sub properties of dc:creator to multiple instances of vra:Work, into an RDF bag of works and a single triple linking the painter as dc:creator to this bag.

- **steps**(*Integer*): limits graph exploration to expand less than *Integer* nodes.
- **threshold**(*0.0..1.0*): cuts off the graph exploration on semantic distance.

For annotation we can use the term search API to suggest terms for a particular annotation field. For example, suppose a user has typed the prefix "pari" in a location annotation field that only allows European locations. We can request matching suggestions by using the URI below, filtering the results to resources that can be reached from `tgn:Europe` using `skos:broader` transitively:

```
/api/autocomplete?query=pari&match=prefix&sort=rdfs:label&
    filter={"reachable":{"relation":"skos:broader","value":"tgn:Europe"}}
```

Parameters specific to the term search API are:

- **match**(`prefix`|`stem`|`exact`): defines how the syntactic matching of literals is performed. Autocompletion, for example, requires `prefix` match.
- **property**(*Property, 0.0..1.0*): is a list of RDF property-score pairs which define the values that are used for literal matching. The score indicates preference of the used literal in case a URI is found by multiple labels. Typically preferred labels are chosen before alternative labels.
- **preferred**(`skos:inScheme`, *URI*): if URIs are smushed the information of the URI from the preferred thesaurus is used for augmentation and organisation.
- **compound**(*Boolean*): if `true`, filter results to those where the query matches the information returned by the `info` parameter. For example, a compound query *paris, texas* can be matched in two parts against a) the label of the place *Paris* and b) the label of the state in which *Paris* is located.

5   Discussion and Conclusion

In this paper we analysed the requirements for searching in large, heterogeneous collections with rich, but formally ill-defined semantics. We presented the ClioPatria software architecture we used to explore this topic. Three characteristics of ClioPatria have proved to be a frequent source of discussion: the non-standard API, the central in-core triple store model and the lack of full OWL DL support.

API standardisation. First, ClioPatria's architecture is based on various client-side JavaScript Web applications around a server-side Prolog-based reasoning engine and triple store. As discussed in this paper, the server functionality required by the Web clients extends that of an off-the-shelf SPARQL endpoint. This makes it hard for Semantic Web developers of other projects to deploy our Web applications on top of their own SPARQL-based triple stores. We acknowledge the need for standardized APIs in this area. We hope that the requirements discussed in this paper provide a good starting point to develop the next generation Semantic Web APIs that go beyond the traditional database-like query functionality currently supported by SPARQL.

Central, in-core storage model. From a data-storage perspective, the current ClioPatria architecture assumes images and other annotated resources to reside on the Web. All metadata being searched, however, is assumed to reside in-core in a central, server-side triple store. We are currently using this setup with a 20M triples dataset, and are confident our current approach will easily scale up to 150M triples on modern hardware (32Gb core). Our central in-core model will not scale, however, to the multi-billion triple sets supported by other state-of-the-art triple stores. For future work, we are planning to investigate to what extent we can move to disk-based or, given the distributed nature of the organisations in our domain, distributed storage strategies without giving up the key search functionalities of our current implementation. Distribution of the entire RDF graph is non-trivial. For example, in the keyword search, the paths in the RDF graph from the matching literals to the target resources tend to be unpredictable, varying highly with the types of the resources associated with the matching literals and the type of the target resources. Implementing a fast, semi-random graph walk in a distributed fashion will likely be a significant challenge. As another example, interface components such as a Web-based autocompletion Widget are based on the assumption that a client Web-application may request autocompletion suggestions from a single server, with response times in the 200ms range. Realizing sufficiently fast responses from this server without the server having a local index of all literals that are potential suggestion candidates will also be challenging. Distributing carefully selected parts of the RDF graph, however, could be a more promising option. In our current datasets for example, the subgraphs with geographical information are both huge and connected to the rest of the graph in a limited and predictable fashion. Shipping such graphs to dedicated servers might be doable with only minor modifications to the search algorithms performed by the main server.

Lack of OWL reasoning. From a reasoning perspective, ClioPatria does not provide traditional OWL DL support. First of all, the heterogeneous and open nature of our metadata repositories ensures that even when the individual data files loaded are in OWL DL, their combination will most likely not be. Typical DL violations in this domain are properties being used as a data property with name strings in one collection, and as an object property with URIs pointing to a biographical thesaurus such as ULAN in the other; or `rdfs:label` properties being used as an annotation property in the schema of one collection and as a data property on the instances of another collection. We believe that OWL DL is a powerful and expressive subset of OWL for closed domains where all data is controlled by a single organisation. It has proved, however, to be unrealistic to use OWL DL for our open, heterogenous Semantic Web application where multiple organisations can independently contribute to the data set.

Secondly, our application requirements require the triple store to be able to flexibly turn on and off certain types of OWL reasoning on a per-query basis. For example, there are multiple URIs in our dataset, from different data sources, representing the Dutch painter *Rembrandt van Rijn*. Ideally, our vocabulary mapping tools have detected this and have all these URIs mapped to one another using `owl:sameAs`. For an end-user interested in viewing all information available on Rembrandt, it is likely beneficial to have the system perform `owl:sameAs` reasoning and present all information related to Rembrandt in a single interface, smushing all different URIs onto one. However, an expert end-user annotating an artwork being painted by Rembrandt will, when selecting the corresponding entry from a biographical thesaurus, be interested into which vocabulary source the URI of the selected concept is pointing, and will also be interested in the other vocabularies define entries about Rembrandt, and how the different entries differ. This requires the system to largely ignore the traditional `owl:sameAs` semantics, present all triples associated with the different URIs separately, along with the associated provenance information. This type of ad-hoc turning on and off of specific OWL reasoning is not supported by most off-the-shelf SPARQL endpoints, but crucial in all realistic multi-thesauri semantic web applications.

Thirdly, we found that our application requirements seldomly rely on extensive subsumption or other typical OWL reasoning. In the weighted graph exploration we basically only consider the graph structure and ignore most of the underlying semantics, with only a few notable exceptions. Results are improved by assigning equivalence relations such as `owl:sameAs` and `skos:exactMatch` the highest weight of 1.0. We search the graph in only one direction, the exception being properties being declared as an `owl:SymmetricProperty`. In case of properties having an `owl:inverseOf`, we traverse the graph as we would have if all "virtual" inverse triples were materialised. Finally, we use a simple form of subsumption reasoning over the property and class hierarchy when presenting results to abstract from the many small differences in the schemas underlying the different search results.

Conclusion. Our conclusion is that knowledge rich domains such as cultural heritage fit well with Semantic Web technology. This is because of a) the clear practical needs this domain has for integrating information from heterogeneous sources, and b) its long tradition with semantic annotations using controlled vocabularies and thesauri. We strongly feel that studying the real application needs of users working in such domains in terms of their search and reasoning requirements will move ahead the state of the art in Semantic Web research significantly.

References

1. Hyvönen, E., Junnila, M., Kettula, S., Mäkelä, E., Saarela, S., Salminen, M., Syreeni, A., Valo, A., Viljanen, K.: MuseumFinland — Finnish museums on the semantic web. Journal of Web Semantics 3, 224–241 (2005)
2. Tordai, A., Omelayenko, B., Schreiber, G.: Semantic excavation of the city of books. In: Proc. Semantic Authoring, Annotation and Knowledge Markup Workshop (SAAKM 2007), CEUR-WS, vol. 289, pp. 39–46 (2007), http://ceur-ws.org/Vol-289
3. Miles, A., Bechhofer, S.: SKOS simple knowledge organization system reference. W3C working draft, World-Wide Web Consortium, Latest version (2008), http://www.w3.org/TR/skos-reference
4. de Boer, V., van Someren, M., Wielinga, B.J.: A redundancy-based method for the extraction of relation instances from the web. International Journal of Human-Computer Studies 65, 816–831 (2007)

708   J. Wielemaker et al.

5. Tummarello, G., Morbidoni, C., Nucci, M.: Enabling Semantic Web communities with DBin: an overview. In: Cruz, I., Decker, S., Allemang, D., Preist, C., Schwabe, D., Mika, P., Uschold, M., Aroyo, L.M. (eds.) ISWC 2006. LNCS, vol. 4273, Springer, Heidelberg (2006)
6. Rocha, C., Schwabe, D., de Aragao, M.: A hybrid approach for searching in the semantic web. In: Proceedings of the 13th International World Wide Web Conference, New York, USA, pp. 374–383 (2004)
7. Ding, L., Finin, T., Joshi, A., Pan, R., Cost, R.S., Peng, Y., Reddivari, P., Doshi, V.C., Sachs, J.: Swoogle: A Search and Metadata Engine for the Semantic Web. In: Proceedings of the Thirteenth ACM Conference on Information and Knowledge Management, Washington, D.C., USA, pp. 652–659 (2004)
8. Tummarello, G., Oren, E., Delbru, R.: Sindice.com: Weaving the open linked data. In: Aberer, K., Choi, K.-S., Noy, N., Allemang, D., Lee, K.-I., Nixon, L., Golbeck, J., Mika, P., Maynard, D., Mizoguchi, R., Schreiber, G., Cudré-Mauroux, P. (eds.) ASWC 2007 and ISWC 2007. LNCS, vol. 4825, pp. 547–560. Springer, Heidelberg (2007)
9. Bast, H., Weber, I.: The CompleteSearch Engine: Interactive, Efficient, and towards IR&DB Integration. In: CIDR 2007, Third Biennial Conference on Innovative Data Systems Research, Asilomar, CA, USA, pp. 88–95 (2007)
10. Wielemaker, J., Huang, Z., van der Meij, L.: SWI-Prolog and the web. Theory and Practice of Logic Programming 8, 363–392 (2008) (accepted for publication)
11. Wielemaker, J., Hildebrand, M., van Ossenbruggen, J.: Using Prolog as the fundament for applications on the semantic web. In: Heymans, S., et al. (eds.) Proceedings of ALPSWS 2007, pp. 84–98 (2007)
12. Wielemaker, J., Schreiber, G., Wielinga, B.: Prolog-based infrastructure for RDF: performance and scalability. In: Fensel, D., Sycara, K.P., Mylopoulos, J. (eds.) ISWC 2003. LNCS, vol. 2870, pp. 644–658. Springer, Heidelberg (2003)
13. Huynh, D., Karger, D., Miller, R.: Exhibit: Lightweight structured data publishing. In: 16th International World Wide Web Conference, Banff, Alberta, Canada, ACM Press, New York (2007)
14. Broekstra, J., Kampman, A., van Harmelen, F.: Sesame: A generic architecture for storing and querying rdf and rdf schema. In: Horrocks, I., Hendler, J. (eds.) ISWC 2002. LNCS, vol. 2342, pp. 54–68. Springer, Heidelberg (2002)
15. Wielemaker, J.: An optimised semantic web query language implementation in prolog. In: Gabbrielli, M., Gupta, G. (eds.) ICLP 2005. LNCS, vol. 3668, pp. 128–142. Springer, Heidelberg (2005)
16. Clark, K.G., Feigenbaum, L., Torres, E.: Serializing sparql query results in json W3C Working Group Note 18 June 2007 (2007)

Deploying Semantic Web Technologies for Work Integrated Learning in Industry - A Comparison: SME vs. Large Sized Company

Conny Christl[1], Chiara Ghidini[2] Joanna Guss[3], Stefanie Lindstaedt[4],
Viktoria Pammer[4], Marco Rospocher[2], Peter Scheir[4], and Luciano Serafini[2]

[1] ISN-Innovation Service Network, Zg. Hajdina 159, 2250 Ptuj, Slovenia
conny.christl@innovation.at
[2] FBK-irst. Via Sommarive, 18. 38100 Trento, Italy
{ghidini,rospocher,serafini}@fbk.eu
[3] EADS France - Innovation Works. 12, Rue Pasteur - BP76 - 92150 Suresnes, France
joanna.guss@eads.net
[4] Know-Center and Graz University of Technology, Knowledge Management Institute
Inffeldgasse 21a, 8010 Graz, Austria
{slind,vpammer,pscheir}@know-center.at

Abstract. Modern businesses operate in a rapidly changing environment. Continuous learning is an essential ingredient in order to stay competitive in such environments. The APOSDLE system utilizes semantic web technologies to create a generic system for supporting knowledge workers in different domains to learnwork. Since APOSDLE relies on three interconnected semantic models to achieve this goal, the question on how to efficiently create high-quality semantic models has become one of the major research challenges. On the basis of two concrete examples-namely deployment of such a learning system at EADS, a large corporation, and deployment at ISN, a network of SMEs-we report in detail the issues a company has to face, when it wants to deploy a modern learning environment relying on semantic web technology.

1 Introduction

In the past years we observed a slow but steady uptake of semantic web technologies in businesses. Increasingly search capabilities, data integration and web service communication enabled by semantic web technologies lead to improved business processes, savings in cost and time and heightened efficiency and competitiveness. However, obtaining the needed semantic models has remained a challenge and an art. This challenge is aggravated in situations where not only one model e.g. an ontology, is needed, but a whole network of models needs to be created and later maintained. This contribution reports about the experiences gained during the creation of three interlinked models (process, domain and learning goal model) at two application organizations (EADS and ISN).

A. Sheth et al. (Eds.): ISWC 2008, LNCS 5318, pp. 709–722, 2008.

These experiences were obtained in the context of the EU-funded integrated project APOSDLE[1] (Advanced Process-Oriented Self-Directed Learning Environment). APOSDLE aims at improving the productivity of knowledge workers by providing learning support during work task execution within the everyday work environment of the user, and by utilizing general organizational resources (documents as well as people) for collaborative learning.

This new learning paradigm of technology-enhanced work-integrated or organizational learning (see [1] for possible scenarios) puts one requirement on technology in the center of attention: Flexibility. In contrast to traditional e-learning systems, it is therefore not desirable to create a system specifically matched to one enterprise and one domain. The developed software system must be as generic as possible. Deployment of the system in a new organization or in a new domain must not require substantial software changes.

Further analysis shows that in a system envisioned as APOSDLE, domain-specific knowledge must exist in some form. The system must know about different users, their competencies and learning needs and about the tasks users perform and which digital resources (text documents, multimedia documents etc.) are helpful in which situation (user/task/learning need).

Within the project APOSDLE environments are employed at four application partners. We focus on two of them that also represent two extremes, EADS a large enterprise and ISN a network of SMEs. Although we describe the experiences related to a specific system, we think that our reports are of interest to a wider audience, as APOSDLE relies on many "standard" semantic web technologies and consequently inherits both their advantages and disadvantages.

2   APOSDLE Semantic Web Technologies

Semantic web technologies enable switching the learning domain of APOSDLE without software changes, and without hand-crafting domain-specific learning material.

Instead, customization to a domain of application happens by the creation of formal models that capture the necessary aspects of the domain of application. These formal models encapsulate the knowledge that otherwise would be implemented within the program code. In APOSDLE Changing the domain of application means just changing the models.

In general, different formalisms for encoding domain-specific knowledge could have been used, that would not count as "semantic web technologies" as understood by the semantic web community. However, at the time of project start of APOSDLE, description logics and OWL seemed the most broadly developed and advanced technology, with a lot of supportive tools such as ontology editors, reasoners, programming frameworks and APIs, and well-understood theory. Learning material is created ad-hoc by analyzing available organizational resources (textmining) and reusing it.

---

[1] APOSDLE-Advanced   Process-Oriented   Self-Directed   Learning   Environment, http://www.aposdle.org

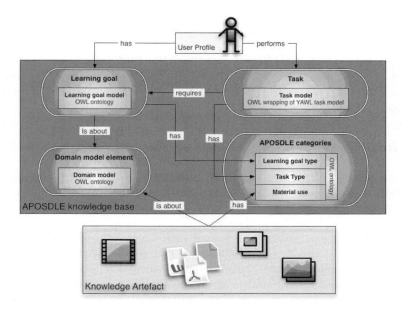

Fig. 1. The APOSDLE Knowledge Base and its relation to other components

In the following we illustrate the realization of this hybrid approach within the APOSDLE system: Firstly, we illustrate the (integrated) schema of semantic models developed inside APOSDLE and proposed as a basis for categorization and retrieval in work-integrated learning, then we illustrate the technology and techniques used to support the construction of these semantic models, and thirdly we illustrate the approach used to connect the semantic models to the resources of the organizational memory (classification).

2.1   The Semantic Models

One of the key problems to solve when trying to build a flexible generic system is to identify the basic knowledge that the system must have in order to deliver the worker with context-sensitive learning events, tailored to her specific learning goals, work situation and learning needs. The model we propose, hereafter called APOSDLE knowledge base, is depicted in Fig. 1. In a nutshell, the APOSDLE Knowledge Base contains all the necessary information about the tasks a user can perform in a certain organization, the learning goals required to perform certain tasks, a description of the domain of affairs (application domain) of the organization, and, finally, specific APOSDLE categories used to classify tasks, learning goals and learning resources. The main idea is that starting from the context-sensitive situation of a user,

which includes her current task(s), APOSDLE is able to determine the learning goals of the user and then use the information contained in the APOSDLE

Knowledge Base to select appropriate resources, so called knowledge artifacts and transform them into learning events[2] proposed to the user for attaining the missing learning goals [2].

A knowledge artifact (KA) is defined in APOSDLE as a document, or part thereof, together with two types of metadata: the learning domain concept addressed /described within the document (piece) and the material use type[3]of the document (piece) [2].

2.2   Building the Semantic Models

Only one of the four models depicted in Fig. 1, the APOSDLE Categories, is an APOSDLE-inherent structure. The others are by nature domain / organization dependent and need to be provided every time the APOSDLE system is configured and deployed for a new organization / domain.

Despite the recent advances of formal modeling and semantic web, we can safely assume–and this is the situation we had to face within the running of the project–that most of the organizations interested in using the APOSDLE system neither have formal models already available, nor all the skills needed to develop them. Therefore, as part of the APOSDLE project we have developed an *Integrated Modeling Methodology (IMM)*[3] as a series of steps, techniques and tools to support the construction of the semantic models depicted in Fig.1. To be effective and tailored to the APOSDLE system, such a methodology had to satisfy some important requirements:
Provide the organization with high level tools useful to specify knowledge in natural language, to minimize the need to become familiar with formal languages. Support an integrated development of the three models (domain, tasks, and learning goals) in order to ensure a conceptual consistency among these models and an easy formal integration. Support the creation of formal models which are described using different languages (YAWL and OWL)Encourage knowledge engineers and domain experts from the organization to work together in a collaborative manner and become the main actors of the modeling phase[4].

The first important semantic technology used to support modeling inside APOSDLE is a semantic wiki[5]. The choice of a semantic wiki as the main tool for *informal modeling* was made for several reasons: first it provides a state of the art collaborative tool which has made possible an active collaboration among all the actors involved in the modeling activities; second it provides a uniform tool for the informal specification of the different models (domain model, task model, learning goal model) using natural language; third, the natural language

---

[2] Learning events are a kind of amalgamation of knowledge artifacts following instructional principles. To read more about learning events, see for instance [2].

[3] Within APOSDLE five material use types have been defined: The material use "Definition", "Example", "Howtodo", "Main points", "More about" [2].

[4] In the event of unavailable knowledge engineers, or knowledge engineers not enough skilled to follow the process in an autonomous manner, we provided some coaching.

[5] We used Semantic MediaWiki (http://www.semantic-mediawiki.org).

descriptions inserted in a semantic wiki can be structured according to predefined templates, and with the help of semantic constructs like attributes and relations. As a consequence the informal descriptions in natural language contain enough structure to be automatically, or semi-automatically, translated in OWL ontologies, thus allowing the reuse of informal descriptions for automatic ontology creation.

The second technology used to support informal modeling (and in particular knowledge elicitation) within APOSDLE were techniques of term extraction to provide lists of candidate concepts. These techniques, illustrated in [4] [2] have also proven useful to reduce the burden of modeling and speed up the process.

The informal models described in the semantic wiki were translated to formal models.

- The domain model was (semi-) automatically translated to an OWL ontology,
- The task model was formalized by using YAWL [5]. This had to be done manually,
- The learning goal model was formally specified with a custom tool, the Task-Competency Tool (TACT for short)[6] by connecting relevant learning goals[7] to tasks.

2.3   Connecting Semantic Models and Resources

To allow for retrieving learning content the formal models have to be connected to the resources of the respective organization (annotation). The annotation process is one of the key challenges in any system that relies on semantic annotations. Semantic annotations in the present system are based on the *Aboutness* of resources [6]. This means that we annotate (parts of) documents with a set of concepts the content of the document is about. Basically, two options exist.

- *Manual annotation*: The resulting annotations are probably of good quality, given that motivated and competent persons perform the annotation. The large drawback is the large effort that manual annotation requires.
- *Automatic annotation*: The resulting annotations are of lower quality than manual annotation given any state-of-the art technology of natural language processing or statistically-based classification. The advantage is of course that it is faster.

In APOSDLE, manual annotation is used to create an initial set of knowledge artifacts of good quality. These are used on the one hand as high-quality learning

---

[6] This Tool is developed by APOSDLE Team [2].

[7] A learning goal is regarded as the combination of a "Learning goal Type" and a "Domain Concept". Learning goal types specify the type and degree of knowledge and skills the person must or typically wishes acquire about this topic. Within APOSDLE Learning Goal Types are "Remember", "Understand", "Create" and "Apply". The domain concept defines the content or topic that the learning goal is about.

material, and on the other hand as training material for automatic classification. Manual annotation can be performed again during usage of APOSDLE, when a user opens a retrieved knowledge artifact and wants to edit it. Automatic annotation is performed at regular intervals. Assignment of domain concepts can be seen as classical text classification problem of assigning documents to a set of predefined classes [7]. A basic algorithm, using Support Vector Machines (SVM) and k-Nearest neighbor algorithms, has been modified to incorporate knowledge about hierarchical relations between the classes, i.e. the domain concepts assigned to pieces of text [8].

Retrieval of learning content can be performed based on the semantic annotations. A detailed description of the approach taken in APOSDLE for retrieval and further references are described by Scheir, Ghidini and Lindstaedt [9].

3   APOSDLE in Application Context

APOSDLE is adapted to four application partner domains. In this section the two most oppositional are compared to each other. On the one site EADS as large company, on the other side ISN, as network of SMEs. The main focus of our attention in this section is on describing shortly the application domains organization, their motivation for using APOSDLE as well as some major activities necessary to deploy APOSDLE.

3.1   Application Domains, Challenges and Motivation for Using APOSDLE

EADS[8] is the largest aerospace company in Europe active in the fields of civil and military aircraft, space, defense systems and services. EADS Innovation Works (IW) decided to implement the APOSDLE system in the *Simulation Domain*. This is due to the growing importance of simulations and the necessity to have quicker operational performances of engineers in this domain. The simulation teams are in charge of designing, developing and achieving numerical simulations of electromagnetic problems or phenomena (Electromagnetic (EM) compatibility, simulations of EM attacks on aeronautical systems or subsystems, etc.). They are composed of engineers who are EM physicists or mathematicians and Soft/Hardware Specialists. The simulation experts are located in different countries and cities and rarely available to help newcomers in learning. Moreover the time allowed to maintain up-to-date worker skills or to acquire new knowledge has to be reduced. The simulation activities require a high level of knowledge and expertise. EADS IW is emphatically interested to introduce an innovative task- and process-oriented learning tool such as APOSDLE to support the simulation development process.

On the other side ISN–Innovation Service Network[9] is a service and research company in the field of innovation- and knowledge management composed of a

---

[8] EADS–European Aeronautic Defense and Space Company EADS N.V., http://www.eads.net

[9] ISN–Innovation Service Network (ISN), http://www.isn.eu

network of SMEs in Slovenia and Austria. ISN is supplemented by more than 40 further partners from universities, competence centers and service companies acting as a pool of experts. All experts are or intend to be specialized in a few, very specific fields such as specific management methods, patent management, creativity methods, etc. In order to stay competitive at ISN each network partner needs to continuously improve her skills in the focus field, rapidly learn aspects of other fields relevant to one customer project, and apply both within the customer's situation. Since the network partners are increasingly involved in partner projects, the time for learning and improvement of competencies becomes more limited. On the other hand, more in-depth and more diverse know-how is asked for by the customers. Thus, within ISN the focus of the APOSDLE domain is on *project processing and management* in order to provide specialized know-how in the area of innovation and knowledge management to their customer organizations.

Currently at both organizations there is no common learning resources database or system that can support the knowledge worker. Learning during work task execution is done by using templates, guidelines, project documentation or internet resources and with few sharing effort to the collective. The initial motivation for implementing APOSDLE at ISN and EADS is therefore to create a common knowledge base which integrates resources from different repositories and backend systems for all knowledge workers in order to improve knowledge acquisition, reuse and sharing.

3.2 Using Semantic Technology–Deriving from an Analysis of Competitive Approaches

At the moment there are several different approaches in the field of knowledge management and e-learning which aim at meeting above mentioned challenges.

Desktop search engines (e.g. Google Desktop) or ontology based search engines are often used for knowledge retrieval within organizations (e.g. KINOA[10] at EADS IW). On the other side knowledge sharing is supported by providing collaboration tools like e-mail, document sharing (e.g. Groove is used at ISN) or instant messaging (e.g. Skype is used at ISN). Even social networking platforms support collaborating and sharing knowledge with each other.

Frequently companies have implemented e-learning systems or learn management systems (LMS) in order to provide employees the possibility to further improve their skills. However knowledge worker in very knowledge intensive domains like at EADS or ISN often do not have the time for explicit learning. They need flexible real-time support within their current work task. Additionally developing an e-learning System is an expensive and time consuming task [1].

By using the potential of semantic web technologies successful approaches of knowledge management and e-learning are combined within APOSDLE. Therefore the essential benefit for ISN and EADS by using APOSDLE is:

---

[10] KINOA–management and sharing of written information content with shared ontologies.

- Learning *during* execution of work tasks.
- *Collaboratively* sharing knowledge within the work environment.
- Learn *from resources* already available within the organization.

4   Customization and Usage of APOSDLE

Both ISN and EADS defined one knowledge engineer (KE), who is responsible for the elicitation of knowledge from the domain experts (DE) and guides the entire modeling process. The DE provides the fundamental knowledge about the domain of the users of APOSDLE and their learning needs. The DE also specifies the pool of resources to be used for knowledge extraction.

Modeling corresponds to customizing the existing APOSDLE system to a specific domain, with a focus on exactly those tasks and competencies that are interesting for the users. Modeling is also very often the "bottleneck" in many semantic web applications, i.e. the applications would be good, or would work if only there were enough / more appropriate models. Therefore we think it of interest to describe the modeling that was necessary at ISN and EADS in order to deploy and use APOSDLE. We followed the *Integrated Modeling Methodology (IMM)* [3]. Below we describe the steps prescribed by the IMM and the corresponding experiences at EADS and ISN.

4.1   Informal Modeling

Informal modeling follows a knowledge elicitation phase, in which a number of relevant "elements'", be it terms, phrases etc. elicited from DE are collected. The goal of the informal modeling phase is that the knowledge engineer enters, after reviewing and filtering them, these first results into a semantic wiki. In a second stage the informal models, i.e. described in natural language, are formalized and connected to each other.

Knowledge elicitation from domain experts. The first stage in APOSDLE models design for Electromagnetic Simulation domain consisted in collecting relevant resources and interviewing experts: At EADS three simulation domain experts (EM physics specialists and mathematics expert) were interviewed separately, during two hours. The discussions focused on simulation process, knowledge needs and difficulties that occur at each simulation development task. As at ISN each domain expert is expert in a very special knowledge field, seven domain experts were interviewed according to a developed guideline asking about tasks they perform, learning needs they have and resources they use for both working and learning. Further the Card Sorting Technique [10] was used to approve the elicited domain knowledge.

Knowledge elicitation from resources. As one way of externalizing domain knowledge is in the form of documents written in natural language, document-based ontology engineering aims at using information available in documents

Fig. 2. Filled template for EADS task

to create formal knowledge representations. We used the discovery tab, a plug-in for Protg OWL, [4] to discover knowledge from text-documents. Grouping documents to form clusters of different topics, gives an overview of the domain covered by the documents. By using statistical and natural language processing methods, relevant terms are extracted from a set of documents. Furthermore, the terms can be grouped by synonymity. Within discovery tab, standard text mining methods have been used. For relevant term extraction, various pre-processing techniques, such as stop-word lists and stemming are employed [4].

Informal modeling with semantic wiki. Using pre-defined templates we have automatically created a page for each one of the elements (concepts or tasks) of the models we wanted to obtain via the wiki. Thus, the domain experts, knowledge engineers and APOSDLE technology partners located in different cities and countries had the possibility to access the models, refine them or comment. The filled template for a task is shown in Fig. 2 and is tailored to the information we needed to obtain for APOSDLE.

As shown the information required in the template is very basic. Nevertheless it allows guiding the modeling team to provide all the information that was needed by the particular application. In addition templates contained also hints to guide the early alignment of models. For instance the template for task description asks also for knowledge required to perform this task, and this has a clear connection with the domain and learning goal models.

Advantages and Difficulties. The experts were in general very busy and tied up in heavy workloads. However their involvement in modeling activity is strongly required. The knowledge elicitation only from documents risks to build up very theoretical and over-complicated knowledge schemes that passes by real knowledge needs. The main difficulty that occurred at this stage of modeling activity consisted in putting the right definition behind the terms used by experts or extracted from documents. Several terms are common to some experts but used to evoke different concepts: e.g. at EADS the term "model" can mean mathematical model, simulation model, data model, numerical model, etc. Some specialists use also different terms to designate the same concepts. Another problem was a lack of documentation, for instance at EADS a large number of simulation knowledge is not really formalized and can be stored in individul's repositories or on desktops.

Currently the Discovery Tab [4] is based on WordNet[11] and Apache's Word-net[12] package . Thus, it was difficult to automatically extract terms from German (ISN) or French (EADS) documents.

The results of the interviews as well as documents analysis allowed the EADS and ISN Knowledge Engineer to create a very first version of the informal models in the Semantic MediaWiki, define some learning scenarios and knowledge / learning needs. Because of both the complexity the domain has and the information needed for the semantic wiki, the informal modeling activity was quite time and resources consuming. However, it turned out that a semantic wiki is very useful for the internal model revision as it allows queries about special relations, shows up all relations a concept/task has to another concepts/task etc.

For the EADS Simulation domain the informal modeling process required about 2.5 person months. Altogether 43 EADS domain concepts and 22 tasks were identified, classified, and described. The modeling process at ISN required 3 person months and 140 concepts and 31 tasks were defined for describing their domain.

4.2   Formal Modeling and Resources Annotation

Based on the KE's semantic descriptions of tasks and concepts in the wiki templates, Semantic MediaWiki generates machine-readable documents in RDF format. These are transformed into an OWL model. The formal domain model could then be viewed and revised by knowledge engineers in the Protégé editor. Basically most of the informally given information was correctly formalized–checking the formal models some specific relation types (e.g. the relationtype *has result*) that could not be translated to OWL had to be defined anew. Additionally some concepts arising from not deleted wiki pages in the ontology had to be deleted. The task model was manually defined by technology partners in YAWL [5] starting from the informal descriptions provided in the semantic wiki. The final step for the task model is the application of formal checks provided by the YAWL editor. These checks result in a debugging of the YAWL specification. This outcome is complemented by a transformation of the YAWL specification to an OWL file containing the complete list of tasks [3].

Once the task and the domain model were formalized knowledge engineers linked with the TACT Tool [2] respective concepts and learning goals to every task. For instance for to the task "Characterize Physical Phenomena to be simulated" (Fig.2) the learning goal "Understand" and the concept "Radiation" was defined (among others).

Finally, the resources annotation was carried out using the annotation tool [2](Fig. 2 shows an example. The annotation tool consists of a pdf-viewer plus the custom functionality to create knowledge artifacts (KA), i.e. to assign domain concepts and material use concepts. The annotation process at EADS and ISN

---

[11] http://wordnet.princeton.edu/

[12] http://lucene.apache.org/java/2_0_0/api/org/apache/lucene/wordnet

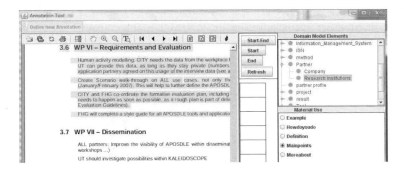

Fig. 3. The Annotation Tool

was composed of 2 main stages. The first one consisted in selecting an initial set of relevant documents (the training corpus). This set of representative resources dealing with the model concepts was manually annotated using APOSDLE annotation tool. For example, at ISN a piece of text within an article containing a definition about the concept "Brainstorming" was annotated with the respective concept and the material use type "Definition". At EADS 60 knowledge artifacts were created from 25 documents. The Knowledge Engineer dedicated 3 days for training corpus building. The results from ISN are very similar. About 100 knowledge artifacts from 50 documents were created within 5 days.

In the second stage the resources annotation process was done automatically. This means that a set of concepts for a document based on a training set of previously annotated documents is suggested. Finally the KE checked these automatic annotations randomly.

Advantages and Difficulties. In total at EADS 150 knowledge artifacts were created from 55 documents. Because of the unavailability of domain experts, the complexity of Simulation Activity and difficulty to collect the right resources the annotation process for APOSDLE Prototype at EADS required about 1.5 person months. At ISN the Annotation process required 1 person moth. And about 200 (KA) were created of 150 documents.

4.3 APOSDLE in Use

Both at EADS and at ISN the prototype of the APOSDLE system was not yet connected to other live systems. Especially at EADS are high security and privacy settings, therefore we decided to first test the system itself, before connecting it to other sources. At EADS the APOSDLE Prototype was used by 10 evaluation users located in Toulouse and Suresnes. They had different levels of knowledge of Simulation Domain and different learning objectives. At ISN installation of APOSDLE Prototype was carried out at the Headquarters in Graz and 10 users were integrated in the evaluation process.

As APOSDLE is a quite complex system we decided to accompany the installation with user trainings. These Trainings have been carried out by the respective knowledge engineers. Generally users are not familiar with the models based and semantic technology approach used in APOSDLE-thus a lot of conceptual questions concerning tasks and related learning goals as well as retrieved resources that depend on task, learning goal and competency arose.

5   Challenges for Semantic Web Technologies in Industry

Although the evaluation of the APOSDLE system is still ongoing and validated assumptions concerning the models and the system can not be made, we figured out the main points to be addressed for future development.

5.1   Models

A very important point concerns the granularity of the models. For this prototype we tried to find out a balance between not too generic and not too specialized models. In consequence it was quite difficult to find a right formulation of a task in order to meet all demands like comprehensibility, feasibility, not too complex etc. and at the same time matching this with a task type (some tasks may cover more than one task type) and a specific concept (required knowledge). For instance we decided to model the task "Select appropriate tools for method". The concept "method" has about 15 sub concepts. As it would have been too much effort to model 15 different tasks and assign the related learning goals and concepts, we decided to follow this quite generic way.

However, after using the real system we identified a need to enrich the domain model in order to allow learners and experts acquire / exchange knowledge on more specific topics (e.g. at EADS measures, different kinds of simulation models and resolution methods). Currently we are working on enabling the KE to represent more complex structures in the formal models.

5.2   Modeling Process

In addition to models, the modeling process also needs to be improved. Firstly we want to integrate more already existing classification like e.g. files structures. Secondly the Discovery Tab could be enriched by new functionalities, e.g. the possibility of associating to the extracted term the snippets of texts and the documents it comes from.

For next model development we will directly use one semantic wiki for each partner-for this prototype we used one semantic wiki for all application domains and encountered some overlapping in concepts and tasks of different domains. The application domains also want to integrate the possibility of a graphical representation of the models and its relations in the semantic wiki.

5.3   Resources Annotation

Based on early results of the evaluation we assume that the quality of the annotations has a huge impact on the users' confidence in APOSDLE system.

The resources provided by APOSDLE have to be relevant to the learner context (domain, task and especially his/her learning objective). Based on our first impressions the manual resource annotations are very good. Assumptions concerning the automatic annotation can not be made at the moment. Therefore it is an open issue, if it would be worth to spend the effort for annotating all documents manually in the case users are not satisfied with automatically provided annotations. The current approach is, that every APOSDLE user can add, edit or delete an annotation to a document–thus the more users give their feedback on the annotations, the better the documents are annotated. During the evaluation phase we started the discussion whether we should follow this approach. Reasons for this are that (a) it is not easy to annotate a document, some people may not do this because it is too complicated. (b) Domain experts mentioned that adding, editing or deleting an annotation is quite subjective, because e.g. domain experts would annotate a document in a different way as others would do this.

6   Conclusion

The paper at hand presented a detailed report of experiences and issues a company has to face, when it wants to deploy a modern learning environment relying on semantic web technologies.

Although the described application domains represent two extremes, both could follow without major differences the same modeling approach in order to deploy a complex system like APOSDLE. Thus, it is not primarily important to which company–whether a large enterprise or a network of SMEs–it is deployed, but to have a clear idea of the domain to be modeled and the users that will use the system. Thus, we will evaluate clear criteria and develop a kind of guideline for which domain and for which granularity of models our modeling approach is suitable.

The IMM approach enables also persons not skilled in knowledge engineering, object-oriented modeling or ontology-building to run through our proposed modeling process. The semantic wiki enabled the knowledge engineer to describe the respective domain in natural language–this information can then (semi-) automatically extracted and translated to OWL. With the development of tools, which can be easily applied in industry–even more business problems could benefit from the potential of semantic web technologies. We will continue to apply and improve the collaborative approach to improve and extend the deployment of semantic technologies in industry.

Acknowledgements. APOSDLE is partially funded under the FP6 of the European Commission within the IST work program 2004 (FP6-IST-2004-027023). The Know-Center is funded within the Austrian COMET Program–Competence Centers for Excellent Technologies–under the auspices of the Austrian Ministry of Transport, Innovation and Technology, the Austrian Ministry of Economics and Labor and by the State of Styria. COMET is managed by the Austrian Research Promotion Agency FFG.

References

1. Lindstaedt, S., Mayer, H.: A Storyboard of the APOSDLE Vision. In: Nejdl, W., Tochtermann, K. (eds.) EC-TEL 2006. LNCS, vol. 4227, pp. 628–633. Springer, Heidelberg (2006)
2. Leemkuil, H., de Hoog, R., Lindstaedt, S., Ley, T., Scheir, P., Ulbrich, A., Beham, G., Klieber, W., Aehnelt, M., Lokaiczyk, R., Zinnen, A., Doering, M., Goertz, M., Leitner, L.S.H., Kump, B., Pammer, V.: Conceptual Framework & Architecture Version 1 (2007),
   http://www.aposdle.tugraz.at/media/multimedia/files/
   conceptual_framework_architecture_version_1
3. Ghidini, C., Rospocher, M., Serafini, L., Kump, B., Pammer, V., Faatz, A., Guss, J.: Integrated Modelling Methodology (IMM) Version 1 (2007),
   http://www.aposdle.tugraz.at/media/multimedia/files/
   integrated_modelling_methodology_version_1
4. Pammer, V., Scheir, P., Lindstaedt, S.: Two Protg plug-ins for supporting document-based ontology engineering and ontological annotation at document-level. In: 10th International Protg Conference (2007)
5. Godehardt, E., Doehring, M., Faatz, A., Goertz, M.: Deploying YAWL for Workflows in Workplace-embedded Learning. In: 7th International Conference on Knowledge Management, pp. 281–288 (2007)
6. Handschuh, S.: Semantic Annotation of Resources in the Semantic Web. In: Semantic Web Services: Concepts, Technologies, and Applications, pp. 135–155. Springer New York, Inc., Secaucus (2007)
7. Sebastiani, F.: Machine learning in automated text categorization. ACM Computing Surveys (CSUR) 34(1), 1–47 (2002)
8. Granitzer, M., Auer, P.: Experiments with Hierarchical Text Classification. In: Proceedings of 9th International Conference on Artificial Intelligence. ACTA Press, Benidorm, Spain, IASTED (2005)
9. Scheir, P., Ghidini, C., Lindstaedt, S.: Improving Search on the Semantic Desktop using Associative Retrieval Techniques. In: Proceedings of I-SEMANTICS 2007, pp. 221–228 (2007)
10. Schilb, S.: Card Sorting-Techniken im Überblick. I-Com 4(1), 49–50 (2005)

Creating and Using Organisational Semantic Webs in Large Networked Organisations

Ravish Bhagdev, Ajay Chakravarthy, Sam Chapman, Fabio Ciravegna,
and Vita Lanfranchi

Department of Computer Science, University of Sheffield, Regent Court,
211 Portobello Street, S1 4DP Sheffield, United Kingdom
{N.Surname}@shef.ac.uk

Abstract. Modern knowledge management is based on the orchestration of dynamic communities that acquire and share knowledge according to customized schemas. However, while independence of ontological views is favoured, these communities must also be able to share their knowledge with the rest of the organization. In this paper we introduce K-Forms and K-Search, a suite of Semantic Web tools for supporting distributed and networked knowledge acquisition, capturing, retrieval and sharing. They enable communities of users to define their own domain views in an intuitive way (automatically translated into formal ontologies) and capture and share knowledge according to them. The tools favour reuse of existing ontologies; reuse creates as side effect a network of (partially) interconnected ontologies that form the basis for knowledge exchange among communities. The suite is under release to support knowledge capture, retrieval and sharing in a large jet engine company.

Keywords: Semantic Web-based Knowledge Management, Knowledge capture, search and retrieval, application of semantic web technologies.

1 Introduction

The classic Knowledge Management environments aim at creating large homogeneous knowledge or document repositories where corporate knowledge is collected and organised according to a single conceptual schema. This schema represents the official agreed view of the organisation with the intent of supporting communication between its different parts and is generally used in an Enterprise Knowledge Portal providing unique standardised access to proprietary knowledge [1]. However, it is a well-known issue that many of these portals are deserted by users, who continue to capture and share knowledge in ways quite different from those provided by the corporate-wide systems. For instance workers use non-official tools such as shared directories, personalised and local databases, etc. [2]. The reason for deserting the central knowledge portals can be summarised as the difficulty in adopting models, schemas and procedures that are unsuitable to specific communities of users. From experience in the manufacturing industry, users tend to organise their cycles of knowledge acquisition and capturing around unstructured (e.g. textual documents) or semi-structured

A. Sheth et al. (Eds.): ISWC 2008, LNCS 5318, pp. 723–736, 2008.

documents (e.g. forms) created with tools like word processors or spreadsheets, rather than using centralized (or even local) databases. Usually this type of document is retrievable via keyword searching and knowledge is individually acquired by reading the document. This problem is amplified by the large amount of unstructured material produced in large organisations, which increases the unmanageability issue. Also, modern organisations are very dynamic and often favour the creation of communities of practice, i.e. more or less formal groups of people with a common interest in some subjects and who collaborate over an extended period to share ideas, mental models, practices, find solutions, and build innovations [4]. As these communities tend to exist across the traditional organisational boundaries, they ill fit pre-determined standard schemas and require flexible customisable knowledge for specific ad-hoc uses.

In summary, modern knowledge management is moving away from the idea of a large centralised schema to suit all situations (*absolute knowledge*) to a more localised approach (*local knowledge*) that "is a partial, approximate, perspectival interpretation of the world" [2].

A similar approach has been adopted by the Semantic Web community, where the concept of small scale distributed interconnected ontologies [3] has replaced the idea of large comprehensive all-encompassing ontologies. The Semantic Web can therefore help such a change in knowledge management with tools and techniques supporting: 1) definition of community-specific views of the world; 2) capture and acquisition of knowledge according to them; 3) integration of captured knowledge with the rest of the organisation's knowledge; 4) sharing of knowledge across communities.

In this paper we introduce K-Forms and K-Search, a suite of tools supporting distributed networked knowledge acquisition, capture, retrieval and sharing. They enable communities of users within or across organisations to define their own views, while at the same time maintaining connections with other communities' views and (if required) with a central schema.

K-Search enables retrieval and sharing of documents and knowledge. It enables accessing multiple repositories (semantic or traditional) using either directly their local reference ontology or other connected ontologies. Queries can return 1) the original documents annotated with the extracted knowledge, or 2) documents generated using metadata for specific user needs, or 3) a summary of the knowledge in the form of graphs or statistics or 4) triples for further elaboration (e.g. by an external Semantic Web Service). Furthermore extracted triples can be exported in CSV or RDF formats for further analysis using external tools.

K-Forms and K-Search enable management modalities that satisfy two main principles claimed by [2] as essential for modern knowledge management:

- *Principle of Autonomy* where each unit is granted a high degree of autonomy to manage their local knowledge;
- *Principle of coordination* where units are enabled to exchange knowledge with other units through a mechanism of mapping other units' context onto their local context.

New communities can be supported by rapidly defining schemas and modalities to capture and share knowledge. As they tend to evolve rapidly, change is supported by enabling easy networked modification of the knowledge schema via a standard browser. Moreover this approach supports the definition and reuse of different views on the same data, particularly useful when communities of users with different tasks and information needs are involved.

In this paper we first introduce a general architecture from a functional point of view, describing how knowledge can be represented, captured, retrieved, shared and visualised using K-Forms and K-Search. Then we will go into the details of system components. An evaluation will show how and in what sense the methodology is appreciated and understood by real users. We will then describe an application developed for an industrial environment to support capture and sharing of knowledge about jet engines. A comparison with the state of the art will follow. Finally we draw conclusions and highlight future work.

2 A Form-Oriented Approach to Knowledge Acquisition and Capture

The architecture is composed of two modules: K-Forms and K-Search. K-Forms supports knowledge acquisition and representation (the definition of the knowledge structures, e.g. the ontology) as well as capture (the creation of instances). K-Search enables searching and sharing of information and knowledge.

Both tools use the concept of forms as the interaction paradigm. Form-based solutions are very familiar to users, as they are very common in knowledge management and in every-day life, therefore they are very easy to understand and use.

2.1 K-Forms: The Final User View

For knowledge acquisition, many organisations adopt forms as a way to capture knowledge, either in the form of static Web forms or as a template in a word processor (e.g. MS Word) or in a spreadsheet. Both solutions have issues.

Static forms are largely unsuitable to dynamic emerging communities as they require a computer expert to be implemented or modified and such expert is rarely available in a community of practice. Therefore Web forms tend to be used mainly in centralised services and changes to the forms are unlikely to happen. This is a serious drawback as it has been noted how users who do not find the appropriate field in a form, start using inappropriate fields for storing the information they need [7], hence causing the consistency of data to be corrupted.

The use of flexible forms (e.g. MS Word) has the opposite disadvantage. Users feel entitled to modify the forms at any time, especially when a centralised consistency check is difficult or impossible. In this case, the change in format causes missing data (e.g. because fields disappear) and/or the insertion of inappropriate data (because these tools do not allow rigid type control on the input). Also the information contained is unstructured and therefore it requires capturing using other means like IE from Text. The extraction is made even more complex when the forms are modified. As noted in [8], in a corpus of 18,000 reports on jet engine diagnostic issues, forms were so disrupted that the application of complex machine learning techniques was required in order to normalise and capture knowledge.

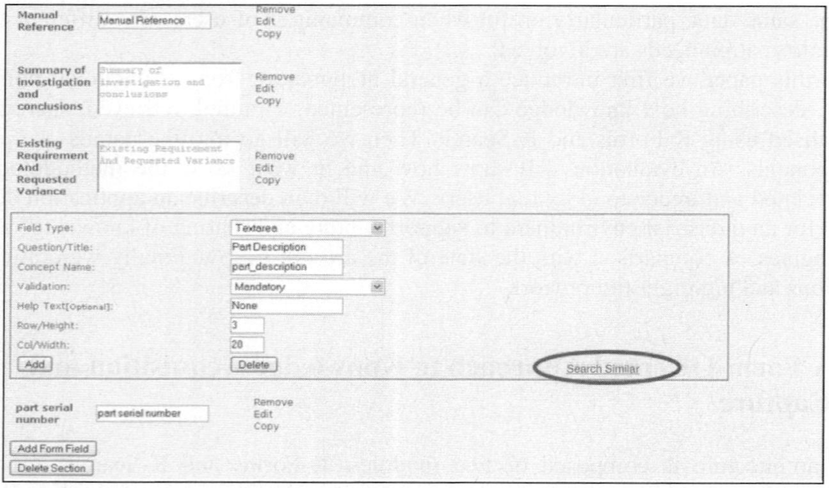

Fig. 1. K-Forms interface for new form creation: highlighted the possibility to search for similar concepts when creating a new one

K-Forms is designed to provide an intermediate semantic solution between rigid web forms and unstructured documents: easily deployable Web forms. K-Forms supports knowledge acquisition, modelling and capture by enabling an easy creation and deployment of Web forms via a graphical user interface. K-Forms typically cannot be modified arbitrarily by whoever fills them but rights can be given to do so should it be appropriate. To encourage semantic interconnections, a functionality for identifying and reusing relevant templates is provided so that the definition of new forms can always start from existing ones (see Figure 1, Figure 2 and Figure 3). This is an important requirement because communities in the same organisation generally tend to need some minor modifications of a similar form within the same domain. For example, components of a jet engine are generally defined via *name*, *model* and *serial number*. A specific community investigating the condition of specific modules during engine overhaul will still need the same information, but will need to add some additional parts (e.g. the name of the module investigated and the engine's number of cycles, etc.). So importing from existing forms not only saves a users time avoiding the redefinition of existing structures but also provides schema linkage between forms and the captured concepts.

K-forms enables easy definition of confidentiality gates, distinguishing between local knowledge (which may be confidential and must not be necessarily shared with others) and shareable knowledge. Knowledge acquired for each domain is stored on a community server. Every user (including external ones) must access the community store in order to access the knowledge. Therefore each community is always in full control of the information and can decide what parts are made available to external users and what parts are not.

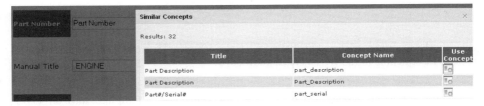

Fig. 2. K-Forms: search and reuse of similar <u>forms</u> before creating a new one

		Similar Concepts		✕
Part Number	Part Number	Results: 32		
		Title	**Concept Name**	**Use Concept**
Manual Title	ENGINE	Part Description	part_description	
		Part Description	Part_Description	
		Part#/Serial#	part_serial	

Fig. 3. K-Forms: search and reuse of similar <u>concepts</u> before creating new ones (here search for concepts similar to part_description is shown)

2.2   Forms: the Technical View

From a technical point of view, K-Forms is a methodology for knowledge acquisition and capture. It is based on semantic technologies (invisible to the users) that provide knowledge sharing and reuse capabilities beyond what is possible with standard technologies.

2.2.1   Forms as Ontologies

When a new form template is created, K-Forms seamlessly translates it into an explicit OWL ontology (see Figure 4); the template concepts and fields and associated constraints are translated into OWL statements. Data inputted into forms based on those templates is transformed into RDF triples that can be searched using standard query languages such as SPARQL and SERQL.

Forms can be divided into sections and fields. Sections can have subsections and fields, e.g. the section designed to hold data about a person will be a concept; a person will have a series of properties e.g. name, address and date of birth. Sections are presented as sub-forms to be filled.

Fields are typed (e.g. text field, integer field, text area, checkbox, option list, etc.). Fields represent meta-properties of the document (e.g. author, date, etc.) or its content (e.g. an issue to be reported). Simple fields will require just the inputting of a value at filling time (a number, a string, a text, a date, etc.). When a field is defined, the user selects a field type and must input appropriate properties for the specific field (e.g. size, validation constraints, default value, help text, etc.). If parts of an existing form are imported, the system automatically imports any associated constraint (e.g. on data types).

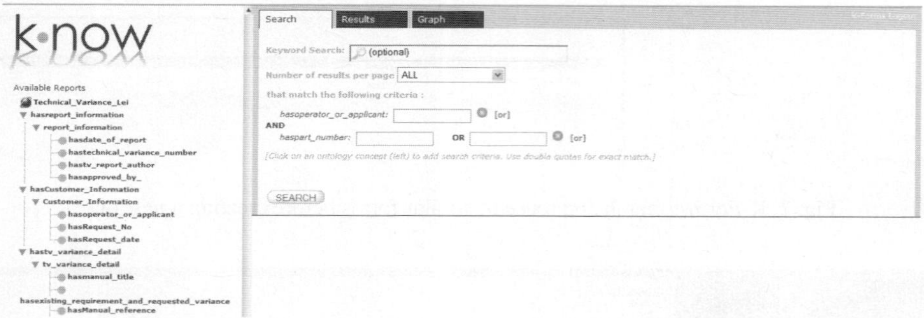

Fig. 4. The Ontology produced by K-Forms after a new form is created can be loaded in K-Search for searching the newly generated knowledge

When a form template is saved, K-Forms translates a template into an OWL Ontology.

The form is represented as an OWL class (`<FormName> Class`) with its own Name-Space to avoid name conflicts. Form sections are represented as OWL classes (`<ConceptName> Class`) that can have subsections (related classes) or individual fields (properties).

Relations among concepts are represented as OWL relations between classes and properties (using owl:equivalentClass and owl:equivalentProperty).

Adding relations between parents (section or entire form) and the contained subsections (object type properties) enables reuse of entire sections. Relational tables can be represented as advanced sections. The domain of some relations may be the overarching `<FormName> Class`. When concepts are introduced at the top level, a relation is formally created domain `<FormName> Class` and range `<ConceptName> Class`.

Individual fields (such as text box, text area, check box) can be added as a property of each section, subsection, or directly in form classes; they are represented as OWL properties. Some of them (as `FormCompilationDate`) are properties of the overarching `<FormName> Class`. Restrictions can be set for the possible values of the individual fields using xml datatype schema (xsd:types) (for example declaring a type to be a positive integer). Again individual fields can be reused between forms and classes.

When concepts are reused across forms, owl:equivalentClass is created. This is due to the fact that users may reuse the concept but change non-conceptual details regarding its visualisation.

2.2.2 Form Reuse as Implicit Ontology Mapping

When (part of) a form is reused, an explicit mapping between (parts of) the two underlying ontologies is established. This means that instead of creating new concepts into the new form namespace, the new ontology will use the concept in the other form's namespace. If the concept is modified, then a link between the two concepts is maintained by automatically creating a meta-class capturing the common parts. This may happen either at the single concept level or at the whole form level.

This creates automatically a network of interconnected ontologies.

The ontology mapping can also be controlled by a system manager who can manually establish further mapping or suggest modifications to the forms in order to make it compatible to other existing forms if possible. This will increase the quality and quantity of interconnections.

Although these connections are far from being complete and ideal, they enable the creation of a web of interconnected ontologies where knowledge can be and shared, much in the way recent Semantic Web developments recommend [5][6].

2.2.3 Knowledge Capture in K-Forms

When a form design is complete (after community validation, if needed), the form can be published for use via Web or Intranet (by simply pressing a "publish" button).

Knowledge capture using the form can then begin (Figure 5). All the inputted values are transformed into RDF statements related to the form ontology.

Fig. 5. The form corresponding to the simple template in Fig. 1

The target users are knowledge workers, for example an engineer diagnosing a fault on a car who has to report about the fault, its symptoms, causes, etc. Form filling is a natural approach to knowledge capture in such environments. The constraints posed on the form (e.g. via selected validation methods and strict data typing) ensure a degree of knowledge quality. Although forms can be modified easily, policies are set to prevent adjustments, as template alterations must be a community decisions and not individual decisions. After data entry, the system generates a document summarising the inputted information. The document can alternatively be sent by email or printed. The format of the summary can be customised with a simple HTML editor during form design. The user can preview the completed form and decide to modify or publish the knowledge. As knowledge capture can occur in places where a Web connection is unavailable, gathered knowledge can be held in a local store and later published to a central server.

2.3   K-Search: Sharing Knowledge Across Communities

Once the knowledge has been thus captured, it can be searched using **K-Search**, a semantic tool for documents and knowledge retrieval and sharing[8]. To guarantee consistency in the framework, K-Search has been designed following the same form-based interaction paradigm, providing an interface that allows users to combine conditions on both metadata and keywords within the same query. K-Search also supports multiple, de-centralised, dynamic knowledge communities by enabling access to knowledge stored in multiple repositories (e.g. triple stores) with multiple ontologies.

The ontology associated to a template is made available to a search mechanism (K-Search) that enables hybrid searches [8]. Hybrid Search (HS) combines the flexibility of keyword-based retrieval with the ontology and its reasoning capabilities, making a synergistic use of both strengths. In HS, users can combine within the same query: (i) ontology-based search; (ii) keyword-based search and (iii) keyword-in-context based search. Keyword-in-context searches for keywords only in the text annotated with a concept in the ontology; in case the document has been created with K-Forms, it searches the content of the form field values associated to the concept.

K-Search interface and modality of interaction is illustrated in Figure 3. The interface works in a standard Web browser, enabling the definition of complex hybrid queries in an intuitive way. Keywords can be inserted into a default form field in a way similar to what required by search engines; Boolean operators AND and OR can be used in their combination. Conditions on the metadata can be added to the query by clicking on the ontology graph (left side of interface in Fig. 3). This creates a form item to insert conditions on the specific concept. Here a unique identifier or a substring of the field can be inserted as condition. AND and OR can be inserted among ontological conditions.

The query output of K-Search is a set of ranked documents displayed as a list on the mid-right panel of the interface; each item in the list is identified by the values in the metadata that satisfy the ontology-based search. Clicking on one item causes the corresponding document to be shown on the bottom right. The document is presented with added annotations via colour highlighting; advanced features or services are associated to annotations [12, 13]: for example right clicking on a concept enables – among other things - query refinement with the selected term. Moreover, K-Search provides ways of inspecting the results of the query as bi-dimensional graphs (pie or bar chart) according to two elements e.g. issue and component); items in the graph (e.g. pie chart slices) are clickable to focus on the associated subset of data (e.g. associated documents or data). The retrieved triples can be exported if needed (in RDF, CSV or XLS formats) for external analysis or automated processing. K-Search is described in details in [8].

2.3.1   Querying Across Ontologies

The Web of interconnected ontologies is exploited in searching. K-Search enables searching multiple repositories at once using one of the available ontologies. This means that a user can decide to:

(i)   Query a specific resource via the original ontology
(ii)   Query a resource using a different ontology interconnected to the original one
(iii) Query multiple repositories using one specific ontology.

When an ontology different from the original is used (cases *ii* and *iii*), the original query is mapped to the original ontology via the formal links. For the common part of the information, there is no issue. For the parts that are not mapped the restrictions on the unmapped parts can be turned into keywords to be matched on the whole document generated from the filled form.

3   Evaluation

An evaluation was carried out to show how using K-Forms users can create a network of interconnected ontologies representing distributed communities and views without any user effort or specialised knowledge. In particular the aims were to evaluate:

- Knowledge reuse: can people look for knowledge they were unaware of and reuse it? Can users from distributed communities create interconnected knowledge in a decentralised way?
- Knowledge conceptualisation: can users conceptualise in terms of forms and use forms as a way to capture knowledge?

6 users tested the system individually. Each session lasted about 90 minutes and users were all academic persons with different degrees of familiarity with semantic technologies and ontology development. The data collected was both objective (logs of the interaction, screen activity) and subjective (questionnaire, interview). Questions were mainly on a 5 points Likert scale measuring the rate of agreement with a specific statement. Upon arrival, users filled in a personal profile questionnaire and were shown a short video (2 minutes) on the system and its use.

Two tasks, written as work task simulations [14] derived from real-world engineering tasks, were then presented to the users. The first task was to analyse a set of technical documents and create a form that could capture the contained knowledge, first on paper and afterwards using K-Forms. The second task was to analyse a different set of documents (that had some semantic intersection with the previous set) and create a form that could capture the knowledge, having as input also the ontology of the previous corpus (this was done in order to test the reuse of concepts). Half of the users received their own ontology (created in Task 1), while half received someone else's ontology (to test the difference between reusing personal knowledge and shared knowledge).

When both tasks were accomplished participants were asked to fill in a user satisfaction questionnaire that covered several aspects of the system, e.g. ease of use, ease of learn, perception of speed and accuracy, etc. The user evaluation ended with an interview aiming at eliciting explanations on participants' behaviour, impressions on the system and personal opinions.

Results for Knowledge Reuse
The ontologies created in task 1 were evaluated to check the degree of similarity when no reuse possibility was offered: users created unrelated ontologies with similar concepts but the similar concepts were not linked in the background.

In average 5 matching concepts between the ontologies were identified. Users created concepts with the same or very similar names to capture similar knowledge but the ontologies being unrelated impairs the possibility of reusing the concept or of searching across the different ontologies and instances.

When given the possibility to reuse an ontology (Task 2) users reused in average 60% of the possible concepts. The users' behaviour had many individual variations, with a peak of 80% and a minimum of 30% concepts reuse. In some cases users reused entire sections, modifying them to better suit their needs.

K-Forms knowledge capture method enabled people to discover new resources they were unaware of (as for ontologies created by other users given as input) and to directly connect what they thought was relevant. This bottom-up decentralized approach created a network of interconnected ontologies with no user effort and allows for a considerable reuse and ability to search across.

From the user interviews more details emerged on the sharing and reuse of concepts.

All the users were happy to reuse their own concepts as they knew not only the name of the concept but also the format they wanted it represented in (i.e. list or textfield) but they were more weary of reusing other people concepts as they may have been differently visualised. The percentage of reuse of own concepts is 68% while the reuse of someone else's concepts is 45%. This leads to the definition of a new requirement: enabling reuse of a concept whilst changing the visualisation manner.

During the interviews reuse strategies emerged: in general users were creating more generic concepts when aiming to reuse them while more specific when they thought they were peculiar to just one document type. Reuse of concepts was appreciated by 90% of the users as it saves time while one user still preferred to type his own concepts. When interviewing the user that did not want to reuse concepts, he commented about not being confident that the concept was exactly matching what he wanted to describe: this issue could be partially solved by showing instances to the users, so he could know whether the concept he is thinking to reuse meets his needs.

Results for Knowledge Conceptualisation

More detailed questions and task observations allowed us to evaluate the conceptualisation feature. When analysing the paper sketches of the corpora provided, all the users managed to easily sketch a form that could capture at least 90% of the existing concepts – some information was ignored as the meaning was not clear or as it was present only in part of the corpus. When translating the paper form into a K-Forms form all the users managed to translate it as they wished. Only 1 user found issues in translating a paper design as a required field type (list) was not available for use inside a table field. The user had to modify the form design to fit with system capabilities.

Overall K-Forms proved successful in supporting Knowledge Conceptualisation: users found easy or very easy (66.7%) to design a form using the system (33.3 % rated it average). Selecting sections, subsections and form fields was considered very clear by 66.6% and clear by 33%.

4   MCR: A K-Forms Application

As a follow-up of this experiment, an experimental application was developed for Rolls-Royce plc for capturing and sharing information about inspections to jet engine components. An existing electronic form requiring the input of some 80 fields has been re-implemented using K-Forms. The form required input of text as well as up-loading and annotation of images. Then a number of existing Module Condition Reports (MCR) have been used to fill the repository. The architecture supports a task where users have to visit overhaul shops around the world where no Intranet connection was available. Therefore they save the filled forms locally and upload them when go back to their offices at a future point. As soon as the triples are uploaded they become available to the rest of the community via K-Search.

The change in upload method from standard word processor documents to K-Forms was seamless and without consequences for the users. The possibility to query the knowledge contained in the forms through K-Search, however, makes the knowledge immediately searchable and reusable, and therefore we expect that many hours will be saved for knowledge workers who no longer need to read documents to access the contained knowledge.

Moreover, the knowledge about the condition of modules can be now correlated and connected to other knowledge contained in other repositories about the same topic. Other important services are now available for querying such as providing images of a module sorted by number of cycles (so to show the relation between condition and the amount of use). All these services where unavailable before the introduction of K-Forms: the only service available was a database allowing document access via some limited metadata (date, engine, component, etc.).

Results in module condition reports can now be compared to the results in two other repositories created for Rolls-Royce containing 18,000 event reports and about 11,000 technical variance documents created using Information Extraction from textual documents [8].

5   Comparison with the State of the Art

A number of previous works are relevant to K-Forms + K-Search. As for acquisition and representation of knowledge, standard technology like ontology editors can be used. However ontology editors like Protégé[1] require specialised knowledge and are generally unsuitable to a community of users who are primarily experts in their domain. The capture phase via Web forms was envisaged in [7] and [9] to capture knowledge and to connect the forms to an ontology. However, they did not envisage the possibility to effectively and efficiently enable the definition of user-defined forms. They still relied on technologies like X-Forms and the manual handcrafting of the forms. This limits the usefulness and flexibility of the tool for supporting distributed dynamic communities because they require the intervention of a computer scientist. Moreover, they do not provide anything like semantic searching mechanism to access information, which has proven very effective in the three applications of K-Search.

---

[1] http://protege.stanford.edu/

Form-based capture is also used in ontology-based knowledge portals [17]. These are system that use the ontology schema and its instances to capture to produce different views on the knowledge (e.g. to enable management of a large distributed project. Knowledge portals presented in literature require complex ontological modelling and therefore it misses the impact of the flexibility of K-Forms.

A technology similar to K-Forms is Semantic Forms within Semantic Media Wiki[2]. These semantic forms can be defined to capture specific semantic knowledge but lack graphical support to development. Knowledge within these semantic forms is captured without optimised storage for hybrid querying of the semantic information contained. Also Semantic Forms do not encourage reuse and interrelation across forms meaning gathered semantic knowledge is mostly unrelated. K-Forms differs by design to create and modify forms easily by users and communities to support emerging and rapidly evolving environments by non-technical users. Annotations are stored in a triple store enabling easy hybrid search. Moreover K-Forms enables to interconnect the underlying ontologies in an easy way in order to support search across form repositories.

The intuitive mixing and reuse of knowledge in existing ontologies has been explored in Potluck [5]. The system enables the mashing up of different pieces of ontological content. Their approach is data oriented, i.e. it involves the correlation of instances for mashing up, not the matching of the ontological schema as we enable in K-Forms. However, the spirit is very similar in the two systems.

A number of works have focused on ontology-based search. Most of the approaches, however, take the view that the user interaction should happen via keyword-based queries that are converted automatically into queries to the ontologies [10, 11, 15, 16]. In our view, the keyword-based approach has a number of issues. First it contributes to the disorientation of the users who do not know what to ask for if they are unfamiliar with the domain. This is visible for example in [16] where in the evaluation they had to remove a number of queries outside the coverage of the ontology. Moreover, keywords may have multiple interpretations (ambiguity) or no semantic interpretation. This requires sometimes sessions of refinements that can be disappointing for the users. Also it is unclear how to mix keywords and ontological queries, as it is difficult for the system to understand if part of the query is out of scope of the ontology or some parts must be interpreted as keyword matching. The use of Boolean operators *OR* and *AND* are difficult in these kind of interfaces, to the point that to our knowledge no one supports them. All the conditions are expected to be in AND. In K-Search, the explicit use of the ontology for query avoids the disorientation as it makes explicit the possible queries. It also enables the sophisticated use of Boolean operators. It enables to clearly define what is to be used as keywords and in what context (the whole document or just the sections annotated with a specific context). This enables very precise queries. Moreover, the possibility to specialize the query by right clicking on information in displayed documents enables a mixture of exploration and browsing which is mostly effective and liked by the users [8].

---

[2] http://www.mediawiki.org/wiki/Extension:Semantic_Forms

6 Conclusions and Future Work

In this paper we have described a suite of tools for knowledge management that enables easy and flexible knowledge acquisition, representation, capturing, retrieval, visualisation and sharing in large distributed organisations. K-Forms enables the intuitive design and deployment of web-based forms that capture semantic information. K-Forms provides a solution to the formalisation of knowledge capture for new cases, but does not provide a solution for legacy data, i.e. Word and Excel forms previously created. In these cases, information extraction from text can be applied; in [8] we have described two applications to two Rolls-Royce datasets that have been extremely successful. Now it is possible to search both new form-based data and legacy documents seamlessly. A user evaluation proved that K-Forms can successfully support users in conceptualising and reusing Knowledge: the users managed to create exactly the form they wanted in 90% of the cases and rated the process as very easy; in some cases they reused up to 80% of already existing concepts. K-Forms uses Semantic Web technologies in two ways: as a way to support the creation of forms (an ontology of form components and constraints guides the template creation) and as a way to create a domain ontology to support search.

K-Search enables accessing multiple repositories using either directly their local reference ontology or portions of other connected ontologies. Queries can either return a document generated with the retrieved knowledge (hence enabling user-specific information presentation), or a summary of knowledge in the form of graphs or statistics. Extracted triples can also be exported to an RDF repository or to a spreadsheet for further elaboration.

K-Forms and K-Search have been used for defining applications in real world environments in the aerospace domain.

Future work will concern further development of the concept of the networked ontologies and their impact on knowledge management. So far, that aspect has been only partially exploited as a way of searching across repositories. However, the creation of partially connected ontologies opens very interesting perspectives for knowledge management that go beyond just searching with a different perspective. Moreover, we need to explore the impact of changes to the existing form schema when some forms have been already filled. At the current point in time, the two versions of the same ontology are treated as different ontologies: as the new version is created by importing parts of the old one, search across the two is possible for the common parts. However, we feel that more sophisticated approach can be taken to address the issue.

Finally, a spin-out company (k-now.co.uk) has been created to commercialise the technology.

Acknowledgments. The work was supported by 1) IPAS, a project jointly funded by the UK DTI (Ref. TP/2/IC/6/I/10292) and Rolls-Royce plc and 2) X-Media, an Integrated Project on large scale knowledge management across media, funded by the European Commission as part of the IST programme (IST-FP6-026978), (www.x-media-project.org).

References

1. Davenport, T.H., Prusak, L.: Working Knowledge: How Organizations Manage What They Know. Harvard Business School Pr, Boston (1997)
2. Bonifacio, M., Bouquet, P., Cuel, R.: Knowledge Nodes: the Building Blocks of a Distributed Approach to Knowledge Management. Journal for Universal Computer Science 8(6), 652–661 (2002)
3. Hendler, J.: Agents and the Semantic Web. IEEE Intelligent Systems 16(2) (March/April 2001)
4. Wenger, E.: Communities of Practice: Learning, Meaning, and Identity. Cambridge University Press, Cambridge (1998)
5. Huynh, D., Miller, R., Karger, D.: Potluck: Data Mash-Up Tool for Casual Users. In: Aberer, K., Choi, K.-S., Noy, N., Allemang, D., Lee, K.-I., Nixon, L., Golbeck, J., Mika, P., Maynard, D., Mizoguchi, R., Schreiber, G., Cudré-Mauroux, P. (eds.) ASWC 2007 and ISWC 2007. LNCS, vol. 4825. Springer, Heidelberg (2007)
6. Sabou, M., Gracia, J., Angeletou, S., D'Aquin, M., Motta, E.: Evaluating the Semantic Web: A Task-based Approach. In: Aberer, K., Choi, K.-S., Noy, N., Allemang, D., Lee, K.-I., Nixon, L., Golbeck, J., Mika, P., Maynard, D., Mizoguchi, R., Schreiber, G., Cudré-Mauroux, P. (eds.) ASWC 2007 and ISWC 2007. LNCS, vol. 4825. Springer, Heidelberg (2007)
7. Gupta, S., Scott Hawker, J., Smith, R.K.: Acquiring and Delivering Lessons Learned for NASA Scientists and Engineers: A Dynamic Approach. In: Proceedings of the 43rd Annual Association of Computing Machinery Southeast Conference, March 2005, Kennesaw, GA (2005)
8. Bhagdev, R., Chapman, S., Ciravegna, F., Lanfranchi, V., Petrelli, D.: Hybrid Search: Effectively Combining Keywords and Semantic Searches. In: Bechhofer, S., Hauswirth, M., Hoffmann, J., Koubarakis, M. (eds.) ESWC 2008. LNCS, vol. 5021. Springer, Heidelberg (2008)
9. Dumas, M., Aldred, L., Heravizadeh, M., Hofstede, A.H.M.: Ontology Markup for Web Forms Generation. In: Proceedongs of the WWW 2002 Workshop on Real-World Applications of RDF and the Semantic Web, Honolulu (May 2002)
10. Manuel Gomez-Perez, J., Blazquez, M., Contreras, J., Jesus Fernandez, M., Paton, D., Rodrigo, L.: An Intelligent Search Engine for Online Access to Municipal Services. In: Proceedings of 1st ESTC, Vienna, May 31, June 1 (2007)
11. Lei, Y., Uren, V., Motta, E.: SemSearch - A Search Engine for the Semantic Web. In: Proceedings of 3rd European Semantic Web Conference, Budva, Montenegro, June 11th - 14th (2006)
12. Dzbor, M., Domingue, J.B., Motta, E.: Magpie - towards a semantic web browser. In: Proceedings of the 2nd International Semantic Web Conference, Sanibel Island, Florida, USA (2003)
13. Lanfranchi, V., Ciravegna, F., Petrelli, D.: Semantic Web-based Document: Editing and Browsing in AktiveDoc. In: Gómez-Pérez, A., Euzenat, J. (eds.) ESWC 2005. LNCS, vol. 3532. Springer, Heidelberg (2005)
14. Borlund, P.: Experimental Components for the Evaluation of Interactive Information Retrieval Systems. Journal of Documentation 56(1), 71–90 (2000)
15. Glover, T., Duke, A., Davies, J.: Squirrel: An Advanced Semantic Search and Browse Facility. In: Proceedings of The 4th European Semantic Web Conference, Innsbruck (May 2007)
16. Tran, T., Cimiano, P., Rudolph, S., Studer, R.: Ontology-based Interpretation of Keywords for Semantic Search. In: Aberer, K., Choi, K.-S., Noy, N., Allemang, D., Lee, K.-I., Nixon, L., Golbeck, J., Mika, P., Maynard, D., Mizoguchi, R., Schreiber, G., Cudré-Mauroux, P. (eds.) ASWC 2007 and ISWC 2007. LNCS, vol. 4825. Springer, Heidelberg (2007)
17. Corcho, O., Gómez-Pérez, A., López-Cima, A., López-García, V., Suárez-Figueroa, M.C.: ODESeW. Automatic Generation of Knowledge Portals for Intranets and Extranets. In: Proc. of the 2nd International Semantic Web Conference, Sanibel island (October 2003)

An Architecture for Semantic Navigation and Reasoning with Patient Data - Experiences of the Health-e-Child Project

Tamás Hauer[1], Dmitry Rogulin[1], Sonja Zillner[2], Andrew Branson[1],
Jetendr Shamdasani[1], Alexey Tsymbal[2], Martin Huber[2],
Tony Solomonides[1], and Richard McClatchey[1]

[1] CCS Research Centre, CEMS Faculty, University of the West of England
Coldharbour Lane, Frenchay, Bristol BS16 1QY, UK
[2] Corporate Technology Division. Siemens AG, Germany
first.last@{cern.ch,siemens.com,uwe.ac.uk}

Abstract. Medical ontologies have become the standard means of
recording and accessing conceptualized biological and medical knowledge. The expressivity of these ontologies goes from simple concept lists
through taxonomies to formal logical theories. In the context of patient
information, their application is primarily annotation of medical (instance) data. To exploit higher expressivity, we propose an architecture
which allows for reasoning on patient data using OWL DL ontologies.
The implementation is carried out as part of the Health-e-Child platform prototype. We discuss the use case where ontologies establish a
hierarchical classification of patients which in turn is used to aid the visualization of patient data. We briefly discuss the treemap-based patient
viewer which has been evaluated in the Health-e-Child project.

1 Introduction

Digitized information management has greatly improved clinical practice during
the past decades. Much patient data from demographic information to lab results
to diagnostic images is now being stored in computerized form. Today, one of the
main challenges for clinical information systems is to find, select and present the
right information to the clinician from the vast amount of data that is available.
This is a daunting task unless effective filtering, classification and visual aids are
available. In this paper we consider an architecture for semantic navigation and
reasoning with patient data, and share our experiences obtained within the EU
FP6 project *Health-e-Child*[1]. The functionality of the architecture hinges on the
patient data stored in a database distributed over the Grid, a domain ontology
with the knowledge relevant for lexicographic classification of patients, and two
key data analysis components, for ontology-based reasoning and visualization.
The focus in this paper is on presenting the later two; reasoning with ontologies

---

[1] http://www.health-e-child.org

A. Sheth et al. (Eds.): ISWC 2008, LNCS 5318, pp. 737–750, 2008.

and ontology-based visualization, which form the backbone of the considered architecture.

There has been lately much work on *ontology visualization* [1,2,3,4] that helps the user display and navigate underlying ontological concepts, see [5] for an extensive survey. In contrast to the mainstream works in the area, the present work proposes not the navigation of the ontology directly, but rather the visualization of instance data with the help of the knowledge that is represented by the ontology, or the deduced knowledge, projecting the respective data onto the ontology of interest. We employ two techniques suitable to ontology-based visualization of projected instance data for that; *facet browsing* and *treemaps* [6]. Each technique has its own benefits and limitations, which complement each other for the two techniques.

Visualization is tightly coupled with the *reasoning* component. Reasoning with ontologies is currently under active study in the semantic web field [7]; with biomedicine being one of the most popular application domains [8,9]. The ability to reason, that is to draw inferences from the existing knowledge to derive new knowledge is an important element for modern systems based on ontologies [7]. In particular, as we demonstrate in this paper using DL reasoning, reasoning with ontologies can help establish a hierarchical classification of patients for their intuitive visualization. By aligning patient data with relevant (fragments of) ontologies and inferring more descriptive patient ontology, improved patient data visualization and comparison can be realized.

The work in our study has been performed as part of the Health-e-Child (HeC) project. HeC is an EU-funded Framework Programme 6 (FP6) project, which was started in 2006, and aims at improving personalized healthcare in selected areas of paediatrics, particularly focusing on integrating medical data across disciplines, modalities, and vertical levels such as molecular, organ, individual and population. The project of 14 academic, industry, and clinical partners aims at developing an integrated healthcare platform for European paediatrics while focusing on some carefully selected representative diseases in three different categories; paediatric heart diseases, inflammatory diseases and brain tumours. The material presented in this paper contributes to the development of decision support facilities within the platform prototype which provide the clinicians with a tool to easily retrieve and navigate patient information and help visualizing interesting patterns and dependencies that may lead, besides personalized decision support concerning appropriate treatment, to establishing new clinical hypotheses and ultimately discovering novel important knowledge.

The paper is organized as follows. In section 2, we look at the requirements we have elicited from collaborating clinicians. In section 3, we review a few medical ontologies which we found useful for our use-cases. Section 4 contains the technical details of our approach of integrating patient data with external knowledge, and section 5 presents the architecture of our prototype platform. In Section 6, we visualize the hierarchical classification using treemap views.

2   Visualization Requirements in Clinical Practice

Clearly arranged visualization of patient data, supports clinicians in their daily tasks of clinical care and medical research. In order to navigate, analyze and visualize the dataset, it is useful to structure information and impose automatic annotation of patient records. In the following sections, we will illustrate how we realized the ontology-based visualization of patient information establishing the backbone of the introduced architecture. We will first describe the particular requirements for visualization of patient data in clinical practice. Bearing in mind the clinical visualization requirements, we will then sketch how we selected relevant medical knowledge sources and how treemaps in combination with our inferred patient ontology can be used for discovering correlations in patient records.

From extensive discussions with clinicians collaborating in Health-e-Child, we have learnt that the clearly arranged presentation of similar patients with respect to the complex and heterogeneous patient data becomes a key requirement for clinical decision support systems. For clinicians, the comparison of similar patients is a valuable source of information in the process of decision making. Therefore clinicians and medical researchers show a particular interest in visual means for comparing and analyzing the heterogeneous data of similar patients that cover demographics, family history, lab results, echocardiograms, MRI, CT, angiograms, ECG, genomic and proteomic samples, history of appointments, and treatments. Our existing data captures for each patient record more than 100 attributes describing the patient history and status data and allows the clinician to analyze patient records at a time. Our requirements elicitation has revealed the following further requirements in aiming for improved patient data visualization:

1. The discovery of patterns and dependencies in patient data. For example, establishing a correlation between the attributes "quality of life" and "tumour location" of brain cancer patients, is a routine task for clinicians. Therefore, the visualization of correlations between selected patient attributes becomes crucial in the clinical decision making process. As patient data attributes are provided in different levels of detail and precision, e.g. tumour location can be specified as "Cerebral Hemisphere" or, more detailed, as "Frontal Left Cerebral Hemisphere", the computation and visualization of data attributes for correlation needs to reflect the variety of detail and precision.
2. The comparison of similar patient records with regard to relevant patient attributes should be supported by browsing facilities over the set of all patient records along context relevant features. Again, the browsing facilities need to reflect the variety of data in detail and precision.

Within traditional applications, users may browse and visualize patient data but little or no help is given when it comes to interpretation because the required semantics are implicit and thus inaccessible to the system. Hence, aiming for the means to enable the easy browsing of patient data and the visualization of complex information e.g. correlations for the establishment of new hypotheses, we are integrating medical ontological knowledge to align patient data with the imposed knowledge structure thereby inferring the correct classification patient records.

3   Identified Medical Ontologies

For a beneficial integration of external semantics, one has to decide which external knowledge sources are appropriate for the purpose in mind, i.e. which external knowledge source captures relevant and helpful knowledge for a particular context. In our case, we aim to provide experts in brain tumour diagnosis and treatment with an improved method for patient data visualization and comparison. More precisely, we aim for the classification of patients with respect to their diagnosed tumour location or WHO tumour classification. We have chosen to use the Foundational Model of Anatomy (FMA) [10] covering the partitive hierarchy of brain regions as relevant and valuable medical knowledge. The coverage of the FMA is very comprehensive, containing approximately 70 thousand distinct anatomical concepts with more than 1.5 million relationships of 170 relationship types. We rely on fragments covering the concepts and relationships relevant to a particular visualization use-case. In our scenario, the established fragment encompasses all anatomical concepts describing possible brain tumour locations hierarchically structured by the regional_part_of relationship.

As second medical knowledge source, we identified the WHO classification of Tumours of the Nervous System establishing a classification and grading of human tumours that is accepted and used worldwide [11]. Its constituted entities establish a hierarchical structuring of histological typed tumours covering a multiplicity of factors, such as the immunohisto-chemistry aspects, genetic profiles, epidemiology, clinical signs and symptoms, imaging, prognosis, and predictive factors. The WHO classification of tumours refers to the ICD-O (International Classification of diseases for oncology) code and includes a WHO-grading scheme that is used for predicting response to therapy and outcome. For improved patient data visualization, we revert to its hierarchical structuring and its grading scheme. Similar to the integration of FMA's knowledge about tumour locations, we use the WHO classification's inherent hierarchical structuring for hierarchically representing patients' data.

4   Patient Record Classification by Reasoning

The Health-e-Child demonstrator is an integrated system that is built on top of a distributed platform, powered by grid technology, which hosts a distributed database and encompasses high-level enabling applications that exploit the intergrated medical data. The medical data is stored according to the Health-e-Child integrated data model and a service-oriented architecture provides access to storage, management and querying of the hosted information.

	Sex	Age at Diagnosis	...	Tumour Site	...
patient1	M	5	...	Cerebellum	...
patient2	F	7	...	Hypothalamus	...
...	...	...	...	...	...

Fig. 1. Simple view of patient data

For the richest possible interpretation, part of the medical data is annotated [12] using selected medical ontologies which allows for integration with external information and in some cases reasoning with expressive ontologies. In particular, when the external knowledge is expressed in some description logic then we can make use of available reasoners to provice enhanced, semantic query answering.

Our example use case is related to the visualization of the hierarchical classification of tumour patients. (Similar use cases have been discussed in [13,6].) We start with a simplified view of the patient database in Fig. 1; the HeC integrated data model is, of course, much more complex but the creation of such simple views is trivial. This database view is "flat" in that all attributes are individual (discrete) labels on the patient without explicitly defined semantics or structuring. Some of the attributes are numeric, others are categorical, taking values in a finite set of predefined concepts, for example tumour location, which refers to an anatomical region of the human body. In order to get correct answers to queries like "does patient x have a tumour in the Hindbrain", the system must have access to and be able to reason with the knowledge about the partitive anatomy of the brain.

The Foundational Model of Anatomy in OWL. The tumour site in the HeC database is annotated [12] with concepts from the FMA which duly encodes the meronomy of human anatomical structures. The FMA is an originally frame-based ontology but there has been effort to convert it to OWL [14,15]. DL Reasoning, unfortunately, does not scale well with the size of ontologies thus in practice one has to identify manageable fragments that are suited both to the use-case and to the data at hand. We have experimented with locality-based fragments [16,17] which are natural choice for reasoning tasks as they guarantee logical consistency but occasionally result in too large fragments and the algorithm is slow. Graph-based fragments (e.g. [18]) have the advantage of performance in terms of speed and fragment size but there is no guarantee about logical completeness. We acknowledge that the selection of such fragments to use (semi-automatic or manual) and the segmentation of the ontology is a difficult task on its own; for the purpose of the description of the architecture in this paper, however, we assume that such fragments are already available. Back to our example, we make use of the `regional_part_of` subontology rooted at the concept `Brain` of [14][2]. The first ingredient to our use-case is the *FMA T-box*:

---

[2] [14] is a stratified ontology with an OWL DL and an OWL full part. Although the `regional_part_of` semantics is encoded in the OWL full part, the fragment we need is purely DL.

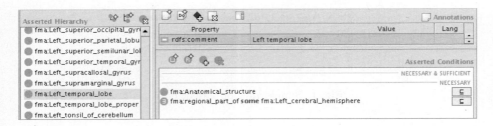

Fig. 2. Brain anatomy as defined in the FMA

$$\text{Tr(regional\_part\_of)} \quad (\textit{Transitive})$$
$$\geq 1 \text{ regional\_part\_of} \sqsubseteq \text{Anatomical\_Structure} \quad (\textit{Domain})$$
$$\top \sqsubseteq \forall \text{regional\_part\_of.Anatomical\_Structure} \quad (\textit{Range})$$
$$\text{Cerebellum} \sqsubseteq \text{Anatomical\_Structure} \sqcap \exists \text{regional\_part\_of.Metencephalon}$$
$$\text{Metencephalon} \sqsubseteq \text{Anatomical\_Structure} \sqcap \exists \text{regional\_part\_of.Hindbrain}$$
$$\text{Hindbrain} \sqsubseteq \text{Anatomical\_Structure} \sqcap \exists \text{regional\_part\_of.Brain}$$
$$\vdots$$

This ontology is our external knowledge that is independent of the Health-e-Child system. In order to make use of it, the information in our database has to be aligned with this ontology. The alignment is in part provided by the *Patient T-box:*

$$\geq 1 \text{ has\_tumour\_location} \sqsubseteq \text{Patient} \quad (\textit{Domain})$$
$$\top \sqsubseteq \forall \text{has\_tumour\_location.Anatomical\_Structure} \quad (\textit{Range})$$

This terminology establishes the semantics for the database records (`Patients`) and provides the alignment with the external ontology through the range of the `has_tumour_location` property. Finally, the relational instance data in Fig. 1 has to be mapped to conformant DL syntax, constituting the *Patient A-box:*

$$\langle \text{patient1, Cerebellum} ,\ldots\rangle$$
$$\langle \text{patient2, Hypothalamus} ,\ldots\rangle$$
$$\cdots$$
$$\Downarrow$$
$$(\text{Patient} \sqcap \exists \text{has\_tumour\_location. Cerebellum })(\text{patient1})$$
$$(\text{Patient} \sqcap \exists \text{has\_tumour\_location. Hypothalamus })(\text{patient2})$$
$$\cdots$$

The queries for our use case are subclasses of the `Patient` class. When establishing the hierarchical classification of patients based on tumour location, these queries make up our *classification T-box* whose defined concepts are the labels for the patient classes:

$$\text{BrainTumourPatient} \equiv \text{Patient} \sqcap \exists \text{has\_tumour\_location.}$$
$$(\text{Brain} \sqcup \exists \text{regional\_part\_of.Brain})$$
$$\text{CerebellumTumourPatient} \equiv \text{Patient} \sqcap \exists \text{has\_tumour\_location.}$$
$$(\text{Cerebellum} \sqcup \exists \text{regional\_part\_of.Cerebellum})$$
$$\vdots$$

In effect, reasoning with the anatomy ontology ensures that patients will be classified/annotated as e.g. *cerebellum tumour patient*s for every case that had originally been annotated with cerebellum or with any known regional part of the cerebellum as the tumour site.

To summarize, we consider the information at hand be represented in terms of DL, where we isolate three parts of the T-Box, a Patient terminology, which gives immediate semantics to the instances, an external knowledge base which is used as annotation and a classification ontology which is simply a collection of meaningful queries and is dictated by the use-case. In turn we arrive at the appropriate semantic classification of patients.

5 Prototype Platform Architecture

Let us now turn to our proposed architecture and implementation. The three core pillars of turning the above theory into a working infrastructure are: an ontology manager that can integrate the different knowledge components; a mapping mechanism between the database schema and the OWL DL A-box; and a reasoner which answers DL queries. These are implemented as services in the HeC platform service layer. Furthermore, these interact with additional components which are dictated by the use case, like a visualization component in our example.

Our implementation is based on the OWL API[3] and some of the design decisions have been influenced by the conformance to it:

- The *OWL-DL Integrator* is a generalized `OWLOntologyManager` which is responsible for importing and managing all the ontology components and loading the knowledge into a reasoner.
- The *DB/OWL DL Mapping component* creates simple views like that of Fig. 1 from the database and maps them to OWL-DL.
- The *Reasoner* uses the set of assertions and knowledge accumulated above and answers semantic queries, in particular creates the inferred patient classification.
- An optional Ontology Transformation Component is used in the hierarchical classification use-case to establish the set of queries or labels (e.g. "Cerebellum Tumour Patients").
- The Interpretation & Visualization Component maps the inferred classification from OWL to the appropriate representation of the user interface.

---

[3] http://owlapi.sourceforge.net/

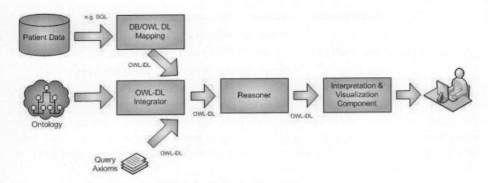

Fig. 3. Prototype Architecture

We briefly discuss these components in turn. The *DB/OWL DL Mapping component* uses the semantic annotations of the patients' data to expose patients' information as OWL DL ontology. First, a flat view is created from the relevant relations which includes the entity identifiers (patient ID), the concept URIs for the hierarchical classification (tumour location) and additional attributes (e.g. status at the end of treatment in later examples) which are of interest but don't contribute to the reasoning. Second, the relevant columns of the relation are translated into DL which is expressed in OWL. This is implemented using a Mapper class which is governed by the Patient terminology and a set of mapping descriptions which bridge the relational and DL schema. When browsing along multiple axes is required, they are all included in the OWL view.

The *OWL-DL Integrator* is a generalized `OWLOntologyManager`. It implements (exposes) multiple ways of accumulating knowledge, including loading OWL from external URI, loading instance data from the database using Mapper instances and adding standalone axioms on the fly. It populates the reasoner with the merged external, patient and classification ontologies and initializes the reasoning.

The *Reasoner* creates the inferred hierarchical classification of patients. We currently use the Pellet 1.5 reasoner engine in the Health-e-Child platform prototype.

The transitive `regional_part_of` property on anatomical concepts induces the subsumption relationship on patient classes (`has_tl` ≡ `has_Tumour_Location`):

$$X \sqsubseteq \exists \texttt{regional\_part\_of}.Y$$
$$\Downarrow$$
$$\exists \texttt{has\_tl}.(\exists \texttt{regional\_part\_of}.X) \sqsubseteq \exists \texttt{has\_tl}.(\exists \texttt{regional\_part\_of}.Y)$$

We can't, however, directly exploit this inference because the classes are not defined: the reasoner only creates inferred subsumption hierarchy for named classes and the visualization also requires the definition of the classes for the lexicographic hierarchy. In other words, we need to supply the set of queries that govern the classification.

To that end, the *ontology transformation component* creates the missing bit: the definition of class names for the hierarchy. In our example this is a set of name definitions based on the corresponding concept names. Creating this set is trivial because we obviously don't use any structure of the ontology.

The *Interpretation & Visualization Component* transforms the inferred OWL into the format conformant to the API of the visualization software. It can also add further attributes from the database which were not part of the reasoning process.

6 Patient Data Visualization Using TreeMaps

The user requirements for visualization patient data revealed the importance of means for discovering correlations in patient data. Therefore, we decided to use treemaps as visualization component. Treemaps [19] are an efficient two-dimensional technique to visualize hierarchical data structures. It is particularly popular for disc storage view of hierarchical filesystems because file size is an aggregate attribute of files. It is a space filling technique, i.e. one that uses the entire screen area by dividing it up between leaf nodes which are subsequently grouped into enclosing rectangles [5]. The image is effectively a rectangular Venn-diagram of nonintersecting sets. Besides the set semantics, there are other attributes such as colour, choice of font and label which can represent attributes of the data beyond that of the hierarchy.

To allow for improved patient visualization and discovery of patient correlation, we represent patients as rectangles of equal size so that the cardinality of patient classes can be easily seen. Figure 4 shows the user interface based on the patient taxonomy that has been inferred from the anatomical meronomy. The medical background information, in our example the hierarchical structured brain tumour locations establishes enclosing and nested areas that are labelled by regions of the brain. Colour may be used to visualize further attributes, in our example the status at the end of treatment for each patient. It is meant to show that space-filling hierarchies can be useful for visualizing correlation, in this example one finds that patients with tumour in the Hindbrain tend to have better prognosis than in other areas (especially the Left Temporal Lobe). In a similar manner, we can correlate and visualize patient record with respect to the "WHO tumour classification" and, for instance their "quality of life" parameter.

A further advantage of treemaps is capability for "binning" the data. An example of Treemap with patient binning would show the patients grouped not only by tumour location but by age as well. It follows visually in Fig. 5 that – in this sample – the end of treatment status of younger patients is worse than that of older patients: there is noticeable correlation which the user can discover by visualizing the data the right way.

While ontology-based visualization using treemaps was already reported, for example to visualize and navigate the Gene Ontology and microarray data [20], we are unaware of its applications in clinical decision support, where patient proximity is visualized by projecting patient data onto existing ontologies.

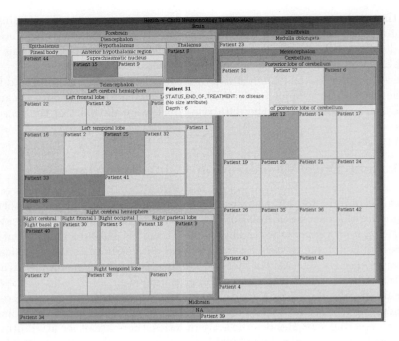

Fig. 4. A full Treemap view of the ontology assisted data representation

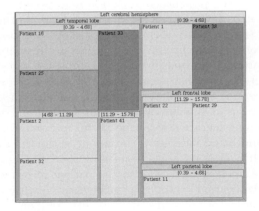

Fig. 5. Binning: patients are grouped both by tumour site as well as age

6.1 Related Approaches in Knowledge Visualization

The work presented in this section relates to visualization for knowledge discovery. The motivating example is that of the physician who believes that an extensive set of patient data would reveal subtle patterns if only it could be visualized in the right way. The right way is here taken to mean analyzed in terms of established or even tentative concepts. Thus the ontology carries what may have already been accepted or adduced as a research hypothesis, and the data is visualized on that basis, as a means of strengthening the evidence or as a means of bolstering the new hypothesis.

There is a significant volume of work on visualization with (and without) ontologies. It may be differentiated from the work presented here on the basis of various criteria, such as: what is visualized, why and how. Visualization may be of data from a homogeneous database in order to display relative volume or from a heterogeneous source to expose some other feature or for exploratory data analysis [21]; it may be of the structure of the data (rather than the data itself) in order to reveal class-level relationships [3]. Visualization has been used to facilitate query formulation or to order threads of data in some schematic way (e.g. temporally); to display a data schema or to inform navigation through the data [22]. In particular cases, ontology-based visualization has been used to support queries based on temporal abstractions [23]; to enrich maps with additional geographic information [24]; to reveal multiple levels of abstraction in decision-tree generation [25] and to assist in information mining [26]; very popularly, to map social networks and communities of common interest [27], [28], [29]. Ontologies have also been used for knowledge discovery without visualization, especially in the integration of heterogeneous scientific repositories ([30]).

For an up to date survey of ontology visualization methods reference must be made to [5]. This paper provides a near exhaustive discussion of methods of classification, of visualization techniques (Euclidean, hyperbolic and spherical space, node-and-link, and other less obviously geometric methods), of representation, overview and focus methods, but it is distinctive in offering no discussion of content. There is a body of work dating back to the mid-'90s on database visualization (see [21]) some of it devised to assist with data mining tasks.

Arguably the work closest to ours in spirit is [31], although their approach is designed more to manage the heterogeneity of web sources in order to summarize the information provided. Their visual method is based on node-and-link representation and resembles mind-mapping. On the other hand, [32] use 'semantic visualization' to support the development process for database-oriented systems, but in the process must tackle questions of data visualization.

7   Conclusions and Future Work

As the biomedical and clinical information available in digitized form is continually increasing, so is the demand for advanced data integration and visualization tools to help clinicians and researchers to explore the information and knowledge at their hands. In this paper we presented a system for patient data visualization which combines reasoning about data using external knowledge with advanced visualization that makes use of the inferred structure. Our approach makes use of existing ontology visualization paradigms with a view to presenting the data whose semantics is governed by the ontology.

We have presented an architecture of a system implementing our framework and demonstrated its functionality using paediatric brain cancer data acquired in Health-e-Child together with anatomical knowledge from the Foundational Model of Anatomy for the purpose of creating annotated treemap views of patient characteristics.

We presented a demonstrator of this framework to clinicians of the Health-e-Child project to evaluate early user experience. We demonstrated the use case of treemap visualization presented in this paper and facet browsing of [6]. In both cases the response was positive: facet browsing helps the clinicians to locate follow-up patients based on incomplete information while the treemap was appreciated for the easy visualization of correlation between clinical attributes. The clinicians also noted that it has potential in education and training.

There are several challenges along the lines we presented. A number of approaches exist for visualizing hierarchies but to really exploit their power, end-users have to be trained so that they can define the visualization which suits their interest and then navigate it. Inferring the hierarchy of the data based on its semantics requires suitable ontologies and alignment with them. In the medical domain, annotation with the UMLS meta-thesaurus is a good choice because it maps to many ontologies. A serious bottleneck is the reasoner as performance scales very badly with the size of the ontologies. We compensated this with fragment extraction but ontology segmentation is also difficult and, especially when one aims for consistent fragment, it is slow and can only be done off-line.

Our future work includes a comparative analysis of how different ontology fragments perform in this approach with real data and we are investigating how this classification scales with the number of patients. We also want to explore the means to give some control over the selection of ontology fragments and definition of classification criteria to the end-users. This is a difficult problem but could prove very interesting.

Acknowledgements

This work has been partially funded by the EU project Health-e-Child (IST 2004-027749). The authors wish to acknowledge support provided by all the members of the Health-e-Child consortium in the preparation of this paper.

References

1. Bosca, A., Bonio, D.: Ontosphere: more than a 3d ontology visualization tool. In: SWAP The 2nd Italian Semantic Web Workshop (2005)
2. Noy, N.F., Fergerson, R.W., Musen, M.A.: The knowledge model of protege-2000: Combining interoperability and flexibility. In: Dieng, R., Corby, O. (eds.) EKAW 2000. LNCS (LNAI), vol. 1937. Springer, Heidelberg (2000)
3. Mutton, P., Golbeck, J.: Visualization of semantic metadata and ontologies. In: IV 2003. Proceedings of the Seventh International Conference on Information Visualization, Washington, DC, USA, p. 300. IEEE Computer Society, Los Alamitos (2003)

4. Pietriga, E.: Isaviz, a visual environment for browsing and authoring rdf models. In: Eleventh International World Wide Web Conference Developer's Day (2002)
5. Katifori, A., et al.: Ontology visualization methodsa survey. In: ACM Comput. Surv., vol. 39, p. 10. ACM, New York (2007)
6. Zillner, S., Hauer, T., Rogulin, D., Tsymbal, A., Huber, M., Solomonides, T.: Semantic Visualization of Patient Information. In: Computer Based Medical Systems (CBMS) (2008)
7. van Harmelen, F., Huang, Z., ten Teije, A.: Reasoning with inconsistent ontologies: framework, prototype, and experiment. In: Studer, R., Davies, J., Warren, P. (eds.) Semantic Web Technologies. Wiley, Chichester (2008)
8. Rubin, D.L., Dameron, O., Musen, M.A.: Use of description logic classification to reason about consequences of penetrating injuries. In: American Medical Informatics Association Conference (AMIA 2005), pp. 649–653 (2005)
9. Dameron, O., Roques, E., Rubin, E., Marquet, G., Burgun, A.: Grading lung tumors using owl-dl based reasoning. In: Proceedings of the 9th International Protege Conference (2006)
10. Rosse, C., Mejino, J.: A reference ontology for bio-medical informatics: the foundational model of anatomy. In: Biomedical Informatics, vol. 36, pp. 478–500 (2003)
11. Louis, D., et al.: The 2007 WHO Classification of Tumours of the Central Nervous System. Acta Neuropathol 114, 97–109 (2007)
12. Rogulin, D., Ruiz, E.J., McClatchey, R., Berlanga, R., Hauer, T., Shamdasani, J., Zillner, S., Branson, A., Nebot, V., Manset, D., Freund, J.: Medical Data Integration and the Semantic Annotation of Medical Protocols. In: Computer Based Medical Systems (CBMS) (2008)
13. Tsymbal, A., Huber, M., Zillner, S., Hauer, T., Zhou, K.: Visualizing patient similarity in clinical decision support. In: Hinneburg, A. (ed.) LWA, pp. 304–311. Martin-Luther-University, Halle-Wittenberg (2007)
14. Noy, N.F., Rubin, D.L.: Translating the foundational model of anatomy into owl. Technical report, Standford Center for Biomedical Research (2007)
15. Golbreich, C., Zhang, S., Bodenreider, O.: Migrating the FMA from Protege to OWL. In: 8th International Protege Conference, Madrid (2005)
16. Grau, B.C., Horrocks, I., Kazakov, Y., Sattler, U.: Just the right amount: extracting modules from ontologies. In: WWW 2007. Proceedings of the 16th international conference on World Wide Web, pp. 717–726. ACM, New York (2007)
17. Sattler, U., Schneider, T., Jiménez-Ruiz, E., Grau, B.C., Berlanga, R.: Safe and economic re-use of ontologies: A logic-based methodology and tool support. In: Bechhofer, S., Hauswirth, M., Hoffmann, J., Koubarakis, M. (eds.) ESWC 2008. LNCS, vol. 5021. Springer, Heidelberg (2008)
18. Jimenez-Ruiz, E., et al.: Ontopath: A language for retrieving ontology fragments. In: OTM Conferences (1), pp. 897–914 (2007)
19. Shneiderman, B.: Tree visualization with tree-maps: 2d space-filling approach. ACM Transactions on Graphs 11, 92–99 (1992)
20. Babaria, K., Baehrecke, E.H., Dang, N., Shneiderman, B.: Visualization and analysis of microarray and gene ontology data with treemaps. BMC Bioinformatics 5(1), 84 (2004)
21. Keim, D.A.: Information visualization and visual data mining. Transactions on Visualization and Computer Graphic 8, 1–8 (2002)
22. Cannon, A., et al.: Ontology-driven automated generation of data entry interfaces to databases. In: Key Technologies for Data Management (LCS). Springer, Heidelberg (2004)

23. Shahar, Y., Cheng, C.: Ontology-driven visualization of temporal abstractions. In: KAW 1998 (1998)
24. Ipfelkofer, F., et al.: Ontology Driven Visualisation of Maps with SVG - An Example for Semantic Programming. In: Conference on Information Visualization, pp. 424–429 (2007)
25. Zhang, J., et al.: Ontology-driven induction of decision trees at multiple levels of abstraction. LNCS (LNAI). Springer, Heidelberg (2002)
26. Castillo, J.A.R., et al.: Information extraction and integration from heterogeneous, distributed, autonomous information sources - a federated ontology-driven query-centric approach. In: Conference On Information Reuse and Integration, IEE International, pp. 183–191 (2003)
27. Mika, P.: Ontologies are us: A unified model of social networks and semantics. In: Gil, Y., Motta, E., Benjamins, V.R., Musen, M.A. (eds.) ISWC 2005. LNCS, vol. 3729, pp. 522–536. Springer, Heidelberg (2005)
28. Zheng, J., Niu, J.: Unified mapping of social networks into 3d space. In: Second International Multi-Symposiums on Computer and Computational Sciences (IMSCCS 2007), pp. 305–311 (2007)
29. Domingue, J., Motta, E.: A knowledge-based news server supporting ontology-driven story enrichment and knowledge retrieval. In: Fensel, D., Studer, R. (eds.) EKAW 1999. LNCS (LNAI), vol. 1621, pp. 103–120. Springer, Heidelberg (1999)
30. Christophides, V., et al.: Ontology-driven integration of scientific repositories. In: 4th International Workshop on Next Generation Information Technologies and Systems, pp. 190–202. Springer, London (1999)
31. Fluit, C., et al.: Ontology-based information visualisation: Towards semantic web applications. In: Geroimenko, V. (ed.) Visualising the Semantic Web, 2nd edn. Springer, Heidelberg (2005)
32. Sheth, A., Avant, D.: Semantic visualization: Interfaces for exploring and exploiting ontology, knowledgebase, heterogeneous content and complex relationships. In: NASA Virtual Iron Bird Workshop (2004)

Requirements Analysis Tool: A Tool for Automatically Analyzing Software Requirements Documents

Kunal Verma and Alex Kass

Accenture Technology Labs, San Jose, CA, USA
{k.verma,alex.kass}@accenture.com

Abstract. We present a tool, called the *Requirements Analysis Tool* that performs a wide range of best practice analyses on software requirements documents. The novelty of our approach is the use of user-defined glossaries to extract structured content, and thus support a broad range of syntactic and semantic analyses, while allowing users to write requirements in the stylized natural language advocated by expert requirements writers. Semantic Web technologies are then leveraged for deeper semantic analysis of the extracted structured content to find various kinds of problems in requirements documents.

Keywords: Requirements analysis, Domain Ontologies, Semantic Analysis, SPARQL.

1 Introduction

Requirements documents for large software systems are very lengthy and complex. Moreover, gathering correct and accurate requirements from customers requires a deep understanding of both the customer's business needs and the technical issues involved, which are often not communicated clearly or at all. These challenges often result in poorly written requirements that are unclear, verbose, and even inconsistent. These poorly written requirements result in extensive rework, which in turn drives up project costs, extends timelines, and decreases customer satisfaction (see, for example, [1]). Many organizations have processes and best practices in place for reviewing requirements documents at various stages of the project lifecycle to detect problems that can lead to extensive rework. This manual review process, however, is painstaking, time consuming and often fails to uncover even the commonly occurring problems in requirements documents. Hence, more automated solutions to assist in the review of requirements documents are needed.

A number of researchers have recognized this need and have proposed a number of tools and techniques to automatically analyze requirements (see, for e.g.,QUARCC [1], QuARS [4], KaOS [3]). There also exist commercial products (e.g. Raven [7]) that provide assistance of this nature. However, these solutions have not achieved a high level of adoption for two reasons. 1) Many solutions (e.g., KaOS [3]) require users to write requirements in formal notations. This restriction is not feasible in practice because requirements documents are written by semi-technical analysts and have

A. Sheth et al. (Eds.): ISWC 2008, LNCS 5318, pp. 751–763, 2008.
© Springer-Verlag Berlin Heidelberg 2008

to be signed-off by business executives. Hence, the communication medium is still natural language. 2) Solutions (e.g. QuARS [4], Raven [7]) that allow users to write requirements in natural language are limited in the kinds of analysis they can perform. For example, QuARS [4] only supports phrasal analysis and Raven [7] is restricted to supporting only use cases. For broad adoption, there is need for a tool that takes a more holistic approach and allows a broad range of analyses over natural language requirements.

In this paper, we report on a tool, called the *Requirements Analysis Tool* (RAT) that automatically performs a wide range of syntactic and semantic analyses on requirements documents based on industry best practices, while allowing the user to write these documents in natural language. RAT encourages users to write in a standardized syntax, a best practice, which results in requirements documents that are easier to read and understand. RAT performs a syntactic analysis by using a set of glossaries to identify syntactic constituents and flag problematic phrases. An experimental version of RAT also performs semantic analysis using domain ontologies and structured content extracted from requirements documents during syntactic analysis. The semantic analysis, which uses semantic Web technologies, detects conflicts, gaps, and inter-dependencies between different sections (corresponding to different subsystems and modules within an overall software system) in a requirements document. Hence, the key contributions of this paper are as follows:

1. We present a tool for enforcing requirements best practices, while allowing users to write requirements in natural language with the help of controlled syntaxes and user-defined glossaries.
2. We show how the structured content extracted from requirements can be used for a deeper semantic analysis with the help of a semantic engine and domain ontologies.

The rest of the paper is organized as follows. Section 2 presents an overview of common problems that occur in requirements documents, and also presents some best practices that RAT supports for avoiding those problems. Section 2 also presents an overview of what RAT does. Sections 3 and 4 present the details of syntactic and semantic analysis performed by RAT. A discussion of the related work is presented in Section 5. Implementation details and an early user evaluation are presented in Section 6. Finally, our conclusions are presented in Section 7.

2   Overview of Common Requirements Problem and Best Practices for Writing Requirements

Despite the challenges associated with writing and analyzing requirements documents, there is general agreement about the kinds of problems that plague these documents. We give an overview of these problems followed by an overview of best practices for writing good requirements.

2.1   Common Requirements Problems

Here is an overview of common problems that occur in a requirements document:

Requirements that are easy to misunderstand. This problem is caused by the use of ambiguous terms, terminological inconsistencies or the use of non-standard syntaxes for documenting requirements.

- *Use of ambiguous terms.* Consider the requirement "The Payroll Database shall respond to all queries quickly". It is unclear what "quickly" means in this context (1 ms, 1 sec, 10 sec, etc.) and this may lead to a mismatch between the expectations of the stakeholder and the interpretation of the developer, ultimately leading to rework and/or customer dissatisfaction.
- *Terminological inconsistency.* Consider the following requirements: 1) "The Order Entry System shall allow users to view all orders placed in a day" and 2) "The Order Processing System shall generate daily reports". In this case, the requirements analyst uses two different terms to refer to the same system, leading to confusion.
- *Use of non-standard syntaxes such as missing agent.* Consider the requirement "Shall have the ability to generate profit reports". It is unclear which agent (e.g., which system or sub-system) shall have the ability to generate the profit reports. In this case, the designers/developers may try to guess which system should have that functionality, potentially leading to rework at a later phase.

Requirements that are inconsistent. This problem is caused by requirements either conflicting with each other or with some policy or business rule. Consider the following two requirements: "The Payroll database shall authenticate all requests using LDAP" and "The payroll database must respond to all requests within 1 millisecond". In this case, it may not be feasible to satisfy this requirement if the LDAP server takes more than 1 millisecond to process each request. If such a defect is not detected early on, then it is possible that the infeasibility is only realized when the LDAP server and the Payroll database have been designed and implemented, leading to large costs in rework.

Requirements that are incomplete. This problem is caused by missing some requirements of a certain type. For example, it is often the case that performance requirements for a system are omitted either due to the lack of knowledge among the stakeholders or because the requirements analysts fail to elicit them. This leaves the technical designers and developers to make design choices about the software system, which may or may not meet the stakeholder's approval.

2.2   Best Practices and Overview of Approach

One of best-known and authoritative sources for software requirements [10] suggests the following best practices for writing requirements to avoid to some of the problems mentioned in the previous sub-section:

1. Write complete sentences that have proper spelling and grammar.
2. Use active voice (For e.g., " The System shall send an e-mail" instead of "E-mail will be sent")

3. Use terms consistently and as defined in the glossary. Do not use different phrases to refer to the same thing. (For e.g., Do not use Order Processing System and Order Entry System to refer to the same system)
4. Write sentences in a consistent fashion starting with the agent/actor, followed by an action verb, followed by an observable result. (For e.g., "The System (agent) shall generate (action verb) a report")
5. Clearly specify the trigger condition that causes the system to perform a certain behavior. (For e.g., "If the user enters the wrongs password , the system shall generate an error message")
6. Avoid the use of ambiguous phrases

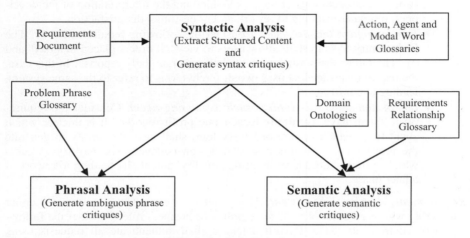

Fig. 1. Analysis Overview

These best practices are designed to deal with syntactic problems such as those caused by the use of ambiguous terms, terminological inconsistencies or the use non-standard syntaxes and can be detected by syntactic analysis of the requirements. However, other problems caused due to inconsistent and incomplete requirements require a more domain specific best practices, where domain knowledge needs to be applied to find the occurrence of such problems.

As shown in Figure 1, our approach starts by syntactically analyzing requirements documents and extracting structured content from the requirements document about each requirement. We leverage a set of controlled syntaxes and user defined glossaries for the syntactic analysis. Then the structured content is leveraged for phrasal and semantic analysis. For the phrasal analysis, RAT uses a problem phrase glossary. For semantic analysis another glossary called the requirements relationship glossary is used. We will discuss the different analysis techniques and the different glossaries in the ensuing sections.

3   Syntactic and Phrasal Analysis

In this section, we will describe our approach for syntactic analysis. We will first cover the set of controlled syntaxes supported by RAT, and then we will discuss the user-defined glossaries, followed by our approach for extracting structured content from requirements documents. Finally we will provide a brief overview of phrasal analysis.

3.1 Controlled Syntaxes for Writing Requirements

RAT supports a set of controlled syntaxes for writing requirements. This set includes the following:

Standard Requirements Syntax. This is the most commonly used syntax for writing requirements and is of the form:

StandardRequirement: <agent> <modal word> <action> <rest>

Where, <agent>, <action>, <modal word> are phrases in their respective glossaries and <rest> is the remainder of the sentence and can consist of agents, actions or any other words and is defined as:

<rest>: [<anyword> | <agent> | <action>]*

Here is an example of an standard requirement: "The system shall generate profit reports." Here "System" is the agent, "shall" is the modal word and "generate" is the action and "profit reports" is the rest of the requirement.

Conditional Requirements Syntax. There are a number of conditional requirements supported by RAT. For brevity, we will only discuss the most common condition syntax:

<"if"> <condition> <"then"> <StandardRequirement>

For example, consider the following requirement: "If the user enters the wrong password, then the system shall send an error message to the user." In the case, "user enters the wrong password" is the condition. The part after "then" is treated like a standard requirement.

Business Rules Syntax. RAT treats all requirements that start with "all", "only" and "exactly" as business rules. An example is: "Only the members of payroll department will be able to access the payroll database".

3.2 User-Defined Glossaries to Parse Documents

RAT uses three types of user glossaries to parse requirements documents: agent glossary, action glossary and modal word glossary. An agent entity is a broad term used to denote systems, sub-systems, interfaces, actors and processes in a requirements document. The agent glossary contains all the valid agent entities for the requirements document. It captures the following information about each agent: name of the agent, immediate class and super-class of the agent and a description of the agent. The class and parent of the class field of the glossary are used to load the glossary into the semantic engine. Table 1 shows some sample entries for an agent glossary.

Table 1. Agent Glossary

Agent Name	Description	Class	Super-Class
Web Server	The Web Server for the project.	System	Agent
Payroll Employee	The payroll dept employee	User	Agent
SAP System	The SAP system for the project	System	Agent
Order Parts Process	The ordering parts process	Process	Agent

The action glossary lists all the valid actions for the requirements document. It has a similar structure to the agent glossary. Examples of actions are "generate", "send" and "allow". The modal word glossary lists all the valid modal words in the requirements document. Examples of modal words are "must" and "shall".

3.3 Parsing and Extracting Structured Content

In this section, we will provide a high level overview of how controlled syntaxes and glossaries are collectively used to parse the requirement. We have a deterministic finite automata based approach to parse the requirements, extract structured content and generate error messages. We shall define the approach informally:

- S is the set of states, and each state denotes a certain stage in the parsing of requirements. s is the start state denoting the start of requirement.
- The entire English language is considered as the alphabet. However only certain members of the alphabet (for e.g., "if", "when", "all", <agent>, <action>, <modal word>) lead to transitions between states.
- G is the set of goal states. Only a transition sequence (for e.g. <agent><modal word><action><".">) that denotes a controlled syntax can lead to a goal state from the start.
- E is the set of error states and each error state has an associated error message. Consider the following requirement: The Messaging System shall allow users to view previously sent messages. If the agent "Messaging System" is not present in the agent glossary, the transition will end in an error state corresponding to missing agent and RAT will mark it with a "Missing Agent" critique.

For brevity, we will not provide more details of the automata here. As a requirement is parsed, structured content is extracted for each requirement. This structured content is used for both the syntactic and semantic analyses. The extracted structured content contains:

- Type of requirement syntax (standard, conditional, business rule)
- All agents, actions and modal words for all the requirements
- Different constituents of conditional requirements.

3.4 Finding the Use of Problematic Phrases

Certain phrases frequently result in requirements that are ambiguous, vague or misleading. The problematic use of such phrases has been well documented in the requirements literature. A classification of problematic phrases is presented in [4]. [10] provides a list of such words and how to correct requirements that use them. We have collected an extensive list of problematic phrases from a number of sources and stored it in a user-extensible glossary called the problem phrase glossary. We believe that the real power of the tool is in creating user-defined, domain-specific glossaries. We have been working with experts in a number of domains to find problematic phrases in their respective domains. Consider this requirement from a financial domain: The Reporting System shall generate new account reports daily. While this requirement seems correct and precise, it turns out that "daily" is ambiguous, since it does not specify what time the reports should be generated (end of day, start of the day, mid-day).

3.5   User Interfaces

The current version of RAT makes the analyses described above available to the user through two distinct user interfaces: First, there is an interactive interface, called the *Requirements Checker*, which is depicted in Figure 2. This interface is similar to Microsoft Word's built-in spelling and grammar checker. The Checker provides an explanation of each problem, one by one, and suggestions for fixing it. For instance, when a problematic phrase is detected, the checker presents the context, explains why the phrase causes problems, and provides a number of suggestions for improving the requirement.

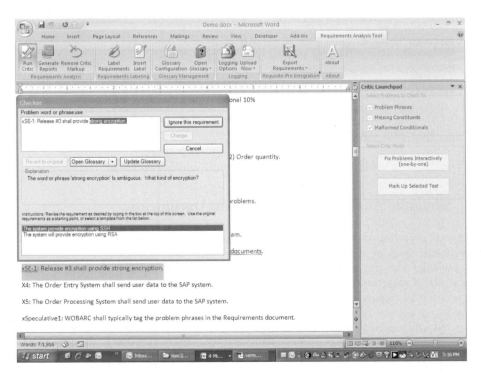

Fig. 2. Interactive Requirements Checker

A second interface, called the *Requirements Tagger*, which is shown in Figure 3, runs through the entire document, and marks it up using Word's margin-note comment feature. It tags all problems it can find with comments that are color-coded to represent the kind of problem detected. The Tagger also highlights the requirements text, underlining the agent and action in each requirement through the use of the corresponding glossaries.

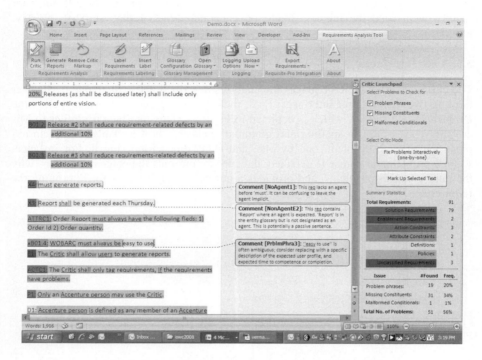

Fig. 3. Comments generated using Requirements Tagger

4   Semantic Analysis of Requirements Documents

Experts use a lot of domain knowledge to study requirements documents for various problems, such as dependencies and conflicts between requirements. Much of this knowledge can be captured using domain-specific ontologies to provide a deeper analysis of requirements document. The crux of our approach is to create a semantic graph for all requirements in the document based on the extracted content. In RAT, users can use the requirements relationship glossary, which is shown in Table 2 to enter domain specific knowledge. The requirements relationship glossary contains a set of requirement classification classes, its super-class, keywords to identify that class and the relationships between the classes. Some samples requirement classes for non-functional requirements and their relationships are shown in Table 2, which essentially states that security based requirements (encryption, authentication) affect time based requirements (query time and response time).

Table 2. Requirements Relationship Glossary

Class	Super-Class	Keywords	Relationships
Security	NonFunctional		Affects:Time
Authentication	Security	Password, token, authentication, Kerberos	Increases:ResponseTime
Encryption	Security	encrypt, SSH, RSA, DSA	Increases: Response Time
Time	NonFunctional		
Query Time	Time	Query time, querytime	
Response Time	Time	response, respond	

For creating the requirements relationship glossary, we have leveraged some of the non-functional classes and relationships presented in QUARCC [1]. For functional requirements, we have been working with experts in various domains such as finance. As is the case with all other glossaries, this glossary is also user editable. Our fundamental belief is that once a requirements document is transformed into a semantic graph (represented as an OWL ontology), users can query for different kinds of relationships that are important to them. Here are the steps that RAT uses to create the semantic graph (graphically depicted in Figure 4) from a requirements document:

1. Load the core requirements ontology in the Semantic Engine. The core requirements ontology is basic requirements ontology with different types of requirements formats (standard, business rule and conditional) and the information that each of them contain.
2. Using the agent and action glossaries to create the agent and action classes and instances of agents and actions.
3. Using the requirement relationship glossary to create requirements classification classes and their relationships.
4. Using the extracted structured content (enhanced by requirement classification information from Table 2) to create instances of requirements.

4.1   Requirements Dependency Analysis

There are many dependencies between the requirements in a document. However, due to the large size of the documents and the fact that different people may have written different sections of the document, it is difficult to identify all these dependencies. Not being able to uncover some of these dependencies can lead to a number of problems such as: 1) Conflicting requirements in the requirements documents that may lead to flawed design and development decisions and 2) Not scheduling some lower priority requirements for development, especially those that may be required for implementing higher priority requirements.

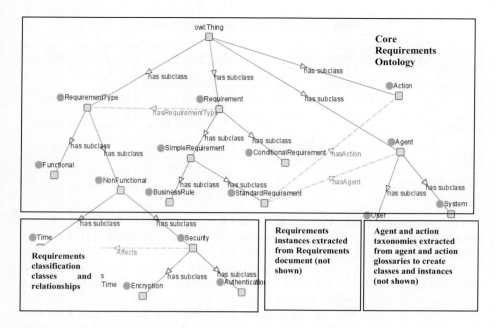

Fig. 4. Core Requirements Ontology and how it is built from different glossaries and requirements document

Let us illustrate a dependency with the help of an example. It is common knowledge among technical architects that increasing the security profile of a system affects it performance. This is modeled in the ontology in Figure 5. Let us see how RAT can leverage this knowledge. Consider the following two requirements: 1) The Web Server shall encrypt all its responses using SSH and 2) The Web Server shall have a response time of 5 milliseconds or less. Once these requirements are entered into Jena, then we can then issue the following SPARQL query to check for any dependencies between them.

```
select ?req1, ?req2 where
{ ?req1 hasRequirementType ?type1 . ?req2 hasRequirementType ?type2 .
 Affects domain ?type1 . Affects range ?type2 .
 ?req2 hasAgent ?agent2 .
 ?req1 hasAgent ?agent1 filter( ?agent1 = ?agent2)})
```

This query returns all requirements (for the same agent) that have requirement-types, which "affect" each other. This is illustrated using Figure 5. In this case, the first requirement (WebServer_EncReq) has requirement-type Encryption, which is a sub-class of Security. Similarly, the second requirement (Web-Server_ResTimeReq) has requirement-type Response Time, which is a sub-class of Time. There is a relationship in the ontology that says that Security affects Time. Thus, the query returns the fact that requirements 1 affects requirement 2.

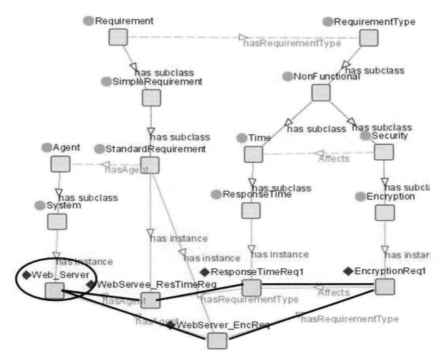

Fig. 5. An ontology snippet (drawn using the Jambalaya plug-in) showing the dependency (using thick lines) between the encryption and response time requirements for the agent Web_Server (circled in figure)

4.2   System and Agent Role Based Analysis

Given the large size of requirements documents, it is useful request reports that highlight various properties and relationships. Using the agent glossary, where we require the users to classify the different agents either as systems, users or any other category, RAT supports the following queries:

- *Systems that are interacting with each other.* We have created a query which returns all requirements with a system as the primary and secondary agent. Consider the following example: "The Web Server shall send the vendor data to the SAP System." In this case, the Web Server is the primary agent and the SAP System is the secondary agent and both of them are classified as systems in the agent glossary, so RAT can deduce that they are interacting with each other.
- *Systems that are missing certain kinds of non-functional requirements.* Non-functional requirements are often overlooked in requirements documents. Hence, we have created a query to identify and return all systems that are missing a non-functional requirement in one or more of following categories: security, performance, reliability, usability, integration and data requirements.

- *Interacting systems that do not have compatible security profile.* This query checks if a system has similar security protocol requirements with another system that it interacts with. Consider the case where one system has a requirement for supporting a certain kind of encryption, while an interacting system does not have any requirement for the same kind of encryption. In this case, RAT points out that there might be a security based incompatibility.

Another important type of analyses is based on the role of the agent. While we do not have any pre-defined queries for this category, we believe that users can enter queries that may be useful in their domains. One way to proceed would be to capture information in the domain ontologies about which agents are allowed to perform which actions (e.g. only purchasing manager may approve new suppliers). Another variation of a similar analysis is "Separation of duty", as outlined in Sarbanes Oxley.

5 Related Work

One approach of analyzing requirements is to treat them as a form of pseudo-code, or even a very high-level language, through which the requirements analyst is essentially beginning to program the solution envisioned by the stakeholders. In such a case, it would be possible to build tools that analyze requirements, just like compilers analyze software programs. This observation has been shared by a number of researchers in this space and a number of tools have been proposed (comprehensive survey of tools is available at [8]). For these tools, the representation of requirements has ranged from free text, use of structured notation, to formal representation using logic. The analyses provided by these tools range from phrasal analysis, use of custom code to analyze small domain models to formal analysis general purpose reasoners. Most of the tools that don't restrict the syntax of requirements are limited to phrasal analysis, while the tools that require formal representation of requirements use general purpose reasoning engines for analyzing the requirements. QuARS [4] presents an approach for phrasal analysis of natural language requirements documents. QUARCC [1] uses a specialized model for identifying conflicts between quality (non-functional) requirements. While we have leveraged both these works (QuARS for classification of problem phrases and QUARCC for creating relationships between non-functional requirements), neither of them provide the broad range of semantic and syntactic analyses that we discussed in this paper. KaOS [3] presents an approach for using Semantic nets and temporal logic for formal analysis of requirements using a goal based approach. While they can perform a broad range of reasoning, their system needs the requirements to be inputted in a formal language.

6 Implementation Details and Early User Evaluation

RAT has currently been created as a plug-in for Microsoft Word 2003/7 using Visual Basic for Applications. The Glossaries are currently also Word documents. The Semantic Engine leverages the reasoning capabilities of Jena [5] and is implemented in Java. We use Protégé to create the OWL [6] ontologies. The Jena semantic engine is used for the reasoning and SPARQL [9] is used for the query language.

Currently, we are piloting an early version of this tool without the full semantic analysis engine, at four client teams. In these pilots, 15 requirements analysts have used RAT to process more than 10,000 requirements. Users have reported that the tool is helping them catch many real world problems. They have reacted positively to the controlled syntaxes and critiques, as the tool helps them follow best practices and comply with a number of industry standards, resulting in much clearer requirements. Some early users requested support for more syntaxes. Users also requested more advanced semantic analysis to detect conflicts, dependencies and missing requirements. The syntactic extensions have been incorporated in the latest release. The semantic analysis discussed in this paper has been implemented in an experimental prototype, and will be made available to users in the next release of RAT.

7 Conclusions and Future Work

In this paper, we have presented RAT: a tool for automatically analyzing requirements documents. We have presented an approach that allows users to write requirement using a set of controlled syntaxes advocated by requirements experts, and user-defined glossaries. This approach allows RAT to extract structured content from the requirements text, which is used for syntactic and semantic analysis using semantic Web technologies. A valuable feature of our approach is that users can extend the analysis by creating user-defined glossaries, creating domain ontologies or writing their own queries. Our future work includes creating a collaborative framework for users to enter domain knowledge.

References

1. Boehm, B., In, H.: Identifying Quality-Requirement Conflicts. IEEE Software 13(2), 25–35 (1996)
2. Boehm, B., Papaccio, P.: Understanding and Controlling Software Costs. IEEE Trans. on Software Eng. 14(10), 1462–1477 (1988)
3. Lamsweerde, A.V., Darimont, R., Letier, E.: Managing Conflicts in Goal-Driven Requirements Engineering. IEEE Trans. Software Eng. 24(11) (1998)
4. Gnesi, S., Lami, G., Trentanni, G.: An automatic tool for the analysis of natural language requirements. CSSE Journal 20(1), 53–62 (2005)
5. Jena – A Semantic Web Framework for Java, http://jena.sourceforge.net/
6. OWL – Web Ontology Language, http://www.w3.org/TR/owl-features/
7. Raven: Requirements Authoring and Validation Environment, http://www.ravenflow.com
8. Robinson, W.N., Pawlowski, S.D., Volkov, V.: Requirements interaction management. ACM Comput. Surv. 35(2), 132–190 (2003)
9. SPARQL Query Language for RDF, http://www.w3.org/TR/rdf-sparql-query/
10. Wiegers, K.: Software Requirements. Microsoft Press (2003)

OntoNaviERP: Ontology-Supported Navigation in ERP Software Documentation

Martin Hepp[1,2] and Andreas Wechselberger[1]

[1] E-Business and Web Science Research Group, Bundeswehr University Munich
[1,2] Semantics in Business Information Systems Group, STI, University of Innsbruck
mhepp@computer.org, andreas.wechselberger@unibw.de

Abstract. The documentation of Enterprise Research Planning (ERP) systems is usually (1) extremely large and (2) combines various views from the business and the technical implementation perspective. Also, a very specific vocabulary has evolved, in particular in the SAP domain (e.g. SAP Solution Maps or SAP software module names). This vocabulary is not clearly mapped to business management terminology and concepts. It is a well-known problem in practice that searching in SAP ERP documentation is difficult, because it requires in-depth knowledge of a large and proprietary terminology. We propose to use ontologies and automatic annotation of such large HTML software documentation in order to improve the usability and accessibility, namely of ERP help files. In order to achieve that, we have developed an ontology and prototype for SAP ERP 6.0. Our approach integrates concepts and lexical resources from (1) business management terminology, (2) SAP business terminology, (3) SAP system terminology, and (4) Wordnet synsets. We use standard GATE/KIM technology to annotate SAP help documentation with respective references to our ontology. Eventually, our approach consolidates the knowledge contained in the SAP help functionality at a conceptual level. This allows users to express their queries using a terminology they are familiar with, e.g. referring to general management terms. Despite a widely automated ontology construction process and a simplistic annotation strategy with minimal human intervention, we experienced convincing results. For an average query linked to an action and a topic, our technology returns more than 3 relevant resources, while a naïve term-based search returns on average only about 0.2 relevant resources.

1 Navigation in ERP Software Documentation

ERP systems like SAP R/3, myERP, or ERP 6.0 are very complex software packages, which makes new users and experienced staff alike largely dependent on online help and other online documentation. At the same time, it is a software category of utmost commercial relevance. Now, due to the broad scope and amount of detail of ERP software, the associated documentation is mostly huge; and the online help and other parts of the documentation combine terminology from business management (e.g. „depreciation"), the various application domains (e.g. „ECR" in the retail sector), and SAP-specific language. With regard to the latter, also, a very specific vocabulary has

A. Sheth et al. (Eds.): ISWC 2008, LNCS 5318, pp. 764–776, 2008.
© Springer-Verlag Berlin Heidelberg 2008

evolved, in particular in the SAP domain (e.g. SAP Solution Maps or SAP module names). This vocabulary is not clearly mapped to business management terminology, as taught at schools and colleges.

In effect, search and navigation in ERP documentation is unsatisfying for many users, since they are unable to express a query using the terminology from their current context or professional background. Instead, they need to be familiar with the particular SAP terminology in order to describe what they are looking for; a skill that imposes a lot of friction on new employees using ERP software. Even for the vendors of respective software, it is extremely difficult to produce and maintain a consistent documentation, in particular due to synonyms and homonyms.

Semantic technology is obviously a promising technology for helping out. However, the enormous size of respective documentation, the ongoing evolution, and pressing business constraints render the creation of perfect ontologies and annotations unfeasible. The particular challenge lies in developing an approach that brings a substantial improvement at little cost, i.e., that minimizes the amount of human labor in the process.

For our evaluation, we have taken the data from a subset of the SAP Logistics branch of functionality, namely the SAP Level II View ("Business Blueprint") regarding the Material Master branch (called "SAP Library" - Material Master ("LO-MD-MM"[1])). The respective part of the documentation consists of only 144 HTML files with a total file size of 1.12 MB. Still, the total number of different words and word groups in this small part exceeds 20,000!

The remainder of the paper is structured as follows: In section 2 we explain the OntoNaviERP approach. In section 3 we summarize the conceptual model for our representation. In section 4 we describe the implementation work carried out. In section 5 we evaluate the technical contribution and discuss the approach in the light of related work. Section 6 summarizes the main points and concludes the paper.

2   OntoNaviERP Approach

Our overall idea is to (1) construct a consolidated set of ontologies covering the general business management domain, the SAP software and solutions domain, and particular industry branch or application domains; (2) integrating those ontologies at a conceptual level, (3) augmenting them with synonyms from Wordnet synsets and other resources, (4) developing a highly automated annotation strategy and infrastructure based on off-the-shelf GATE/KIM technology (see e.g. [1]), and (5) designing a suitable user interface. Figure 1 illustrates our approach.

The main competency question the system should support can be defined as follows:

CQ: *Which [document | part of a document] is relevant as [instruction | term definition | reference] for a software user who wants to [create | modify | retrieve | delete | carry out a certain business function on] a certain business object?*

---

[1] http://help.sap.com/saphelp_erp2005vp/helpdata/en/ff/516a6749d811d182b80000e829fbfe/frameset.htm

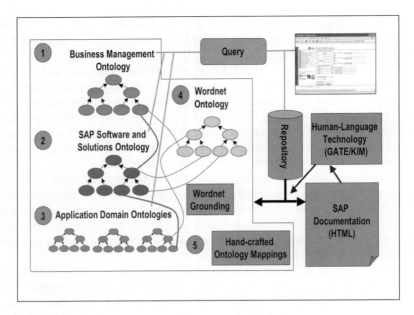

Fig. 1. OntoNaviERP Approach

Of course, the set of competency questions can be extended. However, we would like to stress that we are aiming at a cost-efficient solution that brings a substantial improvement over the state of the art. Given the huge size of both the corpus of text and the vocabulary, a more sophisticated approach is not per se more appropriate. In the long run, we also want to consider the individual usage context and the user's professional background and skills. However, a major problem in using the potential of such extensions is being able to capture respective data without imposing too much additional effort on users.

3   Conceptual Model

Our core conceptual model for supporting search in the SAP software documentation is as follows: First, we assume that a document or part of a document is characterized by (1) whether it offers instructions, explains terminology, or points to further references; (2) which type of action it describes on which type of business object (e.g. tangible or intangible resource or data set). For the type of documents, we use just four classes, a top-level class *TypeOfContent* and three subclasses *Instruction, Term Definition,* and *Reference.*

The topic covered by a document is for us always defined by a pair of an *Action* and a *Business Object.* This could for example be "create new client data set", "change ordering quantity", or "find supplier". Again, this may sound like a rather simple conceptual model, but we will see later that it is sufficient to bring substantial improvement.

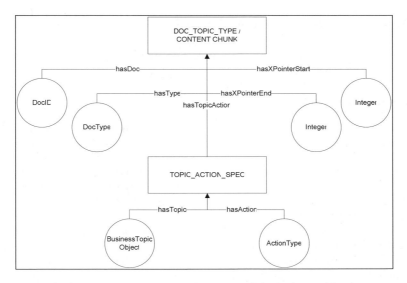

Fig. 2. Representing the type of actions and the business object

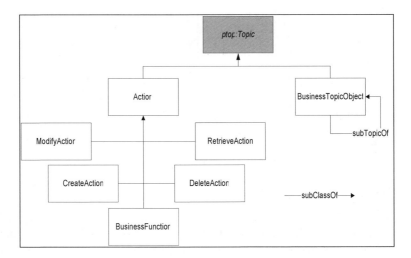

Fig. 3. User Actions and Business Topics as subclasses of Proton `ptop:Topic`

Also, we are dependent on a highly automated annotation process, for which lightweight structures are more promising. One key advantage of this conceptual model is that it reduces the natural-language analysis to spotting the occurrence of named entities representing actions or business objects, which works well with standard GATE/KIM technology without complex linguistic analysis. Figures 2 and 3 illustrate our approach.

4 Implementation: The OntoNaviERP Application

In this section, we describe the implementation of the OntoNaviERP prototype and summarize our experiences.

4.1 System Architecture

For the OntoNaviERP application, we use a very straightforward system architecture, based on mature, mainstream Semantic Web components. In detail, we use Sesame as a repository for the ontologies and the knowledge base, and GATE, KIM, and Lucene for the named entity recognition and other annotation tasks. For controlling the annotation we use KIM directly. For querying the knowledge base, we employ a dedicated GUI implemented as Java Server Pages which access the KIM API.

The KIM platform uses special ".nt" files as input for the named entity recognition. Among other details, they explicitly list the lexical variants of each named entity defined in the ontology. For creating these files from a given OWL ontology augmented by synonyms and other lexical variants, we developed a special converter application. Figure 4 shows the respective architecture.

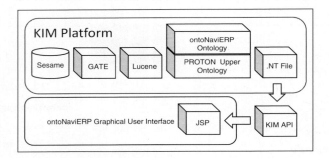

Fig. 4. OntoNaviERP Architecture

4.2 Ontology Engineering

For developing the respective ontologies, we had to meet the following requirements. First, the KIM/GATE infrastructure requires that the PROTON System Module must be present and imported in our own ontology, The `owl:Class` *Entity* must be the superclass of any proprietary ontology class that shall be considered by GATE/KIM for annotating resources. As for the exact location of a domain ontology in the KIM/GATE environment, there are three options: (1) it can be used instead of the PROTON Top Module, (2) instead of the PROTON Upper Module, or (3) in combination with the PROTON Upper Module. The KIM documentation[2] recommends using the domain-independent PROTON Top Module as the basis for any particular domain ontology, and we followed that advice. One positive side-effect of that choice is that the ontologically clean top-level branches "Abstract", "Happening", and

---

[2] http://www.ontotext.com/kim/doc/sys-doc/index.html

"Object'' force us to make good conceptual choices for all of our more specific elements.

For building the OntoNaviERP ontology, we used the following approach: For the *Action* and *Document Type* classes, we simply created respective elements using Protégé, and enriched them by suitable lexical variants of popular synonyms. For all synonyms, we defined a dedicated `owl:annotationProperty` that can later be used to derive the .nt files for KIM/GATE automatically using our tool.

For the *Business Objects* branch, we followed a straightforward approach:

(1) We used the SEO Studio Lite tool (Free Edition, version 2.0.4, build 3452), which is originally a tool for search-engine optimization for Web masters. For us, it returns tables with frequencies of occurrence for all single words or word combinations out of 2 or 3 words. This can be used to get a quick understanding of the active domain vocabulary. Since it was clear that we could not manually engineer an ontology that completely reflects the 20k+ words domain vocabulary, we ordered the resulting lists by descending frequency and considered all single words and 2- or 3-word groups that are used at least ten times in the total text corpus. That cut-of point was mainly determined by practical reasons, i.e., how much time we had available for building the ontology.

(2) We created a term cloud from the concurrency data in order to get a visual aid on the relative importance of certain terms. This step was not really needed, but was perceived a helpful cognitive aid during the ontology engineering process.

(3) Then, we generated a skeleton ontology based on all terms semi-automatically. We mainly applied a script to generate candidate concepts in OWL, consolidated similar concepts, and then manually made them specializations of the PROTON Top Module.

(4) As a last step, we used the Wordnet plug-in for Protégé to augment the concepts by synonyms and lexical variants. We store all synonyms in the ontology using a proprietary `owl:AnnotationProperty` *hasSynonyms.*

In a couple of days, we were able to produce a medium-size ontology for the SAP logistics domain that contains a large amount of synonyms and lexical variants for all entries. One must note that the lexical variants are only necessary because including a stemming engine in the current KIM/GATE package proved burdensome to us, which is why we discarded that option for the moment.

Table 1 summarizes the metrics of the resulting ontology. We can see from the ratio of all concept pairs (action + business objects) vs. the number of concept pairs occurring in at least one document that there is a good fit between the ontology and the document corpus - roughly 60 % of all possible conceptual combinations appear at least once.

We can also see that the 415 pairs of action and business objects on the conceptual level multiply to 27,500 term pairs at the language level, indicating the strength of the consolidation achieved by the ontology.

Note that the ontology needs only a few properties, because we just use very basic recognition of named entities.

Table 1. Metrics of the OntoNaviERP SAP Logistics Ontology

Classes		132
Conceptual level	Action classes	5
	Topic classes	127
	Concept pairs: All	635
	Concept pairs: Subset of pairs that appear in at least one document	415
	Subconcept pairs: Subset of pairs that appear in at least one document	268
Lexical level	Synonyms	383
	Synonyms for actions	126
	Synonyms for topics	257
	Term pairs: All	32382
	Term pairs: Subset for which the respective concept pairs appear in at least one document	27500
Properties	Total	3
	Object	2
	Datatype	0
	Annotation	1

4.3 Annotation

As for the annotation of the corpus itself, we employed the standard KIM/GATE package with existing JAPE rules; we did not modify the named entity recognition nor carried out a linguistic analysis. For putting it to work, we first had to derive an .nt Gazetteer List for the annotation and for the later search. For that, we used a small online tool based on the Jena Semantic Web Framework Java API. It takes as input any OWL ontology and creates from that an .nt file which includes instances of/for the concepts in the OWL ontology, and uses our *hasSynonyms* annotation property to build the *hasMainAlias, hasAlias* and the *subTopicOf* transitive property for the *sub-TopicOf* relation. We will make that tool available for other KIM/GATE users shortly, for we found it quite useful.

Then, we applied a two-stage annotation strategy: First, we used KIM/GATE to store links to all occurrences of known named entities in the repository. Second, we manually decided for each of the 144 HTML documents on the main content type (instruction, term definition, or reference). That took only minimal effort. As a future extension, one could make that distinction individually for action and business object pairs inside the documents.

For populating the KIM/GATE annotation, the following steps are necessary:

(1) Copy the *.owl and the *.nt files into the correct KIM directory. The correct directories are for the OWL files,

> C:\...\kim-platform-1.7.12.15\context\default\kb\owl

and for the .nt files

> C:\...\kim-platform-1.7.12.15\context\default\kb

(2) Edit the file "sesame.inmem.conf" so that our *.owl and *.nt files are included as imports. The file is at

C:\...\kim-platform-1.7.12.15\config

(3) Start KIM, Sesame and Tomcat, and populate the knowledge base. We used a one-time batch annotation run, since the HTML files are static. On-the-fly annotation would work, too, except for the manual step of classifying the type of document. Figure 5 shows the annotation step, and Figure 6 how the recognized entities are highlighted in the generic KIM interface.

Now, with our "brute-force" annotation strategy, we annotated all documents that contain a pair of action and business object anywhere in the text. Since we first thought that was too simple an approach, we added a filtering algorithm that considers a document relevant only if the two words representing the action and the business object respectively are within a range of +/- 25 words, as has been done in traditional information retrieval. However, this extension shows useful only for pure keyword search. As soon as we search at the conceptual level, the impact of that filter becomes limited. Instead, we use a ranking algorithm based on the distance and frequency of occurrence.

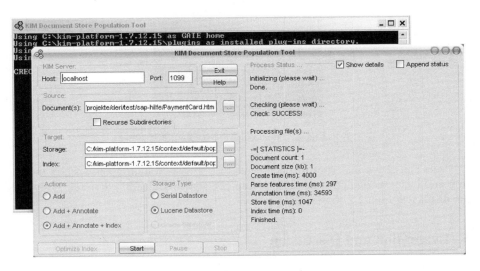

Fig. 5. KIM/GATE Annotation

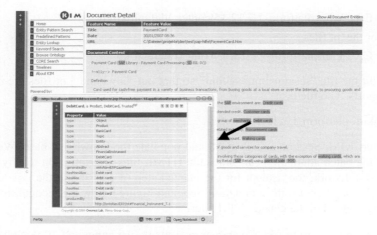

Fig. 6. Recognized entities in the generic KIM interface

4.4 User Interface

We also developed a user interface that hides the ontology-based search behind user-friendly controls.

Fig. 7. OntoNaviERP User Interface

Users can check the types of documents they are interested in and specify the action and business objects. For each chosen conceptual element, all stored synonyms are displayed. Figure 7 shows the interface.

5 Evaluation and Discussion

In the following we summarize our evaluation of the technical contribution of our approach and compare it with the effort for ontology modeling and knowledge base population.

5.1 Contribution of Semantic Technology

From the subset of all term pairs for which the respective conceptual pairs occur at least in one single document, we drew a representative random sample of n=50 (with the random integer generator at http://www.random.org). Then, we determined the number of retrieved documents and the share of truly relevant documents from those documents for the following four techniques:

Technique 1:   Number of documents containing both terms in its exact lexical form (we of course ignore capitalization, since that has been standard in keyword-based retrieval for decades).

Technique 2:   Same as T1, but only those containing both terms within a 50-words range (25 words left and right)

Technique 3:   Number of documents including either combination of a) the given topic term, its synonyms, its subconcepts, or the synonyms of the subconcepts and b) the given action term or its synonyms.

Technique 4:   Same as T3, but only those documents containing the relevant named entities reflecting actions and business objects within a 50-words range.

So in short, techniques 1 and 2 represent the state of the art in simple keyword-based search in ERP documentation, and techniques 3 and 4 are the OntoNaviERP approach. Tables 2 and 3 summarize the results of our evaluation. Note that the precision for techniques 1 and 2 are based on very small return sets, since many search patterns do not appear in this exact lexical form.

Table 2. Impact of OntoNaviERP on retrieved documents and precision

	Technique 1: Term-based			Technique 2: Term-based with 50 words range			Technique 3: OntonaviERP			Technique 4: OntoNaviERP with 50 words range		
	Retrieved	Relevant(*)	Precision(*)	Retrieved	Relevant(*)	Precision(*)	Retrieved	Relevant(*)	Precision(*)	Retrieved	Relevant(*)	Precision(*)
Avg	**0.38**	**0.16**	*0.44*	**0.02**	**0.02**	*1.00*	**11.46**	**3.46**	**0.63**	**5.96**	**2.90**	**0.65**
Min	0.00	0.00	*0.00*	0.00	0.00	*1.00*	1.00	0.00	0.00	0.00	0.00	0.00
Max	6.00	3.00	*1.00*	1.00	1.00	*1.00*	72.00	10.00	1.00	50.00	10.00	1.00
Median	**0.00**	**0.00**	*0.42*	**0.00**	**0.00**	*1.00*	**5.50**	**2.00**	**0.55**	**3.00**	**1.00**	**0.80**
STD	1.10	0.55	*0.43*	0.14	0.14	*n/a*	14.93	3.04	0.33	8.62	3.03	0.37

* Of the first ten results retrieved

The results are very encouraging: Where the mean of retrieved documents in keyword-based search is only 0.38 documents per pair (Technique 1), the ontology-based search (Technique 3) returns, on average, more than 11 documents, and thus almost 30 times as many. Now, one would expect that the simple expansion of a search to synonyms and lexical variants, plus a small subsumption hierarchy would lead to a sharp decrease in precision. However, surprisingly, this is not the case. While the OntoNaviERP approach returns almost 30 times as many documents, more than 60% of the returned documents are relevant, as long as we only look at the top ten documents in our ordered result set. This is the more encouraging as we did not employ any tuning with regard to named entity disambiguation. In other words, the same synonym can be assigned to multiple concepts, and our simplistic annotation counts them for both if found. It seems that the homonyms among the terms are rarely used. There is for sure room for further improvement of the named entity recognition.

Table 3. Statistics on the number of additional, relevant documents

	Effectivity: Number of additional, relevant documents found by OntoNaviERP			
	Additional, relevant documents by technique 3	Comparison: Relevant retrieved documents with technique 1	Additional, relevant documents by technique 4	Comparison: Relevant retrieved documents with technique 2
	T3-T1	T1	T4-T2	T2
Avg	3.30	0.16	2.88	0.02
Min	0.00	0.00	0.00	0.00
Max	10.00	3.00	10.00	1.00
Median	2.00	0.00	1.00	0.00
STD	2.90	0.55	2.99	0.14

While the filtering based on the word distance increases precision in keyword-based search, it has minimal impact on the ontology-based search.

5.2 Discussion: Cost and Benefit

While the technical improvement alone is already very encouraging, it should also be judged in the light of the minimal, straightforward ontology engineering and annotation approach we use. As said, the annotation effort was limited to classifying 144 Web pages according to three branches (instruction, term definition, or reference), and running the out-of-the box named entity recognition of the KIM/GATE platform. Creating the ontology was basically extracting roughly 140 classes, assigning them to PROTON abstractions, and adding a lot of relevant synonyms and lexical variants. By using a dedicated `owl:AnnotationProperty`, such terms could be productively added and maintained directly in standard OWL editors like Protégé. The Gazeteer file was quickly generated from the OWL file using our lightweight conversion tool.

5.3 Related Work

While there is a mature body of literature on the core techniques, like named entity recognition, ontology-supported information retrieval, and ontology learning from text, we found no previous works that apply ontologies for ERP software documentation. This surprised us, because the blend of terminology from multiple spheres, e.g. college textbook general management terminology, vendor-specific business terminology, vendor specific systems terminology, and industry-branch terminology coexists wildly, both in the authoring processes and among the software users.

There is some work on deriving ontologies and populating knowledge bases from software documentation in general, e.g. annotations from APIs etc. Such particular work on ontology learning from software artifacts is described in [2]. The closest works in our direction from the Semantic Web community are [3] and [4], but while both address software documentation, they do not target large Common-Off-The-Shelf ERP packages like SAP solutions. For a general overview on ontology learning and population, see e.g. [5].

The KIM/GATE environment is described in [1]. A conceptual framework for the business process space, which is shaped and reflected by an ERP landscape, is presented in [6] and [7].

In another context, Holger Bast and colleagues [8-10] have worked on completing queries for facilitating search with their CompleteSearch approach; but again, this is not yet applied to ERP documentation. We are considering to using respective techniques for a more intelligent UI, though.

In information systems literature, the problem of modeling activity options for users has been discussed in [11], and the alignment of ERP software documentation and the system configuration has been addressed in [12]

6 Conclusion

We have shown how the navigation in ERP software documentation can be improved substantially by using standard KIM/GATE technology plus rather lightweight ontologies that are massively augmented by synonyms derived from frequent terms in the corpus and a standard Wordnet plug-in for Protégé. We obtained the relevant terminology using readily available search-engine optimization tools.

Despite a mostly automated ontology construction process and a simplistic annotation strategy with minimal human intervention, we experienced convincing results.

It comes as no surprise that ontologies can help improve precision and recall in a large body of text, in particular as long as the effort for creating the ontology and annotating the corpus are not considered and the corpus is stable. Both, however, is not given in ERP software documentation. The sheer size of the documentation and the used terminology makes manual ontology engineering and manual supervision of the annotation unattractive. Thus, we wanted to develop a pragmatic and cheap approach that relies on current semantic technology to tackle a real business problem. Eventually, we were surprised about the substantial improvement our solution shows. As next steps, we will work on more intelligent user interfaces and on trying to consider user skills and backgrounds, the context of a search task, and the customization status of the software for further improving our approach.

Acknowledgements. The work presented in this paper has been supported by a Young Researcher's Grant (Nachwuchsförderung 2005-2006) from the Leopold-Franzens-Universität Innsbruck, and by the European Commission under the project SUPER (FP6-026850). The authors would also like to thank Ontotext for great support with the KIM package.

References

[1] Popov, B., Kiryakov, A., Kirilov, A., Manov, D., Ognyanoff, D., Goranov, M.: KIM - Semantic Annotation Platform. In: Fensel, D., Sycara, K.P., Mylopoulos, J. (eds.) ISWC 2003. LNCS, vol. 2870. Springer, Heidelberg (2003)

[2] Sabou, M.: Extracting Ontologies from Software Documentation: a Semi-Automatic Method and its Evaluation. In: Proceedings of the Workshop on Ontology Learning and Population, ECAI 2004, Valencia, Spain (2004)

[3] Ambrósio, A.P., Santos, D.C.d., Lucena, F.N.d., Silva, J.C.d.: Software Engineering Documentation: an Ontology-based Approach. In: Proceedings of the WebMedia & LA-Web 2004 Joint Conference, Ribeirão Preto-SP, Brazil (2004)

[4] Witte, R., Zhang, Y., Rilling, J.: Empowering Software Maintainers with Semantic Web Technologies. In: Franconi, E., Kifer, M., May, W. (eds.) ESWC 2007. LNCS, vol. 4519. Springer, Heidelberg (2007)

[5] Buitelaar, P., Cimiano, P., Magnini, B.: Ontology Learning from Text: Methods, Evaluation and Applications, vol. 123. IOS Press, Amsterdam, The Netherlands (2005)

[6] Hepp, M., Leymann, F., Domingue, J., Wahler, A., Fensel, D.: Semantic Business Process Management: A Vision Towards Using Semantic Web Services for Business Process Management. In: Proceedings of the IEEE International Conference on e-Business Engineering (ICEBE 2005), Beijing, China (2005)

[7] Hepp, M., Roman, D.: An Ontology Framework for Semantic Business Process Management. In: Proceedings of the 8th International Conference Wirtschaftsinformatik 2007, Karlsruhe (2007)

[8] Bast, H., Chitea, A., Suchanek, F., Weber, I.: ESTER: Efficient Search on Text, Entities, and Relations. In: Proceedings of the 30th Annual International ACM SIGIR Conference on Research and Development in Information Retrieval, Amsterdam, The Netherlands (2007)

[9] Bast, H., Weber, I.: The CompleteSearch Engine: Interactive, Efficient, and Towards IR&DB Integration. In: Proceedings of CIDR 2007, Asilomar, CA, USA (2007)

[10] Bast, H., Majumdar, D., Weber, I.: Efficient Interactive Query Expansion with Complete Search. In: Proceedings of CIKM 2007, Lisboa, Portugal (2007)

[11] Soffer, P., Golany, B., Dori, D.: ERP modeling: a comprehensive approach. Information Systems 28, 673–690 (2003)

[12] Knackstedt, R., Winkelmann, A., Becker, J.: Dynamic Alignment of ERP Systems and their Documentations. An Approach for Documentation Quality Improvement. In: Proceedings of the Americas Conference on Information Systems (AMCIS 2007), Keystone, CO, USA (2007)

Market Blended Insight: Modeling Propensity to Buy with the Semantic Web

Manuel Salvadores[1], Landong Zuo[1], SM Hazzaz Imtiaz[1], John Darlington[1],
Nicholas Gibbins[1], Nigel R Shadbolt[1], and James Dobree[2]

[1] Intelligence, Agents, Multimedia (IAM) Group
School of Electronics and Computer Science
University of Southampton, UK
{ms8,lz,hsmi,jd,nmg,nrs}@ecs.soton.ac.uk
[2] ProspectSpace Ltd
25 Landsdowne Gardens London, SW8 2EQ, UK
james@prospectspace.com

Abstract. Market Blended Insight (MBI) is a project with a clear objective of making a significant performance improvement in UK business to business (B2B) marketing activities in the 5-7 year timeframe. The web has created a rapid expansion of content that can be harnessed by recent advances in Semantic Web technologies and applied to both Media industry provision and company utilization of exploitable business data and content. The project plans to aggregate a broad range of business information, providing unparalleled insight into UK business activity and develop rich semantic search and navigation tools to allow any business to 'place their sales proposition in front of a prospective buyer' confident of the fact that the recipient has a propensity to buy.

1 Introduction

The Market Blended Insight project (DTI Project No: TP/5/DAT/6/I/H0410D) is a three year applied research project funded under the UK Governments Technology Programme.

The innovation challenge for the project is: to overcome the problem that traditional marketing techniques have broad push without knowing if the recipient has a propensity to buy. The project is extending world class Semantic Web research from the EPSRC's "Advanced Knowledge Technologies IRC" [1] and applying it to a large scale collection of real life UK company data sources to understand if an organization's propensity to buy can be discovered within a very large pool of information.

To ensure the research is undertaken in a real world scenario the project has direct involvement from marketing departments within UK businesses. The consortium for this project includes marketing departments of the following companies: ParcelForce Worldwide, British Gas Business, AXA, Clydesdale and York-shire Bank (NAGE), 3M and pH group. Based on their marketing needs the project is developing advanced methods of analysing target markets and the innovation includes:

A. Sheth et al. (Eds.): ISWC 2008, LNCS 5318, pp. 777–789, 2008.

- the anticipated scale of the information source we plan to create, based on the 3.7 million companies that constitute the entire UK economy.
- The complexity of the collection of ontologies, covering a rich depth of information for each company.
- Finally and most difficult, the innovation required to identify within the information the semantic relations and queries required to determine propensity to buy given a sales proposition.

The project is building an industrial scale prototype that aims to provide UK business users with dynamic, relevant insight into their target markets and allows them to make informed decisions on a potential clients propensity to buy.

The first phase performed a needs analysis within the consortium which developed the following classification of marketing processes: strategy, scoping, scanning, data processing and interpreting. Each consortium member has described the different areas and their needs for each in terms of desired outcomes.

A subset of the expected outcomes has been achieved by deploying a number of scenarios in a prototype architecture. This paper discussed two scenarios from the first prototype: *Micro Segmentation* and *Value Chain*. The implementation of both scenarios has tackled technical issues related to Semantic Web such as: semantic annotated data extraction, ontology driven data integration, data generation through onotology reasoning and ontology driven data visualization.

Besides these scenarios and the architecture the following sections also describes related work, the information extraction process and conclusions.

2   Related Work

There is an existing market for Business Intelligence tools and the desire for integration of information services on demand and their descriptive metadata to further improve the performance of corporations in the B2B market is described in [32]. Researchers in the Semantic Web community have been working on exploiting semantic technology for the integration of public information in a wide range of application domains with knowledge held in heterogeneous formats, representations and structures [14]. The research into Semantic Web technologies is diverse, developing emergent middleware frameworks in areas such as service composition [26,27,35], data integration [1,29,31] and data extraction. In following sections it is shown how this project has made use of several data extraction techniques for harvesting the required data and how semantic data integration is one of the basic pilars to build up the scenarios.

The mashing-up of knowledge was previously demonstrated by the AKTivePSI project [2]. The Semantic Web technologies were widely used to integrate both data and metadata of company information with other knowledge from different domains, (for example, spatial knowledge from Ordnance Survey and administrative information from local councils) creating an enhanced view of the market and more efficient access of information from different perspectives [20]. The aggregation of information from structured private data sources such as RDBMs in corporate environments with publicly available sources on the web offers an ability to create an aggregated vision of B2B market analysis where information on the same market is established from a variety of sources.

There has been progress in the extraction of data from public domains like the Internet. SPHINX[30] can be used to develop customized web crawlers. Typical crawling based on keywords has been improved with ontology driven aproaches [25]. Other solutions allow for the extraction of tree data structures and the semantic annotation of the extracted data, manipulation of the tree structures, recording any changes and watching for any modifications [28]. Armadillio [22] uses an alternative strategy dealing with web extraction in a largely automated way that does not require manual annotations. It utilizes information that has been extracted from highly reliable source to train the information extraction.

GATE[24] is an open source architecture that provides an infrastructure to develop software components to process natural language and a GUI development environment. It contains built in components to perform tasks from simple tokenisation to complex semantic tagging and includes a Java Annotation Patterns Engine(JAPE) [23], an annotation comparison tool with performance metrics and plug-in for external parsers etc. Extraction of semantic relations from natural language text can be completed via shallow Jape patterns [34] combined with dependency parsers to identify linguistic triples that can then be mapped to RDF statements using ontology concepts and relations. The entities in the text can be identified by their instantiation in the ontology and the relation between them.

3   Application Architecture

The project architecture is structured with four different components: the web user interface, the Social Network Services (SNS), the Semantic Query and Inference Services (SQIS) and the Common Extraction and Translation Framework (CETF) (see figure 1). The SNS is planned to be built in further prototypes; SQIS , CETF and the Web Portal components are detailed in this section.

The end user application is a web portal developed using Java Server Faces (JSF) [6] which renders in the user's browser XHTML [19]. UI widgets such as trees, tables, lists, maps and network have been developed extending JSF. This set of widgets are able to display the RDF/OWL [11,17] data pulled out from the SQIS. The data is displayed for each scenario by gathering several widgets into one or more web pages. Since it is expected to add more use cases in next iterations, re-utilization of UI compononents gives a flexible and quick method for setting up new data visualizations.

The SQIS component makes wide use of Semantic Web technology to provide the ontology inferencing and data query services to the application. Both reasoning and query functionality are built on top of Jena framework [21] but it has been developed to isolate the application from underlying services in other that other frameworks can be investigated in future prototypes.

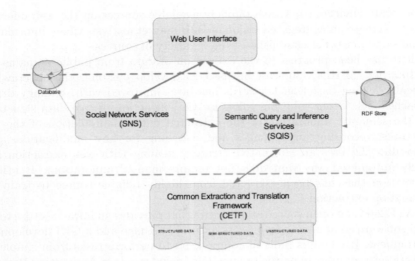

Fig. 1. General architecture diagram

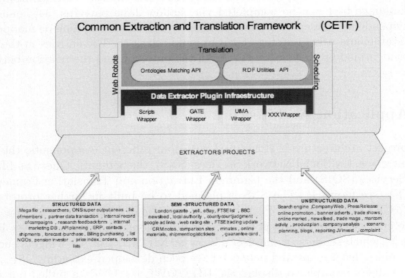

Fig. 2. CETF component

The scenarios deployed in the application make use of the SQIS by defining the queries, in SPARQL [13], and the inference rules. Inference within the SQIS works in two different modes: online and offline. Depending on the specific use case need, inference that requires reasoning over massive amounts of data is performed in offline mode otherwise it is online for results requiring immediate user interaction. Both online and offline reasoning are required for Micro

Segmentation and Value Chain scenarios in order to integrate different data sources as well as to prepared the data to be queried by end user.

The CETF component extracts data and translates it into RDF format. The data sources listed for each of the scenarios are extracted by this component. Currently we classify different data sources in three groups: structured, semi-structured and unstructured data, as it shown in section 4. Each time the CETF extracts a new data source, it notifies the SQIS and the new data source is registered. From that moment the data is avalaible to be used by the inference and query functions implemented in the SQIS. The CETF is built on top of a plugin architecture, see figure 2. Different plugins can be attached to the CETF in order to pull out different information, currently it supports plugins for GATE [24] and UIMA [4]. The CETF also provides crawlers for exploring and scheduling for planning iterative extractions.

4   The Information Extraction Process

To date this project has focused on text based information extraction. It considers "structured data" to refer to the type of data stored in databases with associated metadata reflective of the data schema. Unstructured data does not have a defined structure or schema associated with it for example free text within web pages. Semi-structured is defined as portions of the data have associated structure and meta-data or schema and portions have no meta-data. In our definition the form of structure is not considered i.e. in a HTML web page formatting instructions are helpful for processing the contents of a page but it does not contain the necessary semantics.

The amount of relevant unstructured business data is growing, and will continue to grow in the foreseeable future. According to TDWI [16] estimations of the unstructured and semi-structured data are 53 percent of an organizations overall data. Aware of this opportunity some of the organizations are increasingly feeding unstructured data into their data warehousing and business intelligence processes such as wikis, RSS feeds, instant messaging transcripts, and document management systems, e-mails, office-suite documents, Web pages, and Web logs (or blogs). The amount of e-mail data fed into these processes grew by nearly one-half (47 percent) over the last three years, followed by office-suite documents (35 percent), Web pages (35 percent), and Web logs (27 percent) [16].

The project uses the WebSPHINX [18] web crawler for data extraction which is a implementation of the SPHINX interface [30]. The project has extended the crawler to be switched to a more topic-based focused crawling mode where the crawl is controlled by rating the links based on a high level background knowledge such as an ontology [25], and to direct the crawl for visiting specific links and data patterns. Here the crawl is started by specifying one or several start URLs and an initial ontology.

Once these parameters are specified, the extraction rules extract from the page specific HTML markup any interesting pieces of textual information. In case of PDF documents, this consists of extracting XML markups which comes as an output of the Pdfbox tool [8].

Once the desired pieces of text are extracted, the documents are processed using a GATE [24] pipeline. A GATE pipeline is application dependent but usually consists of stages of tokenisation, gazetteer lookups, pattern matching by JAPE grammars, part of speech tagging and dependency parsing etc. The project has created gazetteers of company names and street names from the company backbone which are used for detecting these entities within the documents. The Jape scripts are selected based on needs of the current extraction and identify items from post codes, phone numbers, emails, URLs etc. to the council planning application numbers and dates.

5   Use Case: Micro Segmentation

The UK standard industrial classification of economic activities SIC(92) [15] has been widely use in marketing analysis due to the fact that its stated objective is to provide the UK with a uniform framework to classify business establishments by the type of economic activity in which they are engaged and also because the information is gathered and published by government.

One form of marketing analysis is to segment, or classify, potential customers by clustering those with common needs. A course segmentation of needs is often defined by creating clusters of organizations with the same business activity. Consortium members have agreed that SIC(92) is not a fine enough classification of business activity, under representing the activity of an organization. For instance it is not possible to find out whether a restaurant is an Italian restaurant or not. The finest SIC(92) classification is "Unlicensed restaurants and cafes", "Take away food shops" and "Licensed restaurants". Many organizations declare more specific details about the activity in which they are engaged in there communication to the market in order to attract the right type of clients. For instance, finding out the type of a restaurant is easily answered by searching the Web or looking at a directory service such as Yellow Pages.

The aim of this scenario is to provide detailed classification, or microsegmentation, by extending SIC(92) classes with sub-classes that are defined by external data sources using Semantic Web techniques.

The data sources involved in this scenario are:

- A backbone of the UK companies within the London boroughs of Lewisham and Camden [9]. This data backbone collects information of more than 83,500 companies in more than 12 million RDF triples.
- SIC(92) hierarchy classification [15], a hierarchy of more than 6k nodes, this information is stored in 61,704 RDF triples.
- Ordnance Survey Point of Interest database (PointX) [7]. PointX dataset contains around 3.9 million geographic and commercial features across Great Britain. The data used in this scenario is the collection related just to the London boroughs of Lewisham and Camden.
- www.mycamden.co.uk [5] is a website provided by the London borough of Camden which publishes a directory, a form of classification, for companies in that specific area. The data extracted from MyCamden contains information

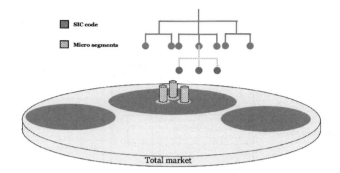

Fig. 3. Micro segmentation process

of 1,824 companies gathered in 85 different classes. This information is stored in 93,989 RDF triples and has been extracted with the information extraction techniques detailed in section 4.

The micro segmentation use case mainly relies on two processes: 1) creating a matching between the companies listed in the additional data sets and their equivalent entity in the backbone listing of UK companies and 2) the extension of the SIC(92) classification by attaching additional sub-classes from the classification structures contained in the additional data sets. These classifications are finer than SIC(92), for instance Mycamden data provides relevant information on whether a restaurant is a Chinese or an Italian restaurant.

Both PointX and Mycamden provide additional data about companies. These data sets need to be attached to the company backbone. The company matching process creates a semantic link between every company extracted from either PointX or Mycamden to the company backbone [9]. In terms of the Semantic Web this link between the backbone and the external data sources is created through a sameAs OWL [17] arc. The inference to see whether two companies in different data sources are the same is achieved by comparing their names and addresses. The rule grammar embeded within thre SQIS for this case is as follows:

```
(?compA rdf:type <mbi:company>) (?compB rdf:type <mbi:company>)
(?compA mbi:hasName ?nameA) (?compB mbi:hasName ?nameB)
equal(?nameA, ?nameB)
(?compA mbi:hasAddress ?addrA) (?compB mbi:hasAddress ?addrB)
equalAddreses(?addrA, ?addrB)
->
 ?compA owl:sameAs ?compB
```

equalAddresses is a built-in primitive developed within Jena to determine with a defined level of confidence if two addresses with different formats are the same.

The finest SIC(92) segments within the restaurant business activity are "unlicensed restaurants and cafes", "Take away food shops" and "licensed restaurants". As it is shown in figure 4, Mycamden and PointX provide finer classification with

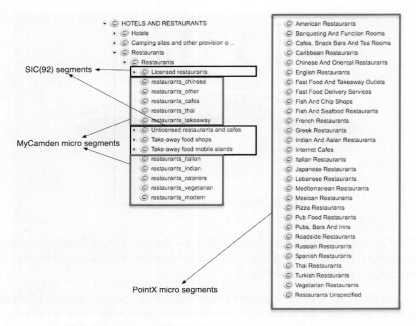

Fig. 4. SIC(92) extension for restaurants business activity

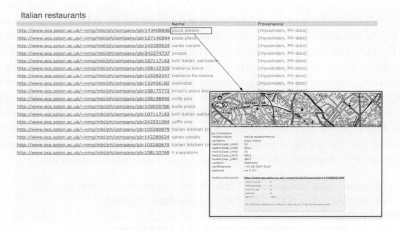

Fig. 5. Micro segmentation query interface

micro segments. In this example, restaurant business activity has been extended with 40 micro segments (30 from PointX and 10 from MyCamden).

SIC(92) is stored in the system in RDF using SKOS [12]. SKOS is a Semantic Web standard that extends RDF with an specific vocabulary for knowledge organization systems and classification schemes. Using SKOS, SIC(92) is extended sub-classes from additional data sets by adding **narrower** arcs.

In our work to date, new information for 5,014 companies (4,406 from PointX and 608 from Mycamden) has been semantically integrated with the companies data backbone, providing 843 micro segments (777 from PointX and 66 from Mycamden). Once the process of creating new micro segments via extending the SIC(92) classification scheme is completed by one user of the system all end users can browse within the web application via the extended hierarchy of micro segments which have finer information about the business activity for the 5,014 companies, see figure 5.

6   Use Case: Value Chain

A Value-Chain is defined by Porter [33] as a series of value-generating activities. Products pass through all activities of the chain in order, and at each activity the product gains some value. The value chain framework quickly made its way to the forefront of management thought as a powerful analysis tool for strategic planning. Clear visualization of value-chain information is one of most desired scenarios from MBI consortium users. They require a tool that allows easy visualisation and manipulation of relationships between companies in order to evaluate their propensity to buy.

The consortium members decided that the initial focus for the prototype would be the relationships within the Building and Construction (B&C) industry. Domain specific data was obtained and a generic solution developed to support ontology-driven data extraction and user-view visualisation based on Semantic Web technology. The value-chain use case is composed of four main functional components, 1) Information Harvesting, 2) Semantic Integration, 3) View Adaptation and 4) Network Visualization (see figure 8).

The Architects Journal [3] is a major web portal providing rich information on the building and Construction Industry including projects, companies and products and all their relations. The prototype harvested information from the web pages and recovered the relating transactions details for 4000 suppliers, 6000 products and 600 projects and all transaction relations using information extraction techniques detailed in section 4.

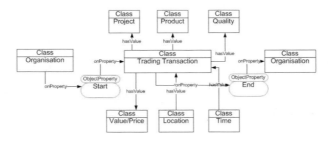

Fig. 6. Pre-inference raw data view

To harvest information the system tries to extract all the specific transaction details in order to form a solid company and project backbone about the construction industry. The data is integrated with relevant information from other sources such as those containing representation of product taxonomy and business activity hierarchy. The integration allows more restricted relations across different domains in order to support filtering out any data that is not of interest in analyzing model of value chain on the users perspective. The network visualisation of value chain is controlled by the user. The user preference is stored in a user ontology and rule syntax that supports the reasoning process necessary to generate a view of value chain model. The generic rule-based inference engine of Jena is deployed to support the knowledge adaptation of raw data model on user demand. The following example shows how a simple user view (see figure 7) is generated from raw data model (see figure 6).

The figure 6 shows a generic schema containing transaction details between any pairs of companies. The relations between companies are classified into types such as, buy, sell, service, offering, shipment, partnership, etc. The conceptual model in figure 7 has addressed a specific user interesting about "Client", "Architect", "Contractor", and "Supplier" in B&C industry, where those concepts and relations may not exist in the raw data model like figure 6. The system has created the following rule syntax to adapt the gap.

```
[Client:  (?n mbi-vc:launch ?t) (?t rdf:type mbi-vc:Investment)
(?n rdf:type aj:Organization)
      -> (?n rdf:type mbi-vc:Client) ]
[Supplier: (?n mbi-vc:launch ?t) (?t rdf:type mbi-vc:Supply)
(?n rdf:type mbi-vc:Architect )
 -> (?n rdf:type mbi-vc:Supplier) ]
 [Supply: (?n1 mbi-vc:launch ?t ) (?t mbi-vc:transactsTo ?n2 )
     (?n1 rdf:type mbi-vc:Supplier) (?n2 rdf:type mbi-vc:Architect) ->
     ( ?n1 mbi-vc:Channel ?n2)]
```

The figure 7 shows the user view model after inferencing processing.

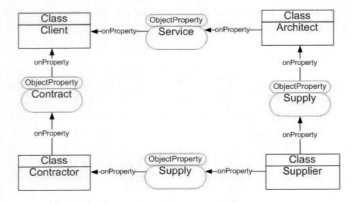

Fig. 7. Post-inference user data view

Fig. 8. Network data representation

By performing the adaptation process, the user does not have to stay at the same granularity level as described in the raw data for example transactions. Instead, his attention of value-chain is more likely addressed in a customised user view over the industry domain. The outcome of view adaption is a new data model that can be well fitted into the SQIS architecture. The SPARQL [13] query is executed against use view model. The query result is visualized in a Java applet in web interface empowered by Prefuse [10], see figure 8.

Our work has shown that a general view of a network of data can be can be constrained by user preferences to show value chains within the data that are of specific interest to the user. The user preference is captured into inference rules that can process the underlying network of data to provide higher level relationships that make the users analysis and decision making much easier.

7   Conclusions

In this paper the MBI project's first prototype has been described in outline and the specific "Micro segmentation" and "Value Chains" use cases explained with real industry data. Both cases demonstrate how with Semantic Web techniques it is possible to extract data from unstructured and semi-structured text from the Web, transform the data into RDF and integrate it with a structured data backbone. Moreover, extra value has been created by applying inference rules to the raw data pulling out new information useful in providing structured views to the end-users. The prototype has demonstrated to the consortium members the facilities Semantic Web technologies can offer when there is a need for data integration. In further work the uses cases presented will be moved to a large scale data backbone encompassing the entire UK data space of businesses and Semantic Web techniques will be included for improved modeling of "propensity to buy".

References

1. AKT research programme (accessed on 04/2008),
   http://www.aktors.org
2. AktivePSI (accessed on 03/2008),
   http://www.aktors.org/interns/2006/aktivepsi/index.php
3. The architects journal (accessed on 04/2008), www.ajspecification.com
4. (accessed on 02/2008), http://incubator.apache.org/uima/
5. (accessed on 05/2008), http://www.mycamden.co.uk/
6. Java server faces (accessed on 05/2008),
   http://java.sun.com/javaee/javaserverfaces/
7. Ordenance survey point of interest database (accessed on 01/2008),
   http://www.ordnancesurvey.co.uk/oswebsite/products/pointsofinterest/
8. Pdfbox - java pdf library (accessed on 04/2008), www.pdfbox.org
9. PH megafile (accessed on 04/2008), http://www.phgroup.com/mfmap.html
10. The prefuse visualization toolkit (accessed on 05/2008),
    http://prefuse.org/
11. Resource description framework (RDF) (accessed on 04/2008),
    http://www.w3.org/rdf/
12. Simple knowledge organization system (SKOS) (accessed on 04/2008),
    http://www.w3.org/2004/02/skos/
13. SPARQL query language for rdf (accessed on 04/2008),
    http://www.w3.org/tr/rdf-sparql-query/
14. SWAD europe deliverable 12.1.1: Semantic web applications - analysis and selection (accessed on 05/2008), http://www.w3.org/2001/sw/europe/reports/
15. The UK standard industrial classification of economic activities uk sic(92) (accessed on 04/2008), http://www.statistics.gov.uk/
16. Unstructured data: Attacking a myth, 9/5/2007, by stephen swoyer (accessed on 04/2008), http://www.tdwi.org/news/display.aspx?id=8577
17. Web ontology language (OWL) (accessed on 04/2008),
    http://www.w3.org/2004/owl/
18. WEBSPHINX: A personal, customizable web crawler (accessed on 04/2008),
    http://www.cs.cmu.edu/~rcm/websphinx/
19. XHTML, http://www.w3.org/tr/xhtml1/
20. Alani, H., Dupplaw, D., Sheridan, J., O'Hara, K., Darlington, J., Shadbolt, N., Tullo, C.: Unlocking the potential of public sector information with semantic web technology. In: ISWC/ASWC, pp. 708–721 (2007)
21. Carroll, J., Dickinson, I., Dollin, C., Reynolds, D., Seaborne, A., Wilkinson, K.: Jena: Implementing the semantic web recommendations. Technical Report HPL-2003 (2003)
22. Ciravegna, F., Chapman, S., Dingli, A., Wilks, Y.: Learning to harvest information for the semantic web (2004)
23. Cunningham, H.: Jape – a java annotation patterns engine. Technical Report, Department of Computer Science, University of Sheffield (1999)
24. Cunningham, H., Maynard, D., Tablan, V., Ursu, C., Bontcheva, K.: Developing language processing components with gate (2001)
25. Ehrig, M., Maedche, A.: Ontology-focused crawling of web documents. In: SAC 2003: Proceedings of the 2003 ACM symposium on Applied computing, pp. 1174–1178. ACM, New York (2003)

26. Sheth, A.P., Verma, K., Sivashanmugam, K., Miller, J.A.: Framework for semantic web process composition. International Journal of Electronic Commerce 9 (2005)

27. Kemper, A., Wiesner, C.: Building scalable electronic market places using hyperquery-based distributed query processing. World Wide Web 8(1), 27–60 (2005)

28. Leonard, T., Glaser, H.: Large scale acquisition and maintenance from the web without source access. In: Handschuh, S., Dieng-Kuntz, R., Staab, S. (eds.) Workshop 4, Knowledge Markup and Semantic Annotation, K-CAP 2001, October 2001, pp. 97–101 (2001)

29. Mena, E., Kashyap, V., Sheth, A.P., Illarramendi, A.: OBSERVER: An approach for query processing in global information systems based on interoperation across pre-existing ontologies. In: Conference on Cooperative Information Systems, pp. 14–25 (1996)

30. Miller, R.C., Bharat, K.: SPHINX: A framework for creating personal, site-specific web crawlers. Computer Networks 30(1-7), 119–130 (1998)

31. Noy, N.F., Musen, M.A.: PROMPT: Algorithm and tool for automated ontology merging and alignment. In: AAAI/IAAI, pp. 450–455 (2000)

32. O'Connell, B.: Building an information on demand enterprise that integrates both operational and strategic business intelligence. In: ICEC 2007: Proceedings of the ninth international conference on Electronic commerce, pp. 85–86. ACM, New York (2007)

33. Porter, M.E.: Competitive Strategy: Techniques for Analyzing Industries and Competitors. Simon and Schuster (1980)

34. Specia, L., Motta, E.: A hybrid approach for relation extraction aimed at the semantic web. In: Larsen, H.L., Pasi, G., Ortiz-Arroyo, D., Andreasen, T., Christiansen, H. (eds.) FQAS 2006. LNCS (LNAI), vol. 4027, pp. 564–576. Springer, Heidelberg (2006)

35. Thompson, P.: Dynamic integration of distributed semantic services: Infrastructure for process queries and question answering. In: HLT-NAACL (2003)

DogOnt - Ontology Modeling for Intelligent Domotic Environments

Dario Bonino and Fulvio Corno

Politecnico di Torino, Torino, Italy
{dario.bonino,fulvio.corno}@polito.it

Abstract. Home automation has recently gained a new momentum thanks to the ever-increasing commercial availability of domotic components. In this context, researchers are working to provide interoperation mechanisms and to add intelligence on top of them. For supporting intelligent behaviors, house modeling is an essential requirement to understand current and future house states and to possibly drive more complex actions. In this paper we propose a new house modeling ontology designed to fit real world domotic system capabilities and to support interoperation between currently available and future solutions. Taking advantage of technologies developed in the context of the Semantic Web, the DogOnt ontology supports device/network independent description of houses, including both "controllable" and architectural elements. States and functionalities are automatically associated to the modeled elements through proper inheritance mechanisms and by means of properly defined SWRL auto-completion rules which ease the modeling process, while automatic device recognition is achieved through classification reasoning.

1 Introduction

Domotic systems, also known as "home automation" systems, have been around on the market for several years, however only few years ago they started to spread over residential buildings, thanks to the increasing availability of low cost devices and driven by new emerging needs on house comfort, energy saving, security, communication and multimedia services.

Current domotic solutions suffer from two main drawbacks: they are produced and distributed by various electric component manufacturers, each having different functional goals and marketing policies; and they are mainly designed as an evolution of traditional electric components (such as switches and relays), thus being unable to natively provide intelligence beyond simple automation scenarios. The first drawback causes an evident interoperation problem that prevents different domotic plants or components to interact with each other, unless specific gateways or adapters are used. While this was acceptable in the first evolution phase, where installations were few and isolated, now it becomes a very strong issue as many large buildings such as hospitals, hotels and universities are mixing different domotic components, possibly realized with different technologies,

A. Sheth et al. (Eds.): ISWC 2008, LNCS 5318, pp. 790–803, 2008.

and need to coordinate them as a single system. On the other hand, the roots of domotic systems in simple electric automation prevent satisfying the current requirements of home inhabitants, who are becoming more and more accustomed to technology, requiring more complex interaction possibilities.

In the literature, solutions to these issues are usually proposed by defining *smart homes*, i.e., homes pervaded by sensors and actuators and equipped with dedicated hardware and software tools that implement intelligent behaviors. Smart homes have been actively researched since the late 90's, pursuing a revolutionary approach to the home concept, from the design phase to the final deployment. Involved costs are very high and prevented, until now, a real diffusion of such systems, that still retain an experimental and futuristic connotation.

The approach proposed in this paper lies somewhat outside the smart home concept, and is based on extending domotic systems, by adding devices and agents for supporting interoperation and intelligence. Our solution takes an evolutionary approach, where commercial domotic systems are extended with a low cost device (embedded PC) allowing interoperation and supporting more sophisticated automation scenarios. In this case, the domotic system in the home evolves into a more powerful integrated system, that we call Intelligent Domotic Environment (IDE), that is able to learn user habits, to provide automatic and proactive security, to implement comfort and energy saving policies and can be immediately exploited, as technologies are low cost and commercially available. IDEs promise to achieve intelligent behaviors comparable to smart homes, at a fraction of the cost, by reusing and exploiting available technology, and providing solutions that may be deployed even today.

A key step towards the definition of IDEs is abstract and formal modeling of domotic device capabilities and functionalities, independently from technology specific aspects. For example a lamp is an object that can be electrically lit and that emits light, independently from the technology with which it is built, provided it is controllable in some way by the domotic system. Abstraction allows to bridge different technologies by associating real devices with their abstract counterparts and by translating low level information into a common, shared language.

This paper introduces DogOnt, a novel modeling language for IDEs, based on Semantic Web technologies. By adopting well known representations such as ontologies and by providing suitable reasoning facilities, DogOnt is able to face interoperation issues allowing to describe:

- where a domotic device is located;
- the set of capabilities of a domotic device;
- the technology-specific features needed to interface the device;
- the possible configurations that the device can assume;
- how the home environment is composed;
- what kind of architectural elements and furniture are placed inside the home.

This information can then be leveraged by inference-based intelligent systems to provide advanced functionality required in Intelligent Domotic Environments. DogOnt is composed of two elements: the *DogOnt ontology*, expressed in OWL, which allows to formalize all the aspects of a IDE, and the *DogOnt rules*, which ease the modeling process by automatically generating proper states and functionalities for domotic devices, and by automatically associating them to the corresponding device instances through semantic relationships. DogOnt is currently adopted to provide house modeling and reasoning capabilities to a domotic gateway called DOG (Domotic OSGi Gateway), which is under development in the authors' research group and that will be distributed as open source toolkit for building IDEs running on low cost PCs. In this context, a third component of DogOnt, namely *DogOnt queries*, not presented in this paper, supports runtime control of the IDE.

The paper is organized as follows: Section 2 introduces relevant related works, while Section 3 describes the DogOnt ontology, starting from the initial assumptions and including the most interesting modeling aspects. Section 4 shows DogOnt rules, i.e., how reasoning mechanisms can be used to ease device modeling and to decouple modeled environments from model evolutions. Section 5 finally provides final remarks and proposes future works.

2   Related Works

Modeling domotic environments through ontologies or taxonomies is an interesting field, but the amount of available literature is very limited. The main contributions are the EHS taxonomy[1] and DomoML [1]. Besides, interesting and complementary works have been done on pervasive and ubiquitous computing modeling [2] and for context representation in ambient intelligence environments [3,4,5].

The EHS taxonomy is a home appliance classification system designed by the EHS (European Home System) consortium (now evolved in the Konnex alliance[2]) that mainly describes so-called white and brown goods located in a domestic environment. It is deployed along four main classes: Meter Reading, which groups all measurement tools, House Keeping, which groups all household appliances and systems, Audio and Video, which encompasses multimedia appliances, and Telecommunication, grouping all tools able to establish a communication. This simple taxonomy has several drawbacks that prevent effective house modeling: first, it takes a somewhat incoherent modeling approach, as overlapping classes are represented as different branches in the taxonomy (thus implying un-existent disjointness unless classification under multiple branches is allowed). Second, it doesn't support non-taxonomic relationships between objects and does not address function and state modeling. Third, it does not deal with representation of appliance functions, capabilities and type of permitted operations, only allowing simple, static description of environments, without any formal notion on operating capabilities of modeled entities.

DomoML [1,6] provides a full, modular ontology for representing household environments. It describes operational and functional aspects together with some

---

[1] The European Home System, http://www.ehsa.com

[2] http://www.konnex.org

preliminary architectural and positioning information and is based on three core ontologies: DomoML-env, DomoML-fun and DomoML-core. DomoML-env provides primitives for the description of all "fixed" elements inside the house such as walls, furniture elements, doors, etc., and also supports the definition of the house layout by means of neighborhood and composition relationships. DomoML-fun provides means for describing the functionalities of each house device, in a technology independent manner. DomoML-core provides support for the correlation of elements described by DomoML-env and DomoML-fun constructs, including the definition of proper physical quantities. DomoML shows some shortcomings when applied to real-world domotic systems. As first, it mixes too different levels of detail in modeling. This implies, on one side, over-specification, e.g., to define that a lamp can be lit, a modeler has to describe the lamp, the attached switch button, down to the single lever. On the other side, it does not address state modeling and doesn't provide facilities to query or auto-complete models, thus requiring a cumbersome modeling effort whenever a new house must be described.

In the context of pervasive computing, the SOUPA ontology [2] provides a modular modeling structure that encompasses vocabularies for representing intelligent agents, time, space, events, user profiles, actions and policies for security and privacy. SOUPA is organized into a core set of vocabularies, and a set of optional extensions. Core vocabularies describe concepts associated with person, agent, belief-desire-intention, action policy, time, space and event. These concepts are expressed as 9 distinct ontologies aligned to well known vocabularies such as FOAF [7], DAML Time [8], OpenCyc [9] and RCC [10]. SOUPA cannot be directly applied to support interoperability and intelligence for domotic systems as many domain-specific concepts are lacking (e.g., no primitives are provided for modeling devices, functionalities, etc.), however it can be useful in a multilayered approach, where DogOnt provides domain-specific, operative knowledge and SOUPA allows modeling high level, pervasive concepts, easing the implementation of intelligent behaviors on top of it.

3   DogOnt Ontology

The DogOnt ontology is designed with a particular focus on interoperation between domotic systems. Base assumptions are directly driven by real-world case studies [11], mainly focusing on device, state and functionality modeling. DogOnt (whose features are reported in Table 1) is deployed along 5 main hierarchy trees (Figure 1):

- *Building Thing*: modeling available things (either controllable or not);
- *Building Environment*: modeling where things are located;
- *State*: modeling the stable configurations that controllable things can assume;
- *Functionality*: modeling what controllable things can do;
- *Domotic Network Component*: modeling features peculiar of each domotic plant (or network).

Table 1. DogOnt ontology statistics

Feature	Value
Expressivity	$ALCHOIN(D)$
Named Classes	167
Number of Siblings per Node (mean)	5
Restrictions	126
Universal restrictions	21
Cardinality restrictions	54
hasValue restrictions	41
Object properties	18
Datatype properties	26

3.1   Environment Modeling

Environment modeling is achieved by means of the DogOnt concepts inheriting from *Building Environment* and from *Building Thing* (Figure 1), both already defined in DomoML [1] but differently formalized in DogOnt to overcome the limitations in IDE modeling described in section 2. Modeling detail is limited to the minimal set of primitives needed to locate domotic components, furniture elements and appliances inside a single flat or living unit. Entire buildings can be represented by extending this section of the ontology through subclassing of *Building Environment* and through the definition of proper relationships (e.g. by introducing principles from spatial modeling and reasoning [12,13]).

The *BuildingEnvironment* tree supports a coarse representation of domestic environments, as whole architectural units, including: several types of *Room*, the *Garage* and the *Garden*. The *BuildingThing* tree, instead, represents all the elements that can be located or that can take part in the definition of a *BuildingEnvironment*. DogOnt defines a clear separation between objects that can be controlled by a domotic system (*Controllable* class) and all the other objects that can be found in a home (*UnControllable* class); they are explicitly modeled as disjoint classes.

Controllable objects can be appliances or can belong to house plants such as the HVAC[3] plant. Appliances are modeled through the homonymous class and are further subdivided in *White Goods* and *Brown Goods*, according to the EHS taxonomy. *House plants* include *HVAC systems*, *electric systems* and *security systems*. They differ from appliances as they are usually installed in fixed positions, and they encompass several components that must be coordinated to reach a specific goal (e.g., delivering electrical power).

Uncontrollable objects are all the home components that cannot be directly controlled by a domotic system. They are mainly subdivided in *Furniture* and *Architectural* elements. *Furniture* models all the elements usually adopted as furniture like chairs, cupboards, desks, etc. Instead, *Architectural* objects model all the elements that define a living environment such as *Walls*, *Floors*, etc. They are mainly grouped in *Vertical* and *Horizontal* elements, which are further

---
[3] Heating, Ventilating, and Air Conditioning

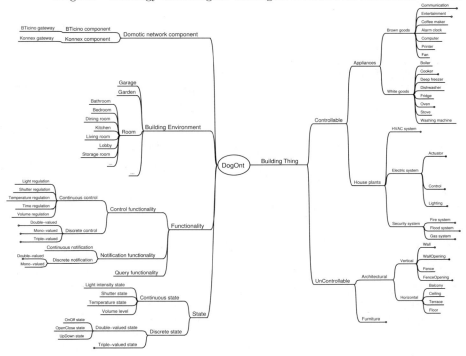

Fig. 1. An overview of the DogOnt ontology

subdivided in subclasses. Architectural modeling is somewhat limited to simple partitions (walls, floors) and openings (windows, doors) and may be extended for implementing advanced modeling, e.g., to support architectural design.

3.2 Device Modeling

All the objects referred to as "device," in this paper, are objects belonging to the *Controllable* sub-tree. A controllable object differs from an uncontrollable one as it must satisfy several restrictions on *Functionalities* and *States*. It must possess at least the functionality of being queried about its operative condition (*QueryFunctionality*) and it must possess a state, intended as the ability of reaching a stable configuration identifiable in some way: a lamp is able to be steadily on or off, a flashing light can be on, off or flashing, a shutter can be moving up, down or being steady, etc.

Example (part 1): We consider a dimmer lamp connected to a KNX network, and located in the living room, as sample device. On the basis of formalization defined until now, our dimmer lamp is an instance of the class *Dimmer Lamp*, which in turn is a *Lamp*, a *Lighting*, an *Electric System*, a *HousePlant* and a *Controllable* object. The corresponding OWL formalization fragment is reported in Figure 2. As can be easily noticed, while the position inside the house is

```
<DimmerLamp rdf:ID="sample_dimmer_lamp">
  <isIn rdf:resource="#sample_living_room"/>
  <rdf:type rdf:resource="#KNXNetworkComponent"/>
  <individualAddress>101</individualAddress>
  <groupAddress>12</groupAddress>
  <hasControl rdf:resource="#switch_sample_dimmer_lamp"/>
</DimmerLamp>
```

Fig. 2. Example (part 1) - representation of a sample DimmerLamp instance

explicitly modeled, as well as the fact that our dimmer lamp is actually connected to a KNX network, and can be controlled by a switch. The inheritance of characteristics from ancestors such as *Lighting* or *Controllable* is left to a simple reasoning step, where the transitive closure of the model is computed and properties are propagated along the ontology *isA* relations.

3.3   Functionality Modeling

Each device class, in DogOnt, is associated to a set of different functionalities, by means of the *hasFunctionality* relationship. Several approaches can be chosen for functionality modeling: a compositional approach, as in DomoML, where the functionalities of a given object derive from the composition of functionalities provided by its components, or a descriptive model, where functionalities are described apart and then associated to the single devices. DogOnt takes this last approach and models functionalities by objectives and by variation modality. This allows to use only one instance per functionality, as device capabilities are modeled independently from device classes.

Each functionality defines the commands to modify a given device property (e.g., light intensity) and the values they can assume. Functionalities are divided in different classes on the basis of their goals: *Control Functionalities* model the ability to control a device or a part of it, e.g., to open up a shutter. *Notification Functionalities* represent the ability of a device to autonomously advertise its internal state and in particular the ability of detecting and signaling state changes. *Query Functionalities* encompass the capabilities of a device to be queried, or polled, about its condition, e.g., failure, internal state values, etc.

Functionalities are also modeled according to the way they modify the internal state of a given device; two main variation families are provided: *Continuous Functionalities* that allow to change device characteristics (e.g., the light intensity) in a continuous manner, between a minimum and a maximum value, and *Discrete Functionalities* that only allow abrupt changes of device properties, e.g., to switch a light on. Most domotic devices such as switches, plugs and lights can be controlled by means of only 2 or 3 different commands (e.g., on-off, open-close, up-down-rest, etc.) while functionalities controlled by more than 3 commands are rare. Reflecting this situation, the DogOnt ontology explicitly models *Discrete Functionalities* as mono-, double- and triple-valued functionalities. Clearly, these 3 subclasses do not define the complete universe of possible discrete functionalities, therefore multi-valued capabilities can be modeled by directly instantiating the *Discrete Functionality* class.

```
<owl:Class rdf:ID="DimmerLamp">
<rdfs:comment rdf:datatype="http://www.w3.org/2001/XMLSchema#string">
  Lamp that varies the level of illumination</rdfs:comment>
<owl:equivalentClass>
  <owl:Class>
    <owl:intersectionOf rdf:parseType="Collection">
      <owl:Restriction>
        <owl:hasValue>
          <LightRegulationFunctionality
            rdf:ID="LightRegulationFunctionalityInstance"/>
        </owl:hasValue>
        <owl:onProperty>
          <owl:ObjectProperty rdf:about="#hasFunctionality"/>
        </owl:onProperty>
      </owl:Restriction>
      <owl:Class rdf:about="#Lamp"/>
    </owl:intersectionOf>
  </owl:Class>
</owl:equivalentClass>
<rdfs:label rdf:datatype="http://www.w3.org/2001/XMLSchema#string">
  DimmerLamp</rdfs:label>
</owl:Class>
```

Fig. 3. The definition of DimmerLamp with associated functionalities

Example (part 2): Let us re-consider the sample dimmer lamp definition started in section 3.2; the capabilities of the lamp can be modeled by means of proper functionalities. Our dimmer lamp, being an instance of the class *Dimmer Lamp*, is connected to the unique *LightRegulationFunctionalityInstance* (continuous) defined in DogOnt, which models the ability to dim the emitted light. As *DimmerLamp* is a subclass of *Lamp*, it inherits the related *OnOffFunctionalityInstance* (discrete) modeling the capability to switch on and off the lamp. Finally, a *Lamp* is a specific subtype of *Controllable* to which is associated the *QueryFunctionalityInstance* representing the capability of the lamp to be queried about its characteristics (e.g., the dimming level). Figure 3 shows the corresponding OWL excerpt.

3.4   State Modeling

States are modeled following the same descriptive approach adopted for functionalities; they must be instantiated for each home device instance since different devices belonging to the same conceptual class, e.g., *Lamp*, can be in different conditions, e.g., on or off.

States are classified according to the kind of values they can assume: continuously changing qualities are modeled as *Continuous States* with an associated *continuousValue* datatype property of type `xsd:float`. Instead, qualities that can only assume discrete values (e.g., On/Off, Up/Down, etc.) are classified as *Discrete States*. Discrete states are subdivided in double- and triple-valued states, while states having more than 3 stable configurations are modeled by directly instantiating the *Discrete State* concept (see Figure 1 for the complete hierarchy).

Discrete states are characterized by the *valueDiscrete* datatype property that describes the current state and by a variable set of *possibleStates* (datatype property) that models all the possible values that *valueDiscrete* can assume for the current state type. Figure 4 reports the definition of the *OnOffState*

```
<owl:Class rdf:about="#OnOffState">
  <rdfs:comment rdf:datatype="http://www.w3.org/2001/XMLSchema#string">
   State: on - off</rdfs:comment>
  <rdfs:label rdf:datatype="http://www.w3.org/2001/XMLSchema#string">
   OnOffState</rdfs:label>
  <rdfs:subClassOf rdf:resource="#DoubleValuedState"/>
  <rdfs:subClassOf>
    <owl:Restriction>
      <owl:onProperty>
        <owl:DatatypeProperty rdf:about="#possibleStates"/>
      </owl:onProperty>
      <owl:hasValue rdf:datatype="http://www.w3.org/2001/XMLSchema#string">
       On</owl:hasValue>
    </owl:Restriction>
  </rdfs:subClassOf>
  <rdfs:subClassOf>
    <owl:Restriction>
      <owl:hasValue rdf:datatype="http://www.w3.org/2001/XMLSchema#string">
       Off</owl:hasValue>
      <owl:onProperty>
        <owl:DatatypeProperty rdf:about="#possibleStates"/>
      </owl:onProperty>
    </owl:Restriction>
  </rdfs:subClassOf>
</owl:Class>
```

Fig. 4. The definition of the OnOffState

typically associated to all *Lamp* instances, the *valueDiscrete* property is defined in the *Discrete State* class and inherited by all subclasses.

Example (part 3): Having defined state modeling, the sample dimmer lamp instance can now be completely defined (Figure 5). As can easily be noticed, many properties of this simple object are not explictly modeled, e.g., its associated functionalities and the relative commands, but need to be deduced by performing a simple reasoning step that computes the transitive closure of the model, turning implicit knowledge into explicit information (Figure 6).

3.5 Network Modeling

Interoperation between domotic systems requires the definition of a technology independent formalization that allows to operate seamlessly with different devices, produced by different manufacturers and operating in plants with different technologies. A minimum set of plant-dependent knowledge must, however, be available for enabling interoperation systems (gateways) to interact with physical devices. DogOnt models such information by means of an ontology branch stemming from the *Domotic Network Component* concept.

```
<DimmerLamp rdf:ID="sample_dimmer_lamp">
  <isIn rdf:resource="#sample_living_room"/>
  <rdf:type rdf:resource="#KNXNetworkComponent"/>
  <individualAddress>101</individualAddress>
  <groupAddress>12</groupAddress>
  <hasControl rdf:resource="#switch_sample_dimmer_lamp"/>
  <hasState>
    <LightIntensityState
     rdf:ID="DimmerLamp_Livingroom1_LightIntensityState"/>
  </hasState>
</DimmerLamp>
```

Fig. 5. The OWL definition of the sample DimmerLamp instance

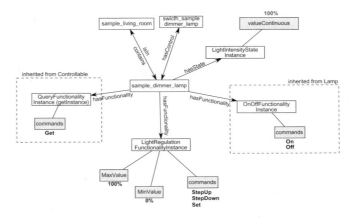

Fig. 6. The complete definition of the sample DimmerLamp instance, with inherited properties

Table 2. Specific features of network-level gateways

Gateway-specific property	BTicino	KNX
connection timeout	x	x
connection trials before failure	x	x
IP address	x	x
port	x	x
sleeptime	x	x
multicast address	–	x
UDP port	–	x
polling interval	–	x

Every controllable object belonging to a domotic plant is modeled as an instance of a specific controllable concept (e.g., a *Lamp*) and, at the same time, as an instance of a proper *Network Component*. Currently 2 network components are already modeled: the *KNXNetworkComponent*, representing KNX-compliant devices and the *BTicinoNetworkComponent*, representing BTicino MyHome devices.

No subclasses are required except for network-level gateways that need more fine grained descriptions to model features such as IP addresses, polling intervals, etc. (see Table 2 for a complete reference). These features are needed by integration systems to interface domotic networks, thus enabling interoperation.

4   DogOnt - Rules

DogOnt provides different reasoning mechanisms responding to different goals: to ease model instantiation, to verify the formal correctness of model instantiations (consistency checking) and to support automatic recognition of device instances from their features (i.e., to support scalability and model evolution):

Model instantiation is a relatively complex task that requires generating device, and state, instances, and to properly link them by means of relationships. Due to the many modeling aspects considered in DogOnt, this process can be quite difficult and error prone. Therefore, a suitable set of auto-completion rules has been defined, which allows, with a single deduction step, to automatically create states and relationships associated to specific type of devices.

Consistency checking allows to verify the formal correctness of generated model instantiations, ensuring correct operation of systems built on top of them.

Classification reasoning is used to automatically recognize device classes, starting from device functional descriptions. This allows to decouple the DogOnt model evolution from model instantiation, thus enabling systems to operate with unknown device classes, on one side, and to automatically re-classify existing devices with respect to new classes defined in forecoming DogOnt model versions, on the other side.

4.1   Rule-Based Model Instantiation

DogOnt represents each device as an object having a given set of functionalities and states (Section 3). Functionalities are automatically added to every device instance by means of suitable restrictions defined at the class level. They are shared by all devices of the same class. On the contrary, states are peculiar of each device.

Manually creating and associating states and devices is absolutely tedious for designers as it is repetitive and can lead to modeling errors and/or inconsistencies (this also happens in the approach taken by DomoML). Thanks to the repetitive nature of the operation, the process can be easily automated. Rule-based reasoning can, in fact, generate the needed instances and links in an automatic and verified manner.

Several rule languages can be applied to the DogOnt ontology; among them, SWRL [14] appears the most suitable solution as it allows to directly embed rules in the DogOnt ontology (SWRL constructs can, in fact, be expressed in OWL). SWRL is a powerful rule language based on Horn-like rules [15], that guarantees decidability in finite time. The most interesting feature of SWRL is the ability to provide/define so-called built-in operators, that can implement non-logic operations such as mathematic calculations, string functions, etc. Built-ins [14] can either be customly built or can belong to standard sets defined in SWRL-B[4] and SWRL-X[5] libraries. SWRL-X, in particular, provides class/instance generation built-ins that, combined with string elaboration built-ins and proper modeling, allow to autocomplete states for device instances. Figure 7 shows a sample SWRL rule for attaching state instances to *SimpleLamp* instances. It must be

---

[4] http://www.daml.org/rules/proposal/builtins.html
[5] http://swrl.stanford.edu/ontologies/built-ins/3.3/swrlx.owl

```
SimpleLamp(?x)^rdfs:label(?x,?y)^
swrlb:stringConcat(?z,?y,"_OnOffState")^
swrlx:createOWLThing(?w,?z)->
OnOffState(?w)^rdfs:label(?w,?z)^hasState(?x,?w)
```

Fig. 7. The auto-completion rule for SimpleLamp states

noticed that to generate human-readable labels and identifiers for SWRL-created states, instances must have a human-readable `rdfs:label`.

SWRL rules can be executed by any SWRL-compliant rule engine. In Protégé [16], for example, the SWRLTab allows to use the Jess[6] rule engine to execute SWRL statements, embedding the newly generated knowledge in the active model. In this way, during house modeling, a domotic designer can instantiate the needed devices without caring of states. Then, she can run the Jess engine with the DogOnt SWRL rules obtaining as result a complete model. Consistency problems are completely avoided for state modeling as rules are predefined and a priori validated.

4.2 Classification Reasoning

Classification reasoning is a type of automated inference that allows to infer the class(es) to which an instance belongs, by checking its properties against the set of necessary and sufficient conditions that define class membership. These conditions are usually adopted in consistency checking to ensure that class instances respect the restrictions defined on properties for individuals belonging to a given class. For example, a *DimmerLamp* instance must have a *LightRegulationFunctionality*, must only possess one *LightIntensityState* and must be a *Lamp*. Every asserted *DimmerLamp* instance respecting these constraints is valid. This kind of reasoning can also be used to discover new class memberships, i.e., to infer instance types: in a sample scenario a given domotic system provides a device able to variate light intensity. The home modeler does not know the existence of the *DimmerLamp* class, and decides to model the device as a *Lamp* instance, with a *LightIntensityState* and a *LightRegulationFunctionality*. Classification reasoning allows to discover that the modeled device is actually a *DimmerLamp*, and can therefore be treated in the way asserted *DimmerLamps* are.

Classification reasoning is a fundamental part of formal modeling of home environments for domotic interoperability. In fact, as manufacturers are always adding new features to their networks and new technologies are emerging too, house models frequently become un-synchronized with the actual environment configuration or capabilities. Being able to automatically discover new classes for already defined instances allows to easily extend/amend the DogOnt ontology without risking to disrupt existing models.

---

[6] The Jess rule engine, `http://www.jessrules.com/`

5   Conclusions

In this paper we introduce DogOnt, a new modeling approach for domotic environments composed of the DogOnt ontology and a set of DogOnt rules. DogOnt provides functionality and state auto-completion, and supports model evolution through classification reasoning. Modeling is done at a detail level that reflects the actual needs of interoperation between real-world domotic systems and supports the development of Intelligent Domotic Environments. Architectural modeling is also provided, although in a very limited, but extensible form. Novelty points include the descriptive modeling approach, more flexible than approaches available in the literature, and the definition of auto-completion mechanisms through reasoning, which eases the house modeling process.

Formal modeling of house environments through ontology-based technologies is a promising research stream for domotic systems and smart homes. Ontologies allow, on one side, to achieve a natural abstraction of networks and devices that can be used for supporting interoperability. On the other hand, well studied reasoning techniques can both ease the modeling process and provide support for the implementation of complex, intelligent behaviors inside domotic homes.

The authors are currently working on the design and implementation of DOG, a domotic gateway based on DogOnt and OSGi, on structural checks, and on verification of safety properties through rule-based reasoning on DogOnt.

References

1. Sommaruga, L., Perri, A., Furfari, F.: DomoML-env: an ontology for Human Home Interaction. In: Proceedings of SWAP 2005, the 2nd Italian Semantic Web Workshop, Trento, Italy. CEUR Workshop Proceedings, December 14-16 (2005)
2. Chenand, H., Finin, T., Joshi, A.: Ontologies for Agents: Theory and Experiences. In: The SOUPA Ontology for Pervasive Computing, pp. 233–258. Birkhäuser Basel, Los Alamitos (2005)
3. Preuveneers, D., Van den Bergh, J., Wagelaar, D., Georges, A., Rigole, P., Clerckx, T., Berbers, Y., Coninx, K., Jonckers, V., De Bosschere, K.: Towards an extensible context ontology for Ambient Intelligence. In: Markopoulos, P., Eggen, B., Aarts, E., Crowley, J.L. (eds.) EUSAI 2004. LNCS, vol. 3295, pp. 148–159. Springer, Heidelberg (2004)
4. Zhang, D., Gu, T., Wang, X.: Enabling Context-aware Smart Home with Semantic Technology. International Journal of Human-friendly Welfare Robotic Systems 6(4) (2005)
5. Kofod-Petersen, A., Aamodt, A.: Contextualised Ambient Intelligence through Case-Based Reasoning. In: Roth-Berghofer, T.R., Göker, M.H., Altay Güvenir, H. (eds.) ECCBR 2006. LNCS (LNAI), vol. 4106, pp. 211–225. Springer, Heidelberg (2006)
6. Furfari, F., Sommaruga, L., Soria, C., Fresco, R.: DomoML: the definition of a standard markup for interoperability of human home interactions. In: EUSAI 2004: Proceedings of the 2nd European Union symposium on Ambient intelligence, pp. 41–44. ACM, New York (2004)
7. Brickley, D., Miller, L.: FOAF Vocabulary Specification 0.91 (November 2007), http://xmlns.com/foaf/spec/

8. Hobbs, J., Pustejovsky, J.: Annotating and Reasoning about Time and Events. In: 2003 AAAI Spring Symposium (2003)
9. Reed, S.L., Lenat, D.B.: Mapping Ontologies into Cyc. In: AAAI 2002 symposium (2002)
10. Li, S., Ying, M.: Region Connection Calculus: its models and composition table. Artificial Intelligence 145(1-2), 121–146 (2003)
11. Pellegrino, P., Bonino, D., Corno, F.: Domotic House Gateway. In: SAC 2006: Proceedings of the 2006 ACM symposium on Applied computing, pp. 1915–1920. ACM, New York (2006)
12. Renz, J., Nebel, B.: On the complexity of qualitative spatial reasoning: A maximal tractable fragment of the Region Connection Calculus. Artificial Intelligence 108, 69–123 (1999)
13. Renz, J., Ligozat, G.: Weak Composition for Qualitative Spatial and Temporal Reasoning. In: van Beek, P. (ed.) CP 2005. LNCS, vol. 3709, pp. 534–548. Springer, Heidelberg (2005)
14. Horrocks, I., Patel-Schneider, P.F., Boley, H., Tabet, S., Grosof, B., Dean, M.: SWRL: A Semantic Web Rule Language Combining OWL and RuleML (November 2003), http://www.daml.org/2003/11/swrl/
15. Horn, A.: On Sentences Which are True of Direct Unions of Algebras. The Journal of Symbolic Logic 16, 14–21 (1951)
16. Gennari, J.H., Musen, M.A., Fergerson, R.W., Grosso, W.E., Crubezy, M., Eriksson, H., Noy, N.F., Tu, S.W.: The evolution of Protégé: an environment for knowledge-based systems development. Int. Journal of Human-Computer Studies 58(1), 89–123 (2003)

Introducing IYOUIT

Sebastian Boehm[1], Johan Koolwaaij[2], Marko Luther[1], Bertrand Souville[1], Matthias Wagner[1], and Martin Wibbels[2]

[1] DoCoMo Euro-Labs, Landsbergerstr. 312, 80687 Munich, Germany
[2] Telematica Instituut, Brouwerijstraat 1, 7523 XC Enschede, The Netherlands

Abstract. We present IYOUIT, a prototype service to pioneer a context-aware mobile digital lifestyle and its reflection on the Web. The application is based on a distributed infrastructure that incorporates Semantic Web technologies in several places to derive qualitative interpretations of a user's digital traces in the real world. Networked components map quantitative sensor data to qualitative abstractions represented in formal ontologies. Subsequent classification processes combine these with formalized domain knowledge to derive meaningful interpretations and to recognize exceptional events in context histories. The application is made available on Nokia Series-60 phones and designed to seamlessly run 24/7.

1 Introduction

In this paper we introduce IYOUIT[1], a mobile application that allows users to automatically collect so-called context information centered on places they visit and people they meet. Context, in a technical sense, is regarded as any piece of information that can be recognized and further processed to adapt the behavior of the application according to a given set of constraints [1]. The application aims at making it easy to automatically collect such data with a standard phone and facilitates an instant and light-hearted sharing of personal experiences within communities.

All data collected by IYOUIT is aggregated into a wealth of context information and made accessible to users on the Web and on a mobile client. For selected context sources, value is added through the transformation of quantitative context information into qualitative statements about a user's given situation. By hooking up to Web 2.0 services like Flickr[2] and Twitter[3], the application allows for sharing personal context with others online. Sharing can be instant, by posting single data items, or through the aggregated contextual experience in potentially lifelong online blogs.

In IYOUIT and its underlying component framework, Semantic Web technology is used in several places to implement key features and to seamlessly connect the application to different services on the Web. IYOUIT is the result of our long-year efforts to leverage the use of context information in mobile applications and has recently been rolled out of our labs as a prototype service for free public use. The remainder of

---

[1] http://www.iyouit.eu
[2] http://www.flickr.com
[3] http://twitter.com

A. Sheth et al. (Eds.): ISWC 2008, LNCS 5318, pp. 804–817, 2008.

Fig. 1. The IYOUIT Mobile Client

this paper is organized as follows: in Section 0 we give a first conceptual overview of IYOUIT and explain its target application domains, which correspond to our general research objectives. In Section 0 we show where Semantic Web technology is embedded into our systems. We do so in concretely discussing selected implementations of the underlying component framework. Section 0 illustrates the practical use of IYOUIT and highlights features that actually reveal semantics to the user. We summarize and conclude with a view on the current IYOUIT user community.

2   A First Look at IYOUIT

IYOUIT has been designed and developed as a "living lab" for our research on mobile community services, context awareness and the smart fusion of both towards the Semantic Web. The idea has always been to spin-off and implement early ideas from research into this living lab for feedback and evaluation within the user community. IYOUIT is the reference implementation of our underlying component framework and comes as a mobile client (a first impression is given in Fig. 1) as well as a Web portal and integrates popular 3rd party Web 2.0 services like Flickr and Twitter.

Since recently, IYOUIT is released to the public as a free service with a wide set of implemented features that have been streamlined towards four complementary, but not mutually exclusive, application domains *Share, Life, Blog* and *Play*. In the spirit of our living lab approach, all features that are showing in the application are fully functional, yet it remains to be seen how well they will be accepted and used by the IYOUIT user community.

In this section we give a first and high-level overview of the IYOUIT application domains as well as the implemented features. We also introduce the underlying context management framework.

2.1   Share, Life, Blog, Play

Share *(community-based context sharing):* application features of the IYOUIT target domain *Share* are concerned with the possible synergies of context-awareness and social networking services. The work builds on results [2] in social networking enhanced through ontology-based reasoning for communities and focuses on the analysis of personal context histories and established relationships to identify possible social network extensions. The social network of users with qualified social relationships is represented using formal ontologies. Through ontology-based reasoning, data consistency can be ensured and additional relationships can be deduced to complement the users' circle of friends as a social portfolio.

Life *(life support through context-aware guidance):* IYOUIT *Life* deals with analyzing user-generated content, such as tags that are manually assigned to photos, and its relation to context over time and space. In the spirit of the Web 2.0, the goal is to extract information from the "wisdom of crowds" through aggregation in the geographic space. As starting point we apply clustering to user-generated tags, which hold location information, to determine regions of interest on maps. These tag clusters – characterized by population, density, position and range – can be used to guide IYOUIT users in their surrounding area. For data sources, IYOUIT Life builds on the services Flickr and Flagr[4] where geo-tagging already enjoys increasing popularity.

Blog *(enhanced contextual blogging):* IYOUIT *Blog* is concerned with enhanced automatic blogging capabilities based on our related work in ontology-based reasoning [3] and a more specialized reasoning [4] to recognize and manage complex contextual events. As most collected context data in IYOUIT is of quantitative nature, abstraction methods and context ontologies haven been introduced to deal with context at a higher level. At the level of these context ontologies, complex conceptual dependencies between context elements are introduced to enrich contextual descriptions and to implement classification-based reasoning about the user's situation, e.g. to describe user places as conceptual abstractions from exact locations ("Office", "Home" or "Business Place").

Play *(playful experience of context-awareness in games):* in IYOUIT *Play* we focus on low-level and quantitative aspects of context gathering, management and similarity detection. Work is concerned with detecting specific context constellations based on low-level context features and describing these constellations in both, a formal and a human readable format. Context histories are observed to discover homogeneous time segments in a higher-dimensional context space and to detect strong correlations between different context dimensions. From the four IYOUIT target domains, Play is at present and by design the one that is the least grounded in Semantic Web technology. It is therefore not further addressed here.

2.2 Context Management

IYOUIT is based on its own Context Management Framework (CMF) to host various services and data sources. Framework components, for instance, track the positions of users via GPS and cell tower information to identify frequently visited places over time. Further context sources include the whereabouts of buddies, scanned Bluetooth and WLAN beacons, local weather, photos, sounds, observed products, books or messages.

---

[4] http://www.flagr.com

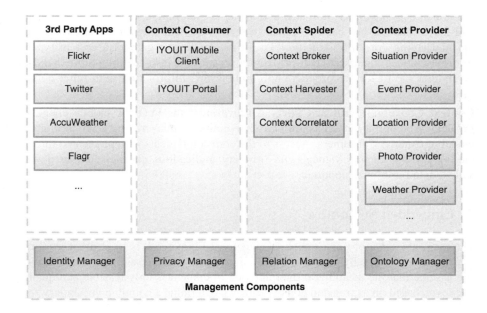

Fig. 2. IYOUIT Context Management Framework

Through distributed CMF components, the gathering of context data can be implemented in a flexible way, to reason about the combination of various context streams. In the following we give an overview of all essential framework components (cf. Fig. 2).

Management Components: the management components are the foundational CMF building blocks. They, for instance, ensure a secure authentication of entities (Identity Manager), enable access control mechanisms to prevent the uncontrolled disclosure of sensitive information (Privacy Manager, see [5]), allow for the usage of domain specific knowledge formalized within a set of core ontologies (Ontology Manager, cf. Section 3.1), or provide means to represent and reason about the social network of users (Relation Manager, cf. Section 3.2).

Context Spiders: the Context Broker, Context Harvester and the Context Correlator form the group of Context Spiders and provide means for easy context lookup, gathering and alignment across components. The Context Broker maintains a complete repository of all registered components and their public schemata to discover appropriate services and to allocate queries to the respective component. The Context Harvester retrieves context data across multiple components independent of where the information is stored physically. The Context Correlator aligns distinct context streams based on their temporal intersection and identifies significant correlations over time.

Context Providers: Context Providers (CP) lie at the very heart of the CMF. They encapsulate the basic context data sources at a quantitative level but can also implement aggregations and abstractions to a qualitative level (cf. Section 3.3 and 3.4). Each CP has been designed to realize domain specific services based on the underlying core infrastructure. To do so, a CP first gathers a certain type of information (e.g. the user's cell-id) from a sensor (for instance the mobile handset) or another CP to further process and combine this information with other context data.

Context Consumers and 3rd Party Applications: the IYOUIT mobile client, the IYOUIT Web portal as well as 3rd party integrations are leveraging CMF through the model of Context Consumers. Access to the core components is implemented through the Privacy and Identity Manager. As 3rd party applications, currently, Flickr, Twitter and Google Earth[5] are supported.

3   Embedded Semantics

The main objective of our Context Management Framework is to abstract from raw sensor data to eventually gain qualitative information about a user in a given situation. We assume that the meaningful interpretation of context is only feasible at a qualitative level, based on aggregated context data. To determine a common vocabulary for a unified interpretation of qualitative context among CMF components we designed a set of specific context ontologies formulated in a decidable fragment of the Web Ontology Language (OWL) [6]. Each Context Provider is responsible to link the quantitative values contained in context elements to qualitative values expressed using this vocabulary. In addition, a Context Provider might also interface with OWL reasoning engines to derive even higher-level of abstractions through the classification of sets of qualitative values using standard Description Logic [7] techniques.

In the following, we give insights into how semantic technology is applied in IYOUIT and highlight the possible alignment of context data to well formed ontologies in a practical application.

3.1   Ontology Manager

To make use of ontology based reasoning mechanisms, we designed a set of core ontologies for distinct application domains including social relationships, location records and weather conditions, amongst others. A total number of twelve interrelated component ontologies defining more than 300 concepts, 200 properties and 300 individuals have been defined. Due to near real-time requirements of most applied reasoning processes, we concentrated on the modeling of highly efficient yet expressive enough ontologies rather than utilizing already existing and widely used ontologies such as DOLCE[6]. Furthermore, these ontologies are not used as the main representation format for all aspects of context, since ontologies are generally weak in handling large amounts of data efficiently [8]. Instead, only distinct higher-level data elements are annotated with ontology references, making them available for further ontology

---

[5] http://earth.google.com
[6] http://www.loa-cnr.it/DOLCE.html

reasoning. This way, the overall scalability is not affected, while at the same time valuable reasoning results can be achieved.

The Ontology Manager provides a gateway within the CMF to simplify the access to OWL ontologies. Given an OWL ontology, it allocates a reasoner resource and instructs it to retrieve the corresponding set of axioms. As foreseen in the OWL 2 draft specification [6], the access to an ontology is accomplished through a physical URI that is given by a mapping from its identifying logical URI. This configurable ontology mapping, the parsing of the concrete OWL syntax and the interfacing with various DL reasoners (at the time of writing we use Pellet [9], FaCT++ [10] and RacerPro [11]) is delegated to a semantic middleware (in our case the OWL API [12]).

After having configured the Ontology Manager appropriately, a CMF component can request structural information about named concepts, properties and individuals. Among the supported queries are requests to retrieve (parts of) the concept and property hierarchy, the (direct) types of individuals, the (direct) individuals of concepts and the relations that hold between two individuals. Each of those requests might also refer to implicit knowledge, which is why a reasoner has to be involved to ensure the completeness of the returned results.

3.2 Relation Manager

All of IYOUIT's community services rely on social networking. Users can establish defined relationships amongst each other, e.g. "friend", "colleague" or "husband", to build social networks and share context data through them. IYOUIT, for instance, supports the concept of buddy lists on its mobile client where users can instantly look up what their friends are doing. Likewise, photos, sounds and other context elements can be shared with others on the Web.

Social networking is implemented through the Relation Manager that exposes a specialized API to allow restricted access to sensitive social context data. The Relation Manager stores social networks as OWL ontologies and enables DL-based reasoning to classify and to combine explicit, user-defined facts about social relationships with the world knowledge encoded for network completion [2]. In short, OWL-based reasoning is crucial for two reasons: for one, the consistency of the provided data can be maintained by dismissing any contradicting or illegitimate definitions; secondly, the applied knowledge discovery techniques can complement the social network of users in deducing implicit relationships.

The entire set of all relationships, including user-defined as well as deduced relationships, is stored within a relational database. The inference process is triggered whenever a relationship has been approved or in case an existing relationship has been removed. All relationships are then transmitted to the inference engine, with the reasoning results again being stored in the database. As a result, the underlying knowledge base does not need to be involved in most cases, because the majority of requests from other CMF components are concerned with retrieving rather than modifying data.

The Relation Manager is used to implement access control policies for privacy protection in IYOUIT: not only is an established relationship between users essentially required to share data, privacy policies can also be defined along relationships

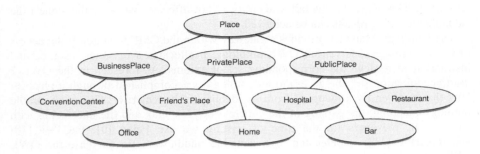

Fig. 3. Snapshot of the Place Ontology

to determine the level of access to personal context. Such access policies can be specified for individuals ("my friend Robert") or types of individuals ("all my friends") and are naturally bound to the underlying social network. The level of access detail is defined per context category, e.g. "show address information only to the level of the city name".

For many common relationships and access policies IYOUIT provides a predefined social ontology and access control directives. To this end, we have modeled the social ontology, in which a hierarchy of more than 50 social relationships has been described. This way sophisticated privacy directives can be expressed to, for instance, specify that colleagues should only know the city of his current location whereas family members may have access to the entire record.

3.3 Location Provider

The main task of the Location Provider is to resolve given location estimations into actual address records, to store location traces and to deduce frequently visited places.

To abstract from exact positions to conceptual places, the Location Provider identifies significant location records of a user by applying profound statistical learning and clustering methods to historic location data [13]. Once established, a place is presented to the user to name and typify it by selecting an appropriate concept from the place ontology, which includes descriptions like "Office", "Home" or "Business Place" (see Figure 3). Staying in a place is from now on recognized automatically by IYOUIT, resulting in qualitative location characteristics also shared among buddies.

3.4 Weather Provider

The Weather Provider enriches a given location record with prevalent meteorological data such as the actual temperature or wind speed. It employs two types of automatic abstractions that link qualitative to quantitative information.

Simple static mappings assign values within certain intervals to the corresponding abstract descriptions. One example is the widely used Beaufort scale [14], which maps wind speeds to 13 qualitative wind conditions, e.g. "light air" or "gentle breeze", that are in turn further characterized within the weather ontology.

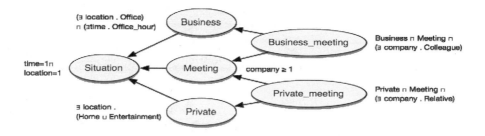

Fig. 4. Snapshot of the Situation Ontology

To categorize quantitative temperature values meaningfully, more complex mappings are needed that take additional context into account. A static mapping as the one described before, might otherwise categorize a given temperature of 5°C as "cool", even though it might have been recorded on an exceptionally warm day somewhere in the north during winter time. Instead, the Weather Provider considers several types of context to dynamically derive a sensible mapping for temperatures and precipitations. For this, historical weather records that provide monthly minimal and maximal mean temperatures and precipitation values are used. In addition, the location, the day of the year and the hour of the day are taken into account to derive a rational categorization of the actual qualitative values such as "low", "moderate", "warm", etc. Note that this mapping is slightly more evolved than the dynamic mapping of places accomplished within the Location Provider as it takes multiple types of context into account and does not involve user interaction for the final semantic categorization.

Mapping of large sets of meteorological data to qualitative abstractions allows for rating an overall weather condition, represented as an individual and linked via object properties to corresponding meteorological abstractions by a standard DL classification process. We defined categories for "bad", "fair", "good" and "splendid" weather conditions based on the ratings for individual weather attributes like temperature, wind, pressure or precipitation and a qualitative health index. This health index is itself derived by classification and expresses to what extend the current weather situation may cause aches and pains. Weather health issues such as chronic pains, aching bones or migraine are again recognized during the classification, based on formalized background knowledge. For instance, an axiom formalizes that weather conditions with low pressure, the passage of a warm front, high temperatures and humidity often cause migraines.

3.5   Situation Provider

The Situation Provider computes an abstract characterization of a user's situation by applying DL classification on several context pieces gathered by the Context Harvester from multiple Context Providers [3]. Abstract situation concepts like "Business Meeting" are formulated w.r.t. the vocabulary of the respective component ontologies (cf. Fig. 4). Each situation individual is assembled of a set of entities representing

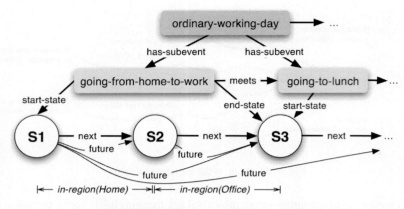

Fig. 5. Example Event of the CMF Event Provider

qualitative context information such as the location (e.g., office), the time (e.g., after-noon) and people in proximity (e.g., friends). Finally, the classification result given by the computed direct individual types is established as a new (derived) context.

3.6   Event Provider

The reasoning within the Situation Provider to classify user context can be seen as static in time as it observes the situation of a user only for one given point in time. To reason about complete context histories over longer periods, specialized reasoning is needed beyond DL-based classifications to detect the line of key situations over time.

   The Event Provider is targeting at such a detection of key events, which are de-fined as temporal and spatial constellations of significant situations in context histo-ries. The primary goal is to use detected events for an optimized composition of online blogs to meaningfully structure blog entries according to their contextual sig-nificance. In [3] we propose an approach based on RacerPro and the semantic query language nRQL [15], which is currently being implemented within the CMF Event Provider. The work exploits complex location concepts, e.g. "my father's house", quantitative concepts of time as well as the linkage of both with other significant context elements. Subsequently, spatio-temporal events like "leaving the home" or "returning from a vacation" can be expressed to indicate a qualitative spatio-temporal change in the user's state or situation.

   Figure 5 gives examples of events as they can be expressed in our model. An event is defined as a time interval having a start state and an end state. It either describes a constancy that holds between states, e.g., like "staying at home", or a certain change that happens, e.g. "going from home to work". Events may naturally depend on each other and can be organized hierarchically. In the model we further distinguish between events as being homogeneous, complex and high-level to allow for the flexible design of large knowledge bases and their reuse in different applica-tion domains.

Fig. 6. Photo Share

4   IYOUIT in Everyday Life

In this section we run through three actual use cases of IYOUIT to illustrate how implemented application features can be bundled to realize the IYOUIT target domains of Share, Life and Blog.

All examples, descriptions and screenshots are taken from the actual IYOUIT service as available today. The mobile client is organized in application tabs, where each tab either displays a certain type of context information (e.g., local weather reports in the Weather tab) or accumulates various pieces of information in a context overview. We encourage all readers who carry a Nokia Series-60 phone to visit our Web site, download IYOUIT and experience the described scenarios in real life.

4.1   Photo Share

Photo Share is the first use case presented in here and reflects the domain of IYOUIT Share. As shown in Figure 6, the mobile client provides access to the phone camera and allows the user to take photos while on the go. Once a picture is captured, IYOUIT automatically compiles all context information available at the given moment and proactively adds it to the photo. Photos are then directly published on Flickr and can be shared with others online.

On Flickr, the automatically added contextual data is made available in photo tags. Semantically enriched bits of information are added in complementary tags, the photo caption or the photo description. Also showing is the derived situation in terms of place, nearby buddies and the prevalent weather condition, qualified through the underlying ontology concepts.

Fig. 7. Buddy Map

The comprehensive tagging capabilities of Photo Share aim at maximizing the use of Flickr for IYOUIT users and allow photos to be organized along the given tags with ease. Photo albums can, for instance, be easily queried for all pictures matching the search for "Pics that I took abroad during lunch on a sunny day?" or "Pics that I took during a private trip to a far destination with my family?".

4.2 Buddy Map

As mentioned above, social networking and buddy-centric context sharing are key features in IYOUIT. Fig. 7 depicts a strong use case that is created in combining the contextual knowledge centered on locations and places with social networking.

The use case of Buddy Map aims at simplifying the daily life of users through keeping track of frequently visited places. Those places add significantly to the semantic value of a user's context, for instance in detecting complex events, but can also be used straightforwardly for basic features: the appearance of buddies in places can trigger small alerts, views to context data can be adjusted based on the type of place, or settings of the phone can be automatically switched based on rules bound to places. As an example, an ad-hoc meeting can be quickly setup in the "Office" place based on the observed fact that all "co-workers" are present in that place. During the actual meeting situation (where again place information is central to detect it) phones can be automatically muted.

The examples in Fig. 7 show a view to the IYOUIT buddy list that is grouping people according to the place they are currently staying in. The right part of the figure depicts actual places on a map revealing their approximated location and size as well as some additional attributes.

4.3 Life Blog

Changes in the context of a user can be hints for significant changes of a current situation and experience. Users may wish to communicate such changes to others in subtle ways, e.g. instead of calling the family at home to say "I just left the office and will be home soon" a small automatic note or a change in the presence settings will indicate that you are now commuting and back at home soon.

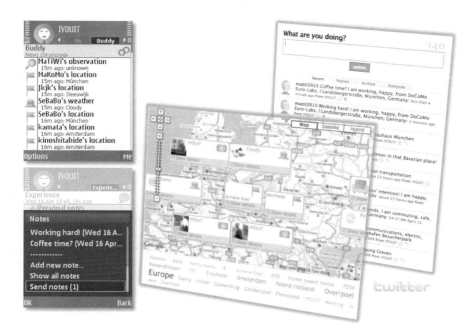

Fig. 8. Life Blog

This is the essential idea of our "Life Blog" use case as depicted in Fig. 8, where changes in the personal context can be automatically posted to others as micro-blog messages. This is done automatically through the IYOUIT mobile client or on the Web. In the example, within the buddy list, recent changes are indicated for six different users in their context data on location, local weather and observed objects. Complementary, and as also shown in Fig. 8, similar contextual notes are posted online on the IYOUIT Web portal and aligned on a map. This way it becomes easy for an observer to follow buddies and to get an idea of what they are up to. In addition to automatic, contextual postings users can submit small text notes to their buddies or the whole user community. These notes are enriched with contextual data and posted to Twitter. In the example, place and location data is automatically added as the message context together with significant presence attributes.

Context histories and micro-blog postings over longer durations are also compiled into permanent blog entries (not shown in the example) to summarize significant periods of time. Such blog entries can, for instance, summarize a trip based on visited locations, experienced weather, buddies in company and the captured photos. Furthermore, the detection of complex events within the observed context history can be used to organize such blogs.

5   Summary and Conclusion

We have introduced IYOUIT, a mobile application to enrich the digital life of people on the go. The application is available through our Web site and can be used on any standard Nokia Series-60 phone. It implements many ideas from the fields of context awareness and community-based sharing while connecting to the established Web 2.0 communities of Flickr and Twitter. As a stable research prototype from our labs, it has been released to the public in June 2008.

We are aware of competing mobile applications like ZoneTag[7], Merkitys-Meaning[8] or Shozu[9] that partially follow similar aims but seem to fall short in the completeness of their key features. A more detailed comparison with these applications is discussed elsewhere [5]. To the best of our knowledge, no applications currently exist that try to leverage Semantic Web technology in similar ways for the mobile domain.

Semantic Web technology is applied in IYOUIT core components, in particular for qualitative context abstraction and reasoning. We have put effort in optimizing our OWL ontologies and focused on efficiently integrating inference techniques based on results gained from extensive evaluation studies [8]. Work in progress is concerned with further extensions to the classification-based reasoning on user situations towards complex events along the spatio-temporal dimensions of context data. However, we also plan to model and publish meta-data of selected public context items with standard vocabularies such as FOAF[10] and SIOC[11] to interlink with other services on the Semantic Web.

With IYOUIT we are taking an agile approach for a rapid development and early release of application features. In the current version, features are manifold and purposely put quite broad. We plan to continuously rectify features, add further functionality or also remove some, based on their popular use as well as the feedback and demands of our users. Albeit early in its development and release, the YOUIT user community is already substantial: to date we support more than 400 users. Amongst other context data items, they have generated more than 250.000 location measurements, took 2.500+ photos, visited over 4.000 cities in 30 different countries where about 12.000 local weather reports were received.

Acknowledgements. We would like to thank all IYOUIT users, especially our friends and co-workers at DoCoMo and Telematica Instituut, for their great feedback and support. We are grateful to our colleagues at DoCoMo i-mode Europe for fruitful discussions on the potential of IYOUIT, compelling real-world use cases and the competitiveness of IYOUIT in the fast-paced world of Web 2.0. Finally, special thanks go to Michael Wessel of Racer Systems for the support and efforts in developing and implementing the spatio-temporal event recognition in context histories.

References

1. Dey, A.K.: Understanding and using context. Personal and Ubiquitous Computing Journal 5(16), 4–7 (2001)
2. Böhm, S., Luther, M., Wagner, M.: Smarter groups – reasoning on qualitative information from your desktop. In: Proc. of the Workshop on The Semantic Desktop, pp. 276–280 (November 2005)

---

[7] http://zonetag.research.yahoo.com
[8] http://meaning.3xi.org
[9] http://www.shozu.com
[10] http://www.foaf-project.org/
[11] http://sioc-project.org/

3. Luther, M., Fukazawa, Y., Wagner, M., Kurakake, S.: Situational reasoning for task-oriented mobile service recommendation. The Knowledge Engineering Review 23, 7–19 (2008)
4. Wessel, M., Luther, M., Wagner, M.: The difference a day makes – recognizing important events in daily context logs. In: Proc. of the 2nd Workshop on Context and Ontologies in conj. with CONTEXT 2007 (August 2007)
5. Böhm, S., Koolwaaij, J., Luther, M.: Share Whatever You Like. In: Proc. of the 1st Int. DisCoTec Workshop on Context-aware Adaptation Mechanisms for Pervasive and Ubiquitous Services (CAMPUS 2008) (June 2008)
6. Motik, B., Patel-Schneider, P., Horrocks, I.: OWL 2 Web Ontology Language: Structural specification and functional-style syntax. In: W3C Working Draft, World Wide Web Consortium (April 2008)
7. Baader, F., Calvanese, D., McGuinness, D., Nardi, D., Patel-Schneider, P.: The Description Logic Handbook – Theory, Implementation and Applications. Cambridge University Press, Cambridge (2003)
8. Weithöner, T., Liebig, T., Luther, M., Böhm, S., von Henke, F.W., Noppens, O.: Real-world reasoning with OWL. In: Franconi, E., Kifer, M., May, W. (eds.) ESWC 2007. LNCS, vol. 4519, pp. 296–310. Springer, Heidelberg (2007)
9. Sirin, E., Parsia, B., Grau, B.C., Kalyanpur, A., Katz, Y.: Pellet: A practical OWL-DL reasoner. Journal of Web Semantics 5, 51–53 (2007)
10. Tsarkov, D., Horrocks, I.: FaCT++ description logic reasoner: System description. In: Furbach, U., Shankar, N. (eds.) IJCAR 2006. LNCS (LNAI), vol. 4130, pp. 292–297. Springer, Heidelberg (2006)
11. Haarslev, V., Möller, R.: Racer: A core inference engine for the Semantic Web Ontology Language (OWL). In: Proc. of the 2nd Int. Workshop on Evaluation of Ontology-based Tools (EON 2003), vol. 87, pp. 27–36, CEUR-WS.org (2003)
12. Horridge, M., Bechhofer, S., Noppens, O.: Igniting the OWL 1.1 touch paper: The OWL API. In: Proc. of the 3rd Int. Workshop on OWL Experiences and Directions (June 2007)
13. Nurmi, P., Koolwaaij, J.: Identifying meaningful locations. In: Proc. of the 3rd Int. Conf. on Mobile and Ubiquitous Systems: Networks and Services (MobiQuitous 2006). IEEE Computer Society, Los Alamitos (2006)
14. Huler, S.: Defining the Wind: The Beaufort Scale. Crown, New York (2004)
15. Wessel, M., Möller, R.: A High Performance Semantic Web Query Answering Engine. In: Proc. Int. Workshop on Description Logics (DL 2005) (July 2005)

A Semantic Data Grid for Satellite Mission Quality Analysis

Reuben Wright[1], Manuel Sánchez-Gestido[1], Asunción Gómez-Pérez[2],
María S. Pérez-Hernández[3], Rafael González-Cabero[2], and Oscar Corcho[2]

[1] Deimos Space, Spain
{reuben.wright,manuel.sanchez}@deimos-space.com
[2] Departamento de Inteligencia Artificial. Facultad de Informática, UPM, Spain
asun@fi.upm.es, rgonza@delicias.dia.fi.upm.es, ocorcho@fi.upm.es
[3] Departamento de Arquitectura y Tecnología de Sistemas Informáticos.
Facultad de Informática, UPM, Spain
mperez@fi.upm.es

Abstract. The combination of Semantic Web and Grid technologies and architectures eases the development of applications that share heterogeneous resources (data and computing elements) that belong to several organisations. The Aerospace domain has an extensive and heterogeneous network of facilities and institutions, with a strong need to share both data and computational resources for complex processing tasks. One such task is monitoring and data analysis for Satellite Missions. This paper presents a Semantic Data Grid for satellite missions, where flexibility, scalability, interoperability, extensibility and efficient development have been considered the key issues to be addressed.

1 Introduction

Earth Observation is the science of getting data about our planet by placing in orbit a Hardware/Software element with several observation instruments, whose main goal is to obtain measurements from the Earth surface or the atmosphere. The instruments on board the satellite act like cameras that can be programmed to take images of specific parts of the Earth at predefined times. This data is sent to Ground Stations and then processed in order to get meaningful scientific information.

Parameters for instrument operations and for the satellite configuration constitute the Mission Plans issued by the Mission Planning System. These plans are issued regularly (e.g., on a weekly basis), and can be modified until they are sent to the satellite. Catastrophic events such as earthquakes, volcanic eruptions, and hurricanes are examples of events that can cause last minute re-planning. These plans and their modifications are sent to the Flight Operation Segment (FOS), which in turn resends that information to a Ground Station and from there to the satellite antenna of the spacecraft. A computer on board the satellite stores the list of MCMD (MacroCommands) that request an instrument or any other part of the satellite to perform an action. These include loading a table, triggering an operation and getting internal status

A. Sheth et al. (Eds.): ISWC 2008, LNCS 5318, pp. 818–832, 2008.
© Springer-Verlag Berlin Heidelberg 2008

information. Images from each of the instruments are stored onboard (in the satellite computer memory) as raw data and when the satellite over-flies the Ground station that data is sent to the Ground Station antenna (Data downlink). Conversion from the raw data to higher level "products" (adding identification labels, geo-location data, etc.) is performed sequentially at the Ground Station and various Payload Data Segment facilities. Fig. 1 shows the overall scenario. A more detailed explanation of the whole system can be found in [1].

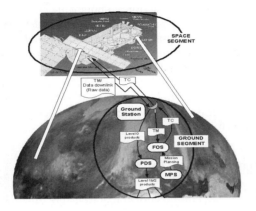

Fig. 1. General overview of an Earth Observation Satellite system (Envisat)

Among the currently active Earth Observation Satellites we find **Envisat**, which monitors the evolution of environmental and climatic changes, and whose data facilitates the development of operational and commercial applications. The satellite carries 10 different instruments and is extensively described in [2]. The work presented in this paper is focused on giving support to this system.

Data circulates within the system as various Plan, MacroCommand and Product Files, with well-defined structures. There can be a variety of hardware or software problems that can occur within the process, hence there is a need for the system to be monitored. **QUARC** is a system that checks off-line the overall data circulation process and in particular the quality of the instrument product files. It checks that the satellite and instrument have performed successfully the measurements (taking images of the Earth), that these images have been stored onboard and transmitted as Raw Data to the Ground station and then processed correctly. QUARC returns reports and plots, which help in the production of new plans. Additionally, the QUARC system is designed to assist in decision making when an instrument or the whole system malfunctions and to detect, in a semi-automated fashion, that something incorrect has occurred in one part of the product generation or data circulation.

The operational QUARC system is located in a single location (ESA-ESRIN, in Italy), which communicates with the archive containing all the products generated from the beginning of the mission and with all the other facilities. The Data Ingestion Modules, one per file type, read the files and convert their contents into parameters that are meaningful to the QUARC data model. The system has been specifically built for this purpose and has bespoke user interfaces. It took several years to build it and there are significant maintenance and development costs as new reports are required and new missions are launched.

Our objective is to replicate some of the features of the QUARC system, namely the comparison between the planned activity and the production of data by the satellite and the further processing of that data in ground systems, demonstrating that we can achieve greater degrees of flexibility, scalability, interoperability and extensibility than the existing system, together with a more efficient development of new functionalities. The existing QUARC system stores implicit metadata about the files that it manages (e.g., hidden in file names) and exposes this metadata through bespoke interfaces. Our approach consists in extracting this metadata to build an explicit semantic representation of the information that is managed, and store it in a way that exposes flexible query interfaces to users and where data is distributed.

In the rest of the paper we look at some detailed use cases for the system, then consider the technical approach and implementation. Finally we summarise the key advantages of the semantic grid approach taken and consider the next steps to be taken in uptake of this approach.

2 Advanced Requirements for Quality Analysis

In addition to functional and non-functional requirements from the existing system we produced the following Use Cases to support incremental, distributed development. These translated directly into Test Cases for evaluation of the system.

Use Case 1: Instrument unavailability. This is a Use Case to ensure our new system is capable of replicating the core functionalities of the existing system. A user needs to find out what planned events and generated products exist for a given time period and instrument, and to plot these results against each other in a timeline. A simple interface is needed, with no underlying complexity exposed to the user.

Use Case 2: Check for the quality of the products in Nominal mode. Certain sorts of products have internal parameters giving a measure of quality of data. The specific situation for this use case at present would be the extraction of one of these quality parameters, over a period of time, for an instrument in a particular mode, being "Nominal". The product files we were to work with did not include this quality data so we expanded this to the more general requirement to be able to extract any piece of metadata from a set of product files. Extracting a new parameter from a file is quite simple for an experienced QUARC user, but it is time consuming and error prone.

Use Case 3: Update of Functionalities with no Software Update. A crucial perceived advantage of the semantic approach was the flexibility with which the system could be adapted. A mission may last 10-15 years and since we are largely investigating anomalous behaviour not all useful queries will be known ahead of time. We needed to know how easily we could develop new queries over our data, and parameterise them for use by ordinary users. This is a complicated process with the current system.

Use Case 4: Data Lifecycle. Satellite plans are not static and the system needed to be able to remove or update metadata from its stores. This needed to be done automatically, and only in the correct circumstances of a new plan covering the same time period and from the provider of the original plan. When querying, the user must be given information about the final, executed, plan. This is managed within an existing QUARC system by virtue of it being centralised, with a single source of data ingestion. In a networked, distributed environment there would be no such facility.

Use Case 5: Modularity of Metadata Service. The desire to be able to change the metadata store comes from wanting flexibility in extending the system. The approach was to design and build a loosely-coupled, service-oriented architecture. In particular we would ensure we could change the metadata store and query engine, but more generally we would use modular components defined by their interfaces. Choices between components can be made on various characteristics including cost, scalability, reliability, and performance. Crucially the user shouldn't have to worry about implementation details. There is no equivalent to this in the existing QUARC system, it is tied to specific versions of the underlying relational database.

3   A Semantically and Grid-Enabled QUARC System

In this section we describe the approach taken for the design and implementation of a semantically and Grid-enabled version of the QUARC system. We start describing the architecture in which the development is founded, and then move to the other ingredients of the development, namely annotation, storage and querying.

3.1   An Architecture for a Semantic Data Grid

We have used the S-OGSA architecture [3] for our development. S-OGSA extends the OGSA model [4], which is commonplace in Grid middleware and applications, and includes two service categories called Semantic Provisioning Services and Semantically Aware Grid Services, as described below.

S-OGSA Information Model. The S-OGSA model identifies three types of entities:

- *Grid Entities* - anything that carries an identity on the Grid, including resources and services [5]. In this system they include planning systems, planning files, satellite instruments, product files and product processing facilities.
- *Knowledge Entities* - Grid Entities that represent or could operate with some form of knowledge. Examples of Knowledge Entities are ontologies, rules, knowledge bases or even free text descriptions that encapsulate knowledge that can be shared. In this system we had a single ontology including classes for times, planning systems, macrocommands, satellite instruments, and the details of the various plan and product file metadata. Ultimately there needs to be a set of ontologies to cover the whole satellite mission domain. In the Satellite Mission Grid an annotation process creates knowledge entities (sets of RDF statements) for the different types of files.

- *Semantic Bindings* - Knowledge Entities that represent the association of a Grid Entity with one or more Knowledge Entities (that is, they represent semantic metadata of a Grid Entity). Existence of such an association transforms the subject Grid entity into a Semantic Grid Entity. Semantic Bindings are first class citizens as they are modeled as Grid resources with an identity and manageability features as well as their own metadata. Grid Entities can acquire and discard associations with knowledge entities through their lifetime. In our system the files are made into Semantic Grid Entities by attaching the created annotations.

Semantic Provisioning Services. These are Grid Services that provision semantic entities. Two major classes of services are identified:

- *Knowledge provisioning services.* They manage Knowledge Entities. Examples of these services are ontology and reasoning services. Ontology services are implemented using the RDF(S) Grid Access Bridge, an implementation of WS-DAIOnt [6], under standardisation at OGF.
- *Semantic Binding provisioning services.* They produce and manage Semantic Bindings. They include annotation services that generate Semantic Bindings from planning and product files, implemented with Grid-KP [7]. They also include a semantic binding storage and querying service [8], which is implemented twice, using the Atlas distributed RDF(S) storage system [9] and Sesame [10].

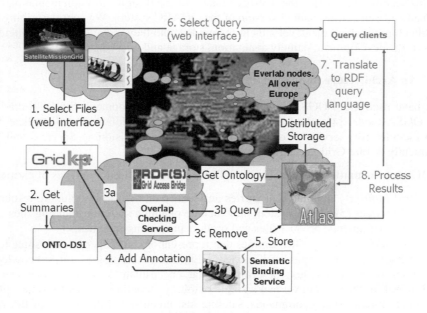

Fig. 2. System architecture

Semantically Aware Grid Services. They are able to exploit semantic technologies to consume Semantic Bindings in order to deliver their functionality. They consume the semantic entities held by Knowledge and Semantic Binding provisioning services, and use their services. The user interface for the Satellite Mission Grid is a Semantically Aware Grid Service, making use of all the aforementioned elements in order to deliver its enhanced functionality.

Figure 2 shows the geographical deployment of the developed system. Software was deployed at 3 sites – Manchester, Madrid and Athens, and Atlas further uses the Everlab cluster of machines throughout Europe. The numbered actions 1-5 and 6-8 show the activity flow for annotating and querying data respectively.

3.2 Annotation: Making Metadata Explicit

Data circulates in the existing systems as files with many common generic features. They are slightly different for planning and product files, and the information about these planned events and generated products is usually bound up with the data involved. Standard ASCII formats encode the information in keyword-value pairs, which are stored as headers for the various files. This is a special format defined for the Envisat mission with an enormous amount of software and documentation generated through years of development. This structure can be simply translated to a fairly flat XML structure. Once this is performed on the planning and product files, the system uses XML software tools.

Product files consist of an ASCII header and a binary part encoded in an ESA proprietary format. The header is just a few Kbs out of an image file size of Gbs. The Onto-DSI [11] component was used to extract and provide just the headers from these files to avoid a large system overhead whilst annotating them.

Much of the metadata is encoded in specific, amalgamated identifiers, with "implicit semantics". For example rules had to be created for product filenames like "RA2_MW__1PNPDK20060201_120535_000000062044_00424_20518_0349.N1". This is decomposed into an Event type (RA2_MW), Processing level (1P) and centre (PDK), a Sensing start time (2006-02-01:12.05.33) and so on. Generic metadata (applied across all captured metadata) and the ontology further add, for example, that the Event type (RA2_MW) is executed by a particular instrument, the Radar Altimeter. A parser extension to Grid-KP carries out the extraction of the relevant file properties.

Another issue was conversion of units. One example of this was converting from date formats, as given above (and given to the users in the webforms) to another standard time format used in space missions, MJD2000. It is the number of seconds (and milliseconds) to have passed since the year 2000, including leap seconds. The conversion routine was wrapped as a Web service using SOAPLAB [12].

Migration of other data to the system would be much simplified by this process and these tools being in place. In addition, the annotation services were deployed in different locations, which supported the distributed nature of the data sources.

3.3 Storage: Managing a (Meta)Data Lifecycle

The Annotation Service was able to use exactly the same mechanisms as the user interface to communicate with the Semantic Binding Service to ask if its current file overlapped with any existing Plan files. This design has two advantages; firstly, no new specific code needed to be written as we already had the query interfaces. Secondly, although the logic needed here was quite simple, we have allowed ourselves full access to the flexibility of RDF querying, which means that if more complex rules are needed in future we will be able to accurately encode them. RDF can be updated using the standard mechanisms provided by metadata stores.

Managing RDF in identifiable, separate Semantic Bindings allows us to better manage the overlaps, and the lifetime of the metadata when several annotations may be created.

3.4 Querying: Exploring Data

We worked with a flexible "Free Querying" interface as we considered how the system would be incrementally improved and developed. This interface simply allowed the user to create queries (in the language of their choice: SPARQL, SeRQL or RQL) and get the results back in a tabular form.

As an example we looked at how we could abstract our queries over the data to a level where new sorts of data could be added to the system and still be identified by our queries. An initial implementation of one of the queries for Use Case 1 was looking for all planned events (DMOP event records) which were using the Radar Altimeter. We matched at the low level of Event identifiers using the implicit metadata that "events with identifiers containing RA are carried out by the Radar Altimeter instrument". The nature of RDF as a web of statements and the existence of an ontology to formalise the existence of different properties made it easy to move these queries to an improved, semantic level.

We were initially searching on the event_id property of the DMOP_er class (DMOP event records), which look like "RA2_IE_00000000002372". It matches REGEX(?EVENT_ID, ".*RA.*") in the SPARQL regular expression syntax. This query works, but we were able to see in the ontology that a better level was possible.

The individual data items about planned events use the event ids, but our system was able to augment that with the knowledge about which event types use which instruments. This was enabled by having an ontology which included instruments and types of events as objects independent of the individual events which they classify. Figure 3, showing part of the Satellite Ontology, specifies that the DMOP_er class (top left) is related to the Plan_Event class by the represents_plan_event property, and that Plan_Event instances have their own identifiers – plan_event_id. They represent the different types of events that can be planned. The next level of abstraction is the Instrument class and the Radar Altimeter is identified as one such task, with instrument_id of "RA".

We translated the WHERE clause of our SPARQL query from

```
?EVENT event_id ?EVENT_ID ;
FILTER ( REGEX(?EVENT_ID,".*RA.*"))
```

to

```
?EVENT event_id ?EVENT_ID ;
       represents_plan_event ?PLAN_EVENT_TYPE .
?PLAN_EVENT_TYPE executed_by_instrument ?INSTRUMENT .
?INSTRUMENT instrument_id "RA"
```

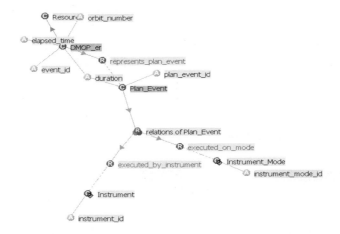

Fig. 3. A section of the Satellite Ontology showing the specific events (DMOP-er), their types (Plan_Event) and the Instruments which carry them out. Diagram from NeOn Toolkit [13]

While this is longer it is both clearer to understand and to implement as a webform where the user will select an instrument. It is also likely to execute more quickly as it is looking for an exact match of `instrument_id` rather than having to rely on regular expression parsing of a string value.

The biggest gain is that it is much more robust in the face of changing data. We can continue to use these "semantic level" queries about instruments even if we add new event types which use this instrument or change the unique identifiers for individual DMOP event records. If further data in the system contained new sorts of events planned and carried out by the Radar Altimeter then our queries would automatically match them. In any of these extended cases a simple statement associates the new event type with an existing instrument or new events with an existing event type. The exact same query (for use of an instrument) will then also report about these new events. We shifted from talking about details of identifiers to the actual objects which the user is concerned about, i.e. we moved to a more semantic level. This process is shown in more detail in an OntoGrid demonstration video [14].

3.5 Accessibility: Providing Tools for End Users

Having developed the queries in a language such as SPARQL we very easily built webforms to allow users to provide the values for any such queries. A web application (running on Tomcat) was developed to serve Java Server Pages which presented interfaces and results to users (see Figure 4), and converted to queries according to RDF data. The results were provided in a simple XML structure and we generated from that either tables or graphs.

The flexibility of the data model, and its level of abstraction from the interface also allowed us to look at different ways of accessing it. A separate tool was developed later to allow the programmatic augmentation of data in the database. As well as

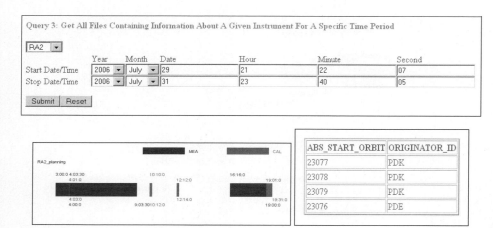

Fig. 4. Webform for user input and timeline and tabular outputs

allowing the adaption of data, for example adding in UTC datetime values as well as the MJD2000 ones supplied in the data, this can help with organising the data. For example, a set of subtypes of instrument types could be created to describe power consumption. Then each of the instruments could be labeled as belonging to one of the subtypes. This would take just a few statements in RDF and the millions of pieces of event information organised by which instrument generates them would now also be connected to this set of subtypes, allowing a user to make queries based on the power consumption.

A webform lets a user define a SPARQL query to identify various RDF nodes, i.e. bound variables in query results. The user can then choose one of these variables as the subject of a new property, and can construct a formula (using another of the variables, and the subset of XPATH expression syntax which does not involve nodes and sequences) for calculating the value of the new property. They also select a name, namespace, and type for the property. For every result the query returns, the system will then calculate the value for the new property and add it to the RDF store.

It does not add the results directly, but to a protected user area ("sandbox"). The user can query, examine or export the contents of their Sandbox to check that the calculated properties are as they expected. They can then choose to add the data to the main RDF store, where it will be available for all queries. This is managed technically with the Sesame concept of "context" for any piece of information. They are put into a user context initially, and can then be moved to the main "core" context.

Another tool allowed the simple creation of reports and graphs (see Figure 5). The idea here was that a web application for report designers could produce from their input a webform which would be used by normal users. These user webforms have public, visible selection fields for the user to use, but also hidden fields detailing the base SPARQL query the designer wants to run and the sort of graphical output they want. The report processor combines the base query with the additional parameter

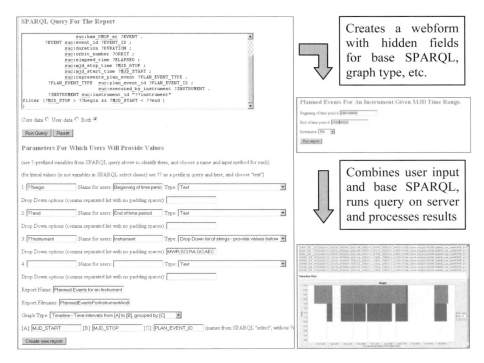

Fig. 5. Process of creating and using one of these auto-generated reports

values provided by the user (e.g. a start and stop date, or instrument type) to create a SPARQL query to run. The user parameters can be used for either specifying bound variables or literal values. The results are then processed according to the report designer's instructions and presented back to the user in the browser.

The user interface was developed using Java Server Pages, Java, XSLT and SPARQL. The Sesame 2 data store and interface were deployed on Apache Tomcat.

A more sophisticated design would combine these two pieces, utilising the idea of processing pipelines. These enable processing to be built up from simple components run one after the other. For example one component could add some data to the "active" dataset, another would run the query, a third would process results to generate a graph, a fourth would generate a summary table, a fifth would generate the results tables and a final stage would collate the outputs of the previous 3 to create the final report as presented to the user.

4   A Grid-Enabled QUARC System

The QUARC system involves a complex process in which distributed data belonging to different organizations must be queried, processed and transferred. The previous version had some clearly defined limitations, namely:

- Data Transfer: It is necessary the transfer of large files among different organization. This was made in a traditional style, through different FTP servers and ``ad hoc'' solutions. Furthermore, the volume of data that is transferred may be decreased, since only a small part of the files is significant for the correct performance of many of the functionalities.
- Security: There was no definition of sophisticated access control mechanisms neither virtual organizations.
- Scalability: In order to deal with a huge number of files, the scalability issue must be addressed.
- Previously the location of the resources and the processing of data were made in a *wired* way, according to filenames and content of these files in an "ad-hoc" format.

All of these limitations have been relieved by means of the intensification of the "griddy" features of the QUARC system. Indeed, the use of a grid framework provides a flexible way of locating required resources and the virtualization of these resources by means of (Semantic) Grid Services.

Several services and functionalities have been developed for tackling these aspects:

- For improving data transfer, we have used GridFTP[15]. We have developed a new Data Storage Interface (DSI) to GridFTP, suitable to deal with Satellite files.
- Regarding security, the involvement of several organizations implies the establishment of different access policies and the definition of virtual organizations. The role of each specific actor within an organization also defines its privileges as a member of the virtual organization. Globus [16] allows us to establish security and define different policies.
- Our system enhances the performance of the data transfer, transferring only the necessary part of the information. This provides a higher scalability.
- Grid provides capabilities for locating services according to metadata.

5 Validation of the System

The implementation shows several advantages compared to a conventional approach, in terms of flexibility, reduction of software running costs, maintainability, expandability, and interoperability.

Legacy Formats and Systems. One of the key issues to bear in mind when implementing a completely new design in a system of this size and complexity is how to manage migration of systems. Some parts of the system (which may belong to other organisations or facilities) are not changed at all, or only partially updated, and there is no simultaneous modernisation of all parts in the system. Legacy systems can be considered in terms of data formats and software.

- **Data Formats.** Although current data files are structured and in well documented formats there remain hundreds of different file types. Much of the metadata is only implicit, such as the information stored within filenames or specialised code systems. For these files we have made this metadata explicit. This makes easier our work in the Use Case 2. More generally, this can help query developers and systems that must manage the lifecycle of metadata. The use of the ontology to store information about properties will make migration of future data simpler. The mapping simply has to be done once between the data and the ontology. This should be especially easy as these existing systems have very strictly defined inputs and outputs and increasingly the formats used are XML based. The process of writing the specific code to extract the information from the datafile and re-encode it as RDF is much simplified, and can be more easily managed.

- **Software.** In the Envisat mission, complex software elements with well-defined interfaces are used in both planning and product generation. Some of these functionalities were incorporated in the prototype (e.g. time conversion utilities) by enabling them to be accessed as Web Services. In a distributed architecture, such as that used in this Semantic Grid, encapsulation allows external systems to interact with individual parts of the system. For example, during a transitional period they might simply make use of a query interface to extract information or an RDF interface to provide some data to an RDF store or annotation component. Use Case 1 ensured we could support existing functionality as well as new techniques.

Flexibility. The system allows a user who is familiar with one of the supported query languages to develop queries directly and iteratively. The process of creating forms which allow other users to provide the specific values for these queries is simple, and made even simpler by the existence of 'common' methods for converting between time formats. In this way we have been able to demonstrate how the system can be enhanced without any significant technical changes being required. This is crucial in an analysis system, where not all relevant queries are known or defined at the beginning of the use of the system.

It is also possible to add in some new relationships or properties without having to change the existing data at all. If a new way of grouping some part of the existing information was required by operators of the system then it could be added into the RDF directly. This aspect of Use Case 3 was explored extensively in the development of user and developer tools as described in the Accessibility section above.

Semantic Technologies and Standards. An Earth observation mission may last 10-15 years and the completely flexible query interface allows exploring of data and development of new queries. This is crucial as anomalies are discovered and examined. Furthermore, new missions may provide information which it would be useful to combine with this existing data. The metadata format and query language defined purposely for the current QUARC implementation, although powerful, cannot be exported and directly used in other systems. The Satellite Mission Grid uses the RDF standard for the storage of the metadata, and the specifics of the format are described by OWL and RDF(S) schemas.

The developed Satellite Ontology has enabled communication, and rapid extension of functionalities as outlined by Use Cases 2 and 3. The addition of "generic" information (such as the linking of event types to instruments) allows us to have semantically rich annotations. This semantic approach to queries where they are moved up to greater levels of abstraction gives us much more flexibility and robustness over time, as we are querying what the users need to know (usage of the instrument) rather than what we traditionally have recorded (the list of codes used for the event types). This shows the general technique of modeling what it is that users wish to search and discuss. As well as making development more efficient it reduces the time to acquaint new users with the system.

RDF allows us to use standard query languages like SPARQL or RQL which are incorporated in tools which have been already tested and proved adequate for re-use in other systems. The use of standard formats also allows different metadata stores to be used, depending on circumstances. For example, we used Atlas and Sesame as two interchangeable components in line with Use Case 5. Another advantage is in not having to train developers in specific new skills but to be able to use what they already know. However, the existence of several query languages can add an overhead in terms of the required skills for developers.

Data Life Cycle. We have shown that controlled updates can be made to stored data. These are automated but only allowed from authorised data sources. This ability supports the fact that data is not static, and that giving access to the correct information can involve removing out-of-date statements as well as adding new ones, as described in Use Case 4. More generally, we hope to be able to integrate data from many sites and missions and reuse the same system across these missions. As such we have created the methodology and tools for adding new data sources. Lightweight annotation components can convert from a legacy system to common RDF which is made available (via the semantic binding service) to the query engines.

There are high volumes of data being produced by Envisat – anticipated to reach a petabyte of data in 10 years of operation [17]. There are also another four Earth Explorer Missions being launched in the next two years. Extraction and management of just metadata (rather than all the data itself) is necessary for any ongoing analytical systems. In the Satellite Mission Grid, we create a single amalgamated dataset from many geographically dispersed data silos. We then store that dataset across many machines and resources, linked by Grid technology. We move from disconnected data islands having to send their data to a central server to having a single virtualised dataset, spread across the machines. Crucially, all the complexity is absent from the user perspective, i.e. they don't have to know about it at all. The resources and computation are distributed but the user has simple, local browser-based interfaces for annotation and querying.

Modularity and Extensibility. The abstraction of components in a loosely coupled system means we have all the advantages of modularity. Interchangeable components can be selected depending on the particular application of the system, as detailed in Use Case 5. The approach allows the users to enjoy a single interface into whichever systems are determined to be best suited to the current scale of data and complexity of query. We also gain in extensibility; it opens up any further development to using "best-of-breed" components, be they new versions of existing ones, or completely new development. Throughout our work we have adopted an incremental development approach as suggested by Use Case 3.

6   Conclusions and Next Steps

A semantic approach, where metadata is created explicitly and managed as a "first class object" gives advantages of flexibility and extensibility. A Grid approach where data and computations are distributed across many sites and machines gives improvements of scalability and robustness. The prototype system has shown itself capable of carrying out the current functionality of mission analysis systems, but across a geographically distributed dataset. It has also shown itself to be easy to extend in capability without significant development effort.

In the Semantic Data Grid community we have helped focus on lightweight protocols and making components more easy to integrate with existing systems. This vision supports a movement towards SOKU – Service Oriented Knowledge Utilities [18]. The next industry steps include the incremental updating and extension of existing systems, where we will add to the metadata we store and make explicit what was formerly implicit.

Acknowledgments

The authors would like to thank the other members of the OntoGrid consortium and European Space Agency (ESA) representatives Olivier Colin (ESA-ESRIN) and Pierre Viau (ESA-ESTEC) for providing access to the actual products and auxiliary tools from the Envisat mission.

References

1. Sánchez Gestido, M.: OntoGrid Business Case and User Requirements Analysis and Test Set Definition For Quality Analysis of Satellite Missions. Deliverable D8.1, OntoGrid (2005), http://www.ontogrid.net
2. ESA bulletin number 106, EnviSat special issue, http://www.esa.int/esapub/pi/bulletinPI.htm
3. Corcho, O., Alper, P., Kotsiopoulos, I., Missier, P., Bechhofer, S., Goble, C.: An overview of S-OGSA: a Reference Semantic Grid Architecture. Journal of Web Semantics 4(2), 102–115 (2006)
4. Foster, I., Kishimoto, H., Savva, A., Berry, D., Djaoui, A., Grimshaw, A., Horn, B., Maciel, F., Siebenlist, F., Subramaniam, R., Treadwell, J., Reich, J.V.: The open grid services architecture, version 1.0. Technical report, Open Grid Services Architecture WG, Global Grid Forum (2005)
5. Estebán-Gutierrez, M., Gómez-Pérez, A., Corcho, O., Muñoz-García, O.: WS-DAIOnt-RDF(s): Ontology Access Provision in Grids. In: The 8th IEEE/ACM International Conference on Grid Computing (2007), http://www.grid2007.org/?m_b_c=papers#2b
6. http://www.isoco.com/innovacion_aplicaciones_kp.htm
7. Corcho, O., Alper, P., Missier, P., Bechhofer, S., Goble, C.: Grid Metadata Management: Requirements and Architecture. In: The 8th IEEE/ACM International Conference on Grid Computing. ACM Press, New York (2007), http://www.grid2007.org/?m_b_c=papers#2b
8. Miliaraki, I., Koubarakis, M., Kaoudi, Z.: Semantic Grid Service Discovery using DHTs. In: 1st CoreGrid WP2 Workshop on Knowledge and Data Management (2005)

9. Sesame, Aduna Open Source project, http://www.openrdf.org

10. ONTO-DSI, OntoGrid, http://www.ontogrid.net

11. Senger, M., Rice, P., Oinn, T.: Soaplab - a unified Sesame door to analysis tools. In: Proceedings of UK e-Science, All Hands Meeting, Nottingham, UK, September 2-4 (2003)

12. NeOn toolkit, http://www.neon-project.org/web-content/

13. Wright, R: Ontogrid Satellite Use Case demonstration video,
    http://www.youtube.com/watch?v=TSbb_8vmKvk

14. Allcock, W., Bester, J., Bresnahan, J., Chervenak, A., Liming, L., Meder, S., Tuecke, S.:
    GridFTP Protocol Specification (September 2002),
    http://www.globus.org/research/papers/GridftpSpec02.doc

15. Globus (accessed 2008), http://www.globus.org

16. European Space Agency Information Note 13,
    http://www.esa.int/esaCP/ESA0MDZ84UC_Protecting_0.html

17. Next Generation Grids Expert Group: Future for European Grids: Grids and Service-Oriented Knowledge Utilities,
    ftp://ftp.cordis.europa.eu/pub/ist/docs/grids/ngg3-report_en.pdf

A Process Catalog for Workflow Generation

Michael Wolverton, David Martin, Ian Harrison, and Jerome Thomere

SRI International
333 Ravenswood Ave
Menlo Park, CA 94025
{Wolverton,Martin,Harrison,Thomere}@ai.sri.com

Abstract. As AI developers increasingly look to workflow technologies to perform complex integrations of individual software components, there is a growing need for the workflow systems to have expressive descriptions of those components. They must know more than just the types of a component's inputs and outputs; instead, they need detailed characterizations that allow them to make fine-grained distinctions between candidate components and between candidate workflows. This paper describes PROCAT, an implemented ontology-based catalog for components, conceptualized as *processes*, that captures and communicates this detailed information. PROCAT is built on a layered representation that allows reasoning about processes at varying levels of abstraction, from qualitative constraints reflecting preconditions and effects, to quantitative predictions about output data and performance. PROCAT employs Semantic Web technologies RDF, OWL, and SPARQL, and builds on Semantic Web services research. We describe PROCAT'S approach to representing and answering queries about processes, discuss some early experiments evaluating the quantitative predictions, and report on our experience using PROCAT in a system producing workflows for intelligence analysis.

Recent research and development in technology for intelligence analysis has produced a large number of tools, each of which addresses some aspect of the *link analysis problem*—the challenge of finding events, entities, and connections of interest in large relational data sets. Software developed in recent projects performs many diverse functions within link analysis, including detecting predefined patterns [1,2,3,4], learning these patterns of interest [5], classifying individuals according to group membership [6] or level of threat [7], resolving aliases for individuals [8], identifying neighborhoods of interest within the data, and others.

While these tools often perform complementary functions within the overall link analysis space, there has been limited success in getting them to work together. One-time integration efforts have been time-consuming to engineer, and lack flexibility. To address this problem, a recent focus of research has been to link these tools together dynamically, through workflow software [9], a blackboard system [10], or some other intelligent system architecture.

One key challenge in building this kind of dynamic link analysis workflow environment is representing the behavior of the individual link analysis processes being composed. In this paper, we describe an implemented Process Catalog software component—PROCAT for short—that serves information about processes that allows

A. Sheth et al. (Eds.): ISWC 2008, LNCS 5318, pp. 833–846, 2008.

a workflow generation component to select, rank, and execute them within a workflow. (Our focus here is on PROCAT'S design and functionality; space constraints allow for only a few brief comments about the characteristics of the larger workflow system.) PROCAT is based on a layered approach to process representation, in which a process is described in terms of both the qualitative features that distinguish it from other processes, and quantitative models that produce predictions of the process's outputs and performance.

PROCAT is implemented and deployed within the TANGRAM workflow architecture, a complex system that generates and executes workflows for intelligence analysis. This deployment requires it to be integrated with several other workflow modules developed by other contractors. PROCAT's current knowledge base encodes a collection of real link analysis tools that perform a variety of functions. It produces quantitative predictions of those tools that early experiments suggest are accurate.

The sections that follow describe PROCAT and its use in more detail. First, we give an overview of its approach and architecture. We then describe the *Capabilities Layer* of the process description, which represents processes at a qualitative level. Next, we cover the *quantitative layers* of the process description, including some experiments evaluating the accuracy of those layers' predictions. We then discuss PROCAT's use within the TANGRAM workflow system in some detail. And finally, we compare this work to other related research, and outline future work and other research issues.

1 Overview

The problem of designing a process characterization language for link analysis presents a number of research challenges, many of which arise because of the need for flexibility in the process description. The representation must be flexible to accommodate heterogeneous processes, multiple possible workflow systems that reason about processes at differing levels of fidelity, and the evolution of the workflow systems' reasoning abilities as they are developed over time.

To meet this need of flexibility, PROCAT is built upon a layered approach to process characterization, where each process is represented at multiple levels of fidelity, and where the workflow system can retrieve and reason about processes at the representation level(s) they can handle. The layers include

- *Capabilities*, which provides a qualitative description of the process's behavior, along with the hard constraints for running it.
- *Data Modification*, which describes how the statistical profile of the data (e.g., the number of nodes and links of each type) changes as the process is run over it.
- *Performance*, which quantitatively describes the performance of processes (e.g., time to complete, maximum amount of memory consumed) given a data set with a particular statistical profile.
- *Accuracy*, which describes how the accuracy of each node and link type in the data changes as the process is run over it.

The Capabilities Layer is described in the next section, while the Data Modification and Performance Layers are described in the subsequent, Quantitative Predictions, section. While PROCAT's current process representation incorporates the Accuracy Layer,

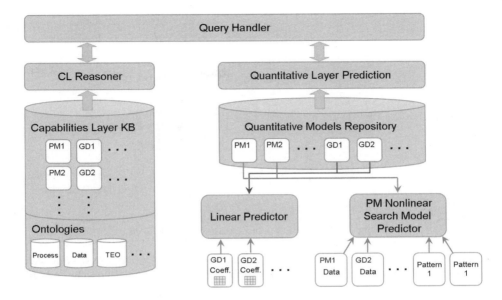

Fig. 1. PROCAT Architecture. The Query Handler accepts a variety of queries about processes from an external Workflow Component (not shown in the figure), and farms these queries out to the appropriate reasoning module.

models for producing accuracy predictions are not currently part of the system and are part of our ongoing work.

The PROCAT architecture is shown in Figure 1. PROCAT feeds information about processes to the workflow system via a set of predefined query types. The system has two mechanisms for answering these queries. Queries that involve the Capabilities Layer are answered by reasoning over a set of ontologies that encode the processes' functionality, resource requirements, invocation details, and so on. Queries that involve the quantitative layers are answered using the processes' models from the Quantitative Models Repository. These two mechanisms are described in more detail in the next two sections.

2   Capabilities Layer

The Capabilities Layer (CL) describes in a qualitative, symbolic manner what a process does, what the requirements are for running it, and how it is invoked. This section explains the representational approach and ontology underlying PROCAT's Capabilities Layer, and the manner in which capabilities queries are handled.

2.1   Capabilities Ontology

PROCAT employs the Resource Description Framework (RDF) [11], the Web Ontology Language (OWL) [12], and the RDF query language SPARQL [13]. Each of these

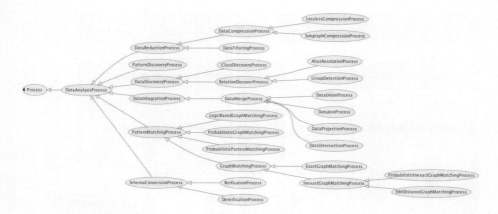

Fig. 2. Class Hierarchy of Processes for TANGRAM

knowledge representation technologies has been standardized at the World Wide Web
Consortium in recent years, as part of the Semantic Web initiative. A number of syntaxes have been defined for OWL, which is layered on top of RDF. PROCAT makes use
of the RDF/XML syntax [14], as discussed below. The internal representation of RDF and
OWL content takes the form of a set of *triples*, which are maintained in a *triple store*.
OWL, a description logic language, is well suited to the Capability Layer's objectives of
describing, classifying, and answering queries about categories of processes, individual
processes, and their characteristics.

In PROCAT, a *process* is any well-defined, reusable procedure, and a *process installation* is an executable embodiment of a process. In the TANGRAM application,
the processes described in PROCAT are data analysis programs, and each process installation is a version of a program installed on a particular machine. The capabilities ontology is organized around the PROCESS class. That class can very naturally
be decomposed into a hierarchy of categories of processes for various purposes. TAN-
GRAM, as shown in Figure 2, employs a hierarchy of data analysis processes. Some
TANGRAM queries quite naturally need to ask for processes belonging to a particular
subclass within this hierarchy, with additional query constraints expressed using properties. We refer to the set of terms defined in the capabilities ontology, along with certain
conventions for its use, as the Process Description Language (PDL).

The core of a capabilities description is a functional characterization of the process,
in terms of its parameters (inputs and outputs), preconditions that must be met to run
it, and postconditions that will hold after running it. The most essential element of
parameter characterization is *type*. The type of each parameter is specified as an OWL
class or an XML datatype. A parameter also has a *role* (e.g., *HypothesisOutputRole*
in Figure 5); roles may be shared across multiple processes. The characterization of
a parameter also includes such things as default values and invocation conventions.
Figure 3 shows the main classes that are found in the process ontology, and relationships
between them. Properties HASINPUT and HASOUTPUT relate a process to instances of
class PARAMETER. In general, those instances will also be instances of data source
classes (classes that indicate the types of parameters). Thus, multiple inheritance is

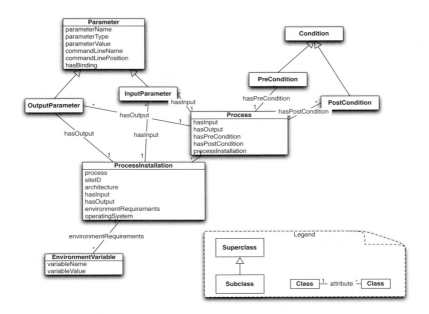

Fig. 3. Key Classes and Relations in the Process Ontology

used to indicate parameter types, which simplifies in some ways the expression of both descriptions and queries. Some conditions may be represented using properties of these data source classes. For instance, some data analysis processes take an analysis pattern as one of their inputs.

A process may be related to zero or more instances of PROCESSINSTALLATION, each representing a specific installation of a process on a particular machine. Various characteristics are specified for process installations, including resource requirements and invocation details. HASINPUT and HASOUTPUT are also present for PROCESSIN-STALLATIONs, to allow for the specification of platform-specific parameters. Information about environment variables required to run the process is also included, as well as details about the physical characteristics of the installed platform, such as machine architecture, operating system, and other details needed to reason about execution requirements and to remotely invoke the process installation.

Preconditions and Postconditions. Preconditions are conditions that must be true in order for a process to successfully occur; postconditions are conditions that will hold true after an occurrence of a process. Preconditions and postconditions in general can be difficult to represent, and can require considerable expressiveness. For one thing, they are not ordinary facts about the process. They are conditions that might or might not hold true when the process is used. The sense of a precondition is that if it evaluates true prior to an *occurrence* (execution) of the process, then (assuming the process executes normally without exceptions) that occurrence will be successful. Similarly, postconditions cannot be understood as ordinary facts about a process. Thus, in a complete logical theory of processes, neither preconditions nor postconditions could be simply asserted, but would need to be reified in some way, and subject to special handling during query answering. Further, as noted just above, preconditions and postconditions apply to process occurrences rather than processes. If both processes and process occurrences are to be explicitly included in a representational framework, it becomes necessary to explicitly capture the relationship between them, as axioms. Such axioms, however, would exceed the expressiveness of RDF and OWL, and thus increase the complexity of reasoning involved in answering queries.

In PROCAT, we have mitigated these difficulties by adopting a simplifying perspective, that a process description is considered to be a *snapshot of an arbitrary successful occurrence of a process*. (Roughly speaking, then, a process description is somewhat like a skolem representative of all possible successful occurrences of the process.) Further, because PROCAT does not store information about *actual* occurrences of a process, there is no inconsistency in including in its description facts that apply to the process itself (rather than to any particular occurrence). Adopting this perspective removes the need to explicitly distinguish between processes and process occurrences. Instead, a process description in the catalog can be viewed as capturing aspects of both at once. Given this perspective, reification is no longer needed and preconditions can be stated as facts about inputs (and postconditions as facts about outputs). For example, Figure 5, in the "PROCAT Implementation and Use" section, shows a query that will match against a process having an output dataset that is *saturatedWith* instances of the class *MoneyLaunderingEvent*. (*saturatedWith* is an ontology term meaning that as many instances as possible have been inferred within a given dataset. In the example, the query uses *?dataVariable5* to stand for the output dataset parameter.) A KB statement matching the query triple (*?dataVariable5 saturatedWith MoneyLaunderingEvent*) would be a simple example of a postcondition.

2.2   Capabilities Layer Functionality

As shown in Figure 1, the Capabilities Layer makes use of the Query Handler component and the CL Reasoning component, which in turn accesses the Capabilities Layer KB. The reasoning component includes both application-specific and general-purpose query processing functionality.

Query Handler. Queries are received by means of a Web service API, as discussed in "PROCAT Implementation and Use" below. To provide catalog services for TANGRAM, PROCAT queries and responses are expressed in a slightly extended form of the RDF/XML syntax. The extension allows for the use of variables, in a manner similar to that of SPARQL. In our syntax, variables can appear in subject, predicate, or object position of any query triple, are named by URIs (and thus belong to a namespace), and are indicated by a question mark as the first character of the name part of the URI. (See Figure 5 for a simple example query using this syntax.) Our experience to date indicates that standard RDF/XML parsers recognize these URIs without difficulty.

SPARQL is used *internally* within PROCAT for accessing the KB, as discussed below. Although we considered using SPARQL as the external query syntax for TANGRAM, we determined that SPARQL was not well suited to meet certain application-specific requirements of TANGRAM. However, we plan to support SPARQL in future versions, as a general-purpose query syntax to supplement the existing, application-specific query conventions.

The Query Handler component is responsible for parsing the RDF/XML-based query syntax and capturing it in an internal format. Because the query syntax is consistent with RDF/XML, we are able to take advantage of standard parsing functionality. Each incoming query is parsed directly into a temporary triple store. In this way, we are able to use triple store queries and triple store manipulations in analyzing and processing the query.

Capabilities Reasoner. An important requirement for PROCAT is to provide flexibility in supporting application-specific query requirements, that is, requirements that cannot be met by a standard query language such as SPARQL. For example, in TANGRAM, one type of query asks PROCAT to formulate a commandline for a particular invocation of a given process installation. Although the KB contains the essential information such as commandline keywords, default values, and proper ordering of commandline arguments, nevertheless the precise formulation of a commandline requires the coding of some procedural knowledge that cannot readily be captured in a KB. ("PROCAT Implementation and Use" below discusses further the types of queries used in TANGRAM.)

PROCAT's architecture allows for the use of arbitrary Lisp code to provide the application-specific query processing. This code, in turn, can use a variety of mechanisms (including SPARQL queries) to examine the temporary triple store containing the parsed form of the incoming query. In most cases, this examination results in the construction of one or more queries (which, again, may be SPARQL queries) to be submitted to the capabilities KB to retrieve the needed information about the process(es) in question. Once that information has been retrieved, Lisp code is called to analyze it and formulate the requested response.

The capabilities KB, including ontologies, is stored within a single triple store. Access to the capabilities KB is provided by a layer of general-purpose (application-independent) utility procedures for triple store update, management, and querying. This includes procedures for formulating and running SPARQL queries.

3   Quantitative Predictions

We designed the quantitative prediction models of PROCAT to meet three criteria:

- The prediction models are *precise*, in that they allow fine-grained predictions of process performance.
- The prediction models are *efficient*, so that predicting a process's performance on a given data set generally takes less time than running the process.
- The individual prediction models are *composable*, so that a workflow component can accurately predict and reason about combinations of processes run in sequence or in parallel.

PROCAT currently produces two types of quantitative predictions: Data Modification and Performance.

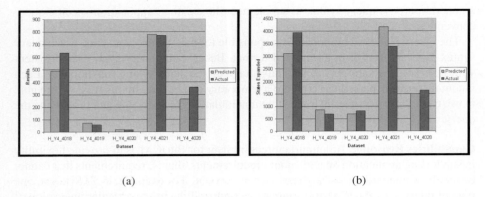

(a)   (b)

Fig. 4. Predicted vs. actual (a) data modification and (b) performance for a pattern matching process

For predictions in the Data Modification Layer, each process is described as a function that maps a problem description and a data model into a data model. The problem description will vary depending on the type of process; for example, a problem for a graph matcher could consist of a pattern graph and various parameters specifying the match criteria. A data model is a statistical description of the data set along whatever parameters are determined to be useful in selecting and composing processes. PRO-CAT's current data model is a statistical description of the data that consists of (1) the number of nodes of each type and (2) the branching factor per node and link type. Here, (2) is the answer to the question: For each entity type T and link type L, given a node of type T, how many expected links of type L are connected to it?[1]

Performance Layer predictions estimate the process's efficiency given the data set and problem. These predictions map a (problem description, data model, resource model) triple into a performance model. The resource model represents the performance characteristics of the hardware available to the process—processor speed, amount of physical and virtual memory, network throughput, database response, and so on. The performance model will represent the predicted time to process the data set, and possibly other measures of performance that we determine are useful for selecting processes. For example, one could imagine building a more complicated model of performance for an anytime process, which includes a tradeoff between execution time and the number of (and completeness of) the results produced.

The quantitative models are represented procedurally—as lisp functions. A model can be custom-built for a particular process by the knowledge engineer populating the catalog, or it can be created by instantiating a preexisting model class. Currently, PRO-CAT has two built-in model classes, shown in the lower right portion of Figure 1. The

---

[1] This level of representation implicitly assumes that link and node type distributions are independent—that is, that the existence of link L_1 attached to node T_1 tells nothing about the likelihood of link L_2 also being attached to T_1. For many processes, especially pattern matching processes, this seems to be a reasonable assumption. However, for other processes, especially relational classification processes, the independence assumption may be too strong.

first is a linear model, where predictions of a quantity characterizing data or performance are derived via a weighted sum of features of the input data.[2] The coefficients for this model can be learned through a regression method, such as least-squares, based on runs of the process being modeled. The second is a nonlinear model that is specific to pattern matchers. This model predicts pattern matcher output and performance by using a recurrence relation to estimate the number of states expanded in the search for a match.

While a detailed description of this latter model is beyond the scope of this paper, we show in Figure 4 the results of some of its predictions for the LAW pattern matcher [4] against actual behavior of the system, to give a sense of the level of accuracy of predictions for one well-understood tool. Figure 4(a) shows the predicted and actual number of results found by LAW matching a relatively simple pattern against five different data sets varying in size, branching factor, and other characteristics. Figure 4(b) shows predicted and actual search states expanded (here we use states expanded as a proxy for runtime because of the ability to get consistent results across machines) for the same runs. The average error of the predictions was 20% for data modification, and 19% for performance. The time taken to run the prediction model was faster than the run of the pattern matcher by over two orders of magnitude.

While this model was built specifically for LAW, the data modification portion of it should be transferable to other pattern matchers. Furthermore, experiments with another pattern matcher, CADRE [3], indicate that its runtimes are roughly proportional to LAW's when matching the same pattern, despite the fact that their pattern-matching approaches are quite different. This suggests that this predictive model may be applicable (with some fitting) to other pattern matchers. Testing that hypothesis is part of our ongoing work.

4   PROCAT Implementation and Use

As discussed above, PROCAT is presently being applied as a module in the TANGRAM system for building and executing workflows for intelligence analysis. To support this application, we worked with the developers of the workflow generation/execution software, WINGS and PEGASUS [9], to design a set of queries that provide the information WINGS/PEGASUS needs to instantiate and execute workflows effectively. These queries are broken into two distinct phases of workflow generation:

- In the *Backward Sweep*, WINGS produces candidate workflows starting with the desired output. Given an output data requirement (a postcondition) and class of processes, PROCAT returns all matching processes that can satisfy that postcondition, along with their input data requirements and other preconditions for running them.
- In the *Forward Sweep*, WINGS and PEGASUS prune the workflow candidates generated in the Backward Sweep, rank them, and map processes to actual grid clusters

---

[2] The linear coefficients also depend on the values of non-data parameters being passed to the process, and, for performance, the hardware on which the process is to be run.

```
<SOAP-ENV:Envelope> <SOAP-ENV:Body>
<pcat:FindInputDataRequirements>
 <pcat:component xsi:type='xsd:string'>
  <rdf:RDF>
   <rdf:Description rdf:about="http://...#?component2">
    <rdf:type rdf:resource="http://.../Process.owl#PatternMatchingProcess"/>
    <pdl:hasOutput rdf:resource="http://...#?dataVariable5"/>
    <pdl:hasInput rdf:resource="http://...#?dataVariable4"/>
    <pdl:hasInput rdf:resource="http://...#?dataVariable3"/>
   </rdf:Description>
  </rdf:RDF>
 </pcat:component>
 <pcat:constraints xsi:type='xsd:string'>
  <rdf:RDF>
   <rdf:Description rdf:about="http://...#?dataVariable5">
    <pdl:hasRole
     rdf:resource="http://.../Process.owl#HypothesisOutputRole"/>
    <rdf:type rdf:resource="http://...#Hypothesis"/>
    <pdl:saturatedWith
     rdf:resource="http://...#MoneyLaunderingEvent"/>
   </rdf:Description>
  </rdf:RDF>
 </pcat:constraints>
</pcat:FindInputDataRequirements> </SOAP-ENV:Body>
</SOAP-ENV:Envelope>
```

Fig. 5. Example PROCAT Query

to run. The queries in this sweep require PROCAT to predict output data characteristics, predict process performance on particular Grid clusters, return actual physical location(s) of the process, return resource requirements for running the process installation, and return the relevant information for building a valid command line for the process installation.

The implementation choices for PROCAT were driven by the requirements described above, together with the fact that PDL is encoded in OWL. We decided to use ALLEGROGRAPH[3], which is a modern, high-performance, persistent, disk-based RDF graph database. ALLEGROGRAPH provides a variety of query and reasoning capabilities over the RDF database, including SPARQL, HTTP, and PROLOG query APIs and built-in RDFS++ reasoning. ALLEGROGRAPH also allows one to define a SOAP Web service API to an ALLEGROGRAPH database, including the ability to generate WSDL files from the code's SOAP server definition, to facilitate the creation of Web service clients.

PROCAT is implemented as an HTTP/SOAP-based Web service with an ALLEGROGRAPH RDF triple store. Component (process) definitions, encoded using PDL, are stored in the triple store, and SOAP services provide the specific information required by WINGS and PEGASUS. Each service is associated with a message handler and response generation code for the SOAP API. No Web service for updating KB content was developed for the initial version of PROCAT, as this capability was not essential to test its ability to provide a useful service as part of a workflow generation/execution experiment. We plan to add this service for the next release of PROCAT.

Figure 5 shows an example of a SOAP query for the Backward Sweep phase. The query specifies a general class of component (in this case, *PatternMatchingProcess*)

---

[3] http://agraph.franz.com/allegrograph/

and a requirement that the component has an output of type *Hypothesis* containing objects of type *MoneyLaunderingEvent*. PROCAT will return any actual components in its repository that match these constraints. It should be noted that all components in the repository are defined as belonging to one or more classes of component, and that these are drawn from the process ontology described above. The results are returned as RDF/XML fragments, one for each matching component instance.

5   Related Work and Discussion

The problem of process characterization and the related problem of process selection to meet a particular set of requirements have been investigated for several decades under various research headings, including program verification, deductive program synthesis, automatic programming, AI planning, agent-based systems, Semantic Web / Grid services, and e-science. Because of space constraints, we can only mention examples from the last three of these areas. For a more extensive summary of related work, see [15].

The field of agent-based systems (ABS) includes a significant body of work on characterizing and reasoning about agent capabilities, which often are conceived as remotely invocable processes. As in earlier work on AI planning, the common denominator of many approaches is the representation of preconditions and effects, often with additional information about the inputs and outputs of the operations that an agent provides. LARKS [16], for example, employs *InConstraints* and *OutConstraints*, expressed as Horn clauses referring to inputs and outputs, respectively, for this purpose. This approach, while flexible, requires special handling for these Horn clauses outside of the description logic framework that underlies LARKS's ontology. PROCAT's approach, in contrast, remains within the representational framework defined by RDF and OWL. Agent systems have also experimented with the use of additional kinds of information, such as quality of service, response time, and other kinds of performance characterization. Generally speaking, however, these have been captured using static, one-size-fits-all characterizations, rather than computed on-the-fly based on the specifics of input datasets and resources in the execution environment, as PROCAT does.

ABS has explored a variety of styles of matchmaking. For example, in the Open Agent Architecture (OAA), [17], the basic capability description is a logic programming predicate structure (which may be partially instantiated), and matchmaking is based on unification of goals with these predicate structures. In addition, both goals and capabilities declarations may be accompanied by a variety of parameters that modify the behavior of the matchmaking routines. Although PROCAT does not make use of unification, it achieves greater flexibility by building on SPARQL, and a more effective means of categorization of capabilities (processes) based on OWL class hierarchies.

Most recently, these same problems have been the focus of inquiry in the context of research on Semantic Web Services (SWS). This field aims to enable the automation of the development and use of Web services. The first challenge in SWS has been the enrichment of service descriptions, which essentially is the same problem as process characterization. OWL for Services (OWL-S) [18], the pioneering effort in this field, introduces the expression of preconditions and effects in a Semantic Web-compatible manner, and also relies on the ability to use OWL to construct class hierarchies of services. PROCAT's ontology is based in part on OWL-S, but goes much further in distinguishing process installations and characterizing their resource requirements and invocation methods. OWL-S also includes a composite process structure model—a set of ontology elements used to formulate a step-by-step representation of the structure of a composite process. PROCAT thus far has had no need for this kind of representation.

The Semantic Web Services Framework (SWSF) [19] builds out from OWL-S by including some additional concepts (especially in the area of messaging); employing first-order logic, which is more expressive than OWL; and drawing on the axiomatization of processes embodied in the Process Specification Language (PSL). The Web Services Modeling Ontology (WSMO) [20], is an EU-funded effort with many of the same objectives and approaches as OWL-S and SWSF. WSMO distinguishes two types of preconditions (called *assumptions* and *preconditions*), and two types of postconditions (called *postconditions* and *effects*). In addition, WSMO associates services with goals that they can satisfy, and models choreography—the pattern of messages flowing between two interacting processes. PROCAT thus far has not encountered requirements for capturing these additional aspects of processes, but could be extended in these directions if needed. On the other hand, PROCAT's expressiveness requirements have deliberately been kept smaller than those of SWSF and WSMO, allowing for relatively lightweight implementation, scalability, and quick response times.

In the field of *e-science*, scientific experiments are managed using distributed workflows. These workflows allow scientists to automate the steps to go from raw datasets to scientific results. The ultimate goal is to allow scientists to compose, execute, monitor, and rerun large-scale data-intensive and compute-intensive scientific workflows. For example, the NSF-funded KEPLER project[4] has developed an open-source scientific workflow system that allows scientists to design scientific workflows and execute them efficiently using emerging Grid-based approaches to distributed computation. Compared to PROCAT, the KEPLER Actor repository can be seen to be a more general-purpose repository for storing workflow components—both the actual software, as well as metadata descriptions of that software. However, compared to PDL, the actor definitions are impoverished, and cover simply I/O parameters.

By and large, process characterization in all these disciplines has been predominantly concerned with what we here call the Capabilities Layer of description. Where quantitative descriptions have been used, they have most often been used to solve very specialized problems. Further, quantitative descriptions have rarely taken advantage of probabilistic methods to characterize data modification and accuracy, or to enable predictions regarding the content and structure of generated data, as has been done in PROCAT. Another significant difference is PROCAT's combined specification of a process's behavioral characterization with the fine-grained characterization of its data products.

Looking ahead, our research directions are focused on four major areas:

- Extending the Capabilities Layer to make more sophisticated use of rules and deduction in finding matching components and inferring the requirements for running them.

---

[4] http://kepler-project.org

- Gathering data and running more experiments to test the accuracy of PROCAT's quantitative predictions, both to evaluate the existing models and to drive the creation of new and better ones.
- Implementing the Accuracy Layer, discussed above. This layer will produce estimates (rough to begin with) of the accuracy of the output data produced by a process.
- Automating some parts of the population of the process repository. For example, we have started to automate some of the experiments needed to create the quantitative models of the Data Modification and Performance layers.

Workflow generation and execution technologies are becoming increasingly important in the building of large integrated systems. More expressive process descriptions, and new kinds of reasoning about them, will play a critical role in achieving this long-term goal. We have described the process representation and reasoning approaches embodied in PROCAT, the rationale behind its current design, its role in a particular integrated system, and research directions under investigation in connection with this work.

Acknowledgments

This research was supported under Air Force Research Laboratory (AFRL) contract number FA8750-06-C-0214. The concept of a process catalog for Tangram is due to Eric Rickard. Thanks also to the builders of the other Tangram modules discussed in this paper, including (but not limited to) Fotis Barlos, Ewa Deelman, Yolanda Gil, Dan Hunter, Jihie Kim, Jeff Kudrick, Sandeep Maripuri, Gaurang Mehta, Scott Morales, Varun Ratnakar, Manoj Srivastava, and Karan Vahi.

References

1. Boner, C.: Novel, complementary technologies for detecting threat activities within massive amounts of transactional data. In: Proceedings of the International Conference on Intelligence Analysis (2005)
2. Coffman, T., Greenblatt, S., Marcus, S.: Graph-based technologies for intelligence analysis. Communications of the ACM 47(3) (2004)
3. Pioch, N.J., Hunter, D., White, J.V., Kao, A., Bostwick, D., Jones, E.K.: Multi-hypothesis abductive reasoning for link discovery. In: Proceedings of KDD 2004 (2004)
4. Wolverton, M., Berry, P., Harrison, I., Lowrance, J., Morley, D., Rodriguez, A., Ruspini, E., Thomere, J.: LAW: A workbench for approximate pattern matching in relational data. In: The Fifteenth Innovative Applications of Artificial Intelligence Conference, IAAI 2003 (2003)
5. Holder, L., Cook, D., Coble, J., Mukherjee, M.: Graph-based relational learning with application to security. Fundamenta Informaticae 66(1–2) (2005)
6. Adibi, J., Chalupsky, H.: Scalable group detection via a mutual information model. In: Proceedings of the First International Conference on Intelligence Analysis, IA 2005 (2005)
7. Macskassy, S.A., Provost, F.: Suspicion scoring based on guilt-by-association, collective inference, and focused data access. In: Proceedings of the NAACSOS Conference (2005)
8. Davis, J., Dutra, I., Page, D., Costa, V.S.: Establishing identity equivalence in multi-relational domains. In: Proceedings of the International Conference on Intelligence Analysis, IA 2005 (2005)

9. Gil, Y., Ratnakar, V., Deelman, E., Mehta, G., Kim, J.: Wings for Pegasus: Creating large-scale scientific applications using semantic representations of computational workflows. In: The Nineteenth Innovative Applications of Artificial Intelligence Conference, IAAI 2007 (2007)
10. Corkill, D.D.: Collaborating software: Blackboard and multi-agent systems and the future. In: Proceedings of the International Lisp Conference (2003)
11. Klyne, G., Carroll, J.J.: Resource description framework (RDF): Concepts and abstract syntax. W3C recommendation, W3C (February 2004), http://www.w3.org/TR/2004/REC-rdf-concepts-20040210/
12. McGuinness, D.L., van Harmelen, F.: Owl web ontology language overview, World Wide Web Consortium (W3C) Recommendation (2004), http://www.w3.org/TR/owl-features/
13. Seaborne, A., Prud'hommeaux, E.: SPARQL query language for RDF. W3C recommendation, W3C (January 2008), http://www.w3.org/TR/2008/REC-rdf-sparql-query-20080115/
14. Beckett, D.: RDF/xml syntax specification (revised). W3C recommendation, W3C (February 2004),
    http://www.w3.org/TR/2004/REC-rdf-syntax-grammar-20040210/
15. Wolverton, M., Harrison, I., Martin, D.: Issues in algorithm characterization for link analysis. In: Papers from the AAAI Fall Symposium on Capturing and Using Patterns for Evidence Detection (2006)
16. Sycara, K., Wido, S., Klusch, M., Lu, J.: LARKS: Dynamic matchmaking among heterogeneous software agents in cyberspace. Journal of Autonomous Agents and Multi-Agent Systems 5(2), 173–203 (2002)
17. Cheyer, A., Martin, D.: The Open Agent Architecture. Journal of Autonomous Agents and Multi-Agent Systems 4(1), 143–148 (2001)
18. Martin, D., Burstein, M., Hobbs, J., Lassila, O., McDermott, D., McIlraith, S., Narayanan, S., Paolucci, M., Parsia, B., Payne, T., Sirin, E., Srinivasan, N., Sycara, K.: Owl-s: Semantic markup for web services (2004) W3C Member Submission 22 (November 2004), http://www.w3.org/Submission/2004/07/
19. Battle, S., Bernstein, A., Boley, H., Grosof, B., Gruninger, M., Hull, R., Kifer, M., Martin, D., McIlraith, S., McGuinness, D., Su, J., Tabet, S.: Semantic web services framework overview (2005) W3C Member Submission (November 22, 2004), http://www.w3.org/Submission/2004/07/
20. Bruijn, J.D., Lausen, H., Polleres, A., Fensel, D.: The web service modeling language WSML: An overview. Technical Report 2005-06-16, DERI (2005), http://www.wsmo.org/wsml/wsml-resources/deri-tr-2005-06-16.pdf
21. Osterweil, L.J., Wisel, A., Clarke, L.A., Ellison, A.M., Hadley, J.L., Boose, E., Foster, D.R.: Process technology to facilitate the conduct of science. In: Unifying the Software Process Spectrum, pp. 403–415. Springer, Heidelberg (2006)

Inference Web in Action:
Lightweight Use of the Proof Markup Language

Paulo Pinheiro da Silva[1], Deborah McGuinness[2]
Nicholas Del Rio[1], and Li Ding[2]

[1] University of Texas at El Paso, El Paso TX 79902, USA
[2] Rensselaer Polytechnic Institute, Troy NY 12180, USA

Abstract. The Inference Web infrastructure for web explanations together with its underlying Proof Markup Language (PML) for encoding justification and provenance information has been used in multiple projects varying from explaining the behavior of cognitive agents to explaining how knowledge is extracted from multiple sources of information in natural language. The PML specification has increased significantly since its inception in 2002 in order to accommodate a rich set of requirements derived from multiple projects, including the ones mentioned above. In this paper, we have a very different goal than the other PML documents: to demonstrate that PML may be effectively used by simple systems (as well as complex systems) and to describe lightweight use of language and its associated Inference Web tools. We show how an exemplar scientific application can use lightweight PML descriptions within the context of an NSF-funded cyberinfrastructure project. The scientific application is used throughout the paper as a use case for the lightweight use of PML and the Inference Web and is meant to be an operational prototype for a class of cyberinfrastructure applications.

1 Introduction

In a question-answering context such as when querying an intelligent agent, answer justifications are used to provide provenance information about the sources and process used by agent (or agents) producing the answers. The Proof Markup Language (PML) is a powerful language for encoding OWL-based justifications that are portable and combinable and that can be distributed over the web [12]. PML also facilitates agents in reusing elements of one justification as part of new justifications, enabling in this way multiple justifications for a single answer to be encoded within a single web artifact. Furthermore, PML design is grounded on proof theory, which enables it to encode formal proofs as justifications. As a consequence of these and many other advanced features of PML, many potential users of PML have not further considered the use of the language and its supporting Inference Web Infrastructure due to perceived complexity issues.

Despite the richness of constructs supporting some of the advanced features mentioned above, PML does not require justifications to be distributed, to be combined, or to be formal proofs. In this paper, we introduce a use case that

A. Sheth et al. (Eds.): ISWC 2008, LNCS 5318, pp. 847–860, 2008.

describes how potential PML users can benefit from a simpler, restricted set of PML constructs to encode very basic justifications as well as to a restricted set of Inference Web tools to perform useful tasks such as the retrieval and browsing of provenance information. We claim that this use case is representative of a set of needs that a broad range of applications face, and further that a broad range of users may similarly use a subset of PML and the Inference Web toolkit to address common problems related to explanation and trust recommendations.

Any simplification of an answer justification may have consequences in terms of missing information that may prove to be critical to support some justification-based tasks such as trust and uncertainty management or even just to verify the justification correctness. For example, in the process of an inference engine encoding a formal proof for a theorem, the omission of discharged assumptions in a single step of the proof may be reason enough for one to consider the entire proof to be unsound. However, such encoding is still a justification capable of identifying the set of axioms used to derive the theorem as well as the collection of information sources asserting the axioms. Thus, it is clear that this unsound proof is still better than no justification at all. With this notion of justification usefulness in mind, the paper describes how lightweight uses of PML have successfully been used to support the inspection of gravity maps as part of the NSF-funded Cyber-ShARE Center of Excellence on Cyber-Infrastructure[1].

The rest of this paper is organized as follows. Section 2 introduces a scientific application as a use case for lightweight use of PML. Section 3 revisits many aspects of PML including its relationship to the Inference Web infrastructure. Section 4 describes some strategies to simplify the process of instrumenting an application to generate PML. This section also highlights how some tools can be easily used in combination with PML. Based on the lightweight use of PML, Section 5 described the results of a user study that identifies the scientists' need of provenance to understand the results of the scientific application introduced in Section 2. Section 6 discusses further strategies to simplify the use of PML as well as describing related work. Section 7 summarizes the main contributions of the paper.

2   A Use Case for Provenance

In this section, we introduce an exemplar scientific use case that has a number of common provenance requirements. Scientists may use gravity data to get a rough idea of the subterranean features that exist within some region. Geoscientists are often only concerned with anomalies, or spikes in the data, which often indicate the presence of a water table or oil reserve. However these anomalies have the potential to be artificial and simply imperfections introduced during the data retrieval process including for instance some data merging and filtering techniques. With the use of provenance, however, one may be able to inspect the process used to retrieve data and this figure out potential sources of imperfections.

---

[1] http://cybershare.utep.edu

This process, which begins by scientists providing a region or interest or footprint, specified in terms of latitude and longitude, is defined by the sequence of tasks below:

1. gather raw point data (possibly from multiple sources) in the region
2. merge point data coming from distinct sources
3. filter raw point data to remove duplicate data
4. create a uniformly distributed dataset by applying a gridding algorithm
5. create a contoured rendering of the uniformly distributed dataset

In a cyber-infrastructure setting, each one of the five tasks above can be realized by a web service. This set of web services is piped or chained together; the output of one service would be forwarded as the input to the next service specified in the *workflow*, such as in [7].

In these types of situations where multiple workflows can satisfy a single type of request, the set of results generated by each workflow are returned to the scientist. As in any question/answer scenario, it is up to the scientist to determine what result to use. However, this situation is no different from how users interact with Web search engines. A single query often yields thousands of results, yet the burden is placed on the user to determine which answer is most appropriate. This is one of the main reasons that applications should be able to explain their results as further discussed in the following section.

3   Inference Web and the Proof Markup Language

The Inference Web [9,10] is a knowledge provenance infrastructure for conclusions derived from inference engines which supports interoperable explanations of sources (i.e. sources published on the Web), assumptions, learned information, and answers associated with derived conclusions, that can provide users with a level of trust regarding those conclusions. The ultimate goal of the Inference Web is the same as the goal of the gravity data scenario which is to provide users with an understanding of how results are derived by providing them with an accurate account of the derivation process and the information sources used (i.e. knowledge provenance [13]).

Inference Web provides PML to encode justification information about basically any kind of response produced by agents. PML is an RDF based language defined by a rich ontology of provenance and justification concepts which describe the various elements of automatically generated proofs. Without getting into the details of the main concepts supporting PML because of obvious reasons, we can say that PML justifications are graphs with the edges always pointing towards the final justification conclusion. PML justifications can also be used to store provenance about associated information sources. PML itself is defined in OWL [1] thus supporting the distribution of proofs throughout the Web. Each PML component, which is not yet defined here, can reside in a uniquely identified document published on the Web separately from the others.

4   Lightweight Use of the Inference Web

Any conclusion may have no justification, one justification, or multiple justifi-cations. In the case of a scientist using a gravity map, it is clear that the map was generated from data provided by some sources, e.g., data points or annota-tions about the region of interest, and by some process, whether the process is computer-based or not. In this case, the map is the conclusion of a justification and the justification is a description of the process used to derive the map. The scientist's knowledge about the map, however, may be restricted to the fact that the map came from Book A. In this case, the scientist can still state that the map is asserted by Book A (or even by the authors of Book A). Both justifica-tions for the map are legitimate and can be encoded in PML. Moreover, PML has been designed to encode all sorts of justifications including the combination of alternate justification for a given conclusion. Because of that, PML has a rich but rather complex set of constructors to encode justifications.

One of the goals of this paper is to demonstrate a lightweight use of PML that relies on three *simplification assumptions* listed below. Please note that lightweight use of PML does not preclude a later enhancement of PML docu-ments that may benefit from the full provenance encoding power of PML.

Simplification Assumption 1 - No use of alternate justifications. The encoding of a single justification for a given conclusion implies that the justification can be considered as a single DAG of **nodes** connected by `hasAntecedent` relationships. In this situation, lightweight use of PML is achieved by considering each node in the DAG to be a single **PML node set** with a single **PML inference step**. The hasAntecedent relationship of the node is the hasAntecedent property of the only inference step inside each node set.

```
<rdf:RDF>
  <NodeSet rdf:about="http://iw.cs.utep.edu//contourMapPS_7355.owl#map">
    <hasConclusion>
      <pmlp:Information>
        <pmlp:hasFormat rdf:resource="http://iw.cs.utep.edu/registry/FMT/ps3.owl#ps3"/>
        <pmlp:hasRawString rdf:datatype="http://www.w3.org/2001/XMLSchema#string">
          (*** THE BASE64 ENCODING OF THE POSTSCRIPT OF THE MAP GOES HERE ***)
        </pmlp:hasRawString>
      </pmlp:Information>
    </hasConclusion>
    <isConsequentOf>
      <InferenceStep>
        <hasIndex rdf:datatype="http://www.w3.org/2001/XMLSchema#int">0</hasIndex>
        <hasAntecedentList>
          <NodeSetList>
            <ds:first rdf:resource="http://iw.cs.utep.edu/griddedData7035.owl#gridmap"/>
          </NodeSetList>
        </hasAntecedentList>
      </InferenceStep>
    </isConsequentOf>
  </NodeSet>
</rdf:RDF>
```

In the example above , we show the last step of our gravity map workflow (step 5 of the use case). The final conclusion of the gravity map workflow is a contour map identified by the URIRef ending on **#map**. In the node, the map itself goes in the **hasRawString** property of the node set. The inference step of the node has that it has been derived from the conclusion of the node identified by the URIRef ending on **#gridmap** (the **#gridmap** URIRef corresponds to step 4 of the use case). The *hasFormat* attribute of a node set is optional as are many other node set and inference step properties. In the case of the contour map node, the **hasFormat** attribute says that the raw string is encoded in PostScript 3. URIs in the fragments of PML documents used in this section have been modified to fit in the paper. Complete PML documents in support of the gravity map use case are available at http://iw.cs.utep.edu:8080/service-output/proofs/.

Simplification Assumption 2 - No knowledge about the inference mechanism used to transform information in a step of a given information manipulation process. In a formal proof, it is expected that one can identify the inference rule used, e.g., resolution, or algorithm, e.g., quick sort, used in each step of the proof. PML can indeed be used for encoding formal proofs but it is often used to encode less structured justifications often called *information manipulation traces*. In concrete terms, it is common that the person instrumenting a process to generate PML may not have full knowledge about the process so that this person can properly document how information is transformed along the process. In the example below, we show how informal but still useful metadata can be added to the process trace of the gravity map. In this case, it is known that the service is a generic service, as identified by the inference rule identified by the URIRef ending on **#genericService** and that the service is named "contour" (as identified by the inference engine identified by the URIRef ending on **#contour**). These URIRef have been created by the example in this paper and they can be reused or new ones can be created on demand. These URIRef are called provenance elements and are also PML documents. In the past, Inference Web used to register these provenance elements in order to facilitate reuse. Currently, Inference Web still incentive the reuse of these documents but also allows these elements to be easily created and stored locally. The IWSearch capability described in Section 4.2 is used instead of the registry to facilitate the location and reuse of PML documents.

```
<rdf:RDF>
   (...)
   <isConsequentOf>
     <InferenceStep>
       <hasIndex rdf:datatype="http://www.w3.org/2001/XMLSchema#int">0</hasIndex>
       <hasInferenceRule
           rdf:resource="http://iw.cs.utep.edu/registry/RUL/GS.owl#genericService"/>
       <hasInferenceEngine
           rdf:resource="http://iw.cs.utep.edu/registry/IE/contour.owl#contour"/>
       <hasAntecedentList>
         <NodeSetList>
           <ds:first rdf:resource="http://iw.cs.utep.edu/griddedData7035.owl#gridded"/>
         </NodeSetList>
       </hasAntecedentList>
     </InferenceStep>
   </isConsequentOf>
   (...)
</rdf:RDF>
```

Simplification Assumption 3 - No knowledge about how information has been asserted from a given source. Conclusions of the leaf nodes in a justification DAG are pieces of information that have been asserted by some source. For example, in the case of the gravity map, the information may correspond to the gravity reading data points that was eventually processed, e.g., gridded, to generate the contour map, where the entire gravity database is the source. In reality, a web service was used to access the database over the web and some additional parameter where required to invoke the service. Leave nodes are called *direct assertions* since they make use of the direct assertion inference rule, i.e., the PML instance identified by the URIRef ending on #told. The example below illustrates the use of a **PML SourceUsage** instance often attached to direct assertions. Source usage is a complex concept since it has a rich set of properties used to specify how exactly a given piece of information was extracted from a given source. In the case of a lightweight use of source usage, however, we restrict its use to the identification of the source that was used without identifying how and when it was used. For instance, the source usage in the example below identifies that the gravity database identified by the URIRef ending on #database was used as a source of the conclusion of the node. It is interesting to note that PML identifies the agent responsible for retrieving the source that is identified by the value of the hasInferenceEngine.

```
<rdf:RDF>
    (...)
    <isConsequentOf>
      <InferenceStep>
        <hasIndex rdf:datatype="http://www.w3.org/2001/XMLSchema#int">0</hasIndex>
        <hasInferenceRule rdf:resource="http://iw.cs.utep.edu/registry/RUL/Told.owl#told"/>
          <hasSourceUsage>
            <pmlp:SourceUsage>
              <pmlp:hasSource
                  rdf:resource="http://iw.cs.utep.edu/registry/PER/GravityDB.owl#database"/>
            </pmlp:SourceUsage>
          </hasSourceUsage>
        <hasInferenceEngine
            rdf:resource="http://iw.cs.utep.edu/registry/IE/AccessDatabase.owl#accessDB"/>
      </InferenceStep>
    </isConsequentOf>
    (...)
</rdf:RDF>
```

4.1 PML Service Wrapper (PSW)

The encoding of lightweight PML is such a straightforward task that we have created a wrapper that can automate the generation of PML justifications for web services [4]. In the case of workflows based on web services, the use of the wrapper may allow PML to be used on the same way provenance is used in workflow-centered infrastructures, as later discussed in Section 6.

The gravity map scenario is realized by a service-oriented workflow composed of five Simple Object Access Protocol (SOAP) services, which gather, merge, filter, grid and render gravity datasets respectively[2]. These Web services are piped

---

[2] The gravity map workflow is available for use at http://iw.cs.utep.edu:8080/service-output/probeit/clientapplet.html

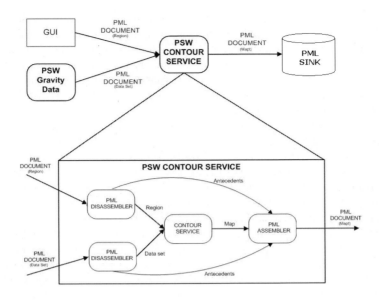

Fig. 1. Example of PSW configured for a contouring service

or chained together; the output of one service is forwarded as the input to the next service specified in the workflow. A workflow director is responsible for managing the inputs/outputs of each service as well as coordinating their execution. Provenance associated with scientific workflows of this nature might include the service's execution sequence as well as each of their respective outputs, which we refer to as *intermediate results.*

PML Service Wrapper (PSW) is a general-purpose Web service wrapper that logs knowledge provenance associated with workflow execution as PML documents. In order to capture knowledge provenance associated with workflows execution, each service composing the workflow has an associated PSW wrapper that is configured to accept and generate PML documents specific to it. Since PML node sets include the *conclusion* element, which is used to store the result of an inference step or Web service, the provenance returned by the wrappers also includes the service output thus workflows can be composed only of these PSWs; this configuration introduces a level of indirection between service consumers (i.e. workflow engine) and the target services that performs the required function. In this sense, PSW can be seen as a server side provenance logger.

The logging capability provided by PSW can be decomposed into three basic tasks: decompose, consume, and compose as illustrated in Figure 1. Upon invocation, the wrapper decomposes the conclusion of an incoming PML document, i.e., extracts the data resident in the PML conclusion using Inference Web's PML API. PSW then consumes the target service, forwarding the extracted data as an input to the target service. The result and associated provenance of the target service is then composed to produce the resultant PML document, the PSW output. For example, a contouring service requires 2D spatial data to map and the region to consider in the mapping therefore a PSW wrapper for this contouring service would require two PML documents, one containing 2D spatial data, coming from some data retrieval service, and the other containing a region, (e.g. specified by latitude and longitude) specified by some user. The output of the contour service is a map, from which a new PML document is created, referencing the two input PML node sets as antecedents.

PSW has been developed in support of scientific workflows able to execute in a distributed environment such as the cyberinfrastructure. In traditional Inference Web applications [11,10], inference engines are instrumented to generate PML. However in a cyberinfrastructure setting, reasoning is not necessarily deductive and is often supported by Web services that can be considered "black boxes" hard to be instrumented at source-code level to generate PML. This is the primary reason why PSW, a sort of external logger, must be deployed to intercept transactions and record events generated by services instead of modifying the services themselves to support logging. Despite this apparent limitation, PSW is still able to record provenance associated with various target systems' important functions. For example, PSW configured for database systems and service oriented workflows can easily record provenance associated with queries and Web service invocations respectively in order to provide a thorough recording of the provenance associated with cyberinfrastructure applications.

4.2   Inference Web Search

IWSearch is developed to facilitate users accessing PML data distributed on the Web. In the case of our use case, the PML data correspond to the values of PML properties such as `hasInferenceRule`, `hasInferenceEngine` and `hasFormat`, as presented in Section 4.

In the past deployments of Inference Web, the provenance metadata are stored in a federated online repository IWBase, which provides both a web user interface and a SOAP web service interface for Inference web users to publish provenance metadata. IWSearch is motivated by the limitations of such provenance data management found in our past practice: (i) IWBase provides limited mechanisms for accessing registered metadata entries, i.e. a user can only browse the type hierarchy of those entries to find entries; and (ii) no service is available to find and reuse PML provenance metadata published on the Web.

IWSearch is implemented as a service in the Inference Web architecture. It provides primitives for discovering, indexing, and searching for PML objects (i.e. instances of PML classes), such as pmlp:Person and pmlj:NodeSet, available on the Web. As shown in Figure 2, IWSearch consists of three groups of services:

1. the discovery services utilize swoogle [5] search results and a focused crawler (searching PML documents in a certain web directory) to discover URLs of PML documents throughout the Web;
2. the index services use the indexer to parse the submitted PML documents and prepare metadata about PML objects for future search, and uses the searcher to serve query calls invoked by users;

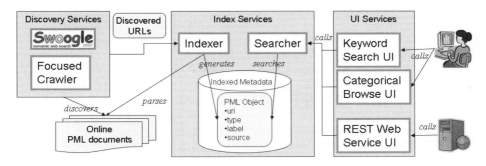

Fig. 2. IWSearch Architecture

3. the UI services offers a keyword search and categorical browsing interface for human or machine users.

On using IWSearch, one may reuse existing PML data that may be expensive to be created. For instance, PML data about a scientific publication may require the creator of the PML data to specify the publication's authors, authors' affiliations and publisher.

4.3   Probe-It!

Probe-It! is a browser suited to graphically rendering provenance information encoded in PML and associated with results coming from inference engines and workflows [3]. Probe-It! consists of four primary views to accommodate the different kinds of provenance information: queries, results, justifications, and provenance, which refer to user queries or requests, final and intermediate data, descriptions of the generation process (i.e., execution traces), and information about the sources respectively.

In a highly collaborative environment such as the cyberinfrastructure, there are often multiple applications published that provide the same or very similar function. A thorough integrative application may consider all the different ways it can generate and present results to users, placing the burden on users to discriminate between high quality and low quality results. This is no different from any question/answer application, including a typical search engine on the Web, which often uses multiple sources and presents thousands of answers back to users. The *query view* visually shows the links between application requests and results of that particular request. The request and each corresponding result is visualized as a node similar to the nodes in the justification view presented later.

Upon accessing one of the answer nodes in the *query view*, Probe-It! switches over to the justification view associated with that particular result. Because users are expected to compare and contrast between different answers in order to determine the *best* result, all views are accessible by a menu tab, allowing users to navigate back to the query view regardless of what view is active.

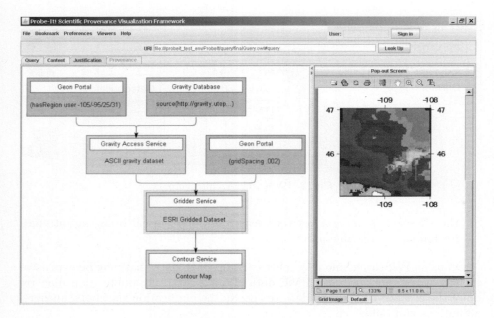

Fig. 3. Probe-It! justification view

The *results view* provides graphical renderings of the final and intermediate results associated with scientific workflows. This view is captured on the right hand side of Figure 3, which presents a visualization of a gridded dataset; this view is initiated by selecting one of the justification nodes, described in the next section. Because there are many different visualizations suited for gridded data and datasets in general, the *results view* is composed of a set of viewers, each implementing a different visualization technique suited to the data being viewed.

The *justification view*, on the other hand, is a complimentary view that contains all the process meta-information associated with the execution trace, such as the functions invoked by the workflow, and the sequencing associated with these invocations. Probe-It! renders this information as a directed acyclic graph (DAG). An example of a workflow execution DAG can be found on the left hand side of Figure 3, which presents the justification of a contour map. From this perspective, Web services and sources (i.e., data sinks) are presented as nodes. Nodes contain a label indicating the name of a source or invoked service, as well as a semantic description of the resulting output data. In the justification view, data flow between services is represented by edges of the DAG; the representation is such that data flows from the leaf nodes towards the root node of the DAG, which represents the final service executed in the workflow. Users can access both provenance meta-information and intermediate results of the sources and services represented by the DAG nodes. In this sense, the justification DAG serves as a medium between provenance meta-information and intermediate results.

The *provenance view*, provides information about sources and some usage information e.g., access time, during the execution of an application or workflow.

For example, upon accessing the node labeled *gravity database*, meta-information about the database, such as the contributing organizations, is displayed in another panel. Similarly, users can access information transformation nodes, and view information about used algorithms, or the hosting organization.

5   Evaluation of Lightweight PML

As MacEachren et al. describe in [8], *provenance* (or lineage, as mentioned in the reference) may be a requirement for understanding uncertainties related to geospatial information. In the case of our use case, map provenance is meta-information about source datasets, services and any other resource used to derive the maps [4]. In this section, we present some results of a user study used to verify whether provenance is needed for scientists to correctly identify and explain the quality of maps, a required condition if scientific communities are going to accept maps from CI-based applications. A comprehensive description of the user study can be found in [2]. It is important to note that this evaluation is part of a more comprehensive and ongoing effort to understand the need for provenance in scientific applications based on cyber-infrastructure resources. Also, the evaluation is not exactly about the lightweight use of the Inference Web. However, the results are significant for this particular paper because the provenance used in the study was *entirely encoded using lightweight PML*.

The user study analyzes how scientists with different levels of expertise on gravity data for geophysics and on GIS can differentiate between contour maps of high quality (e.g., maps with not known imperfections) and maps of low quality (e.g., maps with known imperfections) and to explain the reasons of identified qualities. Two cases *map* and *map+p* have been used, where only a single map M0 has been presented to subjects, thus no averages need to be taken; the scores are equal to the points earned for identifying and explaining the single map.

The hypothesis of concern for this paper is that "Scientists with access to provenance can identify and explain map imperfections more accurately than scientists without access to provenance." There are two types of scores associated with each evaluation case: an *identification score* and an *explanation score*, both of which have been used to assess the validity of our hypothesis. An identification score is computed as the average of points earned for correctly classifying the maps comprising an evaluation case. Similarly, the explanation score is computed as an average of points earned for correctly explaining the map imperfections. Because the measure of identifying and explaining maps is a binary value (e.g., 0 for incorrect answers and 1 for a correct answers), both types of evaluation scores are always between 0 and 1, inclusive.

Table 1. Subjects' average accuracy in identifying and explaining map imperfections

Task	*map*	*map+p*
Imperfection identification	0.10	0.79
Imperfection explanation	0.05	0.78

Significance of the the results collected so far were verified by a single-tail t-test at 95% confidence. The *imperfection identification task* in Table 1 contains the average accuracy of scientists in identifying maps quality. Condition cases *map* versus *map+p* tested whether provenance was needed in order to correctly assess maps; both cases are based on the same map containing the same error with the ability to access provenance in condition *map+p* being the only difference. Prior to the use of provenance, many scientists were unable to determine whether the map contained any imperfections at all, in which case their responses were regarded as unsuccessful earning 0. After the scientists were able to access the provenance, both their accuracy and confidence in determining the quality of the map improved significantly.

6   Discussion and Related Work

The uses of provenance are dictated by the goals of the particular systems; because various dimensions of provenance can be used to achieve various goals, there is no one use that fits all. For instance, one category of provenance systems aim at providing users with a sort of "history of changes" associated with some workflow, thus their view of provenance differs from that of a second category of provenance systems, which aim at providing provenance for use of debugging, or understanding an unexpected result. A third category of provenance systems record events that are well suited for re-executing the workflow it is derived from. From this point of view, PML fits into the second category of provenance systems. Provenance systems representative of these categories are reviewed below.

VisTrails, a provenance and data exploration system, provides an infrastructure for systematically capturing provenance related to the evolution of a workflow [6]. VisTrails users edit workflows while the system records the various modifications being applied. In the context of this system, provenance information refers to the modifications or history of changes made to particular workflow in order to derive a new workflow; modifications include, adding, deleting or replacing workflow processes. VisTrails provides a novel way to render this history of modifications. A treelike structure provides a representation for provenance where nodes represent a version of some workflow and edges represent the modification applied to a workflow. Upon accessing a particular node of the provenance tree, users of Vis-Trails are provided with a rendering of the scientific result which was generated as a result of the workflow associated with the node. In the context of VisTrails, only workflows that generate visualizations are targeted, however the authors describe how this system could be transformed to handle the general case as provided by PML in combination with Probe-It!; to provide a framework that can manage and graphically render any scientific result ranging from processed datasets to complex visualizations.

MyGrid, from the e-science initiative, tracks data and process provenance of workflow executions. Authors of MyGrid draw an analogy between the type of provenance they record for in-silico experiments and the kind of information that a scientist records in a notebook describing where, how and why experimental results were generated [15]. From these recordings, scientists are able to operate in three basic modes: (i) debug, (ii) check validity, and (iii) update mode, which refer to situations when, a result is of low quality and the source of error must be identified, when a result is novel and must be verified for correctness, or when a workflow has been updated and its previous versions are requested. Based on particular user roles, the appropriate dimension of provenance is presented, knowledge, organization, data, or process level [15]. MyGrid is yet another system that supports different tasks or uses of provenance, thus there are multiple "modes" that users can operate in that effectively show only provenance relevant for a particular task. We believe that all levels of provenance are required in order for scientists to identify the quality of complex results.

All these provenance system thus far track provenance related to workflows. Trio is a management system for tracking data resident in a database; provenance is tracked as the data is projected and transformed by queries and operations respectively [14]. provenance related to some function is recorded in a lineage table with various fields such as the derivation-type, how-derived, and lineage-data. Because of the controlled and well understood nature of a database, lineage of some result can many times be derived from the result itself by applying an inversion of the operation that derived it. Additionally, Trio provides the capability of querying the lineage table, thus allowing users to request provenance on demand.

7   Conclusions

The Proof Markup Language has been used in several projects to encode application response justifications. While new justification requirements have been addressed by incremental enhancements of the PML specification, many PML-enabled applications and probably most future PML-enabled applications would need to use just a small subset of the PML concepts and concept properties. Furthermore, most users of these applications would need to use just a small set of tool functionalities to further understand application responses. Through the use of an exemplary scientific application, the paper was demonstrated that lightweight PML has been used by scientists to understand map imperfections. Moreover, the exemplary application has demonstrated the usefulness of IWSearch to support the reuse of PML metadata and of Probe-It! to support the visualization of knowledge provenance encoded with lightweight PML.

Acknowledgments: This work was supported in part by NSF grant HRD-0734825 and by DHS grant 2008-ST-062-000007.

References

1. Dean, M., Schreiber, G.: OWL web ontology language reference. Technical report, W3C (2004)
2. Del Rio, N., Pinheiro da Silva, P.: Identifying and Explaining Map Imperfections Through Knowledge Provenance Visualization. Technical Report UTEP-CS-07-43a, The University of Texas at El Paso (June 2007)

3. Del Rio, N., Pinheiro da Silva, P.: Probe-it! visualization support for provenance. In: Proceedings of the Second International Symposium on Visual Computing (ISVC 2), Lake Tahoe, NV, pp. 732–741. Springer, Heidelberg (2007)
4. Del Rio, N., Pinheiro da Silva, P., Gates, A.Q., Salayandia, L.: Semantic annotation of maps through knowledge provenance. In: Proceedings of the Second International Conference on Geospatial Semantics (GeoS), Mexico City, Mexico. LNCS, pp. 20–35. Springer, Heidelberg (2007)
5. Ding, L., Finin, T., Joshi, A., Pan, R., Cost, R.S., Peng, Y., Reddivari, P., Doshi, V.C., Sachs, J.: Swoogle: A search and metadata engine for the semantic web. In: Proceedings of the 13th CIKM (2004)
6. Freire, J., Silva, C.T., Callahan, S.P., Santos, E., Scheidegger, C.E., Vo, H.T.: Managing Rapidly-Evolving Scientific Workflows. In: Proceedings of the International Provenance and Annotation Workshop (IPAW) (to appear)
7. Ludäscher, B., et al.: Scientific Workflow Management and the Kepler System. In: Concurrency and Computation: Practice & Experience (2005); Special Issue on Scientific Workflows
8. MacEachren, A., Robinson, A., Hopper, S., Gardner, S., Murray, R., Gahegan, M., Hetzler, E.: Visualizing Geospatial Information Uncertainty: What We Know and What We Need to Know. Cartography and Geographic Information Science 32(32), 139–160 (2005)
9. McGuinness, D.L., Pinheiro da Silva, P.: Infrastructure for Web Explanations. In: Fensel, D., Sycara, K., Mylopoulos, J. (eds.) ISWC 2003. LNCS, vol. 2870, pp. 113–129. Springer, Heidelberg (2003)
10. McGuinness, D.L., Pinheiro da Silva, P.: Explaining Answers from the Semantic Web. Journal of Web Semantics 1(4), 397–413 (2004)
11. Murdock, J.W., McGuinness, D.L., Pinheiro da Silva, P., Welty, C., Ferrucci, D.: Explaining Conclusions from Diverse Knowledge Sources. In: Proceedings of the 5th International Semantic Web Conference (ISWC2006), Athens, GA, November 2006, pp. 861–872. Springer, Heidelberg (2006)
12. Pinheiro da Silva, P., McGuinness, D.L., Fikes, R.: A Proof Markup Language for Semantic Web Services. Information Systems 31(4-5), 381–395 (2006)
13. Pinheiro da Silva, P., McGuinness, D.L., McCool, R.: Knowledge Provenance Infrastructure. IEEE Data Engineering Bulletin 25(2), 179–227 (2003)
14. Widom, J.: Trio: A System for Integrated Management of Data, Accuracy, and Lineage. In: Proceedings of the Second Biennial Conference on Innovative Data Systems Research, Asilomar, CA, January 2005, pp. 262–276 (2005)
15. Zhao, J., Wroe, C., Goble, C.,, R.S.: Using Semantic Web Technologies for Representing E-science Provenance. In: Proceedings of the 3rd International Semantic Web Conference, pp. 92–106 (November 2004)

Supporting Ontology-Based Dynamic Property and Classification in WebSphere Metadata Server

Shengping Liu[1], Yang Yang[1], Guotong Xie[1], Chen Wang[1], Feng Cao[1],
Cassio Dos Santos[2], Bob Schloss[3], Yue Pan[1], Kevin Shank[2], and John Colgrave[2]

[1] IBM China Research Laboratory
Building 19A, Zhongguancun Software Park, Beijing, 100094, China
{liusp,yangyy,xieguot,chwang,caofeng}@cn.ibm.com
[2] IBM Software Group
50 Washington Street, Westborough, MA 01581, USA
{scdos,kshank}@us.ibm.com, colgrave@uk.ibm.com
[3] IBM Watson Research Center
P.O. Box 704, Yorktown Heights, NY 10598, USA
rschloss@us.ibm.com

Abstract. Metadata management is an important aspect of today's enterprise information systems. Metadata management systems are growing from tool-specific repositories to enterprise-wide metadata repositories. In this context, one challenge is the management of the evolving metadata whose schema or meta-model itself may evolve, e.g., dynamically-added properties, which are often hard to predict upfront at the initial meta-model design time; another challenge is to organize the metadata by semantically-rich classification schemes. In this paper, we present a practical system which provides support for users to dynamically manage semantically-rich properties and classifications in the IBM WebSphere Metadata Server (MDS) by integrating an OWL ontology repository. To enable the smooth acceptance of Semantic Web technologies for developers of commercial software which must run 24 hours/day, 7 days/week, the system is designed to consist of integrated modeling paradigms, with an integrated query language and runtime repository. Specifically, we propose the modeling of dynamic properties on structured metadata as OWL properties and the modeling of classification schemes as OWL ontologies for metadata classification. We present a natural extension to OQL (Object Query Language)-like query language to embrace dynamic properties and metadata classification. We also observe that hybrid storage, i.e., horizontal tables for structured metadata and vertical triple tables for dynamic properties and classification, is suitable for the storage and query processing of co-existing structured metadata and semantic metadata. We believe that our study and experience are not specific to MDS, but are valuable for the community trying to apply Semantic Web technologies to the structured data management area.

1 Introduction

Metadata is pervasive in large enterprises and can be thought as the "DNA" of enterprise applications. The structured metadata is not only descriptive information about

A. Sheth et al. (Eds.): ISWC 2008, LNCS 5318, pp. 861–874, 2008.

data, but prescriptive information that constrains the structure and content of data. The metadata can be technical metadata, such as relational schemas, XML schemas, schema mappings, UML models and application interface specifications, and can be business metadata, such as business concepts, business rules and business process definitions in an enterprise. The metadata management tool (also known as repository) [15] is crucial for enterprise information management and has become the foundation of Data Warehousing [9], Enterprise Information Integration [8] and Service-Oriented Architecture (SOA).

Recent standards work on MOF/XMI [11] within OMG for metadata representation and interchange has been followed by many vendors, then MOF-based metadata repositories have become the mainstream in industry offerings 5]. Amongst these MOF-based metadata repositories, a common feature is the object-oriented storage strategy where Object-Relational Mapping functionality is used to generate physical schemas for the corresponding MOF meta-models and provide an object-oriented programming interface to the underlying database. One typical example is the IBM WebSphere MetaData Server (MDS), which is a unified metadata services infrastructure within a service-oriented architecture.

Within the enterprise-wide IT environment, metadata management has become more and more challenging because of rapidly-changing business requirements. Metadata repositories are growing from tool-specific, application-specific systems to enterprise-wide, asset-management and architecture decision support systems, in which metadata are shared and integrated across multiple applications or even third party tools [6]. While the metadata and their relationships dramatically grow, it is impossible to design a unified meta-model for all kinds of metadata with all possible attributes and relationships at design stage as the business requirements evolve. Therefore there is a requirement to dynamically add properties for classes in the registered metamodel. For example, after a WSDL meta-model which describes WSDL documents has been registered, a service administrator might add QoS (Quality of Service) metadata to the "WSDLService", such as the "responseTime". Another example is to dynamically build particular relationships across registered meta-models. After the metadata repository has run and collected entries for a period of time, a user needs to create a dynamic relationship "dependsOn" from the class "Activity" in the business process meta-model to the class "WSDLService" in the WSDL meta-model, which later can be used to enable traceability and impact analysis across those models. Moreover, semantic annotations are required to enrich the semantics of dynamic relationships, e.g. annotating "dependsOn" as "transitive". Based on these semantic annotations, ontology reasoning will be made to infer additional information which is not explicitly defined.

In metadata management, a classification scheme is used to classify the metadata objects, such as relational schemas and WSDL definitions in a metadata repository. Examples of classification schemes range from simple tags (keywords), thesauri, taxonomies to formal ontologies. With the growing volumes of metadata in different applications and users of metadata from various business units of enterprises, a flexible and semantic-rich classification scheme is needed to help different users to organize metadata from different viewpoints. This is because: (1)the classification scheme itself needs reasoning on the classifier hierarchy; (2)users need to define rich expressions on the classification scheme to declare dynamic classifiers, in addition, the expression can be defined on dynamic properties. For example, after the dynamic property "dependsOn" is declared, user can define "DataDependentService" as a new classifier, which contains the WSDL services that depend on a "DataService".

With the emergence of the Semantic Web [14], Web metadata markup languages, i.e. RDF (Resource Description Framework) [1] and OWL (Web Ontology Language) [3], have become W3C Recommendations. RDF originated from the W3C Metadata Activity, and is particularly intended for representing metadata about Web resources, such as the title, author, and modification date of a Web page [2]. The most important feature of the RDF data model is that it treats properties as first-class citizens and allows them to be attached to a class dynamically. OWL is a formal logic language to define the vocabularies in RDF documents. It is intended to represent structured metadata ranging from a simple taxonomy, a thesaurus, or to a formal ontology. In practice, OWL is an emerging standard to represent the classification schemes, because of its rich expressivity, formal semantics and reasoning capabilities. Therefore, it is natural to apply the Semantic Web technologies, namely RDF and OWL, to meet the emerging requirements of enterprise-wide metadata management.

In this paper, we propose a practical system which supports (OWL) ontology-based dynamic properties (i.e. dynamic attributes and dynamic relationships) and metadata classifications in the IBM WebSphere Metadata Server (MDS) by integrating an ontology repository. To enable smooth acceptance of Semantic Web technologies for developers of commercial software that must run continuously, the system is designed to consist of integrated modeling paradigm, query language and runtime repository. Our contributions of this work can be summarized as below.

(1) We use the semantic web languages, RDF and OWL in particular, to extend the MOF/XMI based metadata language with more flexibility and semantics. Dynamic attributes and relationships are represented as datatype properties and object properties in OWL respectively, and classification schemes are represented as OWL ontologies.

(2) We extend the object-oriented query language of MDS, XMQL, with additional classification query functions and dynamic property extensions. Users can issue hybrid queries over structured metadata, dynamic properties and semantic classifications simultaneously. The hybrid query follows the language convention of XMQL. MDS users do not need to learn another query language for RDF/OWL, i.e. SPARQL [12]. Ontology reasoning will automatically be conducted when answering queries.

(3) We develop a hybrid storage system by integrating a state-of-the-art ontology repository, namely the SOR Repository [16] to MDS, which is deployed in the same database with the repository of MDS. Specifically, information of dynamic properties and classification are stored in the SOR repository. Hybrid queries will be split to MDS queries and ontology queries, with the SQL fragments translated by MDS and SOR, and subsequently merged and executed by the relational database engine.

The rest of this paper is organized as follows. Section 2 describes the architecture of the integrated system. Section 3 presents the modeling and usage for ontology-based dynamic properties and classifications. Section 4 introduces the metadata query language XMQL with extensions and the query processing engine. Section 5 presents the use study on terminology services implementation. Section 6 discusses related work. Finally Section 7 summarizes the contributions of this paper.

2 System Architecture

To support ontology-based dynamic property and classification in metadata management, one approach is to store all the structured metadata and data on dynamic properties and classifications into an ontology repository. However, most of the current RDF databases (triple-stores) scale poorly since most queries require multiple self-joins on the vertical triples table [21], and the large volumes of structured metadata in enterprise are commonly stored in horizontal tables in a relational database. So we chose the hybrid approach that integrates the MDS with an ontology repository SOR [16], i.e., the structured metadata are still stored in the horizontal tables of MDS and the data on dynamic properties and classifications are stored in the vertical triples table of SOR. This system is called MDS++. Before presenting the detailed system architecture, we give a short introduction to both MDS and SOR.

2.1 WebSphere Metadata Server

The IBM WebSphere Metadata Server (MDS) is a unified metadata services infrastructure that's designed to ease metadata management, access, and sharing within a service-oriented architecture. MDS is available as part of the IBM Information Server platform. It provides metadata management services to products in the IIS platform and is additionally used as a common metadata services infrastructure for metadata products in other IBM software brands.

MDS was built with the Eclipse Modeling Framework[1] (EMF), which is a modeling framework and code generation facility for building tools and other applications based on a structured data model. In general, the design of this product is following the Object-Relational Mapping (ORM) paradigm to manage metadata objects, similar to the well-known ORM tool Hibernate[2]. When a meta-model is registered with the MDS at MDS build time, the CRUD (Create, Read, Update and Delete) API and the relational persistence schema will be automatically generated by EMF tools. Then MDS can support query and persistence of metadata that are instances of this meta-model.

2.2 SOR Repository

SOR (Semantic Object Repository) is a high performance OWL ontology repository built on relational databases. SOR translates OWL semantics into a set of rules which can be easily implemented using SQL statements on RDBMS. SOR supports the RDF data query language SPARQL. In SOR, a SPARQL query is first translated into a

---

[1] Eclipse Modeling Framework□http://www.eclipse.org/emf/
[2] Hibernate: Relational Persistence for Java and .NET. http://www.hibernate.org/

single SQL statement which is evaluated on both explicit assertions and inferred re-sults materialized in the persistent store, benefiting from decades of relational data-base optimization. The following two features of SOR are critical for the integrated system.

(1) The inferred facts are materialized when loading the data, which can improve the response time of queries.

(2) The SPARQL query is translated into a single SQL statement, which can be exe-cuted over the repository directly or embedded as a sub-query of other SQL queries.

2.3 System Architecture of MDS++

To enable smooth acceptance of Semantic Web technologies for developers of com-mercial software that must run continuously, the system is designed to consist of integrated modeling paradigm, query language and runtime repository. Fig. 1 shows the overall system architecture.

For the modeling paradigm, MDS was developed using Model-Driven Architecture and is based on EMF. The SOR architecture follows the model-driven approach for ontology engineering [18]. In this approach, the RDF and OWL is defined based on Ontology Definition Metamodel (ODM)[3]. The ODM specification is implemented by EODM[4], based on the EMF framework with additional inference and model transfor-mation functions. From Fig. 1, we can see that MDS and SOR provide a unified EMF view for users to access the metadata through query and CRUD API.

For the query language, we extend the XMQL query language of MDS to XMQL++, with additional classification query functions and dynamic property exten-sions. Users can simultaneously query over the structured metadata stored in MDS, the information about dynamic properties and classifications stored in SOR.

For the runtime repository, we make the ontology repository tightly-coupled with the MDS repository, i.e., the tables of ontology repository will be deployed in the same database with MDS and are visible to MDS. The two repositories are connected by the unified object identifier in MDS, which is also used as the internal identifier for the RDF resource in SOR. When a hybrid query is issued, the system will translate

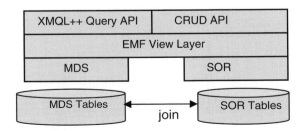

Fig. 1. The system architecture of MDS++

---

[3] Ontology Definition Metamodel Specification, www.omg.org/docs/ad/05-08-01.pdf
[4] EODM homepage on Eclipse, http://www.eclipse.org/modeling/mdt/?project=eodm

the query to one single SQL query which will access the tables of both repositories simultaneously. The advantage of this approach is that it provides high performance because queries are translated to SQL queries which can fully utilize the optimization provided by the underlying DBMS.

3 Dynamic Property and Classification Scheme

In this section, we will show how to model and use the ontology-based dynamic properties and ontology-based classification in MDS++.

3.1 Ontology-Based Dynamic Property

Compared to static EMF properties, the term "dynamic" implies that this kind of property can be declared and attached to the meta-model after the meta-model is registered with the repository. Dynamic properties can be divided into dynamic attributes and dynamic relationships. A dynamic attribute describes some kind of attribute of model elements. The domain of a dynamic attribute must be an EClass. The range of a dynamic attribute can be the supported data types in EMF, such as EString, EInt, etc. A dynamic relationship describes some kind of relationship between model elements. The domain and range of a dynamic relationship must be an EClass. The domain and range constraint on the dynamic properties is an important design consideration to guarantee that the dynamic properties are operated similarly to the static EMF properties.

Because the dynamic properties are treated as properties in OWL ontology, semantic annotations can be further added to enrich their semantics. Currently four kinds of semantic annotations borrowed from the OWL language can be supported: symmetric, transitive, inverseOf and subPropertyOf. After the dynamic properties are declared, the user can fill in the values for these properties. As an example, after the meta-model for WSDL documents, as shown in Fig. 2, is registered with the MDS, a service administrator can create a new dynamic attribute "businessFunction" and declare its domain as the class "WSDLService" in the meta-model and its range as Ecore data type "EString". Similarly, the service administrator can create the dynamic attribute

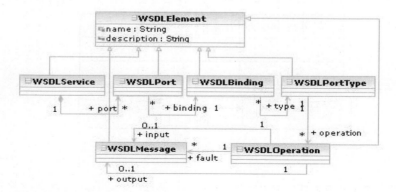

Fig. 2. A simplified meta-model for WSDL Document

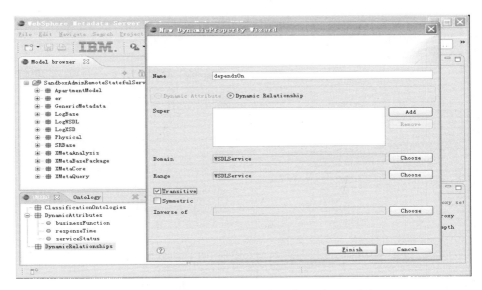

Fig. 3. An eclipse-based UI to declare dynamic properties

"responseTime" and "serviceStatus", and create the dynamic relationship "depend-sOn" to describe the dependency relationships among services, even if this relation-ship is not modeled in the original meta-model. In addition, the user can declare that the "dependsOn" relationship is transitive. The client tool UI to build the dynamic properties is shown in Fig. 3.

3.2 Ontology-Based Classification

In MDS++, a classification scheme is represented by an OWL ontology and the task of classification is supported by the built-in OWL reasoner of SOR. The classification scheme can be created in two ways. One is to load an existing OWL ontology as a classification scheme. In this way, only the named OWL classes inside the ontology will be taken as classifiers. Those anonymous OWL classes or expressions can not be classifiers, because they do not have URIs to get them identified. Another way is to build a classification scheme from scratch by using APIs. A user can create a classi-fier as a named class in the OWL ontology and setup the explicit hierarchy using the OWL construct "subClassOf". In addition, users can define the new classifier using OWL constructs supported in SOR: intersectionOf, and OWL restrictions on dynamic properties: someValuesFrom and hasValue.

After the classification scheme is built, a user can manually classify some metadata to classifiers as shown in Fig. 4. Then OWL reasoning can be applied to find the im-plicit classification information, eg. Find all metadata classified by a high-level class. In addition, based on the semantics of the OWL class expressions and restrictions, automatically classification can be made according to values of dynamic properties by the OWL reasoner. For example, the classifier "DataDependentService" can be de-fined as: *someValuesFrom(dependsOn, DataService)*. If a service s1 dependsOn s2 and s2 is a DataService, then s1 can be automatically classified as "DataDependent-Service" without explicit declaration.

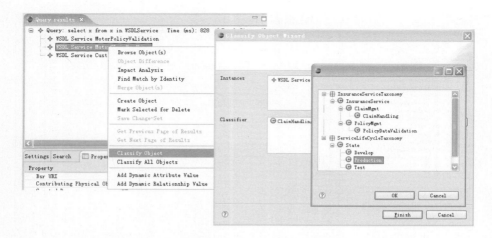

Fig. 4. Classify metadata object to classifier in classification scheme

4   XMQL++ Query Language and Processing

Queries written in the query language XMQL are the main access points to the meta-data stored in MDS. Thus, how to let MDS users access the dynamic properties and classification information though query was a central design problem for MDS++. One alternative we explored was to design a hybrid query language that embeds SPARQL query into the XMQL query language by a predefined function, similar to the *RDF_MATCH* function introduced by Oracle 10g [17]. This design choice was not accepted by the MDS product engineers because it is too complex for MDS users and tool developers to learn two kinds of query languages. Therefore, we designed the query language XMQL++ with a minor extension of XMQL to enable hybrid queries over static metadata, dynamic properties and semantic classifications simultaneously. We will illustrate the extension with simple query examples and introduce the query processing mechanism in this section. The formal specification of XMQL++ and technical details of query processing are omitted due to limited space.

4.1   A Short Introduction to XMQL

XMQL, a subset of ODMG's OQL[5], is the query language of WebSphere Metadata Server (MDS). It is a general purpose SQL-like declarative query language with spe-cial features designed for the efficient retrieval of instances stored in an MDS reposi-tory. A basic structure of an XMQL query is the *select-from-where* clause. The *select* clause defines the structure of the query result, which can be either a collection of objects, or a collection of rows with the selected attribute values. The *from* clause introduces the primary class extent(s) against which the query runs. A class extent is the set of all instances of a given class. A variable needs to be declared to iterate over the instances of each class being queried. The *where* clause introduces a predicate that

---

[5] http://www.odmg.org/

filters the instances of the collections being queried. The XMQL query adopts path expression to denote the traversal of a reference from one object to another, using the "->" operator or the access of an attribute using the "." operator. For example,

```
SELECT e.name
FROM  e IN Employee, p IN e->workForProject
WHERE e->workAt.country="US" AND
      p.name="NewHotel"
```

This query returns a set of rows containing each an employee name for the employees located in the United States that work for the project named "NewHotel".

4.2 Extensions for Dynamic Properties

Dynamic properties play the same role as ordinary EMF properties from user's view, though they are stored independently in separate repositories. To be compatible with the design principal of XMQL, the domain and range of any dynamic property must be explicitly declared in the design stage. Then XMQL query compiler can get the type information of dynamic properties when processing queries. For example, "y in x→*dependsOn*" will be of type "WSDLService".

The following example queries show how dynamic properties are used in XMQL++. The query to get all WSDL services's response time which are dependent on "service1" can be written as (the dynamic properties are in Italic fonts):

```
SELECT  x.name, x.responseTime FROM x IN WSDLService,
        y IN x→dependsOn WHERE y.name ="service1"
```

4.3 Extensions for Classification

Classification functions are provided to enable queries over classification information in XMQL. The basic classification functions are listed as below.

- Object[] *classifiedBy*(URI): It will return a list of objects that are classified to the class represented by the URI argument. This function can be used as an argument of an "IN" predicate in a "WHERE" clause of XMQL.
- URI[] *classifiers*(pathExpression): It will return a list of URIs which are classifiers of the objects denoted by the path expression.

The following example queries show how classification functions are used in XMQL++. The query to get all WSDL services which are classified to "ClaimMgmt" can be written as:

```
SELECT  x FROM x IN WSDLService,
    WHERE x IN classifiedBy("http://foo.org/#ClaimMgmt")
```

The query to get all classifiers for WSDLService "service1" can be written as:
```
SELECT classifiers(x) FROM x IN WSDLService
    WHERE x.name ="Service1"
```

4.4 XMQL++ Query Processing

In MDS, an XMQL query is translated to a single SQL query at compile time, and the results are returned after executing the SQL query by the underlying relational database engine. Similarly, a XMQL++ query which needs to access objects and ontology related information simultaneously will also be translated to one single SQL.

All invocations of dynamic properties and classification functions in XMQL++ will be translated to SPARQL queries during the query translation. Fig. 5 shows the high level workflow of XMQL++ query processing.

To translate a XMQL++ query into SQL query, the XMQL translator firstly will find out all the ontology related invocations. It then will pass these invocations to the *TripleQueryHandler*. The *TripleQueryHandler* will translate the invocations into SPARQL queries and submit them to *SOR SPARQL Engine*. The *SOR SPARQL Engine* will answer those queries by returning SQL sub queries whose results are the answers of the ontology invocations. Finally, the *XMQLTranslator* will merge these sub queries with the SQL query from O-R Mapping.

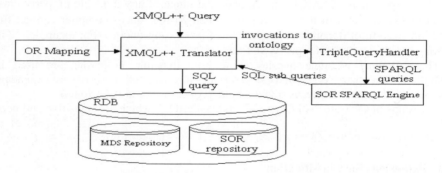

Fig. 5. XMQL++ query translation process

5 Use Study on China Healthcare Solution

Code system (terminology) is an important kind of metadata in healthcare applications, because consistent medical terminology is essential for the sharing, exchange and integration of healthcare information [20]. MDS++ is currently used in the IBM China Healthcare Solution to provide terminology services for healthcare industry. The terminology service is an implementation of HL7 Common Terminology Services (HL7 CTS)[6], which is an Application Programming Interface (API) specification that is intended to describe the basic functionality that will be needed by HL7 Version 3 software implementations to query and access terminological content. CTS API includes two parts of API: a message API that is specific to HL7 software, and vocabulary API, which is general to allow applications to query different terminologies in a consistent, well-defined fashion. The current implementation supports the vocabulary API based on the model[7] shown in Fig. 6.

---

[6] HL7 CTS Specification, http://informatics.mayo.edu/LexGrid/index.php?page=ctsspec
[7] This model does not support "relationship qualifiers" as appeared in the CTS API, because our supported code systems do not have "relationship qualifier" for any relationship.

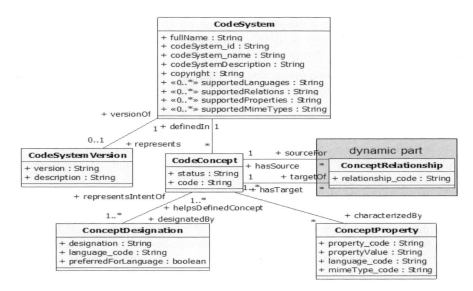

Fig. 6. The reference vocabulary model for CTS implementation

In this model, the class *ConceptRelationship* class represents binary relationships over the set of *CodedConcept*s defined in a single *CodeSystem*. Two of the supported relationships in CTS are "isSubtypeOf" and "isPartOf", which are transitive and need inference support for computing the transitive closure. To register this model to MDS++, we divide the model as two parts: one is static part except the class *ConceptRelationship*; another is the dynamic part that models every "relationship_code" as a dynamic relationship with both domain and range as *CodeConcept*. The instances of the static part will be stored as horizon tables in WMS and the instances of the dynamic part will be stored as vertical triple tables in SOR repository to leverage SOR's reasoning capability.

We have loaded four kinds of healthcare code systems into MDS++: LOINC, ICD-10, TCM(Terminology of Traditional Chinese Medicine) and SNOMED CT [19]. When loading SNOMED CT, we find that the model of SNOMED CT has additional information not captured by the static part of this vocabulary model: the *ConceptDesignation* has three categories: "fully-specified name", "preferred name" and "synonym". To fully keep the information in SNOMED CT, we add one dynamic attribute *designationCategory* with the domain as the class *ConceptDesignation* and the range as an enumeration of the three allowable string values.

We have implemented the CTS Vocabulary APIs by transforming the API calls to XMQL++ queries. For example, for the runtime API call to determine whether two concept codes (e.g. 25064002 and 279001004) in SNOMED CT are related via the relationship "isSubtypeOf", the corresponding XMQL++ query is:

```
SELECT COUNT(x) FROM x IN CodeConcept,z IN x->definedIn,
   y IN x->isSubtypeOf  WHERE  ( x.code="25064002" AND
   y.code="279001004" AND z.codeSystem_name="SNOMED CT" )
```

Based on the experience on implementation of CTS on MDS++, we observe that there are three features essential for the acceptance of the integration of an ontology repository to an industrial-strength metadata repository:

(1) Model flexibility: users/programmers can attach any attributes or relationships to classes in the registered model without redeployment of the model. For example, the *designationCategory* attribute can be added for the class *ConceptDesignation*.

(2) Reasoning support: programmers need not write additional code to handle the complex reasoning problems, such as the transitive closure computing.

(3) User-friendliness: the MDS users write similar APIs and use our modestly-extended query language to handle on dynamic properties and classifications. The underlying Semantic Web technologies are mostly transparent for the MDS users/programmers.

Another encouraging observation is that the hybrid storage pattern, i.e., horizontal tables for the part of structured metadata with no reasoning involved and vertical triple tables for dynamic properties or the part of metadata with reasoning support, is effective and efficient for the storage and query processing of complex metadata that needs reasoning support on part of the metamodel. In a broader sense, the hybrid storage pattern is promising to manage the co-existing structured metadata and semantic metadata.

However, we also observe that this approach can not leverage the full power of Semantic Web technologies, for example, the expressive SPARQL query constructs, such as DESCRIBE query, which is really applicable for the implementation of the CTS API to return a complete description of a coded concept, and also the named graph support, which is useable for modeling a code system as a named graph.

6 Related Work

Metadata management has a long history within the evolving discipline of data management, and the recent focus is on the integration of diverse metadata and MOF-based metadata repositories [5]. Some notable examples are MetaMatrix [10], NetBeans Metadata Repository[8]. As far as we know, both of these metadata repositories do not support the adding of semantic properties to classes in the meta-model at runtime. There is a simple model extension approach that every class has a list of <property, value> pairs each with a name and appropriate value. Another existing approach to add links or relationships between different models is using weaving models [7]. A weaving model conforms to a weaving meta-model, in which a link have multiple link ends that each holding a reference to a model element. For different application scenario, e.g, traceability and schema mapping, the meta-model must be extended and the links can have different semantics. Compared to both approaches, our proposed approach is not just about modeling dynamic properties, but is a system that has unified representation, storage, query and reasoning for dynamic properties. From the modeling perspecitve, OWL is much more expressive and has formal semantics for representation of various kinds of attributes and relationships. From the

---

[8] NetBeans Metadata Repository: http://mdr.netbeans.org/

usage perspective, our approach hides the complexity of management of the <property value> pairs or the weaving links. For the end-user, dynamic properties are accessed similar to the ordinary properites in object-oriented model.

In our approach, a traditional relational database is integrated with a RDF-based triple store to support metadata storage. This kind of hybrid storage is also supported in Oracle 10g [17]. It introduces a SQL table function *RDF_MATCH* that embeds a SPARQL query to query RDF data. Then users can write SQL queries with joins between variables in table function and columns in the relational data. In this implementation, end users have to understand all the complexity of SPARQL query language and write complicated joints by themselves. In contrast, our approach hides the complexity to end users by slightly extending the XMQL language. MDS++ engine will handle joints between static repository and ontology repository because they use the unified object identifier as the shared identifier. The MDS users and MDS tool programmers did not need to learn any query languages specific to RDF.

7 Conclusion

Metadata management systems are growing from tool-specific repositories to enterprise-wide metadata repositories, at the same time, more and more semantic-rich metadata like RDF and OWL ontologies are emerging. The structured metadata and semantic metadata would co-exist in enterprises, and their "marriage" would address some of the key challenges of enterprise-wide metadata management. Traditionally, these two communities, MOF/XMI based structured metadata and RDF/OWL based semantic metadata, have developed their own standards and tools that are incompatible. However, enterprises require a more comprehensive metadata management environment. We believe our work demonstrates a practical architecture moving repositories towards that direction. In our work, the practical system MDS++ provides support for users to manage semantic-rich properties and classifiers in the IBM WebSphere Metadata Server (MDS) by integrating the ontology repository SOR. Our experience can be summarized as this: to enable the smooth acceptance of Semantic Web technologies for commercial metadata product developers with existing products, the integrated system must consist of integrated modeling paradigm, query language and runtime repository. This is essential for the successful application of MDS++ for China Healthcare Solution.

References

1. Manola, F., Miller, E.: RDF primer. W3C recommendation (February 2004)
2. Brickley, D., Guha, R.V.: RDF vocabulary description language 1.0: RDF schema. W3C recommendation (February 2004)
3. Smith, M.K., Welty, C., McGuinness, D.L.: OWL web ontology language guide. W3C recommendation (February 2004)
4. Baker, N.L., Le Goff, J.-M.: Meta Object Facilities and their Role in Distributed Information Management Systems. In: Proc. of the EPS ICALEPCS 1997 (1997)
5. Sen, A.: Metadata management: past, present and future. Decision Support Systems 37(1) (2004)

6. Shankaranarayanan, G., Even, A.: The Metadata Enigma. CACM 49(2) (2006)
7. Stefanov, V., List, B.: Business Metadata for the Data Warehouse: Weaving Enterprise Goals and Multidimensional Models. In: Models for Enterprise Computing 2006 - International Workshop at EDOC 2006 (2006)
8. Thangarathinam, T., Wyant, G., Gibson, J., Simpson, J.: Metadata management: the foundation for enterprise information integration. Intel. Technology Journal 8(4) (2004)
9. Muller, R., Stohr, T., Rahm, E.: An integrative and uniform model for metadata management in data warehousing environments. In: International Workshop on Design and Management of Data Warehouses, pp. 12–28 (1999)
10. Hauch, R., Miller, A., Cardwell, R.: Information intelligence: metadata for information discovery, access, and integration. In: The 2005 ACM SIGMOD, pp. 793–798 (2005)
11. Object Management Group: Meta Object Facility Specification, version 1.4, OMG document formal/02-04-03
12. Prud'hommeaux, E., Seaborne, A.: SPARQL Query Language for RDF. W3C Recommendation (January 15, 2008)
13. Zachman, J.A.: A Framework for Information Systems Architecture. IBM Systems Journal 26(3) (1987)
14. Berners-Lee, T., Handler, J., Lassila, O.: The semantic web. Scientific American 184(5), 34–43 (2001)
15. Marco, D.: Building and Managing the Meta Data Repository: A Full Lifecycle Guide. John Wiley & Sons, Inc., New York (2000)
16. Ma, L., Jing, L., Wang, C., Cao, F., Pan, Y.: Effective and Efficient Semantic Web Data Management over DB2. In: Proc. of the 28th ACM SIGMOD Conference (2008)
17. Chong, E.I., Das, S., Eadon, G., Srinivasan, J.: An Efficient SQL-based RDF Querying Scheme. In: Proc. of VLDB 2005, pp. 1216–1227 (2005)
18. Pan, Y., Xie, G., Ma, L., Yang, Y., Qiu, Z., Lee, J.: Model-driven ontology engineering. Journal of Data Semantics VII, 57–78 (2006)
19. SNOMED Clinical Terms. Northfield, IL: College of American Pathologists (2007)
20. Rector, A.L., Solomon, W.D., Nowlan, W.A., et al.: A Terminology Server for Medical Language and Medical Information Systems. Meth. Inform. Med. 34(1-2), 147–157 (1995)
21. Abadi, D., Marcus, A., et al.: Scalable Semantic Web Data Management Using Vertical Partitioning. In: Proc. of VLDB 2007, pp. 411–422 (2007)

Towards a Multimedia Content Marketplace Implementation Based on Triplespaces

David de Francisco[1], Lyndon JB Nixon[2], and Germán Toro del Valle[1]

[1] Telefónica Research and Development,
Valladolid, Spain
{davidfr,gtv}@tid.es
[2] Free University of Berlin
Berlin, Germany
nixon@inf.fu-berlin.de

Abstract. A Multimedia Content Marketplace can support innovative business models in the telecommunication sector. This marketplace has a strong need for semantics, co-ordination and a service-oriented architecture. Triple Space Computing is an emerging semantic co-ordination paradigm for Web services, for which the marketplace is an ideal implementation scenario. This paper introduces the developed Triple Space platform and our planned evaluation of its value to our telecommunication scenario.

Keywords: telecommunications, co-ordination, marketplace, semantics, triplespaces.

1 Introduction

The Internet, and more specifically Web technologies, have brought about new technological means to explore new business opportunities. In the telecommunications sector, this has been seen as a clear challenge to diversify market coverage of companies. This extended market coverage, as well as the technology adoption of economic models such as specialization and outsourcing (through the Service Oriented Computing paradigm [1]), has meant that companies have began to collaborate to offer more complex and attractive services to their customers (B2B). These collaborative scenarios have led to new technical challenges such as data and process integration of heterogeneous sources (EAI [2]). Therefore, means for transparent communication and integration of data and processes to tackle the inherent requirements of such scenarios have to be defined.

In a previous work [3], Digital Asset Management (DAM) [4] was presented as a strategic collaborative scenario for telecommunication companies. This work identified some key requirements for the communication infrastructure that could be summarized as: (1) arbitrary number of transparent parties involved, (2) the integration of heterogenous data and message formats between them, (3) persistency of the information, (4) reliable and secure access to the information, and last but not least and (5) the support of agile business transactions, where

A. Sheth et al. (Eds.): ISWC 2008, LNCS 5318, pp. 875–888, 2008.

the communication infrastructure could handle the information in an intelligent way, acting consequently.

The aforementioned requirements demand more flexible communication and coordination paradigms beyond the ones used today, such as CORBA or Web Services, characterized by their excessive strictness to the defined interfaces of each participant. One of most innovative communication and coordination paradigms existent today is the one proposed by Triple Space Computing[1] which efficiently combines the highly distributed coordination techniques of Triple Space Computing as a distributed shared memory [5] with the full power of Semantics [6]. Triple Space Computing exposes a simple and standard interface based on the Linda co-ordination language to every client, making communication flexible.

This paradigm has been chosen to explore its applicability to solve the inherent complexity of the DAM scenario, as well as to envision new services and capabilities to be offered to the potential users of this increasingly demanded area of the telecommunication industry. On the other hand, the DAM scenario is also proposed as a validation of the Triple Space Computing paradigm itself. The business scenario is significant as: (1) it acts as a proof of concept of the proposed paradigm, (2) it serves as a way to enable the business exploitation of collaborative scenarios and, last but not least, (3) an evaluation of the designed infrastructure can serve to engender new requirements.

With these objectives in mind, the paper starts outlining the Triple Space Computing paradigm in Section 2 and continues with a presentation of the DAM scenario proposed in Section 3. The definition of the scenario will make emphasis on the relevance to the telecommunication industry and the motivation of choosing Triplespaces as the main back-end technology. Section 4 details the design and implementation of the proposed scenario, whereas Section 5 stresses the evaluation plan to demonstrate the value of the paradigm for collaborative scenarios in the telecommunication sector, stressing the business significance of the evaluation through exploitation models. Conclusions and further work close the article in Section 6.

2   Triple Space Computing and Triplespaces

Triple Space Computing [7] [5] is a novel communication and co-ordination paradigm based on the publication of semantic information (RDF triples [8]). This paradigm combines Tuple Spaces using Linda coordination [9] with Semantic Web principles [6] to provide a time, location, reference and schema decoupling to the communication among any kind of applications, with special emphasis on Web services and semantic Web services [10]. The aforementioned de-coupling aims at the design of asynchronous and transparent communications which permit semantic systems which are more flexible, fault tolerant and unaware of heterogeneity.

---

[1] In this paper we focus on one particular realization of this paradigm as is the one developed in the context of the TripCom FP6 EU project (http://www.tripcom.org)

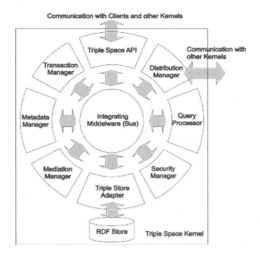

Fig. 1. Triple Space Kernel

Triple Space is the implementation of the Triple Space Computing paradigm. Triple Space is realised as distributed kernels which each host sets of spaces and which are connected by a combination of a Client-Server and a Peer-to-Peer network overlay infrastructure (the former used for direct kernel-to-kernel communication, the latter for network organisation and data look-up).

In Figure 1 the logical view of a single Triple Space kernel [11] with both kernel-internal components (circled around the integrating middleware) and kernel-external clients and other kernels (upper part of the figure) and services that may be connected to it is presented. The components that form a Triple Space kernel communicate over a kernel-internal bus system which is implemented using a tuplespace that allows all components to communicate with each other in a decoupled and co-ordinated manner.

Persistent storage of data at Triple Space kernels is provided through RDF stores which are connected to a kernel by the *Triple Store Adapter* component. The Triple Store adapter (i) abstracts from different RDF stores and their APIs and (ii) enables transparent distribution of data across a number of physical RDF stores. Apart from hosting triplespace data, kernel components can use the RDF store through the Triple Store adapter to persist configuration or runtime data. Security policies are enforced by the *Security Manager* that ensures that all data exchanges across kernel boundaries adhere to specified security policies. The *Metadata Manager* manages and provides access to a kernel's knowledge about itself (e.g. subspaces, triple access statistics) and about the global triplespace infrastructure (e.g. other kernels, clustering and routing information) in the form of semantic (RDF) metadata. The *Query Processor* is responsible for decomposing a query to parts that are satisfiable by the local data store and to parts that must be forwarded to other kernels in order to fulfil the query in its entirety, hence supporting distributed query answering in the Triple Space. The

Distribution Manager connects a single kernel to the global space infrastructure. It implements lookup functionality to find data in other spaces based on a distributed index storage system and carries out communication with remote kernels.

3   A Multimedia Content Marketplace Based on Triplespaces

This section describes the DAM contextualization, motivating why this business model is very promising for the telecommunication sector.A Multimedia Content Marketplace as a specific realization of the more general DAM scenario is presented, pointing out its functional description. Finally, the choice of the Triple Space Computing for its implementation is argued.

3.1   Contextualization

Multimedia content is a collaborative business in which several roles coexist, resulting in a very fragmented value chain (see Figure 2). *Content providers* are entities which own multimedia content (e.g. film producers). This content is usually not commercialized by owners, but by *content brokers* (e.g. film distributors). Content is exploited by *service providers* (e.g. a television company), which offer it to their *customers*. The content is delivered to customers by a *content distributor* (e.g. a telecommunication operator), which can not only deliver it, but also to provide some added value (e.g. QoS features). Content distribution often has to take into account Digital Rights Management (DRM [12]), which is provided by a *DRM provider*.

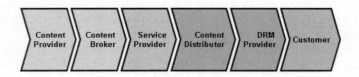

Fig. 2. Multimedia Content Value Chain

For a telecommunication company, such as Telefónica, multimedia content offers many business opportunities. The most obvious is the content distribution, where network and added-value features are already being provided by the company. Furthermore, service provision was handled by the company through the Imagenio product [2].

In the use case we will present below, we focus on the design of services based on multimedia content. The business context presented above is simplified for the business case definition. We consider a content provider as an entity which

---

[2] See http://www.telefonica.es/tol/imagenio.html (in Spanish)

offers multimedia content (being either producer or broker). A service provider is an entity which wants to offer services to their customers, making use of content offered, int he business case. Finally, a customer subscribes to these services and provide feedback about them. These roles will interact through a marketplace infrastructure provided by the marketplace owner. The next section describes the functionality our business case provides.

3.2   Functional Description

The business roles for the business case defined above will interact through a multimedia content marketplace. Telefónica, as a telecommunication operator, will play the marketplace owner role in this business case. Service and content providers will make businesses through the marketplace application. As a result, Telefónica will act as a business mediator among actors. In addition, customers can find, subscribe to and provide feedback about these services by interacting with the marketplace as well. Figure 3 shows an UML business case diagram where the functionality of the marketplace application is summarized.

Content providers can create content catalogues where multimedia content is offered. These catalogues can be consulted by each actor of the business case. Service providers look for content providers which can supply multimedia content, with the aim of assembling services. These searches are published as auctions. The marketplace automatically invites content providers which can supply the content required by auctions. Those which accept the auction invitation compete by offering content they own under certain conditions (e.g. price, QoS, etc.). These conditions constitute a binding bid, which means that bids can neither be withdrawn nor modified after emitted. The auction creator selects the most suitable bid after the auction concludes. The offer is then formalized in a contract

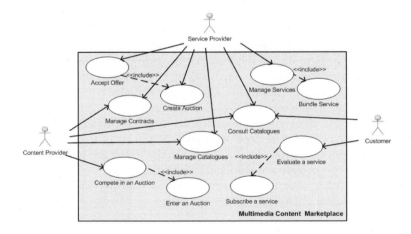

Fig. 3. DAM Marketplace UML business case Diagram

between the service and the content provider. This contract can be managed through the marketplace.

Contracts define the business terms of a service (a service can be defined by several contracts). Services are collaboratively carried out by contract(s) signers, and offered to customers. Customers can look for the most suitable services through service catalogues, and subscribe to them. As customers make use of services, they can evaluate them as well as the multimedia content they offer. Service and content evaluations can be done through the marketplace infrastructure. These evaluations provide valuable feedback to service providers, allowing a better service re-design which takes customer preferences into account.

It's important to note that the aim of the marketplace is not to be a passive business mediator, but instead to make a smart use of the heterogenous information stored. This implies the possibility of actively participate in business transactions (e.g. starting an auction if a content provider resigns). Additionally, the marketplace must mediate between the different information sources.

3.3   Suitability of Triple Space Computing

The current situation of the DAM solutions market is characterized by the existent of highly isolated and proprietary solutions of which ADAM, Apple's Final Cut Server, BrighTech's MediaBeacon, Canto's Cumulus, Chuckwalla, Nstein, OpenEdit, Phrasea, Wave Corp.' MediaBank, Widen and Xinet's WebNative Technology constitutes only a short list[3]. The isolated and proprietary nature of the existent solutions force the involved parties to anchor to some of them and to multiply efforts to support distinct platforms to offer their services.

Although some efforts have been made in the DAM standardization area like, for example, Adobe's Extensible Metadata Platform (XMP)[4], it is clear that there still exists a long way to get a universal DAM solution which efficiently deals with the inherent heterogeneity of these scenarios. The suitability of Triplespaces to implement a marketplace pattern has been motivated in [13]. Our business case is backed on this pattern, therefore we motivate now the applicability of Triplespace technology to its implementation.

As described in Section 2, Triplespaces provide de-coupled communication and co-ordination capabilities. Asynchronous communication is needed in the implementation of our marketplace, since actors might be able to interact in an auction or sign a contract without being online at the same time. Location de-coupling enables the remote communication among actors in a distributed fashion making it possible the decentralized deployment of the marketplace (e.g. with no central point of control or failure). Finally, reference decoupling ensures the transparent and private communication among actors, which do not need to explicitly know other actors to perform business transactions with them. The support of semantics ensure the schema decoupling. Business transactions and

---

[3] Refer to http://www.cmswatch.com/DAM/Vendors/ for a more detailed list and vendors' URLs.

[4] See http://www.adobe.com/products/xmp/

information retrieval can not only infer information not explicitly stated, but also mediate between heterogeneous data schemas used by actors.

In short, Triple Space Computing seems suitable for the implementation of a highly distributed worldwide DAM solution. The aforementioned capabilities ensure a rapidly implementation of a highly functional and open DAM scenario with a minimum consumption of resources, as detailed in the following section.

4   Implementation of the Multimedia Content Marketplace

In this section we detail the design and implementation of the Multimedia Content Marketplace based on Triplespaces. We first stress out the requirements of our system. Then, we describe the high-level architecture of our proposed solution. This architecture will make use of Triplespaces to couple with the identified requirements. An example of how Triplespace functionality is used to address the auction management logic is given. Finally, we describe the ontologies used in our solution, and motivate how semantics tackle the integration and interoperability issues that occur in the scenario.

4.1   Requirements of the Proposed Solution

In this section we list the functional and non-functional requirements needed for the business case implementation. These requirements translate the key requirement identified in [3]. Providing a thorough requirement analysis is out of the scope of this article, but a more detailed analysis can be consulted in [14].

From a functional perspective, publication and retrieval of the semantic information MUST be available. These functionalities SHOULD be blocking, so that information can be queried before it's already published. A destructive consumption of the information SHOULD be also provided in order to allow the information updating. Information MUST be able to be retrieved from different sources via joint operations. Filtering conditions MUST be provided as well. Subscription and notification mechanisms MUST be available in order to coordinate business interactions. The publication of these business transactions MUST be atomic. Consistency of the information published in these cases SHOULD be provided. Actors MUST be able to manage their own information, defining where it should be published and who could access to it.

The information MUST be available and correct for the auction, contract and user management functionalities. It SHOULD be available and correct for catalogue querying. All the information published within the marketplace MUST be kept consistent. Completeness of the information retrieval MUST be ensured to a local data source, and MAY be extended to a distributed data source. Security policies MUST be updatable by actors in order to grant access rights to the information.

From the knowledge specification point of view, the system MUST be able to formally define a taxonomy which enables the classification of multimedia content. Properties and/or rules MUST be defined in order to formalize complex relationships between multimedia assets, actors, and business transactions. All this knowledge MUST be retrieved under the retrieval requirements stated before. The system SHOULD be able to couple with heterogenous information sources, and to actively moderate business transactions by understanding the messages exchanged by actors.

4.2   High-Level Architecture

Figure 4 shows the architecture of the multimedia content marketplace. The architecture is composed of three main components. First, a Front-End (FE) component, which exposes the interfaces used by the distinct actors to interact with the marketplace. Second, a Back-End (BE) component, which abstracts the communication between actors and the Triplespace infrastructure by exposing a Web service interface. The Back-End offers the marketplace functionality (i.e: catalogue creation and queries, etc.) to actors. Finally, the underlying Triplespace infrastructure provides with the storage, communication and coordination and semantic functionality used to implement the marketplace logic.

This architecture reuses the data schemas (XML Schema[15]) being employed in another marketplace implementation. A set of ontologies defines the knowledge needed in this scenario, which will be handled by the Triplespace infrastructure. These ontologies are encoded in OWL[16] (currently OWL Lite version),and will covered in detail later. The marketplace logic is implemented in the Back-End. This logic covers the catalogue, services, contracts, users and auctions management described in Section 3. The Back-End component abstracts the complexity related RDF triples handling and the Triplespace functionality invocation .

The modular nature of the architecture allows actors to interact with the marketplace with three different schemes. First, the usage of the Front-End provided

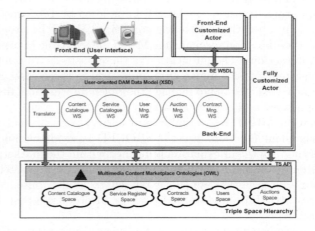

Fig. 4. Multimedia Content Marketplace Architecture

by the marketplace implementation. Second, the invocation of the Web services exposed by the Back-End, allowing actors to develop a custom interface. Finally, the direct access to the Triplespace infrastructure. This access would would be tied to the DAM marketplace data model (the set of ontologies), and the interfaces defined by the Triplespace infrastructure. It is important to point out that the most important functionality resides in the Triplespace infrastructure. Although the figure depicts five spaces, there can be several distributed kernels supporting this logical view.

The architecture described has been defined following two main principles. First, to narrow the logic between the actors and the communication infrastructure in order to fully test the capabilities of the latter. Second, to keep the implementation of the actors as flexible as possible. First principle is achieved by moving almost all business logic to the actors, allowing them to directly interact with the Triplespace infrastructure. Therefore there is no need of neither a central access point to the communication infrastructure nor external synchronization with other agents, following peer-to-peer principles[17]. Second principle has been achieved by the modular definition of the reference components implemented. The marketplace owner can decide the way these components are deployed.

4.3   Integration with the Triplespace Infrastructure

In this section we describe the logic of the auction management in terms of Triplespaces primitives as an example of the marketplace implementation. This sample has been chosen as it better shows the communication and co-ordination functionalities, using Triplespaces primitives.

The Figure 5 depicts the behavior of an auction life cycle in terms of interactions between actor. A *service provider* is looking for content, and starts an auction to get a provider for this content. *Auction participants* will join the auction and perform bids, which might be validated by the auction creator. An *auction creation space* is used to publish all auction calls. Each action is carried in a separate *auction space*. The hierarchy and access policies of the spaces can be seen in the right part of the figure as well.

The storyboard of these interactions is the following:

1. Potential auction participants are subscribed to the auction creation space in order to get notified if any content being searched in any auction can be provided by themselves.
2. A service provider publishes a new auction arrangement in the auction creation space, in order to get a content.
3. Auction participants subscribed to the space whose subscriptions are matched are notified of the new auction.
4. Auction participants interested in the auction can subscribe by publishing a request in the auction creation space.
5. The service provider validates all subscriptions received from auction participants (i.e, checking internal black lists).

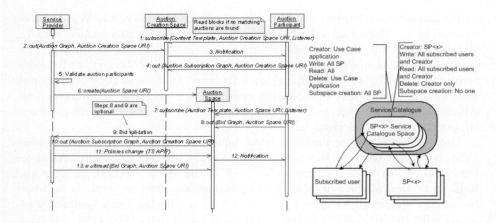

Fig. 5. Auction Management Interactions and Space Hierarchy

6. The service provider creates an Auction Space allowing all validated auction participants to write bids (no bid can be modified or deleted), as well as everyone to consult existing bids in the auction.
7. Subscription to auction management messages to get notified to things like winning bid change or auction end.
8. An auction participant writes a new bid.
9. The bid is validated by the service provider following its own validation logic.
10. If the bid is rejected, it is deleted from the Auction Space by the service provider.
11. Once the auction ends, policies of the auction space are changed, avoiding new bids to be published.
12. Auction participants are notified about the policies change.
13. The service provider reads all the bids published in the space in order to evaluate them.

4.4   Semantics of the Marketplace

Two of the core requirements of the multimedia content marketplace are the heterogeneous information integration and the support for an smart information handling, which can support an intelligent business mediation. The marketplace implementation uses a set of ontologies[5] depicted in the Figure 6, with the purpose of providing a common vocabulary to all actors.

Multimedia content is defined in the *multimedia content ontology*. This ontology defines a taxonomy of assets. Multimedia content is usually retrieved using meta-data annotations. The formalization of semantic descriptions permits the classification of individual assets and the retrieval of inferred information as well. Next code defines a sample property using a TRREE rule:

---

[5] See http://www.tripcom.org/ontologies/dam.php

Fig. 6. Ontology Structure of the Marketplace Implementation

```
x daml:isSimilarToFilm y . [Constraint x != y]
x daml:hasActor a .
y daml:hasActor a .
x daml:hasDirector a .
y daml:hasDirector a .
------------------------
x daml:isSimilarToFilm y .
```

This rule defines that if two film were directed by the same director and some actor played a role in both of them are similar (i.e: The Godfather I and II). As a content provider adds "The Godfather II" asset to the marketplace, the system can infer this information without being explicitly declared. This results in being able to answer next SPARQL query (prefixes are omitted):

```
SELECT ?film_name
WHERE { ?film1 rdf:type daml:Film .
    ?film2 rdf:type dam:Film .
    ?film1 daml:filmIsSimilarTo ?film2 .
    ?film1 daml:hasName "The Godfather" .
    ?film2 daml:hasName ?film_name }
```

Previous query answers the question of "Please, give me titles of films that could be similar to *The Godfather*, which I liked so much". This useful information can't be derived from a non-semantically model without having to explicitly define each film similarity, which is not feasible in a real content catalogue.

The *business transactions ontology* imports concepts from an EDIFACT ontology [18], with the aim of solving the heterogeneity problems that arise when performing a business transaction. Finally, the *marketplace ontology* defines the knowledge related to the marketplace logic, such as actors and auctions related knowledge. Its objective is to facilitate the interoperability between actors and the semi-automatic handling of auctions (i.e: automatically starting an auction when a service is terminated by a content provider).

5   Evaluation of the Marketplace Implementation

In this section we outline the evaluation plan and consequent experiments to measure the suitability of Triplespaces to implement the multimedia content

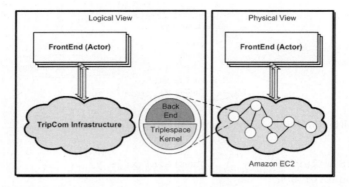

Fig. 7. Tests Architecture to Evaluate the Marketplace Implementation

marketplace. Suitability will be measured in terms of scalability and performance. Both factors are crucial from a business success perspective. Scalability ensures the deployment of a commercial scenario, with the objective of serving as many actors and customers as possible. Performance is crucial in terms of functional capabilities and response time.

We have considered some average figures based on the experiences of current services (e.g. the number of contents offered by a content provider in a real DAM application) in order to define some indicators. Indicators have been divided into scalability indicators (e.g. number of actors interacting within the marketplace) and performance indicators (e.g. content catalogue response time). Success factors have been defined for each indicator, defining expected results to be provided by Triplespaces.

These indicators will be measured through the execution of some experimental tests, whose structure is depicted in Figure 7. The tests will be deployed using the Amazon EC2 service [6]. Each node of the distributed infrastructure is a TripCom kernel [11]. This kernel will offer a marketplace Back-End implementation on its top.

Each experiment will vary the range of users and the number of nodes, and will measure each of the indicators defined. The combination of the aforementioned three coordinates is expected to be classified either as a suitable implementation or not for each individual experiment. The objective of the planned evaluation is to derive conclusions about the suitability of Triplespaces to implement a real scenario, given the expected size and performance with an affordable number of nodes. This can be seen as detecting if Triplespaces cover the technical and economic requirements to implement this real scenario.

6   Conclusions and Future Work

Multimedia content offer a lot of business opportunities that are worth to be explored by a telecommunication company. The collaborative nature of services

---

[6] See http://aws.amazon.com/ec2

based on multimedia content makes transparent communication and integration of data and processes needed. Triple Space Computing is a novel paradigm which efficiently combines the highly distributed coordination techniques of Space Based Computing with semantics. Triplespaces provide an implementation of this paradigm based on tuple spaces, extending the Linda coordination model so that RDF triples can be handled and Web services are supported.

In this article the authors have presented a business case centered on a multimedia content marketplace. Requirements which have arisen in the business case design have been presented. These requirements have motivated the use of Triplespaces and therefore semantics as the underlying communication infrastructure and knowledge representation technology respectively. While coordination capabilities of the Triplespaces have provided a de-coupled and easier implementation, semantics have been crucial to effectively address the heterogeneity of actors and the support of agile business transactions.

The ultimate objective of the prototypical implementation presented in this article is to validate Triplespace suitability for a commercial implementation of a multimedia content marketplace. In this sense, an evaluation plan has been outlined. The authors will carry out this plan with the aim of extracting meaningful conclusions about the scalability and performance indicators achieved by Triplespaces in this context.

Acknowledgements

This work is partially supported by EU funding under the TripCom project (FP6 - 027324).

References

1. Papazoglou, M.P.: Service oriented computing: Concepts, characteristics and directions. In: WISE 2003: Proceedings of the Fourth International Conference on Web Information Systems Engineering, Washington, DC, USA, vol. 3. IEEE Computer Society, Los Alamitos (2003)
2. Linthicum, D.S.: Enterprise application integration. Addison-Wesley Longman Ltd., Essex (2000)
3. de Francisco, D., Pérez, N., Foxvog, D., Harth, A., Martin, D., Wutke, D., Paslaru Bontas Simperl, E.: Towards a digital content services design based on triple space. In: Abramowicz, W. (ed.) BIS 2007. LNCS, vol. 4439, pp. 163–179. Springer, Heidelberg (2007)
4. Austerberry, D.: Digital Asset Management, 2nd edn. Focal Press (2006)
5. Riemer, J., Martín-Recuerda, F., Ding, Y., Murth, M., Sapkota, B., Krummenacher, R., Shafiq, M.O., Fensel, D., Kühn, E.: Triple space computing: Adding semantics to space-based computing. In: Mizoguchi, R., Shi, Z.-Z., Giunchiglia, F. (eds.) ASWC 2006. LNCS, vol. 4185, pp. 300–306. Springer, Heidelberg (2006)
6. Lee, B.T., Hendler, J., Lassila, O.: The semantic web. Scientific American (May 2001)

7. Fensel, D.: Triple-Space Computing: Semantic Web Services Based on Persistent Publication of Information. In: Aagesen, F.A., Anutariya, C., Wuwongse, V. (eds.) INTELLCOMM 2004. LNCS, vol. 3283, pp. 43–53. Springer, Heidelberg (2004)
8. W3C: Rdf specification (February 2004), http://www.w3.org/TR/rdf-primer/
9. Gelernter, D.: Generative communication in linda. ACM Trans. Program. Lang. Syst. 7(1), 80–112 (1985)
10. Fensel, D., Lausen, H., de Bruijn, J., Stollberg, M., Roman, D., Polleres, A.: Enabling Semantic Web Services: The Web Service Modeling Ontology. Springer, Heidelberg (2007)
11. Martin, D., de Francisco, D., Krummenacher, R., Moritsch, H., Wutke, D.: An architecture for a qos-aware application integration middleware. In: Proceedings of the 11th International Conference on Business Information Systems. Lecture Notes in Business Information Processing. Springer, Heidelberg (2008)
12. Uhl, A., Pommer, A.: Image And Video Encryption - From Digital Rights Management To Secured Personal Communication. Springer, Heidelberg (2005)
13. de Francisco, D., Elicegui, J.M., Martin, D., Wutke, D., Murth, M.: Using triple spaces to implement a marketplace pattern. In: Procceedings of the 1st Space Based Computing as Semantic Middleware for Enterprise Application Integration Workshop in 1st European Semantic Technology Conference, Viena, Austria (May 2007), SpaceBasedComputing.org
14. de Francisco Marcos, D., Martin, D., Scheibler, T., Wutke, D., Harth, A., Murth, M., Simperl, E.P.B.: Tripcom requirements analysis and architecture profile for eai applications (April 2007), http://tripcom.org/docs/del/D8B.1.pdf
15. W3C: W3c xml schema definition language (xsdl) 1.1 part 1: Structures (August 2007), http://www.w3.org/TR/xmlschema11-1/
16. W3C: Owl web ontology language overview (February 2004), http://www.w3.org/TR/owl-features/
17. Minar, N., Hedlund, M., Shirky, C., O'Reilly, T., et al.: Peer to Peer: Harnessing the Power of Disruptive Technologies. O'Reilly and Associates Inc., USA (2001)
18. Foxvog, D., Bussler, C.: Ontologizing edi semantics. In: Roddick, J.F., Benjamins, V.R., Si-said Cherfi, S., Chiang, R., Claramunt, C., Elmasri, R.A., Grandi, F., Han, H., Hepp, M., Lytras, M., Mišić, V.B., Poels, G., Song, I.-Y., Trujillo, J., Vangenot, C. (eds.) ER Workshops 2006. LNCS, vol. 4231, pp. 301–311. Springer, Heidelberg (2006)

Semantic Enrichment of Folksonomy Tagspaces

Sofia Angeletou

Knowledge Media Institute (KMi)
The Open University, Milton Keynes, United Kingdom
S.Angeletou@open.ac.uk

Abstract. The usability and the strong social dimension of the Web2.0 applications has encouraged users to create, annotate and share their content thus leading to a rich and content-intensive Web. Despite that, the Web2.0 content lacks the explicit semantics that would allow it to be used in large-scale intelligent applications. At the same time the advances in Semantic Web technologies imply a promising potential for intelligent applications capable to integrate distributed content and knowledge from various heterogeneous resources. We present FLOR a tool that performs semantic enrichment of folksonomy tagspaces by exploiting online ontologies, thesauri and other knowledge sources.

1 Background and Research Problem

The large-scale content annotation and metadata generation has been realised as Web2.0 applications have become very popular. Despite that, **Web2.0 lacks the explicit semantics** that would allow the content to be used in large-scale intelligent applications. At the same time the advances in Semantic Web technologies imply a promising potential for intelligent applications capable to integrate distributed content and knowledge from various heterogeneous resources. There is significant discussion that the combination of Semantic Web and Web2.0 will lead to an interoperable, intelligent Web ([4,8,10]). The goal of this work is to identify methods for the **automatic semantic enrichment** of Web2.0 generated content with a focus on folksonomies.

Folksonomies are Web2.0 systems whose basic elements are **users**, **resources** and **tags**. A resource is a content object depending on the folksonomy (a photo in Flickr[1], a bookmark in Del.icio.us[2], a video in YouTube[3] and so on). A tag can be any sequence of characters a user can attach to a resource. However the semantics of tags, and as a result the semantics of the resources, are not known and are not explicitly stated. This often hampers the resource retrieval within the individual system as well as the integration of resources in cross platform applications. The goal of this work is **to identify the meaning of the tags** attached to a

---

[1] http://www.Flickr.com

[2] http://del.icio.us

[3] http://www.youtube.com

A. Sheth et al. (Eds.): ISWC 2008, LNCS 5318, pp. 889–894, 2008.

resource, **to obtain the formal semantics that correspond to these tags**
and **to attach the formal semantics to the resource**, automatically creating
in that way a semantic layer on top of folksonomy tagspaces. The realisation of
the above raises the following research questions.

- **How can folksonomies' tagspaces be semantically enriched automatically?** This research question can be further analysed into the following
  questions. How to discover automatically the meaning of tags based on their
  context? How can the Semantic Web be exploited for the semantic enrichment of the tags and what other resources are required in case the Semantic
  Web falls short of that task?
- **How can the enriched tagspaces be evaluated in terms of content
  retrieval** against the non enriched tagspaces? What performance measures
  should be established to measure content retrieval in folksonomies before
  and after the semantic enrichment?

In the following we describe the existing research on folksonomies and present
our work.

2   Related Work

Folksonomy research has focused on comprehending the inherent characteristics
of tagging and exploring the semantics that emerge from it. The early works
on folksonomies explore the structure, the types of tags and the user incentives of folksonomies ([7] and [12]). There are also works (see [14] for a detailed
analysis of the specific methods) based on the assumption that frequent co-
occurrence of tags translates to a semantic association among them. These works
use various statistical methods to identify clusters of related tags without defining the exact relations among them. An exception is the work detailed in [14],
where, in addition to clustering the tags, the semantic relations among them are
identified.

More recent research on folksonomies aligns them with knowledge resources
such as WordNet and ontologies. For example, in [9] the authors describe a
method that presents tag clusters as navigable hierarchical structures derived
from WordNet. Using a combination of WordNet based metrics they identify the
possible WordNet sense for each tag. They extract the path of this tag from the
WordNet hierarchy and they integrate it into the hierarchical structure of the
cluster. The TagPlus system [11] uses WordNet to disambiguate the senses of
Flickr tags by performing a two step query. The system returns all the possible
WordNet senses that define a tag and the user selects (disambiguates) which
sense he wishes. Another work aligning folksonomies to a user selected ontology is
described in [1]. The system queries the Web with a variety of linguistic patterns
between the ontological concepts and the tags. Each tag is categorised under the
concept to which it was more related by the Web Search results.

The existing works present methods for tag disambiguation and tag cluster enrichment. Our work aims to address the following additional issues. First, the existing works require some initialising from the user's side (e.g., a priori selecting ontology or knowledge resources for the relevant categories of tags) or they require user contribution to perform the disambiguation of the tags. Our goal is to perform semantic enrichment of folksonomies entirely *automatically* (i.e., without user contribution). Second, we aim to investigate how by using other knowledge resources (e.g., thesauri) and the Semantic Web we can achieve more precise and more complete enrichment of tags compared to the enrichment from single resources (i.e., one ontology, WordNet and so on).

3   FLOR

The goal is to transform a flat folksonomy tagspace into a rich semantic representation. We aim to annotate folksonomy resources with **Semantic Entities** (SEs) rather than raw text tags. However, since tags are the basic description of resources the connection of tags to SEs is the first step prior to connecting the resources to SEs. A SE can ideally be a Semantic Web Entity, SWE (class, relation, instance) defined in an online ontology. Our goal is not just to connect tags to SWEs but also to bring in other knowledge related to these SWEs. In case no SWE exists for a tag the goal is to query other knowledge resources for Semantic Entities.

We present **FLOR**[4], a **FoLksonomy Ontology enRichment** tool, which takes as input a set of tags (either the tagsets of individual resources or clusters derived by the statistical analysis of folksonomies) and automatically relates them to relevant semantic entities (classes, relations, instances) defined in online ontologies. The output of FLOR is a semantically enriched tagset. FLOR performs three basic steps as described in the following.

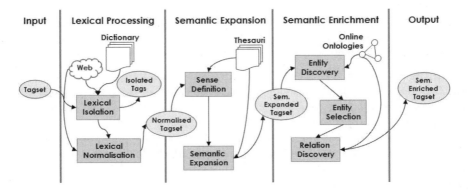

Fig. 1. FLOR Phases

---

[4] http://flor.kmi.open.ac.uk/

3.1   PHASE 1: Lexical Processing

Due to the freedom of tagging as a basic rule of folksonomies, a wide variety of different tag types are in use. Understanding the types of tags is the first step in deciding which of them are meaningful and should be taken into account as a basis of the semantic enrichment. Previous work ([7]) has identified different conceptual as well as syntactic categories of tags. For example, there are tags containing special characters, numbers, concatenated tags or tags with spaces and a big number of non-English[5] tags. The Lexical Processing phase is executed in two steps. The **Lexical Isolation** step uses a set of heuristics to identify which tags will not be further processed by FLOR. The **Lexical Normalisation** step aims to bridge the naming conventions used in folksonomies, ontologies and other knowledge resources by producing a list of possible lexical representations for each tag that will be enriched.

Running Example: Consider the tagset {`buildings`, `corporation`, `bw`, `england`, `road`, `neil101`}. Phase 1 isolates the tags {`bw`, `neil101`} which can't be further processed by FLOR. The Lexical Normalisation step generates the following lexical representations for the tag: `buildings` : {building, buildings}. The rest of the tags are already in a normalised form.

3.2   PHASE 2: Sense Definition and Semantic Expansion

Due to polysemy, the same tag can have different meanings in different contexts. For example, the tag `jaguar` can describe either a car or an animal or an operating system depending on the context in which it appears. The first step of this phase **Sense Definition and Disambiguation** performs sense disambiguation for the tags. The technique described in [2] has been implemented using Word-Net based similarity metrics. Alternative strategies such as [15] and [5] are also considered. Another issue that this phase addresses is the Semantic Web sparseness. While online ontologies might not contain concepts that are syntactically equivalent to a given tag, they might contain concepts that are labeled with one of its synonyms. To overcome this limitation, we perform **Semantic Expansion** for each tag as described in [2]. A combination of thesauri and other knowledge sources is considered in order to achieve an optimal semantic expansion.

Running Example: The Sense Definition step maps a WordNet sense to each of the tags returned from the previous phase. The result in this case is: [`building`: *a structure that has a roof and walls and stands more or less permanently in one place*], [`corporation`: *a business firm whose articles of incorporation have been approved in some state*], [`road`: *an open way (generally public) for travel or transportation*], [`england`: *a division of the United Kingdom*]. Next the Semantic Expansion returns the synonyms and the hypernyms for each tag, i.e.: [`building`: SYNONYMS (edifice) - HYPERNYMS (structure, construction, artefact)], [`corporation`: SYNONYMS (corp) - HYPERNYMS (firm, business, concern)], [`road`: SYNONYMS (route), HYPERNYMS (way, artefact,

---

[5] FLOR deals only with English tags.

object)], [england: SYNONYMS (-), HYPERNYMS (European_Country, European_Nation, land)].

3.3   PHASE 3: Semantic Enrichment

The last phase of FLOR identifies the Semantic Entities that are relevant for each tag by leveraging the results of Phases 1 and 2. The relevant Semantic Web Entities are selected during the **Entity Discovery** step by querying the WATSON Semantic Web Gateway [6], which gives access to all online ontologies. Then in the **Entity Selection** step we filter the SWEs in order to identify the ones that correctly correspond to the tags. Finally the **Relation Discovery** step identifies relations between the SWEs using the SCARLET, semantic relation discovery algorithm [13].

Running Example: The Semantic Enrichment phase links the tags to ontological entities. For example the following semantic information has been attached to the tag: [building: subClassOf (*Infrastructure, Manmade_Structure, HumanShelterConstruction, SpaceInAHOC*) - superClassOf (*Restaurant, RailroadStation*)]. The same happens for the rest of tags.

The output of FLOR for a resource is a semantic layer, containing the definitions of the concepts described in the resource and their relations.

4   Current Status and Outlook

FLOR was designed on the basis of the results presented in [3] where the characteristics of folksonomies versus the characteristics of ontologies were identified. The first functional version of FLOR has been implemented and the results have been manually evaluated. Applying FLOR on a dataset of 250 photos from Flickr with a total of 2819 tags we obtained the results reported in [2]. FLOR enriched approximately the 49% of the tags with enrichment precision of 93%. The main conclusion from this experiment was that folksonomy tagspaces can be automatically enriched with formal semantics extracted from online ontologies and thesauri. Yet, the Relation Discovery (Step 3) of Phase 3 still needs to be implemented. Also the evaluation of FLOR in a large-scale experiment is part of the ongoing work. Additionally the manual evaluation of FLOR revealed the shortcomings of the FLOR algorithm and provided the basis for future work. This is broken down to the following tasks:

- Testing and evaluation of FLOR in a large-scale retrieval task (*M24*)
- Relation Discovery of Phase 3 (*M24*)
- Enhancement of Sense Discovery and Semantic Expansion of Phase 2 (*M30*)
- Testing and evaluation of the improved version of FLOR (*M32*)

Finally the expected contribution of this work is:

- Methodology for semantic enrichment of a set of keywords
- Concept based annotation and retrieval
- Folksonomy content retrieval evaluation strategy.

Acknowledgements

This work was funded by the NeOn project sponsored under EC grant number IST-FF6-027595.

References

1. Abbasi, R., Staab, S., Cimiano, P.: Organizing resources on tagging systems using T-ORG. In: Proc. of the ESWC workshop: Bridging the Gap between Semantic Web and Web 2.0, Innsbruck, Austria, pp. 97–110 (2007)
2. Angeletou, S., Sabou, M., Motta, E.: Semantically enriching folksonomies with FLOR. In: 5th European Semantic Web Conference, Tenerife, Spain (accepted in the Workshop of Collective Semantics, 2008)
3. Angeletou, S., Sabou, M., Specia, L., Motta, E.: Bridging the gap between folksonomies and the semantic web: An experience report. In: Proc. of the ESWC workshop: Bridging the Gap between Semantic Web and Web 2.0, Innsbruck, Austria, pp. 30–43 (2007)
4. Benjamins, R., Davies, J., Baeza-Yates, R., Mika, P., Zaragoza, H., Greaves, M., Gomez-Perez, J., Contreras, J., Domingue, J., Fensel, D.: Near-term prospects for semantic technologies. Intelligent Systems 23, 76–88 (2008)
5. Cilibrasi, R., Vitanyi, P.: The google similarity distance. Transactions on Knowledge and Data Engineering 19(3), 370–383 (2007)
6. d'Aquin, M., Sabou, M., Dzbor, M., Baldassarre, C., Gridinoc, L., Angeletou, S., Motta, E.: Watson: A gateway for the semantic web. In: Poster Session of the 4th ESWC, Innsbruck, Austria (2007)
7. Golder, S., Huberman, B.: Usage patterns of collaborative tagging systems. Journal of Information Science 32(2), 198–208 (2006)
8. Greaves, M.: Semantic Web 2.0. Intelligent Systems 22(2), 94–96 (2007)
9. Laniado, D., Eynard, D., Colombetti, M.: Using WordNet to turn a folksonomy into a hierarchy of concepts. In: Proc.of 4th Italian Semantic Web Workshop, Bari, Italy, pp. 192–201 (2007)
10. Lassila, O., Hendler, J.: Embracing "Web 3.0". Internet Computing 11(3), 90–93 (2007)
11. Lee, S., Yong, H.: TagPlus: A retrieval system using synonym tag in folksonomy. In: Proc. of the Int. Conference on Multimedia and Ubiquitous Engineering, Seoul, Korea, pp. 294–298 (2007)
12. Marlow, C., Naaman, M., Boyd, D., Davis, M.: Position paper, tagging, taxonomy, flickr, article, toread. In: Proc. of the 15th Int. World Wide Web Conference, Edinburgh, Scotland (2006)
13. Sabou, M., d'Aquin, M., Motta, E.: Exploring the semantic web as background knowledge for ontology matching. In: Journal of Data Semantics (accepted for publication, 2008)
14. Specia, L., Motta, E.: Integrating folksonomies with the semantic web. In: Proc. of the 4th ESWC, Innsbruck, Austria, pp. 624–639 (2007)
15. Trillo, R., Gracia, J., Espinoza, M., Mena, E.: Discovering the semantics of user keywords. Journal of Universal Computer Science 13(12), 1908–1935 (2007)

Contracting and Copyright Issues for Composite Semantic Services

Christian Baumann*

SAP AG, SAP Research CEC Karlsruhe
Vincenz-Prienitz-Str. 1, 76131 Karlsruhe, Germany
ch.baumann@sap.com

Abstract. In business webs within the Internet of Services arbitrary services shall be composed to new composite services and therewith creating tradeable goods. A composite service can be part of another composite service and so on. Since business partners can meet just for one transaction not having regular business which justifies frame contracts, ad-hoc automated contracting needs to be established. In addition services have an intangible character and therefore are prone to illegal reproduction. Thus intellectual property rights have to be considered.

Our research approach is to assess the applicability of copyright law for semantic web services and develop a concept for automated contracting. Methodologies to be used are in the field of ontology modeling and reasoning.

Keywords: semantic web service, intellectual property rights, copyright, automated contracting.

1 Research Problem

The interconnectivity utilized by internal networks or the internet and the rise of Service-oriented architectures (SOA) lead to changes in e-business. Reusable components - so called services - can be used to compose individual applications (composite services) fitting for a specific problem. Recently semantic technology is applied to provide semantic services, which allow to more efficiently build composite services out of atomic services. This contributes to the vision of the Internet of Services (IoS) [13] [8], where services become similar to tradeable goods on the internet.

However, the building blocks of the IoS, the services, are more than just technical web services. Services comprise business models, business processes and a technical aspect (see Figure 1) to enable a business web [1]. With the rise of such new business areas participants have substantial interest in the protection of their own (intangible) property, such as copyright. This challenges jurisdictions to provide a legal framework within which business web participants can trade and conclude contracts safely. Typically existing legal regulations are being transfered

* The project was funded by means of the German Federal Ministry of Economy and Technology under the promotional reference "01MQ07012". The authors take the responsibility for the contents.

A. Sheth et al. (Eds.): ISWC 2008, LNCS 5318, pp. 895–900, 2008.

to newly risen business areas, and, if not possible, extended or newly created and passed by the jurisdiction. The advance in information technology, for example, required an adaption in copyright laws worldwide in the past decade, because the costs for information exchange and duplication of digitally recorded work became insignificant [12, pp. 2–3].

The IoS is envisioning a business web with seamless business-to-business (B2B) integration by semantic information exchange and arbitrary reusability of an intangible, tradeable good: the service. To realize the vision, intensive research is currently undertaken. The project TEXO[1] within the research program THESEUS, for example, looks into business webs in the IoS. Within such a business web some

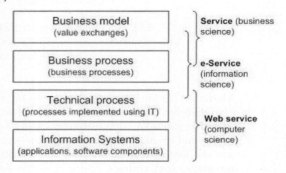

Fig. 1. Service Categorization [1]

services are thought to being composed, e.g. the services a, b, c, to a composite service, e.g. s. These services may all be provided by some legally independent entities, A, B, C and S. The composite service s may also be part of another composite service, let us say z. This short illustration shows the complexity of the IoS and gives an idea on how important representation and control over one's own intellectual property is. Maybe A, B, C do not want to be included in a mashup z or at least receive some payment for this.

Furthermore not only the technology for automated composition of a composite service is required, but also the legal foundation of concluding contracts. Typically frame contracts are being negotiated offline and only afterwards automated negotiation of some service levels is carried out. However, in the IoS this is not sufficient, since business partners may meet just for one transaction, and therefore an explicit offline negotiation of contracts causes too high transaction costs. Hence an automated way for contracting in the IoS is required, with which participants are able to meet on the same eye level, which prevents dominant players.

The example above even shows two different contracts. Let us assume the legal entity Z is the customer of S, integrating s in his own composite service z. Internally A, B, C and S have to conclude a contract, externally S and Z also need to conclude a contract. In this scenario, the intellectual property rights need to be considered and passed through during automated contracting.

This leads us to the **research problem** we want to discuss:

1. In a scenario of collaborative development of new (composite) web services (trade goods), current copyright law is adequate as a legal foundation to protect the exploitation rights of one's own web service.
2. The non-functional property exploitation rights can be represented and formalized with semantic technology for being utilized during automated contracting of service composition. Sound modeling with e.g. ontologies can

---

[1] http://theseus-programm.de/scenarios/en/texo/, visited 05.05.2008.

provide abstractness and flexibility for later extension for the implementation of automated contracting.

Ensuring an adequate handling of exploitation rights can raise trust and transparency in the IoS, therewith boosting its innovation power and give an incentive for participation.

2   Related Work

This section discusses state of the art in the field of copyright and contracting issues for composite semantic services. First, legal aspects are depicted, then related work in the area of semantic web is looked at.

The concept of a web service has taken little notice in the jurisprudence[2] so far. The same is true for the extension with a semantic annotation or the development of composite web services within SOA. The categorization of web services to a special type of contract is an open research question, e.g. [9] is discussing aspects of different contract types. The categorization is important because of the transfer of rights of the contractual object and the further use of this intellectual property. Also collaborative composite service development is not explicitly discussed. For linking and framing of websites copyright law is applicable, e.g. [11], however, this is not settled for composite services and an interesting question for the thesis.

In the research project SESAM[3] self organizing distributed electronic markets in the area of power markets were analyzed. One outcome of the research is the concept of an electronic legal mediator [5] for the process of concluding contracts without any offline steps [4]. This was required, since no central instance was available and contracts had to be concluded hourly. Thus software agents were designed to perform the negotiations between different parties. In contrast to power markets the IoS is, however, not as regulated and has a much lower entry barrier for new participants. As a result the electronic legal mediator is not suitable for on open business web in the IoS, since it only covers a very specific area of contracting in the power market. The IoS, however, requires a flexible framework for automated contracting and the management of intellectual property for the development of composite services.

The research project TrustCoM[4] was put up to build an environment for trust, security and contract management for B2B collaboration based on SOA. The main legal aspect was to enforce and monitor contracts for virtual organizations. This was achieved by implementing contractual terms and conditions as policies within the SOA infrastructure. These common regulations provide transparency for the virtual organizations facilitating trust. Moreover a risk management to reduce uncertainty in regards to the applicable statuary laws is provided. To provide predefined terms for the automated contracting in order to facilitate

---

[2] In this discussion we limit the legal scope to german jurisdiction.

[3] http://www.internetoekonomie.uni-karlsruhe.de/, visited 08.05.2008.

[4] http://www.eu-trustcom.com/, visited 08.05.2008.

trust is a concept we also pursue, however, in TrustCoM "negotiation involves only the SLA, with no reference to the other non-operational aspects of the collaboration that would normally appear in a legal document" [3, p. 29]. These non-operational aspects are the non-functional properties we want to enable, also including negotiation capabilities of intellectual property aspects for composite services.

Automated contracting facilitated by semantic technologies is shown in [10]. Lamparter is using his own core policy ontology, a ontology of bids and a contract ontology, which are based on the foundational ontology DOLCE[5]. Although automated matchmaking and contract formation using semantic annotation is achieved, a pre-negotiated "umbrella contract" [10, p. 123] is set as a precondition. The contracting process is only covering some functional service level agreements, leaving non-functional contractual terms for the offline negotiation of the umbrella contract. Intellectual property rights for service composition is not considered.

In the area of semantic web and law a lot of work is done in the representation of legal knowledge. A key challenge is to give access to existing judgments and represent regulations for educational training of young judges [2]. Also design patterns for legal ontology construction are discussed [6]. Laarshot et al. [14] tried to bridge the gap between legal concepts and actual cases to advice with ontology-based reasoning on possible liability situations. The works mentioned do not discuss automated contracting and representation of intellectual property rights in the semantic environment. However, their work is important for the formalization of legal aspects.

As a foundational ontology in the area of modeling legal aspects, the *Legal Knowledge Interchange Format* (LKIF) is enabling the translation between legal knowledge bases [7]. It is based on the foundational ontology DOLCE. The LKIF consists of a combination of OWL-DL and SWRL. We think LKIF is a good foundation for further formalization in the area of automated contracting and representation of intellectual property rights.

3   Contribution and Evaluation

The results of the research problem as stated in Section 1 will be integrated into the research program THESEUS[6], project TEXO. The legal analysis in copyright law will be reflected in the ontology for the service description framework. The techniques for formalization and reasoning over exploitation rights will make a contribution to the overall context of automated contracting during composition of semantic web services. The methodologies used will be in the field of ontology modeling and reasoning to find possible collisions during negotiation and enabling a management of intellectual property rights during contracting.

The evaluation will also be carried out within the project TEXO. The use case scenario "Eco Calculator" of the project comprises composition of semantic web

---

[5] http://www.loa-cnr.it/DOLCE.html/, visited 08.05.2008.
[6] http://theseus-programm.de/, visited 05.05.2008.

services including different legal entities and an end customer. The scenario will be used to justify the legal analysis and serve as an evaluation for the topic of automated contracting for semantic web services, in particular the exploitation of intellectual property rights.

4 Work Plan

The Ph.D. thesis is located in between the research areas of law and the semantic web. Figure 2 illustrates the interdisciplinary research field. The academic research partners are from Universitt Karlsruhe (TH)[7]: In the area of intellectual property "Institute for Information Law"[8] and for the semantic web domain "Institute of Applied Informatics and Formal Description Methods" (AIFB)[9].

Following different phases of the Ph.D. thesis are outlined. It is differentiated into current status, work in progress and planned work.

Current Status. The overall work time for the Ph.D. thesis is planned to be three years. Three months have passed now. The current status is as outlined in this paper.

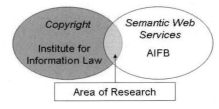

Fig. 2. Interdiciplinary Research Area

Work in Progress. The current work is to *investigate and assess the applicability of copyright law* for collaborative semantic service development. The scope is set to german copyright law, with the option to extend to international copyright acts. In parallel the possibilities of *formalization of such intellectual property rights* for semantic web services is assessed. The next planned step is a paper in 2008 to discuss *copyright issues for collaborative semantic web service development.*

Planned Work. Until end of 2008 the utilization of *semantic modeling of intellectual property rights and semantic reasoning during automated contracting* is planned to be assessed. Until 2009/12 the *legal analysis and the concept for the technical formalization* of intellectual property rights within the context of semantic services shall be finished. Until 2010/12 the *concept of automated contracting and the incorporation of intellectual property rights* into the contracting process shall be ready. The Ph.D. thesis manuscript is planned to be available in 2011/02.

References

1. Baida, Z., Gordijn, J., Omelayenko, B.: A shared service terminology for online service provisioning. In: Janssen, M., Sol, H.G., Wagenaar, R.W. (eds.) ICEC. ACM International Conference Proceeding Series, vol. 60, pp. 1–10. ACM, New York (2004)

---

[7] http://www.uni-karlsruhe.de/, visited 09.05.2008.
[8] http://www.zar.uni-karlsruhe.de/iirdreier/, visited 09.05.2008.
[9] http://www.aifb.uni-karlsruhe.de/english/, visited 09.05.2008.

2. Blázquez-Cívico, M., Pena-Ortiz, R., Contreras-Cino, J., Benjamins, V.R.: Iuris-ervice - Architecture: A Knowledge Management System for Spanish Judicial Domain. In: Casanovas, P., Noriega, P., Bourcier, D., Galindo, F. (eds.) Trends in Legal Knowledge, Florence, Italy, pp. 243–261. European Press Academic Publishing (2007)

3. Cojocarasu, D.I., NRCCL: Final report on legal issues – Enforcing and monitoring of VO Contracts. In: TrustCoM (2005)

4. Conrad, M., Funk, C., Raabe, O., Waldhorst, O.P.: A Lawful Framework For Distributed Electronic Markets. In: Camarinha-Matos, L.M., Afsarmanesh, H., Novais, P., Analide, C. (eds.) Virtual Enterprises and Collaborative Networks. IFIP, vol. 243, pp. 233–240. Springer, Heidelberg (2007)

5. Dietrich, A., Lockemann, P.C., Raabe, O.: Conceptual Modelling in Information Systems Engineering. In: Agent Approach to Online Legal Trade, pp. 177–194. Springer, Heidelberg (2007)

6. Gangemi, A.: Design Patterns for Legal Ontology Construction. In: Casanovas, P., Noriega, P., Bourcier, D., Galindo, F. (eds.) Trends in Legal Knowledge, Florence, Italy, pp. 171–191. European Press Academic Publishing (2007)

7. Hoekstra, R., Breuker, J., Di Bello, M., Boer, A.: The LKIF Core Ontology of Basic Legal Concepts (June 2007)

8. Janiesch, C., Ruggaber, R., Sure, Y.: Eine Infrastruktur für das Internet der Dienste. HMD – Praxis der Wirtschaftsinformatik 261, 71–79 (2008)

9. Koch, F.A.: Web Service als neue IT-Vertragsleistung – Definition und vertragstypologische Einordnung von Web Service-Anwendungenn. IT-Recht kompakt 3, 71–73 (2007)

10. Lamparter, S.: Policy-based contracting in semantic web service markets. Universitätsverlag Karlsruhe (2007)

11. Ott, S.: Urheber- und wettbewerbsrechtliche Probleme von Linking und Framing. Boorberg (2003)

12. Pierson, M., Ahrens, T., Fischer, K.: Recht des geistigen Eigentums. Vahlen (2007)

13. Ruggaber, R.: Internet of Services SAP Research Vision. WETICE, 3 (2007)

14. van Laarshot, R., van Steenbergen, W., Stuckschmidt, H., Lodder, A.R., van Harmelen, F.: The Legal Concepts and the Laymans Terms – Bridging the Gap through Ontology-Based Reasoning about Liability. In: JURIX 2005 (2005)

Parallel Computation Techniques for Ontology Reasoning

Jürgen Bock

FZI Research Center for Information Technologies, Karlsruhe, Germany
bock@fzi.de

Abstract. As current reasoning techniques are not designed for massive parallelisation, usage of parallel computation techniques in reasoning establishes a major research problem. I will propose two possibilities of applying parallel computation techniques to ontology reasoning: parallel processing of independent ontological modules, and tailoring the reasoning algorithms to parallel architectures.

1 Motivation

Scalability is an issue that is subject in many semantic web research discussions. More and more researchers share the awareness that reasoning in its current form will not be able to bear the load of data it is supposed to handle in the near future. A polarising article was published by Fensel and van Harmelen [1], where the authors talk about "10.000 triples just to describe each human, which gives us 100 trillion." Even though this guess may be intensionally provocative, it has in fact been proven several times in the past that even high estimations of growth were beaten in reality often long before they were predicted to eventuate.

An obererervation is, that available state-of-the-art reasoners do not exploit the benefits of parallel computation techniques, as these are not straightforwardly applied for reasoning calculi. Multithreading or other ways of distributed computation cannot easily be taken care of by the operating system. This is a major problem, since computational power at the level of integrated circuits is about to explore its physical limits by being "down to atoms" concerning conductor and transistor size. However, parallel computer architectures emerge, such as grid, peer-to-peer, or multi-core machines even in home-computing environments. This allows for the overall computational power to grow further, provided that software architecture and algorithms respect this computational paradigm shift. Thus parallel architectures do not inherently speed up all kind of computation, as the workload needs to be split into chunks of independent computations. Current reasoning algorithms do *not* naturally decompose into independent computational chunks.

In fact, it remains an open question whether algorithms currently used in reasoning adapt well to the paradigm shift in computer architecture. In particular it is unclear whether well established tableau algorithms, as widely used in state-of-the-art reasoners, can be parallelised.

A. Sheth et al. (Eds.): ISWC 2008, LNCS 5318, pp. 901–906, 2008.

One of the efforts towards making reasoning scalable while still guaranteeing soundness and completeness is the restriction to tractable fragments. While early research in description logics was striving to gain more expressivity, recent trends are going towards limiting that expressivity to those language features that preserve tractability[1]. Since parallelisation in general does not change the computational complexity of reasoning in highly expressive language fragments, provided a fixed number of processors, this research will focus on parallel reasoning in tractable fragments, as the successful outcome will be highly profitable for semantic web applications.

2   State of the Art

Despite the major concerns about scalability, the research area of parallel reasoning in description logics has not received high attention. An early work is the FLEX system [2,3] from 1995, which focuses on subsumption checking, classification, and propagation of rules for ABox reasoning. Due to the fine-grained structural algorithms used for TBox reasoning, certain parallel architectures are not well suited. Another work, and to the best of my knowledge, the only one addressing parallel tableau-based DL reasoning was conducted recently by Liebig and Müller at the University of Ulm [4]. Their method directly gears into the tableau calculus by parallelising nondeterministic branches in the tableau, which are mutually independent computational chunks, and thus can be processed in parallel. A longer history than parallel DL reasoning can be observed in the field of parallel processing of Prolog and Datalog. A summarising survey of many efforts in this direction has been published by Gupta *et al.* [5]. The article highlights Or-Parallelism and And-Parallelism in logic programming.

Another way of applying parallelism to ontology reasoning is the simultaneous processing of different *modules* of an ontology. The problem of ontology modularisation has recently attracted increasing interest. Relevant work in this area addresses the problem of identifying partitions of a large ontology, that are self-contained, *i.e.* one can refer to each partition while still preserving relevant context within this partition. The work of Stuckenschmidt and Klein [6] follows a structure-based approach considering the concept hierarchy of an ontology. A work of Cuenca Grau *et al.* [7] is tackling this shortcoming by introducing a modularisation strategy for OWL DL ontologies. The work, however, distances itself from modularisation for reasoning purposes, and explicitly stresses its contribution for modelling purposes. Similarly the recent work of Doran *et al.* [8] focuses on an improvement of ontology reuse by identifying self-contained modules of ontologies. Other work, which is close to the ontology reasoning problem in terms of ontology modularisation was conducted by Cuenca Grau *et al.* on $\mathcal{E}$-Connections [9] and MacCartney *et al.* [10]. The latter investigates a novel resolution strategy for a first-order theorem prover, by utilising partitioning and

---

[1] Tractability means computing a reasoning request in at most polynomial runtime, *i.e.* not exceeding the complexity class P.

a message-passing algorithm for reasoning on the first-order knowledge base. Recent work on distributed resolution for description logics has been conducted by Schlicht and Stuckenschmidt [11]. The authors delegate clauses to distributed resolution solvers which process them independently.

As mentioned in the previous section, applying parallel computation techniques to highly expressive description logics does not *per se* make them tractable. Hence the focus for large scale scenarios is on the use of tractable fragments, if soundness and completeness of reasoning requests have to be guaranteed. Tractable fragments[2] of OWL 2 that have been identified are DL-Lite, $\mathcal{EL}$, in particular with its more expressive extension $\mathcal{EL}^{++}$, and OWL-R, which is – in its DL version – based on DLP.

Research in description logic reasoning has led to a variety of available reasoners, resulting from the main research groups in the area. In particular these reasoners are FaCT and its successor FaCT++, Pellet, the Racer system, KAON2 and HermiT. While these systems support highly expressive description logics, there are few implementations, which focus on tractable fragments. These are CEL [12] for the fragment $\mathcal{EL}^{+}$, and QuOnto [13] for answering conjunctive queries in DL-Lite. There is currently no effort to investigate established parallel infrastructures such as grid computation as a basis for reasoning algorithms.

3   Contribution

In this research I will strive for bringing the parallel computation paradigm closer to ontology reasoning techniques. I see two possible ways to benefit from parallel computation techniques. The first one is to modularise ontologies and/or queries to process different parts of the ontology (query) in parallel. The second possible way to benefit from parallel compuatation is a low-level redesign of the reasoning algorithm to allow for parallel processing.

3.1   Hypothesis 1 – Independent Ontology Modules

In general, ontologies are not structured as a collection of independent modules. Still it is not clear, whether this general statement is reflected in practice. This means that feasibility of this hypothesis depends on a detailed analysis of real-world ontologies.

The problem in reasoning on separate modules of an ontology or a collection of ontologies is, that in the general case ontologies do not behave modular, *i.e.* all parts of an ontology are connected with each other in some way. The challenge for reasoning on different parts of an ontology in parallel is to identify those modules that do not influence each other in terms of conclusions that can be derived by not considering other modules. While previous work on design issues does not heavily rely on complete independence of the modules considered, this does not hold for reasoning, where any interconnection between modules can potentially require the reasoner to respect the connected modules in total.

---

[2] http://www.w3.org/TR/owl2-profiles/

As this constraint makes modularisation of highly expressive description logics not sound very promising, it might be possible for tractable fragments due to language restrictions.

Respecting the previous work, modularisation in OWL [6,7,9], as well as related first-order theorem proving methods [10] and possible combinations of these two points of view will be taken into account.

3.2   Hypothesis 2 – A Parallel Reasoning Algorithm

Most state-of-the-art reasoning systems are based on tableau algorithms or resolution. Until now it is not clear, whether these methods can be parallelised, in particular since they are highly optimised [14,15] and thus demand a high degree of communication between previously independent chunks of computation (e.g. branches in a tableau). However, modern multi-core, shared-memory machines could overcome this problem by providing computation results of one chunk to other processors via a common repository. Following up the state-of-the-art discussion, there are still unsolved problems in adapting reasoning algorithms to recent developments in multi-core architectures:

1. Parallel tableau algorithms only exploit concurrent processing of nondeterministic branches, which can be computed independently without any interaction [4]. This is insufficient in two main scenarios:
   (a) Some tractable fragments do not allow disjunction or number restrictions and hence cannot benefit from this parallelisation.
   (b) To increase the number of threads, that can be processed in parallel in order to be applicable to an extremely large number of processors, it will be necessary to consider other kinds of (deterministic) branches which need to communicate efficiently with each other.
2. To keep reasoning in tractable fragments in the theoretical complexity range, proprietary algorithms might be required which are not yet parallelised. There is e.g. no implementation for conjunctive query answering in $\mathcal{EL}^{++}$.
3. Extensively studied parallelisation methods for logic programming applied to description logic reasoning is unsatisfactory regarding the following:
   (a) Applying logic programming techniques to description logic reasoning mainly improves ABox reasoning. It is not the preferred method for TBox reasoning.
   (b) Nominals are important in some scenarios and are supported by the tractable fragment $\mathcal{EL}^{++}$. However, it is unclear how nominals can be dealt with in a datalog based description logic reasoner, such as KAON2.

4   Roadmap

Figure 1 shows the high-level roadmap of this research. The first step of benchmarking existing OWL reasoners has been completed and successfully published at the ARea2008 Workshop [16]. In this work we analysed the ontology landscape and identified major expressivity fragments, which are present in real world

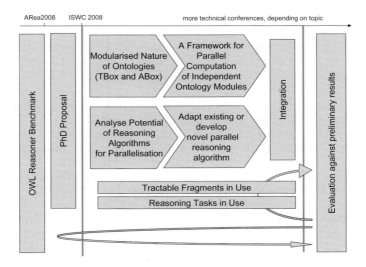

Fig. 1. Roadmap for this PhD research

ontologies. Following this, we identified which reasoner performs best on which fragment. We distinguished between different reasoning tasks (TBox and ABox), and evaluated scalability w.r.t. ABox size.

The second step of narrowing down the topic has also been completed and resulted in the document at hand. While still providing space for two topics, as discussed in Sect. 3, there are two streams, which will be followed simultaneously. Respecting hypothesis 1, the next step will be to analyse real world ontologies in terms of how they naturally decompose into different partitions or modules. This can be built on our previous analysis [16]. In terms of hypothesis 2, the natural next step is to analyse existing work on parallelising tableau algorithms and conduct early experiments based on modified or newly implemented reasoner prototypes, which exploit parallel computation as discussed in Sect. 3.

In the case of hypothesis 1, following steps will include the setup and implementation of a framework for preprocessing, *i.e.* partitioning of ontologies and invoking the reasoner on these independend partitions. As for hypothesis 2, the outcome would be a modified existing, or entirely novel parallel reasoner, based on fundamental parallel tableaux. Finally, I will investigate the possibilities of integrating the two ideas, *e.g.* by checking whether a modularisation can be identified, which enhances the possibilities of parallelising the algorithm, or adapt the ideas of a parallel tableau algorithm to be applicable to distributed ontology modules. Parallel to the two streams, there will be a comprehensive analysis, of how reasonig tasks and tractable tragments are used in partice.

The final step will be an evaluation of the newly developed solutions w.r.t. the initial reasoner benchmark, as well as to the long-term studies on reasoning tasks and tractable fragments in use, to ensure, that the contribution reflects real world needs in terms of optimised performance on important reasoning tasks and frequent language fragments.

References

1. Fensel, D., van Harmelen, F.: Unifying Reasoning and Search to Web Scale. IEEE Internet Computing 11(2), 94–96 (2007)
2. Bergmann, F.W., Quantz, J.J.: Parallelizing Description Logics. In: Wachsmuth, I., Brauer, W., Rollinger, C.-R. (eds.) KI 1995. LNCS, vol. 981. Springer, Heidelberg (1995)
3. Quantz, J.J., Dunker, G., Bergmann, F., Keller, I.: The FLEX System. Technical Report 124, Technische Universität Berlin, Berlin (December 1995)
4. Liebig, T., Müller, F.: Parallelizing Tableaux-Based Description Logic Reasoning. In: Meersman, R., Tari, Z., Herrero, P. (eds.) OTM-WS 2007, Part II. LNCS, vol. 4806, pp. 1135–1144. Springer, Heidelberg (2007)
5. Gupta, G., Pontelli, E., Ali, K.A., Carlsson, M., Hermenegildo, M.V.: Parallel Execution of Prolog Programs: A Survey. ACM Trans. Program. Lang. Syst. 23(4), 472–602 (2001)
6. Stuckenschmidt, H., Klein, M.C.A.: Structure-Based Partitioning of Large Concept Hierarchies. In: McIlraith, S.A., Plexousakis, D., van Harmelen, F. (eds.) ISWC 2004. LNCS, vol. 3298, pp. 289–303. Springer, Heidelberg (2004)
7. Cuenca Grau, B., Parsia, B., Sirin, E., Kalyanpur, A.: Modularizing OWL Ontologies. In: Proceedings of the KCAP-2005 Workshop on Ontology Management. ACM, New York (2005)
8. Doran, P., Tamma, V.A.M., Iannone, L.: Ontology Module Extraction for Ontology Reuse: An Ontology Engineering Perspective. In: Proceedings of the 16th CIKM. ACM, New York (2007)
9. Cuenca Grau, B., Parsia, B., Sirin, E.: Tableau algorithms for $\mathcal{E}$-connections of Description Logics. Technical report, University of Maryland Institute for Advanced Computer Studies (UMIACS) (2004)
10. MacCartney, B., McIlraith, S.A., Amir, E., Uribe, T.E.: Practical Partition-Based Theorem Proving for Large Knowledge Bases. In: Proceedings of the 18th IJCAI. Morgan Kaufmann, San Francisco (2003)
11. Schlicht, A., Stuckenschmidt, H.: Distributed Resolution for ALC. In: Proceedings of DL2008. CEUR Workshop Proceedings, vol. 353 (May 2008)
12. Baader, F., Lutz, C., Suntisrivaraporn, B.: CEL—A Polynomial-time Reasoner for Life Science Ontologies. In: Furbach, U., Shankar, N. (eds.) IJCAR 2006. LNCS (LNAI), vol. 4130. Springer, Heidelberg (2006)
13. Acciarri, A., Calvanese, D., Giacomo, G.D., Lembo, D., Lenzerini, M., Palmieri, M., Rosati, R.: QuOnto: Querying Ontologies. In: Proceedings of the 20th AAAI. AAAI Press, Menlo Park (2005)
14. Horrocks, I.: Optimising Tableaux Decision Procedures for Description Logics. Ph.D thesis, University of Manchester (1997)
15. Motik, B.: Reasoning in Description Logics using Resolution and Deductive Databases. Ph.D thesis, Universität Karlsruhe (TH), Institut AIFB, Karlsruhe, Germany (2006)
16. Bock, J., Haase, P., Ji, Q., Volz, R.: Benchmarking OWL Reasoners. In: Proceedings of the ARea2008 Workshop. CEUR Workshop Proceedings, vol. 350 (June 2008)

Towards Semantic Mapping for Casual Web Users

Colm Conroy

Knowledge and Data Engineering Group,
Trinity College Dublin
coconroy@cs.tcd.ie

1 Introduction

The Semantic Web approach is becoming established in specific application domains, however there has been as yet no uptake within the mainstream internet environment [1]. The reasons for the lack of uptake of the semantic web amongst casual web users can be attributed to technology perception, comprehensibility and ease of use. It is perceived that the creation of ontologies is a top-down and complex process, whereas in reality ontologies can emerge bottom-up and be simple. Ontology technology is based on formal logics that are not understandable for ordinary people. Finally there is significant overhead for a user in the creation of metadata for information resources in accordance with ontologies. To address these three problems, it is proposed that the interfaces to semantic web tools will need to be engineered in such a way that the tools become simplified, disappear into the background, and become more engaging for casual web users. Increasingly techniques from the semantic desktop research community will enable the creation of a personal ontology on behalf of a user. Although the automatic and efficient matching between the personal ontology and the models used by others (for example through the use of collaborative tags, community ontologies) can be achieved through the application of a variety of matching techniques [2], fully automatic derivation of mappings from the resultant set of candidate matches is considered impossible as yet [3]. A mapping can be thought of as the expression of a confirmed correspondence (e.g. equivalence, subclass, some arbitrary formula). The correspondence could be derived perhaps using machine learning approaches but is typically derived by a human. The majority of state of the art tools in the ontology mapping area [4] and the community ontology creation area [5] rely on a classic presentation of the class hierarchy of two ontologies side by side and some means for the user to express the mappings. These approaches predominately assume that the mapping is being undertaken by an expert: who does not require a personalised interface; whose explicit task is to generate a "one size fits all" full mapping (to be used in common by several applications); and who typically undertakes the task during a small number of long sessions. The number of user trials that have taken place have also been small [6] and those that have, have focused purely on the mapping effectiveness and do not address usability issues (an exception recently being that of [7]). In contrast to the semantic web, 'Web 2.0' has seen an explosion in uptake within the mainstream internet environment [8]. Some of the main characteristics of 'Web 2.0' are rich user experience, user participation and collective intelligence [9]. We intend to take user-driven methodologies that exist

A. Sheth et al. (Eds.): ISWC 2008, LNCS 5318, pp. 907–913, 2008.

within 'Web 2.0' to semantic mapping. We propose that the casual web users who will benefit from mappings (through usage by their applications), will undertake themselves partial targeted mappings, gradually and over time, using techniques that address usability issues, support personalization and enable control of the mapping interactions.

2   State of the Art

A widespread issue in making semantic mapping, indeed most semantic, tools more accessible to user is the lack of focus on support and usability for users. PROMPT [10] is a plugin for Protégé which supports managing multiple ontologies including ontology merging and mapping. PROMPT presents a mapping suggestion from a candidate list of mappings to the user which the user verifies or rejects then another suggestion is displayed and the cycle is repeated until the user deems the mapping complete. User evaluations for PROMPT were performed in [11] and the results showed that merging was quite difficult for each user and in particular users found performing any non-automated procedures quite difficult. The paper [12] presents a theoretical framework for cognitive support in ontology mapping. It provides some software tool requirements for each framework principle it suggests, e.g. 'reason for suggesting a mapping'. CoGZ is a mapping tool, built upon PROMPT, which supports this framework using TreeMaps [13] to present an overview of the ontologies and possible mappings. A large usability study to refine the cognitive support framework is planned as future work. In [14] the idea of ontology matching is extended to community-driven ontology matching. Their approach was to partition people into multiple different communities (groups) where alignments can be shared, which extends and preserves the advantages given to communities by the web. A prototype was developed based on their approach and showed the feasibility of acquisition and sharing of ontology mappings among web communities. A side effect of their approach is improving the recall of matching tools via repository of ontology mappings for different domains. Investigating the benefits for human contributors is planned as further work. The paper [15] extends S-Match [16] into a semantic matching system which provides proof and explanations for mappings it has discovered. The matching system uses the Inference Web infrastructure [17] to expose meaningful fragments of S-Match proofs in which users can browse. Proofs are displayed using short, natural language, high level explanations without any technical details which are designed to be intuitive and understandable by ordinary users. Results have been promising and shown the potential to scale for the semantic web. In [18] a proposed formal model for ontology mapping creation is suggested. The formal model is used to get complete correspondence between user's actions and generated alignments. They propose a set of different graphical perspectives that can be linked with the same model, each of them offering different viewpoints on the displayed ontology. By using these different perspectives they hope to hide the complexity of the underlying logical language and give better understanding of the mapping action.

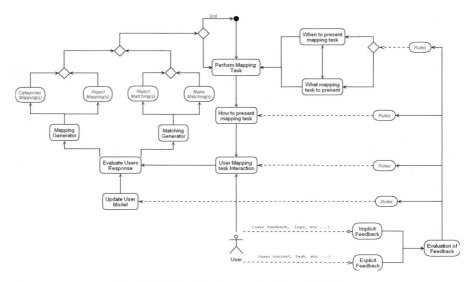

Fig. 1. The Mapping Process for Casual Web Users

3 Research Question

The research question to be addressed is: *what kind of interactions will be acceptable, efficient and effective for casual web users to achieve semantic mappings gradually and over time between conceptual models of interest to the user.* In order to achieve semantic mappings, users will need to undertake mapping tasks as part of the mapping process: e.g. make a mapping based on matching information, categorise a mapping, identify possible mappings, provide corrections for a mapping and possibly withdraw or reject a mapping. Through interaction with the mapping system, the user will undertake the mapping tasks. In our opinion, these interactions need to be adapted so that casual web users see benefit of engaging with the mapping process and are kept usefully engaged over a long period. For example possible available axes of adaptation could be **personalisation, visualisation, collaboration, context** and **the choice of mapping tasks**. These adaptations can occur both at runtime or design time. The adaptations can also be either applied to the mapping process or within the system. In this PhD the primary focus will be on **runtime** adaptation of interactions for an end-user within a single mapping process based on **context**. We prioritise the focus on context-based adaptations of interactions due to the lack of research on this topic within the semantic mapping area. Our definition of context is the current browsing environment the user is within i.e. if the user is using a browser, what web-page and task are they engaged in? In particular the work will:

- *Develop an adaptive mapping process framework to assist casual web users*: There is a need to make the ontology mapping process as unintrusive and as natural as possible, as it is important not to interrupt ordinary users during their daily tasks. This is to ensure that they do not see mapping tasks as inconvenient work but more as something that will be beneficial to them.

- *Determine key context-driven adaptation of interactions with the user and the effect of these*:   The key factors in this problem are determining the most appropriate contexts for people to engage in the mapping process and the appropriate mapping tasks for these contexts.
- *Develop an experimental framework based on the above process for evaluation of adaptive interactions in a long-term mapping process on the web*:   Such an experimental framework for mapping does not currently exist in the state of the art, as most systems implement "one shot" mapping approaches.

An outcome of this research will be a process (Fig. 1), methodology and tools.

4   Evaluation

In this section we outline our current and previous experiments.

4.1   Natural Language

In our initial experiment undertaken early in 2007 we aimed to determine the most practical way of visually displaying the mapping information for different groups of users. Our hypothesis was that *using a "Question & Answer" natural language interface to visually display ontological information helps in making mapping more familiar and accessible and also reduces the complexity of the mapping process for users.* Our experiment investigated the effect and usability of a natural language prototype tool (NL) [19] on three classes of users and tried to determine whether it made the mapping task more user friendly for one group when compared to the others. In addition we wanted to contrast our tool against a current state of the art mapping tool. We chose COMA++ [4] as our state of the art tree-type graph mapping tool. The three different classes of users were: ontologically aware, technology aware and casual web users. The resulting paper [19] describes the experiment in more detail, some key conclusions drawn were:

- On the positive side, results suggested casual web users can map effectively and efficiently even compared to ontologically aware users. Using Natural Language seemed to help people read and understand the information and the Q&A approach helped in navigating through the mapping task.
- On the negative side, casual web users found it very restrictive to be limited to a narrow range of mapping terminology, e.g. "corresponds" and "similar to" when answering mapping questions. In addition, some users were unclear about the benefit in engaging in the mapping task.

4.2   'Tagging' Approach (Finishing August 2008)

In our current experiment we are focusing on whether it is valuable to embed the mapping process within the user environment, designing a user-centric mapping process, and addressing the negative concerns garnered from the previous experiment by allowing the user to be more expressive by allowing them to 'tag' the mapping relation. Our hypothesis is *the mapping task can be simplified, become unintrusive and more engaging by using a 'tagging' approach paradigm, embedding the mapping process within the user environment and showing the benefits of the mapping task.* By using the power of "Web 2.0" through a Firefox extension [20] within our new 'tagging' prototype, we aim to engage the user and display matching tasks at appropriate times within their own work environment, see [21] for more details. We use online questionnaires, interviews, and a log of each user's actions to evaluate the impact of the 'tagging' prototype. In particular through the use of our implementation over the coming months we aim to investigate whether casual web users will be able to use tagging to turn matches into expressive mappings in a straightforward, practical and natural manner. We will also investigate whether embedding the mapping interface inside a browser extension will allow the mapping process to take place over time within a casual web users' work environment in an unobtrusive, sensible, and normal way. The benefits in engaging in the mapping task for the user will be the gain of individually tailored RSS news feed items.

5   Next Step: Context-Driven (September 2008 – June 2009)

A key dilemma is making the mapping process *unintrusive* yet still *engaging* to casual web users. In the next experiment we are going to explore what mapping task should be performed given the context of the user. For example, whether only specific mapping tasks should be performed within a certain context, e.g. when on a site such as del.icio.us [22] only mapping tasks specific to this website should be asked like aligning your 'tags' with the del.icio.us domain ontology. A potential benefit of aligning each user model with the domain ontology is allowing the sharing of information between each user, e.g. alignment on the YouTube [23] site allows sharing of video's via friend of a friend while the user is looking at a video which is similar. The mapping task to be performed should be dependent on both the user context and the web task the user is performing, e.g. if the user is writing an email or document no tasks should be performed while if the user is organizing their bookmarks on the del.icio.us then context-driven mapping tasks might be suggested. The infrastructure setup will involve developing an experimental platform which allows for the evaluation of hundreds of users on the web over a long time period. We would hope to discover via this experiment whether a context-driven mapping task approach will encourage casual web users' to interact with the mapping process.

6   Conclusions

Through our evaluation we intend to weigh up what kind of interactions are necessary for casual web users' to achieve semantic mappings in an acceptable, efficient and effective manner. The thesis write up will occur from July 2009 till October 2009.

References

1. Berners-Lee, T., Shadbolt, N., Hall, W.: The Semantic Web Revisited. IEEE Intelligent Systems 21(3), 96–101 (2005)
2. Shvaiko, P., Euzenat, J.: A Survey of Schema-based Matching Approaches. J. Data Semantics IV, 146–171 (2004)
3. Noy, N.: Semantic Integration: A Survey of Ontology-Based Approaches. ACM SIGMOD Record 33(4), 65–70 (2004)
4. Aumüller, D., Do, H., Massmann, S., Rahm, E.: Schema and Ontology Matching with COMA++. In: ACM SIGMOD international conference on Management of data, pp. 906–908 (2005)
5. Zhdanova, A.: Towards Community-Driven Ontology Matching. In: ACM 3rd International Conference on Knowledge Capture, pp. 221–222 (2005)
6. Jameson, A.: Usability and the Semantic Web. In: Sure, Y., Domingue, J. (eds.) ESWC 2006. LNCS, vol. 4011, p. 3. Springer, Heidelberg (2006)
7. Falconer, S.M., Noy, N.F., Storey, M.: Towards understanding the needs of cognitive support for ontology mapping. In: International Workshop on Ontology Matching, ISWC 2006, CEUR-WS, vol. 225 (2006)
8. Madden, M., Fox, S.: Riding the Waves of 'Web 2.0'. Technical report, Pew Internet & American Life Project (2006)
9. O'Reilly, T.: What is Web 2.0: Design Patterns and Business Models for the Next Generation of Software. J. Communications & Strategies 1, 17 (2007)
10. Noy, N., Musen, M.: The PROMPT suite: interactive tools for ontology merging and mapping. J. Human-Computer Studies 59(6), 983–1024 (2003)
11. Lambrix, P., Edberg, A.: Evaluation of ontology merging tools in bioinformatics. In: Pacific Symposium on Biocomputing 2003, pp. 589–600. World Scientific, Singapore (2003)
12. Falconer, S.M., Storey, M.: A Cognitive Support Framework for Ontology Mapping. In: Aberer, K., Choi, K.-S., Noy, N., Allemang, D., Lee, K.-I., Nixon, L., Golbeck, J., Mika, P., Maynard, D., Mizoguchi, R., Schreiber, G., Cudré-Mauroux, P. (eds.) ASWC 2007 and ISWC 2007. LNCS, vol. 4825, pp. 114–127. Springer, Heidelberg (2007)
13. Shneiderman, B.: Tree visualization with tree-maps: 2-d space filling approach. ACM Transaction on Graphics 11(1), 92–99 (1992)
14. Zhdanova, A., Shvaiko, P.: Community-driven ontology matching. In: Sure, Y., Domingue, J. (eds.) ESWC 2006. LNCS, vol. 4011, pp. 34–49. Springer, Heidelberg (2006)
15. Shvaiko, P., Giunchiglia, F., Pinheiro da Silva, P., McGuinness, D.: Web Explanations for Semantic Heterogeneity Discovery. In: Gómez-Pérez, A., Euzenat, J. (eds.) ESWC 2005. LNCS, vol. 3532, pp. 303–317. Springer, Heidelberg (2005)
16. Giunchiglia, F., Shvaiko, P., Yatskevich, M.: S-Match: an algorithm and an implementation of semantic matching. In: Bussler, C.J., Davies, J., Fensel, D., Studer, R. (eds.) ESWS 2004. LNCS, vol. 3053, pp. 61–75. Springer, Heidelberg (2004)
17. McGuinness, D.L., Pinheiro da Silva, P.: Infrastructure for web explanations. In: Fensel, D., Sycara, K.P., Mylopoulos, J. (eds.) ISWC 2003. LNCS, vol. 2870, pp. 113–129. Springer, Heidelberg (2003)
18. Mocan, A., Cimpian, E., Kerrigan, M.: Formal Model for Ontology Mapping Creation. In: Cruz, I., Decker, S., Allemang, D., Preist, C., Schwabe, D., Mika, P., Uschold, M., Aroyo, L.M. (eds.) ISWC 2006. LNCS, vol. 4273, pp. 459–472. Springer, Heidelberg (2006)

19. Conroy, C., O'Sullivan, D., Lewis, D.: A Tagging Approach to Ontology Mapping. In: 2nd International Workshop on Ontology Mapping, ISWC 2007, CEUR-WS, vol. 304. Springer, Heidelberg (2007)
20. Firefox add-ons, add new functionality to browsing experience,
    http://addons.mozilla.org/
21. Conroy, C., O'Sullivan, D., Lewis, D.: Ontology Mapping Through Tagging. In: International Workshop on Ontology Alignment and Visualisation, CISIS 2008, pp. 886–891 (2008) ISBN 0-7695-3109-1
22. Del.Icio.Us, A social bookmarks manager, http://del.icio.us
23. YouTube, user-generated video site, http://www.youtube.com

Interactive Exploration of Heterogeneous Cultural Heritage Collections

Michiel Hildebrand

CWI, Amsterdam, The Netherlands
michiel.hildebrand@cwi.nl

1 Research Problem

In this research we investigate to what extent explicit semantics can be used to support end users with the exploration of a large heterogeneous collection. In particular we consider cultural heritage, a knowledge-rich domain in which collections are typically described by multiple thesauri. Many institutions have made or are making (parts of) their collections available online. The cultural heritage community has the ambition to make these isolated collections and thesauri interoperable and allow users to explore cultural heritage in a richer environment.

The MultimediaN E-Culture project [1] examines the usability of Semantic Web technology to integrate museum data and to provide effective user interfaces to access this heterogenous data. The project has collected data from multiple museum collections annotated with multiple thesauri. Based on the procedure described in [2] this data is converted to RDFS/OWL and enriched with links across collections and thesauri. The result is represented in a single repository in the form of a large RDF graph. While some of these links have formal semantics as defined by RDFS/OWL, the majority only has "weak" semantics as defined by SKOS[1] and domain specific schemas.

Within the context of the E-Culture project, our research aims at better interfaces and search functionality to support end users with the exploration of large heterogeneous RDF graphs. Here we face two general problems. First, in a heterogeneous graph we have no fixed schema on which we can base the interface design and application functionality. Second, the semantics of our domain are too weak to depend on formal reasoning alone and thus we require alternative strategies that benefit from "weak" semantic structures to provide the required search functionality.

We focus on three types of end user functionality. First, searching for terms within multiple thesauri to support manual annotation. Second, keyword search, as it has become the de-facto standard to access data on the web. Third, faceted browsing as it has become a popular method to interactively explore (image) collections. We investigate the use of explicit semantics to improve support to the user in these three tasks. We propose the following research questions:

---

[1] http://www.w3.org/2004/02/skos/

A. Sheth et al. (Eds.): ISWC 2008, LNCS 5318, pp. 914–919, 2008.

- How can explicit semantics be used in search algorithms to support the user with finding results in a heterogeneous graph?
- How do we organize and visualize the results found in a heterogenous graph to support exploration of this graph?
- How can we evaluate the added value of using explicit semantics in search, result organization and visualization?

2   Approach

We investigate interactive exploration in heterogeneous collections by the implementation and evaluation of three prototype systems on top of large and real world data collections. First, an annotation interface that uses autocompletion to support users with finding terms from multiple thesauri. Second, a search interface that supports the user in exploring museum objects that are semantically related to a keyword query. Third, a facet browser to support the user with the interactive formulation of queries. For all three we investigate how to improve interaction by using explicit semantics in the search algorithm, the result organization and visualization. In the following subsections we describe the details of our approach for each prototype. Due to limited space we do not elaborate on the related work for each. Note, we performed an extensive survey of semantic search applications (see work plan).

For each prototype system we propose an evaluation method. Although several applications use Semantic Web technology to support some form of exploration there are, as yet, no standard metrics to evaluate the solutions. This is probably because the aims, user tasks they intent to support and the use of semantics vary greatly among different applications. Determining appropriate evaluation methods for different forms of semantic search is an intrinsic part of this research.

2.1   Finding Terms within Multiple Thesauri

In an annotation task the user describes an object using terms from domain-specific thesauri. An annotation interface may contain annotation fields for the different properties that should be described. Typically, some form of keyword search makes the thesauri terms accessible. As thesauri become available in an interoperable format, annotators can access terms from multiple sources, including sources provided by other institutions.

At the moment there are several Semantic Web search engines that give access to terms from RDF/OWL documents. Examples of semantic search engines are Swoogle [3] and Sindice [4]. An elaborate analysis of 35 semantic search applications is available in [5]. Generic semantic search engines are not yet suited to support annotators in finding domain specific terms. A search query may return results irrelevant for a particular type of annotation. Furthermore, terms in the result set can be ambiguous in the sense that a naive visualization of these terms would not allow a user to distinguish them from each other.

Could an interface provide large coverage by using multiple thesauri while providing effective organization and visualization of the search results? The user can be supported in the annotation process by presenting only the terms that are appropriate for an annotation field and these terms should be organized and visualized so that they are unambiguous and self explanatory. For example, we can constrain the results of the creator field to persons, and augment the visualization of person names with their birth and death date; an annotation field for the creation site can be constrained to geographical locations, and these locations can be organized in a grouping by the country.

To experiment with different configurations of search, result organization and visualization we implement a configurable term search component. The component uses autocompletion to suggest terms while the user is typing. Based on this autocompletion search component a prototype annotation interface is constructed to support the subject annotation of the Rijksmuseum Amsterdam print collection. The prototype annotation interface will give access to terms from multiple thesauri including thesauri from outside the Rijksmuseum.

Evaluation. The prototype annotation interface is evaluated qualitative with experts at the Rijksmuseum. First, we gather information about the current annotation practices at the Museum. Second, we iteratively design a prototype interface considering the feedback of the experts in each cycle. Third, the professional annotators use the prototype interface in an experiment situated in their own environment. The results of the evaluation consist of observations that impact the practical use of Semantic Web technologies in the search algorithm and result visualization and organization.

2.2 Exploring Heterogeneous Collections through Keyword Search

A common way to start the exploration of the objects in museum collections is through keyword search. The simple "Google like" interface, with a single text-entry box, has become the standard for keyword search interfaces. Using Semantic Web technology we can use keyword search functionality to find museum objects that are semantically related to a query.

Within the E-Culture project we explore graph search algorithms to efficiently perform semantic on a large collection [1]. The results of a graph search are related to the keyword query by a path in the graph, that reflects a possible semantic interpretation of the query. Hollink et. al. showed through statistical analysis that some patterns of graph paths, containing relations from WordNet, performed better than others [6]. In an heterogeneous collection it is, however, difficult to determine in advance all paths that will lead to relevant results.

An interactive interface would let the user choose which interpretation she is interested in. Explicit semantics can be used to organize and visualize the search results to support the user with disambiguation of the search results as well as with the exploration of semantically related objects. We develop a prototype interface that allows interactive exploration of semantic keyword search results. The interface will use the graph-based search algorithms as developed within the E-Culture project. We investigate how explicit semantics can be used to organize the search results. In addition, we explore different visualizations for the presentation of the search results, such as geographical maps and timelines.

Evaluation. Different types of semantic organization techniques will be evaluated in a user study using an interactive exploratory search task. Designing the details of the experimental setup is part of the future work. We strive to use objective measurements, such as click stream data, as well as subjective opinions to determine the satisfaction of discovering (new) museum objects and relations.

2.3 Exploring Heterogeneous Collections with Faceted Browsing

Faceted browsing has become popular as an interface to interactively formulate queries [7,8,9,10]. A single facet highlights a dimension of the underlying data. By visualizing the values of the facets in the user interface, the user can construct multi-faceted queries by navigating through the interface.

Marti Hearst et al. showed that faceted browsing is very well suited to explore a collection of visual resources [7]. They assume a homogeneous collection with a fixed data schema, which allows manual configuration of the facets. Using a prototype implementation we investigate the requirements to apply faceted browsing to a large heterogenous collection. We explore the use of explicit semantics to organize and visualize the many facets of a heterogeneous collection.

Evaluation. A problem with faceted browsing is that large join queries can not always be computed sufficiently fast to support reasonable response times. We evaluate the scalability of faceted browsing and investigate how caching mechanisms can be used to improve response times. We define the theoretical and practical scalability limits, and experimentally, we show the performance statistics of different types of realistic queries on a large real world data set.

3   Contributions

1. Requirements analysis on the semantic data, search algorithms and user interface needed to support annotation using multiple thesauri. Concrete, implementation of an interface along with the underlying algorithms that support efficient term search from multiple thesauri by professional annotators. The Web-based implementation is based on a more generic interface model for term lookup in heterogeneous thesauri. The algorithms provide term search, result organization and visualization and can be configured with domain-specific semantics.
2. Design of an interface and algorithms to support end users with the exploration of a heterogeneous collection. The interface provides keyword search, semantic-based organization of the search results as well as different visualizations. The algorithms provide graph search to efficiently find objects semantically related to a keyword query and result organizing techniques that can be configured with domain specific semantics.
3. Design of a scalable interface and efficient query engine to apply faceted browsing to heterogeneous RDF graphs. The facets can be automatically or manually configured using domain specific semantics. The query engine provides caching mechanism to efficiently support large join queries.

4   Work Plan

Bellow we briefly describe the work that we have achieved so far, that we are currently working on and that is planned for future work. Between brackets we mention to which contribution the work is related.

Results achieved so far

(all) Studied related work in semantic search. In 35 existing systems we analyzed how explicit semantics are used in query construction, the core search process, the presentation of the search results and user feedback on query and results. (To appear as a chapter in the book for the Network of Excellence, K-Space)

(1) Interface design for a configurable autocompletion component. We implemented a client-side autocompletion widget in JavaScript on top of the YAHOO User Interface (YUI) library. We also implemented a server-side algorithm for term search and result organization that can be configured with domain specific semantics. (CWI technical report INS-E0708)

(1) User study on result organization techniques for autocompletion suggestions. In cooperation with Alia Amin we conducted two user studies using web-based interactive surveys to test different methods of grouping term suggestions. (to be submitted)

(2) Design of a search algorithm for semantic search. In cooperation with Jan Wielemaker we implemented a best-first weighted graph search algorithm in Prolog. (Accepted for ISWC 2008)

(2) Initial interface design for organization and visualization of search results. We implemented a client side widget in JavaScript that can visualize a set of results as groups of thumbnails, on a geographical map, timeline or a graph. We also implemented server side algorithms for result organization and visualization of RDF data that can be configured to use domain specific semantics. (Intermediate results are part of the ClioPatria open source toolkit[2])

(3) Interface design for faceted browsing on a heterogeneous RDF graph. The prototype system, /facet, is, for example, used within the E-Culture Demonstrator[3], the K-Space news demonstrator[4]. (Published at ISWC 2006)

Current work

(1) Design and configuration of the interface to support subject annotation of the print collection of the Rijksmuseum Amsterdam.

(1) User study of the prototype annotation interface with professional annotators of the Rijksmuseum Amsterdam.

(3) Evaluation of scalability in /facet. Test caching solutions to improve computation of large join queries.

Planned work

(2) Continuation of interface design to support interactive exploration of search results with semantic clustering.

(2) Experimental design and user study on result clustering for semantic search.

---

[2] http://e-culture.multimedian.nl/software/ClioPatria.shtml
[3] http://e-culture.multimedian.nl/demo/search
[4] http://newsml.cwi.nl/explore/facet

Acknowledgments

I would like to thank Alia Amin, CWI Amsterdam and Jan Wielemaker, UvA Amsterdam for their close cooperation, my supervisors Jacco van Ossenbruggen, CWI Amsterdam and Lynda Hardman, CWI Amsterdam and TU Eindhoven for guiding me in this research trajectory and Guus Schreiber, VU University Amsterdam for his support of my research.

This research was supported by the MultimediaN project funded through the BSIK programme of the Dutch Government and by the European Commission under contract FP6-027026, Knowledge Space of semantic inference for automatic annotation and retrieval of multimedia content — K-Space.

References

1. Schreiber, G., Amin, A., van Assem, M., de Boer, V., Hardman, L., Hildebrand, M., Hollink, L., Huang, Z., van Kersen, J., de Niet, M., Omelayenjko, B., van Ossenbruggen, J., Siebes, R., Taekema, J., Wielemaker, J., Wielinga, B.: MultimediaN E-Culture Demonstrator. In: Cruz, I., Decker, S., Allemang, D., Preist, C., Schwabe, D., Mika, P., Uschold, M., Aroyo, L.M. (eds.) ISWC 2006. LNCS, vol. 4273, pp. 951–958. Springer, Heidelberg (2006)
2. Tordai, A., Omelayenko, B., Schreiber, G.: Thesaurus and metadata alignment for a semantic e-culture application. In: K-CAP 2007. Proceedings of the 4th international conference on Knowledge capture, pp. 199–200. ACM, New York (2007)
3. Ding, L., Pan, R., Finin, T., Joshi, A., Peng, Y., Kolari, P.: Finding and Ranking Knowledge on the Semantic Web. In: Gil, Y., Motta, E., Benjamins, V.R., Musen, M.A. (eds.) ISWC 2005. LNCS, vol. 3729, pp. 156–170. Springer, Heidelberg (2005)
4. Tummarello, G., Oren, E., Delbru, R.: Sindice.com: Weaving the open linked data. In: Aberer, K., Choi, K.-S., Noy, N., Allemang, D., Lee, K.-I., Nixon, L., Golbeck, J., Mika, P., Maynard, D., Mizoguchi, R., Schreiber, G., Cudré-Mauroux, P. (eds.) ASWC 2007 and ISWC 2007. LNCS, vol. 4825, pp. 547–560. Springer, Heidelberg (2007)
5. Hildebrand, M., van Ossenbruggen, J., Hardman, L.: An analysis of search-based user interaction on the Semantic Web. Technical Report INS-E0706, CWI (2007)
6. Hollink, L., Schreiber, G., Wielinga, B.: Patterns of semantic relations to improve image content search. Web Semant 5(3), 195–203 (2007)
7. Yee, K.P., Swearingen, K., Li, K., Hearst, M.: Faceted Metadata for Image Search and Browsing. In: CHI 2003. Proceedings of the SIGCHI conference on Human factors in computing systems, Ft. Lauderdale, Florida, USA, pp. 401–408. ACM Press, New York (2003)
8. Hyvonen, E., Makela, E., Salminen, M., Valo, A., Viljanen, K., Saarela, S., Junnila, M., Kettula, S.: Museumfinland – finnish museums on the semantic web. Journal of Web Semantics 3(2), 25 (2005)
9. Schraefel, M.C., Smith, D.A., Owens, A., Russell, A., Harris, C., Wilson, M.L.: The evolving mSpace platform: leveraging the Semantic Web on the Trail of the Memex. In: Proceedings of Hypertext 2005, Salzburg, pp. 174–183 (2005)
10. Huynh, D., Karger, D., Miller, R.: Exhibit: Lightweight structured data publishing. In: 16th International World Wide Web Conference, Banff, Alberta, Canada. ACM, New York (2007)

End-User Assisted Ontology Evolution in Uncertain Domains

Thomas Scharrenbach

Swiss Federal Institute for Forest, Snow and Landscape Research,
Zürcherstrasse 111, CH-8910 Birmensdorf, Switzerland
thomas.scharrenbach@wsl.ch

Abstract. Learning ontologies from large text corpora is a well understood task while evolving ontologies dynamically from user-input has rarely been adressed so far. Evolution of ontologies has to deal with vague or incomplete information. Accordingly, the formalism used for knowledge representation must be able to handle this kind of information. Classical logical approaches such as description logics are particularly poor in adressing uncertainty. Ontology evolution may benefit from exploring probabilistic or fuzzy approaches to knowledge representation. In this thesis an approach to evolve and update ontologies is developed which uses explicit and implicit user-input and extends probabilistic approaches to ontology engineering.

1 Introduction

The integration of datasources of different origin is quite a difficult task. Even though there exist standard mechanisms for querying like SQL, the underlying schemata may vary significantly. Traditional flat structures are very limited in the representation of the semantics of data. Over the recent couple of years, network based systems like ontologies have become more and more a standard in the semantic representation of data from and within different domains.

Ontologies allow for a sound definition of shared terms within and between different datasources. They are defined by the Web Ontology Language (OWL) recommended by the W3C. Currently, the Ontology layer is the highest layer of sufficient maturity within the Semantic Web [1]. As some sublanguages of OWL directly correspond to Description Logics (DL) traditional rule-based logical reasoning is straightforward and can be seen as state-of-the-art [2].

DL-based systems can model vague information only up to a certain degree by defining, e. g. disjoints, similarity relations etc. However, there are several cases where the explicit modelling of uncertainty is desirable [3]:

- While knowledge and knowledge representation usually are incomplete, there is no sound concept of vague information in DL systems: either something is asserted in the knowledge base or not. And if it is asserted it is true or false, nothing in-between.

A. Sheth et al. (Eds.): ISWC 2008, LNCS 5318, pp. 920–925, 2008.

- In addition to the presence of a logical consequence, it is desired to know how likely a certain event will occur. If e. g. we know that birds can fly with a probability of more than 0.9 and Tweety is a bird then the probability that Tweety has the ability to fly shall be higher than 0.9.

  $Bird(Tweety) \wedge Pr(canFly(Bird)) > 0.9 \rightarrow Pr(canFly(Tweety)) > 0.9$.

- In some cases, contradictions which would violate the consistency of a DL system, must be allowed up to a certain degree. Suppose a simple ontology of birds. Birds can usually fly with the exception of penguins. These are birds that cannot fly. The corresponding DL knowledge base would thus be inconsistent. $\{canFly(Bird), Penguin \sqsubseteq Bird, \neg canFly(Penguin)\}$. The simplest way to overcome this inconsistency is to split the concept of birds into two new concepts of flying and non-flying birds. A more elegant way is to assign the role $canFly$ a probability in which the inconsistency is relaxed such as $Pr(canFly(Bird)) > 0.9$. This states that birds can fly with a probability of more than 0.9 but also allows for non-flying birds without structural changes.

These limitations can be overcome by extending the concept of ontologies with a probabilistic or fuzzy model. The need for such a model is even more acute when the *evolution* of ontologies is considered. The members of a community may want to develop an ontology further. Be it because the ontology is incomplete or because additional knowledge is created which is materialized in new concepts and facts. In such a case these concepts are typically related to the existing ones only in a weakly or undefined way which cannot be put in terms of the primitives of a classical logical formalism. Furthermore, they might introduce inconsistencies into the existing knowledge base which can be relaxated in a probabilistic or fuzzy model. Finally, one of the most interesting question is to what degree the construction or update of ontologies can be automated. This thesis investigates whether and how user-input can be used to automatically construct and/or update application ontologies for heteroneneous data sources by extending probabilistic and fuzzy approaches to description logic reasoning.

2   Related Work

Probabilitstic approaches to ontology engineering can be divided into two groups: approaches directly extending the ontology and approaches where the ontology is transformed into a different representation allowing for probabilistic modelling.

DL are a family of formal languages for structured terminological knowledge representation [2]. On the one hand they can describe the formal concepts of a domain and on the other hand they allow for first-order logic inference. The OWL languages OWL Lite and OWL DL are explicitly based on DL which makes the use of DL for reasoning in ontology based systems straightforward.

While DL provide formal logical representation and inference, uncertainty like "Birds can fly with probability of 0.9" cannot be modelled very well. Lukasiewicz proposed a probabilistic extension called *Probabilistic Description Logics* (PDL) [4]. Individuals can be assigned conditional probability constraints which are asserted to a so-called PABox P_o for every probabilistic individual $o \in \mathbf{I}_P$. Analogously a PTBox $PT = (T, P)$ is defined holding a set of conditional constraints P for the knowledge base $\mathcal{K}$. PDL consist of a set of classical individuals $\mathbf{I}_C$ a set of probabilistic individuals $\mathbf{I}_P$, a PTBox, an ABox, and one PABox for every $o \in \mathbf{I}_P$. This concept allows for modelling quantified uncertainty and first software reasoning tools are available [5].

Classical DL cannot model vague concepts like "Tweety is young". Therefore, Straccia [6] introduced FuzzyOWL. The knowledge base is enriched by fuzzy role inclusion axioms, fuzzy concept inclusion axioms, fuzzy concept assertions and fuzzy role assertions. This induces a fuzzy RBox, a fuzzy ABox, and a fuzzy TBox, respectively. This method enables inferences of "vague rules". While first lacking methods for the reasoning process, recent progress has been made in this area and reasoning tools are available [7]. According to the fuzzy approach, measuring the level of uncertainty is not possible directly [8], at least not as straightforward as in the PDL case.

Although there exist many other approaches for probabilistic ontology based knowledge engineering systems like e. g. BayesOWL [9] the presented ones are considered as the most relevant for the research subject of this thesis. For a more detailed overview the reader is referred to [4].

Work on the update of ontologies has mainly been performed for classical DL based systems [10,11] and for agent systems [12]. The main challenge is to keep the knowledge base consistent which could efficiently only be achieved on the instance level so far. Recently, Haase and Völker proposed a scheme based on finding the minimal inconsistent subontology [13]. According to a confidence measure inconsistent resources are removed until there are no more inconsistencies left. This approachs allowing for the update of arbitrary resources removes contradicting information instead of modelling it.

The aspect of ontology update incorporating inconsistent information using probabilistic or fuzzy approaches has not been addressed yet and will be subject of this thesis.

3   Research Plan

The "Virtual Data Centre" (VDC) of the "Datenzentrum Natur Landschaft" (DNL) project is a collection of several enviromental databases mainly containing data from taxonomies, different land registers, legal documents etc. As such, though there exists no common scheme or explicit references, the data is strongly semantically correlated. The aim of this project is to model the semantics by a multi-lingual eco-ontology on the application level.

3.1   Current State of Research

In a first step, a bilingual eco-ontology was created by expert users from scratch. An open search was realized by an expansion scheme [14] and the reasoning is based on DL. Further research is performed with the objective to extend the knowledge base by a so-called RCCBox representing composition tables for spatial inference based on the Region Connection Calculus (RCC) [15,16]. In late 2008, a first prototype shall be released for a test cycle of selected expert users at the Swiss Federal Office for the Enviroment.

3.2   Future Research

Creating the Baseline. In a first step, the ontology will be extended by a statistical model following both, the PDL and the FuzzyOWL approaches. While there do exist reasoning tools the main challenge lies within the estimation of the parameters for the underlying probability distributions and fuzzy sets. Both approaches will therefore be extended by methods for estimating and automatically updating the corresponding parameters. For this task, classical statistical text-classification approaches will be used like described in [17] which result in a probabilistic and a fuzzy ontology, respectively, acting as the baseline model for this thesis.

Incorporating User-Input. Though the data is semantically connected, in the baseline model these links are not yet established. Furthermore, the baseline model is assumed to be an incomplete representation of the data. Hence, user-input will be incorporated to obtain the required informaton.

One of the main applications of the DNL is the open search that will be used for gathering the desired input during the search process. This step is divided into two parts: Using explicit user-feedback on the one hand and using implicit user-input on the other hand [18]. In this context, the WSL Ontology Webeditor (WOW) is under development allowing for the explicit insertion of new resources into the ontology. While the incorporation of explicit user-input is straightforward, for the implicit input a search context has to be defined. The information of a failed query, i. e. the search terms, will be linked to following search terms and inserted into the knowledge base. In case of a successful query, the confidence of the corresponding resources will be increased.

Along with that, methods for the extension of the ontologies and the corresponding statistical models will be investigated enabling a sound an efficient update.

Handling Inconsistencies. In the first phase of research, the explicit extension will be restricted to the addition of new instances. Later on also the insertion will take place on concept and role level. Inconsistencies will then not be resolved by removing resources like in [19] but modelled explicitly by means of uncertainty. This way, information will not be pruned w. r. t. its relevance but will be kept inside the knowledge base itself. Not only will the proper presentation of the ontology for the insertion of new resources be part of this thesis' research, but also the aspect of how to offer the possibility to let the user specifiy the amount of vagueness for the extension. While the first may be adapted like presented in [20] the latter will be realized in terms of how likely the new individual matches to the actual knowledge base. Within this context it will be interesting to see especially how the insertion will work for geo-spatial data. Particulary, modelling the update of geo-spatial approximations as described in [21] by probabilistic means.

Evaluation. For the evaluation of the performance of the developed methods, a reference dataset will be constructed in cooperation with expert end-users to measure the improvement of precision and recall for the updated ontologies. Since end-user-feedback will be available within the context of the DNL project, this will be used as well for the evaluation of how well the tested methods work for the evolution of the knowledge base.

4   Conclusion and Outlook

The problem of consistent evolution of probabilistic ontologies has not been addressed so far. This thesis investigates how to evolve, i. e. learn and update, a multi-source multi-langual eco-ontology from user-input. Therefore, probabilistic extensions of classical DL knowledge bases will be used with a focus on either Probabilistic Description Logics or FuzzyOWL. These approaches will be extended by an update scheme to incorporate implicit and/or explicit user-input into the knowledge base. Different aspects of how to obtain the desired information from the end-user for the extension of a knowledge base with an underlying statistical model will be investigated. While the extension will be at first restricted to instance level the extension to concept level will be explored based on the gathered results. For systematic evaluation, a reference dataset will be constructed as well as will be used explicit feedback from end-users.

Acknowledgments. I would like to thank Prof. Abraham Bernstein for supervising this thesis as well as Bettina Bauer-Messmer and Rolf Grütter for their support.

References

1. McGuinness, D.L., van Harmelen, F.: OWL Web Ontology Language overview. W3C recommendation, W3C (February 2004),
   http://www.w3.org/TR/2004/REC-owl-features-20040210/
2. Baader, F., Calvanese, D., McGuinness, D.L., Nardi, D., Patel-Schneider, P.F. (eds.): The Description Logic Handbook: Theory, Implementation, and Applications. Cambridge University Press, Cambridge (2003)
3. Bacchus, F.: Representing and Reasoning with Probabilistic Knowledge: a Logical Approach to Probabilities. MIT Press, Cambridge (1990)
4. Lukasiewicz, T.: Probabilistic Description Logics for the Semantic Web. Technical Report, Knowledge-Based Systems Group Tu Vienna (2007)
5. Clark & Parsia: Pronto - a Probabilistic Extension for OWL DL and Pellet,
   http://pellet.owldl.com/pronto

6. Straccia, U.: Towards a Fuzzy Description Logic for the Semantic Web. In: Gómez-Pérez, A., Euzenat, J. (eds.) ESWC 2005. LNCS, vol. 3532. Springer, Heidelberg (2005)

7. Stoilos, G., Stamou, G.: Extending Fuzzy Description Logics for the Semantic Web. In: 3rd International Workshop of OWL: Experiences and Directions, Innsbruck (2007)

8. Stoilos, G., Stamou, G., Tzouvaras, V., Pan, J., Horrocks, I.: Fuzzy owl: Uncertainty and The Semantic Web. In: 21st International Workshop on Description Logics (DL 2008), Galway (2005)

9. Ding, Z., Peng, Y.: A Probabilistic Extension to Ontology Language OWL. In: Proceedings of the 37th Hawaii International Conference On System Sciences (HICSS-37), Big Island, Hawaii (2004)

10. Winslett, M.: Updating Logical Databases. Cambridge University Press, Cambridge (1990)

11. Giacomo, G.D., Lenzerini, M., Poggi, A., Rosati, R.: On the Update of Description Logic Ontologies at the Instance Level. In: AAAI 2006 (2006)

12. McNeill, F., Bundy, A., Walton, C.: Facilitating Agent Communication Through Detecting, Diagnosing and Refining Ontological Mismatch. In: Proceedings of the KR 2004 Doctoral Consortium, AAAI Technical Report (2004)

13. Haase, P., Völker, J.: Ontology Learning and Reasoning - Dealing with Uncertainty and Inconsistency. In: Paulo, C.G., et al. (eds.) Uncertainty Reasoning for the Semantic Web I. Springer, Heidelberg (to appear, 2008)

14. Grütter, R., Bauer-Messmer, B., Frehner, M.: First Experiences with an Ontology-Based Search for Environmental Data. In: Proceedings of the 11th AGILE International Conference on Geographic Information Science (AGILE 2008), Girona, Spain (2008)

15. Grütter, R., Bauer-Messmer, B.: Towards Spatial Reasoning in the Semantic Web: A Hybrid Knowledge Representation System Architecture. In: Proceedings of the 10th AGILE International Conference on Geographic Information Science (AGILE 2007), Aalborg, Denmark (2007)

16. Grütter, R., Bauer-Messmer, B., Hägeli, M.: Extending an Ontology-Based Search with a Formalism for Spatial Reasoning. In: Proceedings of the 23rd Annual ACM Symposium on Applied Computing (ACM SAC 2008), Fortaleza, Brazil (2008)

17. McCallum, A., Nigam, K.: A Comparison of Event Models for Naive Bayes Text Classification. In: Proceedings of the AAAI 1998 Workshop on Learning for Text Categorization, pp. 41–48 (1998)

18. Bauer-Messmer, B., Grütter, R., Scharrenbach, T.: Improving An Environmental Ontology by Incorporating User-Input. In: EnviroInfo 2008 - Environmental Informatics and Industrial Ecology, Lüneburg, Germany (to appear, 2008)

19. Haase, P., Stojanovic, L.: Consistent Evolution of OWL Ontologies. In: Gómez-Pérez, A., Euzenat, J. (eds.) ESWC 2005. LNCS, vol. 3532. Springer, Heidelberg (2005)

20. Bernstein, A., Kaufmann, E.: Gino - a Guided Input Natural Language Ontology Editor. In: Cruz, I., Decker, S., Allemang, D., Preist, C., Schwabe, D., Mika, P., Uschold, M., Aroyo, L.M. (eds.) ISWC 2006. LNCS, vol. 4273, pp. 144–157. Springer, Heidelberg (2006)

21. Grütter, R., Scharrenbach, T., Bauer-Messmer, B.: Improving an RCC-Derived Geospatial Approximation by OWL Axioms. In: Sheth, A., et al. (eds.) ISWC 2008. LNCS, vol. 5318. Springer, Heidelberg (2008)

Learning Methods in Multi-grained Query Answering*

Philipp Sorg

Institute AIFB, University of Karlsruhe
D-76128 Karlsruhe, Germany
sorg@aifb.uni-karlsruhe.de

Abstract. This PhD proposal is about the development of new methods for information access. Two new approaches are proposed: Multi-Grained Query Answering that bridges the gap between Information Retrieval and Question Answering and Learning-Enhanced Query Answering that enables the improvement of retrieval performance based on the experience of previous queries and answers.

1 Introduction

Finding relevant information in the WWW, in knowledge bases of companies, in document repositories or even on personal computers is getting more and more important, as the amount of knowledge contained in these resources continuously increases. In addition, users in a private or professional environment rely heavily on the information. In a professional environment, e.g. as described by Abecker et al [1], building and using Organisational Memories is essential for all companies working in the information sector and reducing the effort of finding information is an important cost factor.

In my PhD research I plan to develop new methods for searching in such heterogeneous knowledge bases. In this PhD proposal I will describe current search methods, identify missing features and present new ideas to improve current search systems.

1.1 Motivation

Current Information Retrieval (IR) and Question Answering (QA) systems traditionally only return answers of a specific granularity. While there are some exceptions, e.g. the search engine Google[1] that implements some heuristics to detect and answer simple factoid questions, IR systems typically return whole documents as answers. On the other end of the spectrum, QA systems try to find an exact answer to the question. They achieve reasonable retrieval results

---

* This work was funded by the Multipla project sponsored by the German Research Foundation (DFG) under grant number 38457858. Many thanks to my PhD supervisor Dr. Philipp Cimiano for his helpful comments.
[1] http://www.google.com

A. Sheth et al. (Eds.): ISWC 2008, LNCS 5318, pp. 926–931, 2008.

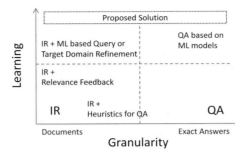

Fig. 1. Current Information Retrieval and Question Answering methods, classified by answer granularity and involved learning

on factoid questions, but are in general not able to answer complex questions, e.g. questions based on a large context or questions for which the answer is not explicitly stated in the text but must be inferred. The gap between IR and QA systems could be filled by systems that handle answers of different granularity ranging from whole documents to exact answers in a flexible way.

Another missing feature of most current IR and QA systems is the missing ability to learn from experience. Intuitively, systems should be able to use information extracted from previous pairs of queries and answers and use this information to improve retrieval results. As far as we know there is no prominent search system that supports this kind of learning [2].

1.2   Current State of My PhD Research

I started my PhD research in October 2007.

First I started to examine different methods of Natural Language Processing (NLP), e.g. relatedness measures on terms and text. A special focus was on defining semantic relatedness measures based on Wikipedia, e.g. by using Explicit Semantic Analysis (ESA) [3] that represents text in a Wikipedia article space. As part of this research I developed a new method to learn new cross-language links in Wikipedia [4], which I used in cross-lingual ESA to define a relatedness measure across languages.

At the current stage I am developing a general framework for accessing and processing unstructured (e.g. plain-text documents) and structured (e.g. ontologies) information sources. Based on this framework I plan to implement the new query answering methods presented in this PhD proposal.

2   Definition of the PhD Topic

In the following section I will define the problem I intend to address during my PhD research and describe state of the art methods that address this problem.

2.1 Definition of the Problem

The problem I want to investigate is to develop new search methods supporting the following features:

− Detect the right granularity of answers and return answers of different granularity. *[Multi-Grained Question Answering]*
− Learn from previous queries and answers to improve retrieval results. *[Learning-Enhanced Question Answering]*

2.2 State of the Art

The problem of Multi-Grained Query Answering is to some extent addressed by commercial Internet search engines like Google or Yahoo where the answer space is mainly defined at the document level but small possibly relevant text snippets are presented as well. Another approach is to use supervised learning methods to learn the ranking function that is used to retrieve document elements [5].

The problem of answering factoid queries based on background knowledge is e.g. solved by matching the query to certain patterns (as implemented in Google) or by finding relevant text by using IR methods on word level and pinpointing the right answer using linguistic analysis on a syntactic/semantic level [6].

A method to improve the retrieval system using previous queries and answers is to use relevance feedback from users [7]. Another approach is to use Machine Learning models trained on query-answer pairs to translate query terms to answer terms for target domain refinement [8].

An example for a QA system based in Machine Learning can be found in [9], where queries and answers are represented as graphs and graph rules mapping queries to answers are learned, which are used for the QA system. Another approach is to learn patterns from question/answer pairs that can be used for QA. A bootstrapping pattern mining approach is e.g. described in [10], where starting from a few hand-crafted examples new patterns are inferred from the Internet using a web search engine.

3 Approach to the Problem

In this section I will first describe how I intend do analyse existing retrieval methods, ranging from Information Retrieval to Question Answering. Then I will present initial ideas for Multi-Grained and Learning-Enhanced Query Answering.

3.1 Analysis of Existing Retrieval Methods

The analysis of existing retrieval methods will focus on IR and QA methods. The expected outcome will be an overview of current retrieval methods and the identification of strengths and weaknesses of those methods.

IR Methods. The analysis of existing IR methods will be mostly concerned with vector space representations of text. The most simple representation is the standard Bag-of-Words model, but there are many systems that extend this model with different weights, by using similarity measures to deal with synonyms or by including background knowledge like annotations of Named Entities. This analysis provides the foundation of purely statistical approaches of Query Answering.

Another important aspect is the use of relevance feedback in IR systems. This is often done by using Machine Learning techniques and will therefore be substantial for the Machine Learning part of my research.

QA Methods. QA systems normally use a more structured representation of text, often based on deep linguistic analysis. A big variety of background knowledge is used in current systems, ranging from patterns matching factoid answers to complex ontologies. This analysis will help to find appropriate representations of text that can be used to develop new retrieval methods.

Many QA systems use IR methods to identify relevant parts of documents. The analysis of these methods will be important as Multi-Grained QA will be based on these existing methods.

3.2   Description of Envisioned Methods

Multi-grained Query Answering. The core of Multi-Grained QA is to develop methods that are able to identify the right granularity of the answers given a query and based on the information sources. One idea is to introduce a measure of Answer Density. The trade-off between completeness of the answer and its length should be modeled in this measure.

The following example shows the advantage of such an Answer Density measure to existing QA systems. For the question

> Why did David Koresh ask the FBI for a word processor?

it is not possible to determine an expected answer type. Users asking this question would expect a short paragraph containing an explanation, like this text snippet of the Wikipedia article "David Koresh":

> ... Communication over the next 51 days included telephone exchanges with various FBI negotiators.
> As the standoff continued, Koresh, who was seriously injured by a gunshot wound, along with his closest male leaders negotiated delays, possibly so he could write religious documents he said he needed to complete before he surrendered. ...

Ideally the Answer Density measure would assign a high value to this snippet. This could e.g. be done by using the semantic relatedness of "word processor" and "write". As the presented snippet is part of the article "David Koresh", contains the term "FBI" and is related to "word processor", the value of the Answer Density is high and could be identified as possible answer.

Learning-Enhanced Query Answering. The problem of learning from previous queries and answers is a problem of unsupervised ML. Supervised learning methods based on user feedback yield good results in improving retrieval performance, but have the problem that feedback is not available in general. As these methods also are widely discussed in the IR research community I intend to focus on unsupervised ML techniques.

One idea is to use clustering techniques to cluster queries and answers. Based on this clustering, query-query, query-answer and answer-answer relations can be extracted. We plan to use syntactic and semantic features of the query for the clustering.

The syntactic features can be used to identify the expected type of answer. E.g. if a question starts with "Who ..." the expected answer will probably be factoid, whereas a more detailed answer is needed for questions starting with "Why ...". Clustering based on Syntactic Tree Kernels [11] is a possible method to use the syntactic features of the queries.

Semantic features express the topic and content of the query. We plan to use Wikipedia as background information source by mapping queries to a space of Wikipedia articles using extracted Named Entities that correspond to Wikipedia articles. This mapping can be used to identify queries with similar topics. Extracted key terms from these queries can then be used for query or answer refinement.

Another application of this mapping to the space of Wikipedia articles is the usage of the categories of these articles. Combined with syntactical information a more abstract representation of queries and answers can be constructed. It is then e.g. possible to use the categories of these articles together with the syntax of the query to find an abstract representation that can be used to cluster similar queries. E.g. the question *"Who wrote Faust?"* with the factoid answer *"Goethe."* could be represented as *"Who wrote {Book}?" "{Person}."*. For new queries assigned to the same cluster the system can then infer that the answer should contain a person.

4   Evaluation

There are different approaches for the planned evaluation of the developed methods. One is automatic evaluation based on existing datasets. As there are no existing datasets for Multi-Grained Query Answering, this evaluation can only be applied to the extrema of Multi-Grained Query Answering by using datasets for the evaluation of IR or QA systems, e.g. datasets provided by TREC[2]. As it will probably not be possible to use this evaluation to compare the new methods with existing IR or QA methods due to the differences in the answer granularity, this evaluation can be mainly used to analyse the benefit of Learning-Enhanced QA. After learning the results should improve on the used datasets. One evaluation step could be the comparison of results of the same query before and after the training phase.

---

[2] http://trec.nist.gov/

To compare the developed system with other IR or QA systems I plan to perform a manual evaluation based on a real world scenario, e.g. a user evaluation involving several people using the system as a desktop search engine.

5 Conclusion

I have presented my PhD research proposal in the field of IR/QA to enhance information access. The main goal is to overcome the rigidity of current systems which either only return full documents or try to pinpoint exact answers. Further, I aim at developing paradigms by which systems can learn from past experience which represents a crucial open problem in the field of information access.

References

1. Abecker, A., Bernardi, A., Hinkelmann, K., Kuhn, O., Sintek, M.: Toward a technology for organizational memories. IEEE Intelligent Systems 13(3), 40–48 (1998)
2. Strzalkowski, T., Harabagiu, S.: Advances in Open Domain Question Answering (Text, Speech and Language Technology). Springer New York, Inc., Secaucus (2006)
3. Gabrilovich, E., Markovitch, S.: Computing Semantic Relatedness using Wikipedia-based Explicit Semantic Analysis. In: Proceedings of the 20th International Joint Conference on Artificial Intelligence (2007)
4. Sorg, P., Cimiano, P.: Enriching the crosslingual link structure of wikipedia - a classification-based approach. In: Proceedings of the AAAI 2008 Workshop on Wikipedia and Artificial Intelligence, Chicago, IL, USA (2008)
5. Vittaut, J.N., Gallinari, P.: Machine learning ranking for structured information retrieval. In: Proceedings of the European Conference on Information Retrieval, pp. 338–349 (2006)
6. Hovy, E.H., Gerber, L., Hermjakob, U., Junk, M., Lin, C.Y.: Question answering in webclopedia. In: Proceedings of the Text Retrieval Conference (2000)
7. Harman, D.: Relevance feedback revisited. In: Belkin, N.J., Ingwersen, P., Pejtersen, A.M. (eds.) Proceedings of the Conference on Research and Development in Information Retrieval, Copenhagen, Denmark, pp. 1–10. ACM, New York (1992)
8. Riezler, S., Vasserman, A., Tsochantaridis, I., Mittal, V., Liu, Y.: Statistical machine translation for query expansion in answer retrieval. In: Proceedings of the 45th Annual Meeting of the Association for Computational Linguistics, Prague, Czech Republic (2007)
9. Molla, D., van Zaanen, M.: Learning of graph rules for question answering. In: Proceedings of the Australasian Language Technology Workshop, Sydney, Australia (2005)
10. Ravichandran, D., Hovy, E.: Learning surface text patterns for a question answering system. In: Proceedings of the 40th Annual Meeting on Association for Computational Linguistics, Morristown, NJ, USA, pp. 41–47. Association for Computational Linguistics (2001)
11. Moschitti, A.: Efficient convolution kernels for dependency and constituent syntactic trees. In: Proceedings of the European Conference on Machine Learning, Berlin, Germany, pp. 318–329. Springer, Heidelberg (2006)

Author Index

Printing: Mercedes-Druck, Berlin
Binding: Stein+Lehmann, Berlin